Gene and its Engineering

Replication • Expression • Cloning • Manipulation

Gene and its Engineering

Replication • Expression • Cloning • Manipulation

H. K. Das
Editor
Formerly Professor of Molecular Biology
Chairman, Centre for Biotechnology
Director, Genetic Engineering Unit
Jawaharlal Nehru University
New Delhi

WILEY

Gene and its Engineering

Editor: H. K. Das

First Edition: 2014

ISBN: 978-81-265-4928-3

www.wileyindia.com
Printed at Shree Maitrey Printech Pvt. Ltd., Noida

Brief Contents

Preface

Why another book on Gene? The design of the experiments which led to the understanding of the nature of the genetic material, its replication, its expression and its engineering was indeed ingenious, and the inferences drawn were highly imaginative. The editor during his long teaching career has assiduously endeavored to share with the students the excitement underlying these experiments. The students enjoyed immensely when they were told how a particular information was generated or derived, rather than just the information itself. The insight imparted to them by this analytical approach has quite often succeeded in inducing them to take up a career in this area and helped them significantly later when they themselves tried to find the answer to an intriguing question.

A drawback of this analytical approach is that when time or space is not unlimited, the nuances of exciting experiments can be presented only at the expense of the amount of information. There is, however, so much information around today that it is impossible to offer everything in a course or a book. One can always refer to innumerable review articles or specialized books to get exhaustive information.

The editor has always emphasized to his students that the understanding of this area of biology is based only on 'strong common sense', and this is the only thing they have to apply to be able to do well in this area. This has been the guiding principle of the editor while working on this book.

Emphasis has been given throughout this book to focus on the significant and the path-breaking experiments that helped in understanding a phenomenon. The presentation of mere facts has been avoided, but attempts have always been made to elucidate how the facts have been unravelled. All pains have been taken to ensure that the readers share fully the excitement of this area. However, the necessity of keeping the reader abreast with the current status has also not been lost sight of.

Wherever possible, the original figures, autoradiograms and photographs have been reproduced to retain the original flavor. The editor is grateful to the scientists and the publishers for giving permission for reproducing images.

There are two parts in the book. These are on

1. Gene, its replication and its expression
2. Engineering of gene

The initial chapter in the first part deals with the evolution of knowledge related to gene as a concept followed by DNA as the genetic material; the constituents of nucleic acids; the structure and important characteristics of DNA; the chromatin structure and its remodelling; DNA replication and its mechanics; the various DNA polymerases; transcription of DNA into RNA, its regulation and RNA processing; replication of RNA as a genetic material; translation of mRNA into protein and regulation of protein synthesis. The second part begins with a chapter on gene cloning – how it all began, including the folk lore involving Stanley Cohen and Herbert Boyer. The following chapters are on the reaction of the scientific community resulting in the historic Asilomar meeting: the guidelines; the tools used in genetic engineering; general strategies for expression of introduced genes in a foreign host; details of various hosts like *Escherichia coli, Bacillus subtilis, Streptomyces, Pichia pastoris,* insect cells using baculovirus as vector, animal cells and plants as host. Subsequent chapters deal with general topics like chemical and enzymatic characterization of a cloned DNA fragment; reverse transcription and cDNA cloning; nucleic acid amplification and its applications; alternative strategies of gene cloning; site-directed *in vitro* mutagenesis and protein engineering; purification and refolding of recombinant proteins; study of regulation of a cloned gene; transgenic and gene knockout technologies; and finally gene therapy.

The chapters of the book have been developed by scientists who teach courses and are involved in active research in this area.

The editor is thankful to Dr. Bushra Rashid and Ms. Vandana Gupta of Wiley India for their painstaking effort and editorial help to improve upon the quality of the book.

H. K. Das
Editor

Contributors

Arun Kumar Sharma
Associate Professor
Department of Plant Molecular Biology
Delhi University, South Campus
New Delhi

Beena Pillai
Scientist
CSIR-Institute of Genomics and Integrative Biology
New Delhi

H. K. Das
Formerly Professor of Molecular Biology
Chairman, Centre for Biotechnology
Director, Genetic Engineering Unit
Jawaharlal Nehru University
New Delhi

Indranil Dasgupta
Professor
Department of Plant Molecular Biology
Delhi University, South Campus
New Delhi

J. Jeya Maheshwari
Scientist
Proteomics Department
Aravind Medical Research Foundation
Aravind Eye Care System
Madurai

Jayanta Kumar Pal
Professor and Head
Department of Biotechnology
University of Pune
Pune

K. Dharmalingam
Director-Research
Aravind Medical Research Foundation
Aravind Eye Care System
Madurai

Krishnaveni Mohareer
SERB Fellow
Dr Reddy's Institute of Life Sciences
University of Hyderabad
Hyderabad

Manoj K. Sharma
Assistant Professor
School of Biotechnology
Jawaharlal Nehru University
New Delhi

Navin Khanna
Group Leader
International Centre for Genetic Engineering and Biotechnology
New Delhi

Pramod C. Rath
Professor
School of Life Sciences
Jawaharlal Nehru University
New Delhi

Pramod Yadava
Professor
School of Life Sciences
Jawaharlal Nehru University
New Delhi

Rohini Muthuswami
Associate Professor
School of Life Sciences
Jawaharlal Nehru University
New Delhi

Sangeeta Chatterjee
DBT-Scientist
Center for Converging Technologies
University of Rajasthan
Jaipur

Seyed Ehtesham Hasnain
Professor
Kusuma School of Biological Sciences
Indian Institute of Technology, New Delhi
New Delhi

Shweta Sharma
Senior Research Fellow
Department of Plant Molecular Biology
Delhi University, South Campus
New Delhi

Shyamal K. Goswami
Professor
School of Life Sciences
Jawaharlal Nehru University
New Delhi

Suman Kundu
Associate Professor
Department of Biochemistry
Delhi University, South Campus
New Delhi

Upasana Arora
Research Associate
International Center for Genetic Engineering and Biotechnology
New Delhi

Uttam Pati
Professor and Former Dean
School of Biotechnology
Jawaharlal Nehru University
New Delhi

Vijay Kumar
Staff Scientist and Group Leader
Virology Group
International Centre for Genetic Engineering and Biotechnology
New Delhi

Vikash Kumar Dubey
Professor
Department of Biotechnology
Indian Institute of Technology, Guwahati
Guwahati

Contents

Part I

Gene, its replication and its expression

1

Concept of a gene

H. K. Das

1.1 Pre-Mendelian ideas and observations

For centuries, heredity was considered as a riddle controlled by some 'mysterious force' (Glass 1957). Artificial pollination of date palm flowers is thought to have been practised by the ancient Babylonians, but heredity, whether in animals or in plants, was an enigma even in the seventeenth century. In 1760, Josef Gottlieb Kolreuter conducted possibly the first systematic study in this area when he crossed two tobacco species, *Nicotiana paniculata* with *Nicotiana rustica*. He conducted a thorough analysis of F_1 (the first filial or the first offspring), F_2 (the second filial or the second offspring) and back-crosses (cross of F_1 or F_2 with the original parent species). He observed that the hybrids (F_1) did not breed true to type, but produced mixtures of the characters of the parents to different degrees. He, however, did not probe any farther (Zirkle 1935). Six decades later, Alexander Seton, John Goss, TA Knight and William Herbert conducted similar experiments with garden pea, *Pisum sativum*. All these workers observed seed colour of only one parent in F_1 and recovery of both parental colours in F_2, but they neither cared to undertake any quantitative analysis nor offered any interpretation (Roberts 1929). In the next four decades, more hybridization studies were reported by Augustin Sageret (Sageret 1826), Karl Friedrich von Gaertner and Charles Naudin (Roberts 1929). Naudin, who has 'most often been referred to as the immediate forerunner of Mendel, had enunciated a "law of disjunction" of two specific "essences" in the pollen and ovules of the hybrid. A hybrid plant is an individual in which are found two different "essences", having their respective modes of development and final direction, which mutually counter one other, and which are essentially in a struggle to "disengage" themselves from one another' (Dunn 1965). Indeed, a firm step forward in understanding heredity.

1.2 Mendel's 'unit factors'

Johann Mendel, who was born to a family of farmers, was ordained to priesthood in the Augustinian monastery at Brunn in Czechoslovakia in 1848 and took the name of Gregor. He studied physics, mathematics and natural sciences in the University of Vienna and became a teacher in Brunn in 1854. In 1857, Mendel started experimenting with garden pea, *Pisum sativum*, in the monastery garden to learn about the ways in which hereditary characters were passed on to the offspring by the parents. (Mendel 1865). He studied seven pairs of characters: stem height – tall or short; pod morphology – axial or terminal, full or constricted, green or yellow; pea seed morphology – round or wrinkled, yellow or green; and flower colour – violet or white. First, he allowed self-pollination for several generations and ensured that the characters were stable (Fig. 1.1). He then carried out crosses between plants displaying alternative forms of the same character, for example, between plants producing round pea seeds and those producing wrinkled pea seeds, or between plants with tall stems and those with short stems. This was done by removing the stamens from the flowers of a plant displaying one form of the character (round seeds) and pollinating with this stamen, the destamenized flower of a plant displaying the other form of the same character (wrinkled seeds). Surprisingly, the pea seeds produced in the cross-fertilized plants (F_1) were all round, none was wrinkled or of intermediate character. When these seeds were sown and the flowers that appeared in the developed plants were allowed to self-pollinate, seeds were generated (F_2) that were of both kinds. Mendel counted 5474 round seeds and 1850 wrinkled seeds; the distribution was roughly 3:1. Thus, the character that disappeared in the F_1 generation was indeed hidden and reappeared in the F_2 generation, albeit at 25% frequency. Mendel christened such a character as *recessive*, as opposed to the other

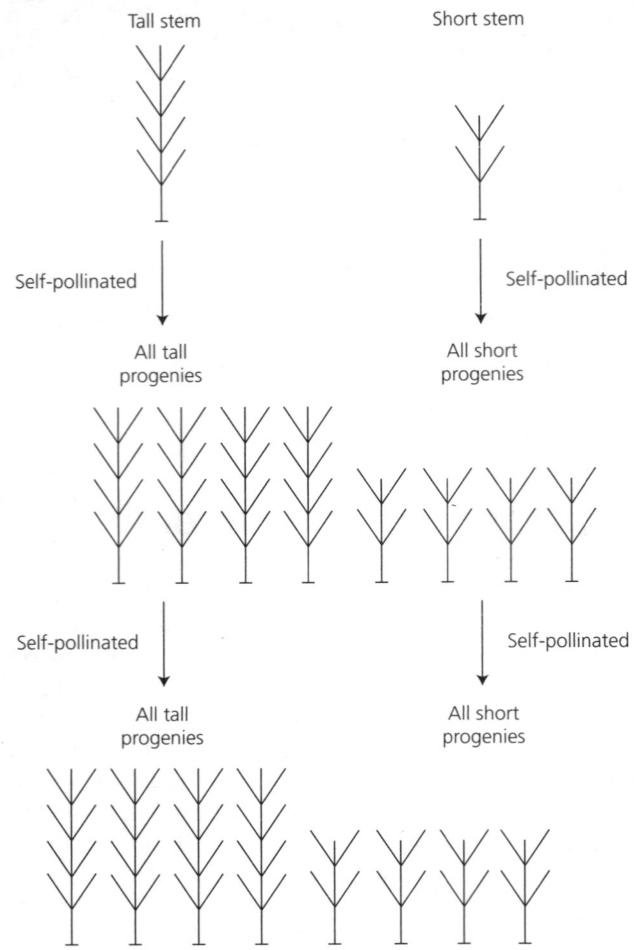

Figure 1.1 Selfing of tall plants or short plants displays stable characters. Self-pollination of tall plants always produces progeny plants that are tall. Similarly, self-pollination of short plants always produces progeny plants that are short.

character, which he called *dominant*. Exactly the same results were obtained irrespective of whether in the first cross fertilization, pollens were used from plants bearing the round seeds or those bearing the wrinkled seeds. Interestingly, Mendel obtained the same kind of results for each of the seven pairs of characters mentioned above. For example, in the experiment in which a plant with tall stem was crossed with a plant with short stem, all F_1 plants were tall, but out of 1064 F_2 plants, 787 plants had tall stems, while 277 had short stems (Fig. 1.2).

When the F_2 plants exhibiting *recessive* characters were allowed to self-pollinate, all the progenies (F_3) were true to type; they all exhibited the same *recessive* characters. This, however, was not true for the F_2 plants exhibiting *dominant* characters. Only 33% of the F_3 offspring of these plants yielded plants, flowers, pods or seeds of exclusively the *dominant* character, while 66% exhibited hybrid characters and yielded plants, flowers, pods or seeds of both kinds in the ratio 3:1 with respect to the *dominant* and *recessive* character, respectively. For example, among 565 plants which were raised from round F_2 seeds and allowed to self-pollinate, 193 yielded round seeds only, while 372 plants yielded both round and wrinkled seeds in the ratio 3:1. Similar results were obtained with respect to the cross between plants having tall stems and short stems (Fig. 1.2).

When round seeds taken from plants exhibiting hybrid characters were sown again and flowers of the

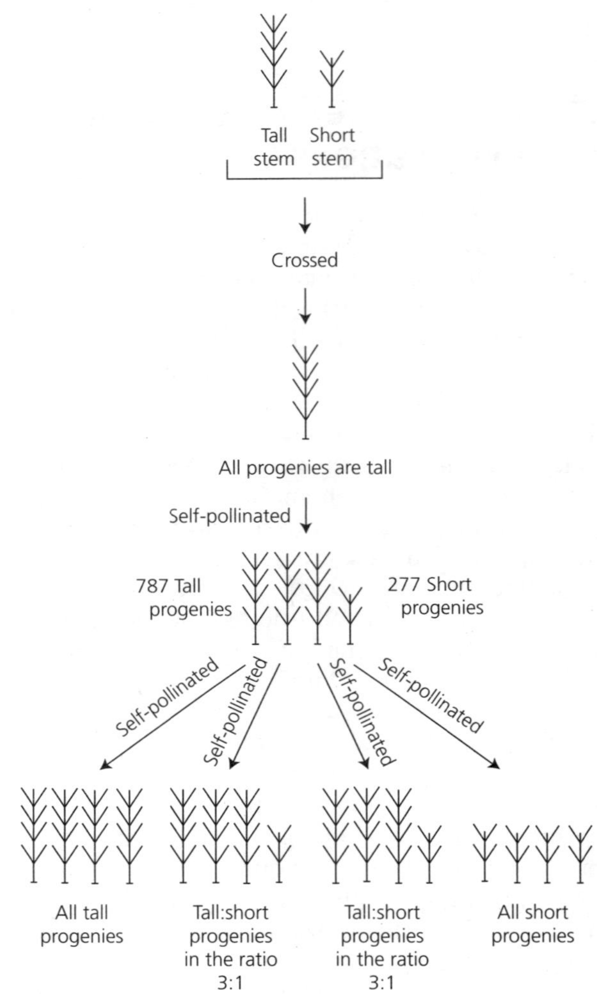

Figure 1.2 Cross between plants with tall stems and short stems. When pollens from flowers of a tall plant are applied to the destamenized flower of a short plant or pollens from flowers of a short plant are applied to the destamenized flower of a tall plant, all progeny plants are tall. When these progeny tall plants (F_1) are allowed to flower and self-pollinate, the next generation of progeny plants are a mixture of tall and short plants and they are in the ratio of 3:1, respectively. The short plants from this generation (F_2), when allowed to flower and self-pollinate, produce all short plants. One-third of the tall plants from this generation (F_2), when allowed to flower and self-pollinate, produce all tall plants, while two-thirds of the tall plants of the F_2 generation, on flowering and self-pollination, produce a mixture of tall and short plants, again in the ratio of 3:1.

developed plants were allowed to self-pollinate, exactly the same distribution of characters was seen in the F_4 generation and similarly in F_5 and F_6 generations. Thus, Mendel observed that the hybrid character resolved in every generation into hybrid and the two constant forms (*dominant* and *recessive*) in the ratio 2:1:1. He wrote, 'since the members of the first generation (F_2) spring directly from the seed of the hybrids (F_1), it is now clear that the hybrids form seeds having one or other of the two differentiating characters, and of these one-half develop again the hybrid form, while the other half yield plants which remain constant and receive the "*dominant*" or the "*recessive*" characters respectively in equal numbers.' This is what is known now as Mendel's law of independent segregation.

Mendel argued that, 'if *A* be taken as denoting one of the two constant characters, for instance the "*dominant*", and *a* the "*recessive*", and *Aa* the hybrid form in which both are conjoined, the expression

$$A + 2Aa + a$$

shows the terms in the series for the progeny of the hybrids of two differentiating characters'. As both *A* and *Aa* would express the *dominant* character, the 3:1 ratio in the F_2 plants was easy to explain.

The concept of 'dominant' and 'recessive' characters and the explanation of appearance of such characters in the progeny on selfing and crossing is described in Fig. 1.3.

Next, Mendel conducted crosses between plants that differed in two separate characters; for example, he fertilized flowers of plants generated from round and yellow seeds (*AB*) with pollens from plants generated from wrinkled and green seeds (*ab*). In the F_3 obtained in a similar manner, the following results were obtained:

38 had round yellow seeds	*AB*
35 had round green seeds	*Ab*
28 had wrinkled yellow seeds	*aB*
30 had wrinkled green seeds	*ab*
65 had round yellow and round green seeds	*ABb*
68 had wrinkled yellow and wrinkled green seeds	*aBb*
60 had round yellow and wrinkled yellow seeds	*AaB*
67 had round green and wrinkled green seeds	*Aab*
138 had round yellow, round green, wrinkled yellow and wrinkled green seeds	*AaBb*

Mendel observed that the different kinds of seeds could be considered to have been obtained approximately in the ratio of 1:1:1:1:2:2:2:2:4 as

$$AB + Ab + aB + ab + 2ABb + 2aBb + 2Aab + 2Aab + 4AaBb$$

which could be resolved as the multiplication product of

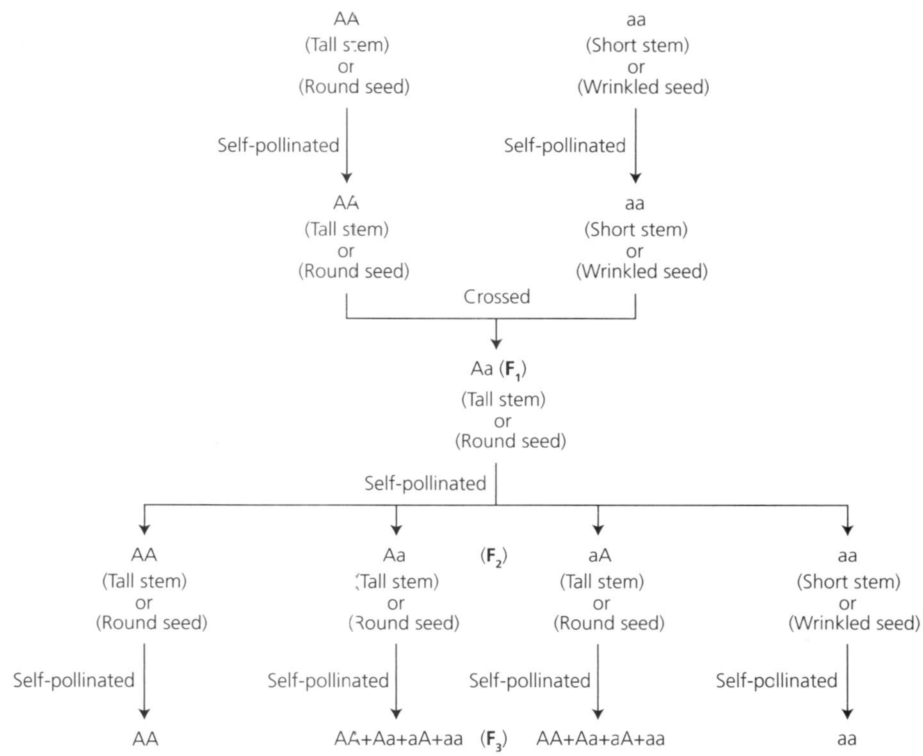

Figure 1.3 The concept of 'dominant' and 'recessive' characters and the explanation of the appearance of such characters in the progeny on selfing and crossing.

$$A + 2Aa + a \quad \text{and}$$
$$B + 2Bb + b$$

Mendel followed up by conducting crosses between plants that differed in three separate characters. Again, the result could be resolved as the multiplication product of

$$A + 2Aa + a \quad \text{and}$$
$$B + 2Bb + b \quad \text{and}$$
$$C + 2Cc + c$$

So, Mendel wrote, 'there is therefore no doubt that for the whole of the characters involved in the experiments, the principle applies that the offspring of the hybrids in which several essentially different characters are combined, exhibit the terms of a series of combinations, in which the developmental series for each pair of differentiating characters are united. It is demonstrated at the same time that the relation of each pair of different characters in hybrid union is independent of the other differences in the two original parental stocks.' This is what is known now as Mendel's law of independent assortment. Mendel's analytical approach and mathematical skills were indeed rare among the biologists of his times, which he possibly had acquired during his early years while studying physics and mathematics at the University of Vienna.

Mendel proposed that each character of the pea plants he studied was controlled by a *factor* (equivalent to the present-day *gene*), which could exist in more than one 'form' (equivalent to the present-day *allele*). Mendel's original paper was published in the *Proceedings of the Brunn Natural History Society* (1865). The paper has been translated into English and titled *Experiments in Plant Hybridization* by the Royal Horticultural Society, London. The English version has been reprinted in the collections of Peters (1959) and Davem (1981), among others.

1.3 Genesis of the term 'gene'

Charles Darwin in his book, *Provisional Hypothesis of Pangenesis*, while dealing with inheritance and the transmission of characters in latent form, wrote of minute granules which were dispersed throughout the whole system and when supplied with proper nutrients, would multiply by self-division. He suggested that these minute granules may be called *gemmules* (Darwin 1868).

Hugo de Vries in his book, *Intracellulare Pangenesis*, postulated self-replicating factors, *pangenes*, while formulating his theory of heredity (de Vries 1889). He borrowed the term from Darwin's *pangenesis*, but attributed to it an entirely different meaning. He wrote in the Introduction, 'if one considers the species characters in the light of the doctrine of descent, it then quickly appears that they are composed of separate more or less independent factors. Almost every one of these is found in numerous species, and their changing combinations and association with rarer factors determine the extraordinary variety of the world of organisms.' 'The existence of a multitude of kinds of invisible "*pangenes*" in all cells, not all alike as in previous conceptions, but representing separate hereditory predispositions and concentrated in the nucleus, from which representative particles entered the cytoplasm to influence reactions there; the properties of self-replication, accompanied by occasional errors like mutation which gave rise to the variety of "*pangenes*" – this sounds very much like a modern description of genes' (Dunn 1965).

The term *genetics* was coined next, before the term *gene* actually came into existence. The English zoologist William Bateson (1861–1926) was deeply involved in plant breeding work to unravel how discontinuous variations were inherited. It was he who first used the term *genetics* in a letter he wrote to his friend Professor Adam Sedgwick on 4 April 1905 on the establishment of a professorship he was aspiring for (Punnet 1928). 'If the Quick Fund were used for the foundation of a professorship relating to Heredity and Variation, the best title would, I think, be The Quick Professorship of "The study of Heredity". No single word in common use quite gives the meaning. Such a word is badly wanted and if it were desirable to coin one, "*Genetics*" might do.' In 1906, Bateson used the term *genetics* in a review he wrote of a book by JP Lotsy (Punnet 1928). Bateson said on 31 July 1906 in his inaugural address on the occasion of the Third Conference on Hybridization and Plant Breeding convened by the Royal Horticultural Society, 'I suggest for the consideration of this congress the term "*Genetics*", which sufficiently indicates that our labours are devoted to the elucidation of the phenomenon of heredity and variation: in other words to the physiology of descent' (Bateson 1907).

Wilhelm Ludwig Johannsen (1857–1927) was the person who finally coined the term *gene*. Johannsen (1909) first defined the concept of *phenotype* as measurable realities, just what can be observed, 'a sum total of the expressed characters which are determined by interplay between inherited predispositions [*Anlage*] and elements of the environment'. He then looked for a single word to replace 'inherited predispositions' and chose the last syllable of de Vries' *pangene* and fused it to *type*, deriving *genotype* as the counterpart of *phenotype*. He added that the '*genotype* consists of elements, "*genes*", which are, in theory, separable but in practice sometimes not separable. The "*gene*" is thus to be used as a kind of accounting or calculating unit [*Rechnungseinheit*]. The conception of the "*gene*" as an organoid, a little body with independent life and similar attributes, in the sense of Darwin's "*gemmules*", is no longer to be considered'. *Gene*, the hypothetical unit of information that was thought to govern inheritance, was still something abstract, but the term acquired a permanent status.

1.4 But how exactly does a 'gene' function?

Naudin's *essence*, Mendel's *factors*, de Vries' *pangenes* and Johannsen's *genes*, all undoubtedly refer to the same thing, but how do these influence heredity? Archibald Garrod, an English physician, was studying some diseases that appeared hereditary. One such condition, alkaptonuria, rendered brown colour to the urine, which on standing in air turned black. Garrod came across five families in which alkaptonuric conditions were inherited and found that the alkaptonuric offsprings were the result of marriages between first cousins. Garrod argued that 'the mating of first cousins gives exactly the conditions most likely to enable a rare and usually recessive character to show itself. If the bearer of such a gamete mate with individuals not bearing it, the character would hardly ever be seen; but first cousins will frequently be bearers of similar gametes, which may in such unions meet each other and thus lead to the manifestation of the peculiar recessive characters in the zygote' (Garrod 1899, 1901). The brown colour of the urine was because of the excretion of alkapton, which was identified to be 2,5-dihydroxyphenylacetic acid or homogentisic acid. Garrod fed the alkaptonurics with the amino acids phenylalanine and tyrosine and observed enhanced excretion of alkapton. He recognized that alkaptonuria must be the consequence of a biochemical defect and concluded that some enzyme responsible for a specific metabolic step was either absent or defective in alkaptonuric persons. He wrote, 'we may further conceive that the splitting of the benzene ring of homogentisic acid in normal metabolism is the work of a special enzyme, that in congenital alkaptonuria this enzyme is wanting' (Garrod 1923). Soon afterwards, it became known that the blood of alkaptonurics was deficient in the enzyme homogentisic acid oxidase. A mutated *determinant of heredity* thus results in a defective enzyme, the first insight into how a *gene* functions.

The pathway of conversion of phenylalanine to homogentisic acid is shown in Fig. 1.4.

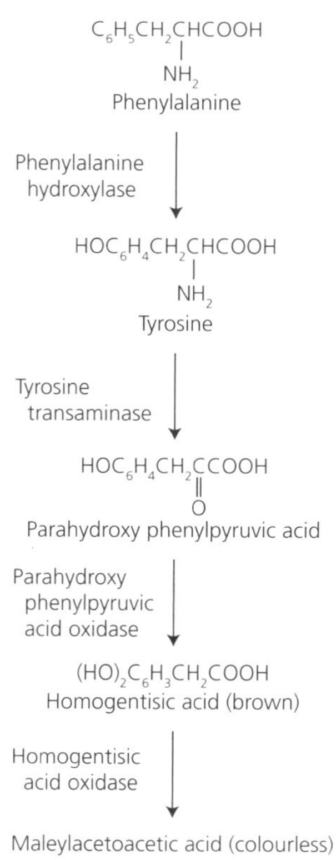

Figure 1.4 The pathway of conversion of phenylalanine to homogentisic acid (brown) and maleylacetoacetic acid (colourless).

1.5 One gene: One enzyme

That biochemical reactions were possibly controlled by specific genes was also suggested from the work on fermentation of sugars by yeast (Winge and Laustsen 1939) and synthesis of anthocyanin pigments in plants (Lawrence and Price 1940). Conclusive evidence in favour of this view came from the work on actinomycetes (bread mould), *Neurospora crassa* (Beadle and Tatum 1941). The approach the authors undertook was 'based on the assumption that X-ray treatment will induce mutations in genes concerned with the control of known specific reactions. If the organism must be able to carry out a certain chemical reaction to survive on a given medium, a mutant unable to do this will obviously be lethal on this medium. Such a mutant can be maintained and studied, however, if it will grow on a medium to which has been added the essential product of the genetically blocked reaction. The experimental procedure based on this reasoning can best be illustrated by considering a hypothetical example. Normal strains of *Neurospora crassa* are able to use sucrose as a carbon source, and are therefore able to carry out the specific and enzymatically controlled reaction involved in the hydrolysis of this sugar. Assuming this reaction to be genetically controlled, it should be possible to induce a gene to mutate to a condition such that the organism could no longer carry out sucrose hydrolysis. A strain carrying this mutant (gene) would then be unable to grow on a medium containing sucrose as a sole carbon source but should be able to grow on a medium containing some other normally utilizable carbon source. In other words, it should be possible to establish and maintain such a mutant strain on a medium containing glucose and detect its inability to utilize sucrose by transferring it to a sucrose medium'. 'Theoretically, any metabolic deficiency can be "bypassed" if the substance lacking can be supplied in the medium and can pass cell walls and protoplasmic membranes' (Beadle and Tatum 1941).

Beadle and Tatum made use of two kinds of media, a 'minimal' medium, that is, one requiring the organism to carry on all the essential syntheses of which it is capable, and a 'complete' medium, that is, one containing as many of the normally synthesized constituents of the organism as is practicable. The 'minimal' medium contained inorganic salts, biotin, a disaccharide and fat, while the 'complete' medium contained inorganic salts, malt extract, yeast extract and glucose. 'Any loss of ability to synthesize an essential substance present in the "complete" medium and absent in the "minimal" medium is indicated by a strain growing on the first and failing to grow on the second medium.' X-ray treatment yielded several mutants that grew normally on the 'complete' medium but failed to grow on the 'minimal' medium. One of these grew normally in the 'minimal' medium when supplemented with vitamin B_6 (pyridoxine), the second mutant grew normally when the 'minimal' medium was supplemented with para-amino benzoic acid and the third one grew normally on supplementation with vitamin B_1 (thiamine).

Crosses between the normal and the pyridoxine-less mutant strain revealed segregation (Fig. 1.5), showing that the defect was indeed inherited and also suggested that the 'inability to synthesize vitamin B_6 is transmitted as it should be if it were differentiated from normal by a single gene'. Beadle and Tatum surmised that by using this approach, 'it should be possible, by finding a number of mutants unable to carry out a particular step in a given synthesis, to determine whether only one gene is ordinarily concerned with the immediate regulation of a given specific chemical reaction'.

In their next paper, the authors (Tatum and Beadle 1942) narrated a similar work with the mutant that could not synthesize para-amino benzoic acid and found that this mutant also was 'differentiated from normal by a single gene'. In this paper, the authors mentioned their 'working hypothesis, a single gene may be considered to be concerned with the primary control of a single specific chemical reaction'. Subsequently, Beadle wrote in the September 1948 issue of the *Scientific American*, 'bread-mold (*Neurospora crassa*) studies have contributed strong support to the hypothesis that each gene controls a single protein' (Beadle 1948).

The formal enunciation of the hypothesis, 'one gene – one enzyme', which is generally attributed to the first *Neurospora* paper (Beadle and Tatum 1941), occurred much later (Beadle 1959). In Beadle's own words, said on the occasion of receiving the Nobel prize, 'it is sometimes thought that the *Neurospora* work was responsible

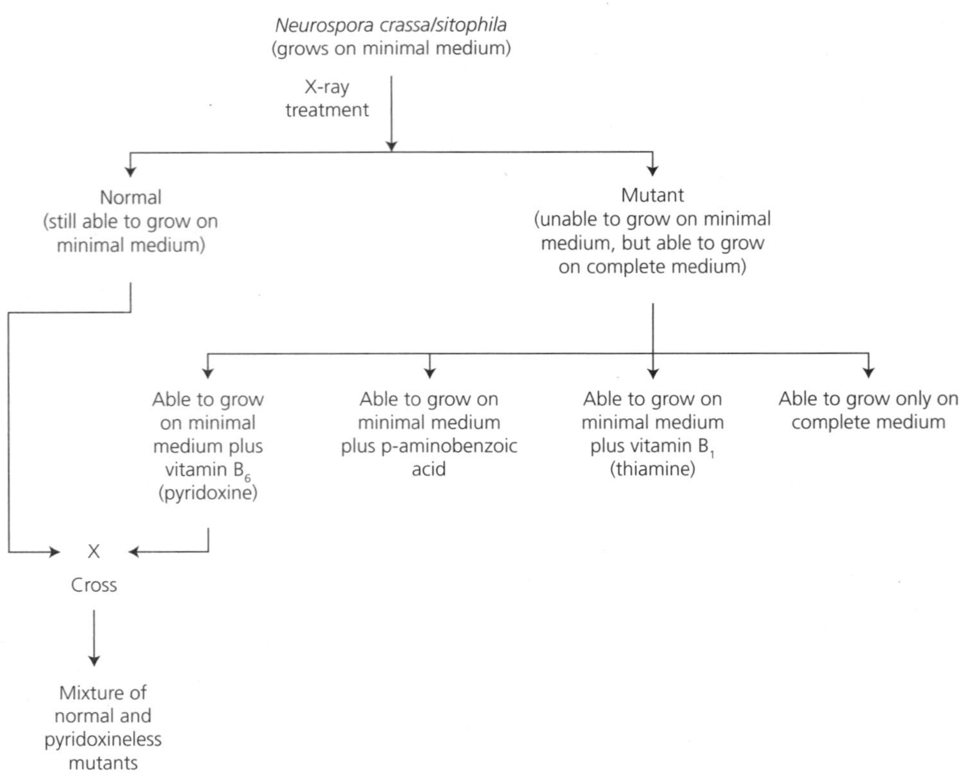

Figure 1.5 X-ray irradiation of *Neurospora crassa/sitophila* and the result of the cross between normal and mutant *Neurospora*. The normal *Neurospora* could grow on the minimal medium, while the mutants could only grow on the complete medium, but not on the minimal medium. Mutants were found that required for their growth vitamin B_6 (pyridoxine) or para-amino benzoic acid or vitamin B_1 (thiamine). There were mutants that could not grow even on addition of all these three nutrients. When pyridoxine-less mutants were crossed with normal *Neurospora*, a mixture of normal and pyridoxine-less mutants were obtained. Similar results were found when the other mutants were used in the crosses.

for the one-gene–one-enzyme hypothesis …'. 'The fact is that it was the other way around – the hypothesis was clearly responsible for the new approach.' 'Regardless of when it was first written down on paper, or in what form, I myself am convinced that the one-gene-one-enzyme concept was the product of gradual evolution, beginning with Garrod and continued to by many, including Moore, Goldschmidt, Troland, Haldane, Wright, Gruneberg, and many others,' 'Although we may not have stated it explicitly, Ephrussi and I had some such concept in mind (when we were earlier working with *Drosophilla* before we started working with *Neurospora*). A more specific form of the hypothesis was suggested by the fact that of all the 26 known eye-color mutants in *Drosophilla,* there was only one that blocked our postulated reactions and one that similarly interrupted the second. Thus, it seemed reasonable to assume that the *total* specificity of a particular enzyme might somehow be derived from a single gene. The finding in *Neurospora* that many nutritionally deficient mutant strains can be repaired by supplying single chemical compounds was a verification of our prediction and, as such, reinforced our belief in the hypothesis, at least in its more general form' (Beadle 1959).

As many enzymes are known today to be composed of multiple dissimilar polypeptide subunits and each subunit is coded for by a separate gene, the present-day version of the concept is 'one gene – one polypeptide'.

References

Bateson W. Facts limiting the theory of heredity (an inaugural address delivered at the Third Conference on Hybridization and Plant Breeding). *Science N.S.* 1907;26:649–660.

Beadle GW. The genes of men and molds. *Scientific American*; 1948. Reprinted in Davern CI, ed. *Genetics: Readings from Scientific American.* San Francisco, CA: WH Freeman & Company; 1981:28–37

Beadle GW. Genes and chemical reactions in *Neurospora. Science* 1959;129:1715–1719.

Beadle GW, Tatum EL. Genetic control of biochemical reactions in *Neurospora. Proc Natl Acad Sci USA* 1941;27:499–506.

Darwin C. *The Variation of Animals and Plants under Domestication.* New York: D Appleton & Company; 1868.

Davem CI, ed. *Genetics: Readings from Scientific American.* San Francisco, CA: WH Freeman and Company; 1981. pp. 8–17.

Dunn LC. *A Short History of Genetics.* New York: McGraw-Hill Book Company; 1965.

Garrod AE. A contribution to the study of alkaptonuria. *Med Chir Trans* 1899;82:369–394.

Garrod AE. About alkaptonuria. *Lancet* 1901;2:1484–1486.

Garrod AE. *Inborn Errors of Metabolism.* New York: Oxford University Press; 1923.

Glass HB. Review of Barthelmess' Verebungswissenschaft. *Quart Rev Biol* 1957;32:376.

Johannsen W. *Elemente der Exakten Erblichkeitslehre* [Elements of the Exact Heredity]. Jena, Germany: Gustav Fischer; 1909.

Lawrence WCJ, Price JR. The genetics and chemistry of flower colour variation. *Cambridge Philos Soc Biol Rev* 1940;15:35–58.

Mendel G. Versuche uber Pflanzenhybriden. *Verh Naturforsch Verein Brunn* 1865;4:3–47.

Peters JA. *Classic Papers in Genetics.* Englewood Cliffs, NJ: Prentice-Hall; 1959.

Punnett RC. *The Scientific Papers of William Bateson.* London: Cambridge University Press; 1928.

Roberts HF. *Plant Hybridization before Mendel.* Princeton, NJ: Princeton University Press; 1929.

Sageret A. Considerations sur la production des hybrids des variantes et des varietes en general, et sur celles des Cucurbitaces en particulier [Considerations on the production of hybrids variants and varieties in general, and in particular those of cucurbits]. *Annales des Sci Nat Prem Serie* 1826;8:294–314.

Tatum EL, Beadle GW. Genetic control of biochemical reactions in *Neurospora*: An 'aminobenzoicless' mutant. *Proc Natl Acad Sci USA* 1942;28:234–243.

de Vries H. *Intracellulare Pangenesis.* Translation in English by Gager CS [1910]. Chicago, IL: Open Court; 1889.

Winge O, Laustsen O. On 14 new yeast types, produced by hybridization. *Compt Rend Lab Carlsberg, Serie Physiol* 1939;22:337–352.

Zirkle C. *The Beginnings of Plant Hybridization.* Philadelphia, PA: University of Pennsylvania Press; 1935.

2

DNA as the genetic material

H. K. Das

2.1 Observations of Frederick Griffith with *Diplococcus pneumoniae*

By the beginning of the twentieth century, Archibald Garrod's study of inheritance of alkaptonuria had more or less convinced the geneticists that a mutated 'gene' resulted in a defective enzyme (Chapter 1). The actual nature of the 'gene' was, however, still shrouded in mystery. Thomas Hunt Morgan, who had mesmerized everybody by his study of inheritance of eye colour in the fruit fly, *Drosophila melanogaster*, said in his acceptance speech of the Nobel Prize, 'At the level at which the genetic experiments lie it does not make the slightest difference whether the gene is a hypothetical unit or whether the gene is a material particle' (Morgan 1910a, 1990b; 1911).

This was the status of the 'gene' in 1923 when a London medical officer, Frederick Griffith, was looking at variation in the different isolates of the bacteria, *Diplococcus pneumoniae*, which caused pneumonia. At that time, three serotypes of *D. pneumoniae* were known, all of these were encapsulated in a slimy coating of polysaccharide around the cell and all of these would cause the disease. The colonies produced by these virulent cells on blood agar plates appeared large, smooth and shiny (Fig. 2.1; colonies at the left). Griffith discovered that each serotype could also exist in a form that lacked the capsule or the shiny coating around the cells and produced small colonies that looked somewhat rough (Fig. 2.1; colonies at the right). The two forms were named 'S' for smooth and 'R' for rough. Griffith also discovered that the 'R' forms were incapable of causing the disease and hence were avirulent. Virulence was tested by injecting subcutaneously a small amount of the bacterial suspension into mice. The mice would contract pneumonia and die when injected with the suspension of the 'S' forms, but would remain healthy when injected with the suspension of the 'R' forms.

Next, Griffith performed the most important experiment. He heated the suspension of the 'S' form to kill all the bacteria, cooled it to room temperature and mixed it with a suspension of the live 'R' form bacteria, which was not heated. This mixture was injected into mice. The mice contracted pneumonia and died. The blood of the dead mice was full of live 'S' form bacteria, which

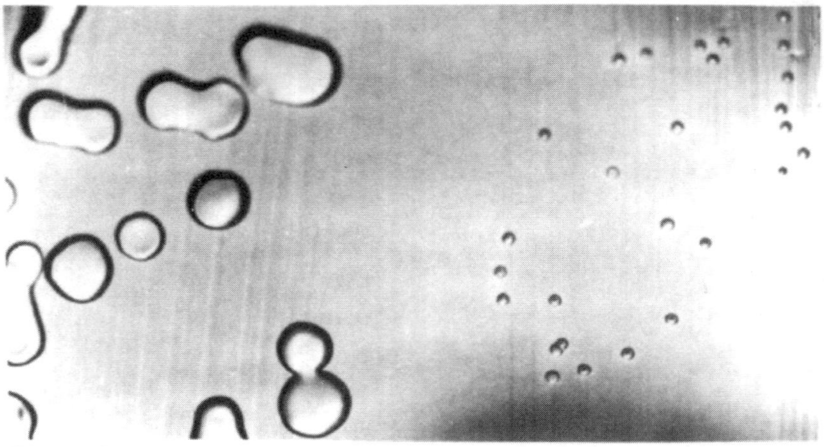

Figure 2.1 Photograph of colonies of R form and S form of *Diplococcus pneumonia*. [Reproduced from Avery, Macleod and McCarty, *J Exp Med* 1944;79:17 by permission of The Rockefeller University Press.]

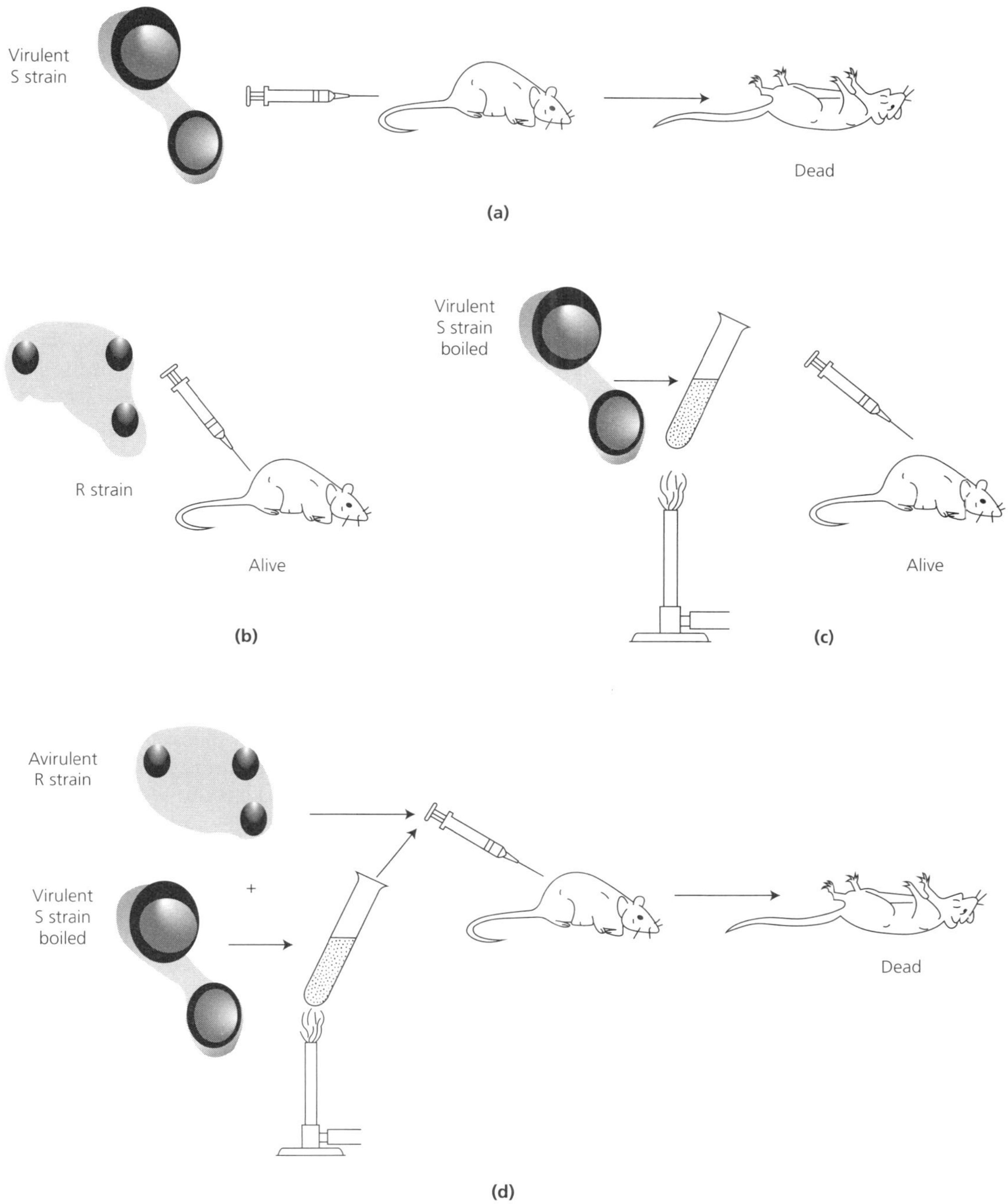

Figure 2.2 Effect of injection of *Diplococcus pneumoniae* into mice – R and S form, and heat-killed S form mixed with R form. The mouse dies when injected with the S form, but is unaffected if injected with the R form or the heat-killed S form. The mouse, however, dies when simultaneously injected with the R form and the heat-killed S form. [Adapted from *Molecular Genetics* by Stent and Calendar, 2nd Edition, by permission of Professor Gunther Stent.]

were highly virulent, as had been found out by injecting these again into mice. This happened even though no live 'S' form bacteria were injected. He performed all the crucial control experiments side by side. When just the heated suspension of the 'S' form bacteria was injected into mice, there was no sign of the disease; the mice were healthy. When the suspension of the 'R' form bacteria that was not heated and not mixed with anything was also injected, the mice were healthy, as expected (Fig. 2.2).

The inference (Griffith 1928) was that the avirulent 'R' form bacteria were 'transformed' into the virulent 'S' form by something coming from the dead 'S' form bacteria. The 'transformation' could not have been done by the polysaccharide from the coat of the dead 'S' form bacteria that were injected, as manifold larger number of live 'S' form bacteria were recovered from the blood of the dead mice. The 'transformed' bacteria had multiplied in the infected mice. The injected live 'R' form bacteria, which were incapable of synthesizing the polysaccharide coat, have now acquired this capability in the form of a new heritable character, which must have been endowed by the 'transforming principle.' The 'transforming principle' was, therefore, something tangible and very likely to be the genetic material, which thus could not be anything abstract or hypothetical.

Subsequently, 'transformation' of the 'R' form of *D. pneumoniae* with heat-killed 'S' form was shown to occur even *in vitro* in the test tube (Dawson and Sia 1931). Later, similar 'transformation' *in vitro* was demonstrated using sterile extracts of 'S' cells from which the cellular debris and other insoluble materials have been removed by Berkefeld filtration (Alloway 1932, 1933).

2.2 The 'transforming principle' of Griffith was nothing but DNA (deoxyribonucleic acid): A discovery that transformed biology

Neither Griffith nor the subsequent workers in the area tried to purify the 'transforming principle', and for almost two decades, the chemical nature of the genetic material remained unknown. Proteins, which were known to be composed of a large number of amino acids, were considered by many biochemists of that time to be a better candidate to serve as the genetic material, which controlled innumerable and diverse characters, than the 'monotonous' DNA or RNA, which were thought to be composed of only four nucleotides in equal proportions. This was the mindset when Oswald Avery (1877–1955) and coworkers published their findings on the purification of the compound that originated from the 'S' form of *D. pneumoniae* and 'transformed' the avirulent 'R' form into the virulent 'S' form (Avery *et al.* 1944).

The 'S' form cells of *D. pneumoniae* Type III were grown in a nutrient broth, sedimented by centrifugation, resuspended in a sodium chloride solution and heated to 65°C for 30 min to kill the cells and to inactivate the enzymes that destroy the transforming principle. Most of the capsular polysaccharides, proteins and ribonucleic acids were removed by washing the dead cells thrice with saline. The cells were then extracted with 0.5% sodium deoxycholate in saline to extract the water-soluble cell components. The insoluble materials were removed by centrifugation and 3–4 volumes of ethanol were added to the supernatant solution. This resulted in the precipitation of the macromolecules, which were collected by centrifugation and dissolved in a sodium chloride solution. The proteins in the solution were removed by extraction with chloroform, and the remaining macromolecules were purified by reprecipitating with ethanol and redissolving in the sodium chloride solution. The residual capsular polysaccharides were removed by digestion with a suitable enzyme preparation. The enzyme in turn was removed by extraction with chloroform. Ethanol was now added to the aqueous solution dropwise till a fibrous substance appeared. 'The precipitate forms a fibrous mass which floats to the surface of the alcohol and can be removed directly by lifting it out with a spatula.' The precipitate was redissolved in saline and 'the solution obtained is usually viscous, opalescent, and somewhat cloudy.' The 'transforming principle' was then further purified by repeated precipitation with alcohol (Fig. 2.3).

As low as 0.003 μg of this material was capable of bringing about 'transformation' *in vitro*.

To say the least, Avery and coworkers' experiments were meticulously thorough and their contribution was, indeed, a classic. The 'transformation' was done *in vitro*. Their assay of 'transformation' was by looking at colony characteristics and morphology, conversion of cells that were forming small, rough colonies (R form) into cells that would form large and glistening mucoid colonies of the S form on blood agar plates (see Fig. 2.1).

The first thing they ensured was that the 'R' form of *D. pneumoniae* they had chosen was not only 'transformable' but was also 'stable' and did not revert to the 'S' form by itself. They used the strain R36A and in their language, 'Strain R36A has become relatively fixed in the R phase and has never spontaneously reverted to the Type II S form. Moreover, repeated attempts to cause it to revert ... have in all instances been unsuccessful.'

To start with, the *in vitro* 'transformation' process was quite irreproducible, possibly because the 'transforming principle' was getting degraded. Hence, they standardized a reproducible 'reaction system' that contained beef heart infusion broth, swine serum and human pleural or ascitic fluid, all preheated at 60–65°C for 30 min. (Fig. 2.4). Serum and the pleural or ascitic fluid were found essential, but these seemed to contain an enzyme that degraded the 'transforming principle.' Heating at 60–65°C destroyed the enzyme.

The purified 'transforming principle' was chemically characterized. 'The purified material gives negative biuret and Millon tests [for proteins]. These tests have been done directly on dry material with negative results. The Dische diphenylamine reaction for desoxyribonucleic acid is strongly positive. The orcinol test (Bial) for ribonucleic acid is weakly positive. However, it has been found that in similar concentrations pure preparations of desoxyribonucleic acid of animal origin prepared by different methods give a Bial reaction of corresponding intensity' (Avery *et al.* 1944). Elemental analysis revealed an average nitrogen:phosphorus ratio of 1.67, 'which is in close agreement with that calculated on the basis of

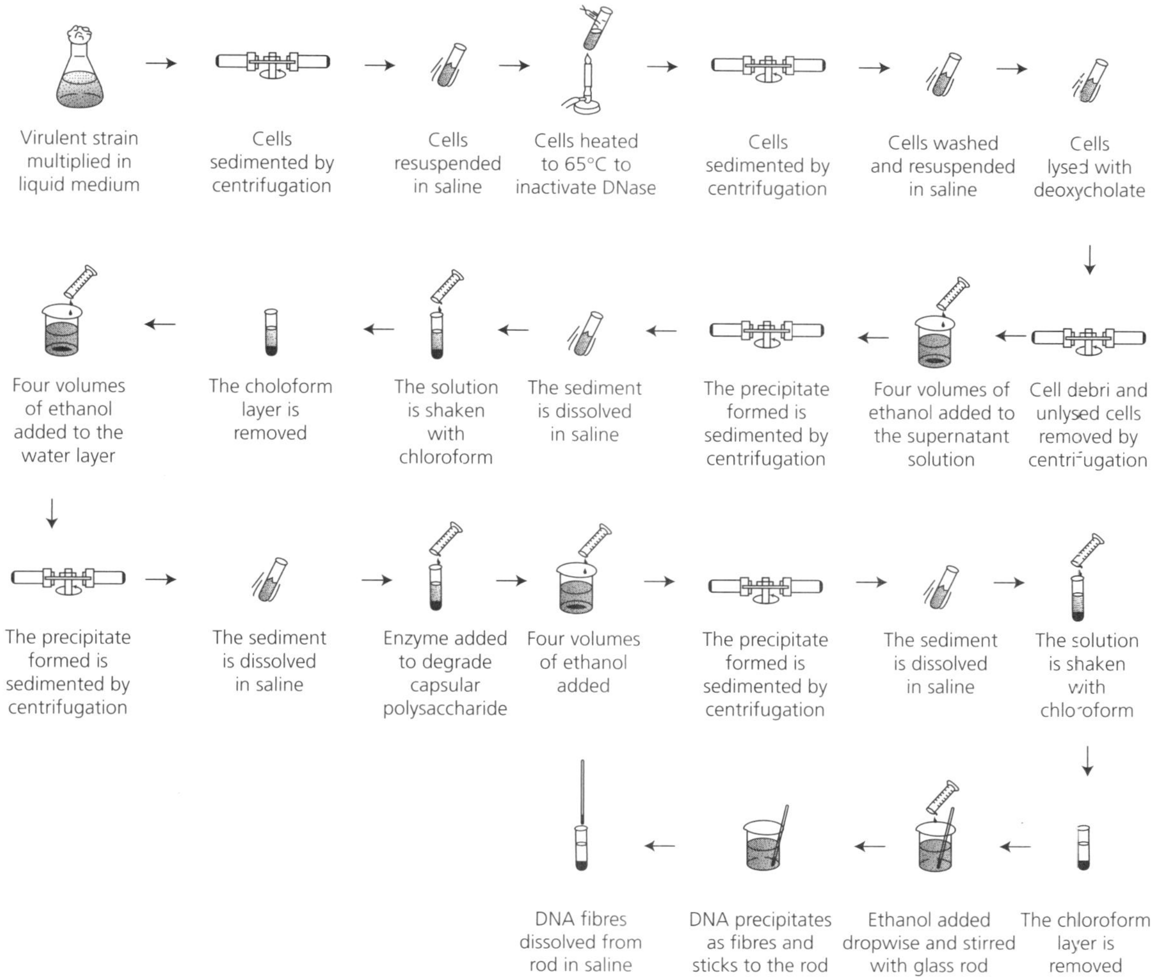

Figure 2.3 Method of isolation of transforming agent. The S form of *Diplococcus pneumoniae* is grown in a nutrient broth, sedimented by centrifugation, resuspended in a sodium chloride solution and heated to 65°C to inactivate the enzymes that destroy the transforming principle. The bacterial cells are then treated with a sodium deoxycholate solution to extract the water-soluble cell components. The insoluble materials are removed by centrifugation and 3–4 volumes of ethanol are added to the supernatant solution. This results in the precipitation of the macromolecules, which are collected by centrifugation and dissolved in a sodium chloride solution. The proteins in the solution are removed by extraction with chloroform, and the remaining macromolecules are purified by reprecipitating with ethanol and redissolving in the sodium chloride solution. The capsular polysaccharides are removed by digestion with a suitable enzyme preparation. The enzyme in turn is removed by extraction with chloroform. Ethanol is now added to the aqueous solution dropwise till a fibrous substance appears, which is collected around a glass rod. This substance works as the transforming principle. [Adapted from *The Mechanics of Inheritance* by FW Stahl by permission of Professor Franklin W Stahl.]

Figure 2.4 *In vitro* transformation of R form into S form by transforming principle. The transforming principle isolated from the S form of *Diplococcus pneumoniae* can, in the presence of serum factors, transform *in vitro* the avirulent R form into the virulent S form of *D. pneumoniae*. [Adapted from *Molecular Genetics* by Stent and Calendar, 2nd Edition, by permission of Professor Gunther Stent.]

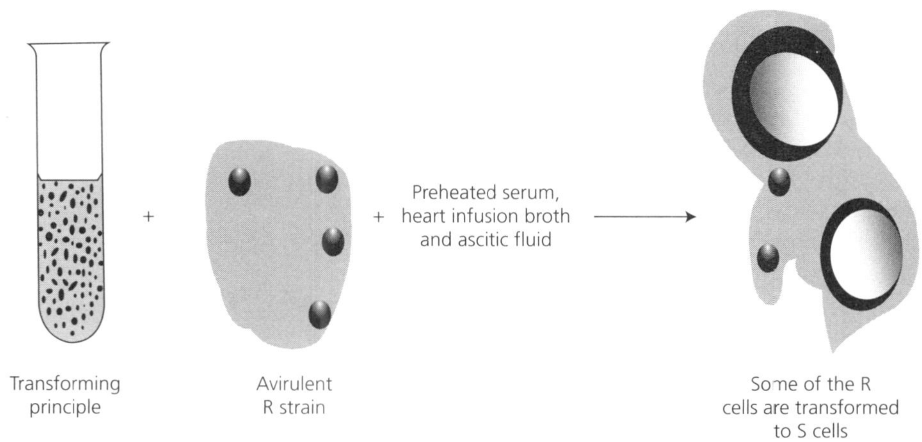

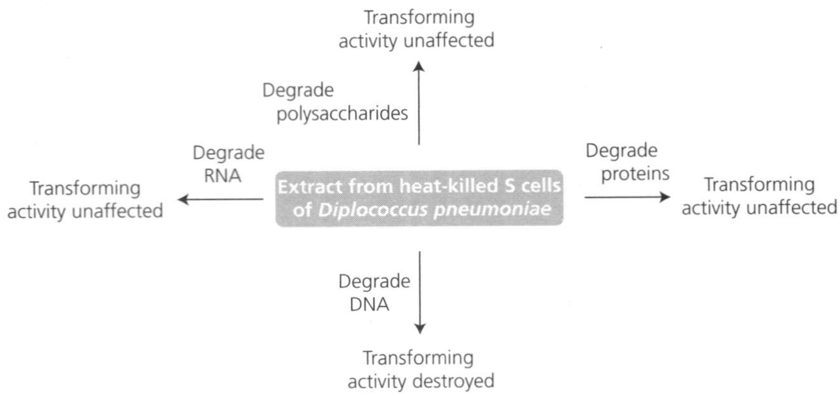

Figure 2.5 Identification of the transforming principle as DNA. The ability of the transforming principle to transform remains unaffected on treatment with the enzyme preparation that degrades polysaccharides, or on treatment with ribonuclease that degrades RNA (ribonucleic acid) or on treatment with trypsin and chymotrypsin that degrade proteins. The ability to transform, however, is severely affected if the transforming principle is treated with deoxyribonuclease that degrades DNA (deoxyribonucleic acid).

the theoretical structure of sodium desoxyribonucleate (tetranucleotide).' 'On the basis of the nitrogen – phosphorus ratio, it would appear that little protein or other substances containing nitrogen or phosphorus are present as impurities since if they were this ratio would be considerably altered.'

Enzymatic analysis of the purified 'transforming principle' confirmed that it was indeed desoxyribonucleic acid. The activity of the 'principle' was unaffected on treatment with crystalline trypsin or chymotrypsin or combinations of both or with crystalline ribonuclease. Purified desoxyribonuclease was not available, but all 'preparations shown to contain an enzyme capable of depolymerizing authentic samples of desoxyribonucleic acid were found to inactivate the "transforming principle".' (Fig. 2.5).

Avery and coworkers (Avery *et al.* 1944) concluded, 'The evidence presented supports the belief that a nucleic acid of the desoxyribose type is the fundamental unit of the transforming principle of *Pneumococcus* Type III.' This finding had indeed transformed biology; molecular biology was born.

2.3 The genetic material of bacteriophage T2 is also desoxyribonucleic acid (DNA)

Viruses that infect and multiply inside bacteria are commonly known as bacteriophages. The bacteriophage T2 infects the bacterium *Escherichia coli,* causing its lysis and consequent release of a large number of progeny bacteriophages. T2 comprises a protein coat and an enclosed desoxyribonucleic acid (DNA) core that could be released by osmotic shock (Herriott 1951). Which of these two components carries the genetic information of the bacteriophage? Electron microscopic studies revealed that the bacteriophage T2 attaches to *E. coli* by its tail and this attachment is preserved during the progress of the infection. (Anderson 1951). Fig. 2.6 shows the electron micrograph of bacteriophage T2 (Brenner *et al* 1959), and Fig. 2.7 depicts a schematic diagram of the process of injection of its genetic material into the host *E. coli*. The whole virus does not enter the bacterium, but only the genetic material is injected inside.

To answer in an unambiguous manner, the question as to whether the DNA is injected or some protein is injected, Hershey and Chase (1952) prepared bacteriophages that had the protein labelled with ^{35}S or the DNA labelled with ^{32}P [proteins are composed of amino acids, of which cysteine and methionine contain sulphur, while the nucleotides in DNA are connected by phosphodiester linkages; the DNA nucleotides do not contain sulphur and the protein amino acids do not contain phosphorus. *E. coli* cells were grown in glycerol-lactate medium, and at least 4 h before infecting with the virus, ^{32}P-labelled orthophosphate

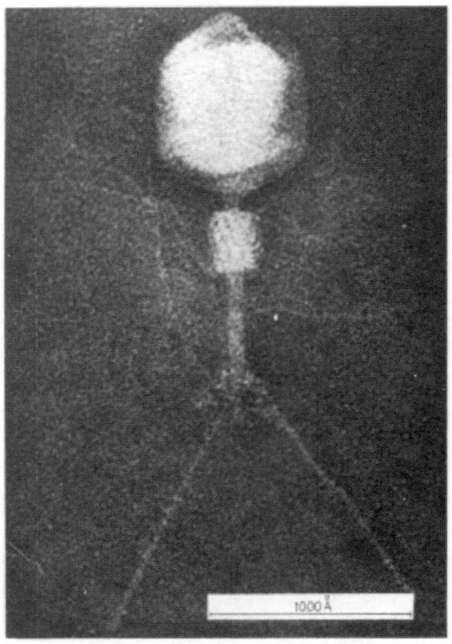

Figure 2.6 Electron micrograph of the bacterial virus T2. [Reproduced from Brenner *et al. J Mol Biol* 1959;1:10 by permission of the Academic Press, London.]

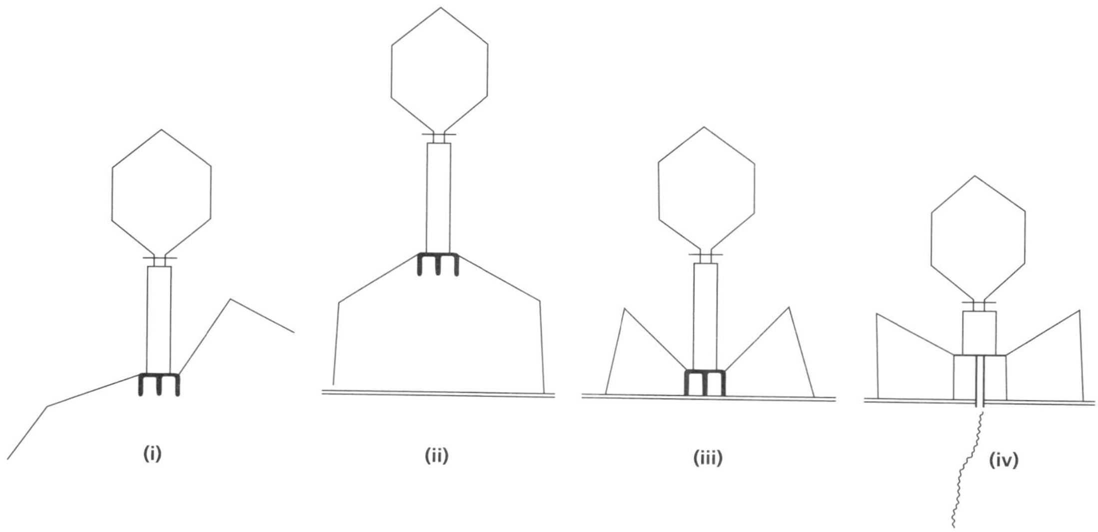

Figure 2.7 Schematic diagram showing injection of genetic material by the bacterial virus T2 into *Escherichia coli*. (i) The yet unattached virus, (ii) the virus attaches to the bacterial cell surface by its tail fibres, (iii) the virus is ready to inject its genetic material into *E. coli* and (iv) the genetic material of the virus is injected into the bacterial cell.

or ^{35}S-labelled sulphate was added. The aerobic incubation was continued overnight to allow the bacteriophages to get adsorbed on to the bacterial surface, inject the genetic material, get multiplied inside the bacteria, form the virus particles and cause lysis of the bacterial cells. The cells and cellular debris were removed by low-speed centrifugation and the bacteriophage particles were recovered by high-speed centrifugation. The bacteriophage particles were purified by repeating the differential centrifugation steps (Fig. 2.8). Radioactivity in the bacteriophage particles was determined in an end-window Geiger counter after drying the samples.

Broth-grown bacteria were mixed with either ^{35}S-labelled or ^{32}P-labelled bacteriophage and time was allowed for the infection to proceed. The bacteria were then sedimented by centrifugation at low speed, leaving the excess bacteriophage particles that were unadsorped on bacterial surface in the supernatant fraction. The sediment was washed, the bacteria were resuspended in an aqueous solution and the radioactivity in the bacteria, which was due to adsorption of the labelled bacteriophage, was determined. This aqueous suspension was next subjected to intense agitation in a Waring blender to dislodge the bacteriophage particles that were adsorped on to the bacterial surface. Samples were withdrawn periodically, centrifuged at low speed and the release of radioactivity into the supernatant fraction was determined. The intactness of the bacteria and the viability of the progeny bacteriophage inside the bacteria in each sample after the agitation were simultaneously monitored by its ability to release bacteriophages

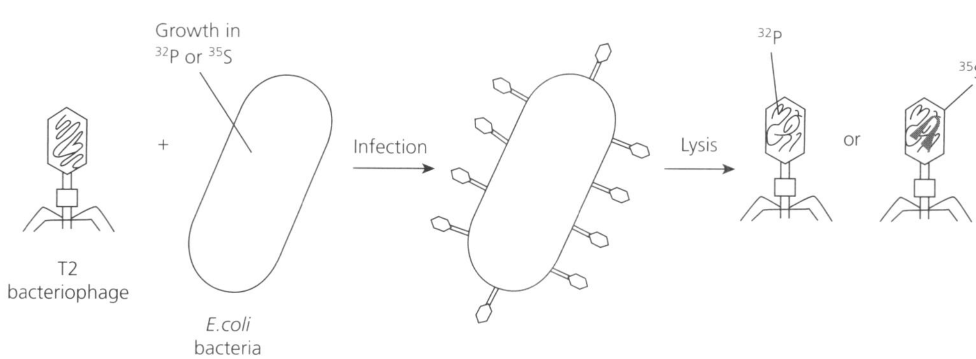

Figure 2.8 Preparation of ^{32}P- or ^{35}S-labelled T2 virus. *Escherichia coli* are grown either in the presence of ^{32}P-orthophosphate or ^{35}S-sulphate. The bacteria are then infected with the T2 virus, which lyse the bacteria after multiplication. The virus thus prepared acquires the ^{32}P or the ^{35}S label. [Adapted from *Genetics* by MW Strickberger, p. 66, Fig. 4-15, The Macmillan Company, New York, 1968, by permission of Professor Monroe W. Strickberger.]

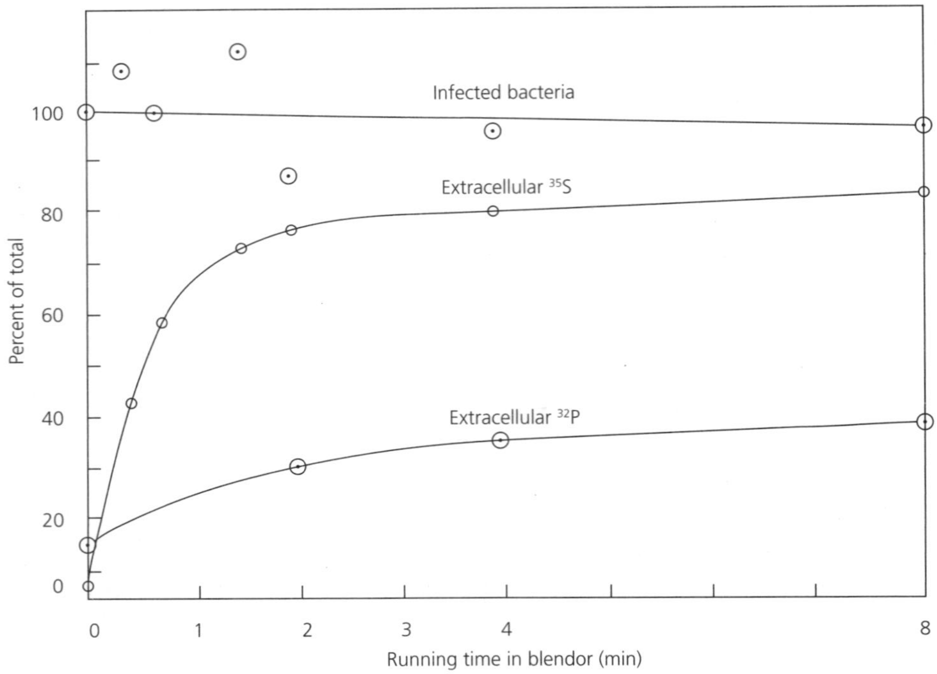

Figure 2.9 Result of the Waring blender agitation of the virus-infected *Escherichia coli* cells – removal of the inactive virus or the viral ghosts still attached on to the bacterial surface. This operation results in the removal of about 80% of the ^{35}S label, but only 30% of the ^{32}P label from the bacteria. [Reproduced from Hershey and Chase, *J Gen Physiol* 1952;32:45 by permission of The Rockefeller University Press.]

on further incubation (using an antiphage serum). The results are shown in Fig. 2.9.

Hershey and Chase (1952) summarized the results as follows:

- '75 to 80 per cent of the phage sulfur can be stripped from the infected cells by violent agitation of the suspension....
- The release of sulfur is accompanied by the release of only 21 to 35 per cent of the phage phosphorus, half of which is given up without any mechanical agitation.
- The treatment does not cause any appreciable inactivation of intracellular phage.
- These facts show that the bulk of the phage sulfur (contained in the protein) remains at the cell surface during infection, and takes no part in the multiplication of intracellular phage. The bulk of the phage DNA (containing the phosphorus), on the other hand, enters the cell soon after adsorption of phage to bacteria.'

What is the fate of the injected ^{32}P- labelled DNA? The infected cells were allowed to undergo lysis and the phage particles were purified as described earlier. As much as 30% or more of the ^{32}P present in the infecting phages could be recovered in the progeny phages. The actual figure would be much higher, because in the procedure followed, a large fraction of the progeny bacteriophages got adsorped to the cell debris and was lost.

When a similar experiment was carried out with ^{35}S-labelled bacteriophages, less than 1% of the parental radioactivity could be recovered in the progeny bacteriophages (Fig. 2.10).

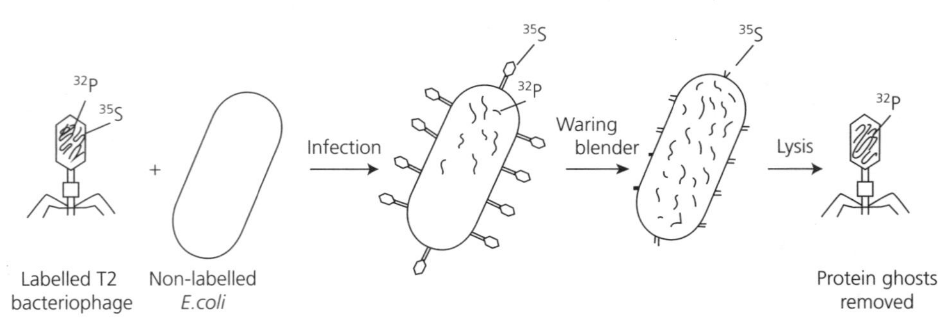

Figure 2.10 Recovery of ^{32}P, but no ^{35}S in the new progeny of the bacterial virus T2. After infection of *Escherichia coli* by the ^{32}P-labelled or ^{35}S-labelled T2 virus, the genetic material of the virus is injected inside the bacteria, while the viral coat is left on the surface. The genetic material of the virus is replicated inside the bacteria and is incorporated into progeny virions. As the bacteria are lysed, the virions are released. These progeny virions have been seen to contain the ^{32}P label, but not the ^{35}S label. [Adapted from Genetics by MW Strickberger, p. 66, Fig. 4-15, The Macmillan Company, New York, 1968, by permission of Professor Monroe W. Strickbeger.]

Hershey and Chase (1952) had stated in a very modest and conservative fashion, 'This protein [that is labelled with ^{35}S] probably has no function in the growth of intracellular phage. The DNA has some function.' The results of their experiments have, nevertheless, been accepted by the scientific community as definite evidence that DNA is the genetic material of T2 bacteriophage.

2.4 Radioisotopes

Chemical properties of an element depend primarily on the number of electrons orbiting in its shells, which is determined by the number of protons present in the nucleus of its atoms. The atomic number is a measure of this. The atomic weight of an element is contributed by the number of both protons and neutrons in its nucleus. Thus, as far as the chemical properties are concerned, it is the atomic number that is important and not the atomic weight of the element. Hence, if there is an extra neutron in the nucleus of an atom, the atomic weight increases, but not the atomic number, and the chemical properties remain practically unchanged. Natural phosphorus has an atomic weight of 31 Da; an addition of a neutron to its nucleus would raise the atomic weight to 32, but the number of protons in the nucleus and hence the atomic number and would remain unchanged at 15. Similarly, if three neutrons are added to the nucleus of the element sulphur, its atomic weight would go up from 32 to 35, but its atomic number would remain unchanged at 16. Atoms of an element (characterized by its atomic number) that have different atomic weights are termed as isotopes of the element. Quite often, the nuclei of the different isotopes of an element are stable; examples are ^{32}S, ^{33}S and ^{34}S, the predominant isotope present in nature being ^{32}S. Sometimes, however, the nucleus becomes unstable and disintegrates, emitting radiation; ^{35}S is an example. Such an isotope is called a radioisotope. ^{32}P is also a radioisotope, ^{31}P being the naturally occurring non-radioactive isotope. Both ^{35}S and ^{32}P emit negatively charged β-particles. Two of the other β-particle–emitting radioisotopes used in molecular biology experiments are ^{3}H and ^{14}C.

What happens to an atomic nucleus after emission of a β-particle? The β-particle is emitted as a consequence of disintegration of a neutron into a proton. This increases the positive charge of the atomic nucleus and hence the atomic number, thus creating an element different from the original one.

$$^{32}P(15) \rightarrow {}^{32}S(16) + \beta\text{-particle}$$
$$^{35}S(16) \rightarrow {}^{35}Cl(17) + \beta\text{-particle}$$
$$^{14}C(6) \rightarrow {}^{14}N(7) + \beta\text{-particle}$$

The figures in parentheses refer to the atomic number.

The energy content of β-particles varies with its source. The rate of disintegration of neutrons also varies considerably with the isotope involved. In addition, the rate of disintegration is always exponential. At any point of time, the rate of disintegration is proportional to the amount of radioactivity present at that point of time. Thus, if r is the radioactivity at any time t, then

$$dr/dt = \lambda r$$

where λ is a constant characteristic of the radioactive element. On integration, one gets

$$r = r_0 e^{-\lambda t}$$

where r_0 is the radioactivity at zero time. As the disintegration of neutrons of the β-ray–emitting radioactive element (or the decay of radioactivity) is exponential, the time taken for complete decay would be infinity. What would be a constant characteristic of the radioactive element is the time taken for the radioactivity to be reduced to half. The laboratory jargon for this characteristic is the 'half-life.' The characteristics of the commonly used β-particle–emitting radioisotopes used in molecular biology are given in Table 2.1.

It must be mentioned here that all radioactive elements do not emit β-particles; there are elements that emit α-particles (two neutrons and two protons) or γ-rays. These are more hazardous to work with.

β-Radioactivity is determined in liquid scintillation spectrometers. The liquid medium into which the radioactive compound is put contains chemicals that fluoresce on being hit with β-particles. The fluorescent emissions are captured in photo cells and quantified in the instrument after amplification.

Table 2.1 Characteristics of β-particle–emitting radioisotopes

Radioisotope	Energy of emission	Half-life
^{3}H	0.08 MeV*	12.3 years
^{14}C	0.115 MeV	5568 years
^{35}S	0.167 MeV	87 days
^{32}P	1.71 MeV	14.2 days

*eV is the unit of energy equal to the work done on an electron in accelerating it through a potential difference of 1 volt; 1 MeV = 96 × 109 J.

References

Alloway JL. The transformation *in vitro* of R *pneumococci* into S forms of different specific types by the use of filtered *pneumococcus* extracts. *J Exp Med* 1932;55:91–99.

Alloway JL. Further observations of the use of *pneumococcus* extracts in effecting transformation of type *in vitro*. *J Exp Med* 1933;57:265–278.

Anderson TF. Techniques for the preservation of three-dimensional structure in preparing specimens for the electron microscope. *Trans New York Acad Sci* 1951;13:130–134.

Avery OT, Macleod CM, McCarty M. Studies on the chemical nature of the substance inducing transformation of *pneumococcal* types. Induction of transformation by a deoxyribonucleic acid fraction isolated from *pneumococcus* type III. *J Exp Med* 1944;79:137–158.

Dawson MH, Sia RHP. *In vitro* transformation of *Pneumococcal* types. I. A technique of inducing transformation of *Pneumococcal* types *in vitro*. *J Exp Med* 1931;54:681–699.

Griffith F. The significance of *pneumococcal* types. *J Hyg* 1928;27:113–159.

Herriott RM. Nucleic acid-free T2 virus 'ghosts' with specific biological action. *J Bacteriol* 1951;61:752–754.

Hershey AD, Chase M. Independent functions of viral protein and nucleic acid in growth of bacteriophage. *J Gen Physiol* 1952;36:39–56.

Morgan TH. Sex-limited inheritance in *Drosophila*. *Science* 1910a; 32:120–122.

Morgan TH. The method of inheritance of two sex-limited characters in the same animal. *Proc Soc Exp Biol Med* 1910b;8:17–19.

Morgan TH. An attempt to analyze the constitution of the chromosomes on the basis of sex limited inheritance in *Drosophila*. *J Exp Zool* 1911;11:365–413.

3

Constituents of nucleic acids

H. K. Das

3.1 History of the term 'nucleic acid'

Much before Avery and coworkers discovered that the 'transforming principle' was a 'nucleic acid of the desoxyribose type' (Chapter 2), research on nucleic acids was initiated by Friedrich Miescher in Tubingen. In 1868, while trying to find out the constituents of human pus cells, he 'removed' the fats by ethanol extraction and the proteins by incubation with a crude preparation of pepsin. The grey precipitate thus obtained contained cell nuclei, which was treated with alkali and then with acid, yielding a material rich in phosphorus. This material, which he termed 'nuclein', was later found in cells of human blood, kidney, liver and testis and also in yeast and salmon sperm. Further analysis revealed that 'nuclein' was strongly acidic and also contained some nitrogenous base. In 1889, Richard Altman found that 'nuclein' was, in fact, a complex of proteins and a phosphorus-rich acidic compound. He could remove the proteins from the complex and named the phosphorus-rich acidic compound as 'nucleic acid'.

3.2 Chemical nature of 'nucleic acid'

As back as 1847, Liebig had isolated a phosphorus-containing acidic compound from beef muscle, which he had termed inosinic acid. Levene and Jacobs found in 1909 that inosinic acid was, in fact, a 5′-phosphoriboside of hypoxanthine. They also found that guanylic acid, which was isolated from beef pancreas, was, indeed, 5′-phosphoriboside of guanine (Fig. 3.1). Similar compounds were subsequently isolated from yeast and were identified as phosphoribosides of adenine, guanine, cytosine and uracil (Fig. 3.2). These compounds were given the general name of 'nucleotides'. The phosphate moiety of these compounds could be removed by alkali, releasing the ribosides of the purine or pyrimidine bases, which were named 'nucleosides' (Fig. 3.3). The structure of the 'nucleotides' was confirmed by total synthesis by Alexander Todd and his coworkers (Hayes *et al.* 1955). The pentose was linked to the purine or pyrimidine base through its 1′-carbon, as shown in Fig. 3.1. The phosphate group in the 'nucleotides' was linked to the 5′-carbon of the pentose, as also shown in Fig. 3.1.

3.3 Existence of two different classes of 'nucleic acids'

'Nucleic acid' isolated from yeast on treatment with aqueous ammonium hydroxide got hydrolysed and yielded the four 'nucleosides', adenosine, guanosine, cytidine and uridine. However, 'nucleic acid' isolated from calf thymus could *not* be hydrolysed by aqueous ammonium hydroxide and, hence, was qualitatively different from the 'nucleic acid' isolated from yeast. In 1929, Levene found that the 'nucleic acid' isolated from calf thymus could be hydrolysed by some enzyme

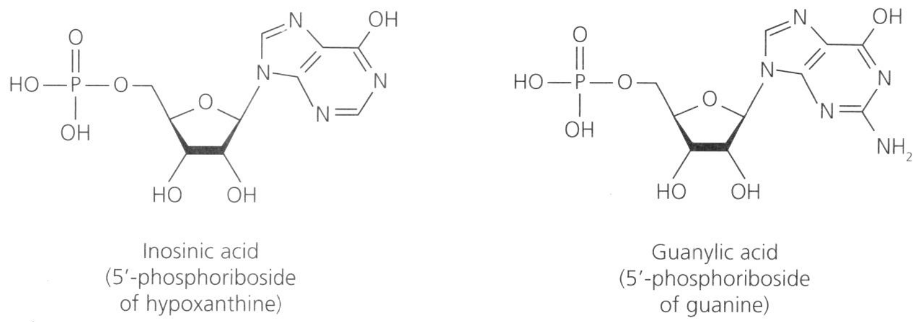

Figure 3.1 Chemical structures of inosinic acid and guanylic acid: the phosphoribosides of hypoxanthine and guanine.

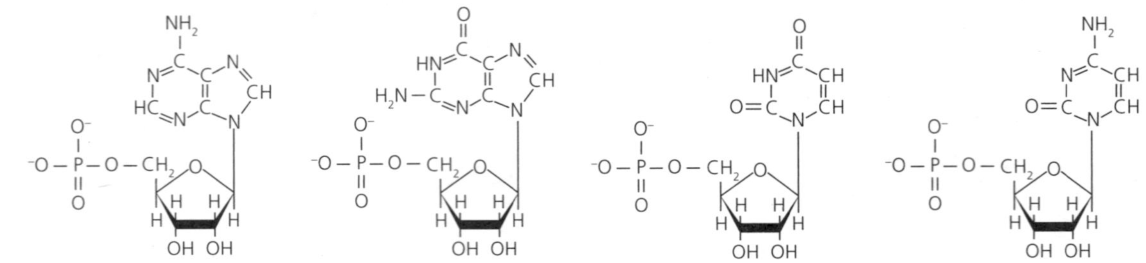

Adenylic acid (5′-phosphoriboside of adenine) | Guanylic acid (5′-phosphoriboside of guanine) | Uridylic acid (5′-phosphoriboside of uracil) | Cytidylic acid (5′-phosphoriboside of cytosine)

Figure 3.2 Chemical structures of the ribonucleotides: the phosphoribosides of adenine, guanine, uracil and cytosine (the monophosphates of the ribonucleosides adenosine, guanosine, uridine and cytidine).

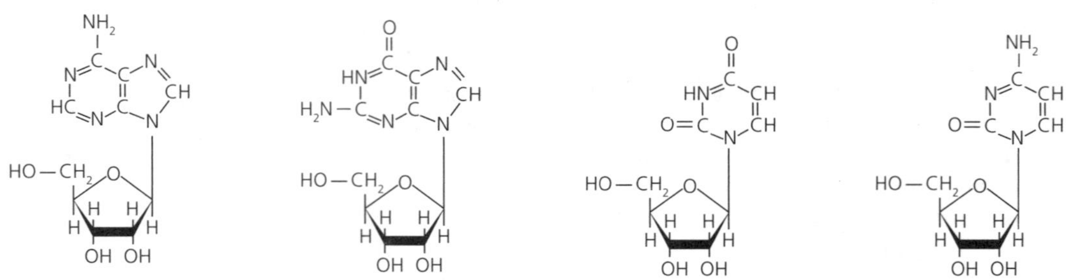

Adenosine (riboside of adenine) | Guanosine (riboside of guanine) | Uridine (riboside of uracil) | Cytidine (riboside of cytosine)

Figure 3.3 Chemical structures of the ribnucleosides: adenosine, guanosine, uridine and cytidine.

preparations, yielding 'nucleotides', which on acid hydrolysis generated 'nucleosides'. It was these 'nucleosides' that were different from the 'nucleosides' isolated from the nucleic acid of yeast. Finally, it turned out that the pentose sugar present in the nucleosides from the calf thymus 'nucleic acid' was not ribose, but 2-deoxyribose (Fig. 3.4). The deoxyribose nucleosides were deoxyadenosine, deoxyguanosine, thymidine and deoxycytidine (Fig. 3.5). Thus, basically two different kinds of 'nucleic acids' existed in nature, ribose nucleic acid (RNA) and deoxyribose nucleic acid (DNA).

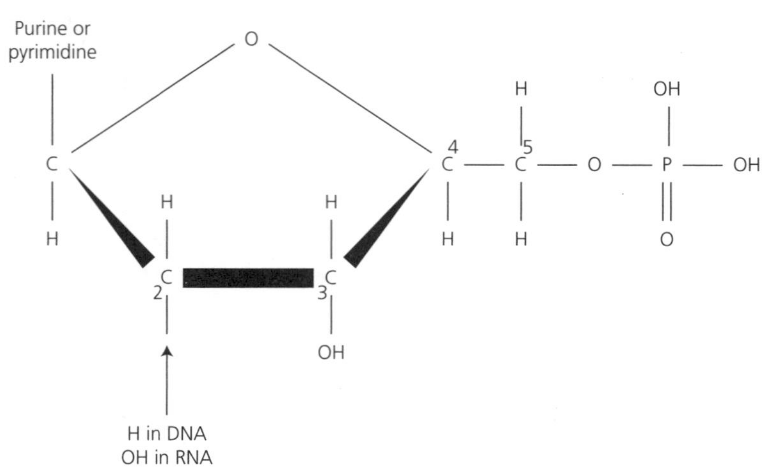

Figure 3.4 The hydroxyl on the 2′-carbon of ribose in RNA is replaced by hydrogen in the pentose (deoxyribose) in DNA.

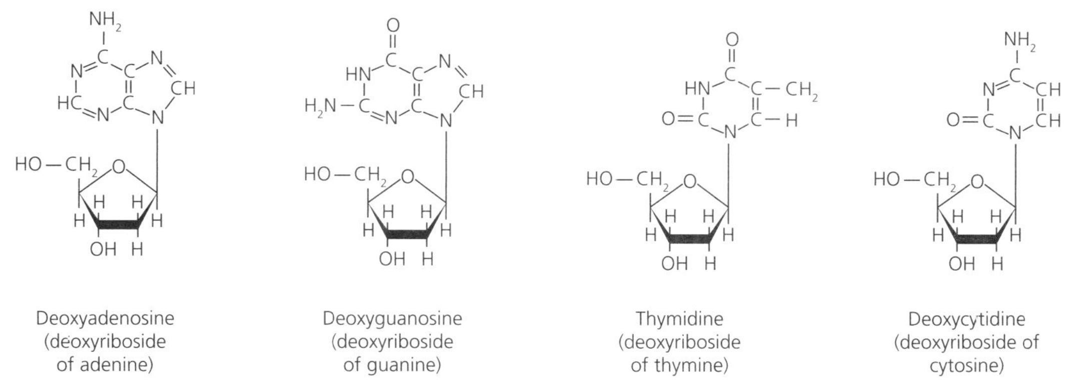

Figure 3.5 Chemical structures of the deoxyribonucleosides: deoxyadenosine, deoxyguanosine, thymidine and deoxycytidine.

3.4 The 'tetranucleotide hypothesis' of the structure of 'nucleic acids'

Till 1940, the erroneous finding that in both kinds of 'nucleic acids' the relative proportion of the four different nucleotides was equal, had prevailed and this had led to the 'tetranucleotide hypothesis' about both kinds of 'nucleic acid'. As no pyrophosphate or phosphomonoester could be detected, Takahashi in 1932 proposed a general cyclic structure for both kinds of 'nucleic acid' (Fig. 3.6). It was this invariant structure that had gone against the candidature of 'nucleic acid' to serve as the genetic material. It was, however, revealed soon that this small cyclic structure could not be correct for the simple reason that the experimentally determined molecular weight of isolated 'nucleic acid' from calf thymus was much higher than that provided by the 'tetranucleotide' model. Flow birefringence studies by Rudolf Signer and ultracentrifugation studies by Levene suggested that molecular weight of deoxyribose nucleic acid could be as high as 1 million daltons. The idea slowly gained ground that 'nucleic acids' could possibly comprise long, linear polynucleotides.

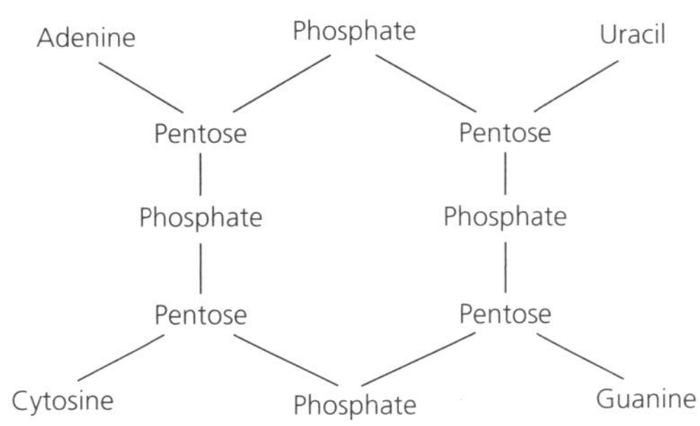

Figure 3.6 Tetranucleotide structure of nucleic acid postulated by Takahashi.

3.5 DNA is a linear macromolecule with a backbone of repeating deoxyribose units linked by 5′–3′ phosphodiester bridges

Electrometric titration of orthophosphoric acid reveals three pK values, as given in Fig. 3.7.

Similar titration of any of the nucleotides elicits only the two pK values in the acidic range, while a 'nucleic acid' shows only one pK, around 1, solely because of primary ionization. Internucleotide linkages in a nucleic acid must, therefore, be of the phosphodiester type. This was inferred early on by Levene and Simms (1925, 1926).

Which carbon atoms of the adjacent pentose molecules in a nucleic acid are involved in this phosphodiester linkage? Enzymatic hydrolysis of DNA with arsenate-treated (for inhibition of phosphomonoesterase) calf intestinal phosphatase produced phosphates of deoxyadenosine, deoxycytidine, deoxyguanosine and thymidine (Klein and Thannhauser 1933, 1934, 1935; Volkin *et al.* 1951). The phosphate group of these nucleotides could be hydrolysed by a specific 5′-nucleotidase (Heppel and Hilmoe 1951), confirming that the first enzymatic digestion had released 5′-phosphates of deoxynucleosides from DNA. The structures of these

Figure 3.7 Three different pK's of orthophosphoric acid.

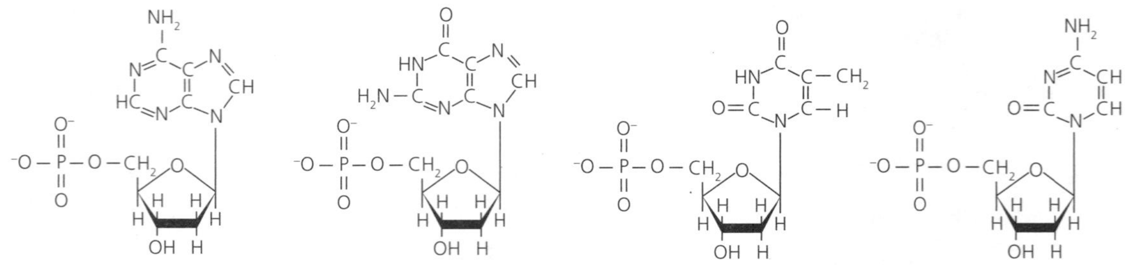

Figure 3.8 Chemical structures of the 5′-deoxyribonucleotides: 5′-phosphates of the deoxyribosides of adenine, guanine, thymine and cytosine.

deoxyribomononucleotides are shown in Fig. 3.8. 5′-Deoxymononucleotides were also obtained when DNA was treated with Russel's viper venom phosphodiesterase and beef pancreas deoxyribonuclease I. That these were indeed 5′-nucleotides was confirmed by comparison with nucleotides synthesized by unambiguous chemical methods (Michelson and Todd 1953, 1954).

Todd and coworkers (Dekker *et al.* 1953) could isolate from acid hydrolysates of herring sperm deoxyribonucleic acid, thymidine-3′,5′-diphosphate and deoxycytidine-3′,5′-diphosphate (Fig. 3.9) and could confirm their exact structure by synthesis. The indication was clear that both the 5′- and the 3′-carbons were involved in the phosphodiester linkage in deoxyribonucleic acid. Indeed, 3′-deoxymononucleotides (structures shown in Fig. 3.10) were obtained if DNA was treated with beef spleen phosphodiesterase and deoxyribonuclease II from spleen [Fig. 3.11(a)] (Sinsheimer and Koerner 1952; Hurst *et al.* 1951). On the other hand, 5′-deoxymononucleotides were obtained if DNA was treated with Russel's viper venom phosphodiesterase [Fig. 3.11(b)]. Thus, the phosphodiester linkage in DNA is indeed 5′–3′. The backbone of the linear macromolucle consists of repeating alternating deoxyribose and phosphate with the purine or pyrimidine bases as side chains attached to deoxyribose.

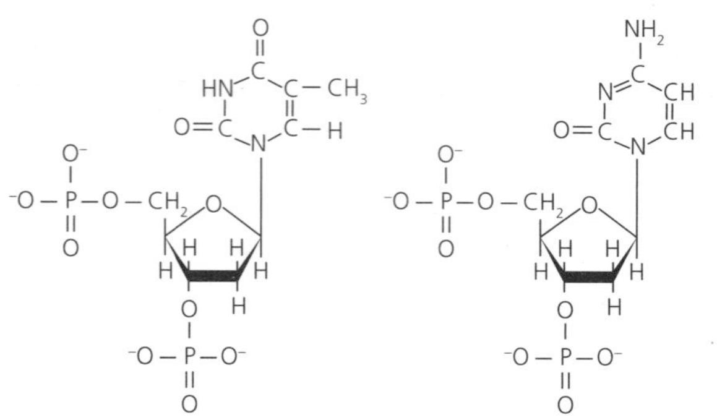

Figure 3.9 Chemical structures of the 3′,5′-diphosphates of thymidine and deoxycytidine.

3.6 RNA: A linear macromolecule with a backbone of ribose units

When ribonucleic acid was treated with pancreatic ribonuclease and arsenate-treated calf intestinal phosphatase, 5′-phosphates of adenosine, guanosine, cytidine and uridine were obtained (Cohn and Volkin 1951) as confirmed by comparison with synthetic ribonucleoside-5′-phosphates (Michelson and Todd 1949). Ribonucleoside-5′-phosphates were also obtained on treatment of RNA with Russel's viper venom phosphodiesterase (Cohn and Volkin 1952). Treatment of RNA with a specific nuclease preparation from beef spleen (Maver and Greco 1949), however, produced ribonucleoside-3′-phosphates (Volkin and Cohn 1954). So, the phosphodiester linkages in RNA are also 5′–3′. The backbone of the macromolecule here comprises alternating ribose and phosphate.

3.7 Direction or polarity in nucleic acids

All linear nucleic acids have two dissimilar ends. At one end, there is a free 5′-carbon of deoxyribose or ribose which may have a hydroxyl or phosphomonoester group (this is termed as the 5′- end), while at the other end, there is a free 3′-carbon of the last deoxyribose or ribose with a hydroxyl or phosphomonoester group (termed

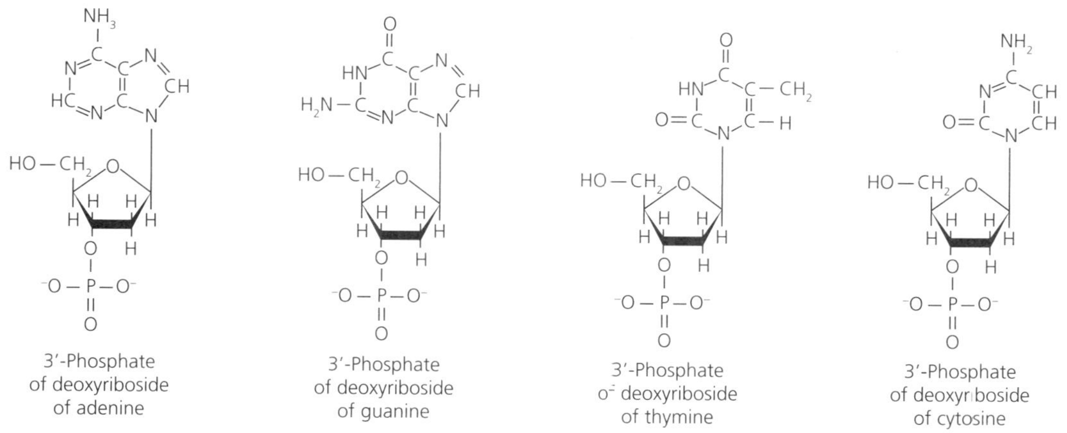

Figure 3.10 Chemical structures of 3′-deoxyribonucleotides: 3′-phosphates of the deoxyribosides of adenine, guanine, thymine and cytosine.

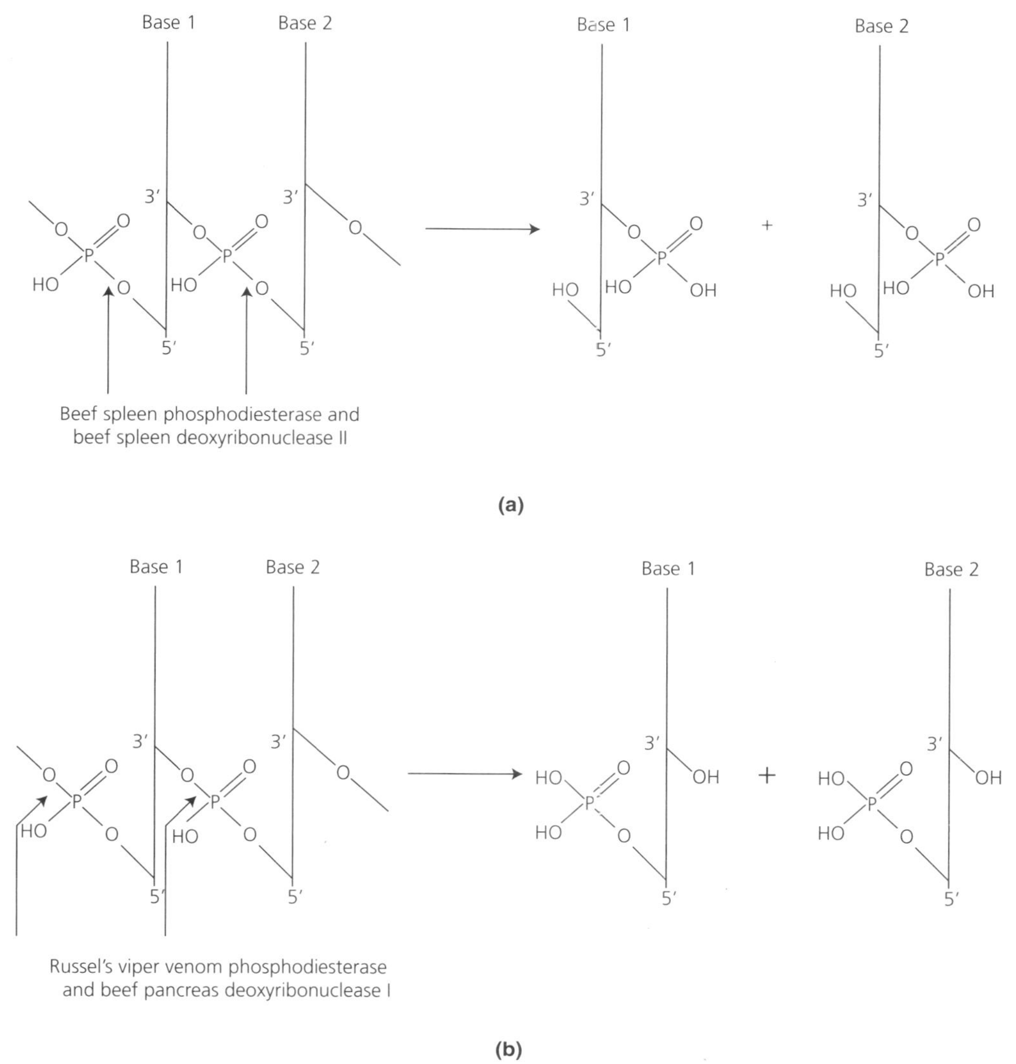

Figure 3.11 Enzymatic cleavage of the phosphodiester linkage of DNA. The deoxyribose sugar has been represented in the figure by a vertical line; the 5′-carbon of the sugar being at the bottom end of the vertical line, while the 1′-carbon being at the top of the line. The purine or pyrimidine base is linked to the 1′-carbon. The successive sugar molecules are linked by phosphodiester linkages. (a) Cleavage by successive treatments with beef spleen phosphodiesterase and beef spleen deoxyribonuclease II releases 3′-deoxymononucleotides. (b) Cleavage by successive treatments with Russel's viper venom phosphodiesterase and beef pancreas deoxyribonuclease I releases 5′-deoxymononucleotides.

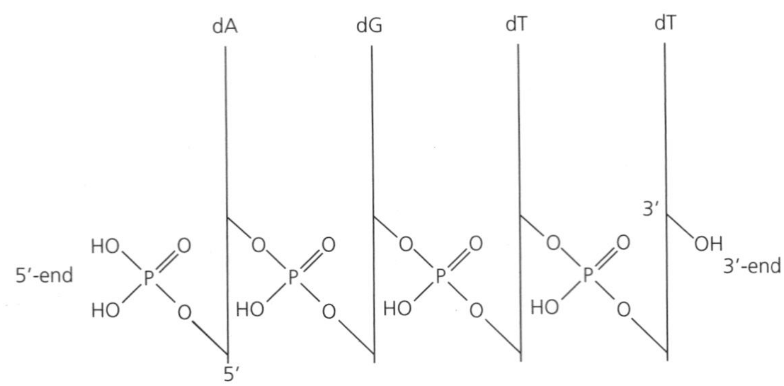

Figure 3.12 Nucleic acids have direction or polarity; one end is called the 5′-end and the other end, the 3′-end. As in Fig. 3.10, here also, the deoxyribose molecules are represented by vertical lines.

as the 3′- end). In between, all the 5′- and the 3′-carbon atoms of the pentoses are involved in phosphodiester linkages. By convention, the 5′- end to the 3′ end of a nucleic acid chain is written from left to right (Fig. 3.12).

3.8 The hydroxyl group on the 2′-carbon of ribose in RNA is responsible for its alkaline hydrolysis

Under alkaline conditions, the free 2′-hydroxyl group of ribose in RNA ionizes and attacks the 5′–3′ phosphodiester linkage, resulting in a cleavage of the bond between the P and the O linked to the 5′-carbon. The liberated P-ion then interacts with the 2′-hydroxyl, forming a cyclic O–P–O structure, which subsequently gives rise to a mixture of 2′- and 3′- phosphates of ribose (Fig. 3.13). This phenomenon did indeed create a lot of confusion early on in deciding as to whether the phosphodiester linkage in RNA was 5′–3′ or 5′– 2′. This hydrolytic effect does not occur in DNA as there is no free hydroxyl group in 2-deoxyribose.

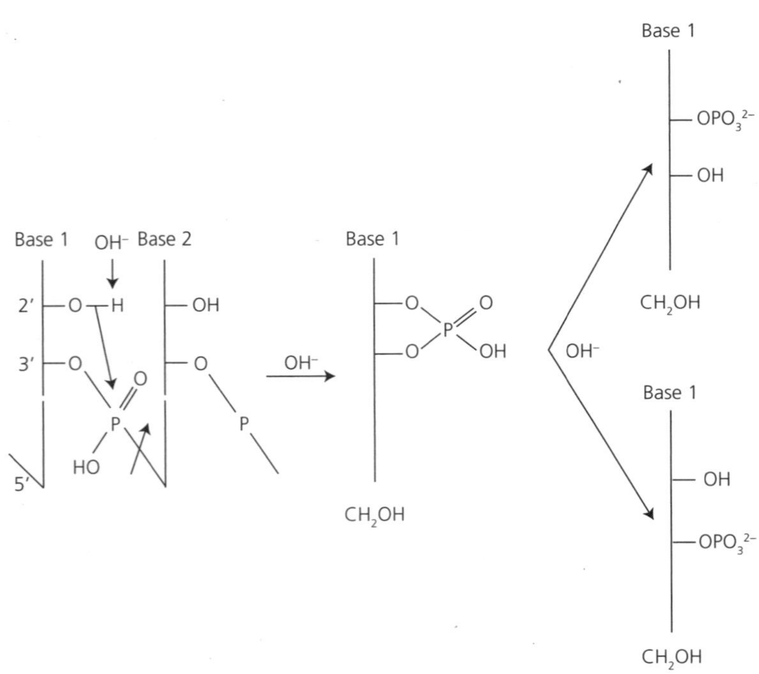

Figure 3.13 Ionization of the 2′-hydroxyl of ribose in RNA and cleavage of the phosphodiester linkage. As in Fig. 3.10, here also, the deoxyribose molecules are represented by vertical lines.

References

Cohn WE, Volkin E. Nucleoside-5′-phosphates from ribonucleic acid. *Nature* 1951;167:483–484.

Cohn WE, Volkin E. Mononucleotides from ribonucleic acid by phosphodiesterase action: The isolation of pyrimidine diphosphates. *Arch Biochem Biophys* 1952;35:465–467.

Dekker CA, Michelson AM, Todd AR. Nucleotides. Part XIX. Pyrimidine deoxyribonucleoside diphosphates. *J Chem Soc* 1953:947–951.

Hayes DH, Michelson AM, Todd AR. Nucleotides. Part XXX. Mononucleotides derived from deoxyadenosine and deoxyguanosine. *J Chem Soc* 1955:808–815.

Heppel LA, Hilmoe RJ. Purification and properties of 5′-nucleotidase. *J Biol Chem* 1951;188:665–676.

Hurst RO, Little JA, Butler GC. The enzymatic degradation of thymonucleic acid. II. The hydrolysis of oligonucleotides. *J Biol Chem* 1951;188:705–715.

Klein W, Thannhauser SJ. Experimentelle Studien über den Nucleinstoffwechsel. XXXII. Über die Ribodesose-Guanylsaüre [Experimental studies on the nuclein metabolism. XXXII. Ribodesose-guanylic acid]. *Z Physiol Chem* 1933;218:173–180.

Klein W, Thannhauser SJ. Experimental studies on the nuclein metabolism. XXXIV. Ribodesoseadenylic acid from thymonucleic acid. *Z Physiol Chem* 1934;224:252–260.

Klein W, Thannhauser SJ. Experimental studies on the nuclein metabolism. XXXV. The pyrimidine nucleotides from thymus nucleic acid. *Z Physiol Chem* 1935;231:96–103.

Levene PA, Simms HS. The dissociation constants of plant nucleotides and nucleosides and their relation to nuclide acid structure. *J Biol Chem* 1925;65:519–534.

Levene PA, Simms HS. Nucleic acid structure as determined by electrometric titration data. *J Biol Chem* 1926;70:327–341.

Maver ME, Greco AE. The nuclease activities of cathepsin preparations from calf spleen and thymus. *J Biol Chem* 1949;181:861–870.

Michelson AM, Todd AR. Nucleotides. Part III. Mononucleotides derived from adenosine, guanosine, cytidine and uridine. *J Chem Soc* 1949:2476–2486.

Michelson AM, Todd AR. Nucleotides. Part XX. Mononucleotides derived from thymidine. Identity of thymidylic acid from natural sources with thymidine 5-phosphate. *J Chem Soc* 1953:951–956.

Michelson AM, Todd AR. Nucleotides. Part XXIII. Mononucleotides derived from deoxycytidine. Note on the structure of cytidilic acids *a* and *b*. *J Chem Soc* 1954:34–40.

Sinsheimer RL, Koerner JF. A purification of venom phosphodiesterase. *J Biol Chem* 1952;198:293–303.

Volkin E, Cohn WE. Estimation of nucleic acids. *Methods Biochem Anal* 1954;1:287–305.

Volkin E, Khym JX, Cohn WE. The preparation of desoxynucleosides. *J Am Chem Soc* 1951;73:1533–1536.

4

Structure of DNA

H. K. Das

4.1 Base composition of DNA

The tetranucleotide hypothesis of nucleic acids (Chapter 3) was based on the belief that both classes of nucleic acids had equal proportion of the four nitrogenous bases. The finding of Avery and coworkers (1944) that the transforming principle of *Diplococcus pneumoniae* was DNA therefore came as a big surprise. How could such a 'simple' organic compound store the hereditary information of such diverse functions? It was at this stage that Erwin Chargaff appeared on the scene and published his painstaking and thorough analysis of the base composition of DNA from several biological sources (Chargaff 1950). In Chargaff's words, 'the old tetranucleotide hypothesis — it should never have been called a theory — was still dominant; and this was characteristic of the enormous sway that the organic chemistry of small molecules held over biochemistry'. After all, all biomolecules, including DNA, are chemical molecules; hence, they would have the same chemical make-up irrespective of the source from which the DNA has been isolated. Chargaff questioned, 'how different must complicated substances be, before we can recognize their difference?' He cautioned, 'there is nothing more dangerous in the natural sciences than to look for harmony, order, regularity, before the proper level is reached'. Referring to the work of Jungner *et al.* (1949), Chargaff reminded, 'the fact remains that the deoxypentose nucleic acids are high-molecular substances which in size resemble, or even surpass, the proteins. It is quite possible that there exists a critical range of molecular weights above which two different cells [of different origin?] will prove unable to synthesize completely identical substances'.

Chargaff extracted the purines from DNA by heating at 100°C for 1 h in 1 N sulphuric acid and precipitated as the hydrochlorides by treatment with dry HCl gas in methanol. The pyrimidines were then extracted with concentrated formic acid at 175°C (under pressure) for 2 h. The individual purines and pyrimidines were separated by paper chromatography using a solvent system containing n-butanol, diethylene glycol and water in an NH_3 atmosphere. The spots were visualized in short-wavelength ultraviolet rays, eluted and then assayed by ultraviolet absorption.

The results can be summarized as follows:

- The proportion of the four nitrogenous bases was *not* the same.
- The base composition of DNA was the same irrespective of the tissue which served as the source as long as the organism was the same. For example, the base composition of DNA isolated from the thymus, spleen or liver of ox was the same.
- The base composition of DNA isolated from man was *different* from that isolated from ox. The composition of DNA from the human sperm, thymus and liver was, however, the same.
- The base compositions of DNA isolated from yeast and avian tubercle bacilli were also different. Thus, the composition is a 'characteristic of the species, but not of the tissue from which they were derived'.

Chargaff (1950) had aptly stated, 'I think there will be no objection to the statement that, as far as chemical possibilities go, they [DNA] could very well serve as one of the agents, or possibly as the agent, concerned with the transmission of inherited properties'. Chargaff had made another far-reaching statement, which he termed as a speculation, 'just as the desoxypentose nucleic acids of the nucleus are species specific and concerned with the maintenance of the species, the pentose nucleic acids of the cytoplasm are organ specific and involved in the important task of differentiation' (Chargaff 1950). This last statement would be referred to again when transcription would be discussed in a later chapter.

In the text of the paper, Chargaff did not mention it anywhere, but it was very obvious from Chargaff's data that whatever the source of DNA, the thymine content is the same as the adenine content and the guanine content is the same as the cytosine content. Cytosine was slightly on the lower side, presumably because of its conversion to uracil during the harsh acid treatment. In summary, the total purine content of DNA is equivalent

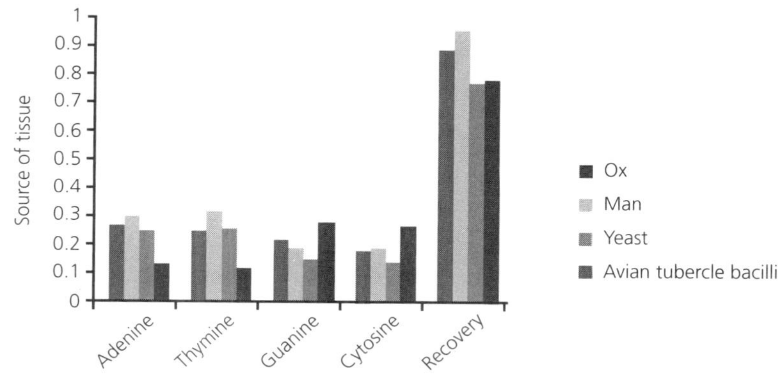

Figure 4.1 Base composition of DNA from several organisms. [compiled from Chargaff (1950).]

to the total pyrimidine content. Base composition of DNA from several organisms is shown in Fig. 4.1 [compiled from Chargaff (1950)]. Similar results were reported subsequently from analysis of DNA isolated from a large number of animals, plant, microorganisms and viruses (Smith and Wyatt 1951; Wyatt 1951, 1952; Zamenhof *et al.* 1952).

4.2 Three-dimensional structure of DNA by Pauling and Corey

Pauling and Corey (1951) had postulated an α-helical configuration of polypeptide chains in proteins. By sheer analogy, they argued, 'it seems not unlikely that [such] a single general operation is also involved in the construction of nucleic acids' which 'would lead to a helical structure. The giant molecule would thus be cylindrical, with approximately circular cross section' (Pauling and Corey 1953b). In support of the postulated circular cross section, they cited the electron micrographs of Williams (1952), which showed fibrils of 15- to 20-Å diameter of apparent circular cross section. Pauling and Corey had at their disposal some X-ray diffraction photographs from their own laboratory and also those published by Astbury and Bell (Astbury and Bell 1938a,b). 'A very strong meridional reflection with spacing about 3.40 Å was seen in these photographs.' Pauling and Corey argued, 'this reflection corresponds to a distance along the fiber axis equal to three times the distance per residue. Accordingly, the reflection is to be attributed to a unit consisting of three residues'. Hence, they proposed that DNA is 'a three-chain structure, each chain being a helix with fundamental translation equal to 3.4 Å', and the three chains 'are coiled about one another'.

The next question they addressed was about 'the nature of the core of the three-chain helical molecule – the part of the molecule closest to the axis'. Pauling and Corey argued that 'because of their varied nature, the purine–pyrimidine groups cannot be packed along the axis of the helix in such a way that suitable bonds can be formed between the sugar residues and the phosphate groups; this choice is accordingly eliminated'. They opined that 'a close-packed core of phosphoric acid residues, HPO_4^-, can easily be constructed'. Pauling and Corey added, 'the purine and pyrimidine groups, on the periphery of the molecule, occupy positions such that their hydrogen-bond forming groups are directed radially. This would permit the nucleic acid molecule to interact vigorously with other molecules'. Thus, the salient features of the model were:

- Each DNA chain is an α-helix.
- The DNA fibril consists of three such α-helices coiled about one another.
- The phosphate groups are at the core of the helix, pointing inside, while the purine and pyrimidine bases are at the periphery, pointing outside.

An artist's impression of the cross sectional view of the nucleic acid structure as imagined by Pauling and Corey is represented in Fig. 4.2 – only three nucleotides of each chain are being shown.

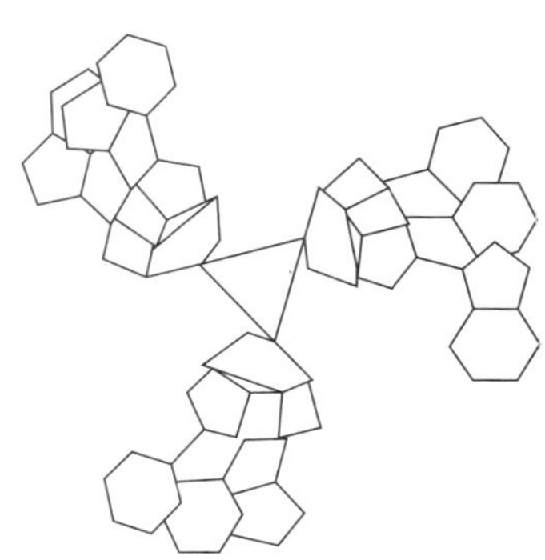

Figure 4.2 An artist's impression of the simplified cross sectional view of the three-chain structure of nucleic acid as imagined by Pauling and Corey. Only three nucleotides of each chain are being shown.

4.3 Watson and Crick's double helix

The announcement of the Pauling and Corey model of DNA first appeared in the 21 February 1953 issue of *Nature* (Pauling and Corey 1953a). James Watson and Francis Crick, who were at that time in Cambridge, England, were not convinced and they had their own view. It is said that the very next Saturday, 28 February 1953, Watson and Crick had described to the lunch-time crowd in the Cambridge pub, The Eagle, their idea of the DNA, 'the secret of life'. Two months later, three papers were published (Watson and Crick 1953; Wilkins *et al.* 1953; Franklin and Gosling 1953) in the 25 April 1953 issue of *Nature*, which together brought about the new concept of the DNA double helix. The first of the three papers presented the model, while the following two papers presented photographs of X-ray diffraction of DNA and their interpretation. Watson and Crick did have access to the photographs and made use of them in designing the model. Watson and Crick also took cognizance of and made use of the base composition data of DNA that appeared between 1950 and 1952 (Chargaff 1950; Smith and Wyatt 1951; Wyatt 1951, 1952; Zamenhof *et al.* 1952). Finally and most important, they appreciated the utmost importance of building scale models incorporating all the available information.

The following were the interpretations of the X-ray diffraction photographs (Figs 4.3 and 4.4; Franklin and Gosling 1953; Wilkins *et al.* 1953):

- The diffraction pattern showed only the periodic or regularly repeated parts of the DNA molecule.
- The diffraction patterns were the same for DNA isolated from diverse species, even though the base composition of the DNA was very much different.
- For the same DNA, the diffraction pattern could assume either of the two forms depending on the moisture content of the sample. Water content of 40% (when relative humidity fell to about 75%) elicited a crystalline pattern, showing that the DNA molecules were aligned regularly in all dimensions (form A; Fig. 4.3). Higher water content (when relative humidity was about 92%) resulted in a paracrystalline form in which the DNA molecules were packed side by side and could presumably slide over one another (form B; Fig. 4.4). This was considered to be the more frequently found physiological form.
- There were two strong 3.4-Å reflexions on the meridian (Fig. 4.4), which corresponded to the internucleotide distance along the fibre axis.
- There were layer lines arranged in the form of an X suggestive of the existence of a helix (Fig. 4.4).
- The 34-Å distance between the layer lines (Fig. 4.4) was, however, thought 'not due to a repeat of a polynucleotide composition, but to chain configuration repeat' (Wilkins *et al.* 1953). Hence, the nucleotides repeated 10 times before the configuration repeated crystallographically. Thus, if the structure is helical, 'there are exactly 10 residues per turn of the helix' (Franklin and Gosling 1953).
- The diameter of the DNA fibre was calculated from the angle formed by the layer lines and was found to be 20 Å.
- 'If ten phosphorus atoms lie on one turn of a helix of radius 10 Å, then distance between neighbouring phosphorus atoms in a molecule is 7.1 Å. This corresponds to the P…P distance in a fully extended molecule, and therefore provides a further indication that the phosphates lie on the outside of the structural unit.' 'The sugar and base groups must accordingly be turned inwards towards the helical axis' (Franklin and Gosling 1953).

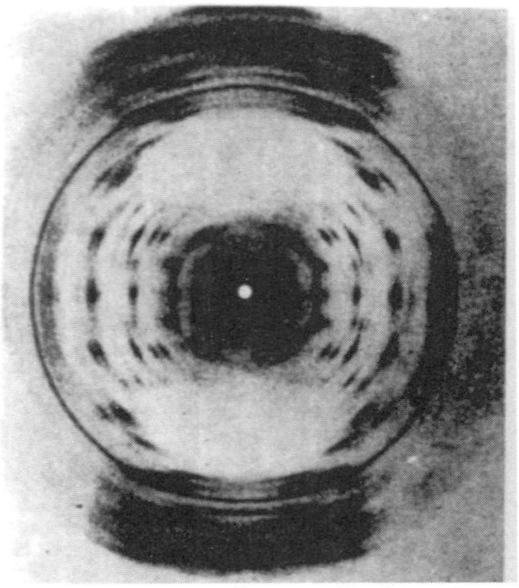

Figure 4.3 X-ray diffraction photograph of DNA, form A. [Reproduced from Wilkins *et al., Nature* 1953;171:738 with permission from *Nature*.]

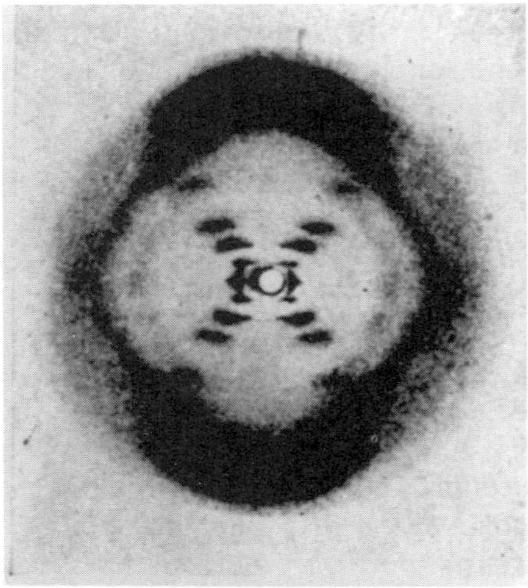

Figure 4.4 X-ray diffraction photograph of DNA, form B. [Reproduced from Franklin and Gosling, *Nature* 1953;171:740 with permission from *Nature*.]

- 'Consideration of density [of the reflexions] show, however, that a cylindrical repeat unit of height 34 Å, and diameter 20 Å, must contain many more than ten nucleotides.' 'From measurements on the X-ray diagram of structure B, it can readily be shown that, whether the number of chains is two or three, the chains are not equally spaced along the fibre axis.' 'Three unequally spaced chains would be crystallographically non-equivalent, and this seems unlikely. It therefore seems probable that there are only two coaxial molecules and that these are unequally spaced along the fibre axis' (Franklin and Gosling 1953).

The helical nature of the DNA structure was also inferred from another consideration. 'The crystallographic repeat did not coincide with the repetition of chemical units in the chain but came at much longer intervals. A possible explanation was that all the links in the chain were the same but the X-rays were seeing every tenth link, say, from the same angle and the others from different angles. What sort of chain might produce this pattern? The answer was easy: the chain might be coiled in a helix. The distance between crystallographic repeats would then correspond to the distance in the chain between one turn of the helix and the next' (Crick 1954).

From considerations of density of the reflexions and from considerations of crystallographic equivalence, Franklin and Gosling (1953) had inferred the presence of two coaxial molecules in a DNA fibre. Watson and Crick (1953) asked the important question as to 'what forces would hold the structure together?' It could not be the negatively charged phosphates near the axis, as postulated by Pauling and Corey (1953a, 1953b), as they would 'repel each other'. Watson and Crick postulated that the purine and pyrimidine bases were inside and hydrogen bonds between them would hold the structure.

While constructing the scale model, Watson and Crick set the following guidelines for themselves:

- The bond angles and the bond distances must be correct.
- Bad van der Waal's contact must be avoided.
- Hydrogen bonds must be allowed wherever possible.
- The model should be space filling.

Their scale model had 'two helical chains each coiled round the same axis'.

'Each chain consists of phosphate diester groups joining β-D-deoxyribofuranose residues with 3′, 5′ linkages. The two chains (but not their bases) are related by a dyad perpendicular to the fibre axis. Both chains follow right-handed helices, but owing to the dyad the sequences of the atoms in the two chains, run in opposite directions' (Fig. 4.5). The chains could not be fitted in the scale model if they were put in the same direction. 'There is a residue on each chain every 3.4 Å, in the z-direction.

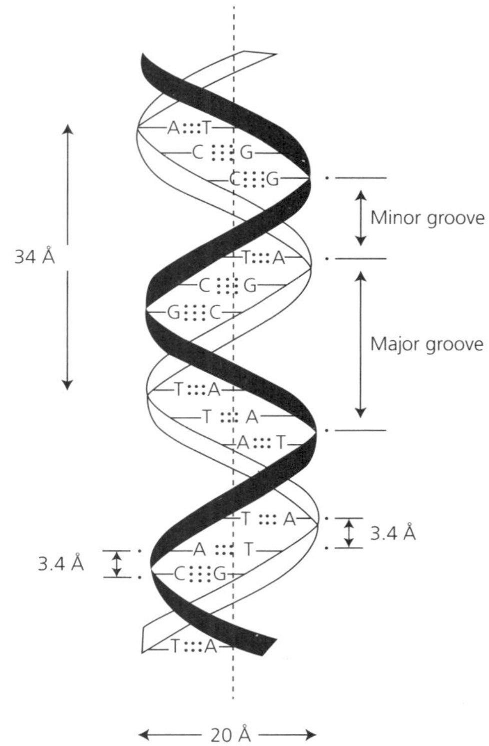

Figure 4.5 The double helical structure of DNA as postulated by Watson and Crick.

An angle of 36° has been assumed between adjacent residues in the same chain, so that the structure repeats after 10 residues on each chain, that is, after 34 Å. The distance of a phosphorus atom from the fibre axis is 10 Å.' 'The planes of the bases are perpendicular to the fibre axis. They are joined together in pairs, a single base from one chain being hydrogen-bonded to a single base from the other chain, so that the two lie side by side with identical z-coordinates. One of the pair must be a purine and the other a pyrimidine for [hydrogen] bonding to occur. The hydrogen bonds are made as follows: purine position 1 to pyrimidine position 1; purine position 6 to pyrimidine position 6.' 'The bases only occur in the structure in the most plausible tautomeric forms (i.e. with the keto rather than the enol configurations).' 'Only specific pairs of bases can bond together. These pairs are: adenine (purine) with thymine (pyrimidine), and guanine (purine) with cytosine (pyrimidine).' It is known now that three hydrogen bonds are possible between guanine and cytosine (Fig. 4.6). In the scale model, adenine could not be paired with cytosine, as hydrogen bonds could not be formed, because two hydrogens appeared in one of the bonding positions, and none at the other. This does not, however, imply that free adenine and cytosine in solution would not be able to form hydrogen bonds. The free bases outside the chain can rotate without any restriction and thus can form hydrogen bonds.

There are two grooves on the surface of the helix: the deep wide one is called the 'major groove', and the

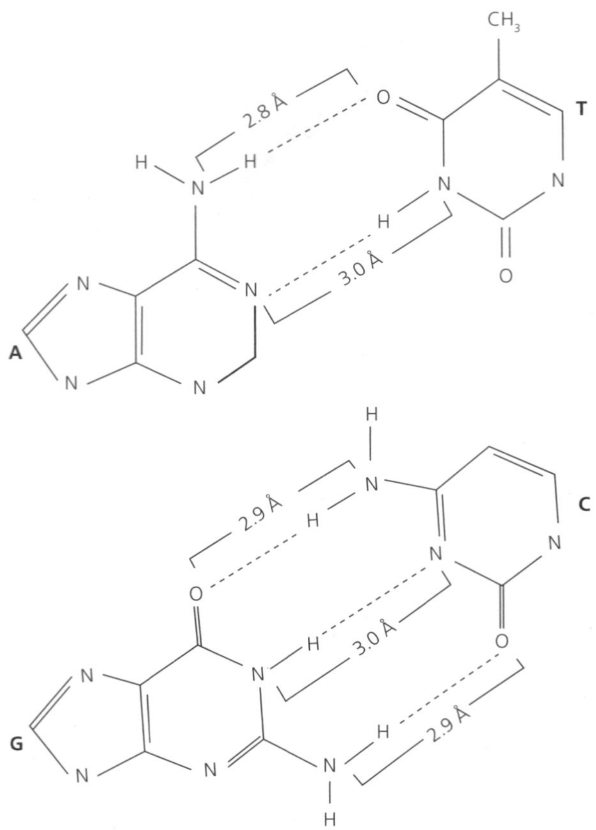

Figure 4.6 Two hydrogen bonds between adenine and thymine and three hydrogen bonds between guanine and cytosine in the DNA double helix.

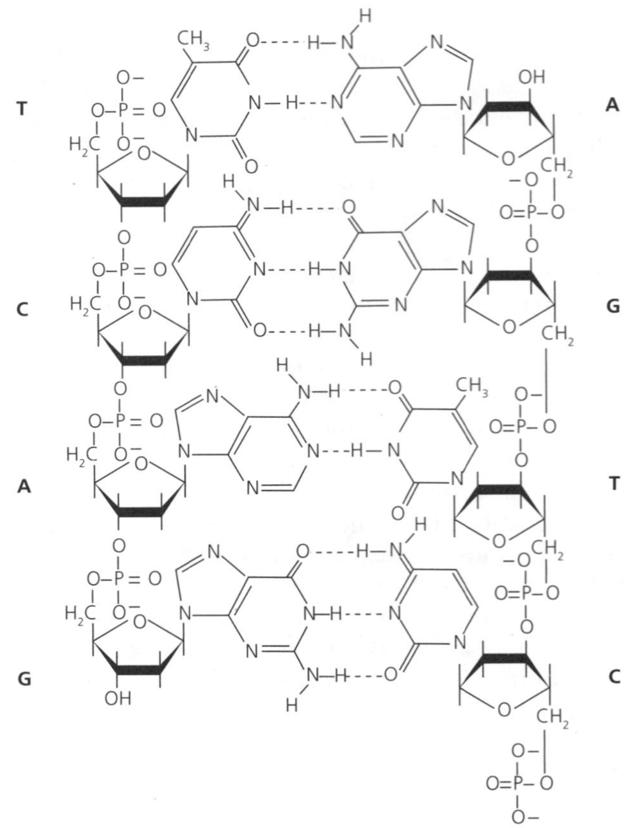

Figure 4.7 Non-helical representation of a hydrogen-bonded tetranucleotide duplex.

shallow narrow one is called the 'minor groove'. Protein molecules can come in contact with the bases inside through these grooves.

Watson and Crick pointed out that 'the sequence of bases on a single chain does not appear to be restricted in any way'. However, 'if the sequence of bases on one chain is given, then the sequence on the other chain is automatically determined'. Fig. 4.7 shows a non-helical representation of a hydrogen-bonded tetranucleotide duplex, the direction of the two strands being antiparallel. Fig. 4.8 is a reproduction of the photograph showing Watson and Crick admiring their scale model of the DNA double helix.

Figure 4.8 James D. Watson and Francis HC Crick with their scale model of DNA. [Reproduced from the *Double Helix*, Atheneum, NY, p. 215 (Copyright 1968 by JD Watson. Photograph by AC Barrington Brown) by permission of Professor JD Watson.]

4.4 Hydrogen bonds

Hydrogen bonds occur between two electronegative atoms. The more is the electronegativity, the stronger is the bond. Elements linked by hydrogen bonds are almost exclusively O, N and F. Sometimes, C can also participate if it is linked to a strong electronegative group. Hydrogen bond is largely electrostatic or ionic in nature. The mean energy required to break a hydrogen bond is about 2–8 kcal per mole, in contrast to

Table 4.1 Approximate lengths of some hydrogen bonds

Nature of bond	Approximate length of bond in Å
N—H-------N	3.10 ± 0.13
N—H-------O	3.04 ± 0.13
N^+—H-------O	2.93 ± 0.10
O—H-------N	2.88 ± 0.13
O—H-------O	2.70 ± 0.10
O—H-------O^-	2.63 ± 0.10

covalent bonds needing about 85 kcal per mole. The approximate lengths of some hydrogen bonds are given in Table 4.1.

4.5 Electronegativity of elements

During bond formation between unlike elements, electrons of the duplet are more likely to be found in the vicinity of one element than the other. The one which attracts the electron more is called the more electronegative element. The order of electron attraction is

$$H < C < N < O < F$$

4.6 Electrometric titration indicating hydrogen bonds between the two strands of DNA

Two main types of titrable groups exist in the purines and pyrimidine bases of DNA: the amino groups in all the bases and –NH–CO– dissociations of guanine and thymine. Gulland *et al.* (1947), however, did not come across any titrable group during forward electrometric titration (employing hydrogen and glass electrodes) of calf thymus DNA between pH 5.0 and pH 11.0. During back titration, nevertheless, the case was different – titration was evident up to pH 8.0 in one side and up to pH 6.0 in the other [Fig. 4.9(a)]. The inference could be drawn that some of the amino groups of the purines and the pyrimidines were possibly blocked because of the formation of hydrogen bonds, and hence, the titrable groups were not evident during forward titration. These groups would be liberated once extreme alkaline or acidic pH values were attained, as hydrogen bonds would have been broken at these pHs. Even ionization of one group would bring about the liberation of both. As a result, titrable groups were encountered during back titration. This inference was borne out by electrometric titration of DNA, which was treated with alkali or acid prior to the titration: titrable groups were indeed evident even during the forward titration [Fig 4.9(b) and (c)].

Titration data alone would not tell whether the hydrogen bonding was intra- or intermolecular. There was, however, marked fall in viscosity at extreme pHs, suggesting the formation of lower-molecular-weight entities. As secondary ionization of phosphate groups of DNA did not increase at these pHs, it was inferred that phosphodiester linkages were not broken. Hence, the reduction of molecular weight must have been because of disruption of intermolecular hydrogen bonds, resulting in the formation of two units of single-stranded DNA from each one unit of the double-stranded DNA.

4.7 Kinetics of enzymatic degradation of DNA

Kinetics of enzymatic degradation of DNA also supported the existence of hydrogen-bonded double-stranded structure.

Bovine pancreatic deoxyribonuclease has been known to cleave phosphodiester linkages of DNA at random. Hence, a single-stranded DNA molecule would be split at each cleavage by the enzyme, but a double-stranded DNA molecule would need at least two cleavages, if not many more, to be split, depending on whether the second cleavage is right across the first cleavage on the other strand or far apart. Hence, the first split in a double-stranded DNA molecule would take place only after several cleavages of phosphodiester linkages. The representation in Fig. 4.10 would explain the situation. Viscosity or the sedimentation coefficient has been used to monitor split in the molecule, while the cleavage of the phosphodiester linkage has been monitored by measuring liberation of the hydrogen ion in a pH-stat. In fact, about 200 phosphodiester linkages had to be broken in order for the viscosity to be reduced to half (Thomas 1956). Schumaker *et al.* (1956) have obtained similar results by both viscosity measurement and determination of the sedimentation coefficient. A schematic representation of the experimental result is given in Fig. 4.11.

4.8 Separation of the two strands of DNA

Electron microscopy has provided visual evidence of separation of the two strands of DNA on disruption of hydrogen bonds. Treatment of DNA with alkali results in the disruption of hydrogen bonds between the two strands of DNA and separation of the strands has been visualized by electron microscopy (Fig. 4.12).

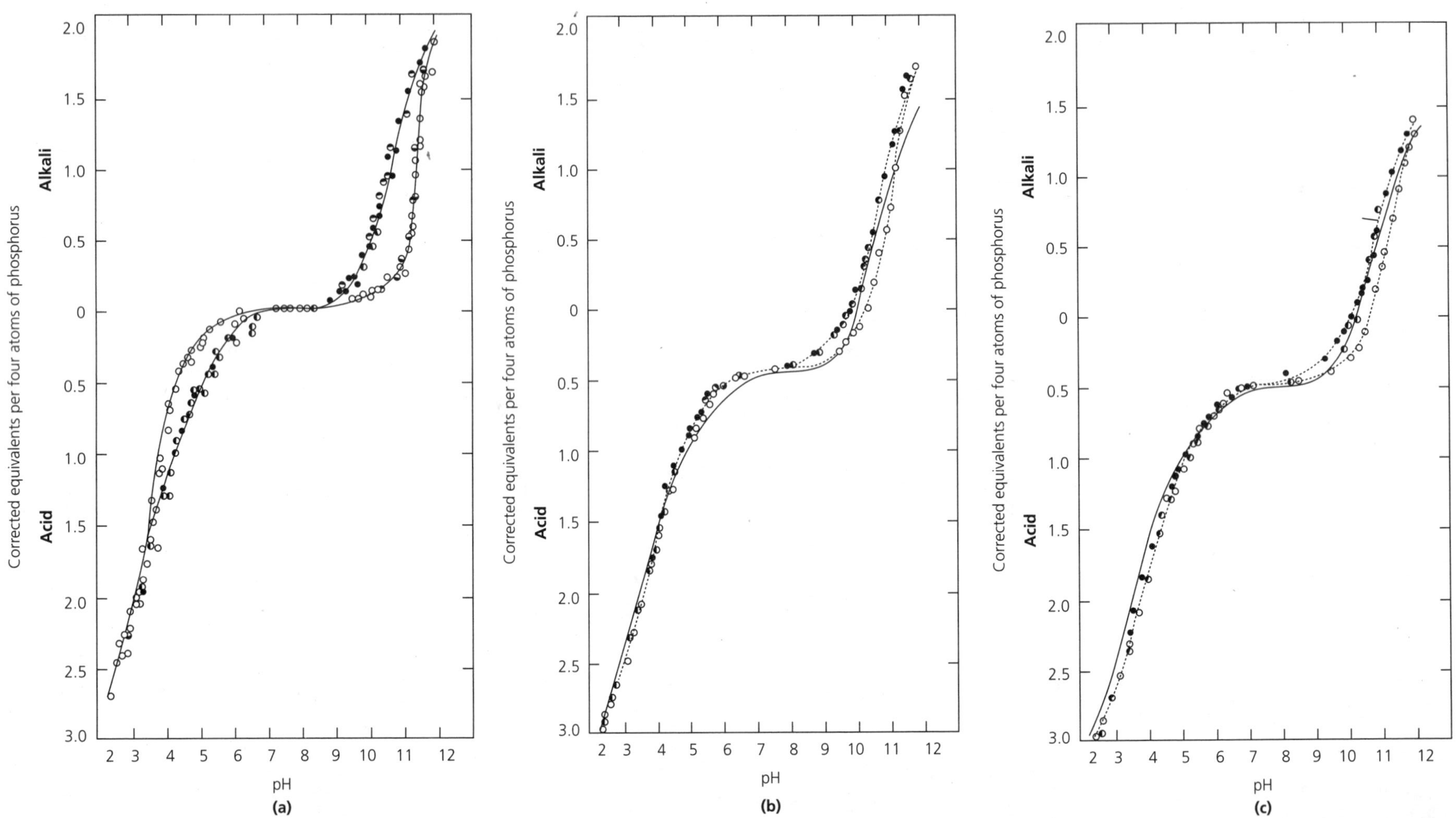

Figure 4.9 Electrometric titration of calf thymus DNA solution. (a) Forward titration with acid or alkali from pH 6.9 (o); back titration with acid from pH 11.4 and with alkali from pH 2.5 (•). It is obvious that the forward titration curve and the back titration curve do not overlap each other. (b) Effect of prior treatment with alkali on electrometric titration of calf thymus DNA. Forward titration with acid or alkali from pH 6.7 (o); back titration with acid from pH 12.0 and with alkali from pH 2.5 (•). Here, the forward titration curve and the back titration curve do overlap. (c) Effect of prior treatment with acid on electrometric titration of calf thymus DNA. Forward titration with acid or alkali from pH 7.1 (o); back titration with acid from pH 12.0 and with alkali from pH 2.3 (•). Here also, the forward titration curve and the back titration curve do overlap. [Reproduced from Gulland *et al.*, *J Chem Soc* 1947:1131–1141 by permission of The Royal Society of Chemistry.]

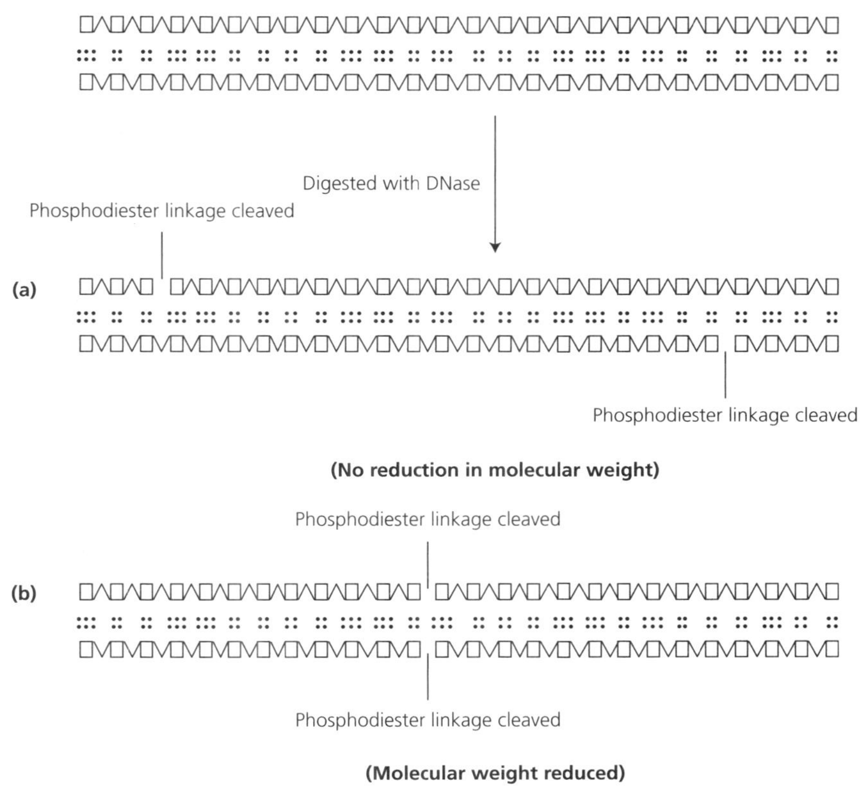

Figure 4.10 Schematic diagram showing cleavage of phosphodiester linkages in a double-stranded DNA and the possibility of split of the molecule. Each small square (□) represents a nucleotide and the (/\) or (\/) a phosphodiester linkage. Each colon (:) between the two DNA bases in the two opposite strands represents a hydrogen bond.

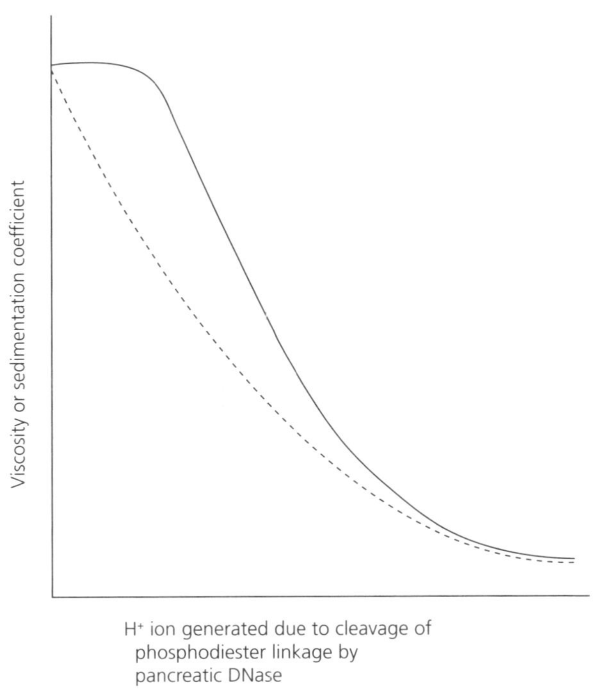

Figure 4.11 Viscosity of DNA solution or the sedimentation coefficient of DNA as a function of the H^+ ion generated due to cleavage of the phosphodiester linkage by bovine pancreatic deoxyribonuclease. Solid line: native double-stranded DNA; broken line: fully denatured single-stranded DNA.

4.9 Opposite polarity of the two strands of DNA

Opposite polarity of the two strands of DNA was confirmed by 'nearest -neighbour frequency' data.

An enzyme had been isolated in the laboratory of Arthur Kornberg from extracts of *Escherichia coli* which could synthesize DNA *in vitro* under the direction of a given DNA template (Lehman *et al.* 1958) (more details in Chapter 9). The enzyme used 5′-triphosphates of deoxyadenosine, deoxyguanosine, deoxycytidine and thymidine as substrates. DNA was synthesized by using one ^{32}P-labelled deoxyribonucleotide at a time and the other three unlabelled deoxyribonucleotide as described below:

Reaction 1: $dATP^{32} + dGTP + dCTP + dTTP$
Reaction 2: $dATP + dGTP^{32} + dCTP + dTTP$
Reaction 3: $dATP + dGTP + dCTP^{32} + dTTP$
Reaction 4: $dATP + dGTP + dCTP + dTTP^{32}$

Phosphodiester linkages (5′→3′) were formed and pyrophosphate was released as shown in Fig. 4.13(a). In the segment of the synthesized polymer that has been shown, dT is at the 3′-end of dC, which is at the 3′-end of dG, which is at the 3′-end of dA and which is at the 3′-end of dT. In fact, all other combinations of sequences would be formed. The synthesized product was purified

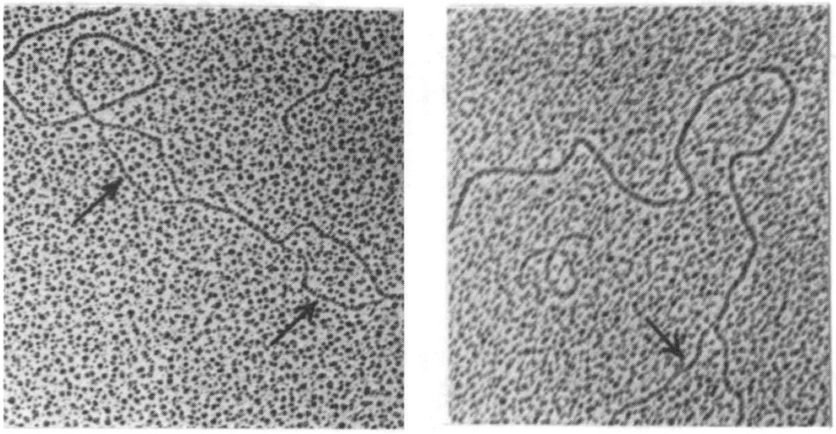

Figure 4.12 Electron micrographs of DNA treated with alkali (magnification ×58,000). Localized separation of the two single-stranded DNA segments can be seen (pointed by arrows). The electron micrographs were taken by Dr Nirupam Roy Choudhury.

from unreacted substrates by repeated precipitation with chilled perchloric acid and dissolution in alkali. The product was successively treated with micrococcal deoxyribonuclease and calf spleen phosphodiesterase. Care was taken to ensure that the enzyme preparations were free from any phosphomonoesterase activity; no ^{32}P-orthophosphate should be detected as reaction product. A mixture of deoxyribonucleotides were obtained with 3′-phosphate groups. Because of the specific site of enzymatic hydrolysis of the P–O bond, the phosphate group that originated in a particular deoxyribonucleotide had been transferred to the deoxyribonucleoside immediately next to it at the 5′-direction [Fig. 4.13(b)].

The individual deoxyribonucleotides were separated by paper electrophoresis and the content of ^{32}P-radioactivity was determined. Though each of the four synthetic reactions was carried out with only one particular 5′-^{32}P-labelled deoxyribonucleotide (plus the other three unlabelled deoxyribonucleotides), radioactivity was now recovered in all the four 3′-^{32}P-labelled deoxyribonucleotides. Thus, the proportion of radioactivity found associated with the four different 3′-^{32}P-labelled deoxyribonucleotides originating from synthesis reaction 1 would be a measure of the proportion in which these four deoxyribonucleotides were present in the synthesis product immediately next in the 5′-direction to deoxyriboadenosine nucleotide. Similarly, the proportion of radioactivity found associated with the four different 3′-^{32}P-labelled deoxyribonucleotides originating from synthesis reaction 2 would be a measure of the proportion in which these four deoxyribonucleotides were present in the synthesis product immediately next in the 5′-direction to deoxyriboguanosine nucleotide. So on for synthesis reactions 3 and 4. The 'nearest-neighbour frequency' data also represent dinucleotide composition of the enzymatically synthesized product. Table 4.2 shows the 'nearest-neighbour frequency' data computed for *Azotobacter vinelandii* DNA as template. The GC content of *A. vinelandii* DNA is about 65%.

Let us consider a double-stranded stretch:

TAGA
ATCT

which may be present in *Azotobacter vinelandii* DNA. The DNA may have opposite polarity

5′ TAGA 3′
3′ ATCT 5′

Table 4.2 'Nearest-neighbour frequency' data computed for *Azotobacter vinelandii* DNA

^{32}P-Labelled deoxyribonucleotide used in the synthesis reaction	Isolated ^{32}P-labelled deoxyribonucleotide after the enzymatic hydrolysis reaction				
	dTMP^{32}	dAMP^{32}	dCMP^{32}	dGMP^{32}	Sums
dATP^{32}	0.018 (TpA)	0.029 (ApA)	0.064 (CpA)	0.064 (GpA)	0.175
dTTP^{32}	0.032 (TpT)	0.037 (ApT)	0.044 (CpT)	0.061 (GpT)	0.174
dGTP^{32}	0.062 (TpG)	0.046 (ApG)	0.134 (CpG)	0.084 (GpG)	0.326
dCTP^{32}	0.064 (TpC)	0.063 (ApC)	0.084 (CpC)	0.115 (GpC)	0.326
Sums	0.176	0.175	0.326	0.324	

* The 'nearest-neighbour frequency data' have been normalized with respect to the *Azotobacter vinelandii* template DNA. Molar proportion of bases: T = 0.174, A = 0.175, C = 0.326 and G = 0.326.

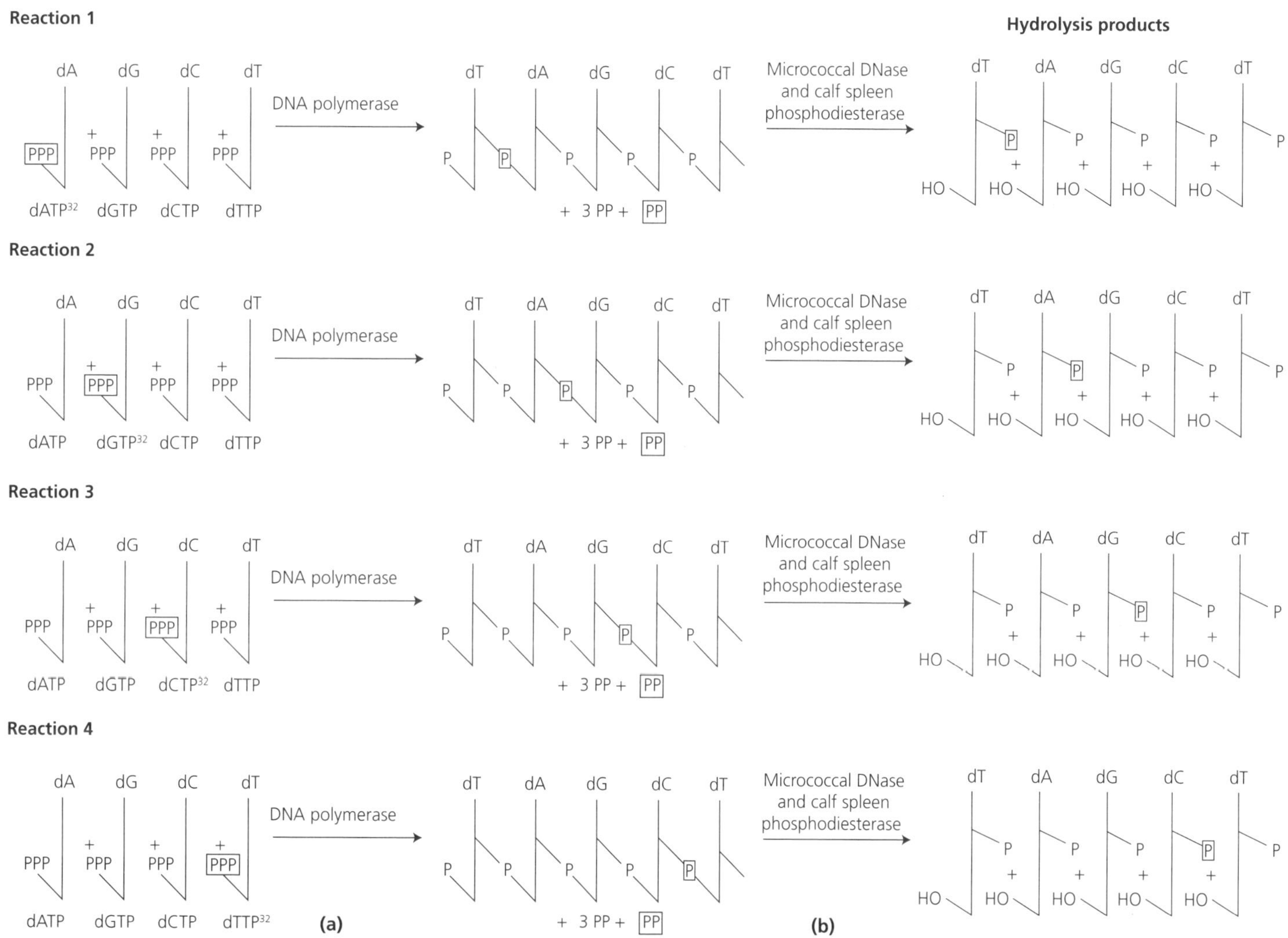

Figure 4.13 (a) Schematic representation of DNA synthesis with ^{32}P- labelled 5′-deoxyribonucleotides. The vertical lines in the figure represent the deoxyribose sugar, the 5′-carbon being located at the bottom end of the line. The phosphate groups are represented in the figure by the letter P; thus, PPP represents triphosphate. The letter P within a box denotes phosphate containing ^{32}P. Only one 5′-deoxyribonucleotide triphosphate is labelled in one reaction, while the other three 5′-deoxyribonucleotides are unlabelled. Four reactions are shown, each using a different ^{32}P-labelled 5′-deoxyribonucleotide. (b) Hydrolysis by consecutive treatment with micrococcal DNase and calf spleen phosphodiesterase of the four DNA molecules synthesized as described in Fig. 4.12(a), yielding mononucleotides, with the phosphate at the 3′-carbon of the deoxyribose (3′-mononucleotides). It is obvious that there has been a shift of ^{32}P to the adjacent nucleoside towards the 5′-end as a result of the enzymatic treatment.

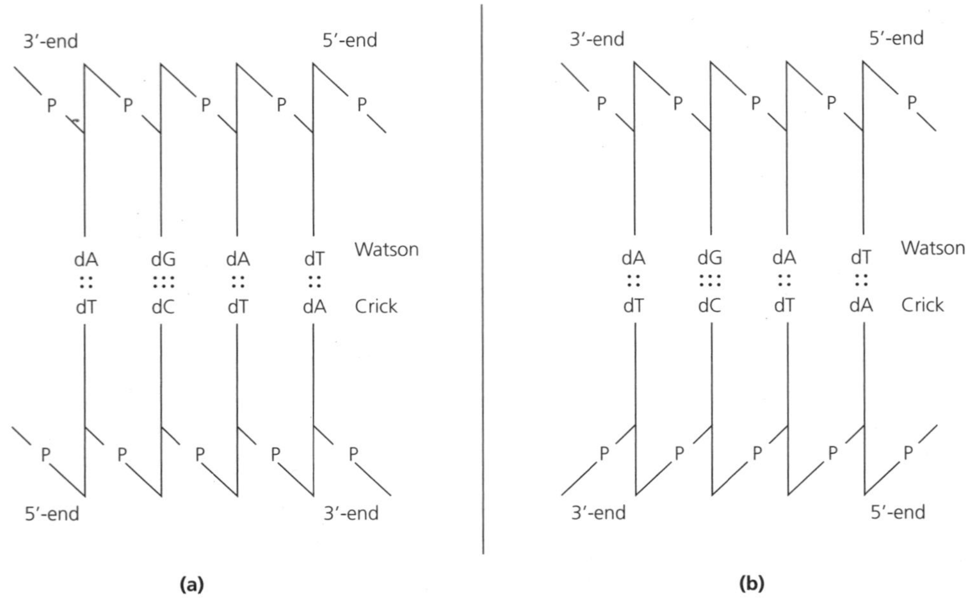

Figure 4.14 Postulated results of nearest-neighbour frequency determination of two double-stranded oligonucleotides, (a) having the strands in opposite polarity and (b) having the strands in similar polarity.

or similar polarity

5′ TAGA 3′
5′ ATCT 3′

Opposite polarity of the two strands of DNA is represented in Fig. 4.14. The nearest-neighbour frequency in this stretch of DNA would depend on the polarity. In the Watson strand, transfer of phosphorus would be from bottom to top in both possibilities (a) and (b). A would transfer its phosphorus to G, G would transfer its phosphorus to A and A would transfer its phosphorus to T. In the Crick strand, transfer of phosphorus would, however, be from top to bottom in the case of opposite polarity [possibility (a)]: A would transfer its phosphorus to T, T would transfer its phosphorus to C and C would transfer its phosphorus to T. However, in the same Crick strand, transfer of phosphorus would be from bottom to top in the case of similar polarity [possibility (b)]: A would transfer its phosphorus to T, T would transfer its phosphorus to C and C would transfer its phosphorus to T. Hence, in the case of similar polarity, the following should hold good:

TpA = ApT
ApG = TpC
GpA = CpT

The data in Table 4.2 reveal that

TpA = 0.018
ApG = 0.046
GpA = 0.064
ApT = 0.037
TpC = 0.064
CpT = 0.044

It is obvious that the above-mentioned sets of equivalence do *not* hold good at all; hence, similar polarity of the two strands of *Azotobacter vinelandii* DNA does *not* appear to be valid.

However, in the case of opposite polarity, the following should hold good:

TpA = TpA
ApG = CpT
GpA = TpC

The data in Table 4.2 reveal that

TpA = 0.018
ApG = 0.044
GpA = 0.064

TpA = 0.018
CpT = 0.044
TpC = 0.064

Thus, the sets of equivalence stipulated for opposite polarity do indeed hold gold and, hence, the 'nearest-neighbour frequency' data from *Azotobacter vimelandii* DNA provide unequivocal evidence for the opposite-polarity model. Josse *et al.* (1961) had examined 16 DNA samples from calf thymus and various bacteria and viruses and have confirmed that the opposite polarity of the two strands of DNA is universally true.

4.10 Base sequencing data

Ultimate confirmation of opposite polarity of the two strands of DNA has come from base sequencing data.

5′ ATGGCATTGCGTCAGTGTGCAATTTACGGCAAGGGT 3′
3′ TACCGTAACGCAGTCACACGTTAAATGCCGTTCCCA 5′

Figure 4.15 Bases in DNA strands of opposite polarity as revealed by sequencing.

Accurate determination of the sequence of bases in individual strands of DNA has been possible by methods involving both specific chemical cleavage of the phosphodiester linkages (Maxam and Gilbert 1977) and enzymatic synthesis of DNA (more details in Chapter 7) with chain-terminating derivatives (Sanger *et al.* 1977). Both kinds of determinations not only reveal the sequence of the bases, but also their polarity in each strand of DNA (Fig. 4.15). This has provided ultimate confirmation of the opposite polarity of the two strands of DNA.

4.11 Determination of DNA base sequence and polarity by chemical cleavage

Treatment with dimethyl sulphate methylates N7 of the guanine residues in DNA and subsequent treatment with piperidine eliminates all the methylated guanines, thereby cleaving the DNA at those points. The reactions are shown in Fig. 4.16. Some adenines also get methylated at N3 on reaction with dimethyl sulphate, but piperidine does not eliminate the methylated adenines. Treatment with formic acid (adjusted to pH 2.0 with pyridine) weakens the glycosidic bonds of both adenine and guanine by protonating the purine-ring nitrogen atoms, which are subsequently opened up by alkali. Piperidine can then eliminate these bases. Hydrazine, on the other hand, attacks at C4 and C6 of the thymine and cytosine residues in the DNA and ultimately forms a new five-membered ring with hydrazine. Piperidine eliminates the thymine and cytosine residues and cleaves the DNA at those points. The sequence of reactions with thymine residues is shown in Fig. 4.17. If molar sodium chloride is present during the hydrazine reaction, only cytosine is affected. These four reactions are, thus, specific for G, G + A, T + C and C.

Prior to the cleavage reaction, either the 5′-end or the 3′-end of the DNA strands is labelled with ^{32}P. Polynucleotide kinase and γ-^{32}P-labelled ATP is used to label the 5′-end. Labelling at the 3′-end is achieved by using DNA polymerase I and α-^{32}P-labelled nucleoside triphosphates.

The rationale for the method is that if the cleavage reaction can be executed in such a way that breaks in the DNA strand (labelled at one end) would take place only at a limited number of sites at random, then one would be very likely to obtain a complete assortment of length of labelled DNA fragments (Fig. 4.18). If all these fragments can be separated by electrophoresis on the basis of length, then the positions of a particular base can be ascertained. The resolution of the electrphoretic separation process has to be such that two fragments of DNA differing only by one base length would also be

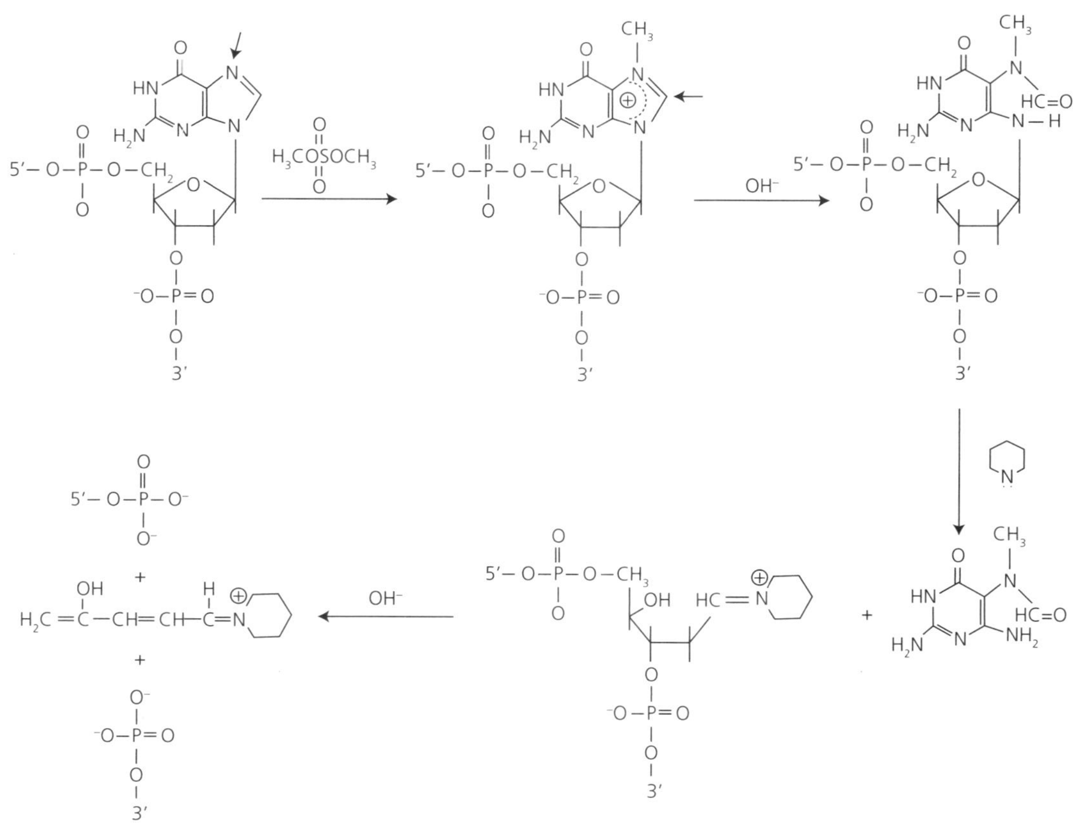

Figure 4.16 Reaction of dimethyl sulphate and piperidine with guanine.

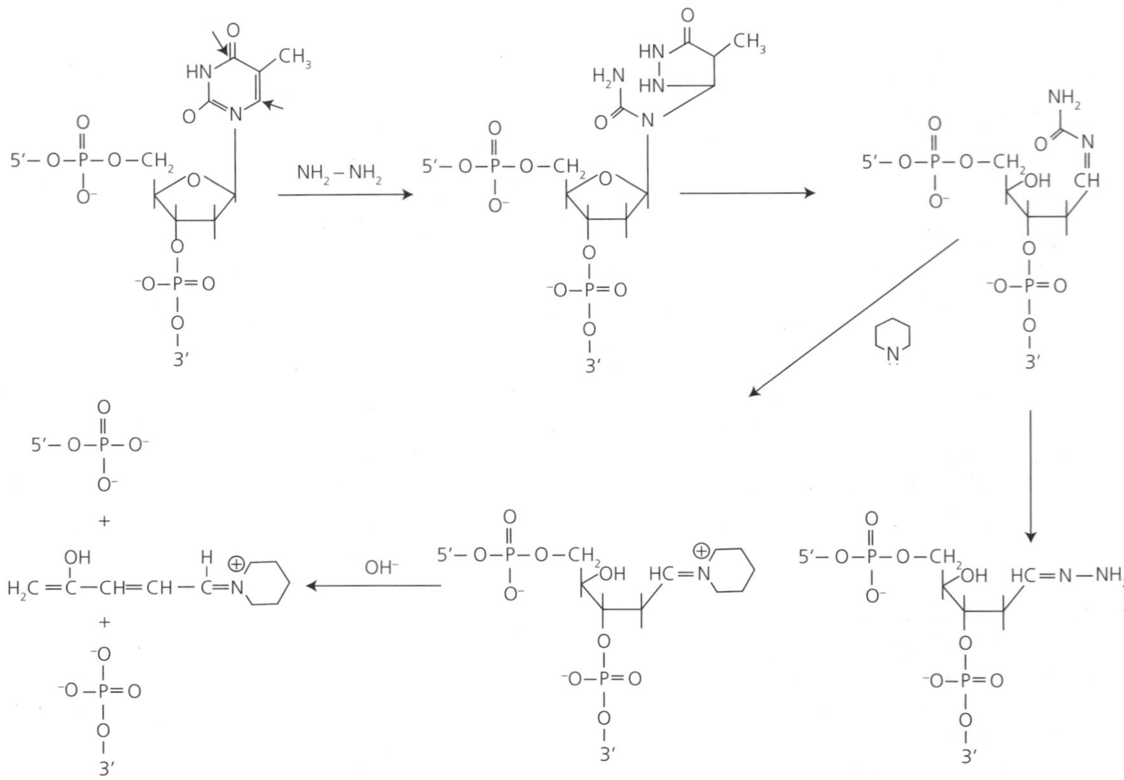

Figure 4.17 Reaction of hydrazine and piperidine with thymine.

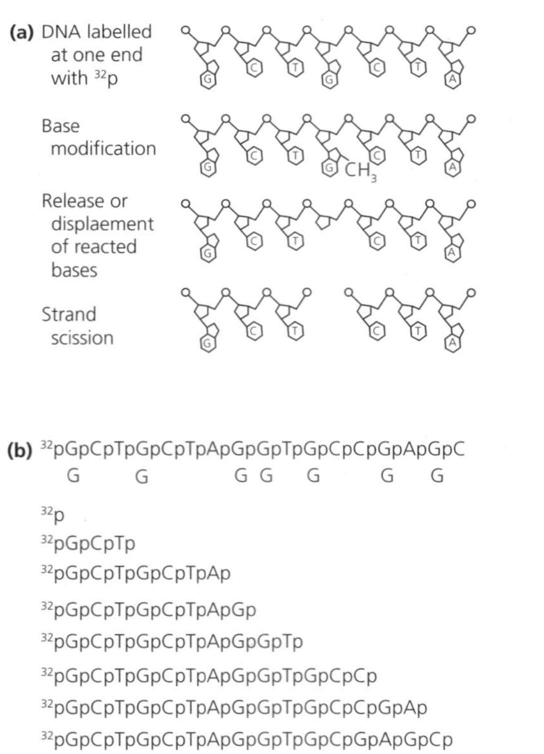

Figure 4.18 Cleavage of guanine residues in DNA (a) and formation of fragments of assorted lengths (b).

resolved. For example, the fragments obtained after treatment with dimethyl sulphate would indicate the positions of guanine residues in the DNA strand. If the products obtained from the four reactions mentioned previously are electrophoresed in four adjacent lanes, the total sequence of DNA can be worked out. This is done by exposure of the gel to an X-ray plate to visualize the ^{32}P-labelled DNA fragments. A sample X-ray plate is shown in Fig. 4.19 (Maxam and Gilbert 1977). As the distance traversed by all the fragments would be according to the length from one specific end (the end which has been labelled), the polarity of the bases would also be automatically determined.

4.12 Determination of DNA base sequence and polarity by enzymatic synthesis

Enzymatic synthesis *in vitro* of DNA complementary to a given DNA template is possible by using DNA polymerase I from *E. coli* and the 5′-triphosphates of deoxyguanosine, deoxyadenosine, deoxycytidine and thymidine (details in Chapter 9). The polymerase catalyzes the formation of 5′–3′. phosphodiester linkages between the 5′-phosphate of one deoxynucleotide and the 3′-hydroxyl of the next one. If the 3′-hydroxyl of one of the four deoxynucleotide triphosphates (say of guanosine) is removed (the compound is called 2′,3′-dideoxy derivative), the chain would be terminated there (at each G residue) as no 5′–3′- phosphodiester linkage

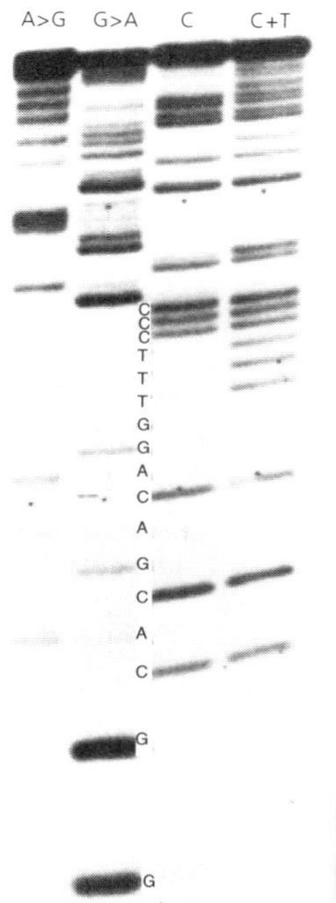

Figure 4.19 A sample X-ray plate showing separation of labelled fragments of DNA for base sequence analysis done by the chemical cleavage method of Maxam and Gilbert. [Reproduced from Maxam and Gilbert, *Proc Natl Acad Sci USA* 1977;74:563 by permission of Professor Walter Gilbert.]

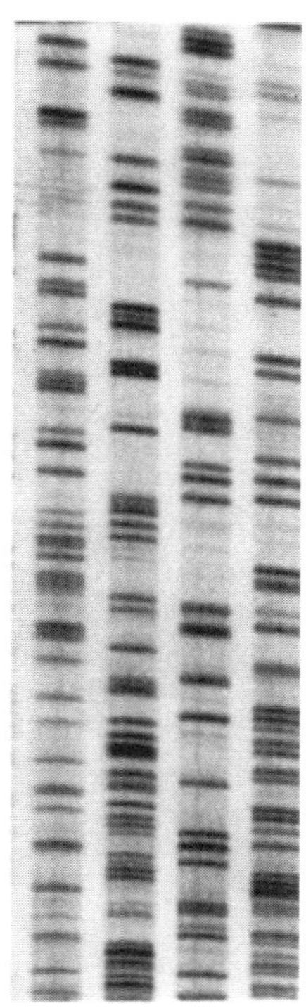

Figure 4.20 A sample X-ray plate showing separation of labelled fragments of DNA for base sequence analysis done by the enzymatic synthesis method of Sanger *et al.* (1977).

can be formed. In practice, a mixture of the deoxynucleotide (2′-monodeoxy) triphosphate and the 2′,3′-dideoxy derivative is used so that termination takes place only at a few G residues at random, thus creating a complete collection of DNA fragments ending with G. Each of the four dideoxy derivatives is used in four separate reactions that are initiated with ^{32}P-labelled primers. The labelled DNA fragments of all sizes are then electrophoresed and the DNA sequence worked is out as described for the chemical cleavage method. A sample X-ray plate is shown in Fig. 4.20.

4.13 Single-crystal X-ray diffraction studies

The existence of two distinct helical structures in isolated DNA fibres (forms B and A) was inferred in 1953 itself from X-ray diffraction photographs (Figs 4.3 and 4.4; Franklin and Gosling 1953; Wilkins *et al.* 1953). Rigorous confirmation of the double helical structure of DNA, at least in crystals, has subsequently been provided from single-crystal X-ray diffraction studies of synthetic oligodeoxyribonucleotide duplexes.

Of course, first real surprise came from the laboratory of Alexander Rich as Andrew H-J Wang solved the structure of the CGCGCG duplex (Wang *et al.* 1979). This hexamer duplex was neither the A form nor the B form. This was a helix all right, but not a right-handed one. It was, in fact, a left-handed helix with a zigzag backbone. It was, hence, named Z-DNA helix. A similar structure was observed for the CGCG duplex by Horace R. Drew in the laboratory of Richard E. Dickerson (Dickerson 1983b).

The CGCGAATTCGCG duplex, however, revealed a classical B-DNA structure and its right-handed helical conformation was confirmed (Wing *et al.* 1980). On the other hand, the GGTATACC duplex (Shakked *et al.* 1981), the GGCCGGCC duplex (Wang *et al.* 1982) and the CCGG duplex (Conner *et al.* 1984) elicited right-handed A-DNA helical structures.

The single-crystal X-ray diffraction studies have also revealed extensive local variation in the helix twist angle (Wing *et al.* 1980). The two bases of each hydrogen-bonded pair did not always lie in precisely the same plane, but elicited a 'propeller twist' that improved the stacking overlap between consecutive bases in the same DNA strand (Dickerson 1983a, 1983b). Unwanted proximity between two purines on opposite backbone strands at adjacent base pairs (bp) as a result of 'propeller twists' were avoided by decreasing the local helix twist angle or by altering the roll angle of the base pairs or by pushing the purines out of the regular stack of bases (Dickerson 1983a, 1983b). The DNA double helix, even in the B or A form, thus is not a regular and monotonous helical structure.

The details of the structural properties of the A, B and Z forms of DNA as derived from X-ray diffraction studies with single crystals of specific oligodeoxyribonucleotides can be found in Dickerson *et al.* (1982).

While the A form is short and broad, the B form is longer and thinner, but the Z form is elongated and slim. The major groove in the A form is extremely narrow, but very deep; however, it is wide and of intermediate depth in the B form. The major groove is flattened out on the helix surface in the Z form. The minor groove in the A form is very broad and shallow, in the B form, it is narrow and of intermediate depth, while in the Z form, it is extremely narrow but very deep.

The existence of a C form of DNA has also been inferred, having ~9 bp per turn, ~38.6° rotation per base pair, ~3.32 Å vertical rise per base pair and a helical diameter of ~19 Å. These parameters for the A, B and Z forms of DNA, respectively, are ~11, ~12 and ~10 base pairs per turn; ~33.6°, 35.9° and 60° (per two base pairs) rotation per base pair; 2.3 Å, 3.32 Å and 3.8 Å vertical rise per base pair; ~25.5 Å, 23.7 Å and 18.4 Å helical diameter.

4.14 Intermediates between A form and B form of DNA

The transition between the A form and the B form of DNA has been mapped through single-crystal structures of the hexamer duplex GGCGCC. This has revealed the existrence of various intermediate states (named E-DNA) between the two forms, which have been trapped by methylating or brominating the cytosine bases (Vargason *et al.* 2000, 2001). The findings have led to the notion that the A form and the B form of DNA are not two isolated conformations.

4.15 Solution structure of the DNA duplex

Solution structures of deoxyoligonucleotide duplexes have now been determined using nuclear magnetic resonance (NMR) and restrained molecular dynamics simulations. The inferred solution structure of d(CCACCGGAAC).(GTTCCGGTGG) was indeed similar to the B form of DNA (Soliva *et al.* 2001).

4.16 Nuclear magnetic resonance spectroscopy

All atomic nuclei have nuclear spin, which may be zero, integral or half integral. The proton has a spin of ½. A spinning nucleus, because of its electric charge, gives rise to a magnetic field. When a magnetic nucleus is subjected to a uniform magnetic field, the magnetic dipole associated with the magnetic moment assumes only a discrete set of orientations, which can be visualized in an NMR spectra.

4.17 Models for non-intertwining side-by-side strands of DNA

Kinetics of enzymatic degradation of DNA (Thomas 1956; Schumaker *et al.* 1956), electrometric titration data (Gulland *et al.* 1947) and electron microscopic evidence, all were in conformity with a hydrogen-bonded double-stranded structure of DNA, but no additional evidence for the double helical structure was forthcoming. Models were, therefore, proposed in which the two strands of DNA of opposite polarity could exist side by side and no double helical intertwining was necessary. According to one model (Rodley *et al.*1976), 'the sugar-phosphate backbone has a twisted strip-like character, yet base-pairing of the Watson-Crick type is still possible'. 'This was achieved by constructing approximately half of the repeat unit in a basically Watson–Crick manner for a right-handed double helix with acceptable structural parameters and then continuing the model in the form of a left-handed double helix' [Fig. 4.21(a) and (b)]. They went on to assert that 'the structure has, on the whole, reasonable stereochemical contacts, and can be shown to produce a fiber diffraction pattern with x-rays not unlike that of the B form of DNA'. A space-filling (Corey–Pauling–Koltun) model in the same general line but slightly different in details was subsequently suggested (Sasisekharan *et al.* 1978). It has been claimed that 'this model is energetically more favourable than the double helix'.

There does not seem to be any doubt that DNA molecules in living cells are mostly double stranded, but whether these molecules really exist as Watson and Crick double helices is still debated by some (Delmonte 1999). In fact, it has been inferred based on symmetry analysis that at least on a cation-coated (amino-silanized) glass surface, as used in microarray technology, 12-mer and 24-mer perfectly complementary oligonucleotides possibly form a non-helical duplex structure (Lemeshko *et al.* 2001).

4.18 Bends and kinks in DNA

Distortion of double helix because of specific nucleotide sequence in DNA can give rise to bends and kinks. Single-crystal X-ray diffraction analysis has revealed that the actual angles of roll at dinucleotide steps and the angles of helical twist are not uniformly same in DNA, in contrast to the idealized Watson and Creek model.

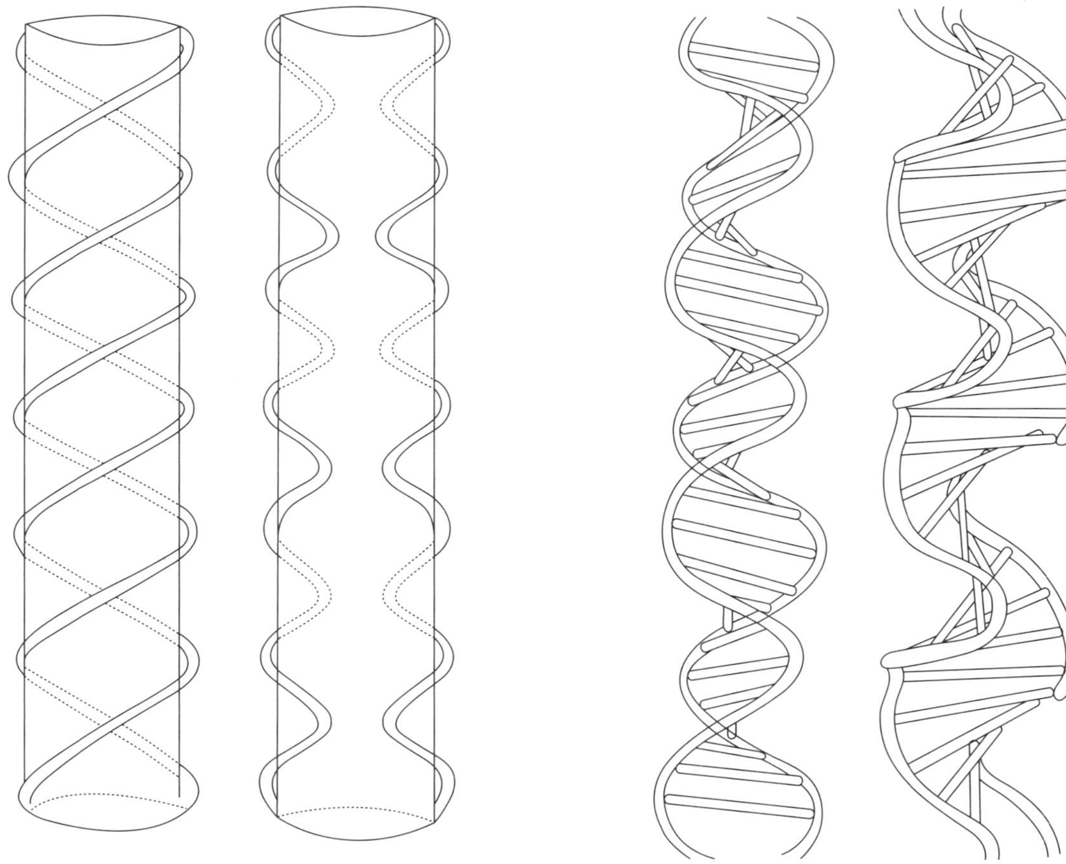

Figure 4.21 (a) Idealized drawings of the double helix (left) and side-by-side (right) structure of DNA. (b) Sketches of the simplified side-by-side structure, side view (left) and angular side view (right). [Reproduced from Rodley *et al.*, *Proc Natl Acad Sci USA* 1976;73:2960–2961 with permission from Professor Robert M. Lewitt.]

It is likely that for some specific sequences, the additive effect of localized distortions could result in bends and kinks in the DNA. Bent DNA was first observed in DNA fragments isolated from the kinetoplast body of trypanosomes (Marini *et al.* 1982) and subsequently from many other sources such as the origin of replication of bacteriophage lambda (Zahn and Blattner 1985), an autonomously replicating sequence of yeast (Snyder *et al.* 1986), upstream of the plasmid promoter P_{CIII} (Perez-Martin and Espinosa 1994) and upstream of the *nifLA* promoter of the nitrogen-fixing bacterium *Klebsiella pneumoniae* (Cheema *et al.* 1999).

Bent DNA molecules exhibit anomalously slow gel electrophoretic mobility (Koo *et al.* 1986), presumably because a bent structure would be able to proceed less quickly through the pores of the gel than a straight structure. Fig. 4.22 shows that a 603-bp long bent DNA fragment containing the *vnf H* promoter of *Azotobacter vinelandii* and its upstream sequence, when electrophoresed in 8% acrylamide gel, migrated like a 704-bp long DNA fragment (Bageshwar *et al.* 1998). The control run was with a 410-bp long straight DNA fragment, which indeed migrated like a 410-bp long fragment. Electron microscopy of the 603-bp long fragment also revealed its bent nature (Fig. 4.23; Bageshwar *et al.* 1998).

4.19 Gel electrophoresis

Separation of charged molecules under an electric field is an age-old technique. Various solid supports such as paper, silica gel, starch gel, agarose gel and polyacrylamide gel have been used. The voltage applied between the two electrodes is the driving force, while the resistance offered by the solid medium is the opposing force. The resistance depends on the pore size of the solid medium, which quite often can be manipulated to some extent. The larger are the molecules, the greater is the resistance and the slower is the mobility. A double-stranded DNA molecule 1-kb long would be expected to have a faster mobility than another double-stranded DNA molecule 2-kb long. On the other hand, a triple helical DNA molecule is likely to have a slower mobility than a double helical DNA molecule of the same length. The shape of the molecule is also a determinant of its mobility. A straight DNA molecule can wriggle

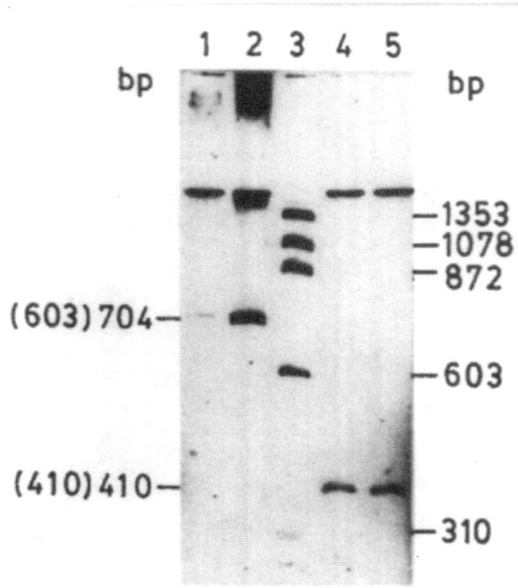

Figure 4.22 Anomalous gel electrophoretic mobility of bent DNA. The bent DNA molecules have reduced mobility compared with a 'straight' DNA molecule of the same length. In lanes 1 and 2, the lowest visible band comprises the DNA fragment containing the *vnf H* promoter of *Azotobacter vinelandii* and its upstream sequence that is 603-bp long but is migrating at a position corresponding to 704 bp. Lane 3 contains DNA fragments obtained by digesting phi X 174 DNA with *Hae* III; the lowest clearly visible band contains DNA that is 603-bp long. Lanes 4 and 5 contain DNA fragments obtained by digesting pBR 322 DNA with *Eco* RI and *Sal* I; the lowest band being 410-bp long and migrates in a position corresponding to 410 bp. [Reproduced from Bageshwar *et al.*, *Can J Microbiol* 1998;44:410 by permission of Professor HK Das.]

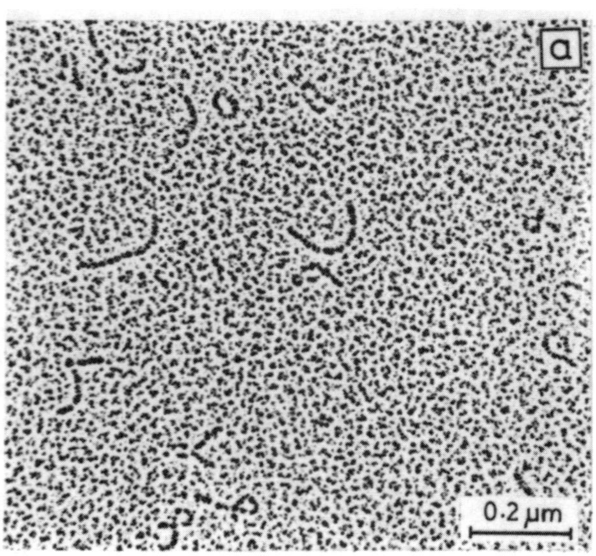

Figure 4.23 Electron micrograph of bent DNA. DNA fragment containing the *vnf H* promoter of *Azotobacter vinelandii* and its upstream sequence that is 603-bp long. The presence of looped molecules may be seen. The photograph was taken by Dr Nirupam Roy Choudhury. [Reproduced from Bageshwar *et al.*, *Can J Microbiol* 1998;44:410 by permission of Professor HK Das.]

through the gel more easily than a bent DNA; hence, bent DNA molecules have slower mobility. For separation of large DNA molecules (0.5–20 kb), agarose gels (0.5–2%) are used as solid support, while for smaller DNA molecules (0.1–1.0 kb), polyacrylamide gels (7–15%) are used. To start with, gels used to be cast inside glass tubes, but now, gels are cast as slabs on plates and samples are applied into wells created during casting by inserting a comb.

It was demonstrated that A/T tracts, each tract being half a helical turn long and repeated in phase with the helix screw, could cause a DNA bend (Koo *et al.* 1986). Koo *et al.* (1986) surmised that 'bending, in the direction of the base-pair tilt rather than roll, occurs at the junctions between the A/T tract and adjacent B-DNA, with a larger angle at the 3′ than at the 5′ end of the A tract'. Calladine *et al.* (1988), however, felt that 'the main effect is due to a difference of about 3° in roll values between AA/TT and other steps, together with a difference of about 1° in the angle of helical twist'. There may be some difference of opinion as to the precise element or combination of elements involved in bending, but a generally accepted view is that the anomaly in the structure is a result of discontinuity at the junction between a random-sequence DNA and oligo A-tracts (Bolshoy *et al.* 1991; Koo and Crothers 1988). Software based on experimentally determined wedge angles of all 16 dinucleotides (Bolshoy *et al.* 1991) and based on dinucleotide parameters such as tilt, roll, twist, shift, slide and rise (Nagaich *et al.* 1994) are available for predicting the existence of bends in DNA from just the base sequence. The 603-bp long DNA fragment (Bageshwar *et al.* 1998) mentioned previously did exhibit a bent structure (Fig. 4.24) when its base sequence was analysed by such a software (Bolshoy *et al.* 1991). A new method for predicting sequence-dependent DNA curvature (Liu and Beveridge 2001), developed by extending the previous applications of dinucleotide step models, has apparently been found to match experimental electrophoretic results better (Hardwidge and Maher 2001).

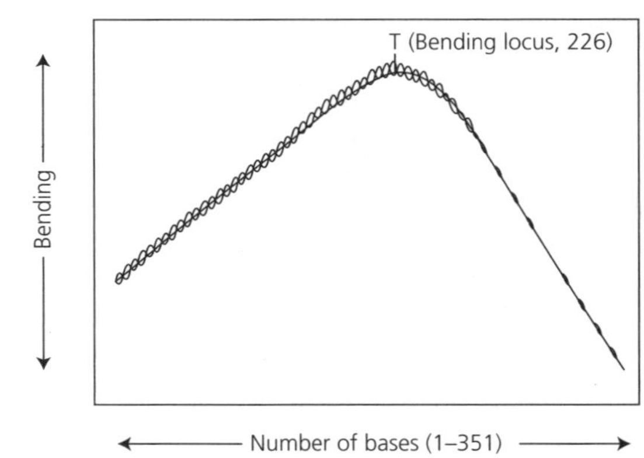

Figure 4.24 Intrinsic bending of the sequence upstream of the *vnf H* promoter of *Azotobacter vinelandii* as predicted by computer-simulated curvature analysis. [Reproduced from Bageshwar *et al.*, *Can J Microbiol* 1998;44:411 by permission of Professor HK Das.]

Some non-A/T DNA sequence elements have also been reported to be bent. X-ray diffraction analysis has revealed that a GGCC element was curved towards the major groove (Goodsell *et al.* 1993). A GGGCCC-containing DNA exhibited anomalous gel electrophoretic mobility, but only in the presence of divalent ions (10 mM Mg^{2+} or Ca^{2+} or 1 mM Zn^{2+}) (Brukner *et al.* 1994).

Kinks (localized bends of average angle 78°) have been observed in axially strained minicircles consisting of tandemly repeated d(A)s and d(GGGCCC) sequences under an atomic force microscope with a magnetically oscillated tip (Hahn *et al.* 1997). The frequency of kinks increased four-fold in the presence of 1 mM Zn^{2+}.

4.20 Triple-stranded DNA

The double-stranded nature of DNA is beyond any doubt. Reports have, meanwhile, appeared of the formation of triple-stranded DNA, particularly between homopolymers, which were, nevertheless, very much different from that postulated in the Pauling and Corey model. When two complementary single strands of DNA are mixed together under appropriate conditions, say in 0.1 M sodium phosphate buffer (pH 7.8), the absorbance at 260 nm decreases (hypochromicity) as a result of the formation of a hydrogen-bonded DNA double helix. This decrease continues till equimolar proportions of the two strands are mixed. Further addition of any one of the single strands results in an increase in the absorbance, signifying the end of double helix formation and the appearance of free single-stranded DNA along with the formed double-stranded DNA. Exactly the same phenomenon could be seen when a solution of dT homopolymer was added to a solution of dA homopolymer in the presence of 0.1 M sodium phosphate buffer (pH 7.8). The absorbance ceased to fall as the dT homopolymer concentration in the mixture exceeded 50% and instead a steady rise ensued [Fig. 4.25(a)]. The dA:dT double helix continued to be formed till the proportion of dA homopolymer and dT homopolymer in the mixture was equal; after that, it was a mixture of dA:dT double helix and dT single-stranded DNA. If, however, the same experiment was done in presence of 1.0 M sodium phosphate buffer (pH 7.8) (or in the presence of 50 mM $MgCl_2$), the decline in absorbance at 260 nm continued till the proportion of dT homopolymer in the mixture reached 66% [Fig. 4.25(b)] (Riley *et al.* 1966). This latter observation was considered to signify the formation of a triple helix consisting of one dA strand and two dT strands. The postulated triple helical structure could be identified as a distinct entity denser than the double helix on caesium

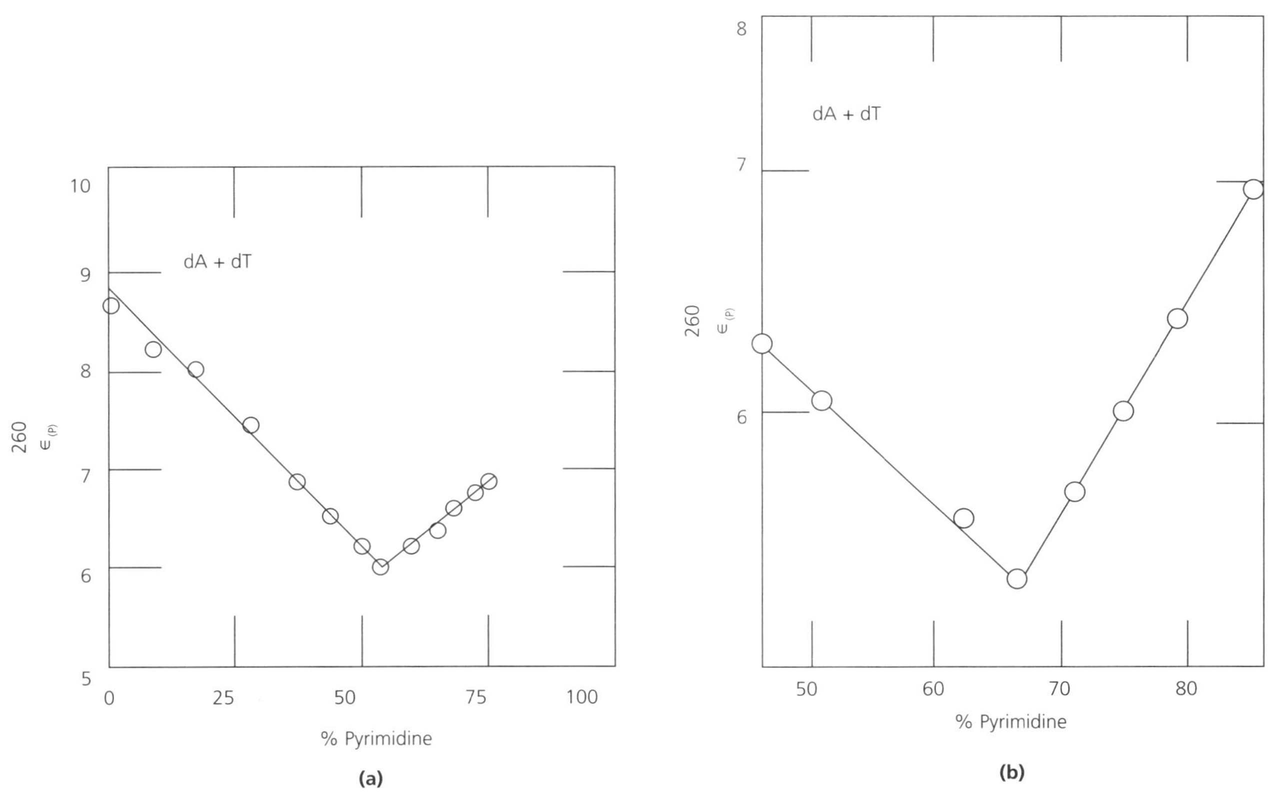

Figure 4.25 (a) Change in absorbance at 260 nm as dT homopolymer is mixed with dA homopolymer in 0.1 M sodium phosphate buffer, pH 7.8. (b) Change in absorbance at 260 nm as dT homopolymer is mixed with dA homopolymer in 1.0 M sodium phosphate buffer, pH 7.8. [Reproduced from Riley *et al.*, *J Mol Biol* 1966;20:367, 376 by permission of the Academic Press, London.]

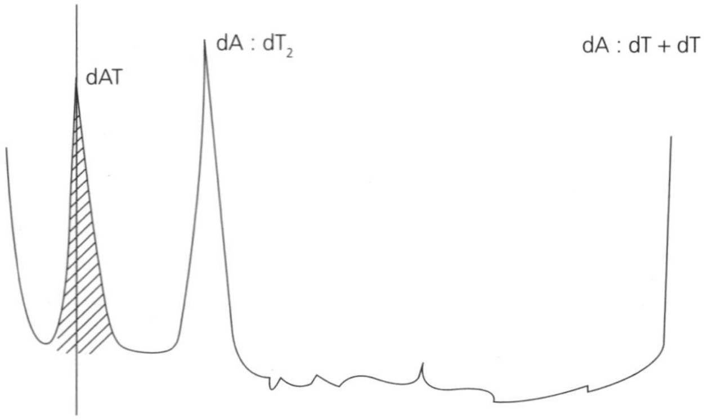

Figure 4.26 Densitometer tracing from ultraviolet photographs of caesium sulphate equilibrium density gradient centrifugation of dA:dT double helix and dA:dT$_2$ triple helix. The dAT copolymer was used as a marker (ρ = 1.426). The double helix has a higher buoyant density than the triple helix. [Reproduced from Riley *et al.*, *J Mol Biol* 1966;20:374 by permission of the Academic Press, London.]

sulphate equilibrium density gradient centrifugation (Fig. 4.26; Riley *et al.* 1966) and also on polyacrylamide gel electrophoresis (Fig. 4.27; Shafer 1998).

4.21 Centrifugal sedimentation

The rate of sedimentation of a particle or a macromolecule can be enhanced manifold by applying centrifugal force. The acceleration imparted to a molecule by centrifugal force is given by ω^2x, where ω is the angular velocity of the centrifuge rotor (the angular velocity of the rotor is equivalent to the product of 2π and the number of revolutions it is undergoing per second) and x being the distance of the molecule from the axis of rotation. The force exerted on the molecule is thus $m\omega^2x$, where m is the mass of the molecule. The mass, of course, would have to be corrected for its buoyancy, which is determined by the density of the liquid medium in which the molecule is present. Both velocity sedimentation and equilibrium sedimentation are employed.

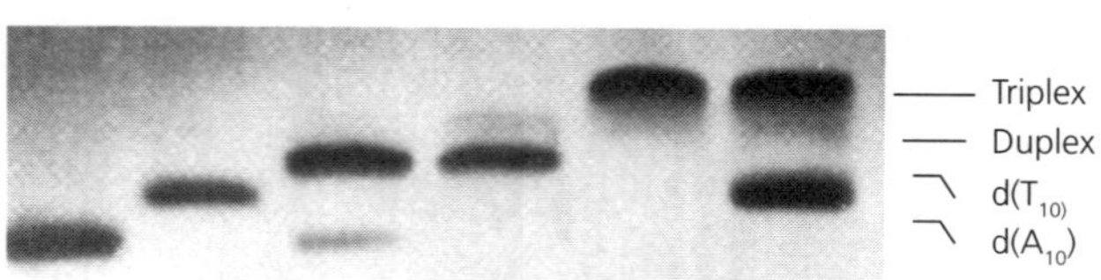

Figure 4.27 Bands of dA, dT, dA:dT and dA:dT2 on polyacrylamide gel after electrophoresis and staining. The single-stranded DNA molecules have the fastest mobility, followed by the duplex DNA, with the triple-stranded DNA being the slowest. [Adapted from Shafer, *Prog Nucleic Acid Res Mol Biol* 1998;59:62 by permission of Professor Richard H. Shafer and Academic Press, Orlando, FL.]

Velocity sedimentation is used when molecules of the same density but different masses are to be separated. An example would be two DNA molecules of same base composition and structure (hence same density) but different chain lengths. As the force exerted on the molecule would be proportional to its mass, the larger DNA molecule would sediment faster and would be present closer to the bottom of the tube, while the smaller molecule would sediment slower and would be away from the bottom. The medium used for centrifugation need to have a density lower than the density of the molecules to be separated. A solution of sucrose or Ficol or a similar substance is used. For the convenience of separating all the sizes of molecules in a small tube without sedimenting the large ones to the bottom and also for getting sharp bands, a density gradient is used. In velocity sedimentation, all the molecules would finally be sedimented to the bottom of the centrifuge tube, some sooner and some later. The centrifugation would, thus, have to be interrupted after a suitable period of time. The separated molecules are collected by puncturing a hole at the bottom of the centrifuge tube and collecting drops in different tubes. Measurement of the optical density of the fractions at 260 μm is used to detect the presence of DNA.

Equilibrium sedimentation is used when molecules of different density but same or different masses are to be separated. An example would be a double-stranded DNA and a triple-stranded DNA irrespective of their chain length. Another example would be two double-stranded DNA: one rich in G/C, another rich in A/T. Here, the medium used for centrifugation need to have a density higher than the density of the molecules to be separated. A concentrated solution of caesium sulphate or caesium chloride is generally used. No density gradient need be made here as the gradient is formed by itself during the centrifugation. The principle behind the operation is that under equilibrium condition the rate at which the molecules sediment down becomes equal to the rate at which they diffuse back because of thermal motion and Brownian movement. It does not matter for how long the centrifugation is done; the molecules stay put at their positions in the centrifuge tube once equilibrium is attained. The position a molecule would occupy depends on the buoyant density of the molecule. Specific bands can be isolated by puncturing, as described earlier. Rotors are also available with quartz windows and the position of the bands can be photographed while the centrifugation is on.

4.22 Nature of bond between the third DNA strand and the DNA duplex

It has been postulated that a new set of hydrogen bonds, referred to as Hoogsteen or Hoogsteen-like bonds, come into play. These are represented in Fig. 4.28 (Shafer 1998). Only the two purine bases have additional hydrogen

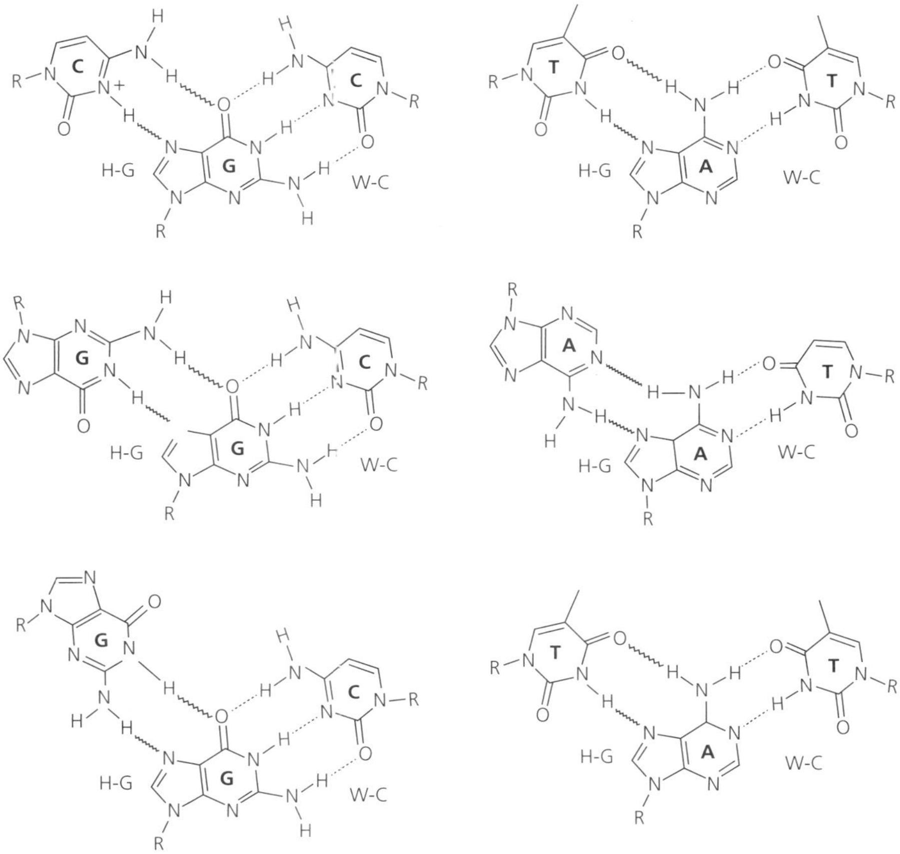

Figure 4.28 Representation of Hoogsteen (H–G) hydrogen bonds (〰〰) that can be formed in addition to Watson–Crick (W–C) hydrogen bonds (- - - - - -), resulting in DNA triplexes. [Reproduced from Shafer, *Prog Nucleic Acid Res Mol Biol* 1998;59:59 by permission of Professor Richard H. Shafer and the Academic Press, Orlando, FL.]

bond–forming capacity and can interact with another base without affecting the Watson–Crick hydrogen bonds in the DNA duplex. The stretch of purine bases would have to be in the same DNA strand in order to be able to form a triple-stranded structure. An adenine in the DNA duplex can form Hoogsteen hydrogen bonds with either a thymine or another adenine, while a guanine in the DNA duplex can form Hoogsteen hydrogen bonds with either a cytosine or another guanine. Thus, a DNA duplex between a purine homopolymer and a pyrimidine homopolymer can form a triplex structure with either a pyrimidine homopolymer (pyr–pur–pyr triplex) or a purine homopolymer (pyr–pur–pur triplex). Because of the postulated T3NH proton that is Hoogsteen hydrogen bonded to A7N (Fig. 4.28), a new imino proton should appear for each triplet, which should be possible to detect in a low-field NMR spectra in water. A mixture of thymine and adenine homopolymers (dT_{10} and dA_{10}) in the molar ratio of 2:1 was taken and the NMR spectra was determined in the presence of increasing concentration of $MgCl_2$ (Pilch *et al.* 1990). A dA–dT double strand and unassociated single-stranded dT would be expected at low Mg ion concentrations, while at higher Mg ion concentrations, the DNA triplex should appear. Indeed, distinct new imino proton resonances did appear at even 10 mM $MgCl_2$ (Fig. 4.29; Pilch *et al.* 1990). Nuclear Overhauser enhancement spectroscopy (NOESY) provided confirmatory evidence that the new imino proton resonances observed were because of the Hoogsteen hydrogen bonds (Pilch *et al.* 1990).

Intramolecular triple-stranded stretches have been inferred to be formed in circular double-stranded DNA (plasmids) containing homopurine–homopyrimidine mirror repeats (Mirkin *et al.* 1987). One of the mirror repeats used is shown in Fig. 4.30.

Two-dimensional (2D) chloroquine agarose gel electrophoresis has indicated the formation of a novel structure (Fig. 4.31; Mirkin *et al.* 1987), very different from double-stranded DNA and presumably containing stretches of triple-stranded DNA (Mirkin *et al.* 1987). Interestingly, formation of this novel structure has rendered the plasmid DNA extremely susceptible to S1 nuclease, which is specific for single-stranded DNA. The S1-susceptible region has been mapped and a partially triple-stranded and partially single-stranded structure (dubbed H-DNA) has been proposed (Fig. 4.32; Mirkin *et al.* 1987). It is thought that a part of the pyrimidine strand folds back onto the intact double-stranded region to form the triple helical region, leaving the complementary region of the purine strand single stranded.

DNA triple-stranded formation at chromosomal translocation sites in human cancer has been inferred by gel electrophoresis, triple-stranded DNA–specific

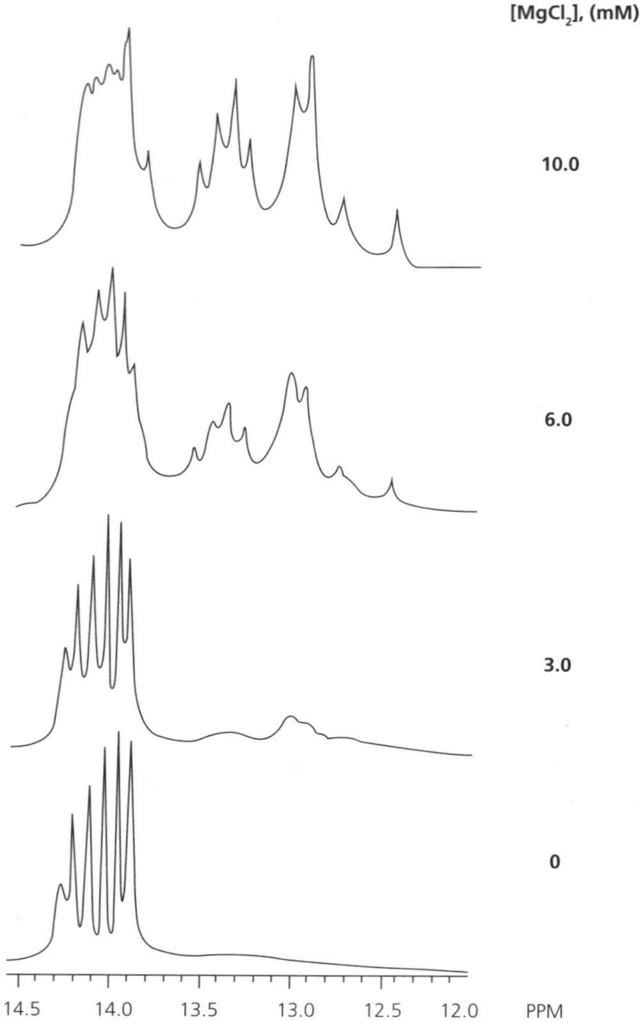

Figure 4.29 Low-field NMR spectra of A_{10}:$2T_{10}$ in water showing the appearance of additional imino proton resonance as a result of an increase in the concentration of $MgCl_2$. [Reproduced from Pilch *et al.*, *Proc Natl Acad Sci USA* 1990;87:1945 by permission of Professor Richard H. Shafer.]

antibody binding and circular dichroism studies (Raghavan *et al.* 2005).

Sites hypersensitive to single-strand–specific S1 nuclease have been detected in many natural DNA segments, and quite often, these DNA segments have homopurine or homopyrimidine mirror repeats (Mirkin *et al.* 1987). Some of the sequences at which S1 hypersensitivity have been observed are: **GGGGA**AGAG G**AGGGG** in chicken *β*-globin gene; **G**GGGTCAGGGG in rat preproinsulin II gene; GGGGAGCCTGGGG in SV40 enhancer; AGAGAGAGAAGAGAAGAGAGAGA, TCTCTC CCACTCTCT and CTCTCTCCACTCTCT in Drosophila heat shock genes hsp26, hsp22 and hsp70, respectively; GGGGGCTATAAAAGGGGG in Adenovirus major late promoter; CCCTCCCCATAAGCGCCCCTCCC in human c-myc gene; CCTCCTCCACCTCCTCC in human *α*2-*α*1 globin intergenic region and TCCCTCCCTTCCTCCCTC CCT in chicken *α*2(1) collagen promoter.

5′-AATTC | AAGGGAGAAGGGG | TATA | GGGGGAAGAGGGAA | GGATC- 3′
3′-TTAAG | T TCCCT CTT CCCC | ATAT | CCCCC TT CT CCCT T | CCTAG- 5′

Figure 4.30 A mirror repeat sequence used to study H-DNA formation. [Reproduced from Mirkin *et al.*, *Nature* 19873;30:496 by permission of *Nature*, London.]

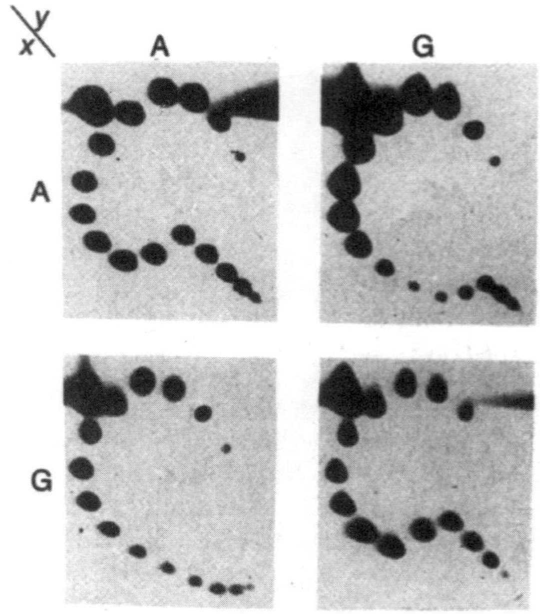

Figure 4.31 Novel structure as revealed by 2D gel electrophoresis, suggesting the formation of H-DNA. A tailored series of plasmids carrying the inserts AAGGGAGAAXGGGGTATAGGGGYAAGAGGGAA, where X and Y may be either A or G, were subjected to 2D gel electrophoresis. The H form of DNA was inferred to have arisen when palindromes were possible, as in the cases where X = Y = G (bottom right) or X = Y = A (top left), but not when palindromes were impossible, as in the cases where X = A, Y = G (top right) or X = G, Y = A (bottom left). Electrophoresis in the first direction (top to bottom) was in 0.2 M sodium citrate buffer (pH 4.3). Chloroquine was added before electrophoresis in the second direction (left to right) to relax superhelical stress. [Reproduced from Mirkin *et al.*, *Nature* 1987;330:496 by permission of *Nature*, London.]

It has, therefore, been suggested that 'the H form is the structural basis for hypersensitivity to S1' (Mirkin *et al.* 1987).

It must be borne in mind that the loop region of a stem-and-loop (also known as a cruciform) structure, which can possibly arise from 'inverted repeats' in DNA, is also single stranded and would be S1 nuclease sensitive. In 'inverted repeats', the copies of the 'repeats' are complementary when read in opposite directions on the same strand and are, thus, fundamentally different from mirror repeats. Fig. 4.33 shows an example of two

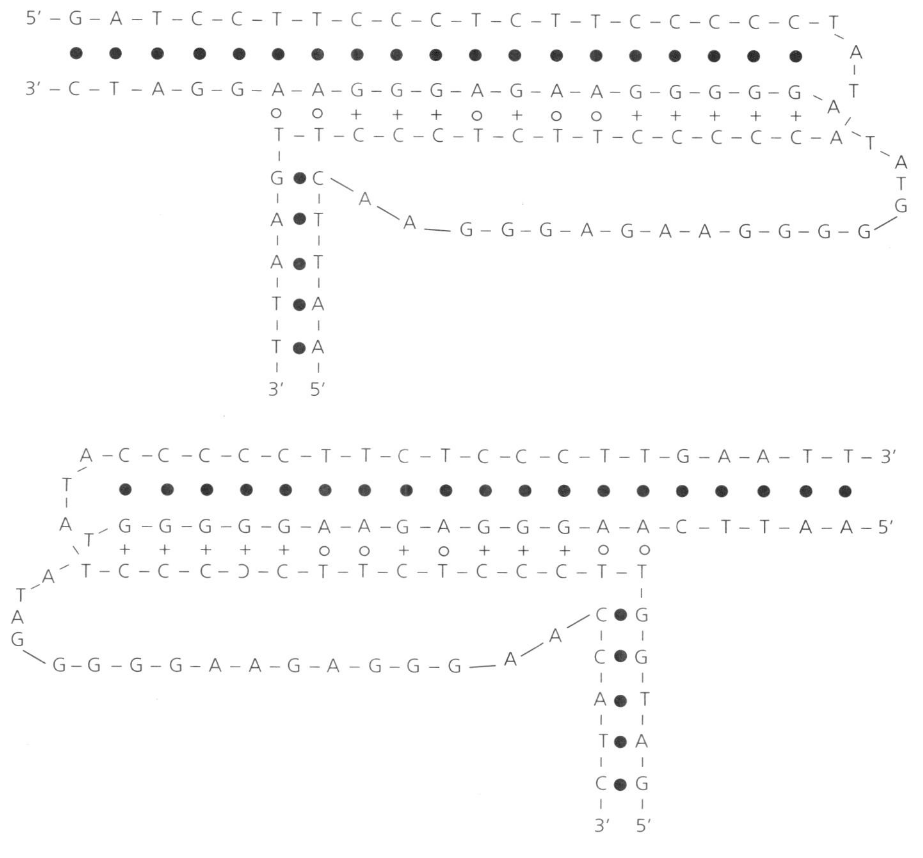

Figure 4.32 The postulated H-DNA structure showing the triple-stranded region and the single-stranded region. [Reproduced from Mirkin *et al.*, *Nature* 1987;330:497 by permission of *Nature*, London.]

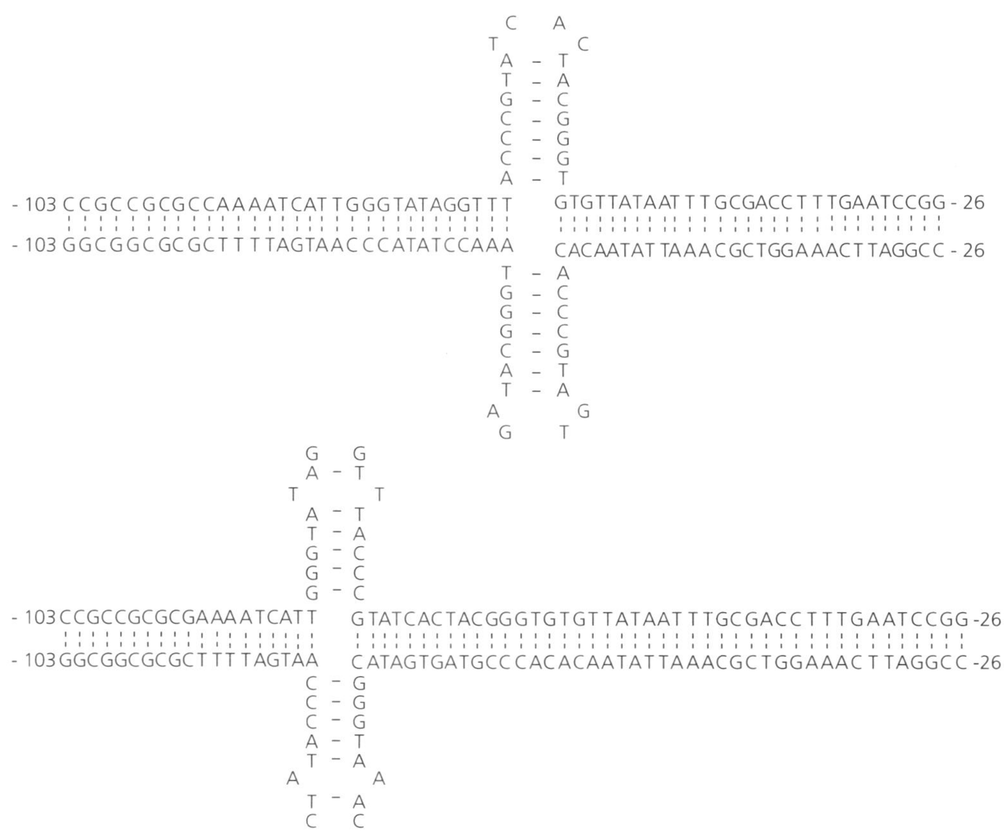

Figure 4.33 Potential stem-and-loop (cruciform) structures as a possible consequence of the presence of 'inverted repeats'. [Reproduced from Dimri and Das, *Nucl Acids Res* 1990;18:155 by permission of Professor HK Das.]

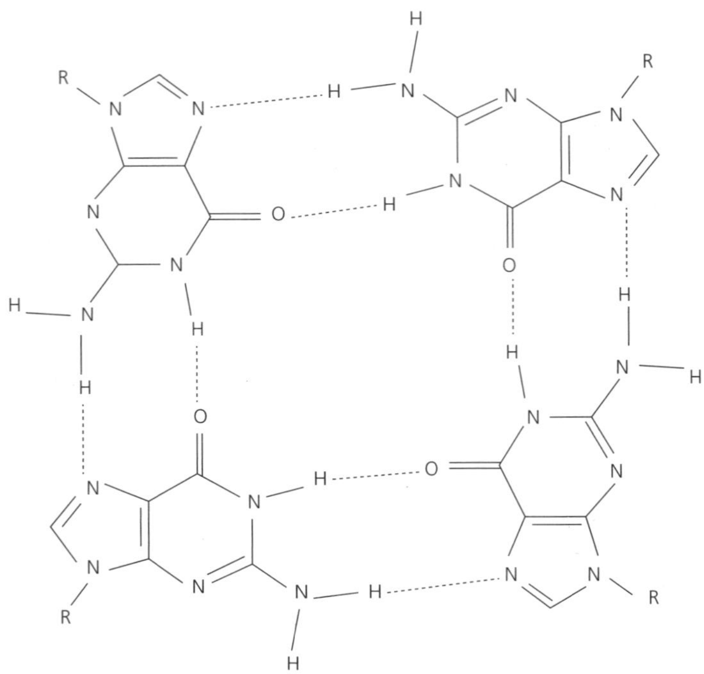

Figure 4.34 Guanine quartet postulated to be formed through Hoogsteen-like hydrogen bonds. [Reproduced from Shafer, *Prog Nucleic acid Res Mol Biol* 1997;59:55–94 by permission of Professor Richard H. Shafer and the Academic Press, Orlando, FL.]

potential stem-and-loop (cruciform) structures as a possible consequence of the presence of 'inverted repeats' in the promoter region of the *gyrA* gene of the nitrogen-fixing bacteria *Klebsiella pneumoniae* (Dimri and Das 1990). Whether cruciform structures are formed *in vivo* is, however, an open question.

4.23 Two-dimensional chloroquine gel electrophoresis

DNA molecules with unusual structures are analysed in agarose gels by electrophoresis in both dimensions one after another. Only one application is made in one corner of the slab gel and the molecules are separated on the basis of native character in one dimension (as explained earlier in the case of electrophoresis in single dimension) and then electrophoresis is done in the other dimension after addition of chloroquine to relax superhelical stress.

4.24 Nucleases

Whatever enzyme causes hydrolytic cleavage of phosphodiester linkages in a nucleic acid or an oligonucleotide is a nuclease. If a nuclease acts on both DNA and RNA, it is just referred to as a nuclease or a phosphodiesterase. Enzymes that are specific for DNA are called deoxyribonucleases, while enzymes that are specific for RNA are called ribonucleases. Enzymes that act on internal phosphodiester linkages, producing polynucleotides or oligonucleotides, are known as endonucleases, while enzymes that act from one end of the nucleic acid, producing a polynucleotide and mononucleotides, are known as exonucleases. There are nucleases specific for single-stranded nucleic acids or single-stranded regions of nucleic acids. Beef pancreatic ribonuclease is an endonuclease specific for RNA, while *E. coli* DNase I is an endonuclease specific for double-stranded DNA. Both Russel's viper venom phosphodiesterase and beef spleen phosphodiesterase are exonucleases that can act on either DNA or RNA, but the former enzyme acts from the 3′-end, releasing 5′-P mononucleotides, while the latter enzyme acts from the 5′-end, releasing 3′-P monnucleotides. Micrococcal nuclease, *Neurospora* nuclease, mung bean nuclease and S1 nuclease are all single-strand–specific endonucleases. *E. coli* exonuclease III and exonuclease V are both double-strand DNA–specific exonucleases from the 3′-end, while bacteriophage λ exonuclease is double-strand DNA–specific exonuclease from the 5′-end. *E. coli* exonuclease I and exonuclease VII are both single-strand DNA specific, but the former acts from the 5′-end, while the latter acts from the 3′-end.

4.25 Quadruple-stranded structures

Guanine residues have the propensity to associate through Hoogsteen-like hydrogen bonds, forming a square planar configuration called the guanine quartet (Fig. 4.34; Shafer 1998). Quadruplex structures comprising stacks of guanine quartets can, thus, be formed if runs of guanines are present in all the four segments, whether these are in four different strands or two strands with hair pins or a single strand with three hair pins.

Repeats of the type GnTm or GnA are known to be present in the telomere or terminal region of linear chromosomes (Blackburn 1991). It has been speculated that such sequences could form hairpin dimers or unimolecular triple hairpins, resulting in guanine quadruplex structures. It has also been speculated that such structures may help in chromosome association and may also provide nuclease-resistant caps at the ends of chromosomes (Shafer 1998).

The human genome contains as many as 376,000 guanine-rich sequences, which have the potential to form guanine quadruplex structures. Working rules have been proposed from theoretical considerations to predict which of these huge number of sequences are likely to form the quadruplex structure (Huppert and Balasubramanian 2005; Todd *et al.* 2005).

References

Astbury WT, Bell FO. Some recent developments in the X-ray study of proteins and related structures. *Cold Spring Harb Symp Quant Biol* 1938a;6:109–121.

Astbury WT, Bell FO. X-ray study of thymonucleic acid. *Nature* 1938b;141:747–748.

Bageshwar UK, Raina R, Choudhury NR, Das HK. Analysis of upstream activation of the vnfH promoter of Azotobacter vinelandii. *Can J Microbiol* 1998;44:405–415.

Blackburn EH. Structure and function of telomeres. *Nature* 1991;350:569–573.

Bolshoy A, McNamara P, Harrington RE, Trifonov EN. Curved DNA without A-A: Experimental estimation of all DNA wedge angles. *Proc Natl Acad Sci USA* 1991;88:2312–2316.

Brukner I, Susic S, Dlakic M, Savic A, Pongor S. Physiological concentration of magnesium ions induces a strong macroscopic curvature in GGGCCC-containing DNA. *J Mol Biol* 1994;236: 26–32.

Calladine CR, Drew HR. McCall MJ. The intrinsic curvature of DNA in solution. *J Mol Biol* 1988;201:127–137.

Chargaff E. Chemical specificity of nucleic acids and mechanism of their enzymatic degradation. *Experientia* 1950;6:201–209.

Cheema AK, Choudhury NR, Das HK. A- and T-tract mediated intrinsic curvature in native DNA between the binding site of the upstream activator NtrC and the *nifLA* promoter of *Klebsiella pneumoniae*. *J Bacteriol* 1999;181:5296–5302.

Conner BN, Yoon C, Dickerson JL, Dickerson RE. Helix geometry and hydration in an A-DNA tetramer CCGG. *J Mol Biol* 1984;174:663–695

Crick FHC. The structure of the hereditary material. *Scientific American*; 1954. Reprinted in Davern CI, ed. *Genetics: Readings from Scientific American*. San Francisco CA: WH Freeman & Company; 1981:49–55.

Delmonte C. *The DNA Structure Puzzles*. Kettering, UK: Clive Delmonte Publications; 2000.

Dickerson RE. Base sequence and helix structure variation in B and A DNA. *J Mol Biol* 1983a;166:419–441.

Dickerson RE. The DNA helix and how it is read. *Sci Am* 1983b;249(6):86–102.

Dickerson RE, Drew HR, Conner BN, Kopka ML, Pjura PE. Helix geometry and hydration in A-DNA, B-DNA and Z-DNA. *Cold Spring Harb Symp Quant Biol* 1982;47:13–24.

Dimri GP, Das HK. Cloning and sequence analysis of *gyrA* gene of *Klebsiella pneumoniae*. *Nucleic Acids Res* 1990;18:151–156.

Franklin RE, Gosling RG. Molecular configuration in sodium thymonucleate. *Nature* 1953;171:740–741.

Goodsell DS, Kopka ML, Cascio D, Dickerson RE. Crystal structure of CATGGCCATG and its implications for A-tract bending models. *Proc Natl Acad Sci USA* 1993;90:2930–2934.

Gulland JM, Jordan DO, Taylor HFW. Deoxypentose nucleic acids. Part II. Electrometric titration of the acidic and the basic groups of the deoxypentose nucleic acid of calf thymus. *J Chem Soc* 1947:1131–1141.

Hahn W, Dlakic M, Zhu YJ, Lindsay SM, Harrington RE. Strained DNA is kinked by low concentrations of Zn. *Proc Natl Acad Sci USA* 1997;94:10565–10570.

Hardwidge PR, Maher LJ. Experimental evaluation of the Liu-Beveridge dinucleotide step model of DNA structure. *Nucleic Acids Res* 2001;29:2619–2625.

Huppert JL, Balasubramanian S. Prevalence of quadruplexes in the human genome. *Nucleic Acids Res* 2005;33:2908–2916

Jose J, Kaiser AD, Kornberg A. Enzymatic synthesis of deoxyribonucleic acid. VIII.

Frequencies of nearest neighbor base sequences in deoxyribonucleic acid. *J Biol Chem* 1961;236:864-875.

Jungner G, Jungner I, Allgén LG. Molecular weight determination of thymonucleic acid compounds by dielectric measurements. *Nature* 1949;163:849–850.

Koo H-S, Crothers DM. Calibration of DNA curvature and a unified description of sequence-directed bending. *Proc Natl Acad Sci USA* 1988;88:1763–1767.

Koo H-S, Wu H-M, Crothers DM. DNA bending at adenine-thymine tracts. *Nature* 1986;320:501–506.

Lemeshko SV, Powdrill T, Belosludtsev YY, Hogan M. Oligonucleotides form a duplex with non-helical properties on a positively charged surface. *Nucleic Acids Res* 2001;29:3051–3058.

Liu Y, Beveridge DL. A refined prediction method for gel retardation of DNA oligonucleotides from dinucleotide step parameters: Reconciliation of DNA bending models with crystal structure data. *J Biomol Struct Dyn* 2001;18:505–526.

Marini JC, Levene SD, Crothers DM, Englund PT. Bent helical structure in kinetoplast DNA. *Proc Natl Acad Sci USA* 1982;79: 7664–7668.

Maxam AM, Gilbert W. A new method of sequencing DNA. *Proc Natl Acad Sci USA* 1977;74:560–564.

Mirkin SM, Lyamichev VI, Drushlyak KN, Dobrynin VN, Filippov SA, Frank-Kamenetskii MD. DNA H form requires a homopurine-homopyrimidine mirror repeat. *Nature* 1987;330:495–497.

Nagaich AK, Bhattacharyya D, Brahmachari SK, Bansal M. CA/TG sequence at the 5′-end of oligo A-tracts strongly modulates DNA curvature. *J Biol Chem* 1994;269:7824–7833.

Pauling L, Corey RB. Atomic coordinates and structure factors for two helical configurations of polypeptide chains. *Proc Natl Acad Sci USA* 1951;37:235–285.

Pauling L, Corey RB. Structure of the nucleic acids. *Nature* 1953a;171:346.

Pauling L, Corey RB. A proposed structure for the nucleic acids. *Proc Natl Acad Sci USA* 1953b;39: 84–97.

Perez-Martin J, Espinosa M. Correlation between DNA bending and transcriptional activation at a plasmid promoter. *J Mol Biol* 1994;241:7–17.

Pilch DS, Levenson C, Shafer RH. Structural analysis of the $(dA)_{10}.2(dT)_{10}$ triple helix. *Proc Natl Acad Sci USA* 1990;87:1 942–1947.

Raghavan SC, Chastain P, Lee JS, Hegde BG, Houston S, Langen R, Hsieh C-L, Haworth IS, Lieber MR. Evidence for a triplex DNA conformation at the bcl-2 major breakpoint region of the t(14;18) translocation. *J Biol Chem* 2005;280:22749–22760.

Riley M, Maling B, Chamberlin MJ. Physical and chemical characterization of two- and three-stranded adenine-thymine and adenine-uracil homopolymer complexes. *J Mol Biol* 1966;20:359–389.

Rodley GA, Scobie RS, Bates RHT. Lewitt RM. A possible conformation for double-stranded polynucleotides. *Proc Natl Acad Sci USA* 1976;73:2959–2963.

Sanger F, Nicklen S, Coulson AR. DNA sequencing with chain-terminating inhibitors. *Proc Natl Acad Sci USA* 1977;74:5463–5467.

Sasisekharan V, Pattabiraman N, Gupta G. Some implications of an alternative structure for DNA. *Proc Natl Acad Sci USA* 1978;75:4092–4096.

Schumaker VN, Richards EG, Schachman HK. A study of the kinetics of the enzymatic digestion of deoxyribonucleic acid. *J Am Chem Soc* 1956;78:4230–4236.

Shafer RH. Stability and structure of model DNA triplexes and quadruplexes and their interaction with small ligands. *Prog Nucleic Acid Res Mol Biol* 1997;59: 55–94.

Shakked Z, Rabinovich D, Cruse WBT, Egert E, Kennard O, Sala G, Salisbury SA, Viswamitra MA. Crystalline A-DNA: The X-ray analysis of the fragment d(G-G-T-A-T-A-C-C). *Proc R Soc London B* 1981;213:479–487.

Smith JD, Wyatt GR. The composition of some microbial deoxypentose nucleic acids. *Biochem J* 1951;49:144–148.

Snyder M, Buchman RA, Davis WR. Bent DNA at a yeast autonomously replicating sequence. *Nature* 1986;324: 87–89.

Soliva R, Monaco V, Gomez-Pinto I, Meeuwenoord NJ, Van der Mare GA, Van Boom JH, Gonzalez C, Orzco M. Solution structure of a DNA duplex with a chiral alkyl phosphonate moiety. *Nucleic Acids Res* 2001;29:2973–2985.

Thomas CA Jr, The enzymatic degradation of desoxyribose nucleic acid. *J Am Chem Soc* 1956;78:1861–1868.

Todd AK, Johnston M, Neidle S. Highly prevalent putative quadruplex sequence motifs in human DNA. *Nucleic Acids Res* 2005;33: 2901–2907.

Vargason JM, Eichman BF, Ho PS. The extended and eccentric E-DNA structure induced by cytosine methylation or bromination. *Nature Struct Biol* 2000;7:758–761.

Vargason JM, Henderson K, Ho PS. A crystallographic map of the transition from B-DNA to A-DNA. *Proc Natl Acad Sci USA* 2001;98:7265–7270.

Wang AH-J, Fujii S, van Boom JH, Rich A. Molecular structure of the octamer d(G-G-C-C-G-G-C-C): Modified A DNA. *Proc Natl Acad Sci USA* 1982;79: 3968–3972.

Wang AH-J, Quigley GJ, Kolpak FJ, Crawford JL, van Boom JH, van der Marel G, Rich A. Molecular structure of a left-handed DNA fragment at atomic resolution. *Nature* 1979;282:680–686.

Watson JD, Crick FHC. Molecular structure of nucleic acids. A structure for deoxyribose nucleic acid. *Nature* 1953;171: 737–738.

Wilkins MHF, Stokes AR, Wilson HR. Molecular structure of deoxypentose nucleic acids. *Nature* 1953;171:738–740.

Williams RC. Electron microscopy of sodium desoxyribonucleate by use of a new freeze-drying method. *Biochim Biophys Acta* 1952;9:237–239.

Wing RM, Drew HR, Takano T, Broka C, Tanaka S, Itakura K, Dickerson RE. Crystal structure analysis of a complete turn of B-DNA. *Nature* 1980;287:755–758.

Wyatt GR. The purine and pyrimidine composition of deoxypentose nucleic acids. *Biochem J* 1951;48:584–590.

Wyatt GR. The nucleic acids of some insect viruses. *J Gen Physiol* 1952;36:201–205.

Zahn K, Blattner RJ. Sequence induced DNA curvature at the bacteriophage lambda origin of replication. *Nature* 1985;317: 451–453.

Zamenhof S, Brawerman G, Chargaff E. On the deoxypentose nucleic acids from several microorganisms. *Biochim Biophys Acta* 1952;9:402–405.

5

Characteristics of DNA

H. K. Das

5.1 Sizes of natural DNA molecules

It has been established in Chapter 4 that DNA is a linear macromolecule consisting of deoxyribonucleotide units joined by 5′–3′ phosphodiester linkages. How long are these molecules? Very long indeed; quite often complete genomes are single molecules. DNA molecules comprising bacterial virus genomes can be 5×10^3 to 10^5 base pairs (bp) long. Bacterial genomes are of the order of 10^6 bp long, while eukaryotic chromosomes are 10^5 to 10^8 bp long. Fig. 5.1 gives a visual impression of the amount of DNA contained in a single cell of the bacterium *Escherichia coli.*

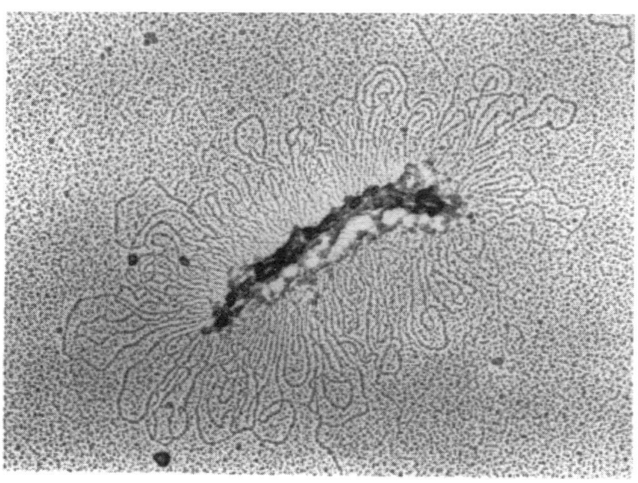

Figure 5.1 Electron micrograph showing DNA released on lysis of a single cell of the bacterium *Escherichia coli*. The remnants of the cell can be seen in the centre, while all the DNA, which is a single molecule, is strewn around it. The preparation was mixed with cytochrome c and formamide and spread on water. The film thus produced was picked up on carbon-coated grids, stained with uranyl acetate and shadowed with Pt-Pd. The micrograph was taken by Dr Nirupam Roy Choudhury.

5.2 Determination of the size of a DNA molecule by electron microscopy and sedimentation analysis

The most difficult step in the determination of the size of a natural DNA molecule is the isolation of the intact DNA molecule. DNA, because of its double helical structure, is a rather rigid molecule and is very prone to shear during isolation. One has to take extra care even for the isolation of an intact DNA molecule 10^5 bp long. Once the DNA is isolated, its molecular weight can be determined by either electron microscopy or sedimentation analysis. In both the methods, the length of the DNA or its molecular weight is determined by comparison with standard isolated DNA molecules. In the method involving electron microscopy, the contour length of the DNA molecules is measured and the number of base pairs in the DNA sample is assessed by comparison of the contour length of the standard DNA. Approximately, 3000 bp constitute 1 μm length of DNA.

In the velocity sedimentation method, the distance d travelled by the sample and the standard DNA are determined and compared. The logarithm of d is proportional to the logarithm of the molecular weight M. The average molecular weight of a base pair is approximately 680 Da.

5.3 Determination of the size of a DNA molecule by pulsed-field gel electrophoresis

Agarose gel electrophoresis has often been used to determine the length of DNA fragments by comparison with DNA of known length, but the resolution is not very good above about 2×10^4 bp under ordinary electrophoresis conditions. Linear DNA molecules are thought to move through agarose matrix by a biased reptation process, and above certain length, the molecules cannot orient themselves properly and hence get entangled. An agarose gel electrophoresis equipment was designed in which large linear DNA molecules, even 2×10^6 bp long, could be resolved. Multiple electrodes were arranged along a closed contour and clamped to a homogeneous electric field that would alternate between two orientations 120° apart (Fig. 5.2) (Chu *et al.* 1986). This presumably allowed the DNA molecules to reorient themselves and reptate smoothly.

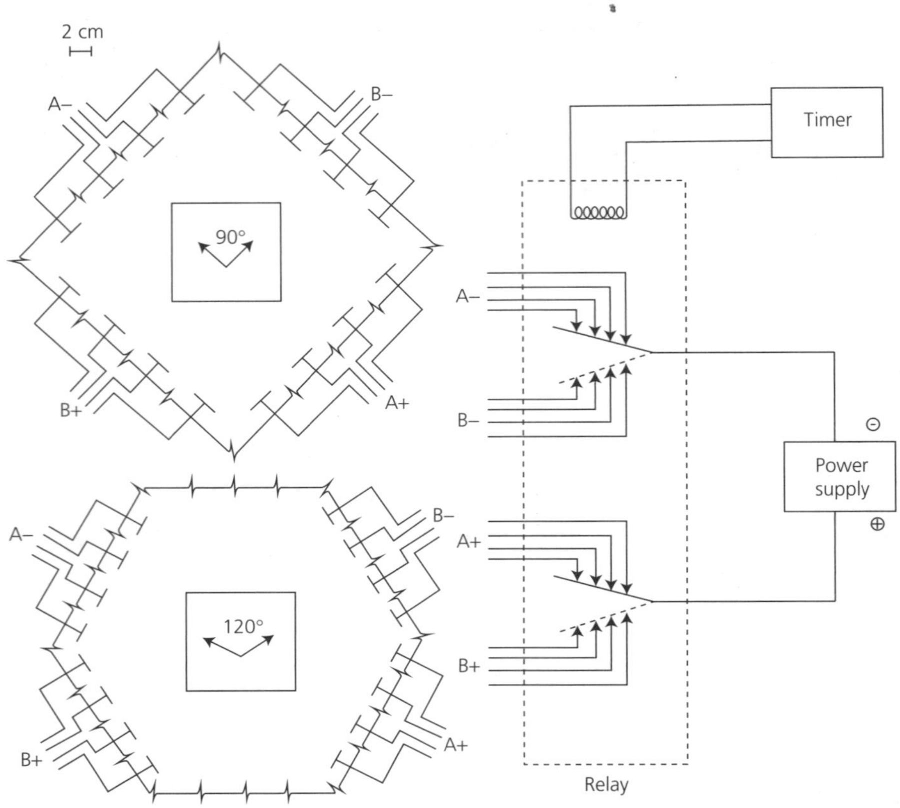

Figure 5.2 Design of an agarose gel electrophoresis apparatus (CHEF) in which multiple electrodes are arranged along a closed contour. A homogeneous electric field can be applied, alternating between two orientations 120° or 90° apart by using a timer (digital timer-intervalometer) that drives the switching relay. [Reproduced from Chu *et al.*, *Science* 1986;234:1583 by permission of the American Association for the Advancement of Science.]

This method turned out to be much superior to the earlier method employing a perpendicularly oriented non-uniform electric field (Schwartz and Cantor 1984). The duration for which the electric field would be allowed to act in one orientation before being switched to the other orientation is called the pulse. The effect of pulse time, the electric field strength, the temperature and the concentration of agarose on the process of separation have been well worked out (Mathew *et al.* 1988a, 1988b) and large linear DNA molecules can now be resolved reproducively and their size determined fairly accurately by comparison with standard molecules. The method has become known as 'pulsed-field gel electrophoresis.' Reptation becomes difficult for large circular molecules and they remain stuck in the well itself. Chromosomal DNA molecules of *Saccharomyces cerevisciae* (baker's yeast), which get separated well by this method (Fig. 5.3) (Chu *et al.* 1986), are used as size standards for large molecules. The yeast chromosomal DNA molecules cannot, however, be isolated intact by standard isolation procedures. A novel method has, therefore, been devised to electrophorese such large molecules. The yeast cells are treated with Zymolyase 5000 and suspended in low-gelling agarose and poured into moulds, which after solidification, form a gel block. The block containing the yeast cells is treated with EDTA (0.5 M) at 37°C overnight and then again with 1% lauroyl sarcosine, proteinase K and EDTA overnight. These successive treatments release the naked DNA from the cells inside the gel block without any shear of DNA. The treated block is then pressed into slots into running gel undergoing electrophoresis (Schwartz and Cantor 1984). This method works equally well with any other cell. The size of a bacterial genome, 2×10^6

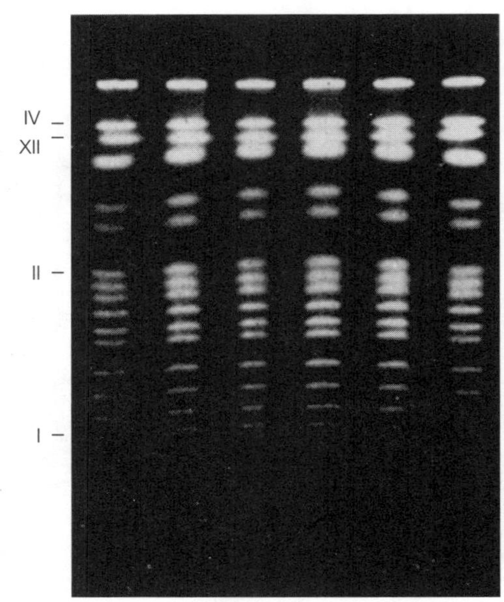

Figure 5.3 Separation of the chromosomes of baker's yeast (*Saccharomyces cerevisiae* strain YNN 295) by pulsed-field agarose gel electrophoresis in a CHEF hexagonal array apparatus. All the lanes contain the same material. The gel was stained with ethidium bromide (0.5 μg/mL) after electrophoresis and visualized under ultraviolet light. The chromosomes I, II, IV and XII are marked. [Reproduced from Chu *et al.*, *Science* 1986;234:1583 by permission of the American Association for the Advancement of Science.]

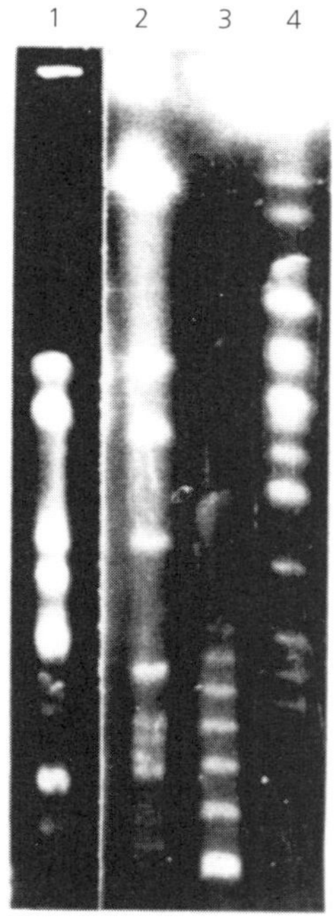

Figure 5.4 Pulsed-field agarose gel electrophoresis of the genome of *Azotobacter chroococcum* M4 after treatment with the restriction endonucleases *Ase*I and *Dra*I. Lanes mentioned here are from left to right. Lane 1, *Ase*I digest; lane 2, *Dra*I digest; lane 3, low-range PFG marker DNA; lane 4, *Saccharomyces cerevisiae* strain YPH 80 chromosomes. The gel was stained with ethidium bromide (0.5 μg/mL) and visualized under ultraviolet light. [Reproduced from Manna and Das, *Microbiology* 1994;140:1238 by permission of AC Manna and HK Das.]

Table 5.1 Recognition sites of some commonly used type II restriction endonucleases that cleave DNA infrequently. The arrows indicate cleavage sites of phosphodiester linkages.

Enzyme	Recognition sequence
*Ase*I	5′-A T↓T A A T-3′ 3′-T A A T↑T A-5′
*Dra*I	5′-T T T↓A A A-3′ 3′-A A A↑T T T-5′
*Swa*I	5′-A T T T↓A A A T-3′ 3′-T A A A↑T T T A-5′
*Nar*I	5′-G G↓C G C C-3′ 3′-C C G↑C G G-5′
*Not*I	5′-G C↓G G C C G C-3′ 3′-C G C C G G↑C G-5′

bp long, can, therefore, be determined this way. In practice, however, the genome is cleaved into several fragments and their sizes is determined. The sum of the sizes of all the fragments gives a more accurate measure of the size of the genome. The treated low-gelling agarose block containing the cells of the bacteria in question is equilibrated with a specific buffer and then treated with the 'type II restriction endonuclease', which would cleave the particular genome infrequently, producing, let us say, 6–10 DNA fragments. Pulsed-field gel electrophoresis is then carried out along with yeast chromosomes and low-range markers (such as bacteriophage λ DNA concatamers) in adjacent lanes and the sizes of the fragments are determined by comparison. Fig. 5.4 shows the results of a pulsed-field gel electrophoresis of the genome of *Azotobacter chroococcum* after treatment with the restriction endonucleases *Ase*I and *Dra*I (Manna and Das 1994). Summation of the sizes of the resolved restriction fragments has yielded the size of the DNA molecule comprising the genome to be around 5.3×10^6 bp.

5.4 Restriction endonucleases

The endonuclease (bovine pancreatic deoxyribonuclease) that has been talked about in Section 4.7, Chapter 4, is very non-specific with respect to the cleavage site. Digestion of DNA with such an endonuclease, therefore, generates mononucleotides and small oligonucleotides. The type II restriction endonucleases, however, cleave DNA at very specific sites, which are generally palindromic. Table 5.1 presents the specific sequences at which some of the more commonly used type II restriction endonucleases cleave DNA infrequently.

5.5 Precise knowledge of the number of base pairs in a DNA molecule from base sequence data

The most precise information on the number of base pairs in a DNA molecule can be derived if the complete sequence of bases of the molecule has been worked out. Initially, only some sequence data were available (Dunham *et al.* 1999; Hattori *et al.* 2000). The technology has indeed advanced over the years. Today, we know the sequences of close to 1500 eukaryotic genomes, including the human genome, of over 25,000 prokaryotic genomes and more than 4000 viral genomes. Table 5.2 lists the precise number of base pairs in some DNA molecules derived from base sequence data.

5.6 DNA molecules: Linear or circular

Eukaryotic chromosomes generally consist of linear DNA molecules. Many bacterial genomes, however, comprise circular DNA molecules. This means that there are no

Table 5.2 Precise number of bases in some DNA molecules as derived from base sequence data.

Source of the natural DNA molecule	Number of base pairs present in the DNA molecule
Escherichia coli phage ϕx 174 genome (Rf)	5,386
Escherichia coli phage T7 genome	39,936
Escherichia coli phage lambda (λ) genome	48,502
Thermotoga maritima genome	498,348
Borrelia burgdorferi genome	910,681
Rickettsia prowazekii genome	1,111,523
Haemophilus influenzae genome	1,830,023
Bacillus subtilis genome	4,214,814
Mycobacterium tuberculosis genome	4,411,529
Escherichia coli genome	4,639,221
Saccharomyces cerevisiae chromosome 1 (smallest yeast chromosome)	230,203
Saccharomyces cerevisiae chromosome 4 (largest yeast chromosome)	1,531,929
Long arm of human chromosome 21 + about 100,000	33,546,361

Source: www.ncbi.nlm.nih.gov

5′-end or 3′-end of these molecules. The genomes of *E. coli* and *Azotobacter vinelandii* are examples. Circularity of the genome of *E. coli* was recognized quite early when the tritium-labelled genome was allowed to leave its impression on a photosensitive film (Fig. 5.5) (Cairns 1963).

5.7 Labelling of the *E. coli* genome with ^{3}H-thymidine

E. coli B3 is a thymidine auxotroph; that is, it cannot synthesize thymidine and hence needs thymidine for its growth. This implies that this strain would not have an internal pool of thymidine, and as such, the ^{3}H-thymidine supplied from outside would not be diluted by any unlabelled thymidine pool pre-existent inside the cell. *E. coli* B3 was grown for several generations by John Cairns (1963) in a glucose salts medium supplemented with ^{3}H-labelled thymidine. This rendered the DNA of the cells labelled with ^{3}H-thymidine. Explosion of the bacterial cells and consequent shear of DNA during cell lysis were avoided by adopting a novel technique. The cells were suspended in 2 M sucrose, 0.02 M NaCl, 0.01 M EDTA pH 8.0 with purified bacteriophage T2 DNA (6 μg/mL) and dialysed in a cylindrical chamber faced on one side with VM Millipore filter (50-μm pore size) against 1% sodium lauryl sulphate (Duponol) in sucrose–NaCl–EDTA. The Duponol was removed by subsequent dialysis and the salt concentration was gradually raised to 4 M NaCl and then lowered to 0.4 M. The cellular proteins were removed from the genomic DNA by the high-salt treatment, which were later bound to the excess carrier T2 DNA during the low-salt treatment. The liquid was drained very slowly by puncturing the Millipore filter. The Millipore filter, which had the released intact DNA bound to it, was stuck to a microscopic slide and overlaid with an autoradiographic stripping film. The *β*-particles from the ^{3}H-labelled DNA released silver grains on the film, which were visualized by developing the film and examining under a light microscope.

The circularity of the DNA molecule comprising the genome of *Azotobacter vinelandii* has been inferred from its failure to leave the well in the sample block even on prolonged pulsed-field gel electrophoresis. *γ*-Irradiation of the DNA, which is known to produce double-strand breaks,

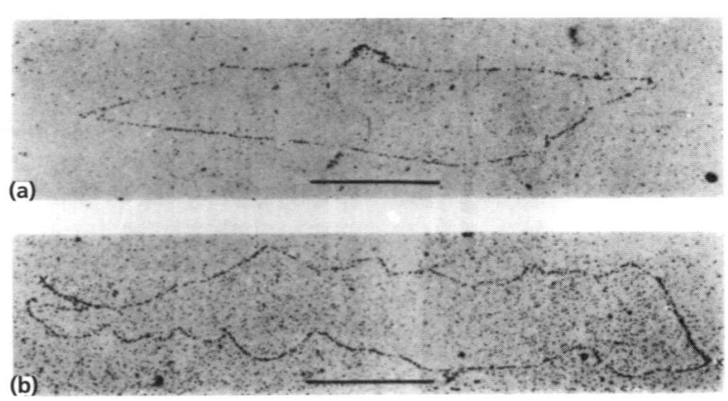

Figure 5.5 *Escherichia coli* genome as a circular DNA molecule as evidenced by the impression on a photosensitive film of the ^{3}H-thymidine-labelled genome of *E. coli* strain B3. The bars represent 100 μm [Reproduced from Cairns, *J Mol Biol* 1963;6:63 by permission of the Academic Press, London.]

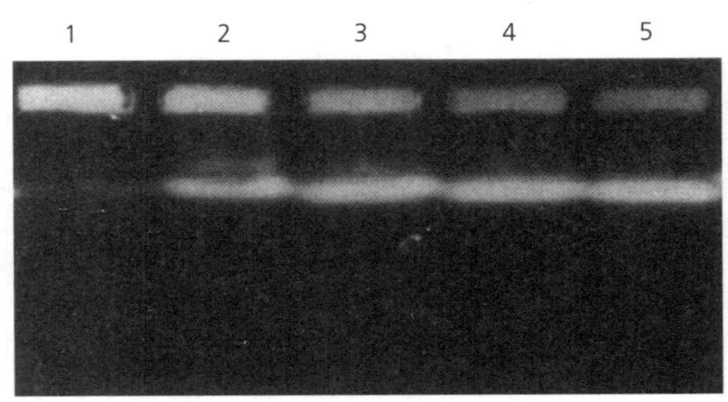

Figure 5.6 *Azotobacter vinelandii* UW genome as a circular DNA molecule as evidenced by its inability to enter agarose gel on pulsed-field electrophoresis. Linearization by *γ*-radiation imparts electrophoretic mobility to it. Agarose plugs containing lysed cells of *A. vinelandii* were gamma-irradiated with 20 Gray (lane 1), 30 Gray (lane 2), 60 Gray (lane 3), 90 Gray (lane 4) and 120 Gray (lane 5). The gel was stained with ethidium bromide after electrophoresis and visualized under ultraviolet light [Reproduced from Manna and Das, *J Genet* 1997;76:59 by permission of AC Manna and HK Das.]

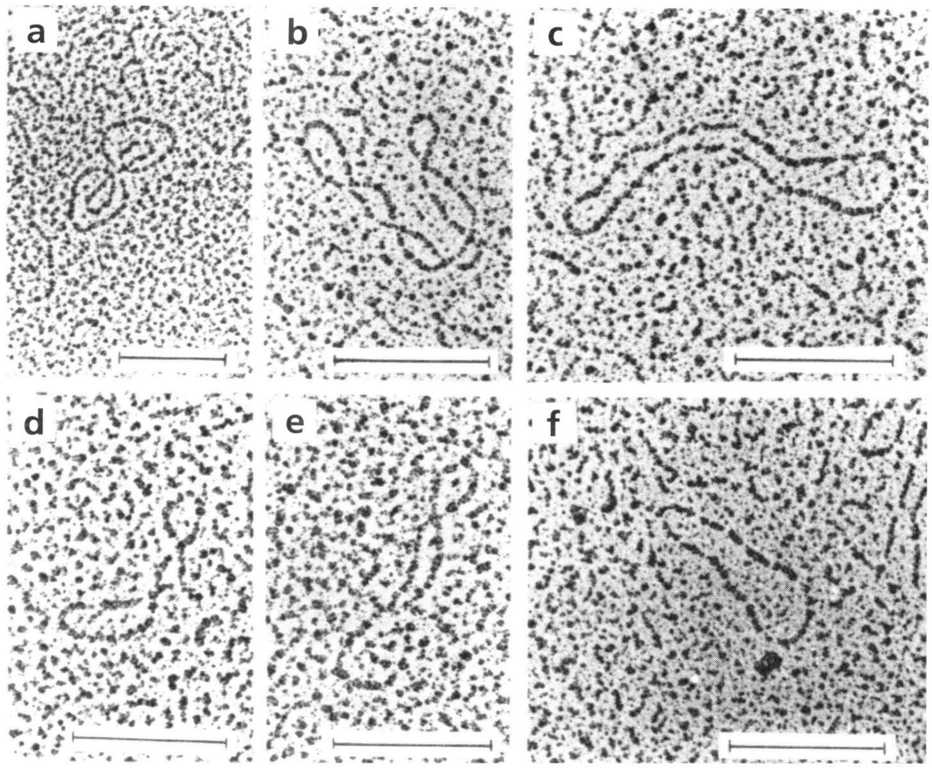

Figure 5.7 Electron micrographs showing plasmids of varying sizes between 1.8 kb and 3.2 kb. The bars represent 0.2 μm. The electron micrographs were taken by Dr Nirupam Roy Choudhury. [Reproduced from the PhD thesis of Ruby Singh by permission of Dr Ruby Singh.]

has enabled it to leave the well and form a band on pulsed-field gel electrophoresis. The dose of irradiation has been so adjusted to yield a significant number of molecules with only one break, converting circular molecules into full-length linear ones (Fig. 5.6) (Manna and Das 1997). Though many bacterial genomes are composed of circular DNA molecules, all need not be so (Ferdows and Barbour 1989; Lin *et al.* 1993). *Agrobacterium tumefaciens* has one linear and one circular chromosome (Allardet-Servent *et al.* 1993).

Many viruses have circular DNA as their genome. Though the genome of bacteriophage lambda (λ), as found in the virion, is linear, the viral DNA as found inside its host *E. coli* is circular. The small *E. coli* virus ϕx 174 also has a circular DNA genome. There are eukaryotic viruses with circular DNA genomes; the monkey virus SV40 and polyoma virus are examples. The *E. coli* viruses T2, T4 and T7, however, have linear DNA as their genome. Eukaryotic chromosomes comprise linear DNA molecules.

Comparatively smaller circular autonomously replicating DNA molecules, known as plasmids, are more or less ubiquitous. Fig. 5.7 shows the electron micrograph of two plasmids, 1.8 and 3.2 kb long. The *E. coli* sex factor F and the various resistance factors (R factors) are all plasmids. Col E1, RK 2 and R 6K are some of the other naturally occurring plasmids found in *E. coli*. The Sym plasmids of *Rhizobium* harbour the nitrogen fixation genes and the symbiotic genes. The ribosomal RNA genes of the human parasitic protozoa, *Entameoba hystolytica*, are located in the plasmid EhR1 and not in its linear chromosomes. Linear plasmids also exist (Barbour and Garon 1987; Crespi *et al.* 1992).

Circular DNA molecules are insensitive to exonucleases because of the absence of any 5′-end or 3′-end. The circular nature of the genome of the bacteriophage ϕx 174 was first inferred because of this property (Fiers and Sinsheimer 1962).

5.8 Supercoiled structure of circular DNA molecules

Polyoma virus genome, which is composed of a circular DNA molecule, was found to exist in two forms, one faster sedimenting (20S) and the other slower sedimenting (16S). Both the forms were infective. The slower sedimenting form could be generated from the faster sedimenting form by treatment with reducing agents such as hydroquinone and $FeCl_2$ (Vinograd *et al.* 1965) and also by mild treatment with beef pancreatic DNase (Dulbecco and Vogt 1963). It was first assumed that the slower sedimenting form was linear, but electron microscopy revealed that these were also circular [Fig. 5.8(b)]. The faster sedimenting form comprised a more compact 'twisted circular form' [Fig. 5.8(a)] (Vinograd *et al.* 1965), which apparently contained tertiary turns. It is in the faster sedimenting form that the DNA helix axis was also coiled, and this higher-order structure is now known as superhelical DNA or closed circular DNA. The slower sedimenting form is known as open circular DNA and

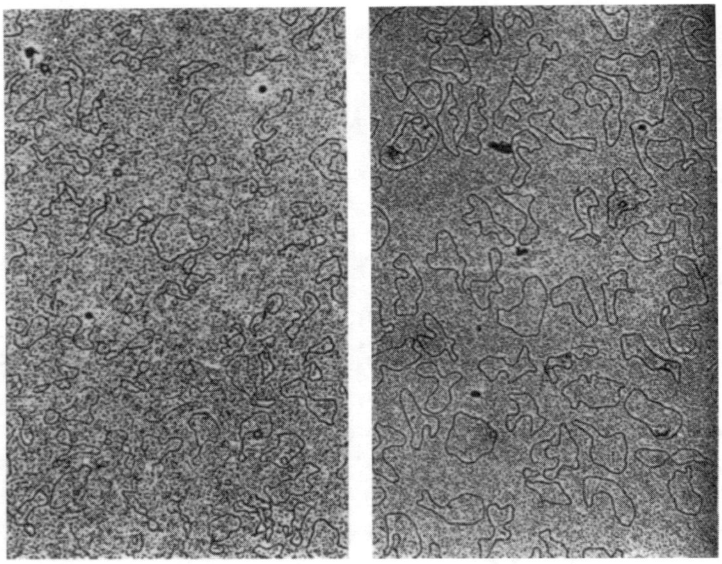

(a) (b)

Figure 5.8 Electron micrograph of supercoiled and open circular polyoma virus DNA. The viral DNA preparation was treated with pancreatic DNase. (a) Sample soon after the treatment with DNase was initiated; most of the circular viral DNA is still supercoiled; (b) sample taken after conversion of the supercoiled to the open circular form was almost complete. [Reproduced from Vinograd *et al.*, *Proc Natl Acad Sci USA* 1965;53:1106 by permission of Professor Philip Laipis.]

has single-strand breaks that releases the tertiary turns (Fig. 5.9). Subsequent observations have indeed confirmed that circular DNA molecules, in general, exist *in vivo* in a supercoiled form irrespective of the organism.

It is, in fact, believed today that virtually all DNA *in vivo*, even the linear ones, are supercoiled because of the presence of distinct topologically constrained domains (Bauer 1978).

5.9 Isolation of supercoiled circular DNA

Supercoiled circular DNA can be isolated in bulk by equilibrium density gradient centrifugation in caesium chloride in the presence of ethidium bromide (2,7-diamino-10-ethyl-9-phenyl-phenanthridinium bromide) (Radloff *et al.* 1967). Ethidium bromide intercalates double-stranded DNA, much more so if the DNA molecule has open ends or single-strand breaks, but not as much if the DNA is a more compact supercoiled closed circular molecule. Intercalation by ethidium bromide reduces the buoyant density of the DNA, and as a result, linear and open circular DNA bands on equilibrium sedimentation at a lower-density position (higher in the centrifuge tube) than the supercoiled DNA (Fig. 5.10). Small-scale or medium-scale purification of supercoiled DNA is possible by simple passage of the cell lysate through cartridges or spin columns containing glass powder or silica membranes or Sepharose or Sephacryl or hydroxyapatite or other matrices. Agarose gel elctrophoresis also separates supercoiled DNA from open circular DNA, the latter migrating slower than the former.

Figure 5.9 Diagrammatic representation of a closed circular (supercoiled) DNA and an open circular DNA. The example shown is the DNA of the polyoma virus. The closed circular DNA (I at the left) sediments at 20S, while the open circular DNA produced by a single-strand break (II at the right) sediments at 16S. [Adapted from Vinograd *et al.*, *Proc Natl Acad Sci USA* 1965;53:1109 by permission of Professor Philip Laipis.]

A critical property of closed circular (superhelical) DNA is its linking number (Lk), which describes the intertwining of the two strands. If the two strands of DNA are lying side by side (no helix) and their ends are joined (basically two separate circles parallel to each other), the linking number would be 0. The linking number would be 1 for a situation where one circle loops through the other just once, 2 for a situation where one circle loops through the other twice and so on. By convention, if the looping is in the right-handed way, the linking number is positive, but if the looping is in the left-handed way, the linking number is negative. Thus, for a molecule of DNA double helix 1000 bp long whose ends are just joined (relaxed circular DNA duplex; its axis lying in one plane), the linking number would be +100, as there are 10 bases for each turn (right handed) of the DNA. The linking number would obviously be higher (say 110) for a circular DNA duplex which has positive superhelical turns and lower (say 90) for the same circular DNA duplex when it has negative superhelical turns. This would be given by the following expression:

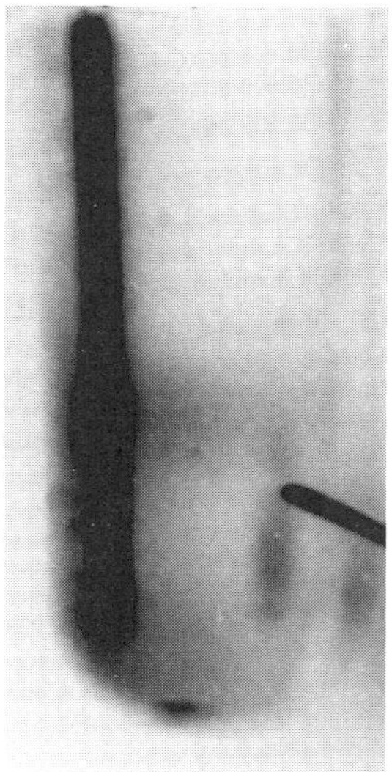

Figure 5.10 Purified supercoiled plasmid DNA in a centrifuge tube after CsCl–ethidium bromide equilibrium centrifugation. Visualization was in ultraviolet light.

$$Lk = Tw + Wr$$

where Tw is the twist and Wr is the writhing number. Tw would be equal to Lk for a relaxed circular DNA duplex (its axis lying in one plane), as under that condition, the value of Wr is 0.

In nature, most closed circular DNA molecules are negatively supercoiled. One can obtain DNA molecules with different degrees of writhing (hence supercoil) by producing nicks in one of the DNA strands and then closing the nick after different periods of time. The resulting DNA topoisomers can be separated by agarose gel electrophoresis (Fig. 5.11). The nicking-closing enzymes, many of which have been characterized now, are called topoisomerases.

5.10 DNA topoisomerases

DNA topoisomerases are ubiquitous enzymes that can carry out interconversions between topological isomers of DNA. Enzymes have been characterized that specifically relax negatively supercoiled DNA, that relax both negatively and positively supercoiled DNA and that introduce negative supercoiling turns into relaxed circular DNA. The first class of enzymes belong to type IA, the second class to type IB and the third class to type II. Type I enzymes transiently cleave only one strand of the double-stranded DNA and then seal it again. No energy is thought to be expended in this operation. Type II enzymes cleave both strands of DNA simultaneously and join them again after negative supercoiling turns have been introduced. This process is coupled to ATP hydrolysis. DNA topoisomerase I and DNA topoisomerase III from both *E. coli* and *S. cerevisiae* belong to type I, while DNA topoisomerase I from mouse and other eukaryotes belong to type IB. On the other hand, DNA gyrase and topoisomerase IV from *E. coli* belong to type II. *E. coli* DNA gyrase can, in fact, convert positive supercoils into negative supercoils. As more and more topoisomerases have been discovered from different sources, subgroups have been conceived within the three types. For example, ATP-dependent topoisomerases capable of introducing positive supercoil in DNA are now known (Wang 1996).

How does supercoiled DNA arise *in vivo*? DNA of the eukaryotic viruses, SV40 and polyoma, are supercoiled presumably because their replication involves wounding on nucleosomes like a solenoid (details in Chapter 7). The replicated DNA assumes superhelical turns as it comes off from the nucleosomes. Nucleosome structures may not be present in prokaryotes, but it is felt that movement of any macromolecular assembly on the double-stranded DNA, as in the case of DNA replication or transcription, would produce supercoil.

5.11 Single-stranded DNA

Some students have the misconception that all DNA molecules, by definition, are double stranded.

All natural DNA molecules are not double stranded. Several viruses have single-stranded DNA as their genetic material. The genomes of the *E. coli* viruses, ϕx174, G4, S3, fd, f1 and M13, are all single-stranded DNA circles. Parvoviruses, which have single-stranded DNA as their genome, can have mouse, cat, dog and

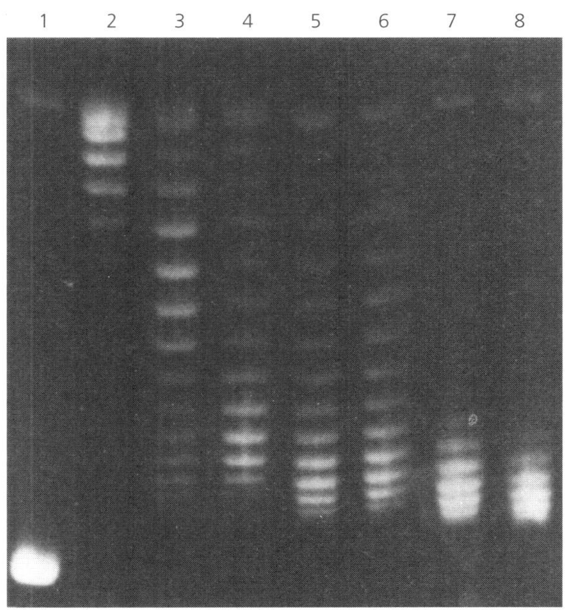

Figure 5.11 Separation of topoisomers of the circular SV40 DNA by agarose gel electrophoresis. Lane 1, the top faint band represents the fully relaxed DNA and the strong bottom band represents the fully supercoiled DNA; lanes 2 to 8 exhibit various DNA topoisomers of SV40 DNA in the order of increasing superhelicity. [Adapted from Rouviere-Yaniv *et al.*, *Cell* 1979;17:268 with permission of Elsevier Science.]

man as their host. The genome of the hepatitis B virus of humans consists of a single circular DNA molecule that is about 75% double stranded and 25% single stranded.

5.12 RNA as genetic material

Some viruses may also have RNA, either single stranded or double stranded, as their genetic material.

The genomes of the *E. coli* viruses MS2, R17, f2 and Qβ are all single-stranded linear RNA molecules, 3600–4200 ribonucleotides long. Nevertheless, it is thought that regions of these linear RNA molecules having complementary sequences fold onto themselves and form base-paired segments, as inferred from the insensitivity of the isolated RNA towards the single-strand-specific ribonuclease T_1 (Fiers *et al.* 1976). A portion of the sequence of the 3569-ribonucleotides long MS2 RNA and postulated base-paired segments is shown in Fig. 5.12 (Fiers *et al.* 1976). These RNA genomes can be directly 'translated' into proteins (see Chapters 14 and 15 for details on 'translation') and hence, these are termed (+) RNA viruses.

Many animal viruses also have RNA as their genome. Examples of single-stranded (+) RNA viruses are yellow fever virus, encephalitis virus, rubella virus, polio virus and foot and mouth disease virus. There are, however, RNA viruses whose genomes cannot be directly 'translated' into proteins. The complementary strands of these RNA molecules would have to be synthesized first, which can only be 'translated' into proteins. These are known as (–) RNA viruses. Examples of single-stranded (–) RNA viruses are influenza virus, mumps virus, measles virus and rabies virus.

Similarly, many plant viruses also have RNA as their genome. The most worked out plant virus is the tobacco mosaic virus, which has a single-stranded (+) RNA genome, only 6395 ribonucleotides long.

Viruses with double-stranded RNA genomes also exist. Examples are Reovirus and rice dwarf virus.

5.13 Buoyant density of DNA: A function of its GC (guanine plus cytosine) content

Buoyant density is determined by equilibrium sedimentation (generally in CsCl) in an ultracentrifuge (see Chapter 4). The buoyant density of DNA has been found to be dependent on the content of guanine and cytosine (GC) in it (Sueoka *et al.* 1959). The GC content (molar percentage) and buoyant density of double-stranded DNA from various sources have been determined experimentally and the following empirical relationship has been arrived at:

$\rho = 1.658 + 0.100\ (GC/100)$ (Rolfe and Messelson 1959)

$\rho = 1.660 + 0.098\ (GC/100)$ (Schildkraut *et al.* 1962)

where (GC) denotes the molar percentage of GC content of DNA. Fig. 5.13 shows a plot of the relationship derived by Schildkraut *et al.* (1962). Thus, a synthetic poly dAT duplex would have a buoyant density of 1.660, and a synthetic poly d(GC) duplex would have a buoyant density of 1.758.

DNA of the *E. coli* phage lambda (λ) has been sheared more or less in the middle and the two have been possible to separate by equilibrium sedimentation because of a marked difference of GC content between the two halves.

5.14 Hyperchromic effect

Nucleic acids absorb maximally around 260 µm. This absorption is the contribution of the purine and pyrimidine bases, but the absorption by the intact nucleic acids is much lower than the summation of the absorptivities of the component mononucleotides. Complete hydrolysis of double-stranded DNA can indeed result in 60–70% increase in absorption at 260 µm. Even dideoxynucleotides possess 5–10% lower absorptivities than the summation of the absorptivities of the constituents. It is thought that even in small polynucleotides, the purine and pyrimidine bases are stacked in layers above each other. The interaction of π-electrons of adjacent rings stabilizes this conformation and, at the same time, forms an essentially new electronic species. The absorption at 260 µm of the new species is the characteristic of the entire molecule and not the characteristics of the component mononucleotides added together.

Interfering with the ionizable groups of the component purine and pyrimidine bases by forming hydrogen bonds also disturbs the π-electron interaction. Besides, restriction of internucleotide rotation because of formation of the double helical structure stabilized by hydrogen bonds should also have a hypochromic effect. Double-stranded DNA on just denaturation (separation of the two strands by breaking hydrogen bonds) without cleavage of the phosphodiester linkage can exhibit 30–40% hyperchromicity.

Denaturation of double-stranded DNA can take place at high temperatures, extreme pH, in low salt concentration or in the presence of formamide. The hyperchromic effect sets in gradually as a function of temperature or pH, and a sigmoidal curve is obtained when absorbance at 260 µm is plotted against temperature or pH. Such curves are known as 'melting curves' and are indeed characteristic of the particular DNA. The melting curves of DNA from some representative sources determined

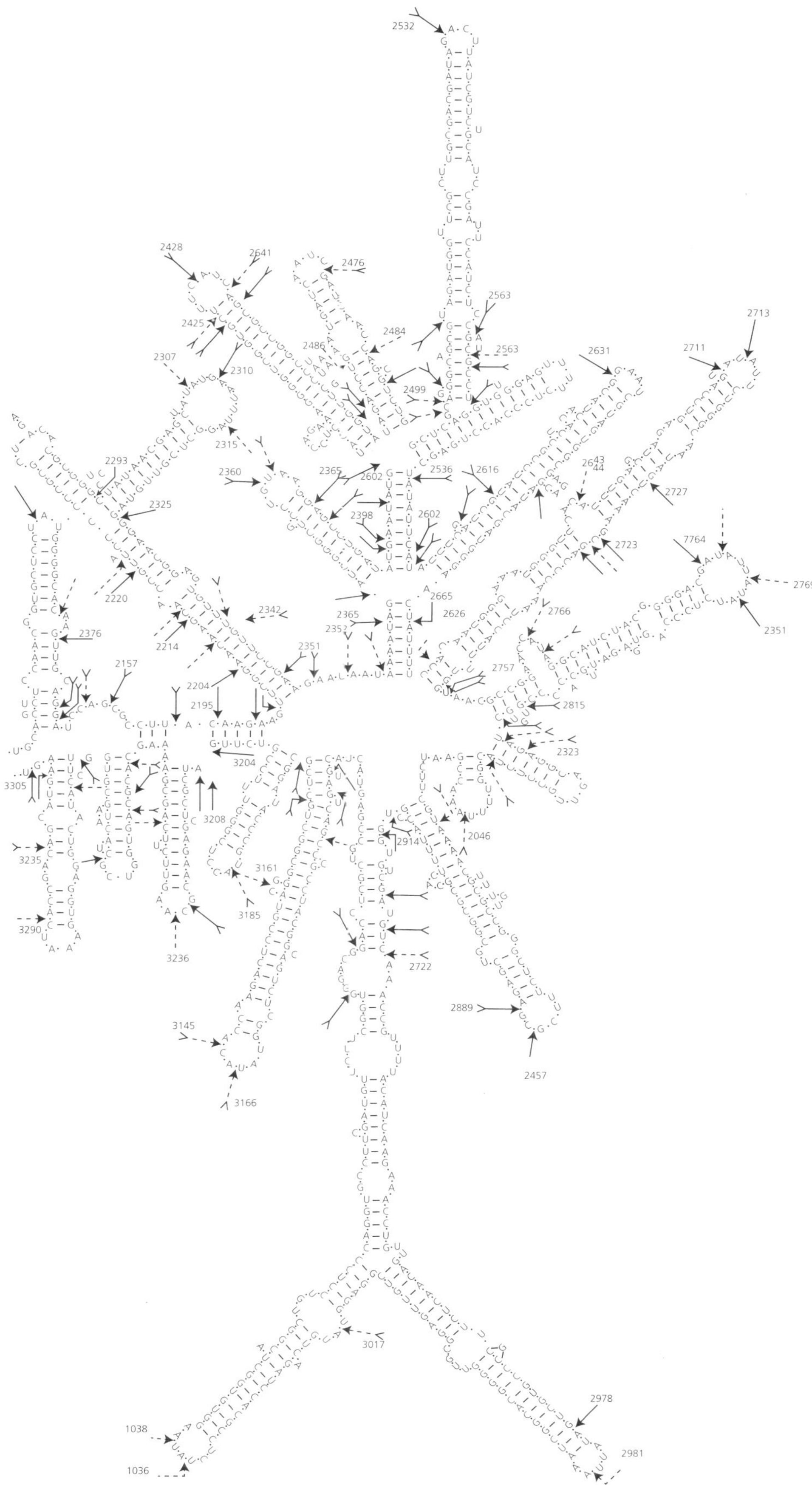

Figure 5.12 Postulated base-paired segments in a portion of the RNA genome of the *Escherichia coli* virus MS2. [Adapted from Fiers *et al.*, *Nature* 1976;260:505 by permission of *Nature*, London.]

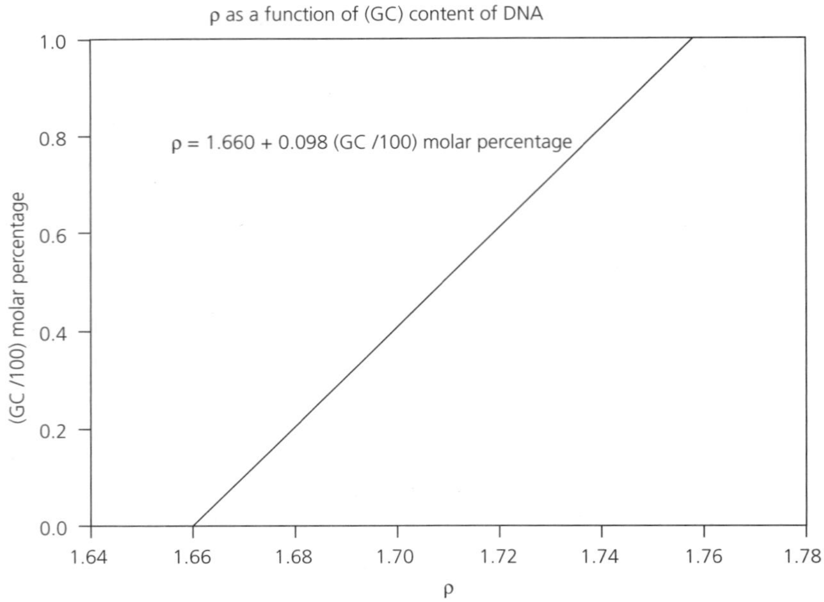

Figure 5.13 Plot of molar percentage of GC in double-stranded DNA as a function of buoyant density. [Adapted from Schildkraut *et al.*, *J Mol Biol* 1962;4:430 by permission of the Academic Press, London.]

in the presence of 20 mM NaCl are shown in Fig. 5.14 (Marmur and Doty 1959). The temperature at which the rise in absorbance at 260 μm is halfway through is called the T_m of the specific DNA. The T_m has been found to be a function of the mole percentage of GC content present in the DNA (GC have three hydrogen bonds between them, while AT have only two) and is given by the following empirical equation

$$T_m = 69.3 + 0.41(GC)$$

where T_m is in °Celsius and (GC) refers to the mole percentage of guanine plus cytosine content of the DNA (Marmur and Doty 1962). Fig. 5.15 shows the plot. It is important to note that the position of the curves is very much dependent on the ionic concentration in the DNA solution, as would be evident from Fig. 5.16, where melting curves of DNA from *E. coli* K12 determined in the presence of various concentrations of KCl have been plotted (Marmur and Doty 1962).

If a double-stranded DNA molecule has stretches of AT-rich regions, preferential separation of strands in those regions is possible under partial denaturation conditions, which can be visualized by electron microscopy (Fig. 5.17).

5.15 Renaturation of denatured DNA

Can the two separated strands (because of denaturation) of a DNA molecule be put together again by proper hydrogen bonding ('renaturation') and is the artificially

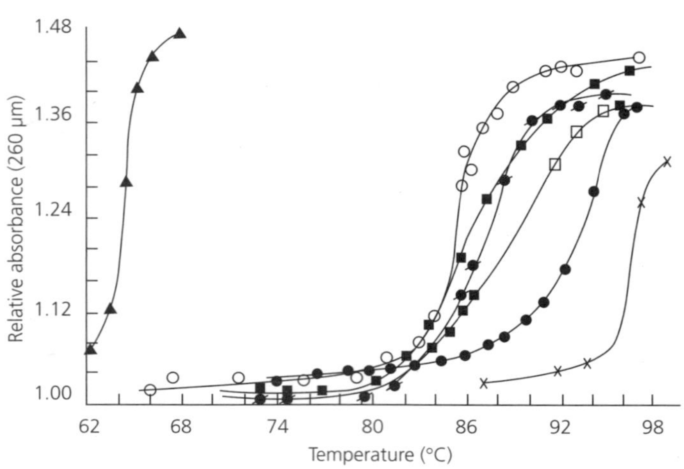

Figure 5.14 Melting curves of double-stranded DNA. Source of DNA: (x) *Mycobacterium phlei* (67% GC), (•) *Escherichia coli*, (o) *Pneumococcus*, (▲) poly dAT (no GC). [Reproduced from Marmur and Doty, *Nature* 1959;183:1427 by the permission of *Nature*, London.]

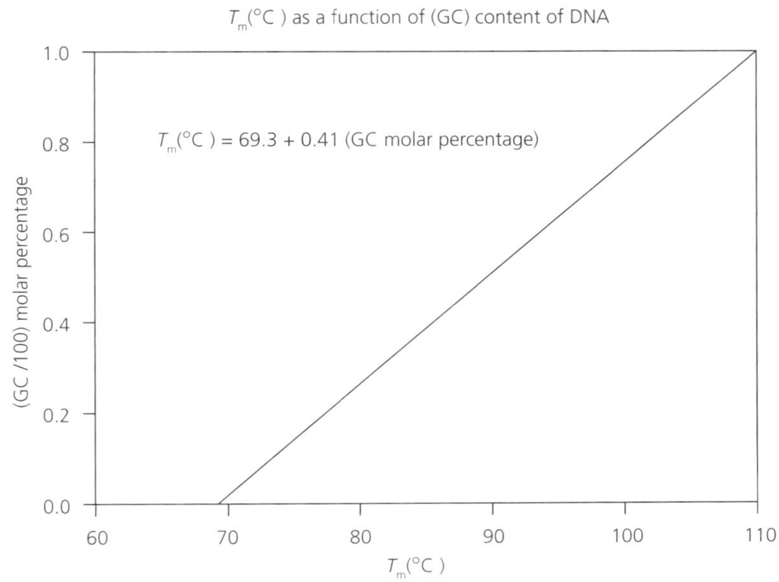

Figure 5.15 Plot of molar percentage of GC in double-stranded DNA as a function of T_m. [Adapted from Marmur and Doty, *J Mol Biol* 1962;5:113 by permission of the Academic Press, London.]

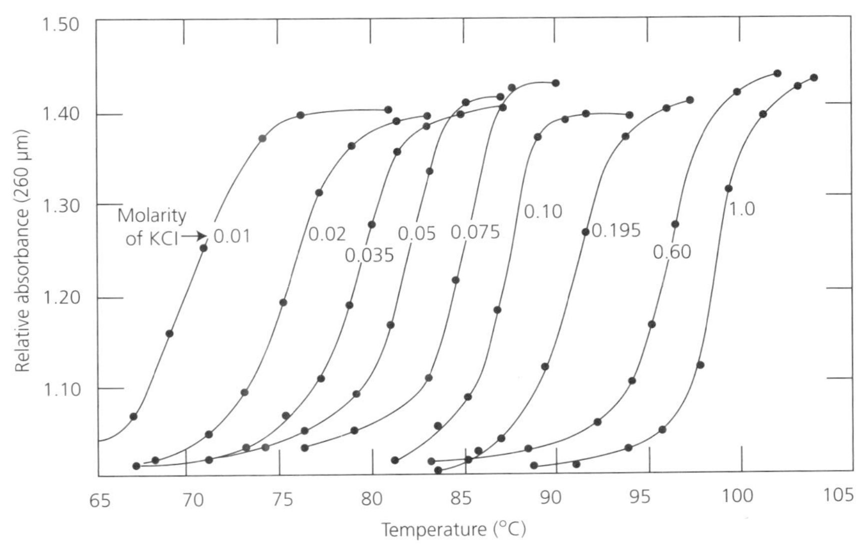

Figure 5.16 Melting curves of DNA of *E. coli* K12 at different concentrations of KCl. [Reproduced from Marmur and Dcty, *J Mol Biol* 1962;5:111 by permission of the Academic Press, London.]

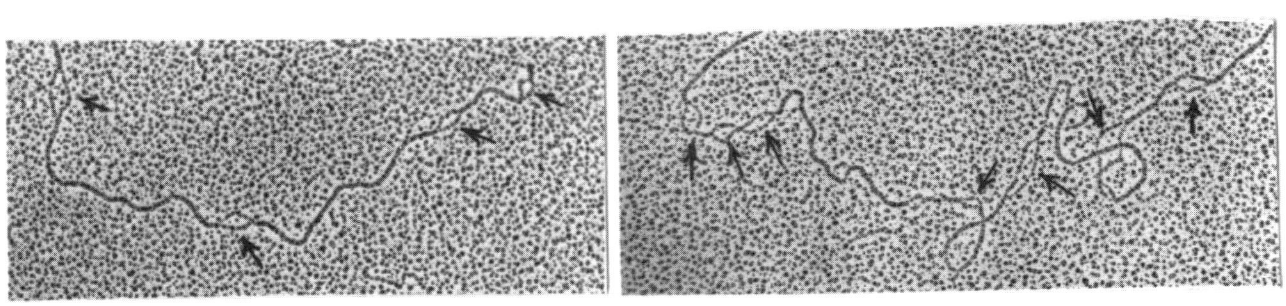

Figure 5.17 Electron micrographs showing preferential separation of DNA strands at AT-rich regions on partial denaturation. The denaturation 'bubbles' are pointed by arrows. The electron micrographs were taken by Dr Nirupam Roy Choudhury.

'renatured' double-stranded DNA biologically active? These questions were asked early on and the answers were in the affirmative. The ability of DNA to carry out genetic transformation (see Chapter 3) was taken as the criterion of biological activity.

Diplococcus pneumoniae was chosen as the experimental microorganism. DNA was isolated from *D. pneumoniae* that was resistant to the antibiotics streptomycin, erythromycin and bryamycin and its ability to transform wild-type (sensitive to the antibiotics) *D. pneumoniae* to the antibiotic-resistant form was checked. The biological activity dropped as the DNA was heat denatured (Marmur and Lane 1960), but when the denatured DNA was slowly cooled, partial recovery of biological activity was evident (Marmur and Lane 1960).

5.16 Other methods of studying renaturation of DNA

An obvious method of studying renaturation is to monitor hypochromicity. One method that has, however, been extensively employed to study renaturation of DNA quantitatively is to determine the appearance of a DNA species that is insensitive towards nucleases, which are specific for single-stranded nucleic acids (see Chapter 4).

Nitrocellulose membranes have been found to specifically bind single-stranded DNA and let double-stranded DNA pass through. Hence, the appearance of a species of DNA in the filtrate is a good indication of renaturation. This method was first developed to isolate RNA complementary to DNA (Nygaard and Hall 1964), but has later been used to detect DNA–DNA hybridization. Similarly, agar gel (Bolton and McCarthy 1962) and cellulose column (Bautz and Hall 1962), which were initially developed for RNA–DNA hybridization, have been successfully used to study DNA–DNA hybridization. Hydroxyapatite columns have also been used to separate double-stranded DNA from single-stranded DNA and have been used to determine the extent of denaturation. A typical elution profile using a phosphate gradient is shown in Fig. 5.18.

5.17 DNA hybridization as a method of determination of homology between different species of DNA

It may be expected that two closely related species would have considerable amount of homology in their DNA sequence. If the DNA from these two species are denatured by heating, mixed together and then renatured by slow cooling, it would be likely that segments of a strand of DNA from one species would 'hybridize' (renature) with segments of the complementary strand of the DNA from the other species. Hence, the extent of cross-hybridization of DNA of two species could be taken as a measure of their relatedness. The important points here are to shear well the DNA from both species before hybridization for effective fishing out of homologous regions and to treat the renatured product with a single-strand–specific nuclease to eliminate artefacts. For ease of detection and quantitation of the hybrid DNA, Schildkraut *et al.* (1961) had density-labelled the DNA from one species with a heavy isotope. For example, *Bacillus subtilis* was grown in a medium containing $^{15}NH_4Cl$ for several generations so that both the strands of DNA were uniformly labelled with ^{15}N. The ^{15}N-labelled DNA was isolated, denatured and then renatured with unlabelled denatured ^{14}N-DNA from *B. natto*. The product was treated with a single-strand–specific nuclease and analysed in a CsCl gradient by equilibrium sedimentation in an ultracentrifuge. In addition to the ^{15}N peak (ρ = 1.744) representing *B. subtilis* DNA and the ^{14}N peak (ρ = 1.705) representing

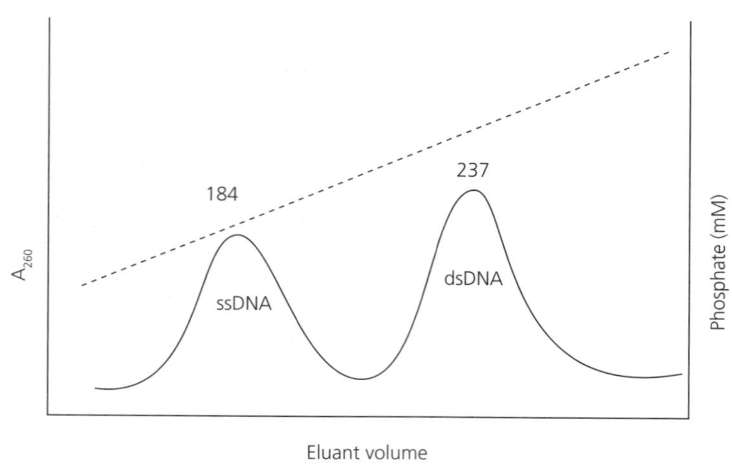

Figure 5.18 Separation of single-stranded DNA from double-stranded DNA by elution with a phosphate gradient in a hydroxyapatite column. The elution of DNA is monitored by following absorbance at 260 µm (A_{260}). The broken line at an angle to the horizontal axis represents the phosphate gradient used for elution. With this particular preparation of hydroxyapatite, the single-stranded (ss) DNA elutes at 184 mM phosphate, while the double-stranded (ds) DNA elutes at 237 mM phosphate.

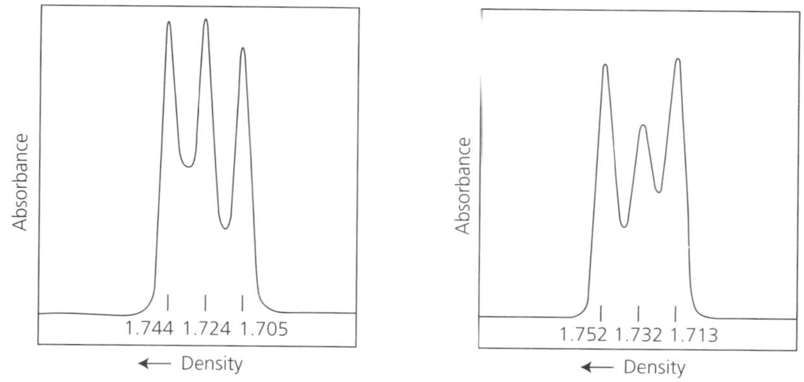

Figure 5.19 Hybridization of DNA from two closely related bacterial species. (Left frame) *Bacillus subtilis* and *B. natto*. *B. subtilis* were grown in a medium containing $^{15}NH_4Cl$ as the sole nitrogen source and D_2O of greater than 99% purity. *B. natto*, however, were grown in a medium containing $^{14}NH_4Cl$ and H_2O. DNA isolated from both species were heated, annealed together and treated with *Escherichia coli* phosphodiesterase. The mixture was then subjected to equilibrium sedimentation by CsCl density gradient centrifugation. The band at the buoyant density of 1.744 represents the heavy *B. subtilis* DNA, the band at the buoyant density of 1.705 represents the light *B. natto* DNA and the band at the buoyant density of 1.724 is due to the hybrid between DNA from *B. subtilis* and *B. natto*. (Right frame) *E. coli* and *Shigella dysenteriae*. Here, *E. coli* were grown in a medium containing $^{15}NH_4Cl$ and D_2O, while *S. dysenteriae* were grown in a medium containing $^{14}NH_4Cl$ and H_2O. DNA isolated from both bacteria were heated, annealed together, treated with *E. coli* phosphodiesterase and subjected to equilibrium sedimentation. The band at the buoyant density 1.752 represents the heavy *E. coli* DNA, the band at the buoyant density of 1.713 represents the light *S. dysenteriae* DNA and the band at the buoyant density of 1.732 is due to the hybrid between DNA from *E. coli* and *S. dysenteriae*. [Reproduced from Schildkraut *et al.*, *J Mol Biol* 1961;3:613 by permission of the Academic Press, London.]

B. natto DNA, an additional peak of hybrid density (ρ = 1.724) appeared, signifying cross-hybridization of DNA [Fig. 5.19(a)] (Schildkraut *et al.* 1961). Similar homology of DNA was observed in the case of *E. coli* and *Shigella dysenteriae* [Fig. 5.19(b)] (Schildkraudt *et al.* 1961). Thus, one could infer that *B. subtilis* and *B. natto* were closely related and so were *E. coli* and *S. dysenteriae*.

5.18 DNA reassociation as a method of detection of 'terminal redundancy' and 'circular permutation' in the genome of bacteriophages

DNA genomes of many bacteriophages, for example, those of *E. coli* phages T3, T5 and T7, have terminal redundancy. This can be illustrated by Fig. 5.20.

The different alphabets denote different genes, while the two strands of the DNA in each gene have been represented as 'w' (Watson) subscript and 'c' (Crick) subscript. As can be seen, the genes A, B, C, D and E are present in both terminii and hence one set is redundant. Now, if the DNA molecules are denatured and then renatured, a few molecules would renature, as in Fig. 5.20(ii). This would then circularize, forming a circle comprising two double-stranded (two regions, both encompassing the genes A, B, C, D and E) and two single-stranded (two regions, both encompassing the genes F, G, H, I, J, K, L, M and N) segments [Fig. 5.20(iii)]. Such structures can be visualized under an electron microscope. The length of the double-stranded regions would thus be a measure of the terminally redundant segment of the genome.

Some genomes, on the other hand, can be 'circularly permuted.' This is illustrated in Fig. 5.21(a). Thus, the individual molecules of bacteriophage DNA are not identical; the genes at the 5′-end and the 3′-end being different as shown in Fig. 5.21(a)(i). The total genomic content, though, is the same in each molecule. If this mixture of different 'circularly permuted' molecules is now denatured and then renatured, there would be shuffling of the complementary strands of DNA and molecules, like the one shown in Fig. 5.21(a)(ii). These molecules would next circularize and fully double-stranded circles with one single-strand break in each strand would be formed [Fig. 5.21(a)(iii)].

DNA genomes of the bacteriophages T2, T4 and P22 are circularly permuted, in addition to having terminal redundancy. This is illustrated in Fig. 5.21(b).

On denaturation and renaturation, here also, there would be shuffling of the complementary strands and molecules, like the one shown in Fig. 5.21(b)(ii). These molecules on circularization would form fully double-stranded circles with two single-stranded tails on opposite strands [Fig. 5.21(b)(iii)]. The tails would represent the terminally redundant portion. Such

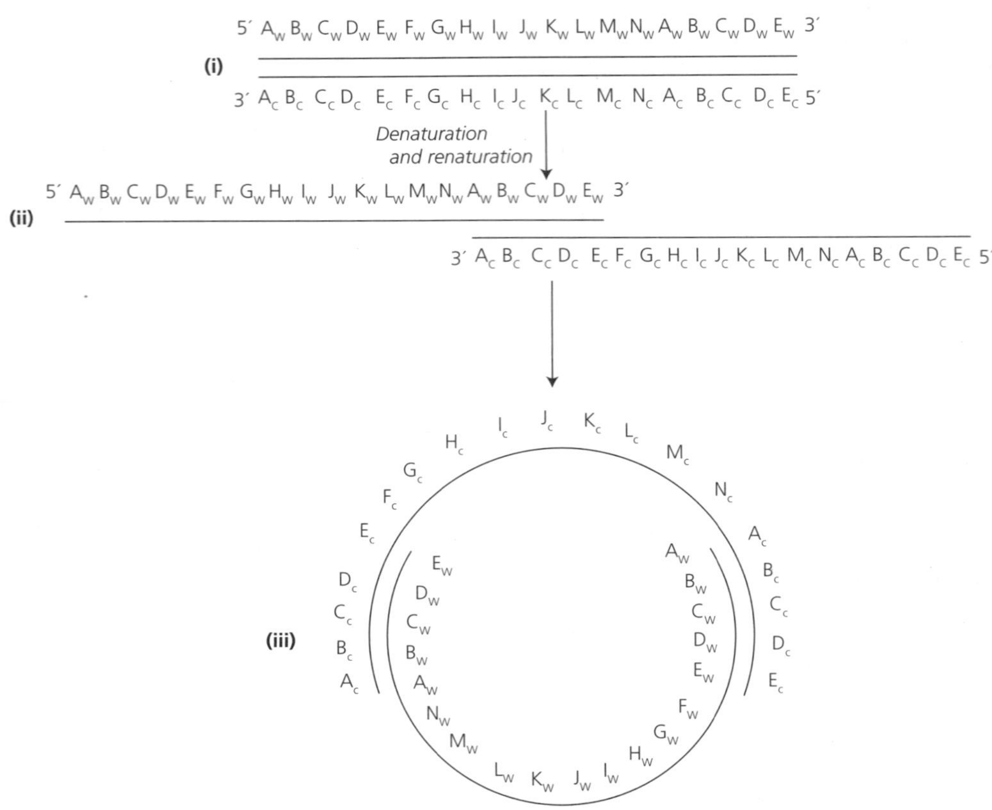

Figure 5.20 Schematic representation of DNA molecules that have terminally redundant regions, and the consequence of denaturation and renaturation.

structures have indeed been visualized under the electron microscope.

5.19 Kinetics of reassociation of denatured strands of DNA as a measure of the complexity of the genome and also as a probe for the presence of repeated sequences

For any reaction which is virtually irreversible, the rate of reaction is proportional to the concentration of the reactants. In DNA reassociation, the concentration of the two strands is the same; hence,

$$dx/dt = k(a - x)^2$$

where a is the concentration of one of the single strands of the particular species of DNA at time t_0, x is the amount of double-stranded DNA formed of the same species at time t_x and k is the rate constant of reassociation. It is obvious, therefore, that the time taken for a particular fraction (t_f) of single-stranded DNA species to be reassociated would be inversely proportional to both the rate constant and the concentration of the single-stranded DNA species:

$$t_f \propto 1/ka$$

The time taken for 50% of the single-stranded DNA to be reassociated is termed as $t_{1/2}$.

5.20 Conditions of efficient reassociation

The conditions of efficient reassociation are as follows (Britten and Kohne 1968):

- An adequate concentration of cations is essential. The reassociation is negligible in less than 0.01 M Na^+ ion; normally, 0.5 M Na^+ is used.
- The temperature at which the reassociation reaction is carried out should be high enough to reduce intra-strand secondary structure. The optimum temperature for reassociation is about 25°C below the T_m.
- The length of incubation and the concentration of DNA have to be adequately high, particularly when reassociation of DNA from a large genome is desired.
- The size of the DNA fragments is critical for proper reassociation. The rate of reassociation is higher with bigger pieces of DNA, but the DNA should be sheared to a conveniently small size to prevent tangling and to achieve specific and reproducible reassociation.

Let us say an organism has 1000 genes, each of which is unique in sequence. Another organism has 10,000 genes,

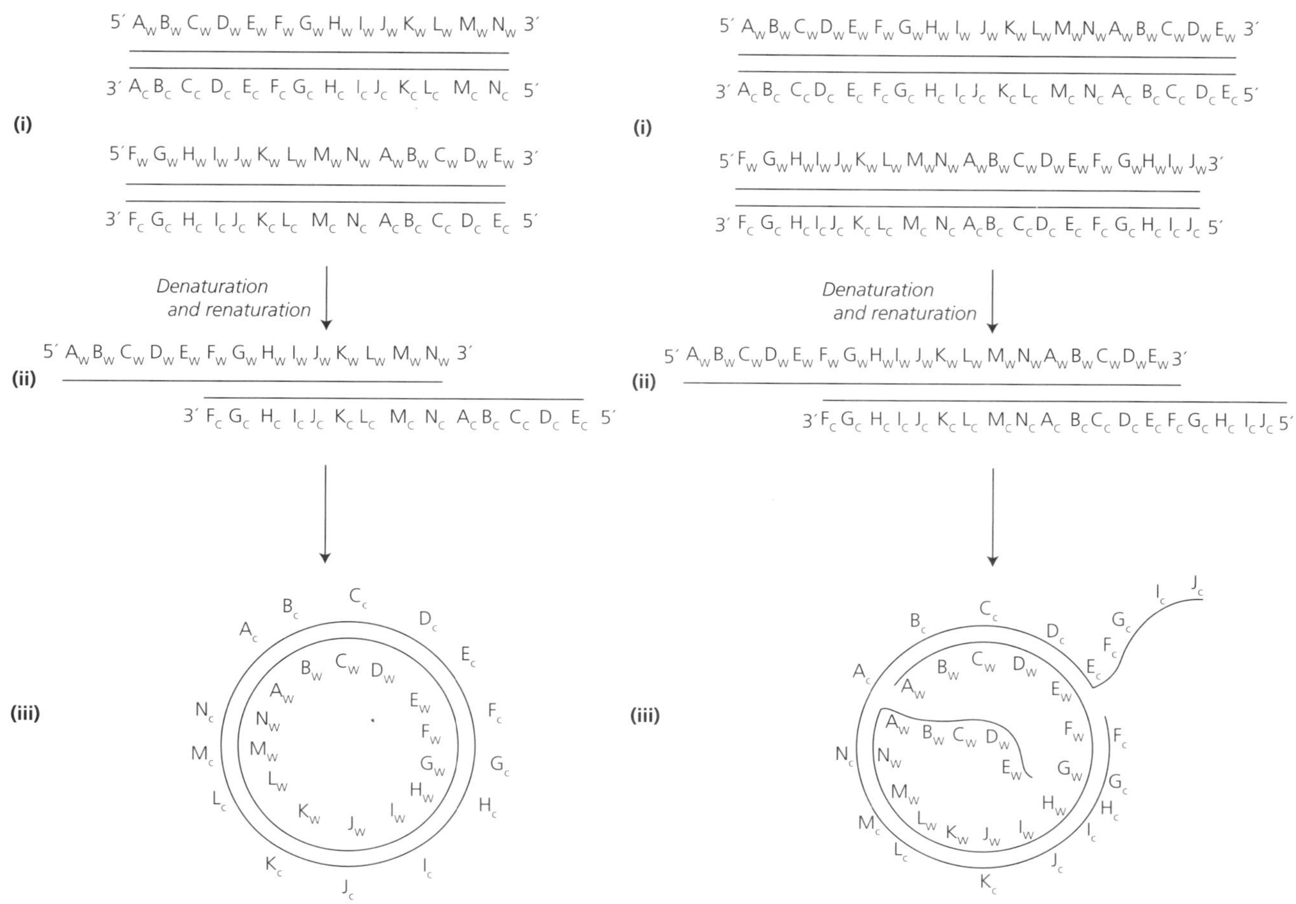

Figure 5.21 (a) Schematic representation of DNA molecules that are circularly permuted, and the consequence of denaturation and renaturation. (b) Schematic representation of DNA molecules that are both circularly permuted and terminally redundant, and the consequence of denaturation and renaturation of such molecules.

again each being unique. The sequence of a particular gene species would obviously be 10 times less concentrated in the latter organism than in the earlier one. Hence, the $t_{1/2}$ of association in the latter case would be 10 times more. An analogous situation would be if a person has to search for his partner in a crowd of 10,000 people rather than in a crowd of 1000 people. This relationship has been found to hold good when DNA from various bacteria or phages was allowed to reassociate; that is, $t_{1/2}$ was proportional to the number of nucleotide pairs (over several orders of magnitude) contained in the DNA of the organism.

In contrast, some DNA fractions from eukaryotic cells very often reassociated with a much lower $t_{1/2}$ than predicted, suggesting the presence of sequences that are repeated many times. In practice, the product of the DNA concentration and the time of incubation is taken as the parameter for the completion of the reaction. This is called Cot, expressed in moles of nucleotides times seconds per litre. The kinetics of reassociation is of the second order. Some Cot curves are shown in Fig. 5.22 (Britten and Kohne 1968). The extent of reassociation can be monitored by determining the fraction of DNA that has become insensitive to a single-strand–specific nuclease, or that passes through a nitrocellulose membrane or that forms a peak separate from the single-strand peak during hydroxyapatite column chromatography. The last method did reveal that about 40% of calf DNA reassociated very fast and required a Cot of 0.03 for half reassociation, while the remaining 60% reassociated much more slowly and required a Cot of 3000. Thus, if the slowly reassociating fraction constitute unique sequences, the fast reassociating fraction can be estimated to contain sequences that may be repeated 1,00,000 times on the average (Britten and Kohne 1968). Sometimes, the DNA from *E. coli*, which contains 4.6×10^6 bp in its genome and is believed to have no repeated sequences, is taken as a standard. The *E. coli* DNA requires a Cot of about 10 for half reassociation. The total mouse genome contains close to 3.0×10^9 bp and 10% of this DNA (i.e. about 3×10^8 bp) reassociates very fast and requires a Cot of only about 0.003 for half reassociation. It can thus be derived that the mouse satellite DNA is repeated to nearly 6,00,000 times and each repeated sequence is less than 500 bp long.

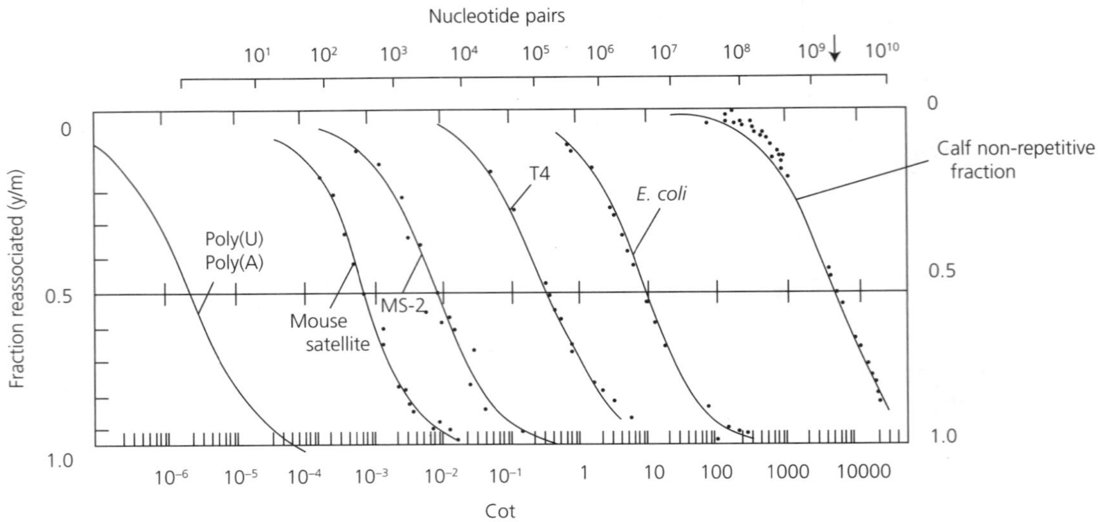

Figure 5.22 Cot curves obtained while reassociating DNA from different sources. The Cot values are written at the bottom, while the numbers of nucleotide pairs in the non-repetitive DNA molecules, which determines the complexity of the DNA, is shown at the top. It is obvious that more complex DNA molecules have higher Cot values. [Reproduced from Britten and Kohne, *Science* 1968;181:530 by permission of the American Association of the Advancement of Science.]

References

Allardet-Servent A, Michaux-Charachon S, Jumas-Bilak E, Karayan L, Ramuz M. Presence of one linear and one circular chromosome in the *Agrobacterium tumefaciens* C58 genome. *J Bacteriol* 1993;175:7869–7874.

Barbour AG, Garon CF. Linear plasmids of the bacterium *Borrelia burgdorferi* have covalently closed ends. *Science* 1987;237: 409–411.

Bauer W. Structure and reactions of closed duplex DNA. *Annu Rev Biophys Bioeng* 1978;7:287–313.

Bautz EKF, Hall BD. The isolation of T4-specific RNA on a DNA-cellulose column. *Proc Natl Acad Sci USA* 1962;48:400–408.

Bolton ET, McCarthy BJ. A general method for the isolation of RNA complementary to DNA. *Proc Natl Acad Sci USA* 1962;48: 1390–1397.

Britten RJ, Kohne DE. Repeated sequences in DNA. *Science* 1968;161:529–540.

Cairns J. The bacterial chromosome and its manner of replication as seen by autoradiography. *J Mol Biol* 1963;6:208–213.

Chu G, Vollrath D, Davis RW. Separation of large DNA molecules by contour clamped homogeneous electric fields. *Science* 1986;234:1528–1585.

Crespi M, Messens E, Caplan AB, Van Montagu M, Desomer J. Faasciation induction of the phytopathogen *Rhodococcus fascians* depends upon a linear plasmid encoding a cytokinin synthase gene. *EMBO J* 1992;11:795–804.

Dulbecco R, Vogt M. Evidence for a ring structure of polyoma virus DNA. *Proc Natl Acad Sci USA* 1963;50:236–243.

Dunham I, Shimizu N, Roe BA, Chissoe S, Hunt AR, Collins JE, *et al.* The DNA sequence of human chromosome 22. *Nature 1999;*402:489–495.

Ferdows MS, Barbour AG. Megabase sized linear DNA in the bacterium *Borrelia burdgorferi*, the Lyme disease agent. *Proc Natl Acad Sci USA* 1989;86:5969–5973.

Fiers W, Contreras R, Duerinck F, Haegemeean G, Iserentant D, Merregaert J, *et al.* Complete nucleotide sequence of bacteriophage MS2 RNA: Primary and secondary structure of the replicase gene. *Nature* 1976;260:500–507.

Fiers W, Sinsheimer RL. The structure of the bacteriophage ϕx 174. I. The action of exopolynucleotidases. *J Mol Biol* 1962;5:408–419.

Hattori M, Fujiyama A, Taylor TD, Watanabe H, Yada T, Park HS, *et al.* The DNA sequence of human chromosome 21. *Nature* 2000;405:311–319.

Lin Y-S, Kieser HM, Hopwood DA, Chen CW. The chromosomal DNA of *Streptomyces lividans* 66 is linear. *Mol Microbiol* 1993;10:923–933.

Manna AC, Das HK. The size of the chromosome of *Azotobacter chroococcum*. *Microbiology* 1994;140:1237–1239.

Manna AC, Das HK. The *Azotobacter vinelandii* chromosome. *J Genet* 1997;76:55–60.

Marmur J, Doty P. Heterogeneity in deoxyribonucleic acids. I. Dependence on composition of the configurational stability of deoxyribonucleic acids. *Nature* 1959;183:1427–1429.

Marmur J, Doty P. Determination of base composition of DNA from its thermal denaturation temperature. *J Mol Biol* 1962;5:109–118.

Marmur J, Lane D. Strand separation and specific recombination in deoxyribonucleic acids: Biological studies. *Proc Natl Acad Sci USA* 1960;46:453–461.

Mathew MK, Smith CL, Cantor CR. High-resolution separation and accurate size determination in pulse-field gel electrophoresis of DNA. 1. DNA size standards and the effect of agarose and temperature. *Biochemistry* 1988a;27:9204–9210.

Mathew MK, Smith CL, Cantor CR. High-resolution separation and accurate size determination in pulse-field gel electrophoresis of DNA. 2. Effect of pulse time and electric field strength and implications for models of the separation process. *Biochemistry* 1988b;27:9210–9216.

Nygaard AP, Hall BD. Formation and properties of RNA-DNA complexes. *J Mol Biol* 1964;9:125–142.

Radloff R, Bauer W, Vinograd J. A dye-buoyant-density method for the detection and isolation of closed circular duplex DNA: the closed circular DNA in HeLa cells. *Proc Natl Acad Sci USA* 1967;57:1541–1521.

Rolfe R, Messelson M. The relative hcmogeneity of microbial DNA. *Proc Natl Acad Sci USA* 1959;45:1039–1043.

Rouviere-Yaniv J, Yaniv M, Germond J-E. *E. coli* DNA binding protein HU forms nucleosome-like structure with circular double-stranded DNA. *Cell* 1979;17:265–274.

Schildkraut CL, Marmur J, Doty P. The formation of hybrid DNA molecules and their use in studies of DNA homologies. *J Mol Biol* 1961;3:595–616.

Schildkraut CL, Marmur J, Doty P. Determination of the base composition of deoxyribonucleic acid from its buoyant density in CsCl. *J Mol Biol* 1962;4:430–443.

Schwartz DC, Cantor CR. Separation of yeast chromosome-sized DNAs by pulsed field gradient gel electrophoresis. *Cell* 1984;37:67–75.

Sueoka N, Marmur J, Doty P. Heterogeneity in deoxyribonucleic acids. II. Dependence of the density of deoxyribonucleic acids on guanine-cytosine content. *Nature* 1959;183:1429–1431.

Vinograd J, Lebowitz J, Radloff R, Watson R, Laipis P. The twisted circular form of polyoma circular DNA. *Proc Natl Acad Sci USA* 1965;53:1104–1111.

Wang JC. DNA topoisomerases. *Annu Rev Biochem* 1996;65: 635–692.

6

Chromatin and its structure

H. K. Das

6.1 How does the DNA reside in a bacterial cell?

The circular DNA molecule comprising the complete genome of the bacterium *Escherichia coli,* which was visualized by John Cairns (1963; see Chapter 5) is about 1300-μm long. The *E. coli* cell is only 1 μm in diameter and about 3 μm in length. It is obvious that the DNA must be highly folded to be accommodated inside a cell and this folding has to be very orderly so that the DNA does not get tangled.

Does the DNA inside a bacterial cell reside as naked DNA without being encumbered with any other macromolecule? To be able to visualize the DNA (labelled with ^{3}H), Cairns (1963) in his classical experiment had treated the *E. coli* cell lysate with 4 M NaCl so that the proteins bound to DNA can be dissociated (see Chapter 5). As DNA is acidic in nature because of the presence of the ionizable hydroxyl groups in the phosphodiester linkages (see Chapter 3), it is but likely that it would attract basic proteins to bind to it.

A compact, fast-sedimenting (sedimentation coefficient 1300–2200S) DNA-containing structure has been found to be released on careful lysis of *E. coli* cells. This has been called 'folded chromosome' (Stonington and Pettijohn 1971; Worcel and Burgi 1972) and has been found to contain 10% RNA and 10% protein in addition to 80% DNA. It was inferred from the response of ethidium bromide and DNase I on the sedimentation rates of 'folded chromosomes' from *E. coli* that these 'chromosomes' contained highly superhelical DNA and these were speculated to be looped around a 'core' (Fig. 6.1; Worcel and Burgi 1972). This 'core' probably contained both RNA and protein as treatment with ribonuclease or trypsin reduced the sedimentation coefficient considerably.

The 'chromosome' released on careful lysis of *E. coli* in physiological salt solution (0.15 M NaCl) directly on the electron microscope–supporting film revealed under the electron microscope, loops of a fibre 120 Å in diameter, with a 130-Å repeating beaded substructure (Fig. 6.2; Griffith 1976). It may be recalled here that the diameter of a naked DNA duplex is only 20 Å (see Chapter 4). It was found that the fibres contained DNA condensed 6.5-fold in blocks of about 250 base pairs (bp) (Griffith 1976).

However, when isolated *E. coli* DNA-binding protein HU (an analogous protein in other bacterial species is known as DNA-binding protein II) was incubated with circular SV40 DNA, condensed circular structures were

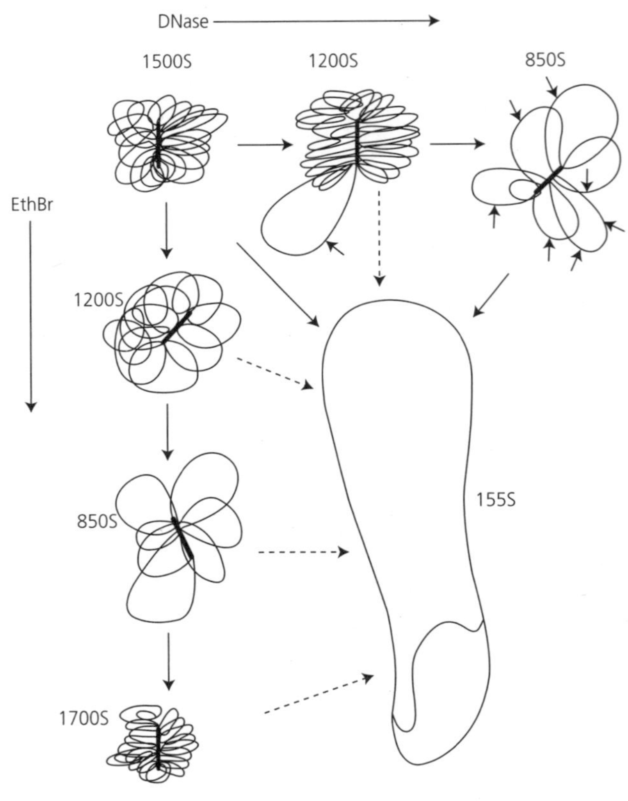

Figure 6.1 Schematic representation of the 'unfolding' of the 'folded' chromosome of *Escherichia coli* as a result of treatment with ethidium bromide or DNase I. The chromosome is thought to be made up of loops. The solid arrows indicate actually observed transitions, while the broken arrows represent postulated transitions. [Reproduced from Worcel and Burgi, *J Mol Biol* 1972;71:143 by permission of the Academic Press, London.]

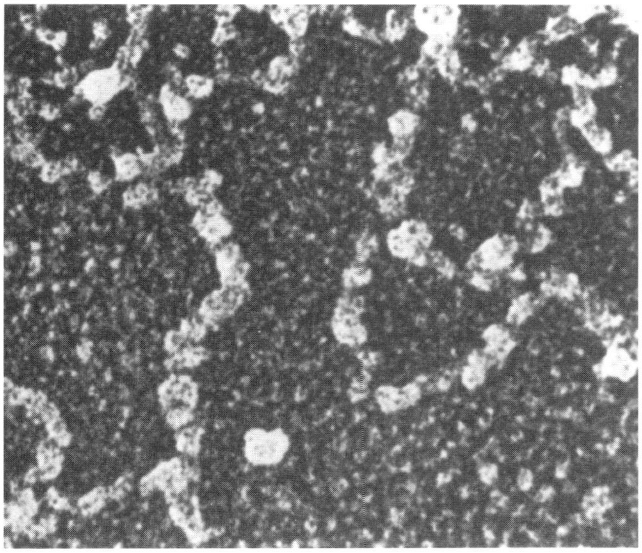

Figure 6.2 An enlarged view of a portion of the released chromosome of *Esherichia coli* Hfr C71 cells, which were lysed carefully on the supporting film laid on the grid of the electron microscope. [Reproduced from Griffith, *Proc Natl Acad Sci USA* 1976;73:566 by permission of Professor Jack Griffith.]

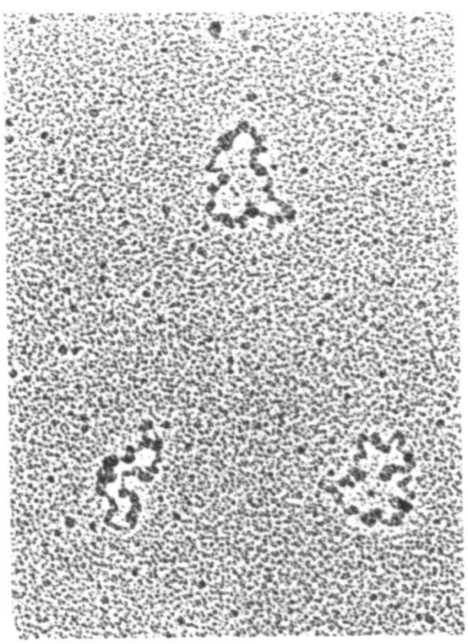

Figure 6.3 Visualization under electron microscope of the beaded and condensed structure of circular SV40 DNA formed as a result of its *in vitro* association with the DNA-binding protein HU isolated from *Escherichia coli*. [Reproduced from Rouviere-Yaniv *et al.*, *Cell* 1979;17:269 by permission of Elsevier Science.]

formed, exhibiting on an average 14 'beads' per SV40 DNA molecule, with a 'bead' diameter of about 180 Å (Fig. 6.3; Rouviere-Yaniv *et al.* 1979). The condensation of DNA in this case of *in vitro* association with the protein HU was to an extent of 2.0–2.5 times compared with naked DNA.

6.2 Proteins

Proteins are linear biopolymers analogous to DNA and RNA, with the basic difference being that here the monomeric units are not nucleotides but amino acids. The amino acids that constitute proteins have the general formula given in Fig. 6.4.

R denotes the side chain, which is different in different amino acids. The amino group is always attached to the α-carbon of the carboxylic acid. In proteins, the amino acids are joined to each other by peptide linkages, in comparison with the phosphodiester linkages joining the mononucleotides in DNA or RNA. The carboxyl group of one amino acid is condensed with the amino group of the next amino acid, thus forming the peptide linkage as shown in Fig. 6.5.

Acid treatment of proteins yields a mixture of amino acids as a result of hydrolysis of the peptide linkages. There are altogether 18 amino acids in natural proteins. The smallest amino acid in proteins is glycine (R = H). The three-letter abbreviation for this amino acid is Gly and the one-letter symbol assigned to it is G. The other aliphatic monocarboxylic amino acids found in proteins are alanine (Ala, A) (R = CH_3), valine (Val, V) (R = C_3H_7), leucine (Leu, L) and isoleucine (Ile, I) (R = C_4H_9), serine (Ser, S) (R = CH_2OH), threonine (Thr, T) (R = C_2H_4OH), cysteine (Cys, C) (R = CH_2SH) and methionine (Met, M) (R = C_3H_6SH). These are all neutral amino acids. The neutral aromatic amino acids are phenylalanine (Phe, F) (R = $C_6H_5CH_2$), tyrosine (Tyr,Y) [R = $C_6H_5(OH)CH_2$] and tryptophan (Trp, W) [R = $C_6H_4(C_2HNH)CH_2$]. Tryptophan has a heterocyclic ring fused to the aromatic ring. Proline (Pro, P) is a neutral heterocyclic amino acid in which the α-carbon and the amino group are a part of the heterocycleic ring (R = C_3H_6). The basic amino acids in proteins are lysine (Lys, K) (R = $C_4H_8NH_2$), arginine (Arg, R)

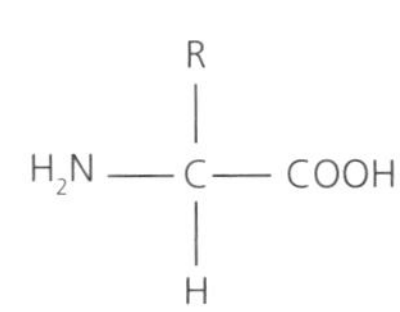

Figure 6.4 General formula of amino acids present in proteins.

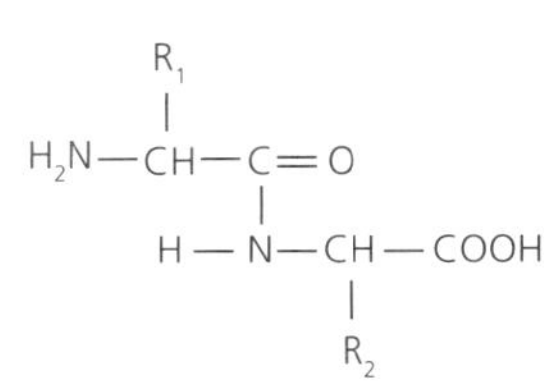

Figure 6.5 Peptide linkage between amino acids in proteins.

```
NH2 - GIVEQCCTSICSLYQLENYCN - COOH                A-chain
           |           |
           S           S
           |           |
           S           S
           |           |
NH2 - FVNQHLCGSHLVEALYLVCGERGFFYTPKT - COOH      B-chain
```

Figure 6.6 Amino acid sequence of human insulin.

[R = $C_3H_6C(NH)_2NH_2$] and histidine (Hys, H) (R = C_4H_4NNH). Histidine is also a heterocyclic amino acid. The two acidic amino acids in proteins are aspartic acid (Asp, D) (R = CH_2COOH) and glutamic acid (Glu, E) (R = C_2H_4COOH), these being dicarboxylic acids. In addition to these 18 amino acids, proteins also contain the two amides, asparagine (Asn, N) (R = CH_2CONH_2) and glutamine (Gln, Q) (R = $C_2H_4CONH_2$), in which the second carboxylic group of aspartic acid and glutamic acid, which is not involved in the peptide linkage, is not free but blocked by an amide group. These two amides cannot be detected in acid hydrolysates of proteins as they get converted into aspartic acid and glutamic acid, respectively.

The sequence of amino acids in a specific protein is fixed, very much similar to the situation in RNA or DNA, where the sequence of bases is fixed. Analogous to the 5′-end and 3′-end of a nucleic acid, proteins have an amino terminal end and a carboxy terminal end. Proteins are small compared with the large DNA molecules. The hormone insulin is composed of two different protein molecules (the A-chain and the B-chain) linked together by –S–S– linkages through the amino acid cysteine in the proteins. The A-chain of insulin has only 21amino acids, while the B-chain has 30 amino acids. The sequence of amino acids in insulin isolated from human pancreas is shown in Fig. 6.6. It would be of interest to note that insulin isolated from the pancreas of other vertebrates has very similar amino acid sequence. For example, the amino acid sequence of the A-chain of bovine insulin has only two differences, alanine at position 8 and valine at position 10 – all other amino acids are identical. The B-chain of bovine insulin has only one difference – alanine at position 31.

The enzymes that function as biological catalysts are mostly proteins. Ribozymes are exceptions. The enzyme DNase I, mentioned earlier, is a protein which hydrolyses phosphodiester linkages in DNA. DNase I isolated from bovine pancreas is composed of 260 amino acids and the sequence is shown in Fig. 6.7.

The enzyme β-galactosidase isolated from *Escherichia coli*, which hydrolyses the disaccharide lactose into glucose and galactose, is a larger protein and has 1023 amino acids. The sequence is presented in Fig. 6.8.

There are many proteins that do not have any enzymatic or hormonal activity. The DNA-binding protein HU, mentioned earlier, is an example. The protein HU isolated from *Escherichia coli* comprises two polypeptide chains, HU-α and HU-β, each having 90 amino acids. The amino acid sequence of these two polypeptides is given in Fig. 6.9.

Proteins can be either fibrous or globular. Fibrous proteins do not have enzymatic or hormonal activity; they are basically involved in the maintenance of structure and are major components of connective tissues, skin, hair, nails, etc. Collagen, elastin, keratins, myosin and actin are examples. Proteins that have enzymatic or hormonal activity are all globular. Proteins that are constituents of membranes and are involved in transport are also globular. The polypeptide chain is folded onto itself in globular proteins, thus forming a compact globular structure. This folding is not random but very specific and involves interaction between specific amino acids in the polypeptide chain.

6.3 DNA-binding proteins

Many proteins have affinity towards DNA. These proteins can be loosely classified into two main groups: the proteins which have affinity for specific sequences in DNA and the proteins which have affinity for DNA irrespective of its sequence. The proteins of the first group are involved in various kinds of DNA transaction processes, which would be described in the subsequent parts of this book. However, many of the proteins found associated with DNA in chromosomes, whether prokaryotic or eukaryotic, belong to the second group. These proteins bind to DNA in stoichiometric amounts. The histones found in eukaryotic chromosomes and the histone-like proteins present in prokaryotic chromosomes are prominent examples of the proteins of the second group. HU is a histone-like protein. Other such proteins are H, H1, H-NS and IHF.

Some proteins are known to prefer single-stranded DNA for binding to double-stranded DNA. The single-strand binding proteins, SSBs as generally called, play important roles in some of the DNA transaction processes. These proteins are also known as DNA unwinding proteins, DNA melting proteins and helix destabilizing proteins.

```
  1 LKIAAFNIRT FGETKMSNAT LASYIVRIVR RYDIVLIQEV RDSHLVAVGK
 51 LLDYLNQDDP NTYHYVVSEP LGRNSYKERY LFLFRPNKVS VLDTYQYDDG
101 CESCGNDSFS REPAVVKFSS HSTKVKEFAI VALHSAPSDA VAEINSLYDV
151 YLDVQQKWHL NDVMLMGDFN ADCSYVTSSQ WSSIRLRTSS TFQWLIPDSA
201 DTTATSTNCA YDRIVVAGSL LQSSVVPGSA APFDFQAAYG LSNEMALAIS
251 DHYPVEVTLT
```

Figure 6.7 Amino acid sequence of the enzyme DNase I isolated from beef pancreas.

1 TMITDSLAVV LQRRDWENPG VTQLNRLAAH PPFASWRNSE EARTDRPSQQ
51 LRSLNGEWRF AWFPAPEAVP ESWLECDLPE ADTVVVPSNW QMHGYDAPIY
101 TNVTTPITVN PPFVPTENPT GCYSLTFNVD ESWLQEGQTR IIFDGVNSAF
151 HLWCNGRWVG YGQDSRLPSE FDLSAFLRAG ENRLAVMVLR WSDGSYLEDQ
201 DMWRMSGIFR DVSLLHKPTT QISDFHVATR FNDDFSRAVL EAEVQMCGEL
251 RDYLRVTVSL WQGETQVASG TAPFGGEIID ERGGYADRVT LRLNVENPKL
301 WSAEIPNLYR AVVELHTADG TLIEAEACDV GFREVRIENG LLLLNGKPLL
351 IRGVNRHEHH PLHGQVMDEQ TMVQDILLMK QNNFNAVRCS HYPNHPLWYT
401 LCDRYGLYVV DEANIETHGM VPMNRLTDDP RWLPAMSERV TRMVQRDRNH
451 PSVIIWSLGN ESGHGANHDA LYRWIKSVDP SRPVQYEGGG ADTTATDIIC
501 PMYARVDEDQ PFPAVPKWSI KKWLSLPGET RPLILCEYAH AMGNSLGGFA
551 KYWQAFRQYP RLQGGFVWDW VDQSLIKYDE NGNPWSAYGG DFGDTPNDRQ
601 FCMNGLVFAD RTPHPALTEA KHQQQFFQFR LSGQTIEVTS EYLFRHSDNE
651 LLHWMVALDG KPLASGEVPL DVAPQGKQLI ELPELPQPES AGQLWLTVRV
701 VQPNATAWSE AGHISAWQQW RLAENLSVTL PAASHAIPHL TTSEMDFCIE
751 LGNKRWQFNR QSGFLSQMWI GDKKQLLTPL RDQFTRAPLD NDIGVSEATR
801 IDPNAWVERW KAAGHYQAEA ALLQCTADTL ADAVLITTAH AWQHQGKTLF
851 ISRKTYRIDG SGQMAITVDV EVASDTPHPA RIGLNQLAQ VAERVNWLGL
901 GPQENYPDRL TAACFDRWDL PLSDMYTPYV FPSENGLRCG TRELNYGPHQ
951 WRGDFQFNIS RYSQQQLMET SHRHLLHAEE GTWLNIDGFH MGIGGDDSWS
1001 PSVSAEFQLS AGRYHYQLVW CQK

Figure 6.8 Amino acid sequence of the enzyme β-galactosidase from *Escherichia coli.*

Amino acid sequence of HU-α

1 MNKTQLIDVI AEKAELSKTQ AKAALESTLA AITESLKEGD AVQLVGFGTF
51 KVNHRAERTG RNPQTGKEIK IAAANVPAFV SGKALKDAVK

Amino acid sequence of HU-β

1 MNKSQLIDKI AAGADISKAA AGRALDAIIA SVTESLKEGD DVALVGFGTF
51 AVKERAARTG RNPQTGKEIT IAAAKVPSFR AGKALKDAVN

Figure 6.9 Amino acid sequence of HU-α and HU-β from *Escherichia coli.*

6.4 Eukaryotic chromosomes

Condensed chromosomes containing DNA have been visualized under the light microscope by staining mitotic cells at metaphase of many eukaryotic organisms. Fig. 6.10 shows such cells stained with the fluorescent dye quinacrine. The cells are ready to divide, and at this stage, each chromosome consists of a pair of 'rods' placed side by side. Each 'rod' (called 'chromatid') is destined to end up in one of the two daughter cells on cell division. Each 'chromatid' contains one continuous linear DNA double helix, a fact that was first demonstrated with chromosomes of the fruit fly *Drosophila melanogaster* (Kavenoff and Zimm 1973; Kavenoff *et al.* 1973). The sizes of the DNA molecules were determined by the viscoelastic technique as sedimentation

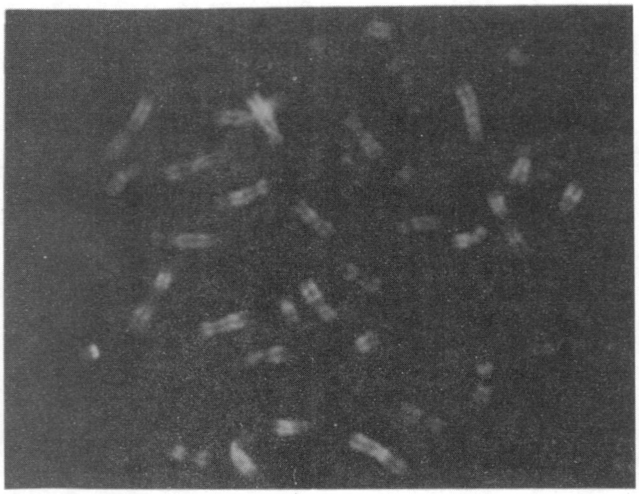

Figure 6.10 Highly condensed chromosomes of human fibroblast cell at metaphase visualized by staining with the fluorescent dye quinacrine.

measurements were not found to be reliable. The viscoelastic relaxation of the DNA molecules was measured by monitoring the recoil decay time after the DNA molecules in the solution have been stretched in a viscometer. The advantage of this method is that it determines the molecular weight of the largest molecule even if it is in a very small minority.

The very condensed visible (after staining) chromatid becomes invisible once the daughter cells are formed from the mother cell and remains so till the daughter cell in its turn is ready to divide. This invisible structure is known as the 'chromatin', but how the 'chromatin' becomes folded and compacted into the 'chromatid' is still an open question. Our knowledge about the structure of 'chromatin', however, is much better. It is more than two decades now that it became evident that 'chromatin' was composed of a consecutive repeated structure of an ordered complex of DNA and proteins, which became known as the 'nucleosome'.

6.5 Concept of the nucleosome

If a molecule of purified DNA is digested with a non-specific endo-deoxyribonuclease, the products would be oligonucleotides of all possible sizes, including some mononucleotides. On gel electrophoresis, these products would appear as a smear, but no specific band would be visible (Hewish and Burgoyne 1973). When, however, the nuclei of liver cells of mouse or rat were incubated in the presence of Ca^{2+} or Mg^{2+}, the endogenous non-specific deoxyribonuclease acted on the chromatins and discrete particle-like complexes were obtained, which sedimented with an approximate minimum mass of 2×10^5 Da. These nuclease-resistant complexes were amenable to digestion with the proteolytic enzyme trypsin (Sahasrabuddhe and van Holde 1974), suggesting the presence of proteins in the complex. If the DNA in these complexes were purified, fragments of about 200 bp or multiples thereof were visible on acrylamide gel electrophoresis (Hewish and Burgoyne 1976; Williamson 1970). Thus, the complexes contained proteins to the extent of more than 50%. Electron microscopic examination of chromatins indeed revealed a structure in which beads of about 10-nm diameter were repeated at uniform distance on a string (Fig. 6.11; Olins and Olins 1974). That the proteins and DNA were associated in chromatins in precise stoichiometry was inferred from chemical cross-linking experiments (Kornberg and Thomas 1974). A model of chromatin incorporating the concept of a regularly repeating structural motif containing specific proteins was proposed (Kornberg 1974), reconciling all these information, including X-ray diffraction patterns. These repeating structures have come to be known as 'nucleosomes'. The specific proteins, which were found to be small, predominantly contained amino acids such as lysine or arginine, which had positively charged side groups. These proteins have come to be known as histones. The histones H1, H2A and H2B are lysine rich, while the histones H3 and H4 are arginine rich – their molecular weights

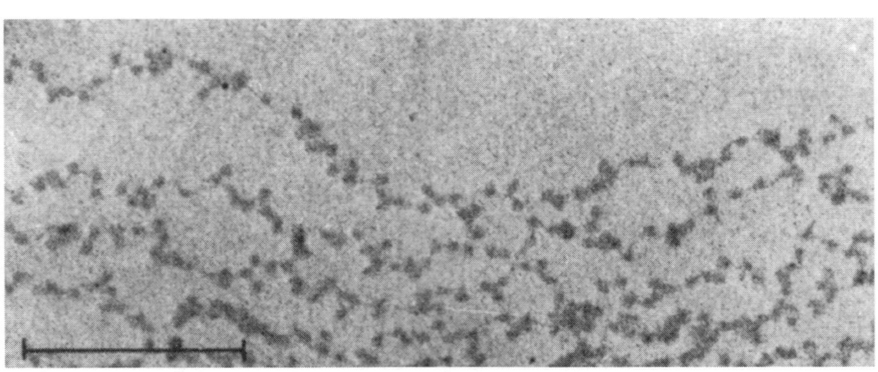

Figure 6.11 Rat thymus chromatin positively stained with phosphotungstic acid. Scale bar: 0.2 μm. [Reproduced from Olins and Olins, *Science* 1974;183:330 by permission from *Science*.]

ranging between 11,000 and 21,000 Da. It is now known that 80% of the nuclear DNA exists as 'nucleosomes' (Noll 1974a).

6.6 Detailed structure of the nucleosome

Controlled treatment of the chromatin with micrococcal nuclease has gone a long way in elucidating some details of the structure of nucleosomes. Treatment for a very short period (15 s) released particulate materials that contained double-stranded DNA molecules about 200-bp long and two molecules each of histones H2A, H2B, H3 and H4, along with a single molecule of histone H1. Longer treatment (10 min) with the nuclease left particles which contained DNA molecules that were only 140-bp long, while histone H1 was lost (Figs 6.12 and 6.13) (Noll and Kornberg 1977). Still further treatment with the nuclease degraded the DNA almost completely and the particulate structure fell apart. It has been inferred that nucleosomes consist of core nucleosome particles linked together by stretches of DNA about 60-bp long bound to histone H1. This part of the structure is referred to as linker DNA and linker histone.

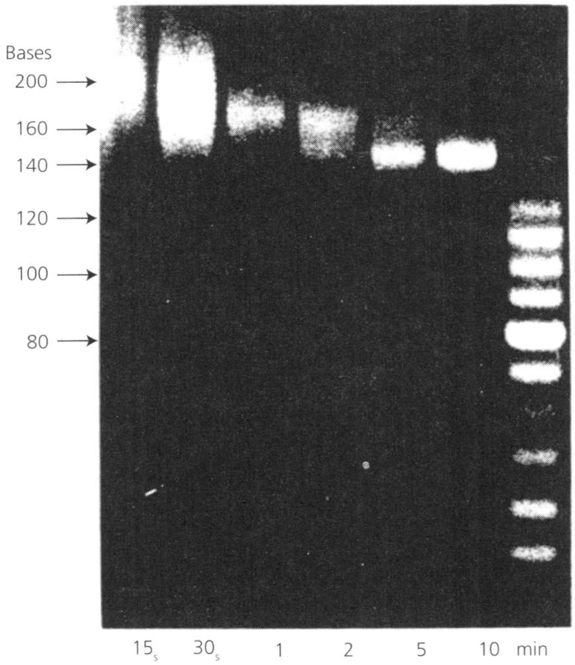

Figure 6.12 Polyacrylamide gel electrophoresis of DNA isolated from rat liver chromatin after progressive treatment with micrococcal nuclease. Length of treatment from left to right: 15 s, 30 s, 1 min, 2 min, 5 min and 10 min. Lengths of the DNA fragments have been indicated on the vertical axis [Reproduced from Noll and Kornberg, *J Mol Biol* 1977;109:398 by permission of the Academic Press, London.]

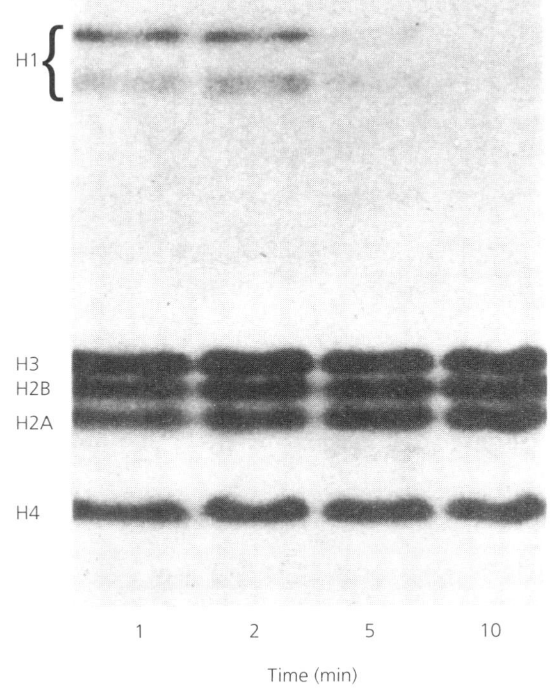

Figure 6.13 Sodium dodecyl sulphate/polyacrylamide gel electrophoresis of histones isolated from rat liver chromatin after progressive treatment with micrococcal nuclease. Identities of the different histone bands have been indicated on the vertical axis, while length of treatment with the nuclease from left to right is: 1 min, 2 min, 5 min and 10 min. [Reproduced from Noll and Kornberg, *J Mol Biol* 1977;109:400 by permission of the Academic Press, London.]

The core particle contained the histone octamer, comprising two molecules each of histones H2A, H2B, H3 and H4 with a 140-bp-long DNA wrapped around the core (Noll and Kornberg 1977). This inference was substantiated by polyacrylamide gel electrophoresis, whereby two complexes could be resolved. There was a larger and slower-migrating complex that contained histone H1 also, while the smaller and faster-migrating complex was devoid of histone H1 (Varshavsky *et al.* 1976).

What is the evidence that DNA is indeed wrapped around the histone octamer core and not embedded in it? When the nucleosome core was treated with a different nuclease, DNase I, cleavages in the phosphodiester bonds were observed every 10–11 bp. The enzyme DNase I is known to recognize the minor groove of the DNA double helix, which is present once in every turn of the helix or, in other words, once in every 10–11 bp. Thus, no part of the entire 140-bp DNA associated with the nucleosome core was hidden or embedded in the histone core and the whole of it was exposed to the action of DNase I. It thus

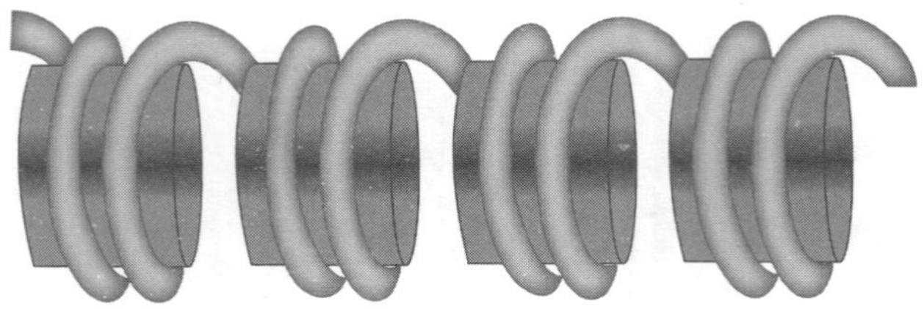

Figure 6.14 An artist's impression of the assembly of four nucleosomes.

has to be on the surface of the histone core (Noll 1974b). X-ray diffraction studies of crystallized nucleosomes have revealed that the histone octamer core assumed a disc-like shape, about 11 nm in diameter and about 5.6 nm in height. The DNA indeed appeared wrapped in 1.75 left-handed superhelical turns around each histone octamer core (Finch *et al.* 1977; Richmond *et al.* 1984). An artist's impression of the nucleosome assembly is presented in Fig. 6.14.

6.7 Does the DNA exist as a regular double helix in the nucleosome or is it distorted?

X-ray diffraction studies of crystallized nucleosomes at 7-Å resolution (Richmond *et al.* 1984) did indicate distortions in the DNA in the form of some very sharp bends. Distortions in the nucleosomal DNA, particularly some unstacking of adjacent bases, have been suggested by probing with singlet oxygen (Hogan *et al.* 1987). However, probing with hydroxyl radicals has also indicated distortions in the nucleosomal DNA because of the existence of 10.7 bp per turn in the central three turns of DNA, while the flanking stretches had the normal 10 bp per turn (Hayes *et al.* 1990, 1991). The presence of exocyclic groups, like the amino group of guanine, in the minor groove of DNA is thought to affect the affinity and positioning of DNA with respect to the histone protein, while the absence of such groups reduces distortion and eases nucleosome formation (Wellenzohn *et al.* 2001).

6.8 How are nucleosomes compacted into chromatins?

It is obvious that an important step in chromatin compaction would be to shield the negative charges of the hydroxyl ions on the phosphodiester linkages of DNA. This has been possible *in vitro* by simply adding appropriate amounts of mono- and divalent cations, and the degree of compaction achieved has been almost the same as that observed *in vivo* (Schwarz and Hansen 1994). Though compaction has been observed even in the absence of the linker histone H1 (Hansen *et al.* 1989; Wolffe *et al.* 1997), it is believed that all the different histone molecules do take part through protein–protein contacts in compaction of nucleosomes (Fletcher and Hansen 1995; Schwarz *et al.* 1996). Electron microscopic studies have added considerably towards the understanding of the compaction of chromatins. In the presence of 0.2 mM EDTA and 1 mM triethanolamine chloride, nucleosomes could be seen attached to a somewhat zigzag fibre [Fig. 6.15(a)]. When the triethanolamine concentration was raised to 5 mM, the structure assumed a flat ribbon-like form, in which the linker DNA also appeared to be coiled between adjacent nucleosome cores [Fig. 6.15(b)]. In the presence of 100 mM NaCl, which is however closer to the physiological condition, one could see 30-nm-wide fibre-like structures that were similar to those observed inside the nucleus [Fig. 6.15(c) and (d); Thoma *et al.* 1979]. A schematic drawing of the chromatin containing histone H1 at different ionic strengths is shown in Fig. 6.16 (Thoma *et al.* 1979).

Folding of DNA into nucleosomes can result in a compaction of only seven-fold, while further seven-fold compaction can be achieved when the nucleosome assembly is folded into the 30-nm chromatin fibres. Conversion of chromatin fibres into the highly condensed metaphase chromosomes, however, possibly needs another 250-fold compaction (Earnshaw 1988). Several models have been proposed to explain the last major compaction – the most favoured one envisaging the formation of loops by the 30-nm chromatin fibres that are arranged radially with respect to the axis of the chromosome (Gasser and Laemmli 1986; Boy de la Tour and Laemmli 1988). When interphase chromosomes of *Drosophila melanogaster* were mildly treated with DNase I, fragments were obtained by velocity sedimentation that contained DNA of approximately 85,000 bp or multiples thereof, suggesting that the chromosomes were arranged in periodic domains (Benyajati and Worcel 1976). Similar results were obtained when the endogenous deoxyribonuclease was made use of and the resulting DNA fragments were separated by pulsed-field gel electrophoresis (Filipski *et al.* 1990). The head of the loops proposed in the model could possibly be the regions most sensitive to DNase I or the endogenous deoxyribonuclease. In another experiment when metaphase chromosomes from HeLa cells were observed under the electron microscope after removal of histones by extraction with high salt, DNA loops that appeared to range between 30,000 and 90,000 bp (10- to 30-μm long) were claimed to have been seen (Figs 6.17 and 6.18; Jackson *et al.* 1990; Paulson and Laemmli 1977).

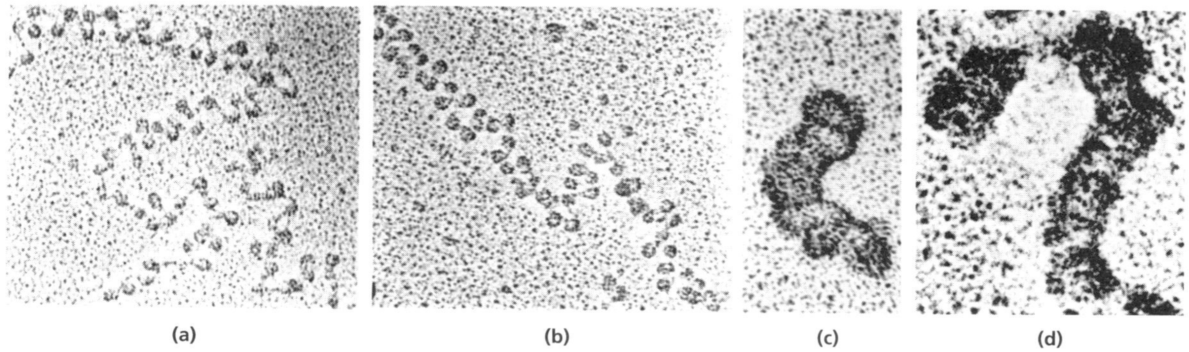

Figure 6.15 Electron micrographs of rat liver chromatin fibres fixed at different ionic strengths. Fixation was performed in (a) 1 mM triethanolaine chloride, 0.2 mM EDTA, (b) 5 mM triethanolamine chloride, 0.2 mM EDTA, and (c) and (d) 100 mM NaCl, 5 mM triethanolamine chloride. [Reproduced from Thoma *et al.*, *J Cell Biol* 1979;83:408 by permission of The Rockefeller University Press.]

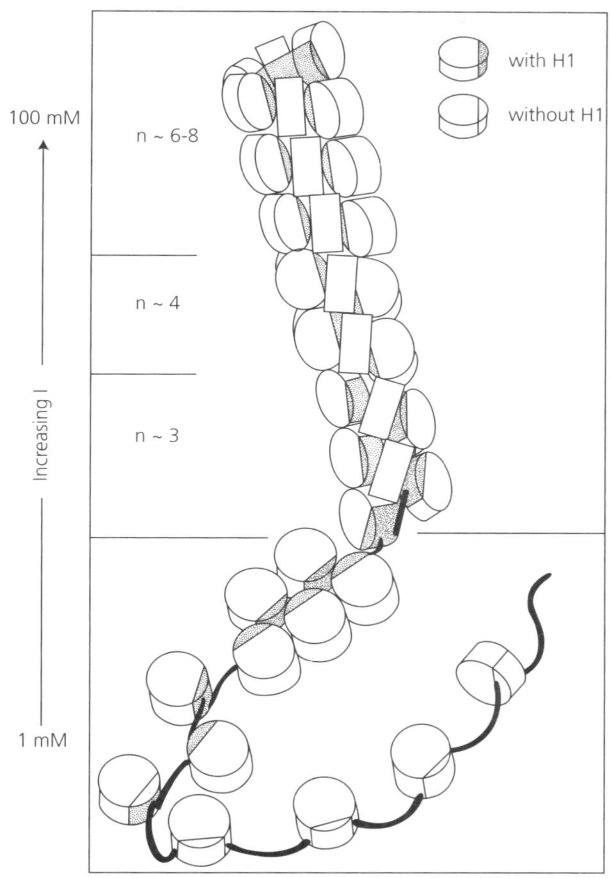

Figure 6.16 Idealized drawing of the chromatin at different ionic strengths. The zigzag pattern of nucleosomes at lower ionic strengths (bottom left) is converted into helices having an increasing number (*n*) of nuclesomes per turn with the increase in the ionic strength. Though in nature the helices are irregular, in the figure, these have been shown as regular. Also, it appears as if the DNA is threaded through the nucleosome cores, but actually the DNA is wrapped around them (see Fig. 6.15) [Reproduced from Thoma *et al.*, *J Cell Biol* 1979;8:419 by permission of The Rockefeller University Press.]

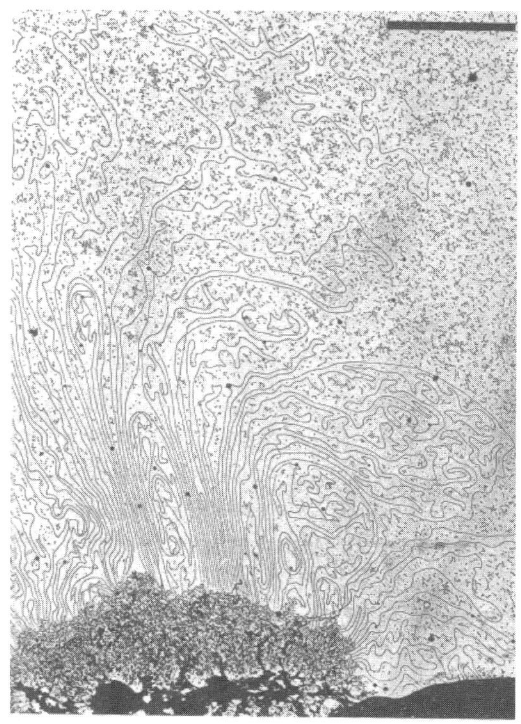

Figure 6.17 Electron micrograph of histone-depleted metaphase chromosome of a HeLa cell. The densely stained central scaffold is at the bottom of the photograph and the DNA can been seen attached to the scaffold in loops, both ends of which appear to emanate from adjacent points in the scaffold. The scale bar is 2 μm. A histogram of the loop lengths of this chromosome is shown in Fig. 6.18. [Reproduced from Paulson and Laemmli, *Cell* 1977;12:823 by permission of Elsevier Science.]

It has been speculated that the loops of the 30-nm chromatin fibres form further complex structures such as loops, rosettes and coils in the path of the assembly of chromosomes, as shown in Fig. 6.19 (Filipski *et al.* 1990).

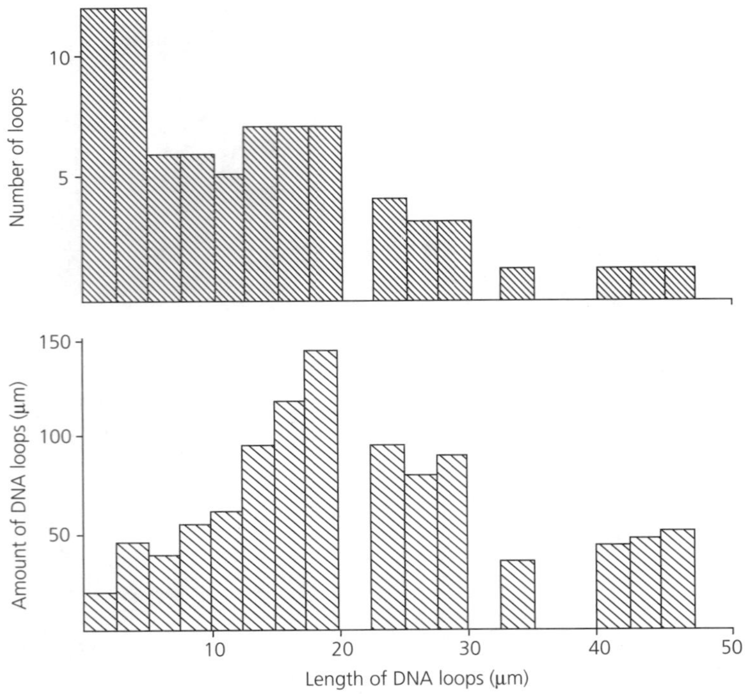

Figure 6.18 Histogram of loop lengths in a histone-depleted metaphase chromosome of a HeLa cell. All loops along one side of the chromatid shown in Fig. 6.17 have been measured. One-micrometre-long DNA comprises approximately 3000 bases. [Reproduced from Paulson and Laemmli, *Cell* 1977;12:824 by permission of Elsevier Science.]

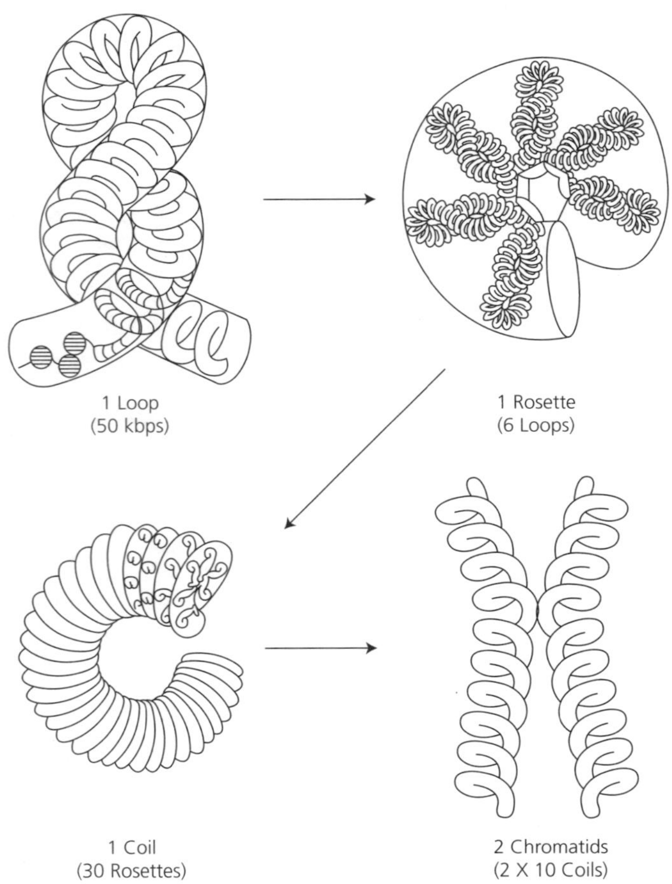

Figure 6.19 Speculated pathway of formation of chromosomes from chromatin fibres. [Reproduced from Filipski *et al*., *EMBO J* 1990;9:125 by permission of *EMBO J*.]

References

Benyajati C, Worcel A. Isolation, characterization and structure of the folded interphase genome of *Drosophila melanogaster*. *Cell* 1976;9:393–407.

Boy de la Tour E, Laemmli UK. The metaphase scaffold is helically folded: Sister chromatids have predominantly opposite helical handedness. *Cell* 1988;55:937–944.

Cairns J. The bacterial chromosome and its manner of replication as seen by autoradiography. *J Mol Biol* 1963;6:208–213.

Earnshaw WC. Mitotic chromosome structure. *BioEssays* 1988;9:147–150.

Filipski J, Leblanc J, Youdale T, Sikorska M, Walker PR. Periodicity of DNA folding in higher-order chromatin structures. *EMBO J* 1990;9:1319–1327.

Finch JT, Lutter LC, Rhodes D, Brown AS, Rushton B, Levitt M, *et al.* Structure of nucleosome core particles of chromatin. *Nature* 1977;269:29–36.

Fletcher TM, Hansen JC. Core histone tail domains mediate oligonucleosome folding and nucleosomal DNA organization through distinct molecular mechanisms. *J Biol Chem* 1995;270:25359–25362.

Gasser SM, Laemmli UK. The organization of Chromatin loops: Characterization of a scaffold attachment site. *EMBO J* 1986;5:511–518.

Griffith JD. Visualization of prokaryotic DNA in a regularly condensed chromatin-like fiber. *Proc Natl Acad Sci USA* 1976;73:563–567.

Hansen JC, Ausio J, Stanik VH, van Holde KE. Homogenous reconstituted oligonucleosomes, evidence for salt-dependent folding in the absence of histone H1. *Biochemistry* 1989;28:9129–9136.

Hayes JJ, Clark DJ, Wolffe AP. Histone contribution to the structure of DNA in a nucleosome. *Proc Natl Acad Sci USA* 1991;88:6829–6833.

Hayes JJ, Tullius TD, Wolffe AP. The structure of DNA in a nucleosome. *Proc Natl Acad Sci USA* 1990;87:7405–7409.

Hewish DR, Burgoyne LA. Chromatin substructure: The digestion of chromatin DNA at regularly spaced sites by a nuclear deoxyribonuclease. *Biochem Biophys Res Commun* 1973;52:504–510.

Hogan ME, Rooney TF, Austin RH. Evidence for kinks in DNA folding in the nucleosome. *Nature* 1987;328:554–557.

Jackson DA, Dickinson P, Cook PR. The size of chromatin loops in HeLa cells. *EMBO J* 1990;9:567–571.

Kavenoff R, Klotz LC, Zimm BH. On the nature of chromosome-sized DNA molecules. *Cold Spring Harb Symp Quant Biol* 1973;38:1–8.

Kavenoff R, Zimm BH. Chromosome-sized DNA molecules from *Drosophila. Chromosoma* 1973:41:1–27.

Kornberg R. Chromatin structure: A repeating unit of histones and DNA. *Science* 1974;184:868–871.

Kornberg R, Thomas JO. Chromatin structure: Oligomers of histones. *Science* 1974;184:865–868.

Noll M. Subunit structure of chromatin. *Nature* 1974a;251: 249–251.

Noll M. Internal structure of the chromatin subunit. *Nucl Acids Res* 1974b;1:1573–1578.

Noll M, Kornberg R. Action of micrococcal nuclease on chromatin and the location of the histone H1. *J Mol Biol* 1977;109: 393–404.

Olins AL, Olins DE. Spheroid chromatin units (V bodies). *Science* 1974;183:330–332.

Paulson JR, Laemmli UK. The structure of histone-depleted metaphase chromosomes. *Cell* 1977;12:817–828.

Richmond TJ, Finch JT, Rushton B, Rhodes D, Klug A. Structure of the nucleosome core particle at 7 Å resolution. *Nature* 1984;311:532–537.

Rouviere-Yaniv J, Yaniv M, Germond J-E. *E. coli* DNA binding protein HU forms nucleosome-like structure with circular double-stranded DNA. *Cell* 1979;17:265–274.

Sahasrabuddhe CG, van Holde KE. The effect of trypsin on nuclease-resistant chromatin fragments. *J Biol Chem* 1974;249: 152–156.

Schwarz PM, Felthauser A, Fletcher TM, Hansen JC. Reversible oligonucleotide self-association: Dependence of divalent cations and core histone tail domains. *Biochemistry* 1996;35: 4009–4015.

Schwarz PM, Hansen JC. Formation and stability of higher order chromatin structure. *J Biol Chem* 1994;269:16284–16298.

Stonington GO, Pettijohn DE. The folded genome of *Escherichia coli* isolated in a protein-DNA-RNA complex. *Proc Natl Acad Sci USA* 1971;68:6–9.

Thoma F, Koller T, Klug A. Involvement of histone H1 in the organization of the nucleosome and the salt dependent superstructures of chromatin. *J Cell Biol* 1979;83:402–427.

Varshavsky AJ, Bakayev VV, Georgiev GP. Heterogeneity of chromatin subunits *in vitro* and location of histone H1. *Nucleic Acids Res* 1976;3:477–492.

Wellenzohn B, Flader W, Winger RH, Hallbrucker A, Mayer E, Liedl KR. Exocyclic groups in the minor groove influence the backbone conformation of DNA. *Nucleic Acids Res* 2001;29:5036–5043.

Williamson R. Properties of rapidly labelled deoxyribonucleic acid fragments isolated from the cytoplasm of primary cultures of embryonic mouse liver cells. *J Mol Biol* 1970;51:157–168.

Wolffe AP, Khochbin S, Dimitrov S. What do linker histones do in chromatin? *BioEssays* 1997;19:249–255.

Worcel A, Burgi E. On the structure of the folded chromosome of *Escherichia coli. J Mol Biol* 1972;71:127–147.

7

Chromatin remodelling

Rohini Muthuswami

7.1 Why chromatin remodelling?

Inside a cell, the DNA is present in a compactly packaged form known as chromatin. The tight packaging is necessitated by the long length of the DNA as compared with the size of the cell. For example, chromosome 1 in humans has approximately 247 million base pairs (bp). If these base pairs were placed end to end, the length of the chromosome would be 84 mm (Box 7.1). A eukaryotic cell is about 10 μm, and the nucleus occupies 1/10th volume of the cell. Thus, the 84-mm-long DNA has to be packed inside a 1-μm cell. The only way this can be achieved is if the DNA is compacted. The compacted DNA is called as chromatin and is composed of DNA, histones and non-histone proteins. However, once the DNA is compacted, it becomes inaccessible to proteins that mediate replication, transcription, repair and recombination. Thus, the DNA present as chromatin has to be made accessible for these processes to occur, and after the process is over, it has to be made inaccessible once more. The reaction that mediates the accessibility and inaccessibility of chromatin is known as chromatin remodelling.

7.2 Chromatin architecture

In a eukaryotic cell, there are three stages of compaction of DNA. The naked double-stranded DNA, in the first stage, is wrapped around a protein complex known as nucleosome (Fig. 7.1). Nucleosomes are composed of four types of histone: H2A, H2B, H3 and H4. Each nucleosome contains two molecules, each of H2A, H2B, H3 and H4, thus forming an octamer. The nucleosome structure shows that 146 bp of DNA is wrapped around each nucleosome (Richmond and Davey 2003). The wrapping of DNA around the nucleosome results in the formation of a left-handed superhelix (Richmond and Davey 2003). The phosphodiester bonds of the DNA make contact with the histone core, while the N-terminus tails of the histones protrude out of the DNA superhelix (Richmond and Davey 2003).

Box 7.1 Calculating the length of chromosomes in a eukaryotic cell

Consider the following DNA sequence:

ATGCTCGAATCG ...

TACGAGCTTAGC ...

Distance between two base pairs	0.34 nm
Number of base pairs in chromosome 1	247×10^6
Length of chromosome 1	$0.34 \times 10^{-9} \times 247 \times 10^6$ m = 84 mm

In the second stage, histone H1 interacts with the nucleosome-wrapped DNA (Thoma 1979; Woodcock *et al.* 2006; Fig. 7.1). This interaction facilitates nucleosome–nucleosome interaction, resulting in the formation of a solenoid structure (Thomas 1984). Information regarding the higher-order structure has remained sketchy. Recently, Schalch and colleagues created a tetranucleosome and showed using X-ray crystallography that the tetranucleosome possessed a zigzag structure (Schalch *et al.* 2005). The linker DNA zigzags between the two nucleosome cores, creating two-start helix structure (Schalch *et al.* 2005).

In the third stage, non-histone proteins associate with the chromatin, modulating its architecture and inducing further compaction (Woodcock and Ghosh 2010). At metaphase, the chromatin has a central scaffold made of structure maintenance chromatin (SMC) proteins (Hirano 1998; Woodcock and Ghosh 2010). Chromatin loops emanate from this central scaffold. At this stage, the chromatin is 10,000-fold compacted with respect to the naked DNA (Woodcock and Ghosh 2010).

Chromatin traditionally has been classified into heterochromatin and euchromatin. The heterochromatin consists of a transcriptionally silent region, while the euchromatin is transcriptionally active. The heterochromatin can further be classified into facultative

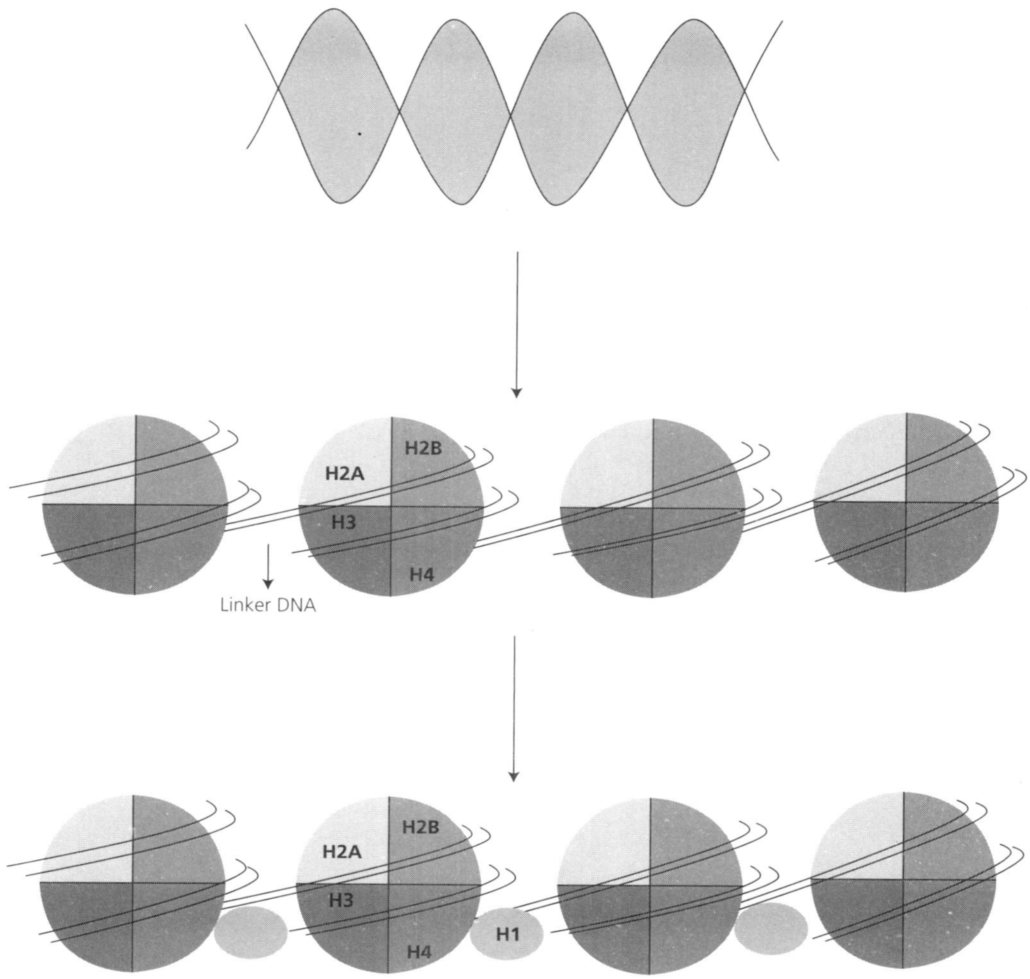

Figure 7.1 Wrapping of the DNA around the nucleosome to create the bead-on-the-string structure. See Colour Plate 1 on page 503.

and constitutive heterochromatin. The facultative heterochromatin can alter between heterochromatin and euchromatin over the course of life of the cell. However, the constitutive heterochromatin does not undergo these changes and is transcriptionally repressed through the course of cell life. For example, the centromeric region is a constitutive heterochromatin region.

7.2.1 DNA metabolic processes

All DNA metabolic processes – replication, transcription, repair and recombination – are mediated by protein complexes. This chapter focuses only on transcription and repair.

DNA transcription minimally requires general transcription factors and an RNA polymerase (Orphanides *et al.* 1996). The eukaryotic RNA polymerase II, which transcribes the mRNA, is a complex of at least 10–12 subunits (Cramer *et al.* 2008; Spahr *et al.* 2009). In addition, transcription often requires additional transcription factors known as activators, a mediator complex and chromatin remodelling complexes (Ptashne 1988; Lin *et al.* 1990; Flanagan *et al.* 1991; Rachez and Freedman 2001; Narlikar *et al.* 2002; Berger 2007; Li B *et al.* 2007). Thus, an enormous protein complex mediates eukaryotic transcription.

DNA repair complexes repair any damage that occurs to the DNA. There are many DNA repair systems operating in a cell. The mismatch repair complexes repair DNA mismatches, the double-strand repair pathway repairs DNA double-strand breaks and the base excision repair pathway repairs base excision damage (Wood 1996).

The double-strand break repair pathway has been most studied with respect to chromatin remodelling (Bao and Shen 2007). Double-strand breaks are potentially dangerous because if they are left unrepaired, they can attach to other chromosomes, causing DNA translocations, one of the major factors of cancer (Khanna and Jackson 2001). The double-strand breaks can be repaired either by homologous recombination or the non-homologous end-joining pathway (Kanaar *et al.* 1998). Both these pathways require Ku proteins, which bind to the DNA ends, DNA protein kinase, chromatin remodelling complexes and DNA ligase (Kanaar *et al.* 1998).

Thus, both transcription and repair are mediated by large protein complexes and these proteins complexes can access DNA only if the chromatin structure is altered or made more accessible.

7.3 Types of chromatin remodelling

The remodelling of chromatin minimally requires alteration of the nucleosome–DNA interactions, followed by nucleosome repositioning. The nucleosome–DNA interactions are altered by introducing covalent modifications in the nucleosome, while nucleosome repositioning is achieved by breaking and reforming the bonds between the nucleosome and the DNA (Fig. 7.2).

In addition to these processes, the core nucleosome can be replaced by histone variants, the DNA itself can be methylated and, recently, small RNA molecules have been shown to participate in remodelling the chromatin (Fig. 7.2).

There is considerable cross-talk between these different processes, and thus, a given section of chromatin is remodelled by the concerted action of at least two or more processes.

7.4 Nucleosome modifications

The four histone molecules can be covalently modified. Histone acetylation was one of the first modifications to be identified (Allfrey *et al.* 1964). Today, many modifications are known – acetylation, phosphorylation, methylation, ubiquitination, SUMOylation and ADP-ribosylation. All these modifications are reversible, which is the one of the major requirements of chromatin remodelling.

7.5 Nucleosome repositioning

Nucleosome repositioning involves physical movement of the nucleosome along the chromatin. The physical movement entails breaking of the existing bonds between the nucleosome and the DNA and re-formation of new bonds between the nucleosome and a new stretch of DNA. As this is a physical work, energy in the form of ATP hydrolysis is required. Thus, the chromatin remodellers mediating nucleosome repositioning are ATPases and are known as ATP-dependent chromatin remodellers (Vignali *et al.* 2000). These remodellers can catalyse both the formation of heterochromatin (where nucleosomes are spaced close together) and euchromatin (where nucleosomes are spaced further apart).

7.6 Histone modifications

The N-terminal tail of histones is post-transcriptionally modified. The range of modifications is enormous and new modifications are continuously being discovered. The earliest post-transcriptional modification to be discovered was acetylation. Now we know that histones can be methylated, phosphorylated, ubiquitinated, SUMOylated, ADP-ribosylated and citrullinated. All these modifications are reversible, thus fulfilling one of the major principles of chromatin remodelling.

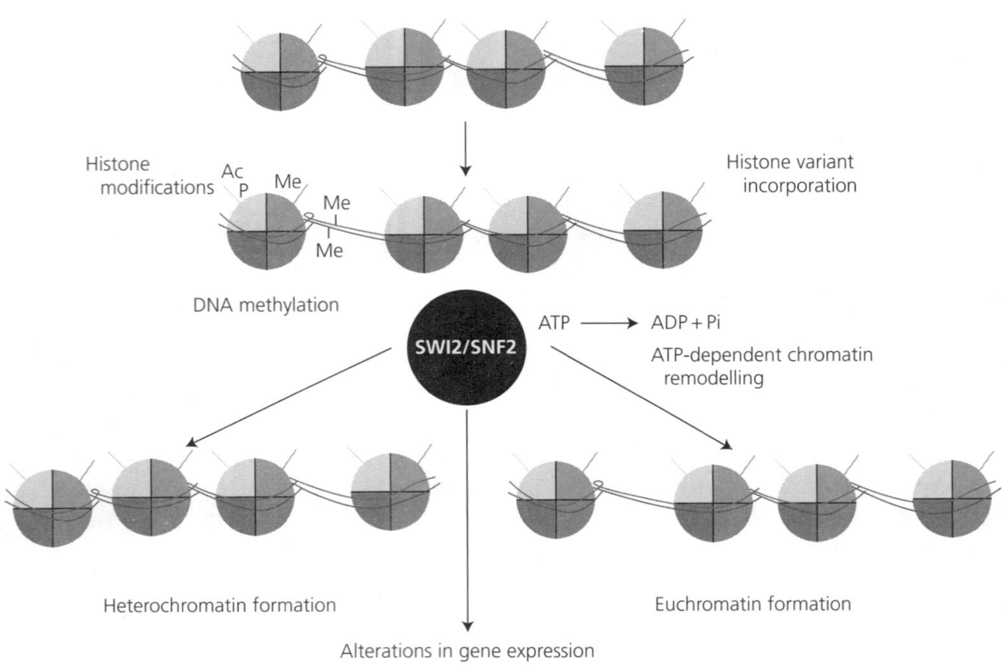

Figure 7.2 Chromatin remodelling involves the concerted action of histone modifications, histone variant incorporation, DNA methylation and ATP-dependent chromatin remodelling, leading to alterations in gene expression. See Colour Plate 2 on page 504.

7.6.1 Histone-modifying reactions

Histone-modifying reactions are mediated by specific enzymes which are subunits of multiprotein complexes. These multiprotein complexes mediate transcription activation/repression, DNA repair, DNA replication and DNA recombination. Thus, for example, a complex known as SAGA is required for transcription activation (Grant *et al.* 1998). The SAGA complex contains, among many proteins, Gcn5p, which is a histone acetyltransferase mediating acetylation (Grant *et al.* 1998). Similarly, a complex known as mSin3A, which mediates transcription repression, contains histone deacetylase as one of its subunits (Pazin and Kadonaga 1997).

7.6.2 Histone acetylation

In 1964, Vincent Allfrey proposed that transcription is regulated by acetylation of histones (Allfrey *et al.* 1964). He studied the incorporation of the acetyl group in isolated HeLa nuclei using ^{14}C-acetate. First, the authors showed that ^{14}C-acetate was specifically incorporated into histones in a time-dependent manner. Next, they studied the incorporation of ^{14}C-acetate in the absence and presence of puromycin, a known translation inhibitor. They found that the incorporation of ^{14}C-acetate was not inhibited in the presence of puromycin, suggesting that the incorporation of the acetate group on histones was a post-translational event. Finally, they measured the rate of RNA synthesis using naked DNA, DNA–histone and DNA-acetylated histone template and found that RNA polymerization was similar for DNA-acetylated histone and naked DNA template, while it was inhibited when DNA–histone template was used. These experiments led Allfrey *et al.* to conclude that histone acetylation was a post-translational event and that this modification favours the transcription process (Allfrey *et al.* 1964).

Subsequently, in 1978, two research papers were published that showed that transcriptionally active chromatin was enriched in acetylated histones. In one of these papers, authored by Sealy and Chalkley, the researchers isolated nuclei from butyrate-treated mammalian HTC cells as butyrate is a known inhibitor of deacetylation (Sealy 1978). These nuclei were subjected to DNase I treatment. After DNase I treatment, the nuclei were separated into soluble and insoluble nucleoprotein fractions. Histones were isolated from both the fractions and analysed using urea–polyacrylamide gel electrophoresis. The experiment clearly showed that acetylated histones were associated with the soluble fraction that was enriched in transcriptionally active chromatin (Sealy 1978).

Histone acetylation is mediated by histone acetyltransferases (HATs) that transfer the acetyl group from acetyl-CoA to the lysine residues present on the N-terminal tail of histones. The first HAT was purified in 1996 by the research group headed by C. David Allis about 30 years after acetylation had been first hypothesized to play a role in transcription (Brownell *et al.* 1996). These researchers cloned the *HAT* gene from *Tetrahymena* and showed that the protein encoded by this gene was homologous to Gcn5p, a known transcriptional activator present in yeast. Further, they demonstrated that both Gcn5p and the protein from *Tetrahymena* possessed histone acetyltransferase activity, thus linking acetylation to transcriptional activation (Brownell *et al.* 1996).

HATs have been classified into five families based on their sequence and structural homologies (Roth *et al.* 2001; Carrozza *et al.* 2003): (a) GNAT family, which comprises Gcn5 and its homologues (Vetting *et al.* 2005); (b) MYST family (for MOZ, Ybf/Sas3,Sas2 and Tip60 (Utley and Côté 2003); (c) p300/CBP (Chan and La Thangue 2001); (d) general transcription factor HAT, which includes the TAF250 subunit of TFIID and (e) nuclear hormone-related HATs.

The GNAT family of HATs plays a critical role in cell growth and development. Knockout of Gcn5 in mice have been shown to be embryonic lethal, suggesting that Gcn5 is critical for development (Xu *et al.* 2000; Yamauchi *et al.* 2000). The MYST family of HATs plays a role in cell cycle (Smith 1998; Clarke *et al.* 1999). Mutations in Esa1, an essential member of this family from *Saccharomyces cerevisiae*, arrests the cell at the G2/M boundary, suggesting a critical role in cell cycle progression (Clarke *et al.* 1999). p300/CBP were independently isolated as a protein (p300) that interacts with the E1A protein of adenovirus (Yang *et al.* 1996) and as a protein that interacts with the phosphorylated form of CBP. Subsequent investigations proved that these two proteins shared a high degree of similarity and in cultured cells behaved interchangeably. Thus, the two proteins are now known as p300/CBP.

7.6.3 Histone deacetylation

Acetylated histones are deacetylated by a specific class of enzymes known as histone deacetylases. The first histone deacetylase was also discovered in 1996 by Taunton *et al.* (Taunton *et al.* 1996). These authors took advantage of the fact that histone deacetylase activity is inhibited by trapoxin, a substrate mimetic. This molecule binds to histone deacetylase, acting as an 'irreversible inhibitor'. The authors, therefore, postulated, that the histone deacetylase activity and the trapoxin-binding activity should co-elute during purification. Purification of proteins from calf thymus using ammonium sulphate precipitation and ion exchange chromatography led to the identification of a nuclear histone deacetylase, HD1, protein. Since then, four classes of histone deacetylases have been identified (de Ruijter *et al.* 2003). Class I, II and IV enzymes are metal-dependent histone deacetylases, while class III enzymes are NAD dependent. Class I enzymes are also known as RPD3-like histone deacetylases. Class II enzymes are known as Hda1-like histone deacetylases, while Class III enzymes constitute the Sir2 family of histone deacetylases (Blander and Guarente 2004).

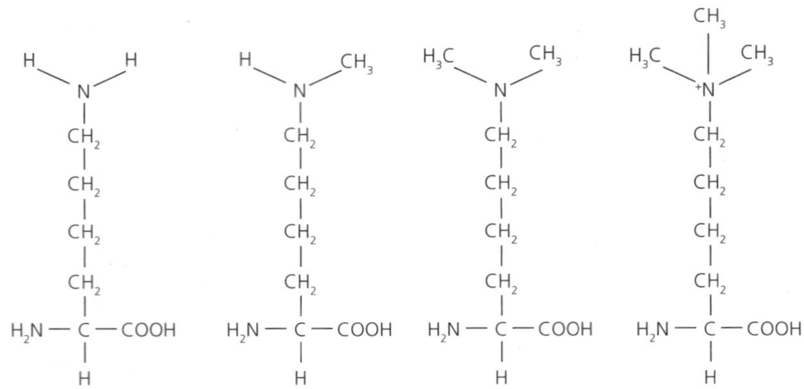

Figure 7.3 Lysine residues present on histone molecules can be monomethylated, dimethylated or trimethylated.

7.6.4 Histone methylation

Histone methylation reaction is mediated by histone methyltransferases, which mediate the transfer of the methyl group from S-adenosylmethionine (SAM) to either lysine or arginine present in the N-terminus tail of histones.

Lysine methylation is mediated by the SET-family of methyltransferases (Dillon *et al.* 2005). These enzymes transfer the methyl group from SAM to the ε-NH2 group of lysine, catalysing the formation of monomethylated, dimethylated or trimethylated lysine (Fig. 7.3). SUV39, which methylates histone H3K9, was the first histone methyltransferase to be discovered (Rea *et al.* 2000). The SET-family of methyltransferases possess the SET domain. The SET domain, which was characterized in 1998, comprises approximately 130 amino acids (Jenuwein *et al.* 1998). In SUV39, the methyltransferase activity resides within the SET domain, and thus, it was hypothesized that the SET domain was necessary for the methyltransferase activity. However, experiments now have shown that methylation occurs only when the SET domain is flanked by two cysteine-rich sequences, known as PRE-SET and POST-SET sequences (Qian and Zhou 2006).

Arginine methylation is mediated by protein arginine methyltransferases (PRMTs) (Wysocka *et al.* 2006). These enzymes transfer the methyl group from SAM to the terminal guanidino nitrogen of arginine, generating monomethyl arginine, symmetric dimethyl arginine or asymmetric dimethyl arginine (Fig. 7.4). Till now, three PRMTs have been identified. CARM1 and PRMT1 are responsible for generating monomethyl arginine and asymmetric dimethyl arginine modifications (Wysocka *et al.* 2006). PRMT5, which is also known as JBP1, modifies arginine to generate symmetric dimethyl arginine derivatives (Branscombe *et al.* 2001).

Unlike histone acetylation, which is strongly associated with transcription activation, histone methylation shows strong context dependence, and thus, histone lysine methylation is associated with both transcription activation and repression (Kouzarides 2002). Therefore, depending on the residue modified and whether the residue is mono-, di- or tri-methylated, methylation signals transcription activation, repression or heterochromatin formation, respectively. Histone arginine methylation, similarly, is associated with both transcription activation and repression (Litt *et al.* 2009). For example, researchers have shown that CARM1 interacts with p300/CBP, and that the synergy between arginine methylation and lysine acetylation results in transcription activation of nuclear receptors (Lee *et al.* 2002).

For a long time, as the search of histone demethylases failed to fructify, it was believed that methylation was an irreversible mark. In a breakthrough research, Shi *et al.* in 2004 reported the identification of the first lysine demethylase (Shi *et al.* 2004). They identified a protein KIAA0601 that possessed sequence homology to amine oxidases and hypothesized that this protein might be capable of oxidizing N-methylated histones. To prove their hypothesis, they expressed the protein in bacteria and purified it. Using the recombinant purified protein, they showed that dimethylated histones H3 K4 and H3 K9 were effectively demethylated. Subsequently, they renamed KIAA0601 as lysine-specific demethylase 1 or LSD1 (Shi *et al.* 2004). Now many demethylases have been identified that specifically remove the methylation mark on H3K4, H3K9, H3K27 and H3K36 (Shi and Whetstine 2007).

7.6.5 Histone phosphorylation

Histones can be phosphorylated on the serine and threonine residues present on the N-terminal tails. Phosphorylation modification is associated with both transcriptional activation and chromatin condensation during mitosis. Interestingly, phosphorylation of histone H3 serine 10 (H3S10) mediates both transcription activation and chromatin condensation (Johansen and Johansen 2006).

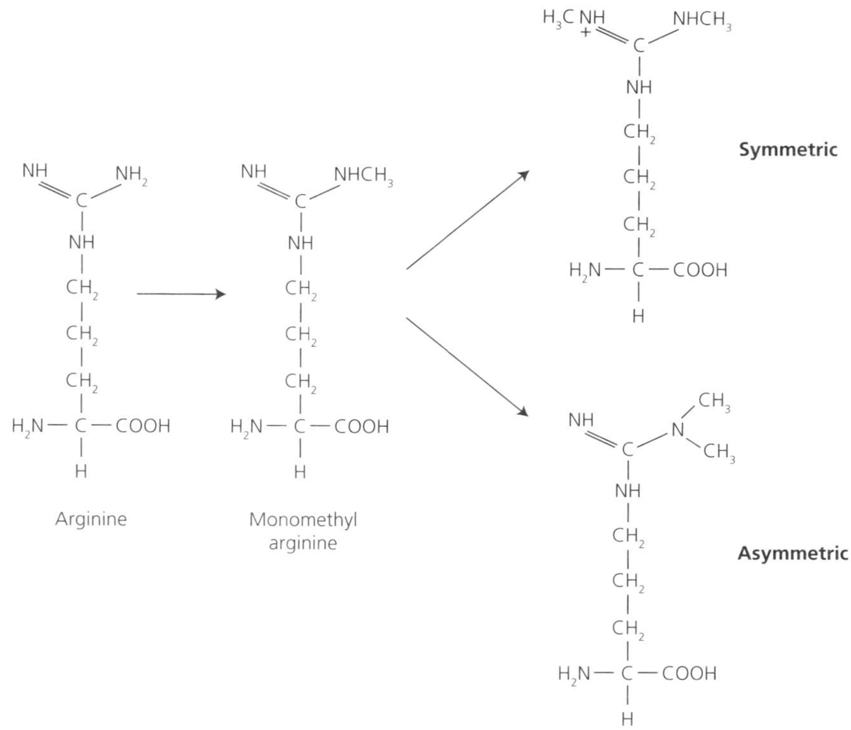

Figure 7.4 Arginine residues on histone molecules are either monomethylated or dimethylated.

Histone phosphorylation was predicted more than 30 years ago. In 1982, Paulson and Taylor reported that histones H3 and H1 were phosphorylated in metaphase chromosomes of HeLa cells (Paulson and Taylor 1982). Subsequently, in 1991, Mahadevan and coworkers showed that when quiescent cells were activated using growth factors such as phorbal esters, early response genes were rapidly induced (Mahadevan *et al.* 1991). They further showed using peptide sequencing that histone H3 was phosphorylated and the phosphorylation could be detected even as early as 5 min after stimulation with growth factor (Mahadevan *et al.* 1991).

Two different kinases that mediate histone phosphorylation have been identified (Nowak and Corces 2004). The MAP kinases phosphorylated histone H3S10 during the G1 phase and this phosphorylation results in transcription activation. During mitosis, aurora kinases phosphorylate histone H3S10 and this phosphorylation mediates chromatin condensation required for cell division to occur (Johansen and Johansen 2006).

Histone phosphorylation is regulated by phosphatases. However, very little information is available about the role of phosphatases. The well-studied example is that of PP1, which has been shown to behave antagonistically to aurora kinases, thus regulating the phosphorylation during mitosis (Hsu *et al.* 2000). These authors showed that histone H3 dephosphorylation was dependent on Glc7p, a PP1 family phosphatase, in *S. cerevisiae*. Mutation in Glc7p resulted in elevated levels of phosphorylated histone H3 and chromosome segregation defects (Hsu *et al.* 2000).

7.6.6 Histone ubiquitination

Ubiquitination of histones H2A and H2B is associated with transcription initiation, and elongation, DNA repair and silencing (Zhang 2003; Weake and Workman 2008). Ubiquitination occurs on the lysine residue of these histone molecules. Ubiquitin is a small protein of 76 amino acid residues that can be conjugated to the lysine residues present on proteins. The ubiquitination reaction is mediated by three separate enzymes: E1 or ubiquitin-activating enzyme, which activates ubiquitin in an ATP-dependent reaction; E2 or ubiquitin-conjugating enzyme, which conjugates ubiquitin via a thioester bond to a cysteine residue present on E2; and E3 or ubiquitin ligase, which transfers the ubiquitin to the lysine residue present on the N-terminal tail of histones. The histone lysines are monoubiquitinated, in contrast to the polyubiquitination of proteins, which targets them to proteosomes for degradation (Fig. 7.5). The monoubiquitination modification of histone lysines does not target them for degradation. However, the monoubiquintation modification results in the recruitment of transcriptional activators, leading to gene activation.

Of all the histone ubiquitination enzymes studied, Rad6, the E2 ubiquitin-conjugating enzyme from *S. cerevisiae*, is the best characterized (Jentsch *et al.* 1987). Rad6 is required for H2BK123 ubiquitination. This

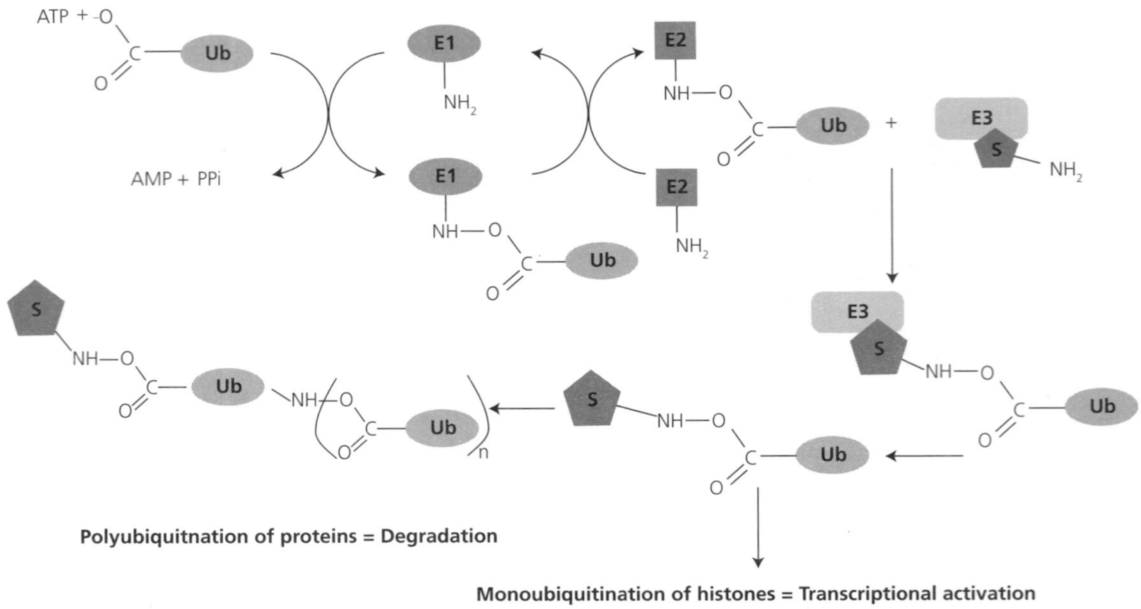

Figure 7.5 Histone ubiquitination mediates transcription activation. See Colour Plate 3 on page 504.

enzyme works in conjunction with Bre1, which is the E3 ubiquitin ligase enzyme.

As with other histone modifications, ubiquitination is also reversible. Deubiquitination is catalysed by ubiquitin-specific proteases. In *S. cerevisiae*, Ubp8 and Ubp10 have been identified as the deubiquitination enzymes (Weake and Workman 2008; Schulze *et al.* 2011). Deletion of Ubp8 and Ubp10 results in increased levels of ubiquitinated H2B in the cell (Weake and Workman 2008). Ubp10 plays a critical role in silencing of telomeric DNA (Osley 2006).

7.6.7 Histone SUMOylation

SUMO or small ubiquitin-like modifier protein shows significant sequence homology to ubiquitin. Like ubiquitination, SUMOylation is also catalysed by three separate enzymes: E1, E2 and E3. The E2 enzymes, also known as Ubc9, directly interact with substrates to catalyse the formation of an isopeptide bond between the C-terminus of SUMO protein and the ε-amino group of lysine present on histone tails.

Modification of histone H4 lysine by SUMO is associated with transcription repression. Shiio and Eisenman FLAG-tagged the N-terminus of histones and examined the status of SUMOylation after transfecting the constructs into HEK293T cells using anti-SUMO antibody (Shiio and Eisenman 2003). These experiments revealed that histone H4 is primarily SUMOylated and this modification leads to transcription repression. Further experiments by Nathan *et al.* showed that in *S. cerevisiae,* when Ubc9 is down-regulated, global acetylation is upregulated, leading to the concept that SUMOyation and acetylation are antagonistic within the chromatin remodelling context (Nathan 2006).

SUMOylation is reversible and specific isopeptidases have been identified that cleave the isopeptide bond.

7.7 Role of histone acetylation in transcription regulation

Histone modifications regulate DNA transcription, repair, replication as well as recombination. Of these processes, the role of histone modification, especially histone acetylation, has been the most studied and has been consistently correlated with transcription activation (MacDonald and Howe 2009). In contrast, histone deacetylases mediate transcription repression. Both histone acetyltransferases and histone deacetylases exist as multi-subunit protein complexes.

The identification of Gcn5p in *Tetrahymena* provided the impetus to study the acetylation status at individual promoters (Brownell *et al.* 1996). Experimental evidence has shown that Gcn5p is the catalytic component of the SAGA complex(Grant *et al.* 1997). SAGA (Spt-Ada-Gcn5 acetyltransferase) is a 2-MDa complex that functions as a transcriptional co-activator. This complex, through its auxiliary subunits, can interact with multiple transcriptional activators and therefore be recruited to the promoter (Qiu *et al.* 2004). Recruitment of the SAGA complex to the promoter facilitates binding of the TATA-binding protein (TBP) (Bhaumik and Green 2002). These authors first identified the promoters that required the SAGA activity. Analysis of TBP recruitment to these promoters in *spt7-*, *ada1-* and *spt20*-deleted yeast cell lines showed that the recruitment was dependent on the presence of the factors, leading them to conclude that the SAGA complex facilitates recruitment of TBP to the promoter (Bhaumik and Green 2002).

Recently, researchers have begun to map global histone modification patterns in genomes. These mapping studies have shown that histones are predominantly acetylated in the transcriptionally active regions of the genome (Li B *et al.* 2007).

One of the intriguing questions is whether histone modifications are stable or whether there is a high rate of turnover. An elegant experiment done in 1975 by Jackson *et al.* showed that acetylated histones are rapidly turned over (Jackson *et al.* 1975). These researchers used radioactive [^{3}H] acetate to label the acetylated fraction of histones in a time-dependent manner. They found that the acetylation reaction was rapid and approximately 70% of the histones were acetylated when a short pulse was given (Jackson *et al.* 1975). They further found that the deacetylation reaction was equally rapid, and that most of the metabolically active acetate was removed with a half-life of 30–40 min (Jackson *et al.* 1975). These experiments were further extended and showed that within the cell, many different pools of acetylated histones exist with characteristic turnover rates. For example, one of these pools has a turnover rate of 1–5 min (Jackson *et al.* 1975). Thus, the histones are continuously acetylated and deacetylated, and recent technological advances have enabled researchers to conclude that the fast turnover dynamics rather than the stable histone acetylation levels are associated with transcription activation (Clayton *et al.* 2006).

7.8 Alteration in nucleosomal structure

The nucleosomal structure was deduced using unmodified histones (Richmond and Davey 2003). Does modification alter the basic nucleosomal structure? Many of the modifications, especially on K115 and K122 of histone H3, are located on the DNA–histone dyad interface and, therefore, have the potential to alter the histone–DNA interactions. In addition, many of the modifications that occur on lysine residues have the ability to reduce the positive charge on the nucleosome and again, therefore, alter the histone–DNA interactions. In 2009, Manohar and colleagues showed using semi-synthetic histone H3 acetylated at lysine 115 that acetylation did not alter the nucleosome position but weakened the nucleosome–DNA interaction, thus suggesting that histone modifications possess the possibility of altering the nucleosomal structure (Manohar *et al.* 2009). However, the nucleosomal structure of modified histones has not been delineated as yet, and therefore, it is not evident whether histone modifications alter the nucleosomal structure.

7.9 Histone code

The pattern of histone modifications on nucleosomes constitutes the histone code (Strahl and Allis 2000). This code determines the transcriptional status of a gene and thereby decides how much of the genomic information should be utilized in a given cell in a given time. Similarly, this code signals whether the DNA is damaged, and if so, which pathway should be involved in repairing it. Thus, the code supplements the genetic information encoded within the DNA. The histone modification pattern is transmitted faithfully during cell division, and therefore, the daughter cells possess the same histone modification pattern as the mother cell.

The histone code has three components: (a) a writer that comprises a histone-modifying enzyme responsible for modifying the histones, (b) a reader that can read the modification and (c) an eraser comprising enzymes that can remove the histone modifications (Cosgrove and Wolberger 2005; Fig. 7.6). Thus, within this system, histone acetyltransferase is a writer, whereas the histone deacetylase is an eraser. What are the readers or how is the histone code read?

The histone modifications constitute a platform for protein modules to interact and, therefore, recruit other proteins required for mediating the process. For example, a protein module known as bromodomain recognizes and binds acetylated lysines (Zeng and Zhou 2002). Similarly, a protein module known as chromodomain recognizes methylated lysines (Blus *et al.* 2011). Bromodomain are generally present on transcriptional activators, while chromodomains are present on transcriptional repressors. Thus, acetylated lysines recruit bromodomain-containing transcriptional activators and thereby facilitate gene expression. The methylated lysines recruit chromodomain-containing transcriptional repressors and thereby facilitate gene repression. The bromodomain- and chromodomain-containing proteins, therefore, are examples of readers of the histone code.

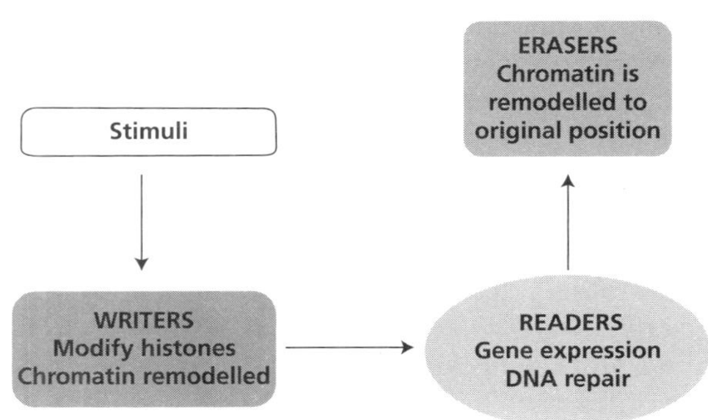

Figure 7.6 Histone code involves three components: writers, readers, and erasers. See Colour Plate 4 on page 505.

The histone code is complicated because a histone molecule can contain multiple modifications. Thus, utilization of the epigenetic information is a sum of the modifications present on the nucleosome. It is difficult to predict *a priori* what the histone modifications imply; the histone code can only be interpreted through experimental results.

7.10 Histone variants

Histone genes are among the slowest evolving genes. However, non-allelic forms of core histones do exist, and these are known as histone variants. The sequence differences between core and variant histones can range from single amino acid to significant divergence in the primary sequence. The biophysical characteristics of histone variants can also be considerably different from that of core histones (Kamakaka and Biggins 2005).

Unlike the genes encoding core histones, histone variant genes are not found in a cluster in the genome. Furthermore, the histone mRNA is not polyadenylated, while the variant histone mRNA is polyadeneylated. Finally, core histones are expressed only during the S phase, but variant histones are expressed throughout the cell cycle (Kamakaka and Biggins 2005). Incorporation of histone variants instead of core histones plays a critical role in heterochromatin formation, DNA repair and transcription.

7.11 How many histone variants?

All core histones except H4 have histone variants. Histone H2A and H3 have the largest number of histone variants identified, while histone H2B has the least.

7.11.1 Variants of histone H1

Eleven variants have been identified in mouse and humans that are expressed in a spatio-temporal manner (Happel and Doenecke 2009). H1.1, H1.5 and H1t in humans are incorporated in a replication-dependent manner and are expressed exclusively during the S phase, while H1.0 is incorporated in a replication-independent manner and is expressed through the cell cycle (Happel and Doenecke 2009). Some of the histone variants, for example, H1.2, are expressed in all cell types, whereas others like sperm-specific and testis-specific H1 variants are expressed only in particular tissues (Happel and Doenecke 2009).

The differences between core H1 and variant H1 are limited mainly to the non-globular N-and C-terminal tails.

7.11.2 Variants of histone H2A

First described by West and Bonner, the largest number of variants has been identified for H2A (West and Bonner 1980; Kamakaka and Biggins 2005) . As in the case of H1 variants, H2A variants too show spatio-temporal differences in expression. H2A-Bbd, for example, localizes on the active X-chromosome, while macroH2A localizes on the inactive X-chromosome in mammals (Chadwick and Willard 2001a, 2001b; Mietton *et al.* 2008). The variants also show differences in conservation during evolution. H2AZ is evolutionarily conserved, while macroH2A as well as H2A-Bbd are present only in vertebrates or mammals.

7.11.3 Variants of histone H2B

In 2002, Zalensky *et al.* cloned and characterized the only well-studied testis/sperm-specific variant of histone H2B (Zalensky *et al.* 2002). This variant, known as TH2B, has been shown to possess telomere-binding activity (Zalensky *et al.* 2002).

7.11.4 Variants of histone H3

Three main classes of histone H3 genes have been identified in metazoans. One class of genes encode for core histone H3, the second class encodes for the replication-independent histone H3.3 and the third class encodes for the centromeric histone H3 known as CENP-A. In addition to these variants, mammalian cells also possess a variant known as H3.1, which is incorporated in a replication-dependent manner.

Ahmed and Henikoff showed in *Drosophila* using green fluorescent protein (GFP)–tagged H3.3 that this histone variant was deposited in the rDNA regions that were actively transcribing rRNA, thus marking out the transcribing regions of the genome (Ahmad and Henikoff 2002). CENP-A is exclusively present in the centromeric regions, while H3.1 and H3.2 have been shown to localize in the heterochromatin. H3.1 and H3.2 differ by a single amino acid; however, the two variants undergo different post-translational modifications, suggesting that the two variants play different roles in the cell (Garcia *et al.* 2005).

7.12 Histone variants and DNA repair

The damage to DNA has to be repaired such that the genetic and epigenetic information is maintained without any alteration. Smerdon in 1991 and later Green and Almouzni in 2002 put forward the access–repair–restore model to explain how DNA repair takes place within the context of the chromatin (Smerdon 1991; Green and Almouzni 2002). According to this model, the first step is to make the damaged DNA accessible to the repair machinery. In the second step, the damage is repaired by DNA repair enzymes. And in the final step, the epigenetic modifications are restored to the original status. Section 7.13 focuses only on how histone variants are involved in making the DNA damage accessible to the DNA repair machinery. In particular, it focuses on the role of H2AX.

7.13 Role of H2AX in DNA repair

H2AX, a highly conserved variant of histone H2A, differs from the core histone primarily in the carboxyl terminus (Fernandez-Capetillo *et al.* 2004). Sequence analysis of H2AX from various species revealed that the protein contained a conserved SQ motif in the carboxyl terminus, which was absent in the core H2A protein. Rogaku *et al.* showed using two-dimensional gel electrophoresis that the serine of the conserved SQ motif is rapidly phosphorylated in response to double-strand DNA breaks induced by ionizing radiation (Rogakou *et al.* 1998). The phosphorylated H2AX is denoted as γH2AX.

Mutation studies have underscored the importance of phosphorylation of H2AX in DNA repair. For example, Celeste *et al.* (2002) showed by targeted disruption of H2AX in mice that H2AX was not necessary for survival. However, $H2AX^{-/-}$ mice were growth retarded and highly sensitive to ionizing radiation. Further, these mice showed increased genomic instability (Celeste *et al.* 2002).

γH2AX helps in recruiting DNA repair proteins to the site of DNA damage. Using specific antibodies, researchers have shown that various proteins such as BRCA1, Nbs1 and ATM1 co-localize with γH2AX at the DNA damage site (Paull *et al.* 2000).

7.14 Nucleosomal structure of histone variants

Does the incorporation of histone variants alter the structure of nucleosomes? Many researchers have investigated this question. The crystal structure of a nucleosome containing H2AZ, a histone H2A variant, has been resolved by Suto *et al.* (2000). These researchers have shown that the nuclesosome particle containing H2AZ variant is structurally similar to the nucleosome particle containing core histones. However, the interaction between (H2AZ.H2B) dimer and (H3.H4) tetramer is destabilized, implying that the nucleosome particle containing H2AZ is inherently less stable than the nucleosome particle containing core histones (Suto *et al.* 2000). Bao *et al.* (2004) have studied the structure of the nucleosome particle containing H2A.BbD, another H2A variant, using two different experimental approaches. In one methodology, these researchers used fluorescence resonance energy transfer (FRET) to measure the distance between the ends of a DNA molecule wound across a nucleosome particle containing the H2A.BbD variant (Bao *et al.* 2004). In another experiment, they digested a nucleosome particle containing the H2A.BbD variant using micrococcal nuclease (Bao *et al.* 2004). Both these experimental strategies showed that only 118 ± 2 bp DNA was wound across the nucleosome particle containing the H2A.BbD variant (Bao *et al.* 2004). Thus, the structural evidence points that histone variants maybe regulating the chromatin dynamics by altering the nucleosomal structure (Ausió 2006).

7.15 Histone barcode

Histone variants are not only incorporated into the nucleosome but are also post transcriptionally modified. Thus, the chromatin consists of nucleosome particles made of either core or variant histones that are further post-transcriptionally modified in myriad ways. The post-transcriptional modification on core as well as variant histone particles thus defines the epigenetic status of a cell not only during the cell cycle but also during cellular differentiation and development. Hake and Allis in 2006 proposed the H3 histone barcode hypothesis to explain how the post-transcriptional modifications on core and variant H3 histones alter the chromatin dynamics and thus the epigenetic status during cellular differentiation and development (Hake and Allis 2006). According to this hypothesis, histones index the chromatin into transcriptionally active and inactive regions by non-random distribution of histone variants (Hake and Allis 2006). Thus, H3.3 is deposited preferentially onto regions that are transcriptionally active, while H2.2 is deposited into silent regions that can be reversibly activated. Therefore, each chromosome has a distinct pattern of post-transcriptionally modified variant histones. These patterns are inherited during cell division. The pattern can undergo changes during cell differentiation, and thus, each cell lineage will have a distinct pattern or a barcode of histone deposition. They further speculated that a cell contains three codes – a genetic code, a post-transcriptional histone code and a histone barcode – that regulate gene expression in the cell (Hake and Allis 2006).

7.16 ATP-dependent chromatin remodelling

Histone modifications are now believed to act as markers or protein scaffolds for other proteins, known as readers, to bind. One such reader is the ATP-dependent chromatin remodelling factor, also known as SWI/SNF proteins, which couples the energy released by ATP hydrolysis to physically move nucleosomes. A mammalian cell contains 24 such ATP-dependent chromatin remodelling factors (Flaus *et al.* 2006).

7.17 SWI/SNF family of ATPases

SWI/SNF proteins were discovered in *S. cerevisiae*. Neigeborn and Carlson (1984) isolated 10 revertants of sucrose non-fermenting mutants (Snf) by plating them on a rich medium containing sucrose, galactose, glycerol or raffinose as the carbon source. Simultaneously, Stern *et al.* (1984) screened for mating-type switch mutants and identified five genes that were required for the expression of the *HO* gene in *S. cerevisiae*. Subsequently, it was discovered that one of the Snf genes (Snf2) and one of the Swi genes (Swi2) encoded an identical product and this was named as the Swi2/Snf2 protein. The Swi2/Snf2 protein was found to possess ATPase activity, and sequence analysis showed that the protein possesses seven conserved motifs (Gorbalenya and Koonin 1993). These motifs are also present in helicases, and therefore, the SWI/SNF family of ATPases is classified as helicases (Gorbalenya and Koonin 1993). However, it needs to be remembered that the SWI/SNF family of proteins does not possess the DNA unwinding property. The energy released by ATP hydrolysis is coupled to nucleosomal movement.

SWI/SNF proteins are present in all organisms, including bacteria, though they are more abundant in eukaryotes than in Archaea and bacteria. Flaus *et al.* (2006), using bioinformatics tools, have further divided these proteins into 24 distinct subfamilies.

7.17.1 SWI/SNF ATPases as catalysts

Nucleosomal positions can be altered such that the chromatin either becomes more open (increased space between nucleosomes) or becomes more closed (decreased space between nucleosomes). The movement of nucleosomes involves breaking of existing bonds between the nucleosome particle and the DNA and making new bonds between nucleosomes and the DNA at a different site from the previously existing position. The breaking and making of nucleosome–DNA bonds requires a certain amount of energy. SWI/SNF proteins act as catalysts, lowering the free energy required for this process and thus facilitating chromatin fluidity (Kingston and Narlikar 1999; Fig. 7.7). It needs to be emphasized that SWI/SNF proteins do not alter the thermodynamic

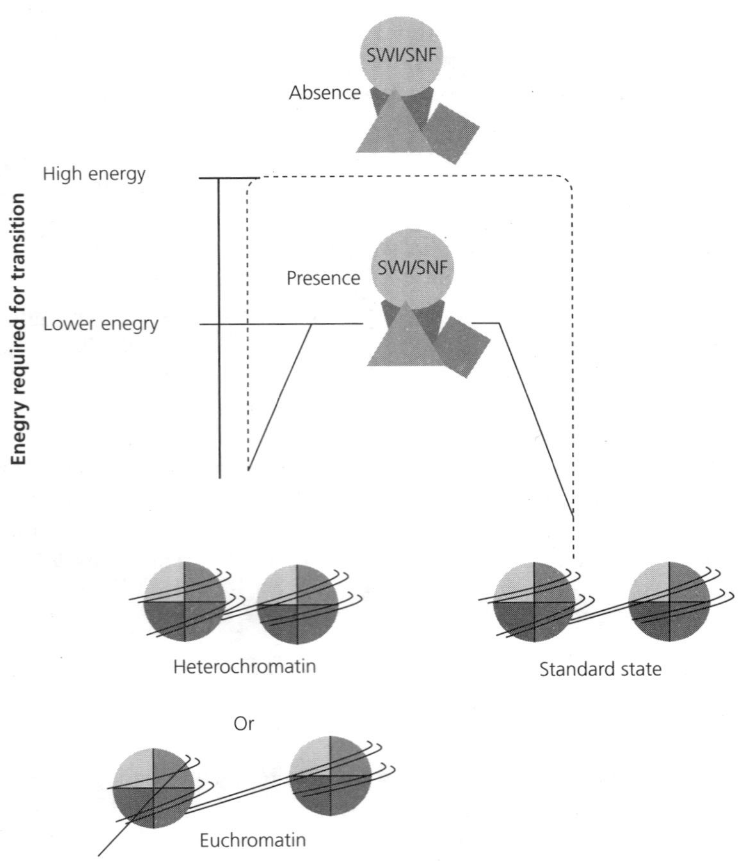

Figure 7.7 ATP-dependent chromatin remodellers catalyse the formation of both the euchromatin and the heterochromatin state. See Colour Plate 5 on page 505.

stability of the nucleosome–DNA interaction. Rather, these proteins facilitate the interconversion between different states (heterochromatin–euchromatin intercoversion) by increasing the rate of interconversion.

7.18 Nucleosomal movement

Chromatin fluidity, as discussed previously, requires movement of the nucleosome along the length of the chromosome. How do SWI/SNF proteins facilitate this movement? Havas *et al.* (2000) showed that SWI/SNF proteins alter the topological constraints of the DNA. This group of scientists used the formation of a cruciform DNA as a measure of superhelical tension. Cruciform DNA is formed when inverted repeats are present in a DNA molecule. However, the formation of a cruciform DNA by inverted repeats is energetically expensive. When the DNA is negatively supercoiled, the formation of a stable cruciform DNA is favoured. Protein molecules that induce negative supercoils in DNA favour the formation of the cruciform. Havas *et al.* (2000) found that the formation of a cruciform DNA was favoured in the presence of SWI/SNF proteins and concluded that these proteins induce negative supercoil in the DNA, thus inducing topological stress in the nucleosomal particle (Havas *et al.* 2000). The superhelical tension destabilizes nuclesomal–DNA interactions.

This notion was challenged by other researchers as they believed that the generation of superhelical tension was not sufficient to break histone–DNA bonds. Kassabov *et al.* (2002) mapped the contacts between histones and the DNA using a photochemical method. Through their experiments, they showed that SWI/SNF proteins break the contacts between histone molecules and the DNA, unwrapping about 50 bp of DNA from the nucleosomal edge (Kassabov *et al.* 2002). The histone octamer is translocated beyond the DNA ends. The process involves the formation of a bulge as the DNA between the nucleosomal entry and exit points are looped out (Kassabov *et al.* 2002). The mechanism of nucleosomal movement is still not resolved. It is, of course, possible that different SWI/SNF proteins favour different mechanisms for moving nucleosomes.

7.19 SWI/SNF proteins and transcription

The role of SWI/SNF proteins in transcription has been extensively studied in yeast, *Drosophila* and mammals. These proteins can activate as well as repress transcription. Snf2, the first protein of this family to be identified, is a transcriptional activator. Snf2 contains helicase motifs as well as a bromodomain. The bromodomain specifically recognizes acetylated lysines, and thus, Snf2 interacts with acetylated histones, leading to transcriptional activation. Snf2 in yeast exists as a multiprotein complex of approximately 11 MDa. The protein complex was purified from *S. cerevisiae* and the components of the complex were analysed by Kornberg's group (Cairns *et al.* 1994). In mammalian cells, two homologues of Snf2 have been identified – Brgl and Brm. These two proteins are constituents of many different protein complexes.

7.20 SWI/SNF proteins and repair

The access–repair–restore model proposes that the chromatin is remodelled during DNA repair. Many of the SWI/SNF proteins have been shown to play a role in DNA repair (Osley *et al.* 2007). For example, Rad54 has been shown to be important for DNA double-strand break repair. Human Rad54 has been shown to rescue γ-ray and mitomycin C sensitivity in Rad54 knocked-out mouse embryonic stem cells (Swagemakers 1998). Similarly, in *S. cerevisiae*, Sth1, a member of the SWI/SNF family of proteins, forms a multi-subunit complex known as RSC which is required for repairing double-strand breaks via the homologous recombination pathway (Chai 2005).

The ATP-dependent chromatin remodelling process is intimately connected to histone modifications. Thus, researchers have shown that Ino80, a SWI/SNF protein, interacts with γH2AX at the site of DNA damage. Ino80 forms a multi-subunit complex known as the INO80 complex. Morrison *et al.* (2004) showed that the complex localizes to double-strand breaks induced in *S. cerevisae*. Further, they were able to purify the INO80 complex from cells which had been treated with methylmalonate sulphonate (MMS) to induce double-strand breaks. They found that the complex purified along with γH2AX and the interaction with γH2AX was important for recruitment of the INO80 complex to the site of double-strand break (Morrison *et al.* 2004).

7.21 DNA methylation

DNA methylation occurs in both prokaryotes and eukaryotes. In prokaryotes, both adenine and cytosine are methylated and constitute the restriction modification system. In eukaryotes, cytosine present as 5´-CpG-3´ is specifically modified by DNA methyltransferases at the C-5 position, generating 5-methylcytosine. Methylation of cytosine results in transcription repression; thus, DNA methylation controls gene expression. DNA methylation is important for development; mice that lack the enzymes mediating DNA methylation show embryonic lethality. The modification is also important for X chromosome inactivation as well as for establishment and maintenance of genome imprinting.

7.21.1 Types of DNA methylation

DNA methylation is catalysed by DNA methyltransferases (DNMT) (Bestor 2000). DNA methylation has been classified into *de novo* methylation and maintenance methylation. *De novo* methylation occurs during embryogenesis, and maintenance methylation occurs during DNA replication, leading to conservation of the methylation pattern during cell division. *De novo* methylation is mediated by DNMT3a and DNMT3b (Okano *et al.* 1999), while maintenance methylation is mediated by DNMT1 and DNMT2 (Bestor 2000).

7.21.2 *De novo* methylation

During mouse embryo development, methylation marks are erased on the onset of gastrulation. During gastrulation, primordial germ cells emerge from the epiblast, proliferate, and migrate towards the genital ridge, where gonads eventually develop. At this stage, the methylation marks are erased such that by embryonic day E13.5, only 10% of methylation marks remain. Primordial germ cells establish the gonads, thus determining the sex of the embryo. At this stage, the methylation marks are re-established depending upon the sex of the embryo. Thus, the sperm and the mature oocyte have distinct methylation profiles. This process is known as *de novo* methylation. Okano *et al.* showed in 1999 that DNMT3a and DNMT3b are essential for *de novo* methylation by targeted disruption of genes (Okano *et al.* 1999). Genes were disrupted in embryonic stem (ES) cells. These cells were infected with MMLV retrovirus (lacking DNA methylation) and the ability of these cells to methylate proviral DNA *de novo* was examined. They found that $Dnmt3a^{-/-}$ as well as $Dnmt3b^{-/-}$ ES cells were able to methylate proviral DNA, but $Dnmt3a^{-/-}$, $Dnmt3b^{-/-}$ double-mutant ES cells were unable to methylate proviral DNA, suggesting that these enzymes are critical for *de novo* methylation. They further showed that $Dnmt3a^{+/-}$ heterozygous mice were normal, but $Dnmt3a^{-/-}$ homozygous pups died at 4 weeks of age. Similarly, $Dnmt3b^{+/-}$ were normal, but in contrast to $Dnmt3a^{-/-}$ pups, no viable $Dnmt3b^{+/-}$ pups were born (Okano *et al.* 1999).

7.21.3 Maintenance methylation

The DNA methylation pattern needs to be maintained in somatic cells through cell division. DNMT1 plays a major role in faithfully maintaining the genome-wide DNA methylation profile. In contrast to DNMT3a and DNMT3b, which use unmethylated DNA as substrate, DNMT1 uses hemi-methylated DNA, which appears during replication or repair as substrate. Thus, it uses the template of methylation pattern present on one DNA strand and replicates it on the new DNA strand synthesized in the S phase. Li E *et al.* (1992) have shown that while $Dnmt1^{+/-}$ mice behaved like $Dnmt1^{+/+}$, the $Dnmt1^{-/-}$ mutant embryos showed severely stunted growth and gross developmental abnormalities. Furthermore, these scientists also investigated the levels of methylation in cells and found that there was no discernible difference in methylation levels when only one copy of the *Dnmt1* gene was disrupted, whereas the levels of methylation decreased by one-third when both copies of the *Dnmt1* gene were disrupted (Li E *et al.* 1992).

Unlike DNMT1, DNMT2 is localized to the cytoplasm (Goll 2006). Goll *et al.* showed that DNMT2 can methylate $tRNA^{Asp}$ in *Drosophila*, *Arabidopsis* and mouse (Goll 2006).

7.22 DNA methylation and transcription

It has been known for over 20 years that methylation of DNA results in transcription repression. How does DNA methylation mediate transcription? We began to understand the mechanism of transcription repression mediated by DNA methylation when Adrian Bird's group isolated and characterized methyl-binding proteins (Meehan *et al.* 1992; Hendrich and Bird 1998). These methyl-binding proteins recognize and bind specifically to 5-methyl cytosine. Since then, other methyl-binding proteins have been identified (Ballestar and Wolffe 2001). These methyl-binding proteins are classified into five families – MeCP2, MBD1, MBD2, MBD3 and MBD4. All these proteins contain a methyl-binding motif which is about 70 amino acids long. This motif is necessary and sufficient to bind to 5-methyl cytosine–containing DNA.

Methyl-binding proteins are necessary for mediating transcription repression. Nan *et al.* (1998) showed that MeCP2 contains a transcription repression domain (TRD). Further, they immunoprecipitated MeCP2 and showed that it interacted with HDAC1 as well as HDAC2, thus linking histone modification with DNA methylation (Nan *et al.* 1998). A similar approach has been used to show that other methyl-binding proteins too interact with histone deacetylases. Thus, methyl-binding proteins recognize and bind methylated DNA. These proteins, in turn, recruit proteins that promote heterochromatinization of genes, thus repressing transcription.

7.23 RNA-mediated chromatin remodelling

Eukaryotic cells transcribe a plethora of small RNA molecules. These small RNA molecules, about 20–22 nt long, generated by the action of Dicer enzyme, regulate gene expression both at transcriptional and post-transcriptional stages. However, experimental evidence has now shown that these small RNA molecules are also involved in chromatin remodelling. This section focuses on two phenomena. The first phenomenon involves heterochromatin formation in *Schizosaccharomyces pombe*, which is mediated by siRNAs (small interfering RNAs). The second process occurs exclusively in *Tetrahymena*, where, in an extreme case of chromatin remodelling, DNA sequences are removed from the genome in a process mediated by siRNAs.

7.23.1 siRNAs and heterochromatin formation in *Sch. pombe*

Emil Heitz in 1920 classified chromatin as euchromatin and heterochromatin. Euchromatin is traditionally associated with transcriptionally active genes, while heterochromatin was believed to be transcriptionally silent. Heterochromatin can further be classified either as facultative or as constitutive. The centromere present on chromosomes is an example of constitutive heterochromatin, and for many years, it was believed that the centromere is transcriptionally silent throughout the cell cycle. In 2002, Volpe *et al.* challenged this view (Volpe *et al.* 2002). These researchers showed that the centromere in *Sch. pombe* transcribes small RNA molecules. The centromere in *Sch. pombe* is structurally similar to that in higher eukaryotes. The central region of the centromere consists of a kinetochore-binding site. The binding site is flanked by inverted repeats containing transfer RNA (tRNA) genes. Volpe *et al.* (2002) integrated the *ura4* gene into the centromeric region. In the wild-type strains, they found that the *ura4* transcript was not made, but in mutant strains that lacked Dicer (*dcr*$^-$), the *ura4* transcript was present, suggesting that centromeric silencing required Dicer. They further investigated the role of Argonaute2 and RNA-dependent RNA polymerase and found that when the gene encoding these proteins was deleted, centromeric silencing was also relieved. Thus, the authors concluded that centromeric silencing requires Dicer, Argonaute2 and RNA-dependent RNA polymerase (Volpe *et al.* 2002).

7.24 DNA elimination in *Tetrahymena*

Ciliates have a dimorphic nucleus. The micronucleus is usually diploid and transcriptionally silent during the vegetative phase of growth, serving as repository of the genetic material, while the macronucleus is usually polyploid and transcriptionally active during this phase. The number of micronuclei and macronuclei varies among ciliates. For example, *Tetrahymena* sp. has only one micronucleus and one macronucleus, while *Paramecium* sp. has two micronuclei and one macronucleus.

The micronucleus and the macronucleus differ in their genomic content. The macronucleus lacks many genetic sequences that are present in the micronucleus. The deleted sequences are known as internal eliminated sequences (IES), which are present in both coding and non-coding regions of the micronuclear genome. These sequences are present between two short direct repeats. The IES are primarily repetitive sequences and transposons, and thus, the transcriptionally active nucleus lacks these sequences. *Tetrahymena* contains about 5000–6000 repetitive sequences, constituting about 20% of the micronucleus genome. Nearly all of these sequences are absent in the macronucleus.

During sexual reproduction, two different strains of *Tetrahymena* fuse. The micronucleus present in each cell undergoes meiosis to produce four nuclei. Three of the nuclei disintegrate, while one survives in each cell. The surviving nuclei undergo one round of mitotic division and create a zygotic nucleus. This zygotic nucleus eventually gives rise to a micronucleus and a macronucleus.

In 2001, Chalker and Yao showed that during nuclear development, many non-coding RNA molecules were transcribed from the micronucleus (Chalker and Yao 2001). These authors showed that the elimination of an element known as M element from the macronucleus was dependent upon the transcription of non-coding RNA molecules from this element in the micronucleus early during the conjugation process. The next step in understanding the phenomenon of DNA elimination was the discovery of a protein called Twi1p. Mochizuki *et al.* (2002) cloned *TWI1* from early mating cells of *Tetrahymena thermophila* using reverse transcription polymerase chain reaction (RT-PCR) technique. They showed that Twi1p was a 780-amino acid–containing protein that was not essential for vegetative growth but for the formation of viable progenies, thus indicating a role in the developmental process. The authors constructed a haemagglutinin (HA)-tagged Twi1p protein and showed using the immunofluorescent technique that the protein translocated from the old to the new macronucleus (Mochizuki *et al.* 2002).

Based on these experimental evidence, Mochizuki and Gavorosky (2004) have proposed a model to explain DNA elimination in ciliates. According to their model, the IES present in the micronucleus are transcribed bidirectionally early during the conjugation process. These double-stranded RNA molecules are processed by the RNAi machinery, generating 21- to 22-nt non-coding RNA molecules known as scan RNAs (scnRNAs). scnRNAs associate with Twi1p in the old macronucleus. As their name suggests, scnRNAs scan the old macronucleus for homologous sequences. Those scnRNAs that possess homologous sequences in the old macronucleus are degraded, and the remaining translocate to the new macronucleus. In the new macronucleus, scnRNAs identify the homologous sequences and eliminate them by base pairing (Mochizuki and Gorovsky 2004).

7.25 Chromatin remodelling and diseases

The epigenome of the cell is defined by histone modifications, histone variants, ATP-dependent chromatin remodelling factors and non-coding RNA molecules. Alterations in any of these factors would inevitably lead to alteration in the epigenome and, thus, changes in the gene expression pattern. Thus, diseases such as cancer, which results due to an aberrant gene expression pattern, are expected to have an altered epigenome. Further, mutations in any of the proteins involved in defining the epigenome are expected to lead to aberrant gene expression and, thus, disease conditions. Sections 7.26 and 7.27 discuss Rett syndrome and cancer caused due to aberrations in chromatin remodelling factors, respectively.

7.26 Rett syndrome

Rett syndrome was first described by Dr Andreas Rett in 1966. Rett syndrome is a post-natal progressive neurodevelopmental disorder that manifests predominantly in girls. The patients develop normally for 6–8 months after birth. Subsequently, development stagnates, leading to loss in hand skills, speech and social interaction. The patients also exhibit autistic behaviour and eventually develop Parkinsonian features (Samaco and Neul 2011).

Though the predominant manifestation in girls suggests an X-linked dominant disorder, more than 99% of the cases appear to be sporadic with no obvious familial inheritance. In 1999, Amir *et al.*, using information from rare familial cases, identified mutations in *MECP2* as the leading cause of Rett syndrome (Amir *et al.* 1999). *MECP2* maps to the Xq28 chromosomal location and encodes for MeCP2, the methyl CpG–binding protein.

MeCP2 is a global transcriptional regulator that can function both as a transcription repressor and as a transcription activator. Cohen *et al.* (2003) showed that the level of MeCP2 is developmentally regulated and increased post-natally in neurons constituting the central nervous system (CNS). This increase correlated with neuronal maturation and synaptogenesis, suggesting a role in CNS development (Cohen *et al.* 2003). Chen *et al.* (2001) subsequently showed that the deletion of MeCP2 in mouse models led to the development of Rett syndrome–like symptoms, providing a correlation between MeCP2 and Rett syndrome. These authors created a conditional allele of MeCP2 using the *Cre–LoxP* recombination system and disrupted the exon 3 of the gene that encodes for the methyl-binding domain. They compared the RNA as well as protein levels from the knockout mice with the normal wild-type mice and found that both RNA and protein levels were downregulated in the CNS of the knockout mice. As expected of Rett syndrome patients, the MeCP2 knockout pups developed normally, but later developed the classical symptoms characterizing the syndrome, including nervousness and body trembling (Chen *et al.* 2001).

At present, no treatment is available for patients suffering from Rett syndrome. Therapeutic interventions currently focus on identifying the genes regulated by MeCP2. One such gene is *BDNF*, which triggers neuronal maturation as well as synaptogenesis. *BDNF* is also controlled by insulin growth factor-1 (IGF-1), and indeed, administration of IGF-1 has been shown to partially reverse the effect of Rett syndrome (Tropea *et al.* 2009).

7.27 Cancer

The transformation of a normal cell into a malignant cell is driven by alterations in the expression of genes involved in cell cycle control, differentiation, DNA repair, cell death and maintenance of genomic integrity. Alterations/dysregulation in gene expression implies perforce an alteration in the epigenome, and therefore, investigations have focused on understanding the role of chromatin remodelling factors in driving the transformation of a normal cell into a malignant cell.

7.27.1 Role of histone modifications in cancer

Fraga *et al.* (2005) compared histone modifications in normal lymphocytes with those present in leukaemia cell lines as well as primary tumours and found that H4 was deacetylated in transformed cell lines as well as in primary tumours as compared with normal cells. Concomitantly, they found that H4 showed 75% loss in trimethylation in transformed cells and primary tumours as compared with normal cells (Fraga *et al.* 2005). Further analysis led them to identify that the loss of acetylation as well as trimethylation occurred on the repetitive sequences (Fraga *et al.* 2005).

7.27.2 Role of ATP-dependent chromatin remodelling factors in cancer

Wong *et al.* (2000) analysed 92 tumour cell lines and found that *Brg1* was mutated in many of these cell lines. Importantly, as a result of these mutations, *Brg1* expression was not observed in these tumour cell lines. A similar analysis was performed by DeCristofaro *et al.* (2001), who too reported loss of *Brg1* expression in many breast cancer cell lines, adenocarcinomas and carcinomas. Further, Reisman *et al.* (2003) showed loss of expression of Brg1 and Brm in 30% of non-small lung cancer cell lines. An analysis of primary lung adenocarcinomas and primary lung squamous carcinomas showed loss in expression of both *Brg1* and *Brm* in 10% of cases. This loss was directly correlated to poor prognosis, suggesting a role for these proteins in cancer progression (Reisman *et al.* 2003).

Experimental evidence suggests that *Brg1* functions as a tumour suppressor gene. Dunaief *et al.* in 1994 showed using a yeast two-hybrid system that the retinoblastoma (Rb) protein interacted with *Brg1* (Dunaief *et al.* 1994). They further showed that the interaction of *Brg1* interacted specifically with the hypophosphorylated form of Rb. Finally, using an *in vivo* assay, they showed that *Brg1* can act as a tumour suppressor gene (Dunaief *et al.* 1994).

Brg1 also interacts with *BRCA1*, a transcription activator. Mutations in *BRCA1* have been implicated in familial breast cancer. Bochar *et al.* (2000) showed that *BRCA1* co-purified with *Brg1* from HeLa cells, and that these two proteins form a stable complex. They further showed that transient transfection of *BRCA1* into a *Brg1*-expressing cell line resulted in three-fold stimulation of transcription. However, transcription stimulation was not observed when mutated *BRCA1* was transiently transfected into the *Brg1*-expressing cell line. Finally, the authors showed that the cancer-causing mutation of exon 11 in *BRCA1* was unable to stimulate transcription when cells were expressing dominant-negative ATPase-deficient Brg1 protein, leading to the conclusion that the ATPase activity of *Brg1* was essential for the transcription stimulation mediated by *BRCA1* (Bochar *et al.* 2000; Roberts and Orkin 2004).

7.27.3 Role of DNA methylation in cancer

As early as 1988, Feinberg *et al.* showed that DNA methylation is altered in cancer cells (Feinberg *et al.* 1988). Using high-performance liquid chromatography (HPLC) to measure the amount of 5-methyl cytosine, this group of scientists showed that there was an average 8–10% reduction of this modification in tumour cells as compared with normal cells, indicating that tumour cells are hypomethylated (Feinberg *et al.* 1988). They further analysed benign and malignant tumours and found that in both of them, DNA was hypomethylated, indicating that DNA hypomethylation was a malignant transformation (Feinberg *et al.* 1988).

Analysis of the tumour genome has shown that intergenic regions are specifically hypomethylated, while promoters of tumour suppressor genes are hypermethylated. Two consequences are immediately evident: one, transposable elements, which are present in the intergenic regions, become active and become the major contributor to genomic instability; second, tumour suppressor genes, due to hypermethylation of the promoter region, become transcriptionally silent.

Studies have also indicated that aberrations in DNA methylation pattern are one of the earliest events in the progression of a normal cell into a tumour cell. Issa *et al.* showed in 1994 that DNA methylation–mediated inactivation of the oestrogen receptor in colorectal mucosa could be one of the earliest events that lead to colorectal tumours (Issa *et al.* 1994).

As DNA methylates during development, what causes aberrant DNA methylation in somatic cells during tumourigenesis? To address this question, Vertino *et al.* (1996) sought to understand whether increased levels of *DNMT1* can lead to aberrant methylation. They overexpressed human *DNMT1* in human fibroblasts and showed that increased levels of *DNMT1* lead to hypermethylation (Vertino *et al.* 1996). However, bi-allelic loss of *DNMT1*, surprisingly, did not lead to hypomethylation, suggesting that the aberrant hypermethylation of cytosines during tumourigenesis cannot be attributed solely to upregulation of *DNMT1* (Vertino *et al.* 1996). Tumour cells also show upregulation of *de novo* methyltransferases, and hence, it is possible that a combination of DNA methyltransferases is responsible for aberrant methylation in tumour cells (Vertino *et al.* 1996).

Recommended textbooks

- *Molecular Biology of the Cell* (5th Edition) by Alberts *et al.* New York: Garland Science; 2008.
- *Molecular Cell Biology* (5th Edition) by Lodish *et al.* New York: WH Freeman; 2004.
- *Epigenetics* by C. David Allis *et al.* Cold Spring Harbor, NY: Cold Spring Harbor Laboratory Press; 2007.

References

Ahmad K, Henikoff S. The histone variant H3.3 marks active chromatin by replication-independent nucleosome assembly. *Mol Cell* 2002;9:1191–1200.

Allfrey VG, Faulkner R, Mirsky AE. Acetylation and methylation of histones and their possible role in the regulation of RNA synthesis. *Proc Natl Acad Sci USA* 1964;51:786–794.

Amir RE, Van den Veyver IB, Wan M, Tran CQ, Francke U, Zoghbi HY. Rett syndrome is caused by mutations in X-linked MECP2, encoding methyl-CpG-binding protein 2. *Nat Genet* 1999;23:185–188.

Ausió J. Histone variants – the structure behind the function. *Brief Funct Genomic Proteomic* 2006;5:228–243.

Ballestar E, Wolffe AP. Methyl-CpG-binding proteins. Targeting specific gene repression. *Eur. J. Biochem.* 2001;268:1–6.

Bao Y, Konesky K, Park Y-J, Rosu S, Dyer PN, Rangasamy D, *et al.* Nucleosomes containing the histone variant H2A.Bbd organize only 118 base pairs of DNA. *EMBO J* 2004;23:3314–3324.

Bao Y, Shen X. Chromatin remodeling in DNA double-strand break repair. *Curr Opin Cell Biol* 2007;17:126–131.

Berger SL. The complex language of chromatin regulation during transcription. *Nature* 2007;447:407–412.

Bestor TH. The DNA methyltransferases of mammals. *Human Mol Gen* 2000;9:2395–2402.

Bhaumik SR, Green MR. Differential requirement of SAGA components for recruitment of TATA-box-binding protein to promoters *in vivo*. *Mol Cell Biol* 2002;22:7365–7371.

Blander G, Guarente L. The Sir2 family of protein deacetylases. *Annu Rev Biochem* 2004;73:417–435.

Blus BJ, Wiggins K, Khorasanizadeh S. Epigenetic virtues of chromodomains. *Crit Rev Biochem Mol Biol* 2011;46:507–526.

Bochar DA, Wang L, Beniya H, Kinev A, Xue Y, Lane WS, *et al.* BRCA1 is associated with a human SWI/SNF-related complex: Linking chromatin remodeling to breast cancer. *Cell* 2000;102:257–265.

Branscombe TL, Frankel A, Lee JH, Cook JR, Yang Z, Pestka S, *et al.* PRMT5 (Janus kinase-binding protein 1) catalyzes the formation of symmetric dimethylarginine residues in proteins. *J Biol Chem* 2001;276:32971–32976.

Brownell JE, Zhou J, Ranalli T, Kobayashi R, Edmondson DG, Roth SY, *et al.* Tetrahymena histone acetyltransferase A: A homolog to yeast Gcn5p linking histone acetylation to gene activation. *Cell* 1996;84:843–851.

Cairns BR, Kim YJ, Sayre MH, Laurent BC, Kornberg RD. A multisubunit complex containing the SWI1/ADR6, SWI2/SNF2, SWI3, SNF5, and SNF6 gene products isolated from yeast. *Proc Natl Acad Sci USA* 1994;91:1950–1954.

Carrozza MJ, Utley RT, Workman JL, Côté J. The diverse functions of histone acetyltransferase complexes. *Trends Gen* 2003;19:321–329.

Celeste A, Petersen S, Romanienko PJ, Fernandez-Capetillo O, Chen HT, Sedelnikova OA, *et al.* Genomic instability in mice lacking histone H2AX. *Science* 2002;296:922–927.

Chadwick BP, Willard HF. A novel chromatin protein, distantly related to histone H2A, is largely excluded from the inactive X chromosome. *J Cell Biol* 2001a;152:375–384.

Chadwick BP, Willard HF. Histone H2A variants and the inactive X chromosome: Identification of a second macroH2A variant. *Hum Mol Genet* 2001b;10:1101–1113.

Chai B. Distinct roles for the RSC and Swi/Snf ATP-dependent chromatin remodelers in DNA double-strand break repair. *Genes Dev* 2005;19:1656–1661.

Chalker DL, Yao MC. Nongenic, bidirectional transcription precedes and may promote developmental DNA deletion in tetrahymena thermophila. *Genes Dev* 2001;15:1287–1298.

Chan HM, La Thangue NB. p300/CBP proteins: HATs for transcriptional bridges and scaffolds. *J Cell Sci* 2001;114:2363–2373.

Chen RZ, Akbarian S, Tudor M, Jaenisch R. Deficiency of methyl-CpG binding protein-2 in CNS neurons results in a Rett-like phenotype in mice. *Nat Genet* 2001;27:327–331.

Clarke AS, Lowell JE, Jacobson SJ, Pillus L. Esa1p is an essential histone acetyltransferase required for cell cycle progression. *Mol Cell Biol* 1999;19:2515–2526.

Clayton AL, Hazzalin CA, Mahadevan LC. Enhanced histone acetylation and transcription: A dynamic perspective. *Mol Cell* 2006;23:289–296.

Cohen DRS, Matarazzo V, Palmer AM, Tu Y, Jeon O-H, Pevsner J, *et al.* Expression of MeCP2 in olfactory receptor neurons is developmentally regulated and occurs before synaptogenesis. *Mol Cell Neurosci* 2003;22:417–429.

Cosgrove MS, Wolberger C. How does the histone code work? *Biochem Cell Biol* 2005;83:468–476.

Cramer P, Armache K-J, Baumli S, Benkert S, Brueckner F, Buchen C, *et al.* Structure of eukaryotic RNA polymerases. *Annu Rev Biophys* 2008;37:337–352.

de Ruijter AJM, van Gennip AH, Caron HN, Kemp S, van Kuilenburg ABP. Histone deacetylases (HDACs): Characterization of the classical HDAC family. *Biochem J* 2003;370:737–749.

Decristofaro MF, Betz BL, Rorie CJ, Reisman DN, Wang W, Weissman BE. Characterization of SWI/SNF protein expression in human breast cancer cell lines and other malignancies. *J Cell Physiol* 2001;186:136–145.

Dillon SC, Zhang X, Trievel RC, Cheng X. The SET-domain protein superfamily: Protein lysine methyltransferases. *Genome Biol* 2005;6:227.

Dunaief JL, Strober BE, Guha S, Khavari PA, Alin K, Luban J, *et al.* The retinoblastoma protein and BRG1 form a complex and cooperate to induce cell cycle arrest. *Cell* 1994;79:119–130.

Feinberg AP, Gehrke CW, Kuo KC, Ehrlich M. Reduced genomic 5-methylcytosine content in human colonic neoplasia. *Cancer Res* 1988;48:1159–1161.

Fernandez-Capetillo O, Lee A, Nussenzweig M, Nussenzweig A. H2AX: The histone guardian of the genome. DNA Repair (Amst.) 2004;3:959–967.

Flanagan PM, Kelleher RJ III, Sayre MH, Tschochner H, Kornberg RD. A mediator required for activation of RNA polymerase II transcription *in vitro*. *Nature* 1991;350:436–438.

Flaus A, Martin DMA, Barton GJ, Owen-Hughes T. Identification of multiple distinct Snf2 subfamilies with conserved structural motifs. *Nucleic Acids Res* 2006;34:2887–2905.

Fraga MF, Ballestar E, Villar-Garea A, Boix-Chornet M, Espada J, Schotta G, *et al.* Loss of acetylation at Lys16 and trimethylation at Lys20 of histone H4 is a common hallmark of human cancer. *Nat Gen* 2005;37:391–400.

Garcia BA, Barber CM, Hake SB, Ptak C, Turner FB, Busby SA, *et al.* Modifications of human histone H3 variants during mitosis. *Biochemistry* 2005;44:13202–13213.

Goll MG. Methylation of tRNAAsp by the DNA methyltransferase homolog Dnmt2. *Science* 2006;311:395–398.

Gorbalenya AE, Koonin EV. Helicases: Amino acid sequence comparisons and structure-function relationships. *Curr Opin Cell Biol* 1993;3:419–429.

Grant PA, Duggan L, Côté J, Roberts SM, Brownell JE, Candau R, *et al.* Yeast Gcn5 functions in two multisubunit complexes to acetylate nucleosomal histones: Characterization of an Ada complex and the SAGA (Spt/Ada) complex. *Genes Dev* 1997;11:1640–1650.

Grant PA, Sterner DE, Duggan LJ, Workman JL, Berger SL. The SAGA unfolds: Convergence of transcription regulators in chromatin-modifying complexes. *Trends Cell Biol* 1998;8:193–197.

Green CM, Almouzni G. When repair meets chromatin. First in series on chromatin dynamics. EMBO Rep. 2002;3:28–33.

Hake SB, Allis CD. Histone H3 variants and their potential role in indexing mammalian genomes: The 'H3 barcode hypothesis'. *Proc Natl Acad Sci USA* 2006;103:6428–6435.

Happel N, Doenecke D. Histone H1 and its isoforms: Contribution to chromatin structure and function. *Gene* 2009;431:1–12.

Havas K, Flaus A, Phelan M, Kingston R, Wade PA, Lilley DM, *et al.* Generation of superhelical torsion by ATP-dependent chromatin remodeling activities. *Cell* 2000;103:1133–1142.

Hendrich B, Bird A. Identification and characterization of a family of mammalian methyl-CpG binding proteins. *Mol Cell Biol* 1998;18:6538–6547.

Hirano T. SMC protein complexes and higher-order chromosome dynamics. *Curr Opin Cell Biol* 1998;10:317–322.

Hsu JY, Sun ZW, Li X, Reuben M, Tatchell K, Bishop DK, *et al.* Mitotic phosphorylation of histone H3 is governed by Ipl1/aurora kinase and Glc7/PP1 phosphatase in budding yeast and nematodes. *Cell* 2000;102:279–291.

Issa JP, Ottaviano YL, Celano P, Hamilton SR, Davidson NE, Baylin SB. Methylation of the oestrogen receptor CpG island links ageing and neoplasia in human colon. *Nat Genet* 1994;7:536–540.

Jackson V, Shires A, Chalkley R, Granner DK. Studies on highly metabolically active acetylation and phosphorylation of histones. *J Biol Chem* 1975;250:4856–4863.

Jentsch S, McGrath JP, Varshavsky A. The yeast DNA repair gene RAD6 encodes a ubiquitin-conjugating enzyme. *Nature* 1987;329;131–134.

Jenuwein T, Laible G, Dorn R, Reuter G. SET domain proteins modulate chromatin domains in eu- and heterochromatin. *Cell Mol Life Sci* 1998;54:80–93.

Johansen KM, Johansen J. Regulation of chromatin structure by histone H3S10 phosphorylation. *Chromosome Res* 2006;14:393–404.

Kamakaka RT, Biggins S. Histone variants: deviants? *Genes Dev* 2005;19:295–310.

Kanaar R, Hoeijmakers JH, van Gent DC. Molecular mechanisms of DNA double-strand break repair. *Trends Cell Biol* 1998;8:483–489.

Kassabov SR, Henry NM, Zofall M, Tsukiyama T, Bartholomew B. High-resolution mapping of changes in histone-DNA contacts of nucleosomes remodeled by ISW2. *Mol Cell Biol* 2002;22:7524–7534.

Khanna KK, Jackson SP. DNA double-strand breaks: Signaling, repair and the cancer connection. *Nat Genet* 2001;27:247–254.

Kingston RE, Narlikar GJ. ATP-dependent remodeling and acetylation as regulators of chromatin fluidity. *Genes Dev* 1999;13:2339–2352.

Kouzarides T. Histone methylation in transcriptional control. *Curr Opin Genet Dev* 2002;12:198–209.

Lee Y-H, Koh SS, Zhang X, Cheng X, Stallcup MR. Synergy among nuclear receptor coactivators: Selective requirement for protein methyltransferase and acetyltransferase activities. *Mol Cell Biol* 2002;22:3621–3632.

Li B, Carey M, Workman JL. The role of chromatin during transcription. *Cell* 2007;128:707–719.

Li E, Bestor TH, Jaenisch R. Targeted mutation of the DNA methyltransferase gene results in embryonic lethality. *Cell* 1992;69: 915–926.

Lin Y-S, Carey M, Ptashne M, Green MR. How different eukaryotic transcriptional activators can cooperate promiscuously. *Nature* 1990;345:359–361.

Litt M, Qiu Y, Huang S. Histone arginine methylations: Their roles in chromatin dynamics and transcriptional regulation. *Biosci Rep* 2009;29:131–141.

MacDonald VE, Howe LJ. Histone acetylation: Where to go and how to get there. *Epigenetics* 2009;4:139–143.

Mahadevan LC, Willis AC, Barratt MJ. Rapid histone H3 phosphorylation in response to growth factors, phorbol esters, okadaic acid, and protein synthesis inhibitors. *Cell* 1991;65:775–783.

Manohar M, Mooney AM, North JA, Nakkula RJ, Picking JW, Edon A, *et al.* Acetylation of histone H3 at the nucleosome dyad alters DNA-histone binding. *J Biol Chem* 2009;284:23312–23321.

Meehan RR, Lewis JD, Bird AP. Characterization of MeCP2, a vertebrate DNA binding protein with affinity for methylated DNA. *Nucleic Acids Res* 2009;20:5085–5092.

Mietton F, Sengupta AK, Molla A, Picchi G, Barral S, Heliot L, *et al.* Weak but uniform enrichment of the histone variant macroH2A1 along the inactive X chromosome. *Mol Cell Biol* 2008;29:150–156.

Mochizuki K, Fine NA, Fujisawa T, Gorovsky MA. Analysis of a piwi-related gene implicates small RNAs in genome rearrangement in tetrahymena. *Cell* 2002;110:689–699.

Mochizuki K, Gorovsky MA. Small RNAs in genome rearrangement in Tetrahymena. *Curr Opin Genet Dev* 2004;14:181–187.

Morrison AJ, Highland J, Krogan NJ, Arbel-Eden A, Greenblatt JF, Haber JE, *et al.* INO80 and gamma-H2AX interaction links ATP-dependent chromatin remodeling to DNA damage repair. *Cell* 2004;119:767–775.

Nan X, Ng HH, Johnson CA, Laherty CD, Turner BM, Eisenman RN, *et al.* Transcriptional repression by the methyl-CpG-binding protein MeCP2 involves a histone deacetylase complex. *Nature* 1998;393:386–389.

Narlikar GJ, Fan H-Y, Kingston RE. Cooperation between complexes that regulate chromatin structure and transcription. *Cell* 2002.108:475–487.

Nathan D. Histone sumoylation is a negative regulator in Saccharomyces cerevisiae and shows dynamic interplay with positive-acting histone modifications. *Genes Dev* 2006;20:966–976.

Neigeborn L, Carlson M. Genes affecting the regulation of SUC2 gene expression by glucose repression in Saccharomyces cerevisiae. *Genetics* 1984;108,845–858.

Nowak SJ, Corces VG. Phosphorylation of histone H3: A balancing act between chromosome condensation and transcriptional activation. *Trends Genet* 2004;20:214–220.

Okano M, Bell DW, Haber DA, Li E. DNA methyltransferases Dnmt3a and Dnmt3b are essential for de novo methylation and mammalian development. *Cell* 1999;99:247–257.

Orphanides G, Lagrange T, Reinberg D. The general transcription factors of RNA polymerase II. *Genes Dev* 1996;10:2657–2683.

Osley MA. Regulation of histone H2A and H2B ubiquitylation. *Brief Funct Genomic Proteomic* 2006;5:179–189.

Osley MA, Tsukuda T, Nickoloff JA. ATP-dependent chromatin remodeling factors and DNA damage repair. *Mutat Res* 2007;618:65–80.

Paull TT, Rogakou EP, Yamazaki V, Kirchgessner CU, Gellert M, Bonner WM. A critical role for histone H2AX in recruitment of repair factors to nuclear foci after DNA damage. *Curr Biol* 2000;10:886–895.

Paulson JR, Taylor SS. Phosphorylation of histones 1 and 3 and nonhistone high mobility group 14 by an endogenous kinase in HeLa metaphase chromosomes. *J Biol Chem* 1982;257:6064–6072.

Pazin MJ, Kadonaga JT. What's up and down with histone deacetylation and transcription? *Cell* 1997;89:325–328.

Ptashne M. How eukaryotic transcriptional activators work. *Nature* 1988;335:683–689.

Qian C, Zhou M-M. SET domain protein lysine methyltransferases: Structure, specificity and catalysis. *Cell Mol Life Sci* 2006;63:2755–2763.

Qiu H, Hu C, Yoon S, Natarajan K, Swanson MJ, Hinnebusch AG. An array of coactivators is required for optimal recruitment of TATA binding protein and RNA polymerase II by promoter-bound Gcn4p. *Mol Cell Biol* 2004;24:4104–4117.

Rachez C, Freedman LP. Mediator complexes and transcription. *Curr Opin Cell Biol* 2001;13:274–280.

Rea S, Eisenhaber F, O'Carroll D, Strahl BD, Sun ZW, Schmid M, *et al.* Regulation of chromatin structure by site-specific histone H3 methyltransferases. *Nature* 2000;406:593–599.

Reisman DN, Sciarrotta J, Wang W, Funkhouser WK, Weissman BE. Loss of BRG1/BRM in human lung cancer cell lines and primary lung cancers: Correlation with poor prognosis. *Cancer Res* 2003;63:560–566.

Richmond TJ, Davey CA. The structure of DNA in the nucleosome core. *Nature* 2003;423:145–150.

Roberts CWM, Orkin SH. The SWI/SNF complex — chromatin and cancer. *Nat Rev Cancer* 2004;4:133–142.

Rogakou EP, Pilch DR, Orr AH, Ivanova VS, Bonner WM. DNA double-stranded breaks induce histone H2AX phosphorylation on serine 139. *J Biol Chem* 1998;273:5858–5868.

Roth SY, Denu JM, Allis CD. Histone acetyltransferases. *Annu Rev Biochem* 2001;70:81–120.

Samaco RC, Neul JL. Complexities of Rett syndrome and MeCP2. *J Neurosci* 2011;31:7951–7959.

Schalch T, Duda S, Sargent DF, Richmond TJ. X-ray structure of a tetranucleosome and its implications for the chromatin fibre. *Nature* 2005;436:138–141.

Schulze JM, Hentrich T, Nakanishi S, Gupta A, Emberly E, Shilatifard A, *et al.* Splitting the task: Ubp8 and Ubp10 deubiquitinate different cellular pools of H2BK123. *Genes Dev* 2011;25:2242–2247.

Sealy L. The effect of sodium butyrate on histone modification. *Cell* 1978;14:115–121.

Shi Y, Lan F, Matson C, Mulligan P, Whetstine JR, Cole PA, *et al.* Histone demethylation mediated by the nuclear amine oxidase homolog LSD1. *Cell* 2004;119:941–953.

Shi Y, Whetstine JR. Dynamic regulation of histone lysine methylation by demethylases. *Mol Cell* 2007;25:1–14.

Shiio Y, Eisenman RN. Histone sumoylation is associated with transcriptional repression. *Proc Natl Acad Sci USA* 2003;100: 13225–13230.

Smerdon MJ. DNA repair and the role of chromatin structure. *Curr Opin Cell Biol* 1991;3:422–428.

Smith ER. ESA1 is a histone acetyltransferase that is essential for growth in yeast. *Proc Natl Acad. Sci. USA* 1998;95:3561–3565.

Spahr H, Calero G, Bushnell DA, Kornberg RD. Schizosacharomyces pombe RNA polymerase II at 3.6-A resolution. *Proc Natl Acad Sci USA* 2009;106:9185–9190.

Stern M, Jensen R, Herskowitz I. Five SWI genes are required for expression of the HO gene in yeast. *J Mol Biol* 1984;178:853–868.

Strahl BD, Allis CD. The language of covalent histone modifications. *Nature* 2000;403:41–45.

Suto RK, Clarkson MJ, Tremethick DJ, Luger K. Crystal structure of a nucleosome core particle containing the variant histone H2A. *Z Nat Struct Biol* 2000;7:1121–1124.

Swagemakers SMA. The human Rad54 recombinational DNA repair protein is a double-stranded DNA-dependent ATPase. *J Biol Chem* 1998;273:28292–28297.

Taunton J, Hassig CA, Schreiber SL. A mammalian histone deacetylase related to the yeast transcriptional regulator Rpd3p. *Science* 1996;272:408–411.

Thoma F. Involvement of histone H1 in the organization of the nucleosome and of the salt-dependent superstructures of chromatin. *J Cell Biol* 1976;83:403–427.

Thomas JO. The higher order structure of chromatin and histone H1. *J Cell Sci Suppl* 1984;1:1–20.

Tropea D, Giacometti E, Wilson NR, Beard C, McCurry C, Fu DD, *et al.* Partial reversal of Rett Syndrome-like symptoms in MeCP2 mutant mice. *Proc Natl Acad Sci USA* 2009;106:2029–2034.

Utley RT, Côté J. The MYST family of histone acetyltransferases. *Curr Top Microbiol Immunol* 2003;274:203–236.

Vertino PM, Yen RW, Gao J, Baylin SB. De novo methylation of CpG island sequences in human fibroblasts overexpressing DNA (cytosine-5-)-methyltransferase. *Mol Cell Biol* 1996;16: 4555–4565.

Vetting MW, S de Carvalho LP, Yu M, Hegde SS, Magnet S, Roderick SL, *et al.* Structure and functions of the GNAT superfamily of acetyltransferases. *Arch Biochem Biophys* 2005;433:212–226.

Vignali M, Hassan AH, Neely KE, Workman JL. ATP-dependent chromatin-remodeling complexes. *Mol Cell Biol* 2000;20:1899–1910.

Volpe TA, Kidner C, Hall IM, Teng G, Grewal SIS, Martienssen RA. Regulation of heterochromatic silencing and histone H3 lysine-9 methylation by RNAi. *Science* 2002;297:1833–1837.

Weake VM, Workman JL. Histone ubiquitination: Triggering gene activity. *Mol Cell* 2008a;29:653–663.

West MH, Bonner WM. Histone 2A, a heteromorphous family of eight protein species. *Biochem.* 1980;19:3238–3245.

Wong AK, Shanahan F, Chen Y, Lian L, Ha P, Hendricks K, *et al.* BRG1, a component of the SWI-SNF complex, is mutated in multiple human tumor cell lines. *Cancer Res* 2000;60:6171–6177.

Wood RD. DNA repair in eukaryotes. *Annu Rev Biochem* 1996;65:135–167.

Woodcock CL, Ghosh RP. Chromatin higher-order structure and dynamics. *Cold Spring Harbor Perspect Biol* 2010;2:a000596.

Woodcock CL, Skoultchi AI, Fan Y. Role of linker histone in chromatin structure and function: H1 stoichiometry and nucleosome repeat length. *Chromosome Res* 2006;14:17–25.

Wysocka J, Allis CD, Coonrod S. Histone arginine methylation and its dynamic regulation. *Front Biosci* 2006;11:344–355.

Xu W, Edmondson DG, Evrard YA, Wakamiya M, Behringer RR, Roth SY. Loss of Gcn5l2 leads to increased apoptosis and mesodermal defects during mouse development. *Nat Genet* 2000;26:229–232.

Yamauchi T, Yamauchi J, Kuwata T, Tamura T, Yamashita T, Bae N, *et al.* Distinct but overlapping roles of histone acetylase PCAF and of the closely related PCAF-B/GCN5 in mouse embryogenesis. *Proc Natl Acad Sci USA* 2000;97:11303–11306.

Yang X-J, Ogryzko VV, Nishikawa J, Howard BH, Nakatani Y. A p300/CBP-associated factor that competes with the adenoviral oncoprotein E1A. *Nature* 1996;382:319–324.

Zalensky AO, Siino JS, Gineitis AA, Zalenskaya IA, Tomilin NV, Yau P, *et al.* Human testis/sperm-specific histone H2B (hTSH2B). Molecular cloning and characterization. *J Biol Chem* 2002;277:43474–43480.

Zeng L, Zhou MM. Bromodomain: An acetyl-lysine binding domain. *FEBS Lett* 2002;513:124–128.

Zhang Y. Transcriptional regulation by histone ubiquitination and deubiquitination. *Genes Dev* 2003;17:2733–2740.

8

Classic experiments on DNA replication *in vivo*

H. K. Das

8.1 Mechanism of replication apparent to Watson and Crick

Replication is the minimum attribute of the genetic material. How does DNA replicate itself? In the very first paper in which Watson and Crick proposed 'a structure for deoxyribose nucleic acid' (Watson and Crick 1953a), they wrote, 'It has not escaped our notice that the specific base pairing [of DNA] we have postulated immediately suggests a possible copying mechanism for the genetic material'. In their second paper, they continued, 'We imagine that prior to duplication the hydrogen bonds [in the double stranded DNA] are broken, and the two chains unwind and separate. Each chain then acts as a template for the formation onto itself of a new [complementary] companion chain [following the Watson and Crick base pairing rule], so that eventually we shall have two pairs of chains, where we only had one before. Moreover, the sequence of the pairs of bases will have been duplicated exactly' (Watson and Crick 1953b). 'We imagine that at this stage in the life of the cell, free nucleotides, strictly polynucleotide precursors, are available in quantity. From time to time the base of a free nucleotide will join up by hydrogen bonds to one of the bases on the chain already formed. We now postulate that the polymerization of these monomers to form a new chain is only possible ...' (Watson and Crick 1953b). A fairly precise postulation, indeed.

8.2 Experimental confirmation by Meselson and Stahl of the mechanism of replication of DNA postulated by Watson and Crick

Five long years had elapsed since the postulation by Watson and Crick of the mechanism of replication of DNA, but no experimental confirmation was forthcoming. This was mainly because of the unavailability of a method by which the parental DNA strands could be differentiated from the newly synthesized ones. To this end, Meselson and Stahl (1958) developed a method based on the use of the density-labelled isotope N^{15} and demonstrated that labelled DNA molecules could be separated by equilibrium sedimentation (see Chapter 4) in caesium chloride (CsCl) density gradients, a technique that they had described a year back (Meselson *et al.* 1957). Separation was possible even when the density difference was as little as 0.3%. *Escherichia coli* were grown in glucose salts medium containing NH_4Cl as the sole nitrogen source. In one experiment, the NH_4Cl containing the natural N^{14} isotope was replaced by NH_4Cl containing the heavy isotope N^{15} of 96.5% isotope purity. The cells were allowed to grow in the N^{15}-containing medium for 14 generations so that all the DNA was labelled uniformly

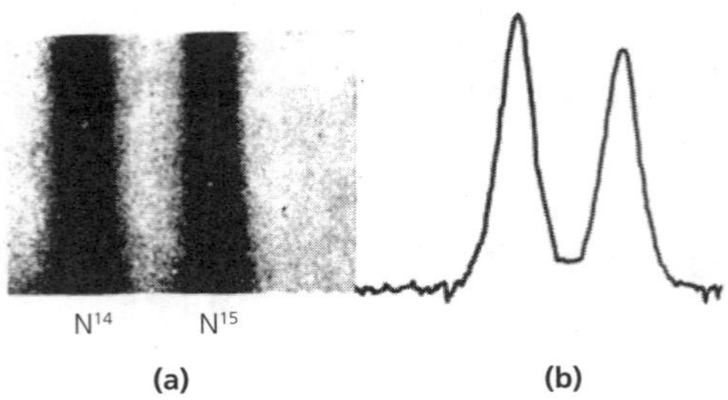

Figure 8.1 Resolution by equilibrium sedimentation of DNA containing N^{14} and DNA containing N^{15}. *Escherichia coli* were grown in a medium containing N^{14} and in a medium containing N^{15}, lysed, mixed together and subjected to equilibrium sedimentation by CsCl density gradient centrifugation. (a) Ultraviolet absorption photograph showing the resolved DNA bands, (b) a microdensitometer tracing showing the distribution of DNA in the region of the two bands. [Reproduced from Meselson and Stahl, *Proc Natl Acad Sci USA* 1958;44:673 by permission of Professor Matthew Meselson.]

with the heavy isotope N^{15}. It was seen that DNA isolated from cells grown in the medium containing N^{14} and from cells grown in the medium containing N^{15} could be resolved very clearly by equilibrium sedimentation in a CsCl gradient (Fig. 8.1) (Meselson and Stahl 1958).

In their classic experiment, Meselson and Stahl (1958) added 10-fold excess of NH_4Cl containing the natural isotope of N^{14} to the *E. coli* cells that had grown for 14 generations in the N^{15}-containing medium, thus essentially bringing about an abrupt replacement of the N^{15}-containing medium by the N^{14}-containing medium. This would ensure that all new DNA synthesis from this point onwards would be with N^{14}-containing precursors. Samples of the cells were then withdrawn periodically, lysed with sodium dodecyl sulphate (SDS) and the DNA was analysed by equilibrium sedimentation. The results of this experiment are presented in Fig. 8.2 (Meselson and Stahl 1958). The top of the centrifuge tube, where the molecules with lighter density would be, is at the left of the photograph, while the bottom, where the molecules with heavier density would band, is at the right. If each of the two complementary strands of DNA can serve as a template for the formation onto itself of a new complementary strand, then after one generation time, a duplex DNA containing N^{15} would be converted into DNA with hybrid density (because onto an existing N^{15}-containing strand, a new N^{14}-containing strand would be formed, producing the hybrid duplex DNA). The position in the photograph of the DNA band at zero time represents the DNA with both N^{15}-containing strands. It can indeed be seen that after one generation, the position of the DNA band shifted towards the lighter side, but it was still a single band, presumably composed of one parental strand containing N^{15} and the other newly synthesized strand containing N^{14}. The presumption was confirmed by

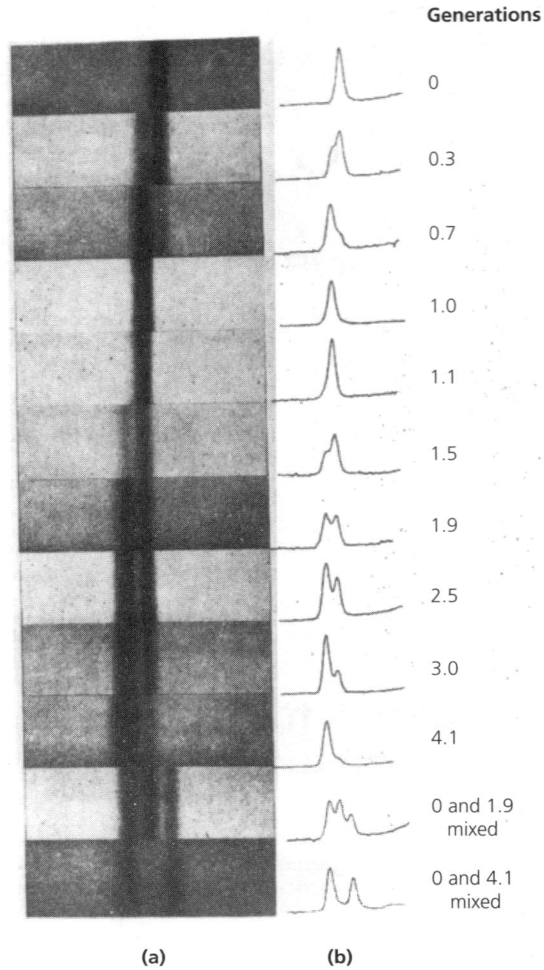

Figure 8.2 Ultraviolet absorption photographs showing DNA bands in centrifuge tubes. *Escherichia coli* were grown in a medium containing N^{15} NH_4Cl for 14 generations and then 10-fold excess of N^{14} NH_4Cl was added. Samples of cells (about 10^8) were withdrawn periodically, beginning with the addition of N^{14} NH_4Cl (zero generation). One generation is equivalent to the time taken for the cell number to double. The cell samples were lysed and subjected to equilibrium sedimentation by CsCl density gradient centrifugation. [Reproduced from Meselson and Stahl, *Proc Natl Acad Sci USA* 1958;44:675 by permission of Professor Matthew Meselson.]

heat denaturation of this DNA and recentrifugation, which revealed two bands, one corresponding to N^{14}-containing single-stranded DNA and the other corresponding to N^{15}-containing single-stranded DNA (Fig. 8.3) (Meselson and Stahl 1958). On the other hand, if the parental DNA had been fully conserved and both the strands of the progeny DNA were newly synthesized, comprising N^{14}, we would have seen two bands instead of one after one generation in the experiment presented in Fig. 8.2, one containing DNA with both strands comprising N^{15} and the other containing DNA with both strands comprising N^{14}.

After two generations, however, two DNA bands of almost equal intensity could be seen, one appearing

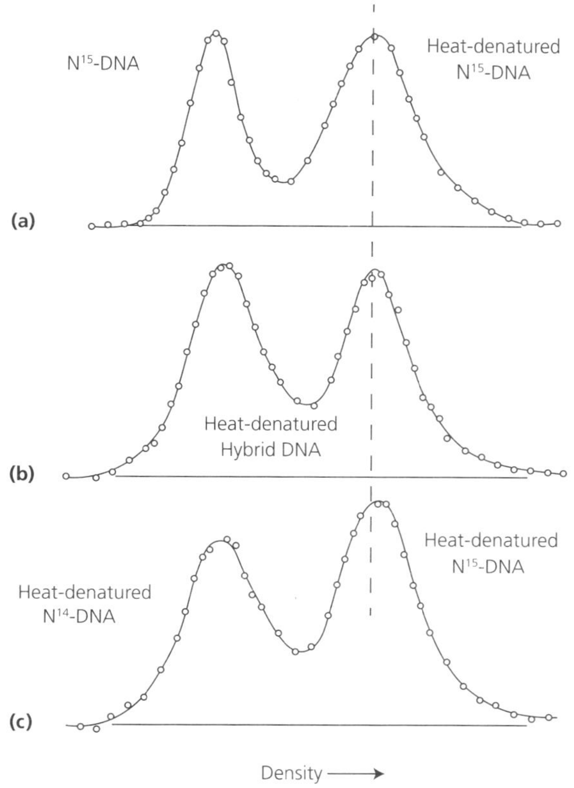

Figure 8.3 Microdensitometer tracing of bands of single-stranded DNA resolved by equilibrium sedimentation. (a) Top frame: mixture of unheated N^{15} DNA (double stranded) (left peak) and heated N^{15} DNA (single stranded) (right peak) shown for comparison. Single-stranded DNA has a higher density than double-stranded DNA, but half the molecular weight; (b) middle frame: DNA isolated one generation after addition of N^{14} NH_4Cl in the experiment presented in Fig. 8.2 was denatured and resolved by equilibrium sedimentation; (c) bottom frame: double-stranded N^{14}-containing DNA and double-stranded N^{15}-containing DNA mixed together, denatured and subjected to equilibrium sedimentation. [Reproduced from Meselson and Stahl, *Proc Natl Acad Sci USA* 1958;44:680 by permission of Professor Matthew Meselson.]

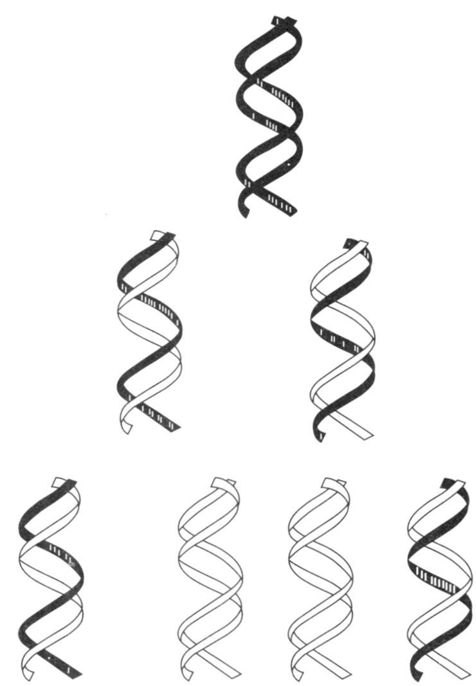

Figure 8.4 Diagrammatic representation of replication of DNA as inferred by Meselson and Stahl. The double helix at the top represents the original parent molecule with both strands containing N^{15}. The two double helices just below the parent molecule represent the first-generation daughter molecules, which have one strand containing N^{15} and the other strand containing N^{14}. The four double helices at the bottom represent the second-generation daughter molecules. Two of these molecules have one N^{15} strand and one N^{14} strand, while the other two molecules have both N^{14} strands. [Reproduced from Meselson and Stahl, *Proc Natl Acad Sci USA* 1958;44:678 by permission of Professor Matthew Meselson.]

exactly at the same position as the hybrid N^{15}–N^{14} band seen after one generation, and the other lighter than that. The lighter DNA band on denaturation and recentrifugation yielded a homogeneous band of single-stranded DNA that matched with N^{14} DNA, while the heavier band revealed the presence of an N^{15} single strand and an N^{14} single strand. The same two DNA bands persisted after the third and fourth generations, but the proportion of the N^{15}–N^{14} hybrid band progressively decreased. When the DNA samples from zero time and the second generation were mixed, all the three DNA bands representing N^{15}–N^{15}, N^{15}–N^{14} and N^{14}–N^{14} DNA could be seen, while the N^{15}–N^{14} band was barely visible in the mixture of zero time and the fourth generation.

The findings of Meselson and Stahl support the view that DNA is replicated by the 'semiconservative' mechanism, and this has been summarized diagrammatically in Fig. 8.4.

The perspective on the Meselson–Stahl experiment written by Philip Hanawalt (2004) is a 'must read' to really appreciate the elegance, the far-reaching implications and the subsequent developments that were based on the principle underlying the experiment.

The article by Tinsley Davis (2004) reveals the human aspects behind the great experiment.

8.3 The fork mechanism of DNA replication as visualized in autoradiographs by John Cairns

The experiments of Meselson and Stahl established that a new strand of DNA is synthesized on an old one and the daughter duplex is composed of an old and a new one. This novel experiment did not, however, shed any light on the mechanistic details of this process. Do the two strands of the duplex DNA first get unwound completely, disrupting all the hydrogen bonds and the resulting two single strands serve as separate templates for the sequential synthesis of the two complementary strands? Besides, is the direction of synthesis the same or opposite for both the new strands? These two modes of synthesis are represented in options I and II in Fig. 8.5. Alternatively, the synthesis of the complementary strands could start at a small localized denatured region and the hydrogen bonds between the two strands of the parent duplex could break gradually with the progress of the synthesis. In this alternative also, there could be two different modes represented by options III and IV in Fig. 8.5.

John Cairns attempted to distinguish between these options by labelling the newly synthesized DNA strands in *E. coli* B3 with ^{3}H-labelled thymidine and allowing the DNA to generate silver grains on a radiosensitive film. Some details of the method of labelling of DNA, lysis of cells and autoradiography have been mentioned in Chapter 5.

Density of silver grains was determined for uniformly labelled double-stranded DNA after labelling for several generations. The density of grains for single-stranded labelled DNA was also obtained by denaturation of the DNA after cell lysis. Henceforth, counting the number of grains per unit length would indicate whether the radioactive label was in both strands of DNA or in just one strand.

The *E. coli* cells were allowed to incorporate ^{3}H-thymidine (9 c/m-mole) into their DNA for just a few minutes.

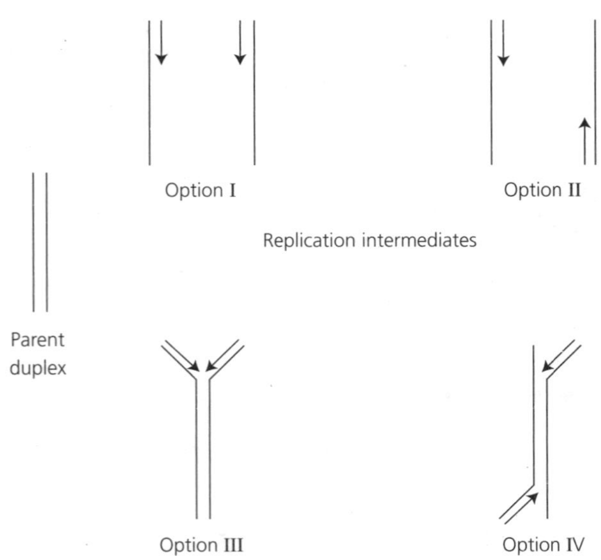

Figure 8.5 Options of replication of DNA. The arrows indicate the ongoing synthesis of the new complementary DNA strands and the lines indicate the parental DNA template strands. In options I and II, the parental DNA strands are first completely separated, while in options III and IV, the separation of the parental DNA strands is concomitant to the synthesis process.

This was achieved by diluting the bacterial culture after the desired period of labelling, 250-fold into chilled 0.15 M NaCl, containing 0.01 M KCN and 0.002% bovine serum albumin. The laboratory jargon for this operation is 'pulse labelling'. A 1-min pulse produced a 30-μm-long label, while a 2-min pulse produced a 60-μm-long label. This can be taken as evidence that DNA is synthesized sequentially in *E. coli.* and also as a measure of the rate of replication in these cells. Sometimes, the pulse was terminated by diluting 50-fold into a medium containing 20 μg/mL of unlabelled thymidine. This is known as 'pulse and chase'.

An artist's impression of some of the autoradiograms obtained by John Cairns (1963a) are shown in Fig. 8.6(a) and (b). Fig. 8.6(a) is the impression of the autoradiogram after pulse labelling for 3 min. Here, the grain density represents a double-stranded DNA or both labelled strands of DNA still in close association. The result of this experiment clearly preclude option I and II and also option IV, as these options would have shown label only on one strand of DNA in one place. It is important to remember that the visible grains are only from the newly synthesized labelled DNA and the parental unlabelled DNA cannot be seen in this experiment. Thus, in the impression of the autoradiogram shown in Fig. 8.6(a), one can imagine the presence of four DNA strands in close association at the site of synthesis, two parental strands of the original duplex that are not labelled with ^{3}H-thymidine and two newly synthesized ^{3}H-thymidine labelled DNA strands produced by the semiconservative mode of synthesis. It is not possible to infer from this autoradiogram as to whether the origin of replication is at one of the fork ends, while the replication head is at the other fork end, or the replication origin is at the centre and replication is proceeding simultaneously in both the directions with two replication heads at the two fork ends. The impression of the autoradiogram shown in Fig. 8.6(b) has been obtained after a 15-min chase subsequent to the 3-min pulse labelling. Here, the grain density reveals the presence of two separated ^{3}H-labelled single strands, implying that the daughter duplexes containing one parental unlabelled DNA strand and one labelled newly synthesized DNA strand, which have separated completely as a result of the chase. The forks cannot be seen as they are unlabelled and have moved away on both sides. Subsequently, John Cairns published his classic autoradiogram, showing the complete circular chromosome of *E. coli* undergoing replication (Cairns 1963b). An artist's impression of this autoradiogram is shown in Fig. 8.6(c).

These results were the basis of the 'fork mechanism' of DNA synthesis proposed by John Cairns. Both the strands of DNA are replicated simultaneously at the same site and separation of the parental strands of the DNA duplex follows the sequential synthesis of DNA, resulting in a fork-like structure. Electron micrographs of actual forked DNA molecules (stained with uranyl acetate) from a different experiment are shown in Fig. 8.7(a) and (b). As the two strands of DNA

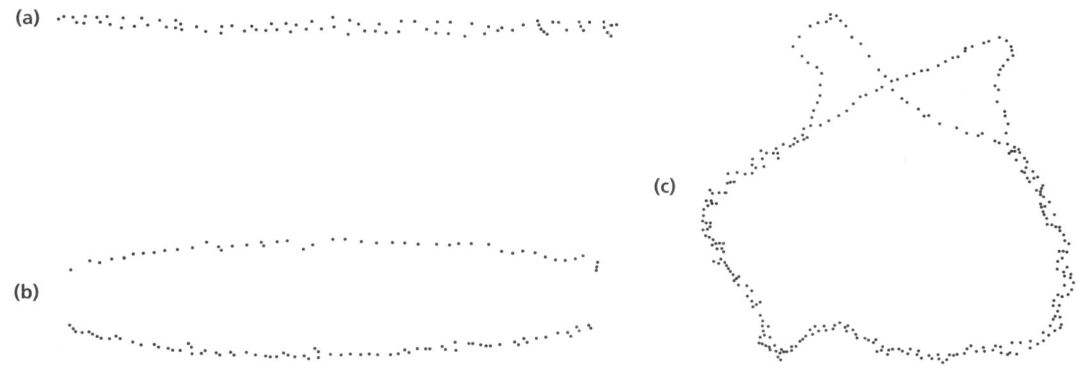

Figure 8.6 An artist's impressions of autoradiograms of John Cairns. Impressions of segments of DNA in *Escherichia coli* labelled with ^{3}H-thymidine. (a) Immediately after 3-min pulse labelling, (b) 3-min pulse labelling followed by 15-min chase. (c) impression of the ^{3}H-thymidine–labelled complete circular replicating chromosome of *E. coli*. The segments with both the DNA strands labelled can be distinguished from the segments with only one strand labelled on the basis of grain density. The two forks are visible. [An artist's impressions of autoradiograms of Cairns, *J Mol Biol* 1963;6:81.]

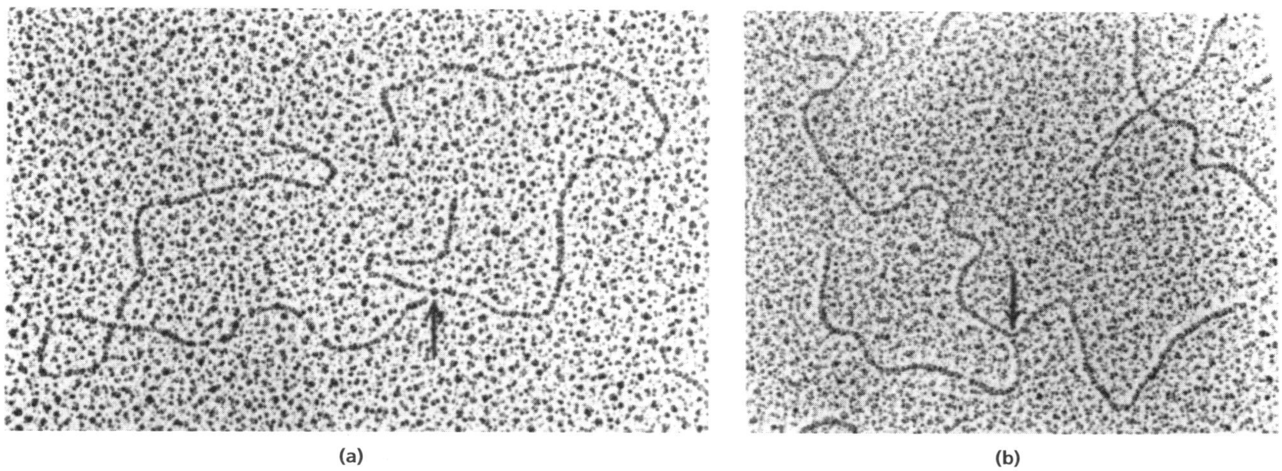

Figure 8.7 Electron micrographs of forked DNA molecules (stained with uranyl acetate). The positions of the forks have been marked by arrows. Magnification: (a) ×90,000; (b) ×1,00,000. The electron micrographs shown in Fig. 8.7 (a) and (b) have been taken by Dr Nirupam Roy Choudhury in the laboratory of Professor HK Das.

are 'antiparallel', an important implication is that if the synthesis of one strand of DNA is in the 5′→ 3′ direction, the net direction of synthesis of the other strand would have to be in the 3′→5′ direction. (See DNA replication of the 5′ strand in Chapter 9, Sections 9.20– 9.25, for clarification on this point.)

8.4 Concept of unique origin of replication of DNA in the bacterium *E. coli*

John Cairns had assumed that one of the two forks in Fig. 8.6(c) was the origin of replication while the replication was in progress at the other fork. As a corollary, he postulated that *E. coli* had a unique and fixed replication origin and the replication was unidirectional, as represented diagrammatically in Fig. 8.8 (Cairns 1963b). Though the first half of the postulate was proved to be correct and the unique replication origin has now been precisely located, the second half was later found to be incorrect (see Section 8.6).

8.5 Tandem initiation of replication of DNA in Chinese hamster cells

Chromosomes of higher organisms are much larger than the genomes of prokaryotes such as *E. coli* (see Table 5.2, Part 1). Does replication still get initiated in such large chromosomes at a unique replication origin?

Joel A. Huberman and Arthur D. Riggs (1968) checked on replication of DNA in Chinese hamster cells to answer this question by using basically the same approach as John Cairns. When hamster fibroblast cells were pulse-labelled with ^{3}H-thymidine (1.8 c/mM), tandem arrays of labelled DNA were seen on the autoradiogram [an artist's impression in Fig. 8.9(a)], suggesting simultaneous initiation of replication at multiple origins of replication along the same DNA molecule. Here, the length of pulse labelling was 30 min; this was because the doubling time of the cells was 17 h in contrast to the doubling time of 30 min for *E. coli* cells.

Next, the 30-min pulse labelling was followed by a 45-min chase. As the Chinese hamster cells were not auxotrophic for thymidine and the internal pool of thymidine is large, the chase was not expected to be instantaneous as was seen for *E. coli*, but was supposed to be gradual. In other words, the silver grains would not be terminated instantly as a result of the chase, but the density of the silver grains along the line of synthesis would be reduced gradually. This was exactly what was seen, but the gradual decrease in grain density was observed to be taking place on both ends of the initial labelled region (Huberman and Riggs 1968). An artist's impression of the autoradiogram is presented in Fig. 8.9(b). This suggested strongly that at least in the case of Chinese hamster cells, replication of DNA proceeded simultaneously in both the directions from the origin of replication. This was again in contrast to the postulate of John Cairns that DNA synthesis in *E. coli* took place in only one direction.

Is the fork mechanism of replication of DNA operative in this case also? Huberman and Riggs did obtain autoradiograms in some of their pulse and chase experiments which clearly gave an affirmative answer (Huberman and Riggs 1968). An artist's impression of such an autoradiogram is presented in Fig. 8.10. A schematic summary of the results on replication of DNA in the Chinese hamster cells, as has been proposed by Huberman and Riggs (1968), is presented in Fig. 8.11.

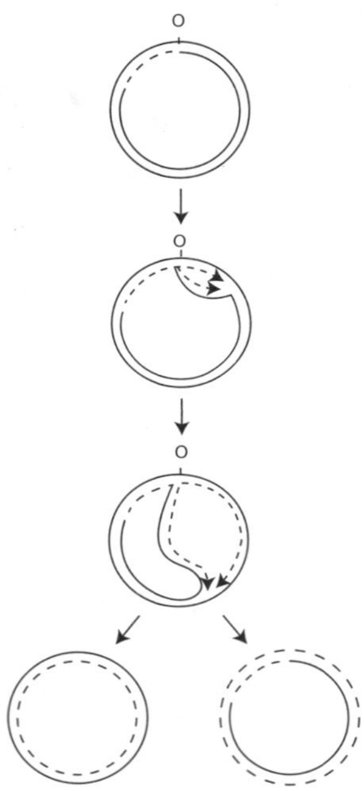

Figure 8.8 Schematic diagram of the mechanism of replication of the circular genome of *Escherichia coli* as postulated by John Cairns. The replication was thought to have initiated at a unique replication origin, marked O at the top of the circle. The solid lines represent the pre-existing DNA and the broken lines, the newly synthesized segments. The arrow heads indicate the direction of synthesis. Both the strands of the DNA were replicated simultaneously, but the replication fork advanced from the replication origin in only one direction. (The second half of the postulate has been found to be incorrect later.) Two daughter DNA duplexes were thought to have been formed when the fork reached the terminus, that is, the origin itself or was close to the origin.

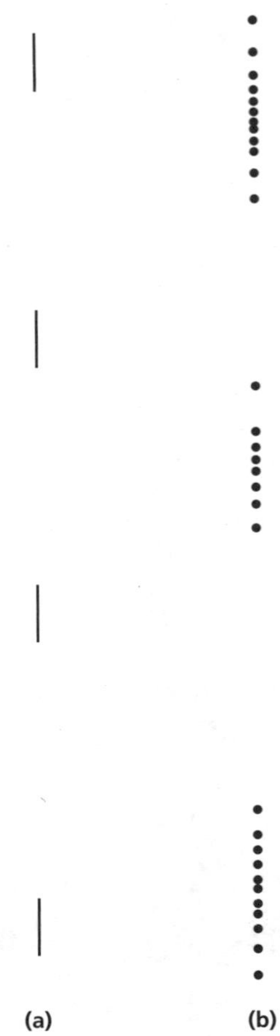

Figure 8.9 (a) An artist's impression of autoradiogram of tandem arrays of labelled DNA observed by Huberman and Riggs (1968) on pulse labelling for 30 min of Chinese hamster cells with ^{3}H-thymidine. (b) An artist's impression of autoradiogram of labelled DNA on pulse labelling for 30 min with ^{3}H-thymidine, followed by a chase for 45 min with unlabelled thymidine. The gradual decrease in grain density at both ends of the labelled stretches is obvious, suggesting that the replication was progressing at both the ends.

Figure 8.10 An artist's impression of an autoradiogram obtained by Huberman and Riggs (1968), revealing replication forks subsequent to the separation of labelled DNA strands after pulse and chase of Chinese hamster cells. The initial tandem arrays of labelled DNA with decreasing grain density at the ends can still be seen.

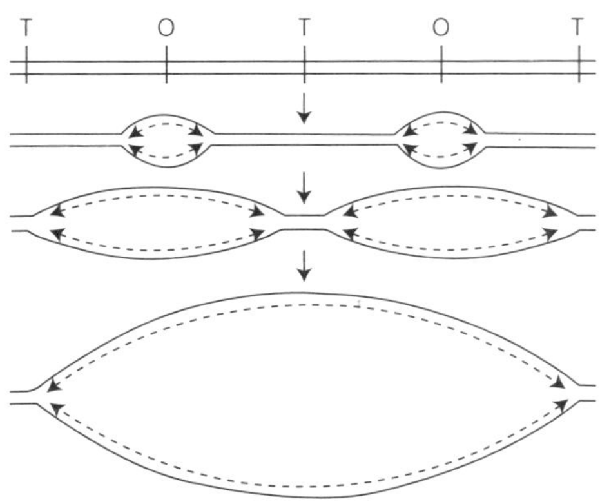

Figure 8.11 An artist's impression of the schematic summary of the autoradiograms arising out of labelling of DNA of Chinese hamster cells [Huberman and Riggs (1968)]. Tandem initiation of DNA replication at multiple replication origins, advance of replication forks at both sides of the replication origins, forming replication bubbles, and coalescence of the small bubbles into large bubbles can be seen. The parental DNA strands have been represented by continuous lines, while the newly synthesized DNA strands have been represented by broken lines. The arrow heads indicate the direction of replication. The replication origins have been marked by the letter O and the replication terminii situated at both sides of the origins have been marked by the letter T.

8.6 Bidirectional replication is also the mode in *E. coli*

The experimental approach adopted by DM Prescott and PU Kuempel (1972) to determine whether the replication of DNA in *E. coli* was unidirectional or bidirectional was basically similar to that adopted by Huberman and Riggs (1968) for Chinese hamster cells, except that instead of applying a chase with unlabelled thymidine, Prescott and Kuempel (1972) added ^{3}H-thymine plus ^{3}H-thymidine of high specific activity after labelling with ^{3}H-thymine of moderate specific activity. The increase in the density of the silver grains would, thus, be expected after the application of the high specific radioactivity ^{3}H-thymidine.

The important point to remember here, however, is that replication is supposed to be initiated in *E. coli* from a unique origin in the whole chromosome. This is very much unlike the situation in Chinese hamster cells, where a large number of tandem replication origins not far from each other exist and initiation of replication is always taking place in one or another origin. An unambiguous inference would thus be impossible unless labelling can be initiated at the unique origin of replication of *E. coli*. How does one lay a finger on to those *E. coli* cells that are about to initiate replication?

KG Lark had inferred early on (1966) that if an amino acid auxotroph of *E. coli* (a strain that has a mutation in the pathway of the biosynthesis of that particular amino acid) was starved of the required amino acid, only the ongoing round of DNA replication was completed and no new round of replication was initiated. The evidence in favour of this inference was the following. If ^{3}H-thymidine was added to an asynchronously dividing culture of an amino acid auxotroph of *E. coli* starving for the required amino acid, incorporation of radioactivity into the DNA slowly tapered off and finally came to a halt after about one doubling time. If ^{3}H-thymidine was added at the point when the cells had just started starving, the total amount of radioactivity incorporated into the DNA of all the starving cells in the culture was just about half the total amount of radioactivity incorporated into the cells of a control culture to which the amino acid was added back along with ^{3}H-thymidine and allowed to incorporate the labelled precursor into the DNA for a period of one doubling time. In an asynchronously dividing culture of cells, the replication fork in each individual cell would be at any point between the replication origin and the replication terminus. In some cells, the replication of DNA would have just begun; in some cells, the replication fork(s) would be about to reach the terminus; while in others, the forks would be at various positions between the origin and the terminus. A schematic representation of the situation is given in Fig. 8.12(a). For simplicity's sake, replication of DNA has been assumed to proceed in only one direction. It can, thus, be assumed that the total distance covered by the replication fork(s) of all the cells taken together just before ^{3}H-thymidine was added would be equivalent to the same as if all the forks have covered a distance halfway to the terminus from the origin [Fig. 8.12(b)]. Incorporation of ^{3}H-thymidine into the DNA would now begin at the arrow heads. If in amino acid–starved cells,

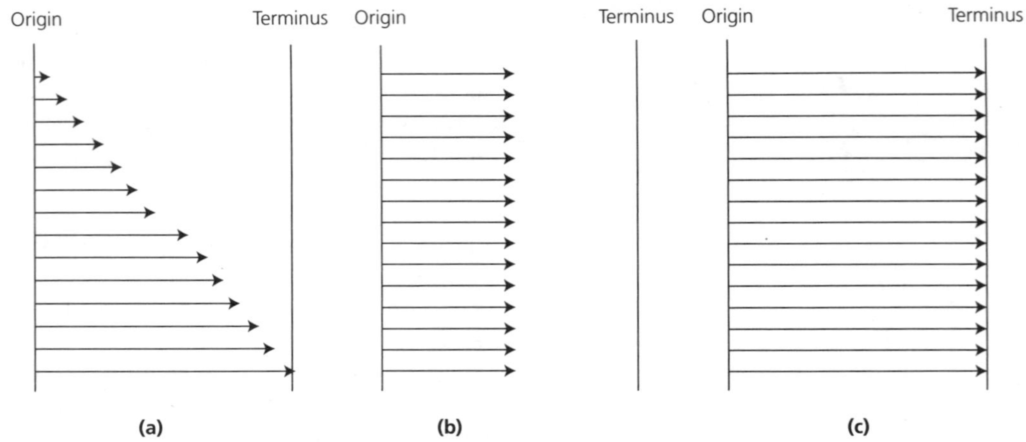

Figure 8.12 Schematic representation of the positions of the replication forks with respect to the replication origin and the terminus in individual cells of *Escherichia coli*. The arrow heads represent the positions of the replication forks, the lines behind the arrow heads represent the already replicated portions, while the gaps in front of the replication forks represent the unreplicated portions of DNA. (a) Actual positions of the replication forks at any given point of time in individual cells of an asynchronous culture. (b) On an average, the replication forks in all the cells can be considered to have progressed up to the halfway mark between the replication origin and the terminus. The summation of the distance covered by all the forks in Fig. 8.12(a) would be equivalent to the summation of the distance covered by all the forks in Fig. 8.12(b). (c) In amino acid–starved cells, the replication forks in all the cells can reach the terminus, but are unable to initiate a new round of replication.

just the ongoing round of replication is completed, but no new round is initiated, all the forks would reach the terminus and stay put there [Fig. 8.12(c)]. The total incorporation in this case would obviously be only half as much as would have been incorporated during one doubling time.

If the required amino acid is added to the medium after the starvation has been allowed for more than one doubling time, replication of the DNA would be expected to get initiated at the origin of replication in all the cells and a synchronous replication of DNA would ensue, resulting in synchronous division of cells. This has indeed been verified by monitoring the number of cells in the culture during growth (Matney and Suit 1966). In an asynchronous culture, the number of cells increases continuously and exponentially with time. If synchrony is achieved, the number of cells remains constant for one doubling time and then almost instantly becomes double. This phenomenon is repeated every doubling time. Instead of an exponential curve, one gets a step curve with alternating horizontal and vertical components (Fig. 8.13).

Ptrescott and Kuempel (1972), therefore, took amino acid–starved *E. coli* cells and added back the required amino acid along with ^{3}H-thymine of moderate specific radioactivity (10 Ci/mM). As they found in preliminary experiments that it took anywhere between 5 to 10 min to get the replication initiated, they waited for about 15 min and then again added ^{3}H-thymidine of higher specific radioactivity (52 Ci/mM). The incubation was continued for another 2.5 min and the reaction was stopped with EDTA and NaCN. The cells were lysed with SDS, spread on microscopic slides and exposed to a radiosensitive film. The tracks of silver grains observed in the autoradiograms are shown in Fig. 8.14. It is obvious that the increase in the density of grains as a result

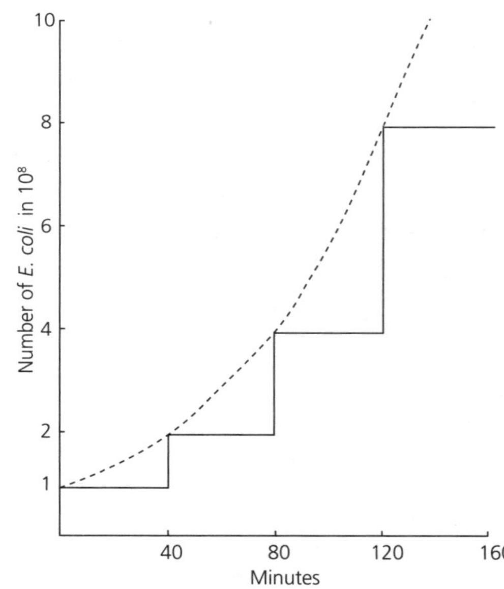

Figure 8.13 Simulation of the number of *Escherichia coli* cells during growth in liquid shake cultures. The cells are auxotrophic for an amino acid. The broken line represents the growth of asynchronous cells in the complete medium (containing the amino acid); the doubling time being 40 min. The solid line represents the stepwise increase in cell numbers on replenishment of the amino acid in the culture medium after the cells were starved of the amino acid for several doubling times.

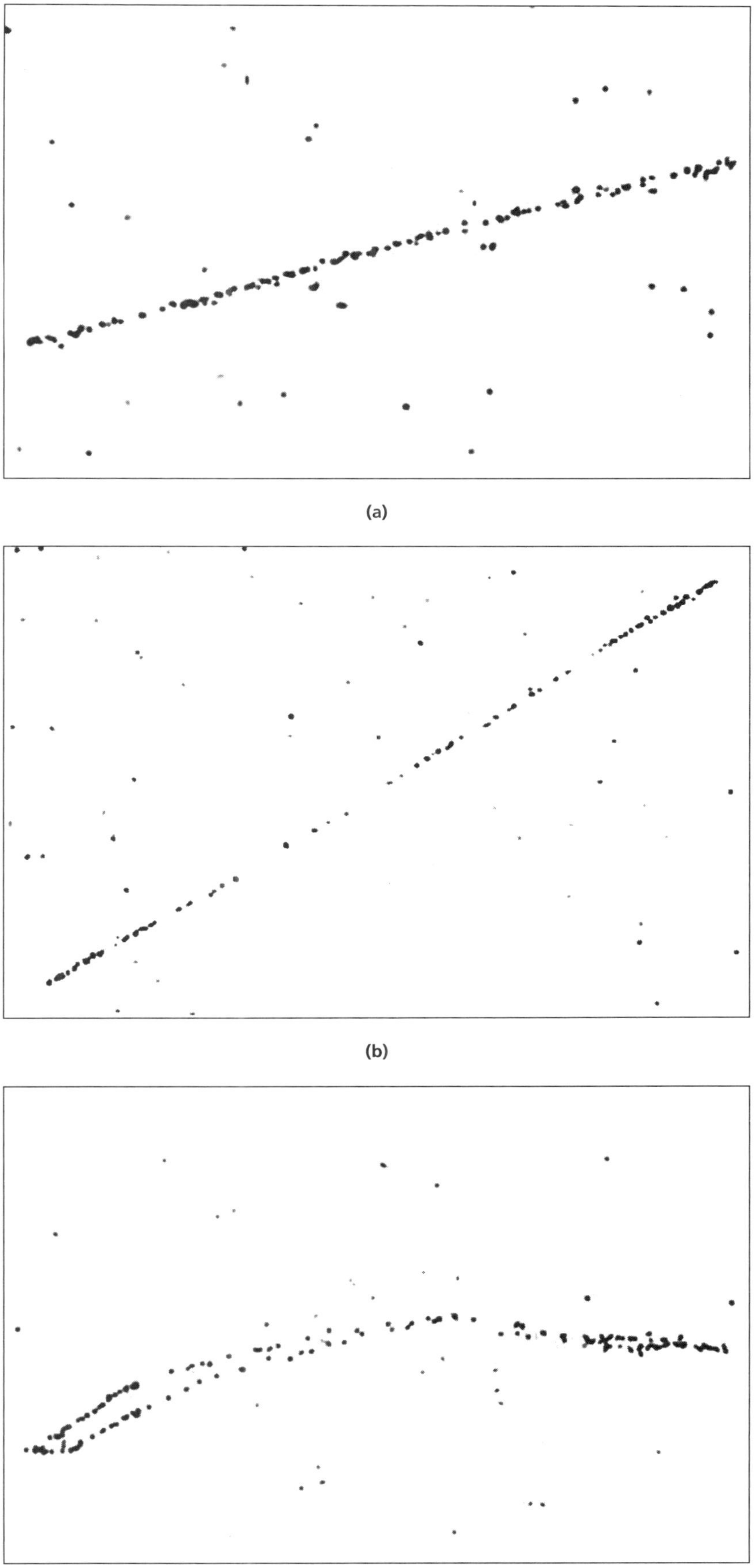

(a)

(b)

(c)

Figure 8.14 Autoradiograms showing tracks of silver grains as a result of incorporation of first low specific radioactivity ^{3}H-thymine and then high specific radioactivity ^{3}H-thymidine at the origin of replication of DNA of *Escherichia coli* cells synchronized by amino acid starvation. The starvation has been relieved by the addition of the required amino acid when the ^{3}H-label has also been added. (a) Control experiment. Track of silver grains developed from a cell fed with ^{3}H-thymidine (52 Ci/mM) along with the required amino acid and incubated for 16 min. The uniformly high density of silver grains can be seen throughout the track. (b) Track of silver grains developed from a cell incubated with ^{3}H-thymine (10 Ci/mM) for 19 min (added along with the required amino acid) and for additional 2.5 min with ^{3}H-thymidine (52 Ci/mM). Higher density of silver grains can be seen on both sides at the two ends and the low density stretch in the middle. (c) An autoradiogram obtained from an experiment similar to that described in Fig. 8.14(b), where the two daughter duplexes are separated and the two replication forks are visible. [Reproduced from Prescott and Kuempel, *Proc Natl Acad Sci USA* 1972;69:2844 with permission from Professor PL Kuempel.]

of the addition of a high specific activity label is on both sides of the low-density central stretch. The inference was, therefore, drawn that replication of the DNA in *E. coli* was bidirectional. Similar inference had also been drawn from genetic transduction experiments (Masters and Broda 1971).

8.7 Electron microscopic visualization of replication of DNA of the bacterial virus lambda (κ): Bidirectional replication from a unique origin is the predominant mode

The genome of the virus lambda (λ), as isolated from the virion, is a linear double-stranded DNA about 17.5-μm long. The DNA, however, is found as a circular molecule inside its host *E. coli*. Intact replicating lambda (λ) DNA has been isolated from inside *E. coli* and has been visualized under the electron microscope (Schnos and Inman 1970). An artist's impression of intact replicating lambda (λ) is presented in Fig. 8.15. The replication bubble can be seen between the two replication forks marked with arrows in Fig. 8.15. The questions Maria Schnos and Ross Inman asked were:

- Is there a unique origin of replication of lambda (λ) DNA? If so, what is the physical location of the replication origin?
- Is replication unidirectional or bidirectional?

The strategy decided on was to isolate a large number of lambda DNA molecules from inside the infected *E. coli* cells and examine the replication bubbles in them under the electron microscope. The problem, however, is that a circle does not have any reference point. How does one determine the position of the replication bubble or the replication forks on a circle and relate it to the bubble or forks on another circle in the absence of a reference point? If the replicating DNA had been linear, things would have been straightforward, the two ends could have served as reference points. One has to appreciate that no restriction endonuclease was known in those days and there was no way of cleaving a circular DNA molecule at a unique site. Maria Schnos and Ross Inman, nevertheless, devised a novel approach to obtain reference points. They isolated linear DNA from the mature virions of lambda (λ), partially denatured it by holding it for 10 min at around pH 11 at room temperature and examined the individual DNA molecules under the electron microscope. Quite a number of denaturation bubbles (bubbles comprising single-stranded arms) could be seen, whose positions and sizes were found to be invariant from molecule to molecule. Temperature denaturation was not as reproducible as alkaline pH denaturation. Schnos and Inman constructed a 'partial denaturation map' of linear lambda DNA. When they subjected circular lambda DNA to the same conditions of denaturation, they obtained an exactly same denaturation map with the same sequence of the individual denaturation bubbles at the same positions. This allowed them to assign on the circular DNA the location of the site at which a cleavage would have generated the ends of the linear DNA isolated from the virion. Hence, they had the reference point they were looking for (Schnos and Inman

Figure 8.15 An artist's impression of an electron micrograph of replicating DNA of the virus lambda (λ) isolated from inside its host *Escherichia coli*. The replication bubble in the circular double-stranded DNA is seen between the two replication forks marked with arrows.

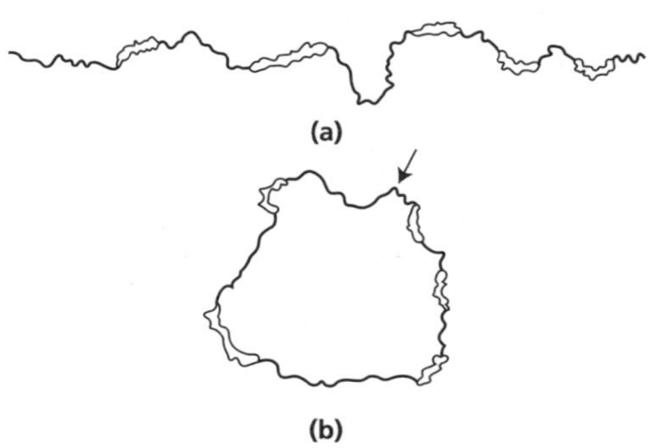

Figure 8.16 Mini models of lambda DNA, both the linear (a) and the circular form (b), showing denaturation bubbles obtained on treatment at pH 11.05 for 10 min at room temperature. Only five denaturation bubbles are shown in the mini model. The arrow in the circular DNA indicates the site of cleavage for linearization before packaging in the virion.

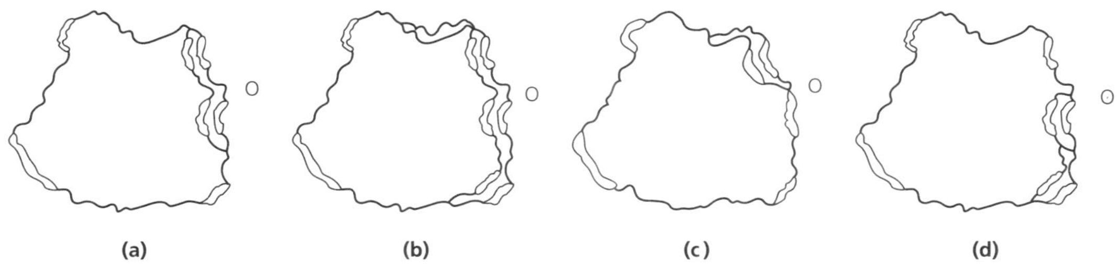

Figure 8.17 Mini models of lambda DNA showing a replication bubble and also the denaturation bubbles in circular lambda DNA. (a) Bidirectional replication. (b) Bidirectional replication with a larger replication bubble. (c) Unidirectional replication. (d) Unidirectional replication in the opposite direction.

1970). Mini models of lambda DNA, both the circular and the linear form, containing only five denaturation bubbles is presented in Fig. 8.16.

Schnos and Inman next isolated a large number of replicating circular DNA molecules from inside *E. coli* cells, subjected these to partial denaturation conditions and examined these under the electron microscope. The analysis of 68 circular lambda DNA molecules has revealed that 46 of these molecules have been replicating bidirectionally, while only 22 molecules have been replicating unidirectionally. Of these 22 molecules, 10 were replicating in one direction, while 12 were replicating in the other direction. The replication origin of all the 68 molecules appeared to be located at around 14.3 μm from the left end of the linear lambda DNA molecule (Schnos and Inman 1970). Fig. 8.17 shows mini models of four such replicating lambda DNA molecules: DNA molecules in Fig. 8.17(a) and 8.17(b) replicating bidirectionally, while DNA molecules in Fig. 8.17(c) and 8.17(d) replicating unidirectionally but in opposite directions.

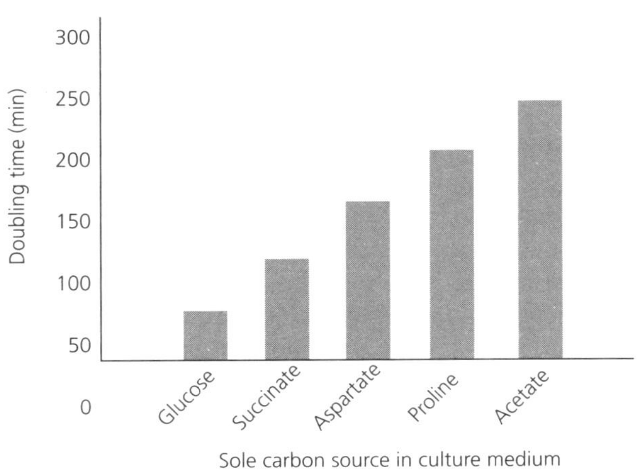

Figure 8.18 Effect of various sole carbon sources on the doubling time of *Escherichia coli.*

8.8 Does the speed of advancement of the replication fork change with the change in the time of doubling of the number of cells of an organism?

The time of doubling of the number of cells in a culture of *E. coli* varies greatly with the nature of nutrients in the culture medium (Fig. 8.18; Lark 1966). The basic assumption is that the division of a cell takes place only after the DNA replication fork has reached the replication terminus, so that the two replicated DNA molecules can partition into the two daughter cells. If that is so, then the reduction in the time taken for doubling of the number of cells from let us say 120 min to 40 min can take place in either of the following two alternative ways:

1. The speed of movement of the replication fork has been enhanced three fold because of the availability of a better carbon source in the culture medium.
2. The replication fork had already covered considerable distance between the replication origin and the terminus when the last round of cell division had taken place.

Growing cells of *E. coli* were pulse-labelled with ^{3}H- thymidine, lysed gently, the lysate fixed on a glass slide and a radiosensitive film was layered on the slide similar to the way it was done by John Cains (Fig. 8.6). The film was developed, revealing a stretch of silver grains which represented the newly synthesized labelled DNA. The length of this stretch of grains would be a measure of the distance traveled by the replication fork during the period of pulse labelling. This was indeed found to be more or less the same irrespective of whether aspartate or glucose was used in the culture medium as the sole carbon source. This observation would tend to suggest that the rate of movement of the replication fork was practically independent of the sole carbon source used in the culture medium. In other words, as Cooper and Helmstetter (1968) have stated, 'the time for a complete round of DNA replication in *E. coli* is constant and independent of the growth rate'. Then, the alternative (ii), in which it has been postulated that the replication fork had already covered considerable distance between the replication origin and the terminus, when the last round of cell division had taken place, was possibly the correct one. One way for this to happen is to imagine multiple replication forks in a fast dividing cell (Fig. 8.19). During cell division, replication fork 1 would be 40 min away from the replication terminus at T, assuming that the replication fork takes 120 min to travel from the replication origin to the terminus. This fork was, hence, initiated 80 min earlier. Replication forks 2 and 3 are 80 min away from the terminus and were initiated 40 min earlier. Four new replication forks would now be initiated at the replication origin. During the next cell division, replication fork 1 would reach the terminus, giving rise to two DNA molecules, each of which would look exactly the same as shown in Fig. 8.19. These two identical DNA molecules, one containing replication fork 2 and the other containing replication fork 3, would end up in the two daughter cells and would have by this time moved to the position where replication fork 1 was at the earlier cell division. This cycle would be repeated. The cell doubling time would, however, be dependent on the medium.

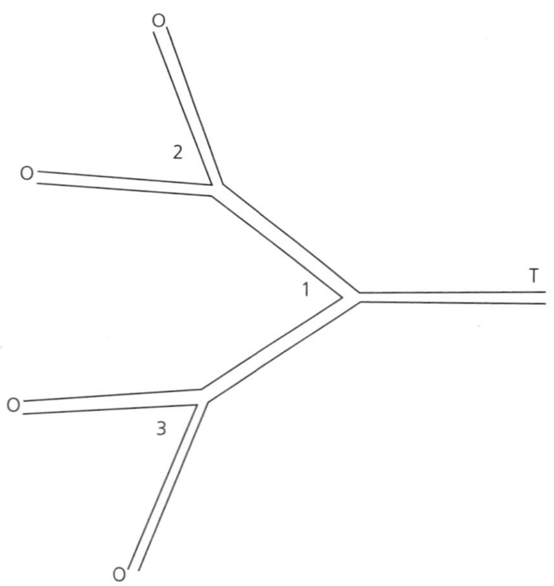

Figure 8.19 Postulated presence of multiple DNA replication forks in *Escherichia coli* growing in a medium containing glucose with a doubling time of 40 min. O stands for the replication origin and T for the terminus. The assumption is that replication fork 1 has already covered about two-third the distance between the replication origin and the terminus when the cell division has taken place. Replication fork 1 and replication fork 2 have been initiated 40 min early and have covered one-third the distance at cell division.

8.9 Experimental evidence in favour of the existence of multiple DNA replication forks

It was Noboru Sueoka and his associates who first demonstrated the existence of multiple DNA replication forks in the gram-positive bacterium *Bacillus subtilis* (Oishi *et al.* 1964). When spores of *B. subtilis* are germinated, DNA replication is simultaneously initiated at the replication origin of almost all the cells, thus generating a synchronous culture.

B. subtilis were first grown for several generations in D_2O and allowed to sporulate. The washed deuterium-labelled spores were then put in a germination medium rich in nutrients and containing H_2O and ^{32}P-labelled orthophosphate. Cell samples were withdrawn periodically, the DNA isolated, sheared and analysed by CsCl density gradient centrifugation.

Two DNA bands were first observed, a heavy-light band containing radioactivity and a heavy-heavy band containing no radioactivity. Transformation experiments with auxotrophic mutants of *B. subtilis* suggested that the heavy-light radioactive DNA contained the adenine marker which was earlier demonstrated to be close to the replication origin, while the heavy-heavy non-radioactive DNA contained the leucine and methionine markers that were closer to the replication terminus. Normalized transformation frequency data also suggested the presence of twice as many DNA fragments containing the adenine marker in the heavy-light fraction compared with the DNA fragments containing the leucine or methionine marker in the heavy-heavy fraction. The inference was that there was a single DNA replication fork which had yet not gone past the leucine marker (Oishi *et al.* 1964). Soon enough, however, and much before the time for cell division, a third band containing light-light radioactive DNA appeared in the CsCl gradient. The DNA fragments in this band contained the adenine marker and none of the threonine, or leucine, or methionine markers. The total amount of adenine marker present at this point in the combined light-light and light-heavy fractions was four times the methionine marker in the heavy-heavy band. The leucine marker was present at this stage in the heavy-light band and was found to be twice in amount compared with the methionine marker (Oishi *et al.* 1964). A schematic diagram representing the successive stages of DNA replication in *B. subtilis*, when its spores were allowed to germinate in a medium rich in nutrients, is presented in Fig. 8.20.

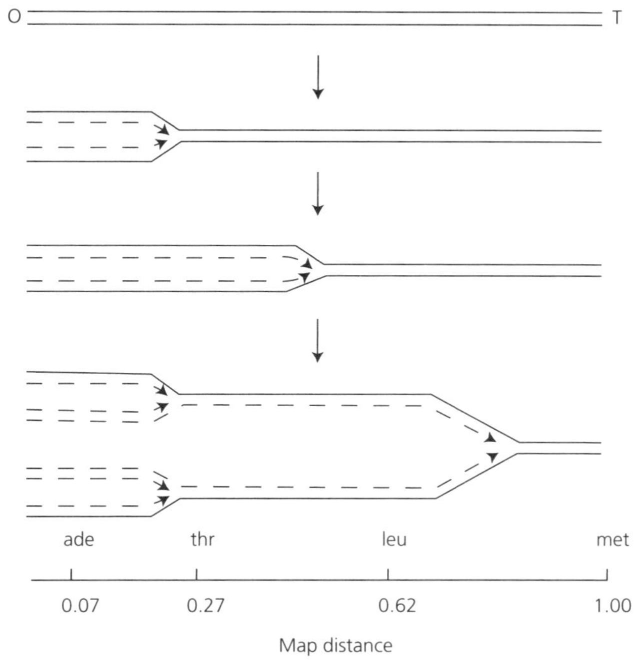

Figure 8.20 A schematic diagram representing the successive stages of DNA replication in *Bacillus subtilis* when its spores were allowed to germinate in a medium rich in nutrients. Deuterium-labelled spores of *B. subtilis* were washed and allowed to germinate in a rich medium containing H_2O and ^{32}P-labelled orthophosphate [Oishi *et al.* (1964)]. The topmost double line represents DNA in the spores where replication has not been initiated. O stands for the replication origin and T for the terminus. The next stage (immediately below) shows that DNA replication has been initiated and the fork has moved only about 0.2 map distance, covering the *ade* marker, but not the *thr* marker. So the *ade* marker is in two copies, while all the other markers are still in single copy. The solid lines represent the parental DNA strands that are deuterium heavy, and the broken lines represent the newly synthesized light and radioactive DNA strands. The arrow heads indicate the direction of synthesis. At the next stage (immediately below the second stage), the fork has moved to about 0.5 map distance, covering the *thr* marker also, but not the *leu* marker. So at this stage, both the *ade* and the *thr* markers are in two copies, while the *leu* and the *met* markers are still in single copy. At the last stage (bottom most), the initial fork has moved to about 0.8 map distance, covering the *leu* marker also, but not the *met* marker. Meanwhile, two new forks have been initiated, which have covered about 0.2 map distance, covering the *ade* marker. Thus, at this stage, the *ade* marker is in four copies, the *thr* and *leu* markers are in two copies and the *met* marker is still in single copy.

The inference was unambiguous; additional DNA replication forks were initiated at the DNA replication origin of *B. subtilis* even before the primary replication fork reached the terminus, if the spores were germinated in rich medium. This mode of DNA replication was called 'dichotomous'. Dichotomous replication of DNA has also been observed in rapidly growing *B. subtilis* independent of sporulation or spore germination (Yoshikawa *et al.* 1964). Symmetric multifork (dichotomous) DNA replication takes place in rapidly growing *E. coli* also (Cooper and Helmstetter 1968; Fritsch and Worcel 1971; Helmstetter and Cooper 1968).

Cooper and Helmstetter (1968) have found that the time for a replication fork to travel from the origin of replication to the terminus of *E. coli* B/r is about 40 min. They have also found that there was only one chromosome in cells dividing in 60 min (in nutritionally very poor medium). Replication was initiated immediately after the cell division and the replication fork took 40 min to reach the terminus. The next cell division happened after a gap of another 20 min and during this period, the two daughter chromosomes remained in the same cell, with no initiation of a new round of replication. In cells dividing in 50 min (in nutritionally slightly richer medium), the replication was initiated 10 min before the last cell division took place and the replication fork took another 30 min to reach the terminus. The two daughter chromosomes remained dormant in the same cell for the next 10 min and then replication was initiated in both the chromosomes. The cell division and partitioning of the two chromosomes took place another 10 min later. In cells dividing in 35 min (in nutritionally rich medium), the replication was initiated 30 min before the last cell division took place and the replication fork took only 5 more min to reach the terminus. DNA replication was initiated immediately in both the chromosomes while still in the same cell and the replication continued for another 30 min when the cell division took place. Dichotomous replication (multiple replication forks in the same chromosome) occurred only when the doubling time was less than 30 min. In cells dividing in 20 min (in nutritionally very rich medium), two chromosomes existed even at the time of cell division, and in both these chromosomes, the replication fork had already advanced halfway. In another 5 min, replication was again initiated at the four origins in the two chromosomes, leading to dichotomous replication. During the next cell division, there were four chromosomes in the cell, and in each of them, the replication fork had advanced halfway. The inferences were drawn on the basis of incorporation of 3H-thymidine in each individual cell after pulse labelling, as determined by counting the number of silver grains on the radiosensitive film per cell at different periods after cell division (Cooper and Helmstetter 1968).

References

Cairns J. The bacterial chromosome and its manner of replication as seen by autoradiography. *J Mol Biol* 1963a;6:208–213.

Cairns J. The chromosome of *Escherichia coli. Cold Spring Harb Symp Quant Biol* 1963b;28:43–46.

Cooper S. Helmstetter CE. Chromosome replication and the division cycle of *Escherichia coli* B/r. *J Mol Biol* 1968;31:519–540.

Davis TH. Meselson and Stahl: The art of DNA replication. *Proc Natl Acad Sci USA* 2004;101:17895–17896.

Fritsch A, Worcel A. Symmetric multifork chromosome replication in fast growing *Escherichia coli. J Mol Biol* 1971;59:207–211.

Hanawalt PC. Density matters: The semiconservative replication of DNA. *Proc Natl Acad Sci USA* 2004;101:17889–17894.

Helmstetter CE, Cooper S. DNA synthesis during the division cycle of rapidly growing *Escherichia coli* B/r. *J Mol Biol* 1968;31:507–518.

Huberman JA, Riggs AD. On the mechanism of DNA replication in mammalian chromosomes. *J Mol Biol* 1968;32:327–341.

Lark KG. Regulation of chromosome replication and segregation in bacteria. *Bacteriol Rev* 1966;30:3–32.

Masters M, Broda P. Evidence for the bidirectional replication of the *Escherichia coli* chromosome. *Nature New Biol* 1971;232:137–140.

Matney TS, Suit JC. Synchronously dividing bacterial cultures. 1. Synchrony following depletion and resupplementation of a required amino acid in *Escherichia coli. J Bacteriol* 1966;92:960–966.

Meselson M, Stahl FW. The replication of DNA in *Escherichia coli. Proc Natl Acad Sci USA* 1958;44:671–682.

Meselson M, Stahl FW, Vinograd J. Equilibrium sedimentation of macromolecules in density gradients. *Proc Natl Acad Sci USA* 1957;43:581–588. Oishi M, Yoshikawa H, Sueoka N. Synchronous and dichotomous replications of the *Bacillus subtilis* chromosome during spore germination. *Nature* 1964;204: 1069–1073.

Prescott DM, Kuempel PL. Bidirectional replication of the chromosome in *Escherichia coli. Proc Natl Acad Sci USA* 1972;69:2842–2845.

Schnos M, Inman RB. Position of branch points in replicating λ DNA. *J Mol Biol* 1970;51:61–73.

Watson JD, Crick FHC. Molecular structure of nucleic acids. A structure for deoxyribose nucleic acid. *Nature* 1953a;171: 737–738.

Watson JD, Crick FHC. Genetical implications of the structure of deoxyribonucleic acid. *Nature* 1953b;171:964–967.

9

DNA polymerases

H. K. Das

9.1 Is any enzyme at all necessary for the replication of the genetic material?

In their first paper on the proposed structure of DNA, James Watson and Francis Crick wrote, 'It has not escaped our notice that the specific pairing we have postulated immediately suggests a possible copying mechanism for the genetic material' (Watson and Crick 1953a). A loaded statement that avoids spelling out much detail. In their second paper, of course, Watson and Crick came out with some detailed ideas, 'We imagine that prior to duplication the hydrogen bonds are broken, and the two chains unwind and separate. Each chain then acts as a template for the formation on to itself of a new companion chain, so that eventually we shall have two pairs of chains, where we only had one before'. 'From time to time the base of a free nucleotide will join up by hydrogen bonds to one of the bases on the chain already formed. We now postulate that the polymerization of these monomers to form a new chain is only possible if the resulting chain can form the proposed structure'. 'Whether a special enzyme is required to carry out the polymerization, or whether the single helical chain already formed acts effectively as an enzyme, remains to be seen' (Watson and Crick 1953b). Thus, if at all any enzyme was involved, its role was envisaged to only carry out a 'zipper' action.

It was Arthur Kornberg who, however, was convinced that 'all biosynthetic and degradative events are catalysed by enzymes, making possible refinements of control and specificity and rapid rates of reaction' (mentioned in Kornberg 1980). Does not matter how complex the process is involving four different nucleotides and a template DNA that has to be copied faithfully, as long as it is a biological process, it has to be catalysed by an enzyme. An enzyme was indeed isolated from *Escherichia coli* which appeared to be synthesizing DNA (Lehman *et al.* 1958). The enzyme was called DNA polymerase, but since subsequently other enzymes were found, we would call it DNA polymerase I.

9.2 DNA polymerase I from *E. coli*

The following lists what is so special about DNA polymerase.

- **Substrate:** Before one goes about exploring for an enzyme, one must be able to decide on the correct substrate(s). DNA is a polymer of four different deoxyribonucleoside monophosphates linked by 5′–3′ phosphodiester linkages (see Chapter 3). Lehman *et al.* (1958) were guided by what was known then on the synthesis of other nucleotide derivatives such as coenzymes. These were synthesized by a mononucleotidyl transfer from a nucleoside 5′-triphosphate. So, they used ^{32}P-labelled 5′-triphosphates of thymidine, deoxyadenosine, deoxyguanosine and deoxycytidine as substrates. With a natural DNA template, addition of all the four deoxyribonucleoside 5′-triphosphates was found essential for the polymerization reaction catalysed by the putative DNA polymerase. The diphosphates or the 3′-triphosphates did not serve as substrates. Dinucleotides or longer oligonucleotides also did not serve as subsrates. The ribonucleoside triphosphates were not good substrates either (Lehman *et al.* 1958).
- **Enzyme:** The enzyme was initially purified more than 2000 fold from the soluble fraction of a lysate of *E. coli* (Lehman *et al.* 1958). The enzyme has subsequently been purified to homogeneity (more than 5000-fold purification) and characterized thoroughly (Richardson *et al.* 1964). The enzyme consists of a single polypeptide containing 911 amino acid residues and has a molecular weight of 102,000 Da (Kornberg 1980; Kornberg and Baker 1992). The isoleucine residue at position 709 has been found to be involved in sugar discrimination in *in vitro* polymerization

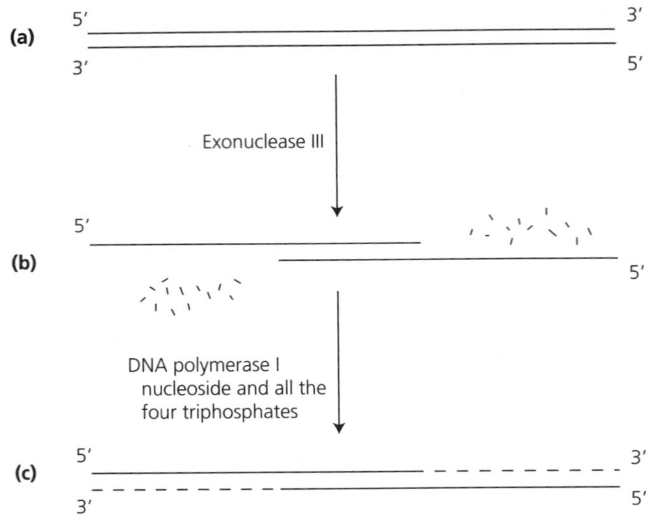

Figure 9.1 A double-stranded DNA molecule with 5′-overhangs as template for DNA polymerase I. *Escherichia coli* Exonuclease III acts on the 3′-ends of the strands of the double-stranded DNA (a), thus producing a molecule with 5′-overhangs (b). Addition of DNA polymerase I along with the four ^{32}P-labelled nucleoside triphosphates results in addition of mononucleotides to the recessed 3′-ends of DNA using the overhanging 5′-ends of the other strand as the template, thus restoring the double-stranded DNA (c). A part of the acid-soluble ^{32}P-radioactivity in the nucleoside triphosphates gets converted into acid-insoluble radioactivity. The newly synthesized portion of DNA is shown by broken lines.

reactions. Site-directed mutagnesis of the gene for DNA polymerase I has been used to replace isoleucine at position 709 of the enzyme protein by methionine, asparagine, phenylalanine or alanine. Sugar selectivity has been tested in an *in vitro* DNA-synthesizing system by substituting dGTP with rGTP. All the four mutants had poor sugar selectivity, the enzyme with phenylalanine at position 709 being the poorest (Shinkai *et al.* 2001).

- **DNA template:** A DNA template was necessary for the reaction. In the absence of a template, deoxypolynucleotides with a repetitive base sequence were formed *de novo* after a long lag. This activity has been attributed to oligonucleotide impurity in enzyme preparation and reiterative replication using these oligonucleotides as template (Kornberg *et al.* 1964).
- **DNA primer:** A DNA primer onto which mononucleotides would actually be added by the formation of a phosphodiester linkage was also found to be a strong requirement. A good template–primer combination was a DNA molecule that had 5′-overhangs [Fig. 9.1(b)]. Such molecules could be obtained by partial treatment of double-stranded DNA with exonuclease III or any other exonuclease which specifically cleaved from the 3′-end of double-stranded DNA [Fig. 9.1(a) and (b)]. A single-stranded DNA molecule could also serve as a good template–primer combination if self-complementarity of bases was possible at the 3′-end. Single-stranded dAT copolymer is an example. The flexibility of single-stranded DNA allows the 3′-end of the molecule to bend on itself and form a hydrogen-bonded hairpin-like structure. This end then serves as a primer on to which monnucleotides are added using the other end of the DNA as template (Fig. 9.2).
- **Divalent cation:** Addition of Mg^{++} was necessary, but Mn^{++} could also be used under certain conditions.
- **Measure of polymerization:** The reaction can be terminated by addition of chilled perchloric acid. Conversion of acid-soluble radioactivity in ^{32}P-labelled deoxyribonucleoside triphosphates into acid insoluble radioactivity was taken as a measure of DNA synthesis by the isolated enzyme (Lehman *et al.* 1958). While mononucleotides have been found to be soluble in chilled perchloric acid or trichloroacetic acid solution, polynucleotides of 10 or more residues have been found to be insoluble in these acids.
- **Reaction:** A polymer of nucleoside monophosphates is formed from nucleoside triphosphate substrates by addition of nucleoside monophosphates to the DNA primer as a consequence of the nucleophilic attack of the free hydroxyl group in the primer on the nucleotidyl phosphorus in the substrate with the elimination of inorganic pyrophosphate (PP). Pyrophosphorolysis catalysed by DNA polymerase I or reversal of the

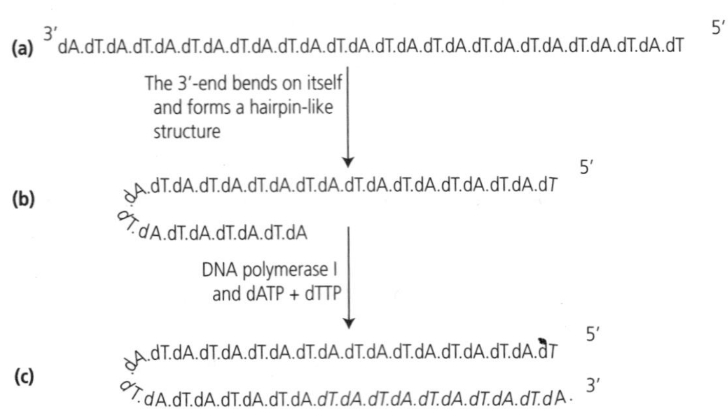

Figure 9.2 Single-stranded dAT copolymer can serve as a template–primer combination for synthesis of DNA-like polymer by DNA polymerase I. (a) The single-stranded dAT copolymer. (b) The 3′-end of the molecule bends on itself and forms a hydrogen-bonded hairpin-like structure, which serves as a template primer combination. (c) Addition of dA and dT to the 3′-end when acted upon by DNA polymerase I and dATP and dTTP. The newly synthesized portion is shown in italics.

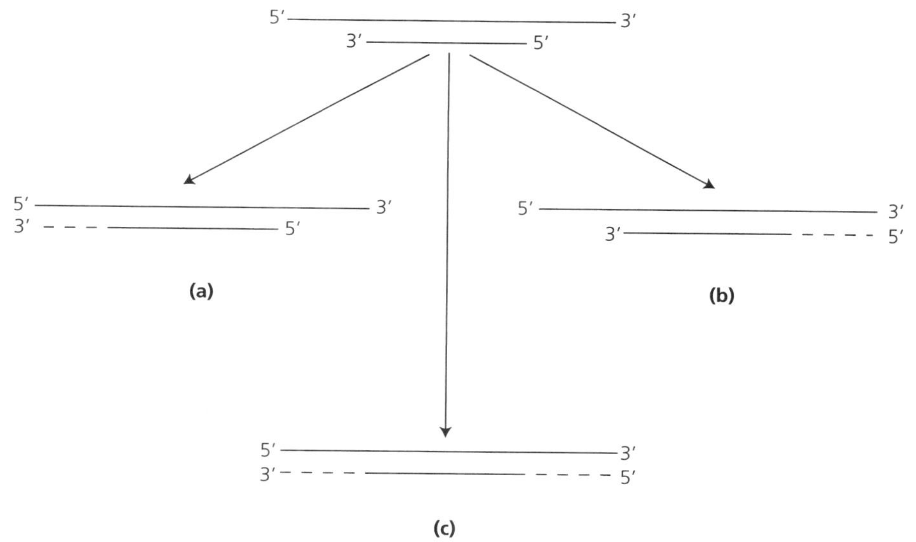

Figure 9.3 Possibilities of addition of mononucleotides to the primer strand when the template strand has both a 5′-overhang and a 3′-overhang. The newly synthesized portion of DNA is shown by broken lines. (a) Addition only to the recessed 3′-end of the primer strand. (b) Addition only to the recessed 5′ -end of the primer strand. (c) Addition to both the recessed 3′-end and the recessed 5′-end of the primer strand.

synthetic reaction can indeed be demonstrated in the presence of inorganic pyrophosphate at concentrations much higher than that of nucleoside triphosphate substrates (Deutscher and Kornberg 1969). The synthetic reaction can be rendered irreversible if an enzyme that cleaves inorganic pyrophosphate to inorganic phosphate (pyrophosphatase) is added to the reaction mixture.

- **Site of addition of mononucleotides:** DNA polymerase I adds on by phospghodiester linkage mononucleotides to the primer at the recessed 3′-ends [Fig. 9.1(b)] using the 5′-overhang of the other strand as template by following the complementary base pairing rule of Watson and Crick [Fig. 9.1(b) and (c)]. A completely double-stranded DNA molecule is thus restored. That during the polymerization reaction, nucleotides are added only to the 3′-end and never to the 5′-end of the primer has been confirmed. The polymerization reaction was carried out by using ^{32}P-labelled deoxyribonucleoside triphosphates. The product of polymerization was purified free of radioactive mononucleotides by gel filtration and treated separately with exonucleases specific for the 3′-end (Exonuclease III from *E. coli*) and specific for the 5′-end (*Bacillus subtilis* exonuclease). The template–primer combination used for this experiment had both 5′- and 3′-overhang, as shown in Fig. 9.3. There are three possibilities, the polymerization reaction results in the addition of mononucleotides to the primer at the 3′-end only (a), at the 5′-end only (b) or at both the 3′-end and the 5′-end (c). The newly synthesized portion of DNA has been shown by the broken line [Fig. 9.3(a)–(c)]. The course of release of acid-soluble radioactivity on treatment of the product with the 3′-end–specific Exonuclease III from *E. coli* was as shown in Fig. 9.4(a), while the course of release of acid-soluble radioactivity on treatment with the 5′-end–specific *B. subtilis* exonuclease was as shown in Fig. 9.4(b). These results furnished conclusive evidence that in the reaction catalysed by DNA polymerase I, nucleotides were added only to the 3′-end of the primer. Addition to the 5′-end would have yielded opposite results. Addition to both the 3′-end and the 5′-end would have resulted in a situation shown in Fig. 9.4(c), irrespective of whether a 3′-exonuclease or a 5′-exonuclease was used. This was never seen.

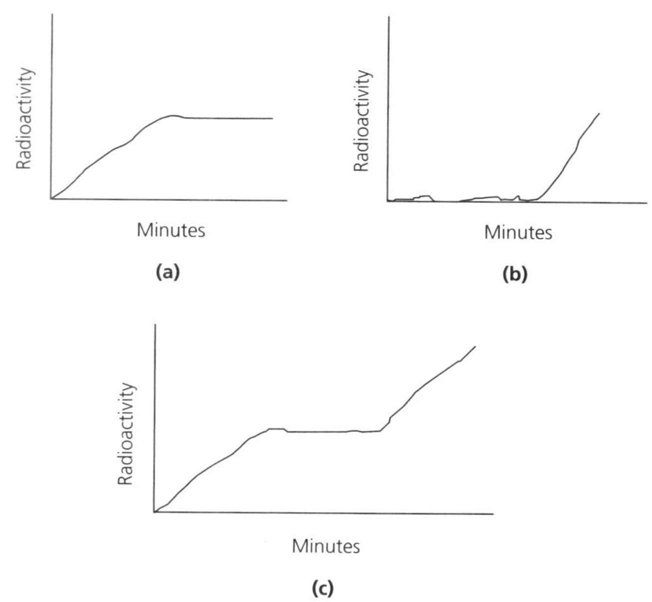

Figure 9.4 Release of acid-soluble radioactivity from labelled DNA on treatment with the 3′-end-specific *Escherichia coli* Exonuclease III and the 5′-end–specific *Bacillus subtilis* exonuclease. ^{32}P-Labelled deoxyribonucleoside triphosphates were used to carry out the polymerization reaction catalysed by DNA polymerase I. The template used was the same as shown in Fig. 9.3. The product of synthesis was purified free of any acid-soluble mononucleotide by gel filtration and then treated with the specific exonuclease. (a) Synthesis product treated with 3′-end–specific *E. coli* Exonuclease III. (b) Synthesis product treated with 5′-end–specific *B. subtilis* exonuclease. (c) Hypothetical situation when addition of mononucleotides had taken place at both the 3′-end and the 5′-end of the primer strand and either of the 3′-end–specific or the 5′-end–specific exonuclease had been used.

- **Product:** The product produced by DNA polymerase I exactly resembles DNA, more particularly the template–primer combination used in the reaction, as revealed by the analysis of base composition and nearest-neighbour frequency.
- **Fidelity:** When poly dAT was used as template–primer and ^{32}P-labelled deoxyguanosine triphospate was added in equimolar proportion along with deoxyadenosine triphosphate and thymidine triphosphate as substrates, incorporation of the guanine nucleotide could not be detected even when 10^5 molecules of adenine nucleotide and thymidine nucleotides were incorporated (Kornberg 1980).
- **Processivity:** Processivity value of a polymerase is a measure of the number of nucleotides added to the primer before dissociating from the template–primer. This value for DNA polymerase I is 11–13 when an oligo(dT):poly(A) primer–template system [as in Fig. 9.9(b)] is used, 170–200 in a poly dAT system [Fig. 9.2(b)] and 45–55 if a plasmid (ColE1) with a single-strand gap is used as a template–primer (Kornberg and Baker 1992).

9.3 *In vitro* synthesis of infectious virus ϕx174 DNA using DNA polymerase I

The final proof of fidelity of DNA polymerase I would be its ability to synthesize *in vitro* a biologically active DNA molecule. This has been achieved. The 5386-nucleotide-long single-stranded circular DNA molecule, which is the genome of the bacterial virus ϕx174, has been completely synthesized *in vitro* and this DNA molecule has been found to be able to infect its host *E. coli* (Goulian and Kornberg 1967; Goulian *et al.* 1967). The ϕx174 genome is known to first form a double-stranded DNA molecule by synthesizing the complementary circular strand inside the host. This double-stranded DNA is known as the replicative form or RF. The viral DNA is the (+) strand and the complementary DNA the (–) strand. More + strands are subsequently formed *in vivo* and packaged into the viral coat.

The following steps, schematically described in Fig. 9.5 (Goulian *et al.* 1967), were executed for the *in vitro* synthesis:

1. *E. coli* was infected with the ϕx174 virus in a medium containing ^{3}H-labelled thymidine, deoxyadenosine, deoxyguanosine and deoxycytidine. The virus particles were purified and the ^{3}H-labelled viral DNA (+ strand) was isolated.
2. The ^{3}H-labelled viral DNA (+ strand) was then incubated with DNA polymerase I, triphosphates of deoxyadenosine, deoxyguanosine and deoxybromouracil and ^{32}P-labelled triphosphate of deoxycytidine. Deoxybromouracil can base-pair with deoxyadenosine and has been used instead of thymidine to give the product a density label. A boiled extract of *E. coli* was included to supply fragments of DNA that could serve as the primer for the polymerization reaction. A polynucleotide-joining enzyme was added as well to join the ends of the newly synthesized (–) strand and form a circle. Formation of double-stranded circular DNA was confirmed by analysis in an alkaline sucrose density gradient when these molecules sedimented faster than the (+) strand of the viral DNA and also by electron microscopy (Goulian and Kornberg 1967).
3. The polymerized product was purified by gel filtration and subjected to a mild treatment with bovine pancreatic deoxyribonuclease to an extent sufficient to have single cleavage in just one of the strands in about 50% of the molecules. Thus, about half of the double-stranded circles would have a nick in one of the strands, while the other strand would still remain circular.
4. The DNA molecules were then heat denatured. Strand separation would take place in those circular duplexes that had a scission giving rise to a mixture of single-stranded DNA molecules comprising circular (+) strands, linear (+) strands, circular (–) strands and linear (–) strands.
5. The DNA molecules were next fractionated by CsCl equilibrium density gradient centrifugation to separate (a) the thymidine-containing circular and linear (+) single strands that had a buoyant density of 1.722, (b) the bromouracil-containing circular and linear (–) single strands that had a buoyant density of 1.809 and (c) the hybrid double-stranded circular molecules that had a buoyant density of 1.747. The centrifugation was conducted long enough and it was ensured by direct determination of radioactivity that the ^{32}P-labelled (–) strands were not contaminated by the ^{3}H-labelled (+) strands or the hybrids.
6. The (–) strands were collected, and using these as template, the (+) strands were synthesized the same way as in step (2), but ^{32}P-labelled deoxybromouracil triphosphate was replaced by thymidine triphosphate and the labelled substrate was ^{3}H-labelled dCTP. Completely *in vitro* synthesized ϕx174 RF DNA duplexes, some linear and some circular, were thus obtained. The presence of circular DNA duplexes was confirmed, as usual, by analysis in an alkaline sucrose gradient, when the circular duplexes sedimented faster.
7. *In vitro* synthesized ^{3}H-labelled circular (+) strands were obtained from the *in vitro* synthesized ϕx174 RF DNA duplexes, as described in steps (3), (4) and (5).
8. The (+) circular DNA strands thus obtained were indeed found to be capable of infecting *E. coli* and of producing ϕx174 virus as evidenced by the formation of plaques. The (+) linear DNA molecules are not infective.

The results of this experiment would tend to strengthen the candidature of DNA polymerase I as the enzyme responsible for DNA replication *in vivo*, though it does not really establish it. In the words of Arthur Kornberg, 'The fact that *E. coli* DNA polymerase can synthesize biologically active DNA does not establish its function in the replication of the bacterial chromosome. However, the effectiveness of the combined action of the polymerase and the polynucleotide-joining enzyme in forming infective DNA may have considerable significance for chromosomal replication' (Goulian *et al.* 1967).

9.4 Nuclease activities as integral parts of DNA polymerase I

DNA polymerase I has two nuclease activities as its integral parts.

9.4.1 A 3′→5′ exonuclease activity

Completely purified homogeneous DNA polymerase I from *E. coli* contains as an integral part of the enzyme a 3′→5′ exonuclease activity (Lehman and Richardson 1964). This activity can be detected in the following manner. With a suitable template–primer combination, like that shown in Fig. 9.1(b), ^{32}P-labelled deoxyriboside triphosphates are used as substrates for the synthetic reaction and the newly synthesized DNA is purified free of the acid-soluble nucleotides by gel filtration. DNA polymerase I can now release labelled acid-soluble nucleoside monophosphates from the 3′-ends of the synthesized DNA strands, if applied in the absence of deoxynucleoside triphosphates (Fig. 9.6). This exonuclease activity releases only mononucleotides and degrades single-stranded DNA as well.

9.4.2 A 5′→3′ exonuclease activity

Purified DNA polymerase I from *E. coli* also contains as an integral part of the enzyme a 5′→3′ exonuclease activity (Klett *et al.* 1968), which has been detected in the following manner. A single-stranded polydAT copolymer was first synthesized from ^{32}P-labelled dATP and dTTP using DNA polymerase I from *E. coli* in the presence of small synthetic repeating dAdT oligomers. Incorporation of ^{32}P-labelled radioactivity in acid-insoluble product ensured polymerization. The polymer was purified thoroughly from any ^{32}P-labelled dATP and dTTP by gel filtration. DNA polymerase I was again added, but this time along with ^{3}H-labelled d[DAP]TP (deoxyribose 2,6-diaminopurine triphosphate) and dTTP (Klett *et al.*

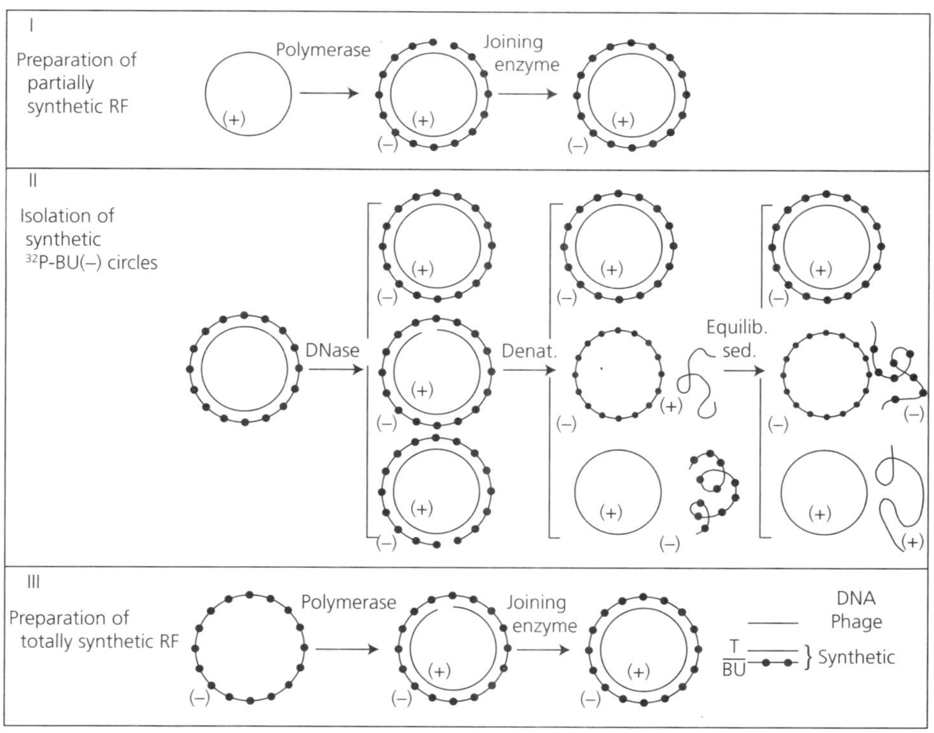

Figure 9.5 Schematic representation of the *in vitro* synthesis of circular ϕx174 DNA(−) strands and RF DNA duplexes by using DNA polymerase I from *Escherichia coli*. Frame I in the figure shows step (2) as described in the text, frame II shows steps (3), (4) and (5); while frame III of the figure shows step (6) of the text. The thin circle in frame I is the ^{3}H-labelled (+) DNA isolated from the virus particles, the beaded thick circle is the bromouracil containing ^{32}P-labelled (−) DNA synthesized *in vitro*. The thick plain circle in frame III is the thymidine-containing ^{3}H-labelled (+) DNA synthesized *in vitro*. [Reproduced from Goulian *et al.*, *Proc Natl Acad Sci USA* 1967;58:2323 with permission from Professor Mehran Goulian.]

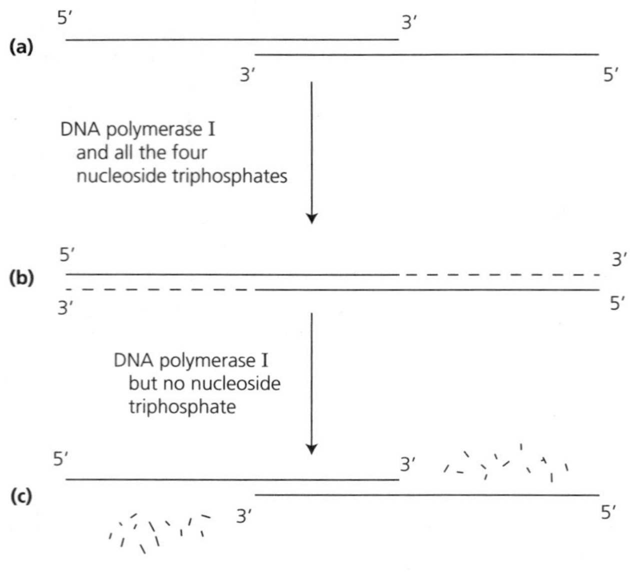

Figure 9.6 Schematic representation of a 3′→5′ exonuclease activity as an integral part of *Escherichia coli* DNA polymerase I. (a) Shown in solid lines is the template–primer combination used, which is similar to molecule (b) in Fig 9.1. (b) Completely double-stranded molecule obtained by employing DNA polymerase I and all the four ^{32}P-labelled nucleoside triphosphates. A part of the acid-soluble ^{32}P-radioactivity in the nucleoside triphosphates gets converted into acid-insoluble radioactivity. The newly synthesized ^{32}P-labelled portion of DNA is shown by broken lines. (c) Result of the 3′→5′ exonuclease activity of DNA polymerase I, which can be apparent when the nucleoside triphosphates are withheld. A lot of acid-insoluble radioactivity is converted into acid-soluble radioactivity.

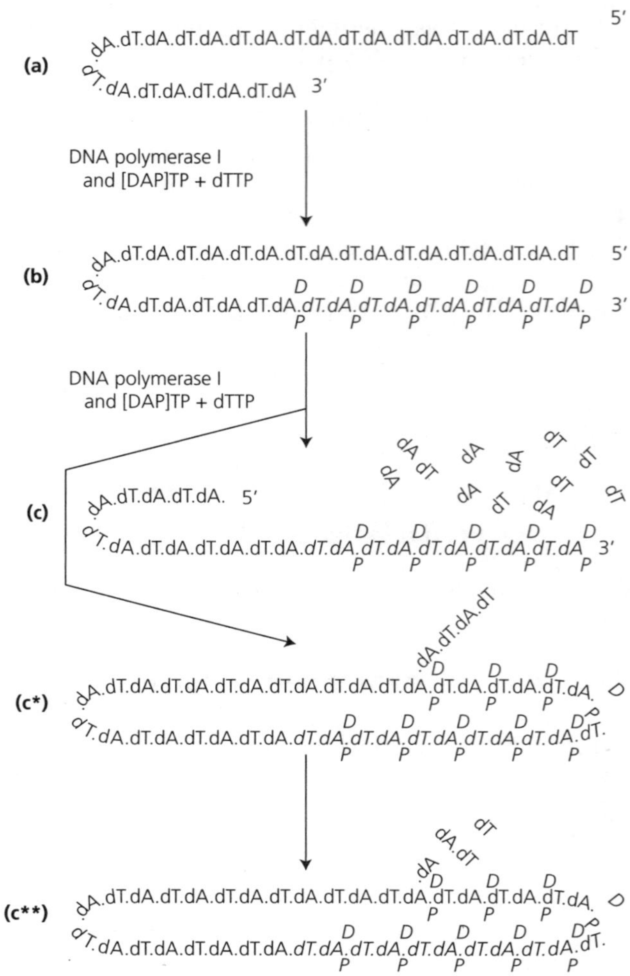

Figure 9.7 Schematic representation of a 5′→3′ exonucleolytic activity as an integral part of *Escherichia coli* DNA polymerase I. (a) A dAT copolymer, the 3′-end of which has bent on itself and formed a hydrogen-bonded hairpin-like structure, thus serving as a template–primer combination. (b) Addition of d[DAP]MP and dTMP to the 3′-end, forming a hybrid hair pin containing a double-stranded molecule. This molecule has dA-dT at the 5′-end and d[DAP]-dT at the 3′-end. (c) The dAMP and dTMP residues have been released from the 5′-end of the hybrid molecule due to the 5′-exonucleolytic activity of DNA polymerase I.

1968). Because of the possibility of the 3′-end of the dAT copolymer bending on to itself and forming a template primer complex [Fig. 9.7(a)], a hair pin containing a hybrid double-stranded DNA product, as schematically shown in Fig. 9.7(b), was expected. This product was detected and purified by CsCl density gradient centrifugation, taking advantage of the difference in the buoyant density imparted by the incorporation of d[DAP]P. The buoyant density of pure dAT copolymer was found to be 1.679, while that of pure d[DAP]T copolymer was 1.718. The hybrid molecule banded with an intermediate buoyant density depending on the proportion of dA and d[DAP] in the product (Klett *et al.* 1968). While the dAT copolymer [Fig. 9.7(a)] contained ^{32}P-radioactivity, the hybrid dAT-d[DAP]T copolymer containd both ^{32}P and ^{3}H-radioactivity. If this hybrid molecule was treated again with DNA polymerase I along with d[DAP]TP and dTTP, then it was exonucleolytically degraded from the 5′-end, apparently in a manner shown in Fig. 9.7(c). ^{32}P-labelled dAMP and dTMP were released in the acid-soluble fraction and the buoyant density of the residual polymer gradually increased till it finally attained the value 1.718. The result of an actual experiment is shown in Fig. 9.8 (Klett *et al.* 1968). It is obvious that DNA polymerase I has a nucleolytic activity that cleaves the DNA from the 5′-dAT end. This nuclease activity has been found to not only release mononucleotides but also small oligonucleotides.

The significance of certain observations was not clear when the 5′→3′ exonuclease activity was first discovered. Why is the addition of d[DAP]TP and dTTP necessary along with the polymerase to elicit exonuclease activity? Also, only double-stranded DNA molecules that are amenable to the polymerase reaction are attacked by the 5′→3′ exonuclease and single-stranded DNA that cannot fold on itself and cannot form intrastrand hydrogen bonds serving as template–primer for DNA polymerase is not acted upon by this nuclease. It has now been revealed that there is close coordination

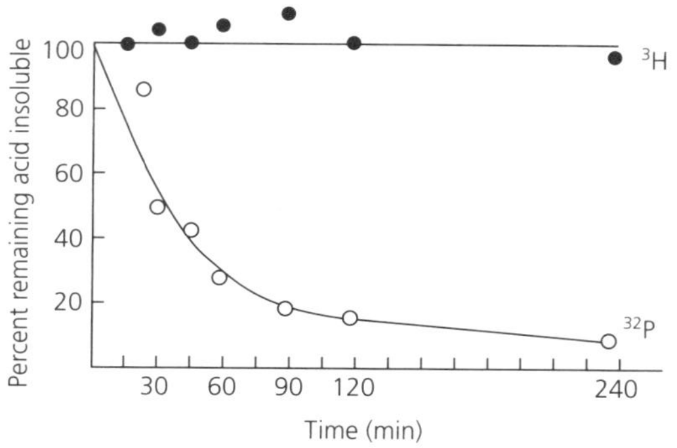

Figure 9.8 Action of DNA polymerase I on 5′-dAdT-P^{32}/3′-d[DAP] dT-H^3 hybrid in the presence of d[DAP]TP and dTTP. [Adapted from Klett *et al.*, *Proc Natl Acad Sci USA* 1968;60: 947 with permission from Professor Edward Reich.]

between the polymerase activity and the 5′→3′ exonuclease activity, even though the two active sites are physically separate and operate independently. In fact, the preferred substrate for the 5′→3′ exonuclease is a 'double-flap' structure having a base or bases at the primer terminus overlapping the displaced strand that is to be cleaved by the exonuclease (Xu *et al.* 2000). The actual situation is more like what is represented in Fig. 9.7(c*) and (c**) rather than in Fig. 9.1(c).

9.5 Significance of nuclease activities associated with DNA polymerase I

In vitro experiments with synthetic DNA polymers have revealed that the 3′→5′ exonuclease activity associated with DNA polymerase I was specifically directed against mispaired or unpaired bases at the 3′-terminus, particularly in the presence of the relevant deoxytriphosphates. Addition of new bases at the 3′-end began only after all mispaired bases were first removed and a properly paired base was reached (Brutlag and Kornberg 1972), suggesting that the function of this activity is to 'proofread' the incorporation of the correct nucleotide by DNA polymerase I (Brutlag and Kornberg 1972). A schematic representation is given in Fig. 9.9. In case of chance incorporation of a nucleotide that does not obey the base pairing rule of Watson and Crick and hence cannot form proper hydrogen bonds, the 3′→5′ exonuclease activity would first remove it and then the right nucleotide would be incorporated by the polymerase. This is thus a contributing factor towards the high fidelity of DNA polymerase I.

In vivo experiments have also confirmed the contribution of the proofreading 3′→5′ exonuclease activity towards the fidelity of DNA polymerase I. It has been revealed in subsequent sections of this chapter that DNA polymerase I is not necessary for the replication of the chromosome of *E. coli*. Nevertheless, it has been found that it is this DNA polymerase I that is essential for the replication of the plasmids which have been derived from Col E1 (Kingsbury and Helinski 1970), even though the host for these plasmids is *E. coli* itself. The frequency of reversion of a point mutation in the

β-Lactamase gene contained in a Col E1–derived plasmid has been determined as a measure of fidelity of replication of the plasmid, ostensibly by DNA polymerase I. Higher reversion would imply lower fidelity of DNA replication. Two strains of *E. coli* have been used as host of the plasmid, one having the wild-type DNA polymerase I and wild-type 3′→5′ exonuclease function and the other having the wild-type DNA polymerase I but lacking the 3′→5′ exonuclease function, arising out of a mutation in the relevant domain. A four-fold enhancement in the reversion frequency was observed in the strain in which the 3′→5′ exonuclease function was abolished (Shinkai and Loeb 2001). The reversion frequency was monitored by scoring colonies that reverted to carbenicillin resistance from the initial carbenicillin sensitivity because of the original mutation in the β-lactamase gene. It was, however, pointed out that the reversion frequency was enhanced 5–23 fold if the residue Ile[709] in the DNA polymerase I active site motif

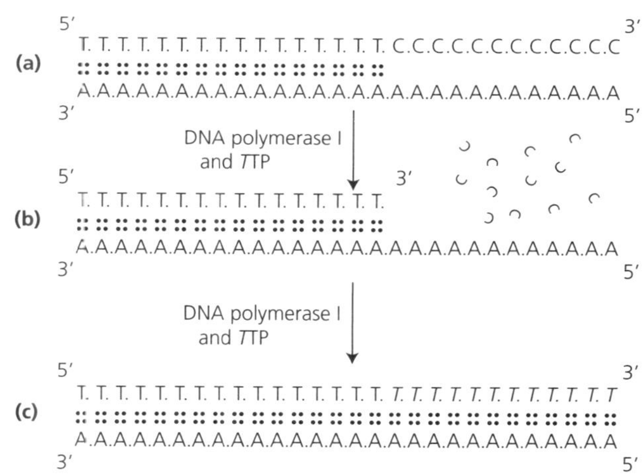

Figure 9.9 Schematic representation of the experiment by Brutlag and Kornberg (1972) on the proofreading function of the 3′→5′ exonuclease activity associated with DNA polymerase I. (a) A segment of dA/dT duplex with a stretch of mismatched dCs at the 3′-end of the dT strand. The phosphodiester linkages between the dTs had ^{32}P and the dCs contained ^{3}H label. (b) Removal of the mismatched dCs by the proofreading function of the 3′→5′ exonuclease activity. The ^{3}H-label was rendered acid soluble and paper chromatography has revealed that the ^{3}H-labelled dCs were removed as mononucleotides. Very few dTs (containing ^{32}P) were cleaved in the presence of dTTP. (c) Incorporation of dTs from the ^{14}C-labelled dTTP (acid soluble) added along with DNA polymerase I. The newly incorporated dTs have been shown in italics. A part of the ^{14}C label was converted into an acid-insoluble form.

A was mutated to Asn or Phe or Ala, in spite of the fact that the 3′→5′ exonuclease function was left unaltered (Shinkai and Loeb 2001). The combination of mutation in Ile^{709} and in the 3′→5′ exonuclease domain, of course, enhanced the reversion frequency by 400 fold. Thus, though the 3′→5′ exonuclease function contributes a lot to the fidelity of replication, accurate discrimination of the incoming nucleotide at the active centre of DNA polymerase I is also very important for the fidelity of polymerization *in vivo* (Shinkai and Loeb 2001).

The 5′→3′ exonuclease activity of DNA polymerase I has been found to be capable of excising thymine dimers and other mismatched sequences in DNA (Kelly *et al.* 1969). This is in spite of the fact that the 5′→3′ exonuclease activity is otherwise specific for hydrogen-bonded double-stranded regions.

9.6 Thymine dimers

Thymine dimers are formed when cells get exposed to ultraviolet (UV) rays (wave length 200–400 nm). Purine or pyrimidine bases may be transformed into an excited state by UV

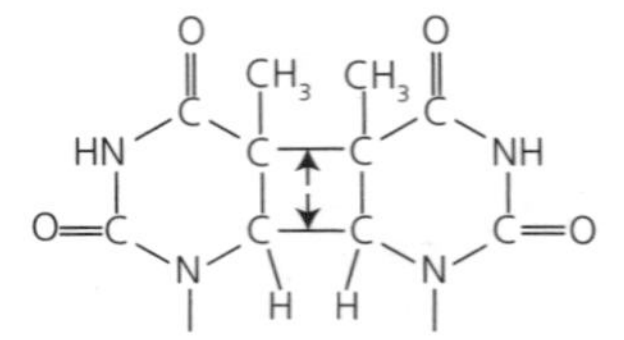

Figure 9.10 A thymine dimer formed from two adjacent thymines in the same DNA strand by the formation of a cyclobutane ring through two new C—C covalent bonds (shown by arrows) as a result of irradiation by ultraviolet rays.

radiation. Adjacent excited thymine residues in a DNA strand have often been found to interact with each other, forming a cyclobutane ring, by covalently connecting the two thymines through two new C—C bonds (shown by arrows; Fig. 9.10). Formation of thymine dimers distorts the double helix and the hydrogen bonds between these thymines and the adenines on the other strand are disrupted.

Kelly *et al.* (1969) constructed a DNA duplex with a mismatched region. A penta-dC was labelled at the 5′-end with polynucleotide kinase and γ ^{32}P-labelled ATP. About 200 residues of ^{3}H-labelled dT residues were added to the 3′-end of

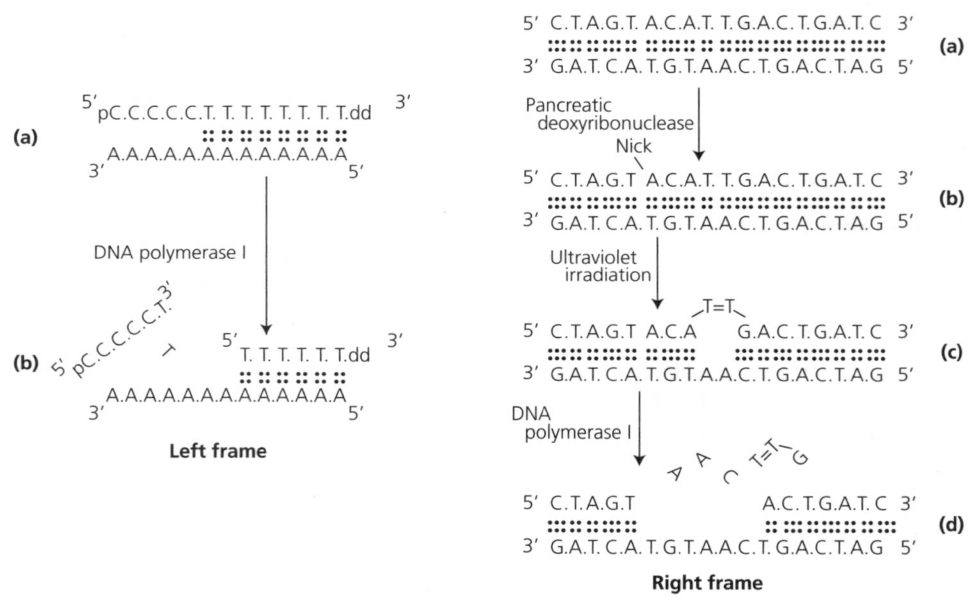

Figure 9.11 Schematic representation of the experiment by Kelly *et al.* (1969) on excision *in vitro* of mismatched segments and thymine dimers from DNA by the 5′→3′ exonucleolytic activity of DNA polymerase I. Left frame: (a) The p at the 5′-end of penta-dC of the top strand represents a ^{32}P and the Tdd at the 3′-end of oligo-dT represents dideoxy thymine. In the actual experiment, there were about 200 dT residues. All the dTs were labelled with ^{3}H. (b) pdCdCdCdCdCdT has been cleaved off as a block from the 5′-end of the top strand along with a few mono dTs by the 5′→3′ exonucleolytic activity of DNA polymerase I. These have been detected by paper chromatography. No cleavage has taken place from the 3′-end of the top strand by the 3′→5′ exonucleolytic activity, because of the presence of the dideoxy thymine there. Cleavage from the 5′-end and the 3′-end of the bottom strand must have also taken place, but has not been shown here as this was not detectable because of the absence of any label in the bottom strand. Right frame: (a) A segment of *Escherichia coli* DNA containing two adjacent thymines. All the thymines in the DNA have been labelled with ^{3}H. (b) The DNA has been subjected to mild treatment with pancreatic deoxyribonuclease for producing single-strand nicks at random to allow the 5′→3′ exonucleolytic activity of DNA polymerase I to initiate action. (c) The DNA was irradiated with ultraviolet rays until about 6% of its thymines have been converted into thymine dimers. (d) The DNA was then treated with DNA polymerase I when 5′→3′ exonucleolytic degradation was initiated at the nick and the thymine dimer was removed along with a few other bases. The excised ^{3}H-labelled thymine dimer was detected by paper chromatography.

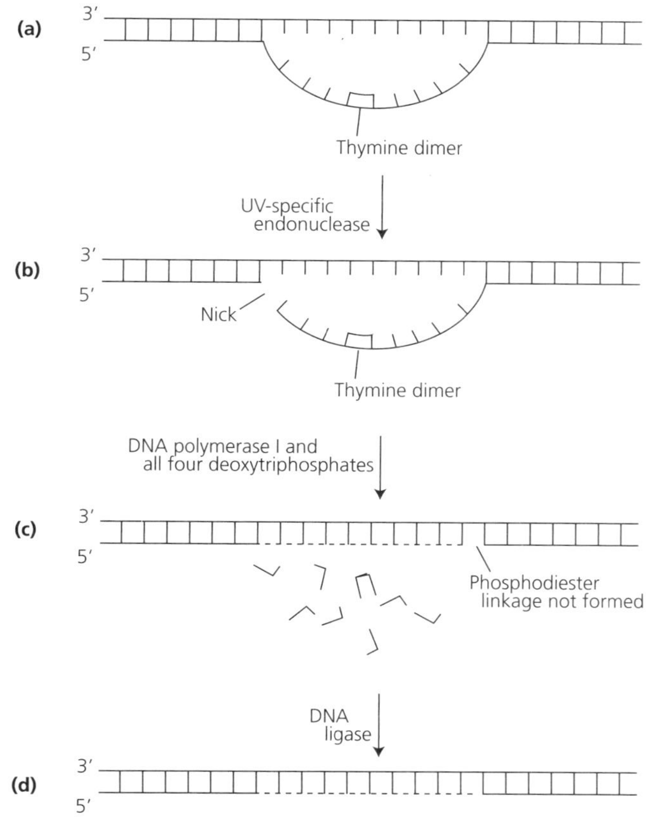

Figure 9.12 Schematic representation of *in vivo* excision of a thymine dimer and repair synthesis by DNA polymerase I. (a) A segment of a distorted DNA duplex because of the presence of a thymine dimer is shown at the top. (b) Cleavage of the phosphodiester bond close to and 5′ to the distorted region by ultraviolet-specific endonuclease. (c) Excision of the thymine dimer by the 5′→3′ exonucleolytic activity of DNA polymerase I and repair synthesis by DNA polymerase I. The broken line represents the newly repair synthesized DNA segment. The last phosphodiester linkage is yet to be formed. (d) Formation of the last phosphodiester linkage by ligation (joining) of the DNA ends by the joining enzyme, DNA ligase.

the penta dC from ^{3}H-dTTP using terminal deoxynucleotidyl transferase and this was terminated with a dideoxy thymidine. This strand was hybridized with unlabelled poly dA. The product is schematically shown in the left frame at (a) of Fig. 9.11. The dideoxy thymidine at the 3′-end would prevent any cleavage and release of ^{3}H-dT from that end by the 3′→5′ exonuclease activity. Treatment of this construct with DNA polymerase I released ^{32}P-labelled penta-dCdT and some ^{3}H-labelled mono dTs from the 5′-end because of the 5′→3′ exonuclease activity. The released hexaanucleotide and mononucleotides were detected by paper chromatography. No mono-, di-, tri-, or tetra-dC was released.

For the excision of thymine dimers *in vitro, E. coli* DNA labelled with ^{3}H-thymidine was treated mildly with bovine pancreatic deoxyribonuclease to produce single-strand nicks. The DNA was then irradiated with UV rays till about 6% of the thymine residues were converted into thymine dimers. DNA polymerase I, when added to this DNA, initiated the 5′→3′ exonucleolytic activity at the nicks and excised out the thymine dimers (schematically shown in Fig. 9.11, right frame), as detected by paper chromatography. The gap thus formed could be filled up if all the four deoxytriphosphates were added along with DNA polymerase I. The ultimate phosphodiester linkage could also be formed if a joining enzyme (DNA ligase) was included. Inside the cell, however, the nicking of irradiated DNA close to the thymine dimer is done by a UV-specific endonuclease, while excision and subsequent repair synthesis is carried out by DNA polymerase I (schematic representation in Fig. 9.12).

9.7 DNA ligase: The joining enzyme

DNA polymerase I can form a phosphodiester linkage between a mononucleotide and a polynucleotide (primer), but never between two polynucleotides. The joining enzyme,

Figure 9.13 Circularization of the linear form of DNA of the bacterial virus lambda (λ) by the joining enzyme, DNA ligase. (a) Linear form of DNA of lambda (λ) virus. The double-stranded DNA contains 48,502 base pairs with single-strand overhangs at the two 5′-ends. The base sequences of these two overhangs are complementary to each other such that these bases can form hydrogen bonds when annealed. (b) Annealed lambda (λ) DNA. The DNA is circularized as a result of the formation of hydrogen bonds between the bases at the 5′-overhangs. The individual circles are not covalently linked as the phosphodiester linkages are missing between the 5′- and the 3′-end of the strands. (c) Formation of phosphodiester linkages and conversion into covalently linked circles mediated by the joining enzyme, DNA ligase.

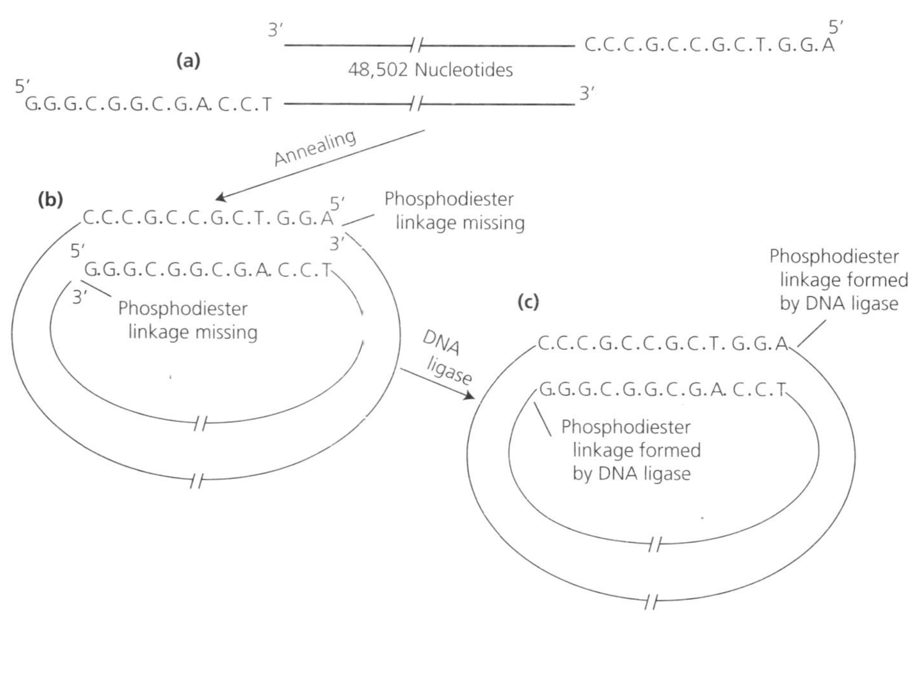

DNA ligase, can join two polynucleotides by the formation of a phosphodiester linkage. The substrate chosen for the search of the joining enzyme was the DNA of the bacterial virus lambda (λ), which has two 5′-end overhangs of single-stranded regions comprising 12 bases each that are exactly complementary to each other (Fig. 9.13). When this DNA molecule is annealed, a double-stranded circular molecule is formed through the formation of hydrogen bonds between the two complementary 5′-end overhangs (Fig. 9.13). DNA ligase can convert this 'hydrogen-bonded circle' into a 'covalently closed circle'. The DNA ligase isolated from *E. coli* (Gellert 1967; Olivera and Lehman 1967a; Olivera and Lehman 1967b; Zimmerman *et al.* 1967) needs NAD (nicotinamide adenine dinucleotide) in addition to ATP. The requirement for the ligation (joining) reaction is a phosphate group at the 5′-end and a hydroxyl group at the 3′-end. DNA ligase cannot add mononucleotides. This enzyme can form a phosphodiester linkage between two DNA single strands only if the linkage across in the other strand of the DNA duplex is intact (Fig. 9.14). The joining enzyme (also called sealase) isolated from the T4 virus–infected *E. coli* (Weiss and Richardson 1967), however, needs only ATP, and no NAD, and forms a phosphodiester linkage between two DNA strands even though the linkage across in the other strand is also missing (Fig. 9.14). This enzyme can thus carry out what is commonly known as 'blunt-end ligation'. An enzyme with similar properties was also described in virus T7–infected *E. coli.* (Becker *et al.* 1967).

A second NAD-dependent DNA ligase of similar amino acid sequence and similar enzymatic properties has now been demostrated in *E. coli* and has been named LigB (Sriskanda and Shuman 2001). Both the NAD-dependent DNA ligase (LigA) and the ATP-dependent DNA ligase (LigD) have now been found in many bacterial species (Gong *et al.* 2004). The LigD enzymes are thought to have a DNA-modifying function catalysed through their ATP-dependent DNA ligase domain fused to a nuclease and/or a DNA polymerase/primase domain (Weller and Doherty 2001; Zhu and Shuman 2005). DNA ligases exist in mammalian cells, too (Lindahl and Barnes 1992).

DNA ligase has also been dealt with in Chapters 17 and 20.

9.8 Terminal deoxynucleotidyl transferase

The 58,000-Da enzyme can be purified from calf thymus gland and bone marrow tissue. The 32,000-Da product of partial proteolysis is fully active. The enzyme uses deoxynucleoside triphosphates as substrates and adds on mononucleotides to the 3′-end of a DNA primer. Neither any template strand is necessary, nor the addition of nucleotides is directed by any template. Even the intact enzyme is not associated with any exonucleolytic activity. One of the isoforms of terminal deoxynucleotidyl transferase of mouse, which happens to be the isoform conserved across the vertebrate phylum, has been found to be unable to discriminate between ribonucleotides and deoxyribonucleotides for incorporation (Boule *et al.* 2001). The amino acid sequence of human terminal deoxynucleotidyl transferase has considerable

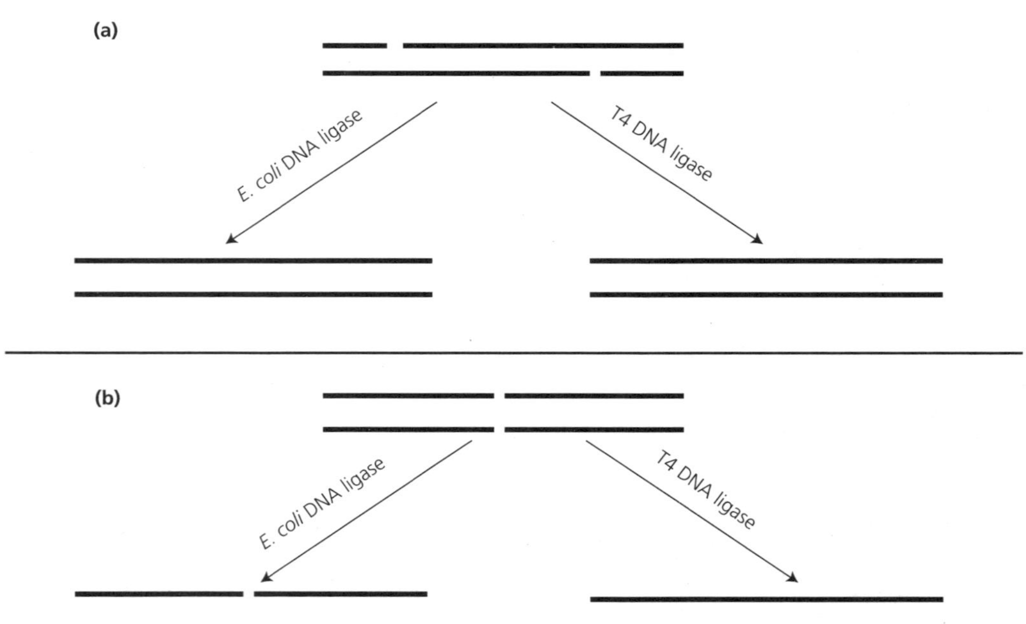

Figure 9.14 Schematic representation of the distinctive functions of *E. coli* DNA ligase and T4 DNA ligase. (a) Ligation of DNA with single-strand breaks. Across the single-strand break in the other DNA strand, the phosphodiester linkage is intact. Both the DNA ligases can join the DNA strands in this situation. (b) Ligation of DNA with double-strand breaks or 'flush ends'. Only T4 DNA ligase can join the DNA molecules in this situation. The efficiency of ligation is, however, less than in the case of single-strand breaks.

homology with human DNA polymerase μ (Aoufouchi *et al.* 2000).

What is the physiological function of terminal deoxynucleotidyl transferase (TdT) in a mammalian cell? There is good evidence to suggest that TdT is involved in creating a diversity of immunoglobulins and antigen receptors. It has been found that lymphocytes isolated from mice whose gene for TdT has been knocked out had no N regions in their variable-region genes (Gilfilan *et al.* 1993; Komori *et al.* 1993). The N region, which adds to the diversity in the germline-encoded repertoire, has been thought to be the outcome of template-independent addition of nucleotides at gene segment junctions during rearrangement of the variable (V), diversity (D) and joining (J) segments prior to the assembly of immunoglobulin and T-cell receptor variable-region genes.

9.9 Polynucleotide kinase

A dimer or tetramer of an identical 34,000-Da protein, polynucleotide kinase, can be isolated from *E. coli* infected with T even viruses. The enzyme phosphorylates the free 5′-OH end of DNA, RNA or oligonucleotides. The phosphate group can be donated by ATP or any other NTP. DNA kinases and RNA kinases have been identified from eukaryotes.

9.10 Klenow fragment

Proteolytic cleavage of DNA polymerase I yields two fragments, the larger one carries out the polymerase activity and the 3′→5′ exonuclease activity, while the smaller fragment harbours the 5′→3′ exonuclease activity.

Arthur Kornberg and his associates had observed that if a pure preparation of DNA polymerase I from *E. coli* was stored at 0°C for 4 weeks, it lost the 5′→3′ exonuclease activity, though the polymerase activity and the 3′→5′ exonuclease activity were unaffected (Brutlag *et al.* 1969). They also reported that if the freshly isolated enzyme was treated with an extract of *B. subtilis*, two fragments of 76,000 Da and 36,000 Da could be recovered, the larger fragment containing only the polymerase activity and the 3′→5′ exonuclease activity (Brutlag *et al.* 1969). Soon afterwards, Klenow and Henningsen (1970) also reported the elimination of the 5′→3′ exonuclease activity by limited proteolysis with subtilisin. The larger fragment is now popularly known as the Klenow fragment. Detailed characterization of the two fragments has subsequently been done by Kornberg and his coworkers (Setlow *et al.* 1972; Setlow and Kornberg 1972). Cleavage with subtilisin in the presence of DNA allowed them to recover the two fragments with full respective activities. All the three functions of the intact enzyme, the polymerization of the deoxyribonucleotides, the 3′→5′ exonuclease activity and the 5′→3′ exonuclease activity, could be restored by combining the two fragments (Setlow and Kornberg 1972).

The Klenow fragment comprises two domains. The ~200-residue N-terminal domain is the seat of the 3′→5′ exonuclease activity, while the ~400-residue C-terminal domain, which contains a deep crevice of the appropriate shape and size for binding double stranded B-DNA, catalyses the polymerase activity (Ollis *et al.* 1985; Beese *et al.* 1993). It is inside this crevice, in the so called J-helix, that amino acid residues exist which regulate both the polymerase and the 3′→5′ exonuclease activity. The asparagine residue at position 678, if replaced by alanine, enhances mismatch-directed 3′→5′ exonuclease activity, leaving the polymerase activity unchanged. Replacement of the glutamine residue at position 677 by alanine, on the other hand, affects the polymerase activity, but does not interfere with the 3′→5′ exonuclease activity. The proline residue at position 680 seems to regulate the processivity. Replacement of proline by either glycine or glutamine abolishes processivity of the Klenow fragment (Tuske *et al.* 2000).

9.11 Proteolytic enzymes

Proteolytic enzymes cleave peptide linkages in proteins by adding a molecule of water. Subtilisin is a proteolytic enzyme isolated from *B. licheniformis*. Other commonly known proteolytic enzymes are trypsin, chymotrypsin, pepsin and papain.

9.12 Reappraisal of the template–primer combinations that are useful for DNA synthesis by DNA polymerase I

A very important characteristic of enzymatic synthesis of DNA is the essentiality of a template and a primer. Some inaccurate notions had, however, crept in at the early stages of study because of inadequate availability of methods of purification of the template, primer, substrates and the enzyme. It soon appeared that an intact double-stranded DNA was unable to serve as a template for DNA synthesis. Such a DNA molecule could serve as a template–primer combination only because of the existence of nicks (broken phosphodiester linkages) in the DNA. The rate of incorporation of labelled nucleotides was indeed proportional to the number of nicks in the DNA (Kelly *et al.* 1970). DNA polymerase I adds on nucleotides to the free 3′-end at a nick, and at the same time, the 5′-exonuclease activity of DNA polymerase initiates degradation from the free 5′-end at that nick. The synthetic reaction is thus dependent on the progressive degradation. On the other hand, the degradation from the 5′-end at the nick was enhanced considerably by the presence of the deoxynucleoside triphosphate substrates of the synthetic reaction (Kelly *et al.* 1970). An exactly similar phenomenon was apparent with double-stranded circular DNA which could serve as a template–primer combination only if nicks existed. As has been mentioned earlier, a double-stranded DNA with 5′-overhangs and a single-stranded DNA whose 3′-end can bend on itself and form hydrogen bonds are good template–primer combinations. A single-stranded circular DNA serves as a good template when a small primer is around (Kelly *et al.* 1970) (schematic representation in Fig. 9.15).

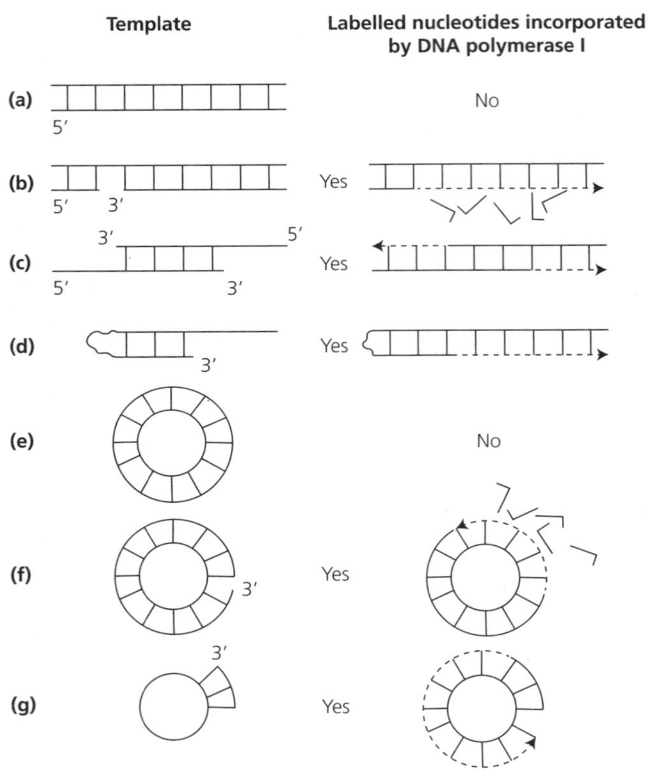

Figure 9.15 Schematic representation of various DNA structures and their ability to serve as templates for DNA synthesis by DNA polymerase I. The solid lines represent the preexisting DNA and the broken lines, the newly synthesized segments. The arrow heads indicate the direction of synthesis. (a) Intact double-stranded DNA is unable to serve as a template. (b) Double-stranded DNA with a nick serves well as a template–primer combination. (c) Double-stranded DNA with 5′-overhangs serves well as a template–primer combination. (d) Single-stranded DNA, the 3′-end of which can bend onto itself, forming hydrogen bonds, functions as a good template–primer combination. (e). Intact double-stranded circular DNA fails to serve as a template. (f) Double-stranded circular DNA with a nick functions as a good template–primer combination. (g) A single-stranded circular DNA with a DNA primer hydrogen bonded to it serves well as a template.

9.13 Temperature-sensitive mutants of *E. coli* that are unable to duplicate their DNA at the non-permissive temperature have active DNA polymerase I in normal amounts at the same non-permissive temperature

A temperature-sensitive mutant of *E. coli* which could not replicate its DNA *in vivo* at 42°C but contained fully active DNA polymerase I in normal amounts at 42°C has been found (Ogawa *et al.* 1968). The mutant also contained normal amounts of all the deoxyriboside triphosphates at 42°C. When single-stranded ϕx 174 DNA was introduced into a similar temperature-sensitive mutant strain, it could not convert the viral DNA into the double-stranded replicative form at 42°C (Steinberg and Denhardt 1968). DNA repair synthesis after ultraviolet irradiation was, however, normal in the mutant strains even at 42°C. These results had cast serious doubts about whether DNA polymerase I was indeed the enzyme responsible for DNA replication *in vivo*. Interestingly, DNA repair synthesis after UV irradiation and normal DNA replication have been observed to differ considerably in the extent to which these two processes discriminate between 5-bromouracil and thymine as the acceptable substrate (Kanner and Hanawalt 1968), thus strengthening the view that if DNA polymerase I was responsible for repair synthesis, some other enzyme was carrying out DNA replication.

9.14 An *E. coli* mutant that lacks DNA polymerase I but can grow normally and replicates its chromosomal DNA perfectly well

A group of summer students in the laboratory of John Cairns have finally set at naught the prospect of DNA polymerase I to serve as the enzyme responsible for chromosomal DNA replication *in vivo*. They tested indiscriminately several thousand bacterial colonies of *E. coli* that were mutaginized with N-methyl-N′-nitroso-N-nitrosoguanidine (1mg/mL) for 30 min. Each colony was separately grown in liquid medium, and a standard protocol was followed for the purification of DNA polymerase I. The presence of the enzyme was checked *in vitro* by the ability of the purified preparation to polymerize deoxyriboside triphosphates (monitored with ^{3}H-labelled TTP) into the acid-insoluble form in the presence of sonicated calf thymus DNA as the template–primer combination. A mutant was detected which had less than 1% of the normal level of DNA polymerase I (de Lucia and Cairns 1969). This mutant grew normally in minimum or complete medium at all temperatures between 25°C and 42°C. It was found later that the mutant strain also replicated its DNA normally, as measured by *in vivo* incorporation of labelled thymidine into its DNA. This mutant, which has come to be known as *pol A1,* was, however, markedly more sensitive to UV rays. This mutant could not support or maintain the replication of plasmids derived from the plasmid Col E1 (Kingsbury and Helinski 1970).

9.15 Determination of sensitivity of bacterial cells towards UV rays

A small amount of bacterial suspension in a petri plate is held at a fixed distance from a short-wavelength ultraviolet lamp and exposed for a fixed time. The suspension is chilled and held in the dark for about half an hour. The bacteria in the suspension are then diluted and a fixed volume is plated on agar in a nutrient medium. The plates are incubated for growth of the bacteria and the number of colonies counted. Fewer colonies would appear if the strain is UV sensitive. An actual experiment is shown in Fig. 9.16.

Another widely used procedure is to streak the suspension on an agar plate containing a growth medium and expose the streak to UV rays for different time periods. The plate is then incubated for growth of the bacteria. Resistant strains would survive through longer exposure, while the sensitive strains would only survive through shorter exposure (Maniatis *et al.* 1982).

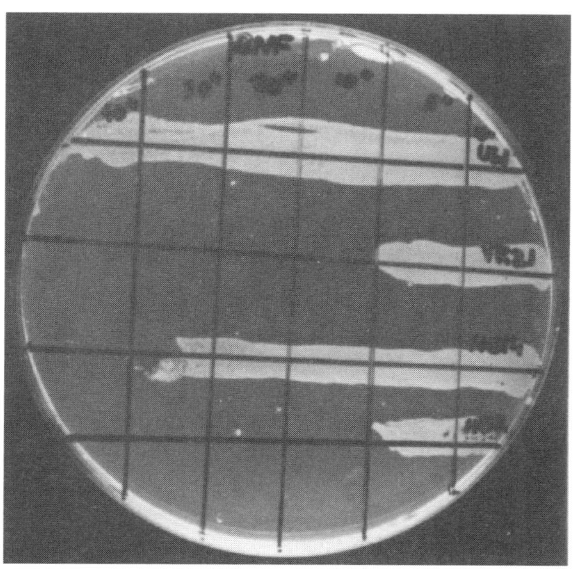

(a)

(b)

(c)

(d)

Figure 9.16 Sensitivity of different bacterial strains towards ultraviolet (UV) rays, as determined by the UV streak test (Maniatis *et al.* 1982). (a) The least sensitive strain – has withstood 50 s of exposure. (b) and (d) Strains that are quite sensitive – these have withstood only 10 s of exposure. (c) Moderately sensitive strain – has withstood 35 s of exposure.

9.16 Discovery of DNA polymerase II from *E. coli*

Several groups initiated search for an alternative DNA polymerase in *E. coli* once it became obvious that the cells can replicate their DNA without DNA polymerase I. All these groups had worked with the *polA1* mutant of de Lucia and Cairns (1969). A second DNA polymerase was soon found in extracts of *E. coli* (Knippers 1970; Kornberg and Gefter 1970; Knippers and Stratling 1970; Moses and Richardson 1970a; Smith *et al.* 1970) and named DNA polymerase II. DNA polymerase II could also be isolated from wild-type *E. coli.* Phosphocellulose chromatography would separate DNA polymerase II from DNA polymerase I (Moses and Richardson 1970b). The enzyme has been purified to homogeneity and consists of a single polypeptide of 90,000 Da molecular weight. DNA polymerase II had many properties similar to DNA polymerase I; it needed all the four deoxyribonucleoside triphosphates, Mg^{++} and native DNA for maximal activity. The product was covalently linked to the primer. DNA synthesis was semiconservative and proceeded from 5′ to 3′. The enzyme did have a 3′→5′ exonucleolytic activity. The distinct properties of DNA polymerase II were its requirement of NH_4^+, its sensitivity to high ionic strength, to sulphydril-blocking agents such as parachloro mercuribenzoate and insensitivity to antibody raised against DNA polymerase I (Moses and Richardson 1970a; Kornberg and Gefter 1971). DNA polymerase II did not have any 5′→3′ exonucleolytic activity (Gefter *et al.* 1972).

9.17 Is DNA polymerase II the enzyme responsible for DNA replication *in vivo*?

In 1970 itself, Moses and Richardson (1970b) had found out that in two independent *E. coli* mutants (CR43-26 and CR266-26) temperature sensitive for DNA replication at 42°C, not only was DNA polymerase I fully active at 42°C, but so also was DNA polymerase II. Finally, Campbell *et al.* (1972) were able to isolate a mutant strain of *E. coli* by treating the *polA1* strain with N-methyl-N'-nitro-N-nitroso guanidine which contained less than 0.5% of normal levels of DNA polymerase II. This mutant strain grew normally at 25° and 42°C and could support the growth of bacterial viruses T4, T7, lambda (λ), ϕx174, P2 and fd. This sealed the prospect of DNA polymerase II being the enzyme responsible for DNA replication.

9.18 Isolation of DNA polymerase III from *E. coli*

A third DNA polymerase was also detected in *E. coli* (Kornberg and Gefter 1971), purified and characterized (Kornberg and Gefter 1972). DNA polymerase III required all four deoxyribonucleoside 5′-triphosphates, Mg^{++}, mercaptoethanol or dithio threitol and native DNA for maximal activity. Ethanol was stimulatory. DNA synthesis was inhibited by sulphydryl reagents such as N-ethylmaleimide. DNA polymerase III required a primer to initiate DNA synthesis and polymerization proceeded in the 5′→3′ direction by covalent extension of the primer. Very much unlike with DNA polymerase I, DNA with single-strand scissions did not serve as template

DNA polymerase III holoenzyme has a very complex structure and consists of at least 10 different polypeptides. These polypeptides differ in the tightness of their association in the holoenzyme complex, as becomes evident on dilution of the holoenzyme. Polypeptides α, ε and θ are the most tightly associated and comprise the DNA polymerase III core (Maki *et al.* 1988). Polypeptide α is the DNA polymerase; ε, the 3′→5′ exonuclease, while no function is known for θ (Maki and Kornberg1985). The purified DNA polymerase III core is devoid of processivity; that is, it is unable to replicate single-stranded circular viral DNA molecules when these are several thousand nucleotides long, even when a proper primer is provided (Wickner and Kornberg 1974). The auxiliary polypeptides, τ, γ, δ, δ', χ, ψ and β, present in DNA polymerase III holoenzyme in 1:1 stoichiometry impart high processivity to the enzyme and allow it to synthesize long stretches of DNA without dissociating from the DNA template by providing some kind of clamping function (Wickner *et al.* 1973). The molecular weights of the polypeptides α, ε, θ, τ, γ, δ, δ', χ, ψ and β are 130, 27.5, 10, 71, 47.5, 35, 33, 15, 2 and 40.6 kDa, respectively (McHenry and Kornberg 1977; Maki *et al.* 1988). The complete holoenzyme behaves like a 800-kDa particle on gel filtration and apparently contains two each of the individual polypeptides. Polypeptide β, also termed as the copolymerase III, can be retained in the holoenzyme only in the presence of ATP and Mg++ (Maki *et al.* 1988). It has been possible to assemble DNA polymerase III holoenzyme *in vitro* from the 10 different subunits (Pritchard and McHenry 2001).

The properties of DNA polymerases I, II and III of *E. coli* can be summarized as follows: The polymerization direction of all the three DNA polymerases is 5′→3′ and the 3′→5′ exonuclease activity is present in all of these; the 5′→3′ exonuclease activity is, however, present in only DNA polymerase I; none of the three polymerases can use intact double-stranded DNA as template–primer; DNA polymerase I can use double-stranded DNA with single-strand scissions as template–primer, but the other two polymerases can not; the activity of DNA polymerase II and III is inhibited by sulphydryl (–SH)-blocking agents, but not of DNA polymerase I (Kornberg and Baker 1992).

9.19 Mutants lacking DNA polymerase III?

So far, no *E. coli* mutant has been discovered which does not have the normal amount of DNA polymerase III but nevertheless has normal DNA replication ability. Gefter *et al.* (1971) found out that if the temperature-sensitive mutation was in the *dnaE* locus, DNA polymerase III was inactive at 42°C and no DNA replication was also possible *in vivo* at 42°C. They looked at four independent mutations in the *dnaE* locus. Three of these were phenotypically similar: no DNA replication at 42°C and no DNA polymerase III activity *in vitro* at 42°C. DNA polymerase II was fully active at 42°C in these mutants. In the fourth mutant (*dnaE* 511), they observed slow cessation of DNA replication *in vivo* at 42°C, and the isolated DNA polymerase III from this mutant was slightly temperature sensitive at 42°C.

9.20 Any DNA polymerase that can replicate DNA in the 3′→5′ direction?

The fork mechanism of DNA replication as observed in *E. coli in vivo* (Cairns 1963) (Chapter 7) demands that both the 5′-strand and the 3′-strand of the DNA be replicated simultaneously in the same general direction. This implies that if the 3′-strand of the template DNA is replicated in the 5′→3′ direction, the 5′-strand would have to be replicated in the 3′→5′ direction. None of the three DNA polymerases discovered in *E. coli* was capable of carrying out DNA replication in the 3′→5′ direction. Concerted efforts had been made in several laboratories to isolate a DNA polymerase that could replicate DNA in the 3′→5′ direction, but with no success.

9.21 Is there any way of replication of the 5′-strand using the existing DNA polymerases?

A phenomenon reported by Mitra and Kornberg (1966) was of interest in this context. It was common knowledge that when DNA polymerase I was allowed to replicate double-stranded DNA *in vitro*, the product was hard to denature. Electron microscopic examination of this product had revealed multiple branched structures in the DNA. What Mitra and Kornberg (1966) found out was that the product could be made devoid of any branched structure and easily denaturable if *Neurospora* nuclease was added to the reaction mixture along with DNA polymerase I. *Neurospora* nuclease specifically degrades single-stranded DNA. Mitra and Kornberg (1966) offered the following explanation for this phenomenon:

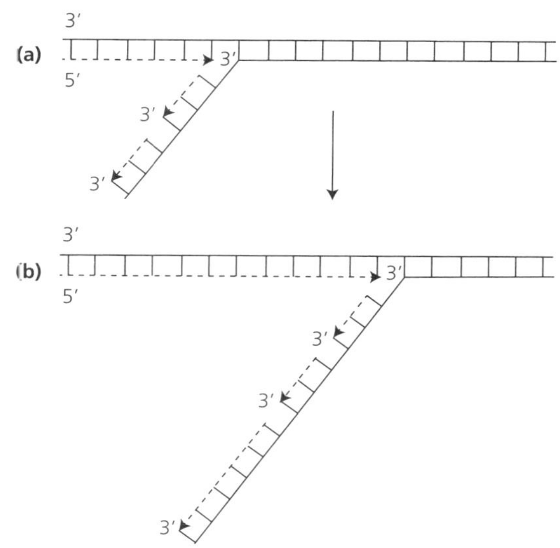

Figure 9.18 A speculative scheme of simultaneous replication of both strands of a DNA duplex effectively in the same general direction by a DNA polymerase that can only synthesize in the 5′→3′ direction. The solid lines represent the preexisting DNA and the broken lines, the newly synthesized segments. The arrow heads indicate the direction of synthesis. (a) The replication of the 5′-strand had to be discontinuous and (b) the replicated DNA fragments thus produced had to be joined later.

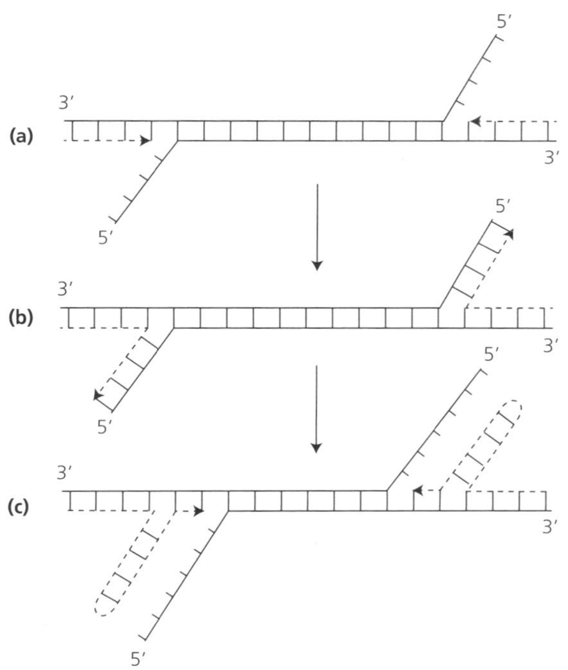

Figure 9.17 Schematic representation of the explanation offered for the appearance of branched structures in the product of *in vitro* replication of double-stranded DNA by DNA polymerase I. The solid lines represent the preexisting DNA and the broken lines, the newly synthesized segments. The arrow heads indicate the direction of synthesis. (a) DNA polymerase replicates the 3′-strand in the 5′→3′ direction, displacing the 5′-strand, which dangles as a single strand. (b) DNA polymerase switches over to the 5′-strand and replicates it in the 5′→3′ direction. (c) DNA polymerase turns 180° and replicates itself, again displacing the 5′-strand.

- The 3′-strand of the DNA served as the template and replication was initiated in the 5′→3′ direction on this strand progressively displacing the 5′-strand [schematic representation in Fig. 9.17(a)].
- The 5′-strand of the template DNA remained unreplicated and the portion of it displaced as a result of the movement of the DNA polymerase on the 3′-strand, kept dangling as a single strand [schematic representation in Fig. 9.17(a)].
- At this stage, the polymerase switched over to the dangling 5′-strand and proceeded to replicate it, but of course, in the 5′→3′ direction [schematic representation in Fig. 9.17(b)]. This replication continued till the polymerase reached the end of the 5′-strand when it turned around 180° and started replicating itself again, displacing the template 5′-strand [schematic representation in Fig. 9.17(c)]. These steps were repeated.
- *Neurospora* nuclease had degraded the displaced and dangling end of the 5′-strand of the DNA because it was single stranded and thus prevented the formation of the branched structures.

Could the cue to the mechanism of replication of the 5′-strand be there in Fig. 9.17? It was indeed speculated in a subsequent publication (Mitra *et al.* 1967) that 'a stretch of 5′-strand, freed by replication of the 3′-strand, would be replicated from some higher point along the strand down to the terminus [schematic representation in Fig. 9.18(a)]; subsequent repetitions of this process, and connections of terminating with earlier starting points, would then account for an over-all simultaneous, unidirectional replication of a helical chain' [schematic representation in Fig. 9.18(b)]. The essential feature of this speculative model was that the 5′-strand would have to be replicated in a discontinuous fashion, hence the replication products of the 5′-strand would initially be short DNA fragments.

9.22 Okazaki fragments

Sakabe and Okazaki did report about a year back (1966) of the appearance of 'slowly sedimenting small molecular DNA' in the 'labelled nascent portion of the chromosome' when *E. coli* cells were pulse-labelled with ^{3}H-deoxythymidine (Sakabe and Okazaki 1966). They also reported recovering the labelled DNA in the large molecular DNA fraction when a chase with unlabelled thymidine was applied after the pulse labelling. Sakabe and Okazaki did not seem to have appreciated the full significance of their finding at that point of time. They wrote, 'these results strongly suggest the presence of a unique structure at the growing point of the bacterial chromosome, one possibility being that there are breaks between the replicating region and the remaining portion of the chromosome' and they left it at that.

It was only 2 years later that they extended their earlier experiments and reinterpreted the results and finally came out with their model of discontinuous replication of DNA (Okazaki *et al.* 1968). *E. coli* B cells were grown at 20°C to slow down DNA synthesis and pulse-labelled with ^{3}H-thymidine for different lengths of pulse. The incorporation was then stopped by pouring the cells into crushed ice containing 0.02 M KCN. The DNA was isolated by a procedure that minimized shear and treated with NaOH–EDTA to denature the DNA and release single strands. The DNA was then analysed by centrifugation in alkaline sucrose gradient. During the first 10 s, most of the labelled DNA was of the fast-sedimenting type, the sedimentation coefficient being 11S. This would possibly imply that both the 3′-strand and the 5′-strand were being replicated in a discontinuous fashion. Labelling for 30 s led to the appearance of a larger-sized DNA peak along with the 11S peak. The 11S peak represented only a small fraction of the labelled DNA among the mostly large-sized ones when the labelling was for 60 s or more (Fig. 9.19; Okazaki *et al.* 1968). Of the three possible models of DNA synthesis (Fig. 9.20; adapted from Okazaki *et al.* 1968), the results obviously would be more in line with model C. This view, that both strands of DNA were being synthesized discontinuously,

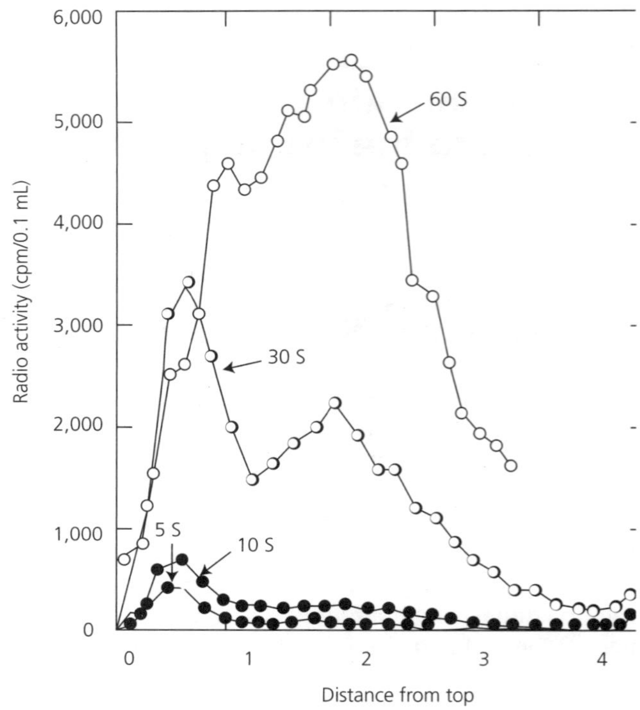

Figure 9.19 Sedimentation analysis of denatured radioactive DNA of *Escherichia coli* after pulse labelling with ^{3}H-thymidine. Pulse labelling of *E. coli* culture was done at 20°C. DNA was isolated by a procedure that minimized shear and denatured by treatment with 0.1 N NaOH and 1 mM EDTA. Alkaline sucrose density gradient centrifugation was used for the sedimentation analysis. The smallest DNA fragments are closest to the top. Most of the radioactivity after a 5- or 10-s pulse was in short DNA fragments, while most of it after a 60-s pulse was in long DNA fragments. [Adapted from Okazaki *et al.*, *Proc Natl Acad Sci* 1968;59:600 with permission from Professor T. Okazaki.]

appeared to be corroborated by subsequent observation that the T4 nascent short DNA chains isolated from T4-infected *E. coli*, annealed equally with the separated complementary strands of T4 DNA (Sugimoto *et al.* 1969).

Discontinuous synthesis seemed to be established, but was the 5′-strand also replicated in the 5′→3′ direction? Okazaki and Okazaki (1969) grew *E. coli* in ^{14}C-thymidine and infected with T4 virus in the presence of ^{14}C-thymidine. Pulse labelling with ^{3}H-thymidine was then done and the nascent short DNA strands were isolated. If indeed the replication of both the 3′-strand and the 5′-strand was proceeding in the 5′→3′ direction, all the nascent short DNA fragments would have ^{3}H-label at the 3′-end, but the 5′-end would have ^{14}C-label. Okazaki and Okazaki (1969) separated the nascent short DNA fragments on the basis of hybridization with the two complementary strands of T4 DNA and treated them separately with *E. coli* exonuclease I that degrades specifically in the 3′ to 5′ direction and with *B. subtilis* exonuclease that degrades specifically in the 5′ to 3′ direction. The released soluble radioactivity was

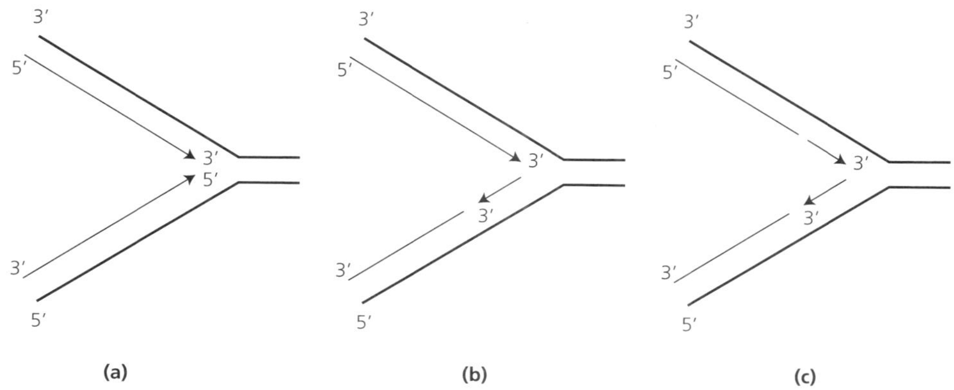

Figure 9.20 Models of DNA replication *in vivo*. (a) The 3′-strand of the double-stranded DNA is being replicated in the 5′→3′ direction, while the 5′-strand is being replicated in the 3′→5′ direction. (b) Both strands of the double-stranded DNA are being replicated in the 5′→3′ direction: replication of the 3′-strand is continuous, while replication of the 5′-strand is discontinuous. (c) Replication of both strands of the double-stranded DNA is discontinuous. [Adapted from Okazaki *et al.*, *Proc Natl Acad Sci USA* 1968;59:599 with permission from Professor T. Okazaki.]

then analysed for ^{14}C and ^{3}H. Plots of ^{14}C-radioactivity released versus ^{3}H-radioactivity released are presented in Fig. 9.21(Okazaki and Okazaki 1969). It is obvious that initially more ^{3}H-radioactivity was released in comparison with ^{14}C-radioactivity, but subsequently, more ^{14}C-radioactivity was released when the DNA fragments were treated with the 3′-end–specific *E. coli* exonuclease I. On the other hand, initially more ^{14}C-radioactivity was released in comparison with ^{3}H-radioactivity, but subsequently, more ^{3}H-radioactivity was released when the DNA fragments were treated with the 5′-end–specific *B. subtilis* exonuclease. These results would definitely suggest that both the DNA strands were being replicated in the 5′→3′ direction.

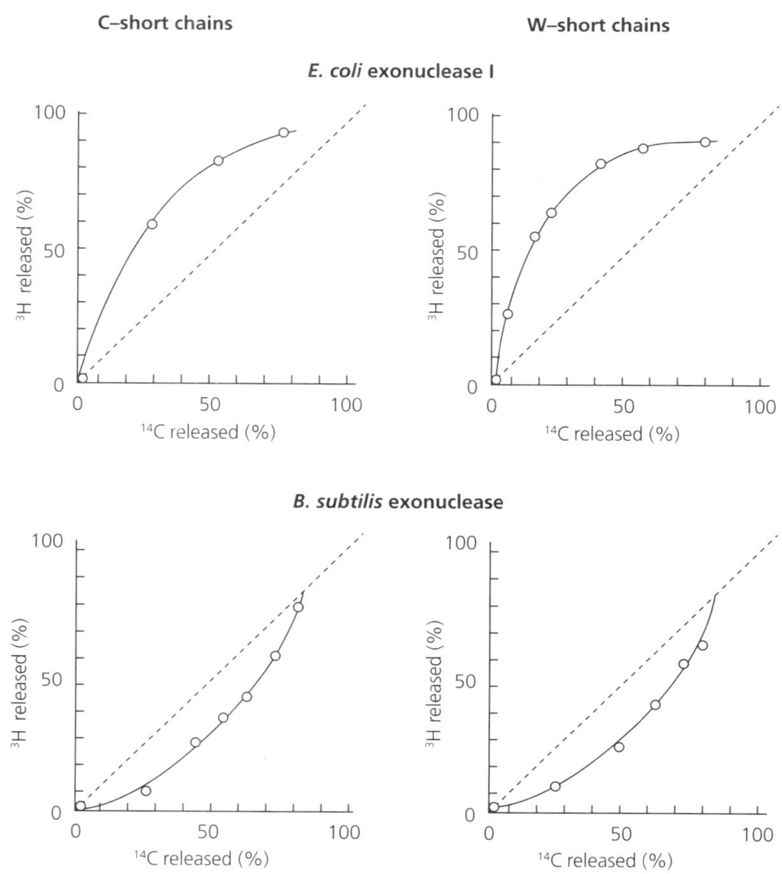

Figure 9.21 Release of radioactivity by exonucleolytic digestion of the Okazaki fragments. The DNA of a replicating bacteriophage T4 was first uniformly labelled with ^{3}H-thymidine and then pulse-labelled with ^{14}C-thymidine. The Okazaki fragments were isolated and separated on the basis of hybridization with the C-strand and the W-strand of bacteriophage T4. The Okazaki fragments were then separately treated with exonuclease I from *Escherichia coli* that cleaves from the 3′-end towards the 5′-end and with exonuclease from *Bacillus subtilis* that cleaves from the 5′-end towards the 3′-end. Samples were withdrawn periodically and ^{3}H- and ^{14}C-radioactivity release (acid-soluble radioactivity) was determined. From both the DNA strands, comparatively more ^{3}H radioactivity was released initially on treatment with exonuclease I from *E. coli*. However, comparatively more ^{14}C radioactivity was released initially on treatment with exonuclease from *B. subtilis*. [Reproduced from Okazaki and Okazaki, *Proc Natl Acad Sci USA* 1969;64:1247 with permission from Professor T. Okazaki.]

9.23 Why is it necessary to replicate the 3′-strand discontinuously?

There has been no good answer to this question for quite some time. Ultimately, some insight into the situation appeared from an unrelated area. The size of Okazaki fragments obtained after pulse labelling with ^{3}H-thymidine was substantially smaller (3-6.4S) if the *E. coli* cells were defective in dUTPase (*dut* mutation) (Konrad and Lehman 1975; Tye *et al.* 1977). The enzyme dUTPase hydrolyses dUTP, which would otherwise compete with dTTP as a substrate during DNA replication and dU would tend to get incorporated across dA in place of dT. Does this imply that higher incorporation of dU in DNA is correlated with the smaller size of the Okazaki fragments? As much as 30% of the thymine in the DNA of the bacteriophage T4 got replaced with uracil if the host *E. coli* was deficient in dUTPase (*dut-*) and in another enzyme uracil-DNA glycosylase (*ung-*) (Warner and Duncan 1978). Uracil-DNA glycosylase is known to catalyse the cleavage of the uracil–deoxyribose linkage in DNA (Lindahl 1974), creating apyrimidinic sites in the DNA. Subsequent endonuclease action at these sites is thought to bring about single-strand scissions that finally lead to excision of that region of the DNA backbone by exonuclease action. Thus, if a cell, in spite of it being *dut+*, is not 100% efficient in hydrolysing all the dUTP, some dU would still be incorporated into DNA and in a *ung+* situation, single-strand scissions would take place at these dU residues.

In an *in vitro* system comprising a cell lysate of *E. coli*, when no dUTP was added, about half of the newly synthesized DNA was of high molecular weight that sedimented along with the chromosomal DNA, while the other half was of small molecular weight (Fig. 9.22, top frame; Olivera 1978). Small DNA fragments formed were enhanced in amount but reduced in size by addition of dUTP, and finally, at 100-μM dUTP, no high-molecular-weight DNA was left (Fig. 9.22, bottom frame). This effect was further augmented if the lysate was from cells that were deffective in dUTPase (Olivera *et al.* 1979).

The inference is that DNA replication of the 3′-strand is continuous, but of the 5′-strand is discontinuous. The continuously replicated 3′-strand is fragmented later

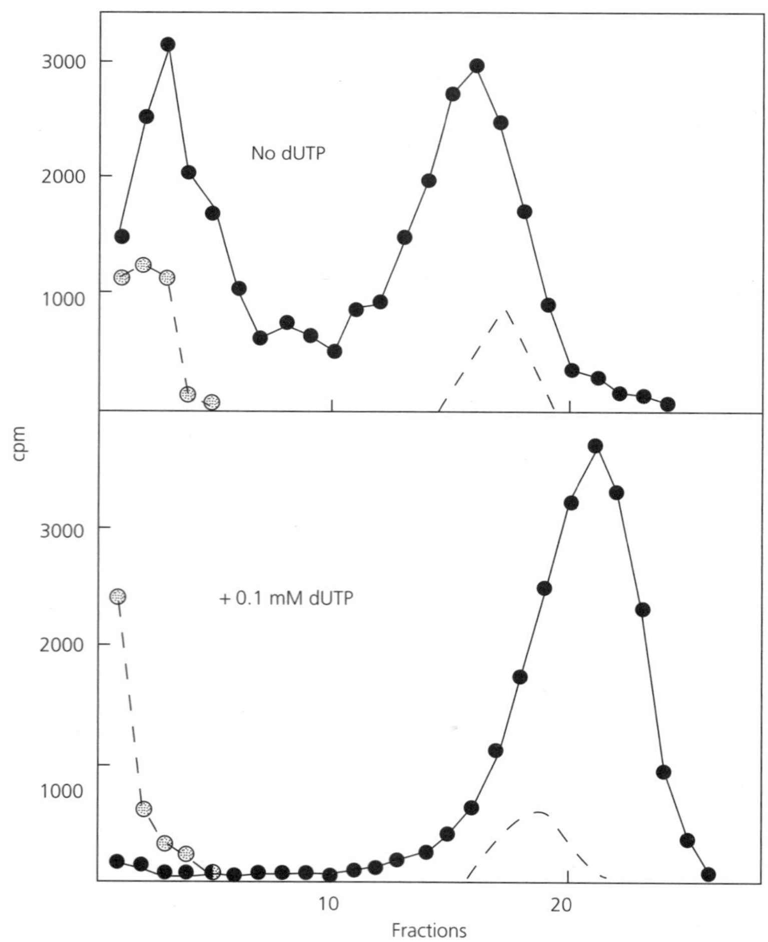

Figure 9.22 Effect of dUTP on the size and relative amount of Okazaki fragments formed in an *in vitro* DNA replication system from *Escherichia coli*. *E. coli* cells were grown in ^{3}H-thymidine and a lysate was prepared. The lysate was incubated in an *in vitro* DNA replication system with ^{32}P-dTTP and NMN for 15 min. The denatured DNA was analysed by alkaline sucrose density gradient centrifugation. Fractions were collected from the bottom of the centrifuge tubes. (•—•) ^{32}P radioactivity representing DNA synthesized *in vitro*; (o—o) ^{3}H radioactivity representing chromosomal DNA synthesized *in vivo*; (- - - -) DNA size marker of S = 9.5. Top frame: no dUTP added in the *in vitro* system; bottom frame: 0.1 mM dUTP added. [Reproduced from Olivera, *Proc Natl Acad Sci USA* 1978;75:239 with permission from Professor Baldomero Olivera.]

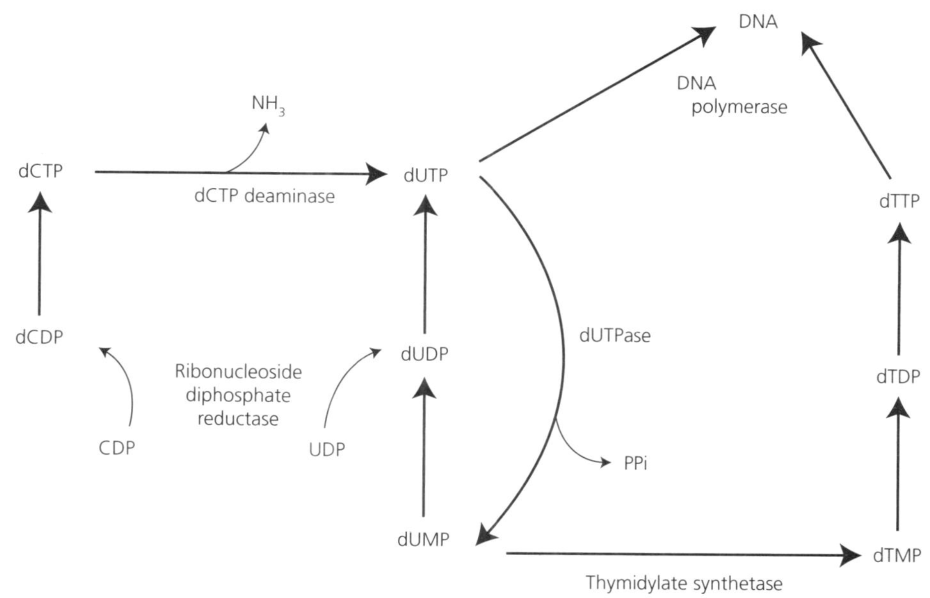

Figure 9.23 Key role of dUTPase in elimination of dUTP and in *de novo* synthesis of dTTP. [Reproduced from Tye *et al.*, *Proc Natl Acad Sci USA* 1977;74:154 with permission from Professor Bik-Kwoon Tye.]

because of the incorporation of dU during replication and the consequent process of the removal of the dU's initiated by uracil-DNA glycosylase.

9.24 Role of dUTPase in minimizing incorporation of dU in DNA

Fig. 9.23 (Tye *et al.* 1977) illustrates the key position of dUTPase among the various steps of interconversion between nucleotides. The enzyme dUTPase not only reduces the level of dUTP but the dUMP produced is also converted to dTMP through thymidylate synthase en route to dTTP.

9.25 Involvement of DNA ligase in joining the DNA fragments

What is the enzyme that joins the DNA fragments to produce continuous DNA strands? The conditional lethal mutant of bacteriophage T4, *ts* A80, produces an active T4 DNA ligase at 25°C, but an inactive enzyme at 37°C. Newman and Hanawalt (1968) observed that the proportion of short labelled DNA fragments was much higher if *E. coli* cells infected with T4 ts A80 were pulse-labelled with ^{3}H-labelled precursor at 43°C instead of at 23°C. Cells infected with the wild-type T4 bacteriophage when pulse-labelled at 43°C had a proportionately negligible amount of short labelled DNA fragments. It could thus be inferred that DNA ligase was involved in joining the nascent short DNA fragments.

9.26 Discovery of DNA polymerase IV from *E. coli*

The discovery of DNA polymerase IV from *E. coli* was attained under very interesting circumstances. It has been known for quite some time that normal DNA replication in *E. coli* by DNA polymerase III gets blocked when the DNA is damaged by UV irradiation or acted upon by chemical agents such as mytomycin c or nalidixic acid (Walker 1984). It has also been known that in response to this adverse situation, expression of a number of genes, including the genes *dinB*, *umuC* and *umuD*, is induced to resurrect the synthesis of DNA (Walker 1984). This is commonly known as the SOS response, which has also been found to result in an increased mutation rate (Walker 1984). It has been assumed that mutation in SOS-induced cells is the consequence of translesion DNA synthesis that is inherently error prone. The mutations could, however, be both targeted (at the site of the lesion in DNA of irradiated bacteria or virus) or untargeted (at undamaged DNA sites; Caillet-Fauquet *et al.* 1977).

If unirradiated λ phage was allowed to infect UV-irradiated *E. coli* cells, the frequency of mutation in the progeny λ phage DNA was one to three orders of magnitude more than when unirradiated *E. coli* cells were infected (Brotcorne-Lannoye and Maenhaut-Michel 1986). This is an example of untargeted mutation. Interestingly, mutation in the *dinB* gene of *E. coli* suppressed untargeted mutation in phage λ DNA (Brotcorne-Lannoye and Maenhaut-Michel 1986), thus implicating the *dinB* gene product in untargeted mutation. The frequency of this class of mutation was unaffected by mutations in the *umu* loci (Maenhaut-Michel and Caillet-Fauquet 1984; Wood and Hutchinson 1984). The *dinB* gene has been subsequently shown to be identical to the *dinP* gene and overexpression of the cloned *dinB/dinP* gene, achieved through recombinant DNA technology, resulted in an enhancement of untargeted mutation in F'*lac* plasmids by two to three orders of magnitude, without UV irradiation or any other DNA-damaging treatment (Kim *et al.* 1997). Enhancement of this class of untargeted mutation could not be reversed by mutations in the *umu* loci. The *dinB/dinP* gene product by itself thus elicits potent mutagenic activity (Kim *et al.* 1997). This was the background when characterization of the purified protein product of the overexpressed *dinB* gene revealed that it was indeed a novel template-dependent *E. coli* DNA polymerase capable of elongating a DNA primer *in vitro* (Wagner *et al.* 1999). This new polymerase was named DNA polymerase IV. No polymerase activity could be observed if *E. coli* cells with a mutation in the *dinB* gene was used as the source material for isolation of the enzyme (Wagner *et al.* 1999).

It is obvious from the foregoing that DNA polymerase IV is induced as a consequence of the SOS response to DNA damage. In addition, it has been observed that DNA polymerase IV is induced under the starvation stress in stationary phase culture under the positive control of RpoS, the sigma factor (Layton and Foster 2003). It has now been claimed that induction of DNA polymerase IV is regulated by the heat shock chaperone GroE by a process that is independent of LexA (Layton and Foster 2005). No direct interaction between GroE and DNA polymerase could, however, be shown.

9.27 DNA polymerase IV has a very low processivity and does not have any associated 3′→5′ exonuclease activity

The template used to study processivity of DNA polymerase IV was a single-stranded circular DNA derived from a natural plasmid and the primer was a 24-mer deoxyoligonucleotide complementary to the template. The primer was hybridized to the template after labelling it at the 5′-end with ^{32}P by using polynucleotide kinase and γ-^{32}P–labelled ATP. The reaction mixture contained all the four deoxyribonucleotide triphosphates, the template/primer at 50 nM concentration and different concentrations of the purified DNA polymerase IV. The reaction product was analysed by electrophoresis on a 12% denaturing polyacrylamide gel (Fig. 9.24). The low processivity of the enzyme was obvious from the fact that the reaction product was only one nucleotide longer than the primer when equimolar amounts of the enzyme and the substrates were present in the reaction mixture (Wagner *et al.* 1999).

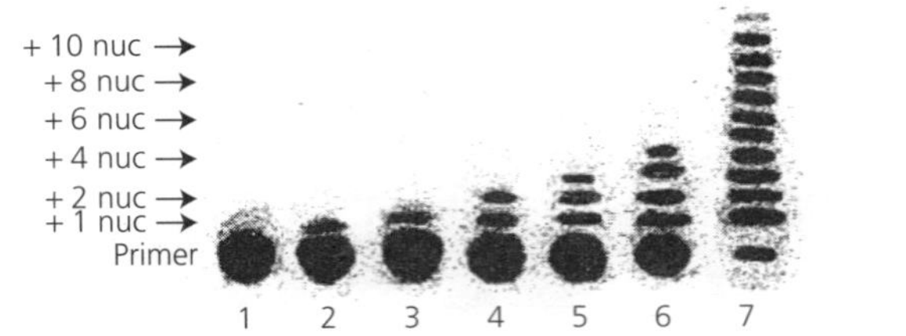

Figure 9.24 An artist's impression of the radioautogram showing low processivity of DNA polymerase IV. The template used for the study was a single-stranded circular DNA. The primer was a 24-mer deoxyoligonucleotide complementary to the template, which was labelled at the 5′-end with ^{32}P by employing γ-^{32}P–labelled ATP and T4 polynucleotide kinase. All the four deoxynucleotide triphosphates were added in excess along with a fixed amount of the template–primer (50 nM) and different amounts of purified DNA polymerase IV. The reaction product was electrophoresed on a 12% denaturing polyacrylamide gel and radioautographed. Lane 1: no DNA polymerase IV added, only the primer can be seen; lane 2: 5 nM DNA polymerase IV added, faint appearance of the product containing one additional nucleotide can be seen; lane 3: 10 nM DNA polymerase IV added, one clear product band containing one additional nucleotide can be seen; lane 4: 20 nM DNA polymerase IV added, two product bands, one containing one additional nucleotide and the other containing two additional nucleotides, can be seen; lane 5: 50 nM DNA polymerase IV added, three product bands containing one additional, two additional and three additional nucleotides can be seen; lane 6: 100 nM DNA polymerase IV added, four product bands containing one additional, two additional, three additional and four additional nucleotides can be seen; lane 7: 300 nM DNA polymerase IV added, a ladder of 11 product bands can be seen–the largest product band contains 11 additional nucleotides.

The presence of any associated 3′→5′ exonuclease activity would be expected to be manifested by a reduction in the size of the primer, when the specific deoxyribonucleotide triphosphate complementary to the next base in the template would not be available in the reaction mixture. An experiment done with a template and a primer of defined sequence where no deoxyribonucleotide triphosphate or a single deoxyribonucleotide triphosphate was added as substrate unambiguously excluded the possibility of the presence of any 3′→5′ exonuclease activity associated with DNA polymerase IV (Wagner *et al.* 1999). The length of the primer was increased by one nucleotide as a result of the polymerase reaction when the substrate added (in this case dATP) was complementary to the next base in the template. The enzyme did not cause any reduction in the size of the primer on addition of no nucleotide or on addition of the nucleotides (T, G or C)

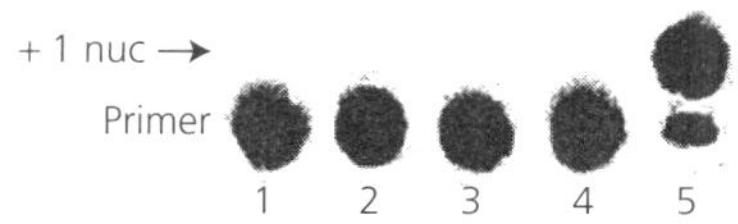

Figure 9.25 An artist's impression of the autoradiogram showing that DNA polymerase IV lacks any associated 3′→5′ exonuclease activity. The template used in the reaction was a 44-mer deoxyoligonucleotide. The primer was a complementary 24-mer deoxyoligonucleotide ^{32}P-labelled at the 5′-end, which was annealed to the template. The enzyme added to each reaction mixture was purified DNA polymerase IV. The reaction product was electrophoresed and radioautographed as mentioned in Fig. 9.24. Lane 1: no deoxyribose triphosphate substrate was added; lane 2: only one nucleotide, dTTP was added as substrate; lane 3: only one nucleotide, dGTP was added as substrate; lane 4: only one nucleotide, dCTP was added as substrate; lane 5: only one nucleotide, dATP was added as substrate. It can thus be seen that addition of dATP increased the length of the primer by one nucleotide, while additon of either of the other three triphosphates or no addition of a triphosphate did not reduce the size of the primer by enzyme action.

that were not complementary to the next base in the template (Fig. 9.25).

9.28 Replication by DNA polymerase IV results in mutation, predominantly of the single-base or two-base deletion type

Overexpression of DNA polymerase IV in *E. coli* caused enhanced mutagenesis *in vivo* even in the absence of any exogenous treatment to damage DNA, and this was predominantly of the single-base or two-base deletion type (frameshift mutation) along with some base substitution (Wood and Hutchinson 1984; Kim *et al.* 1997). When a synthetic template/primer combination was designed such that the base at the 3′-end of the primer was not complementary to the template, replication *in vitro* by DNA polymerase IV produced a sizable amount of product that had one nucleotide less than the template. It was as if a nucleotide in the template bulged out till a match was found with the primer at the 3′-end. This was in contrast to the products obtained by using the Klenow fragment of DNA polymerase I (Wagner *et al.* 1999). Base sequencing of the largest polymerization products obtained by using DNA polymerase IV has revealed that the mismatched base in the template has indeed been skipped. This appeared to be the basis of single- or double-base deletion mutation observed earlier *in vivo*. The Klenow fragment that was devoid of 3′-exonuclease could not correct the mismatch, but the Klenow fragment that did have the 3′-exonuclease could carry out the correction (Wagner *et al.* 1999).

9.29 Discovery of DNA polymerase V from *E. coli*

The genes *umuC* and *umuD* are also induced as a part of the SOS response consequent to DNA damage. Mutations in the *umu* loci nullified the increase in the rate of the targeted mutation in *E. coli* DNA arising out of UV irradiation of *E. coli* (Kato and Shinoura 1977). The proteins UmuC and UmuD were thus termed as mutagenesis proteins (Burckhardt *et al.* 1988; Woodgate *et al.* 1989). It was demonstrated that RecA, the protein involved in recombination, mediated cleavage of UmuD and the resulting UmuD' was the active form (Burckhardt *et al.* 1988). It was thought that the proteins UmuC, UmuD' and the activated RecA (RecA*) got bound to DNA polymerase III and produced a modified DNA polymerase III complex that could replicate past the damaged DNA block, but only in an error-prone fashion (Woodgate *et al.* 1989; Rajagopalan *et al.* 1992). This view persisted till 1998, when it was recognized that UmuC and UmuD' formed the complex UmuD'$_2$UmuC, which was termed as the 'mutagenic complex' that was thought to serve as an elongation factor involved in the extension of DNA synthesis by DNA polymerase III past a site of lesion, albeit by misincorporation of bases (Tang *et al.* 1998). The very next year, it was possible to purify the UmuD'$_2$UmuC complex absolutely free from any contaminating DNA polymerase and still the complex exhibited DNA polymerase activity. No addition of any DNA polymerase III was necessary for the polymerization reaction. The UmuD'$_2$UmuC complex has been named as DNA polymerase V (Tang *et al.* 1999). The DNA polymerase activity of the UmuD'–UmuC complex has also been reported by Reuven *et al.* (1999).

9.30 Lesion bypass by DNA polymerase V

DNA polymerase V was found to be capable of bypassing targeted mutations (lesion bypass). Tang *et al.* (1999) found that addition of RecA*, single-strand DNA binding protein (SSB) and the auxiliary factors β and γ, which impart processivity to the core DNA polymerase III, was necessary for lesion bypass. Reuven *et al.* (1999) too found the necessity of adding RecA* and SSB for the *in vitro* lesion bypassing polymerase activity, but not of the auxiliary factors β and γ.

Schlacher *et al.* (2005) have found that a complex composed of just polymerase V and two molecules of RecA was capable of carrying out translesion DNA synthesis.

The lesion bypass by the UmuD'$_2$UmuC complex was, as expected, very error prone, but was 100- to 150-fold more efficient than by DNA polymerase III. The common DNA template lesions are thymine–thymine pyrimidine (6-4) pyrimidone photoproducts, thymine–thymine *cis-syn* cyclobutane photodimers caused by UV radiation, and apurinic/apyrimidinic abasic sites resulting from the loss of a DNA base spotaneously or because of action of alkylating agents. All these lesions were found to be efficiently bypassed within 30 s by DNA polymerase V (Tang *et al.* 2000). There was no detectable bypass of these lesions by either DNA polymerase III or DNA polymerase IV on this time scale. DNA polymerase V would invariably incorporate a guanine opposite the 3′-thymine of a TT (6-4) photoprotuct (Tang *et al.* 2000). The processivity of DNA polymerase V was, however, only 6–8 nucleotides (Tang *et al.* 2000).

9.31 Untargeted mutagenesis by DNA polymerase V

A view had gained ground that the two error-prone DNA polymerases of *E. coli*, pol IV and pol V, had complementary roles in the cell: pol IV being involved in untargeted mutagenesis, while pol V carried out translesion DNA synthesis or targeted mutagenesis (Tang *et al.* 2000). It has, however, been demonstrated that DNA polymerase V was highly mutagenic even when it carried out gap-filling DNA replication *in vitro* in association of RecA* and SSB protein, there being no apparent DNA lesion anywhere (Maor-Shoshani *et al.* 2000). The template–primer used was a gapped plasmid with a single-stranded DNA region of about 350 nucleotides long. Replication by DNA polymerase V produced point mutations at a frequency of 2.1×10^{-4} per nucleotide, which was more than 70-fold higher than that produced by DNA polymerase III holoenzyme. DNA base sequencing of the replication product has revealed that 53% of the mutations were transversions (Maor-Shoshani *et al.* 2000).

Fuchs *et al.* (2004) have reviewed the properties and functions of *E. coli* DNA polymerase IV and DNA polymerase V.

9.32 RecA protein

RecA protein (~37,000 Da) is essential for all pathways of homologous DNA recombination and has a number of enzymatic activities. It can catalyse the following related reactions: (a) DNA-dependent hydrolysis of ATP, (b) ATP-stimulated renaturation of complementary single-stranded DNA (ss) DNA molecules, (c) ATP-dependent reciprocal exchange of single strands of DNA between homologous double-stranded DNA molecules and (d) cleavage of LexA protein and lambda repressor proteins in an ATP- and DNA-dependent reaction. RecA* is the activated form of RecA protein.

9.33 Single-stranded DNA binding (SSB) protein

SSB protein (~19,500 Da) binds single-stranded DNA selectively and cooperatively facilitates RecA protein binding to single-stranded DNA and consequent RecA-promoted reactions. SSB protein stimulates DNA polymerases II, III, IV and V and favours melting of DNA.

9.34 SOS response

Exposure of *E. coli* to agents that damage DNA (ultraviolet radiation) or interfere with DNA replication (mitomycin C and nalidixic acid) results in the induction of some 20 or more genes, including the *din* (damage inducible) genes, the *umu* (ultraviolet mutagenesis) genes and the *uvr* (ultraviolet reactivation) genes. The SOS genes, which are repressed by LexA, are induced when RecA* (activated RecA) binds to LexA and promotes its self-cleavage. The signal for SOS response is the appearance of single-stranded DNA at a stalled DNA replication fork. The RecA protein gets activated when it comes in contact with single-stranded DNA.

9.35 Mutations: Different kinds

Mutation refers to any structural alteration of DNA. A mutation may or may not elicit a changed phenotype. A point mutation signifies a single nucleotide change, which could be the substitution of a nucleotide. Substitution of a purine by a purine or of a pyrimidine by a pyrimidine is called nucleotide transition. On the other hand, substitution of a purine by a pyrimidine or of a pyrimidine by a purine is known as nucleotide transversion (Fig. 9.26). The chemical mutagen nitrous acid (HNO_2) causes deamination of nucleotide bases. Thus, adenine is converted into hypoxanthine, guanine is converted into xanthine and cytosine is converted into uracil [Fig. 9.27 (a)]. During pairing with thymine by hydrogen bonding, the NH_2 group of adenine serves as the H-donor [Fig. 9.27 (b)]. In the absence of the NH_2 group, hypoxanthine cannot play the same role, but its OH group serves as the hydrogen receiver. During replication, hypoxanthine, thus, pairs with cytosine, which during the subsequent replication pairs with guanine. Thus, an AT pair gets converted into a GC pair, a case of transition (Fig. 9.28). Similarly, the converted uracil pairs with adenine, which during the next replication pairs with thymine. Here, a GC pair is converted into

Wild type
TGCAATGGAC

Single nucleotide substitution
TGCAGTGGAC — Transition
TGCACTGGAC — Transversion

Single nucleotide deletion or insertion
TGCATGGAC
TGCAAGTGGAC

Multiple nucleotide deletion or insertion
TGCGAC
TGCAA ---------- TGGAC

Nucleotide rearrangement
TGCGTAAGAC

Figure 9.26 Different kinds of mutations. The substituted and the inserted bases(s) have been underlined. The site of deletion has also been marked by underlining between bases.

an AT pair, another case of transition (Fig. 9.28). Xanthine, on the other hand, though devoid of one amino group, can still pair with cytosine through two hydrogen bonds. Hence, no change takes place.

The structures of some of the chemical mutagenic agents are shown in Fig. 9.29. The alkylating agents methyl methane sulphonate and ethyl methane sulphonate add on an alkyl group to N-7 of purine, primarily guanine. One effect of this alkylation is labilysation of the *N*-glycosylic bond, resulting in depurination. It must be noted that this is not a case of deletion of a purine nucleotide. The phosphodiester backbone of the DNA remains unaffected, only a purine base is knocked off. DNA polymerase V can replicate through such an abasic site, but any of the four nucleotides can be inserted opposite the depurinated site, thus potentially creating both nucleotide transition and nucleotide transversion.

Insertion or deletion of a single nucleotide pair can also be considered a point mutation. The intercalating dye acridine orange, if present during DNA replication, engineers addition of a nucleotide pair and sometimes deletion of a nucleotide pair. This kind of mutation is also known as frameshift mutation. An amino acid is coded for by three consecutive nucleotides; consecutive amino acids in a polypeptide are coded for by consecutive nucleotide triplets of a polynucleotide. An insertion or deletion of a single nucleotide pair would, therefore, shift the frame of triplets and hence the whole sequencr of amino acids would be changed (Fig. 9.30). DNA polymerase IV is known to produce frameshift mutation.

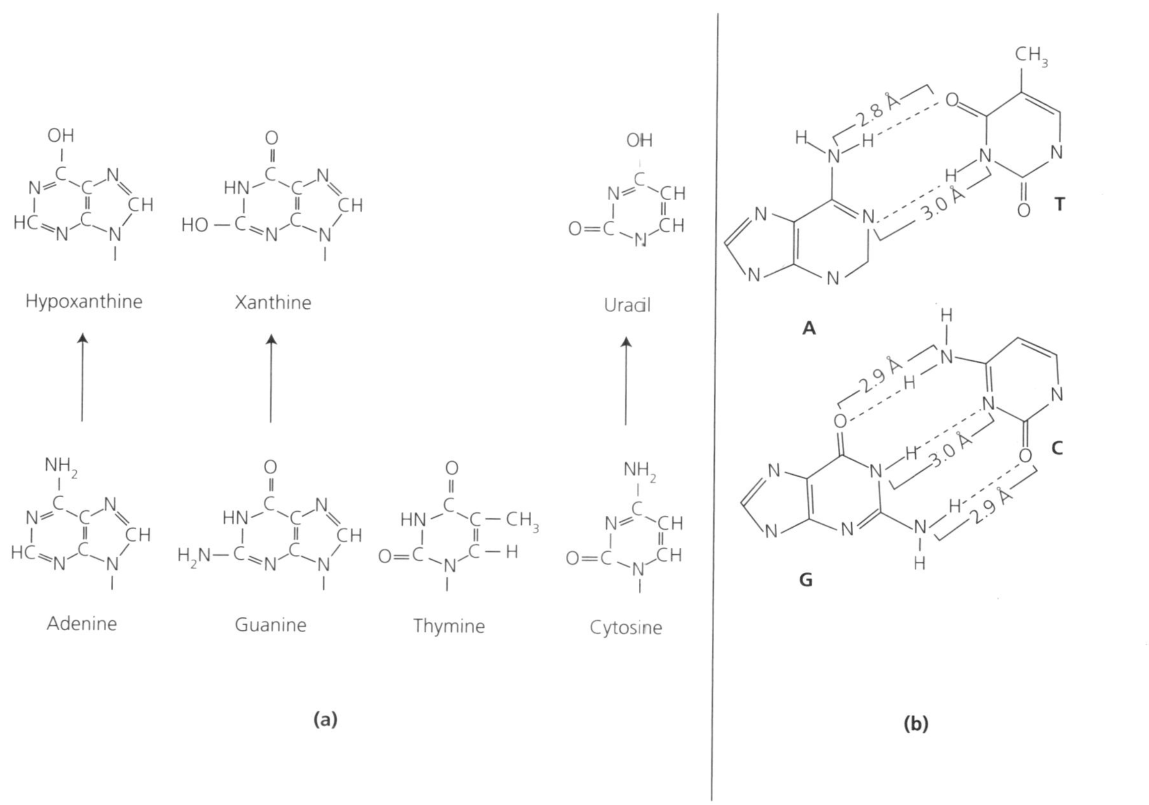

Figure 9.27 Deamination of nucleotide bases by nitrous acid (HNO_2). (a) Adenine is deaminated to hypoxanthine, guanine to xanthine and cytosine to uracil. Thymine has no amino group; hence, it cannot be deaminated. (b) Involvement of amino groups as donor for hydrogen during hydrogen bonding between adenine (A) and thymine (T) and between guanine (G) and cytosine (C).

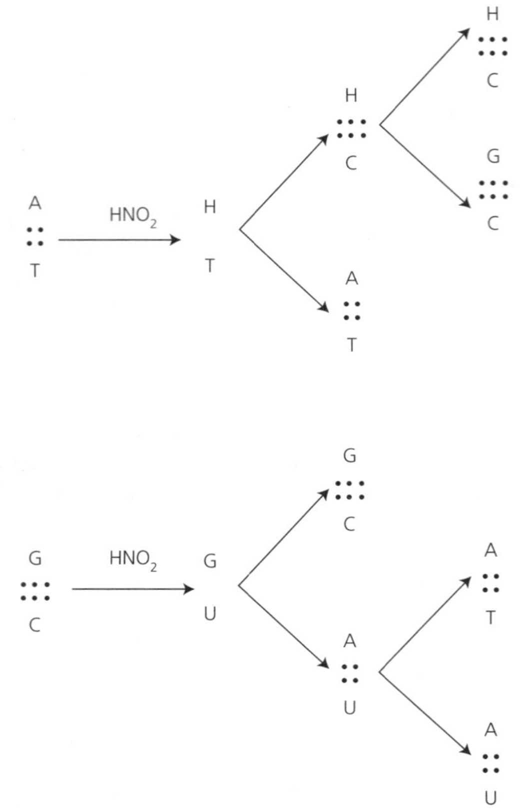

Figure 9.28 Schematic representation of nucleotide transition as a result of deamination by HNO_2. The adenine of an A:T pair is deaminated to hypoxanthine. In the next replication round, cytosine is put across hypoxanthine, and during the subsequent replication, guanine is put across cytosine. Thus an A:T pair is converted into a G:C pair. Similarly, cytosine of a G:C pair is deaminated to uracil and finally a G:C pair is converted into an A:T pair.

(a) CCA ACG GAC CAC CGA GAC AGC
pro thr asp his arg asp ser

A ↑ (*−1 nucleotide*)

(b) CCA ACG GCC ACC GAC ACA GCG
pro thr ala thr glu thr ala

G ↓ (*+1 nucleotide*)

(c) CCA ACG GGA CCA CCG AGA CAG
pro thr gly pro pro arg gln

Figure 9.30 Frameshift mutation due to single-base deletion and single-base insertion. (a) Triplets in wild-type nucleotide sequence CCA-ACG-GAC-CAC-CGA-GAC-AGC coding for the polypeptide pro-thr-asp-his- arg-asp-ser. (b) Nucleotide dATP has been deleted from codon GAC, converting this into codon GCC using nucleotide C from the first letter of the next codon CAC. The new triplets in this mutated sequence are CCA-ACG-*GCC-ACC-GAG-ACA-GCG* and would code for the polypeptide pro-thr-*ala-thr-glu-thr-ala*, the amino acids after threonine being completely different as a result of a backward shift in the reading frame. (c) Nucleotide G has been inserted after G in codon GAC. The new triplets in this mutated sequence are CCA-ACG-*GGA-CCA-CCG-AGA-CAG*, which would now code for the polypeptide pro-thr-*gly-pro-pro-arg-gln*, again the amino acids after threonine being completely different as a result of a forward shift in the reading frame. It is obvious that frame shift would also take place in the case of deletion or addition of two bases because of the triplet nature of the genetic code.

Irradiation with UV rays (wavelength 200–400 nm) can transform purine or pyrimidine bases into an excited state. Adjacent excited thymine residues in a DNA strand have often been found to interact with each other, forming a cyclobutane ring by covalently connecting the two thymines through two new C—C bonds (shown by arrows; Fig. 9.31). Formation of thymine dimers distorts the double helix and the hydrogen bonds between these thymines and the adenines on the other strand are disrupted. Though DNA polymerase III cannot continue replication through thymine dimers, DNA polymerase V can, but guanine would invariably be incorporated instead of adenine.

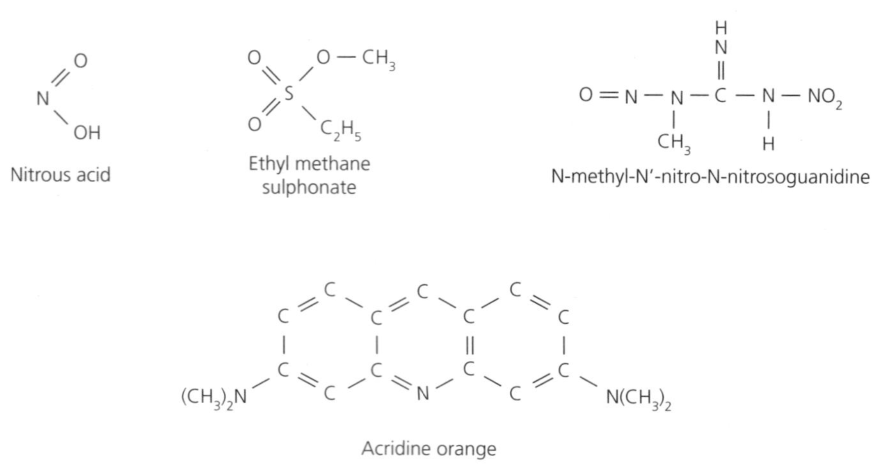

Figure 9.29 Chemical structures of the mutagenic agents nitrous acid, ethyl methane sulphonate (EMS), *N*-methyl-*N'*-nitro-*N*-nitrosoguanidine and acridine orange.

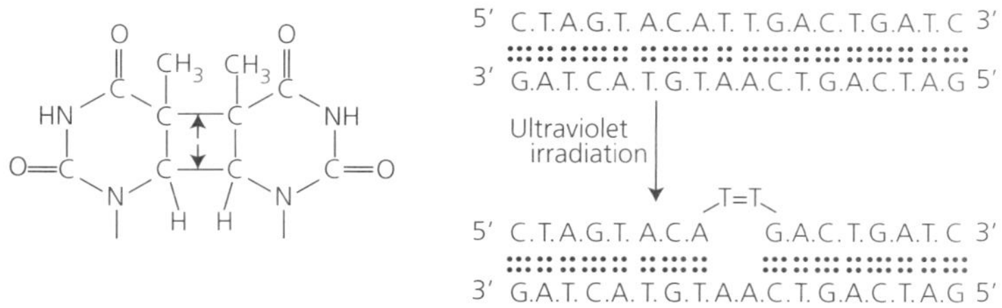

(a) (b)

Figure 9.31 Formation of thymine dimer as a result of ultraviolet irradiation of DNA. (a) The structure of the thymine–thymine *cis-syn* cyclobutane photo dimer. (b) Distortion in the DNA duplex structure because of thymine dimer formation.

9.36 DNA polymerases from *B. subtilis*

DNA polymerases have been isolated and characterized from a large number of bacteria, both gram positive and gram negative. Though the gram-positive bacterium *B. subtilis* is phylogenetically very distantly related to the gram-negative *E. coli*, it has three DNA polymerases of very analogous properties. Here also, DNA polymerase III is the enzyme responsible for DNA replication *in vivo* (Karmata and Gross 1970; Cozzarelli and Low 1973). The holoenzyme has a processive 5′→3′ polymerization activity and non-processive 3′→5′ exonuclease activity (Low *et al.* 1976), but there is no separate ε subunit. The 3′→5′ proofreading exonuclease activity is an integral part of the 1335 amino acid–long α-subunit (Sanjanwala and Ganesan 1989).

9.37 Thermostable DNA polymerases

The DNA polymerase from the thermophilic eubacterium *Thermus aquaticus* is 832 amino acid long and is significantly similar in properties to DNA polymerase I of *E. coli* (Lawyer *et al.* 1989). One property of this enzyme, however, has made it unique: the enzyme is stable at 95°C, the half-life at this temperature being 45 min (Lawyer *et al.* 1989). As most DNA would completely denature at 95°C, this enzyme has been extensively used for polymerase chain reaction (PCR) as the necessity of addition of the polymerase after each denaturation step is obviated. An important drawback of the *T. aquaticus* enzyme is the lack of the proofreading 3′→5′ exonuclease activity in it, which could make the synthesis error prone (Dunning *et al.* 1988; Tindall and Kunkel 1988). The fidelity of DNA synthesis, however, has been found to be quite high. Under defined reaction conditions, the error rate per nucleotide polymerized at 70°C has been found to be as low as 10^{-5} for base substitution errors and 10^{-6} for frameshift errors (Eckert and Kunkel 1990).

DNA polymerase from the hyperthermophilic marine archaebacterium *Thermococcus litoralis* (Belkin and Jannasch 1985; Neuner *et al.* 1990), also known as 'Vent' polymerase, is even more stable at high temperatures. The half-life of the enzyme at 95°C is 8 h, and at 100°C, it is 2 h (Kong *et al.* 1993). The other advantage with this enzyme is that it does have the proofreading 3′→5′ exonuclease activity (Mattila *et al.* 1991; Kong *et al.* 1993), thus resulting in a frequency of mutation for base substitution during replication that is 5–10 times lower than that with the *T. aquaticus* enzyme. The amino acid sequence of this enzyme bears greater homology with DNA polymerase II of *E. coli* as also the DNA polymerases from most other archaebacterium classified as Euryarchaeotic DNA polymerases I.

9.38 Euryarchaeotic DNA polymerases

A new class, Euryarchaeotic DNA polymerases II, has been reported which are altogether different from Euryarchaeotic DNA polymerases I (Cann and Ishino 1999). DNA polymerase II from the hyperthermophilic archaeon *Pyrococcus furiosus* contains two subunits of estimated molecular mass of 69 kDa and 143 Da, both subunits being essential for *in vitro* DNA polymerase activity and 3′→5′ exonuclease activity (Cann and Ishino 1999). Published genome sequences have revealed the conservation of euryarchaeotic DNA polymerase II among several archaebacterium, but not among other prokaryotes or eukaryotes (Cann and Ishino 1999).

9.39 Polymerase chain reaction

Polymerase chain reaction is the name given to a procedure by which one can amplify a very small amount of a specific DNA sequence, may be as small an amount as that present in one single cell, to a very large amount (Saiki *et al.* 1985; Mullis and Faloona 1987). After 20 cycles of the reaction, one can indeed expect an amplification of 200,000 fold. The protocol is very straightforward. Two oligodeoxynucleotide primers that have sequence complementary to the flanking 3′-regions of the two strands of the DNA sequence to be amplified are synthesized. The DNA is heat denatured and the oligodeoxynucleotides, which are in large excess, are annealed to the DNA by slow cooling. The temperature is allowed to fall to 37°C and DNA polymerase I from *E. coli* is added along with all the four deoxyribonucleotides and DNA replication is allowed to proceed at 37°C. The reaction is initiated by the addition of mononucleotides at the 3′-ends of the two primers and both the strands of DNA are replicated in the 5′→3′ direction. This completes the first cycle. The mixture is heated to denature the two newly formed DNA duplexes and then cooled so that the primers, which are in excess, can again anneal, but this time to both the two original DNA strands and also to the two newly synthesized ones. Once the mixture is cooled to 37°C, DNA polymerase I from *E. coli* is again added and replication is allowed to proceed. The second cycle is now completed. The steps are repeated for the third cycle and subsequent cycles. At the end of 20 cycles, one can thus theoretically expect an amplification of 2^{20} fold, which is equivalent to more than 10^6. The logic of the amplification steps is schematically represented in Fig. 9.32.

The discovery of the thermostable DNA polymerase from *Thermus aquaticus* has rendered the amplification procedure simpler (Saiki *et al.* 1988). It is not necessary any more to add fresh polymerase in the beginning of every cycle and the reaction mixture does not have to be cooled to 37°C for the polymerization reaction. After the annealing is carried out at 40°C or near about, the temperature is raised to 70°C, the optimum temperature for polymerization. At this temperature, all poorly matched primer–template hybrids dissociate and only highly

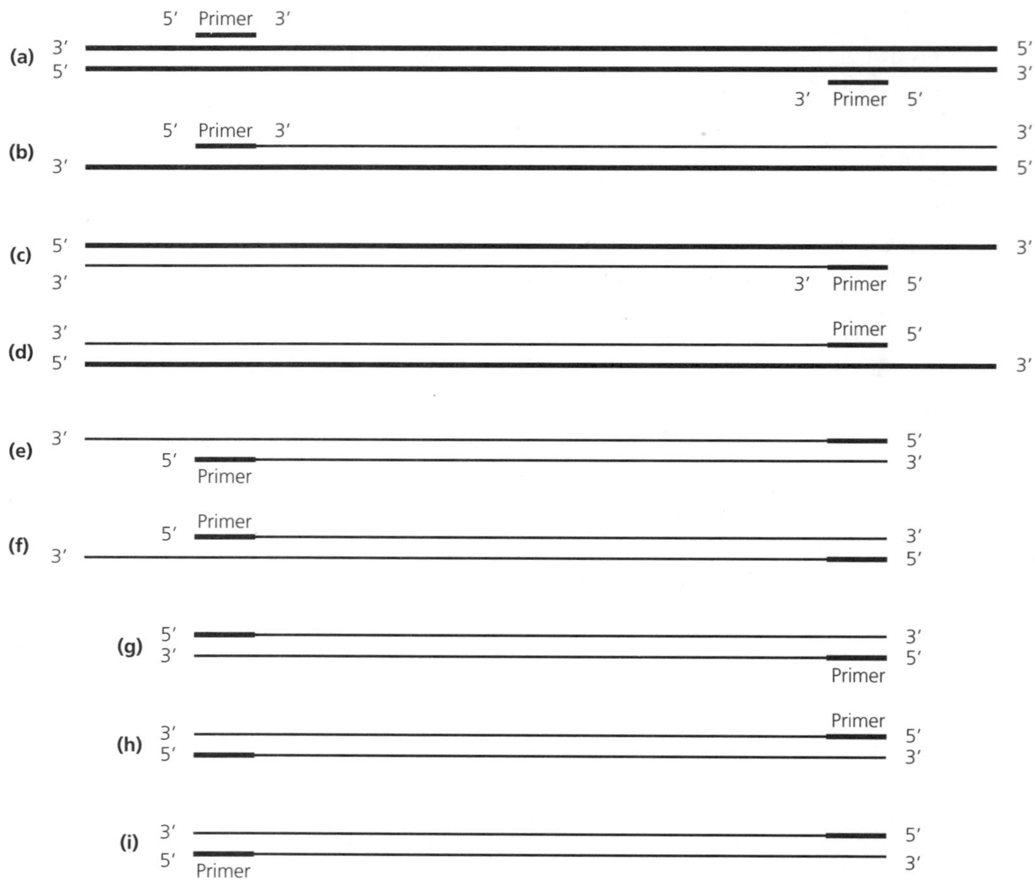

Figure 9.32 Schematic representation of polymerase chain reaction. (a) The DNA to be amplified with two small pieces of complementary DNA annealed to the two strands. The annealed small pieces of DNA would serve as primers for the replication process. The original DNA and the primers are shown by thick lines. The two strands of the original DNA get separated and two daughter DNA duplexes are formed as a result of DNA replication by the semiconservative mechanism. The newly synthesized DNA strands are shown by thin lines. In the next cycle, the primers are annealed to both the newly synthesized duplexes (b) and (c) and the replication process is repeated with either fresh *E. coli* polymerase or thermostable polymerase. In the figure, only replication of the duplex (c) has been shown. The two strands of (c) get separated as the two new daughter duplexes (d and e) are formed. In the next cycle, all the steps are repeated. Again, the replication of only the duplex (e) has been shown. The daughter duplexes (f) and (g) are formed. In the next cycle, the duplex (g) has been shown in the figure, forming the daughter duplexes (h) and (i). It may be noted that the duplexes (g) onwards are all of the same length bounded by the two primers. The PCR product would thus be of a fixed size, exhibiting a specific band of DNA on electrophoresis.

complementary hybrids survive, thus imparting strict specificity and hence purer polymerized products. The reaction rate is very high and hence greater yield is obtained in much less time.

9.40 DNA polymerases of prokaryotic viruses

Viruses are generally known to mobilize the host machinery for their multiplication.

Many viruses, however, code for their own DNA polymerase. The bacteriophage T2 is one such virus. The rate of DNA synthesis in *E. coli* is enhanced 10-fold after infection with bacteriophage T2 because of the elaboration of the T2-coded DNA polymerase (Aposhian and Kornberg 1962). The T4 bacteriophage–coded DNA polymerase has also been purified, which functions as a holoenzyme of the core polymerase and three other phage-coded proteins (Goulian *et al.* 1968). The enzyme requires a template, a primer and deoxyribonucleoside 5′-triphosphates for replication that proceeds in the 5′→3′ direction. The T4 DNA polymerase has 3′→5′ exonuclease activity (Bedinger and Alberts 1983), which possibly contributes to its fidelity (Sinha 1987). The exonuclease did remove *in vitro* 50–86% of alkyl mispairs (Khare and Eckert 2001). T4 DNA polymerase apparently lacks the 5′→3′ exonuclease activity. The DNA polymerase from the bacteriophage T7 has been purified and its crystal structure has been elucidated (Doublie *et al.* 1998). The T7 DNA polymerase is similar to the Klenow fragment of the *E. coli* DNA polymerase I, in having an amino terminal 3′→5′ exonuclease domain and a carboxy terminal polymerase domain. The polymerase domain of the T7 DNA polymerase, the Klenow fragment of *E. coli* DNA polymerase I, the *Thermus aquaticus* DNA polymerase I and the *B. stearothermophilus* DNA polymerase I, all have a shape of a human right hand in which the fingers, thumb and palm subdomains form a DNA-binding groove that leads to the polymerase active site. On the basis of the amino acid sequence, these DNA polymerases have been classified to belong to family A (Blanco *et al.* 1991; Shinkai *et al.* 2001).

9.41 DNA polymerase α from eukaryotes

Quite a number of DNA polymerases have by now been isolated from eukaryotic cells. The name DNA polymerase α was given to the enzyme that was isolated from calf thymus (Yoneda and Bollum 1965), the first DNA polymerase from a eukaryote. The name was given in a conference held in Asilomar, CA, convened to formalize the nomenclature of eukaryotic DNA polymerases (Weissbach *et al.* 1975). DNA polymerase α was then characterized as the enzyme that was an acidic protein with a molecular weight higher than 100,000 Da which was inhibited by sulphydryl reagents. All eukaryotes have a DNA polymerase analogous to the calf thymus enzyme. DNA polymerase α, irrespective of its eukaryotic source, is characterized by its nuclear location and inhibition *in vitro* by the sulphydryl inhibitor N-ethylmaleimide (Kaguni and Lehman 1988) and the antibiotic aphidicolin (Ohashi *et al.* 1978) and insensitivity to dideoxynucleoside triphosphates (Edenberg *et al.* 1978). DNA polymease α from human KB cells was found to utilize 'activated' duplex DNA template–primers prepared by treatment of Salmon sperm DNA with beef pancreatic DNase till 25% of it became acid soluble (Schlabach *et al.* 1971). The enzyme, however, would work much better on a template–primer that had single-strand gaps of 30–60 nucleotides (structure 'd' in Fig. 9.33) but lacked activity on double-stranded DNA that had just nicks (structure 'b' in Fig. 9.33), or had gaps of less than 30 nucleotides (structure 'c' in Fig. 9.33) or on polydA:oligodT template–primer (failed to convert structure 'b' to structure 'c', as shown in Fig. 9.9; Fisher *et al.* 1979; Wang and Korn 1980). DNA polymerase I isolated from *Saccharomyces cerevisiae* is similar to DNA polymerase α of higher eukaryotes (Morrison *et al.* 1990). Common structural features exist in DNA polymerase α isolated from various eukaryotes. These polymerases are composed of four distinct tightly associated structural units: one large unit comprising a 180-kDa polypeptide or a group of polypeptides 140–185 kDa in size, a polypeptide of around 70 kDa and two smaller polypeptides of around 60 kDa and 50 kDa (Lehman and Kaguni 1989).

DNA polymerase α is thought to be the enzyme directly involved in the replication of nuclear DNA for the following reasons (Lehman and Kaguni 1989):

- The level of DNA polymerase α is highest in cells that are dividing rapidly, but is comparatively low in non-dividing cells.
- Cellular DNA synthesis *in vivo* is drastically inhibited by aphidicolin, the antibiotic that inhibits DNA polymerase α *in vitro*.
- Monoclonal antibodies raised against DNA polymerase α inhibit DNA replication in permeabilized nuclei and also in cultured mammalian cells when the antibodies are directly injected into the cells.
- Mutant mouse cells that contain temperature-sensitive DNA polymerase α are unable to replicate DNA at the non-permissive temperature.
- Disruption of the gene for the largest subunit of DNA polymerase α of *S. cerevisiae* is lethal.

DNA polymerase α, whatever its source may be, replicates DNA in the 5′→3′ direction and this property has been ascribed to the largest polypeptide in the polymerase complex (Kaguni and Lehman 1988). The smallest

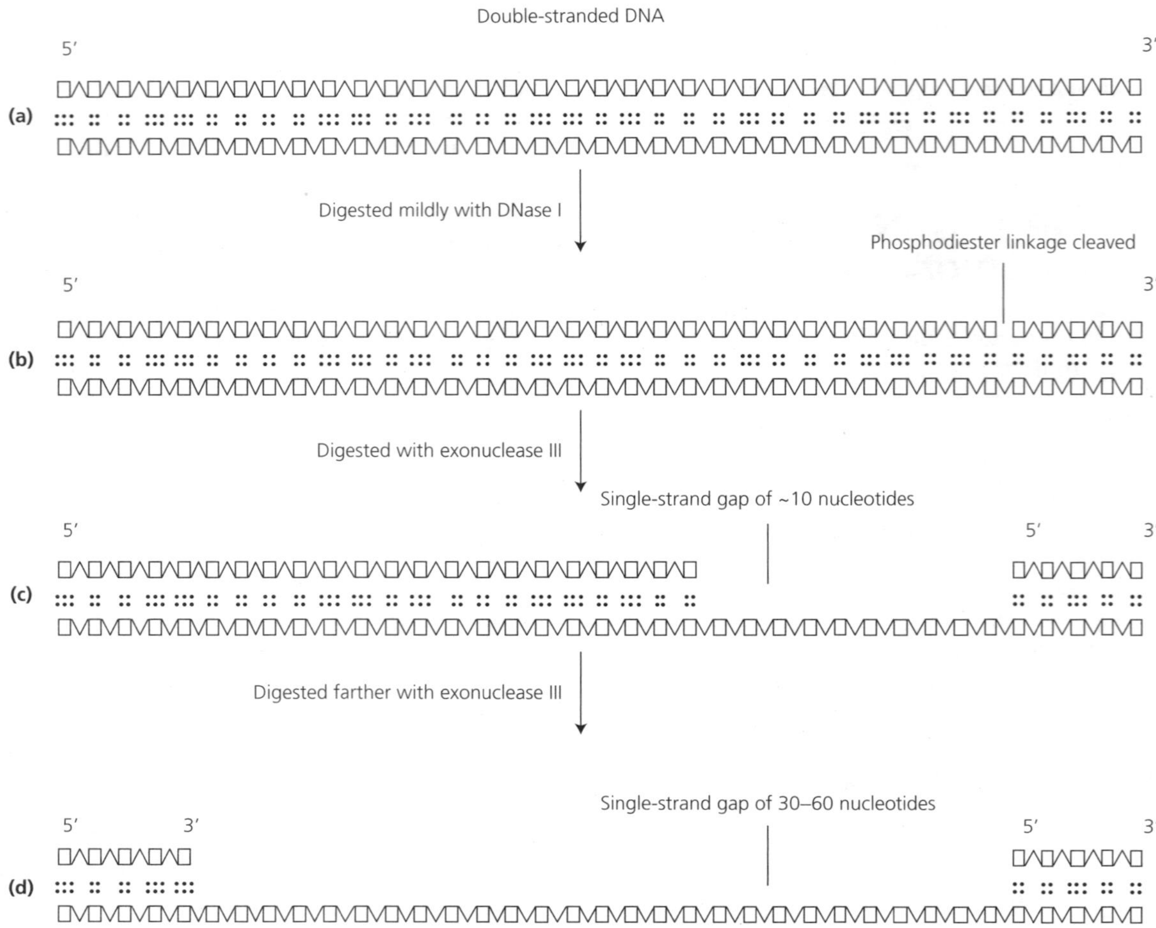

Figure 9.33 Different forms of DNA template–primer. Each small square (□) represents a nucleotide or base and the (/\) or (\/), a phosphodiester linkage. Each colon (:) between the two DNA bases in the two opposite strands represents a hydrogen bond. (a) Double-stranded DNA with no nick or gap in it. (b) Double-stranded DNA with a nick (a phosphodiester bond is cleaved, but no base is deleted) produced by mild treatment of DNA with DNase I from *Escherichia coli* or beef pancreatic DNase. (c) Double-stranded DNA with a single strand gap of about 10 nucleotides produced by treatment of the nicked DNA with exonuclease III from *E. coli.* (d) Double-stranded DNA with a single strand gap of 30–60 nucleotides produced by treatment of the nicked DNA with exonuclease III from *E. coli.*

two polypeptides are thought to provide the primase component which is essential for initiating the Okazaki fragments (Burgers 1989). Except for the *Drosophila* enzyme, which may have some cryptic activity, the purified whole DNA polymerase α complex has been found to lack the proofreading 3′→5′ exonuclease activity. It does not have any associated 5′→3′ exonuclease activity either (Chang and Bollum 1973; So and Downey 1988).

9.42 DNA polymerase δ from eukaryotes

DNA polymerase δ is also located in the nucleus and its activity is also inhibited *in vitro* by aphidicolin (Goscin and Byrnes 1982; Hammond *et al.* 1987; Goulian *et al.* 1990), but the important distinguishing features are that this enzyme has a proofreading 3′→5′ exonuclease activity (Byrnes *et al.* 1976; Goscin and Byrnes 1982: Crute *et al.* 1986), and does not have any primase activity (Lehman and Kaguni 1989). DNA polymerase δ can also be distinguished from DNA polymerase α by its chromatographic behaviour and preference for template–primer combination. DNA polymerase δ is more active with the poly(dA):oligo(dT) system (structure 'b' in Fig. 9.9), while DNA polymerase α is more active with the DNase-treated double-stranded DNA system (structure 'd' in Fig. 9.33; So and Downey 1988). DNA polymerase δ has been found to be hundred times less sensitive than DNA polymerase α towards the inhibitors butylphenyl-dGTP and butylanilino-dATP (Hammond *et al.* 1987). Polyclonal antibodies raised against DNA polymerase δ from human placenta or calf thymus did not inhibit DNA polymerase α from the same species. In addition, monoclonal antibodies to KB cell DNA polymerase α, though neutralized DNA polymerase α isolated from calf thymus, human placenta and rabbit bone marrow, did not inhibit DNA polymerase δ from the same sources (So and Downey 1988; Wong *et al.* 1989). Two-dimensional tryptic peptide mapping of DNA polymerase α and DNA polymerase δ did not reveal any relatedness between the two enzymes (Wong *et al.* 1989). Nevertheless, levels of DNA polymerase δ in different tissues parallel that of DNA polymerase α. DNA polymerase III isolated from *S. cerevisiae* is similar to DNA polymerase δ from higher eukaryotes (Morrison *et al.* 1990). The number of constituent polypeptides of DNA polymerase δ of different eukaryotic sources varies between 2 aned 3 and their sizes range between 40 and 225 kDa (Burgers 1989; Johansson *et al.* 2001; Goulian *et al.* 1990).

Is DNA polymerase δ essential for DNA replication *in vivo*? Following evidences would suggest that it is indeed so. Concentrations of butylphenyl-dGTP, which totally inhibited DNA polymerase α *in vitro* but not DNA polymerase δ, were only partially effective in reducing DNA replication *in vivo* in permeabilized CV-1 cells (Hammond *et al.* 1987). Besides, monoclonal antibody that specifically blocked DNA polymerase α but did not interfere with DNA polymerase δ activity *in vitro* could abolish only about 50% of DNA replication *in vivo* in permeabilized CV-1 cells (Hammond *et al.* 1987). Levels of DNA polymerase α and DNA polymerase δ declined in parallel during maturation and terminal differentiation of neonatal heart of rat. (Zhang and Lee 1987).

DNA polymerase δ appeared relatively non-processive when tested on the poly(dA):oligo(dT) template–primer system with a high-template–low-primer ratio. Addition of 'proliferating cell nuclear antigen' (PCNA) could, however, largely restore the processivity (Goulian *et al.* 1990; Syvaoja *et al.* 1990). It became known later that at least two separate domains of polymerase δ bind tightly to at least two domains of PCNA. In addition, it is thought that during replication, two molecules of PCNA interact with each molecule of polymerase δ (Johansson *et al.* 2004).

9.43 DNA polymerase ε from eukaryotes

A new class of DNA polymerase was encountered in the nucleus, which very much like DNA polymerase δ, was highly active on poly(dA):oligo(dT) template–primer (Crute *et al.* 1986), was inhibited by aphidicolin (Wahl *et al.* 1986; Goulian *et al.* 1990), had tightly associated 3′→5′ exonuclease activity (Wahl *et al.* 1986), which removed mismatched nucleotides but not matched nucleotides from the 3′-terminus of oligo(dT) annealed to poly(dA) (Syvaoja and Linn 1989) and did not have any associated primase activity. This enzyme was even more insensitive to the inhibitors butylphenyl-dGTP and butylanilino-dATP (Wahl *et al.* 1986). This enzyme, however, very much unlike DNA polymerase δ, exhibited high processivity on the poly(dA):oligo(dT) template–primer system with a high-template–low-primer ratio. Addition of PCNA to this new polymerase did not elicit any further stimulatory effect (Syvaoja and Linn 1989; Goulian *et al.* 1990; Syvaoja *et al.* 1990). Monoclonal antibody against PCNA had no inhibitory effect on this DNA polymerase, and the antibody did not recognize any peptide from this DNA polymerase in immunoblots (Syvaoja and Linn 1989). This enzyme was initially given the name DNA polymerase $δ_2$ or δII, while the original non-processive DNA polymerase δ was termed DNA polymerase $δ_1$ or δI. DNA polymerase ε is the new name of the PCNA-independent DNA polymerase $δ_2$ or δII (Syvaoja *et al.* 1990). DNA polymerase ε from HeLa cells was found to be different from both DNA polymerase α and DNA polymerase δ, as was evident from partial peptide mapping and lack of immunological cross-reaction (Syvaoja *et al.*1990). DNA polymerase ε was, however, inhibited by aphidicolin (Wahl *et al.* 1986). DNA polymerase II isolated from *S. cerevisiae* has been found to be similar to DNA polymerase ε from higher eukaryotes (Hamatake *et al.* 1990).

DNA polymerase ε (then known as DNA polymerase δII) has been known to be capable of DNA repair synthesis in UV-irradiated permeabilized human diploid fibroblasts. In contrast, DNA polymerase α could not promote any DNA repair synthesis. Monoclonal antibody raised against DNA polymerase α could not inhibit the repair synthesis catalysed by DNA polymerase ε (Nishida *et al.* 1988).

The gene for DNA polymerase II of *S. cerevisiae* has been cloned, and on the basis of base sequence data, the molecular weight of 2,55,649 Da has been assigned to the protein. Disruption of the gene results in the disappearance of the enzyme and loss of viability of the cells (Morrison *et al.* 1990). DNA polymerase II and by analogy DNA polymerase ε can thus be inferred as essential for DNA replication. Sizes of DNA polymerase ε from different eukaryotic sources range between 122 and 290 kDa (Goscin and Byrnes 1982; Crute *et al.* 1986; Nishida *et al.* 1988; Syvaoja and Linn 1989; Goulian *et al.* 1990; Morrison *et al.* 1990).

9.44 DNA polymerase β from eukaryotes

DNA polymerse β, a single polypeptide enzyme that is the smallest among all eukaryotic DNA polymerases, is also present in the nucleus. The rat enzyme is 39 kDa (Matsukage *et al.* 1987; Kumar *et al.* 1990), the mouse enzyme is 40 kDa (Tanabe *et al.* 1979), the human enzyme is 39 kDa (Wilson *et al.* 1988) and the enzyme from *Drosophila* is 110 kDa (Sakaguchi and Boyd 1985).

DNA polymerase β, however, is insensitive to aphidicolin or sulphydryl-blocking inhibitors such as N-ethyl maleimide, but is quite sensitive to dideoxy-NTPs and only somewhat sensitive to butylphenyl dGTP (Sakaguchi and Boyd 1985; Wilson *et al.* 1988). DNA polymerase β is capable of both DNA-dependent and RNA-dependent DNA synthesis, as is the case with DNA polymerase I from *E. coli* (Kumar *et al.* 1990), but does not have any associated 3′→5′ or 5′→3′ exonuclease activity (Tanabe *et al.* 1979). The processivity of DNA polymerase β has been found to be low. Only one nucleotide is added per polymerization cycle by human DNA polymease β in the presence of Mg^{++}, while 4–6 nucleotides are added in the presence of Mn^{++} (Wang and Korn 1982).

DNA polymerase β from human KB cells was active on DNase-nicked double-stranded DNA (structure 'b' in Fig. 9.33), incorporating ~15 nucleotides at each nick by strand displacement. DNA polymerase β, however, was more active with primer–templates having short gaps of 10–20 nucleotides (structure 'c' in Fig. 9.33; Wang and Korn 1980; Mosbaugh and Linn 1983), thus suggesting its role in DNA repair. Involvement of DNA polymerase β in DNA repair has also been inferred from other lines of evidence. The specific activity of this enzyme was twofold higher in skin fibroblast cell lines from patients with ataxia telangiectasia, which were more X-ray sensitive, than fibroblast cell lines from normal people (Mitchell *et al.* 1985). *In vitro* DNA excision repair of UV-induced thymine dimers was possible by HeLa DNA polymease β after removal of the dimers by HeLa DNaseV (Mosbaugh and Linn 1983; Evans and Linn 1984). DNA polymerase β has been identified to be the enzyme involved in DNA repair *in vivo* as inhibitors such as dideoxythymidine triphosphate, which specifically affect polymerase β *in vitro*, prevent DNA repair. This has been found to be particularly so when the extent of DNA damage has been low. Various DNA damaging agents such as bleomycin, neocarzinostatin, alkylating agents and UV ray have been used on HeLa cells and diploid fibroblasts. Involvement of DNA polymerase α in addition to DNA polymerase β has been inferred when the repair process was affected by inhibitors such as aphidicolin and sulphydryl reagents that specifically block DNA polymerase α *in vitro*, but not DNA polymerase β. Both DNA polymerase α and β were necessary when the extent of damage was high (Miller and Chinault 1982; Dresler and Liberman 1983; Yamada *et al.* 1985).

DNA polymerase β has now been shown to play an important role in the repair of both single-nucleotide base excision repair and long-patch base excision repair. For single-nucleotide base excision, polymerase β makes use of its constituent 5′-deoxyribosephosphate (dRP) lyase activity to remove the unmodified 5′-deoxyribose group (Matsumoto and Kim 1995) to facilitate ligation. For removal of modified 5′-deoxyribose group or for long-patch base excision repair by polymerase β, a separate enzyme, the flap endonuclease 1 is necessary for the removal of the DNA flap structure. It is now known that such repair takes place by alternating short gap production by flap endonuclease 1 and gap filling by polymerase β (Liu *et al.* 2005).

DNA polymerase β has also been implicated as the major polymerase involved in repair of oxidative base lesions. A single 8-oxoguanine or 5,6-dihydrouracil has indeed been found to be replaced by DNA polymerase β through the same constituent 5′-deoxyribosephosphate (dRP) lyase activity (Allinson *et al.* 2001).

Levels of DNA polymerase β in cultured animal cells were found to be unaltered irrespective of whether the cells were stationary or were growing vigorously and dividing (Chang *et al.* 1973; Mitchell *et al.* 1985). This has led to the belief that this enzyme could not be necessary for DNA replication.

9.45 DNA polymerase λ from eukaryotes

DNA polymerase β–deficient human cells have been found to be more sensitive to DNA-alkylating agents such as ethyl methane sulphonate (EMS) and methyl methane sulphonate (MMS), but not to UV radiation. This suggested that other yet unidentified polymerase β such as DNA polymerase might exist. Degenerate primers were designed on the basis of sequences homologous in human and some yeast DNA polymerase β and a human cDNA library was screened by PCR. A human cDNA was thus isolated which coded for a DNA polymerase that was similar to polymerase β but was distinct from it. This new polymerase was named DNA polymerase β2 (Nagasawa *et al.* 2000), but has now been renamed DNA polymerase λ (Burgers *et al.* 2001). An analogous gene (POLL) was identified in a genomic clone from chromosome 19 of mouse (*Mus musculus*) by computer (BLAST) search of the 'high-throughput genome sequence' (HTGS) database using different conserved sequences of the catalytic core of various DNA polymerase β and allied enzymes as probes (Garcia-Diaz *et al.* 2000). A sequence of segments of DNA from the genomic clone was chosen on the basis of the amino acid sequence of DNA polymerase β and was employed to search the expressed sequence tags (ests) database using the GAPPED BLAST programme. One cDNA clone that contained the complete coding sequence of the new protein was overexpressed in *E. coli* and the protein was found to have DNA polymerase activity. This new polymerase was named DNA polymerase λ. This is the agreed new name of this and other homologous enzymes (Burgers *et al.* 2001). The murine DNA polymerase λ consists of 573 amino acid residues and has 32% identity to murine DNA polymerase β. Isolation of mouse and human DNA polymerase λ has also been reported by Aoufouchi *et al.* (2000) by similar screening of sequence databases and searching of the ESTs. Physical clones were finally obtained from cDNA libraries and the protein which had DNA polymerase activity was obtained by overexpression. DNA polymrase λ appeared to be strictly distributive and lacked any 3′→5′ exonuclease activity. Interestingly, DNA polymerase λ was downregulated by UV rays (Aoufouchi *et al.* 2000).

High levels of DNA polymerase λ had been detected in mouse testis and was then speculated to be involved in DNA repair synthesis associated with meiosis (Garcia-Diaz *et al.* 2000). Subsequent studies have indeed confirmed its involvement in DNA repair processes (Hubscher *et al.* 2002; Braithwaite *et al.* 2005). Polymerase λ has, therefore, been classified as a member of the pol X family along with polymerase β and polymerase μ.

Polymerase λ has been thought to be involved in a number of DNA repair processes such as abasic site translesion DNA synthesis, base excision repair and double-strand break repair (Garcia-Diaz *et al.* 2002; Lee *et al.* 2004). Polymerase λ tends to produce −1 frameshifts as a result of microhomology base pairing between two DNA strands and this possibly facilitates abasic site bypass (Blanca *et al.* 2004; Garcia-Diaz *et al.* 2004).

A feature of DNA polymerase λ that is worth mentioning is that it can elongate RNA primers (Villani *et al.* 2003).

An interesting finding on DNA polymerase λ has been that it has in the same polymerase β-like core domain in the C-terminal region, a terminal deoxynucleotidal transferase activity (see Section 9.8), in addition to the DNA polymerase activity (Villani *et al.* 2003; Ramadan *et al.* 2004). The terminal deoxynucleotidal activity of polymerase λ is capable of adding several nucleotides in a template-independent manner to the 3′-OH end of a single-stranded DNA or a double-stranded DNA that contains a 3′-protruding single-stranded DNA tract 9–12 nucleotides long (Maga *et al.* 2005).

PCNA has been found to stabilze binding of DNA polymerase λ to the 3′-OH end of DNA (Maga *et al.* 2002). Interestingly, PCNA appears to stimulate the terminal deoxynucleotidal transferase activity rather than the DNA polymerase activity of polymerase λ. On the other hand, replication protein A preferentially stimulates DNA polymerase activity of polymerase λ in preference to its terminal transferase activity (Maga *et al.* 2005).

9.46 DNA polymerase μ from eukaryotes

A database search using the protein sequence of human terminal deoxynucleotidyl transferase has enabled the retrieval of the sequence of yet another mammalian DNA polymerase. The protein has been obtained by overexpression of the cDNA clone and has indeed been found to have template-dependent DNA polymerase activity with no associated 3′→5′ exonuclease activity. This new polymerase has been named DNA polymerase μ and was found to be more processive compared with DNA polymerase λ (Aoufouchi *et al.* 2000). This polymerase was downregulated by UV rays (Aoufouchi *et al.* 2000). Subsequent evidence have suggestsed that DNA polymerase μ is involved in non-homologous end joining of double-stranded DNA breaks. An interesting feature of mammalian DNA polymerase μ is that it incorporates rNTPs into DNA with high efficiency.

DNA polymerase μ was also found to exhibit deoxynucleotidal transferase activity (Dominguez *et al.* 2000), but in contrast to polymerase λ, it needs a recessed 3′-end for this activity (Covo *et al.* 2004).

9.47 DNA polymerase σ from eukaryotes

The gene *TRF4*, together with its close homolog *TRF5*, was identified in the yeast *S. cerevisiae* as essential in chromosome segregation (Castano *et al.* 1996a; Castano *et al.* 1996b) and repair of DNA damaged by some chemical agents (Walowsky *et al.* 1999) and was required for sister chromatid cohesion (Wang *et al.* 2000). Computer methods involving iterative database search, multiple alignment, motif analysis and structural modelling have led Aravind and Koonin (1999) to conclude that the *TRF4* and *TRF5* gene products belong to the DNA polymerase β-like nucleotidyltransferase superfamily. The cloned *TRF4* cDNA was overexpressed in *E. coli* and the protein was found to have template-dependent DNA polymerase activity (Wang *et al.* 2000). This new polymerase was initially named DNA polymerase κ (Wang *et al.* 2000), but later, it was decided to be named as DNA polymerase σ according to the revised nomenclature (Burgers *et al.* 2001). Very much like DNA polymerase β, DNA polymerase σ (initially called DNA polymerase κ) was found to be distributive in character in an oligo dT primer–poly dT template system, was resistant to aphidicolin and N-ethylmaleimide, but was inhibited by dideoxythymidine triphosphate. In contrast, the enzyme was quite processive when tried with a complex template–primer (Wang *et al.* 2000). Genes analogous to *TRF4* and *TRF5* have been identified in *S. pombe* (the *cyd* genes) and also in human cells.

9.48 DNA polymerase IV from *S. cerevisiae*

DNA polymerase IV is the sole representative of DNA repair polymerase of *S. cerevisiae* in contrast to the multiple DNA repair polymerases of mammalian cells. Yeast polymerase IV has constituent deoxyribose 5-phosphate lyase activity similar to mammalian DNA polymerase β and polymerase λ. Yeast DNA polymerase IV is capable of filling short gaps in DNA, the processivity of the enzyme being quite low. The fidelity of yeast DNA polymerase is very low, the errors generated are single-base deletions and T to C and A to G substitutions. Yeast DNA polymerase has a very low ability to discriminate against rNTPS (Bebenek *et al.* 2005).

9.49 DNA polymerase γ from eukaryotes

Yet another DNA polymerase has been isolated from mammalian cells, such as HeLa cells, chick embryo cells and rat liver cells, that has been named polymerase γ. It turned out that this polymerase was identical with the mitochondrial DNA polymerase described earlier (Bolden *et al.* 1977; Tanaka and Koike 1977; Tanaka and Koike 1978; Yamaguchi *et al.* 1980). DNA polymerase γ from chick embryo has been reported to be a tetramer of a single polypeptide of 47 kDa (Yamaguchi *et al.* 1980), from *Drosophila* embryo to consist of two polypeptides of 35 kDa and 125 kDa (Wernette and Kaguni 1986), and from *Xenopus laevis* to comprise a catalytic subunit of 140 kDa plus a 40-kDa and a 30-kDa polypeptide that copurify and cosediment with the catalytic sbunit (Insdorf and Bogenhagen 1989). DNA polymerase γ is resistant to aphidicolin, but is sensitive to N-ethylmaleimide and dideoxy TTP (Wernette and Kaguni 1986; Insdorf and Bogenhagen 1989). The rate of DNA synthesis with the enzyme from rat liver and *Drosophila melanogaster* is highest with the homopolymeric template–primer polydA-oligodT (would convert at a high rate structure 'b' to structure 'c', as shown in Fig. 9.9), the processivity being about 30 (Tanaka and Koike 1987; Wernette *et al.* 1988). The average size of dT product formed on an oligodT primer has been found to be ~4,100 nucleotides with the enzyme from chick embryo (Yamaguchi *et al.* 1980). DNA polymerase γ is unable to carry out displacement synthesis; nicked DNA (structure 'b' in Fig. 9.33) is not active as a template–primer. 'Activated' DNA, prepared by treatment of Salmon sperm DNA with beef pancreatic DNase till 25% of it became acid soluble (Schlabach *et al.* 1971), is a satisfactory template–primer (Tanaka and Koike 1977). This polymerase also contains a 3′→5′ exonuclease activity that copurifies and cosediments with the polymerizing activity, but lacks any 5′→3′ exonuclease activity (Kunkel and Soni 1988; Insdorf and Bogenhagen 1989; Kaguni and Olson 1989). In the presence of deoxynucleotide triphosphates, the enzyme excises *in vitro* mismatched nucleotides in the 3′-end of the primer with greater than 98% efficiency, prior to subsequent matched nucleotide addition [analogous to DNA polymerases from *E. coli*, as shown in Fig. 9.9 (a)–(c)]. DNA polymerase γ from *Drosophila* mitochondria has high fidelity and commits less than one error for every 2,60,000 nucleotides incorporated (Kunkel and Soni 1988: Wernette *et al.* 1988). The human mitochondrial DNA polymerase γ also has a high fidelity of one error in 2,80,000 base pairs and a high processivity of 1850 (Johnson and Johnson 2001a). With functional 3′→5′ exonucleolytic proofreading activity, the fidelity coud be as high as one error in 1 million base pairs (Johnson and Johnson 2001b).

9.50 DNA polymerase ζ from eukaryotes

DNA damage–induced mutagenesis in *S. cerevisiae* is dependent on the products of the genes *REV1, REV3 and REV7.* The polypeptide products of the overexpressed cloned genes *REV3* and *REV7* of the yeast *S. cerevisiae* associate to form a DNA polymerase, which has been named polymerase ζ (zeta) (Nelson *et al.* 1996a). The interesting property of this enzyme is that, though other polymerases get stalled, this one can replicate past a thymine–thymine *cis-syn* cyclobutane dimer without repairing it. Two adenines are correctly inserted opposite the thymine dimer. The efficiency of bypass *in vitro* is ~10% with polymerase ζ, while it is less than 1% with polymerase α. DNA polymerase ζ has a low processivity and lacks any 3′→5′ exonuclease activity. It is insensitive to aphidicolin and dideoxy NTP, but is moderately sensitive to butyl phenyl guanosine triphosphate (Nelson *et al.* 1996a). PCNA substantially stimulates translesion synthesis by polymerase ζ of UV-damaged DNA, even though neither of the subunits of the polymerase has the consensus PCNA-binding motif QxxLxxFF (Garg *et al.* 2005).

9.51 *S. cerevisiae REV1* gene codes for a deoxycytidyl transferase

The cloned *REV1* gene has been overexpressed in yeast and the ~116-kDa protein was found to have a template-dependent deoxynucleotidyl transferase activity highly specific for dCTP (Nelson *et al.* 1996b). No other nucleotide was added and no other dNTP served as a substrate. One C residue was added at the 3′-end of the primer across a G residue in the template, while two consecutive C residues were added if two consecutive G residues were present across in the template. Nothing was added if the first template nucleotide was a C or T. One C was, however, added to the primer if the first nucleotide in the template was A, even though at about half the efficiency. No more C after the first one was added even if a series of As were there in the template (Nelson *et al.* 1996b). A template oligonucleotide containing a dUMP residue was constructed and treated with uracil-*N*-glycosylase, thus creating an abasic site. The *REV1* protein added C across this abasic site five times more efficiently than across a G in the template (Nelson *et al.* 1996b). In a prallel experiment, it has been found that the efficiency of synthesis by DNA polymerase ζ fell to 7%, when a dU in the template was converted into an abasic site. This efficiency was enhanced fivefold in the presence of the *REV1* protein (Nelson *et al.* 1996b).

It is relevant to mention here that preferential insertion of C across an abasic site has been observed to occur in *S. cerevisiae in vivo* also. *S. cerevisiae* cells were transformed with an otherwise duplex plasmid vector with a single abasic site specifically located within a 28-nucleotide single-stranded region. The plasmid was reisolated after several generations and base sequencig done of the specific region. Base C was found to have appeared across the abasic site in more than 80% of the cases (Gibbs and Lawrence 1995). It is possible that the *REV1* protein is instrumental in the preferential insertion of C across abasic sites in *S. cerevisiae.*

9.52 DNA polymerase η from eukaryotes

The *RAD30* gene of *S. cerevisiae* codes for yet another DNA polymerase that has been named polymerase η (Johnson *et al.* 1999a, Johnson *et al.* 1999b). DNA polymerase η can replicate and bypass with more than 80% effeciency a thymine–thymine *cis-syn* cyclobutane dimer, incorporating two adenines opposite the dimer even when incubated *in vitro* with all the four deoxynucleotide triphosphates. DNA polymerase η is resistant to aphidicolin and N-ethylmaleimide, but is modestly sensitive to dideoxy NTPs. Deletions in the genes for DNA polymerase ζ and DNA polymerase β in the chromosome of *S. cerevisiae* have no effect on the activity of DNA polymerase η (Johnson *et al.* 1999a).

Yeast polymerase η can also incorporate the correct nucleotide C opposite 7,8-dihydro-8-oxoguanine, the binding efficiency of pol η to dCTP being the same opposite for both G and 8-oxoG. It has been inferred that yeast pol η replicates DNA through 8-oxoG as if the lesion has offered no barriers (Carlson and Washington 2005).

The human counterpart of yeast *RAD30* gene is *hRAD30A*, which codes for human DNA polymerase η. The gene *hRAD30A* has been found to harbour nonsense or frameshift mutation in cell lines derived from the skin of Xeroderma pigmentosum patients who suffer from an autosomal recessive disease that makes them prone to skin cancer (Johnson *et al.* 1999c). Human DNA polymerae η has been purified. Its fidelity, however, is low, and it misincorporates every 10^{-2} to 10^{-3} bases. Nevertheless, the misincorporation frequency across a thymine–thymine *cis-syn* cyclobutane dimer is not any higher than the misincorporation frequency across two adjacent normal thymines (Johnson *et al.* 2000a).

The ability of DNA polymerase η to bypass other damaged or modified bases in DNA has been assessed. Some of the guanine residues in DNA are converted into O^6-methyl guanine (m6G) on treatment with alkylating agents such as N-methyl-N'-nitro-N-nitrosoguanidine and this modified base presents a block to DNA polymerases. Both yeast and human DNA polymease η are able to bypass m6G lesions in DNA by inserting C opposite it (Haracska *et al.* 2000a). The adduct 7,8-dihydro-8-oxoguanine is formed by the action of oxygen-free radicals on DNA. Other DNA polymerases incorporate A opposite this adduct, but DNA polymerase η is able to incorporate C correctly (Haracska *et al.* 2000b). DNA polymerase η can correctly bypass a (6-4)TT photoproduct, provided, of course, DNA polymerase ζ is also around (Johnson *et al.* 2001). DNA polymerase η, whether from yeast or from human, is however very inefficient in both inserting a nucleotide opposite an abasic site and extending from the nucleotide, if at all inserted (Haracska *et al.* 2001a). PCNA, together with replication factor C (RFC) and replication protein A (RPA), stimulates ~12 fold the activity of human DNA polymerase η (Haracska *et al.* 2001c).

9.53 DNA polymerase κ from eukaryotes

While searching the database of expressed sequence tags (Dbest) for human cDNAs that share homology with the yeast *RAD30* gene, a second cDNA that was similar but distinct from the *hRAD30A* gene was identified. As this clone shared a high degree of homology with the *E. coli dinB* gene, this was named *hDINB1*. The protein obtained on overexpression of this gene was found to be a DNA polymerase. This new DNA polymerase was first named DNA polymerase θ (Johnson *et al.* 2000b), but has now been renamed DNA polymerase κ (Burgers *et al.* 2001). Human DNA polymerase κ has been found to be unable to bypass a *cis-syn* thymine–thymine dimer or a (6-4) photoproduct or an abasic site (Johnson *et al.* 2000b). The fidelity of DNA polymerase κ on an undamaged template has been found to be low, frequency of misincorporation of deoxyribonucleotides being 10^{-4} to10^{-3} (Johnson *et al.* 2000b). PCNA, together with RFC and RPA, stimulates 50–200 fold the activity of DNA polymerase κ (Haracska *et al.* 2002). DNA polymerase κ has now been found to be essential for recovery from S-phase checkpoint arrest induced by the carcinogen benzo[α]pyrene-dihydrodiolepoxide in human lung carcinoma H1299 cells (Bi *et al.* 2005).

9.54 DNA polymerase ι from eukaryotes

Among the human cDNAs that shared homology with the yeast *RAD30* gene, a third cDNA that was similar but distinct from the *hRAD30A* gene was identified and named *hRAD30B* (Johnson *et al.* 2000c). This gene coded for a 81-kDa DNA polymerase, which has been named DNA polymerase ι. A counterpart of this polymerase has not been found in yeast. DNA polymerase ι, in contrast to DNA polymerase η, is unable to replicate through a *cis-syn* T–T dimer, and cannot even insert a nucleotide across the 3′-T of the dimer. DNA polymerase ι, however, is capable of incorporating nucleotides opposite the 3′-T of the (6-4) T–T photoproduct and opposite an abasic site (Johnson *et al.* 2000c). DNA polymerase ζ, the efficient extender of mismatched base pairs, readily extends from the nucleotides inserted by DNA polymerase ι across the lesion sitees (Johnson *et al.* 2000c). Human DNA polymerase ι is a low-fidelity copier of even undamaged DNA, frequently incorporating G against T rather than A (Vaisman *et al.* 2001). PCNA, together with RFC and RPA, has also been found to stimulate 80–150 fold the activity of DNA polymerase ι (Haracska *et al.* 2001b). The processivity of polymerase ι [and also of other class Y polymerases] is, however, not enhanced by binding to PCNA. Though polymerase ι contains three potential PCNA-binding sites, the binding actually occurs only at one site, unlike the situation with polymerase δ (Haracska *et al.* 2005).

Goodman (2002) has reviewed the error-prone repair DNA polymerases in both prokaryotes and eukaryotes.

9.55 DNA polymerase θ from eukaryotes

The gene *mus38* in *Drosophila melanogaster* was identified as one that conferred resistance to some chemical agents such as nitrogen mustard, which damaged DNA (Boyd *et al.* 1990) The gene *mus38* was subsequently cloned and shown to have extensive homology with the prokaryotic DNA polymerase I genes (Harris *et al.* 1996). While searching (BLAST) the database of ESTs for human foetal spleen cDNAs that share homology with the *Drosophila mus38* gene, a cDNA was indeed detected with extensive similarity to the *mus38* gene. Similar clones were also present in a HeLa cell cDNA library, a human KB cell cDNA library and a human cerebellum cDNA library, but not in any *S. cerevisiae* library. The putative protein product has a C-terminal region that contains the canonical DNA polymerase motifs A, B and C found in DNA polymerase I of *E. coli* and an N-terminal region that contains an ATP-binding domain. The putative protein product has been named DNA polymerase θ, which would have 1762 amino acids and a calculated molecular mass of 198 kDa (Sharief *et al.* 1999).

9.56 Classification of DNA polymerases

DNA polymerases have been classified into six main groups based upon phylogenetic relationships with *E. coli* DNA polymerase I (class A), *E. coli* DNA polymerase II (class B), *E. coli* DNA polymerase III (class C), Euryarchaeotic DNA polymerases II (class D), human DNA polymerase β (class X) and *E. coli* UmuC/DinB and *S. cerevisiae RAD30*/human Xeroderma pigmentosum variant (class Y; Braithwaite and Ito 1993; Burgers *et al.* 2001; Ohmori *et al.* 2001). No eukaryotic homologue of class C or class D DNA polymerases has so far been detected. The eukaryotic DNA polymerases and their main functions have been reviewed by Burgers *et al.* (2001), while Bebenek and Kunkel (2004) have reviewed the functions of DNA polymerases from both prokaryotes and eukaryotes.

References

Allinson SL, Dianova II, Dianov GL. DNA polymerase β is the major dRP lyase involved in repair of oxidative base lesions in DNA by mammalian cell extracts. *EMBO J* 2001;20:6919–6926.

Aoufouchi S, Flatter E, Dahan A, Faili A, Bertocci B, Storck S, *et al.* Two novel human and mouse DNA polymerases of the polX family. *Nucleic Acids Res* 2000;28:3684–3693.

Aposhian V, Kornberg A. Enzymatic synthesis of deoxyribonucleic acid. IX. The polymerase formed after T2 bacteriophage infection of *Escherichia coli*: A new enzyme. *J Biol Chem* 1962;237:519–525.

Aravind L, Koonin EV. DNA polymerase β-like nucleotidyltransferase superfamily: Identification of three new families, classification and evolutionary history. *Nucleic Acids Res* 1999;27:1609–1618.

Bebenek K, Garcia-Diaz M, Patishall SR, Kunkel TA. Biochemical properties of *Saccharomyces cerevisiae* DNA polymerase IV. *J Biol Chem* 2005;280:20051–20058.

Bebenek K, Kunkel TA. Functions of DNA polymerases. Adv Protein Chem 2004;9:137–165.

Becker A, Lyn G, Gefter M, Hurwitz J. The enzymatic repair of DNA. II. Characterization of phage induced sealase. *Proc Natl Acad Sci USA* 1967;58:1996–2003.

Bedinger P, Alberts BM. The 3′–5′ proofreading exonuclease of bacteriophage T4 DNA polymerase is stimulated by other T4 DNA replication proteins. *J Biol Chem* 1983;258:9649–9656.

Beese LS, Derbyshire V, Steitz TA. Structure of DNA polymerase I Klenow fragment bound to duplex DNA. Science 1993;260:352–355.

Belkin S, Jannasch HW. A new extremely thermophilic, sulfur-reducing heterotrophic, marine bacterium. *Arch Microbiol* 1985;141:181–186.

Bi X, Slater DM, Ohmori H, Vaziri C. DNA polymerase κ is specifically required for recovery from the benzo[α]pyrene-dihydrodiol epoxide

(BPDE)-induced S-phase checkpoint. *J Biol Chem* 2005;280: 22343–22355.

Blanca G, Villani G, Shevelev I, Ramadan K, Spadari S, Hubscher U, *et al.* Human DNA polymerase λ and β show different efficiencies of translesion DNA synthesis past abasic sites and alternative mechanisms for frameshift generation. Biochemistry 2004;43:11605–11615.

Blanco L, Bernard A, Blasco MA, Salas M. A general structure for DNA-dependent DNA polymerases. *Gene* 1991;100:27–38.

Bolden A. Noy GP, Weissbach A. DNA polymerase of mitochondia is a γ polymerase. *J Biol Chem* 1977;252:3351–3356.

Boule J-B, Rougeon F, Papanicolaou C. Terminal deoxynucleotidyl transferase indiscriminately incorporates ribonucleotides and deoxyribonucleotides. *J Biol Chem* 2001;276:31388–31393.

Boyd JB, Sakaguchi K, Harris PV. *mus308* mutants of *Drosophila* exhibit hypersensitivity to DNA cross-linking agents and are defective in a deoxyribonuclease. *Genetics* 1990;125:813–819.

Braithwaite DK, Ito J. Compilation, alignment, and phylogenetic relationships of DNA polymerases. *Nucleic Acids Res* 1993;21: 787–802.

Braithwaite EK, Prasad R, Shock DD, Hou EW, Beard WA, Wilson SH. DNA polymerase λ mediates a back-up base excision repair activity in extracts of mouse embryonic fibroblasts. *J Biol Chem* 2005;280:18469–18475.

Brotcorne-Lannoye A, Maenhaut-Michel G. Role of RecA protein in untargeted UV mutagenesis of bacteriophage λ: Evidence for the requirement for the *dinB* gene. *Proc Natl Acad Sci USA* 1986;83:3904–3908.

Brutlag D, Atkinson MA, Setlow P, Kornberg A. An active fragment of DNA polymerase produced by proteolytic cleavage. *Biochem Biophys Res Commun* 1969;37:982–989.

Brutlag D, Kornberg A. Enzymatic synthesis of deoxyribonucleic acid. XXVI. A proofreading function for the 3′–5′ exonuclease activity in deoxyribonucleic acid polymerases. *J Biol Chem* 1972;247:241–248.

Burckhardt SE, Woodgate R, Scheuermann RH, Echols H. UmuD mutagenesis protein of *Escherichia coli*: Overproduction, purification, and cleavage by RecA. *Proc Natl Acad Sci USA* 1988;85:1811–1815.

Burgers PMJ. Eukaryotic DNA polymerases α and δ: Properties and interactions, from yeast to mammalian cells. *Prog Nucleic Acid Res Mol Biol* 1989;37: 235–280.

Burgers PMJ, Koonin EV, Bruford E, Blanco L, Burtis KC, Christman MF, *et al.* Eukaryotic DNA polymerases: Proposal for a revised nomenclature. *J Biol Chem* 2001;276: 43487–43490.

Byrnes JJ, Downey KM, Black VL, So AG. A new mammalian DNA polymerase with 3′–5′ exonuclease activity: DNA polymerase δ. *Biochemistry* 1976;15:2817–2823.

Caillet-Fauquet P, Defais M, Radman M. Molecular mechanisms of induced mutagenesis. Replication *in vivo* of bacteriophage fx174 single-stranded, ultraviolet light irradiated DNA in intact and irradiated host cells. *J Mol Biol* 1977;117:95–112.

Cairns J. The bacterial chromosome and its manner of replication as seen by autoradiography. *J Mol Biol* 1963;6:208–213.

Campbell JL, Soll L, Richardson CC. Isolation and partial characterization of a mutant of *Escherichia coli* deficient in DNA polymerase II. Proc Natl Acad Sci USA 1972;69:2090–2094.

Cann IKO, Ishino Y. Archaeal DNA replication: Identifying the pieces to solve a puzzle. Genetics 1999;152:1249–1267.

Carlson KD, Washington MT. Mechanism of efficient and accurate nucleotide incorporation opposite 7,8-dihydro-8-oxoguanine by *Saccharomyces cerevisiae* DNA polymerase η. *Mol Cell Biol* 2005;25:2169–2176.

Castano IB, Brzoska PM, Sadoff BU, Chen H, Christman MF. Mitotic chromosome condensation into the rDNA requires *TRF4* and DNA topoisomerase I in *Saccharomyces cerevisiae. Genes Dev* 1996a;10:2564–2576.

Castano IB, Heath-Pagliuso S, Sadoff BU, Fitzhugh DJ, Christman MF. A novel family of *TRF* (DNA topoisomerase I-related function) genes required for proper nuclear segregation. *Nucleic Acids Res* 1996b;24:2404–2410.

Chang LMS, Bollum FJ. A comparison of associated enzyme activities in various deoxyribonucleic acid polymerases. *J Biol Chem* 1973;248:3398–3404.

Chang LMS, Brown M, Bollum FJ. Induction of DNA polymerase in mouse L cells. *J Mol Biol* 1973;74:1–8.

Covo S, Blanco L, Livneh Z. Lesion bypass by human DNA polymerase μ reveals a template-dependent, sequence-independent nucleotidyl transferase activity. *J Biol Chem* 2004;279:859–865.

Cozzarelli NR, Low RL. Mutational alteration of *Bacillus subtilis* DNA polymerase III to hydroxyphenylazopyrimidine resistance: Polymerase III is necessary for DNA replication. *Biochem Biophys Res Commun* 1973;51:151–157.

de Lucia P, Cairns J. Isolation of an *E. coli* strain with a mutation affecting DNA polymerase. *Nature* 1969;224:1164–1166.

Deutscher M, Kornberg A. Enzymatic synthesis of deoxyribonucleic acid: XXVIII. The pyrophosphate exchange and pyrophosphorolysis reactions of deoxyribonucleic acid polymerase. *J Biol Chem* 1969;244:3019–3028.

Dominguez O, Ruiz JF, Lain de Lera T, Garcia-Diaz M, Gonzalez MA, Kirchoff T, *et al.* DNA polymerase mu (Pol μ), homologous to TdT, could act as a DNA mutator in eukaryotic cells. *EMBO J* 2000;19:1731–1742.

Doublie S, Tabor S, Long AM, Richardson CC, Ellenberger T. Crystal structure of a bacteriophage T7 DNA eplication complex 2.2 Ao resolution. *Nature* 1998;391:251–258.

Dresler SL, Liberman MW. Identification of DNA polymerases involved in DNA excision repair in diploid human fibroblasts. *J Biol Chem* 1983;258:9990–9994.

Dunning AM, Talmud P, Humphries SE. Errors in polymerase chain reaction. *Nucl Acids Res* 1988;16:10393–10396.

Eckert KA, Kunkel TA. High fidelity DNA synthesis by the *Thermus aquaticus* DNA polymerase. *Nucl Acids Res* 1990;18:3739–3744.

Edenberg HJ, Anderson S, DePamphilis ML. Involvement of DNA polymerae α in simian virus 40 DNA replication. *J Biol Chem* 1978;253:3273–3280.

Evans DH, Linn S. Excision repair of pyrimidine dimers from simian virus 40 minichromosomes *in vitro*. *J Biol Chem* 1984;259: 10252–10259.

Fisher PA, Wang TS-F, Korn D. Enzymological characterization of DNA polymerase α. Basic catalytic properties, processivity, and gap utilization of the homogeneous enzyme from human KB cells. *J Biol Chem* 1979;254:6128–6137.

Fuchs RP, Fujii S, Wagner J. Properties and functions of *Escherichia coli* Pol IV and Pol V. *Adv Protein Chem* 2004;69:229–264.

Garcia-Diaz M, Bebenek K, Krahn JM, Blanco L, Kunkel TA, Pedersen LC. A structural solution for the DNA polymerase λ-dependent repair of DNA gaps with minimal homology. *Mol Cell* 2004;13:561–572.

Garcia-Diaz M, Bebenek K, Sabariegos R, Dominguez O, Rodriguez J, Kirchoff T, *et al.* DNA polymerase λ, a novel DNA repair enzyme in human cells. *J Biol Chem* 2002;277:13184–13191.

Garcia-Diaz M, Dominguez O, Lopez-Fernandez LA, de Lera LT, Saniger ML, Ruiz JF, *et al.* DNA polymerase lambda (Pol λ), a novel eukaryotic DNA polymerase with a potential role in meiosis. *J Mol Biol* 2000;301:851–867.

Garg P, Stith CM, Majka J. Burgers PMJ Proliferating cell nuclear antigen promotes translesion synthesis by DNA polymerase ζ. *J Biol Chem* 2005;280:23446–23450.

Gefter ML, Hirota Y, Kornberg T, Wechsler JA, Barnoux C. Analysis of DNA polymerase II and III in mutants of *Escherichia coli* thermosensititive for DNA synthesis. *Proc Natl Acad Sci USA* 1971;68:3150–3153.

Gefter ML, Molineux IJ, Kornberg T, Khorana HG. Deoxyribonucleic acid synthesis in cell-free extracts. III. Catalytic properties of deoxyribonucleic acid polymerase II. *J Biol Chem* 1972;247: 3321–3326.

Gellert M. Formation of covalent circles of lambda DNA by *E. coli* extracts. *Proc Natl Acad Sci USA* 1967;57:148–155.

Gibbs PEM, Lawrence CW. Novel mutagenic properties of abasic sites in *Saccharomyces cerevisiae. J Mol Biol* 1995;251: 229–236.

Gilfilan S, Dierich A, Lemeur M, Benoist C, Maltes D. Mice lacking TdT: Mature animals with an immature lymphocyte repertoire. *Science* 1993;261:1175–1178.

Gong C, Martins A, Bongiorno P, Glickman M, Shuman S. Biochemical and genetic analysis of the four DNA ligases of mycobacteria. *J Biol Chem* 2004;279:20594-20606.

Goodman MF. Error-prone repair DNA polymerases in prokaryotes and eukaryotes. *Annu Rev Biochem* 2002;71:17–50.

Goscin LP, Byrnes JJ DNA polymerase δ: One polypeptide, two activities. Biochemistry 1982;21:2513–2518.

Goulian M, Herrmann SM, Sackett JW, Grimm SL. Two forms of DNA polymerase δ from mouse cells. Purification and properties. *J Biol Chem* 1990;265:16402–16411.

Goulian M, Kornberg A. Enzymatic synthesis of DNA. XXIII. Synthesis of circular replicative form of phage ϕx 174 DNA. *Proc Natl Acad Sci USA* 1967;58:1723–1730.

Goulian M, Kornberg A, Sinsheimer RL. Enzymatic synthesis of DNA. XXIV. Synthesis of infectious phage phix 174 DNA. *Proc Natl Acad Sci USA* 1967;58:2321–2328.

Goulian M, Lucas ZJ, Kornberg A. Enzymatic synthesis of DNA. XXV. Purification of properties of deoxyribonucleic acid polymerase induced by infection with phage T4. *J Biol Chem* 1968;243:627–638.

Hamatake RK, Hasegawa H, Clark AB, Bebenek K, Kunkel TA, Sugino A. Purification and characterization of DNA polymerase II from the yeast *Saccharomyces ceevisiae.* Identification of the catalytic core and a possible holoenzyme form of the enzyme. *J Biol Chem* 1990;265:4072–4083.

Hammond RA, Byrnes JJ, Miller MR. Identification of DNA polymerase δ in CV-1 cells: Studies implicating both DNA polymerase δ and DNA polymerase α in DNA replication. *Biochemistry* 1987;26:6817–6824.

Haracska L, Acharya N, Unk I, Johnson RE, Hurwitz J, Prakash L, *et al.* A single domain in human DNA polymerase ι mediates interaction with PCNA: Implication for translesion DNA synthesis. *Mol Cell Biol* 2005;25:1183–1190.

Haracska L, Johnson RE, Unk I, Phillips BB, Hurwitz J, Prakash L, Prakash S. Physical and functional interactions of human DNA polymerase η with PCNA. *Mol Cell Biol* 2001c;21:7199–7206.

Haracska L, Johnson RE, Unk I, Phillips BB, Hurwitz J, Prakash L, *et al.* Targeting of human DNA polymerase ι to the replication machinery via interaction with PCNA. *Proc Natl Acad Sci USA* 2001b;98:14256–14261.

Haracska L, Prakash S, Prakash L. Replication past O6-methylguanine by yeast and human DNA polymerase η. *Mol Cell Biol* 2000a;20:8001–8007.

Haracska L, Unk I, Johnson RE, Phillips BB, Hurwitz J, Prakash L, Prakash S. Stimulation of DNA synthesis activity of human DNA polymerase κ by PCNA. Mol Cell Biol 2002;22:784–791.

Haracska L, Washington T, Prakash S, Prakash L. Inefficient bypass of an abasic site by DNA polymerase η. *J Biol Chem* 2001a;276:6861–6866.

Haracska L, Yu S-L, Johnson RE, Prakash L, *et al.* Efficient and accurate replication in the presence of 7,8-dihydro-8-oxoguanine by DNA polymerase η. *Nat Genet* 2000b;25:458–461.

Harris PV, Mazina OM, Leonhardt EA, Case RB, Boyd JB, Burtis KC. Molecular cloning of *Drosophila mus308*, a gene involved in DNA cross-link repair with homology to prokaryotic DNA polymeraseI genes. *Mol Cell Biol* 1996;16:5764–5771.

Hubscher U, Maga G. Spadari S. Eukaryotic DNA polymerases. *Annu Rev Biochem* 2002;71:133–163.

Insdorf NF, Bogenhagen DF. DNA polymerase γ from Xenopus laevis. II. A 3′–5′ exonuclease is tightly associatd with the DNA polymerase activity. *J Biol Chem* 1989;264:21498–21503.

Johansson E, Garg P, Burgers PMJ. The pol 32 subunit of DNA polymerase δ contains separable domains for processive replication and proliferating cell nuclear antigen (PCNA) binding. *J Biol Chem* 2004;279:1907–1915.

Johansson E, Majka J, Burgers PMJ. Structure of DNA polymerase δ from *Saccharomyces cerevisiae. J Biol Chem* 2001;276: 43824–43828.

Johnson AA, Johnson KA. Fidelity of nucleotide incorporation by human mitochondrial DNA polymerase. *J Biol Chem* 2001a;276:38090–38096.

Johnson AA, Johnson KA. Exonuclease proofreading by human mitochondrial DNA polymerase. *J Biol Chem* 2001b;276:38097–38107.

Johnson RE, Haracska L, Prakash S, Prakash L. Role of DNA polymerase η in the bypass of a (6-4)TT photoproduct. *Mol Cell Biol* 2001;21:3558–3563.

Johnson RE, Kondratick CM, Prakash S, Prakash L. *hRAD30* mutations in the variant form of Xeroderma pigmentosum. *Science* 1999c;285:263–265.

Johnson RE, Prakash S, Prakash L. Efficient bypass of a thymine-thymine dimer by yeast DNA polymease, pol η. *Science* 1999a;283:1001–1004.

Johnson RE, Prakash S, Prakash L. Requirement of DNA polymerase activity of yeast Rad 30 protein for its biological function. *J Biol Chem* 1999b;274:15975–15977.

Johnson RE, Prakash S, Prakash L. The human DINB1 gene encodes the DNA polymerase Pol θ. *Proc Natl Acad Sci USA* 2000;97:3838–3843.

Johnson RE, Washington MT, Haracska L, Prakash S, Prakash L. Eukaryotic polymerases ι and ξ act sequentially to bypass DNA lesions. *Nature* 2000c;406:1015–1019.

Johnson RE, Washington MT, Prakash S, Prakash L. Fidelity of human DNA polymerase η. *J Biol Chem* 2000;275:7447–7450.

Kaguni LS, Lehman IR. Eukaryotic DNA polymerase-primase: Structure, mechanism and function. *Biochim Biophys Acta* 1988;950:87–101.

Kaguni LS, Olson MW. Mismatch-specific 3′–5′ exonuclease associated with the mitochondrial DNA polymerase from *Drosophila* embryos. *Proc Natl Acad Sci USA* 1989;86:6469–6473.

Kanner L, Hanawalt P. Efficiency of utilization of thymine and 5-bromouracil for normal and repair DNA synthesis in bacteria. *Biochim Biophys Acta* 1968;157:532–545.

Karmata D, Gross JD. Isolation and genetic analysis of temperature sensitive mutants of *Bacillus subtilis* defective in DNA synthesis. *Mol Gen Genet* 1970;108: 277–287.

Kato T, Shinoura Y. Isolation and characterization of mutations by ultraviolet light. *Mol Gen Genet* 1977;156:12–131.

Kelly RB, Atkinson MR, Huberman JL, Kornberg A. Excision of thymine dimers and other mismatched sequences by DNA polymerase of *Escherichia coli*. *Nature* 1969;224:495–501.

Kelly RB, Cozzarelli NR, Deutscher MP, Lehman IR, Kornberg A. Enzymatic synthesis of deoxyribonucleic acid. XXXII. Replication of duplex deoxyribonucleic acid by polymerase at a single strand break. *J Biol Chem* 1970;245:39–45.

Khare V, Eckert KA. The 3′–5′ exonuclease of T4 DNA polymerase removes premutagenic alkyl mispairs and contrbutes to futile cycling at O^6-methylguanine lesions. *J Biol Chem* 2001;276:24286–24292.

Kim S-R, Maenhaut-Michel G, Yamada M, Yamamoto Y, Matsui K, Sofuni T, *et al*. Multiple pathways for SOS-induced mutagenesis in *Escherichia coli*: An overexpression of *dinB/dinP* results in strongly enhancing mutagenesis in the absence of any exogenous treatment to damage DNA. *Proc Natl Acad Sci USA* 1997;94:13792–13797.

Kingsbury DT, Helinski DR. DNA polymrase as a requirement for the maintenance of the bacterial plasmid colicinogenic factor E1. *Biochem Biophys Res Commun* 1970;41:1538–1544.

Klenow H, Henningsen I. Selective elimination of the exonuclease activity of the deoxyribonucleic acid polymerase from *Escherichia coli* B by limited proteolysis. Proc. Natl. Acad. Sci. USA 1970;65:168–175.

Klett RP, Cerami A, Reich E. Exonuclease VI. A new nuclease activity associated with *E. coli* DNA polymerase. *Proc Natl Acad Sci USA* 1968;60:943–950.

Knippers R. DNA polymerase II. *Nature* 1970;228:1050–1053.

Knippers R, Stratling W. The DNA replicating capacity of isolated *E. coli* cell wall-membrane complexes. *Nature* 1970;226:713–717.

Komori T, Okada A, Stewart V, Alt FW. Lack of N regions in antigen receptor variable region genes of TdT-deficient lymphocytes. *Science* 1993;61:1171–1175.

Kong H, Kucera RB, Jack WE. Characterization of a DNA polymerase from the hyperthermophile archaea *Thermococcus litoralis*. Vent DNA polymerase, steady state kinetics, thermal stability, processivity, strand displacement and exonuclease activities. *J Biol Chem* 1993;268:1965–1975.

Konrad EB, Lehman IR. Novel mutants of *Escherichia coli* that accumulate very small DNA replicative intermediates. *Proc Natl Acad Sci USA* 1975;72:2150–2154.

Kornberg A. *DNA Replication*. San Francisco, CA: WH Freeman & Company;1980:87–88.

Kornberg A, Baker TA. *DNA Replication* (2nd edition). San Francisco, CA: WH Freeman & Company;1992.

Kornberg A, Bertsch LL, Jackson JF, Khorana HG. Enzymatic synthesis of deoxyribonucleic acid. XVI. Oligonucleotides as templates and the mechanism of their replication. *Proc Natl Acad Sci USA* 1964;51:315–323.

Kornberg T, Gefter ML. DNA synthesis in cell-free extracts of a DNA polymerase-defective mutant. *Biochem Biophys Res Commun* 1970;40:1348–1355.

Kornberg T, Gefter ML. Purification and DNA synthesis in cell-free extracts:Properties of DNA polymerase II. *Proc Natl Acad Sci USA* 1971;68:761–764.

Kornberg T, Gefter ML. Deoxyribonucleic acid synthesis in cell-free extracts. IV. Purification and catalytic properties of deoxyribonucleic acid polymerase III. *J Biol Chem* 1972;247:5369–5375.

Kumar A, Abbotts J, Karawya EM, Wilson SH. Identification and properties of the catalytic domain of mammalian DNA polymerase β. *Biochemistry* 1990;29:7156–7159.

Kunkel TA, Soni A. Exonucleolytic proofreading enhances the fidelity of DNA synthesis by chick embryo DNA polymerase γ. *J Biol Chem* 1988;263:450–4459.

Layton JC, Foster PL. Error-prone DNA polymerase IV is controlled by the stress-response sigma factor, RpoS, in *Escherichia coli*. *Mol Microbiol* 2003;50:549–561.

Layton JC, Foster PL. Error-prone DNA polymerase IV is regulated by the heat shock chaperone GroE in *Escherichia coli*. *J Bact* 2005;187:449–457.

Lee JW, Blanco L, Zhou T, Garcia-Diaz M, Bebenek K, Kunkel TA, *et al*. Implication of DNA polymerase λ in alignment-based gap filling for nonhomologous DNA end joining in human nuclear extracts. *J Biol Chem* 2004;279:805–811.

Lehman IR, Bessman MJ, Simms ES, Kornberg A. Enzymatic synthesis of deoxyribonucleic acid. I. Preparation of substrates and partial purification of an enzyme from *Escherichia coli*. *J Biol Chem* 1958;233:163–170.

Lehman IR, Kaguni LS. DNA polymerase α. *J Biol Chem* 1989;264:4265–4268.

Lehman IR, Richardson CC. The DNases of *Escherichia coli*. IV. An exonuclease activity present in purified preparations of DNA polymerase. *J Biol Chem* 1964;239: 233–241.

Lindahl T. An *N*-glycosidase from *Escherichia coli* that releases free uracil from DNA containing deaminated cytosine residues. *Proc Natl Acad Sci USA* 1974;71:3649–3653.

Lindahl T, Barnes DE. Mammalian DNA ligases. *Annu Rev Biochem* 1992;61:251–281.

Liu Y, Beard WA, Shock DD, Prasad R, Hou EW, Wilson SH. DNA polymerase β and flap endonuclease 1 enzymatic specificities sustain DNA synthesis for long patch base excision repair. *J Biol Chem* 2005;280:3665–3674.

Low RL, Rashbaum SA, Cozzarelli NR. Purification and characterization of DNA polymerase III from *Bacillus subtilis*. *J Biol Chem* 1976;251:1311–1325.

Maenhaut-Michel G, Caillet-Fauquet P. Effect of umuC mutations on targeted and untargeted ultraviolet mutagenesis in bacteriophage λ. *J Mol Biol* 1984;177:181–187.

Maga G, Ramadan K, Locatelli GA, Shevelev I, Spadari S, Hubscher U. DNA elongation by the human DNA polymerase λ and terminal transferase activities are differentially coordinated by proliferating cell nuclear antigen and replication protein A. *J Biol Chem* 2005;280:1971–1981.

Maga G, Villani G, Ramadan K, Shevelev I, LeGac NT, Blanco L, *et al*. Human DNA polymerase λ functionally and physically interacts with proliferating cell nuclear antigen in normal and translesion DNA synthesis. *J Biol Chem* 2002;277: 48434–48440.

Maki H, Kornberg A. The polymerase subunit of DNA polymerase III of *Escherichia coli*. II. Purification of the α subunit, devoid of nuclease activity. *J Biol Chem* 1985;260:12987–12992.

Maki H, Maki S, Kornberg A. DNA polymerase III holoenzyme of *Escherichia coli*. IV. The holoenzyme is an asymmetric dimer with twin active sites. *J Biol Chem* 1988;263: 6570–6578.

Maniatis T, Fritsch EF, Sambrook J. *Molecular cloning: A laboratory manual*. Cold Spring Harbor NY: Cold Spring Harbor Laboratory;1982.

Maor-Shoshani A, Reuven NB, Tomer G, Livneh Z. Highly mutagenic replication by DNA polymerase V (Umu C) provides a mechanistic basis for SOS untargeted mutagenesis. *Proc Natl Acad Sci USA* 2000;97:565–570.

Matsukage A, Nishikawa K, Ooi T, Sato Y, Yamaguchi M. Homology between mammalian DNA polymerase β and terminal deoxynucleotidyl transferase. *J Biol Chem* 1987;262: 8960–8962.

Matsumoto Y, Kim K. Excision of deoxyribose phosphate residues by DNA polymerase β during DNA repair. *Science* 1995;269: 699–702.

Mattila P, Korpela J, Tenkanen T, Pitkanen K. Fidelity of DNA synthesis by the *Thermococcus litoralis* DNA polymerase – an extremely heat stable enzyme with proof reading activity. *Nucleic Acids Res* 1991;18:4967–4973.

McHenry C, Kornberg A. DNA polymerase III holoenzyme of *Escherichia coli*. Purification and resolution into subunits. *J Biol Chem* 1977;252:6478–6484.

Miller MR, Chinault DN. The roles of DNA polymerases α, β and γ in DNA repair synthesis induced in hamster and human cells by diferent DNA damaging agents. *J Biol Chem* 1982;257:10204–10209.

Mitchell JB, Karawya E, Kinsella TJ, Wilson SH. Measurement of DNA polymerase β in skin fibroblast cell lines from patients with ataxia telangiectasia. *Mutation Res* 1985;146:295–300.

Mitra S, Kornberg A. Enzymatic mechanisms of DNA replication. *J Gen Physiol* 1966;49 (Part 2):59–79.

Mitra S, Reichard P, Inman RB, Bertsch LL, Kornberg A. Enzymic synthesis of deoxyribonucleic acid. XXIL. Replication of a circular single-stranded DNA template by DNA polymerase of *Escherichia coli*. *J Mol Biol* 1967;24:429–447.

Morrison A, Araki H, Clark AB, Hamatake RK, Sugino A. A third essential DNA polymerase in *S. cerevisiae*. *Cell* 1990;62:1143–151.

Mosbaugh DW, Linn S. Excision repair and DNA synthesis with a combination of HeLa DNA polymerase β and DNase V. *J Biol Chem* 1983;258:108–118.

Moses RE, Richardson CC. A new DNA polymerase activity of *Escherichia coli*. I. Purification and properties of the activity present in *E. coli* polA1. *Biochem Biophys Res Commun* 1970a;41: 1555–1564.

Moses RE, Richardson CC. A new DNA polymerase activity of *Escherichia coli*. II. Properties of the enzyme purified from wild-type *E. coli* and DNA_{ts} mutants. *Biochem Biophys Res Commun* 1970b;41:1565–1571.

Mullis KB, Faloona FA. Specific synthesis of DNA *in vitro* via a polymerized catalyzed chain reaction. *Methods Enzymology* 1987;155:335–350.

Nagasawa K-I, Kitamura K, Yasui A, Nimura Y, Ikeda K, Hirai M, *et al*. Identification and characterization of human DNA polymerase β2, a DNA polymerase β-related enzyme. *J Biol Chem* 2000;275:31233–31238.

Nelson JR, Lawence CW, Hinkle DC. Thymine-thymine dimer bypass by yeast DNA polymerase ζ. *Science* 1996a;272:1646–1649.

Nelson JR, Lawence CW, Hinkle DC. Deoxycytidyl transferase activity of yeast REV1 protein. *Nature* 1996b;382:729–731.

Neuner A, Jannasch HW, Belkin S, Stetter KO. *Thermococcus litoralis* sp.nov.: A new species of extremely thermophilic marine bacteria. *Arch Microbiol* 1990;153:205–207.

Newman J, Hanawalt PC. Role of T4 DNA ligase in in T4 DNA replication. *J Mol Biol* 1968;35:639–642.

Nishida C, Reinhard P, Linn S. DNA repair synthesis in human fibroblasts requires DNA polymease δ. *J Biol Chem* 1988;263: 501–510.

Ogawa H, Shimada K, Tomizawa J. Studies on radiation sensitive mutants of *E. coli*. *Mol Gen Genet* 1968;101:227–244.

Ohashi M, Taguchi T, Ikegami S. Aphidocolin: A specific inhibitor of DNA polymerases. *Biochem Biophys Res Commun* 1978;82:1084–1090.

Ohmori H, Friedberg EC, Fuchs RPP, Goodman MF, Hnaoka F, Hinkle D, *et al*.. The Y-family of DNA polymerases. Mol Cell 2001;8:7–8.

Okazaki R, Okazaki T, Sakabe K, Sugimoto K, Sugino A. Mechanism of DNA chain growth. I. Possible discontinuity and unusual secondary structure of newly synthesized chains. *Proc Natl Acad Sci USA* 1968;59:598–605.

Okazaki T, Okazaki R. Mechanism of DNA chain growth. IV. Detection of synthesis of T4 short DNA chains as revealed by exonucleolytic degradation. *Proc Natl Acad Sci USA* 1969;64:1242–1248.

Olivera BM. DNA intermediates at the *Escherichia coli* replication fork: Effect of dUTP. *Proc Natl Acad Sci USA* 1978;75:238–242.

Olivera BM, Lehman IR. Linkage of polynucleotides through phosphodiester bonds by an enzyme from *Escherichia coli*. *Proc Natl Acad Sci USA* 1967a;57:1426–1433.

Olivera BM, Lehman IR. Diphosphopyridine nucleotide: A cofactor for the polynucleotide joining enzyme from *Escherichia coli*. *Proc Natl Acad Sci USA* 1967b;57:1700–1704.

Olivera BM, Manlapaz-Ramos P, Warner HR, Duncan BK. DNA intermediates at the *Escherichia coli* replication fork. II. Studies using *dut* and *ung* mutants *in vitro*. *J Mol Biol* 1979;128:265–275.

Ollis DL, Brick P, Hamlin R, Xuong NG, Steitz TA.Structure of large fragment of *Escherichia coli* DNA polymerase I complexed with dTMP. *Nature*1985;313:762–766.

Pritchard AE, McHenry CS. Assembly of DNA polymerase III holoenzyme. Coassembly of gamma and tau is inhibited by DnaX complex accessory proteins but stimulated by DNA polymrase III core. *J Biol Chem* 2001;276: 35217–35222.

Rajagopalan M, Lu C, Woodgate R, O'Donnell M, Goodman MF, Echols H. Activity of the purified mutagenesis proteins UmuC, UmuD', and RecA in replicative bypass of an abasic DNA lesion by DNA polymerase III. *Proc Natl Acad Sci* 1992;89:10777–10781.

Ramadan K, Shevelev IV, Maga G, Hubscher U. *de novo* DNA synthesis by human DNA polymerase λ, DNA polymerase μ and terminal deoxyribonucleotidyl transferase. *J Mol Biol* 2004;339:395–404.

Reuven NB, Arad G, Maor-Shoshani A, Livneh Z. The mutagenesis protein UmuC is a DNA polymerase activated by UmuD', RecA, and SSB and is specialized for translesion replication. *J Biol Chem* 1999;274:31763–3766.

Richardson CC, Schildkraut CL, Aposhian HV, Kornberg A. Enzymatic synthesis of deoxyribonucleic acid. XIV. Further purification and properties of DNA polymerase of *Escherichia coli*. *J Biol Chem* 1964;239:222–231.

Saiki RK, Gelfand DH, Stoffel S, Scharf SJ, Higuchi R, Horn GT, *et al*. Primer-directed enzymatic amplification of DNA with a thermostable DNA polymerase. 1988;239:487–491.

Saiki RK, Scharf S, Faloona F, Mullis KB, Horn GT, Erlich A, *et al*. Enzymatic amplification of β-globin genomic sequences and restriction site analysis for diagnosis of sickle cell anemia. *Science* 1985;230:1350-1354

Sakabe K, Okazaki R. A unique property of the replicating origin of chromosomal DNA. *Biochim Biophys Acta* 1966;129: 651–654.

Sakaguchi K, Boyd JB. Purification and characterization of a DNA polymerase β from *Drosophila*. *J Biol Chem* 1985;260: 10406–10411.

Sanjanwala B, Ganesan AT. DNA polymerase III gene of *Bacillus subtilis*. *Proc Natl Acad Sci USA* 1989;86:4421–4424.

Schlabach A, Fridlender B, Bolden A, Weissbach A. DNA-dependent DNA polymerases from HeLa cell nuclei. II. Template and substrate utilization. *Biochem Biophys Res Commun* 1971;44:879–885.

Schlacher K, Leslie K, Wyman C, Woodgate R, Cox MM, Goodman MF. DNA polymerase V and RecA protein, a minimal mutasome. *Mol Cell* 2005;17:561–572.

Setlow P, Brutlag D, Kornberg A. Deoxyribonucleic acid polymerase: Two distinct enzymes in one polypeptide. I. A proteolytic fragment containing the polymerase and 3′–5′ exonuclease functions. *J Biol Chem* 1972;247:224–231.

Setlow P, Kornberg A. Deoxyribonucleic acid polymerase: Two distinct enzymes in one polypeptide. II. A proteolytic fragment containing the 3′–5′ exonuclease function. Restoration of intact enzyme functions from the proteolytic fragments. *J Biol Chem* 1972;247:232–240.

Sharief FS, Vojta PJ, Ropp PA, Copeland WC. Cloning and Chromosomal mapping of the human DNA polymerase θ (*pol Q*), The eighth human DNA polymerase. *Genomics* 1999;59:90–96.

Shinkai A, Loeb LA. *In vivo* mutagenesis by *Escherichia coli* DNA polymerase I. Ile[709] in motif A functions in base selection. *J Biol Chem* 2001;276:46759–46764.

Shinkai A, Patel PH, Loeb LA. The conserved active site motif A of *Escherichia coli* DNA polymerase I is highly mutable. *J Biol Chem* 2001;276:18836–18842.

Sinha NK. Specificity and efficiency of editing of mismatches involved in the formation of base-substitutions mutations by the 3′–5′ exonuclease activity of phage T4 DNA polymerase. *Proc Natl Acad Sci USA* 1987;84:915–919.

Smith DW, Schaller HE, Bonhoeffer FJ. DNA synthesis *in vitro*. Nature 1970;226:711–713.

So AG, Downey KM. Mammalian DNA polymerases α and δ: Current status in DNA replication. *Biochemistry* 1988;27:4591–4595.

Sriskanda V, Shuman S. A second NAD$^+$-dependent DNA ligase (LigB) in *Escherichia coli*. *Nucleic Acids Res* 2001;29:4930–4934.

Steinberg RA, Denhardt DT. Inhibition of synthesis of ϕx174 DNA in a mutant host thermosensitive for DNA synthesis. *J Mol Biol* 1968;37:525–528.

Sugimoto K, Okazaki T, Imae Y, Okazaki R. Mechanism of DNA chain growth. III. Equal annealing of T4 nascent short DNA chains with the separated complementary strands of the phage DNA. *Proc Natl Acad Sci USA* 1969;63:1343–1350.

Syvaoja J, Linn S. Characerization of a large form of DNA polymease δ from HeLa cells that is insensitive to proliferating cell nuclear antigen. *J Biol Chem* 1989;264: 2489–2497.

Syvaoja J, Suomensaari S, Nishida C, Goldsmith JS, Chui GSJ, Jain S, *et al.* DNA polymerases α, δ and ε: Three distinct enzymes from HeLa cells. *Proc Natl Acad Sci USA* 1990;87:6664–6668.

Tanabe K, Bohn EW, Wilson SH. Steady state kinetics of mouse DNA polymerase β. *Biochemistry* 1979;18:3401–3406.

Tanaka S, Koike K. Template specificity of rat mitochondrial DNA polymerase. *Biochim Biophys Acta* 1977;479:290–299.

Tanaka S, Koike K. DNA polymerase γ is localized in mitochondria. Biochem Biophys Res Commun 1978;81:791–797.

Tang M, Bruck I, Eritja R, Turner J, Frank EG, Woodgate R, *et al.* Biochemical basis of SOS-induced mutagenesis in *Escherichia coli*: Reconstitution of *in vitro* lesion bypass dependent on the UmuD'2C mutagenic complex and RecA protein. *Proc Natl Acad Sci USA* 1998;95:9755–9760.

Tang M, Pham P, Shen X, Taylor J-S, O'Donnell M, Woodgate R, *et al.* Roles of *E. coli* DNA polymerases IV and V in lesion-targeted and untargeted SOS mutagenesis. *Nature* 2000;404:1014–1018.

Tang M, Shen X, Frank EG, O'Donnell M, Woodgate R, Goodman MF. UmuD'2C is an error-prone DNA polymerase, *Escherichia coli* pol V. *Proc Natl Acad Sci USA* 1999;96:8919–8924.

Tindall KR, Kunkel TA. Fidelity of DNA synthesis by the *Thermus aquaticus* DNA polymerase. *Biochemistry* 1988;27:6008–6013.

Tuske S, Singh K, Kaushik N, Modak MJ. The J-helix of *Esherichia coli* DNA polymerase I (Klenow fragment) regulates polymerase and 3′–5′ exonuclease function. *J Biol Chem* 2000;275:23759–23768.

Tye B-K, Nyman P-O, Lehman IR, Hochhauser S, Weiss B. Transient accumulation of Okazaki fragments as a result of uracil incorporation into nascent DNA. *Proc Natl Acad Sci USA* 1977;74:154–157.

Vaisman A, Tissier A, Frank EG, Goodman MF, Woodgate R. Human DNA polymerase ι promiscuous mismatch extension. *J Biol Chem* 2001;276:30615–30622.

Villani G, Blanco L, Hubscher U. Human DNA polymerase λ possesses terminal deoxyribonucleotidyl transferase activity and can elongate RNA primers: Implications for novel functions. *J Mol Biol* 2003;328:63–72.

Wagner J, Gruz P, Kim S-R, Yamada M, Matsui K, Fuchs RPP, *et al.* The *dinB* gene encodes a novel *E. coli* DNA polymerase, DNA pol IV, involved in mutagenesis. Mol Cell 1999;4:281–286.

Wahl AF, Crute JC, Sabatino RD, Bodner JB, Marraccino RL, Harwell LW, *et al.* Properties of two forms of DNA polymerase δ from calf thymus. Biochemistry 1986;25:7821–7827.

Walker GC. Mutagenesis and inducible responses to deoxyribonucleic acid damage in *Escherichia coli*. *Microbiol Rev* 1984;48:60–93.

Walowsky C, Fitzhugh DJ, Castano IB, Ju JY, Levin NA, Christman MF. The topoisomerase-related function gene *TRF4* affects cellular sensitivity to the antitumor agent campothecin. *J Biol Chem* 1999;274:7302–7308.

Wang TS-F, Korn D. Reactivity of KB cell deoxyribonucleic acid polymerases α and β with nicked and gapped deoxyribonucleic acid. Biochemistry 1980;19:1782–1790.

Wang TS-F, Korn D. Specificity of the catalytic interaction of human DNA polymerase β with nucleic acid substrates. *Biochemistry* 1982;21:1597–1608.

Wang Z, Castano IB, de Las Penas A, Adams C, Christman MF. Pol κ: A DNA polymerase required for sister chromatid cohesion. *Science* 2000;289:774–779.

Warner HR, Duncan BK. *In vivo* synthesis and properties of uracil-containing DNA. *Nature* 1978;272:32–34.

Watson JD, Crick FHC. Molecular structure of nucleic acids. A structure for deoxyribose nucleic acid. *Nature* 1953a;171:737–738.

Watson JD, Crick FHC. Genetical implications of the structure of deoxyribonucleic acid. *Nature* 1953b;171:964–967.

Weiss B, Richardson CC. Enzymatic breakage and joining of deoxyribonucleic acid. I. Repair of single strand breaks in DNA by an enzyme system from *Escherichia coli* infected with T4 bacteriophage. *Proc Natl Acad Sci USA* 1967;57: 1021-1028.

Weissbach A, Baltimore D, Bollum F, Gallo R, Korn D. Nomenclature of eukaryotic DNA polymerases. *Science* 1975;190:401–402.

Weller GR, Doherty AJ. A family of DNA repair ligases in bacteria? *FEBS Lett* 2001;505:340–342.

Wernette CM, Conway MC, Kaguni LS. Mitochondrial DNA polymerase from *Drosophila melanogaster* embryos: Kinetics, processivity and fidelity of DNA polymerization. *Biochemistry* 1988;27:6046–6054.

Wernette CM, Kaguni LS. A mitochondril DNA polymerase from embryos of *Drosophila melanogaster*. Purification subunit structure and partial characterization. *J Biol Chem* 1986;261:14764–14770.

Wickner W, Kornberg A. A holoenzyme form of deoxyribonucleic acid polymerase III. Isolation and properties. *J Biol Chem* 1974;249:6244–6249.

Wickner W, Schekman R, Geider K, Kornberg A. A new form of DNA polymerase III and a copolymerase replicate a long single-stranded primer-template. *Proc Natl Acad Sci USA* 1973;70:1764–1767.

Wilson SH, Abotts J, Widen S. Progress toward molecular biology of DNA polymerase β. *Biochim Biophys Acta* 1988;949:149–157.

Wong SW, Syvaoja J, Tan C-K, Downey KM, So AG, Linn S, *et al.* DNA polymerases α and δ are immunologically and structurally distinct. *J Biol Chem* 1989;264:5924–5928.

Wood RD, Hutchinson F. Non-targeted mutagenesis of unirradiated lambda phage in *Escherichia coli* host cells irradiated with ultraviolet light. *J Mol Biol* 1984;173: 293-305.

Woodgate R, Rajagopalon M, Lu C, Echols H. UmuC mutagenesis protein of *Escherichia coli*: Purification and interaction with UmuD and UmuD'. *Proc Natl Acad Sci USA*. 1989;86:7301–7305.

Xu Y, Grindley NDF, Joyce CM. Coordination between the polymerase and 5′-nuclease components of DNA polymerase I of *Escherichia coli*. *J Biol Chem* 2000;275:20949–20955.

Yamada K, Hanaoka F, Yamada M. Effects of aphidicolin and/ or 2′–3′-ideoxythymidine on DNA repair induced in HeLa cells by four types of DNA-damaging agents. *J Biol Chem* 1985;260: 10412–10417.

Yamaguchi M, Matsukage A, Takahashi T. Continuous synthesis of long DNA chains by chick embryo DNA polymerase γ. *Nature* 1980;285:45–47.

Yoneda M, Bollum FJ. Deoxynucleotide-polymerizing enzymes of calf thymus gland. I. Large scale purification of terminal and replicative deoxynucleotidyl transferases. *J Biol Chem* 1965;240:3385–391.

Zhang SJ, Lee MYWT. Biochemical characterization and development of DNA polymerases α and δ in the neonatal rat heart. *Arch Biochem Biophys* 1987;252: 24–31.

Zhu H, Shuman S. A primer dependent polymerase function of *Pseudomonas aeruginosa* ATP-dependent DNA ligase (LigD). *J Biol Chem* 2005;280:418–427.

Zimmerman SB, Little JW, Oshinsky CK, Gellert M. Enzymatic joining of DNA strands: A novel reaction of diphosphopyridine nucleotide. *Proc Natl Acad Sci USA* 1967;57:1841–1848.

10

Mechanics of DNA replication

Pramod C. Rath

10.1 Prokaryotic DNA replication

DNA is the blueprint of all known forms of cellular life on this planet. The nucleotide sequence of DNA carries the genetic information (Hartwell *et al.* 2006). Before any cell divides to duplicate itself, its DNA has to replicate faithfully and completely (Kornberg 1960). DNA replication is the fundamental principle of continuity of life. Therefore, both the processes and the mechanisms of DNA replication are very well conserved during evolution (Lewin 2004). DNA replication occurs in three steps: initiation, elongation and termination. The mechanisms of DNA replication have been mostly studied and described in bacteria (Synder and Champness 2003), yeast (Lodish *et al.* 2004) and mouse, as well as in human cells (Alberts *et al.* 2002).

DNA replication occurs by a semiconservative mechanism (Meselson and Stahl 1958). Each DNA strand is used as a template to synthesize the complementary strand. During DNA replication, one out of the two strands of a duplex DNA in a chromosome undergoes discontinuous synthesis [Fig. 10.1(a)–(f)] (Watson *et al.* 2004). When a linear duplex DNA opens up at a point to separate the two DNA strands, this is called a 'replication fork', a Y-shaped DNA molecule. One of the strands has a free 3′-end and the other has a free 5′-end at the separated end. From the free 3′-end, a primer can start synthesizing the new strand (leading strand) from the 5′- to the 3′-end up to the other end of the fork [Fig. 10.1(a)]; whereas, the other strand can start synthesizing the new strand (lagging strand) from the opposite end of the fork from the 5′- to the 3′-end. Both the leading and the lagging strands are synthesized first by making a small RNA primer complementary to the DNA template and then extended by DNA polymerase III in a template-dependent manner. RNA primer for the lagging strand is synthesized by primase. As soon as the lagging strand is synthesized, the fork opens up in the opposite direction towards the closed end of the fork [as shown in Fig. 10.1(b)]. This creates a gap between the closed end of the fork and the 5′-end of the lagging strand. This gap is filled by synthesizing again another RNA primer by primase from the closed end of the fork and then extended by DNA polymerase III. Thus, as shown in Fig. 10.1(c), the strand with a free 5′-end of the fork can synthesize its complementary strand in many discontinuous DNA strands. Therefore, it is called the lagging strand. These newly synthesized discontinuous DNA strands form the lagging strand, which contains a piece of RNA primer and a piece of DNA, and are called 'Okazaki fragments' (Okazaki *et al.* 1968). Two more enzymatic activities are necessary for converting the Okazaki fragments into a single continuous DNA strand. RNaseH is required to digest the RNA primer, and DNA polymerase I is required to replace it with the corresponding DNA sequence. The single nucleotide gap between the two DNA fragments is finally sealed by establishing a phosphodiester bond by DNA ligase at the expense of ATP. These steps are shown in Fig. 10.1(d)–(f). If a circular DNA duplex opens up at a given region by dissociating hydrogen bonds between the base pairs (bp), it creates a 'bubble'. A bubble actually contains two replication forks. DNA replication is therefore bidirectional at a bubble (Brown 2002).

From the lagging strand, short Okazaki DNA fragments are synthesized from an RNA primer by primase followed by DNA polymerase III. Synthesis of each Okazaki fragment needs a new RNA primer about 10-12 nucleotides long. DnaG primase makes these RNA primers at a rate of once for every 2 kilobases in *Escherichia coli*. It recognizes the sequence 3′GTC5′ and begins synthesis opposite to the T. DNA polymerase III utilizes these RNA primers to synthesize the DNA until it encounters another such piece being synthesized (Fig. 10.2). By using RNA rather than DNA as the primer for DNA synthesis, the cell lowers the rate of mistake in incorporation of nucleotides into the newly made DNA strand.

RNase H removes the RNA primers from the DNA–RNA hybrid. DNA polymerase I utilizes its 5′ exonuclease and 5′ to 3′ polymerase activities to replace the RNA

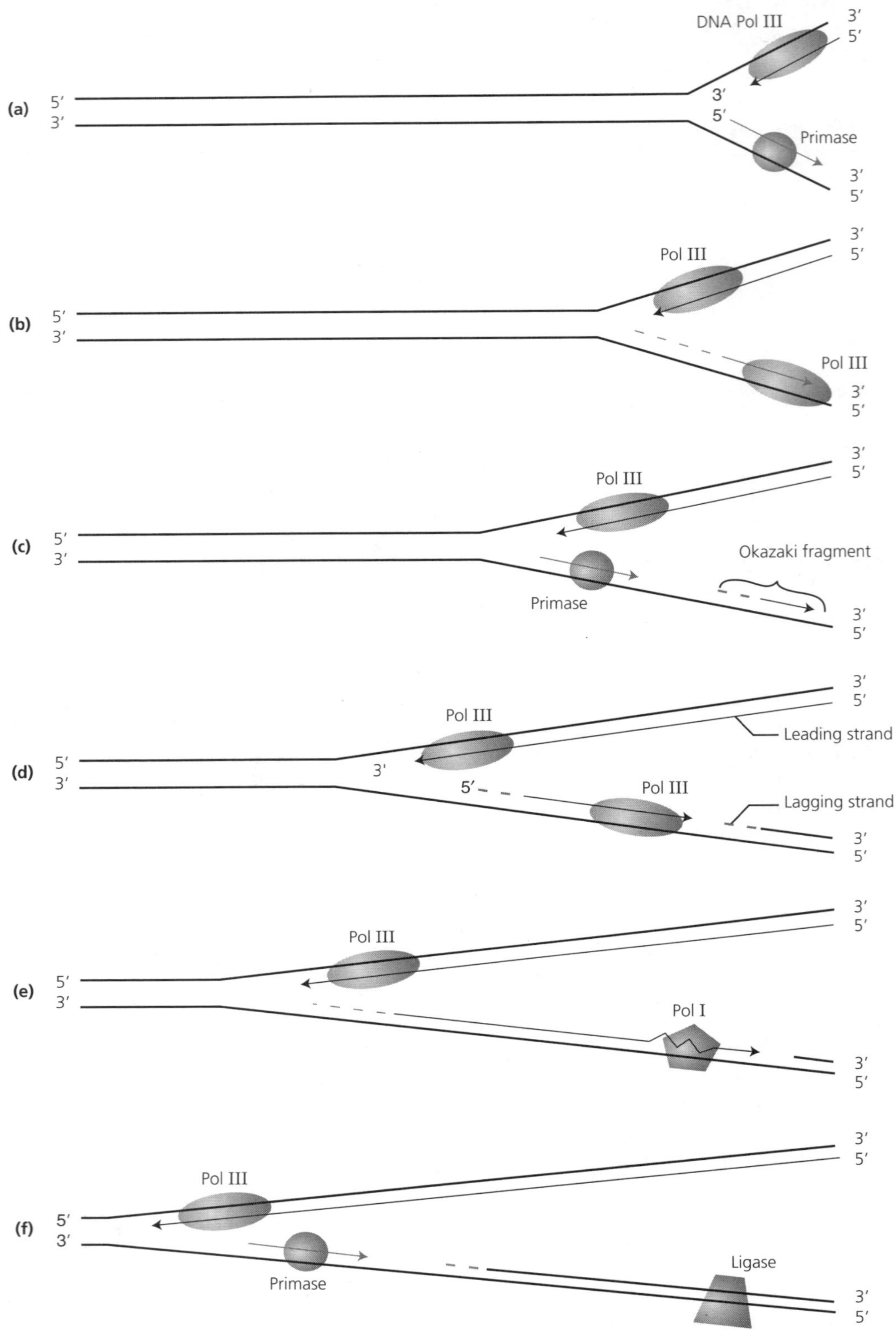

Figure 10.1 Discontinuous synthesis of one strand out of two strands of DNA in the chromosome during replication. Primase, DNA polymerase I and ligase functions are shown in case of the lagging strand synthesis in the replication fork. The short RNA primers are removed from the lagging strand and re-synthesized as DNA using upstream DNA as primer.

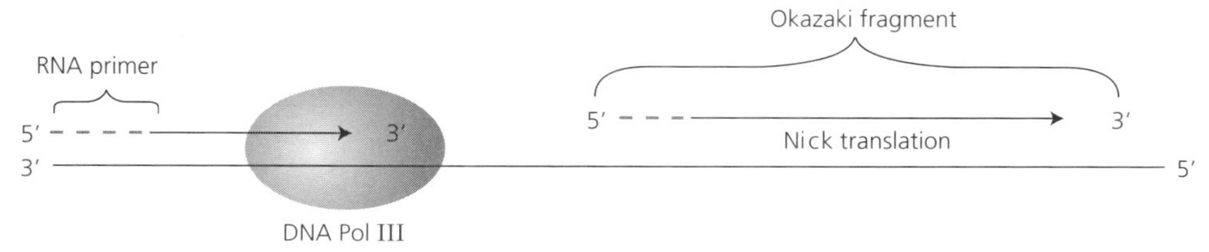

Figure 10.2 Synthesis of Okazaki DNA fragments from RNA primer.

primers with appropriate DNA sequence [Fig. 10.3(a)–(c)]. This process is called 'nick-translation'. Thus, although DNA polymerase III is the actual enzyme for DNA synthesis, DNA polymerase I plays an important role in converting the discontinuous DNA fragments into a continuous DNA strand.

The *E. coli* chromosome is a circular duplex DNA of about 4.6×10^6 bp. It has a topological organization of

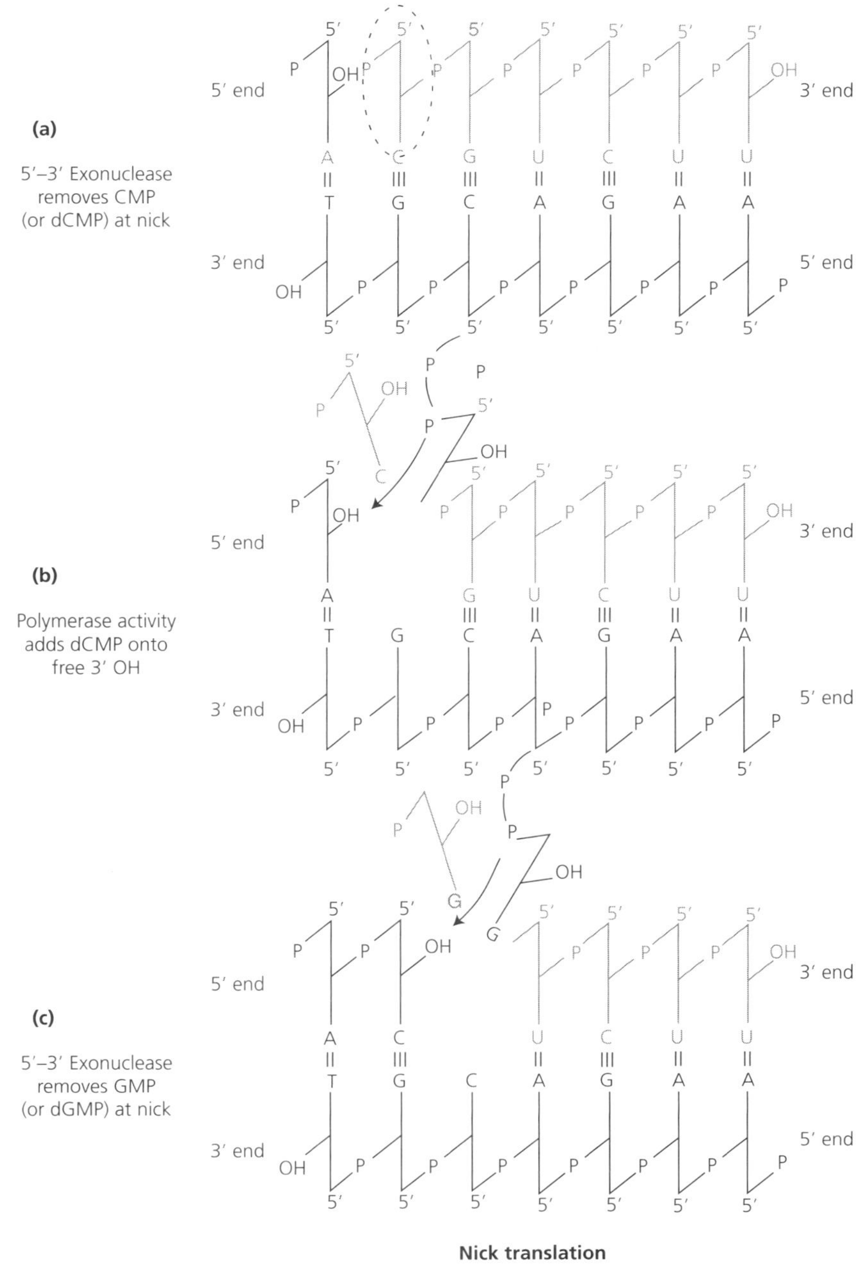

Figure 10.3 DNA polymerase I removes nucleotides of the RNA primer by 'nick translation'. (a) A nick is created by DNA polymerase III while incorporating the last deoxynucleotide in front of the RNA primer. 5′ to 3′ exonuclease activity removes CMP (or dCMP) at nick. (b) DNA polymerase activity incorporates a dCMP onto the free 3′-OH. (c) Nick translation process continues and the nick moves in 5′ to 3′ direction.

supercoils or superhelices of DNA. Topology of a DNA molecule refers to its position in space. As the circular DNA is very long and its strands are wrapped around each other, pulling the two strands apart exerts stress into other regions of the DNA in terms of supercoiling. The two DNA strands are not free to rotate around each other. Therefore, what actually happens at a replication fork is far more complex. Topoisomerases are the enzymes that relieve or undo the supercoiling ahead of the replication fork. DNA gyrase is the enzyme which introduces supercoiling into the DNA (Wang 2002). In Fig. 10.1, it is shown that the two DNA strands are independently replicated. However, instead of replicating separately, the leading and the lagging strands are replicated simultaneously. The two DNA polymerase III molecules are joined to each other through their τ subunits. Table 10.1 shows various proteins involved in DNA replication in *E. coli.* A 'trombone' model for simultaneous replication of the leading strand and the lagging strand is shown in Fig. 10.4. The lagging strand DNA loops out as the Okazaki fragment is synthesized. This allows the two DNA polymerase III molecules to move in opposite directions while they are still joined to each other. DnaB helicase unwinds the DNA ahead of DNA polymerase III to open the replication fork. The lagging strand loop is then relaxed as the sliding clamp releases the lagging strand polymerase; this allows the DNA polymerase to hop ahead to the next RNA primer to begin synthesizing the next Okazaki fragment. This has been described as the replication clamp-loading machine at work (Indiani and O'Dannell 2006). This formation and contraction of the loops at the replication fork is the basis for calling it a trombone model. Thus, DNA replication consists of many protein machines (Alberts 1998).

Table 10.1 Proteins involved in initiation and elongation during DNA replication in *E. coli*

Protein	Gene	Function
DnaA	*dnaA*	Initiator protein primosome (priming complex) formation
DnaB	*dnaB*	DNA helicase
DnaC	*dnaC*	Delivers DnaB to the replication complex
SSB	*ssb*	Single-stranded DNA binding
Primase	*dnaG*	RNA primer synthesis
DNA pol I	*polA*	Primer removal; gap filling
DNA pol III (holoenzyme):		
α	*dnaE*	Polymerization
ε	*dnaQ*	3′ ⟶ 5′+ editing
RNaseH	*rnhA*	Removes RNA primers
θ	*holE*	Present in the core (αεθ)
β	*dnaN*	Sliding clamp
τ[a]	*dnaX*	Organizes the complex; joins leading and lagging DNA pol III
γ[b]	*dnaX*	Binds clamp loaders and SSB
δ	*holA*	Clamp loading
δ'	*holB*	Clamp loading
χ	*holC*	Binds SSB
ψ	*holD*	Binds SSB
DNA ligase	*lig*	Seals DNA nicks
DNA gyrase:		
α	*gyrA*	Nick closing
β	*gyrB*	ATPase

[a]Full-length product of the *dnaX* gene.

[b]Shorter product of the *dnaX* gene produced by translational frameshifting. For details, see Snyder and Champness (2003).

10.1.1 Proofreading and editing

DNA replication should be error-free to maintain genomic stability of a species. If the sequence of DNA is altered during replication, then altered bases cause 'mutations' (Fig. 10.5). Mutations are passed from generation to generation. Mutations could be harmful because they may affect gene function. To avoid this, misincorporation of bases during DNA replication is kept to a minimal level. For this purpose, DNA polymerase has a proofreading (editing) function. In Fig. 10.5, a T is put against a G by mistake during replication. This leads to an A:T base pair replacing a G:C base pair in the newly synthesized DNA. This may represent a mutation. In Fig. 10.6(a), an A is put against a G by mistake during replication. DNA polymerase stops at this misincorporation, the G is removed and it is replaced by a T to correct the misincorporation before DNA replication is allowed to proceed further [Fig. 10.6(b) and (c)]. The DNA repair mechanisms have evolved as protective responses against DNA damage (Friedberg 2000).

The editing function of DNA polymerase reduces misincorporation of bases during replication. Proteins which are part of DNA polymerase or accessory proteins perform the editing function. If due to misincorporation of a base a mismatch is created, then DNA polymerization will stop and the editing function will rectify the mismatch by removing the incorrect base and replacing it with the correct base. As a DNA chain grows from 5′ to 3′, the editing function removes the incorrect base by 3′ to 5′ exonuclease activity, which is called the proofreading activity. A mismatched base pair creates a minor distortion in the DNA structure, which is possibly the basis of identification of a misincorporation by the editing activity. The 3′ exonuclease activity is indeed a part of the DNA polymerase I activity. Several accessory proteins in *E. coli* actually travel along the DNA with the

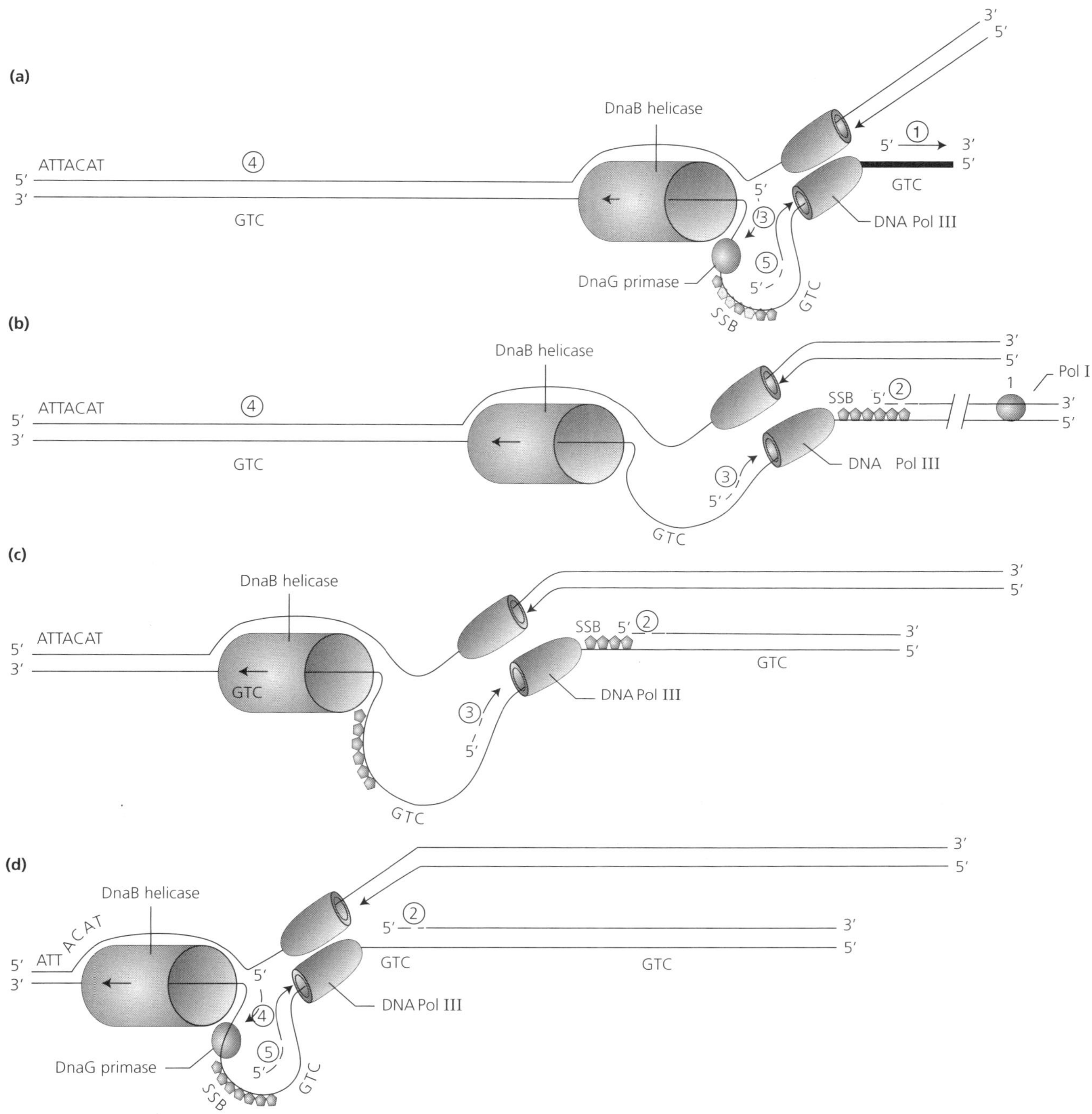

Figure 10.4 A 'trombone' model for simultaneous replication of both the leading and lagging strands at the replication fork. SSB, single-stranded DNA-binding protein; RNA primer. (a) DNA pol III holoenzyme synthesizing lagging strand DNA from primer 2 and just run into primer 1. (b) Lagging strand pol III released from the template at primer 1 and hopped ahead, reassembling on DNA at primer 3 to synthesize an Okazaki fragment. Both leading and lagging strand pol III enzymes remain bound to each other during release and reassembly processes. (c) Pol III encounters lagging strand synthesis from primer 3. (d) Pol III completes Okazaki fragment and runs into primer 2. Pol III will now dissociate from DNA and reassemble at primer 4. 5′ATTACAT3′ arbitrary sequence shows progress of replication fork. 3′GTC5′ shows initiation of primer synthesis.

DNA polymerase for carrying out the editing function. The 3′ exonuclease is coded by the *dnaQ* gene in *E. coli*. An RNA primer is linked to the editing function. It allows better editing function because when DNA synthesis begins at a replication fork, which has just reopened, the DNA distortion due to misincorporation-mediated mismatch is not so easy to be detected by the editing function. Instead, the use of an RNA primer allows its

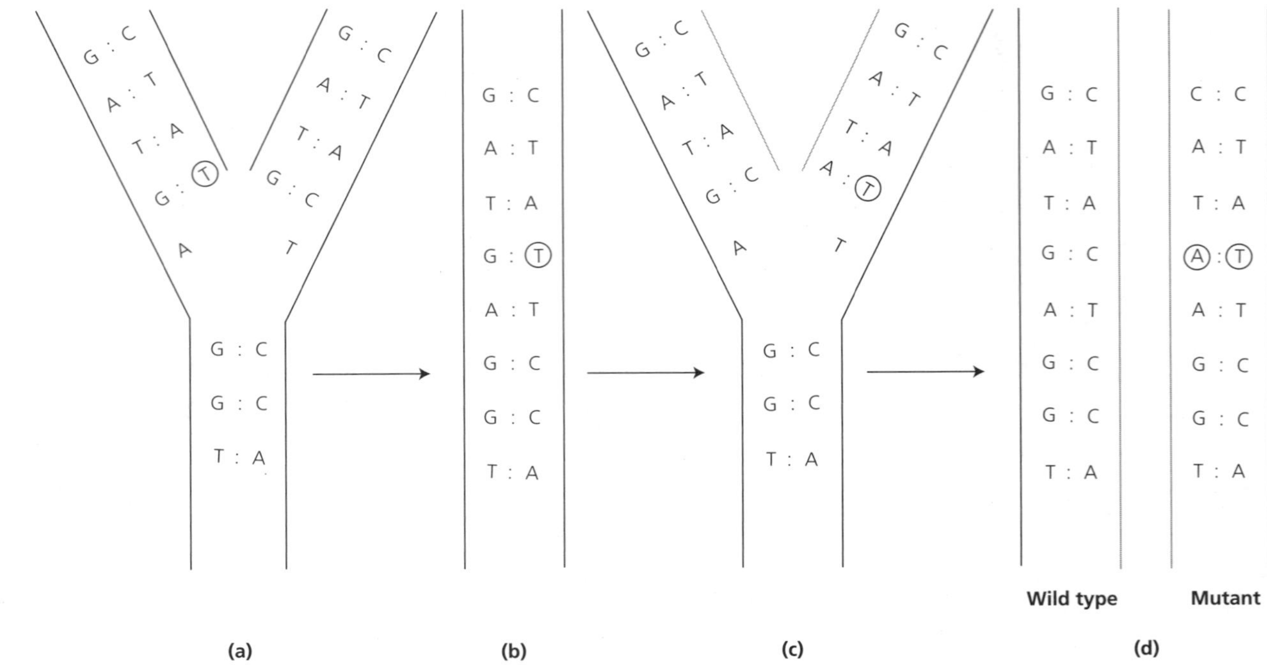

Figure 10.5 Mistakes in base pairing lead to changes in DNA sequence shown as 'mutations'. For example, T put opposite to G during replication (a) leads to an A:T base pair replacing a G:C base pair in the progeny, DNA (b)–(d).

complete replacement by a DNA sequence later, thus removing if there would have been any misincorporation in the primer.

10.1.2 Mismatch repair

It so happens that sometimes a wrong base is misincorporated into the DNA in spite of the editing process. The mismatch repair system, another error-correcting mechanism of the cell, can recognize this and removes the DNA around the mismatch (Modrich and Lahue 1996). This leaves a gap in the DNA, which is refilled by DNA polymerase, which inserts the correct nucleotides. The methylation status of the DNA allows the newly formed DNA strand to be distinguished from the parent strand in *E. coli*. The A in GATC/CTAG is methylated at the 6-position. Deoxyadenosine methylase (Dam methylase) adds a methyl group at the 6-position of the A. The newly synthesized DNA will remain temporarily hemimethylated at these positions. This helps in identifying the new strand, and it is called the methyl-directed mismatch repair system (Fig. 10.7). The *E. coli* DNA replication process probably makes a mistake about once every 10^{10} times it incorporates a nucleotide. Taking 4.7×10^6 bases per *E. coli* chromosomal DNA strand, it will make 4.7×10^{-4} mistake per DNA replication. As the entire bacterial chromosome should replicate once the cell divides, this will amount up to 1 in every 2000 daughter *E. coli* cells having a mistake in its DNA. Such lower mistake rates are tolerated and may even contribute to the diversity in population and speed up evolution. The fidelity of DNA replication is thus related to editing and mismatch repair of the DNA.

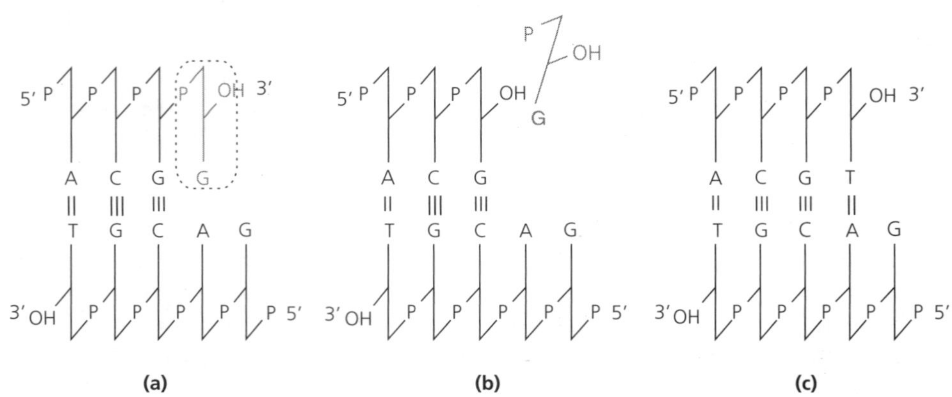

Figure 10.6 DNA polymerase has proofreading (editing) function.

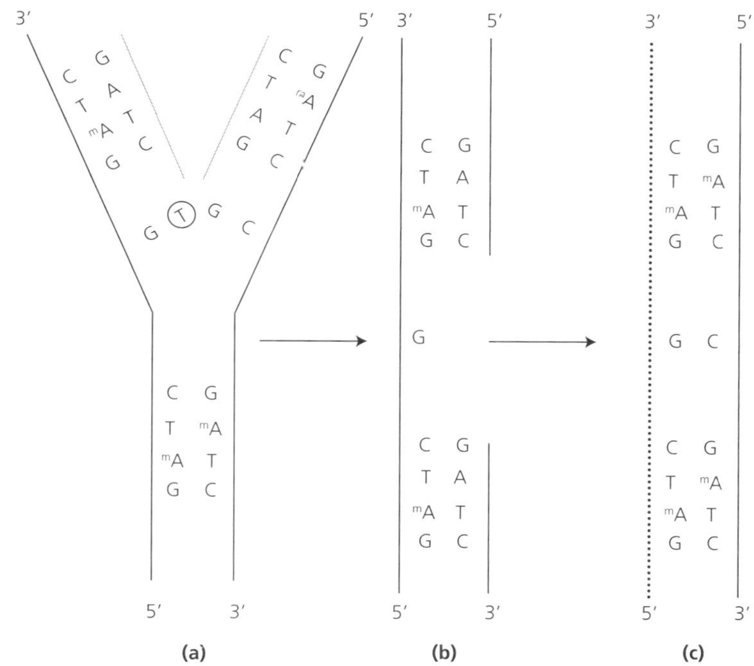

Figure 10.7 The methyl-directed mismatch repair system.

10.1.3 The *oriC* region of *E. coli*

Initiation of DNA replication in *E. coli* involves specific sites or DNA sequences at which proteins act to initiate DNA replication. These DNA sequences are called *cis*-acting sites, and the protein factors are called *trans*-acting factors. The *oriC* site is a well-defined conserved DNA sequence of less than 250-bp DNA with AT-rich sequences (Fig. 10.8). Within *oriC*, there are 9-bp DnaA boxes repeated four times. A 13-bp AT-rich region is repeated three times. These two DNA sequences are important for initiation of *E. coli* DNA replication.

10.2 Enzymes and accessory proteins involved in DNA replication

Many transacting proteins are required for initiation of DNA replication in *E. coli* at the *oriC* site such as DnaA, DnaB and DnaC proteins. Dna A recognizes *oriC*, DnaB is a helicase to unwind the DNA and DnaC delivers DnaB. Replicative helicases have been described as a staircase with a twist (Dillingham 2006).

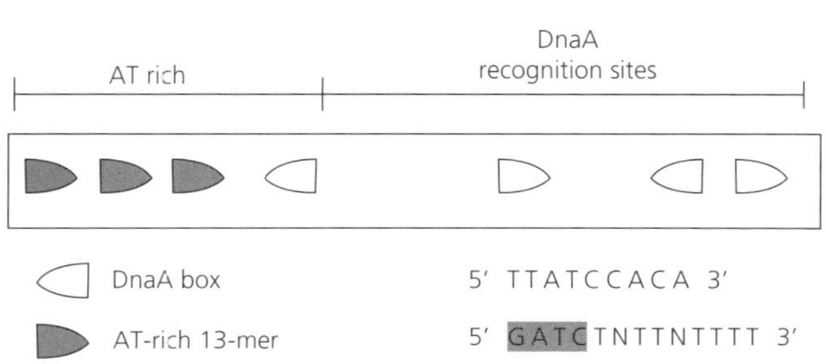

Figure 10.8 Structure of the *oriC* region of *E. coli*.

10.2.1 Initiation of prokaryotic DNA replication

Initiation of DNA replication has been investigated in *E. coli* (Mott and Berger 2007), whereby both the mechanisms and the regulations have been described. The architecture and the dynamics of the replication complex (replisome) have been described in detail (O'Donnell 2006). In brief, DnaA is needed for initiation, but DnaB and DnaC are also needed for primer synthesis once DNA synthesis is underway. Among others, primase (DnaG) and the normal RNA polymerase of *E. coli,* which synthesizes most of the RNA are also needed (Fig. 10.9). First, 10–12 molecules of DnaA protein bind to DnaA boxes in the *oriC* site; this wraps up the DNA and bends it around the aggregated DnaA so that the DNA strands separate in the region of the bend. Then, DnaB and DnaC proteins bind to the *oriC* site. DnaB is a helicase which opens the strands further for priming and replication. The helix-destabilizing proteins or single-stranded DNA-binding proteins (SSBs) prevent the double helix from reforming. The DnaC protein may now leave. This association of *oriC* with DnaA, DnaB and other proteins is called a primosome, which may help the DnaG primase or another RNA polymerase to synthesize an RNA primer to start replication. Although an RNA primer is necessary for initiation of DNA replication, it is not clear which RNA polymerase synthesizes the RNA primer for the leading strand. It is possible that when RNA polymerase moves through the *oriC* site to open the two

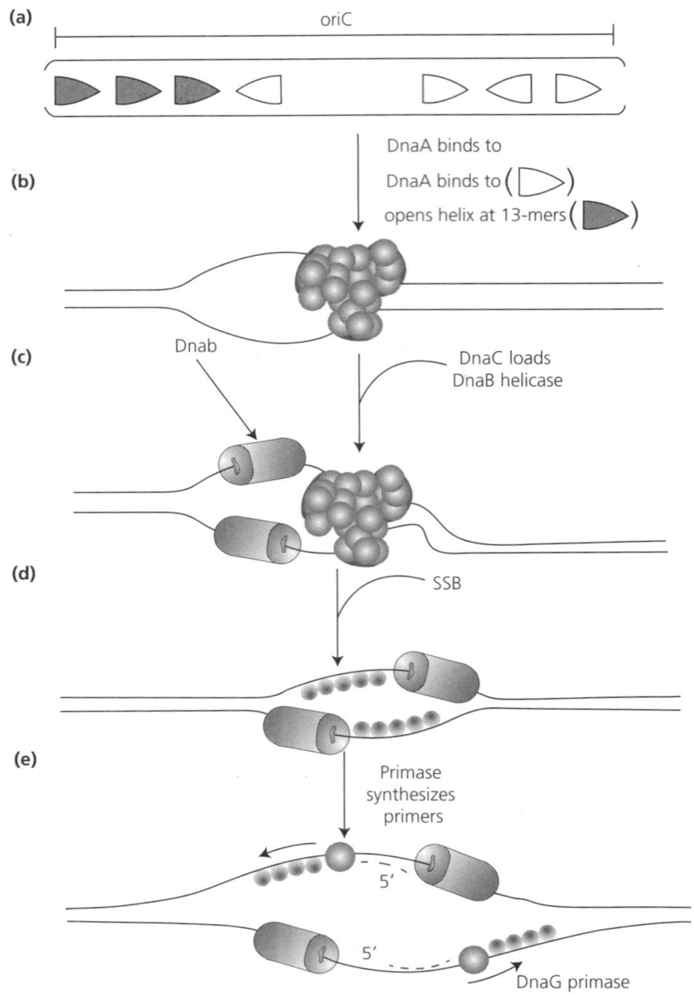

Figure 10.9 Initiation of DNA replication in *E. coli*.

strands in order to facilitate subsequent binding of DnaA protein, this RNA polymerase also makes the primer for the leading strand. Alternatively, DnaG primase makes the RNA primer for the leading strand as well. A list of proteins involved in initiation and elongation during DNA replication in *E. coli* has been given in Table 10.1.

10.2.2 Termination of prokaryotic DNA replication

Initiation of DNA replication at *oriC* allows replication of the *E. coli* chromosome to continue in both directions along the circular chromosome. After this bidirectional replication continues for a considerable period of time, the two replication forks must meet somewhere on the other side of the chromosome and the two daughter chromosomes must separate. Termination of replication occurs in a certain region, which is not a well-defined unique site. The termination region, called *ter*, contains a cluster of *ter* sequences. These 22-bp *ter* sequences allow the replication fork to pass through in one direction but not in the other, like a one-way vehicular traffic regulation (Fig. 10.10). Two *ter* site clusters *terA* and *terB* bracket the termination region. The replication fork can pass the *terA* site in the clockwise direction only; *terB* permits the opposite; that is, it allows the replication fork to pass only in the counterclockwise direction. Thus, when the two replication forks meet at *terA* or *terB* or in a region between the two, the two forks will terminate replication and the two daughter DNAs will be released. Actually, each replication fork passes through a number of *ter* sites, each time the replication rate slows down and finally it stops. While one replication fork is slowing down, the other replication fork has time to move around the chromosome to meet it at a point more or less opposite to *oriC*. There are specific proteins called terminus utilization substance (Tus) in *E. coli* and replication termination protein (RTP) in *B. subtilis*, which bind to *ter* sites and stop the replicating helicase (DnaB in *E. coli*) that separates DNA strands ahead of the fork. Such proteins also help in the separation of daughter DNAs without further 3′-extension.

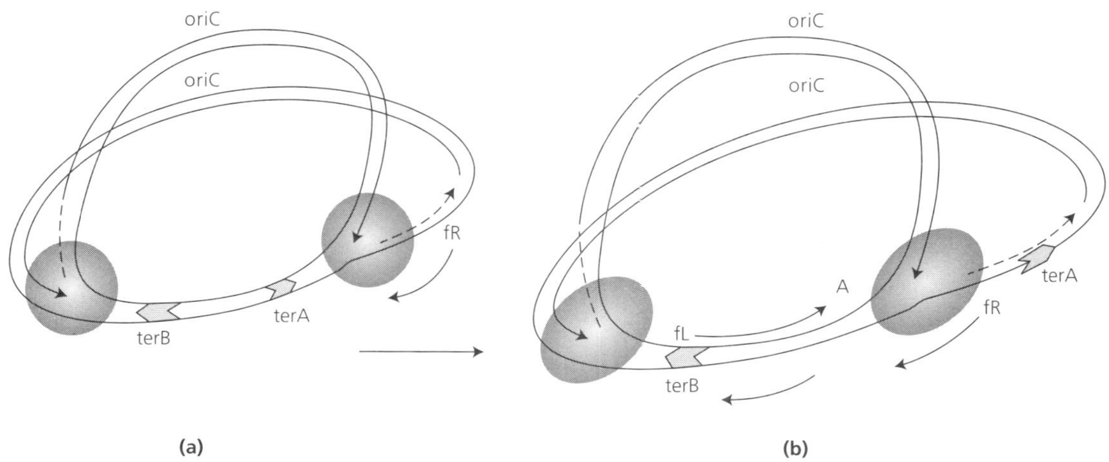

Figure 10.10 Termination of DNA replication in *E. coli*. fR, fork right; fL, fork left.

10.3 Eukaryotic DNA replication

The process of DNA replication in eukaryotes is far more complex than that in prokaryotes.

10.3.1 Major differences between prokaryotic and eukaryotic DNA replication

The major differences between prokaryotic and eukaryotic DNA replication are as follows:

- Bacterial chromosomes have a single origin of DNA replication, but eukaryotic chromosomes contain multiple origins of replication (Alberts *et al.* 2002).
- In eukaryotes, DNA replication takes place during only one part of the cell cycle, that is, 'DNA synthesis phase' or 'S phase'. Different regions on the same chromosome replicate at distinct times in the S phase (Lodish *et al.* 2004).
- Eukaryotic genomic DNA is organized into a nucleoprotein complex called the 'chromatin'. Chromatin contains genomic DNA, histones and non-histone chromosomal proteins as well as other proteins. Nucleosome is the fundamental unit of the eukaryotic chromatin. All transactions of the genomic DNA in eukaryotes occur in the chromatin (Lodish *et al.* 2004). The chromatin challenges during DNA replication and repair have been a matter of intense research (Groth *et al.* 2007).
- Eukaryotic chromatin remains condensed at one place and decondensed or relaxed at another place in chromosomes. Highly condensed regions of the chromatin replicate their DNA late, while genes in less condensed regions of the chromatin tend to replicate their DNA early during the S phase (Alberts *et al.* 2002).
- In the budding yeast (*Saccharomyces cerevisiae*), a simple eukaryote, very well-defined DNA sequences have been identified, which function as replication origins. A large multi-protein complex binds to eukaryotic origins of replication. This is called the 'origin recognition complex' or 'ORC'. It has been difficult to identify the DNA sequences that specify the initiation of DNA replication in mammalian cells. They may be several thousand nucleotide pairs in length. Similar to the ORC of the yeast, there is a human ORC required for the initiation of DNA replication. Soon after the DNA is replicated during the S phase, new nucleosomes are assembled behind the replication fork (Lodish *et al.* 2004).
- There are telomeres at the ends of eukaryotic chromosomes. An RNA–protein complex enzyme called telomerase replicates the telomeric sequences at the ends of chromosomes (Watson *et al.* 2004).

10.3.2 Unique features of eukaryotic DNA replication

The unique features of eukaryotic DNA replication are discussed in the following subsections.

10.3.2.1 DNA polymerases

There are distinct forms of DNA polymerases, which play different roles at the replication forks in prokaryotic and eukaryotic cells. DNA polymerase III is the main enzyme for replication in *E. coli*, which synthesizes both the leading strand and Okazaki fragments from RNA primers. There are three distinct DNA polymerases in eukaryotes: DNA polymerases α, δ and ε for nuclear DNA replication. The study of DNA replication of animal viruses and yeast cells has provided knowledge about eukaryotic DNA polymerases. Experiments involving DNA replication of the Simian Virus 40 (SV40) in cell-free extracts prepared from animal cells have been used to analyse biochemical activities of the three different forms of DNA polymerases (Li and Kelly 1984). Similarly, yeast cells and their powerful genetic analysis (Raghuraman *et al.* 2001) have shown cellular functions of eukaryotic DNA polymerases. Mammalian cell fusion studies have been used to investigate regulation of DNA synthesis during mitosis (Rao and Johnson 1970). DNA polymerase α, δ and ε are present in yeast as well as in human cells. First, polymerase α and primase form a complex, which synthesizes short RNA–DNA fragments at origins of replication and during lagging strand synthesis. Then, polymerase δ and ε act and take part in the synthesis of lagging and leading strands, respectively (Fig. 10.11).

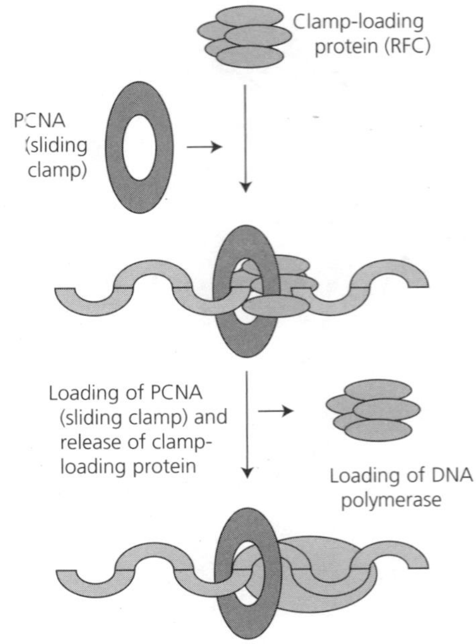

Figure 10.12 Role of additional proteins of DNA polymerase for DNA replication. The clamp-loading protein (RFC) and the sliding-clamp protein (PCNA) of mammalian cells form a complex, which binds the RNA primer–template DNA junction. The PCNA is loaded onto the DNA, RFC is released and DNA polymerase then binds to PCNA.

10.3.2.2 Additional proteins involved in replication

Mutant strains of *E. coli* and yeast with defects in DNA replication helped in identifying additional proteins acting at the replication fork and involved in DNA replication. Such proteins were also purified from mammalian cells and biochemically tested for *in vitro* cell-free DNA replication of SV40 virus DNA. These proteins are as follows. The sliding clamp protein [proliferating cell nuclear antigen (PCNA) in eukaryotes] binds to both *E. coli* DNA polymerase III and eukaryotic DNA polymerases δ and ε, loads the polymerase onto the primer and template, increases the activity of the polymerases and helps them to remain stably bound to the template DNA to continue synthesizing the new DNA strand (Fig. 10.12). The clamp-loading protein [replication factor C (RFC)] loads the sliding clamp protein (PCNA) onto the RNA primer and template DNA junction in mammalian cells. The clamp-loading protein opens the sliding clamp by utilizing the energy produced by ATP hydrolysis. Then, the clamp-loading protein is released and the sliding clamp protein remains bound as a ring around the DNA and loads DNA polymerase onto the DNA and the complex is maintained during DNA synthesis. The role of clamp loaders during initiation of DNA replication has been reviewed (O'Donnell and Kurien 2005).

During DNA replication, it is necessary to unwind the DNA and maintain the single-stranded DNA regions. Helicases enzymatically catalyse the unwinding of double-stranded DNA ahead of the replication fork by utilizing the energy from ATP hydrolysis. Single-stranded DNA-binding proteins (SSBs) stabilize the single-stranded DNA. Due to unwinding, the DNA ahead of the replication fork rotates; therefore, topoisomerases catalyse the breakage and reunion of the DNA to avoid twisting of the DNA due to unwinding,

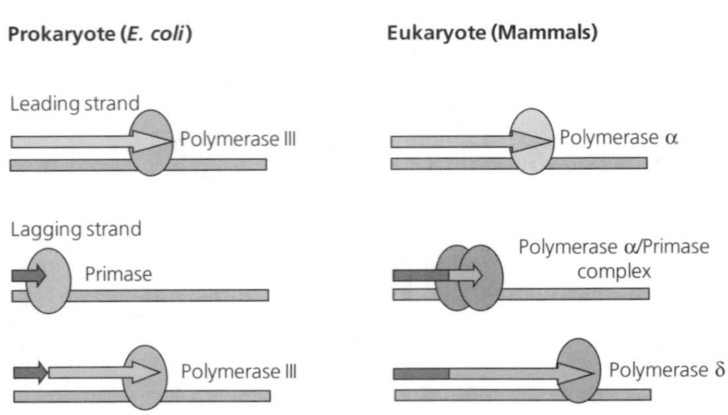

Figure 10.11 Comparison of the DNA polymerases of prokaryotes (*E. coli*) and eukaryotes (mammals) in terms of their biochemical activities with respect to synthesis of the leading and the lagging strands. The leading strand synthesis is carried out by polymerase III in *E. coli* and polymerase δ in mammals. The lagging strand synthesis is initiated by primase to synthesize the RNA primers, which are elongated by polymerase III in *E. coli*. However, the lagging strand is initiated by a complex of primase and polymerase α to synthesize short RNA–DNA hybrids, which are then elongated by polymerase ε in mammals.

which could otherwise stop DNA polymerase. The type I topoisomerases make single-stranded DNA breaks, and type II topoisomerases make double-stranded DNA breaks. Both topoisomerase I and topoisomerase II introduce breaks into the double-stranded DNA to allow it to unwind without twisting itself. This is possible by rotating the two strands of DNA freely around each other so that the template DNA is replicated. Topoisomerase II also helps to separate the two daughter chromatids after DNA replication is completed during mitosis in eukaryotes. In eukaryotic chromatin DNA, nucleosomes are disrupted during DNA replication and the newly synthesized DNA reassembles the nucleosomes from both the existing nucleosomes and nucleosomes assembled from freshly synthesized histones, which needs nucleosome or chromatin assembly factors.

10.3.2.3 Origin of eukaryotic DNA replication

Several animal viruses such as SV40 in animal cells, yeast (*S. cerevisiae* and *Schizosaccharomyces pombe*) and the ribosomal DNA of mammalian cells have been studied for understanding the origin and initiation of DNA replication in eukaryotes. SV40 virus has a 64-bp single origin of replication, which functions in mammalian cells and cell-free extracts. SV40 T antigen (T-Ag) protein, synthesized from the viral genome, initiates DNA replication from this origin. T-Ag also functions like a helicase and unwinds the DNA at the origin. An SSB stabilizes the single-stranded DNA template and the cellular DNA polymerase α–primase complex initiates DNA synthesis. As eukaryotic chromosomal DNA is large in size and DNA synthesis has to be completed within a short period of time, it utilizes multiple origins of replication, unlike the single origin in *E. coli* and viral genome. The size of the mammalian (mouse or human) genome (haploid set of chromosomes) is very large, that is, about 3×10^9 bp, compared withthe small *E. coli* genome (4×10^6 bp); therefore, if *E. coli* chromosome is replicated in 30 min, at the same rate, mammalian chromosomes would need about 30,000 min (3 weeks) if replicated from a single origin of replication. In addition to this, the mammalian genome is packaged into chromatin; therefore, the rate of replication in mammalian cells is about 10-fold slower than in *E. coli*. In spite of all these, mammalian chromosomes are replicated within a few hours due to the presence of thousands of origins of replication (Fig. 10.13).

The replication origins in eukaryotes have been described to possess elusive determinants (Costa and Blow 2007). Replication origins in the yeast genome are about 40 kb apart from each other, and in the case of human genome, the average distance between the origins is about 50–300 kb, resulting into around 30,000 origins. Yeast DNA elements called autonomously replicating sequences or ARS have been isolated and biochemically tested for their function for replication of plasmids as well as yeast chromosomes. ARS elements are 100-bp elements with an 11-bp core sequence, which is common to many ARS elements and is essential for the origin function. A six-subunit protein complex called the origin recognition complex (ORC), which binds to the core 11-bp sequence, supports initiation of DNA synthesis by recruiting the minichromosome maintenance (MCM) DNA helicase and other proteins by a mechanism similar to *E. coli* and eukaryotic viruses. ARS element and ORC proteins have conserved functions in eukaryotes. The *Schizosaccharomyces pombe* (fission yeast) has ARS sequences spread over 0.5- to 1-kb DNA with A/T-rich repeats for binding of ORC proteins, but the *Drosophila* and mammalian ORC proteins do not bind specific DNA sequences. In higher eukaryotes, the chromatin structure may determine ORC binding sites and initiation of DNA synthesis.

The ends of eukaryotic chromosomes are replicated by a different mechanism. The ends contain specialized DNA structures called telomeres and the telomeric sequences are several thousands of continuous stretch of repeats of the 5′TTAGGG3′ conserved sequence

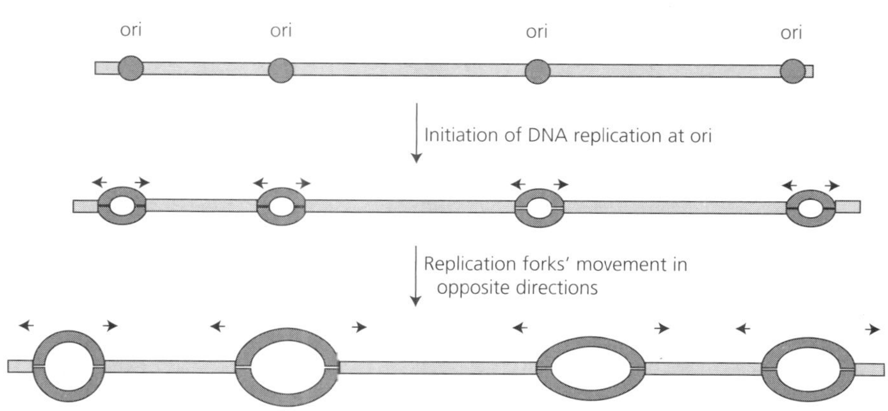

Figure 10.13 Origin and initiation of DNA replication in large eukaryotic chromosomes. The chromosomal DNA (a single duplex of DNA) melts into the two individual strands at the origins of replication (ori) occurring at multiple sites. The melting spreads in opposite directions from the ori sites and simultaneously the newly synthesized DNA is produced in opposite directions. As a result, two chromatids are formed from the parental chromosome, which ultimately separates into two individual chromosomes.

(Chan and Blackburn 2004). These repeats form a four-stranded G4-DNA structure by folding of the double-stranded DNA and the reverse transcriptase. Telomerase enzyme utilizes the telomeric RNA as a template to copy and synthesize the telomeric DNA repeats, which are added at the ends of the chromosomes. Many cellular proteins bind to telomeric repeats, and telomerase is regulated by telomeric proteins (Smogorzewska and Lange 2004). The ends protect the chromosomes from damage and maintain genomic stability. Defective and insufficient telomeric repeats at the ends of the chromosomes in human cells are associated with aging, and excessive telomeric repeats are linked to cancer.

10.3.3 Regulation of eukaryotic DNA replication

The mechanics of eukaryotic DNA replication is intimately associated with its regulation. Regulation of eukaryotic DNA replication may be understood in three different aspects by asking three different questions: Where does replication start? When does replication start? How does replication start? A recent review on these aspects has been published (Masai *et al.* 2010). DNA replication in metazoans has been described to occur at the right place, right time and only once (Machida *et al.* 2005). The origins of DNA replication have been intensely studied (Robinson and Bell 2005), and the processes as well as the mechanisms of initiation of cellular DNA replication have been recently reviewed (Costa *et al.* 2013). In addition to this, DNA replication has a high level of plasticity in its regulation. There are many potential replication origins (pre-RC) in the eukaryotic genome in excess of the need for complete genome replication. Some initiation sites are selectively used from the pre-RCs. If the rate of the replication fork formation is fast, the initiation is less frequent. The initiation occurs more frequently when the rate of fork formation is slower (Fig. 10.14).

The process of assembly and activation of pre-RC is complex. It involves a large number of proteins and a sequential coordination in a stepwise manner. First, the ATP–ORC complex binds to potential origin (ori) sequences. Then,the ATP–Cdc6 complex binds to the ORC. Then, the MCM complex+Cdt1 together binds to it. After this, the MCM complex associates with the ori, Cdc6 gets dissociated from the ori and ATP on the ORC is hydrolysed and this completes pre-RC assembly. Alternatively, the MCM complex moves away and repeats MCM loading on the ori. Thus, there is either formation of an unperturbed replication fork or a stalled replication fork, which produces a longer single-stranded DNA coated with replication protein A (RPA) and ataxia telangiectasia–mutated proteins and Rad3-related inhibitory protein (ATRIP) (Fig. 10.15). Table 10.2 shows the regulatory proteins and their complexes involved in the process of assembly and activation of the pre-RC complex. Many

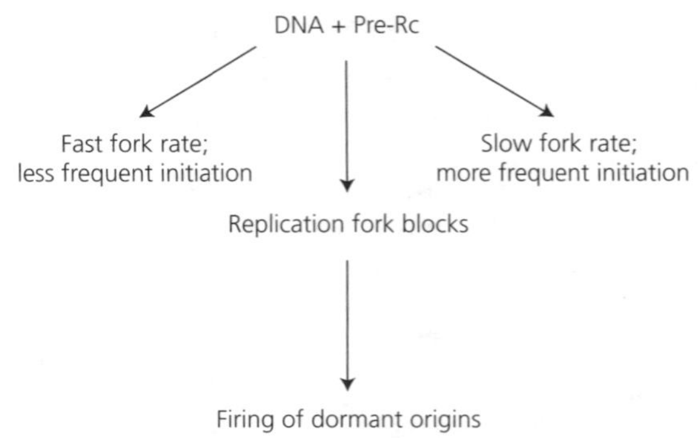

Figure 10.14 DNA replication shows high level of plasticity in eukaryotes. There are many potential replication origins (pre-RC), some of which are used for initiation. The rate of fork formation and the frequency of initiation are inversely related to each other. Blocking of replication fork may lead to activity of the dormant origin.

molecular details of the mechanisms of action of these protein complexes are still under investigation.

Genome-wide DNA replication studies have mainly focused on two aspects: (a) determination of replication origin mapping and binding sites of various factors, and (b) replication timing mapping. With respect to the first aspect, the following methods were adopted. Replicated DNA as heavy-light and unreplicated DNA as light-light fractions were separated at different time points after labelling the cells with bromodeoxyuridine (BrdU) during the S phase and were fractionated by density gradient centrifugation. Copy number changes in the DNA were also analysed. Single-stranded DNAs generated were also measured. The ORC–MCM binding sites on the DNA were also determined by the chromatin immunoprecipitation-chip (ChIP-chip) assay technique, where the antibody against the protein isolated the DNA bound to the protein and this DNA was then detected by an array of DNA sequences. The BrdU incorporation was also analysed by the ChIP-chip assay. The findings from these studies were as follows. The pre-RC sites (ORC–MCM binding) and the actual initiation sites coincided in budding yeast. But in fission yeast, the pre-RC was found once in 10 kb, that is, with an average inter origin length of 10 kb. Forty percent of the total pre-RC sites were used for initiation. The origins did not produce any consensus sequence. In more than 70% cases in fission yeast, AT-rich intergenic sequences were used as potential origins. In *Drosophila*, ORC-binding sites overlapped with RNA polymerase II–binding sites, suggesting a role for transcription factors in ORC binding. In human cells, 50% of origins were abundantly found in or near the CpG islands of promoter regions. Origins were strongly associated with transcriptional regulatory elements of the promoters, for example, c-Jun and c-Fos binding sites. Mammalian origins were also found to be conserved along evolution of the species. In mouse, the origin specification and its firing efficiency were maintained across different cell types.

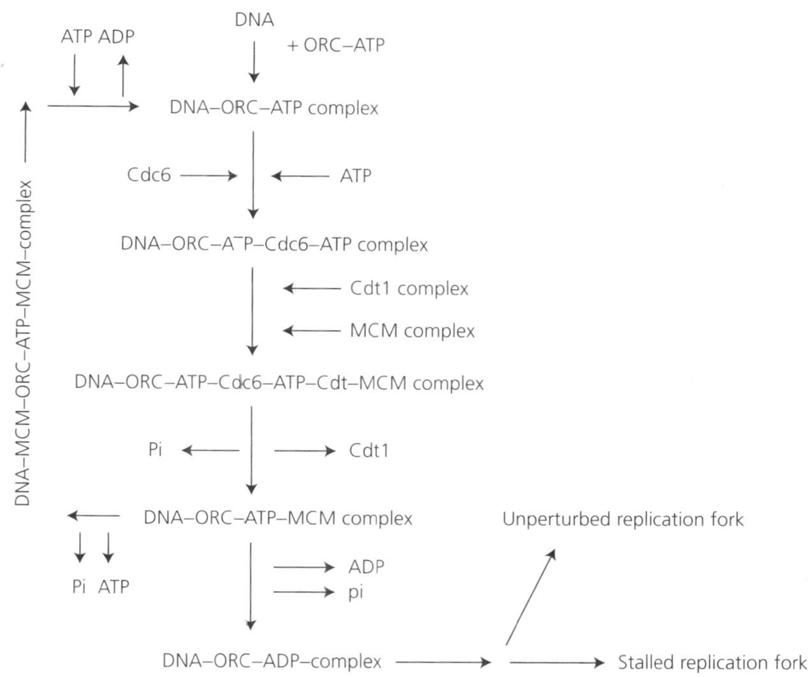

Figure 10.15 The processes of assembly and activation of the pre-RC complex in the eukaryotic genome occurs in a sequential and stepwise manner. The DNA–ORC–ATP complex is formed at the first step. Then, the Cdc6–ATP binds to it to form the DNA–ORC–ATP–Cdc6–ATP complex. The Cdt1 complex and the MCM complex bind to it in the next step to form the DNA–ORC–ATP–Cdc–ATP–Cdt–MCM complex. This is followed by ATP hydrolysis and dissociation of Cdt1. In the next step, the DNA–ORC–ATP–MCM complex is formed, which gets converted to the DNA–ORC–ADP complex by ATP hydrolysis. This is followed by one of the two fates: the unperturbed replication fork or the stalled replication fork. The DNA–MCM–ORC–ATP–MCM complex may be recycled to the initial DNA–ORC–ATP complex.

Table 10.2 Examples of some regulatory proteins involved in the process of assembly and activation of the potential replication origin (pre-RC) complex in mammalian cells (with homologues in other eukaryotes)

ATR-ATRIP	A complex of ataxia telangiectasia–mutated and Rad3-related (a phosphatidylinositol 3-kinase playing a major role in DNA replication stress checkpoint response) and ATR interacting protein
Asf1	Anti-silencing function 1 (histone H3–H4 chaperone)
Cdc45	An essential protein for a functional helicase at the fork
Ctf4	A conserved replication fork factor
Dpb 11	DNA polymerase B11 required for initiation and checkpoint response
GINS	Go-Ichi-Ni-San complex, essential for a functional helicase at the fork
Mcm 10	An essential factor for replication
Mcr 1	A conserved factor required for stable maintenance of the replication fork
PCNA-RFC-RFC2~5	Proliferating nuclear antigen-replication factor C (clamp and clamp loader for DNA chain elongation)
Polα-Pri	DNA polymerase alpha-primase
9-1-1-Rad 17-RFC 2~5	Rad 9-Rad1-Hus1 complex-Rad 17-replication factor C2~5 (clamp and clamp loader for DNA damage response)
RPA	Replication protein A
Sld2 or Sld3	Synthetically lethal with dpb11-1 mutant 2 or 3 (required for assembly of the initiation complex and a target of CDK phosphorylation)

For details, see Masai *et al.* (2010).

Origins used for normal replication and re-replication were different in budding yeast. With respect to the second aspect, cells were fractionated into early and late S-phase populations and BrdU incorporation into the DNA was measured as well as round mitotic cells were collected and measured from synchronized cell populations after pulse labelling, washed and incubated for different times. The findings from these studies revealed that eukaryotic genomes have different domains with different replication timings, each domain may contain from one to tens of replication origins, which fire in a coordinated fashion. No correlation between the replication timing and transcription or the GC content was observed in budding yeast. In metazoan genomes, the reverse was true. Here, the replication timing, transcription or GC content showed correlation. During complex cellular processes such as development and differentiation in multicellular organisms, the replication timing domains usually undergo substantial reorganization. Mutations in different diseases and single-nucleotide polymorphisms frequently occur in replication timing transition regions of the genome.

Thus, eukaryotic DNA replication is unique and complex. Its basic mechanism is synchronized with regulations in cellular processes. In metazoans, ORC binds DNA with less specificity, generally in AT-rich DNA sequences, whereas transcription factors and chromatin structures and their remodelling may provide ORC–DNA binding specificity. The pre-RC assembly process selects actual initiation sites on the chromosomal DNA. DNA replication shows high plasticity. Origin selection and fork rate control each other. Global chromatin context may regulate the replication timing and coordinate it with transcription. There are two fundamental principles of DNA replication. They are as follows: the pre-RC assembly and its activation occur during mutually exclusive periods of cell cycle, and there is only one round of DNA replication per each cell cycle. The stochastic and opportunistic initiation processes within a given replication domain may occur. Replication stress induces a checkpoint, which represses both progression of the fork and firing of a new origin. A large RPC has been isolated from cells, which contains helicase, fork stabilizers and histone chaperones, with the MCM complex being the central helicase of RPC. The CMG complex may be assembled as an active helicase with the help of RPC, GINS and Cdc 45.

We still need to fully understand the site-specific initiation of DNA replication in eukaryotes. Mechanism of timing of origin firing, coordination of fork rate and new origin firing during the S phase, molecular details of the eukaryotic replication fork, its regulation, link between DNA replication defects and diseases, and metabolic regulation of DNA replication are some more aspects to be studied and fully understood.

References

Alberts B. The cell as a collection of protein machines: Preparing the next generation of molecular biologists. *Cell* 1998;92:291–294.

Alberts B, Johnson A, Lewis J, Raff M, Roberts K, Walter P. *Molecular Biology of the Cell*, 4th Edition. New York: Taylor & Francis Group; 2002.

Brown TA. *Genomes 2*, 2nd Edition. New York: Wiley-Liss; 2002.

Chan SR, Blackburn EH. Telomeres and telomerase. *Philos Trans R Soc Lond B Bio Sci* 2004;359:109–121.

Costa A, Hood IV, Berger JM. Mechanisms for initiating cellular DNA replication. *Ann Rev Biochem* 2013;82:25–54.

Costa S, Blow JJ. The elusive determinants of replication origins. *EMBO Rep* 2007;8:332–334.

Dillingham MS. Replicative helicases: A staircase with a twist. *Curr Biol* 2006;16: R844–R847.

Friedberg EC. Biological response to DNA damage. *Cold Spring Harb Symp Quant Biol* 2000;65:593–602.

Groth A, Rocha W, Almouzni G. Chromatin challenges during DNA replication and repair. *Cell* 2005;128:721–733.

Hartwell L, Hood L, Goldberg ML, Reynolds AE, Silver LM. *Genetics: From Genes to Genomes*.Boston: McGraw-Hill; 2006.

Indiani C, O'Donnell M. The replication clamp-loading machine at work in the three domains of life. *Nat Rev Mol Cell Biol* 2006;7:751–761.

Kornberg A. Biological synthesis of DNA. *Science* 1960;131: 1503–1508.

Lewin B. *Genes VIII*. Upper Saddle River, NJ: Prentice-Hall; 2004.

Li JJ, Kelly TJ. SV40 DNA replication *in vitro*. *Proc Natl Acad Sci USA* 1984;81:6973.

Lodish H, Berk A, Zipursky SL, Matsudaira P, Baltimore D, Darnell J. *Molecular Cell Biology*, 5th Edition. New York: WH Freeman & Co.; 2004.

Machida YJ, Hamlin JL, Dutta A. Right place, right time and only once replication initiation in metazoans. *Cell* 2005;123:13–24.

Masai H, Matsumoto S, You Z,Yoshizawa-Sugata N, Oda M. Eukaryotic chromosome DNA replication: Where, when and how? *Ann Rev Biochem* 2010;79:89–130.

Meselson M, Stahl FW. The replication of DNA in *E. coli*. *Proc Natl Acad Sci USA* 1958;44:671–682.

Modrich P, Lahue R. Mismatch repair in replication fidelity, genetic recombination and cancer biology. *Ann Rev Biochem* 1996;65:101–133.

Mott ML, Berger JM. DNA replication initiation: Mechanisms and regulation in bacteria. *Nat Rev Microbiol* 2007;5:343–354.

O'Donnell M. Replisome architecture and dynamics in *E. coli*. *J Biol Chem* 2006;281:10653–10656.

O'Donnell M, Kuriyan J. Clamp loaders and replication initiation. *Curr Opin Struct Biol* 2005;16: 35–41.

Okazaki R, Okazaki T, Sakabe K, Sugimoto K, Sugino A. Mechanism of DNA chain growth. I. Possible discontinuity and unusual secondary structure of newly synthesized chains. *Proc Natl Acad Sci USA* 1968;59:598–605.

Raghuraman MK, Winzeler EA, Collingwood D, Hunt S, Wodicka L, Conwat A, *et al*. Replication dynamics of the yeast genome. *Science* 2001;294:115–121.

Rao PN, Johnson RT. Mammalian cell fusion: Studies on the regulation of DNA synthesis and mitosis. *Nature* 1970;225:159.

Robinson NP, Bell SD. Origins of DNA replication in the three domains of life. *FEBS J* 2005;272:3757–3766.

Smogorzewska A, de Lange T. Regulation of telomerase by telomeric proteins. *Ann Rev Biochem* 2004;73:177–208.

Synder L, Champness W. *Molecular Genetics of Bacteria*, 2nd Edition. Washington, DC: ASM Press; 2003.

Wang JC. Cellular roles of DNA topoisomerases: A molecular perspective. *Nature Rev Mol Cell Biol* 2002;3:430–440.

Watson JD, Baker TA, Bell SP, Gann A, Levine M, Losick R. *Molecular Biology of the Gene*, 5th Edition.Harlow, UK: Pearson Education; 2004.

11

RNA, its transcription and processing

Shyamal Goswami

11.1 Discovery of RNA and the emergence of the central dogma of molecular biology

In the late nineteenth century, laboratories in Europe and North America were trying to understand the chemical constitution of cells. In those days, organic chemistry had developed as a discipline, but biochemistry was not on the horizon. As modern-day purification procedures were non-existent, isolation of a particular type of biomolecule from cell extracts was not possible. People used to study only those biomolecules which were in abundance in certain cell types. One example is isolation and crystallization of haemoglobin from red blood cells. Friedrich Miescher, while working in the University of Tübingen, Germany, first isolated 'nuclein' (DNA with associated proteins) from the nucleus of pus cells (lymphoid cells). He then demonstrated that it was made up of hydrogen, oxygen, nitrogen and phosphorus and there was a unique ratio of phosphorus to nitrogen (suggesting a defined molecular constitution). He subsequently continued his interest in nucleic acids and isolated it from salmon sperm (till date, it is used as a source of DNA in molecular biology research). However, until early twentieth century, existence of two distinct types of nucleic acids, that is, RNA and DNA, with discrete biological functions was not known. They were rather named after the source from which they were commonly isolated; RNA was 'yeast nucleic acid' and DNA was 'pancreas nucleic acid'. When carbohydrate chemists showed that the two nucleic acids are characterized by different sugars, they were named accordingly. It was then shown that except containing uracil instead of thymine, RNA contains same bases as DNA.

Involvement of RNA in protein synthesis was evident as early as late 1930s (Allfrey *et al.* 1953). In 1950s, the American scientist Severo Ochoa (born in Spain) discovered how RNA is synthesized (Grunberg-Mano *et al.* 1956) and was awarded the Nobel Prize in Medicine in 1955.

Following the discovery by Oswald Avery, Colin MacLeod and Maclyn McCarty that DNA is the active principle of genetic transformation in bacteria and hence is the genetic material (Avery *et al.* 1944), it was further confirmed by Alfred Hershey and Martha Chase through the bacteriophage lambda experiments (Hershey and Chase 1952). The central role of DNA in storage and transmission of genetic information was finally established through the discovery of its double helical structure by James D. Watson and Francis Crick in 1953 (Watson and Crick 1953; Martin *et al.* 1962).

In subsequent years, accumulated knowledge on the structure and synthesis of DNA, RNA and proteins led to the emergence of the 'central dogma of molecular biology' (Crick 1970), which deals with the detailed residue-by-residue transfer of sequential information from DNA to protein ('DNA makes RNA makes protein', Marshal Nirenberg 1961; Crick 1970). It was also accepted that such process of information transfer cannot be backward, that is, from protein to nucleic acid.

In the last 50 years, the world of RNA has further expanded, and till date, RNA research has fetched about 30 Nobel prizes, of which some significant winners are:

- Robert William Holley, Nobel Prize in Physiology or Medicine 1968 (together with Har Gobind Khorana and Marshall Warren Nirenberg) for describing the structure of alanine transfer RNA (Apgar and Holley 1964)
- David Baltimore, Howard Temin and Renato Dulbecco, Nobel Prize for Physiology or Medicine 1975 for their discoveries concerning the interaction between tumour viruses and the genetic material of the cell (discovery of reverse transcriptase) (Temin and Baltimore 1972)

- Sidney Altman and Thomas R. Cech, Nobel Prize in Chemistry 1989 for their work on the catalytic properties of RNA (North 1989)
- Richard John Roberts and Phillip Allen Sharp, Nobel Prize in Physiology or Medicine 1993 for the discovery of introns in eukaryotic DNA and the mechanism of gene splicing (Berget *et al.* 1977)
- Andrew Zachary Fire and Craig C. Mello, Nobel Prize for Physiology or Medicine 1998 for the discovery of RNA interference (RNAi) (Ebhardt 2007)
- Roger David Kornberg, Nobel Prize in Chemistry 2006 for discovering the molecular basis of eukaryotic transcription (Kornberg 2007)
- Venkatraman Ramakrishnan, Thomas A. Steitz and Ada E. Yonath, Nobel Prize in Chemistry 2009 for studies of the structure and function of the ribosome (Ogle *et al.* 2003)

11.2 Types of RNA and their functions

As described in the previous section, understanding of the biology of RNA has co-evolved with that of molecular biology, leading to many discoveries that are interconnected. Further, with the progress of time, the world of RNA has become more and more diverse, highlighting the complex role of RNA in the propagation and sustenance of the life process.

In late 1940s and early 1950s when DNA was established as the genetic material that leads to the synthesis of proteins, it was also demonstrated that RNA is synthesized in the nucleus and moves to the cytoplasm (while some RNAs always remain in the nucleus; Zalokar 1959). However, the connection between the synthesis of RNA and that of proteins was not known. The first evidence in this regard came from the observation that ribosomes, a large assembly of RNA and proteins located in the cytoplasm, are essential for protein synthesis (Brenner 1961). Therefore, at one point, it was thought that these RNAs are produced from the DNA and carry the necessary information to the cytoplasm and get translated into proteins therein. However, if it is so, as proteins present in a cell are of many different types, the composition of ribosomal RNAs would also be very different from one another. However, upon analysis, RNA content in ribosomes was found to be very similar under different conditions (Brenner *et al.* 1961). Also, ribosomes could be separated into two parts, each of which contained specific types of RNAs. To explain these observations, François Jacob and Jacques Monod put forward the 'mRNA hypothesis' in 1961, suggesting that a special class of RNA named messenger RNA (mRNA) is synthesized from the DNA (with base sequences complementary to the gene sequence) and transported to the ribosomes for protein synthesis (Gros *et al.* 1961). They also suggested that after protein synthesis, the mRNA leaves the ribosome, making way for other mRNAs to be translated. This hypothesis was subsequently confirmed by Sydney Brenner, François Jacob and Matthew Meselson by showing that when bacteriophage T4 infects an *E. coli* cell, a virus-specific RNA is made that rapidly but only transiently associates with bacterial ribosomes (Brenner *et al.* 1961). Thereafter, transfer RNAs (tRNAs) were discovered that cargo specific amino acids to the growing polypeptide chains in the ribosome. Every cell has a repertoire of tRNAs, and each species is specific for one particular amino acid. tRNAs are small (~75 nucleotides), and each has one anti-codon sequence complementary to the triplet codon in the mRNA, ensuring the incorporation of corresponding amino acid in the growing polypeptide chain (Allen *et al.* 1960) (Fig. 11.1). In 1965, Robert W. Holley determined the primary structure of tRNA, and based upon that, proposed the cloverleaf structure of tRNA (Holley *et al.* 1965). Methods of sequencing of RNA were developed more than a decade before those for DNA, but they became obsolete thereafter as they were much more labour intensive.

The presence of a population of small RNAs in the nucleus was observed as early as the 1950s (Weinberg and

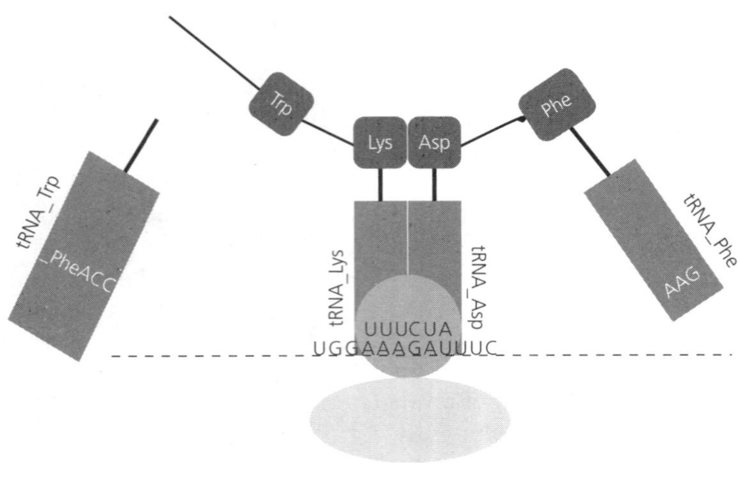

Figure 11.1 Function of tRNA as adapters of amino acids to the growing polypeptide chain. The mRNA is shown as dotted line wherein the nucleotides are partially shown. Each tRNA carries a complementary triplet sequence (anticodon) that base-pairs with the codon. Two subunits of ribosomes harbouring the translating codons (UUU and CUA) are shown as spheres. Actively engaged tRNA, namely tRNA_Lys and tRNA_Asp, are shown. The incoming tRNA_Phe and outgoing tRNA_Trp are also shown on the left and right, respectively.

Penman 1968). However, their function in RNA metabolism became apparent only after the discovery of splicing and polyadenylation in the 1980s (Rogers and Wall 1980; Richard John Roberts and Phillip Allen Sharp, Nobel Prize in Physiology or Medicine 1993).

Research on RNA processing in the 1980s and 1990s led to another astonishing discovery that some RNA molecules can cleave themselves in the absence of any proteins. Such function of RNA as an enzyme led to the term 'ribozyme', that is, RNA acting as an enzyme (Cech and Bass 1986). However, unlike the wide occurrence of protein enzymes, ribozymes are rare in nature, though a number of them with designer catalytic functions have been developed *in vitro* (Koizumi *et al.* 1989). Two other novel classes of small RNAs, namely microRNA (miRNA) and small interfering RNA (siRNA), were discovered in the 1990s (Ambros 2001). They are generated in cells as large precursor transcripts and then processed through distinct pathways (Cullen 2004). Many distinct types of miRNAs have been identified in eukaryotic cells and they play key regulatory roles in gene expression (He and Hannon 2004). Aberrant expressions of these miRNAs are often associated with various human diseases (Cronce and Calin 2005). Eukaryotic cells also contain long non-coding RNAs (ncRNA), which also modulate gene expression through interaction with various proteins (Goodrich and Kugel 2006).

11.3 Only one strand of DNA is transcribed to produce one particular type of RNA species

Following the establishment of the central dogma of molecular biology that 'DNA is the genetic material' and 'DNA makes RNA makes protein', exhaustive efforts were made to understand the mechanisms involved. In the 1960s, three independent lines of research thus emerged, that is, understanding how genetic information is propagated (i.e. DNA replication), how genetic information is transcribed into mRNA and how mRNAs are translated into proteins?

In transcription research, one important question to be answered was, 'Do both the strands of DNA get transcribed?' Although early *in vitro* studies suggested that both strands of DNA can make RNA, in 1963, a group of researchers led by S. Spiegelman at the University of Illinois answered this question by studying bacteriophage ϕX174, which has a double-stranded genome while replicating in the host, but upon maturation carries only one strand as its genetic material. By studying hybridization of single-stranded genomic DNA and its RNA, they demonstrated that the viral RNA is complementary to only one strand (Hayashi *et al.* 1963). Subsequent studies were done by synthesizing various cDNAs, cloning them in plasmid vectors and sequencing. Upon comparing cDNA sequences to those of their genes, it was established that *in vivo*, only one strand of a gene is transcribed into RNA (Hahn *et al.* 1977; Fig. 11.2). Conventionally, during transcription, the strand that is used as template is called the antisense strand as it harbours a sequence complementary to the RNA. This strand is also called the negative or the non-coding strand. The other strand that harbours the same sequence as that of the RNA (except containing thymidine instead of uracil) is called the sense strand or the positive strand. Therefore, in real sense, the non-coding/antisense strand actually carries the genetic code that is transcribed into the mRNA. Also, it has to be remembered that in a large DNA molecule harbouring many different genes (as in chromosomes), the sense sequence of a particular gene can be in one strand (called the Watson strand), while the sense sequence of another gene can be in the other strand (called the Crick strand). In other words, both the strands can

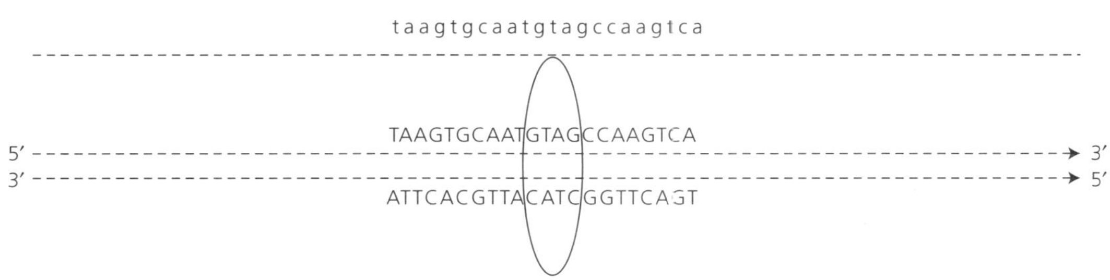

Figure 11.2 RNA polymerase transcribing only one of the two strands of the DNA (gene). Both the strands are shown as dotted lines and their relative orientation (3′ and 5′) are shown). RNA polymerase locally melts the double-stranded DNA and transcribes only one strand, using it as a template (lower strand in above picture, going in 3′ and 5′ direction as shown by an arrow). The resultant mRNAs thus have the complementary sequence that is same as that of the other strand (upper strand in this picture) of the DNA. In higher eukaryotes, generally one fragment of DNA contains only one gene and only one strand of the DNA is transcribed into an mRNA. Some viral genes are unique in nature as both the strands code for mRNAs and they are transcribed independently by the RNA polymerase. However, more recent studies show that in many genes in higher eukaryotes, both the strands are actually transcribed but only one produces full-length mRNA, while the other gives RNAs (non-coding) of various sizes which are called antisense RNAs (taking the mRNA as the sense RNA). The mechanisms of their synthesis and the functions of the antisense RNA are not understood as yet.

harbour sense and antisense sequences of different genes, which are independently transcribed in a gene-specific manner (Fig. 11.2). Noticeably, in the past decade, technologies have emerged where almost every RNA molecule isolated from a cell can be sequenced (Tang *et al.* 2010). It has now been shown that *in vivo*, the entire genome is transcribed from both the strands, producing numerous small and large transcripts, of which only a subset correspond to the coding strands of the annotated genes (encoding rRNAs, tRNAs and mRNAs; Jacquier 2009). So, this discovery has completely changed the earlier concept of sense and antisense strands of DNA (and the gene). Nevertheless, despite such pervasive transcription of the genome, the relative abundance of these RNA is much less compared with those transcribed from annotated genes.

11.4 Prokaryotic and eukaryotic RNA polymerases

Transcription is the first step of gene expression when the antisense strand of the double-stranded DNA encompassing the gene is copied to the functional RNA. This is done by RNA polymerase in association with numerous other core and auxiliary transcription factors. RNA polymerase copies the DNA strand from the 3′ to 5′ direction so that the resultant RNA strand is synthesized from the 5′- to 3′-end. RNA polymerase was discovered in 1960 by two independent groups of researchers, one led by Jerard Hurwitz and the other by Samuel B. Weiss, Audrey Stevens and James Bonner (Hurwitz 1968–1969). Although Nobel Prize in Medicine had already been awarded to Severo Ochoa in 1959 for discovering RNA polymerase ('Ochoa enzyme', as it was called in those days), later it was found to be polynucleotide phosphorylase having both DNA-independent oligonucleotide polymerase activity and exoribonuclease activity. Nevertheless, the contribution made by this enzyme was enormous as it was used in synthesizing RNA templates *in vitro*, which were then used for deciphering the genetic code. RNA polymerases are involved in the synthesis of mRNA, tRNA, rRNA, miRNA, catalytic RNA (Ribozyme) and siRNA etc.

In prokaryotes, transcription machinery is relatively less complex and involves only one RNA polymerase that synthesizes both mRNAs and other non-coding RNAs. The core enzyme has five subunits, namely β′, β, I &II and ω (~400 kDa), while the complete holoenzyme has another subunit named sigma (σ) (Burgess 1971).

In eukaryotes, however, there are five RNA polymerases, namely RNA polymerase I–V (Wild and Cramer 2012). RNA polymerase I synthesizes 45S pre-rRNA (in yeast, it is 35S) that upon maturation gives 28S, 18S and 5.8S rRNAs. RNA polymerase II is functionally most versatile as it transcribes the precursors of mRNAs, most of the small nuclear RNAs and miRNAs. RNA polymerase III synthesizes tRNAs, 5S rRNAs and several small RNAs. RNA polymerase IV and V are recently discovered in plants and they, respectively, synthesize siRNA and RNAs involved in siRNA-directed heterochromatin formation. Another completely different type of RNA polymerases are those using RNA as the template (RNA-dependent RNA polymerase). They are encoded by RNA viruses and are required for the replication of the viral genome. Among all these RNA polymerases, RNA polymerase II has been widely investigated. It has 10–12 subunits (RPB 1–12) and the total complex is ~550 kDa. The largest subunit II is RPB 1 and has an unique 'carboxy terminal domain' (CTD) comprising 26 (yeast) to 52 (human) heptapeptide (Tyr-Ser-Pro-Thr-Ser-Pro-Ser) repeats, which play a crucial role in the polymerase II activity (Egloff and Murphy 2008) (see Section 11.5 for further details). However, eukaryotic RNA polymerases on their own do not recognize the target promoters and initiate transcription. They require a number of general transcription factors for precise identification and assembly on the transcription start site on the target promoters. The number of such general transcription factors is six for RNA polymerase II (Smale and Kadonaga 2003) (see Section 11.5 for details).

11.5 Eukaryotic transcription machinery is an assembly of a large number of proteins

As transcription is the first step in decoding the genetic information, it needs to be tightly regulated. Although RNA polymerase is the enzyme that transcribes DNA into RNA, it alone cannot do it in the cellular milieu. During evolution, nature has created several other associated factors that work together with the RNA polymerases, ensuring synthesis of the right transcript at the right time in the right type of cells. Transcription factors that perform such functions are called 'general transcription factors' (Smale and Kadonaga 2003). The reason for using the term 'general' in describing them is, there is another category of transcription factors that also play a critical role in transcription. They are called 'gene-specific transcription factors' as their role is more in the context of specific type of cell and the gene they help in transcribing (Amoutzias *et al.* 2008). As an example, transcription factor NF-κB plays a nodal role in transcribing genes mediating immune response. General transcription factors are highly conserved from

Saccharomyces cerevisiae to human. They were named in the sequential order as they were discovered by biochemical analysis and thereafter validated by genetic and other methods. General transcription factors associated with RNA polymerase II (Pol II) are better characterized, and there are six such factors comprising about 30 subunits (TFII- A, B, D, E, F and H). Although the initial model of RNA polymerase II assembly suggested that they are sequentially engaged in the transcription process, later studies suggested that they also can engage in transcription as a preassembled complex (holoenzyme) with the RNA polymerase (Smale and Kadonaga 2003). Further, as among three eukaryotic RNA polymerases, Pol II–mediated transcription is most versatile, the general transcription factors associated with it are more in number and their functions are better understood. Nevertheless, many of these factors are also involved in the initiation of transcription by other two RNA polymerases. The following is a list of the major factors in this category:

- **TFIID:** It is a multiprotein complex containing the TATA box–binding protein (TBP) and several other proteins known as TBP-associated factors (TAFs). TBP is the first protein recruited to the promoter, which induces a bend in the DNA. TBP participates in the function of TFIID even in TATA-less promoters. The TAFs have histone-like structures and some have serine kinase and histone acetyltransferase activities.
- **TFIIB:** It associates with TBP and the TFIIB–TBP–DNA ternary complex is formed. TFIIB physically links TFIID at the promoter with the Pol II/TFIIF complex.
- **TFIIA:** It is required for the activation of transcription.
- **TFIIE:** It recruits TFIIH to the promoter and is involved in promoter melting.
- **TFIIF:** It escorts the RNA polymerase II to the promoter. It is the eukaryotic homolog of bacterial sigma subunit. Without the involvement TFIIF, Pol II cannot stably associate with the TFIID and TFIIB assembly at the promoter.
- **TFIIH:** It is a multi-subunit protein with 3′-5′ helicase and RNA Pol II C-terminal domain kinase activity.
- **P-TEFb:** It is also a kinase that targets the C-terminal domain of RNA polymerase II and prevents promoter-proximal pausing and premature termination, thus promoting the entry of the polymerase into productive elongation.
- **TFIIS:** It is critical for the release of stalled RNA Pol II, promoting elongation and decreasing pausing.

The sequence of events for the formation of a pre-initiation complex and transcription by Pol II is (Shilatifard *et al.* 2003; Smale and Kadonaga 2003):

- Recognition of core promoter elements by TFIID (containing TBP and TAFs).
- Recognition of and binding of TFIIB to the TFIID–promoter complex.
- Recruitment of a TFIIE/Pol II complex.
- Promoter melting and formation of an 'open' initiation complex.
- Synthesis of the first phosphodiester bond of the nascent mRNA transcript
- Release of Pol II from the promoter (promoter clearance)
- Elongation of the RNA transcript

11.6 How do RNA polymerases know where to start and where to end the transcription?

Once it was established that RNA polymerase transcribes only one strand of DNA, next question was how does it recognize which strand to copy, and where to start and where to end transcription? Two major discoveries in molecular biology research facilitated understanding of the transcription process. In early 1970s, David Baltimore and Howard Temin (Nobel prize in Physiology and Medicine 1975) discovered the enzyme reverse transcriptase (Verma *et al.* 1974). These enzymes are used by retroviruses to copy their RNA genome into DNA (pro viral), which is then integrated into the host genome, followed by latency. Another discovery was the development of the methodology for sequencing DNA (Maxam and Gilbert 1977; Walter Gilbert and Frederick Sanger, Nobel prize in Chemistry 1980). These two discoveries, together, made it easier for the researchers to determine the sequence of an mRNA (first converting it into a cDNA and then cloning and sequencing it) and identifying the gene from which the mRNA was transcribed. Studies with numerous prokaryotic and eukaryotic genes led to the realization that transcription of each gene starts at a particular nucleotide (called transcription start site; Juven-Gershon and Kadonaga 2010). Conventionally, the portion of the gene further upstream of the start site is called the promoter (that regulates its transcription), while the sequence downstream of the start site is called the coding region. However, the actual protein-coding region starts a few hundred nucleotides further downstream from the transcription start site. These few hundred nucleotides are copied as the 5′ untranslated region of the resulting mRNA. Also, the promoter region is further divided into two sections, namely the proximal and the distal promoter. The proximal one generally is spread up to 200–250 nucleotides upstream of the start site, while the distal promoter may spread up to 2000–3000 nucleotides further upstream (Uptain *et al.* 1997). While RNA polymerase and various other general transcription factors bind to the proximal promoter and initiate transcription, various gene-specific transcription factors bind to the distal promoter and communicate with the RNA polymerase, boosting the transcriptional output (Tan and Richmond 1998). In this regard, one has to remember that DNA-binding proteins, including RNA polymerases, contact DNA at small but specific stretches of nucleotides that can be identified by studying DNA protein interactions *in vitro*. Initial studies in early 1980s revealed that a small stretch of T- and A-rich sequence (5′-TATAAA-3′ or a variant), called TATA sequence, is located ~25 base pairs upstream of the transcription site (Smale and Kadonaga 2003). It is conserved in eukaryotic evolution and is bound by the TBP, a constituent of TFIID–RNA polymerase complex (Vannini and Cramer 2012). Following the discovery of TATA element in the promoters of various protein-coding genes, once it was thought that they are present in all mRNA-encoding genes and it is the only requirement for RNA polymerase to bind the promoter and initiate transcription (at a point ~25 nucleotides further downstream). However, in 1989, another core promoter element with functions similar to those of a TATA box was identified and named 'initiator element (Inr)' (Thomas and Chiang 2006). The Inr elements harbour the transcription start within the consensus sequence $P_yP_yA^{+1}N^{T/}{}_{A}P_yP_y$ (P_y denotes pyrimidine base and +1 denotes transcription initiation site, N denotes any nucleotide base), and like TATA box, they also bind the TBP (TFIID complex). The Inr element is found in many TATA-less genes, where it decides the transcription start site, as well as in TATA-containing genes, where it influences the efficiency of transcription initiation by the TATA sequence. Another core promoter element named downstream promoter element (DPE) was first (1996) discovered in *Drosophila melanogaster* and subsequently found in other metazoans, including humans (Kadonaga 2002). It has been observed that the functions of DPE and Inr in basal transcription are interdependent. The DPE sequence is located at about 28–33 nucleotides downstream of the transcription start site and the consensus sequence is RGWYV(T) [R: pu**R**ine i.e. A/G; W: Weak i.e. A/T; which form only two hydrogen bonds; Y: (p**Y**rimidine, i.e. C/T; V: A/C/G. DPE is found in both TATA-containing and TATA-less promoters and is required for the binding of TFIID to TATA-less promoters. The fourth core promoter element that was discovered in 1998 is the 'B recognition element (BRE)' found in many eukaryotic genes either immediate upstream (BRE_u: consensus sequences ${}^{G}/{}_{C}{}^{G}/{}_{C}{}^{G}/{}_{A}$CGCC) or downstream ($BRE_d$: consensus sequences ${}^{G}/{}_{A}$T-T/G/A-${}^{T}/{}_{G}{}^{G}/{}_{T}{}^{T}/{}_{G}{}^{T}/{}_{G}$) of the TATA box (Juven-Gershon *et al.* 2008). Two other core promoter elements named MTE (motif ten element) and DCE (downstream core element) have also been described in various eukaryotic promoters (Juven-Gershon *et al.* 2008). The consensus sequence for MTE is $C^{G}/{}_{C}{}^{G}/{}_{C}A^{A}/{}_{G}C^{G}/{}_{C}{}^{G}/{}_{C}$AAC, located at positions +18 to +27 relative to the transcription start site, and its function depends on the Inr element; although it is independent of the TATA box and DPE, it shows synergy with them (Juven-Gershon *et al.* 2008). Genome-wide promoter sequence analysis has suggested that TATA-containing promoters are less in vertebrates (~15%) and more in *Drosophila* (~38%). Rather, Inr elements are more frequent in both vertebrates and fly (55% and 70%, respectively). DPE is also found in abundance in human and fly promoters (48% and 40%, respectively). BRE is more prevalent in vertebrates (12–62% depending on species), while MTE is found primarily in fly (~8.5%) (Juven-Gershon *et al.* 2008). Once it was thought that core promoters are least diverse and just function as the assembly point of RNA polymerase II. However, following the discovery of multiple core promoter elements, it now appears that core promoters are quite diverse in nature and play an active role in regulating RNA polymerase II assembly in multiple ways, actively regulating gene expression (Sandelin *et al.* 2007).

In recent years, comprehensive sequencing of total cellular RNA has led to another surprising discovery. It has been shown that initiation of transcription of many mammalian promoters start at various points spanning over about hundred nucleotides in the proximal promoter. These genes generally code for proteins that are constitutively expressed (generally called housekeeping genes; like those involved in metabolism). The promoters of these genes are called disperse promoters, while those having only one transcription start site are called focused promoters (Seila *et al.* 2009). They contain repeats of CpG sequences called CpG islands. Therefore, it now appears that two independent mechanisms have evolved in regulating genes which are constitutively expressed and those which are modulated under different conditions. Although both uses RNA polymerase II for their transcription, the mechanisms of its assembly is totally different for the two classes of promoters (Kulaeva *et al.* 2013; Fig. 11.3).

11.7 Transcription initiation of eukaryotes requires extensive coordination between chromatin modifiers and upstream activators

The total length of DNA strands in a single human cell is several metres, while the diameter of a nucleus is only a few micrometres. To accommodate itself therein, genomic DNA thus condenses by winding around histones, forming coils (nucleosomes) and kinks, known as chromatin. However, in this process, DNA also wraps itself with chromatin proteins, becoming inaccessible to various regulatory proteins. Such constraints are further compounded by numerous other structural proteins in the nucleus. Eukaryotic genes are

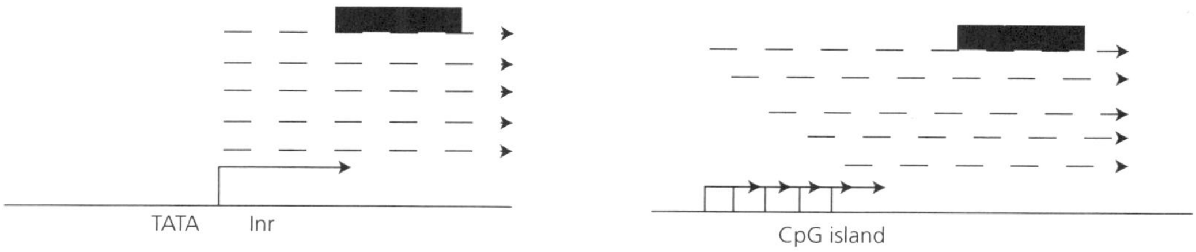

Figure 11.3 Two different modes of transcription initiation by RNA polymerase II. (Left panel) Transcription of genes which contain defined TATA and Inr elements is generally initiated at a specific nucleotide in the promoter and the resulting transcripts (dotted arrows with the coding region marked as a grey box) have the same length. (Right panel) Transcription of genes having CpG island in their promoters is generally initiated at multiple points spreading over about 100–150 nucleotides. These transcripts have (dotted arrows with the coding region marked as a grey box) variable lengths. Generally, genes which are regulated by various signals are transcribed by the first mechanism, while those which are constitutively expressed follow the second mechanism. The mechanistic difference by which RNA polymerase II can initiate transcription in two different ways is not understood as yet.

not transcribed until and unless they are opened up, accessing RNA polymerase and transcription factors (both general and gene specific). In fact, the opening of chromatin is a highly coordinated process, playing a major role in the regulation of gene expression (and other DNA-related activities such as replication). Chromatin organization and remodelling is a dynamic process involving several chromatin modifiers and histone variants. Over the past two decades, a large number of chromatin-modifying activities have been characterized. Broadly, they are of two types: (a) ATP-dependent chromatin-remodelling complexes which move, eject or restructure nucleosomes and (b) covalent histone-modifying enzymes (Flaus and Owen-Hughes 2011; Kulaeva *et al.* 2013). At least five families of ATP-dependent chromatin remodellers, namely SWI/SNF, ISWI, NURD/Mi-2/CHD, INO80 and SWR1, have been characterized, of which the first two are better understood, especially in yeast. Several histone-modifying complexes catalysing addition or removal of chemical groups on histones have been reported. Commonly found histone modifications include acetylation, methylation, phosphorylation and ubiquitination at the N-terminal region. Such modifications loosen or tighten the DNA wrapped around histones, altering their interactions (Rando 2007). As an example, while methylation of certain lysine residues in histone 3 (H3) enhances the condensation of DNA, inhibiting transcription, methylation of certain other lysine (and arginine) residues (Lys 4 and 36) results in transcriptional activation. Similarly, while acetylation of histone by histone acetyl transferase (HAT) relaxes the condensed DNA and makes it accessible to gene regulatory proteins, increasing gene expression, deacetylation of histones by histone deacetylase (HDAC) enhances the packing of DNA, causing repression. Taken together, chromatin remodelling and histone modifications are very essential components of the eukaryotic gene-expression machinery. In recent years, chromatin remodelling has been extensively associated with many human diseases such as cancer and various remodelling proteins are being studied as therapeutic targets (O'Sullivan and Karlseder 2012).

11.8 Eukaryotic mRNAs are extensively processed by capping, polyadenylation and splicing

Most of the early studies on the mechanism of gene transcription were done in *E. coli*, which showed that once a particular segment of the DNA is transcribed into mRNA, it is simultaneously translated into proteins (Cozzone 1980). However, in early 1970s, it was first reported that in sea urchin eggs, between fertilization and the two-cell stage, the cellular content of polyadenylic acid almost doubles and it remains associated with the ribosome–polyribosome fraction (Wilt 1973). It was then established that polyadenylation occurs at the 3′-ends of a pool of preexisting mRNAs, facilitating their translation (Lizardi *et al.* 1975). Subsequently, it was also established that all eukaryotic mRNAs (with the exception of a few histone mRNAs) are polyadenylated immediately after their transcription. Normally, the poly(A) tail is 200–300 bases long and it plays a key role in the export of mRNA from the nucleus, its translation and stability (Wells 2012). Polyadenylation of mRNAs is independent of template DNA and is catalysed by an enzyme named Poly(A) polymerase. Transcription by RNA polymerase starts from the 3′-end of the gene and the 5′-end of the RNA is first to synthesize. Once RNA polymerase II completes the transcription process and the 3′-untranslated region (UTR) of the mRNA comes out of the RNA polymerase II complex, nascent RNA is cleaved by a multiprotein complex called CPSF at about 10–30 nucleotides downstream of a specific sequence (5′-AAUAAA-3′) called polyadenylation signal.

Poly(A) polymerase then starts adding A residues at the free 3′ hydroxyl group of the mRNA. Many eukaryotic non-coding RNAs are also polyadenylated, while a subset of them, such as microRNAs, do not carry the Poly(A) tail in their mature form (Tuck and Tollervey 2011). Many mRNAs carry multiple polyadenylation signals in their 3′-UTR and thus can be polyadenylated at any one of those, resulting in multiple subspecies of the same mRNA (with varying 3′-UTR) (Di Giammartino *et al.* 2011). As 3′-UTRs often carry binding sites for many RNA-binding proteins and micro RNAs, such differential polyadenylation with different UTR lengths plays a major role in regulating their functions (such translation, stability, transport and cellular localization). In many bacteria, mRNAs and non-coding RNAs are also polyadenylated, but the poly(A) tail is about 30 nucleotides long and it promotes their degradation rather than stability (Kushner 2004). In metazoans and some parasites, mitochondrial RNAs might undergo both stabilizing and destabilizing polyadenylations (Rorbach and Minczuk 2012).

Almost immediately after the discovery of polyadenylation of eukaryotic mRNAs, people working on mRNAs from various mammalian DNA viruses and their hosts noticed that their 5′-ends do not have a free phosphate group as in bacterial mRNAs (Shatkin 1976). Further study revealed that they are actually blocked by a methyl guanosine residue through a 5′-5′ triphosphate linkage, hence named 'Cap' (Mizumoto and Lipmann 1979). Capping of eukaryotic mRNAs is linked to their efficient translation, while uncapped mRNAs are unstable and are rapidly degraded. Like polyadenylation, capping is also an enzymatic process occurring immediately after the initiation of transcription (i.e. during early elongation). More recent studies have shown that eukaryotic cells have large pools of small RNAs which are also capped but they are not translated (Shimba *et al.* 1992). Also, the mechanism of their capping is different from that of mRNAs. So, it appears that in eukaryotes, capping plays a wider role other than only translation (Darnell 2013).

In mid-1970s, researchers analysing sequences of mRNAs from DNA viruses (such as adenovirus and SV40) revealed another astonishing phenomenon. They noticed that the mRNAs were not the uninterrupted copy of the corresponding gene, but rather, they had certain segments of the gene sequence missing. It was then established that following transcription, the transcript (called pre-mRNA) undergoes a processing wherein certain intervening sequences (called introns) are excised and the remaining sequences (called exons) are ligated back in the same order, producing the mature mRNA harbouring an open reading fame preceded by the 5′-UTR and followed by the 3′-UTR (Crick 1979). Thus, most eukaryotic genes are discontinuous, carrying sequences that are present in mature mRNA (exons), in addition to those (introns) that are removed after transcription. Such removal of introns and joining of exons in a specific order is called splicing (Darnell 2013). Splicing is a highly precise and coordinated process as any mistake resulting in the deletion or addition of even an extra nucleotide will result in faulty mRNAs with an altered or disrupted open reading frame. Splicing is carried out by a family of proteins that correctly recognize the exon–intron boundary, cut it and then ligate back the exons in the same order (Sharp 1985). Splicing was among the fundamental discoveries in molecular biology research (Phillip A. Sharp, Nobel Prize in Physiology or Medicine 1993). Many pre-mRNA species are in fact spliced in multiple ways so that several subspecies of mRNAs are generated, which then produce proteins with partially same and partially different amino acid sequences (Kelemen *et al.* 2013). Till date, the record number of splicing variation is held by the DSCAM gene in *Drosophila*, which produces a mind-boggling 38,016 dscam protein isoforms arising out of alternative splicing (Zipursky *et al.* 2006) (every *Drosophila* neuron has a unique set of dscam proteins).

Another integral part of gene expression is mRNA turnover. Processing bodies (or P-bodies) present in the cytoplasm play a key role in general mRNA decay, nonsense-mediated mRNA decay (faulty mRNAs are identified and degraded by a specific mechanism), AU-rich element-mediated mRNA decay (UTRs containing AU-rich sequences are targeted by certain proteins, followed by the decay) and microRNA-induced mRNA silencing (Franks and Lykke-Andersen 2008). P-bodies are distinctly localized in the cytoplasm and contain several mRNA-metabolizing enzymes. Apart from mRNA degradation, these also store mRNA prior to translation (Parker and Sheth 2007).

11.9 Transcription and mRNA processing are intricately coupled

Although transcription and mRNA processing (capping, splicing and polyadenylation) were studied independently, later studies revealed that they all are well coordinated and occur concurrently (Hsin and Manley 2012). The CTD of the largest subunit (RPB1) of RNA polymerase II plays a critical role in this process. Once the pre-intiation complex is formed by RNA polymerase II and associated transcription factors (TFIIA, B, D, E, F, H), transcription is initiated. Thereafter, the serine residues in the CTD domain of the RPB1 subunit of RNA polymerase II gets phosphorylated by TFIIH, resulting in dissociation of most of those basal transcription factors from the complex (promoter clearance) and RNA polymerase goes into the elongation mode (Hirose and Ohkuma 2007). Apart from promoter clearance, phosphorylation of the CTD has much wider roles as well. It actually acts as a platform for various capping, splicing and polyadenylation enzymes while the elongation process is still on (Phatnani and Greenleaf 2006). Due to the presence of repeat sequences, many different combinations of serine and threonine residues in the CTD can be phosphorylated and their phosphorylation status rapidly changes during the progress of transcription, hence recruiting different modifying enzymes at different stages of elongation (Egloff and Murphy 2008) Among the two serine residues, phosphorylation of serine 5 is primarily involved in the initiation process, while that of serine 2 is important for mRNA elongation and processing. As an evidence of such co-transcriptional mRNA processing, researchers have studied the expression of dystrophin gene. In human, *dystrophin* is the longest gene known [i.e. about 2.2 megabases (2.2×10^6 bases, ~0.07% of the entire genome)]. The primary transcript measures about 2400 kilobases, with 79 exons that takes about 16 h to transcribe. The mature mRNA is about 14.0 kilobases, encoding a protein of over 3500 amino acid residues (Ahn and Kunkel 1993). Experiments including electron microscopic analysis have shown that dystrophin mRNA is spliced while the transcription elongation is still going on (Tennyson *et al.* 1995). However, there are also examples where certain splicing event occurs only after transcription is complete and the mRNA is transported to the cytoplasm (Meshorer and Misteli 2005).

11.10 RNA editing causes change in amino acid sequence originally encoded in the mRNA

In 1986, Rob Benne and coworkers in the University of Amsterdam, the Netherlands, noticed that in *Trypanosoma brucei* (a parasite), the mRNA for the subunit 2 of cytochrome c oxidase (a mitochondrial protein) contains four uridylate (U) residues that do not correspond to any A residues in its gene (Benne *et al.* 1983). They coined the term 'RNA editing', implying those four residues were inserted into the RNA after its transcription. Such incorporation of U residues into the mRNA alters the genetic information by causing a frame shift in protein-coding sequence in the genome. Subsequent study showed that such insertions of U occur quite frequently in trypanosomal mitochondrial RNAs, and in some cases, more than 50% of the protein-coding sequences are generated through insertions (Shaw *et al.* 1988). There are also examples where U residues are deleted from the transcript, causing changes in the reading frame (Shaw *et al.* 1988). Studies on the mechanism of editing suggest that first the unedited primary transcript base pairs with a guide RNA-containing complementary sequences. The double-stranded RNA is then enveloped by a protein complex called editosome (same way, the protein complex that catalyses splicing is called splicosome). Editosome cleaves the transcript at a point where there is a mismatch in nucleotide (between the guide RNA and the transcript) and starts adding uridine residues from UTP [in a manner poly(A) polymerase adds A residues to the 3′ end of mRNAs]. The inserted uridines base-pair with the guide RNA and insertion continues as long as there are A residues in it (Simpson *et al.* 2003). Although transcript editing is quite frequent in protozoan parasites, later it was found to be occurring in many other eukaryotes through different mechanisms (Farajollahi and Maas 2010). Another mechanism of mRNA editing is by the enzyme cytidine deaminase that converts a C residue into a U residue. Such C-to-U conversion occurs in human apolipoprotein B mRNA (Anant *et al.* 2003). In liver, the protein Apo B100 is derived from the unedited transcript carrying the CAA sequence. In the intestine, it is edited to a stop codon (UAA), resulting in a smaller protein apo B48. A third mechanism of editing, called 'A-to-I editing', is catalysed by the enzyme adenosine deaminases acting on RNA (ADARs). It acts on double-stranded RNA (duplex), causing deamination of adenosine to inosine. Conversion of A to I can be very selective, where a single adenosine is targeted within a small stretch of duplex RNA (like an intron sequence base-paired with an exon). A-I editing can also be more extensive when as much as up to 50% of the adenosine residues present in larger duplexes are modified. Such type of editing occurs during the replication of RNA virus genome (that forms a duplex with the antigenome during replication; Hundley and Bass 2010). Subsequent to conversion from A to I, it behaves like G, both during translation and formation of secondary structures. A to I conversion thus affects coding capacity, interactions of miRNA/siRNA to their targets (mRNA), nuclear and cytoplasmic localization of RNA and splicing (Nishikura 1988). RNA editing occurs more extensively in the brain and many neuronal diseases are associated with faulty editing (Mattick *et al.* 2008).

11.11 Genome-wide transcription is far more pervasive than initially thought

Studies on gene expression analysis done in 1970s and 1980s were primarily based on the analysis of cDNAs and characterization of corresponding genes (i.e. those encoding mRNA, rRNA, tRNA and snRNAs). Based on those results, it was believed that each organism is defined by the tightly regulated expression of certain number of genes, while the complexity of the process increases from prokaryotes to higher eukaryotes. It was also observed that eukaryotic cells from yeast to human contain a large amount of DNA that does not correspond to any mRNA or other known RNAs. As an example, early studies suggested that only 1.2% of mammalian genome encodes for mRNAs. Thus, the remaining DNA that was once considered to be 'junk' accumulated during evolution

(Wright and Bruford 2011). When the genome sequence of many organisms, including humans, were known, a majority of those regions were found harbouring repeat sequences and pseudo-genes (dysfunctional genes that do not express. They might be impaired due to the insertion of premature stop codons, frame shifts, or a lack of promoters.) (López-Flores and Garrido-Ramos 2010). However, with the emergence of microarray and deep sequencing technologies, where almost each and every RNA species can be sequenced and matched with the genome, it became apparent that a large part of the genome, including the regions beyond the annotated genes, are indeed transcribed (Birney 2007). Such phenomenon of extensive transcription of the genome was termed 'pervasive transcription' (Birney *et al.* 2007; Jacquier 2009). Since then, genome biologists have made extensive efforts to elucidate which part of the genome is transcribed and which part is not. It has now been proposed that genome-wide transcription is neither as pervasive as claimed, nor it is as restrictive as believed earlier. Available genome-wide data now suggest that in mouse, a large segment of the genome is transcribed, producing thousands of novel protein-coding RNAs as well and more than 30,000 long non-coding RNAs generated from the intron and intergenic (in between two genes) regions (Berretta and Morillon 2009). Also, many segments of the genome are transcribed from both the strands, producing sense and antisense transcripts (Sonmez and Dean 2012). These results have thus opened several key questions regarding the role of RNA in cellular processes, that is: (a) What are the roles of these RNAs, especially those which do not code for any protein? (b) How they are being transcribed as their transcription is not initiated from bonafide promoters (like those generated from introns)? (c) Whether or not there are other yet unknown mechanisms of their regulation? Taken together, it appears that the research on RNA biology has entered into an exciting phase where newer functions of RNA in regulation of gene expression might be discovered and the central dogma of molecular biology, that is, 'DNA makes RNA makes protein' might be revised.

11.12 Many non-coding RNAs of unknown functions are now being identified

Non-coding RNAs are those which do not encode a protein. Thus, the earliest known non-coding RNAs were transfer and ribosomal RNAs identified in the 1950s and 1960s. In subsequent years, several other non-coding RNAs such as small nucleolar RNAs (snoRNAs), long non-coding RNA, siRNA, piRNA and micro RNA were discovered (Sabin *et al.* 2013). With the advent of technologies that virtually identify almost each and every RNA in a cell, many other small and large non-coding RNAs of unknown functions have been identified. Noticeably, some of the recently identified small (~100–200 nucleotides) non-coding RNAs are even polyadenylated (Derti *et al.* 2012). Non-coding RNAs are uniquely characterized by their ability to interact with proteins, DNA and RNA, regulating their functions. In view of the recently discovered diverse types of RNAs, it appears that their role in cell biology is much more complex than previously anticipated. Some of the interesting functions of non-coding RNAs are as follows (Wassarman 2007; Aalto and Pasquinelli 2012; Guenzl *et al.* 2012; Guil and Esteller 2012; Kortmann and Narberhaus 2012; Lee 2012; Letonqueze *et al.* 2012; Natoli and Andrau 2012; Rinn and Chang 2012):

- snoRNAs guide the covalent modifications (maturation) of ribosomal, transfer and small nuclear RNAs.
- RNase P helps in generating mature 5′-ends of tRNAs by cleaving the 5′-leader sequence of the precursors.
- Small nuclear RNAs (snRNA) are a type of non-coding RNAs which associate with protein factors, forming small nuclear ribonucleoproteins (snRNP). They are involved in splicing of eukaryotic pre-mRNAs and hence named spliceosomes.
- Micro RNAs (miRNAs) are small non-coding RNAs (~22 nucleotides) involved in regulation of gene expression. One single miRNA generally has partially or fully complementary sequences in the 3′-UTRs of many mRNAs, where they bind and downregulate their expression. They are synthesized from specific genes as long polyadenylated precursors, which are then processed through a pathway, leading to their maturation into small RNA molecules.
- 7SK is a small nuclear RNA that forms a ribonucleoprotein complex (snRNP) that inhibits the function of transcription elongation factor P-TEFb.
- In *E. coli*, 6S RNA binds to the RNA polymerase holoenzyme containing the σ70 factor, repressing the expression of σ70-dependent genes.
- Under oxidative stress, bacterial non-coding RNA OxyS binds to the Shine–Dalgarno sequences of mRNAs, blocking the binding of ribosomes and preventing their translation.
- Several long non-coding RNAs have been associated with chromatin remodelling, intracellular protein transport and regulation of gene expression via interaction with specific target proteins.

11.13 Nucleus is structurally organized to support transcription and RNA processing

In eukaryotic cells, the nucleus is the largest organelle encapsulating the genome (mitochondria are the other organelle which also contain their own genome, which are much smaller). The nucleus is enclosed by a double membrane called the nuclear envelope, which regulates the entry and exit of proteins, RNAs and other metabolites (Palazzo and Akef 2012). Nucleoskeleton is a meshlike interconnected assembly of proteins that provides the structural support for the nucleus (like cytoskeleton provides the support for the cytoplasm; Butin-Israeli *et al.* 2012). The function of the nucleus is to maintain the integrity of the genome, to control its expression and to replicate it whenever required. As genome is the blueprint for the sustenance and propagation of all living cells, nuclear events are tightly regulated (Capelson *et al.* 2010).

Cell biological and genetic studies have shown that rather than being evenly distributed in the nucleoplasm, various proteins, RNA and DNA are organized in the three-dimensional space within the nucleus (Van Bortle and Corces 2012). Such organization plays a key role in gene regulation. In metazoan nuclei, individual chromosomes are positioned at distinct locations, called chromosome territories (Cremer *et al.* 2006). Also, gene-rich regions of the chromosomes are placed towards the interior and gene-poor regions are oriented towards the exterior. It has been found that certain gene loci move to the nuclear interior prior to transcription, but certain others remain in the periphery while being transcribed (Heard and Bickmore 2007). Interestingly, unlike in mammalian cells, in *Saccharomyces cerevisiae,* transcribing genes move in the opposite direction, that is, from the nuclear interior to the periphery (Taddei and Gasser 2012). So, it is likely that in yeast and in mammalian cells, while a subset of genes are transcribed in the interior of the nucleus, others are transcribed in the peripheral region. Such dynamic rearrangements of genes in the three-dimensional space within the nucleus play a critical role in regulation of gene expressions during major cellular events such as the progression of cell cycle and cell differentiation (Hernandez-Verdun 2011). Another fascinating aspect of regulation of gene expression by three-dimensional genome organization is the co-localization of genes present in different chromosomes (Chuang and Belmont 2005). It has been suggested that to be efficiently transcribed, genes must reach the 'transcription factories' rich in RNA polymerase II and other transcription factors. As transcription factories are limited in numbers, a group of genes expressing under a given condition come together by repositioning of their chromosomes and share the factory (Edelman and Fraser 2012).

Apart from the three-dimensional organization of the nucleus, there are certain sub-nuclear bodies containing specialized proteins and RNAs. One such subnuclear body is the nucleolus, where ribosomal RNA are synthesized, matured, assembled into ribosomes and then transported to the cytoplasm (Hernandez-Verdun 2011). Sub-nuclear structures called speckles are rich in splicing factors and snRNPs (Spector and Lamond 2011). Speckles are very dynamic in nature as certain proteins and RNAs continuously cycle between them and the transcription factories. The nucleus of metabolically active or proliferating cells contains another sub-nuclear body named Cajal body, after its original discoverer Santiago Ramón y Cajal (Brasch and Ochs 1992). Cajal bodies have been implicated in RNA metabolism such as biogenesis of snRNPs and maturation of histone mRNAs.

11.14 Conclusion

The past century has seen a magnificent elucidation of the biology of nucleic acids. Starting as an innocuous entity, both DNA and RNA have established themselves as the sole mediator of storage and propagation of the cellular information. Although the core dogma of molecular biology remained untouched and rather strengthened over the decades, the intricacies of storage and decoding of genetic information have turned out to be amazingly complex. It now appears that a functional circuitry comprising DNA, RNA and proteins creates an information highway that ensures almost flawless sustenance of the life process. Although it is still widely accepted that among these three macromolecules, proteins are the most versatile in structure and function, recent studies have shown that RNAs are equally versatile in this regard.

It is thus expected that in coming years, more astonishing discoveries on RNA structure and function will give us a better understanding of the fascinating world of nucleic acid biology.

References

Aalto AP, Pasquinelli AE. Small non-coding RNAs mount a silent revolution in gene expression. *Curr Opin Cell Biol* 2012;24: 333–40.

Ahn AH, Kunkel LM. The structural and functional diversity of dystrophin. *Nat Genet* 1993;3:283–291.

Allen EH, Glassman E, Cordes E, Schweet RS. Incorporation of amino acids into ribonucleic acid. II. Amino acid transfer ribonucleic acid. *J Biol Chem* 1960;235:1068–1074.

Allfrey V, Daly MM, Mirsky AE. Synthesis of protein in the pancreas. II. The role of ribonucleoprotein in protein synthesis. *J Gen Physiol* 1953;20;37:157–175.

Ambros V. microRNAs: Tiny regulators with great potential. *Cell* 2001;107:823–826.

Amoutzias GD, Robertson DL, Van de Peer Y, Oliver SG. Choose your partners: Dimerization in eukaryotic transcription factors. *Trends Biochem Sci* 2008;33:220–229.

Anant S, Blanc V, Davidson NO. Molecular regulation, evolutionary, and functional adaptations associated with C to U editing of mammalian apolipoproteinB mRNA. *Prog Nucleic Acid Res Mol Biol* 2003;75:1–41.

Apgar J, Holley RW. Separation of *E. coli* leucine-acceptor RNA's by countercurrent distribution. *Biochem Biophys Res Commun* 1964;16:121–123.

Avery OT, Macleod CM, McCarty M. Studies on the chemical nature of the substance inducing transformation of pneumococcal types: Induction of transformation by a desoxyribonucleic acid fraction isolated from pneumococcus type iii. *J Exp Med* 1944;79:137–158.

Benne R, De Vries BF, Van den Burg J, Klaver B. The nucleotide sequence of a segment of Trypanosoma brucei mitochondrial maxi-circle DNA that contains the gene for apocytochrome b and some unusual unassigned reading frames. *Nucleic Acids Res* 1983;11:6925–6941.

Berget SM, Moore C, Sharp PA. Spliced segments at the 5′ terminus of adenovirus 2 late mRNA. *Proc Natl Acad Sci USA* 1977;74:3171–3175.

Berretta J, Morillon A. Pervasive transcription constitutes a new level of eukaryotic genome regulation. *EMBO Rep* 2009;10:973–982.

Birney E, *et al.* (ENCODE Project Consortium). Identification and analysis of functional elements in 1% of the human genome by the ENCODE pilot project. *Nature* 2007;447:799–816.

Brasch K, Ochs RL. Nuclear bodies (NBs): A newly "rediscovered" organelle. *Exp Cell Res* 1992;202:211–223.

Brenner S. RNA, ribosomes, and protein synthesis. *Cold Spring Harb Symp Quant Biol* 1961;26:101–110.

Brenner S, Jacob F, Meselson M. An unstable intermediate carrying information from genes to ribosomes for protein synthesis. *Nature* 1961;190:576–581.

Burgess RR. RNA polymerase. *Annu Rev Biochem* 1971;40: 711–740.

Butin-Israeli V, Adam SA, Goldman AE, Goldman RD, Nuclear lamin functions and disease. *Trends Genet* 2012;28:464–471.

Capelson M, Doucet C, Hetzer MW. Nuclear pore complexes: Guardians of the nuclear genome. *Cold Spring Harb Symp Quant Biol* 2010;75:585–597.

Cech TR, Bass BL. Biological catalysis by RNA. *Annu Rev Biochem.* 1986;55:599–629.

Chuang CH, Belmont AS. Close encounters between active genes in the nucleus. *Genome Biol* 2005;6:237–249.

Cozzone AJ. Stringent control and protein synthesis in bacteria. *Biochimie* 1980;62:647–664.

Cremer T, Cremer M, Dietzel S, Müller S, Solovei I, Fakan S. Chromosome territories – a functional nuclear landscape. *Curr Opin Cell Biol* 2006;18:307–316.

Crick F. Central dogma of molecular biology. *Nature* 1970;227: 561–563.

Crick F. Split genes and RNA splicing. *Science* 1979;204:264–271.

Croce CM, Calin GA. miRNAs, cancer, and stem cell division. *Cell* 2005;122:6–7.

Cullen BR. Transcription and processing of human microRNA precursors. *Mol Cell* 2004;16:861–865.

Darnell JE Jr. Reflections on the history of pre-mRNA processing and highlights of current knowledge: A unified picture. *RNA* 2013;19:443–460.

Derti A, Garrett-Engele P, Macisaac KD, Stevens RC, Sriram S, Chen R, *et al.* A quantitative atlas of polyadenylation in five mammals. Genome Res. 2012;22:1173–1183.

Di Giammartino DC, Nishida K, Manley JL. Mechanisms and consequences of alternative polyadenylation. *Mol Cell* 2011;43:853–866.

Ebhardt HA. RNA interference (RNAi) patents and human health related applications of RNAi. *Recent Pat DNA Gene Seq* 2007;1:128–133.

Edelman LB, Fraser P. Transcription factories: Genetic programming in three dimensions. Curr Opin Genet Dev 2012;22:110–114.

Egloff S, Murphy S. Cracking the RNA polymerase II CTD code. *Trends Genet* 2008;24:280–288.

Farajollahi S, Maas S. Molecular diversity through RNA editing: A balancing act. *Trends Genet* 2010;26:221–230.

Flaus A, Owen-Hughes T. Mechanisms for ATP-dependent chromatin remodelling: The means to the end. *FEBS J* 2011;278:3579–3595.

Franks TM, Lykke-Andersen J. The control of mRNA decapping and P-body formation. *Mol Cell* 2008;32:605–615.

Goodrich JA, Kugel JF. Non-coding-RNA regulators of RNA polymerase II transcription. *Nat Rev Mol Cell Biol* 2006;7:612–616.

Gros F, Gilbert W, Hiatt HH, Attardi G, Spahr PF, Watson JD. Molecular and biological characterization of messenger RNA. *Cold Spring Harb Symp Quant Biol* 1961;26:111–132.

Grunberg-Mano M, Ortiz PJ, Ochoa S. Enzymic synthesis of polynucleotides. I. Polynucleotide phosphorylase of azotobacter vinelandii. *Biochim Biophys Acta* 1956;20:269–285.

Guenzl PM, Barlow DP. Macro lncRNAs: A new layer of *cis*-regulatory information in the mammalian genome. *RNA Biol* 2012;9:731–741.

Guil S, Esteller M. *Cis*-acting noncoding RNAs: Friends and foes. *Nat Struct Mol Biol* 2012;19:1068–1075.

Hahn WE, Pettijohn DE, Van Ness J. One strand equivalent of the *Escherichia coli* genome is transcribed: Complexity and abundance classes of mRNA. *Science* 1977;197:582–585.

Hayashi M, Hayashi MN, Spiegelman S. Restriction of *in vivo* genetic transcription to one of the complementary strands of DNA. *Proc Natl Acad Sci USA* 1963;50:664–672.

He L, Hannon GJ. MicroRNAs: Small RNAs with a big role in gene regulation. *Nat Rev Genet* 2004;5:522–531.

Heard E, Bickmore W. The ins and outs of gene regulation and chromosome territory organisation. *Curr Opin Cell Biol* 2007;19:311–316.

Hernandez-Verdun D. Assembly and disassembly of the nucleolus during the cell cycle. *Nucleus* 2011;2:189–194.

Hershey AD, Chase M. Independent functions of viral protein and nucleic acid in growth of bacteriophage. *J Gen Physiol* 1952;36:39–56.

Hirose Y, Ohkuma Y. Phosphorylation of the C-terminal domain of RNA polymerase II plays central roles in the integrated events of eucaryotic gene expression. *J Biochem* 2007;141:601–608.

Holley RW, Apgar J, Everett GA, Madison JT, Marquisee M, Merrill SH, *et al.* Structure of a ribonucleic acid. *Science* 1965;147: 1462–1465.

Hsin JP, Manley JL. The RNA polymerase II CTD coordinates transcription and RNA processing. *Genes Dev* 2012;26:2119–2137.

Hundley HA, Bass BL. ADAR editing in double-stranded UTRs and other noncoding RNA sequences. *Trends Biochem Sci* 2010;35:377–383.

Hurwitz J. Studies on the DNA-dependent synthesis of RNA with RNA polymerase. *Harvey Lect* 1968–1969;64:157–178.

Jacquier A. The complex eukaryotic transcriptome: Unexpected pervasive transcription and novel small RNAs. *Nat Rev Genet* 2009;10:833–844.

Juven-Gershon T, Hsu JY, Theisen JW, Kadonaga JT. The RNA polymerase II core promoter – the gateway to transcription. *Curr Opin Cell Biol* 2008;20:253–259.

Juven-Gershon T, Kadonaga JT. Regulation of gene expression via the core promoter and the basal transcriptional machinery. *Dev Biol* 2010;339:225–229.

Kadonaga JT. The DPE, a core promoter element for transcription by RNA polymerase II. *Exp Mol Med*. 2002;34:259–264.

Kelemen O, Convertini P, Zhang Z, Wen Y, Shen M, Falaleeva M, *et al.* Function of alternative splicing. Gene 2013;514:1–30.

Koizumi M, Hayase Y, Iwai S, Kamiya H, Inoue H, Ohtsuka E. Design of RNA enzymes for sequence-dependent cleavage of RNA. *Nucleic Acids Symp Ser* 1989:107–108.

Kornberg RD. The molecular basis of eukaryotic transcription. *Proc Natl Acad Sci USA* 2007;104:12955–12961.

Kortmann J, Narberhaus F. Bacterial RNA thermometers: Molecular zippers and switches. *Nat Rev Microbiol* 2012;10:255–265.

Kulaeva OI, Hsieh FK, Chang HW, Luse DS, Studitsky VM. Mechanism of transcription through a nucleosome by RNA polymerase II. *Biochim Biophys Acta* 2013;1829:76–83.

Kushner SR. mRNA decay in prokaryotes and eukaryotes: Different approaches to a similar problem. *IUBMB Life* 2004;56:585–594.

Lee JT. Epigenetic regulation by long noncoding RNAs. *Science* 2012;338:1435–1439.

Letonqueze O, Lee J, Vasudevan S. MicroRNA-mediated post-transcriptional mechanisms of gene expression in proliferating and quiescent cancer cells. *RNA Biol* 2012;9:871–880.

Lizardi PM, Williamson R, Brown DD. The size of fibroin messenger RNA and its polyadenylic acid content. *Cell* 1975;4:199–205.

López-Flores I, Garrido-Ramos MA. The repetitive DNA content of eukaryotic genomes. *Genome Dyn* 2010;7:1–28.

Martin RG, Matthaei JH, Jones OW, Nirenberg MW. Ribonucleotide composition of the genetic code. *Biochem Biophys Res Commun* 1962;6:410–414.

Mattick JS, Mehler MF. RNA editing, DNA recoding and the evolution of human cognition. *Trends Neurosci* 2008;31:227–233.

Maxam AM, Gilbert W. A new method for sequencing DNA. *Proc Natl Acad Sci USA* 1977;74:560–564.

Meshorer E, Misteli T. Splicing misplaced. *Cell* 2005;122:317–318.

Mizumoto K, Lipmann F. Transmethylation and transguanylylation in 5′-RNA capping system isolated from rat liver nuclei. *Proc Natl Acad Sci USA* 1979;76:4961–4965.

Natoli G, Andrau JC. Noncoding transcription at enhancers: General principles and functional models. *Annu Rev Genet* 2012;46:1–19.

Nishikura K. Functions and regulation of RNA editing by ADAR deaminases. *Annu Rev Biochem* 2010;79:321–349.

North G. Nobel prizes: Chemistry. RNA's catalytic role. *Nature* 1989;341(6243):556.

Ogle JM, Carter AP, Ramakrishnan V. Insights into the decoding mechanism from recent ribosome structures. *Trends Biochem Sci* 2003;28:259–266.

O'Sullivan RJ, Karlseder J. The great unravelling: chromatin as a modulator of the aging process. Trends Biochem Sci. 2012;37:466–476.

Palazzo AF, Akef A., Nuclear export as a key arbiter of "mRNA identity" in eukaryotes. Biochim Biophys Acta. 2012;1819:566–577.

Parker R, Sheth U. P bodies and the control of mRNA translation and degradation. *Mol Cell* 2007;25:635–646.

Phatnani HP, Greenleaf AL. Phosphorylation and functions of the RNA polymerase II CTD. *Genes Dev* 2006;20:2922–2936.

Rando OJ. Global patterns of histone modifications. *Curr Opin Genet Dev* 2007;17:94–99.

Rinn JL, Chang HY. Genome regulation by long noncoding RNAs. *Annu Rev Biochem* 2012;81:145–166.

Rogers J, Wall R. A mechanism for RNA splicing. *Proc Natl Acad Sci USA* 1980;77:1877–1879.

Rorbach J, Minczuk M. The post-transcriptional life of mammalian mitochondrial RNA. *Biochem J* 2012;444:357–373.

Sabin LR, Delás MJ, Hannon GJ. Dogma derailed: The many influences of RNA on the genome. *Mol Cell* 2013;49:783–794.

Sandelin A, Carninci P, Lenhard B, Ponjavic J, Hayashizaki Y, Hume DA. Mammalian RNA polymerase II core promoters: Insights from genome-wide studies. *Nat Rev Genet* 2007;8:424–436.

Seila AC, Core LJ, Lis JT, Sharp PA. Divergent transcription: A new feature of active promoters. *Cell Cycle* 2009;8:2557–2564.

Sharp PA. On the origin of RNA splicing and introns. Cell 1985;42:397–400.

Shatkin AJ. Capping of eucaryotic mRNAs. *Cell* 1976;9(Part 2):645–653.

Shaw JM, Feagin JE, Stuart K, Simpson L. Editing of kinetoplastid mitochondrial mRNAs by uridine addition and deletion generates conserved amino acid sequences and AUG initiation codons. *Cell* 1988;53:401–411.

Shilatifard A, Conaway RC, Conaway JW. The RNA polymerase II elongation complex. *Annu Rev Biochem* 2003;72:693–715.

Shimba S, Buckley B, Reddy R, Kiss T, Filipowicz W. Cap structure of U3 small nucleolar RNA in animal and plant cells is different. gamma-Monomethyl phosphate cap structure in plant RNA. *J Biol Chem* 1992;267:13772–13777.

Simpson L, Sbicego S, Aphasizhev R. Uridine insertion/deletion RNA editing in trypanosome mitochondria: A complex business. RNA 2003;9:265–276.

Smale ST, Kadonaga JT. The RNA polymerase II core promoter. *Annu Rev Biochem* 2003;72:449–479.

Sonmez C, Dean C. Transcription beyond borders has downstream consequences. *RNA Biol* 2012;9:143–147.

Spector DL, Lamond AI. Nuclear speckles. *Cold Spring Harb Perspect Biol*. 2011;3.

Taddei A, Gasser SM. Structure and function in the budding yeast nucleus. *Genetics* 2012;192:107–129.

Tan S, Richmond TJ. Eukaryotic transcription factors. *Curr Opin Struct Biol*. 1998;8(1):41–48.

Tang F, Barbacioru C, Nordman E, Li B, Xu N, Bashkirov VI, *et al.* RNA-Seq analysis to capture the transcriptome landscape of a single cell. *Nat Protoc* 2010;5:516–535.

Temin HM, Baltimore D. RNA-directed DNA synthesis and RNA tumor viruses. *Adv Virus Res* 1972;17:129–186.

Tennyson CN, Klamut HJ, Worton RG. The human dystrophin gene requires 16 hours to be transcribed and is cotranscriptionally spliced. *Nat Genet* 1995;9:184–190.

Thomas MC, Chiang CM. The general transcription machinery and general cofactors. *Crit Rev Biochem Mol Biol* 2006;41:105–178.

Tuck AC, Tollervey D. RNA in pieces. *Trends Genet* 2011;27:422–432.

Uptain SM, Kane CM, Chamberlin MJ. Basic mechanisms of transcript elongation and its regulation. *Annu Rev Biochem* 1997;66:117–172.

Van Bortle K, Corces VG. Nuclear organization and genome function. *Annu Rev Cell Dev Biol* 2012;28:163–187.

Vannini A, Cramer P. Conservation between the RNA polymerase I, II, and III transcription initiation machineries. *Mol Cell* 2012;45:439–446.

Verma IM, Firtel RA, Lodish HF, Baltimore D. Synthesis of DNA complementary to cellular slime mold messenger RNA by reverse transcriptase. *Biochemistry* 1974;13:3917–3922.

Wassarman KM. 6S RNA: A small RNA regulator of transcription. *Curr Opin Microbiol* 2007;10:164–168.

Watson JD, Crick FH. Molecular structure of nucleic acids; a structure for deoxyribose nucleic acid. *Nature* 1953; 171:737–738.

Weinberg RA, Penman S. Small molecular weight monodisperse nuclear RNA. *J Mol Biol* 1968;38:289–304.

Wells DG. mRNA translation: Regulating an out of soma experience. *Curr Opin Cell Biol* 2012;24:554–557.

Wild T, Cramer P. Biogenesis of multisubunit RNA polymerases. *Trends Biochem Sci* 2012;37:99–105.

Wilt FH. Polyadenylation of maternal RNA of sea urchin eggs after fertilization. *Proc Natl Acad Sci USA* 1973;70:2345–2349.

Wright MW, Bruford EA. Naming 'junk': Human non-protein coding RNA (ncRNA) gene nomenclature. *Hum Genomics* 2011;5: 90–98.

Zalokar M. Nuclear origin of ribonucleic acid, *Nature* 1959;183:1330.

Zipursky SL, Wojtowicz WM, Hattori D. Got diversity? Wiring the fly brain with Dscam. *Trends Biochem Sci* 2006;31: 581–588.

12

How transcription of genes is switched on and off

Shyamal Goswami

12.1 Tryptophan and lac operons as paradigms of activating and repressing gene expression

Identifying and utilizing available nutrients is the key to the survival of every organism. When glucose is in abundance, *Escherichia coli* consume it as the primary source of carbon even if other sugars are available. When glucose is depleted, enzymes metabolizing other sugars such as lactose are expressed. The mechanism by which *E. coli* switches from glucose to lactose metabolism was first explained by Francois Jacob and Jacques Monod in 1961 and they won the Nobel Prize in Physiology and Medicine in 1965 for this discovery (Cohen 1995). The three genes, namely *lacZ, lacY* and *lacA,* required for lactose utilization are located in tandem in the *E. coli* genome and are called the 'lac operon' (originally named the 'lactose operon'; Loomis and Magasanik 1965). The first gene *lacZ* codes for β-galactosidase that hydrolyses lactose to glucose and galactose. The second one, *lacY,* encodes a permease that transports extracellular lactose inside the cell. *LacA* codes for a transacetylase that transfers an acetyl group from acetyl-CoA to β-galactosides. Though it is necessary for the survival of *E. coli* cells, its function in lactose metabolism is not fully understood as yet. Since its discovery, the lac operon has long been studied as a model for understanding how genes are regulated (turned 'on' and 'off') in response to metabolic stimuli. It is also studied for understanding the basic mechanisms of gene regulation in prokaryotes (the other widely studied model is the replication of bacteriophage λ in infected *E. coli*; Kaiser 2006; Wilson *et al.* 2007).

The three lactose-metabolizing genes *lacZ, lacY* and *lacA* are transcribed as a single polycistronic mRNA, which is then translated to three separate proteins (synthesis of polycistronic mRNA followed by translation into the respective protein is a common phenomenon in bacterial cells). Transcription of the lac operon is regulated by its promoter and the operator. Under normal conditions, when glucose is present, a regulatory protein named lactose repressor inhibits the transcription of the operon. The repressor is a product of the *LacI* gene located just upstream of the operon in the genome (Wilson *et al.* 2007). The operon has three sites (operator) where LacI repressor can bind. One of them is the primary operator (O1), while the other two are auxiliary or pseudo-operators (O2 and O3, located in the *lacZ* and *lacI* gene, respectively). All three sites have similar sequences (Table 12.1). These auxiliary operators are necessary for maximum repression. A tetramer of the lac repressor binds two operators simultaneously through a looping of the target DNA that enhances and stabilizes the LacI–operator complexes. Binding of LacI to the operator hinders the binding of RNA polymerase to the promoter and the transcription of the lac operon is substantially inhibited (Reznikoff 1992; Fig. 12.1). When lactose is available, minimally available lac permease brings it into the cell and β-galactosidase converts a small fraction of lactose into 1,6-allolactose (besides galactosidase, it is another activity of this enzyme). LacI has high affinity for 1,6-allolactose and the LacI–1,6-allolactose complex has a much lower affinity for the operator sequence. Hence, in the presence of lactose, the inhibition of transcription of the

Table 12.1 Nucleotide sequence of the three operator sites

O1	AATTGTGAGCGGATAACAATT
O2	AAATGTGAGCGAGTAACAACC
O3	GGCAGTGAGCGCAACGCAATT

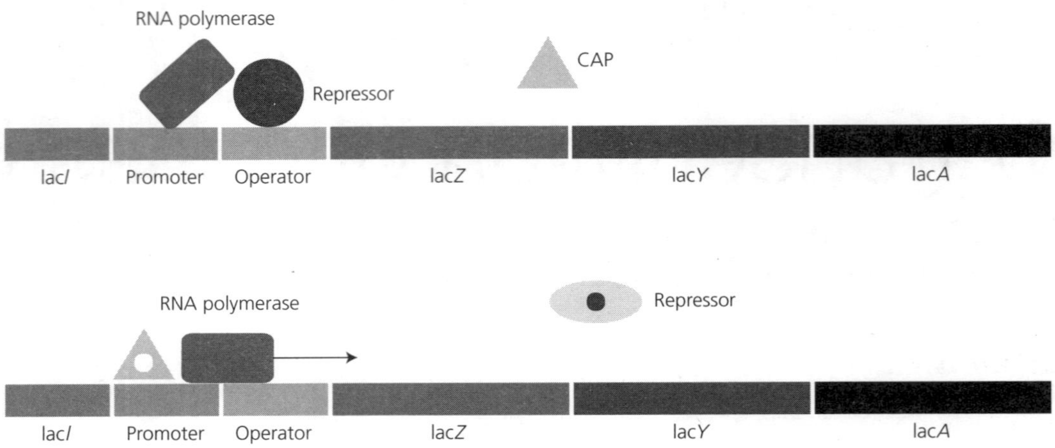

Figure 12.1 The lac operon and its regulation. The three genes, namely *lacZ, lacY* and *lacA*, are marked. The gene encoding the lac repressor, that is, LacI, is located upstream. Binding of lac repressor (LacI) to the operator hinders the binding of RNA polymerase to the promoter and the transcription of the lac operon is substantially inhibited. When lactose (small black dot) is present and glucose is absent, lactose binds to LacI, inducing a change in conformation in LacI and it does not bind to the operator, allowing RNA polymerase to bind to the promoter and initiate transcription. The absence of glucose also leads to the increase in cAMP (white dot) that binds to CAP, inducing a change in the conformation. CAP–cAMP complex then binds to the promoter, boosting the activity of RNA polymerase.

lac operon is withdrawn (Juers *et al.* 2012). This is an example of how the activity of a protein can be regulated by an allosteric regulator. However, this is not the only mechanism through which the transcription of *lacZYA* is regulated. When both lactose and glucose are available in the medium, transcription of the lac operon is not induced due to another mechanism that is glucose dependent. The activation of the lac operon also needs a protein named catabolite activator protein (CAP). CAP can activate transcription only when bound to cyclic AMP (cAMP) (Reznikoff 1992). When glucose level is high, cAMP level is low. When the glucose level decreases, the activity of adenyl cyclase increases, which then generates more cAMP from ATP. A homodimer of CAP then binds to cAMP and the complex then binds to a 22-bp sequence at –62 position of the lac

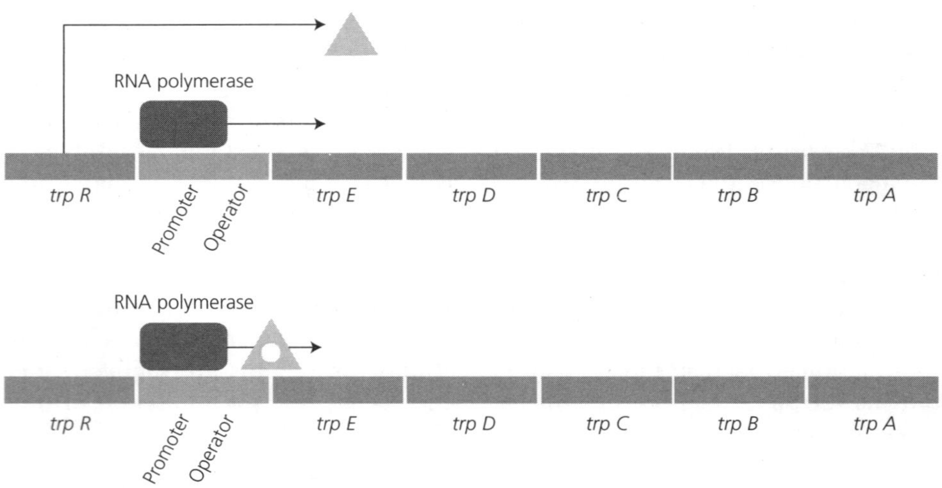

Figure 12.2 Tryptophan operon and its regulation. It contains five genes, namely *trp E, trp D, trp C, trp B* and *trp A*, encoding enzymes required for tryptophan synthesis. In the presence of tryptophan, their expression is inhibited. Another gene named *trp R* located upstream encodes for a repressor (TrpR: trp repressor) that binds to tryptophan and when bound to tryptophan, it has high affinity for the operator. When bound to the operator, it strongly blocks the transcription from the promoter. In the absence of tryptophan, the repressor does not bind to the operator, allowing transcription, resulting in the synthesis of enzymes necessary for tryptophan biosynthesis. Once produced, tryptophan then shuts down its own synthesis.

promoter and bends the DNA by 80°. Promoter-bound CAP then recruits RNA polymerase by interacting with its carboxyl terminal domain (CTD). CAP increases the affinity of RNA polymerase to the promoter, stimulates the formation of the open complex and facilitates promoter escape, thereby boosting the transcription rate by ~50-fold (Reznikoff 1992; Fig. 12.1). This is also another example of allosteric regulation. Therefore, CAP acts positively and the lac repressor acts negatively by controlling the rate and strength at which RNA polymerase attaches to the *lac* promoter. Interplay between repression (by LacI) and activation (by CAP) thus determines the extent of synthesis of carbon-metabolizing enzymes in *E. coli*. Many other metabolic genes are also regulated by such mechanisms. Detailed structural analysis of the interactions between the lac promoter-operator, RNA polymerase and CAP has given excellent insight into transcription activation in prokaryotes (Reznikoff 1992).

Another cluster of genes that has been extensively studied as a model for understanding gene regulation in prokaryotes is the tryptophan operon. There are five genes which encode for the enzymes required for the synthesis of tryptophan. They are adjacent to one another and the entire arrangement is called the trypophan operon. When tryptophan is absent, the repressor is in an inactive conformation and does not bind the operator, resulting in constitutive transcription. When tryptophan is present, it binds to the repressor and induces a conformational change. It then binds to the operator and prevents RNA polymerase to bind the promoter, inhibiting transcription (Youderian and Arvidson 1994; Fig. 12.2).

12.2 Role of enhancers, insulators and silencers in metazoan gene expression

12.2.1 Enhancers

Unlike in prokaryotic cells where variability of gene expression is primarily due to altered nutritional status, gene regulation in metazoan cells is highly complex. In multicellular organisms, different tissue types are generated by differential expression of the same genome. Enhancers are the key components of the eukaryotic genome that play a complex role in such differential gene expression. Enhancers were first described in 1981 in the polyoma virus SV40 genome (Khoury and Gruss 1983). Subsequent studies demonstrated that enhancers are generally 30–100 base pair long DNA element harbouring multiple transcription factor binding sites that can substantially increase the transcription efficiency of a neighbouring gene. Enhancers can function independent of their position relative to the gene and their orientation. Also, they may exert their effects even from a location far away from the transcription start site. Studies over the past 25 years have identified numerous enhancers associated with the expression of many mammalian genes, including immunoglobulin, insulin and β-globin (Harmston and Lenhard 2013). Enhancers can be both inducible and constitutive. Examples of constitutive enhancers are those found in viruses such as adenovirus and SV40. In mammalian cells, most of the enhancers are tissue specific and inducible. One example of an inducible enhancer is that present in human γ-interferon (IFNγ) gene, activating its expression in virally infected cells (Buecker and Wysocka 2012). Understanding the function of enhancers present in muscle creatine kinase (MCK) gene has significantly contributed to our knowledge of muscle cell differentiation and development. There are instances where enhancers were located several hundred kilobases away from the target promoter (Kulaeva *et al.* 2012). Also, some enhancers have been found within the introns of their target genes (Pound *et al.* 2011). While enhancers are generally located on the same chromosome where its target gene is present (*cis*-acting), there are examples where enhancers can act from a different chromosome (*trans*-acting; Bateman *et al.* 2012). Each enhancer is targeted by a number of transcription factors, forming an assembly on their target DNA elements (enhanceosome), which then interact with RNA polymerase II via the mediator complex. Such interaction between RNA polymerase II located on the transcription start site and the enhanceosome occurs through the bending (looping) of the intervening DNA that brings the two regulatory units (enhancer and the promoter) in proximity (Kulaeva *et al.* 2012; Fig. 12.3). Enhancers play a major role in metazoan tissue development and differentiation. They turn on genes in specific tissues while remaining ineffective in another. Studies on *Drosophila* embryos as a model system have revealed the role of enhancers in establishing the three germ layers and body pattern (Sanson 2001). In more recent studies using comparative genomics approach, sequences from multiple species are aligned and conserved regions are computationally identified as potential enhancers. It now appears that there might be as many as 1,10,000 enhancers in the human genome (Ong and Corces 2012). Chromatin immunoprecipitation is an advanced technique wherein antibody against a specific transcription factor (or other DNA binding proteins) is used for co-precipitating the bound DNA, followed by its identification by sequencing. Using this technique, many novel enhancers involved in tissue type–specific (erythroid, muscle, etc.) gene expression are now being discovered (Su *et al.* 2013). Specific epigenetic modification such as monomethylation of histone H3 lysine 4 (H3K4me1) has also been found to be the hallmark of active enhancers (Pekowska *et al.* 2011). Criticality of enhancers in

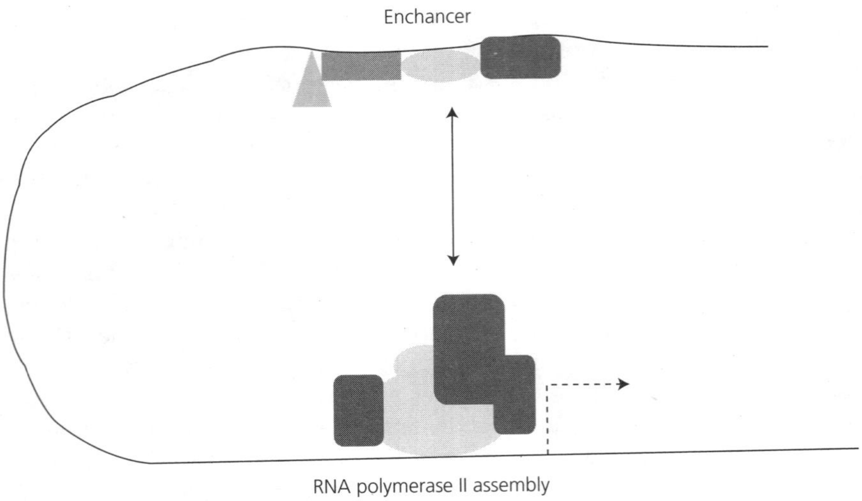

Figure 12.3 Mechanism of enhancer function. Enhancers are targeted by a number of transcription factors forming an assembly (enhanceosome) on their target DNA, which then interact with the RNA polymerase II assembly via the mediator complex (shown by the two-way arrow). Such interaction between the RNA polymerase II assembly on the promoter and the enhanceosome occurs through the bending (looping) of the intervening DNA that brings the two regulatory units (enhancer and the promoter) in proximity.

embryonic development and tissue differentiation is further highlighted by the observation that certain genes involved in critical developmental processes contain multiple enhancers, ensuring tight regulation of their expression (Perry *et al.* 2011). Certain transcription factors such as Oct4, Sox2 and Nanog are involved in the genesis and maintenance of pluripotent stem cells (they are also called master transcription factors). Recently, it has been shown that in embryonic stem cells, these master transcription factors occupy several enhancer domains that are required for the control of the pluripotent state. These enhancers are called super-enhancers, consisting of clusters of enhancers occupied by a high density of transcription factors. Super-enhancers are now believed to have important roles in the control of cell identity in mammals (Whyte *et al.* 2013).

12.2.2 Insulators

As enhancers activate their target genes from a distant location, one question that needs to be explained is how do they maintain their specificity? This is achieved through the functions of another regulatory element named insulator. These are the elements where several protein factors assemble and create a genetic boundary that physically blocks the non-specific interaction between enhancers and nearby promoters (Valenzuela and Kamakaka 2006). Insulators thus determine the set of genes an enhancer can regulate. Such role of insulators becomes critical when two adjacent genes are differentially expressed in two different tissues (Fig. 12.4).

12.2.3 Silencers

Some genetic elements have the same characteristics as enhancers, except that they repress rather than activate certain genes. They are called silencers. Silencers are thus functionally opposite to enhancers. They bind to repressors, form a repression complex and inhibit gene expression (Dhillon and Kamakaka 2002). Neural restrictive silencer factor, NRSF (also called REST), binds to a silencer element, ensuring the repression of neuron-specific genes in non-neuronal cells (Murai *et al.* 2004).

12.3 Certain unique features of gene expression learned from viruses

Viruses are uniquely placed in the living world where they can survive and propagate in their host with minimal genetic material. In fact, most of the viruses replicate

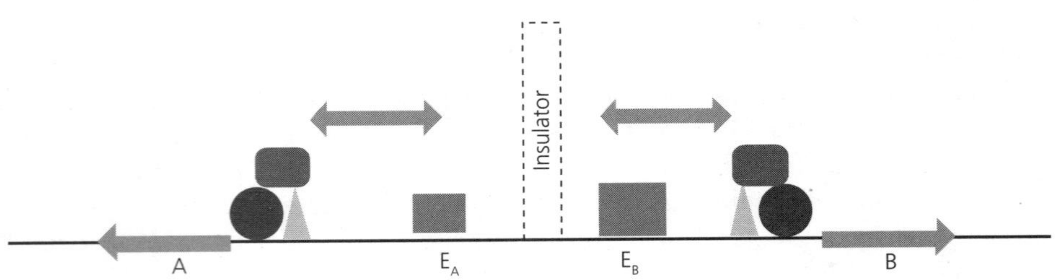

Figure 12.4 The specificity of enhancer function. Two genes A and B are located in proximity. They also have two enhancers nearby named E_A and E_B which activate the expression of gene A and B, respectively. The insulator element located in between the two genes does not allow the E_A assembly to interact with polymerase II assembled on gene B and the E_B assembly to act on Pol II assembled on gene A.

using the host machinery and carry genes that only supplements that need. Further, viruses have the uncanny ability to generate maximum output from a smaller genome. This is achieved through certain features present in their genome sequence. Replication of viruses has thus been studied for the understanding of certain key aspects of transcription and translation. A few of those are discussed next to have an insight into these processes.

Polioviruses contain RNA as their genetic material. The special feature of this RNA is that it is an uncapped mRNA coding for viral proteins (while the host mRNAs are capped). For many years, it was not understood how uncapped viral mRNAs are so efficiently translated in host cells that are equipped for translating capped mRNAs. In 1988, it was discovered that translation of the poliovirus mRNA is mediated by a unique structure which allows assembly of the ribosomal machinery at a position close to or directly on the initiation codon (Sonenberg and Meerovitch 1990). This finding broke one of the fundamental concepts of translation initiation, that is, 'eukaryotic ribosomes can bind to mRNA only at the 5′ end and then proceed towards the initiation codon by scanning it' (Malys and McCarthy 2011). The sequence where the ribosomes assemble on the poliovirus mRNA is called internal ribosomal entry sequence or IRES (Fig. 12.5). In recent years, some mammalian mRNAs (e.g. p53 mRNA) have also been shown to have IRES. Generally, they are translated from the first AUG codon using the cap structure following the conventional mechanism of ribosomal engagement and scanning (Kozak 2002). However, under certain conditions, translation starts from an internal in-frame AUG synthesizing a shorter polypeptide devoid of a part of the N-terminal sequence present in the full-length protein. Another type of internal ribosomal entry has also been described in mammalian cells. In this case, the mRNA is bicistronic (means one mRNA codes for two proteins), where the first coding sequence is followed by a second one and there is an intervening sequence containing IRES. While in certain cells, the first coding sequence is translated, in others, the second one is translated only through the IRES. Certain viral mRNAs have overlapping open reading frames (ORFs) and translation from both is ensured by another process called leaky scanning, thereby economizing on the coding space (Kozak 2002). Leaky scanning has been found in several plant viruses and mammalian retroviruses such as HIV (Firth and Brierley 2012). In HIV-1, the envelope protein (Env) is translated from an mRNA that also contains an upstream ORF encoding an accessory protein Vpu in a different reading frame. Interestingly, the Vpu initiation site is suboptimum (it means, the nucleotide sequences are not optimum to be recognized by the ribosomes as a start codon). Hence, when the ribosomes reach that

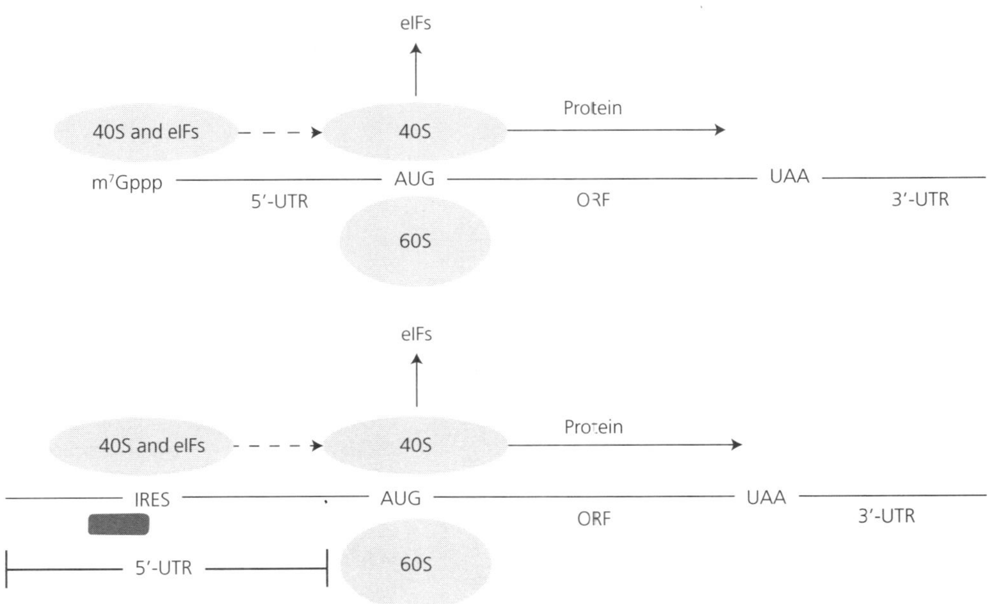

Figure 12.5 Translation by internal ribosomal entry. Upper panel: Translation of capped mRNA, wherein the 40S ribosome in association with a number of initiation factors (eIFs) recognizes the cap structure (m7Gppp). It then starts scanning the 5′-UTR till it reaches the first AUG, wherein it binds the 60S subunit of ribosome and forms the 80S complex. Thereafter, translation starts. Lower panel: 40S ribosome recognizes the IRES and scans the mRNA until it reaches the first AUG. The rectangular box bound to IRES represents certain protein factors that are necessary for the recognition of IRES by the 40S ribosome. Besides IRES-dependent translation of uncapped mRNAs (such as poliovirus mRNA), certain capped mRNAs (host) can also be translated by this mechanism where translation from the first AUG occurs by capped-dependent mechanism while that from internal AUG starts by IRES-dependent mechanism.

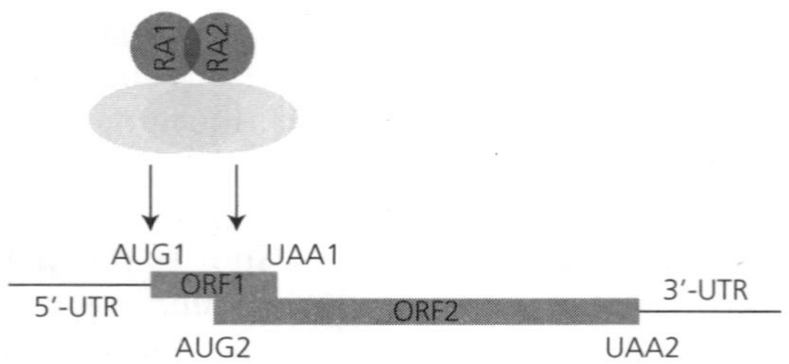

Figure 12.6 Leaky scanning. Certain viral mRNAs have two overlapping ORFs. They are translated by leaky scanning, where only a fraction of ribosomes (RA1) upon reaching AUG (AUG1) of first ORF translates it. The remaining fraction (RA2) does not start translation there and continues scanning the mRNA till it reaches the AUG of the second ORF (AUG2), where translation starts. By this mechanism, two proteins are generated from the same mRNA, economizing the genetic output. An example of such leaky scanning is the expression of vpu and env proteins of HIV. Noticeably, the sequences around the first AUG of the first ORF is so that the fraction of ribosome skipping it is controlled and the relative yield of the first and second proteins remains constant.

sequence, only a small fraction recognizes the initiation codon while the majority bypasses it and starts translation from the initiation codon for the Env coding sequence (Fig. 12.6). Interestingly, through this mechanism, a proportionate amount of both Vpu and Env proteins are produced from the same mRNA. The simian virus 40 (SV40) genome is only 5.2 kb. A 400-bp regulatory region harbours the origin of replication (ori), the promoters and enhancers that control replication and transcription. The remaining regions produce three proteins named TAg, tAg and 17 kT from one strand, while three others named VP1, 2 and 3 are produced from the opposite strand. Two transcripts are then alternatively spliced, producing functional mRNAs encoding different proteins. Such generation of mRNAs from the same DNA by the transcription of the two complementary strands is called bidirectional transcription. This is another example how a DNA virus uses a minimal DNA sequence to produce so many different proteins (Fig. 12.7).

12.4 How genes respond to external cues, modulating their expression?

It is necessary for all organisms to adapt to the surroundings for its survival and propagation. Also, in multicellular organisms, different cell types communicate with each other so that they can perform their functions in a harmonious way (as in the case of immune cells). As discussed in the previous section, the first evidence on regulation of gene expression by external stimuli came from studies on the lac operon in *E. coli*. Since then, monumental progress has been made in delineating the mechanisms of how external factors modulate gene activities. External modulators of gene activities can be simple metabolites such as lactose, diffusible molecules such as steroid hormones, growth factors such as nerve growth factor, cytokines such as TNF-α, and other stimuli such as bacterial lipopolysaccharides (LPS). These modulators often bind to cell surface

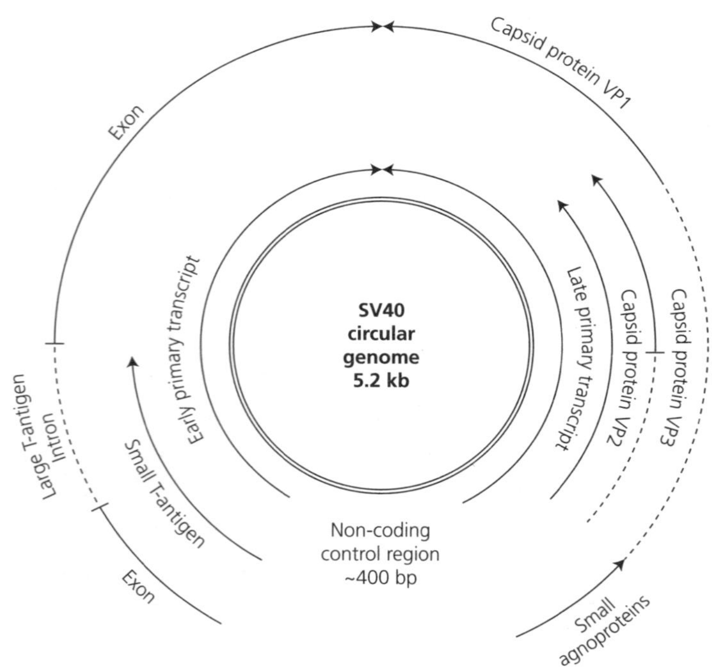

Figure 12.7 Schematic representation of Simian Virus 40 (SV40) genome (5.2 kb, double-stranded DNA) and its expression. The small (~400 bp) non-coding control region (NCCR) contains the origin of replication (ori), the early and late promoters and enhancers that control its expression. The early and late primary transcripts (shown by the two innermost arrows in the opposite direction) are generated from the two different strands of the genome. The outermost arrow on the left side represents the spliced mRNA of the early protein, the large T-antigen (solid line: exon; dotted line: intron). The arrow in the middle on the left side shows the mRNA of the other early protein, the small T-antigen. The arrows on the right side represent the mRNAs of the late capsid proteins VP1, VP2 and VP3 and the small agnoproteins. The early proteins are expressed soon after the virus enters a cell. The late proteins are expressed only after viral DNA is replicated.

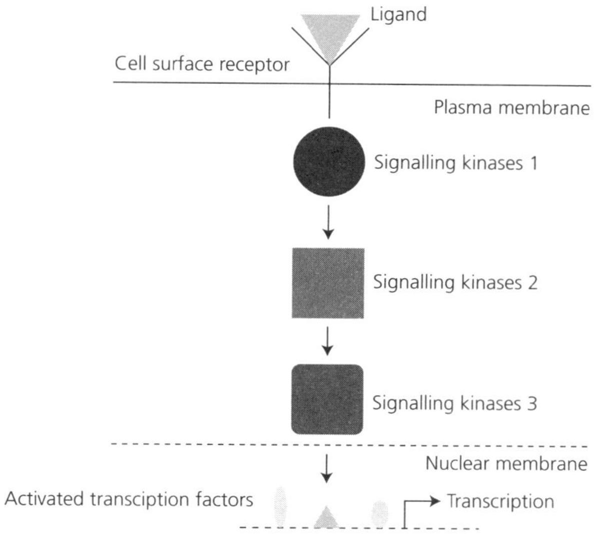

Figure 12.8 Schematic representation of signal transduction in eukaryotic cells. When ligands bind to the cell surface receptors, a cascade of events is initiated which is triggered by the generation of second messengers, which in turn activate a series of kinases (as in the case of G-protein–coupled receptors activating adenylate cyclase–cAMP–protein kinase A). Alternatively, certain kinases can be directly activated by receptor stimulation as in the case of the activation receptor tyrosine kinases. Subsequently, the upstream kinases activate the downstream kinases, and through that mechanism, the signal is passed on to the end-point kinases that phosphorylate various transcription factors, regulating their activities. One well-studied example of such signal transduction is through a family of kinases known as MAP Kinase kinase kinase (MAP3 Kinase)–MAP Kinase kinase (MAP 2 Kinase)–MAP kinase. The best studied MAP Kinases are ERKs.

receptors (growth factors, cytokines, LPS, etc.) and activate signal transduction cascades that transmit the signal to the nucleus through stepwise phosphorylation–dephosphorylation of various regulatory proteins. Eventually, such signals lead to activation of various transcription factors that in turn activate their target genes (Fig. 12.8). There are thousands of examples how various genes are activated by such signal transduction processes. One widely studied example of activation of gene expression is by the transcription factor NFκB mediating in immune response. NFκB is a dimer of two similar proteins named p50 and p65. In unstimulated cells, p65 is sequestered in the cytoplasm by another protein called IκB (inhibitor of NFκB). When cells sense the presence of infectious agents, a specific signal induces a kinase that phosphorylates IκB. Phosphorylated IκB is immediately degraded by cellular proteasomes and p65 moves to the nucleus. Its counterpart is also sent to the nucleus by a similar mechanism. The two together then form a dimer and activate target genes (Pereira and Oakley 2012; Fig. 12.9). Another transcription factor FoxO remains in the nucleus under stress conditions (such as nutritional stress) and activates stress-responsive and pro-apoptotic genes. When cells are stimulated by pro-survival hormones such as insulin, it activates a kinase named Akt that goes to the nucleus and phosphorylates FoxO. In the phosphorylated form, its cytosolic localization signal gets exposed and FoxO cannot stay in the nucleus. It then moves to the cytoplasm and expression of pro-apoptotic genes is repressed (Tzivion *et al.* 2011). Many steroid hormones have their receptors in the nucleus. These are transcription factors involved in the upregulation of many steroid response genes. Normally, they remain in the nucleus but do not bind their targets. In the presence of specific steroid hormones such as glucocorticoids or oestrogen, the hormone–receptor complex becomes functional (as in the case of CAP–cAMP complex in *E. coli*) and specifically binds to their target sites, activating responsive genes. How specific signals stimulate gene expression is a vast area of research in understanding embryonic development, cell differentiation and certain diseases such as cancer. In the past quarter of a century, many such pathways have been extensively characterized, and one among them is the Hedgehog signalling involved in embryonic development. This pathway is named from an intercellular polypeptide signaling molecule called Hedgehog (Hh), first found in *Drosophila*. It is involved in establishing the basis of the fly body plan. Mammals have three Hedgehog homologues, of which Sonic is the best studied. The pathway is involved in brain, skeleton, musculature, gastrointestinal tract and lung development. This pathway has also been implicated in the development of some cancers (Briscoe and Thérond 2013).

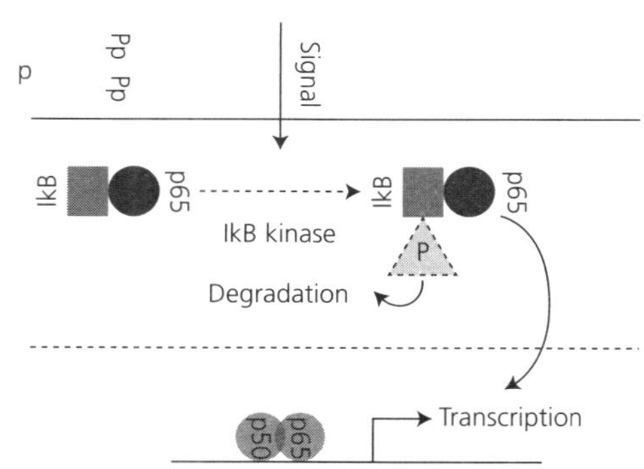

Figure 12.9 Schematic representation of activation of transcription factor NF-κB. Under normal condition, the p65 subunit of NF-κB is sequestered in the cytoplasm by its binding to the inhibitory factor IκB. When stimulated by specific stimuli such as cytokines, viruses etc., a kinase named IκB kinase phosphorylates IκB. Phosphorylated IκB then dissociates from p65 and is degraded by the proteasome. Free p65 moves to the nucleus and dimerizes with its counterpart p50 (also activated by a similar mechanism) and activates cognate genes.

References

Bateman JR, Johnson JE, Locke MN. Comparing enhancer action in cis and in trans. *Genetics*. 2012;191:1143–1155.

Briscoe J, Thérond PP. The mechanisms of Hedgehog signalling and its roles in development and disease. *Nat Rev Mol Cell Biol* 2013;14:416–429.

Buecker C, Wysocka J. Enhancers as information integration hubs in development: Lessons from genomics. *Trends Genet* 2012;28: 276–284.

Cohen GN. Thirty years since the award of the Nobel Prize to Lwoff, Monod, and Jacob. *FASEB J* 1995;9:981–982.

Dhillon N, Kamakaka RT. Breaking through to the other side: Silencers and barriers. *Curr Opin Genet Dev* 2002;12:188–192.

Firth AE, Brierley I. Non-canonical translation in RNA viruses. *J Gen Virol* 2012;93(Part 7):1385–1409.

Harmston N, Lenhard B. Chromatin and epigenetic features of long-range gene regulation. *Nucleic Acids Res* 2013;41:7185–7199.

Juers DH, Matthews BW, Huber RE. LacZ β-galactosidase: Structure and function of an enzyme of historical and molecular biological importance. *Protein Sci* 2012;21:1792–1807.

Kaiser D. A microbial genetic journey. *Annu Rev Microbiol* 2006;60:1–25.

Khoury G, Gruss P. Enhancer elements. *Cell* 1983;33:313–314.

Kozak M. Pushing the limits of the scanning mechanism for initiation of translation. *Gene* 2002;299:1–34.

Kulaeva OI, Nizovtseva EV, Polikanov YS, Ulianov SV, Studitsky VM. Distant activation of transcription: Mechanisms of enhancer action. *Mol Cell Biol* 2012;32:4892–4897.

Loomis WF Jr, Magasanik B. Genetic control of catabolite repression of the lac operon in *Escherichia coli*. *Biochem Biophys Res Commun* 1965;20:230–234.

Malys N, McCarthy JE. Translation initiation: Variations in the mechanism can be anticipated. *Cell Mol Life Sci* 2011;68:991–1003.

Murai K, Naruse Y, Shaul Y, Agata Y, Mori N. Direct interaction of NRSF with TBP: Chromatin reorganization and core promoter repression for neuron-specific gene transcription. *Nucleic Acids Res* 2004;32:3180–3189.

Ong CT, Corces VG. Enhancers: Emerging roles in cell fate specification. *EMBO Rep* 2012;13:423–430.

Pekowska A, Benoukraf T, Zacarias-Cabeza J, Belhocine M, Koch F, Holota H, *et al.* H3K4 tri-methylation provides an epigenetic signature of active enhancers. *EMBO J.* 2011;30:4198–4210.

Pereira SG, Oakley F. Nuclear factor-kappaB1: Regulation and function. *Int J Biochem Cell Biol* 2008;40:1425–1430.

Perry MW, Boettiger AN, Levine M. Multiple enhancers ensure precision of gap gene-expression patterns in the *Drosophila* embryo. *Proc Natl Acad Sci USA* 2011;108:13570–13575.

Pound LD, Sarkar SA, Cauchi S, Wang Y, Oeser JK, Lee CE, *et al.* Characterization of the human SLC30A8 promoter and intronic enhancer. *J Mol Endocrinol* 2011;47:251–259.

Reznikoff WS. The lactose operon-controlling elements: A complex paradigm. *Mol Microbiol* 1992;6:2419–2422.

Sanson B. Generating patterns from fields of cells. Examples from *Drosophila* segmentation. *EMBO Rep* 2001;2:1083–1088.

Sonenberg N, Meerovitch K. Translation of poliovirus mRNA. *Enzyme* 1990;44:278–291.

Su MY, Steiner LA, Bogardus H, Mishra T, Schulz VP, Hardison RC, *et al.* Identification of biologically relevant enhancers in human erythroid cells. *J Biol Chem* 2013;288:8433–8444.

Tzivion G, Dobson M, Ramakrishnan G. FoxO transcription factors: Regulation by AKT and 14-3-3 proteins. *Biochim Biophys Acta* 2011;1813:1938–1945.

Valenzuela L, Kamakaka RT. Chromatin insulators. *Annu Rev Genet* 2006;40:107–138.

Whyte WA, Orlando DA, Hnisz D, Abraham BJ, Lin CY, Kagey MH, *et al.* Master transcription factors and mediator establish super-enhancers at key cell identity genes. *Cell* 2013;153:307–319.

Wilson CJ, Zhan H, Swint-Kruse L, Matthews KS. The lactose repressor system: Paradigms for regulation, allosteric behavior and protein folding. *Cell Mol Life Sci* 2007;64:3–16.

Youderian P, Arvidson DN. Direct recognition of the trp operator by the trp holorepressor—a review. *Gene* 1994;150:1–8.

13

Structure and replication of RNA

Pramod Yadava

13.1 Structure of RNA

The discovery of the acidic content of nuclei of leucocytes by Friedrich Miescher in 1869 was followed by the isolation of similar fractions from other organisms and cells. Much later, a preparation from yeast was shown to be more alkali labile and, later, chemical analyses showed that it had ribose in place of deoxyribose. This was designated ribonucleic acid. It was realized only around 1940 that all cells have both DNA and RNA. Gradually, more was learnt about the unique features of the two kinds of nucleic acid.

13.1.1 RNA does not produce diffraction patterns typical of the double helical structure

Encouraged by the success of X-ray crystallography of DNA fibres, scientists, including James Watson, tried producing similar pictures also for RNA but got no resolute patterns for double helical organization. In some preparations, there did appear clusters of spots, suggesting segmental double strandedness. Also, RNA was known not to abide by Chargoff's rules (A + G = T + C) as much as DNA did. In line with these observations, RNA showed a much lower hyperchromic/hypochromic factor than DNA when subjected to denaturing conditions. It was increasingly being clear that RNA in general remained single stranded. Later, it was shown that complementary sequences of RNA did anneal to form a double helix with its coordinates similar to the A-helices formed by DNA under special conditions (see Table 13.1).

13.1.2 The small chemical difference between DNA and RNA makes a lot of difference

Like DNA, RNA consists of four kinds of nucleotides, with two of them being purines (A and G) and two pyrimidines (U and C); thymine occurring in DNA is replaced by uracil (Fig. 13.1). Also, the sugar in RNA is ribose sugar, which has a hydroxyl group at the 2′ position. This is the cause of a lot of intra- and inter-molecular reactivity of RNA as compared with DNA. For example, abstraction of a proton from the hydroxyl group can initiate a cascade of electron transfers to re-equilibrate the charge micro-environment. This may culminate into 2′–3′ cyclization through phosphoester bonds, followed by snapping of the phosphodiester bond, causing cleavage of the RNA molecule. Stretches of the complementary sequence of RNA do form a double helical structure. However, rather than making a compact B helix, they form an A helix. As an intra-molecular motif, this is commonly referred to as hairpin with or without bulges. In replicating RNA, however, longer double-stranded intermediates do form transiently. Thus, an RNA molecule, though considered to be single stranded, does assume a double helical structure. The A-helical conformation of RNA also has to be due to the seemingly minor difference in the chemical constituents of RNA when compared with DNA. The coordinates of the three helical forms assumed by nucleic acids are summarized in Table 13.1.

As in the case of DNA, five dihedral bonds make for the backbone of each nucleotide, beginning with the α characterizing the link between 5′ phosphorus and oxygen of the hydroxyl group, β between oxygen of the 5′ OH and the 5′ carbon itself, γ between 5′C and 4′C of ribose,

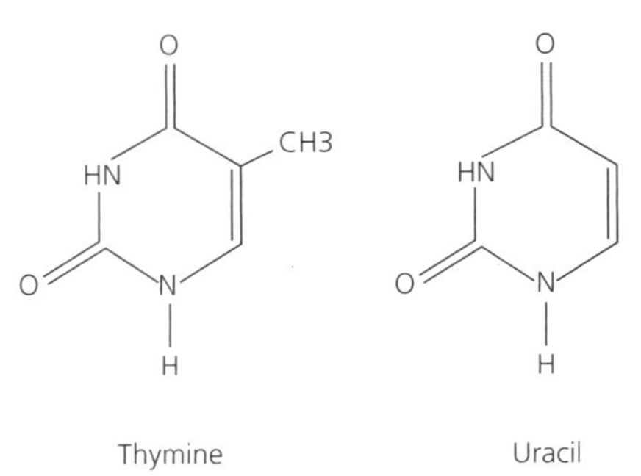

Figure 13.1 Uracil monophosphate replaces thymidine monophosphate in RNA.

Table 13.1 Coordinates and conformations of common helical forms of nucleic acids

Property	A-helix	B-helix	Z-helix
Diameter	23.7 Å	25.5 Å	18.4 Å
Pitch of the helix	24.6 Å	33.2 Å	45.6 Å
Base pairs per pitch	11	10	12
Rise per base	2.3 Å	3.32 Å	3.8 Å
Mean rotation per base pair	33.6°	35.9°	30°
Base pair tilt from perpendicular	+19°	−1.2 ± 4.1°	−9°
Mean propeller twist	+18°	116 ± 7°	0°
Helix axis location	Major groove	Through base pair	Minor groove
Major groove	Narrow but deep	Wide but with intermediate depth	Flattened on helix surface
Minor groove	Broad and shallow	Narrow with intermediate depth	Extremely narrow but quite deep
Glycosyl bond conformation	Anti	Anti	Anti at C, syn at G
Handedness	Right handed	Right handed	Left handed

δ between 4′C and 3′C of ribose, ε between 3′C and oxygen of the hydroxyl group and ζ between oxygen of the 3′ hydroxyl and phosphorus of the 3′ phosphate group.

A ubiquitous structural motif of RNA is minor-groove base triplexes resulting from interactions of the 2′ hydroxyl group of ribose.

13.2 Replication of RNA

It had been established that viruses had either DNA or RNA in the infectious particles even before DNA was experimentally proven to be the material basis of inheritance. Tobacco mosaic virus was known to contain RNA (Stanley 1935). Later, bacteriophages specifically infecting bacteria with F pilus were isolated (the f1 series of viruses; Zinder *et al.* 1963). Many male-specific phage isolates (i.e. those infecting through F pili) turned out to be RNA bacteriophages (see Davis *et al.* 1961). It was obvious that the RNA viruses must employ an RNA-dependent copying mechanism for their propagation. The characterization of DNA-dependent polynucleotide phosphorylase by Sevro Ochawa's group and subsequently DNA-dependent RNA polymerase by Hurwitz (cf. Hurvitz 2005) and DNA-dependent DNA polymerase by Arthur Kornberg's group (Lehman *et al.* 1958) in late 1950s probably inspired biochemists to look for RNA-dependent polymerases during early 1960s.

13.2.1 Phage RNA can initiate infection cycle

Male-specific RNA bacteriophages MS2 and Qβ were isolated in the 1960s (Davis *et al.* 1961). The infection cycle of these bacteriophages could be initiated by their genomic RNA, suggesting that the genome had a plus-strand polarity and could undergo translation. Full-length cDNA corresponding to the phage genome could also initiate intracellular phage cycle and yield infectious virions. Mutation and complementation studies suggested that the phage genome coded for at least four proteins, including coat, maturation, lysis and a replicase protein. The genomes of these RNA bacteriophages were typically in the range of 3000–4000 nucleotides by size. A diagrammatic representation of the genome of the phage MS2 is given in Fig. 13.2.

13.2.2 RNA replicase from phage Qα has been studied in detail

For phage Qβ, it was soon possible to purify and characterize the replicase protein for their *in vitro* reactions and assess the *in vivo* efficacy of their variants. The replicase holoenzyme in the case of Qβ was shown to consist of four subunits α, β, γ and δ, of which only β-subunit was coded for by the phage replicase gene, while α consists of the *E. coli* ribosomal protein S1, γ-subunit is bacterial translation elongation factor TU and δ-subunit is elongation factor TS. In addition to the replicative RNA-dependent RNA polymerase (RdRp) function, the enzyme was also shown to promote the incorporation of nucleotides in a template-independent manner. A diagrammatic representation of the Qβ replicase is given in Fig. 13.3.

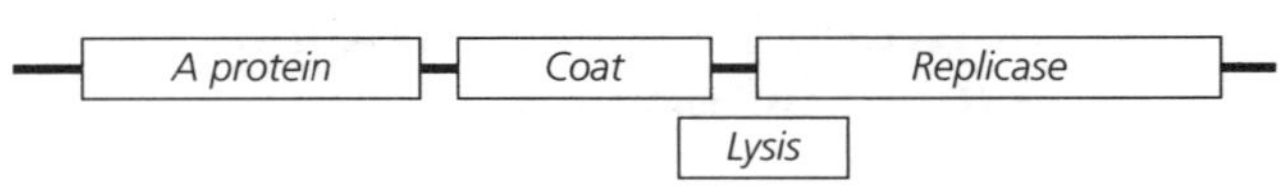

Figure 13.2 RNA genome of phage MS2.

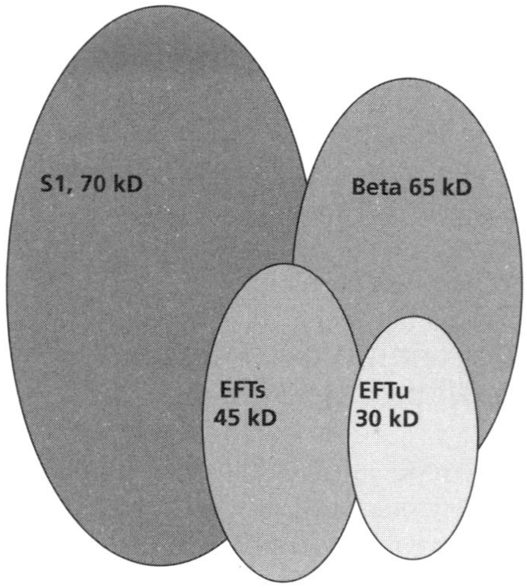

Figure 13.3 The Qβ RNA replicase, consisting of four subunits, of which only the β-subunit is coded by the virus.

13.2.3 RNA replication is quite error prone

It was realized that RNA replication was error prone and any replicating pool of phage genome consisted of variants which could be represented by a common consensus sequence. The latter could be referred to as quasi-species. Tadatsugu Taniguchi and Walter Schaffner in Zurich (Taniguchi *et al.* 1978) had characterized phage genome by 1980 and demonstrated the presence of alternative and much smaller RNA to be accepted as a template by the phage replicase. These RNAs were designated midivariants, minivariants and nanovariants. It was clear that the replicating phage RNA yielded a collection of RNA sequences that probably competed for replicase and those found suitable were amplified. This made a good system to study evolutionary selection at the molecular level. Manfred Eigen and his group in Gottingen (Biebricher *et al.* 1981) adopted this to develop an *in vitro* pool of replicons interacting with Qβ RNA replicase. Eigen could project the entire pool of RNA as an evolving milieu, as a sequence space that would give rise to variants, which would then be subjected to selection for their fitness as template (Eigen and Biebricher 1988).

13.2.4 Genome of phage MS2 was the first to be sequenced

Characterization of RNA sequences during the early 1970s was done by cleaving them with different ribonucleases and following chromatographic patterns of the wandering spots. Walter Fiers and his group could stitch together such information into the complete sequence (3569 nucleotides) of phage MS2 RNA by 1976. This was the first genome of any organism or virus to be sequenced (Jou *et al.* 1972; Fiers *et al.* 1976).

13.2.5 Phage RNA replicase is specific for phage RNA

Bacteriophage MS2 isolated by Davis *et al.* (1961) had no DNA and it was very keenly studied for how it replicated and expressed itself. It turned out that the isolated RNA of the bacteriophage could initiate infection, resulting in plaque formation in susceptible bacterial hosts. Bacteria were not known to have the mechanism of making RNA copy of RNA templates. Hence, it was increasingly accepted that phage genome itself must start off making the required enzyme or, alternatively, it ought to carry the enzyme as part of the virion. Even a DNA copy of the viral RNA, a mechanism later found to work for many tumour viruses, was initially proposed to serve as the template for viral gene expression. The only possibility left was to look for an RNA-dependent RNA replication mechanism. It would be necessary for the viral replication machinery to be able to specifically recognize viral RNA among a much larger number of host RNA, else the delivered viral RNA would be lost. It was soon established that there was indeed an RdRp in phage-infected cells. This could be isolated and purified and was demonstrated to specifically replicate phage RNA and remained unmoved with host RNA. This was a deviation because the nucleic acid polymerases characterized till then were not restricted to particular templates. Cells infected with the related phage Qβ yielded a replicase showing better solubility and reaction in *trans* than the enzyme from MS2-infected bacteria. Complementarity at termini and secondary structure assumed by the phage RNA determine preferential translation of different cistrons in phage genome. The RdRps distinguish themselves from other nucleic acid polymerases (viz. DNA-dependent DNA polymerase and DNA-dependent RNA polymerase) in that structural elements of the template are important for RNA replication while other polymerases primarily copy any sequence incorporated in the template.

13.2.6 *In vitro* replicated phage RNA retained biological properties

Spiegelman (Doi and Spiegelman 1963) could demonstrate amplification of phage RNA with the RNA replicase *in vitro*. The resulting mass of RNA shared not only physicochemical but also biological properties and could initiate infection cycle, producing phage progeny indistinguishable from those of the natural infection process. This marks the earliest demonstration of making a biologically active molecule and an autonomous entity largely *in vitro* as being ventured currently under synthetic biology.

13.2.7 Organization and life cycle of phages MS2 and Qα

Both MS2 and Qβ phage have plus-strand genomic RNA of 3569 and 4218 nucleotides, respectively, and code for four proteins, namely maturation, coat, lysis and replicase protein. Both the viruses have an icosahedral symmetry (an icosahedron has 20 faces and 12 vertices). Each virion carries a single maturation protein which binds to its receptor on the F pilus. The incoming phage RNA acts as mRNA to produce the first batch of phage proteins, including replicase. The latter in turn makes complementary negative-strand antigenomic RNA which acts as a template for synthesis of more positive strands. Phage RNA has extensive secondary structure resulting from self-folding. It should be appreciated that no specific mechanism is yet established for separation of the complementary strands from the replication intermediate and the elaborate secondary structure may be helping the phage in keeping the copy strands separate and accessible to ribosomes or to the RNA replicase. The secondary structures also ensure that the coat gene is translated most efficiently, probably using an internal ribosome entry mechanism, while maturation and lysis proteins are made in lower abundance in proportion with their need for completion of the phage infection cycle. Binding of the maturation protein to phage RNA might signal initiation of phage assembly in the cell. The coat proteins can self-assemble to form icosahedra, yet binding of the coat protein to a hairpin-like sequence element on phage genome is shown to nucleate the assembly of phage particles. As the concentration of lysis proteins builds up, holes are created in the membrane of infected cells that would cause loss of osmotic control and rupture of cell wall, releasing the progeny phage.

13.2.8 Higher-order structure of phage RNA may be the discerning basis for replication, transcription and translation

The secondary and perhaps higher-order structures of genomic RNA of Qβ as well as MS2 determine the preferred sites of interaction of replicase with them. These structures also offer the basis for interaction with host factors that may steer replication in favour of producing positive strands. The coat protein may bind in regions upstream of replicase gene and may influence its translation. Phage MS2 infects *via* the F pilus of the male bacteria. Maturation protein contacts a cognate receptor on the side of the pilus. Access of the ribosomes to the phage replicase open reading frame, which is otherwise occluded due to the strong secondary structure surrounding it, probably follows a transient opening of the secondary structure under the influence of momentum of the moving ribosome complex as the latter translates coat gene sequence. The tapering of translation of replicase follows the accumulation of coat protein molecules in the infected cell. The initial event upon arrival of phage genome in host cell is translation of the plus strand to produce some molecules of replicase which produces minus strands. When minus strands are used for copying, more of plus strands are produced. The phage replicase has some sensing mechanism for the sequence and associated structure of termini of the genome. In the case of phage Qβ, the 5′ and 3′ termini have small strings of Gs and Cs. The terminal sequences may also anneal with each other, mimicking a primer–template complex.

13.2.9 Darwinian evolution in replicating RNA

The phage Qβ replication system was adopted as a model to study Darwinian evolution of replicating entities *in vitro* as the enormous number of the phage and a very rapid proliferation compressed the time scale for such evolutionary processes within humanly manageable limits. By growing the phage under varied environmental conditions, one could demonstrate the emergence of variants with enhanced fitness. A natural offshoot of phage replication is the smaller templates that could efficiently interact with phage replicase and get amplified. This also represented a system where evolutionary processes could occur outside cells – something of relevance to pre-cellular evolution. Considering the mutual fitness of replicase and the RNA template, Manfred Eigen (Eigen and Biebricher 1988) postulated sequence space (cumulative complexity of all variants created during replication) as a positive determinant of evolution. Interestingly, phage Qβ replicase could synthesize a self-propagating RNA, much smaller than the phage genome, upon being incubated with ribonucleotides. At least, the initial reactions would be template independent as substantiated by the great diversity among the resulting sequences (Spiegelman *et al.* 1965; Oehlenschlager and Eigen 1997).

13.2.10 RNA-dependent RNA polymerases

The enzyme required for making the complementary strand of a given RNA template was discovered in RNA bacteriophages MS2 and Qβ and from the RNA virus known to be causal of poliomyelitis in mammals around the same time in early 1960s. As RNA polymerase had by then become the designation of DNA-dependent RNA polymerase involved in transcription, the RdRps were given the alternative name RNA replicase. In contrast to DNA replication, initiation of replication by RNA replicase at the 3′-end of the template does not require nucleic acid primer, though some viral enzymes use a viral protein as a genome-linked (VPg) primer (Gu *et al.* 2009). The same enzyme mediates transcription (production of mRNA) and replication (production of copies of the genome). In addition to their presence in RNA viruses, RdRps are also known to amplify interfering RNA in many eukaryotes (cf. Iyer *et al.* 2003). Evolutionarily, the RdRps are quite conserved across phyla and have their similarity with other RNA-dependent enzymes such as the telomerase reverse transcriptase.

The VPg primer is covalently bound to the 5′-end of the virus genome. This is a small protein and undergoes uridylylation at the tyrosine residue at position 3 and binds with RdRp. A secondary structure assumed by the viral RNA has been characterized to be a *cis*-acting replication element (CRE) which directs uridylylation of VPg. Mutations in CRE and in VPg cause impairment in RdRp function, particularly with reference to positive-strand synthesis (Murray and Barton 2003).

13.2.11 RNA viruses exhibit diverse genome organization and correspondingly diverse replication strategies

RNA replication thus seems to have many unique features to it while it is essentially guided by base complementarity for synthesis of the copy strand. RNA replicases usually are specific for their cognate template unlike DNA-dependent polymerases, which can act on any given template once the essential elements (such as ori for DNA polymerase and promoter elements for RNA polymerase) are provided. The higher-order structure of the template seems to play an important role in offering access to the RdRp enzyme (Gu *et al.* 2009), which is quite error prone due to lack of corrective activities such as proofreading. A summary of genome organization and replication strategies applied by various RNA viruses is presented in Table 13.1.

13.2.12 Single-stranded positive-sense eukaryotic RNA viruses

This group includes polio and yellow fever viruses, among many others. RNA of these viruses can initiate the infection process, suggesting their translatability to gene products. The translation products get processed into all functional and structural constituents of the virus. In the case of poliovirus, for example, the translation product is a polyprotein which undergoes processing to produce all the unit viral proteins. A small viral protein VpG is covalently linked to poliovirus RNA and helps in initiating RNA replication. Polio virus is not known to produce any subgenomic transcripts, while alphaviruses, also single-stranded positive-sense eukaryotic RNA viruses, produce subgenomic transcripts from the negative-strand RNA template (Suopanki *et al.* 1998).

13.2.13 Single-stranded non-segmented negative-sense eukaryotic RNA viruses

This group is represented by measles, mumps, rubella and rabies viruses. All negative-stranded RNA viruses are enveloped. As the genomic RNA is negative sense, it needs to be copied into corresponding plus strands for undergoing translation to produce viral gene products. Hence, all negative-sense RNA viruses carry RdRp in the virion. Hence, the polymerase in these viruses is to be treated as the structural protein of the virus. The viral transcripts get modified to look like the host mRNAs. The capping and polyadenylation functions also reside in the viral RdRp. Production of full-length copies is considered replication, while that of subgenomic transcripts is referred to as transcription (Conzelmann 1998).

13.2.14 Single-stranded segmented negative-sense eukaryotic RNA viruses

The most familiar virus of this class is influenza virus (Orthomyxoviridae), which undergoes uncoating in the cytosol and enters the nucleus for undergoing transcription. The viral polymerase does not have capping and polyadenylation activities. The viral transcripts snatch the cap region of the host mRNA and are sent out for translation in the cytosol. As the viral genome consists of several pieces of RNA, each carrying a coding sequence for viral proteins, the virus can produce new combinations of these pieces, particularly following infection at high multiplicity (i.e. several virions entering and replicating in the same host cell). This enhances the variability of influenza viruses and is referred to as genetic shift.

13.2.15 Cap-snatching is a strategy of influenza virus to make its RNA transcripts translatable

Influenza virus gets into the nucleus for initiating its intracellular cycle. Like other RNA viruses, it employs its own RNA polymerase. However, as its polymerase does not cap the mRNAs, it has acquired another strategy of cleaving a piece from the 5′ region of host mRNA and extends it on its own nucleocapsid to produce capped viral mRNAs. It may be noted that negative-stranded RNA viruses that complete their life cycle in the cytosol (e.g. paramyxoviruses) have capping and polyadenylation activities associated with their RNA polymerase. A diagrammatic representation of the phenomenon of cap-snatching is given in Fig. 13.4.

13.2.16 RNA-dependent RNA polymerase may amplify interfering RNA in some eukaryotes

The RNA interference phenomenon in plants seems to far exceed the extents of stoichiometric reactions between interfering RNA and its target. The observed

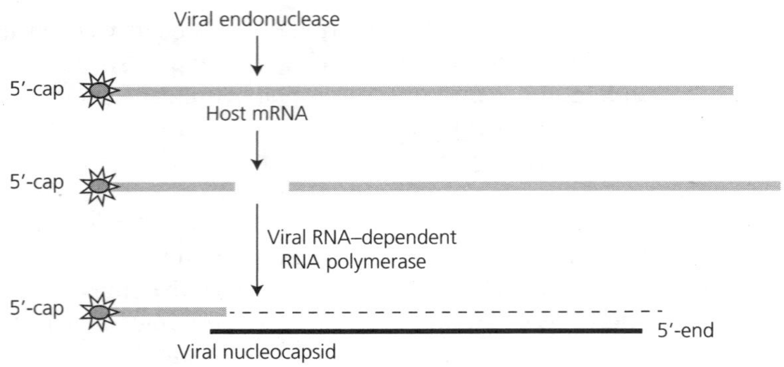

Figure 13.4 Cap-snatching is a strategy of influenza virus to make its RNA transcripts translatable.

effects can be accounted for if the interfering RNA is amplified replicatively with the help of RdRp.

13.2.17 Retroviruses are single-stranded positive-sense eukaryotic RNA viruses with a DNA phase

RNA of this class of viruses, even though positive-sense single stranded, is not directly adaptable for translation by ribosomes. Instead, the virions carry an RNA-dependent DNA polymerase that produces a cDNA copy of the viral RNA which is integrated into the host cell genome for transcription and translation, like one of the natural genes of host.

In addition to the above, several viral genomes are ambivalent and code for viral gene products from either the genomic or the antigenomic strand. Yet another variant is exhibited by Reoviruses that have several double-stranded RNA, which together form the viral genome. All RNA viruses have to have their own RNA-dependent enzyme and, in many cases, this enzyme is even carried by the virions.

13.2.18 Darwinian selection of replicating RNA (Spiegelman's experiment)

RdRp reaction products (midivariants, minivariants, etc.) when diluted and offered fresh enzyme showed emergence of diverse small replicons, showing conservation of structure rather than sequence. Left to compete in the wild, such RNA molecules will not have maximal survival potential as they do not code for the complete enzyme. However, being small, they get replicated more efficiently than full-length templates would.

13.2.19 Poliovirus employs a covalently linked protein (VpG) for initiation of replication

VpG consisting of 22 amino acid proteins covalently linked via a phosphodiester bond to the 5′-end of viral RNA serves as a primer in poliovirus. This follows uridylylation at the 3′-terminal of VpG.

Various expression and replication mechanisms of RNA viruses have been summarized in Table 13.2.

Table 13.2 Expression and replication mechanisms of RNA viruses

Nature of viral genome	Examples	Expression/ Replication mechanism
Single-stranded positive-sense RNA	Picornaviruses Flaviviruses Caliciviruses Astroviruses Togaviruses Coronaviruses Arteriviruses Nodaviruses Tetraviruses	Genomic RNA is translatable and can initiate infection cycle once inside the host cell. Replication employs viral RNA-dependent RNA polymerase (RdRp).
Single-stranded negative-sense RNA	Rhabdoviruses Filoviruses Bornaviruses Paramyxoviruses Orthomyxoviruses Deltaviruses	Viral RdRp Cellular RNA Pol II produces mRNA.
Single-stranded positive-sense RNA (with a DNA intermediate)	Retroviruses	Viral reverse transcriptase makes a cDNA copy of the genome, which is integrated in host cell genome and acts as a template for RNA Pol II to produce mRNA.
Single-stranded ambisense	Arenaviruses	Viral RdRp
Double-stranded RNA	Reoviruses Birnavirus	Viral RdRp
Ambisense RNA		

References

Biebricher CK, Eigen M, Luce R. Product analysis of RNA generated de novo by Qbeta replicase. *J Mol Biol* 1981;148: 369–390.

Conzelmann KK. Nonsegmented negative-strand RNA viruses: Genetics and manipulation of viral genomes. *Ann Rev Genet* 1998;32:123–162.

Davis JE, Strauss JH, Sinsheimer RL. Bacteriophage MS2: Another RNA phage. *Science* 1961;134:1427.

Doi RH, Spiegelman S. Conservation of a viral RNA genome during replication and translation. *Proc Natl Sci Acad USA* 1963;49:353–360.

Eigen M, Biebricher CK. Sequence space and quasispecies distribution. In: Domingo E, Holland JJ, Ahlquist P, eds. *RNA Genetic Vol. III: Variability of RNA Genomes*. Boca Raton, FL: CRC Press; 1988:211–245.

Fiers W, Contreras R, Duerinck F, Haegeman G, Iserentant D, Merregaert J, *et al.* Complete nucleotide sequence of bacteriophage MS2 RNA: Primary and secondary structure of the replicase gene. *Nature* 1976;260(5551):500–507.

Gu C, Zeng T, Li Y, Xu Z, Mo Z, Zheng C. Structure-function analysis of mutant RNA-dependent RNA polymerase complexes with VPg. *Biochemistry Mosc* 2009;74:1132–1141.

Hurwitz J. The discovery of RNA polymerase, *J Biol Chem* 280:42477–42485.

Iyer LM, Koonin EV, Aravind L. Evolutionary connection between the catalytic subunits of DNA-dependent RNA polymerases and eukaryotic RNA-dependent RNA polymerases and the origin of RNA polymerases. *BMC Struct Biol* 2003;3:1.

Jou WM, Haegeman G, Ysbaert M, Fiers W. Nucleotide sequence for gene coding for bacteriophage MS2 coat protein. *Nature* 1972;237: 82–88.

Lehman IR, Bessman MJ, Simms ES, Kornberg A. Enzymatic synthesis of deoxyribonucleic acid. I. Preparation of substrates and partial purification of an enzyme from *Escherichia coli*. *J Biol Chem* 1958;233:163–170.

Murray KE Barton DJ. Poliovirus CRE dependent VpG uridylylation is required for positive strand RNA synthesis but not for negative strand RNA synthesis. *J Virol* 2003;77:4739–4750.

Oehlenschläger F, Eigen M. Thirty Years Later – A new approach to Sol Spiegelman's and Leslie Orgel's *in vitro* evolutionary studies. Dedicated to Leslie Orgel on the occasion of his 70th birthday. *Orig Life Evol Biosph* 1997;27:437–457.

Spiegelman S, Haruna I, Holland IB, Beaudreau G, Mills D. The synthesis of a self-propagating and infectious nucleic acid with a purified enzyme. *Proc Natl Acad Sci USA* 1965;54:919–927.

Stanley WM. Isolation of a crystalline protein possessing properties of tobacco mosaic virus. *Science* 1935;81:644–645.

Suopanki J, Sawicki DL, Sawicki SG, Kaariainen L. Regulation of alphavirus 26S mRNA transcription by replicase component nsP2. *J Gen Virol* 1998;79:309–319.

Taniguchi T, Palmieri M, Weissmann C. Qβ DNA-containing plasmids giving rise to Qb phage formation in the bacterial host. *Nature* 1978;274:223–228.

Zinder ND, Valentine RC, Roger M, Stoeckenius W. f1, a rod-shaped male-specific bacteriophage that contains DNA. *Virology* 1963;20:638–640.

14

Translation of mRNA into protein: Classic and recent experiments

Jayanta Kumar Pal and Sangeeta Chatterjee

14.1 Flow of genetic information and the 'central dogma of molecular biology'

Various exciting early discoveries in the first half of the 20th century which culminated in the 1950s proved beyond doubt that the genetic information of an organism/cell is embedded/contained in the deoxyribonucleic acid (DNA). The very obvious question at that point was how is this genetic information in the DNA of an organism/cell actually disseminated in the phenotype of the organism; that is, how does the genetic information flow from the DNA to a protein phenotype/character of the organism. Based on various fragmented information available, towards the end of 1950s, Francis HC Crick, being an authority in this search, had proposed a scheme of genetic information flow which is popularly called the 'central dogma of molecular biology' (Crick 1958). This he elaborated further in a *Nature* paper in 1970 (Crick 1970). According to the central dogma, the genetic information contained in the DNA (gene) flows through an RNA intermediate to a protein phenotype. The two steps in the information flow is called transcription

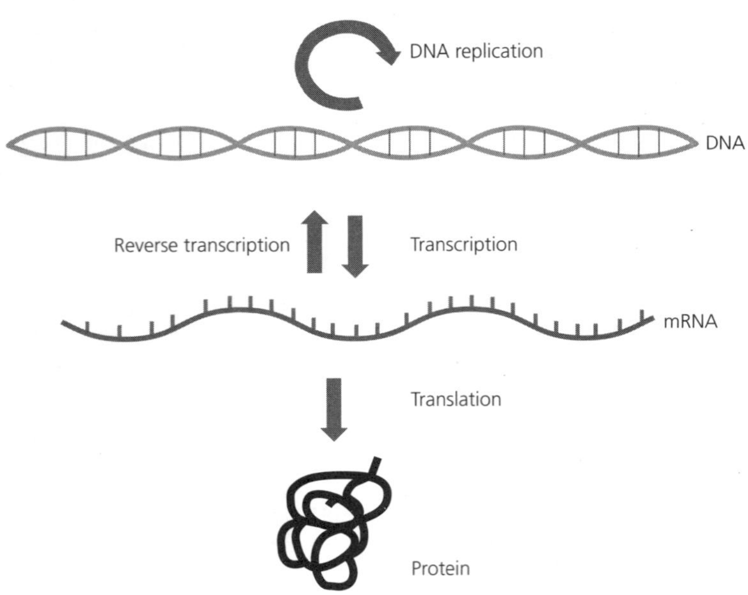

Figure 14.1 Central dogma of molecular biology: Schematic representation. The genetic information flows from DNA to RNA to protein, with the exception of RNA viruses. In such viruses that contain RNA as the genetic material, RNA is reverse transcribed to DNA and then it follows the general scheme. See Colour Plate 6 on page 506.

(generating mRNA) and translation (synthesis of protein from mRNA), respectively. However, at a later date, the seminal independent discovery of David Baltimore (1970) and Howard Temin and S Mizutani (1970), that many viruses (RNA tumour viruses) begin the information flow in a reverse direction, that is, from RNA to DNA and then follows the proposed direction, led to the revision of the central dogma, which is depicted in Fig. 14.1. For this outstanding discovery, they shared the Nobel Prize in 1973. Essentially, the central dogma served as the guiding principle for various other important discoveries in the 1960s and thereafter led to the current understanding on the detailed mechanism of genetic information flow in cells.

This chapter deals with the second important part of the central dogma, that is, translation (mRNA to protein), in detail.

14.2 Discovery of mRNA as the intermediate between genetic information in DNA and protein phenotype

The seminal discovery by Macleod, Avery and McCarty, 1944 (Avery *et al.* 1944), that DNA was the genetic material in an organism and the elucidation of the three-dimensional (3D) structure of DNA by James Watson and Francis HC Crick, 1952–53 (Watson and Crick 1953), shook the imagination of the scientific community at large. These discoveries stimulated researchers to resolve the primary question as to how the genetic information contained in DNA was conveyed from the genes to the cytoplasm to produce the proteins required for cellular functions and hence the phenotypic trait of an organism. However, they were confronted with the following problem – DNA resides in the nucleus and proteins were already shown to be synthesized in the cytoplasm in eukaryotic cells (McQuillen *et al.* 1959). Then, how is this information transferred from the nucleus to the cytoplasm? The discovery of mRNA (messenger RNA) as the intermediate between gene and protein is a matter of experiment and collective insight. Many theories were propounded; among them contributions of Crick, Volkin, Astrachan, Brenner, Jacob and Meselson were significant. The biologists François Jacob, Andre M Lwoff and Jacques Monod received Nobel Prize in Physiology and Medicine for their part in this research and discovery of operons in 1965. The following is an account of the further details of their significant contributions.

14.2.1 Contribution of Francis HC Crick

Crick and Watson (1957) proposed RNA as the intermediate carrier of genetic information. By this time, ribosomes were already discovered and were shown to contain RNA (which was termed ribosomal RNA or rRNA). Further, it was already known that numerous ribosomes exist in a cell. This information led to the establishment of a school of thought (led by Crick) which believed that rRNAs in the ribosomes were directly responsible for the synthesis of proteins. They assumed that a single gene passed its information to a single rRNA in a ribosome; thus, there is a specialized ribosome for each protein corresponding to a gene. However, this hypothesis failed and was shown to have many loopholes. If each rRNA (thus each ribosome) corresponded to a single gene, then its product, that is, protein, should have similar diversity in population as ribosomes. All ribosomes were shown to contain the same quantity of rRNA, while proteins were already known to have diverse molecular weights. Moreover, there should be a distinct correlation between the DNA and the rRNA base compositions. Comparative studies in different microorganisms revealed that no such correlation exists.

14.2.2 Contribution of Elliot Volkin and Lazarus Astrachan

Elliot 'Ken' Volkin and Lazarus Astrachan (1956), while working in Oak Ridge National Laboratory, TN, provided important evidence regarding the existence of mRNA (messenger RNA). They infected *Escherichia coli* with T2 bacteriophage in the presence of ^{32}P-labelled nucleotides. Later, they isolated the nucleic acid, hydrolysed it with sodium hydroxide and separated the alkaline products by ion exchange chromatography. It was already known that when T2 bacteriophage infects the bacteria, it halts bacterial protein synthesis and initiates viral protein synthesis. While analysing their results, they came across a minor species of RNA (among total RNA) that had a short half-life and whose base composition had a direct correlation to the DNA base composition of T2 bacteriophage and not that of *E. coli.* Volkin and Astrachan called this variant RNA 'DNA-like-RNA'. However, they did not understand the implication of their finding. Based on their research, other researchers later were able to decipher that these 'DNA-like-RNA' were the molecules that encoded genetic information from the DNA and were messengers that carried information for protein formation. Thus, these molecules came to be known as messenger RNA or mRNA.

14.2.3 Experiments of Arthur Pardee, François Jacob and Jacques Monod

Arthur Pardee, François Jacob and Jacques Monod, working at Institute Pasteur, Paris, conducted the

'PaJaMo' (Pa from Pardee, Ja from Jacob and Mo from Monod) experiment in 1958 (Pardee *et al.* 1958). Monod in his previous experiments had devised a system for genetically manipulating the β-galactosidase pathway for sugar metabolism in *E. coli*. In this experiment, they took two types of bacteria (male and female) and allowed conjugation to take place. Female bacteria were mutants for β-galactosidase pathway, while male bacteria were wild type. They observed that in female bacteria, the β-galactosidase pathway was restored within minutes. By this time, it was already established that recombination event between male and female bacteria takes place after 60–90 min after injection of the chromosome into the female (Wollman *et al.* 1956). Thus, they concluded that the male bacteria were sending a cytoplasmic message to the female bacteria which was restoring its function. This experiment though did not lead to direct discovery of mRNA, it laid the foundation of Jacob's later experiment, which would lead to its discovery. Jacob later also gave mRNA its name.

14.2.4 Experiments of Sydney Brenner, François Jacob and Matthew Meselson

Sydney Brenner, François Jacob and Matthew Meselson (Brenner *et al.* 1961) experimentally disproved Crick's hypothesis that rRNA was the carrier of genetic information. The key to this experiment was density gradient centrifugation that could separate 'heavy' and 'light' macromolecules according to their heavy isotope content. *E. coli* cells grown on synthetic medium (labelled with heavy isotopes ^{15}N and ^{13}C) were infected with bacteriophage T2. These cells contained all components (including ribosomes) uniformly labelled with heavy isotopes. These ribosomes were termed 'old'. After infection with T2, these cells were grown for some time in a light medium (containing ^{14}N and ^{12}C) supplemented with ^{32}P. It was already known that T2 only contains DNA and protein and no RNA, so after infection, newly made phage RNA would be labelled with ^{32}P. So, if Crick's hypothesis were true, then phage-specific 'new' ribosomes, which were light, would be complexed with ^{32}P-labelled newly made phage RNA. Otherwise, old (heavy) ribosomes complexed with ^{32}P-labelled newly made phage RNA would be observed. Brenner and colleagues took the infected *E. coli* cells, lysed them and separated the lysate in a CsCl density gradient. Newly made ^{32}P-labelled phage RNA was found to be associated with old ribosomes. Thus, T2 phage utilized old bacterial ribosome for new viral protein synthesis (Fig. 14.2). Clearly, this old bacterial rRNA cannot carry any phage genetic information; by extension, it is unlikely that it carries bacterial genetic information either. Thus, they proposed that messengers were independent RNAs that were translated by non-specialized ribosomes.

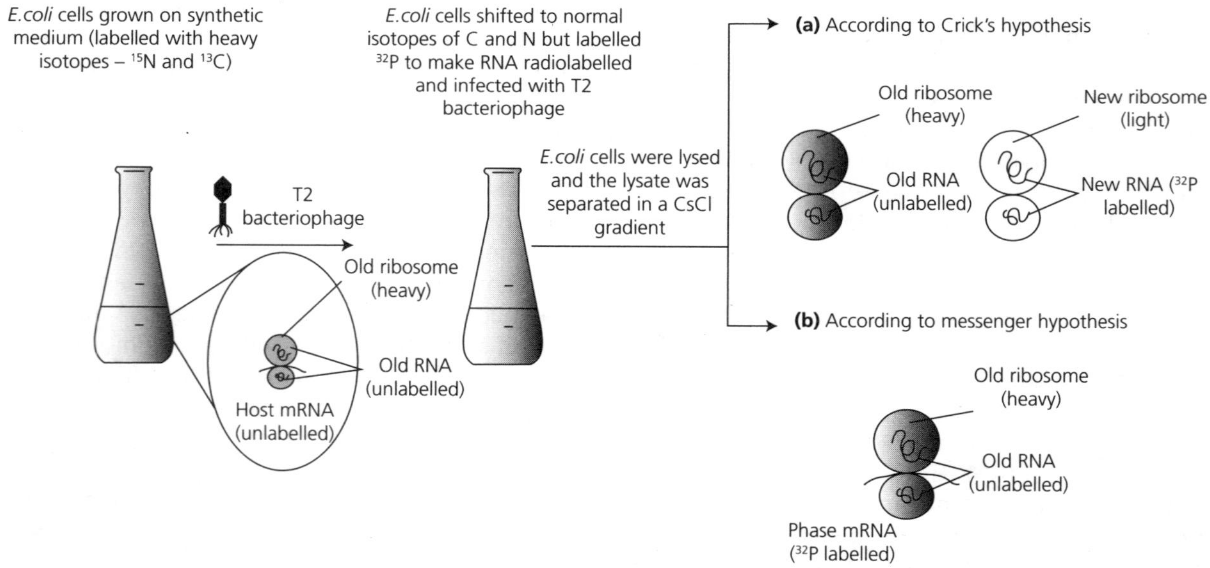

Figure 14.2 Schematic representation of the experimental test for messenger hypothesis. *E. coli* cells grown on a synthetic medium (labelled with heavy isotopes: ^{15}N and ^{13}C) were infected with bacteriophage T2 and shifted to a lighter medium containing labelled ^{32}P. Infected *E. coli* cells were lysed and separated in a CsCl density gradient. (a) According to Crick's hypothesis, ribosomal RNA carried the message for making proteins; therefore, whole new ribosomes with phage-specific ribosomal RNA would have been made after phage infection. (b) According to messenger hypothesis by Jacob and colleagues, a messenger RNA (mRNA) carried genetic information to the ribosomes. Therefore, bacteriophage T2 infection would cause the synthesis of new phage-specific mRNAs that would be ^{32}P-labelled. These would then associate with old, heavy ribosomes.

14.2.5 World of mRNA

Since the discovery of mRNAs, a lot has been achieved with respect to its characterization and determination of its role in translation. Elucidation of regulatory elements present at the 5′- and 3′-ends of the coding sequence [called the 5′-and 3′-untranslated regions (UTRs)] and their interaction with *trans*-acting elements, including proteins, has opened up a new area of research, with particular reference to gene expression. Genetic variations such as point mutations, truncations and single nucleotide polymorphisms (SNPs) in the 5′- and 3′-UTR are associated with a number of human diseases and increased susceptibility to diseases, such as X-linked Charcot–Marie–Tooth (CMTX) disease, multiple myeloma, hereditary hyperferritinaemia/cataract syndrome (HHCS), breast cancer and fragile X syndrome. A lot of interest is currently focused on this region (UTRs) for development of drugs that specifically target this region, bringing specific up- or downregulation of genes. The importance of this region will be discussed in more detail in the subsequent sections.

14.3 Ribosomes: Structure–function updated with the exciting recent discoveries

Ribosomes, the prime subcellular organelles that sponsor and catalyse protein synthesis in a cell, were discovered in the 1940s. However, their structure could be observed only in 1950s in an electron microscope as dense particles by a Romania-born American cell biologist, George Palade (1955). At that time, he worked in the laboratory of Albert Claude, at the Rockefeller Institute in New York. Later, both Palade and Claude along with Christian de Duve jointly shared the 1974 Nobel Prize in Physiology and Medicine for their work on ribosomes, mitochondria and endoplasmic reticulum. Subsequent to Palade's discovery in 1955, with the understanding of the nature and link of these dense particles with protein synthesis, they were named ribosomes (ribonucleic acid, i.e. ribosomal RNA + Greek 'soma' = body) by Richard B Roberts in 1958.

With the advancement of technology as well as refinement of classical techniques such as electron microscopy, centrifugation, etc., many aspects of the ribosome structure–function relationship in the context of protein synthesis were understood through painstaking efforts of many scientists over a few decades. A crucial understanding that came in view was that ribosomes are molecular machines and are the sites of protein synthesis. These RNA–protein complexes bring in the template mRNA and the aminoacyl-tRNAs with the respective cognate amino acids in them. This way, they catalyse the process of translation of mRNA with their intrinsic enzymatic activities, including the one that forms peptide bonds between amino acids in a polypeptide, called the peptidyl transferase activity.

Structurally, ribosomes are highly conserved RNA–protein complexes, although their sizes vary in prokaryotes and eukaryotes. Ribosomes are composed of two subunits, large (50S in prokaryotes; 60S in eukaryotes) and small (30S in prokaryotes; 40S in eukaryotes) subunits. In *E. coli*, ribosomes are composed roughly of two-thirds of RNAs and one-third of proteins; large and small subunits contain 23S and 16S rRNAs and 34 and 21 proteins, respectively. In eukaryotes, the 60S subunit is composed of three rRNAs, namely 5S, 5.8S and 28S rRNAs, and about 45 proteins. The 40S subunit is composed of one 18S rRNA molecule and 35 proteins.

The major work that elucidated the structure–function relationship of ribosomes was completed during the early part of molecular biology research, predominantly based on bacterial systems. It was apparent that the small subunit is primarily responsible for recognition and reading of the mRNA, while the large subunit appeared to provide the main platform for the process of protein synthesis per se. Later, it was realized that actually both the subunits are required to assemble the three sites, P, A and E, required for tRNA–ribosome interaction. P and A sites are meant for binding of peptidyl-tRNA and aminoacyl-tRNA, respectively. The discovery that there exists an E site where the de-acetylated tRNA gets discharged from the ribosome took place later (Rheinberger *et al.* 1981). Further details on these are given in the context of discussion on antibiotics and protein synthesis.

The characterization of the ribosome structure and elucidation of its structure–function relationship took a long way, starting from its discovery in the 1950s, and only in the early 2000, the major left-out puzzle, that is, the source of the peptidyl transferase activity involved in peptide bond formation, was solved. This horrendous task was undertaken by three independent groups from three different countries, headed by Ada Yonath from Weizman Institute (Israel), Thomas Steitz from Yale (the United States) and Venkat Ramakrishnan from Cambridge (the United Kingdom). Naturally, these three stalwarts were awarded the Nobel Prize in Chemistry in 2009. Their success was essentially achieved through improvement on the crystallography techniques, which are the best methods available for determining the atomic structure of macromolecular complexes such as ribosomes. For the details on the development of technological competence and the involvement of many other scientists in this endeavour, readers are referred to an excellent review by Steitz and Moore (2003), in addition to a few seminal publications (Ban *et al.* 1998, 2000; Cate *et al.* 1999; Nissen *et al.* 2000, 2001; Hansen *et al.* 2002; Schmeing *et al.* 2002).

In summary, ribosomes turned out to be one of the few amazing macromolecular machines that work smoothly and accurately to synthesize the pivotal macromolecules, the proteins, that keep the cells alive and make them capable of what they like to do or like to be. The precision in their performance/functioning lies in their precise structure, and this has been well experienced by the fact that small molecules such as a number of antibiotics can jeopardize their activity and make the cell go out of order within no time. Because of the achievement of obtaining high-resolution (2.4Å) crystal structure of the ribosomal subunits, many of the ill-understood processes became clearer. For example, Ramakrishnan's group could demonstrate how some of the antibiotics work in obliterating the function of ribosomes. According to them, some of the antibiotics actually alter the opening and closing behaviour of the 30S subunit and thus interfere with ribosome functioning. Further details on these are given in the context of discussion on antibiotics and protein synthesis.

Eventually, these advances in our understanding lend credential to design and develop newer generations of antibiotics/drugs. In parallel, Steitz's group, who devoted their attention to the 50S subunit, discovered the involvement of a highly ordered set of water molecules and hydrogen bonds in the functioning of this subunit. It is also predicted that the protein synthesis is driven by changes in entropy rather than enthalpy (http://pipeline.corante.com/archives/2009/10/07/a_nobel_for_ribosome_structure.php). Further research on the role of other non-ribosomal proteins that bind to ribosomes transiently through the structural biology approach will help in unravelling many mysteries still prevailing in the structure–function relationship of ribosomes.

14.4 Adaptor hypothesis of Crick and discovery of tRNA by Zamecnik and Hoagland fitting into the concept

Francis HC Crick was a person with many ideas. Ideas so elegant and spirited that they often formed the tenets of molecular biology as we know it today. Early 1950s was abuzz with excitement all around in the scientific world, with each individual wanting to immortalize his name in living memory by answering some of the key questions about how a cell functions. By this time, the DNA double helix structure was already deciphered and the existence of ribosome and RNA was already known. However, mRNA and genetic code were still to be unravelled. The missing links were being worked upon by several groups with feverish haste. The missing links were – How was the genetic information in DNA passed onto the proteins? Was there an intermediate between DNA and protein? How was the code on the DNA read? What was the mechanism by which the amino acids were assembled into proteins? One of the popular notions at that time was the folding of RNA into definite pockets into which amino acids would get accommodated and thus the template would be read. However, this notion had the following fallacies – First, RNA had side chains which were hydrophilic; thus, these could never accommodate amino acids with hydrophobic side chains. Second, RNA was too simple a molecule to discriminate between two chemically similar amino acids which differed by only a single methyl group (e.g. Gly and Ala). The contributions of the three key scientists (Francis HC Crick, Paul Zamecnik and Mahlon Hoagland) who helped in answering some of these questions are discussed next.

14.4.1 Contributions of Francis HC Crick

Crick, a member of the RNA tie club, in early 1955 wrote a paper on degenerate template and adaptor hypothesis for general circulation among the club members (Crick 1955). This paper is the most significant unpublished work of any author till date. He postulated, in this paper, the existence of small adaptor molecules (preferably RNA) that would attach to specific amino acids on one hand and recognize the RNA template on the other hand. These adaptor molecules would thus function as mediators between DNA and protein. He even predicted that 2–3 bases of the adaptor molecule would pair with the codons, though at this time, nobody had an exact idea what they actually were. Though the existence of adaptor molecules was correctly predicted, the task of proving their existence ultimately fell upon Paul C Zamecnik and Mahlon B Hoagland in 1957.

14.4.2 Contributions of Paul C Zamecnik and Mahlon B Hoagland

Paul C Zamecnik (1952), a trained medical doctor, was inspired to become a scientist while performing an autopsy of a morbidly obese woman. He found that she had very little muscle and protein and only vast amount of fat. This made him question how proteins were actually synthesized. To dissect out the individual steps in protein synthesis, he started to develop a cell-free translation system, a system that would be able to synthesize proteins *in vitro*. He was partially successful (1952) in making a cell-free lysate from rat liver (Note: the details of development of cell-free protein synthesis system will be discussed in Section 14.7). He later perfected this system and synthesized proteins from radiolabelled amino acids (Zamecnik and Keller 1954; Hoagland *et al.* 1956). Radiolabelling is the method of incorporating a radioactive element into a compound to investigate its metabolism, fate and utilization. Amino acids are labelled on the amine group, carboxyl group or on sulphur (in the case of cysteine and methionine) with radioactive isotopes such as ^{3}H, ^{14}C and ^{35}S. One of the components of this system was the so-called pH 5 enzyme fraction which was capable of translating added mRNA into protein. This fraction mostly contained soluble proteins. Zamecnik and coworkers, to their surprise, found a small RNA in this fraction. This RNA could be coupled to amino acids on addition of ATP in the presence of the pH 5 fraction. They further proved that this RNA was capable of transferring the coupled amino acid to a growing protein chain. The exact experimental details are as follows. Zamecnik and colleagues added increasing amount of labelled [^{14}C]leucine to pH 5 fractions in the presence of ATP and took samples at different time points. They then plotted binding of [^{14}C]leucine as a function of labelled [^{14}C]leucine added. They observed that with an increase in the amount of [^{14}C]leucine, there was an increase in its retention in the RNA fraction. RNA was separated from proteins by phenol extraction. This RNA was named as soluble RNA (sRNA) by Zamecnik. This was so called because most of it was found in the supernatant fraction while preparing a cell-free protein synthesis system at 1,05,000 *g* centrifugation step (Fig. 14.3). Mahlon B Hoagland, who was also a trained medical doctor and was working as a post-doctoral fellow in Zamecnik's lab, showed that this soluble RNA was capable of transferring the coupled amino acid to growing protein chains. Hoagland took [^{14}C]leucine-charged pH 5 RNAs and mixed them with microsomes. Microsomes are small vesicular artefacts that formed during isolation of the endoplasmic reticulum and these contain ribosomes. He observed that there was a near-perfect correlation between the loss of radioactive leucine from the pH 5 RNA and the gain of the radiolabelled leucine by the protein in the microsomes. This represented the incorporation of leucine from $tRNA^{leucyl}$ into nascent polypeptides on ribosomes (Hoagland *et al.* 1958). Later, sRNA was renamed as transfer RNA (tRNA) with reference to its role in transferring amino acids to nascent polypeptide chains.

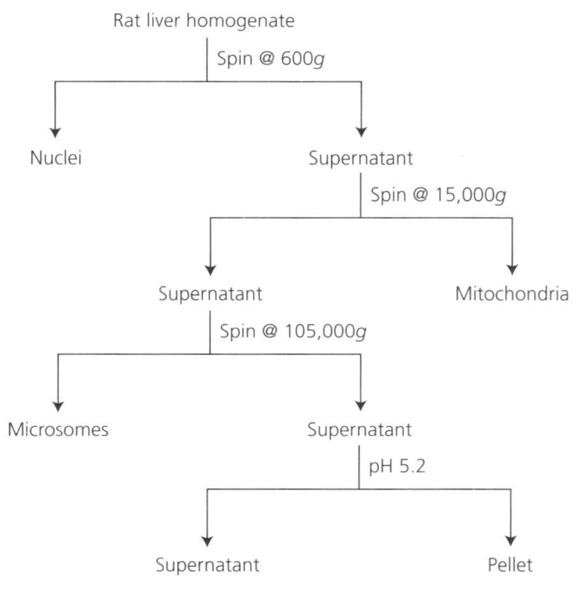

Figure 14.3 Schematic representation of the various steps in preparation of a cell-free lysate from rat liver. The crude rat liver homogenate was sequentially centrifuged at different speeds, namely 600*g*, 15,000*g* and 1,05,000*g* to obtain nuclear, mitochondrial and microsomal fractions, respectively. The microsomal fractions contained the ribosomes. The supernatant obtained at this stage when precipitated at pH 5 yielded the so-called 'pH 5 enzymes' (aminoacyl-tRNA synthetases) and the sRNA (tRNA) in the supernatant.

14.4.3 How did the discovery of tRNA reinforce the genetic code?

Many groups (Berg, Lipmann, Novelli, Schweet and Zamecnik) were involved in discovering aminoacyl-tRNA synthetases, the enzymes responsible for charging of tRNAs with their cognate amino acids (Berg and Ofengand 1958; Lipmann *et al.* 1959) and working out the finer details of the reaction (Raj Bhandary and Köhrer 2006). Holley and coworkers (Holley *et al.* 1965) in 1965 sequenced the first tRNA (yeast $tRNA^{ala}$) and proposed three possible secondary structures, one of which was the cloverleaf structure, which has been experimentally proven both by X-ray diffraction and nuclear magnetic resonance (NMR) analyses. Availability of several tRNA sequences (1966–1967) and the knowledge of their folding into a similar kind of cloverleaf structure led to the establishment of the cloverleaf structure as the secondary structure of tRNA and identification of anticodon sequences in tRNA (Raj Bhandary and Köhrer 2006). Crick (1966) proposed the Wobble hypothesis (a set of rules for interaction between anticodon and codon) in the light of identification of anticodon sequences of tRNA and discovery of genetic code. There are 64 possible codons, out of which three code for stop codons. Thus, there are 61 codons which would require 61 specific tRNAs for their recognition by canonical base pairing (G≡C and A≡T). However, in nature, there are only 45 tRNAs. To explain this inconsistency, Crick proposed that some of the 45 tRNAs must recognize more than one codon. He suggested that the 5′ base on the anticodon, which binds to the 3′ base on the mRNA, was not as spatially confined as the other two bases were, and could thus, have non-standard base pairing (wobble base pairing). This hypothesis was experimentally proven by Söll and Raj Bhandary (1967) using yeast $tRNA^{phe}$ with the anticodon sequence GmAA that could be read either of the phenylalanine codons UUU and UUC in an *in vitro* protein synthesis system (Söll and Raj Bhandary 1967). Thus, through all the above experiments, it was unequivocally proved that an RNA adaptor molecule existed which paired through its anticodon with the codon in mRNA and, consequently, this is how the 45 tRNAs read 61codons.

14.5 Discovery and deciphering of the genetic code

After elucidation of the double helical structure of DNA by Watson and Crick in 1953, the whole scientific community went agog with the possibilities that it would entail. One of the interesting questions at that time was how genetic information coded in the DNA was translated into protein? Scientists already knew that there were four bases that made DNA (guanine, cytosine, adenine and thymine). So, if there were codons made of only two bases, then there would only be 16 unique combinations. However, in nature, there were 20 different amino acids. Alternatively, if the code was a triplet, then there could be 64 unique combinations. It was indeed found to be true; there are 64 codons, out of which 61 code for 20 amino acids and the three remaining are stop codons (Fig. 14.4). The possibility that the code was a quadruplet was ruled out through the law of parsimony, which states that the simplest solution is often right. If a quadruplet code were to exist, then there would be 256 unique combinations available. The salient features of the genetic code such as that it is a triplet, non-overlapping, degenerate and universal were later established through several meticulous experiments by some of the brightest minds in science. Among them, Marshall W Nirenberg, Har Gobind Khorana and Robert W Holley received Nobel Prize in 1968 'for their interpretation of the genetic code and its function in protein synthesis'. The details of their contributions are described later in this chapter.

14.5.1 Genetic code is in triplet: Contribution of George Gamow

George Gamow, a Russian physicist, had come to collect his post when he discovered the May (1953) issue of *Nature* describing Watson and Crick's famous double helix model of DNA in his mailbox. He read the entire paper while standing near the mailbox with one arm on it. He immediately thought that three bases in DNA corresponded to one amino acid considering that there are four kinds of bases in DNA, and 20 kinds of amino acids in protein. Taking them three at a time, there are 64 possible combinations of the three bases. He later communicated this postulation in a letter to Watson and Crick, on July 1953 (http://www.dnalc.org/view/15348-Gamov-s-conclusion-that-three-DNA-bases-link-to-one-amino-acid-Marshall-Nirenberg.html). George Gamow later established the 'RNA Tie Club' with a purpose to crack the genetic code. It had 20 members, each code-named as one of the 20 amino acids (Crick, Watson, Feynman and Brenner were its eminent members). However, it was Marshall Nirenberg, a non-member of the club, who pioneered in cracking the genetic code.

14.5.2 Indirect evidence of triplet code by Francis Crick and Sydney Brenner

Francis Crick, Sydney Brenner, Leslie Barnett and Watts-Tobin in a series of intricate genetic experiments in late

	U	C	A	G	
U	UUU Phe	UCU Ser	UAU Tyr	UGU Cys	U
	UUC	UCC	UAC	UGC	C
	UUA Leu	UCA	UAA Stop	UGA Stop	A
	UUG	UCG	UAG	UGG Trp	G
C	CUU Leu	CCU Pro	CAU His	CGU Arg	U
	CUC	CCC	CAC	CGC	C
	CUA	CCA	CAA Gln	CGA	A
	CUG	CCG	CAG	CGG	G
A	AUU Ile	ACU Thr	AAU Asn	AGU Ser	U
	AUC	ACC	AAC	AGC	C
	AUA	ACA	AAA Lys	AGA Arg	A
	AUG Met	ACG	AAG	AGG	G
G	GUU Val	GCU Ala	GAU Asp	GGU Gly	U
	GUC	GCC	GAC	GGC	C
	GUA	GCA	GAA Glu	GGA	A
	GUG	GCG	GAG	GGG	G

Figure 14.4 The standard genetic code. The genetic code was found to be a triplet. Thus, there could be 64 unique combinations. Out of these 64 codons, 61 code for 20 amino acids and the remaining 3 are stop codons. As can be seen, there is a degeneracy of codes, that is, a single amino acid can be coded by multiple codons.

1950s in T4 bacteriophage showed that the genetic code is a triplet, non-overlapping and read from a starting point. They induced mutations in one of its essential genes of T4 bacteriophage (B cistron of r_{II} gene) through proflavine. Proflavine, an acriflavine derivative, causes base pair deletions or base pair insertions and not base substitutions. They showed that two or four mutations (either additions or deletions) inactivated the gene, while three mutations (either additions or deletions) restored the activity of the gene. While working with FC0 (a r_{II} B mutant), a proflavine-induced mutant, they found that this mutant was produced by insertion (they assumed it was insertion) of an additional base in the wild-type sequence. Wild-type T4 bacteriophage grows on both *E. coli* B strain and K12 strain, while r_{II} B mutants grow exclusively on *E. coli* B strain. When *E. coli* B strain was infected with FC0 mutant strain and exposed to proflavine, it was observed that some of them reverted to 'wild' phenotype. As this second exposure is likely to add or delete a base (they assumed it was deletion), Crick concluded that the message must be read from a point (reading frame) and this exposure restored the reading frame (Fig. 14.5). This kind of mutation was called suppressor mutation (FC1) and designated as – (arbitrarily considered base deletion) and original FC0 mutant as + (arbitrarily considered base addition). When they made crosses between the same group (+ and +, and – and –), they found no revertants. Crosses were made by infecting *E. coli* with different strains of T4 bacteriophage at the same time. The fact that the cross between two members of the same group does not produce a revertant suggests that the code is not a doublet. However, when + was crossed with –, then revertants were obtained. These were called double mutants. They concluded that when both FC0 and FC1 mutations were present on the same piece of DNA, some part of gene (between FC0 and FC1) would be misread but reading frame would be restored for rest of the gene. Hence, a pseudo-wild-type protein would be obtained. The pseudo-wild-type strain would infect *E. coli* K12 strain; however, the plaque produced would be quite different from wild-type plaque. Next, they tried to make a triple mutant by crossing members of the same group. They found that revertants were obtained. These revertants also had pseudo-

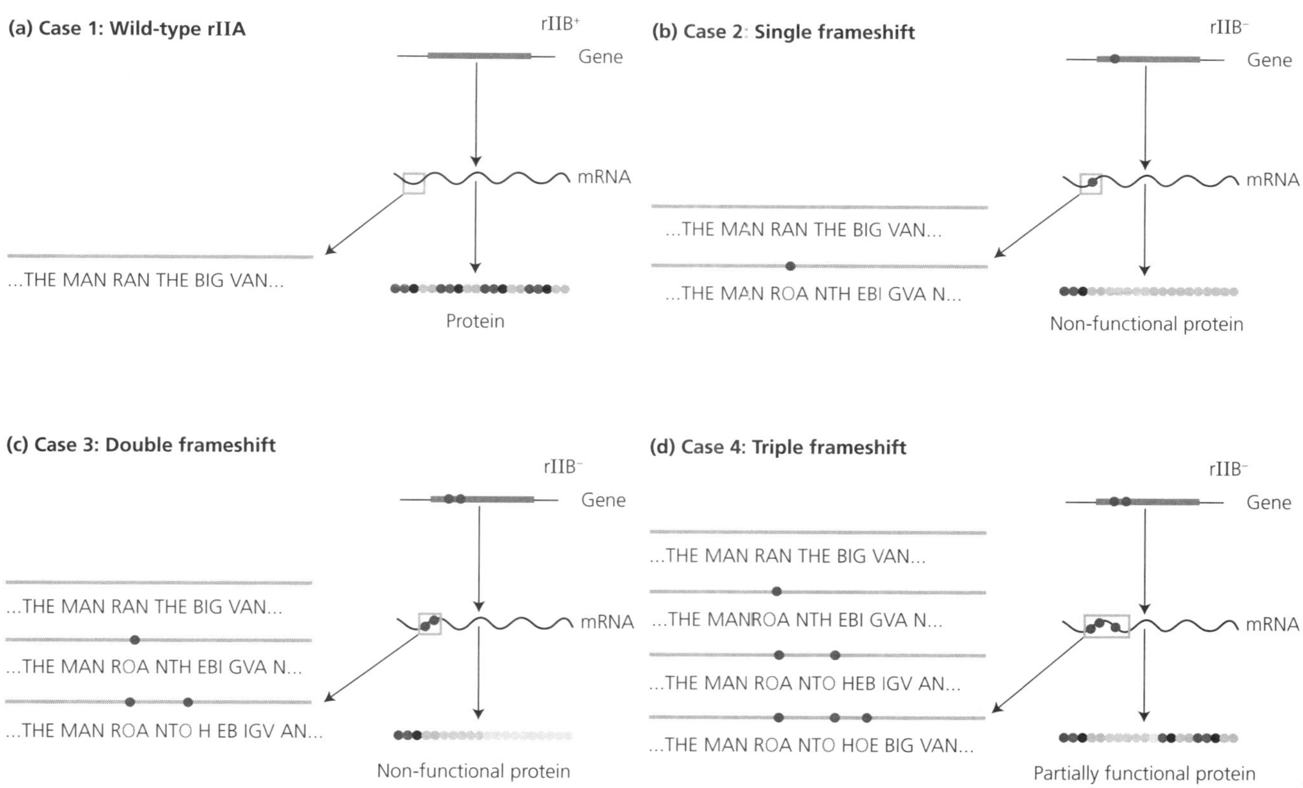

Figure 14.5 Graphical representation of proflavine-induced genetic mutation experiments in T4 bacteriophage that led to the concept of reading frame and triplet codons. (a) Case 1: Wild-type T4 bacteriohage with rII B gene. In this case, the functional protein is produced from the mRNA. (For better understanding, the message in the mRNA is denoted by the phrase 'THE MAN RAN THE BIG VAN', with each word representing a codon.) (b) Case 2: T4 bacteriophage with rII B mutant (denoted as rII B–) with a single frameshift mutation leads to the message being misread as 'THE MAN ROA NTH EBI GVA N' and thus a non-functional protein is produced. (c) Case 3: T4 bacteriophage with rII B mutant with a double frameshift mutation. Second mutation in the above rII B mutant results in the message being misread as 'THE MAN ROA NTO HEB IGV AN'. (d) Case 4: T4 bacteriophage with rII B mutant with a triple frameshift mutation. The third mutation in the above rII B mutant results in the message being partially misread as 'THE MAN ROA NTO HEO BIG VAN'. Thus, a partially functional protein is produced. See Colour Plate 7 page 506.

wild-type phenotype. Crick concluded as three same kinds of mutations (as triplets were obtained by crossing three times in the same group) restored the reading frame and produced a pseudo-wild-type protein, the code must be a triplet. Crick describes the realization that the genetic code is a triplet in his book *What a Mad Pursuit* through a very telling remark to Leslie Barnett, coauthor of his 1961 *Nature* paper, where he presented these results as, 'I looked across at Leslie. "Do you realize," I said, "that you and I are the only people in the world who know it's a triplet code?"' (Crick 1988).

14.5.3 Cracking of the code: Contributions of Marshall Nirenberg and Heinrich Mathaei

In 1961, Marshall Nirenberg and his post-doctoral student Heinrich Matthaei (Nirenberg and Mathaei 1961) synthesized homopolymers of nucleic acid using polynucleotide phosphorylase (originally discovered by Grunberg-Manago, Ortiz and Ochoa, for which Severo Ochoa was later awarded Nobel Prize in Physiology and Medicine in the year 1959). The first homopolymer to be used was poly(U). This homopolymer was incubated in a cell-free, DNAse-treated *E. coli* extract fortified with GTP and other factors in 20 different tubes. In each tube, there was a different amino acid that was radioactively labelled. They found radioactively labelled polypeptide in one tube only. This tube contained radioactively labelled phenylalanine; thus, they concluded that poly(U) codes for polyphenylalanine (Fig. 14.6). To prove that polypeptide was being synthesized, they used TCA (trichloroacetic acid), which is known to precipitate polypeptides. The precipitate was shown to contain radioactivity. To irrevocably prove that it was polyphenylalanine, he had to first characterize the precipitate. He wanted to show that the peptide polyphenylalanine had the solubility expected of authentic polyphenylalanine, but he did not know its solubility. For this, Nirenberg took the help of his fellow scientist Michael Sela, who suggested that he should dissolve it in 15% concentrated acetic acid dissolved in hydrobromic acid. Nirenberg was astounded on hearing of such an unusual solvent; however, he tried just the same. Later, he found that Sela was the only person in the world at that time who knew what would dissolve polyphenyalanine. It was by serendipity that Sela discovered the solvent for polyphenyalanine. Earlier, while working in his lab, he had mistakenly added polyphenyalanine in a solvent for the C-terminal analysis of proteins and it had dissolved in it. Thus, Sela suggested the same to Nirenberg, and Nirenberg thus could prove that UUU codes for phenylalanine (http://history.nih.gov/archives/downloads/Nirenberg%20oral%20history%20Chap%203a-%20%20Recognition%20Moscow,%20MIT.pdf). Later, using the same approach, they proved that poly(C) codes for proline and poly(A) codes for lysine. No code could be found for poly(G) as under denaturing conditions, it assumes what is believed to be a four-stranded helix.

Two groups, Ochoa and his associates and Nirenberg and his colleagues, simultaneously tried deducing the code for heteropolymeric nucleic acids. They made short artificial RNAs with defined proportions of two or three bases. This was possible as they already knew that polynucleotide phosphorylase joins nucleotides randomly, so these oligonucleotides would contain bases in the proportion of bases originally used. Thus, they could speculate about the potential codon for a particular amino acid. This was a speculation because a triplet with 2 G and 1 C could either be of these three

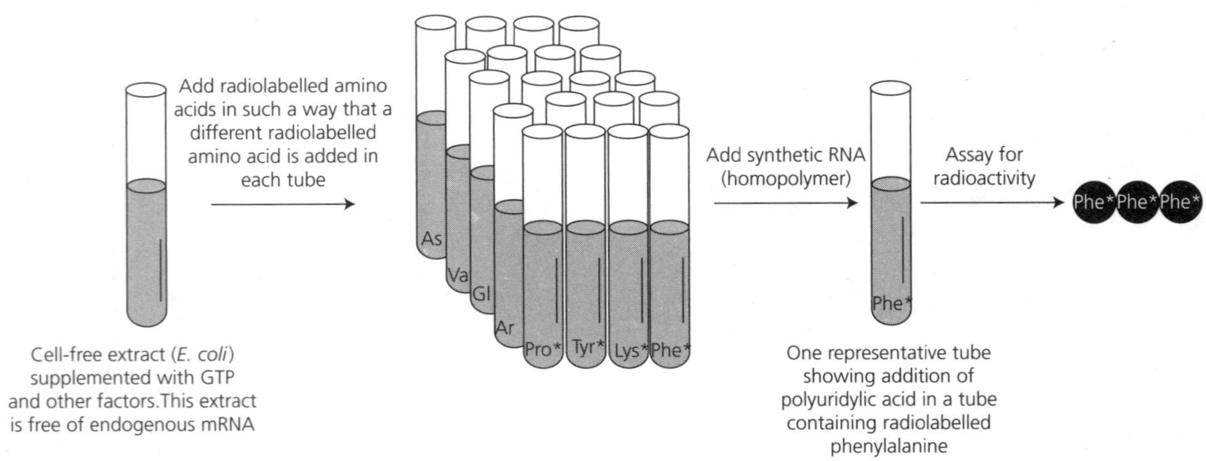

Figure 14.6 Graphical representation of the experiment done by Nirenberg and Matthaei (1962). Cell-free and endogenous nucleic acid–free extract of *E. coli* is incubated with a mixture of 20 amino acids in 20 different tubes in such a way that a different radiolabelled amino acid is added in each tube. Artificially synthesized homopolymer of ribonucleotide is added to the tubes and scored for radioactive polypeptide formation.

combinations, namely GGC, GCG or CGG. Using this approach, they determined putative 50 codons for various amino acids (Speyer *et al.* 1963; Nirenberg *et al.* 1963).

14.5.4 Cracking of the code: Contributions of Nirenberg and Phil Leder

Nirenberg extended his work of deciphering codons from heteropolymeric nucleic acids with his post-doctoral student Phil Leder. By this time, Holley had already predicted the structure of Ala-tRNA from yeast (Holley *et al.* 1965). In 1964, Nirenberg and Leder found that trinucleotides promoted the binding of specific aminoacyl-tRNAs by ribosomes if GTP was absent in the medium and when $MgCl_2$ concentration was 0.02 M. Leder thus made short RNA chains that were three nucleotides long and incubated ribosomes with a mixture of aminoacyl-tRNAs and a single kind of template in the absence of GTP and with $MgCl_2$ concentration 0.02 M. There were different such tubes, with each tube having a different aminoacyl-tRNA, which was radiolabelled. Contents of each assay tube were poured on a cellulose nitrate filter and the filter was assessed for radioactivity (it was already known that the filter would retain complex of ribosome, template and aminoacyl-tRNAs and not aminoacyl-tRNAs alone). For each triplet used, they found only a single radiolabelled amino acid on the filter (Fig. 14.7). Later, using a cell-free extract of *E. coli*, they added the corresponding radiolabelled amino acid and synthesized the polypeptide (Nirenberg and Leder 1964; Nirenberg 1972). This assay is also called the filter binding assay.

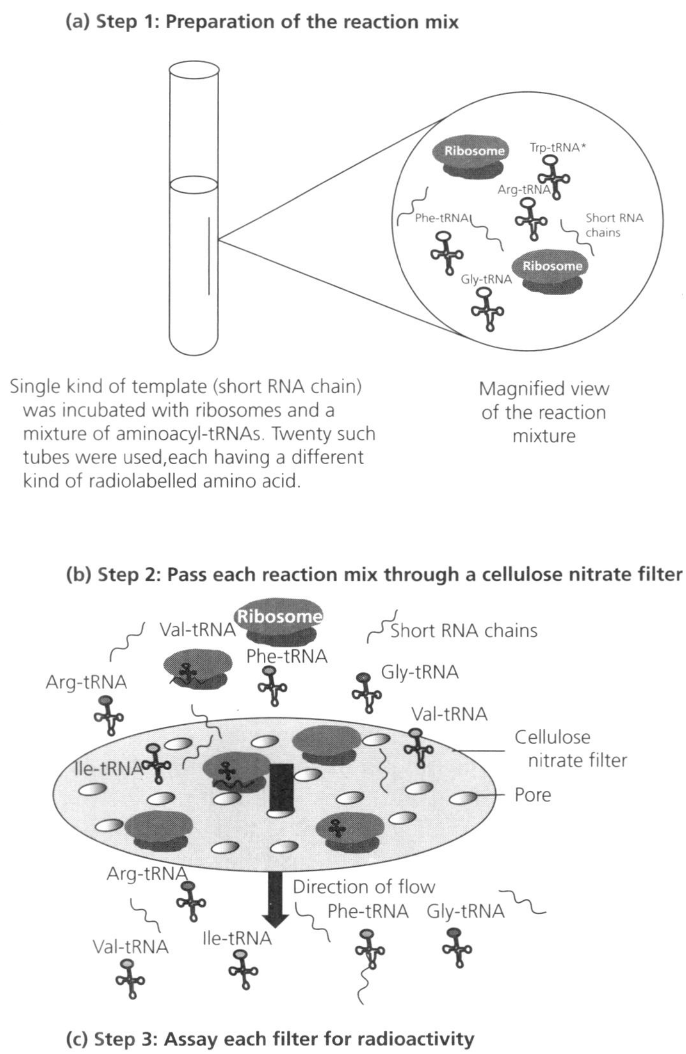

Figure 14.7 Graphical representation of the filter-binding assay done by Nirenberg and Leder (1964). (a) Step 1: Preparation of the reaction mix – short RNA chains were incubated with ribosomes, a mixture of aminoacyl-tRNAs and a single-kind template. Twenty such tubes were used, each having a different kind of radiolabelled amino acid. (b) Step 2: Pass each reaction mix through a cellulose nitrate filter. (c) Step 3: Assay each filter for radioactivity. See Colour Plate 8 on page 507.

14.5.5 Cracking of the code: Contributions of Har Gobind Khorana

In 1965, Har Gobind Khorana and his colleagues chemically synthesized short RNA molecules of known repeating sequences. For example, UCUCUCUCUCUC sequence would be read as UCU and CUC; thus, this sequence should contain two alternate amino acids. Acid hydrolysis of this polypeptide resulted in a mixture of Ser–Leu dipeptides. This exercise was based on the knowledge that peptide bonds involving the amino group of serine were sensitive to acid hydrolysis. He further used repeats of trinucleotides for similar experiments. If UUCUUCUUCUUCUUC was used, there could be three ways this transcript can be read, namely $(UUC)_n$, $(UCU)_n$ and $(CUU)_n$. Thus, three different homopolypeptides would be obtained. Khorana using the earlier transcript in a cell-free protein synthesizing system obtained homopolypeptides of Phe, Ser and Leu. Similar results were obtained when using repeats of tetranucleotides (Khorana 1965, 1972). Important observations from these experiments were that genetic code was triplet, non-overlapping, degenerate and without any commas or punctuations.

14.5.6 Genetic code is degenerate

There is more than one triplet code for an amino acid. This is called degeneracy of the genetic code. Degeneracy of the genetic code was elucidated from separate experiments done by Nirenberg and Khorana, as indicated in the earlier section. Let us understand this better by taking an example. GGC, GGU, GGA and GGG code for glycine; thus, there are four triplet codes for glycine, and likewise there are two triplet codes for phenylalanine, six for leucine and so on (Fig. 14.4). Redundancy of the genetic code is beneficial as replacement of one base by another in DNA often does not result in the replacement of one amino acid by another in protein. Many mutations, therefore, are silent ones.

14.5.7 Genetic code is universal

Earlier experiments led to the belief that the genetic code is universal; meaning that same code exists from *E. coli* to elephant. Synthesis of rabbit hemoglobin in a cell-free extract of *E. coli* suggested that it may be so. Experiments by Marshall *et al.* (1967) further supported the idea that different forms of life use the same genetic language. Hinegardner and Engelberg (1963) suggested that assignment of specific codon to amino acids was a 'historical accident' which was carried over across generations along different phyla. Their reasoning was simple; there was nothing inherently beneficial in allotment of any specific code to any specific amino acid. Thus, UUU, which codes for phenylalanine, could very well have been assigned to any other amino acid, namely leucine, during the course of evolution. Any fundamental change in this system would be lethal as the occurrence of two events would be a rare phenomenon. What are these two events? The first is the aminoacyl-tRNA that recognizes a specific codon should incorporate the corresponding change in its anticodon and the second event is replacement of old codon with new codon in important (if not all) sequences in an organism's genome. Let us understand this better with the help of an example: tRNA for serine recognizes UCU, but now UUU codes for serine. Thus, there should be two events for this to happen: first, the anticodon of the serine–tRNA has to undergo a specific set of mutation to recognize the UUU; second, in the genome wherever the codon for important serine exists, it has to be replaced with UUU. Thus, the occurrence of two such events would be rare and it would mean lethality if any of the events occurred separately.

It is remarkable considering the earlier arguments that the genetic code is not so universal. DNA from mitochondria, mycoplasma, *Candida albicans* and some ciliates such as *Paramecium* show deviations from the universal genetic code (Table 14.1). It is postulated that these genomes, being very small and coding for a limited number of proteins, have more autonomy to change than nuclear genomes; thus, the genetic code is 'nearly' universal, with exceptions few and far between.

14.5.8 How genetic code helped in understanding human genome?

The enormous amount of data generated from the Human Genome Project has led to a spurt of research in various areas. One of the thrust areas was bioinformatics, which includes cataloguing genes and its product, be it RNA and/or protein. Validation of data generated from the Human Genome Project with the data already published was a daunting task. However, scientists were successful in doing so. With the genetic code being known, they could translate (*in silico*) hundreds of putative protein-coding nucleotide sequences and assign corresponding information to the gene. This information is of relevance as many of the proteins are yet to be discovered and studied in detail, be it in humans, whose genome is already sequenced, or other organisms, whose DNA sequencing is in progress or is yet to be done.

14.6 Learning details of steps of translation through the use of antibiotics

The term 'antibiotic' (Greek: *anti*-against; *bios*-life) was coined by Selman Waksman to denote substances produced by a microorganism that are detrimental

Table 14.1 Deviations from standard genetic code. 'N' stands for any base

Source organism	Codon	Usual meaning	New meaning
Mitochondrial genomes			
Mammals	UGA	Stop	Trp
	AGA, AGG	Arg	Stop
	AUA	Ile	Met
Drosophila	UGA	Stop	Trp
	AGA	Arg	Ser
	AUA	Ile	Met
Saccharomyces cerevisiae	UGA	Stop	Trp
	CUN	Leu	Thr
	AUA	Ile	Met
Fungi	UGA	Stop	Trp
Maize	CGG	Arg	Trp
Prokaryotic genomes			
Several protozoa	UAA, UAG	Stop	Gln
Candida cylindracea	CUG	Leu	Ser
Micrococcus sp.	AGA	Arg	Stop
	AUA	Ile	Stop
Mycoplasma sp.	UGA	Stop	Trp
	CGG	Arg	Stop

to the growth of other microorganisms. Some of the popular antibiotics used for basic research are penicillin, streptomycin, neomycin, puromycin, paromomycin, rifampicin, rapamycin and chloramphenicol. Among these, we will be discussing three major antibiotics (streptomycin, chloramphenicol and puromycin) which have contributed to our understanding of the mechanism of translation.

14.6.1 Streptomycin

Few antibiotics have stirred the collective imagination of the scientific world as much as penicillin and streptomycin. Penicillin, discovered by Alexander Flemming, earned its distinction of being the first antibiotic (also referred to as 'the wonder drug' then) that could kill a variety of gram-positive bacteria, while streptomycin for its ability to act against gram-negative bacteria. including tuberculosis (TB), and for the controversy surrounding its discovery. Streptomycin also has the privilege of being associated with two Nobel Prizes; let us discuss these discoveries in detail.

14.6.1.1 Streptomycin: The discovery, controversy and the first Nobel Prize

Dr Selman Waksman (Rutgers University), a soil microbiologist and world authority on *Actinomycetes* species, is credited with the discovery of as many as 20 antibiotics such as neomycin, actinomycin, streptothricin and streptomycin. Dr Waksman was awarded Nobel Prize in 1952 for discovery of streptomycin (he claimed to be the sole discoverer of the same), the first antibiotic to cure TB. However, this claim was furiously contested by his graduate student Albert Schwatz, who asserted that it was he who originally isolated two strains of *Actinomyces grieus*. The first one was a sample from heavily manured farmyard soil near his laboratory and the other sample was from the petri dish of his fellow graduate student Doris Jones, where she had streaked a swab from the throat of a healthy chicken. In his doctoral thesis, Schatz (1945) describes how he proved that streptomycin acted against *E. coli* and also on non-virulent and virulent strains of *Mycobacterium tuberculosis*. His preparation of streptomycin was used by

William Feldman and Corwin Hindshaw at Mayo Clinic to prove its activity against TB in guinea pig and human patients (Schatz and Waksman 1944, 1945; Schatz *et al.* 1944). Waksman, along with Rutgers University, had industrial collaboration with Merck laboratories, where this drug was finally developed for human consumption. Schatz was the first author of two major papers related to streptomycin and he, along with Selman, was granted patent for the same. In 1948, Selman duped Schatz of his share of royalty received from the sale of the drug and under false impression made Schatz sign his share of royalty to Rutgers University, saying that if he did not do so, it will 'kill his job chances' and he himself had done so. In 1949, Schatz sued him for his share of royalty and later Waksman made an out-of-court settlement with him. In 1952, the Nobel committee awarded Nobel Prize for Physiology and Medicine to Waksman as under the terms of the will of Alfred Nobel, Nobel Prize could be shared at the maximum by three people. Discovery and development of streptomycin involved contributions of more than three people (Waksman, Schatz, scientists at Merck and doctors at Mayo Clinic); hence, influenced by the active propaganda of Waksman and personal relations department at Rutgers University (which sought to minimize the contributions of others), the committee awarded the Nobel Prize to Waksman (Pringle 2012). Schatz died in 2005 as an embittered old man with the reputation of a trouble maker who could never find a job in the scientific community. However, the drug that he and others discovered and developed was used by the scientific community and common man at large for betterment of quality of human life. We shall now discuss how the mechanism of action of this drug shed light on the finer details of translation.

14.6.1.2 Streptomycin: The mechanism of action and the second Nobel Prize

Streptomycin increases error during translation, with the ribosomes making mistakes while incorporating amino acids in the polypeptide chain. It interferes with initial codon–anticodon recognition and proofreading by binding very near to the A site (where decoding of the codons occurs) of the 30S ribosomal subunit. It stabilizes the *ram* site (ribosome ambiguity site), a site which allows easy base-pairing between a codon and non-cognate aminoacyl-tRNAs. This site lies very close to the helix H27 of the 16S rRNA component of the ribosome. Helix H27 has two conformations during translation – *ram* state and restrictive state. Restrictive state is required for accurate proofreading; if a ribosome is locked in restrictive state, it seldom makes mistakes. Thus, this site is said to be hyper-accurate. Streptomycin acts in two ways: first, by reducing the fidelity of translation by allowing non-cognate aminoacyl-tRNAs to bind to the A site and, second, by preventing the shift to restrictive state that is essential for proofreading. This study was a piece of a large part of work done by Venky Ramkrishnan and colleagues on the 30S ribosomal subunit from *Thermus thermophilus.* This study yielded the complete atomic model (refined to 3Å resolution) of the 30S ribosome (Wimberly *et al.* 2000). They obtained crystal structure of the 30S ribosomal subunit bound to three different antibiotics: spectinomycin (inhibits translocation), streptomycin (causes errors in translation) and paromomycin (increases the error rate) and illustrated the finer details about the mechanism of translation. As stated before, Ramakrishnan, Yonath and Steitz were jointly awarded Nobel Prize in Chemistry in 2009 for unravelling the structure of ribosome.

14.6.2 Puromycin and chloramphenicol

Puromycin is an aminonucleoside first isolated from *Streptomyces alboniger*. Aminonucleoside is an amino acid linked to an adenosine analogue. It closely resembles the aminoacyl adenosine at the end of an aminoacyl-tRNA. Thus, it binds to the A site in the ribosome and substitutes for the peptidyl–tRNA in peptidyl transferase activity. Puromycin once bound to the polypeptide chain diffuses out of the A site, as in comparison with the tRNA, it is quite small; hence, the ribosome is unable to retain it and thus premature chain termination occurs. Puromycin affects both eukaryotic and prokaryotic ribosomes. Puromycin and chloramphenicol have been very effectively utilized to study the mechanism of translation in detail.

14.6.2.1 Two-site model of ribosome

Bretscher and Marcker, working at the Medical Research Council Laboratory of Molecular Biology, Cambridge, in 1966 used puromycin to prove that the ribosome has two sites 'A' site and 'P' site (Bretscher and Marcker 1966). 'A' stands for acceptor site and 'P' stands for peptidyl site. They postulated that if aminoacyl-tRNA is in P site, then puromycin at the A site will be transferred to the growing polypeptide chain and chain termination will occur. This complex being acid soluble can be easily isolated. However, if P site is already occupied by aminoacyl-tRNA, then puromycin will not be able to bind. Such aminoacyl-tRNA bound to ribosome is acid insoluble. This is called a puromycin release assay [Fig. 14.8(a)]. Bretscher and Marcker incubated triplet AUG with [^{35}S]fMet-tRNA$_f$, ribosomes and puromycin [Fig. 14.8(b)(i)]. They found the presence of fmet–puromycin while precipitating the reaction mix with perchloric acid. They further confirmed this observation by replacing [^{35}S]fMet-tRNA$_f$ with [^{35}S] Met-tRNA$_{met}$ [Fig. 14.8(b)(ii)]. In this experiment, no acid-soluble precipitate was obtained as Met-tRNA$_{met}$ occupies A site and thus puromycin is not able to enter. Thus, there are two states in a ribosome – a puromycin reactive state and a puromycin unreactive state, thus envisaging for at least two binding sites in the ribosome, which were thus called A site and P site. Later, Knud Nierhaus and colleagues in 1981 presented evidence for a third site, the E site or exit site (Rheinberger *et al.* 1981).

14.6.3 50S Ribosomal subunit contributes to the peptidyl transferase activity

The peptidyl transferase assay (Fig. 14.9) was devised by Robert Traut and Robert Monro (1964) and later was modified by Harry Noller and colleagues (Noller *et al.* 1992) and was called fragment reaction. Both these assays were effectively utilized to prove that 50S ribosomal subunit carries the peptidyl transferase activity (capacity to form peptide bonds) and wherein it resides. Traut and Monro incubated poly(U) mRNA coding for poly(phe) with ribosomes, labelled phenylalanine and $tRNA^{phe}$. Later, they removed the 30S subunit by incubating in a buffer containing low Mg^{2+} ion concentration, followed by ultracentrifugation. Pure 50S subunit complexed with poly(phe)–tRNA in P site was isolated by repeatedly washing the fraction with salt solutions to remove traces of initiation or elongation factors in it. Such primed 50S subunit was utilized for puromycin release assay. They incubated these 50S subunit with puromycin and found that poly(phe) puromycin was being formed. This reaction they tried with and without ATP and found that in both conditions, puromycin was releasing the peptide from the 50S subunit, a true characteristic of the peptidyl transferase activity. But where does this peptidyl transferase activity in 50S subunit reside? Harry Noller, Mankin, Thomas Steitz and their associates working in three different labs tried to solve this riddle.

14.6.3.1 23S RNA also contributes to the peptidyl transferase activity

By 1980s, the pioneering work by Thomas Cech and colleagues (Cech 2000) had shown that RNA can also act as enzymes (ribozymes). This led Harry Noller and colleagues to think that 23S rRNA might play a role. Noller and colleagues in 1992 (Noller *et al.* 1992) used a modified version of peptidyl transferase assay (called the fragment reaction) to determine wherein 50S ribosome peptidyl transferase activity resides – RNA or protein or both? They utilized labelled CCA-fMet or CAACCA-fMet (fragment of tRNA), which resembles the fMet-$tRNA_{f\ Met}$ enough to bind to the P site. This they incubated with necessary components of puromycin release assay and 50S subunit, which was robbed of all its protein content by pre-treatment with either harsh reagents such as SDS, proteinase K or phenol. They found that treatment with SDS and

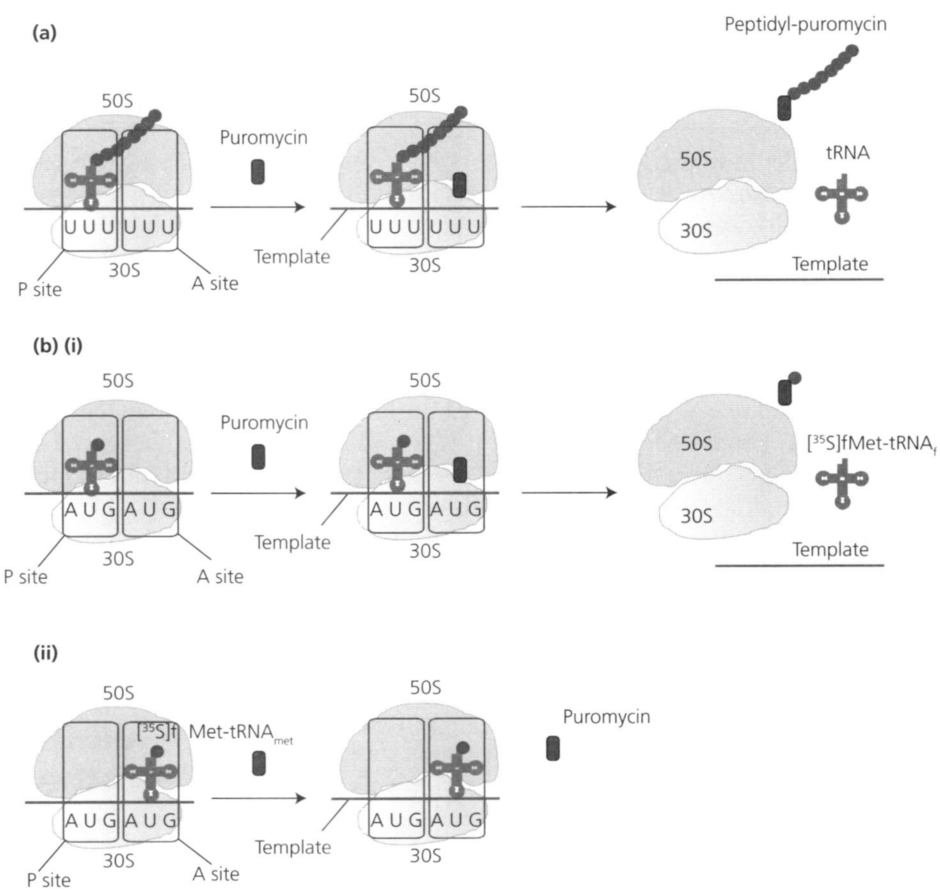

Figure 14.8 (a) Puromycin release assay. A synthetic template such as poly(U) is incubated with ribosomes, tRNA and other buffer components. Such ribosomes are isolated and incubated with puromycin. Puromycin attaches to the growing peptide chain and causes the release of the peptide from the ribosome. This peptidyl puromycin is acid soluble; hence, it can be isolated and characterized. (b) Puromycin release assay as employed by Bretscher and Marcker. They utilized an AUG triplet template in the above reaction along with (i) [35S]fMet-tRNAf or (ii) [35S] Met-tRNAmet. In the first reaction, [35S]fMet-tRNAf occupies the P site; hence, puromycin is able to enter the A site and terminate the peptide chain. In the second case, [35S] Met-tRNAmet occupies the A site; hence, puromycin is not able to enter. As a result, chain termination does not occur. See Colour Plate 9 on page 508.

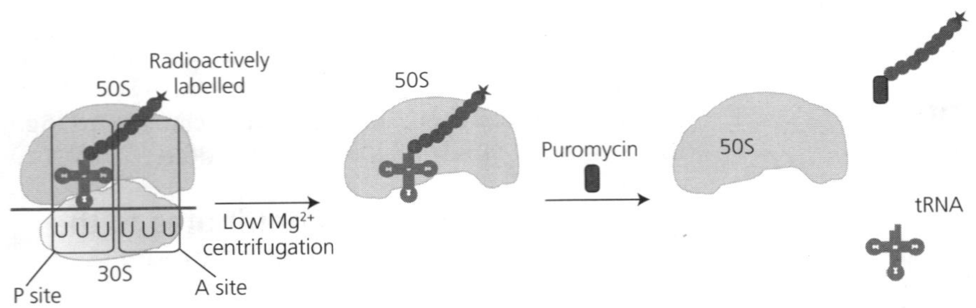

Figure 14.9 Peptidyl transferase assay. A synthetic poly(U) template is incubated with ribosomes, tRNA and other buffer components. Such ribosomes are isolated and centrifuged under low Mg^{2+} concentration. Pure 50S subunit complexed with poly(phe)-tRNA in the P site is isolated by repeatedly washing the fraction with salt solutions to remove traces of initiation or elongation factors in it. Such primed 50S subunit was utilized for puromycin release assay [see Fig. 14.8(a)]. See Colour Plate 10 on page 508.

proteinase K did not hamper puromycin activity; however, phenol did. They postulated that phenol treatment may be disrupting the higher-order structure of RNA; hence, it was unreactive. They envisaged that if 50S ribosome isolated from *Thermus aquaticus* was used in the above reaction, then it should be able to withstand harsh treatment with phenol and still retain its activity as the organism itself is an inhabitant of hot thermal springs. As expected, 50S ribosome from *T. aqauticus* was able to retain its activity under treatment with all the three reagents. To further strengthen this observation, they repeated the experiment in the presence of chloramphenicol, a known inhibitor of the peptidyl transferase activity. Chloramphenicol, as expected, inhibited the release of poly(phe)$_n$ by puromycin. They abolished this activity by treating the 50S subunit by RNAse. Thus, they proved that RNA is an essential component of 50S subunit necessary for peptidyl transferase activity. Later, Noller, in collaboration with Alexander Mankin (Khaitovich *et al.* 1999), proved that the minimum necessary components for peptidyl transferase activity (*in vitro*) are 23S rRNA and proteins L2 and L3. X-ray crystallography studies by Thomas Steitz and colleagues (2000) showed that 23S rRNA is at the catalytic centre of peptidyl transferase and thus has peptidyl transferase activity (Nissen *et al.* 2000).

Chloramphenicol was originally isolated from *Streptomyces venezuelae* in 1947, but like most antibiotics, it is presently produced synthetically. It was reported by Hahn and colleagues (Gale and Folkes 1953; Wisseman *et al.* 1954) that the addition of chloramphenicol to growing bacterial cultures results in an almost immediate cessation of protein biosynthesis. Avram Goldstein and colleagues working at Stanford University School of Medicine tried to dissect out the mode of its action by using *E. coli* cultures grown at 0°C (Das *et al.* 1966). Growing cultures at very low temperatures effectively slowed down metabolism and could be utilized to study slowed-down assembly of protein synthesis machinery, which was not possible at physiological temperatures (a protein of molecular weight 5×10^4 takes 5 s to assemble). They proposed that chloramphenicol might act by competing with the carboxy-terminal amino acid of the growing chain of nascent protein (attached to tRNA) for a stereospecific locus associated with one of the tRNA binding sites on the ribosome (Das *et al.* 1966). Detailed analyses by Coutsogeorgopoulos, Nierhaus and others indicated that it inhibited protein synthesis by binding to the 50S ribosomal subunit and blocking the peptidyl transferase reaction (Coutsogeorgopoulos 1966; Nierhaus and Nierhaus 1973). X-ray crystallographic studies by Steitz and colleagues (2003) have further mapped two binding sites of chloramphenicol in 50S ribosome. One site lies in a crevice in the peptidyl transferase centre and another in a hydrophobic crevice at the entrance of the peptide exit tunnel (Hansen *et al.* 2003).

14.6.4 What about other antibiotics?

A host of antibiotics have been utilized by both doctors and researchers alike to treat human diseases and learn the mechanisms of action of drugs. Tetracyclin, doxycline, azithromycin, paromomycin, erythromycin and cycloserine have been either utilized to treat human diseases or in stages of modification so that they can be suitable for human consumption in diseases such as filariasis, malaria, amoebiasis, trachoma and leishmaniasis. These antibiotics are effective against bacterial or protozoan infections because of their ability to target specifically prokaryotic or protozoan translation machinery, and in doing so, they tell us much about the finer details of translation. Many other antibiotics such as cycloheximide, sulphonamides, fusidic acid and others have been utilized to study protein synthesis; however, they have not been discussed in detail in this chapter as it is beyond the scope of the chapter. The readers may refer to the reviews (Blanchard *et al.* 2010; Guo *et al.* 2012) for more information.

14.7 Cell-free protein synthesis

The demonstration that protein synthesis does not require intact cells and can be carried on disintegrated cells was one of the major breakthroughs that led to the birth of molecular biology of protein synthesis. Establishment of the genetic code by Nirenberg, discovery of tRNA and intermediate steps of protein translations are some of the early discoveries that were achieved using this system. Original observations were made with homogenates from animal tissues or fractionates thereof and this system was later perfected with prokaryotic (*E. coli*) and plant- (wheat germ lysate; WGL) based systems. We discuss here some of the pioneering discoveries that led to the use of cell-free protein synthesis systems as a tool for understanding how mRNAs are translated into functional polypeptides.

14.7.1 Contribution of Paul C Zamecnik's laboratory

Paul C Zamecnik had earlier worked with Fritz Lipmann and had proved that proteins were made of amino acids in a reaction requiring ATP (http://www.dnaftb.org/21/bio.html). He became further curious to dissect out the sequence of anabolic events that occur during protein synthesis. Zamecnik had earlier utilized rat liver slices to incorporate exogenously supplied radioactive amino acids into proteins. However, this system was not suitable for studying intricate steps of protein synthesis. Thus, he set out to develop a cell-free protein synthesis system, a truly daunting task that took almost 5 years of research. In 1951, Philip Siekevitz and Zamecnik homogenized minced rat liver with a glass pestle, fitted tightly into a test tube with a rounded ground glass bottom. The liver tissue was further ground briefly by a rotating motor maintained at 4°C in an icebath. The resulting homogenate was utilized for further reactions. However, the rate of protein synthesis observed in this case was found to be low. One day, Zamecnik loaned his homogenizer pestle and ground glass acceptor tube, which served as a mortar with a looser fit than their usual one, to his colleague Nancy Bucher, who was studying cholesterol synthesis in the rat liver. Bucher found that 'loose homogenization' yielded better evidence for cholesterol synthesis. Zamecnik reasoned that harsher homogenization ruptured cellular compartments which stored ATPases; hence, ATP was not available for protein synthesis as they were being gradually destroyed by the liberated ATPases. Thus 'loose homogenization' became the standard practice after this observation. The crude homogenate so obtained was sequentially centrifuged at different speeds, namely 600*g*, 15,000*g* and 1,05,000*g*, to obtain nuclear, mitochondrial and microsomal fractions, respectively (Fig. 14.3). The microsomal fractions contained the ribosomes (Zamecnik 2005). The supernatant obtained at this stage when precipitated at pH 5 yielded the so called 'pH 5 enzymes' (amino acid synthetases) and the sRNA (tRNA) in the supernatant. This supernatant was utilized by Zamecnik's lab to discover tRNA and aminoacyl tRNA synthetases (Hoagland *et al.* 1958). Thus, Zamecnik and his colleagues (Littlefield *et al.* 1955) made a real cell-free ribosomal system for protein synthesis based on mitochondria-free cytoplasmic extracts of rat liver.

In 1957, Marvin Lamborg from Zamecnik's lab attempted to make an *E. coli*–based cell-free translation system (earlier Zamecnik had tried doing so but had failed) and finally succeeded in making it by 1960. Live *E. coli* cells were always found to be a contaminant in their cell-free system and thus a constant source of worry as they synthesized protein at a speed many times greater than the cell-free system itself. Moreover, they found that only early log phase *E. coli* cells were suitable for making the lysate as it had a low level of ATPase activity while mid- and late log phase cells had high activity, which easily overwhelmed the capacity of an ATP generating system (phosphopyruvate and pyruvate kinase, or phosphocreatine and creatine kinase) to maintain the steady-state level of ATP essential to fuel amino acid activation. Lamborg empirically determined 10^5 live cells/mL of cell-free preparation were an acceptable degree of contamination for a 15-min incubation at 37°C and scaled up the culture to gallons to obtain early log phase *E. coli* cells (Lamborg and Zamecnik 1960). This *E. coli*–based cell-free translation system was later lent to Marshall Nirenberg for his experiments with which he achieved a breakthrough in genetic code in 1962 (Nirenberg and Matthaei 1961).

14.7.2 Contribution of Nirenberg and Matthaei

E. coli S30 extracts were used as the first prokaryotic cell-free protein-synthesizing system. *E. coli* S30 extracts contain supernatant obtained after centrifuging *E. coli* cell lysate at 30,000*g* which is supplemented with amino acids, ATP and GTP. This system (and the earlier ones also) was inefficient in translating exogenous RNA as endogenous RNA got translated in addition to exogenous RNA under the reaction conditions (Tissieres *et al.* 1960). Moreover, it was also observed that DNA was a constant contaminant in this system. Nirenberg and Matthaei used this system more effectively by pre-incubating the extract at physiological temperature, thus allowing mRNAs (endogenous) already attached to the ribosomes to complete translation and thus leave the ribosomes vacant to accept the next ribosome-free mRNA. They also used DNAse to prevent DNA contamination. This simple modification allowed translation of exogenous natural mRNAs or synthetic polyribonucleotides (Nirenberg and Matthaei 1961) more efficiently.

14.7.3 Other cell-free translation systems

Rabbit reticulocyte lysate, wheat germ extract or combinations of different types of lysates are the other cell-free translation systems that are commercially available for research. All these systems contain crude extracts with macromolecular components (70S or 80S ribosomes, tRNAs, aminoacyl-tRNA synthetases, initiation, elongation and termination factors, etc.) required for translation of exogenous RNA. To ensure efficient translation, each extract is supplemented with amino acids, energy sources (ATP, GTP), energy-regenerating systems (creatine phosphate and creatine phosphokinase for eukaryotic systems, and phosphoenol pyruvate and pyruvate kinase for the *E. coli* lysate) and other cofactors (Mg^{2+}, K^+ and others). These systems suffer from contaminations from NTPases and intensive degradation of NTP due to NTP-dependent metabolic processes, which break down added ATP (required for activation of amino acid) and GTP (for the function of ribosomes). Therefore, to maintain adequate levels of ATP and GTP, they are supplemented with energy-regenerating systems.

14.7.4 Rabbit reticulocyte lysate

Unfractionated rabbit reticulocyte lysate (RRL) allows significant translation of exogenously added message; however, it has to compete with more abundantly present endogenous globin mRNA (Hunt and Jackson 1974). Reticulocytes are immature erythrocytes that are primarily responsible for the synthesis of haemoglobin, which represents more than 90% of the proteins made in the reticulocyte. Reticulocytes lack nuclei; however, they have abundant RNA and complete translation machinery. This type of lysate is typically used for studying the translation machinery, for example, studying the effects of inhibitors on globin translation. Pelham and Jackson (1976) used Ca^{2+} dependent micrococcal ribonuclease for destruction of endogenous RNA, which was subsequently inactivated with chelating agents such as EGTA or EDTA. This type of lysate is called nuclease-treated rabbit reticulocyte lysate and is used for synthesis of microgram quantities of proteins from capped or uncapped mRNAs (Pelham and Jackson 1976).

14.7.5 Wheat germ lysate

Nakano and colleagues in 1996 were successful in generating a plant-based cell-free translation system (Nakano *et al.* 1996). They highly concentrated the WGL by polyethylene glycol precipitation and further treated it with Cu^{2+} to inhibit inherent nuclease and phosphatase activity. WGL did not require pre-treatment to remove endogenous RNA because it had intrinsically low levels of mRNA. Removal of endosperm from wheat embryos (before preparing the extract) eliminates translation inhibitors, such as tritin (ribosome-inactivating protein of wheat), thionins and RNases effectively, increasing its stability and productivity (Madin *et al.* 2000). WGL efficiently translates exogenous RNA from a variety of different organisms, from viruses and yeast to higher plants and mammals.

14.7.6 Coupled transcription and translation

In prokaryotes, translation begins as soon as the mRNA is transcribed by the RNA polymerase; hence, here the term 'coupled' transcription and translation is used. Avram Goldstein and colleagues working with *E. coli* maintained at 0°C demonstrated that ribosomes attach to nascent mRNAs while transcription was still going on. As discussed earlier, *E. coli* grown at 0°C has a lower metabolic rate; thus, many of the metabolic processes such as transcription and translation can be studied with ease at this temperature. The average rate of nucleotide addition at the 3′ end of an mRNA falls down to five nucleotides/min (Goldstein *et al.* 1965) at 0°C. *E. coli* cells starved of uridine at 37°C when grown at 0°C in media supplemented with radioactive uridine incorporates uridine in their mRNAs. Goldstein and colleagues used these cells to study whether transcription and translation was coupled in *E. coli*. They postulated that if ribosomes were attaching to nascent mRNA while transcription was still going on, then the label would appear mostly in ribosome fraction, and if the transcription and translation were not coupled, then the label would be predominately in the RNA fraction. To experimentally prove this hypothesis, they did a number of experiments where uridine exposure to cells was limited to 2.5–6 min. This duration allowed only 12–30 residues of uridine being incorporated at the 3′ end of mRNA. They assumed if the average size of mRNA is 1000 nucleotides, then during the duration of the experiment, the fraction of RNA that completed transcription would not exceed 3% of the total RNA population. Implying that if transcription and translation were two separate processes, then only 3% of the RNA (only those will be labelled which are on the stage of completion of transcription) would have the chance of binding to the ribosomes. (Please note that here only radioactively labelled RNAs synthesized during the course of the experiment are considered). However, if the reverse were true, then most of the label would be associated with the ribosome fraction. Goldstein and colleagues found that nearly half of the label was associated with the ribosomes. When they confirmed their observation with *E. coli* grown at 37°C and incorporation duration of 10 s, they concluded that coupled transcription and translation occurs in *E. coli* (Das *et al.* 1967).

This discovery of coupled transcription and translation in *E. coli* has been utilized for developing a prokaryotic cell-free translation system. In general, mRNA species are added to the cell-free translation systems to obtain proteins. However, one may add plasmids, an isolated gene, or a synthetic DNA fragment to the lysate and obtain resulting proteins. However, these DNA species should have a prokaryotic promoter and the Shine–Dalgarno sequence for proper recognition by prokaryotic RNA polymerase and ribosomes, respectively. The

E. coli extracts utilized for this purpose are devoid of endogenous DNA and mRNA (Zubay 1973; DeVries and Zubay 1967).

14.7.7 Combined transcription and translation

Unlike prokaryotes, coupled transcription and translation does not take place in eukaryotes due to compartmentalization of the nucleus from the cytoplasm. Transcription takes place in the nucleus and translation in the cytoplasm; hence, the mRNA has to be transported out of the nucleus before it is acted upon by the ribosomes for protein synthesis. RRL and WGL can be utilized for synthesis of proteins directly from DNA template (including PCR products with cognate promoter); however, these systems lack inherent RNA polymerases. Therefore, they have to be supplemented with SP6, T3 or T7 bacteriophage RNA polymerases. The DNA templates used for combined transcription and translation have SP6, T3 or T7 promoters for recognition by the RNA polymerase used. Here, the prefix applied for transcription and translation is 'combined' and not 'coupled' as the viral RNA polymerases have high transcript turnover rate; hence, the transcript is made much before it is acted upon by translation machinery (Spirin 1991; Craig *et al.* 1992; Baranov and Spirin 1993).

14.7.8 Continuous-flow and continuous-exchange cell-free systems

The major drawback of the above process of *in vitro* protein synthesis is the scale of the setup, which is small. Moreover, the reaction stops after 20–60 min due to depletion of substrate or accumulation of inhibitory by-products. To do away with such limitations, Spirin in 1988, devised the continuous-flow cell-free system (CFCF; Spirin *et al.* 1988). In this system the consumable substrates were continuously supplied to the reaction mixture (amino acids, NTPs and energy-regenerating system) and reaction by-products (inorganic phosphates and NMPs) were continuously removed with the help of a peristaltic pump. Here, the higher-molecular-weight translational machinery is separated from the low-molecular-weight substrates and reaction by-products by a porous ultrafiltration membrane with molecular-weight cut-offs of 10, 20 or 30 kDa depending upon the size of the product formed. The term 'molecular-weight cut-off' of number 'X' means that only molecules higher than the molecular weight 'X' will be retained by the membrane. Thus, Spirin created a novel reactor (Fig. 14.10) which was capable of sustained protein synthesis till 20–40 h and yielded milligram quantity of proteins. This system has been utilized for synthesis of functionally active DHFR, CAT, GFP, IL-2 and IL-6, as well as virus coat proteins, globin and enzymes (Spirin 2002; Spirin and Swartz 2008).

Continuous-exchange cell-free systems use passive diffusional force for exchange of substrate and by-products across the dialysis membrane or hollow fibres. The system may be rejuvenated by addition of fresh mRNA every 24–48 h and maintained even up to 2 weeks (Sawasaki *et al.* 2002). Milligram quantities of protein can be obtained by this process using *E. coli* extract, RRL or WGL (Kigawa *et al.* 1999).

14.7.9 Future of cell-free protein synthesis

Cell-free protein synthesis offers several advantages over conventional cell-based overexpression methods, including high yield, favourable conditions for soluble protein formation, decreased sensitivity to product toxicity and suitability for high-throughput screening (Katzen *et al.* 2005). *Drosophila* embryo extract and wheat germ extracts are now being widely utilized for studying the RNA silencing machinery because of the less restrictive nature of the biochemical framework in this system (Novina and Sharp 2004). Recent advances in technology and sequencing of genomes from organisms across phyla have generated a lot of interest in using this system for expression of proteins from putative open reading frames, expression of proteins with labelled or unnatural amino acids, *in vitro* protein engineering and screening of engineered and theoretically designed proteins. The research in the field for development and application of cell-free based systems is sure to expand and progress in coming years.

14.8 Molecular mechanism of the orchestrated act of translation

As discussed above, from various pioneering discoveries, it became apparent that for protein synthesis, the following principal components are involved: (a) mRNA, the template; (b) tRNA, the adaptor molecule; and (c) ribosomes, the work bench. mRNA carries the genetic information in the form of genetic code, tRNA interprets the trinucleotide codons in the form of amino acids and ribosomes sponsor the formation of polypeptides by assembling and supporting all the necessary factors, both intrinsic and extrinsic, that catalyse the act of translation. For detailed characteristics of these components, the readers may refer to Pal and Ghaskadbi (2009).

For the convenience of our understanding the molecular mechanism of protein synthesis, it can be divided into three major steps: initiation, elongation and termination. These major steps are conserved

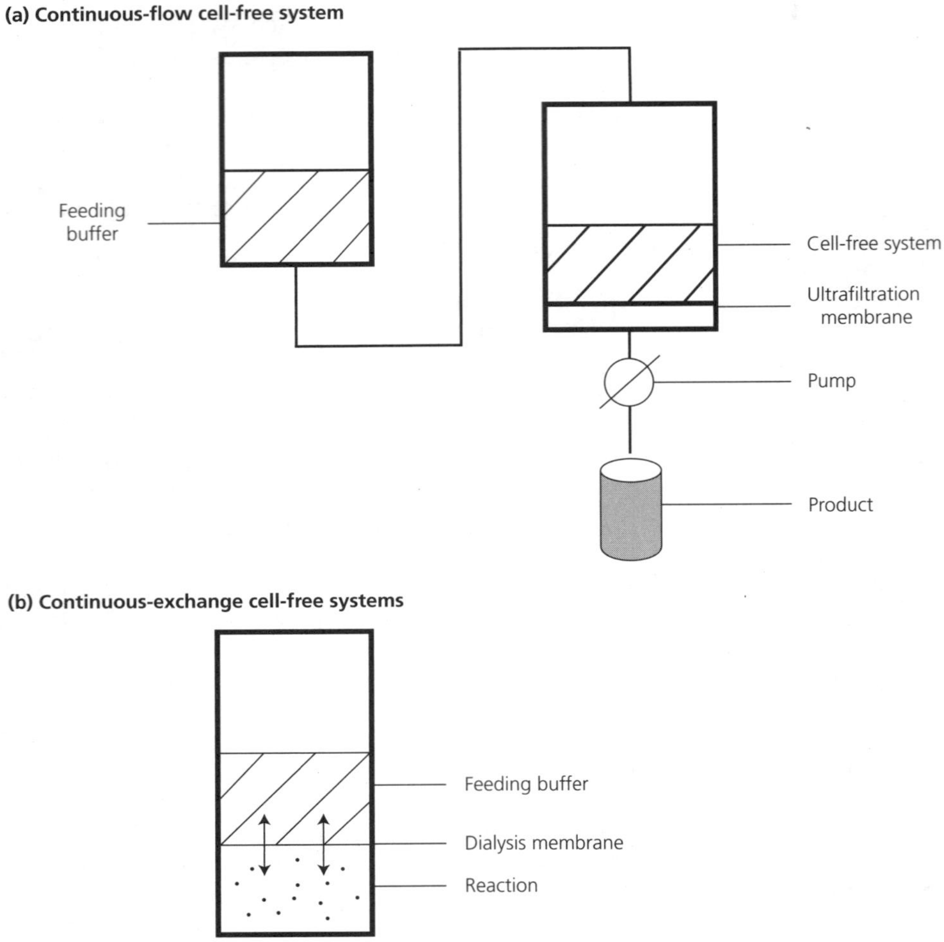

Figure 14.10 Schematic representation of (a) the continuous-flow cell-free system and (b) continuous-exchange cell-free system. The feeding buffer includes amino acids, NTP, energy components, co-factors and other accessory components, while the reaction component includes ribosomes, translation factors, tRNA, aminoacyl tRNA synthetases, template (RNA or DNA) and RNA polymerase (if required).

between prokaryotes and eukaryotes. The mechanism described here is of eukaryotes; however, the variations that occur in prokaryotes are appropriately addressed. Each of these three steps is catalysed in a cascade manner by a number of soluble protein factors that associate with and dissociate from ribosomes in a dynamic and orchestrated manner. They are called initiation, elongation and termination factors.

The following are certain prerequisites for translation of mRNAs: (a) ribosomes and their individual dissociated subunits are needed to be in equilibrium in the cytoplasm, (b) tRNAs need to be charged with the cognate amino acids by amino acid-specific tRNA synthetases and (c) all the necessary factors need to be functional and adequate in the cytoplasm.

14.8.1 Initiation

Initiation of protein synthesis refers to the formation of a complex of the fully formed ribosome with the mRNA and initiator tRNA on it such that the anticodon of the initiator (methionine)-tRNA is aligned with the initiation codon (AUG) of mRNA. This complex is formed through a series of steps, each of which is catalysed by a group of initiation factors. The entire process of initiation, which is a highly ordered phenomenon, is schematically shown in Fig. 14.11. For a detailed description, the readers are suggested to refer to a recent review (Jackson *et al.* 2010). In a summarized form, the various steps involved in canonical initiation (cap-dependent) are as follows. In some eukaryotic mRNA and a number of viral mRNAs, initiation is cap-independent and internal ribosome entry site (IRES)-mediated. The details of this process are dealt under the regulation of translation later:

- Step 1: Prevention of the small subunit of ribosome to associate with the large subunit and preparation of the small subunit of ribosome for its binding with the initiator methionine-tRNA. The initiation factors that catalyse these events are eukaryotic initiation factor-1 (eIF-1), eIF-1A and eIF-3.
- Step 2: Interaction of the small subunit of ribosome with the preformed initiator methionine-$tRNA_i$ in the form of a ternary complex, eIF-2-GTP-Met-$tRNA_i$; the involved intiation factor is eIF-2. This results in the formation of a 43S pre-initiation complex (PIC).

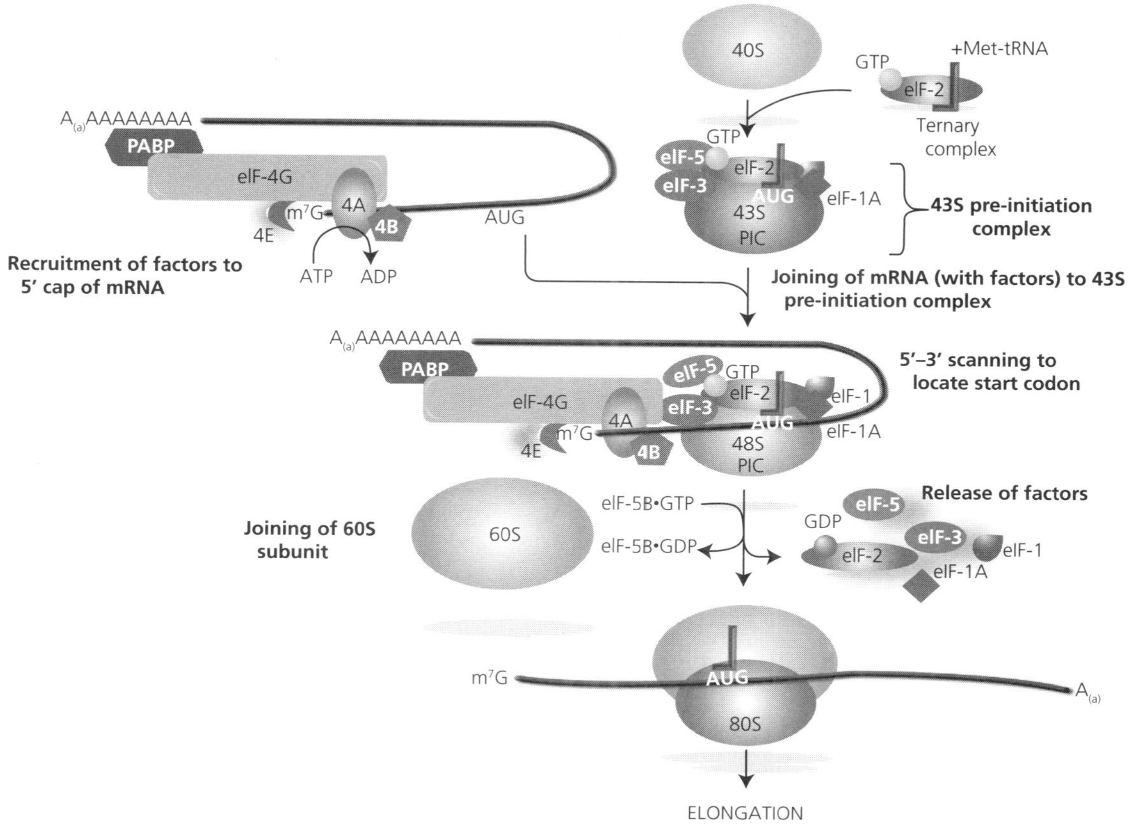

Figure 14.11 Summary of initiation of eukaryotic protein synthesis. There are four major steps: (1) Formation of the pre-initiation complex (PIC) containing 40S ribosome, eIF2.GTP.Met-tRNA and other initiation factors (eIF1A, -3 and -5). (2) Preparation of mRNA with several initiation factors, including (eIF-4E) (cap-binding protein), -4A, -4B, -4G and possibly PABP [poly(A) binding protein], which links 5′- and 3′-ends of mRNA. (3) Interaction of PIC with such mRNA, followed by scanning for AUG by the 40S ribosome. (4) Joining of the 60S ribosome to form the 80S initiation complex along with eIF5. See Colour Plate 11 on page 509.

- Step 3: PIC interacts with activated mRNA at or near the cap. These processes, that is, activation of mRNA and binding of 43S PIC to it, are catalysed by a group of initiation factors called eIF-4 group of polypeptides. The various protein factors that are part of the group are eIF-4E (cap-binding protein), eIF-4A, eIF-4B (RNA helicases) and eIF-4G (the adapter protein that links the 5′ end and 3′ end of mRNA through eIF-4E and poly A binding protein, PABP, respectively). This results in the circularization of mRNA. eIF-4E, -4A and -4G are also collectively referred as eIF-4F.
- Step 4: The 43S PIC scans through the 5′ untranslated region (UTR) of the mRNA for the initiation codon, AUG in Kozak context/consensus (RccAUGG: R is a purine and it is located three bases upstream of AUG, which is flanked by a G; Kozak 1984). This has originally been hypothesized by Marylene Kozak, and it is referred as Kozak's scanning hypothesis. Upon reaching the initiation codon and formation of 48S PIC, the scanning complex gets converted to a closed conformation and leads to the release of eIF-1 and simultaneous eIF-5-mediated hydrolysis of GTP in the eIF-2. GTP complex.
- Step 5: The 60S subunit of ribosome now joins the 48S PIC to form a 80S initiation complex, and as a consequence, eIF-2-GDP binary complex and other bound initiation factors, eIF-1, eIF-3, eIF-4B, eIF-4F and eIF-5, are displaced. The release of these factors is mediated by eIF-5B. The released initiation factors recycle as such or get converted to active form prior to recycle, as is the case for eIF-2-eIF-2.GDP complex gets converted to an active eIF-2-GTP complex by the action of a guanine nucleotide exchange factor called eIF-2B, and eIF-2-GTP can now interact with methionine initiator tRNA and thus eIF-2 recycles.
- Step 6: The 80S initiation complex gains competence to initiate elongation by the following further events: GTP hydrolysis by eIF-5B; release of eIF-1A and eIF-5B-GDP.

In the case of prokaryotes, the process of initiation is almost the same. A few deviations are: (a) As most of the prokaryotic mRNAs are without a cap structure,

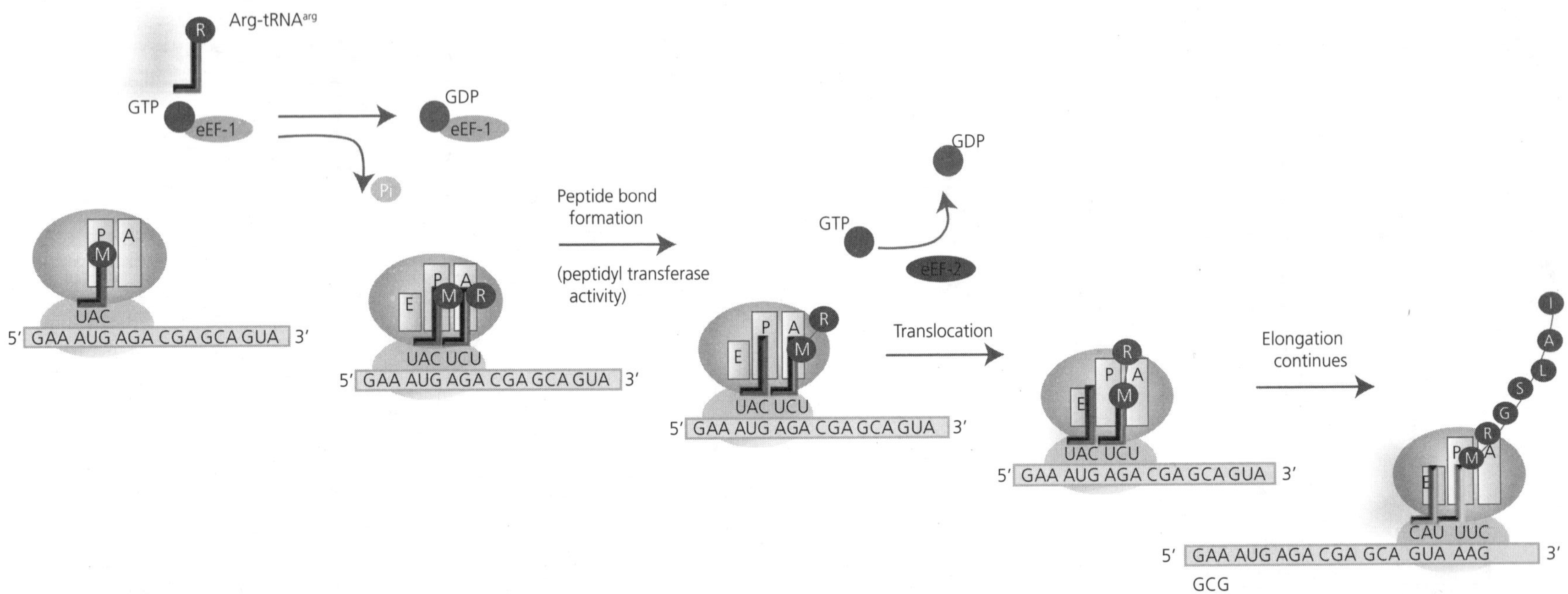

Figure 14.12 Elongation during translation. First, the initiator tRNA binds at the P site in the ribosome. In the second step, the next aminoacyl tRNA binds at the A site, followed by peptide bond formation. Translocation of the ribosomes shifts the vacant tRNA to the E site and the aminoacyl tRNA to the P site. Then, the next aminoacyl-tRNA comes at the A site and the process is repeated till the termination codon is reached. See Colour Plate 12 on page 510.

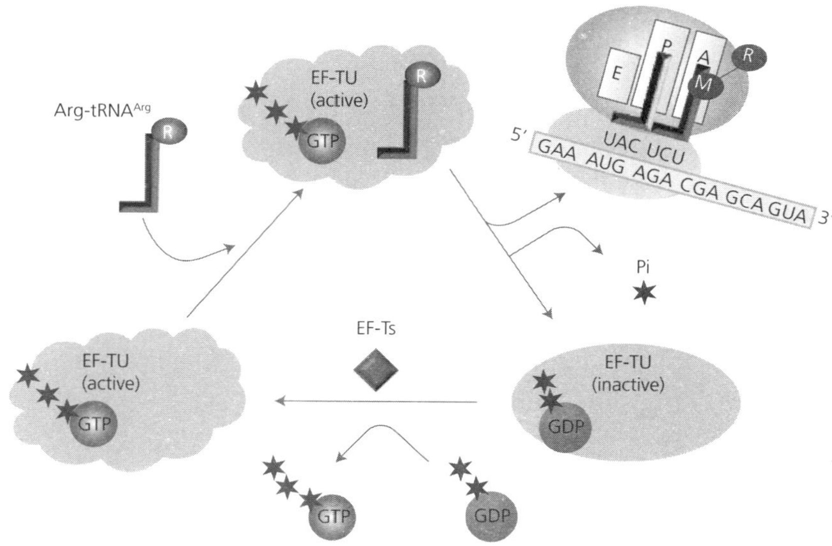

Figure 14.13 Recycling of EF-Tu/GTP during elongation of translation. During elongation, EF-Tu-GTP binds to the aminoacyl-tRNA. On binding of the correct aminoacyl-tRNA to the codon, EF-Tu-GTP gets hydrolysed and EF-Tu-GDP is released. The inactive EF-Tu-GDP complex is converted to EF-Tu-GTP with the help of EF-Ts, making the EF-Tu-GTP ready for binding to the next aminoacyl-tRNA. See Colour Plate 13 on page 511.

ribosome (small subunit) binds to a ribosome-binding sequence called the Shine–Dalgarno sequence, which is located at –7 position in the UTR (Fig. 14.13). (b) The number of initiation factors that are involved in this process is less in prokaryotes.

14.8.2 Elongation

Elongation refers to the process through which step-wise addition of amino acids to the existing methionine (held onto the peptidyl-tRNA–binding site on the 80S ribosome) takes place, resulting in the elongation of the polypeptide (Fig. 14.12). This process is much simpler and involves aminoacyl-tRNAs and two major elongation factors called eEF-1 and eEF-2. In bacteria, these are called EF-TU, EF-Ts and EF-G, respectively. For further details of these elongation factors and their role in protein biosynthesis, the readers are referred to an excellent review by Anderson *et al.* (2003). There are three distinct steps in this process, which are as follows:

- Step 1: Aminoacyl-tRNA forms a ternary complex with eEF-1 and GTP and binds to the aminoacyl-tRNA binding site (A) on the ribosome whose P site is occupied by the initiator methionine-tRNA. This step is catalysed by hydrolysis of GTP bound to eEF-1, which is liberated into the cytosol as an inactive binary complex eEF-1-GDP. As eEF-1 is rate limiting, it needs to recycle for activating aminoacyl-tRNAs. This process has been well studied in prokaryotes and the details are as follows. EF-TU-GDP binds to EF-TS, which is a guanine nucleotide exchange factor, and due to exchange of GDP for GTP, it gets activated and interacts with an aminoacyl-tRNA that gets activated to interact with the ribosome (Fig. 14.13).
- Step 2: Peptide bond (between the two amino acids) formation necessitates transfer of the initiator methionine from the peptidyl-tRNA to the amino acid attached to the aminoacyl-tRNA bound to the A site. This is catalysed by an intrinsic activity of ribosome called the peptidyl transferase activity. As

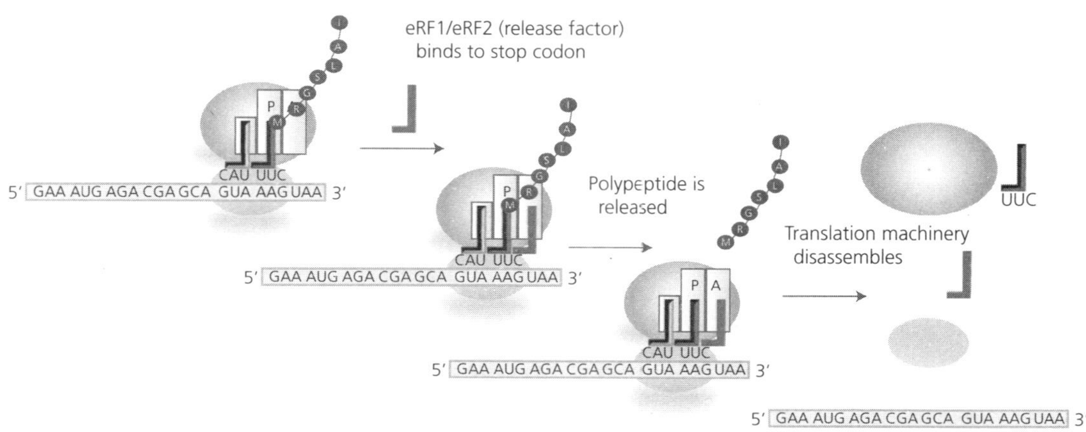

Figure 14.14 Termination of translation. On reaching the stop codon, the release factor binds to it (UAA, UGA and UAG). After the binding of the release factor, the translational machinery disassembles as the ribosome dissociates and the polypeptide and the tRNA are released. See Colour Plate 14 on page 511.

discussed before, this is rather a recent discovery that this activity is an autocatalytic activity of ribosomal RNAs.

- Step 3: After completion of step 2, P site contains an empty (discharged) tRNA and A site is occupied with the peptidyl-tRNA holding the dipeptide. To add the next amino acid to the peptide by an aminoacyl-tRNA, A site needs to be emptied to allow the incumbent aminoacyl-tRNA, and this is accomplished by a process referred as translocation. In this, the ribosome moves through the mRNA by a triplet codon, resulting in the transfer of the discharged tRNA to E site and peptidyl-tRNA with the elongated peptide to P site, thus making an empty A site available for the next aminoacyl-tRNA. This process is catalysed by the elongation factor eEF-2 (EF-G in bacteria), and it requires GTP. Following GTP hydrolysis, eEF-2 is released from the ribosome. This process continues until the ribosome reaches the termination codon, thereby completing the full peptide.

14.8.3 Termination

Elongation of the peptide by the ribosome as per the genetic codes in the mRNA continues until the ribosome encounters the stop codon (UGA, UAA or UAG). Upon reaching the stop codon, due to lack of any tRNA to recognize the stop codon, the ribosome pauses and with the help of release factor called eRF-1/eRF-2, the translation terminates. These release factors require GTP for binding to the ribosome. They are released after GTP hydrolysis. As a consequence, the fully formed polypeptide is released from the last tRNA, the tRNA is ejected from the ribosome and the ribosome dissociates from the mRNA (Fig. 14.14). The molecular details regarding dissociation of ribosomes from the mRNA are still unclear. However, in prokaryotes, it is postulated that this process is catalysed by another protein called the ribosome recycling factor (RRF). The RRF possesses a tRNA-like structure and by entering P or A site, it catalyses the dissociation of the ribosome into its subunits. However, no RRF-like protein has yet been identified in eukaryotes.

In prokaryotes, there are two release factors, RF-1 and RF-2. RF-1 recognizes UAA and UAG, while RF-2 recognizes UGA and UAA. There is another release factor, RF-3, which does not recognize any specific termination codon and seems to act along with RF-1 and RF-2.

14.9 Summary

The discovery of DNA structure by James Watson and Francis Crick in 1953 and the subsequent proposition of the central dogma of molecular biology in 1957 by Francis Crick led the foundation of search to unravel the mechanism of gene regulation. Over a period of two decades (1950–1970), many scientists through their extremely innovative and dedicated search established the molecular details of the central dogma, thereby establishing the concept of molecular biology. In this chapter, we have emphasized on the exciting discoveries that resulted in building our concept on the cytoplasmic contribution of gene regulation, that is, translation of mRNA to protein in eukaryotes.

In the first half of the chapter, we described the exciting pioneering experiments of many biologists to discover each of the major components of the protein-synthesizing machinery, namely messenger RNA, ribosomes and tRNA. The pioneering work of the following biologists led to the discovery of mRNA: Elliot 'Ken' Volkin, Lazarus Astrachan, Francis Crick, Sydney Brenner, François Jacob and Matthew Meselson. Credit for the discovery of ribosome structure and its involvement in protein synthesis goes to George Palade and Richard Roberts. However, for a comprehensive understanding on the structure–function relationship of ribosomes, the recent outstanding contributions of Ada Yonath, Thomas Steitz and Venkat Ramakrishnan have been recognized by awarding them with Nobel Prize in 2009. Further, these advances in our understanding of ribosome structure lend credential to design and develop newer generation of antibiotics/drugs. Francis Crick's adaptor hypothesis culminated in the discovery of tRNAs as the translator of nucleotide codes into amino acids. The scientists who pioneered in this discovery were Paul C Zamecnik, Mahlon B Hoagland, Francis Crick and Uttam Raj Bhandary. Another horrendous task was cracking the genetic code. The original and pioneering contributions of Marshall Nirenberg, Phil Leder and Har Gobind Khorana are phenomenal.

In the second half, we discussed the assembly of the above-mentioned major components and a large number of cytoplasmic protein factors for an orchestrated act of translation. We begin with the contributions of two approaches, namely antibiotics in perturbation of protein synthesis and *in vitro* protein synthesis systems, in our understanding the importance of various factors and the process of protein synthesis per se. We then summarized the different steps of protein synthesis, namely initiation, elongation and termination, and the detailed molecular mechanisms involved thereof.

References

Anderson GR, Nissen P, Nyborg J. Elongation factors in protein biosynthesis. *Trends Biochem Sci* 2003;28:434–441.

Avery OT, MacLeod CM, McCarty M. Studies on the chemical nature of the substance inducing transformation of pneumococcal types: Induction of transformation by a desoxyribonucleic acid fraction isolated from *Pneumococcus* type III. *J Exp Med* 1944;79:137–158.

Baltimore D. Viral RNA-dependent DNA polymerase. *Nature* 1970; 226:1209–1211.

Ban N, Freeborn B, Nissen P, Penczek P, Grassucci RA, Sweet R, *et al.* A 9Å resolution X-ray crystallographic map of the large ribosomal subunit. *Cell* 1998;93:1105–1115.

Ban N, Nissen P, Hansen J, Moore PB, Steitz TA. The complete atomic structure of the large ribosomal subunit at 2.4 Å resolution. *Science* 2000;289:905–920.

Baranov VI, Spirin AS. Gene expression in cell-free systems on preparative scale. *Methods Enzymol* 1993;217:123–142.

Berg P, Ofengand EJ. An enzymatic mechanism for linking amino acids to RNA. *Proc Natl Acad Sci USA* 1958;44:78–86.

Blanchard SC, Cooperman BS, Wilson DN. Probing translation with small-molecule inhibitors. *Chem Biol* 2010;17:633–645.

Brenner S, Jacob F, Meselson M. An unstable intermediate carrying information from genes to ribosomes for protein synthesis. *Nature* 1961;190:576–581.

Bretscher MS, Marcker KA. Peptidyl-sRibonucleic acid and amino-acyl-sRibonucleic acid binding sites on ribosomes. *Nature* 1966;211:382–383.

Cate JH, Yusupov MM, Yusupova GZh, Earnest TN, Noller NF. X-ray crystal structures of 70S ribosome functional complexes. *Science* 1999;285:2095–2104.

Cech TR. The ribosome is a ribozyme. *Science* 2000;289: 878–879.

Coutsogeorgopoulos C. On the mechanism of action of chloramphenicol in protein synthesis. *Biochim Biophys Acta* 1966;129:214–217.

Craig D, Howell MT, Gibbs CL, Hunt T, Jackson RJ. Plasmid cDNA-directed synthesis in a coupled eukaryotic *in vitro* transcription-translation system. *Nucleic Acids Res* 1992;20:4987–4995.

Crick FHC. On degenerated templates and adaptor hypothesis. Unpublished observations. 1955. Available at: http://genome.wellcome.ac.uk/assets/wtx030893.pdf.

Crick FHC. On protein synthesis. *Symp Soc Exp Biol* 1958;12: 138–163.

Crick FHC. Codon-anticodon pairing: The wobble hypothesis. *J Mol Biol* 1966;19: 548–555.

Crick FHC. The central dogma of molecular biology. *Nature* 1970;227:561–563.

Crick FHC. *What a Mad Pursuit: A Personal View of Scientific Discovery.* New York: Basic Books; 1988. p. 122, 128, 133.

Crick FHC, Watson JD (1957) Virus structure: General principles. In: Wolstenholme GEW, Millar ECP, eds. *Ciba Foundation Symposium on the Nature of Viruses.* London: J & A Churchill; 1957. pp. 5–13.

Das HK, Goldstein A, Kanner LC. Inhibition by chloramphenicol growth of nascent protein chains in *Escherichia coli. Mol Pharmacol* 1966;2:158–170.

Das HK, Goldstein A, Lowney LI. Attachment of ribosomes to nascent messenger RNA in *Escherichia coli. J Mol Biol.* 1967;24: 231–245.

DeVries JK, Zubay G. DNA-directed peptide synthesis. II. The synthesis of the a-fragment of the enzyme α-galactosidase. *Proc Natl Acad Sci USA* 1967;57:1010–1012.

Gale EF, Folkes JP. The assimilation of amino-acids by bacteria. 15. Actions of antibiotics on nucleic acid and protein synthesis in *Staphylococcus aureus. Biochem J* 1953;53:493–498.

Goldstein A, Kirschbaum JB, Roman A. Direction of synthesis of messenger RNA in cells of *Escherichia coli. Proc Natl Acad Sci USA* 1965;54:1669–1675.

Guo L, Wan Y, Wang X, Wang PG, Zhao W. Development of aminoglycoside antibiotics by carbohydrate chemistry. *Med Chem* 2012 [Epub ahead of print].

Hansen JL, Moore PB, Steitz TA. Structures of five antibiotics bound at the peptidyl transferase center of the large ribosomal subunit. *J Mol Biol* 2003;330:1061–1075.

Hansen JL, Schmeing T, Moore PB, Steitz TA. Structural insights into peptide bond formation. *Proc Natl Acad Sci USA* 2002;99:11670–11675.

Hinegardner RT, Engelberg J. Rationale for a universal genetic code. *Science* 1963;142:1083–1085.

Hoagland MB, Keller EB, Zamecnik PC. Enzymatic carboxyl activation of amino acids. *J Biol Chem* 1956;218:345–358.

Hoagland MB, Stephenson ML, Scott JF, Hecht LI, Zamecnik PC. A soluble ribonucleic acid intermediate in protein synthesis. *J Biol Chem* 1958;231:241–257.

Holley RW, Apgar J, Everett GA, Madison JT, Marquisee M, Merrill SH, *et al.* Structure of a ribonucleic acid. *Science* 1965;147: 1462–1465.

Hunt T, Jackson RJ. *Modern Trends in Human Leukuemia.* In: Neth R, Gallo RC, Spiegelrnan S, Stohlman F, eds. Munich: JF Lehrnanns Verlag; 1974. pp. 300–307.

Jackson RJ, Hellen CUT, Pestova TV. The mechanism of eukaryotic translation initiation and principles of its regulation. *Nature Rev Mol Cell Biol* 2010;10:113–127.

Katzen F, Chang G, Kudlicki W. The past, present and future of cell-free protein synthesis. *Trends Biotech* 2005;23:150–156.

Khaitovich P, Mankin AS, Green R, Lancaster L, Noller HF. Characterization of functionally active subribosomal particles from *Thermus aquaticus. Proc Natl Acad Sci USA* 1999;96: 85–90.

Khorana HG. Polynucleotide synthesis and the genetic code. *Fed Proc* 1965;24:1473–1487.

Khorana HG. Nucleic acid synthesis in the study of the genetic code. In: *Nobel Lectures, Physiology and Medicine 1963-1970.* Amsterdam: Elsevier Publishing Company; 1972.

Kigawa T, Yabuki T, Yoshida Y, Tsutsui M, Ito Y, Shibata T, *et al.* Cell-free production and stable-isotope labelling of milligram quantities of proteins. *FEBS Lett* 1999;442:15–19.

Kozak M. Point mutations close to the AUG initiator codon affect the efficiency of translation of rat preproinsulin *in vivo. Nature* 1984;308:241–246.

Lamborg M, Zamecnik PC. Amino acid incorporation into protein by extracts of *E. coli. Biochim Biophys Acta* 1960;42:206–211.

Lipmann F, Hülsmann WC, Hartmann G, Boman HG, Acs G. Amino acid activation and protein synthesis. *J Cell Comp Physiol* 1959;54:75–88.

Littlefield JW, Keller EB, Gross J, Zamecnik PC. Studies on cytoplasmic ribonucleoprotein particles from the liver of the rat. *J Biol Chem* 1955;217:111–123.

Madin K, Sawasaki T, Ogasawara T, Endo Y. A highly efficient and robust cell-free protein synthesis system prepared from wheat embryos: Plants apparently contain a suicide system directed at ribosomes. *Proc Natl Acad Sci USA* 2000;97: 559–564.

Marshall RE, Caskey CT, Nirenberg MW. Fine structure of RNA codewords recognized by bacterial, amphibian, and mammalian transfer RNA. *Science* 1967;155:820–826.

McQuillen K, Roberts RB, Britten RJ. Synthesis of nascent protein by ribosomes in *Escherichia coli. Proc Natl Acad Sci USA* 1959;45:1437–1447.

Nakano H, Tanaka T, Kawarasaki Y, Yamane T. Highly productive cell-free protein synthesis system using condensed wheat-germ extract. *J Biotech* 1996;46:275–282.

Nierhaus D, Nierhaus KH. Identification of the chloramphenicol-binding protein in *Escherichia coli* ribosomes by partial reconstitution. *Proc Natl Acad Sci USA* 1973;70:2224–2228.

Nirenberg MW. The genetic code. In: *Nobel Lectures, Physiology or Medicine 1963-1970*. Amsterdam: Elsevier Publishing Company; 1972.

Nirenberg MW, Jones OW, Leder P, Clark BFC, Sly WS, Pestka S. On coding of genetic information. *Cold Spring Harbor Symp Quant Biol* 1963;28:549–557.

Nirenberg MW, Leder P. RNA codewords and protein synthesis. The effect of trinucleotides upon the binding of sRNA to ribosomes. *Science* 1964;145:1399–1407.

Nirenberg MW, Matthaei JH. The dependence of cell-free protein synthesis in *E. coli* upon naturally occurring or synthetic polyribonucleotides. *Proc Natl Acad Sci USA* 1961;47: 1588–1602.

Nissen P, Hansen J, Ban N, Moore PB, Steitz TA. The structural basis of ribosome activity in peptide bond synthesis. *Science* 2000;289:920–930.

Nissen P, Ippolito JA, Ban N, Moore PB, Steitz TA.RNA tertiary interactions in the large ribosomal subunit: The A-minor motif. *Proc Natl Acad Sci USA* 2001;98:4899–4903.

Noller HF, Hoffarth V, Zimniak L. Unusual resistance of peptidyl transferase to protein extraction procedures. *Science* 1992;256:1416–1419.

Novina CD, Sharp PA. The RNAi revolution. *Nature* 2004;430:161–164.

Schatz A. Streptomycin, an antibiotic produced by *Actinomyces griseus*. PhD dissertation. New Brunswick, NJ: Rutgers University; 1945.

Pal JK, Ghaskadbi SS. *Fundamentals of Molecular Biology*. New Delhi: Oxford University Press; 2009. pp.1–374.

Palade GE. A small particulate component of the cytoplasm. *J Biophys Biochem Cytol* 1955;1: 59–68.

Pardee AB, Jacob F, Monod J. [The role of the inducible alleles and the constitutive alleles in the synthesis of beta-galactosidase in zygotes of *Escherichia coli*]. *CR Hebd Seances Acad Sci* 1958;246:3125–3128.

Pelham HRB, Jackson RJ. An efficient mRNA-dependent translation system from reticulocyte lysates. *Eur J Biochem* 1976;67:247–256.

Pringle P. *Experiment Eleven: The Dark Secrets Behind the Discovery of a Wonder Drug*. New York: Walker & Co.; 2012.

Raj Bhandary UL, Köhrer C. Early days of tRNA research: Discovery, function, purification and sequence analysis. *J Biosci* 2006;31:439–451.

Rheinberger H-J, Sternbach H, Nierhaus KH. Three tRNA binding sites on *Escherichia coli* ribosomes. *Proc Nat Acad Sci USA* 1981;78:5310–5314.

Sawasaki T, Ogasawara T, Morishita R, Endo Y. A cell-free protein synthesis system for high throughput proteomics. *Proc Natl Acad Sci USA* 2002;99:14652–14657.

Schatz A, Bugie E, Waksman SA. Streptomycin, a substance exhibiting antibiotic activity against gram-positive and gram-negative bacteria. *Proc Soc Exptl Biol Med* 1944; 55:66–69.

Schatz A, Waksman SA. Effect of streptomycin upon *Mycobacterium tuberculosis* and related organisms. *Proc Soc Exptl Biol Med* 1944;57:244–248.

Schatz A, Waksman SA. Strain variation and production of antibiotic substances. IV. Variations among actinomycetes with special reference to *Actinomyces griseus*. *Proc Natl Acad Sci USA* 1945;31:129–137.

Schmeing TM, Seila AC, Hansen JL, Freeborn B, Soukup JK, Scaringe SA, *et al.* A pretranslocational intermediate in protein synthesis observed in crystals of enzymatically active 50S subunits. *Nat Struct Biol* 2002;9:225–230.

Söll D, Raj Bhandary UL. Studies on polynucleotides. LXXVI. Specificity of transfer RNA for codon recognition as studied by amino acid incorporation. *J Mol Biol* 1967;29:113–124.

Speyer JF, Lengyel P, Basilio C, Wahba AJ, Gardner RS, Ochoa S. Synthetic polynucleotides and the amino acid code. *Cold Spring Harbor Symp Quant Biol* 1963;28:559–567.

Spirin AS. Cell-free protein synthesis bioreactor. In: Todd P, Sikdar SK, Beer M, eds. *Frontiers in Bioprocessing II*. Washington, DC: American Chemical Society; 1991. pp. 31–43.

Spirin AS, ed. *Cell-Free Translation Systems*. Berlin; Springer Verlag; 2002.

Spirin AS, Baranov VI, Ryabova LA, Ovodov SY, Alakhov YB. A continuous cell-free translation system capable of producing polypeptides in high yield. *Science* 1988;242:1162–1164.

Spirin AS, Swartz JR. Cell-free protein synthesis: Historical landmarks, classification and general methods. In: Spirin AS, Swartz JR, eds. *Cell-free Protein Synthesis: Methods and Protocols*. Weinheim, Germany: Wiley-VCH; 2008. pp. 1–34.

Steitz TA, Moore PB. RNA, the first macromolecular catalyst: The ribosome is a ribozyme. *Trends Biochem Sci* 2003;28:411–418.

Temin HM, Mizutani S. Viral RNA-dependent DNA polymerase. *Nature* 1970;226:1211–1213.

Tissieres A, Schlessinger AD, Gros F. Amino acid incorporation into proteins by *Escherichia coli* ribosomes. *Proc Natl Acad Sci USA* 1960;46:1450–1463.

Traut RR, Monro RE. The puromycin reaction and its relation to protein synthesis. *J Mol Biol* 1964;10:63–72.

Volkin E, Astrachan L. Phosphorus incorporation in *Escherichia coli* ribo-nucleic acid after infection with bacteriophage T2. *Virology* 1956;2:149–161.

Watson JD, Crick FHC. A structure for deoxyribose nucleic acid. *Nature* 1953;171:737–738.

Wimberly BT, Brodersen DE, Clemons WM Jr, Morgan-Warren RJ, Carter AP, Vonrhein C, *et al.* Structure of the 30S ribosomal subunit. *Nature* 2000;407:327–339.

Wisseman CL Jr, Smadel JE, Hahn FE, Hopps HE. Mode of action of chloramphenicol. I. Action of chloramphenicol on assimilation of ammonia and on synthesis of proteins and nucleic acids in *Escherichia coli*. *J Bacteriol* 1954;67:662–673.

Wollman EI, Jacob F, Hayes W. Conjugation and genetic recombination in *Escherichia coli* K-12. *Cold Spring Harb Symp Quant Biol* 1956;21:141–162.

Zamecnik PC. From protein synthesis to genetic insertion. *Annu Rev Biochem* 2005;74:1–28.

Zamecnik PC, Keller EB. Relation between phosphate energy donors and incorporation of labelled amino acids into proteins. *J Biol Chem* 1954;209:337–354.

Zubay G. *In vitro* synthesis of protein in microbial systems. *Annu Rev Genet* 1973;7:267–287.

15

Translation regulation of gene expression and human diseases

Jayanta Kumar Pal and Sangeeta Chatterjee

15.1 Definition and importance of regulation of protein synthesis

Cell structure and function are determined by expression of genes in a spatio-temporal manner. This phenomenon at the organism level is particularly evident during embryonic development. Gene expression is controlled at multiple levels, namely transcription, post-transcription (mRNA processing, transport, localization, half-life) and translation levels. Among these, expression at the level of transcription and translation is widely studied. From various studies, it appears that gene expression is regulated both at the level of transcription and at the level of translation. However, some genes are regulated preferentially at the level of transcription, while some others are regulated predominantly at the level of translation. Indeed, often it is observed that under normal physiological condition vis-à-vis certain specific circumstances, although the transcript level for a gene remains unchanged, the protein level changes significantly. Such a phenomenon where gene expression is regulated at the level of translation is referred to as translation regulation of gene expression.

As far as the importance of translation regulation is concerned, the following advantages are envisaged: (a) In eukaryotes, in particular, the entire process of gene expression is rather lengthy and time-consuming. Therefore, under conditions such as stress, where quick response is warranted, the provision of translation regulation is explored by cells. (b) For spatial regulation, as mRNAs are differentially localized in the cytoplasm, their translation only can provide the spatial pattern of proteins often required by the cells. This is particularly important during embryonic development and in neurons for localized protein production. (c) Specific cell types such as mammalian reticulocytes lack nuclei, and therefore, the only option left for them is differential regulation of translation of mRNAs. For further details, the readers are referred to reviews by Mathews *et al.* (2007) and Lackner and Bahler (2008). The various physiological processes and conditions that employ this strategy are embryonic development (Thompson *et al.* 2007) and various stress conditions, including toxicity, apoptosis (Morley and Coldwell 2007), nutrient starvation and cancer (Schneider and Sonenberg 2007). The importance of this phenomenon is further evident in a recent estimate indicating that in yeast, over 70% of the genes are regulated at the level of mRNA through mRNA-directed mechanisms (Lu *et al.* 2007).

Translation regulation can be a global approach when a cell has to modulate the global translation activity, that is, translation of almost the entire repertoire of cellular mRNA. For example, during environmental stress, global translation activity of a cell is substantially decreased until the cell recovers from stress. Alternatively, translation regulation can be mRNA-specific (Gebauer and Hentze 2004). This approach is prevalent during early embryonic development when differential protein synthesis from the stored maternal mRNAs is quickly carried out to link it with rapid cleavage, leading to extensive cell proliferation. In the past 10 years, the molecular mystery behind these two types of translation regulation has been unravelled substantially and a brief account of it is given next.

15.2 Global regulation by translation initiation factors

It has been well established that translation regulation takes place predominantly at the initiation step of protein synthesis. As discussed in Chapter 14, the initiation step is very elaborate, consisting of multiple substeps, which are catalysed by over 20 different protein factors called the initiation factors. Due to these inherent multiple substeps, the initiation process offers ample opportunities for regulation. As discussed next, the essence of this regulation is primarily through modulation of activities of the initiation factors, leading to activation/suppression of protein synthesis.

15.2.1 Modulation of eIF-2, influencing the formation of eIF-2-GTP-Met tRNA ternary complex

As described in Chapter 14, in the first step of initiation, binding of the initiator Met $tRNA_i$ with the small subunit of ribosome is catalysed by eIF-2. eIF-2 along with GTP binds to Met $tRNA_i$ to form a ternary complex (eIF-2-GTP-Met $tRNA_i$), which in its turn binds to the 40S ribosomal subunit. Towards the end of initiation, eIF-2-bound GTP gets hydrolysed and eIF-2 is ejected out of the initiation complex in the form of eIF-2-GDP, an inactive form. Cells need to recycle the eIF-2 by converting the inactive eIF-2-GDP form to the functional eIF-2-GTP form. This reversal or exchange reaction is carried out by another initiation factor, eIF-2B (previously called as the reversing factor, RF), also called the guanine nucleotide exchange factor (Fig. 15.1). This conversion process is perturbed when the α-subunit of the trimeric eIF-2 gets phosphorylated. Phosphorylated eIF-2α-GDP cannot be converted to its functional form by eIF-2B, and further eIF-2B gets sequestered while it forms a complex with eIF-2αP-GDP in the reversal reaction. This results in perturbation of recycling of eIF-2B, which is a rate-limiting enzyme, thus causing inhibition of initiation right at its very first step (Fig. 15.1).

Phosphorylation of eIF-2α is caused by a family of kinases called the eIF-2α kinases. In mammals, there are four such kinases, namely the haem-regulated eIF-2α kinase (also called the haem-regulated inhibitor, HRI), double-stranded RNA-dependent eIF-2α kinase (PKR), PKR-like endoplasmic reticulum (ER) stress kinase (PERK) and GCN2-like eIF-2α kinase. Although these kinases phosphorylate the α-subunit of eIF2 at Ser51 site, they are regulated differently (Fig. 15.2). For a detailed description, including the initial discoveries, the readers may refer to a few excellent reviews (London *et al.* 1987; Chen and London 1995; Pal *et al.* 1996, 1999; Pal and Ghaskadbi 2009). In summary, these kinases are expressed as well as activated during a variety of stress conditions such as haem-deficiency (in erythroid cells), viral infections, ER stress and amino acid starvation and thus act as the first line of global regulator of protein synthesis (Dever *et al.* 1992; Gebauer and Hentze 2004; Holcik and Sonenberg 2005; Ron and Harding 2007).

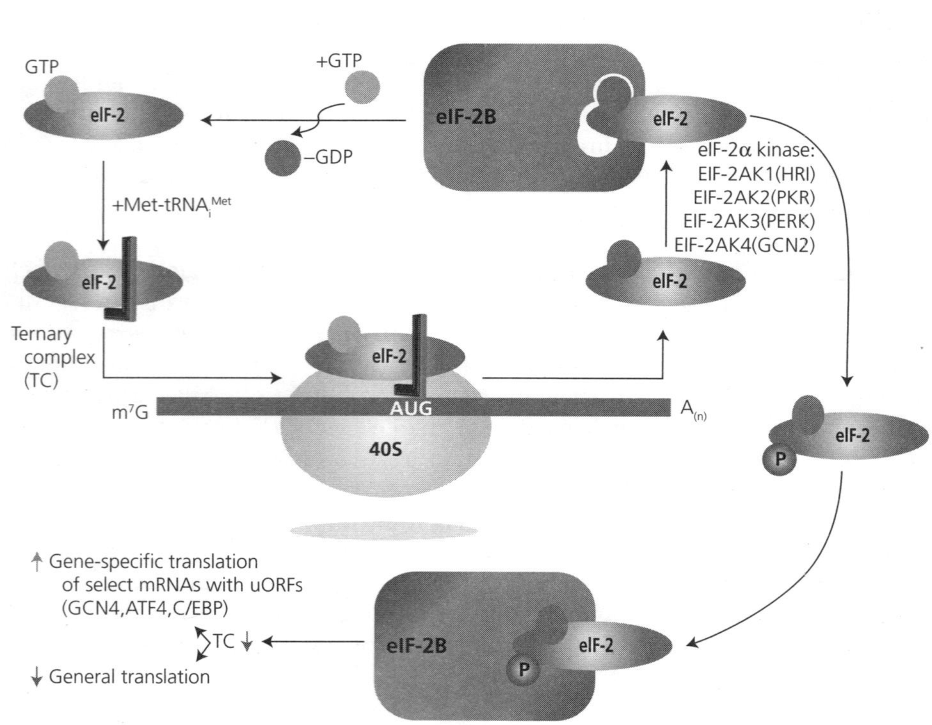

Figure 15.1 Schematic diagram showing the role of eIF-2α kinases in modulating initiation of protein synthesis through phosphorylation of eIF-2α. eIF-2 is rate limiting in a cell and thus is required to be recycled for continuation of initiation of protein synthesis. eIF-2 is liberated after one round of initiation as a non-functional eIF2.GDP binary complex. This is acted upon by eIF-2B, a guanine nucleotide exchange factor which converts eIF-2-GDP into a functional complex eIF-2-GTP. eIF-2-GTP binds to Met $tRNA_i$ and forms a ternary complex eIF-2-GTP-Met $tRNA_i$, which is utilized for initiation. If a member of the family of eIF-2α kinases (HRI, PKR, PERK and GCN2) gets activated, it phosphorylates eIF-2α. Phosphorylated eIF-2α cannot be recycled by eIF-2B, resulting in the inhibition of initiation of protein synthesis. However, in contrast to this, mRNAs of some specific genes such as *GCN4*, *ATF4* and *C/EBP* with uORFs are favourably translated under this condition. See Colour Plate 15 on page 512.

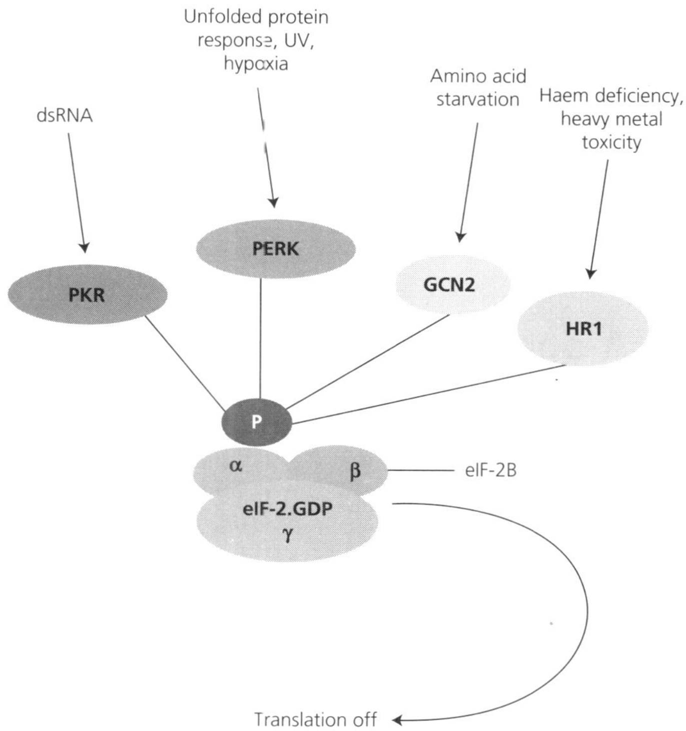

Figure 15.2 Differential regulation of eIF-2α kinases in mammals. Although each of these eIF-2α kinases phosphorylates the same Ser 51 residue of the substrate eIF-2α, each gets activated differently. The major modes of activation are indicated on the top of each kinase. See Colour Plate 16 on page 513.

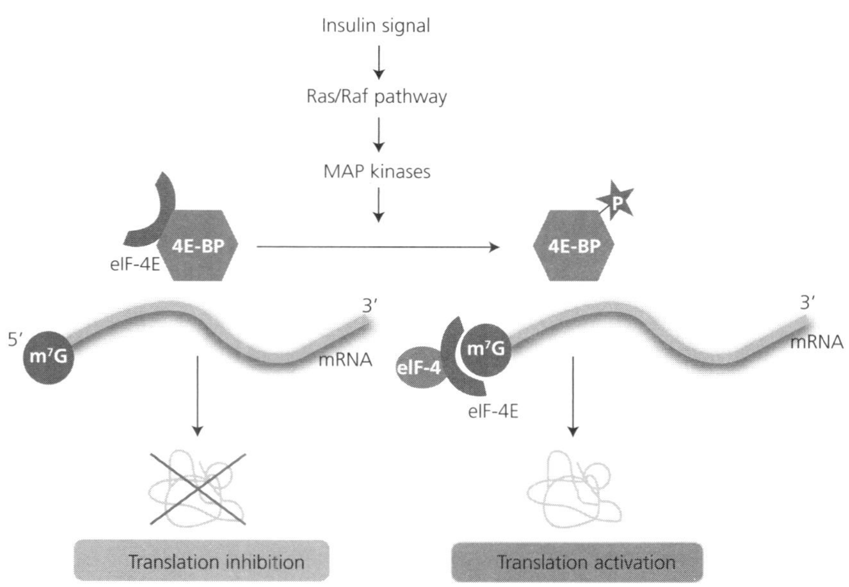

Figure 15.3 Schematic diagram showing stimulation of translation initiation (exposed to insulin) with the availability of increased quantity of eIF-4E. The cap-binding protein – eIF-4E – is crucial for cap-dependent initiation of translation. eIF-4E-BPs binds to eIF-4E and inhibits its cap-binding activity. During insulin signalling, MAP kinases phosphorylate eIF-4E-BPs. The phosphorylated eIF-4E-BPs fall off eIF-4E and thus more free eIF-4E is available for translation. See Colour Plate 17 on page 513.

15.2.2 Modulation of eIF-4E, influencing cap-dependent ribosome binding to mRNA

The role of eIF-4E, the cap-binding protein, in cap-dependent translation of mRNAs, in particular, is well established. eIF-4F complex, which includes eIF-4E, -4G,-4A and -4B, regulates the interaction of mRNA with the ribosomal complex. Binding of eIF-4E to mRNA cap structure and eIF-4G is regulated by either of two protein factors called the eIF-4E-binding proteins, eIF-4E-BP1 and eIF-4E-BP2 (Fig. 15.3; Richter and Sonenberg 2005). These, when abundant in the cytoplasm, are hypophosphorylated and can bind to all available eIF-4E molecules and thus sequester them, resulting in perturbation of their interaction with eIF-4G and thus an mRNA–ribosomal complex interaction, leading to inhibition of initiation (Lackner and Bahler 2008). Interestingly, for a positive regulation of initiation or enhanced initiation such as during growth factor stimulation, these eIF-4E-BPs get hyperphosphorylated through the MAP kinase pathway, resulting in their dissociation from the eIF-4E, leading to enhanced binding of eIF-4E with eIF-4G and promotion of cap-dependent initiation (Richter and Sonenberg 2005). Indeed, 4E-BPs are known to regulate a number of vital processes such as development or cell growth and inhibition of tumour formation (Richter and Sonenberg 2005).

15.3 mRNA-specific regulation

The main target of this regulation is the mRNA itself. The elements in the mRNA involved in this regulation, that is, the *cis* regulatory elements, reside at the 5′- and 3′-untranslated regions (UTRs). Interaction of *trans*-acting cytoplasmic proteins [mRNA-binding proteins (messenger ribonucleoproteins; mRNPs)] with these *cis* elements brings about regulation of translation of the mRNA (Fig. 15.4). This approach is the primary basis of translation regulation of a selective set of mRNAs (e.g. maternal mRNAs) that are responsible for temporal and spatial regulation of gene expression often encountered during early embryogenesis, environmental stress, viral infection, etc. The details of these regulations with specific examples of mRNAs are described next. Deregulation of translation often leads to various pathological disorders in humans (Chatterjee and Pal 2009).

15.3.1 Approaches often employed during oogenesis and embryonic development

The oogenesis and embryonic development are unique and specialized processes which demand robust and highly dynamic regulation of gene expression. Interestingly, in many species, gene regulation during these two processes occurs predominantly at the level of translation. These processes use multiple approaches and modes of translation regulation, and some of the major ones are described next.

15.3.1.1 Regulation by modulating the length of poly(A) tail

Polyadenylation of mRNAs is a co-transcriptional modification of most of the eukaryotic mRNAs. A minimum length of the poly(A) tail of mRNAs is crucial for their stability as well as for translation. This property has been exploited by organisms particularly during oogenesis and early embryonic development. There are dynamic changes in the poly(A) tail of maternal mRNAs by the cytoplasmic poly(A) polymerases (PAPs), leading to their extensive translation regulation and stability (Read and Norbury 2002; Stevenson and Norbury 2006). A direct correlation of the length of poly(A) tail of mRNAs with their rate of translation was previously documented for a number of mRNAs for the proteins that are crucial for developmental regulation (Richter 1995). This phenomenon has been well studied in the case of maternal mRNAs in *Xenopus* oocytes. In immature oocytes, a large number of maternal mRNAs after their transport to the cytoplasm remains stable and translationally repressed with short poly(A) tails, while others with long poly(A) tails undergo translation into proteins (Fig. 15.5). However, during maturation of oocytes, those repressed mRNAs undergo further polyadenylation by cytoplasmic PAP and thereby get activated and translated (Richter 1995, 2007; Mendez and Richter 2001). Similarly, the active mRNAs with long poly(A) tails (in immature oocytes) may undergo deadenylation during oocyte maturation. Thus, it appears that there exists a dynamic change in the poly(A) tail executed by two cytoplasmic enzymes, PAP and deadenylase, resulting in promotion or suppression of translation

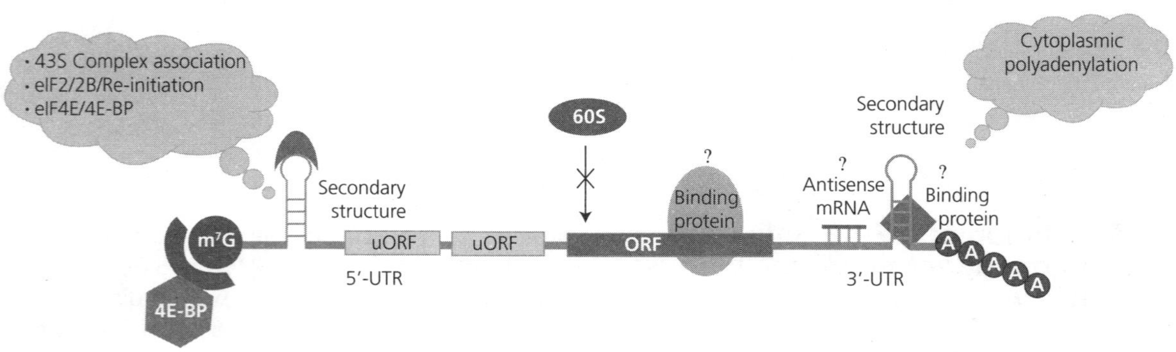

Figure 15.4 Summarized scheme of translation regulation in eukaryotic cells. Interaction of various *cis* elements present at the 5′- and 3′-UTR (untranslated regions) in the mRNA with specific *trans*-acting factors present in the cytoplasm determines the regulation of its translation. Some possible such *cis*-*trans* regulations that are still to be unravelled are indicated with notes of interrogation. See Colour Plate 18 on page 514.

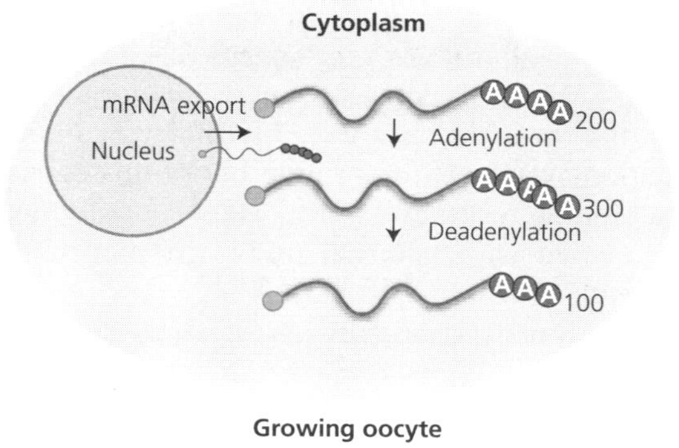

Figure 15.5 Translation regulation by changing the length of the poly(A) tail of mRNAs. In oocytes, the length of the poly(A) tail is altered by adenylation or deadenylation by cytoplasmic enzymes, resulting in the regulation of its translation. See Colour Plate 19 on page 514.

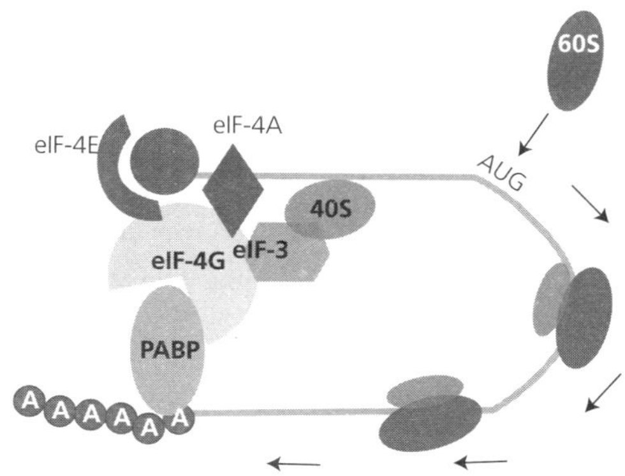

Figure 15.6 Translation regulation by eIF-4G and poly(A) binding protein (PABP) through circularization of mRNA. The circular form generated due to the interaction of eIF-4G (as a part of the cap-binding complex at the 5′-end of an mRNA) and PABP at the 3′-end is stabilized by proteins which are bound to the 5′- and 3′-end of the mRNA. The ribosome which dissociates from the 3′-end of mRNA can easily find the neighbouring ribosomal subunits for reassembling at the 5′-end. This further ensures translation of only full-length mRNAs. See Colour Plate 20 on page 515.

activity of maternal mRNAs. Similarly, maternal mRNAs of a large number of proteins, such as Toll and torso in *Drosophila* [reviewed in Curtis *et al.* (1995)], c-mos in frog and mouse (Sheets *et al.* 1994, 1995; Gebauer *et al.* 1994), etc., are subjected to translation regulation during oogenesis and early embryogenesis using this approach.

Large poly(A) tail often attributes to the positive translation activity through enhanced initiation, particularly of capped mRNAs. An mRNP called poly(A)-binding protein (PABP) that binds to poly(A) tail (Burd *et al.* 1991; Tarun and Sachs 1995; Imataka *et al.* 1998; Piron *et al.* 1998) has been demonstrated to integrate the 3′-end and the 5′-end of an mRNA through an adapter protein called the eIF-4G that is a part of the eIF-4F complex at the 5′-end of mRNA during the initiation step (Fig. 15.6). This results in an enhanced translation efficiency due to shunting of the ribosomes from the 3′-end (at the end of termination step) to the 5′-end for re-initiation. There are a few other proteins that assist in this process as well. Indeed, such circular mRNAs in the form of circular polysomes have been detected by electron microscopy (Christensen *et al.* 1987; Visa *et al.* 1996). This observation thus clearly establishes the previous indirect observations of Gallie and Tanguay (1994) that the purified initiation factors formed a complex with poly(A).

15.3.1.2 Regulation by interaction of *cis* elements in the 5′- and 3′-UTRs with *trans*-acting mRNPs

Eukaryotic mRNAs contain additional significant sequences other than the coding sequences that are actually translated into the cognate proteins. These sequences that reside in the mRNA flanking the coding sequence both in the 5′ and in the 3′ direction are not translated and are termed as the 5′-untranslated region (5′-UTR)- and 3′-untranslated region (3′-UTR), respectively. These UTRs regulate translation of the coding sequence in the mRNA. These contain mRNA-specific sequences called *cis* sequences/elements which most often form complex secondary structures [reviewed in Chatterjee and Pal (2009)]. These sequences are recognized by specific *trans*-acting factors that have RNA-binding motifs/domains (RBD) or RNA recognition motifs (RRM), and they are generally referred to as mRNA-binding proteins (mRNPs).

mRNPs are inseparably associated with eukaryotic mRNAs at various levels from the time they are synthesized and also immediately after their synthesis in the nucleus. They direct and precisely regulate various functions of the mRNAs, namely post-transcriptional/co-transcriptional modifications of mRNAs, including capping, polyadenylation and splicing; transport to the cytoplasm; localization; stability and translation into proteins (Dreyfuss *et al.* 2002; Hieronymus and Silver 2004). Here, both the stability and the translation of specific mRNAs by various mRNPs in the context of embryonic development are dealt with, as the other aspects are beyond the scope of the present discussion.

In eukaryotes, the fate of the mRNAs that are transported from the nucleus to the cytoplasm in the form of mRNPs is determined primarily by mRNA-associated proteins. The pool of mRNPs for a particular mRNA is often distributed into two different populations, namely ribosome-associated actively translating mRNPs and translationally inactive ribosome-free cytoplasmic mRNPs. The latter population can be either irreversibly repressed or reversibly repressed [reviewed in Scherrer (1980)]. These observations were primarily made in duck erythroblasts, the precursors of erythrocytes. Interestingly, the type of proteins that associate with these two populations of mRNAs were not the same; some were common and the others were specific to their function, that is, active or inactive. Indeed, it was demonstrated that such inactive mRNPs when deproteinized by phenol extraction (extracted mRNA) could be translated *in vitro* (Vincent *et al.* 1983).

The initial observations on the existence of such translationally inactive mRNPs in the embryos of sea urchin and fish were made during the late 1960s by Spirin (1966, 1969), and these were called informosomes. However, it took many years to characterize them and assess their biological significance in the context of development. These were universally detected in unfertilized eggs, spermatocytes and embryos of a variety of species. These were thus speculated to play a key role in the translation regulation of gene expression during gametogenesis, cell cycle, development and differentiation [reviewed in Pal *et al.* (1999)]. The search for these inhibitory mRNPs in oocytes, and embryos in particular, and their precise role in developmental regulation took a rapid pace in the 1980 and 1990s. Some of the initial findings are described next. In mature eggs of *Xenopus laevis*, Darnbrough and Ford (1981) identified two oocyte-specific proteins, mRNP3 and mRNP4, that were associated with translationally inactive mRNPs and that inhibited translation. mRNP4 was subsequently found to be identical with a previously identified Y-box transcription factor, FRGY2 (Deschamps *et al.* 1992; Murray *et al.* 1992; Wolffe *et al.* 1992). Further, Ranjan *et al.* (1993) demonstrated that FRGY2 mRNA, which is expressed exclusively during oogenesis, if expressed ectopically in somatic cells, stimulates transcription from a promoter containing a Y-box, and at the same time, it inhibits translation of all mRNAs non-specifically. These results supported the earlier observation of Richter and Smith (1984) that these proteins were inhibitory in an *in vitro* translation system. Another member of this conserved Y-box–binding protein family, MYS1, was shown to be associated with paternal mRNA in spermatocytes (Tafuri *et al.* 1993).

In mammalian cells, mRNAs of two proteins that maintain iron homeostasis, namely ferritin and transferrin receptor, undergo translation regulation through such *cis–trans* regulation. The *trans*-acting proteins that interact with the 5′- and 3′-UTR of ferritin mRNA and transferrin receptor mRNA, respectively, are iron regulatory proteins (IRPs). The *cis* sequences to which these proteins bind are folded into secondary structures and are called iron responsive elements (IREs). Binding of IRP to the IRE located at the 5′-UTR of ferritin mRNA inhibits its translation. However, binding of IRPs to the IREs located at the 3′-UTR of transferrin receptor mRNA leads to its stabilization and upregulation of translation. However, under conditions of excess iron in a cell, iron traps the IRPs and prevents their binding to the ferritin mRNA IRE, leading to upregulation of ferritin mRNA translation; whereas lack of binding of IRPs to the 3′-UTR of transferrin receptor mRNA leads to its destabilization and degradation (Fig. 15.7). Contrary to this, under iron-deficient conditions, the free IRPs can bind to the ferritin mRNA IRE, leading to its inhibition; whereas binding of free IRPs to the transferrin receptor

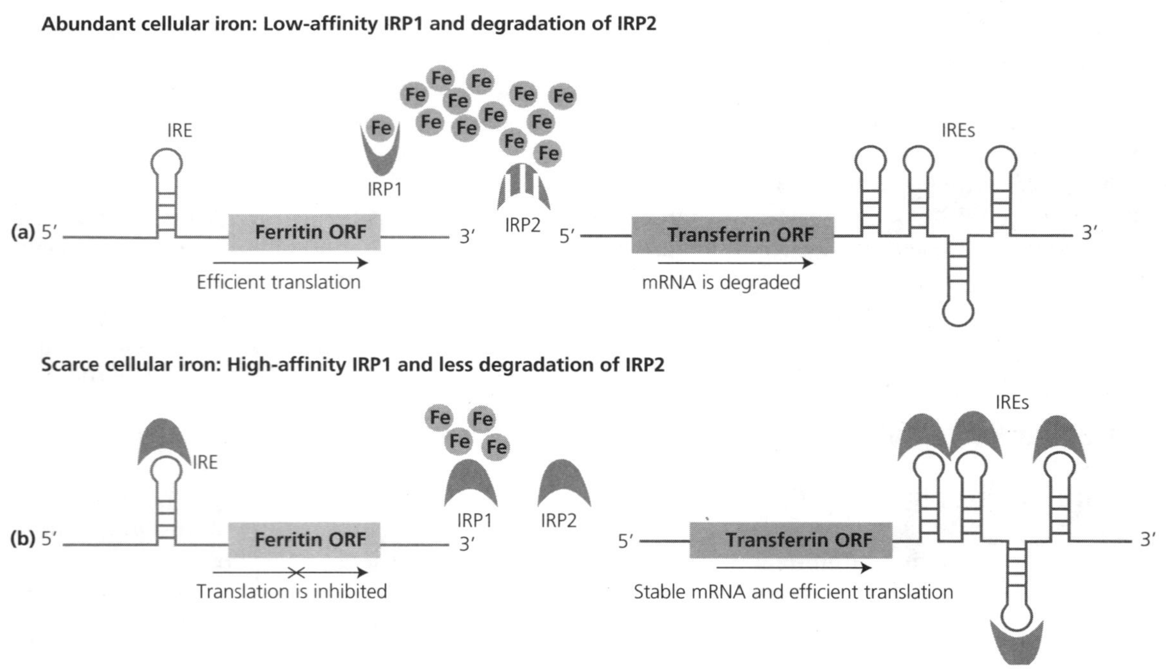

Figure 15.7 Translation regulation of ferritin and transferrin receptor mRNAs. 5′-UTR of ferritin mRNAs have IREs (iron-responsive elements). The IRP (iron regulatory protein) binds to IRE and inhibits translation of the mRNA. When iron is present in excess, iron binds to IRP; thus, IRP cannot bind to IRE and translation of ferritin mRNA takes place (a). The reverse is true in the case of iron-deficient condition (b). 3′-UTR of transferrin receptor mRNA also contains IREs. During iron deficiency, IRPs stabilize the mRNA by binding to it and thus facilitate its efficient translation (b). But when iron is present in excess, IRPs do not bind to IREs and the transferrin mRNA gets degraded (a). See Colour Plate 21 on page 515.

mRNA IREs leads to its upreguation of translation. In this manner, translation regulation of these two protein mRNAs regulates iron homeostasis in a cell. Any perturbation in this mechanism may result in pathological consequences.

15.3.1.3 Regulation by miRNAs

The first microRNA (miRNA) to be discovered was lin-4 in *Caenorhabditis elegans* (a nematode). Since then, a large number of miRNAs have been discovered across species in metazoans. miRNAs are 18–25 nt long and found in all metazoans studied so far (Bartel 2004). They display considerable evolutionary conservation, both in the sequence and in the genomic organization of the encoding regions of their precursors. miRNAs are derived from larger hair-pin–forming precursors (~70 nt), and on binding to mRNA, either trigger mRNA cleavage and decay or lead to translation inhibition without degradation (Lee and Ambros 2001). Some even stimulate translation (Vasudevan *et al.* 2007). miRNAs play decisive roles in many key biological processes such as cell growth, tissue differentiation, embryonic development, cell proliferation and apoptosis (Esquela-Kerscher and Slack 2006). Deregulation of miRNA genes (downregulation or upregulation) typically results in deleterious effects, especially relating to the development of the organism. miRNAs have also been implicated in either susceptibility or progression of diseases in animals, suggesting its overall significance in biology (Lopez *et al.* 2007; Latronico and Condorelli 2009).

Discovery of miRNA

Mutation studies involving the *lin-4* gene advocated its importance in the development of *C. elegans* as, on being targeted, the development of the worm was impaired. Victor Ambros and colleagues (Lee *et al.* 1993) cloned and analysed *lin-4* and found that it did not code for protein; rather, it coded for a 21-nt-long RNA which had antisense complementarity to lin-14 mRNA. lin-14 mRNA codes for LIN-14 protein, which regulates *C. elegans* development. LIN-14 levels are high during the first larval stage, where it helps to determine cell fate. During the second larval stage, LIN-14 expression falls due to imperfect base pairing of lin-4 miRNA with several target sequences in the 3′-UTR of lin-14 mRNA. This event ultimately leads to translation repression of lin-14 mRNA (Ha *et al.* 1996; Olsen and Ambros 1999). Gary Ruvkun and his colleagues showed that 3′-UTR of lin-14 mRNA was essential for lin-4's LIN-14 suppressive activity. Seven years later, Ruvkun and colleagues discovered that let-7, a 22-nt-long miRNA, also regulates developmental timing in worms (Reinhart *et al.* 2000). These discoveries heralded the age of miRNA in molecular biology. With time, an increasing number of miRNAs were discovered in thousands of genes in different animals and their targets and their effects were analysed.

Functional significance of miRNA in translation

As discussed above, miRNAs are known to cause translational repression, translation stimulation or mRNA cleavage. It is postulated that different miRNAs may adopt different functions or the same miRNA may adopt different functions depending on the cellular context. Let us discuss how it does so. miRNA causes translation repression by inhibiting translation initiation. When Filipowicz (2005) and colleagues added miRNA with target mRNA sequence in a reaction mix, they found that the miRNA and their target mRNA localized in mRNPs and not with polysomes. mRNPs usually contain non-translating mRNAs, while polysomes contain actively translating mRNAs. They further found that the influence of miRNA on target mRNA was abolished if the mRNA contained IRES (internal ribosome entry site) but not a cap. Such capless mRNAs would initiate in a cap-independent manner and were found in polysomes fraction, thus reinforcing the idea that translation initiation was the step being targeted by miRNA. miRNAs are also known to collaborate with Argonaute (AGO) proteins to compete with translation initiation factors for attaching to mRNA caps, thereby blocking initiation. AGO proteins are essential catalytic component of the RNA-induced silencing complex (RISC), the protein complex responsible for the gene silencing phenomenon known as RNA interference (RNAi). Contradictory evidence regarding at which step of translation miRNA acts was presented by Philip Sharp and colleagues (Zamore *et al.* 2000). They suggested that miRNA did not target translation initiation but a much later step, which might be elongation or termination. Whatever might be the step that miRNA targets, it essentially causes translation repression or decay of mRNA. Imperfect base pairing between target sequence and miRNA causes translation repression, while perfect base pairing with the target sequence causes cleavage of mRNA. This scenario pertains to animal mRNAs, while in plants, miRNAs tend to base-pair perfectly or near-perfectly with their cognate mRNAs and cause cleavage of these mRNAs. However, exceptions exist wherein translation blockage is also known to occur in plants (Weaver 2012).

Joan Steitz and colleagues (Vasudevan *et al.* 2007), while working on ARE (AU-rich elements) in TNF-α (tumor necrosis factor α) mRNA in serum-starved cells, demonstrated that miRNAs were capable of translation stimulation as well. They had in an earlier experiment reported that TNF-α ARE was transformed into a translation activation signal during serum starvation. The association of TNF-α ARE with the AGO2 and fragile X mental retardation-related protein 1 (FXR1) was proved to be essential for this translation activation (Vasudevan and Steitz 2007). Earlier studies had already indicated the involvement of miRNA in recruitment of AGO proteins which ultimately resulted in translation repression or mRNA cleavage (Valencia-Sanchez *et al.* 2006). Steitz and colleagues hypothesized that an miRNA must be involved in TNF-α ARE–mediated translation activation during serum starvation. Using bioinformatics screen, they searched the human genome for miRNAs with seed sequences complementary to the TNF-α ARE. (Seed sequences are minimum essential sequences required for miRNA activity). The screening yielded six candidate miRNAs, out of which one (miR-16) was already reported to reduce TNF-α mRNA level. The effect of all the other five miRNAs was tested on a synthetic construct bearing a firefly luciferase reporter gene attached to a TNF-α ARE under serum starvation. Only one miRNA (miR369-3) was found to have a direct correlation with TNF-α ARE–mediated translation activation during serum starvation. miR369-3 levels increased with serum starvation, and this increase could be reversed with a specific siRNA (single-stranded RNA known to cause RNA silencing) targeting the loop of miR369-3. They further strengthened their observation by repeating the experiment in a variety of conditions, where they manipulated the presence or absence of serum and siRNA and examined the effect of miR369-3 on translation activation. They also mutated seed regions in AREs of the mRNA used and observed that miR369-3 did not cause translational activation. However, when complementary mutations were made in seed regions in miR369-3, then its translation activation property was restored. Previous studies had already shown that AGO2 and FXR1 were essential for translational activation. Using aptamer-tagged mRNAs, they measured the recruitment of these two proteins to the ribonucleoprotein (RNP) complex. They showed that miR369-3, with the help of AGO2 and FXR1, activates translation of TNF-α mRNA in serum-starved cells. However, miR369-3, with the help of AGO2, represses translation of the mRNA in synchronized cells growing in serum. Thus, the function of an miRNA might actually depend on its cellular context.

Future of miRNAs

The fact that miRNAs are capable of influencing gene expression by causing translation stimulation, repression or mRNA cleavage opens interesting new avenues, particularly with reference to human health. Mutations in miRNAs and their cognate target sequences have been predicted to influence the risk of diseases as well as influence an individual's response to drugs. Such mutations either abolish or weaken existing binding sites or produce a new illegitimate binding site altogether. The resultant variation in miRNA binding due to the generated mutation may conclude in a related escalation or decline in translation. A host of diseases such as Parkinson's disease, several types of cancer, diabetes, hypertension and cardiovascular disease, schizophrenia, irritable bowel syndrome and others have demonstrated the role of miRNAs in disease susceptibility and progression (Chatterjee and Pal 2009). The strong association of miRNAs with diseases does not necessarily entail that drugs targeting them will be entirely successful, as many diseases are polygenic and are hugely influenced by aetiological and environmental factors. Moreover, targeting an miRNA may altogether have a contradictory effect as miRNAs are capable of influencing the functioning of many mRNAs and their functions may differ with the cellular context. The call of the day is to use large, well-characterized population studies and stringent statistical evaluation while establishing drug efficacies. Having said this, let us not forget that though research in this area is relatively new, it holds many promises for the elucidation of finer molecular mechanisms of gene expression.

15.3.2 Approaches discovered in viruses

It is well known that viruses utilize cellular machinery of their hosts in many of their molecular processes, including transcription, post-transcriptional processing of mRNA and translation. Indeed, mechanisms of many of these molecular processes, including splicing, that occur in eukaryotes have been discovered in viruses. One such unique approach in the process of translation regulation in viral mRNAs is described next.

15.3.2.1 Internal ribosome entry sites (IRES)

Translation is the key determinant of proteome of a cell. The magnitude of protein synthesis varies in a spatio-temporal manner in such a way so that the cell maintains homeostasis and is protected from cellular insult. The phenomenon of translation is severely compromised during viral infection. Viruses, as we all know, are a group of organisms which contain DNA or RNA as their genetic material and hijack the cellular machinery of their host for their own metabolism. On viral infection, the host protein synthesis is shut down and viruses utilize the host translation machinery for their protein synthesis. The fact that the host protein synthesis machinery is capable of supporting translation of the viral RNA and not of itself signifies that the host and the virus might have different requirements for protein synthesis. When we study and compare virus and eukaryotic cells, we find that they differ widely in the manner in which they recruit the translation machinery to synthesize proteins. Most of the eukaryotic mRNAs translate through a cap-dependent mode, while viral RNA translates through a cap-independent mode. The cap-dependent and cap-independent modes of translation vary in their requirement of the 5′-end of the mRNA, which usually bears a cap. During the cap-dependent mode of translation, the translation initiation takes place through the capped end of the mRNA and the ribosome scans the mRNA from the 5′-end till it reaches an AUG in Kozak's context and starts protein synthesis. In the case of the cap-independent mode of translation, the ribosome directly lands on or near an

IRES and mediates protein synthesis. IRES is also called as ribosome landing pad (RLP). This distinctive way of ribosome landing does away with the requirement of the cap and many of the factors required for cap recognition and binding. Moreover, the ribosome by landing internally is no longer required to make a lengthy and energy-inefficient journey through the long and highly complex UTR, which may be dotted with stable secondary structures.

IRESs were first discovered in viruses and later were found in some cellular mRNAs also. These cellular mRNAs usually follow cap-dependent translation, and during certain conditions of stress, they activate the IRES-dependent mode of translation. It appears that almost 10% of all cellular mRNAs undergo IRES-mediated translation (Stoneley and Willis 2004). mRNAs with cellular IRES are generally involved in cell growth (c-Myc), or apoptosis [X-linked inhibitor of apoptosis (XIAP) and apoptotic protease-activating factor 1 (Apaf-1)] or stress conditions such as hypoxia [vascular endothelial growth factor (VEGF)] or heat shock [fibroblast growth factor 2 (FGF-2); Sonenberg 2006].

Discovery of viral IRES: A new beginning

The seminal discovery by Pelletier and Sonenberg (1988) that naturally uncapped RNA of poliovirus initiated translation internally, that is, through IRES, heralded a new chapter in molecular biology. This finding was followed by the discovery of IRES in encephalomyocarditis virus (EMCV) by Wimmer and coworkers (Jang *et al.* 1988). Poliovirus belongs to the family picornaviridae (the etymology of which means small RNA viruses), which includes among others, cardiovirus, for example, EMCV and rhinovirus. Picornaviruses infect animal cells and cause a variety of diseases, from common cold to poliomyelitis. It was observed that a poliovirus-infected cell stopped cellular protein synthesis, while the RNA from poliovirus was efficiently translated. On looking more closely at the polioviral RNA, it was noted that like other members of its family, its UTR was very long (length varies from 600 to 1200 nucleotides), structured and encumbered by a number of upstream AUGs. All these features are contradictory to efficient cap-dependent translation. Moreover, poliovirus induces the cleavage of eIF-4G, an essential component of the cap-binding complex and thereby inhibits cellular cap-dependent protein synthesis. This led Sonenberg and Pelletier to ponder whether poliovirus was using an altogether different mechanism to get its protein synthesized. They ultimately cloned the 5′-UTR of polioviral RNA in a dicistronic vector and proved that polioviral RNA initiated internally.

A dicistronic vector has two cistrons which are separated by an intergenic region. The upstream and downstream cistrons are usually reporter genes such as chloramphenicol acetyltransferase, firefly and *Renilla* luciferase whose activity can be easily measured. The first cistron is translated by the cap-dependent mode, whereas the downstream reporter is translated if the intergenic region contains a functional IRES element; that is, it is able to direct internal entry of a ribosome. The vector is so constructed that there is no leaky translation from the first cistron into the second cistron. This is made possible by the introduction of a very strong secondary structure at the end of the first cistron which impedes further movement of the ribosome. Therefore, translation from the second cistron is possible only if ribosomes initiate internally (Fig. 15.8).

Over a period of years, it was subsequently proved that all members of picornaviridae have IRES and it is also present in some of the non-picornaviridae members such as hepatitis C virus (HCV), plant viruses [tobacco mosaic virus (TMV)] and insect viruses [cricket paralysis virus (CrPV)]. Following the discovery of IRES in viruses, it was predicted that cellular mRNAs may also possess such a feature because of the fact that about 15–20% of the total mRNA population have long and structured UTR. It is estimated through polysome association and DNA microarray studies that less than 3% of total cellular mRNAs are capable of translation during poliovirus infection (Sonenberg 2006), thus indicating that cellular mNRAs also bear IRES.

Cellular IRES

BiP (immunoglobulin heavy-chain-binding protein) mRNA has the distinction of being the first cellular IRES to be discovered. The majority of cellular IRESs are involved in important functions of the cell such as growth control, mitosis, cell cycle, hypoxia and apoptosis, such as cellular myelocytomatosis oncogene (c-Myc), apaf-1 and ornithine decarboxylase (ODC). Under the circumstances that impair cap-dependent translation, IRES-mediated translation provides an alternative method for the translationa of those mRNAs which are implicated in decisions concerning the fate of the cell (Holcik and Sonenberg 2005; Pickering and Willis 2005). The loss of IRES-mediated translation regulation is implicated in the pathophysiology of various diseases such as X-linked Charcot–Marie–Tooth disease (CMTX), multiple myeloma and fragile X syndrome (FXS; Chatterjee *et al.* 2010).

According to IRESite (http://iresite.org/), an IRES-related database which catalogues information regarding all known IRES (viral, cellular and synthetic), there exist 68 viral IRES and 115 cellular IRES till the time the database was last updated in 2011. However, how many of these actually function as IRES is a question for debate. It is argued that the so-called 'gold standard' for testing an IRES, that is, a dicistronic assay, is not capable of distinguishing an IRES element from a non-IRES element. The dicistronic vectors often give false-positive results due to the presence of cryptic promoters and splice variants in the putative element that can cause translation from the second cistron independent of the first cistron (Kozak

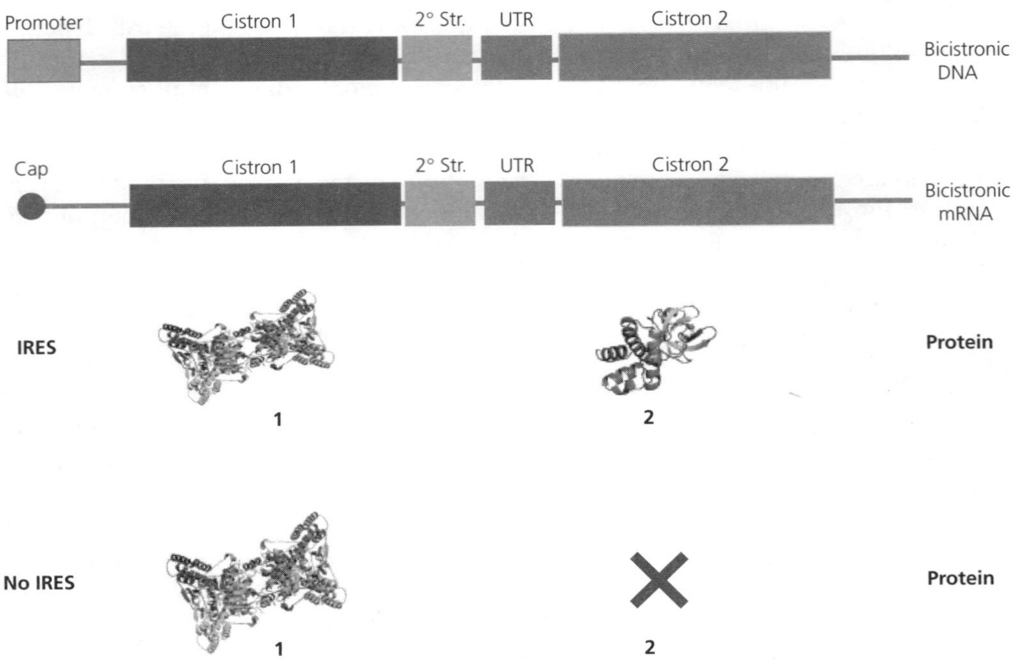

Figure 15.8 Schematic representation of bicistronic assay. A bicistronic construct is made in such a way that the transcribed mRNA from the first cistron is translated in a cap-dependent manner, while the mRNA of the second cistron is translated in a cap-independent manner. The 5′-UTR of the prospective IRES-containing mRNA (to be tested) is cloned next to a highly structured sequence in the UTR of the second cistron. If this prospective UTR contains an IRES, due to internal ribosome entry, it will be translated, as can be observed by expression of the specific reporter gene (second cistron). See Colour Plate 22 on page 516.

2005). Reports suggest that pRF (a dicistronic vector), which was used for detection of VEGF, c-Myc, XIAP, vascular endothelial growth factor receptor-1 (VEGFR-1), early growth response 1 (Egr-1) or hypoxia-inducible factor-1α (HIF-1α) IRES may not be reliable as pRF plasmids containing 5′-UTRs of candidate genes produce downstream luciferase activity primarily as a result of activation of cryptic promoter activity by downstream simian virus 40 (SV40) enhancer elements (Bert *et al.* 2006). Many such controversies surround this field; however, it cannot be denied that IRES does provide an additional mechanism of mRNA-specific translation regulation.

Can IRES be defined and classified?

The answer to the above question is a complicated one. Yes, it can be classified, but to frame a definition that will tell all about IRES is tricky. The picornaviral IRES elements, on the basis of their primary sequences, predicted secondary structures, and requirements for efficient translational initiation have been assigned to three groups: the enterovirus and rhinovirus group (type I IRES, e.g. poliovirus), the cardio- and aphtovirus (type II IRES, e.g. EMCV), and the hepatitis A virus IRES (Jackson and Kaminski 1995). Merrick defined IRES as an RNA sequence/structure that promotes the binding of 40S ribosomal subunits to an internal portion of mRNA (Merrick 2004). There is no consensus sequence or structure common to all viral or cellular IRESs. A pseudo-knot structure has been shown to be required absolutely for IRES activity in HCV. A Y-shaped structural motif is claimed to be common to IRESs of immunoglobulin heavy-chain-binding protein (Bip), Antennaepedia (Antp) and FGF-2. However, to date, there exists an ambiguity about their structure (Baird *et al.* 2006). It is also alleged that cellular IRES elements can undergo transitions that can substantially influence their activity (Komar and Hatzoglous 2005). IRES seem to be a scaffold that contains multiple sites for interaction with other components of translation apparatus.

Moreover, IRESs exhibit flexibility in their requirement of canonical factors and non-canonical IRES *trans*-acting factors (ITAFs). ITAFs play a variety of functions in cells other than modulating IRES-mediated translation. In the context of IRES, it is postulated that they may help to recruit the 40S ribosomal subunit or promote/stabilize specific active conformation of IRES. EMCV requires eIF-4G C-terminal (eIF4G-Ct), eIF-4A, eIF-3, eIF-2 and polypyrimidine tract-binding protein (PTB; Pestova *et al.* 1991; Kolupaeva *et al.* 1998), while HCV requires eIF-3, eIF-2, PTB, poly rC–binding protein 2 (PCBP2) and La (Pestova *et al.* 1998). Surprisingly, CrPV does not require any canonical factors, and *trans*-acting factors required for it are yet to be found (Wilson *et al.* 2000). A rule of thumb does not exist with regard to the requirement of

ITAFs and canonical translation factors. The diversity in length, secondary and tertiary structures, sequence and requirement of *trans*-acting factors of IRESs suggests that they may have evolved separately over a period of time and were positively selected because of their ability to provide alternative routes of translation under less favourable conditions.

Perspective

It has been 25 years since the discovery of IRES, but a lot remains to be unearthed. Along the years, a number of cellular and viral IRESs have been reported across kingdoms. However, the foremost question that still remains unanswered is the molecular mechanism of IRES-mediated translation in the context of cellular IRES. How do cellular IRESs recruit the 40S ribosomes to initiate translation? It is widely assumed that IRES elements possess complex secondary and tertiary structures, which allow for multiple interactions with the components of the translational machinery (canonical initiation factors, ITAFs and 40S ribosomes). It should be taken into account that there is no nucleotide sequence similarity among cellular IRES elements (Komar and Hatzoglous 2005).

There is no doubt that elucidation of this kind of structure–function relationship will enrich the field of molecular biology and translational medicine. Translational medicine is integrative application of knowledge obtained from research from various fields in a manner such that there is easy transition 'from bench to bedside' medicine formulation. IRES or IRES-related machinery may be targeted in cases where downregulation or upregulation of IRES activity is the known cause of disease.

15.3.3 Approaches employed under environmental stress

The lower eukaryotes have evolved various strategies to cope up with environmental stress. This appears to be particularly evident in the regulation of gene expression. It is known that stress-mediated regulation of gene expression occurs predominantly at the level of translation, and one such regulation is described next.

15.3.3.1 Regulation through multiple AUGs and upstream ORFs at the 5′-UTR

Many eukaryotic mRNAs that are tightly regulated contain multiple AUGs and open reading frames (ORFs) at their 5′-UTR. These generally precede the actual AUG and actual ORF, respectively. These *cis* sequences are instrumental in controlling the rate of translation of mRNAs whose products need to be correlated with the their precise quantitative requirement by the cells. In this context, the example of the mRNA of an amino acid biosynthetic gene, *GCN4* (general growth control non-derepressible), in yeast and its translation during amino acid starvation is exemplary. GCN4 mRNA contains four additional initiation codons located 5′ to the actual initiation codon, called upstream AUGs (uAUGs) and thus results in four upstream ORFs (uORFs). During normal physiological condition (Fig. 15.9), when the cells do not require GCN4 protein because of the presence of ample concentration of amino acids in the medium, these uAUGs play a central role (Hinnebusch and Natarajan 2002). With the available ternary complex (eIF-2-GTP-Met tRNA),

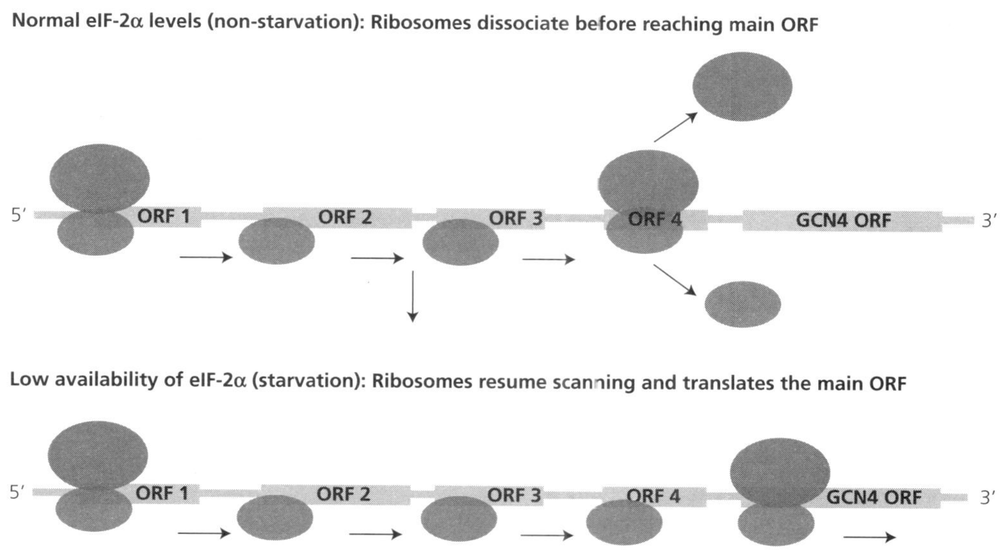

Figure 15.9 Translation regulation of GCN4 mRNA in yeast. GCN4 mRNA contains 4 uORFs apart from the actual ORF. Sequences downstream of the ORF 1 allow the ribosomes to resume scanning and to initiate at downstream ORFs. During normal conditions, ribosomes are released after translating one of these ORFs. On amino acid starvation, the ribosomes resume scanning and re-initiate at the actual ORF (coding sequence), thus allowing its translation. See Colour Plate 23 on page 516.

translation begins at uORF 1, the ribosome dissociates and resumes scanning to translate uORF 2, uORF 3 and uORF 4. Ribosomes however fail to re-initiate translation of the actual ORF of GCN4 after termination of translation at these latter uORFs (Fig. 15.9). On the contrary, during amino acid starvation, when yeast eIF-2 (GCN2) gets phosphorylated, resulting in limiting the concentration of the functional ternary complex, ribosomes do not resume scanning at the uORFs and initiate translation of the actual ORF. This results in the upregulation of GCN4, leading to stimulation of amino acid biosynthesis. This amino acid starvation response in yeast leading to translation regulation of GCN2 eIF-2α kinase appears to be evolutionarily conserved up to mammals. For example, in mouse brain, GCN2 protein kinase activity appears essential for a restricted intake of diets devoid of essential amino acids (Hao *et al.* 2005; Maurin *et al.* 2005).

15.4 Deregulation of translation leading to human pathologies

Gene expression is so organized that higher-order regulatory switches exist to realize tight control of RNA and protein synthesis. Dynamic reprogramming of ongoing protein synthesis is a major part of the stress-related response programme. This reprogramming can be at global and/or mRNA-specific level. Global control takes place through phosphorylation/dephosphorylation of initiation and elongation factors and other components of the translation machinery (Yamasaki and Anderson 2008), while mRNA-specific regulation takes place through their UTRs. Cells using these stretches at their 5′ or 3′ (containing highly complex secondary structures) inhibit or promote protein synthesis by binding to mRNA-specific and/or global *trans*-acting factors. UTRs are primarily involved in translation regulation through manipulation of mRNA stability, expression and localization; any error in these acts results in perturbation in normal cellular functions. Errors introduced through point mutations or truncations in coding or UTRs lead to an atypical protein profile. Even slight change in levels of protein, particularly those involved in cell differentiation, cell cycle, stress response and apoptosis, may lead to tilting of the physiological balance from healthy to the diseased state, resulting in conditions such as hereditary thrombocythaemia, pancreatitis, breast cancer, FXS, bipolar affective disorder and Alzheimer's disease.

15.4.1 Effect of mutations on effectors of global translation regulation

There are many effectors of global translation regulation which determine the proteome of an organism. We discuss here in particular three effectors (PEK, eIF-2B and HRI), mutations in which lead to serious pathological disorders. Pancreatic eIF-2α kinase (PEK), also called PKR-like endoplasmic reticulum kinase (PERK), is an important eIF-2α kinase. It is activated by an unfolded protein response, and on being activated, it inhibits protein synthesis by phosphorylation of the α-subunit of eIF-2, an essential component of the translation initiation machinery. PERK mutations in humans are associated with the Wolcott–Rallison syndrome of infantile diabetes. Mutations that prevent the α-subunit of eIF-2 from being phosphorylated by PERK block β-cell development and impair gluconeogenesis (Harding and Ron 2002).

eIF-2B, an important guanine nucleotide exchange factor, is involved in recycling eIF-2. eIF-2B has five subunits, each encoded by a different gene; mutations in these genes result in severe autosomal recessive neurodegenerative disease, described in young children as childhood ataxia with central nervous system hypomyelination (CACH)/leucoencephalopathy with vanishing white matter (VWM) syndrome. eIF-2B mutations are associated with a wide spectrum of symptoms, from cataract, ovarian failure, pancreatitis, hepatosplenomegaly to renal hypoplasia (Fogli *et al.* 2003; Van der Knaap *et al.* 2003).

The haem-regulated eIF-2α kinase is also called haem-regulated inhibitor (HRI) or haem-controlled repressor (HCR). The primary function of HRI is thought to be the coordination of the synthesis of globin chains in reticulocytes with the availability of haem, which is required for the assembly of α- and β-globin chains into haemoglobin. Thus, a deficiency of haem induces the activation of HRI, which subsequently phosphorylates the α-subunit of eIF-2, leading to the inhibition of protein synthesis. This prevents the accumulation of unassembled globin chains, the aggregation of which is toxic to red blood cells (Chen 2007). Studies in $HRI^{-/-}$ mice show that HRI or eIF-2AK1 is involved in three types of anaemias, namely iron deficiency anaemia, erythropoietic protoporphyria (EPP) and α-thalassaemia intermediates (Chen 2007).

15.4.2 Effect of mutations on effectors of mRNA-specific translation regulation

Mutations in secondary structures in 5′-and 3′-UTRs, multiple ORFs, multiple uAUGs, IRESs and motifs such as IREs effect translational regulation. Mutations that lead to the creation of the premature termination codon (PTC) and delayed stop codon modify the length of the polypeptide chain and hence non-functional proteins of atypical molecular weights are formed. Mutations may also abolish the canonical polyadenylation signal and hamper efficient transcription termination and polyadenylation of mRNA. Mutations in the 3′-UTR alter the binding sites of interacting miRNAs or proteins. AU-rich element–associated mutations may lead to atypical stabilization of mRNA. Impaired translation thus results in the production of non-functional or partially functional proteins, which ultimately culminates in a diseased state or susceptibility to a disease. We discuss here, in brief, associations of mutations in different motifs of UTRs and disease. Interested readers are requested to look at the following reviews for comprehensive information with reference to the role of UTRs in human diseases: Chatterjee and Pal (2009), Chatterjee *et al.* (2010) and Pal *et al.* (2010).

15.4.3 Mutations and 5′-UTR

Mutations in uORFs of mRNAs of thrombopoietin (TPO), β-site APP (amyloid precursor protein) cleavage enzyme 1 (BACE), 5-hydroxytryptamine receptor 3A (5-HT3A), transforming growth factor β3 (TGF-β3) and cyclin-dependent kinase inhibitor 2A (CDKN2A) result in diseases such as hereditary thrombocythaemia, Alzheimer's disease, bipolar affective disorder, arrhythmogenic right ventricular cardiomyopathy/dysplasia and melanoma, respectively. The loss of IRES-mediated translation regulation is implicated in the pathophysiology of various diseases such as CMTX, multiple myeloma and FXS; while mutations in ferritin L-chain mRNA IRE result in hereditary hyperferritinaemia/cataract syndrome (HHCS).

15.4.4 Mutations and 3′-UTR

Diseases such as cystic fibrosis, Duchenne muscular dystrophy (DMD), haemophilia, aniridia, lysosomal disorders and epidermolysis bullosa simplex (EBS) are caused due to PTCs. Delayed stop codons are associated with aniridia, EBS and Fabry's disease. Haemoglobin H (HbH) disease, and immune dysfunction, polyendocrinopathy, and enteropathy, X-linked (IPEX) are caused due to the disruption of the polyadenylation signal. Mutations associated with the disruption of the secondary structure of 3′-UTR is associated with selenoprotein-related myopathies and congenital heart disease (CHD). miRNA target sequence–related mutations are implicated in Tourette's syndrome, hereditary spastic paraplegia (HSP), papillary thyroid carcinoma, Parkinson's disease and colorectal cancer.

15.4.5 Human pathologies and implication of translation studies

The huge amount of data generated through genome sequencing and molecular profiling of metabolic pathways has inspired many to look more closely at the molecular basis of diseases. Many diseases are documented whose origin, cause of susceptibility and progression at the molecular level is unknown. Studies documenting mutations in coding and UTRs of genes will yield valuable information relating to their respective importance in disease susceptibility and progression. It has to be taken into account that most diseases per se are not monogenic, and most of them are controlled by a variable number of parameters. These parameters change in a spatial and temporal manner and are also highly influenced by aetiological factors. In such a scenario, the task of studying the molecular basis of diseases may seem daunting, but careful experimental designing, rigorous statistical analysis and large sample population may help arriving at a more definite conclusion. In the long run, these studies will herald genome-based drug targeting and personalized medicine. However, the application of this knowledge in routine diagnosis practiced in third-world countries will still remain a challenge. The need of the hour is technical breakthroughs which will take care of the prohibitive cost factor in the context of application of high-end technologies such as DNA sequencing.

15.5 Summary

Regulation of gene expression in eukaryotic cells is tightly linked with various cell physiologies, basic metabolic activity and specific activity in time and space. In the past, research towards understanding the regulation of gene expression was primarily focused at the transcription level, generating mRNAs. However, since the past two decades, it has been increasingly realized that a major portion of the regulation lies at the post-transcriptional level, translation being the major contributor. Cells adopt the global regulation approach, particularly under adverse environmental conditions. On the other hand, mRNA-specific diverse approaches have been discovered during a variety of cell functions, including gametogenesis, development and differentiation, including abnormal differentiation leading to cancer, etc. In this chapter, we have documented some such regulations in a comprehensive manner. Although we have described global regulation separately from mRNA-specific regulation, often these two may overlap, as exemplified by the regulation of GCN4 mRNA translation, where both the components of global regulation (GCN2 eIF-2α kinase) and mRNA-specific regulation (5′-UTR, uAUGs and uORFs) work together under environmental stress.

Various non-coding RNAs, particularly miRNAs, appear to have a tremendous potential for translation regulation. A large amount of data on this area are being continuously added to the literature. This area offers a great promise to unravel many unsolved puzzles in the context of regulation of gene expression, particularly in human pathologies.

There appears to be a paradigm shift in the approach of investigation in this area. In the past, investigations were restricted to single-gene levels. Now, with the advent of modern technologies, the genome-wide analysis approach is gaining momentum and it will unravel the coordinated regulation of gene cassettes rather than single genes involved in related cell functions.

References

Baird SD, Turcotte M, Korneluk RG, Holcik M. Searching for IRES. *RNA* 2006;12:1755–1785.

Bartel DP. MicroRNAs: Genomics, biogenesis, mechanism, and function. *Cell* 2004;116:281–297.

Bert AG, Grépin R, Vadas MA, Goodall GJ. Assessing IRES activity in the HIF-1α and other cellular 5′ UTRs. *RNA* 2006;12:1074–1083.

Burd CG, Matunis EL, Dreyfuss G. The multiple RNA binding domains of the mRNA poly(A)-binding protein have different RNA-binding activities. *Mol Cell Biol* 1991;11:3419–3424.

Chatterjee S, Pal JK. Role of 5′- and 3′-untranslated regions of mRNAs in human diseases. *Biol Cell* 2009;101:251–262.

Chatterjee S, Pal JK, Berwal SK. Pathological variations in 5′-untranslated regions of human genes. In: *Encyclopaedia of Life Sciences*. Chichester: John Wiley & Sons; 2010. pp. 1–8.

Chen J-J. Regulation of protein synthesis by the heme-regulated eIF2α kinase: Relevance to anemias. *Blood* 2007;109:2693–2699.

Chen J-J, London IM. Regulation of protein synthesis by heme-regulated eIF-2 alpha kinase. *Trends Biochem Sci* 1995;20:105–108.

Curtis D, Lehmann R, Zamore PD. Translational regulation in development. *Cell* 1995;81:171–178.

Darnbrough CH, Ford PJ. Identification in *Xenopus laevis* of a class of oocyte-specific proteins bound to messenger RNA. *Eur J Biochem* 1981;113:415–424.

Deschamps S, Viel A, Garrigos M, Denis H, Le Maire M. mRNP4, a major mRNA-binding protein from *Xenopus* oocytes is identical to transcription factor FRG Y2. *J Biol Chem* 1992;267: 13799–13802.

Dever TE, Feng L, Wek RC, Cigan AM, Donahue TF, Hinnebusch AG. Phosphorylation of initiation factor 2 alpha by protein kinase GCN2 mediates gene-specific translational control of GCN4 in yeast. *Cell* 1992;68:585–596.

Dreyfuss G, Kim VN, Kataoka N. Messenger-RNA-binding proteins and the messages they carry. *Nat Rev Mol Cell Biol* 2002;3:195–205.

Esquela-Kerscher A, Slack FJ. Oncomirs – microRNAs with a role in cancer. *Nat Rev Cancer* 2006;6:259–269.

Filipowicz W. RNAi: The nuts and bolts of the RISC machine. *Cell* 2005;122:17–20.

Fogli A, Rodriguez D, Eymard-Pierre E, Bouhour F, Labauge P, Meaney BF, *et al.* Ovarian failure related to eukaryotic initiation factor 2B mutations. *Am J Hum Genet* 2003;72:1544–1550.

Gallie DR, Tanguay R. Poly(A) binds to initiation factors and increases cap-dependent translation *in vitro*. *J Biol Chem* 1994;269:17166–17173.

Gebauer F, Hentze MW. Molecular mechanisms of translational control. *Nat Rev Mol Cell Biol* 2004;5:827–835.

Gebauer F, Xu W, Cooper GM, Richter JD. Translational control by cytoplasmic polyadenylation of c-mos mRNA is necessary for oocyte maturation in the mouse. *EMBO J* 1994;13:5712–5720.

Ha I, Wightman B, Ruvkun G. A bulged lin-4/lin-14 RNA duplex is sufficient for *Caenorhabditis elegans* lin-14 temporal gradient formation. *Genes Dev* 1996;10:3041–3050.

Hao S, Sharp JW, Ross-Inta CM, McDaniel BJ, Anthony TG, Wek RC, *et al.* Uncharged tRNA and sensing of amino acid deficiency in mammalian piriform cortex. *Science* 2005;307:1776–1778.

Harding HP, Ron D. Endoplasmic reticulum stress and the development of diabetes: A review. *Diabetes* 2002;51:S455–S461.

Hieronymus H, Silver PA. A systems view of mRNP biology. *Genes Dev* 2004;18:2845–2860.

Hinnebusch AG, Natarajan K. Gcn4p, a master regulator of gene expression, is controlled at multiple levels by diverse signals of starvation and stress. *Eukaryot Cell* 2002;1:22–32.

Holcik M, Sonenberg N. Translational control in stress and apoptosis. *Nat Rev Mol Cell Biol* 2005;6:318–327.

Imataka H, Gradi A, Sonenberg N. A newly identified N-terminal amino acid sequence of human eIF4G binds poly(A)-binding protein and functions in poly(A)-dependent translation. *EMBO J* 1998;17:7480–7489.

Jackson RJ, Kaminski A. Internal initiation of translation in eukaryotes: The picornavirus paradigm and beyond. *RNA* 1995;1:985–1000.

Jang SK, Krausslich HG, Nicklin MJ, Duke GM, Palmenberg AC, Wimmer E. A segment of the 5′ nontranslated region of encephalomyocarditis virus RNA directs internal entry of ribosomes during *in vitro* translation. *J Virol* 1988;62:2636–2643.

Kolupaeva VG, Pestova TV, Hellen CU, Shatsky IN. Translation eukaryotic initiation factor 4G recognizes a specific structural element within the internal ribosome entry site of encephalomyocarditis virus RNA. *J Biol Chem* 1998;273:18599–18604.

Komar AA, Hatzoglou M. Internal ribosome entry sites in cellular mRNAs: Mysteries of their existence. *J Biol Chem* 2005;280:23425–23428.

Kozak M. Regulation of translation via mRNA structure in prokaryotes and eukaryotes. *Gene* 2005;361:13–37.

Lackner DH, Bahler J. Translational control of gene expression: From transcripts to transcriptomes. *Int Rev Cell Mol Biol* 2008;271:199–251.

Latronico MV, Condorelli G. MicroRNAs and cardiac pathology. *Nat Rev Cardiol* 2009;6:419–429.

Lee RC, Ambros V. An extensive class of small RNAs in *Caenorhabditis elegans*. *Science* 2001;294:862–864.

Lee RC, Feinbaum RL, Ambros V. The *C. elegans* heterochronic gene lin-4 encodes small RNAs with antisense complementarity to lin-14. *Cell* 1993;75:843–854.

London IM, Levin DH, Matts RL, Thomas NSB, Petryshyn R, Chen J-J. Regulation of protein synthesis. In: Boyer PD, Krebs EG, eds. *The Enzymes* (3rd Edition). New York: Academic Press; 1987. pp. 359–380.

López de Silanes I, Quesada MP, Esteller M. Aberrant regulation of messenger RNA 3′-untranslated region in human cancer. *Cell Oncol* 2007;29:1–17.

Lu P, Vogel C, Wang R, Yao X, Marcotte EM. Absolute protein expression profiling estimates the relative contributions of transcriptional and translational regulation. *Nat Biotech* 2007; 25:117–124.

Mathews MB, Sonenberg N, Hershey JWB. Origins and principles of translational control. In: Mathews MB, Sonenberg N, Hershey JWB, eds. *Translational Control in Biology and Medicine*. New York: Cold Spring Harbor Laboratory Press; 2007. pp. 1–40.

Maurin AC, Jousse C, Averous J, Parry L, Bruhat A, Cherasse Y, *et al.* The GCN2 kinase biases feeding behavior to maintain amino acid homeostasis in omnivores. *Cell Metab* 2005;1:273–277.

Mendez R, Richter JD. Translational control by CPEB: A means to the end. *Nat Rev Mol Cell Biol* 2001;2:521–529.

Merrick WC. Cap-dependent and cap independent translation in eukaryotic systems. *Gene* 2004;332:1–11.

Morley SJ, Coldwell MJ. Matters of life and death: Translational initiation during apoptosis. In: Mathews MB, Sonenberg N, Hershey JWB, eds. *Translational Control in Biology and Medicine*. New York: Cold Spring Harbor Laboratory Press; 2007. pp. 433–458.

Murray MT, Schiller DL, Franke WW. Sequence analysis of cytoplasmic mRNA-binding proteins of *Xenopus* oocytes identifies a family of RNA-binding proteins. *Proc Natl Acad Sci USA* 1992;89:11–15.

Olsen PH, Ambros V. The lin-4 regulatory RNA controls developmental timing in *Caenorhabditis elegans* by blocking LIN-14 protein synthesis after the initiation of translation. *Dev Biol* 1999;216:671–680.

Pal JK, Anand S, Joseph J. Association of hsp90 with the heme-regulated eukaryotic initiation factor 2α kinase – a collaboration for regulating protein synthesis. *J Biosci* 1996;21:191–205.

Pal JK, Chatterjee S, Berwal SK. Pathological variations in 3′-untranslated regions of human genes. In: *Encyclopaedia of Life Sciences*. Chichester: John Wiley & Sons, Ltd.; 2010. pp. 1–9.

Pal JK, Ghaskadbi SS. *Fundamentals of Molecular Biology*. Oxford University Press; 2009.

Pal JK, Kar A, More NS. Translational regulation of gene expression during early embryonic development. In: Gakhar SK, Mishra SN, eds. *Recent Trends in Developmental Biology*. Mumbai: Himalaya Publishing House; 1999. pp. 27–46.

Pelletier J, Sonenberg N. Internal initiation of translation of eukaryotic mRNA directed by a sequence derived from poliovirus RNA. *Nature* 1988;334:320–325.

Pestova TV, Hellen CUT, Wimmer E. Translation of poliovirus RNA: role of an essential cis-acting oligopyrimidine element within the 5′ nontranslated region and involvement of a cellular 57-kilodalton protein. *J Virol* 1991;65:6194–6204.

Pestova TV, Shastsky IN, Fletcher SP, Jackson RJ, Hellen CUT. A prokaryotic-like mode of cytoplasmic eukaryotic binding to the initiation codon during internal translation initiation in hepatitis C and classic swine fever virus RNAs. *Genes Dev* 1998;12:67–83.

Pickering BM, Willis AE. The implications of structured 5′ untranslated regions on translation and disease. *Semin Cell Dev Biol* 2005;16:39–47.

Piron M, Vende P, Cohen J, Poncet D. Rotavirus RNA binding protein NSP3 interacts with eIF4GI and evicts the poly(A) binding protein from eIF4F. *EMBO J* 1998;17:5811–5821.

Ranjan M, Tafuri SR, Wolffe AP. Masking mRNA from translation in somatic cells. *Genes Dev* 1993;7:1725–1736.

Read RL, Norbury CJ. Roles for cytoplasmic polyadenylation in cell cycle regulation. *J Cell Biochem* 2002;87:258–265.

Reinhart BJ, Slack FJ, Basson M, Pasquinelli AE, Bettinger JC, Rougvie AE, *et al.* The 21-nucleotide let-7 RNA regulates developmental timing in *Caenorhabditis elegans*. *Nature* 2000;403:901–906.

Richter JD. *Translational Control*. New York: Cold Spring Harbor Lab Press; 1995.

Richter JD. CPEB: A life in translation. *Trends Biochem Sci* 2007;32:279–285.

Richter JD, Smith LD. Reversible inhibition of translation by *Xenopus* oocyte-specific proteins. *Nature* 1984;309:378–380.

Richter JD, Sonenberg N. Regulation of cap-dependent translation by eIF4E inhibitory proteins. *Nature* 2005;433:477–480.

Ron D, Harding HB. eIF2α phosphorylation in cellular stress response and disease. In: Mathews MB, Sonenberg N, Hershey JWB, eds. *Translational Control in Biology and Medicine*. New York: Cold Spring Harbor Laboratory Press; 2007. pp. 345–368.

Scherrer K. In: Kolodny GM, ed. *Eukaryotic Gene Regulation* (Volume 1). Boca Raton, FL: CRC Press; 1980. pp. 57–129.

Schneider RJ, Sonenberg N. Translational control in cancer development and progression. In: Mathews MB, Sonenberg N, Hershey JWB, eds. *Translational Control in Biology and Medicine*. New York: Cold Spring Harbor Laboratory Press; 2007. pp. 401–431.

Sheets MD, Fox CA, Hunt T, Vande Woude G, Wickens M. The 3′-untranslated regions of c-mos and cyclin mRNAs stimulate translation by regulating cytoplasmic polyadenylation. *Genes Dev* 1994;8:926–938.

Sheets MD, Wu M, Wickens M. Polyadenylation of c-mos mRNA as a control point in *Xenopus* meiotic maturation. *Nature* 1995;374:511–516.

Sonenberg N. Protein synthesis in eukaryotes: IRES-mediated internal initiation. In: *Encyclopaedia of Life Sciences*. Chichester: John Wiley & Sons; 2006.

Spirin AS. 'Masked' forms of mRNA. *Curr Top Dev Biol* 1966;1:1–38.

Spirin AS. Informosomes. *Eur J Biochem* 1969;10:20–35.

Stevenson AL, Norbury CJ. The Cid1 family of non-canonical poly(A) polymerases. *Yeast* 2006;23:991–1000.

Stoneley M, Willis AE. Cellular internal ribosome entry segments: Structures, trans-acting factors and regulation of gene expression. *Oncogene* 2004;23:3200–3207.

Tafuri SR, Familari M, Wolffe AP. A mouse Y box protein, MSY1, is associated with paternal mRNA in spermatocytes. *J Biol Chem* 1993;268:12213–12220.

Tarun SZ Jr, Sachs AB. A common function for mRNA 5′ and 3′ ends in translation initiation in yeast. *Genes Dev* 1995;9:2997–3007.

Thompson B, Wickens M, Kimble J. Translational control in development. In: Mathews MB, Sonenberg N, Hershey JWB, eds. *Translational Control in Biology and Medicine*. New York: Cold Spring Harbor Laboratory Press; 2007. pp. 507–544.

Valencia-Sanchez MA, Liu J, Hannon GJ, Parker R. Control of translation and mRNA degradation by miRNAs and siRNAs. *Genes Dev* 2006;20:515–524.

Van der Knaap MS, Van Berkel CGM, Herms J, van Coster R, Baethmann M, Naidu S, *et al.* eIF2B-related disorders: Antenatal onset and involvement of multiple organs. *Am J Hum Genet* 2003;73:1199–1207.

Vasudevan S, Steitz JA. AU-rich element-mediated upregulation of translation by FXR1 and Argonaute2. *Cell* 2007;128:1105–1181.

Vasudevan S, Tong Y, Steitz JA. Switching from repression to activation: MicroRNAs can up-regulate translation. *Science* 2007;318:1931–1934.

Vincent A, Akhayat O, Goldenberg S, Scherrer K. Differential repression of specific mRNA in erythroblast cytoplasm: A possible role for free mRNP proteins. *EMBO J* 1983;2:1869–1876.

Visa N, Alzhanova-Ericsson AT, Sun X, Kiseleva E, Björkroth B, Wurtz T, *et al.* A pre-mRNA-binding protein accompanies the RNA from the gene through the nuclear pores and into polysomes. *Cell* 1996;84:253–264.

Weaver RF. *Molecular Biology* (5th Edition). New York: McGraw-Hill; 2012.

Wilson JE, Pestova TV, Hellen CUT, Sarnow P. Initiation of protein synthesis from A site of ribosome. *Cell* 2000;102:511–520.

Wolffe AP, Tafuri SR, Ranjan M, Familari M. The Y-box factors: A family of nucleic acid binding proteins conserved from *Escherichia coli* to man. *New Biol* 1992;4:290–298.

Yamasaki S, Anderson P. Reprogramming mRNA translation during stress. *Curr Opin Cell Biol* 2008;20:222–226.

Zamore PD, Tuschl T, Sharp PA, Bartel DP. RNAi: Double-stranded RNA directs the ATP-dependent cleavage of mRNA at 21 to 23 nucleotide intervals. *Cell* 2000;101:25–33.

16

Non-coding RNA

Beena Pillai

16.1 Introduction: Central dogma of molecular biology

Molecular biology was built on the basic tenet that the DNA in our cells is the source of all information that is decoded and eventually forms the basis on which proteins are made. Proteins were considered the foot soldiers of cell biology, carrying out various metabolic, regulatory and structural functions. Genomic DNA, the repository of genetic information, is faithfully replicated at high fidelity. Between the single-copy, informational DNA molecule and the large milieu of proteins playing functional roles were the intermediary RNA molecules (Crick 1970). The total RNA of a cell, when purified, can be visualized by gel electrophoresis. It appears as two major bands of ribosomal RNA easily visible due to the high abundance of rRNA, along with some smaller RNAs. However, messenger RNAs (mRNAs) that carry the code for producing protein products are quite scarce. mRNAs are a large and diverse group consisting of individual transcripts of sizes as variable as the genes they are produced from. Hence, each mRNA is not abundant enough to be visualized by gel electrophoresis. As a group, mRNAs are also not of a certain size uniform enough to be visualized as a single band. The mRNA in the cell is captured by ribosomes – a ribonucleoprotein machinery containing the ribosomal RNA and associated proteins – and translated into protein with the assistance of tRNA. Thus, the ribosomal RNA and tRNA were the main representatives of the non-coding RNA (ncRNA), while the large number of mRNA in the cell was labelled 'coding'. As our knowledge of eukaryotic systems grew, it became clear that other RNAs, especially the small nucleolar RNAs (snoRNAs), could form important structural components of ribonucleoprotein complexes involved in splicing. The telomerase, an enzyme consisting of an RNA and several proteins, provides another example of an ncRNA that plays an important role in the cell. In spite of the knowledge about a few ncRNAs, their diversity, complexity and functional capabilities were not thought to be comparable to that of proteins.

In the 1960s, a golden era for molecular biology, it was widely held that RNA was a prime candidate for the 'primordial molecule', the original self-replicating molecule (Woese 1967). This idea is captured by the phrase, 'the RNA world', first used by Walter Gilbert in 1986 (Gilbert 1986). Although time and again, the basic similarity of RNA to DNA has supported the idea that it can play a central role in cellular information transfer (Box 16.1), proteins were thought to carry out the more mundane cellular functions. Specifically, gene regulation, that is, the spatial and time-dependent expression of genes that controls the developmental programme of a cell and orchestrates cellular response to external stimuli was thought to be largely dependent on transcription factors. Transcription factors are proteins that typically bind to specific sequences in the DNA and recruit the RNA polymerase through diverse mechanisms to increase the rate of mRNA expression of specific genes. Some transcription factors such as the retinoic acid receptor can both bind to external ligands and localize to the cell nucleus to then function as transcription factors, thus combining the sensor and the effector in the same protein. Other transcription factors may be modified by kinases in the cytoplasm, to localize them to the nucleus. Whatever be the mechanism, transcription factors are central to deciding which gene should be expressed where and how. Thus, cells were thought to express a specific gene expression profile – the relative expression of a set of genes that defined the cells and are usually related to the specific functions of the cell – depending on their developmental state. For instance, a terminally differentiated neuronal cell would express relatively high levels of certain ion channels but may shut down certain metabolic capabilities. The sequential expression of transcription factors was thought to be the driving factor for this 'programmed gene expression'. In eukaryotes, nucleosomal organization and histone modifications reinforce this expression

Box 16.1 RNA Tie Club

The RNA Tie Club: A club formed by George Gamow and James D. Watson in the early, heady days of molecular biology has been fondly recounted by both Watson and Crick. The club had 20 members for the 20 amino acids and four honorary members for the four nucleotides.

The famous Adaptor hypothesis, accounting theoretically for a tRNA before it was shown to exist, was postulated by Crick and first informally presented as a paper to the RNA Tie club. The RNA tie club is often used as an illustration of how speculation and discussion, even incorrect ones can shape scientific thinking and provide hypothesis to test, since some of the theories discussed at the club were later proven wrong.

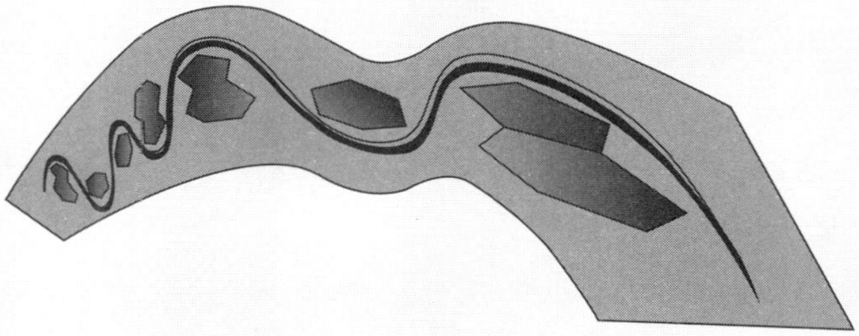

pattern by facilitating the expression of certain genes. Besides this, external cues such as the availability of nutrients and physical stresses such as changing temperature or the presence of an antigen in the body can be sensed by cell surface receptors. These signals can then be transmitted through conformational changes of a series of proteins, post-translational modifications of proteins and even changed localization, finally activating the transcription factors required to mount a 'dynamic gene expression change' encompassing hundreds or even thousands of genes that may be upregulated or downregulated. The T-cell activation response, for instance, results in the induction of a few thousand protein-coding genes. While the transcription factors, all proteins, re-programmed the cell, the ncRNAs, rRNA, tRNA and snoRNAs helped to express the required proteins by organizing the splicing machinery in the nucleus and the translation machinery in the cytoplasm. Thus, in the view of molecular biology of the 1990s, the ncRNAs, abundant, highly conserved, composed of a few species and expressed almost constitutively in all cells, had a largely supportive role to play in the expression of genes. Even in these roles, they worked in unison with a few protein partners, suggesting that the proteins formed the cell and performed its functions.

But all that was set to change soon. During the late 1990s and early 2000s, three parallel lines of investigations led us to new knowledge, suggesting that the RNA world is not just a primordial, life-giving environment, but hidden within cells today, there exists a second 'RNA world' teeming with ncRNAs, interacting with each other, proteins and DNA, and participating in a host of normal cellular functions (Eddy 2001). This chapter describes the major experiments that led to this new level of understanding, organizing the field along certain lines to help a reader navigate the overwhelming amount of research that has made it possible to unmask the 'RNA world' that lies within our cells. Craig C. Mello refers to this as the RNAi world in his Nobel Prize lecture on the discovery of RNA interference. In the interest of brevity and conceptual understanding, only prototypical examples have been included, while a lot of emerging complexity has been brushed aside. Some ncRNAs hold promise of applications: therapeutics, diagnostics and tools for studying genes. This chapter also shares the excitement of this relatively novel area of molecular biology, speculating on some applications we may see in the future.

16.2 Discovery of a functional non-coding RNA

Three main lines of evidence were instrumental in disrupting the protein-centred view of molecular biology. Time and again, laboratories working on the genetics of unconnected phenomena and across various model systems encountered mutants that were difficult to explain (Sen and Blau 2006). Molecular biologists found 'activities' for which the responsible protein could not be isolated (Crick 1970; Gunnery and Datta 1987). A typical example of this was the study of heterochronism in the laboratory of Robert Horvitz. Two young researchers working on the phenomenon of developmental expression timing of genes encountered a mutant with a dramatic phenotype: The female of *Caenorhabditis elegans*, a tiny worm made up of transparent cells that greatly facilitate observation of developmental changes, lays mature eggs through the vulva – openings in the middle of their body. However, the mutant worm, called lin-4, would accumulate unlaid eggs because the vulval opening was never properly formed (Fig. 16.1). A mutation in another gene, lin-14, could suppress the lin-4 mutant, allowing the first connections between these genes to be made. Try as they might, with different researchers making a go for it over many years, the protein for lin-4 could not be identified. The laboratories of Victor Ambrose and Gary Ruvkun could not find the protein responsible for the proper formation of the vulva. Working on the mutant again in the late 1980s, through

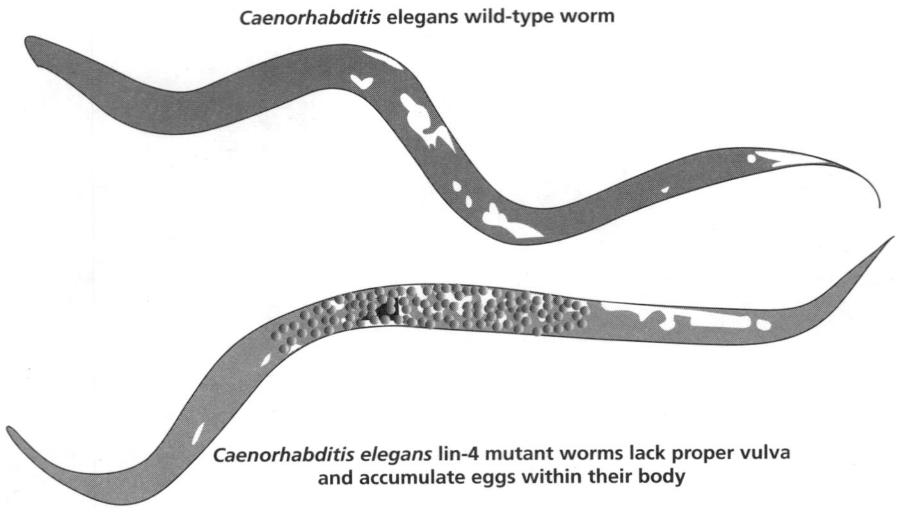

Figure 16.1 Schematic representation of the lin-4 mutant phenotype that led to the discovery of microRNAs by Ambros (2008).

systematic mutations and positional cloning, they found that it may be a small RNA, only 21 nt in length, that was the elusive culprit (Lee *et al.* 2004). Ruvkun, once a colleague of Ambrose and later a collaborator, while setting up his own new laboratory, had chosen lin-14 to work on. Looking closely at the sequences of the two genes, the presence of some mutually complementary regions attracted the attention of Ruvkun. They went on to show that the small RNA lin-4 could bind to lin-14 mRNA and control its expression. Even then, the small regulatory RNA Ruvkun and Ambrose discovered, then called stRNA for spatiotemporal RNA, was thought to be an oddity, perhaps of the *C. elegans* system (Lee *et al.* 1993; Wightman *et al.* 1993). In another 7 years, only one more such RNA was discovered, again in *C. elegans* (Reinhart *et al.* 2000). However, the second one, let-7, unlike lin-4, was highly conserved. It seemed to be present in practically all metazoan genomes and now we know that it is expressed in several tissues. The discovery of let-7 opened the floodgates for small RNAs that showed partial complementarity to regions of protein-coding mRNAs (Pasquinelli *et al.* 2000). These were called microRNAs (miRNAs) and have now become the best-studied class of small regulatory RNA.

16.3 Discovery of RNAi pathway

Around the same time, Craig C. Mello made an interesting observation that eventually turned out to have a direct bearing on the discovery of small RNAs. Although he was studying events during *C. elegans* development, it came as a surprise when he injected double-stranded RNA (dsRNA) into *C. elegans* only to find that it led to repression of specific genes. Eventually, with Andrew Fire, he discovered the RNAi pathway, that is, an endogenous pathway for dealing with dsRNA that can be used

Figure 16.2 Artist's representation of petunia flowers that led to the discovery of co-suppression. See Colour Plate 24 on page 517.

to regulate gene expression (Fire *et al.* 1998). They went on to win the Nobel Prize for Medicine in 2006, for work that they did in the late 1990s. The artificial small RNAs that they used for knocking down genes were called siRNAs (small interfering RNAs).

Several groups working with plants also knew that plants had special mechanisms for dealing with dsRNA (Hamilton and Baulcombe 1999), perhaps as a defence response to viral infections. One such observation also came up serendipitously when Napoli and Jorgensen were studying the pigmentation of petunia (Fig. 16.2). The anthacyanin pathway was responsible for the purple colouration in petunia. Chalcone synthase (CS) was thought to be the key enzyme to enhance the purple colouration as it catalyses the rate-limiting step in the pathway. But when Napoli and Jorgensen tried to over-express the CS gene, they got white flowers. Apparently, the CS gene, both endogenous and the externally expressed copy, was downregulated, prompting the cell to invoke the 'co-suppression' phenomenon (Napoli *et al.* 1990). Shortly thereafter, another group showed a similar phenomenon in *Neurospora crassa* and called it quelling (Romano and Macino 1992). Guo's laboratory, working on *C. elegans,* described a similar phenomenon in *C. elegans* (Guo and Kemphues 1995). Guo's intriguing observation was that both sense and antisense RNA resulted in the suppression of the expression of endogenous genes.

When Craig C. Mello and Andrew Fire made their observations, they not only showed a mechanistic basis of RNA degradation triggered by dsRNA but also integrated these previous observations and suggested that degradation of mRNAs triggered by complimentary small RNA derived from exogenous dsRNA, variously called co-suppression and quelling, was the same phenomenon. RNA interference, as it is known now, was immediately useful as it came at a time when scientists were grappling with the need to rapidly assign function to the large number of genes that had been discovered through genome sequencing. Knocking down the gene through the expression of dsRNA was an appealing alternative to the conventional and tedious route of mutations, screening and complementation.

16.4 Ultra-conserved non-coding regions in the genome

Genome sequencing had not just thrown up a large number of genes, it had also revealed secrets of the non-geneic regions of the genome. A cornerstone of biology is that if something is conserved, it must be important. With more and more genomes becoming available, this guiding principle was put to test and found to largely conform to expectations as exons were conserved better than introns and many essential genes were indeed highly conserved. The discovery of let-7 was quickly followed by the realization that miRNAs were a universal phenomenon because of their high conservation. However, not all regions of the genome showed expected levels of conservation. In the midst of introns and gene deserts were sometimes large and small stretches of ultra-conserved regions (Bejerano *et al.* 2004), implying that they must be important enough as nature did not allow it to be altered through the ever-shaping force of mutations (Fig. 16.3). Some of them were miRNAs and many others turned out to be other types of ncRNA. Ultra-conserved regions include enhancers and regions that harbour regulatory signals also, but a significant subset is indeed ncRNA.

Disparate lines of evidence emerging from different model organisms all led to the same finding: that the eukaryotic cell expresses a previously unexpected number of ncRNAs. They were often highly conserved (Diederichs 2014) and tissue specific in expression (Lagos-Quintana *et al.* 2002), and mutations in them, although rare due to their small size, could result in

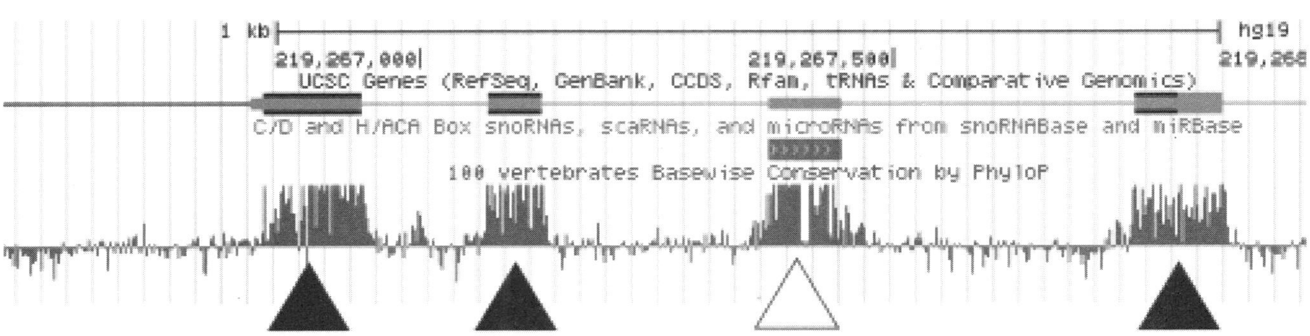

Figure 16.3 A region of human chromosome 2 that harbours the protein-coding gene *CTDSP1* and within its intron, the microRNA (miRNA) hsa-miR-26b as displayed in the UCSC Genome Browser (www.genome.ucsc.edu). The black arrowheads highlight the conservation in protein-coding exons, while the white arrowhead shows the comparable conservation in the miRNA region. See Colour Plate 25 on page 517.

strong phenotypes. Introduction of ncRNA into the cell could also result in the suppression of similar genes, giving a convenient tool to study genes.

16.5 Definition and types of non-coding RNA

Today, with the rapid growth of sequencing technology, it has become relatively straightforward to identify all the RNA molecules expressed in the cell. As the depth of sequencing increases, we can dig for RNAs that are expressed at lesser and lesser abundance and compare them with highly expressed genes. Most techniques based on hybridization could report accurately about three orders of magnitude, while RNA sequencing can routinely report 5–7 orders of magnitude. Once it became clear that the cell expresses small RNA, it became easier to identify small RNAs by size fractionating total RNA and sequencing various fractions. Because of this trend, today, ncRNAs are classified mainly based on their size. However, as we learn more about their functions, a new type of classification based on their functional roles is likely to emerge. For now, the classification based on size also serves a purpose because RNAs of different sizes seem to have different types of functions.

16.5.1 microRNAs

The best-studied class of small ncRNA, 19–21 nt in length, is a single-stranded (ss) RNA used by the protein complex to repress protein expression from mRNAs that have regions complementary to the miRNA (Bartel 2004). Most miRNAs are highly conserved, and several are expressed in specific cell types or during specific developmental phases (Dugas and Bartel 2004; He and Hannon 2004; Pasquinelli *et al.* 2005; Pillai 2005; Vella and Slack 2005). Loss of certain miRNAs can lead to cancer as they are involved in regulation of cell division, while others are elevated in cancers (Calin and Croce 2006a, 2006b). Their deep involvement in various diseases has attracted attention, and efforts are on to explore their therapeutic potential. One miRNA-based lead, miR-34, is already in clinical trials for hepatocellular carcinoma.

16.5.2 siRNAs

siRNAs are the synthetic equivalent of the endogenous miRNA. Although about the same size, they are derived from artificial dsRNA (He and Hannon 2004). Some of the cellular enzymes involved in the biogenesis of miRNA are also involved in the processing of the dsRNA to siRNA. Both siRNAs and miRNAs cause target repression, but a major difference is that siRNAs are usually completely complementary to their target sites, while miRNAs seem to need only partial complementarity. Plant small RNAs are notable exceptions, as they are endogenously produced but bind to targets with complete complementarity (Rhoades *et al.* 2002).

16.5.3 piRNAs

While sequencing small RNAs prepared by size fractionation of total RNA, it was found that a class of small RNAs were slightly longer than miRNAs, ranging from 25 to 35nt. Their name is derived from their association with a class of proteins called PIWI proteins. They have specific modifications at their ends and a distinct biogenesis pathway from miRNAs, and are often involved in the silencing of transposons (Aravin *et al.* 2006; Girard *et al.* 2006; Grivna *et al.* 2006; Kim 2006; Watanabe *et al.* 2006). Although there have been isolated reports of piRNAs expressed in somatic tissues, they are most abundant in germline tissue. There are large clusters of piRNA in the genome, suggesting that they are processed from long precursors. The repeat-associated siRNAs (rasiRNAs) are now considered a subset of piRNAs.

16.5.4 lncRNAs

lncRNAs or long ncRNAs are defined as RNAs longer than 200nt but with no apparent open reading frame (ORF), that is, a run of codons uninterrupted by stop codons (Fatica and Bozzoni 2014). Increased sequencing prowess has allowed the discovery of large numbers of such RNAs. Assigning function to them is a major challenge because of their large numbers (Rinn and Chang 2012). Although small RNAs are a recently added feature of eukaryotic gene regulation, some lncRNAs such as Xist have been known for a long time and studied extensively for their effects on chromatin organization and gene silencing (Pontier and Gribnau 2011).

16.5.5 lincRNAs

The long intergenic ncRNAs (lincRNAs) are very long and a distinct subclass of the lncRNAs defined by their genomic context (Khalil *et al.* 2009; Guttman *et al.* 2011). Many lincRNA genes bear histone modifications that support a robust expression, such as trimethylation of the fourth lysine of histone (H3K4me3) at the promoter and H3K36me3 modifications all along the transcribed length. Some typical lincRNAs are known to drive the epigenetic modification of the loci they are expressed from, thus influencing the expression of genes in their vicinity. A well-studied lincRNA, Hotair, is described in greater detail later in this chapter.

A cursory examination of the types of ncRNA listed in Table 16.1 shows that there are thousands of ncRNA, several times more than the protein-coding genes of

Table 16.1 Classes and protypes of regulatory RNAs

Name	Subtypes/ Examples	Size	Key features/Function	Organism
RNAI		108 nt	ColE1 replication blocker	Bacteria
pOUT (promoter-Tn10 transposon)		70 nt	Repress translation of transposase RNA	Bacteria
MicF		174 nt	Repress translation of OmpF RNA	
Riboswitches			Present within the 5′ UTR of protein-coding transcripts	Eubacteria, Archaebacteria
	Temperature regulated			
	Metabolite regulated		Most metabolite bound riboswitch conformations are repressive	
sRNA		Many sRNAs need the Hfq RNA chaperone	Can mask the ribosome-binding site; some can trigger RNaseE-mediated cleavage of complementary RNA	Mainly gram-negative bacteria
	CsrB		Regulates carbon usage and bacterial motility by binding to CsrA protein	
	6S		Mimics open promoter; binds to sigma factor	
	GlmY	~160 nt	Activates GlmS mRNA by interfering with another sRNA, GlmZ	Eubacteria
	IsrR		Represses *cis*-encoded IsiA transcript	Cyanobacteria
	Qrr		Quorum sensing	*Vibrio cholerae*
	RyhB		Iron sulphur cluster regulation	*Escherichia coli*
CRISPR		~50 nt	Phage immunity; has several applications such as oligotyping and genome editing	Eubacteria, Archaebacteria
miRNA		19–21 nt	Dicer dependent; thousands of miRNAs are known now, each can target many mRNAs; usually repressors of translation or post-transcriptional mRNA stability	Metazoans
miRtrons		19–21 nt	Derived from introns, different from intronic miRNA because they are not dependent on Drosha	
piRNA		24–36 nt	Abundant in germline tissue	
rasiRNA			Repeat-associated siRNA; now considered a subclass of piRNA	
eRNA			Enhancer RNA, derived from near the enhancer sites; includes two classes: 1D eRNAs are long polyadenylated transcripts reflecting the nearby gene and often incorporating an additional first exon; 2DeRNAs are shorter, bidirectional products	
Circular RNA			One circular RNA, with several binding sites for microRNA miR-7, is a sponge for the neuronal miRNA; covalently joined ends make these circular RNAs highly stable	
lncRNA		>200 nt		
lincRNA			A subclass of lncRNA defined by their genomic location, found in 'intergenic regions'	
Endo-siRNA/ esiRNA			Derived from endogenous dsRNA precursors; involved in transposon control; enriched in germline tissue	
21URNA		21 nt	Precisely 21 nt long, starts with a uridine residue	*C. elegans*
Guide RNA		Variable	RNA editing: insertion or deletion of uridine residues into mitochondrial mRNAs in kinetoplastid protists	Trypanosomes

the genome. The large numbers in fact prompted a transient view that perhaps transcription is error prone, producing a large number of transcripts from all over the genome in a poorly controlled process. However, as we dig deeper, each lncRNA seems to offer a new discovery, a new mechanism by which it acts. Thus, the currently popular view is that, unbelievable though it may seem, there is indeed a swarm of RNAs of various sizes, forming the scaffold of multi-protein complexes, or binding to complementary RNA, folding into structures that can be tweaked by the binding of metabolites, small molecules and even physical stresses such as heat to organize the cell, execute its functions and ensure its survival. The conventional view to understanding the classes of ncRNAs described above is to explore in detail the proteins that are required for their expression, processing and modification, followed by the mechanisms by which they may act and the impact they have on the cells' functions. However, in this chapter, we take a detour at this point to revise and understand the basics of RNA structure and the principles of RNA–RNA interaction before we come back to the story of ncRNAs.

16.6 RNA–RNA interactions

It is now clear that RNA has two key features of proteins and DNA, the other major macromolecules involved in information transfer of the cell. With DNA, it shares the ability to bond with complementary nucleic acids in a sequence-specific manner. This ability to 'zip-up' can be used to identify sequence motifs in the DNA or other RNA molecules. However, ssRNA also shares an important property of proteins – the ability to fold back on itself to form complex three-dimensional structures that allow binding to other RNA, proteins and small molecules by navigating a shape and charge contour map. The repetitive nature of the backbone provides a charged, uniform and generic way of interacting with proteins. Thus, the combination of backbone charge, sequence information encoded in the bases and the three-dimensional structure make RNA molecules complex, versatile mediators of intermolecular interactions (Fig. 16.4).

Let us examine examples of each type of interaction. When siRNAs and miRNAs interact with their targets,

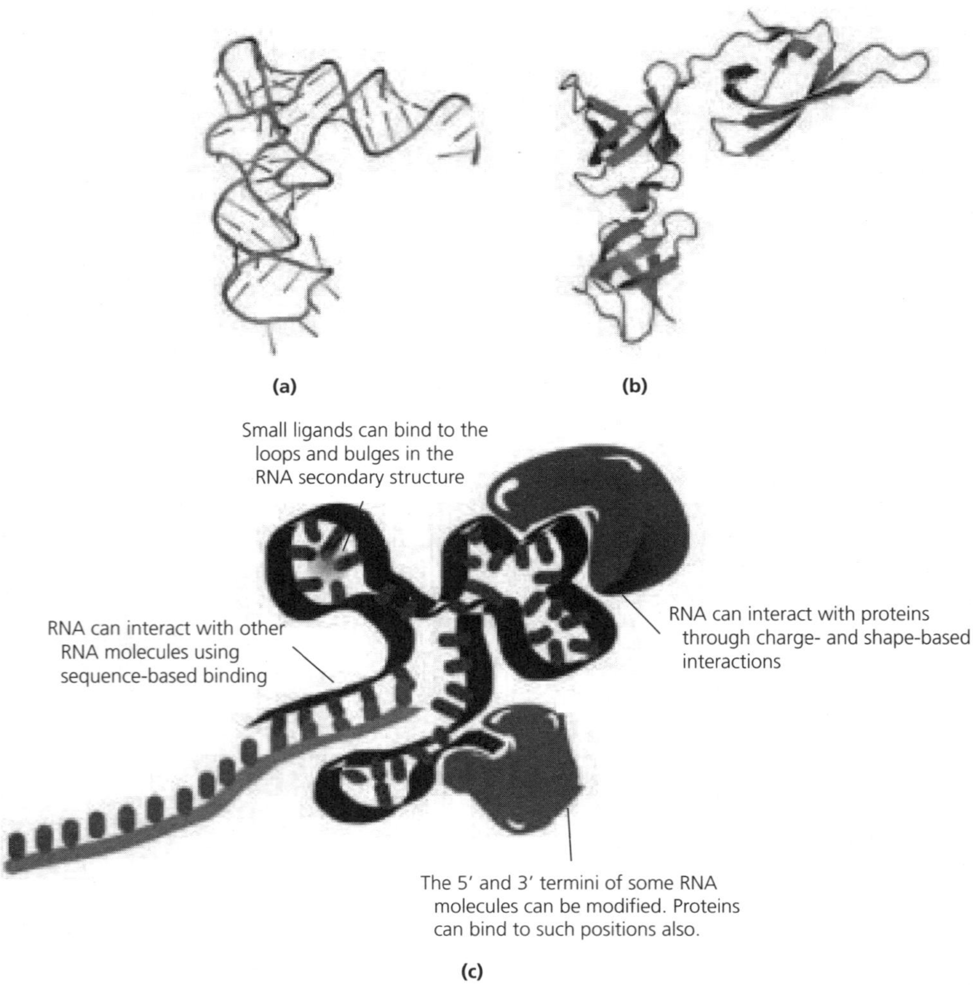

Figure 16.4 The three-dimensional L-shaped structure of the non-coding transfer RNA (tRNA) (a) is mimicked by some proteins (b) that modulate translation. The tRNA coordinates from chain R of PDB ID 1ASY and EF-P coordinates from chain D of PDB ID. 3A5Z were used to make figures in Pymol (Image by Bhupesh Taneja, CSIR–Institute of Genomics and Integrative Biology, New Delhi). Schematic representation of the types of interactions RNA can show with small molecules, other RNA, DNA and proteins (c); see Colour Plate 26 on page 518.

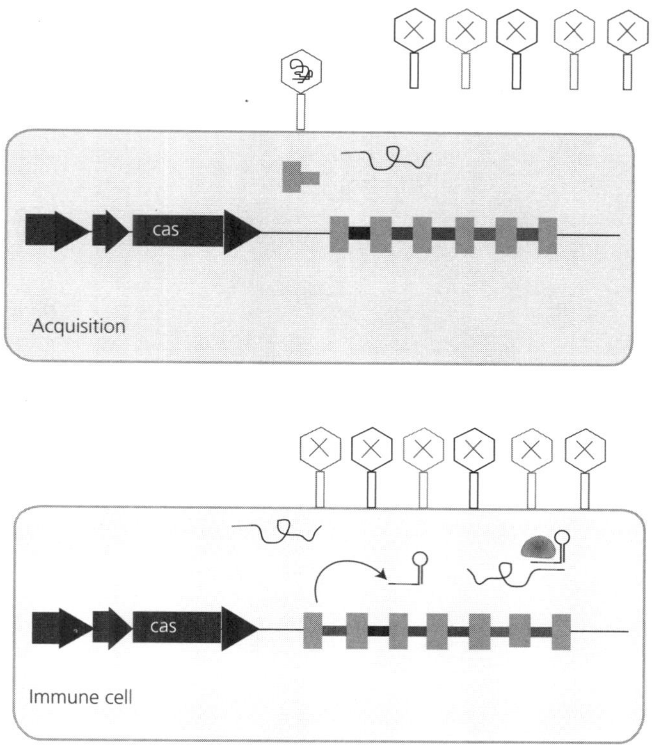

Figure 16.5 The CRISPR locus consists of the CAS genes, followed by a leader sequence and several repeats (vertical boxes) separated by spacers (horizontal boxes). A new spacer, with an additional repeat, is acquired when a phage infects the bacterial cell. The transcripts from the CRISPR locus are processed (curved arrow) to produce the ncRNA, which binds to the infecting phage genetic material during subsequent infections and directs its cleavage by protein factors. This provides acquired immunity against the phage. The hexagons represent phages. Phages against which the cell is immune are represented with a cross within the hexagon. See Colour Plate 27 on page 519.

they primarily use sequence-based interaction. The seed region of the miRNA, that is, the 7-nt stretch from the second to the ninth position, in the miRNA is an example of the importance of this type of interaction. miRNA interaction with target sites in mRNA is greatly favoured by perfect interaction between the seed region and its complementary region in the target site (Box 16.2). However, the interaction of the miRNA with proteins that bring about target repression is mediated through backbone interactions. The 2'-OH group, which forms the key difference in the DNA and RNA backbone, provides a uniformly repeated potential site for hydrogen bonding.

Like the hydroxyl group at the C2 position, the presence of uracil in place of thymine is a distinctive feature of RNA. Uracil plays an important role in RNA structure by allowing the formation of G:U base pairs (bp). Interestingly, some classes of small RNAs are uracil rich, specially some of the spliceosomal RNAs important in the recognition of introns. Uracil is also found at specific positions in small regulatory RNA. For instance, it is frequently found at the first position of piRNAs and in a small, little known class of ncRNA called 21U RNAs in *C. elegans* marked by their uniform 21-nt length.

Riboswitches (Winkler *et al.* 2002; Cheah *et al.* 2007) offer a great example of three-dimensional folded structures in RNA. One of the best-studied riboswitches is the SAM riboswitch that binds to the methyl donor S-adenosyl methionine. After the methyl transfer, a toxic by-product, S-adenosyl homocysteine (SAH), is produced. SAH and SAM are structurally similar small molecules, but riboswitches can accurately distinguish between them (Montange *et al.* 2010). High-resolution structures of the SAM riboswitches found in different organisms have revealed that the same structural fold is not responsible for the individual SAM-binding riboswitches (Roth and Breaker 2009).

RNA footprinting is used widely to map the secondary structure of RNA. The basic principle is that the bases that are engaged in the secondary structure formation are protected from enzymatic or chemical modification; their differential susceptibility can then be mapped by endonucleolytic cleavage. Hydroxyl radicals can cleave at solvent-exposed regions, while specific RNAses such as RNAse T1 and RNAse V1 have preferences for ssRNA and dsRNA, respectively. Compared with structure prediction *in silico*, structural mapping has the advantage that the *in vitro* transcribed RNA is folded in solution.

In the famous words of Francis Crick, 'If you want to understand function, study structure'. This adage is indeed true for non-coding RNAs today. New methods of mapping RNA structure in high throughput have been developed recently. One such method, SHAPES,

Box 16.2 Calculation of ΔG and effect of mismatch on binding affinity. What would be the effect of a single mutation in an miRNA–target interaction? Even ignoring the effect of protein binding and other contextual information, the effect of change in ΔG by 1 kcal/mol is about a fivefold drop in the equilibrium binding constant

$$\Delta G_{wt} = -RT\ln(K_{eq_{wt}})$$

$$\Delta G_{mut} = -RT\ln(K_{eq_{mut}})$$

$$\frac{K_{eq_{wt}}}{K_{eq_{mut}}} = \frac{e^{\frac{\Delta G_{wt}}{-RT}}}{e^{\frac{\Delta G_{mut}}{-RT}}}$$

$$= e^{\left[\left(\frac{\Delta G_{wt}}{-RT}\right)-\left(\frac{\Delta G_{mut}}{-RT}\right)\right]}$$

$$= e^{\left[\frac{(\Delta G_{wt} - \Delta G_{mut})}{-RT}\right]}$$

For a change in ΔG from –20 kcal/mol to –10 kcal/mol

$$\frac{K_{eq_{wt}}}{K_{eq_{mut}}} = e^{\frac{-20-10}{-RT}} = e^{\frac{-10}{-RT}} = e^{\frac{10}{RT}}$$

$$e^{\frac{10}{0.615}} = e^{16.26}$$

For a change in ΔG by one unit

$$\frac{K_{eq_{wt}}}{K_{eq_{mut}}} = e^{\frac{1}{0.615}} = e^{1.626} = 5.08$$

abbreviation for selective 2'- hydroxyl acylation analysed by primer extension, relies on the chemical N-methylisotoic anhydride (NMIA), which forms adducts at the 2'-OH (Roth and Breaker 2009). These adducts block the extension by polymerases. Therefore, in a primer extension following treatment with the NMIA, each termination event is an indication of an accessible 2'-OH site. Data is collected at almost every position in the RNA, but the relative accessibility of single- and double-stranded regions gives signals of different intensities. SHAPE has been used to solve the structures of entire viral RNA genomes. *In vivo* folding maps of transcriptomes are now being generated. With the growing emphasis on RNA structure, perhaps it is a matter of time before we have knowledge of RNA folds like we today appreciate the nature of the alpha helix or beta-sheet for proteins.

In the cell, RNA structure is likely to be influenced by the binding of protein factors and other complementary RNAs. How does one study RNA–protein interactions? A significant portion of the protein-coding genes in any mammalian genome is composed of RNA-binding proteins. The specificity of an antibody binding to a protein is used to immunoprecipitate the RNA-binding protein. All the RNAs associated with the protein are co-immunoprecipitated and then separated from the protein through the action of proteases. The RNA is then subjected to RNA-seq or a microarray analysis to identify all the RNAs bound to the protein. These approaches are called RIP-Seq or RIP-ChIP (Keene *et al.* 2006).

16.7 Evolution of ncRNA

Besides structure, a source of insight into the biogenesis and function of ncRNA comes from their comparative genomics, both the conservation of miRNA sequences and their organization in the genome. Let us first examine which classes of organisms have what types of ncRNA and then look at the genomic organization of ncRNA genes.

16.7.1 Classes of ncRNA in different branches of the tree of life

Bacteria were initially thought to contain only the rRNA and tRNA – abundant, translation-related ncRNA. Later, sRNAs (small RNAs) produced by certain bacteria to regulate the expression of target genes were discovered to be analogous in function to miRNAs, although they have a different biogenesis pathway (Waters and Storz 2009).

More recently, a kind of ncRNA employed for the defence of bacterial cells from attacking bacteriophages has been discovered. CRISPRs (clustered regularly interspaced short palindromic repeats) quite accurately describe what they are. Short, almost palindromic sequences are found evenly spaced from each other in clusters in the bacterial genome (Sorek *et al.* 2008). The number of repeats in a cluster varies widely from a couple to a few hundreds. The locus usually has an AT-rich leader sequence. The spacer sequences and the repeats are about 23 nt to 50 nt each. Almost at the same time, three groups discovered that the spacer sequences showed homology to different viruses or plasmids (Bolotin *et al.* 2005; Mojica *et al.* 2005; Pourcel *et al.* 2005). Eventually, this led to the suggestion that when bacteria are infected by phages, small portions of their genome may be clipped and inserted into the host microbial genome at the CRISPR locus. Thus, the bacteria carry a catalogue of previous infections at the CRISPR locus. On subsequent infections, small RNAs derived from the CRISPR locus can bind to the invading virus and reduce the susceptibility of the host. Thus, in concept, CRISPRs are analogous to antibody-mediated immunity, wherein previous exposure results in a rapid eviction of the pathogen (Fig. 16.5). CRISPR systems of various bacteria vary widely, but nearly 40% of all bacteria explored so far contain them.

The baker's yeast *Saccharomyces cerevisiae* remains exceptional in terms of RNAi because it does not seem to have any of the protein components required for RNAi. The RNAi genes and Dicer, Drosha and Argonaute proteins do not seem to have clear homologues in this yeast. A concerted effort to identify distant homologues did show the presence of small RNAs and components of the RNAi machinery in the closely related *Saccharomyces castelli* (Drinnenberg *et al.* 2009). In spite of this discovery a few years back, the RNAi pathway of *Saccahromyces cerevisiae* is still a matter of speculation. *Monosiga brevicollis*, a marine, single-celled colony-forming organism, also seems to have no homologue of miRNAs or miRNA-processing genes. But one cannot readily assume that the miRNA pathway evolved only in multicellular eukaryotes because many other single-celled eukaryotes and even some common ancestors have miRNAs or even miRNA-processing genes. Thus, the predominant view is that certain organisms seem to have lost the miRNA pathway after it evolved once very early on in eukaryotes (Berezikov 2011). The small RNAs of the starlet sea anemone *Nematostella vectensis* and the sponge *Amphimedon queenslandica* have been sequenced to explore the early miRNAs. They contained about 40 and 8 miRNAs each and several proteins, homologous to the metazoan miRNA biogenesis pathway. miR-100 is one of the earliest evolved and most highly conserved miRNAs found from sponges to humans (Chen and Rajewsky 2007). Besides the miRNAs, another small RNA of ancient origin is the piRNA.

In thinking about plant small ncRNAs, there are two very important factors to consider. In many vertebrates, miRNAs are a major class of small RNA. However, in plants, the pool of small RNA is relatively more complex and miRNAs may even be a minor class. Further, plants have a special RNA polymerase called RNA polymerase IV/V that is involved in producing endogenous siRNA (see above for differences between miRNA and siRNA).

The human genome consists of more than 1000 miRNAs according to the miRbase, the online repository of miRNA sequences (Griffiths-Jones 2006; Kozomara and Griffiths-Jones 2014). However, a recent attempt at listing about 1800 'high-confidence' miRNAs includes only 20% of these. Thus, the human sequence data seems to be prone to misannotations, leading to an overestimate of miRNAs. Barring such annotation issues, it is still perhaps acceptable to think of miRNAs as a diverse and major class of regulatory RNAs. In spite of reports of a few somatic ones, piRNAs are largely restricted to germline tissue, where they are very abundant.

Looking at the 'tree of life' from the point of ncRNA origin and evolution, three general principles emerge (Fig. 16.6). piRNAs and miRNAs have very early origins but have diversified in bilaterans, keeping up with the increasing complexity of animals. The protein machinery required for the generation of miRNAs and piRNAs are also quite ancient, although there are differences in the exact complement. For instance, Pasha, the interacting partner of Drosha, is known to be used only for miRNA biogenesis and is absent along with miRNAs in some organisms. However, these organisms continue to have the AGO proteins or Drosha, presumably to produce siRNAs or carry out some other RNA processing in their cells. In some organisms, there is a sudden spurt of the number of homologues of some RNAi machinery components. For instance, *Arabidopsis* has 10 homologues of the argonaute protein, each one of which seems to preferentially bind to one type of small RNA (Mi *et al.* 2008).

lncRNAs, with their poor conservation, heterogenous mix of possible mechanisms of action and the lack of defined RNA or protein partners, cannot be easily visualized in Fig. 16.6. Nevertheless, an intuitively appealing view is that lncRNAs contain small conserved stretches that are key to their function and highly conserved promoters that hint at a conservation in expression levels and patterns. Some lncRNAs are simply precursors of miRNAs and other known small RNAs, while others may contain small ORFs that give rise to peptides. But even after accounting for these 'conventionally explainable lncRNAs', there remains an overwhelming number with low expression and conservation to be studied.

16.7.2 Genomic organization

ncRNAs are found in almost all genomic contexts. lncRNAs may overlap protein-coding genes in the sense or antisense strand. They may also be bidirectional, transcribed outwards from a shared promoter region, usually less than 1000 bp apart. Some lncRNAs are embedded within larger transcription units and others are apparently far away from other genes (Mercer *et al.* 2009). miRNAs also arise from diverse locations. About half of all human miRNAs arise from introns of protein-coding genes. But it would be an over-simplification to assume that they also share the host genes functions and expression domains. Some do, but it is far from a universal principle. Other miRNAs arise from their own precursors, which resemble mRNAs in being capped and polyadenylated, but are relatively less abundant, perhaps due to their role as a transient intermediate in the production of stable miRNAs. Some distinct miRNAs arise from a common precursor RNA, even from the same intron. There are certain loci in the human and mouse genomes that give rise to a large number of miRNAs. The *HOX* gene locus is an interesting example illustrating the positional conservation of the genomic locus of one miRNA (Fig. 16.7) (Yekta *et al.* 2004), while the *C19MC* locus is an amazing example of a large miRNA cluster.

The human genome has four *Hox* gene loci, from which arise miRNAs belonging to two families. Two closely related members of the miR-10 family, miR-10a and miR-10b, arise from two *HOX* gene loci, while miR-196a and miR-196 arise from two other *HOX* gene loci. By comparing the organization of the *HOX* gene loci and the embedded miRNA genes, it is immediately obvious that their genomic context is conserved. Interestingly, tagging a reporter RNA with the miR-196 target site showed that the cleavage of HOXB8 is brought about by miR-196a, thus restricting its expression in the appropriate area during development (Yekta *et al.* 2004). In

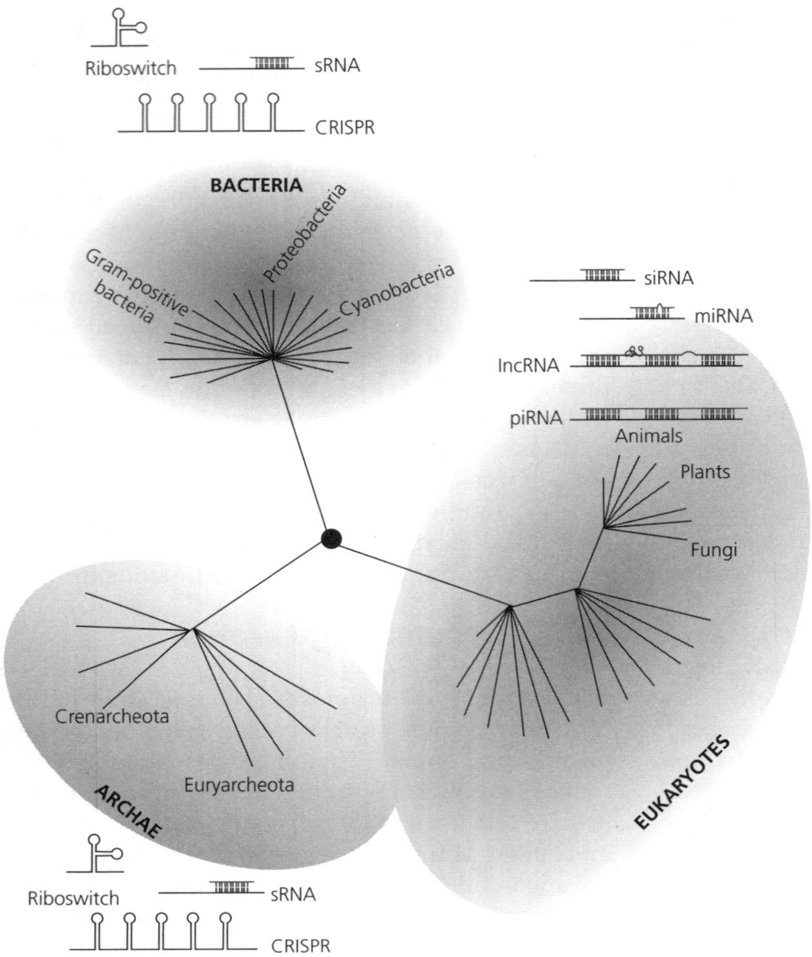

Figure 16.6 The major types of non-coding RNA found in different classes of organisms. See Colour Plate 28 on page 520.

insects, a similar function of the miR-iab-4 locus in repressing an upstream *hox* gene in the Antennapedia locus has been shown by Eric Lai and coworkers. This locus expresses two miRNAs from the sense and the antisense strands, both of which have multiple binding sites in target *hox* gene RNAs. Misexpression of miR-iab-4 antisense miRNA transforms halteres to wings, thus unequivocally demonstrating that the miRNA qualifies to be a *hox* locus member through its ability to cause homeotic transformations.

The chromosome 19 miRNA cluster *C19MC* locus is the largest known cluster of miRNAs in the human genome (Bortolin-Cavaille *et al.* 2009; Noguer-Dance *et al.* 2010). In sharp contrast to the *HOX* cluster miRNAs, the *C19MC* cluster is seen only in primates and is strongly imprinted, ensuring that only the paternal copy is expressed during pre-natal developmental stages. In adult tissues though, both alleles seem to be methylated and repressed. This cluster of miRNAs, containing about 59 miRNAs from a region spanning about 100 kb, is one of the extreme examples of miRNA gene clustering in the genome. Recently, a genomic fusion of the C19MC locus, placing it under the nearby brain-specific *TTYH* gene, allowed the aberrant expression of the miRNAs in infants (Kleinman *et al.* 2014). The miRNAs then set in motion a neural developmental programme that should normally be expressed only in early development, leading to fatal paediatric tumours. At least in this case, the genomic position of the miRNA is certainly critical for its correct spatiotemporal expression.

Thinking more deeply about genomic positions, miRNAs offer an interesting conundrum. Some miRNA genes are expressed from multiple genes in different locations in the genome. They produce identical end products, such as in the case of the miR-196a-1 and -2 in the *HOXB* and *HOXC* loci. On the other hand, functionally different miRNAs and protein-coding genes seem to be expressed from the same primary transcript in many other cases. Thus, in analysing miRNAs and lncRNAs, their genomic location has not offered any general insights into their biogenesis, but at least in some cases, it is linked to their function.

The genomic location of piRNAs by contrast offered the first clues about their biogenesis. The piRNAs arise from clusters that coincide with transposable elements in the genome. They map to both strands of the genomic

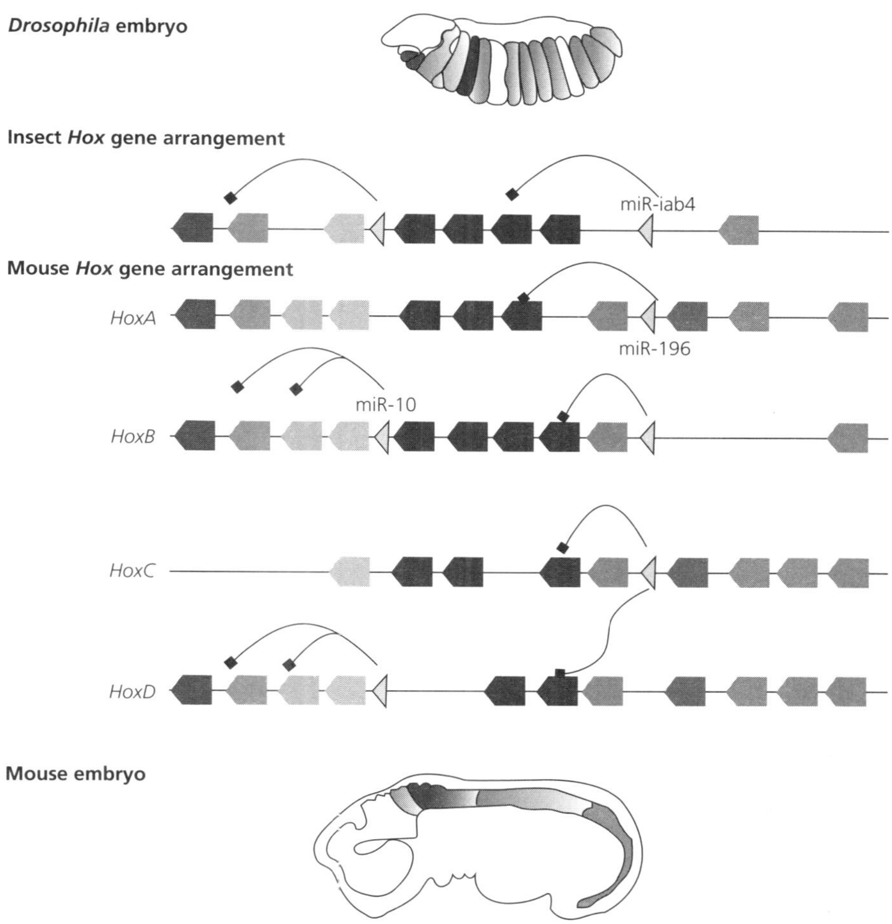

Figure 16.7 The mammalian and fly Hox clusters contain genes that are arranged in the spatial order of their expression along the anterior–posterior axis (Yekta *et al.* 2004). Two microRNAs found within these clusters can repress the more anterior *Hox* genes in the cluster. As microRNAs act in *trans*, they can target *Hox* genes from distant clusters. See Colour Plate 29 on page 521.

DNA in a specific pattern that led to the current ping-pong model of piRNA biogenesis. Once long primary transcripts are made from transposons, they are processed by the protein Zuchini and bound to PIWI proteins. These primary piRNAs guide their PIWI protein partners to slice antisense transcripts, thus generating a new antisense piRNA. This cycle goes back and forth, asymmetrically, to produce abundant antisense piRNAs that eventually destroy the transcripts from the transposable elements (Kleinman *et al.* 2014). This primitive pathway seems critical for genome integrity because loss of the PIWI proteins in *Drosophila* and mouse lead to genome instability.

16.8 miRNA biogenesis and action

Of the ncRNAs, the biogenesis of miRNAs is the best understood. miRNAs arise from long precursor transcripts called pri-miRNA. These primary transcripts resemble mRNAs in that they are capped, polyadenylated and form stem-loop secondary structures corresponding to inverted repeats (Bartel 2004). The stem-loop structure is processed by the 'microprocessor' complex. The Drosha–DGCR8 (called Pasha in *Drosophila*) complex is the functional core of the microprocessor complex. These proteins of the microprocessor complex are critical in chopping the stem-loop structure at the base. There are two alternative routes to forming miRNAs. mirtrons are miRNA stem-loop structures that are released as introns are spliced out (Berezikov *et al.* 2007; Okamura *et al.* 2007). A less prevalent pathway is the production of stem-loops from snoRNAs that contain such secondary structures. Ultimately, the stem-loop known as the pre-miRNA is exported into the cytoplasm by exportin, and in the cytoplasm, a complex containing the highly conserved and functionally versatile Dicer protein cuts the loop region off. This leaves a dsRNA intermediate, whose fate is similar but not identical to injected dsRNAs used in RNAi technology. In most dsRNAs, one of the arms is preferentially loaded into a protein complex called the RNA-induced silencing complex. The fate of the less abundant strand has been debatable. In most miRNAs, the degradation of one of the strands results in a selective miRNA retention from one of the arms of the stem-loop structure. In several miRNAs though, both strands are retained, and in some rare cases, the predominant arm may even switch between tissues or between organisms. Many pre-miRNAs give rise to a more abundant miRNA and a less abundant one from the opposite strand. The less abundant strand is usually called miR*. miR-9 and miR-9* make a pair of miRNAs that arise from the same precursor from either arm of the same hairpin and seem to regulate the same target. The picture of miRNA processing described above is only a thematic overview. Today, we have convincing structural insights into how the miRNA biogenesis pathway members cleave their targets. We also know that many ancillary proteins work closely in these complexes to modulate the processing and stability of the miRNA. 3′-end modification through adenylation, uridinylation or 2-O-methylation can all impact the stability of miRNAs (Li *et al.* 2005). In general, miRNAs are thought to be highly stable. But some miRNAs have a rapid turnover rate that can be important for their function (Krol *et al.* 2010).

A major component of the miRISC complex is the Argonaute protein. There are four AGO protein members in mammals, several more in plants, while a single AGO seems sufficient in *Drosophila* (Tolia and Joshua-Tor 2007). The many attempts to elucidate rules for which miRNAs prefer which AGO proteins are beyond the scope of this discussion. A general theme is that mammalian AGO2, the only AGO that can directly cleave the target mRNA, generally binds to miRNA–target pairs that are completely complementary – a situation reminiscent of siRNAs. In fact, the siRNA-induced silencing complex called siRISC is analogous to miRISC. The other AGO proteins seem to modulate the translation from the target mRNA by inhibiting translation initiation or promoting mRNA degradation by deadenylation. Close interactions between members of miRISC-like GW182 protein with the translation initiation complex and the CCR4–CAF1 deadenylation complex suggest that miRNAs impact target gene expression through multiple, parallel pathways (Eulalio *et al.* 2008). The basis for the apparent preference for one mechanism over another in certain contexts is not understood (Valencia-Sanchez *et al.* 2006).

16.9 How do miRNAs regulate functions?

There is a general consensus now that, a few recent reports of activating small RNAs notwithstanding, miRNAs are generally repressors of gene expression. What sort of genes do they repress and what sort of effects can this trigger? In the case of transcription factors, we have learned several years back, through several natural mutations and directed mutational studies, that their loss can lead to the shutdown of entire programmes of gene expression and cells can be directed to alternative fates such as apoptotic cell death, developmental arrest and metabolic auxotrophy by the loss of transcription factors. More recently, as few as four transcription factors called the Yamanaka factors have been shown to reprogramme differentiated cells into induced pluripotent states and many factors have been shown to trans-differentiate one cell type from another. In contrast, the expression of individual mammalian miRNAs does not usually result in such large swathes of gene expression changes. This seems to be both due to the small magnitude of change they cause on their direct targets and due to their positioning within gene regulatory networks. Many miRNAs function in 'tuning' gene expression, that is, in tweaking the expression of a gene rapidly and slightly to respond to changes in the environment (Bartel and Chen, 2004; Bartel 2009). These changes maybe then reinforced through a more robust transcriptional response. However, some miRNAs can impact entire developmental programmes. In *C. elegans*, a cascade of gene expression is involved in establishing the identity of two bilaterally asymmetric neurons using miRNAs (Bartel 2009). In the mammalian brain, the expression of miR-124 directs the expression of several neuronal splice variants and can lead to neuronal differentiation (Makeyev *et al.* 2007). Similarly, the expression of miR-125 in the haematopoetic lineage is thought to 'cause' leukaemia by affecting the differentiation of white blood cells in mice (Bousquet *et al.* 2010).

16.10 Functions of lncRNAs

Unlike miRNAs, lncRNAs are a large and heterogenous class of ncRNAs that may impact gene expression in diverse ways (Mercer *et al.* 2009; Fatica and Bozzoni 2014). In fact, it is still debatable whether all of them will be functional. For instance, recently, some lncRNAs have been reannotated as they are now proven to have peptide-coding capability. The other reservation about lncRNAs is that they are poorly conserved and individually expressed at low levels. However, the high conservation of short stretches within them and in their promoters suggests that we may need new ways of studying function. In the absence of one unifying principle, several prototypical lncRNAs are described in Table 16.2. Their modes of action offer frameworks to think about other lncRNAs. But like proteins, each lncRNA may well reveal a character of its own. Furthermore, some really long lncRNAs may include subdomains that use different strategies to work together, much like multi-domain proteins.

As a general principle, lncRNAs have certain inherent features that allow them to perform specialized functions. lncRNAs are associated with the locus from which they are transcribed, either by direct DNA–RNA

Table 16.2 Some well-known lncRNAs and their roles

lncRNA	Name	Process	Phenotype (loss of function)	Reference
Xist	X (inactive)-specific transcript	Dosage compensation	Embryonic lethal	Reviewed in Fatica and Bozzoni (2014)
Tsix	X (inactive)-specific transcript	Dosage compensation	Embryonic lethal	
Kcnq1ot1	Kcnq1 overlapping transcript 1	Genomic imprinting	Growth defects	
Airn	Antisense Igf2r (insulin-like growth factor 2 receptor) RNA	Genomic imprinting	Growth defects	
Fendrr	Foxf1 adjacent non-coding developmental regulatory RNA	Gene expression regulation in mesoderm	Embryonic lethal	
Hotair	HoxA transcript antisense RNA	*Hox* gene regulation	Defects in skeletal system development	
Dlx1os	Distal-less homeobox 1	Homeodomain transcription factor regulation in developing forebrain	Skull and neurological defects	
Dlx6os1	Dlx6 opposite strand transcript 1	Homeodomain transcription factor regulation in developing forebrain	Altered GABAergic interneuron development	
Malat1	Metastasis-associated lung adenocarcinoma transcript 1	Tumourigenesis		
Miat	Myocardial infarction-associated transcript (also known as Rncr2)	Retina development	Defects in specification of retina cell types	
Six3os1	Six3 opposite strand transcript 1	Retina development	Defects in specification of retina cell types	
Tug1	Taurine upregulated gene 1	Retina development	Defects in differentiation of photoreceptor progenitor cells	
Vax2os	Vax2 opposite strand transcript	Retina development	Defects in differentiation of photoreceptor progenitor cells	
Cyrano	Cyrano	Zebrafish embryogenesis	Developmental defects	
Megamind	Megamind	Zebrafish embryogenesis	Defects in brain morphogenesis and eye development	
HOTTIP	HOXA distal transcript antisense RNA	HOXA regulation	Altered limb morphology in chicken	
Bace1AS	Bace1 antisense	Neuronal survival	Transcribed from antisense strand of the Bace1 locus; implicated in Alzheimer's disease; modulates hippocampal neurogenesis	Faghihi *et al.* (2008)
hsr omega	Heat shock response omega	*Drosophila* heat shock response	Associates with RNA processing factors; involved in the formation of omega speckles during stress response	Jolly and Lakhotia (2006)
ciRS-7	Circular RNA sponge for miR-7	Human and mouse brain development	Quenches the neuronal miRNA, miR-7	Hansen *et al.* (2013)

interactions or even by virtue of high local concentration at the site of transcription. The allele-specific association makes them ideal for nucleating chromatin-mediated effects at the locus of transcription. This has led to the view that even the 'act of transcription of an lncRNA' can lead to chromatin modification at the site, even when the RNA product may serve no direct role. This feature also allows a memory function if the RNA has a slow turnover rate. Further, the DNA–RNA interactions allow tethering at the DNA and recruitment of protein complexes at the other end. The modular DNA-binding domains and activation domains also follow the same principle but are limited by the small footprints of proteins on DNA and the necessary loss of genomic context information as proteins are translated in the cytoplasm. As a generalization, one may argue that proteins can mediate regulation in *trans* by tethering to short DNA motifs all over the genome and recruit other proteins, while lncRNA can mediate *cis* effects by tethering to long stretches of DNA and recruiting regulatory factors. Perhaps due to these features, many well-studied lncRNAs have roles in epigenetic phenomena, chromosomal silencing and *cis*-regulation of gene expression.

Another inherent feature of lncRNAs is that when they arise from geneic sites, they are complementary to protein-coding genes. This is perhaps the reason why many lncRNAs function as antisense regulators of protein-coding genes. They may either target the mRNA for degradation or have a finer control by masking splice sites and thus altering the relative abundance of transcript isoforms through alternative splicing.

Lastly, lncRNAs serve as decoys by offering sites to miRNAs and RNA-binding proteins, sequestering them and keeping them away from their legitimate targets. This decoy function can be useful to interfere with the function of protein-coding genes in both the nucleus and the cytoplasm. The assumption that decoys can only play the role of general disruptor is in fact simplistic. They may also act as scaffolds to bring together proteins of diverse functions, thus increasing local concentrations; a strategy used often by biological systems to overcome the concentration requirements for certain reactions to be feasible. Until we know the impact of decoys in the turnover of the factors they bind, we cannot be sure if they are disruptors or if they could also stabilize the binding partner, holding it until a later time when its function may again be needed. Indeed, this would convert them from disruptors to 'depots'.

The recent spurt of information on ncRNAs suggests that there is much to be learned from closely studying the cellular roles of these molecules. As new methods to look at structure and function of RNAs emerge, our view of the role and importance of ncRNAs is likely to further change. In parallel, several applications that rely on this emerging knowledge are taking shape. Many ncRNAs are emerging as therapeutic targets and biomarkers for the development of novel diagnostics. Further, ncRNAs are also offering new avenues in the development of genome engineering tools for synthetic biology and biotechnological applications.

Web Resources

- http://www.hhmi.org/biointeractive/rna-interference
- http://www.nobelprize.org/nobel_prizes/medicine/laureates/2006/mello-telephone.html
- http://www.nature.com/nm/journal/v14/n10/pdf/nm1008-1036.pdf
- http://www.plosgenetics.org/article/info%3Adoi%2F10.1371%2Fjournal.pgen.1000853
- http://www.stanford.edu/group/blau/pdfs/Sen-2006-A%20brief%20history%20of%20R.pdf
- http://www.mirnarx.com/___Pipeline/Pipeline_MRX34.aspx
- http://www.nature.com/nrg/posters/regulation/regulation_microRNA.pdf
- http://www.webofstories.com/play/francis.crick/84;jsessionid=B06881E84E6A585821699E9F40BACAE7

References

Ambros V. The evolution of our thinking about microRNAs. *Nat Med* 2008;14:1036–1040.

Aravin A, Gaidatzis D, Pfeffer S, Lagos-Quintana M, Landgraf P, Iovino N, *et al.* A novel class of small RNAs bind to MILI protein in mouse testes. *Nature* 2006;442:203–207.

Bartel DP. MicroRNAs: Genomics, biogenesis, mechanism, and function. *Cell* 2004;116:281–297.

Bartel DP. MicroRNAs: Target recognition and regulatory functions. *Cell* 2009;136:215–233.

Bartel DP, Chen CZ. Micromanagers of gene expression: The potentially widespread influence of metazoan microRNAs. *Nat Rev Genet* 2004;5:396–400.

Bejerano G, Pheasant M, Makunin I, Stephen S, Kent WJ, Mattick JS, *et al.* Ultraconserved elements in the human genome. *Science* 2004;304:1321–1325.

Berezikov E. Evolution of microRNA diversity and regulation in animals. *Nat Rev Genet* 2011;12:846–860.

Berezikov E, Chung WJ, Willis J, Cuppen E, Lai EC. Mammalian mirtron genes. *Mol Cell* 2007;28:328–336.

Bolotin A, Quinquis B, Sorokin A, Ehrlich SD. Clustered regularly interspaced short palindrome repeats (CRISPRs) have spacers of extrachromosomal origin. *Microbiology* 2005;151:2551–2561.

Bortolin-Cavaille ML, Dance M, Weber M, Cavaille J. C19MC microRNAs are processed from introns of large Pol-II, non-protein-coding transcripts. *Nucleic Acids Res* 2009;37:3464–3473.

Bousquet M, Harris MH, Zhou B, Lodish HF. MicroRNA miR-125b causes leukemia. *Proc Natl Acad Sci USA* 2010;107: 21558–21563.

Calin GA, Croce CM. MicroRNA signatures in human cancers. *Nat Rev Cancer* 2006a;6:857–866.

Calin GA, Croce CM. MicroRNA-cancer connection: The beginning of a new tale. *Cancer Res* 2006b;66:7390–7394.

Cheah MT, Wachter A, Sudarsan N, Breaker RR. Control of alternative RNA splicing and gene expression by eukaryotic riboswitches. *Nature* 2007;447:497–500.

Chen K, Rajewsky N. The evolution of gene regulation by transcription factors and microRNAs. *Nat Rev Genet* 2007;8:93–103.

Crick F. Central dogma of molecular biology. *Nature* 1970;227: 561–563.

Diederichs S. The four dimensions of noncoding RNA conservation. *Trends Genet* 2014;30:121–123.

Drinnenberg IA, Weinberg DE, Xie KT, Mower JP, Wolfe KH, Fink GR, *et al.* RNAi in budding yeast. *Science* 2009;326:544–550.

Dugas DV, Bartel B. MicroRNA regulation of gene expression in plants. *Curr Opin Plant Biol* 2004;7:512–520.

Eddy SR. Non-coding RNA genes and the modern RNA world. *Nat Rev Genet* 2001;2:919–929.

Eulalio A, Huntzinger E, Izaurralde E. Getting to the root of miRNA-mediated gene silencing. *Cell* 2008;132:9–14.

Faghihi MA, Modarresi F, Khalil AM, Wood DE, Sahagan BG, Morgan TM, *et al.* Expression of a noncoding RNA is elevated in Alzheimer's disease and drives rapid feed-forward regulation of beta-secretase. *Nat Med* 2008;14:723–730.

Fatica A, Bozzoni I. Long non-coding RNAs: New players in cell differentiation and development. *Nat. Rev. Genet.* 2014;15:7–21.

Fire A, Xu S, Montgomery MK, Kostas SA, Driver SE, Mello CC. Potent and specific genetic interference by double-stranded RNA in *Caenorhabditis elegans*. *Nature* 1998;391:806–811.

Gilbert W. Origin of life: The RNA world. *Nature* 1966;319:618.

Girard A, Sachidanandam R, Hannon GJ, Carmell MA. A germline-specific class of small RNAs binds mammalian Piwi proteins. *Nature* 2006;442:199–202.

Griffiths-Jones S. miRBase: The microRNA sequence database. *Methods Mol Biol* 2006;342:129–138.

Grivna ST, Beyret E, Wang Z, Lin H. A novel class of small RNAs in mouse spermatogenic cells. *Genes Dev* 2006;20:1709–1714.

Gunnery S, Datta A. An inhibitor RNA of translation from barley embryo. *Biochem Biophys Res Commun* 1987;142:383–388.

Guo S, Kemphues KJ. par-1, a gene required for establishing polarity in *C. elegans* embryos, encodes a putative Ser/Thr kinase that is asymmetrically distributed. *Cell* 1995;81:611–620.

Guttman M, Donaghey J, Carey BW, Garber M, Grenier JK, Munson G, *et al.* lincRNAs act in the circuitry controlling pluripotency and differentiation. *Nature* 2011;477:295–300.

Hamilton AJ, Baulcombe DC. A species of small antisense RNA in posttranscriptional gene silencing in plants. *Science* 1999;286:950–952.

Hansen TB, Jensen TI, Clausen BH, Bramsen JB, Finsen B, Damgaard CK, *et al.* Natural RNA circles function as efficient microRNA sponges. *Nature* 2013;495:384–388.

He L, Hannon GJ. MicroRNAs: Small RNAs with a big role in gene regulation. *Nat Rev Genet* 2004;5:522–531.

Jolly C, Lakhotia SC. Human sat III and Drosophila hsr omega transcripts: A common paradigm for regulation of nuclear RNA processing in stressed cells. *Nucleic Acids Res* 2006;34:5508–5514.

Keene JD, Komisarow JM, Friedersdorf MB. RIP-Chip: The isolation and identification of mRNAs, microRNAs and protein components of ribonucleoprotein complexes from cell extracts. *Nat Protoc* 2006;1:302–307.

Khalil AM, Guttman M, Huarte M, Garber M, Raj A, Rivea MD, *et al.* Many human large intergenic noncoding RNAs associate with chromatin-modifying complexes and affect gene expression. *Proc Natl Acad Sci USA* 2009;106:11667–11672.

Kim VN. Small RNAs just got bigger: Piwi-interacting RNAs (piRNAs) in mammalian testes. *Genes Dev* 2006;20:1993–1997.

Kleinman CL, Gerges N, Papillon-Cavanagh S, Sin-Chan P, Pramatarova A, Quang DA, *et al.* Fusion of TTYH1 with the C19MC microRNA cluster drives expression of a brain-specific DNMT3B isoform in the embryonal brain tumor ETMR. *Nat Genet* 2014;46:39–44.

Kozomara A, Griffiths-Jones S. miRBase: Annotating high confidence microRNAs using deep sequencing data. *Nucleic Acids Res* 2014;42:D68–D73.

Krol J, Busskamp V, Markiewicz I, Stadler MB, Ribi S, Richter J, *et al.* Characterizing light-regulated retinal microRNAs reveals rapid turnover as a common property of neuronal microRNAs. *Cell* 2010;141:618–631.

Lagos-QuintanaM, Rauhut R, Yalcin A, Meyer J, Lendeckel W, Tuschl T. Identification of tissue-specific microRNAs from mouse. *Curr Biol* 2002;12:735–739.

Lee RC, Feinbaum RL, Ambros V. The *C. elegans* heterochronic gene lin-4 encodes small RNAs with antisense complementarity to lin-14. *Cell* 1993;75:843–854.

Lee RC, Feinbaum R, Ambros V. A short history of a short RNA. *Cell* 2004;116:S89–S92.

Li J, Yang Z, Yu B, Liu J, Chen X. Methylation protects miRNAs and siRNAs from a 3′-end uridylation activity in Arabidopsis. *Curr Biol* 2005;15:1501–1507.

Makeyev EV, Zhang J, Carrasco MA, Maniatis T. The MicroRNA miR-124 promotes neuronal differentiation by triggering brain-specific alternative pre-mRNA splicing. *Mol Cell* 2007;27:435–448.

Mello CC. Return to the RNAi world: Rethinking gene expression and evolution. Nobel Lecture, 8 December 2006. Available from: http://www.nobelprize.org/mediaplayer/index.php?id=87

Mercer TR, Dinger ME, Mattick JS. Long non-coding RNAs: Insights into functions. *Nat Rev Genet* 2009;10:155–159.

Mi S, Cai T, Hu Y, Chen Y, Hodges E, Ni F, *et al.* Sorting of small RNAs into Arabidopsis argonaute complexes is directed by the 5′ terminal nucleotide. *Cell* 2008;133:116–127.

Mojica FJ, ez-Villasenor C, Garcia-Martinez J, Soria E. Intervening sequences of regularly spaced prokaryotic repeats derive from foreign genetic elements. *J Mol Evol* 2005;60:174–182.

Montange RK, Mondragon E, van TD, Garst AD, Ceres P, Batey RT. Discrimination between closely related cellular metabolites by the SAM-I riboswitch. *J Mol Biol* 2010;396:761–772.

Napoli C, Lemieux C, Jorgensen R. Introduction of a chimeric chalcone synthase gene into petunia results in reversible co-suppression of homologous genes in trans. *Plant Cell* 1990; 2:279–289.

Noguer-Dance M, bu-Amero S, Al-Khtib M, Lefevre A, Coullin P, Moore GE, *et al.* The primate-specific microRNA gene cluster (C19MC) is imprinted in the placenta. *Hum Mol Genet* 2010;19:3566–3582.

Okamura K, Hagen JW, Duan H, Tyler DM, Lai EC. The mirtron pathway generates microRNA-class regulatory RNAs in Drosophila. *Cell* 2007;130:89–100.

Pasquinelli AE, Hunter S, Bracht J. MicroRNAs: A developing story. *Curr Opin Genet Dev* 2005;15:200–205.

Pasquinelli AE, Reinhart BJ, Slack F, Martindale MQ, Kuroda MI, Maller B, *et al.* Conservation of the sequence and temporal expression of let-7 heterochronic regulatory RNA. *Nature* 2000;408:86–89.

Pillai RS. MicroRNA function: Multiple mechanisms for a tiny RNA? *RNA* 2005;11:1753–1761.

Pontier DB, Gribnau J. Xist regulation and function explored. *Hum Genet* 2011;130:223–236.

Pourcel C, Salvignol G, Vergnaud G. CRISPR elements in Yersinia pestis acquire new repeats by preferential uptake of bacteriophage DNA, and provide additional tools for evolutionary studies. *Microbiology* 2005;151:653–663.

Reinhart BJ, Slack FJ, Basson M, Pasquinelli AE, Bettinger JC, Rougvie AE, *et al.* The 21-nucleotide let-7 RNA regulates developmental timing in *Caenorhabditis elegans*. *Nature* 2000;403:901–906.

Rhoades MW, Reinhart BJ, Lim LP, Burge CB, Bartel B, Bartel DP. Prediction of plant microRNA targets. *Cell* 2002;110:513–520.

Rinn JL, Chang HY. Genome regulation by long noncoding RNAs. *Annu Rev Biochem* 2012;81:145–166.

Romano N, Macino G. Quelling: Transient inactivation of gene expression in *Neurospora crassa* by transformation with homologous sequences. *Mol Microbiol* 1992;6:3343–3353.

Roth A, Breaker RR. The structural and functional diversity of metabolite-binding riboswitches. *Annu Rev Biochem* 2009;78:305–334.

Sen GL, Blau HM. A brief history of RNAi: The silence of the genes. *FASEB J* 2006;20:1293–1299.

Sorek R, Kunin V, Hugenholtz P. CRISPR—a widespread system that provides acquired resistance against phages in bacteria and archaea. *Nat Rev Microbiol* 2008;6:181–186.

Tolia NH, Joshua-Tor L. Slicer and the argonautes. *Nat Chem Biol* 2007;3:36–43.

Valencia-Sanchez MA, Liu J, Hannon GJ, Parker R. Control of translation and mRNA degradation by miRNAs and siRNAs. *Genes Dev* 2006;20:515–524.

Vella MC, Slack FJ. C. elegans microRNAs. *WormBook*. 21 September 2005:1–9.

Watanabe T, Takeda A, Tsukiyama T, Mise K, Okuno T, Sasaki H, *et al.* Identification and characterization of two novel classes of small RNAs in the mouse germline: Retrotransposon-derived siRNAs in oocytes and germline small RNAs in testes. *Genes Dev* 2006;20:1732–1743.

Waters LS, Storz G. Regulatory RNAs in bacteria. *Cell* 2009;136:615–628.

Wightman B, Ha I, Ruvkun G. Posttranscriptional regulation of the heterochronic gene lin-14 by lin-4 mediates temporal pattern formation in C. elegans. *Cell* 1993;75:855–862.

Winkler W, Nahvi A, Breaker RR. Thiamine derivatives bind messenger RNAs directly to regulate bacterial gene expression. *Nature* 2002;419:952–956.

Woese CR. *The Genetic Code: The Molecular Basis for Genetic Expression*. New York: Harper & Row. p. 186.

Yekta S, Shih IH, Bartel DP. MicroRNA-directed cleavage of HOXB8 mRNA. *Science* 2004;304:594–596.

Part II

Engineering of Gene

17

Cloning of gene: How it all began

H. K. Das

The discovery of the technique to clone the gene has, to say the least, revolutionized biology. How did it all begin? As Stanley Cohen, one of the architects of this discovery, has said, 'Science isn't carried out in a vacuum. The discovery of methods for cloning DNA was dependent not just on insights that Boyer and I had about plasmids and restriction enzymes, but also on the fruits of years of effort in dozens of other laboratories—as is the case for most discoveries' (Cohen 2004).

Put simply, gene cloning involves cleaving of DNA and joining the desired DNA fragment into an autonomously replicating entity such as a plasmid or a viral DNA. Task 1 in gene cloning, therefore, is to cleave the DNA.

17.1 Hydrodynamic shear of DNA

DNA can be cleaved by just hydrodynamic shear. The double helical structure imparts rigidity to the DNA fibre and this leads to shear cleavage of the phosphodiester linkage in DNA even during laboratory manipulation. This was the main reason why in the mid-1950s, there was quite a confusion about the size of DNA, as several workers reported different values for the size of DNA of both the bacteriophages T2 and T4. The sizes were determined by measuring the contour length of the autoradiograph of the DNA and also by measuring the sedimentation coefficient of the DNA during velocity centrifugation (see Chapter 5, Section 5.2). The first systematic study of the effect of hydrodynamic shear on the size of DNA was reported by Peter F Davison (1959). He demonstrated that the size of the DNA of the bacteriophage T2, as measured by its sedimentation coefficient, was reduced by simply shaking the DNA solution in a test tube [Fig. 17.1(a)]. Drastic reduction of size occurred when the DNA solution was forced through a hypodermic needle [Fig. 17.1(b)]. This has since become the standard technique of shearing DNA. In fact, one can obtain DNA fragments of nearly any specific size by this method by varying the time taken to force a given quantity of the DNA solution through the hypodermic needle. The only problem here is that the cleavage can take place anywhere and there is no way of choosing the cleavage site.

17.2 Enzymatic cleavage of the phosphodiester linkage of DNA

Cleavage of DNA can be achieved by enzymatically hydrolysing the phosphodiester linkage between deoxyribonucleotides.

17.2.1 Non-specific enzymatic cleavage

An enzyme capable of non-specific cleavage of phosphodiester linkage was detected in gastrointestinal juices as far back as 1911 (Levene and Medigreceanu 1911), which was isolated from beef pancreas and crystallized later (Kunitz 1950). The problem with this enzyme and with other similar enzymes purified subsequently is that cleavage takes place almost at random and the product of cleavage is generally small in size, far smaller than the size of an intact gene, even a prokaryotic gene.

17.2.2 Discovery of restriction endonucleases: The enzymes that cleave DNA at specific sites

A suitable biochemical cleaving tool would, however, be an enzyme which would not cleave just any phosphodiester linkage or next to one particular base (e.g. T1 RNase cleaves next to guanine), but would recognize and cleave only at a specific sequence of several (let us say six) nucleotides. The statistical frequency of coming across a specific six-base sequence in a segment of DNA where all the four bases are randomly distributed would be $(1/4)^6 = 1/4096$. This would be a fairly long

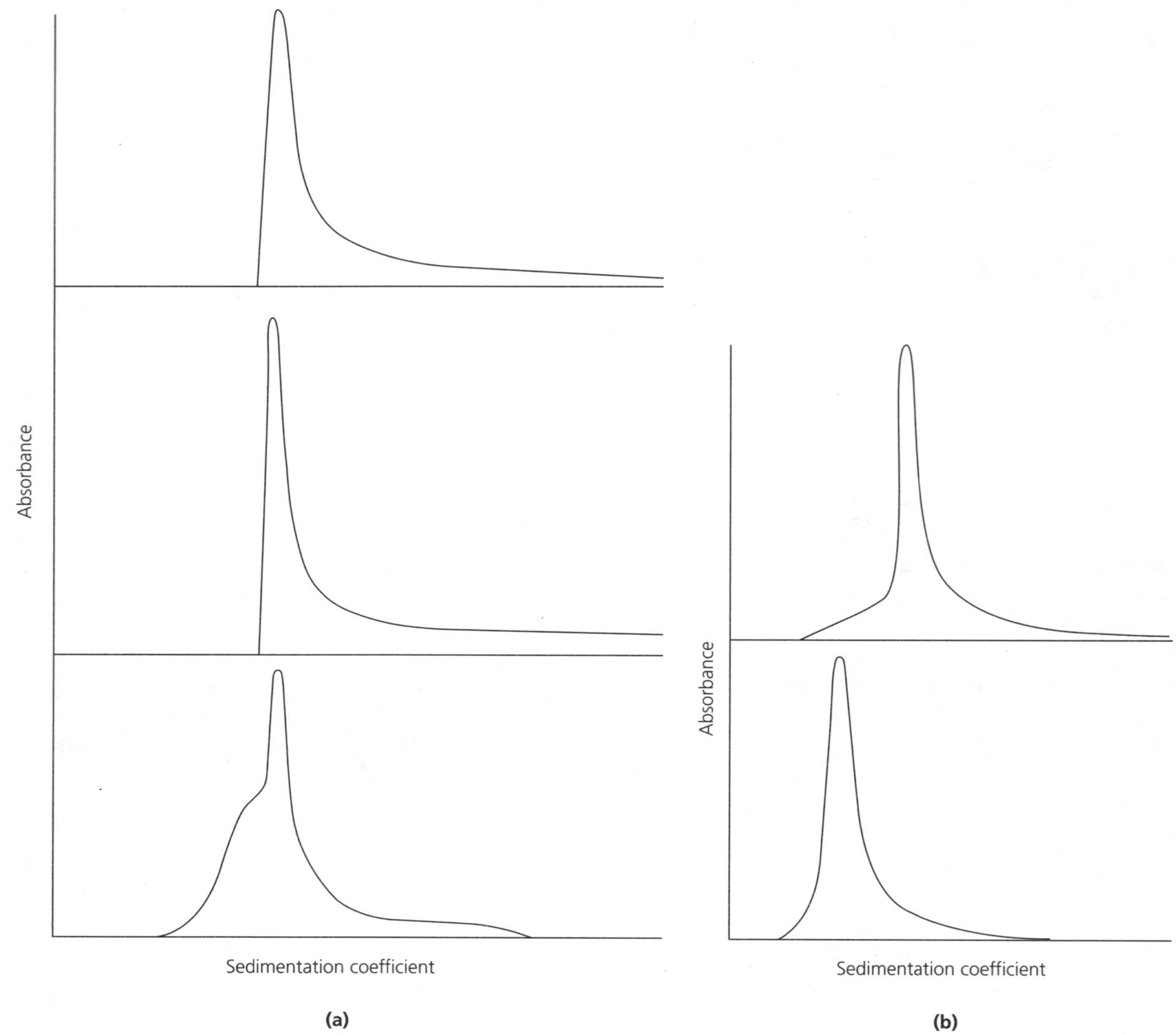

Figure 17.1 (a) Effect of shaking in a test tube T2 DNA solution (0.001 %) in 0.2 M phosphate buffer, pH 8.0 on the size of DNA as monitored by its sedimentation coefficient. Top frame: untreated DNA solution; middle frame: shaken 30 times; bottom frame: shaken 30 times with equal volume of chloroform. (b) Effect of forced shear on T2 DNA (0.002% solution) as evidenced by its sedimentation coefficient. Top frame: untreated DNA solution; bottom frame: the DNA solution was forced through a No. 27 hypodermic needle at the rate of 1 mL in 23.5 s.

stretch and the chances of accommodating a gene (at least a prokaryotic gene) in such a stretch of DNA would be quite high. It would be possible to obtain larger fragments of DNA by carrying out partial digestion. Such enzymes have been found and are known as restriction endonucleases. The observation, 'Science isn't carried out in a vacuum' is very much valid here also.

17.2.2.1 The phenomenon of restriction *in vivo*

Existence of some kind of a restriction phenomenon that was induced by the host was suspected quite early, when it was observed that a bacteriophage would fail to reproduce in a bacterial cell unless it had originated from a similar type of cell (Luria and Human 1952; Bertani and Weigle 1953). It was assumed that the similar type of cell possessed the specific modification allele *m*, which was responsible for modifying the phage DNA and making it immune to restriction. On the other hand, the dissimilar cells did not have this specific allele. The failure of the phage to reproduce was later correlated with the appearance of the radioactivity of the viral DNA in the growth medium, suggesting degradation of DNA (Lederberg 1957). Subsequent studies confirmed the relationship between restriction of reproduction of the phage and degradation of the invading phage DNA (Arber and Dussoix 1962; Dussoix and Arber 1962). The degradation was thought to be the function of the allele *r* in the cells, which coded for the restriction enzyme. Arber (1965) inferred that the extensive degradation of the restricted phage DNA was very likely the consequence of some initial cleavages 'by a mechanism of recognition of certain base sequences' followed by general non-specific hydrolysis of phosphodiester linkages at random. Arber (1965) went on further, 'If this last idea should be correct, one may further speculate that a restriction enzyme might provide a tool for the sequence specific cleavage of DNA'. What a profound prediction!

17.2.2.2 Restriction *in vitro*

Arber himself subsequently got indication of the presence of such an enzyme activity in the extract of *Escherichia coli B* (Linn and Arber 1968). Infectivity of the replicative form (RF) of the DNA of bacteriophage fd was lost if incubated with the *E. coli* extract. Electron microscopy of the incubation product revealed that the (RF) fd DNA had suffered only limited double-strand breaks. The dialysed extract needed the addition of Mg^{2+}, ATP and S-adenosylmethionine for acting on the (RF) fd DNA. This loss of infectivity of the (RF) fd DNA occurred only when the extract was obtained from an *E. coli* strain that would restrict the *in vivo* growth of the bacteriophage. An extract from *E. coli* K or *E. coli* C, the strains that could not restrict the bacteriophage *in vivo,* did not affect the infectivity of the bacteriophage (RF) DNA.

17.2.2.3 Purification of a restriction endonuclease from *Escherichia coli*

In 1968 itself, Meselson and Yuan purified a restriction enzyme from *Escherichia coli* strain K (m_K+, r_K+) and found that it was selective in its action and produced only a limited number of double-strand cleavages in the DNA of the bacteriophage λ which has been grown in an mK-strain of *Escherichia coli* . The inference was based on sedimentation profiles of the λ DNA before and after limit treatment with the enzyme. This enzyme needed for its activity adenosine triphosphate and S-adenosyl methionine, in addition to Mg^{2+} ion (Meselson and Yuan 1968). The recognition sequence of the enzyme and the exact nature of cleavage were not determined then. As it became known later, there is one drawback of this restriction enzyme: the exact site of cleavage is variable and could be 1,000 base pairs away from the recognition sequence, which is fixed. Such enzymes are now classified as Type I restriction endonucleases. The enzyme involved in the *in vitro* restriction phenomenon observed by Linn and Arber (1968) has also been purified later from the *E. coli* B strain and this restriction enzyme too has been found to belong to Type I.

17.2.2.4 Purification of a restriction endonuclease from *Haemophilus influenzae*

In 1970, Hamilton Smith of Johns Hopkins University School of Medicine at Baltimore discovered a similar endonuclease from the extract of the bacteria *Haemophilus* influenzae Rd. and this enzyme was called *Hind*II (Smith and Wilcox 1970). This enzyme needs for its activity only Mg^{2+} and no ATP or S-adenosyl methionine is needed. Hamilton Smith has described his discovery as 'chance discovery'. The chance finding that initiated the whole study was that radioactively labelled DNA of the *Salmonella* phage P22 got degraded, as seen by disappearance of the original band in a caesium chloride (CsCl) density sedimentation experiment, when incubated with *Haemophilus influenzae* cells. The observation that really clinched the issue was that the viscosity of the DNA from *Haemophilus influenzae* was unaltered when incubated with a cell extract of *Haemophilus influenzae,* while the viscosity of the *Salmonella* phage P22 DNA progressively decreased under the same conditions [Fig. 17.2(a)]. Zone centrifugation in sucrose density gradients after incubation of the *E. coli* bacteriophage T7 DNA with the 200-fold purified enzyme confirmed degradation of this DNA, while *Haemophilus influenzae* DNA was again unaffected [Fig. 17.2(b)]. Zone centrifugation also revealed that only double-stranded T7 DNA was cleaved by the enzyme; no reduction in size took place if denatured single-stranded T7 DNA was incubated with the enzyme.

The cleavage of the phosphodiester linkage by the enzyme yielded a 5′-phosphoryl group and a 3′-hydroxyl group. This was inferred because polynucleotide kinase could not transfer radioactivity from γ-^{32}P–labelled ATP to the cleavage products unless prior treatment with bacterial alkaline phosphatase was done.

17.2.2.5 Characterization of the specific nucleotide sequence at which *Hind*II cleaves

Limit digestion of double-stranded T7 DNA with the enzyme produced DNA fragments of an average length of about 1000 nucleotide pairs, thus indicating that only about 0.1% of the total available phosphodiester linkages were available for cleavage by the enzyme. This was suggestive of the possibility that the enzyme *Hind*II could cleave only when it encountered a specific nucleotide sequence in the DNA. This specific nucleotide sequence was determined (Kelly and Smith 1970) as follows:

Step 1. The limit digestion product by *Hind*II of the uniformly ^{33}P-labelled phage T7 DNA was treated with bacterial alkaline phosphatase and then incubated with γ-^{32}P–labelled ATP and polynucleotide kinase (see Chapter 9, Section 9.9), thereby ^{32}P-labelling all the 5′-ends of the cleaved DNA fragments.

Step 2. The 5′-end ^{32}P-labelled DNA fragments from *Hind*II digested T7 DNA were treated with Russel's viper venom phosphodiesterase and beef

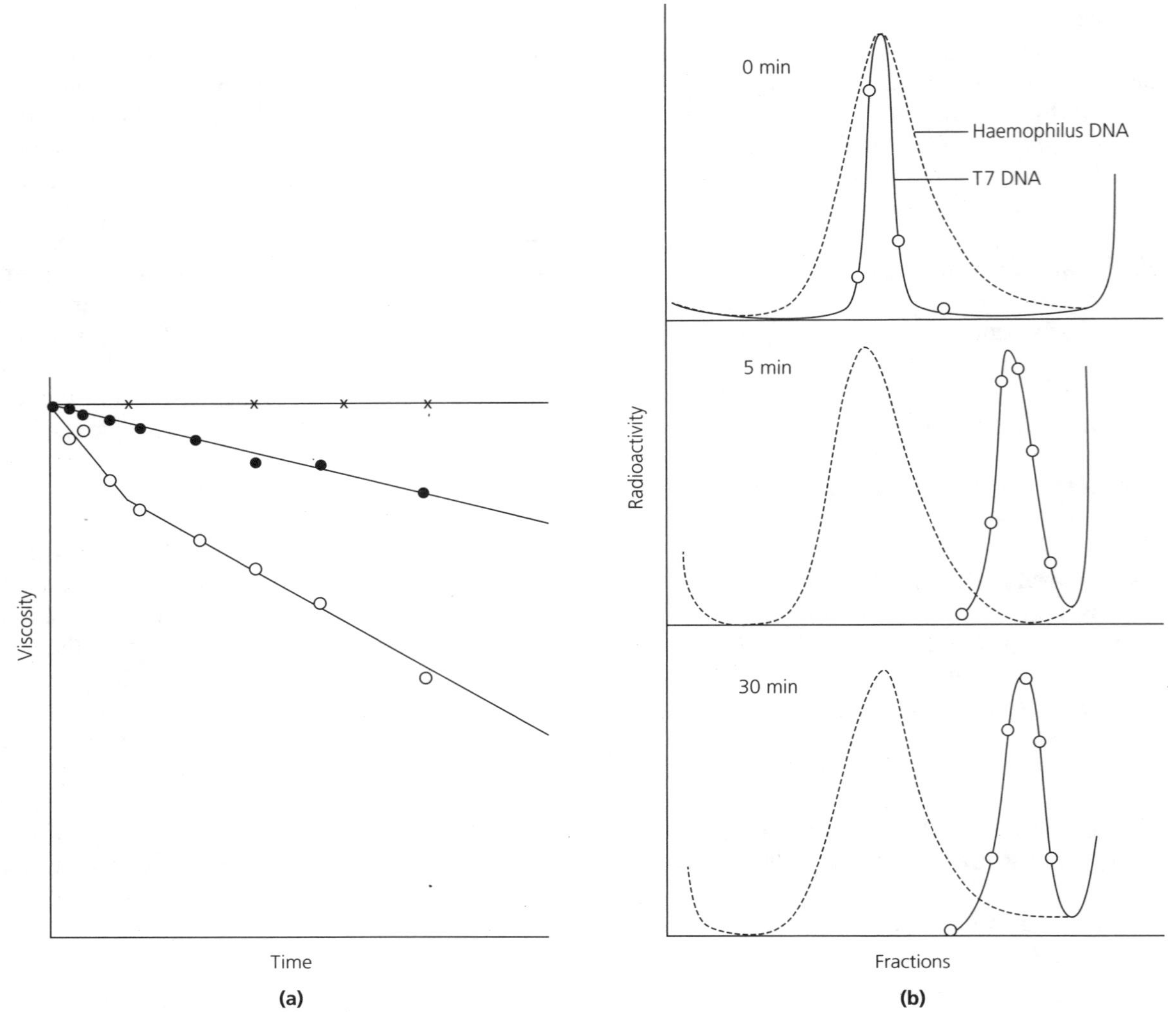

Figure 17.2 Action of the restriction endonuclease from *Haemophilus influenzae.* (a) DNA from *Haemophilus influenzae* and bacteriophage P22 were acted upon by cell extract of *Haemophilus influenzae* and specific viscosity of the DNA samples, which was a measure of the size of the DNA, was determined periodically. (x) Viscosity of *Haemophilus influenzae* DNA as affected by 20 μL of cell extract; (•) viscosity of P22 DNA as affected by 10 μL of cell extract; (o) viscosity of P22 DNA as affected by 50 μL of cell extract. (b) Equal amounts of radioactive DNA from *Haemophilus influenzae* (^{3}H-labelled) and bacteriophage T7 (^{32}P-labelled) were mixed and acted upon by purified restriction endonuclease from *Haemophilus influenzae*, samples were withdrawn at 0 min, 5 min and 30 min of incubation and zone sedimentation of the DNA was done in neutral sucrose gradients. Fractions were collected from the bottom of the centrifuge tubes, the bottom of the centrifuge tubes containing larger DNA molecules is at the left and the top of the centrifuge tubes containing smaller DNA molecules is at the right of the figure. (....) ^{3}H-labelled *Haemophilus influenzae* DNA; (o) ^{32}P-labelled bacteriophage T7 DNA.

pancreas deoxyribonuclease I [see Fig. 3.11(b)] and the resulting mononucleotides were analysed by electrophoresis on thin-layer cellulose. It was found that the dAMP spot contained about 63% of ^{32}P-radioactivity and the dGMP spot contained 37% ^{32}P-radioactivity, while the dCMP spot and the dTMP spots contained no ^{32}P-radioactivity.

Step 3. Steps 1 and 2 were then performed again, but with the phage P22 DNA. It was now found that the dAMP spot contained about 85% of ^{32}P-radioactivity and the dGMP spot contained 15% ^{32}P-radioactivity, while the dCMP spot and the dTMP spot again contained no ^{32}P-radioactivity.

As the ratio of radioactivity in dAMP and dGMP varied depending on the source of DNA and hence the base composition of it, it was inferred that the enzyme could not discriminate between the two purines and cleaved at the 5′-end of any of the two purines.

Step 4. The 5′-end ^{32}P-labelled DNA fragments from *Hin*dII digested T7 DNA was next treated with only beef pancreas deoxyribonuclease I, yielding a mixture of oligonucleotides. The oligonucleotides were then denatured and digested with exonuclease I, which is known to carry out stepwise degradation from the 3′-OH end, leaving the 5′-dinucleotide intact. Separation of the dinucleotides and mononucleotides by electrophoresis on thin-layer cellulose revealed the presence of 62% of the radioactivity in pApA and 38% in pGpA. Thus, the inference was drawn that at the cleavage site that the first two nucleotides from the 5′-end were pPuA.

Step 5. The 5′-end ^{32}P-labelled DNA fragments from *Hin*dII digested T7 DNA was again treated with only beef pancreas deoxyribonuclease I, yielding a mixture of oligonucleotides. These oligonucleotides were then subjected to partial digestion with Russel's viper venom phosphodiesterase such that only a few nucleotides were removed from the 3′-OH end of the oligonucleotides. The smaller oligonucleotides thus obtained were separated on the basis of chain length by DEAE-cellulose chromatography in the presence of 7 M urea. The trinucleotide fraction was collected and analysed by electrophoresis on thin-layer cellulose. Two spots containing ^{32}P were detected: one had dAMP and dCMP in the ratio of 2:1, while the other had dGMP, dAMP and dCMP in equal proportions. Combining the information obtained in step 4, it was inferred that the two ^{32}P -containing trinucleotides had the sequence d(pApApC) and d(pGpApC). The sequence of nucleotides at the 5′-end of the DNA fragments obtained by cleavage with *Hin*dII was, therefore, d(pPupApC).

Next, the tetranucleotide fraction was collected from the DEAE-cellulose chromatography eluate and analysed by electrophoresis on thin-layer chromatography. Altogether, six distinct spots containing ^{32}P were detected. The obvious inference was that the fourth nucleotide from the 5′-end of the *Hin*dII cleaved fragments was not unique. Ideally, eight ^{32}P-containing tetranucleotides should have been detected, but possibly, the other two tetranucleotides were in very small quantities and escaped detection.

Step 6. Nucleotide sequence at the 3′-end of the *Hin*dII-cleaved fragments was next determined. Bacteriophage T7 DNA, which was uniformly labelled with ^{32}P, was limit digested with *Hin*dII and the resulting DNA fragments were incubated exhaustively with micrococcal nuclease. This is an endonuclease that cleaves the phosphodiester linkage, releasing 3′-phosphates [similar to the beef spleen enzymes, as shown in Fig. 3.11(a)], but produces mononucleoside 3′-phosphates (Xp) and dinucleoside 3′-phosphates (XpYp) in about equal quantities. The 5′-end of the DNA would thus yield pXp and pXpYp, while the 3′-end would yield XpY and Y. The uniqueness in size and charge of XpY would enable one to separate this specific class of dinucleotides from all other mononucleotides and oligonucleotides by DEAE-cellulose chromatography. The XpYs thus obtained were fractionated by electrophoresis on thin-layer cellulose. Two radioactive spots were detected. Spot 1 produced the mononucleotide pC when digested with Russel's viper venom phosphodiesterase and the mononucleotide Tp when digested with beef spleen phosphodiesterase, thus indicating that spot 1 contained TpC. Spot 2 when digested with Russel's viper venom phosphodiesterase released the mononucleotide pT, and on digestion with spleen phosphodiesterase, released the mononucleotide Tp, hence allowing the inference that spot 2 contained TpT.

The two dinucleotides TpC and TpT at the 3′-end of the *Hin*dII cleavage site were thus exactly complementary to the dinucleotides pGpA and pApA at the 5′-end of the *Hin*dII cleavage site. The enzyme *Hin*dII, therefore, produces a 'flush' cut or an 'even' cut, resulting in a 'blunt'-ended DNA fragment [Fig 17.3(a)], rather than producing a 'staggered' cut, resulting in a 5′-overhang or a 3′-overhang [Fig. 17.3(b) and (c)].

It was inferred that *Hin*dII exclusively recognized in DNA the sequence given below and also cleaved right in the middle of this sequence and nowhere else:

$$\begin{array}{c}\downarrow\\ 5'\ \text{GTPyPuAC}\ 3'\\ 3'\ \text{CAPuPyTG}\ 5'\\ \uparrow\end{array}$$

Py and Pu denote a pyrimidine and a purine, respectively. The arrows indicate the sites for cleavage of the phosphodiester linkages in the two strands of DNA (Kelly and Smith 1970). Such enzymes are now classified as Type II restriction endonucleases.

17.2.2.6 A type II restriction endonuclease from *E. coli* RY13

RN Yoshimori, a PhD student in the laboratory of Herbert Boyer at the University of California at San Francisco, CA, isolated the *Escherichia coli* strain RY 13 from a hospital patient and discovered and purified a Type II restriction endonuclease from this strain, which was named *Eco*R I (Yoshimori 1971).

EcoRI would cleave superhelical circular (form I) SV40 DNA at a single specific site. This was evident from analysis of partial denaturation maps obtained by alkaline denaturation of unit-length linear molecules formed by cleavage (Mulder and Delius 1972) and also from the

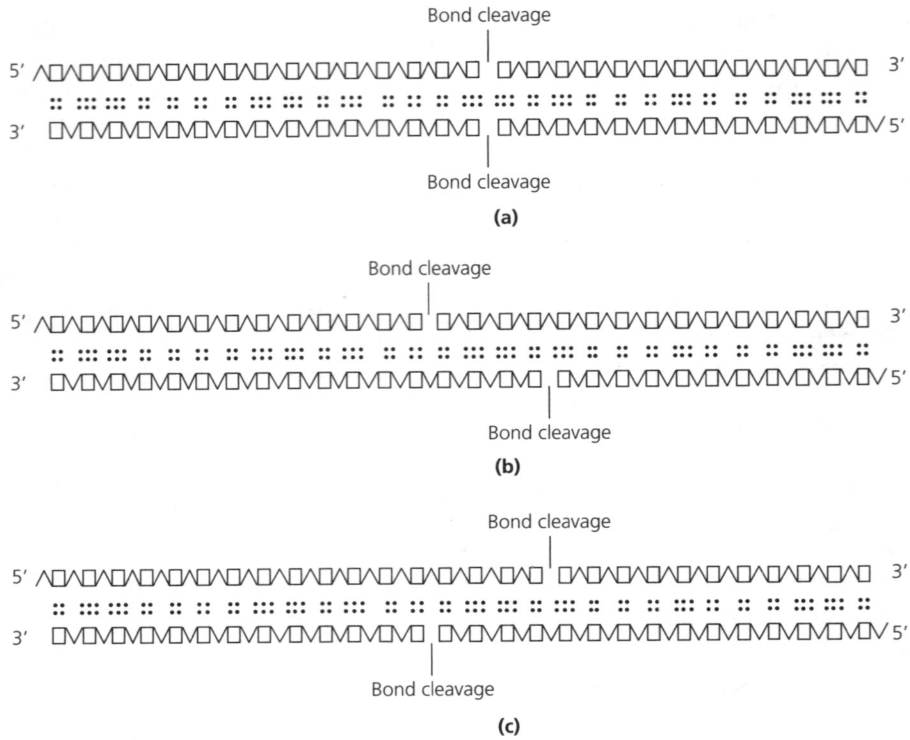

Figure 17.3 Three possible ways a restriction endonuclease can cleave a double-stranded DNA. (a) A 'flush' cut or an 'even' cut resulting in a 'blunt'-ended DNA fragment; (b) a 'staggered' cut resulting in a 5′-overhang; (c) a 'staggered' cut resulting in a 3′-overhang.

observation that denaturation and renaturation of all these linear molecules together yielded only linear but no circular molecules (Morrow and Berg 1972).

17.2.2.7 *Eco*RI makes a 'staggered' cut in DNA strands, generating 'cohesive' ends

A subsequent interesting finding was that even after circular SV40 form I DNA was linearized to the extent of 99.9% by *Eco*RI, its infectivity to the extent of 10% still persisted (Mertz and Davis 1972). The belief at that time was that only the circular form of the DNA was capable of replication *in vivo*. The obvious inference was that the SV40 form I DNA linearized by *Eco*RI *in vitro* was getting recircularized *in vivo*. Indeed, the linearized DNA could be recircularized, as visualized by electron microscopy, even *in vitro* when proper annealing conditions were adopted. Almost all the molecules were linear at 25°C in 50% formamide, but about 50% of the molecules were circular at 6°C in 0.5 M ammonium acetate (Mertz and Davis 1972). Very few circular molecules remained circular if the temperature was elevated by 3°C. The conclusion clearly was that circularization had taken place by hydrogen bonding between two short stretches of single-stranded DNA, which could be possible only if *Eco*RI had generated 'cohesive' ends through 'staggered' cuts. This was confirmed by the production of fully infective covalently linked circular DNA molecules by incubation of the hydrogen-bonded circular molecules with DNA ligase (Mertz and Davis 1972). These molecules could not be linearized by elevation of temperature, but treatment again with *Eco*RI could form linear molecules. If the cohesive ends generated by *Eco*RI were always the same irrespective of the DNA used, such cohesive ends from different DNA molecules should be joined by DNA ligase, forming a covalently linked joint molecule. This was indeed found to be so; SV40 DNA and λdv-120 DNA could be joined together, which banded together at intermediate density on equilibrium centrifugation (Mertz and Davis 1972).

17.2.2.8 Characterization of the specific nucleotide sequence at which *Eco*RI cleaves

The cleavage of the phosphodiester linkage by *Eco*RI also yielded a 5′-phosphoryl group and a 3′-hydroxyl group, as was inferred because polynucleotide kinase could not transfer radioactivity from γ-^{32}P–labelled ATP to the *Eco*RI cleavage products of λ DNA above the background value observed with uncleaved λ DNA unless prior treatment with bacterial alkaline phosphatase was done. The 5′-^{32}P–labelled λ DNA fragments were successively digested with beef pancreatic DNaseI and Russel's viper venom phosphodiesterase. Analysis by paper electrophoresis of the mononucleotides thus produced revealed that pA was the nucleotide at the 5′-end of the cleaved DNA fragments (Hedgpeth *et al.* 1972).

The 5′-end ^{32}P-labelled DNA fragments from *Eco*RI-digested λ DNA were next mixed with uniformly 33P-labelled λ DNA for ease of tracking. The mixture was first digested exhaustively with beef pancreatic DNase I and then briefly with Russel's viper venom phosphodiesterase to yield 5′-^{32}P–labelled oligonucleotides of average chain length of 3–5 residues. The oligonucleotides were first separated according to chain length and then each size class was analysed by two-dimensional high-voltage paper electrophoresis. The base composition of the ^{32}P-containing oligonucleotides was determined by digestion with Russel's viper venom phosphodiesterase and re-electrophoresis. The deduced 5′-terminal sequences were pA, pApA, pApApT, pApApTpT, pApApTpTpC, pApApTpTpCpA and pApApTpTpCpT, respectively, for the base composition of ^{32}P-containing oligonucleotides 1A; 2As; 2As, 1T; 2As, 2Ts; 2 As, 2 Ts, 1 C; 3 As, 2 Ts, 1 C; and 2 As, 3 Ts, 1 C, respectively (Hedgpeth *et al.* 1972).

The sequence of nucleotides at the 5′-end of the DNA fragments obtained by cleavage with *Eco*R I was, hence, d(pApApTpTpCpA/T); the sixth position could be either A or T.

As it became known that *Eco*RI generated cohesive ends by staggered cuts (Section 17.2.2.7), what remained to be ascertained was whether a 5′-overhang or a 3′-overhang was being produced [see Fig. 17.3(b) and (c)]. A 5′-overhang would have a 3′-recess, which should be possible to fill up by the correct deoxynucleotides catalysed by a DNA polymerase [see Fig. 9.1(b)]. Rous Sarcoma Virus DNA polymerase was used as, unlike *E. coli* DNA polymerase I (see Chapter 9, Sections 9.2 and 9.4), it did not have any associated nuclease activity. Various combinations of α-^{32}P deoxynucleoside triphosphates and non-radioactive deoxynucleoside triphosphates were used for incorporation into λ DNA cleaved with *Eco*R I. Significant incorporation was indeed detected in *Eco*R I-cleaved DNA over and above the incorporation into uncleaved λ DNA, indicating that the 'staggered' cut by *Eco*R I did indeed produce 5′-overhangs (Hedgpeth *et al.* 1972).

The sequence of the 5′-overhangs was confirmed by nearest-neighbour transfer analysis [see Chapter 4, Section 4.9 and Figs 4.12(a) and (b)] to be pApApTpT. This was done by conducting four consecutive incorporation reactions using one of the four ^{32}P-labelled deoxyribonucleotides at a time and the other three unlabelled deoxyribonucleotides. No additional incorporation over and above that with the uncleaved λ DNA could be detected with ^{32}P-CTP or ^{32}P-GTP. The level of incorporation from ^{32}P-ATP and ^{32}P-TTP was practically unaltered even if unlabelled GTP and CTP were excluded from the reaction.

The sequence at the cleavage site was, therefore, inferred (Hedgpeth *et al.* 1972) to be

↓

5′ ....A/TpG pApApTpTpCpT/A ....3′
3′ ....T/ApC pTpTpApApGpA/T ....5′

↑

where the arrows indicate the precise site for phosphodiester bond cleavage.

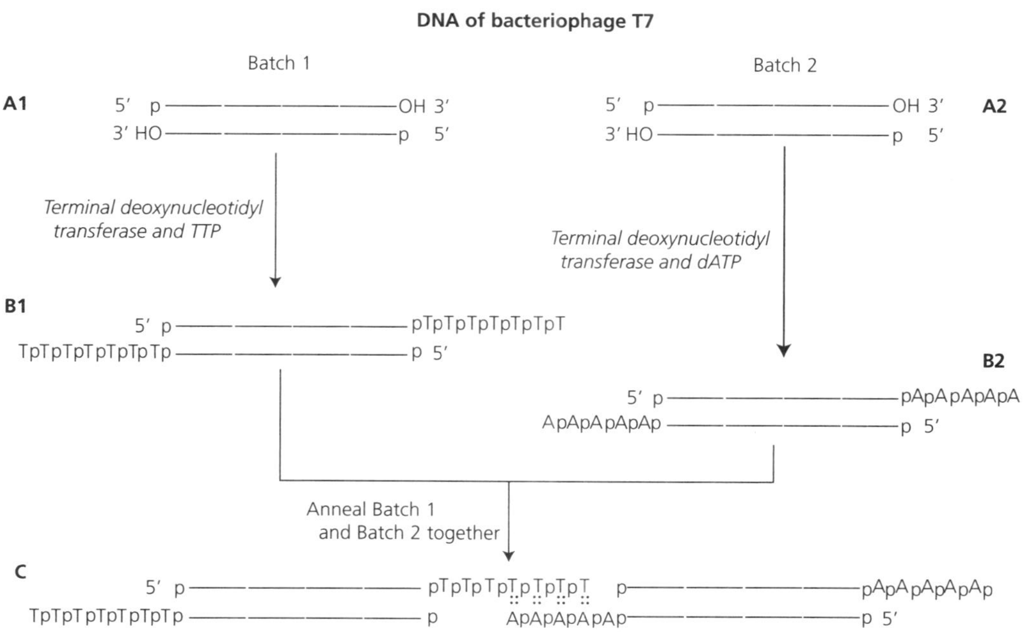

Figure 17.4 Diagrammatic representation of the polydA–polydT tailing method of forming 'cohesive' ends on 'blunt'-ended DNA and forming hydrogen-bonded catenanes by annealing. DNA ligase cannot form covalently linked catenanes in this case because of the presence of gaps between the ends of the two tails.

17.2.2.9 Enzymatic addition of 'cohesive' ends to 'blunt' ends

'Cohesive' ends can be added to 'blunt' ends by the use of terminal deoxynucleotidyl transferase isolated from calf thymus (see Chapter 9, Section 9.8). This enzyme catalyses the successive addition of mononucleotides to the 3′-end of DNA from nucleoside triphosphates. Jensen *et al.* (1971) separately introduced several thymidylate residues to the 3′-ends of one batch of T7 DNA and deoxyadenylate residues to another batch of the same DNA. Equimolar amounts of the two batches of DNA were mixed and annealed. Catenanes measuring two or three times the length of normal T7 DNA were formed, as visualized by electron microscopy. The inference was that the polydA and polydT tails have served as 'cohesive' ends and hydrogen bonds have been formed between the complementary tails during annealing. DNA ligase (see Section 17.3.1) could not, however, join the hydrogen-bonded single strands of these catenanes. This was obviously because the lengths of the polydA and the polydT chains were not exactly the same; hence, there were gaps between the ends of the two tails after annealing (see Fig. 17.4 for a diagrammatic representation of the situation).

17.3 Joining of two polydeoxynucleotides by restoration of phosphodiester linkage

Task 2 in gene cloning is to join the desired DNA fragment thus isolated to an autonomously replicating DNA molecule.

17.3.1 DNA ligase: The joining enzyme

The enzymes that join polynucleotides are called DNA ligases. Such enzymes were of great interest as DNA strand breaking and joining was thought to be of central importance in genetic recombination. Five papers appeared in quick succession in 1967 in Volume 57 of the Proceedings of the National Academy of Sciences of the United States of America that described purification and characterization of an enzyme from *E. coli* (Gellert 1967; Olivera and Lehman 1967a, 1967b; Zimmerman *et al.* 1967) and another from virus T4–infected *E. coli* (Weiss and Richardson 1967) that could join two polynucleotides by the formation of a phosphodiester linkage. In the same year in Wolume 58 of the same journal, a third enzyme with similar properties was described in virus T7–infected *E. coli* (Becker *et al.* 1967). Yet another paper on DNA ligase also appeared in the same year (Cozzarelli *et al.* 1967).

The substrate chosen for the search of the joining enzyme was the DNA of the bacterial virus lambda (λ). The DNA, as isolated from the purified virus particles, is a linear double-stranded molecule containing 48,502 base pairs. The interesting thing about this DNA is that it has two 5′-end overhangs of single-stranded regions comprising 12 bases each that are exactly complementary to each other [Fig. 17.5(a)]. These ends are commonly known as 'cohesive ends'. When this DNA molecule is annealed, a double-stranded circular molecule is formed through the formation of hydrogen bonds between the two complementary 5′-end overhangs [Fig. 17.5(b)]. This molecule was termed by Martin Gellert as 'hydrogen-bonded circle'. This circle, however, has no phosphodiester linkage between the 5′-end and the 3′-end of the individual DNA strands and is reconverted into a linear DNA on mild heating.

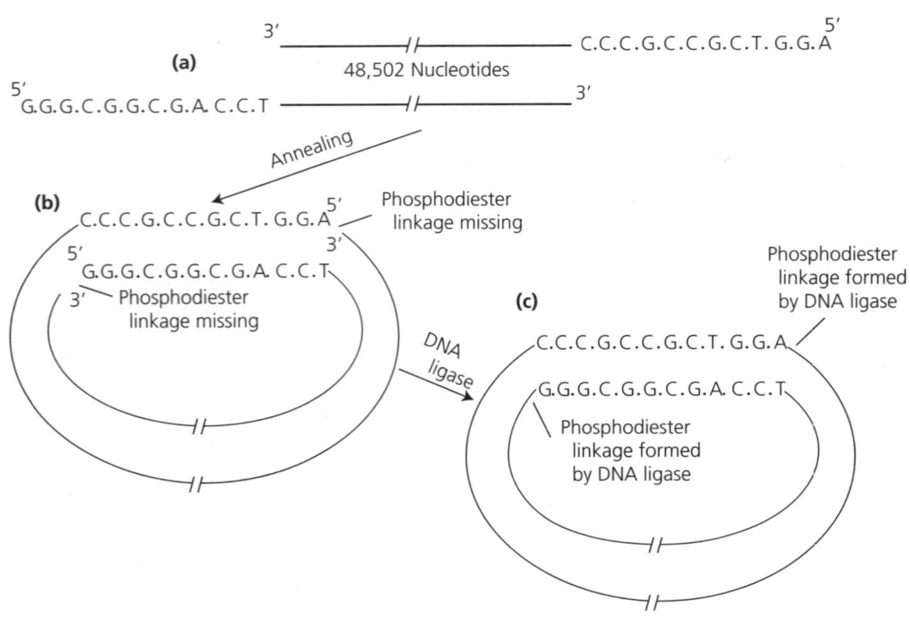

Figure 17.5 (a) Linear DNA of the bacteriophage lambda (λ); (b) Hydrogen bonded circular form of the DNA of bacteriophage lambda (λ) produced by annealing. This circular form would be reconverted into the linear form by mild denaturation; (c) DNA ligase has converted the hydrogen-bonded circular form (see Chapter 9, Fig. 9.5l) into covalently linked circular DNA. Denaturation cannot convert this into the linear form.

Gellert (1967) could get this 'hydrogen-bonded circle' converted into a 'covalently closed circle' by treating with an extract of *E. coli*, which is the natural host of the lambda (λ) virus. The putative joining enzyme in the extract has mediated the formation of the two phosphodiester linkages between the ends of the two DNA strands [Fig. 17.5(c)]. The 'covalently closed circle' cannot be linearized by heating. The enzyme has subsequently been purified to homogeneity and is now known as *E. coli* DNA ligase. The enzyme needs NAD (nicotinamide adenine dinucleotide) in addition to ATP. The requirement for the ligation (joining) reaction is a phosphate group at the 5′-end and a hydroxyl group at the 3′-end. This enzyme cannot add mononucleotides. This enzyme can form phosphodiester linkage between two DNA single strands only if the linkage across in the other strand of the DNA duplex is intact [Fig. 17.6(a)]. The joining enzyme (also used to be called sealase) isolated from the T4 virus–infected *E. coli*, however, needs only ATP but no NAD. An important feature of the T4 DNA ligase is that it forms a phosphodiester linkage between two DNA strands even though the linkage across in the other strand is also missing [Fig. 17.6(b)]. The only requirement, as was first found out by using synthetic deoxyoligonucleotides, is that there must be a phosphate at the 5′-end and a hydroxyl group at the 3′-end (Sgaramella *et al.* 1970). DNA of the *Salmonella* bacteriophage P22, which has fully base-paired termini, has subsequently been converted into dimers, trimers and even higher oligomers by T4 DNA ligase, as has been visualized by electron microscopy and monitored by velocity centrifugation (Sgaramella 1972). This enzyme can thus carry out what is commonly known as 'blunt-end ligation'.

17.4 Autonomously replicating DNA molecules that were used for gene cloning

An autonomously replicating DNA molecule is necessary to act as the vehicle.

17.4.1 Simian Virus 40 DNA and the first ever exercise towards gene cloning

The first DNA virus that was used as a gene cloning vehicle was the Simian Virus 40 (SV40). In fact, the first ever exercise towards gene cloning was reported from the laboratory of Paul Berg at the Department of Biochemistry in the Stanford University Medical Center (Jackson *et al.* 1972) and was described as a 'biochemical method for inserting new genetic information into DNA of Simian Virus 40 DNA molecules containing lambda phage genes and the galactose operon of *Escherichia coli*'.

SV40 was first discovered in the late 1950s as a contaminant in a batch of polio virus vaccine which was being prepared in rhesus monkey cells. SV40 has later been found to be able to grow in many mammalian cell

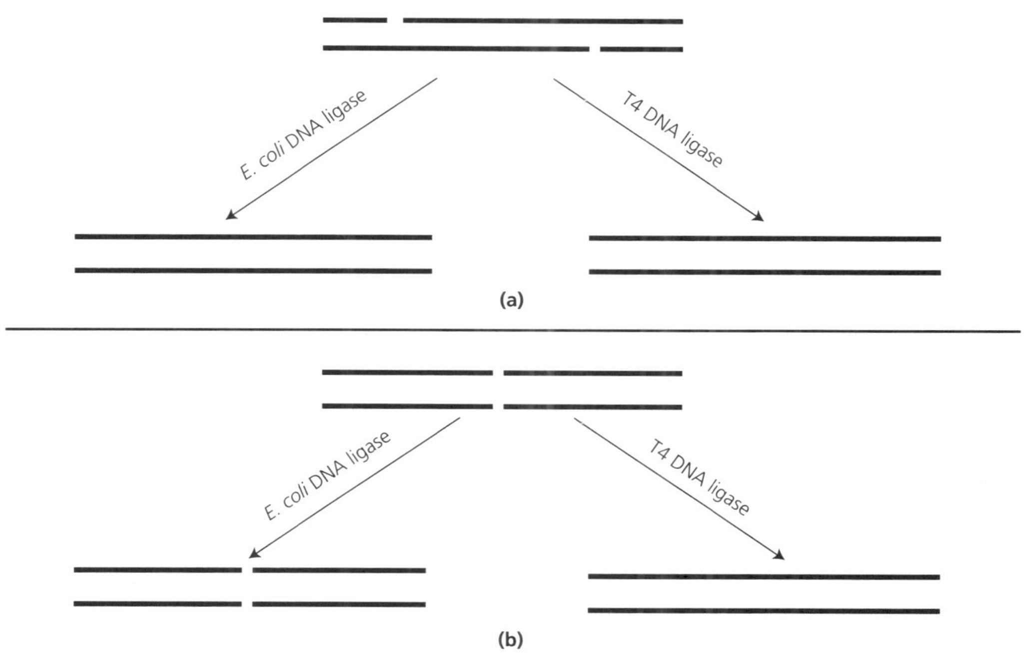

Figure 17.6 Schematic representation of the distinctive functions of *E. coli* DNA ligase and T4 DNA ligase. (a) Ligation of DNA with single-strand breaks. Across the single-strand break in the other DNA strand, the phosphodiester linkage is intact. Both the DNA ligases can join the DNA strands in this situation. (b) Ligation of DNA with double-strand breaks or 'flush ends'. Only the T4 DNA ligase can join the DNA molecules in this situation. The efficiency of ligation is, however, less than in case of single-strand breaks.

types and has been observed to produce tumours in rodents. Its genome is a double-stranded circular DNA of 5,243 base pairs. Though much less efficiently, isolated SV40 DNA can be made to infect competent mammalian cells.

As Donald S Fredrickson wrote about Paul Berg, 'His curiosity about whether a virus might be used to transfer a foreign gene into eukaryotic cell cultures led him to become familiar with the virus called simian virus 40 (SV40). Berg considered the relationship between phage and bacteria to be closely analogous to that of SV40 and eukaryotic cells, and he wondered if the virus might work more efficiently as a vector for a bacterial gene. Berg enlisted two coworkers, David Jackson and R. H. Symons, to determine if they could insert a bacterial gene held by a modified lambda phage into the SV40 genome' (Fredrickson 2001).

It turned out that the restriction endonuclease *Eco*RI cleaved the SV40 DNA at a single site, producing a full-length linear DNA (Morrow and Berg 1972; Mulder and Delius 1972). Apparently, Jackson *et al.* did not have the information that *Eco*RI produced a staggered cut, creating cohesive ends, so they took recourse to homopolymer tailing (see Section 17.2.2.9). They briefly treated the *Eco*RI-cleaved linearized SV40 DNA with lambda (λ) exonuclease to create 3′-overhangs to facilitate subsequent dA addition catalysed by terminal transferase. They then isolated the galactose operon of *E. coli* by treatment of λdvgal with *Eco*RI. The latter DNA was tailed with dT after treatment with lambda (λ) exonuclease. These two species of DNA molecules were then annealed together. At this point, they recognized that the failure of DNA ligase to join covalently the hydrogen-bonded single strands of polydA and polydT to the original double strands in the experiment conducted by Jensen *et al.* (1971) (see Section 17.2.2.9) was because of the presence of gaps, consequent upon the unequal lengths of the polydA and polydT stretch. So, after annealing the dA-tailed SV40 DNA and the dT-tailed λdvgal DNA molecules, they filled up the gaps by incorporating deoxynucleotides catalysed by *E. coli* DNA polymerase I. The DNA ligase could now carry out covalent joining of the ends. *E. coli* exonuclease III was used prior to DNA polymerase I addition to remove 3′-phosphoryl residues if present at any nick produced inadvertently during manipulation. This involved procedure for dA/dT tailing was actually devised by Peter E Lobban and his PhD supervisor AD Kaiser, who had advised Jackson *et al.* in this matter. The details about the development of this procedure were published subsequently (Lobban and Kaiser 1973). A diagrammatic representation of the steps is given in Fig. 17.7. Thus, Jackson *et al.*

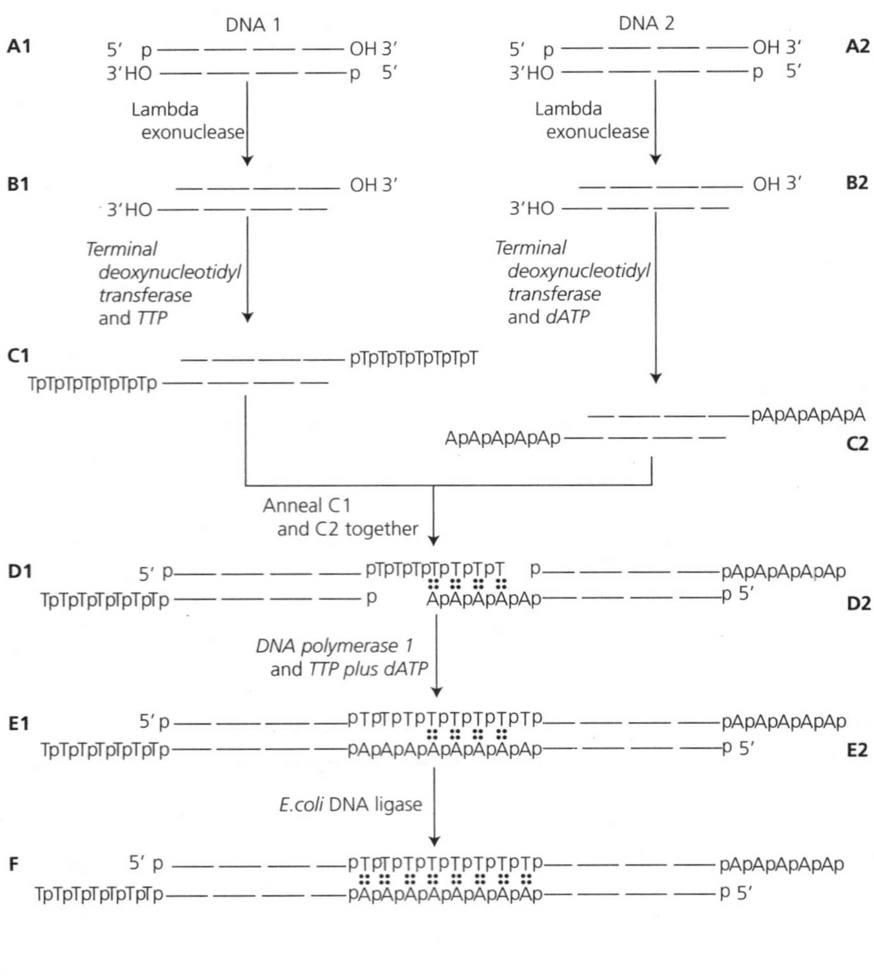

Figure 17.7 Diagrammatic representation of the steps used (Jackson *et al.* 1972; Lobban and Kaiser 1973) for covalently joining two DNA molecules by the dA–dT tailing method. A1 and A2: the two DNA molecules to be joined; B1 and B2: the two DNA molecules with 3′-overhangs as the result of brief treatment with lambda (λ) exonuclease; C1: treatment with terminal deoxyribonucleotidyl transferase plus TTP has added dT tail at the 3′-ends of B1; C2: treatment with terminal deoxyribonucleotidyl transferase plus dATP has added dA tail at the 3′-ends of B2; D1 and D2: annealed form of C1 and C2; E1 and E2: gaps at the end of the dT and dA tails have been filled up by treatment with *E. coli* DNA polymerase I plus TTP and dATP; F: E1 and E2 have been covalently linked by *E. coli* DNA ligase.

(1972) were successful in covalently inserting *in vitro* the galactose operon of *E. coli* along with some lambda genes into the SV40 genome. Jackson *et al.* did not try to introduce their novel construct into mammalian cells or bacterial cells because of their serious concern about the unknown but potential hazards of propagating the tumour-causing SV40 with its artificially linked components of some of the genes of the bacterial virus lambda (λ) and the *E. coli gal* operon.

17.4.2 Plasmids

Plasmids are autonomously replicating DNA vehicles that are not infective as viruses.

17.4.2.1 Sex factor in *E. coli*

The first evidence of the presence of an extrachromosomal, autonomously replicating and transmissible genetic element in the bacteria *E. coli* K 12 was obtained in the form of sex factor or F factor. The F^+ cells were the donors of the genetic element, while the F^- cells were the recipients. The F^- cells became F^+ after receiving the F factor. The sex factor could proliferate autonomously and more rapidly than the chromosome of the recipient cell (Lederberg *et al.* 1952; Hayes 1953; Cavalli-Sforza *et al.* 1953). Josua Lederberg proposed the term 'plasmid' for these extranuclear genetic elements that were able to replicate in an autonomous state (Lederberg 1952).

17.4.2.2 Colicinogenic genetic elements

Colicins are polypeptide toxins elaborated by some strains of *Enterobacteriaceae (Escherichia, Shigella, Salmonella)* that kill, but do not lyse, other strains of this family. Analogous to sex factors, the determinants of certain colicins can be transmitted from one bacteria (col^+) to another (col^-) by cellular contact, and the recipient cell (originally col^-) now becomes col^+ (Fredericq 1957; Clowes 1961). Genetic evidence suggested that the determinants of these colicins were indeed extrachromosomal genetic elements that replicated autonomously (Fredericq 1958). These elements, therefore, qualified as plasmids.

17.4.2.3 Resistance transfer factors

Since 1957, quite a large number of *Shigella* strains isolated from the stool of bacillary dysentery patients in Japan were found to be resistant to multiple antibiotics and sulpha drugs. It was then discovered that this multiple resistance character could be transmitted *en bloc* to other strains and even other bacteria, including *E. coli*, by physical contact or mixed cultivation (Akiba *et al.* 1960; Mitsuhashi *et al.* 1960). Various genetic evidence had led to the conclusion that the resistance transfer factors were also extrachromosomal genetic elements that replicated autonomously and thus qualified as plasmids (Hayes 1968). The similarity of genetic properties between the R factors and the Col elements has been conclusively established (Frederick 1969).

A good account of the history of plasmids has been written by Donald R Helinski of the Division of Biology of the University of California at San Diego (Helinski 2004).

17.4.2.4 Chemical nature of plasmids

The DNA nature of the F plasmids was inferred when these were found to be labelled *in vivo* by ^{32}P and subsequently inactivated as the ^{32}P decayed. The actual experiment was as follows. (a) An F^+ *E. coli* strain was cultured in a medium containing ^{32}P-orthophosphate, thereby labelling all the nucleic acids. The cells were washed thoroughly and mixed with an F^- strain of *E. coli* in a medium containing no labelled orthophosphate. The cells of the F^- strain that have become F^+, because of the transfer of the F factor from the original F^+ strain, were selected for and found to have acquired ^{32}P-radioactivity. The recipient cells now containing the ^{32}P-labelled F factor could again transfer these into another F^- strain. (b) In a parallel experiment, mitomycin C was added along with ^{32}P-orthophosphate to the culture medium of the original F^+ strain to inhibit DNA replication. These cells were then mixed with the F^- strain. The cells of the F^- strain that have become F^+ did not, however, acquire any ^{32}P-radioactivity (Lavalle and Jacob 1961). In experiment (a), if the recipient cells that had acquired the ^{32}P-labelled F factor were held for some months, allowing the ^{32}P to decay to ^{32}S (half-life 14 days; see Chapter 2, Section 2.4), they failed to transfer the F factor to the second F^- strain (Driskell and Adelberg 1961).

17.4.2.5 Physical nature of plasmids

The colicin plasmid ColE1 was found to be a circular DNA molecule when viewed under the electron microscope (Roth and Helinski 1967). The bacterial cells were lysed in alkaline buffer (0.1 M Tris, pH 8.0) using lysozyme and the ionic detergent sodium dodecyl sulphate (1%). The lysate was extracted with phenol-chloroform saturated with the pH 8.0 buffer and DNA was precipitated from the aqueous layer with ethanol. The ColE1 DNA, which was only 0.2–0.3% of the total DNA, was purified essentially free from chromosomal DNA by fractionation on a methylated albumin kieselghur (MAK) column. Electron microscopy revealed the presence of typical double-stranded DNA, mostly in the form of supercoiled circles and about 10–20% in the open circular form (double-stranded circular DNA with one or more single-strand scissions). The contour length of the circular DNA was about 2.3 μm, with some being 4.7 μm. Electron micrographs of some supercoiled and open circular plasmid DNA molecules can be seen in Figs 5.7 and 5.8.

17.4.2.6 The dye-boyant-density procedure for purification of plasmid DNA

A couple of months ahead of the Helinski paper (Roth and Helinski 1967), there appeared a paper from the laboratory of Jerome Vinograd of the California Institute of Technology in Pasadena, CA, which described a method for the purification of supercoiled circular DNA using the DNA of the polyoma virus as an example (Radloff *et al.* 1967). This method still continues to be the method of choice for purification of all supercoiled circular DNA. The binding of intercalating dyes such as ethidium bromide had just been shown to result in partial unwinding of double-stranded DNA. In the case of superhelical circular DNA, as pointed out by Radloff *et al.* (1967), 'any unwinding of the duplex causes a change in the number of superhelical turns, so that the total number of turns in the molecule remains constant. A small and critical amount of dye-binding reduces the number of superhelical turns to zero. Further dye binding results in the formation of superhelices of the opposite sign or handedness. The creation of these new superhelices introduces mechanical stresses into the duplex and a more ordered conformation into the molecule. These effects increase the free energy of formation of the DNA–dye complex. The maximum amount of dye that can be bound by the closed molecule is therefore *smaller* than by the linear or nicked circular molecule. Correspondingly, since the buoyant density of the DNA–dye complex is inversely related to the amount of dye bound, the buoyant density of the closed circular DNA–dye complex at saturation is *greater* than that of the linear or nicked circular DNA–dye complex. The above effect results in a buoyant density difference of approximately 0.04 gm/mL in CsCl containing saturating amounts of ethidium bromide'. The method has been applied for isolation of closed circular DNA from HeLa cells which has been found to band lower than the other DNA . Fig. 17.8 is a representative illustration of a CsCl–ethidium bromide equilibrium centrifugation gradient showing the relative positions of proteins, linear DNA or nicked plasmid DNA, supercoiled plasmid DNA and single-stranded RNA.

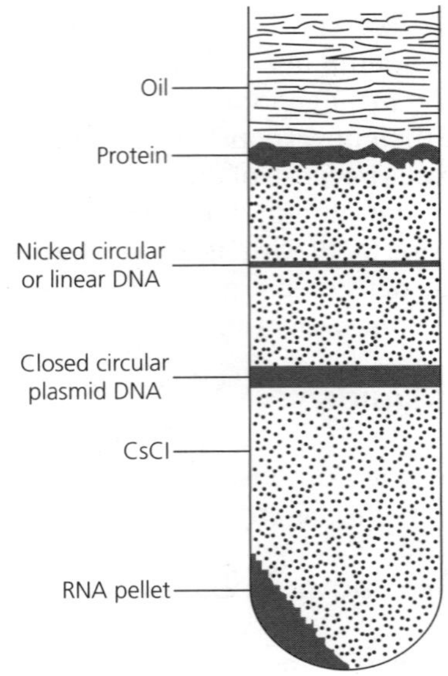

Figure 17.8 Equilibrium centrifugation of cell lysate of a strain of *Escherichia coli* containing a plasmid on a caesium chloride (CsCl)–ethidium bromide gradient. The bands have been visualized under ultraviolet ray. The centrifuge tube has been topped up with mineral oil. The relative positions of proteins, linear DNA or nicked plasmid DNA, supercoiled plasmid DNA and single-stranded RNA are shown.

17.4.2.7 The construction of plasmid pSC101

The transmissible resistance transfer factor R6 is a circular plasmid that has been isolated from both *E. coli* and *Proteus mirabilis* by CsCl–ethidium bromide equilibrium centrifugation and has been shown to have a molecular weight of approximately 64×10^6 Da (Cohen and Miller 1970). When *E. coli* C600 was transformed with this plasmid by the calcium chloride procedure (Mandel and Higa 1970), the strain acquired resistance towards tetracycline, chloramphenicol, kanamycin, neomycin, streptomycin and sulphonamide (Cohen and Miller 1970). The R6 plasmid was too large for routine manipulation, and it would have been nice to have a smaller derivative of this natural plasmid. The isolated R6 plasmid was, therefore, sheared for 30 min by a stirrer rotating at 2000 rpm and *E. coli* C600 was transformed with the sheared DNA fragments (Cohen and Chang 1973). The rationale of this step was the following. Chances were that some of the linear fragments, once inside the cell, would circularize through the action of the cellular DNA ligase. Only those circularized DNA fragments that would have the replication origin of the plasmid in them would, however, replicate and survive. This putative smaller plasmids would, of course, be cryptic unless one of the antibiotic resistance markers was also present in it, besides the replication origin. The transformed cells were, therefore, plated on a medium containing each of the antibiotics separately. Transformants did indeed appear on plates containing only tetracycline and no other antibiotic. A plasmid, which could be isolated from the transformed cells, could again transform *E. coli* C600 to yield tetracycline-resistant bacteria, but the plasmid was not self-transmissible. Electron microscopy of the plasmid revealed that it was circular and was only 5.8×10^6 Da in molecular weight, which was less than 10% of the size of the natural plasmid R6. The plasmid was then named Tc6-5, but was later renamed as pSC101 (schematic representation in Fig. 17.9).

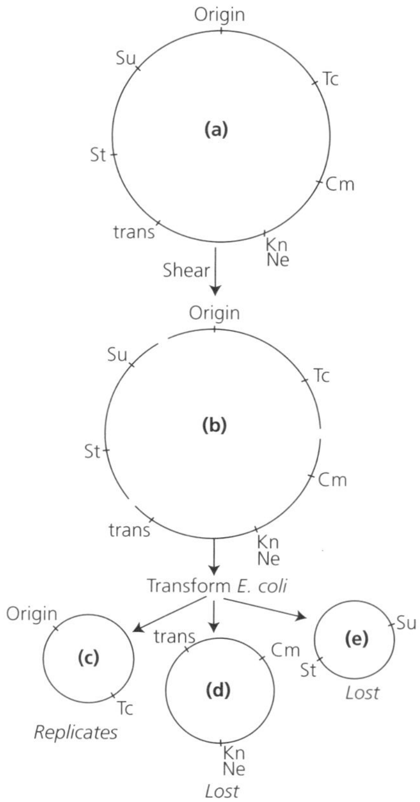

Figure 17.9 Principle behind construction of the plasmid pSC101. (a) The *E. coli* transmissible resistance factor plasmid R 6. Origin signifies the origin of replication of the plasmid,; Tc, Cm, Kn, Ne, St and Su stand for genes that impart resistance towards tetracycline, chloramphenicol, kanamycin, neomycin, streptomycin and sulphonamide; and trans denotes the transfer function responsible for self-transmissibility (the sequence of the genes is arbitrary); (b) The plasmid R6 has been sheared, for the sake of convenience of representation, say into three linear fragments. One fragment contains the origin and the tetracycline resistance genes (Tc). The second fragment contains the resistance gene for chloramphenicol (Cm) and that for kanamycin (Kn) and neomycin (Ne) and also the transfer function (trans), but no replication origin. The third fragment contains the resistance genes for streptomycin (St) and sulphonamide (Su), but again no replication origin. (c)–(e) *Escherichia coli* cells are transformed with the three fragments of DNA, wherein these are circularized by ligation *in vivo*. Only (c), which has the origin, replicates, while (d) and (e) cannot and are ultimately lost. The plasmid C is notionally pSC101, which imparts tetracycline resistance, but is not self-transmissible.

17.5 The meeting in Honolulu: The initiation of collaboration between Stanley Cohen and Herbert Boyer

It was 1970 and Tsutomu Watanabe of Japan, who had done exhaustive work on resistance transfer factors, contacted Donald Helinski of the United States, who discovered the circular nature of the Col plasmids, to assess Helinski's interest in organizing a joint US–Japan conference on plasmids under the auspices of the US National Science Foundation as a part of their programme for the support of US–Japan scientific exchange. The conference finally got scheduled for November 1973 to be held at the East West Center in Honolulu and Stanley Cohen of Stanford University, CA, subsequently joined the duo as the third co-organizer. Less than 2 weeks before the conference, Tsutomu Watanabe passed away, but the conference took place as scheduled. Thirty scientists from Japan and the United States deliberated on various molecular aspects of plasmid biology. Stanley Cohen presented his work on the construction of the small plasmid pSC101 and its ability to transform *E. coli*, thereby endowing tetracycline resistance to the bacteria. Herbert Boyer described the purification of the restriction endonuclease *Eco*RI and elucidation of the base sequence at the cohesive end it generated on cleavage of DNA. As Stanley Cohen has said, 'I listened with excitement to Herbert Boyer describe the enzymatic actions of the restriction enzyme *Eco*RI, which he and his colleagues had purified from bacteria' (Cohen 2004).

In the evening, some of the participants, including Stanley Cohen and Herbert Boyer, walked for an evening snack to a delicatessen off the Waikiki beach. The folklore is that it was there that Cohen and Boyer agreed on a collaborative proposal over hot pastrami and corned beef sandwich. Boyer's *Eco*RI would be used to linearize Cohen's pSC101, which would then be used as a vehicle to ligate other DNA fragments also cleaved with the same *Eco*RI.

17.6 The outcome of the Cohen–Boyer collaboration

In Cohen's language, 'Herb and I began our collaborative experiments shortly after returning from Hawaii. Plasmids isolated at Stanford were transported to Boyer's lab in San Francisco for cutting by *Eco*RI and then transported back to Stanford, where randomly joined fragments were introduced into *E. coli* by the transformation procedure. Bacterial colonies expressing only some of the R6-5 antibiotic resistance genes were isolated and then analysed by electrophoresis in gels at UCSF (University of California at San Francisco) and by centrifugation and electron microscopy at Stanford. Annie Chang, then a research technician in my laboratory, isolated plasmids and carried out bacterial transformations; Bob Helling, who was on sabbatical leave in Boyer's lab, did many of the gel analyses. The strategy that Boyer and I had devised in Hawaii worked even better than we had hoped, and the months in early 1973 were a period of almost unbelievable excitement for all of us' (Cohen 2004).

17.6.1 The first collaborative communication

The first communication containing the results of the collaboration was reported in July 1973 and was published in November the same year (Cohen *et al.* 1973). The title of the paper was 'Construction of biologically functional bacterial plasmids *in vitro*'.

The abstract of the paper read as follows: 'The construction of new plasmid DNA species by *in vitro* joining of restriction endonuclease-generated fragments of separate plasmids is described. Newly constructed plasmids that are inserted into *Escherichia coli* by transformation are shown to be biologically functional replicons that possess genetic properties and nucleotide base sequences from both of the parent DNA molecules. Functional plasmids can be obtained by reassociation of endonuclease generated fragments of larger replicons, as well as by joining of plasmid DNA molecules of entirely different origins'.

Initially, they cleaved the large plasmid R6-5 with *Eco*RI into multiple fragments and introduced the fragments singly or in combination into *E. coli*, thereby imparting newer antibiotic resistance to the bacteria and subsequently allowing isolation of newer smaller plasmids (pSC102, pSC105) from the transformed *E. coli*. This experiment established that a DNA fragment that had an antibiotic resistance marker in it, but did not have the replication origin, could be ligated *in vivo* to a fragment that had the replication origin in it, to form a stable antibiotic-resistant plasmid.

A fortuitous but very important finding was that the plasmid pSC101, which was earlier constructed *in vitro* (Cohen and Chang 1973), had only one *Eco*RI cleavage site. The plasmid pSC101 was only linearized on digestion with the restriction endonuclease, but no portion of the plasmid was lost. Besides, the cleavage by *Eco*RI did not appear to interrupt the replication function or the tetracycline resistance function. Thus, pSC101 qualified to be used as a vector for cloning by *in vitro* ligation of other *Eco*RI-cleaved DNA fragments not having any replication origin,.

The most significant contribution reported in the paper was the incorporation *in vitro* into pSC101 of genes conferring resistance to streptomycin and sulphonamide from the plasmid RSF1010, which was originally isolated from the bacteria *Salmonella typhimurium*. The plasmid RSF1010 also had only one cleavage site for *Eco*RI. The plasmids pSC101 and RSF1010 were linearized by digestion with *Eco*RI, mixed together and treated with *E. coli* DNA ligase. *E. coli* was transformed with the ligation product and plated on a medium containing both tetracycline and streptomycin. Colonies did indeed appear that contained a single molecular circular DNA species (designated pSC109) which sedimented at 33.5S, corresponding to an approximate molecular weight of 11.5 × 106 Da, which was the sum total of the molecular weights of pSC101 and RSF1010. When isolated pSC109 was digested with *Eco*RI and analysed by agarose gel electrophoresis, it revealed two DNA fragments that were indistinguishable from *Eco*RI-treated pSC101 and RSF1010. The presence in pSC109 of both pSC101 and RSF1010 in covalent linkage was confirmed by heteroduplex formation between pSC109 and RSF1010 and also between pSC109 and pSC101, as seen under the electron microscope.

Subsequently, the plasmid p1258 from *Staphylococcus aureus* was used as a source for the penicillin resistance gene for cloning into pSC101 (Chang and Cohen 1974). The plasmid p1258 was cleaved with *Eco*RI to generate four different fragments. The plasmid pSC101, linearized by digestion with *Eco*RI, was mixed with all the four *Eco*RI-cleaved fragments of the plasmid p1258, treated with *E. coli* DNA ligase and *E. coli* C600 transformed with the ligated mixture. The transformed *E. coli* cells were plated on a medium containing both tetracycline and penicillin. Colonies resistant to both the antibiotics did appear. No penicillin-resistant colony appeared when *E. coli* C600 was transformed with the intact plasmid p1258 or the *Eco*RI-treated p1258 or pSC101. Plasmid was isolated from the cells that were resistant to both tetracycline and penicillin (designated pSC112). The presence in pSC112 of both pSC101 and a fragment from p1258 in covalent linkage was confirmed by heteroduplex formation between pSC 112 and p1258 and also between pSC112 and pSC101, as seen under the electron microscope.

17.6.2 The second collaborative communication

The second collaborative paper entitled 'Replication and transcription of eukaryotic DNA in *Escherichia coli*' was communicated in January 1974 and was published in May (Morrow *et al.* 1974). Cloning of the genes for ribosomal RNA of the eukaryote *Xenopus laevis* in the plasmid pSC101 has been described in this paper. The restriction endonuclease *Eco*RI has been used to cleave both the DNA of pSC101 and the DNA for Xenopus ribosomal RNA. While the pSC101 DNA produced a single linear DNA species (5.8×10^6 Da molecular weight), the Xenopus DNA gave rise to three major fragments (molecular weights 3.1×10^6, 4.3×10^6 and 5.1×10^6 Da). The linear pSC101 DNA was mixed with all the three fragments of *Xenopus* DNA together, incubated with *E. coli* DNA ligase and *E. coli* C600 was transformed with the ligated mixture. The transformed cells were plated on a medium containing tetracycline. Plasmids were isolated from each colony, digested with *Eco*RI and analysed by agarose gel electrophoresis. All the plasmid samples showed the 5.8×10^6 Da pSC101 band. In addition to the pSC101 band, some of the plasmid samples (class I) revealed the presence of the 3.1×10^6 Da band corresponding to one of the *Xenopus* DNA fragments, while some others (class II) showed the 4.3×10^6 Da band corresponding to another *Xenopus* DNA fragment. A few of the plasmid samples (class II) exhibited all the three, the 5.8×106 Da band corresponding to pSC101, the 3.1×10^6 Da *Xenopus* DNA band and also the 4.3×10^6 Da Xenopus band. The identity of the DNA bands was confirmed by comparing their buoyant densities and observing heteroduplex formation under the electron microscope with the authentic DNA samples. Finally, separate hybridization with ^{32}P-labelled 18S and 28S *Xenopus* rRNAs revealed the presence of both 18S and 28S rDNAs in class III plasmids, while class I and class II plasmids contained only one of the rDNAs predominantly.

Next, the possibility of transcription in *E. coli* cells of the *Xenopus* DNA cloned in pSC101 was ascertained by transforming a minicell producing *E. coli* strain with each class of plasmid. Minicells, which bud off from such *E. coli* strains, do not contain any chromosomal DNA, but could contain the plasmids that were present in the parent cell. As the minicells have all the machinery for macromolecular synthesis, transcription of DNA in the plasmids can be studied in isolation and free of interference from chromosomal DNA. Minicells from the transformed cells were incubated with ^{3}H-uridine and RNA purified from the minicells was hybridized with *Xenopus* rDNA immobilized on nitrocellulose membranes. RNA produced by minicells containing each of class I, class II and class III plasmids hybridized with the relevant immobilized *Xenopus* rDNA, but RNA produced by minicells containing only pSC101 did not hybridize. Thus, the DNA of the eukaryote *Xenopus laevis* linked to the plasmid pSC101 not only was replicated in *E. coli*, but also was transcribed properly.

17.7 The gene cloning technique has been patented

An US patent has been granted to the trustees of the Stanford University for a 'process for producing biologically functional molecular chimeras'. The inventors of the process, as appears in the patent, are Stanley Cohen and Herbert Boyer. The abstract of the process is as follows.

'Method and compositions are provided for replication and expression of exogenous genes in microorganisms. Plasmids or virus DNA are cleaved to provide linear DNA having ligatable termini to which is inserted a gene having complementary termini, to provide a biologically functional replicon with a desired phenotypical property. The replicon is inserted into a microorganism cell by transformation. Isolation of the transformants provides cells for replication and expression of the DNA molecules present in the modified plasmid. The method provides a convenient and efficient way to introduce genetic capability into microorganisms for the production of nucleic acids and proteins, such as medically or commercially useful enzymes, which may have direct usefulness, or may find expression in the production of drugs, such as hormones, antibiotics, or the like, fixation of nitrogen, fermentation, utilization of specific feedstocks, or the like.' The full text of the patent granted is given in Appendix 1.

17.8 The terms 'genetic engineering' and 'recombinant DNA technology'

The terms 'genetic engineering' and 'recombinant DNA technology' have become synonymous today and are being used in an interchangeable manner. The genesis of the two terms, however, is not similar.

17.8.1 The term 'genetic engineering'

The term 'genetic engineering' was, in all probability, first used in 1965 by Rollin Hotchkiss, as he speculated on the possibility of future intervention for enhancing of mental and physical characteristics of the human kind (Hotchkiss 1965). On 17 August 1965, Hotchkiss delivered the second Wilhelmine E. Key lecture entitled 'Portents for a genetic engineering' at the annual meetings of the American Institute of Biological Sciences, University of Illinois, Urbana. He said, 'The new potential program, "genetic engineering", which is raised by the exploits of molecular biology, is a genetic intervention that could be practiced.... on individual genes'. Hotchkiss acknowledged, 'My own specialty, that of genetic transformation of bacteria by administration of DNA from related strains, probably has been the principal stimulus for recent speculation along these lines'. He conjectured, 'A more effective mode of introduction of DNA would be through viruses or episomes'. Hotchkiss further said, 'One can imagine farms of cultured human cells producing desired markers and delivering them up to viruses that can then profusely replicate them'. He also predicted, 'I believe the art of DNA identification and fractionation would soon bypass the laborious genetic crosses'. Hotchkiss did not have even the faintest inkling of what unfolded during 1972–1973; nevertheless, his vision, though rather hazy, was definitely not wild.

The term 'genetic engineering' had indeed been used in the past as a broad term implying genetic manipulation of any sort (Shapiro *et al.* 1969; Kass 1971). Let us also look at the first patent of a life form filed in June 1972 by Ananda Chakrabarty and General Electric Company (patent granted in 1981). The very first line of the abstract of the patent mentions about the use of genetic engineering techniques, though perusal of the abstract quoted below would reveal that no recombinant DNA technique was applied. 'Unique microorganisms have been developed by the application of genetic engineering techniques. These microorganisms contain at least two stable (compatible) energy-generating plasmids, these plasmids specifying separate degradative pathways. The techniques for preparing such multi-plasmid strains from bacteria of the genus Pseudomonas are described. Living cultures of two strains of Pseudomonas (*P. aeruginosa* [NRRL B-5472] and *P. putida* [NRRL B-5473]) have been deposited with the United States Department of Agriculture, Agricultural Research Service, Northern Marketing and Nutrient Research Division, Peoria, Ill. The *P. aeruginosa* NRRL B-5472 was derived from *Pseudomonas aeruginosa* strain 1c by the genetic transfer thereto, and containment therein, of camphor, octane, salicylate and naphthalene degradative pathways in the form of plasmids. The P. putida NRRL B-5473 was derived from *Pseudomonas putida* strain PpG1 by genetic transfer thereto, and containment therein, of camphor, octane, salicylate and naphthalene degradative pathways and drug resistance factor RP-1, all in the form of plasmids'.

Pushpa Mitra Bhargava, of the Centre for Cellular and Molecular Biology, Hyderabad, had mentioned genetic engineering in his article entitled 'Fantastic strides in Biochemistry' published in the weekly newspaper *The Motherland* in its New Delhi (India) edition of Sunday, 11 March 1973. The relevant paragraph is reproduced here. 'Some day in the future it should be possible for us to carry out "genetic engineering" to control at will – though admittedly, at least in the beginning, in a limited way – the genetic potential of a living organism, including man. It is already possible for us to "construct" new micro-organisms from existing ones, which will behave in a predictable manner'. The Cohen–Boyer paper was still a few months away, so Bhargava also was apparently referring to genetic manipulation of any sort.

17.8.2 The term 'recombinant DNA technology'

In all probability, the term 'recombinant DNA' first appeared as the subject matter of the letter, 'Potential biohazards of recombinant DNA molecules', which was published in the 26 July 1974 issue of *Science* (Berg *et al.* 1974). The first line of the letter stated, 'Recent advances in techniques for the isolation and rejoining of segments of DNA now permit construction of biologically active recombinant DNA molecules *in vitro*'. The letter was signed by as many as 11 eminent molecular biologists, including Paul Berg, Stanley Cohen and Herbert Boyer, and urged all scientists throughout the world to voluntarily deferring experiments aimed at construction of new recombinant DNA molecules till the potential hazards of such recombinant molecules were ascertained. The technology for producing biologically active recombinant DNA molecules *in vitro* has later been termed recombinant DNA technology.

References

Akiba T, Koyame K, Ishiki Y, Kumira S. On the mechanism of the development of multiple drug-resistant clones of Shigella. *Japan J Microbiol* 1960;4:219–225.

Arber W. Host-controlled modification of bacteriophage. *Annu Rev Microbiol*1965;19:365–378.

Arber W, Dussoix D. Host specificity of DNA produced by *Escherichia coli*. I. Host controlled modification of bacteriophage λ. *J Mol Biol* 1962;5:18–36.

Becker A, Lyn G, Gefter M, Hurwitz J. The enzymatic repair of DNA. II. Characterization of phage induced sealase. *Proc Natl Acad Sci USA* 1967;58:1996–2003.

Berg P, Baltimore D, Boyer HW, Cohen SN, Hogness DS, Nathns D, *et al.* Potential biohazards of recombinant DNA molecules. *Science* 1974;185:303.

Bertani G, Weigle JJ. Host controlled variation in bacterial viruses. *J Bacteriol* 1953;65:113–121.

Cavalli-Sforza LL, Lederberg J, Lederberg EM. An infective factor controlling sex compatibility in *Escherichia coli*. *J Gen Microbiol* 1953;8:89–95.

Chang ACY, Cohen SN. Genome construction between bacterial species *in vitro*: Replication and expression of *Staphylococcus* plasmid genes in *Escherichia coli*. *Proc Natl Acad Sci USA* 1974;71:1030–1034.

Clows RC. Colicin factors as fertility factors in bacteria: *Escherichia coli*. *Nature* 1961;190:986–992.

Cohen SN. DNA cloning. Available at: www.ergito.com; 2004.

Cohen SN, Chang ACY. Recircularization and autonomous replication of a sheared R-factor DNA segment in *Escherichia coli* transformants. *Proc Natl Acad Sci USA* 1973;70: 1293–1297.

Cohen SN, Chang ACY, Boyer HW, Helling RB. Construction of biologically functional bacterial plasmids *in vitro*. *Proc Natl Acad Sci USA* 1973;70:3240–3244.

Cohen SN, Miller CA. Non-chromosomal antibiotic resistance in bacteria. II. Molecular nature of R-factors isolated from *Proteus mirabilis* and *Escherichia coli*. *J Mol Biol* 1970;50:671–687.

Cozzarelli NR, Melechen NE, Jovin TM, Kornberg A. Polynucleotide cellulose as a substrate for a polynucleotide ligase induced by phage T4. *Biochem Biophys Res Commun* 1967;28:578–586.

Davison PF. The effect of hydrodynamic shear on the deoxy ribonucleic acid from T2 and T4 bacteriophages. *Proc Natl Acad Sci USA* 1959;45:1560–1568.

Driskell PJ, Adelberg EA. Inactivation of the sex factor of *Escherichia coli* K12 by the decay of incorporated radiophosphorus. *Bacteriol Proc* 1961;186:647–653.

Dussoix D, Arber W. Host specificity of DNA produced by *Escherichia coli*. II. Control over acceptance of DNA from infecting phage λ. *J Mol Biol* 1962;5:37–49.

Fredrickson DS. *The Recombinant DNA Controversy: A Memoir - Science, Politics, and the Public Interest 1974-1981*. Washington, DC: ASM Press; 2001. p. 8.

Fredericq P. Colicins. *Annu Rev Microbiol* 1957;11:7–28.

Fredericq P. Colicins and colicinogenic factors. *Symp Soc Exp Biol* 1958;12:104.

Fredericq P. The recombination of colicinogenic factors with other episomes and plasmids. In: Wolstenholme GEW, O'Connor M, eds. *Bacterial Episomes and Plasmids*. London: J & A Churchill; 1969. pp. 163–178.

Gellert M. Formation of covalent circles of lambda DNA by *E. coli* extracts. *Proc Natl Acad Sci USA* 1967;57:148–155.

Hayes W. The mechanism of genetic recombination in *Escherichia coli*. *Cold Spring Harbor Symp Quant Biol* 1953;18:75.

Hayes W. *The Genetics of Bacteria and Their Viruses, Second Edition*. New York: John Wiley & Sons; 1968.

Hedgpeth J, Goodman HM, Boyer HW. DNA nucleotide sequence restricted by the RI endonuclease. *Proc Natl Acad Sci USA* 1972 ;69:3448–3452.

Helinski DR. Plasmids: A selective view of their history. In: Funnel BE, Philips GJ, eds. *Plasmid Biology*. Washington, DC: ASM Press; 2004. pp. 1–21.

Hotchkiss RD. Portents for a genetic engineering. *J Heredity* 1965;56:197–202.

Jackson DA, Symons RH, Berg P. Biochemical method for inserting new genetic information into DNA of Simian Virus 40 DNA molecules containing lambda phage genes and the galactose operon of *Escherichia coli*. Proc *Natl Acad Sci USA* 1972;69:2904–2909.

Jensen RH, Wodzinski RJ, Rogoff MH. Enzymatic addition of cohesive ends to T7 DNA. *Biochem Biophys Res Com* 1971;43:384–392.

Kelly TJ Jr, Smith HO. A restriction enzyme from *Haemophilus influenzae*. II. Base sequence of the recognition site. *J Mol Biol* 1970;51:393–409.

-Kass LR. The new biology: What price relieving man's estate. *Science* 1971;174:779–788.

Kunitz M. Crystalline deoxyribonuclease. I. Isolation and general properties. *J Gen Physiol* 1950;33:349–362.

Lavalle R, Jacob F. Sur la sensibilite des episomes sexuels et colicinogenes d'*E. coli* K12 a la desintegration du radiophosphore. *C R Acad Sci Paris* 1961; 252:1678–1682.

Lederberg J. Cell genetics and hereditary symbiosis. *Physiol Rev* 1952;32:403.

Lederberg J, Cavalli-Sforza LL, Lederberg EM. Sex compatibility in *E. coli*. *Genetics* 1952;37:720–726.

Lederberg S. Suppression of the multiplication of heterologous bacteriophages in lysogenic bacteria. *Virology* 1957;3:496–513.

Levene PA, Medigreceanu F. The action of gastro-intestinal juices on nucleic acids. *J Biol Chem* 1911;9:375–403.

Linn S, Arber W. Host specificity of DNA produced by *Escherichia coli*. X. *In vitro* restriction of phage fd replicative form. *Proc Natl Acad Sci USA* 1968;59:1300–1306.

Lobban PE, Kaiser AD. Enzymatic end-to-end joining of DNA molecu es. *J Mol Biol* 1973;78:453–471.

Luria SE, Human ML. A nonhereditary, host-induced variation in bacterial viruses. *J Bacteriol* 1952;64:557–569.

Mandel M, Higa A. Calcium-dependent bacteriophage DNA infection. *J Mol Biol* 1970;53:159–162.

Mertz JE, Davis RW. Cleavage of DNA by R1 restriction endonuclease generates cohesive ends. *Proc Natl Acad Sci USA* 1972;69:3370–3374.Meselson M, Yuan R. DNA restriction enzyme from *E. coli*. *Nature* 1968;217:1110–1114.

Mitsuhashi S, Harada K, Hashimoto H. Multiple resistance of bacteria and transmission of drug-resistance to other strains by mixed cultivation. *Japanese J Exp Med* 1960;30:179–185.

Morrow JF, Berg P. Cleavage of Simian Virus 40 DNA at a unique site by a bacterial restriction enzyme. *Proc Natl Acad Sci USA* 1972;69:3365–3369.

Morrow JF, Cohen SN, Chang ACY, Boyer HW, Goodman HM, Helling RB. Replication and transcription of eukaryotic DNA in *Escherichia coli*. *Proc Natl Acad Sci USA* 1971;71:1743–1747.

Mulder C, Delius H. Specificity of the break produced by restricting endonuclease R_1 in Simian Virus 40 DNA, as revealed by partial denaturation mapping. *Proc Natl Acad Sci USA* 1972;69: 3215–3219.

Olivera BM, Lehman IR. Linkage of polynucleotides through phosphodiester bonds by an enzyme from *Escherichia coli*. *Proc Natl Acad Sci USA* 1967a;57:1426–1433.

Olivera BM, Lehman IR. Diphosphopyridine nucleotide: A cofactor for the polynucleotide joining enzyme from *Escherichia coli*. *Proc Natl Acad Sci USA* 1967b;57:1700–1704.

Radloff R, Bauer W, Vinograd J. A dye-buoyant-density method for the detection and isolation of closed circular duplex DNA: The closed circular DNA in HeLa cells. *Proc Natl Acad Sci USA* 1967;57:1514–1521.

Roth TF, Helinski DR. Evidence for circular DNA forms of a bacterial plasmid. *Proc Natl Acad Sci USA* 1967;58:650–657.

Sgaramella V. Enzymatic oligomerization of bacteriophage P22 DNA and of linear Simiun Virus 40 DNA. *Proc Natl Acad Sci USA* 1972;69:3389–3393.

Sgaramella V, van de Sande JH, Khorana JH. Studies on polynucleotides, C. *Proc Natl Acad Sci USA* 1970;67:1468–1475.

Shapiro J, Eron L, Beckwith J. More alarums and excursions. *Nature* 1969;224:1337.

Sriskanda V, Shuman S. A second NAD^+-dependent DNA ligase (LigB) in *Escherichia coli*. *Nucleic Acids Res* 2001;29:4930–4934.

Smith HO, Wilcox KW. A restriction enzyme from *Haemophilus influenzae*. I. Purification and general properties. *J Mol Biol* 1970;51:379–391.

Weiss B, Richardson CC. Enzymatic breakage and joining of deoxyribonucleic acid. I. Repair of single strand breaks in DNA by an enzyme system from *Escherichia coli* infected with T4 bacteriophage. *Proc Natl Acad Sci USA* 1967;57:1021–1028.

Wright S. *Molecular Politics: Developing American and British Regulatory Policy for Genetic Engineering, 1972-1982*. Chicago, IL: University of Chicago Press; 1994. p. 124.

Yoshimori RN. PhD dissertation. University of California School of Medicine, San Francisco, CA; 1971.

Zimmerman SB, Little JW, Oshinsky CK, Gellert M. Enzymatic joining of DNA strands: A novel reaction of diphosphopyridine nucleotide. *Proc Natl Acad Sci USA* 1967;57:1841–1848.

18

Reaction of the scientific community after successful cloning of a gene

H. K. Das

18.1 Apprehensions even in the pre-gene–cloning years

Much before the actual recombinant DNA experiments were contemplated, there were concerns about the possible misuse of the explosive possibilities that had been opened up by the achievements in molecular genetics of bacteria and their viruses, as several scientists extrapolated and speculated on the possibility of future genetic intervention for enhancing the mental and physical characteristics of humans (Hotchkiss 1965). Salvador Luria, who was awarded the Nobel Prize in 1969 for his work on the molecular genetics of bacterial viruses, remarked in 1963, 'What has been foremost in my mind has not been a feeling of optimism, but one of tremendous fear of the potential dangers that genetic surgery (apparently referred to molecular genetic manipulation of humans), once it becomes feasible, can create if misapplied' (Wright 1994).

In 1969, the isolation of pure lac operon DNA from *Escherichia coli* was reported from the laboratory of Jon Beckwith of the Harvard Medical School (Shapiro, MacHattie *et al.* 1969). This was indeed a remarkable molecular genetic fit and was hailed widely both by biologists and the media. Three prominent members of this group had on this occasion stated, 'progress in the field of molecular genetics in the last few years has been extraordinary. ... the isolation of pure *lac* operon DNA was a graphic, useful and easily understood example of that progress'. 'As to the specific issue of genetic engineering (apparently meant molecular genetic experiments), we cannot predict the future. But who in 1896 (the year radioactivity was discovered by Becquerel) could have foreseen the weapons of mass destruction which now threatens us all?' They went on to state, 'scientists are obligated to inform the public about what is happening in their secluded fields of research so that people can demand control over decisions which profoundly affect their lives. If our arguments mean that that the progress of science itself may be interrupted, that is an unfortunate consequence we will have to accept' (Shapiro, Eron *et al.* 1969).

The fear of imminent extension of molecular genetic manipulations to humans and its possible misapplication was all pervading those days. In 1971, Leon R Kass, then the executive secretary of the Committee on the Life Sciences and Social Policy, National Research Council, US National Academy of Sciences, analysed the situation in exhaustive details in an article, 'The New Biology: what price relieving man's estate?' (Kass 1971). The article started with the sentence, 'Recent advances in biology and medicine suggest that we may be rapidly acquiring the power to modify and control the capacities and activities of men by direct intervention and manipulation of their bodies and minds'. Kass continued, 'These developments have been viewed both with enthusiasm and alarm'. Kass pleaded for restraint and even for abstention. He said, 'we can be wise enough to know that we are not wise enough. When we lack sufficient wisdom to do, wisdom consists in not doing. Caution, restraint, delay, abstention are what is second best (and, perhaps, only) wisdom dictates with respect to the technology of human engineering'. 'If we can recognize that biomedical advances carry significant social costs, we may be willing to adopt a less permissive, more critical stance toward new developments'. Kass had indeed advocated 'the need for institutional controls'. He cautioned that, 'If attempts are not made early to detect and diminish the social costs of biomedical advances by intelligent institutional regulation, the society is likely to react later with more sweeping, immoderate, and throttling controls'. Kass asserted, 'Clearly, the question is not control versus no control, but rather what kind of control, when, by whom, and for what purpose'. Kass also advocated, 'Means for achieving international regulation and control need to be devised. Biomedical technology can be no nation's monopoly'. He emphasized, 'The probable or possible harmful effects of the widespread use of a new technique should be anticipated and introduced as "costs" to be weighed in about the *first* use. The regulatory institutions should be encouraged to exercise restraint and to formulate the grounds for saying "no". We must all get used to the idea that biomedical technology makes possible many things we should never do'.

18.2 Apprehensions generated on the news of joining of SV40 DNA with the lambda DNA in the laboratory of Paul Berg

The scientific community reacted with great expectations and apprehensions as the news of the experiments in the laboratory of Paul Berg leaked out. The experiments were initiated in the spring of 1971. As David S Fredrickson has narrated (Fredrickson 2001), 'Janet Mertz, a graduate student newly arrived (in Paul Berg's laboratory at Stanford) from the Massachusetts Institute of Technology, was intrigued by the possibility that SV40 chromosomes could be reproduced in bacteria'. Fredrickson also mentioned about 'Mertz's growing ambivalence about such an experiment'. Though Mertz was not directly involved in the project, she discussed about the experiments when she attended a workshop at Cold Spring Harbor Laboratory in June 1971. On 28 June 1971, Paul Berg got a telephone call from Robert Pollack, a microbiologist in the research staff of the Cold Spring Harbor Laboratory. 'Pollack told Berg in effect that he should "put genes into a phage that does not grow in a bug that grows in your gut" and reminded him that SV40 is a small animal tumor virus and transforms human cells in culture, making them look like tumor cells. Pollack later described the idea as a pre-Hiroshima condition' (Fredrickson 2001). 'After Pollack's call Berg undertook further sampling of the opinions of his peers', which included Maxine Singer, Leon Kass and Andrew Lewis, and 'in the fall of 1971, he informed his coworkers that they should postpone that part of their proposed experiment that would transfect the SV40–lambda hybrid into *E. coli*' (Fredrickson 2001). The report of the truncated work from the laboratory of Paul Berg was communicated on 31 July 1972 and appeared in print in October 1972 (Jackson *et al.* 1972).

18.3 A two-edged sword

The Cell Biology correspondent of *Nature* wrote his comments on the paper from the laboratory of Paul Berg in the 10 November 1972 issue: 'Both the rationale of these experiments and their execution, as far as it has been carried to date, are, of course, impeccable, but there is a fly in the ointment which no doubt Berg and his colleagues are well aware of and which may cause them misgivings about proceeding further and could result in their reagent being left permanently in the deep freeze'. 'The dilemma is obvious enough; if the SV40 moiety of this hybrid DNA can act as a vector for the insertion of the bacterial DNA into chromosomes of animal cells, then no doubt the bacterial and phage moieties can act as vectors for the insertion of the SV40 DNA into *E. coli* chromosomes'. '... but *E. coli* and its relatives happen to constitute the human gut flora, SV40 transforms some human cells in culture and a virus virtually indistinguishable from SV40 has recently been isolated from brain tissue of people suffering from multifocal eucoencephalopathy'.

'What would be the consequences if the reagent Berg and his colleagues have made somehow infected and lysogenized *E. coli* in someone's gut as the result of an accident? This possibility, remote though it may seem to be, can hardly be ignored, and it would be most interesting to learn what criteria the group adopts when it decides whether or not the scientific information that might be obtained by continuing the experiment justifies the risk. Perhaps those involved would decide that the game is not worth the candle' (Cell Biology correspondent 1972).

18.4 Asilomar I

At that point in time, it appeared that an important ingredient of Berg's experiment that generated the serious apprehension was the use of SV40 DNA, as SV40 was suspected to have the ability to cause cancer. Berg mooted the idea of a meeting on hazards and containment issues related to work on tumour viruses. The Human Cell Biology Steering Committee accepted the idea and convened a meeting of its members, which was held on 22–24 January 1973 in Asilomar Conference Center in Pacific Grove, CA. Many have dubbed this meeting as Asilomar I. Around 100 scientists attended, almost all of them from the United States. Among them were Paul Berg, David Baltimore, Andrew Lewis, Wallace Rowe, James Watson and Norton Zinder. Various scientists presented their findings on many tumour viruses. Nobody reported that there was any evidence of any of the known viruses causing human cancer. Nevertheless, safety precautions were prescribed, but only for those who were searching for a virus that could cause human cancer. Berg stated that 'prudence demands caution and some serious efforts to define the limits of whatever potential hazards exist' (Fredrickson 2001). A larger number of the scientists, however, felt that fear being spread about presumed hazards was unwarranted and expressed concern and unhappiness about it. The proceedings of the meeting did not have any mention of the experiments that would later be referred to as recombinant DNA experiments. Susan Wright has commented on the outcome of the meeting, 'The scientific community's inaction with respect to the hazards of cancer virus research demonstrates the competitive pressures that inhibited researchers from even trying to control widely acknowledged hazards. The safety of research with cancer viruses was handled internally and quietly, and moves toward external regulation were headed off' (Wright 1994).

18.5 The EMBO workshop in Basel

The EMBO workshop held in Basel in September 1972 was on DNA restriction and modification and was attended by about 50 scientists from 12 countries. In this workshop, discussions were also held on the use of restriction endonucleases for the construction of hybrid DNA molecules and as a tool in genetic engineering and also the potential biohazards.

18.6 The Gordon Conference on Nucleic Acids, New Hampton, NH, 13 June 1973

The fourth day of the conference was kept for discussion on the use of restriction endonucleases for the analysis of DNA. It was on this day that 'Herbert Boyer had shared with the conferees information about the capabilities of the restriction enzyme *Eco*RI to splice DNAs of different origin, and how two plasmids bearing genes specifying resistance to two different antibiotics had been joined' (Fredrickson 2001). [The results of the work was communicated in July and was published in November, the same year (Cohen *et al.* 1973)]. 'It was after Boyer's comments that someone loudly sounded the excited comment that "Now we can combine any DNA". Other reaction to this hint that biology was approaching something akin to the nuclear physicists' chilling arrival at "critical mass" was delayed until late afternoon, when two researchers at the Cambridge Molecular Biology Laboratory, Edward Ziff and Paul Sedat, sought out the two chairpersons of the conference. These scientists, Maxine Singer of NIH and Ditter Soll of Yale, listened to the urging of Ziff and Sedat to have the conference discuss the potential hazards disclosed in the afternoon's session. Left with only a half day's conference to go, Singer and Soll nevertheless agreed to arrange for the matter to be taken up at the beginning of the Friday morning session. Within about a half hour, approximately three-quarters of the enrollees at the conference who were still on hand reached a consensus that a communication should be sent to the presidents of the National Academy of Sciences (NAS) and the Institute of Medicine' (Fredrickson 2001). Edward Ziff later followed it up, expressing his concern in an article that appeared in the *New Scientist* (Ziff 1973).

18.7 The Singer–Soll letter

The letter, as desired by the enrollees of the Gordon Conference, was sent on 17 July 1973 to Philip Handler, president of the National Academy of Science and to John R Hogness, president of the National Institute of Medicine. The following are the excerpts of the letter: 'We are writing to you, on behalf of a number of scientists, to communicate a matter of deep concern. .... we presently have the technical ability to join together, covalently, DNA molecules from diverse sources. .... This technique could be used, for example, to combine DNA from animal viruses with bacterial DNA, or DNA's of different viral origin might be so joined. In this way new kinds of hybrid plasmids or viruses, with biological activity of unpredictable nature, may eventually be created. .... Certain such hybrid molecules may prove hazardous to laboratory workers and to the public. .... The conferees suggested that the Academies establish a study committee to consider this problem and to recommend specific actions or guidelines, should that seem appropriate ....' (Singer and Soll 1973).

18.8 Response of the National Academy of Science: Formation of a study committee

Philip Handler consulted the members of the Council of the National Academy of Science and referred the Singer–Soll letter to the Executive Committee of the Assembly of Life Sciences with the suggestion that a study committee be formed to assess the hazards involved. The Executive Committee of the Assembly of Life Sciences gave a personal hearing to Maxine Singer in September 1973, and on her advice, requested Paul

Berg to head the study committee. The members of the study committee, as selected by Berg, were David Baltimore, Daniel Nathans, Richard Roblin, Maxine Singer, James Watson, Sherman Weissman and Norton Zinder. The one-day meeting of the Committee was held on 7 April 1974 in the Massachusetts Institute of Technology. Berg subsequently co-opted Herbert Boyer, Stanley Cohen, Ronald Davis and David Hogness as members of the study committee that evolved into the Committee on Recombinant DNA.

18.9 Report of the Committee on Recombinant DNA and the appeal to the world scientific community

The report of the Committee on Recombinant DNA was released 3 months later by the National Academy of Science on 18 July 1974 and was published in *Science* as a letter to the world scientific community (Berg *et al.* 1974). The excerpts of the report are as follows. 'Recent advances in techniques for the isolation and rejoining of segments of DNA now permit construction of biologically active recombinant DNA molecules *in vitro*.' '... Although such experiments are likely to facilitate the solution of important theoretical and practical biological problems, they would also result in the creation of novel types of infectious DNA elements whose biological properties cannot be completely predicted in advance. There is serious concern that some of these artificial recombinant DNA molecules could prove biologically hazardous'. '... Thus new DNA elements introduced into *E. coli* might possibly become widely disseminated among human, bacterial, plant, or animal populations with unpredictable effects'. The members of the committee proposed: 'First, and most important, that until the potential hazards of such recombinant DNA molecules have been better evaluated or until adequate methods are developed for preventing their spread, scientists throughout the world join with the members of this committee in voluntarily deferring the following types of experiments.' Apparently, it was Norton Zinder who had said during the meeting, 'If we had any guts at all, we'd tell people not to do these experiments'. The types of experiments that were to be deferred were, '*Type 1*: Construction of new, autonomously replicating bacterial plasmids that might result in the introduction of genetic determinants for antibiotic resistance or bacterial toxin formation.... *Type 2*: Linkage of all or segments of the DNA's from oncogenic or other animal viruses to autonomously replicating DNA elements such as bacterial plasmids or other viral DNA's.'

'Second, plans to link fragments of animal DNA's to bacterial plasmid DNA or bacteriophage DNA should be carefully weighed in light of the fact that many types of animal cell DNA's contain sequences common to RNA tumor viruses'.

'Third, the Director of the National Institutes of Health is requested to give immediate consideration to establishing an advisory committee charged with (i) overseeing an experimental program to evaluate the potential biological and ecological hazards of the above types of recombinant DNA molecules; (ii) developing procedures which will minimize the spread of such molecules within human and other populations; and (iii) devising guidelines to be followed by investigators working with potentially hazardous recombinant DNA molecules'.

'Fourth, an international meeting of involved scientists from all over the world should be convened early in the coming year to review scientific progress in this area and to discuss appropriate ways to deal with the potential biohazards of recombinant DNA molecules'. Apparently, it was James Watson who had insisted on a meeting of scientists from all over the world, and this very wise move had definitely 'materially helped to buffer a possible inference that a gang of seven (or was it 11?) American scientists had impulsively doused the boiler of what arguably would become the most powerful scientific engine of the century' (Fredrickson 2001).

The letter ended with the repetition of the appeal for restraint, '... our concern for the possible unfortunate consequences of indiscriminate application of these techniques motivates us to urge all scientists working in this area to join us in agreeing not to initiate experiments of the types 1 and 2 above until attempts have been made to evaluate the hazards and some resolution of the outstanding questions has been achieved'.

18.10 Preparations for the international meeting

The organizing committee consisting of chairman Paul Berg, David Baltimore, Richard Roblin and Maxine Singer met on 10 September 1974 at the Massachusetts Institute of Technology. Other people that joined were Sherman Weissman, Norton Zinder, Donald Brown, Richard Novick and Aaron Shatkin. William Gartland attended on behalf of the National Institutes of Health, who had agreed to fund the conference. Herman Lewis represented the Human Cell Biology Steering Committee. Sydney Brenner of the Molecular Biology Laboratory at Cambridge, England, and Neils Jerne of the EMBO Council were invited to join the organizing committee as representatives from outside the United States. Four discussion panels were envisaged, one on ecology and potential dissemination of *E. coli* (to be chaired by Stanley Falkow), one on potential biohazards associated with experimentation involving genetically altered microorganisms, with special reference to bacterial plasmids and phages (to be chaired by Richard Novick), one on guidelines for animal virus–plasmid recombinants (to be chaired by Aaron Shatkin) and one on guidelines for plasmid–eukaryotic cell DNA recombinant experiments (to be chaired by Donald Brown). A format for a three-and-a-half day meeting, from 24–27 February 1975, was prepared. It was decided that there would be no publication of the conference as this would require manpower and time. The proceedings of the conference would be tape recorded and maintained as an archive. A total of about 90 scientists from the United States and 60 scientists from 12 other countries were to be invited. A few selected reporters of the press would be allowed, but they would be prohibited from reporting anything before the conclusion of the conference.

The invitees were told that the purpose of the international meeting was to identify (a) the kinds of experiments one would like to do with the recombinant molecules, (b) the important information that could be obtained from these experiments, (c) the potential risks involved for the investigator and others, and (d) the measures that can be employed to test for and minimize the biohazards so that the work could go on.

18.11 International Conference on Recombinant DNA Molecules, Asilomar Conference Center, Pacific Grove, CA, 24–27 February 1975

The International Conference on Recombinant DNA Molecules was sponsored by the National Institutes of Health contract no. NO1-OD-2103 and the National Science Foundation grant no. GB,S 75-05293.

The introductory speech, 'History, Aim of Meeting, Outline of Task' was delivered by David Baltimore on Monday, February 24, 1975, at 8:30 a.m. He said, 'a new technique of molecular biology appears to allow us to outdo the standard events of evolution by making combinations of genes which could be unique in natural history' (Wright 1994). 'He explained that the meeting had been convened to lay out the existing technology and what had been done to find answers to questions of what (experiments) should be done or not done, and what should be done before an experiment was undertaken. Baltimore emphasized that the balance of risks and benefits would be considered, but he stated that discussion of the hazards was a subject more important than either the benefits or molecular biology *per se*. His summary of the program ended with the observation that on the last morning, the organizing committee expected to present a summary statement including general guidelines for discussion and consensus. Baltimore reminded the audience that if it could not reach consensus there was no one else to turn to' (Fredrickson 2001).

Then, Paul Berg spoke on 'Assessment of present recombinant DNA technology, joining methods and vectors'. Stanley Falkow next talked on 'Ubiquity of microorganisms, natural history, plasmid ecology and introduction to normal bowel flora'. Talks on 'General problem of *Salmonella* infection in man and domestic animals and the use and misuse of antibiotics in medicine and agriculture, enhancement of R plasmid emergence', 'Plasmids which contribute to virulence or colonization of enteric species' and 'Gene transfer *in vivo* of both enteric organisms and *Staphylococcus*' by Ephraim Anderson, H Williams Smith and Mark Richmond, respectively, followed. The first panel discussion on 'Ecology and potential dissemination of *E. coli* strains' was held next.

Paul Berg in the beginning of his talk in the morning had 'mused aloud that the writers of the original letter to the world scientific community had not anticipated how it would affect the scientific community and that the organizing group was not prepared or experienced in how one should arrive at a decision. He said therefore that a panel of lawyers arranged by Dan Singer would present views on public policy issues' (Fredrickson 2001). Harold Green of the George Washington University National Law Center presented his talk on 'A public policy perspective' immediately after lunch of the first day's session. 'Green told the scientists that the conference and its unique moratorium on research – for which he gave them high praise – would serve as a moral precedent and a model of how science should proceed to deal with such issues. He was asked several questions about how the responsibilities for risk, or the framework for proceeding with experimentation, would be determined, and he offered his opinion that the government would ultimately determine the public policy' (Fredrickson 2001).

Late in the afternoon, talks were delivered on 'Molecular biology of bacterial plasmids', 'Molecular biology of bacterial conjugation and conjugative mobilization of plasmid and other DNAs', 'Phenotypic expression of plasmids in bacteria' and 'Some features of DNA uptake in bacteria' by Richard Novick, Roy Curtiss III, Julian Davies and Maurice Fox, respectively. After dinner, three very important talks were held; Herbert Boyer spoke on 'Genetics and biochemistry of DNA restriction', Stanley Cohen spoke on 'Molecular cloning of DNA as a tool for the study of plasmids and eukaryotic biology' and Kenneth Murray spoke on 'Development of bacteriophage systems on cloning vehicles'.

An observer's perception of the proceedings of the first day of the conference can be had from the writings of Janet H Weinberg (Weinberg 1975). 'On the first day of the conference, they presented each other with technical information on the ecology of the organisms used for gene transplant research. The beckoning of an exquisite Monterey spring did not reach those inside the chapel. With shades drawn and slide projector humming, they pursued their cerebral gray obsession, their slightly dangerous and exciting biotool for molding evolution. The unique language and concepts of molecular biology seemed a shelter from moral decision making'. 'One scientist grumbled, "People are only presenting the work that made them famous – they just can't resist with such an important audience".'

The next day, Tuesday, February 25, the proceedings got initiated at 8:30 a.m. with the panel discussion on 'Potential biohazards associated with experimentation involving genetically altered microorganisms, with special reference to bacterial plasmids and phages', which went on till 12 noon. The discussion was based on the exhaustive (35 single-spaced pages) and very well-organized report prepared by the working party comprising Royston Clowes, Stanley Cohen, Roy Curtiss III and Stanley Falkow, with Richard Novick as the chairman. The report was most conservative and forthright. For example, under the section of 'experimental systems', it recommended that 'where it is judged that the escape of even a small number of experimental organisms would constitute a serious biohazard, the experiment should not be attempted'. In the same section, it was suggested that 'where evaluation of biohazard potential is difficult or impossible, the undesirable alternative of simply accepting the best available guess and acting accordingly should be circumvented by developing an experimental organism with very low potential for survival or transfer of its genetic material upon escape'. All experiments have been classified from class I to class VI depending on the extent of possible biohazard and containment levels have been prescribed for each class. Class VI experiments have been prohibited even if the most stringent containment conditions were available. Class I experiments could be done under standard aseptic conditions, while class V experiments called for the most stringent containment measures, but these were all physical containment and no specific mention of biological containment was there in the containment section, even though there was such a suggestion in the section of 'experimental systems'.

It was Sydney Brenner who, in the early afternoon of Tuesday, strongly advocated the desirability of 'biological containment'. 'Brenner chaired a series of well-attended sessions on precisely how a series of mutations might be produced in *E. coli* to ensure that it would die as required outside the test tube. And the challenge of producing the strains was taken up by Roy Curtiss III, who announced in a memo to the organizing committee that he was hopeful that his objective could be met in a matter of months' (Wright 1994). Simultaneously, discussions were on for 'the designing of plasmids, phage, or other vectors that could not survive in a new ecological niche outside the laboratory.... That night a group of 'lambda people' worked late and by morning had a design on paper of a potentially safer phage vector' (Fredrickson 2001).

Tuesday late afternoon and evening were devoted to animal viruses. Daniel Nathans spoke on 'Design, construction and value of new kinds of molecules using SV40 (deletion mutants, transducing viruses, plasmid–viral gene recombinants)', Joe Sambrook discussed 'Natural history of viruses; especially, are viral genes attached to plasmids a special case?', Andrew Lewis talked about 'Adeno-SV40 hybrid DNAs as a case history; including: (a) NIH distribution policy, (b) conditions required for use in lab, (c) results of risk assessment experiments', Duard Walker drew 'Comparison of SV40 and human papova virus: pathogenicity, cross reactivity and oncogenicity', Michael Bishop narrated 'Isolation and study of Rous sarcoma proviral DNA presence of "virogenes" in "normal" cells' and Bernard Roizman described 'Possible experiments with herpes viruses DNA fragments'.

Finally, Aaron Shatkin initiated panel discussion on 'Preparation and use of recombinant molecules involving animal virus genomes' based on the report prepared by the working party comprising Michael Bishop, David Jackson, Daniel Nathans, Bernard Roizman, Joe Sambrook and Duard Walker, with Aaron Shatkin as the chairman. The report was, however, amended subsequently in the light of suggestions offered during the discussion. Experiments involving animal virus genomes were divided into two categories. The use of low-risk animal viruses as vectors in animal cells could be carried out only under physical containment measures. Experiments involving viral genomes, viral genome segments or eukaryotic DNA recombined with prokaryotic vectors and introduced into prokaryotic cells must be done under more stringent containment conditions using both physical and biological barriers. Andrew Lewis, one of the members of the working party disagreed with certain aspects of the report. He asserted, 'I believe that the risks associated with the widespread, semi-contained use of this procedure exceed the rewards from the information to be obtained'. 'The initial step in the reduction of these risks is the development and testing of new and theoretically safe vectors'.

The following are some of the observations of reporter Janet H Weinberg (Weinberg 1975) on the proceedings of the second day of the conference. 'Tuesday morning, the first working group presented its paper on how the experiments in that area could be done safely, and the first signs of confrontation appeared'. Objections to the letter and the spirit of the paper came pouring from the previously silent audience. Several young researchers (described by one bemused British as young colts chafing at the starting line for their Nobel Prizes) were obviously anxious to resume their work they had deferred since the previous in compliance with the ban. They wanted to know how their own research interests would be affected by the tentative restrictions in the working paper. Nobel Laureate Joshua Lederberg of Stanford feared that a specific document would be crystallized into law by suspicious legislators hardly qualified to make decisions on highly technical problems such as containment of risky organisms. Harvard Nobel laureate James D Watson, looking tousled and myopic, insisted that risks from gene transplant research could not be measured at this early stage. And even if there are great dangers, he said, 'there should be no illusions that regulation is possible'. One outspoken young researcher, David Botstein, countered Watson, 'I would like to make a very simple argument for guidelines in this field. I'm not omniscient. My experiments usually go wrong, then I learn from them and correct my mistakes. There may be a danger of guidelines becoming rigid laws that cut into academic freedom, but the prize-winning scientists of the world are not the only ones who will do the experiments. Ten years from now, genetic engineering kits might be standard equipment in high-school classes'. Throughout the uncomfortable exchanges, Paul Berg's craggy face remained downcast and apprehensive. The consensus decision to which his reputation would be linked was slipping away like the pure white Asilomar sand through dry fingers – lost in a barrage of unyielding, self-indulgent and conflicting attitudes. 'Halfway through the discussion, Sydney Brenner made an impassioned statement to steer the discussion from barren territory. In sonorous tones, Brenner, a five-and-a-half-foot scientific puck, chastened the group for its parochial concerns. The group is here to make a unique, broad policy decision about the future of genetic engineering – shall the field proceed, how, when, by whom and for what purpose?' 'Subsequently, Brenner led a massive brainstorming session on ways to construct safe experimental organisms that will self-destruct outside of the test tube, thereby reducing the chances of toxic agents escaping the laboratory. As if recoiling from the confrontation, the scientists retreated to the safety of pure science. The flexibility and creativity that many lacked toward making policy that morning literally blossomed in the light of a forward-looking scientific challenge. They spoke for hours on plasmids and phages and reagents and culture methods as the afternoon light changed from white to amber. As if in a Carlos Casteneda fantasy, one could almost see them connected by strands of crystalline energy, relaying sparks from the electric mass of ideas.'

The morning session of Wednesday, February 26, was devoted to plasmid–eukaryotic cell DNA recombinants. David Hogness, Donald Brown and Dana Carroll presented 'What has already been done in linking eukaryotic cell and plasmid DNA'. Then, Sydney Brenner, Peter Day and Raymond Valentine gave a glimpse of what they thought were the 'Major problems which can be solved by making eukaryotic cell DNA–plasmid recombinant'. Next, was the panel discussion on guidelines for plasmid–eukaryotic cell DNA recombinant experiments based on the report prepared by Sydney Brenner, Robert Burris, Dana Carroll, Ronald Davis, David Hogness, Kenneth Murray and Raymond Valentine, with Donald Brown as the chairman. It was envisaged that there could be 'three ways that recombinant eukaryotic DNA molecules could be hazardous. (a) A gene could function in the bacteria in which it is cloned and produce a toxic product. (b) A DNA component could in some way enhance the virulence or change the ecological range of the bacterium in which it is cloned. (c) A DNA component could infect some plant or animal, integrate into its genome, or replicate, or by its expression could produce a modification of the cells of the organism'. 'Mammalian DNAs, primates in particular, were rated high in biohazard because DNAs from these sources are more likely to contain infectious agents pathogenic for humans. Shotgun experiments involve DNA of unknown function, and introduce into the bacterium new sequences with unpredictable consequences. These considerations place all shotgun experiments in class 3 or higher regardless of the source of DNA. Shotgun experiments with primate DNA with animal vectors were placed in class 5. Cloning of high-risk animal virus nucleic acid or purified genes of known toxic products in animal or plant or prokaryotic vectors was also placed in class 5. An addendum was issued in response to the suggestions received during the discussion in which some low-risk experiments were identified. 'In the low-risk category are experiments involving fusion of prokaryotic vectors with DNA from all eukaryotes except warm-blooded vertebrates. Purified DNAs whose functions are non-toxic may be recombined under low-risk conditions'.

The after dinner session of Wednesday was reserved for the lawyers to address the scientists. Daniel Singer, Vice President, Institute of Society, Ethics and Life Sciences and husband of Maxine Singer, was the chairman and the first speaker. This session was arranged 'in hopes of strengthening the framework for the final discussion in the following morning. Singer began by complimenting the scientists on the exercise of public responsibility which he perceived in their undertaking. He also wished to remind them that he saw the issues as being not only scientific, but the benefits and risks were social issues, and the public which was paying for the research would have to have its say' (Fredrickson 2001). The next speaker was Alexander Capron, professor in the University of Pennsylvania School of Law. Capron said, 'Many people have talked about this research under the banner of "academic freedom". By overstating their case they risk provoking greater restriction. Freedom of thought does not encompass freedom to cause physical injury to others, and "prior restraint" as one person has termed it, is absolutely an appropriate response where irreversible harm is threatened. The public has a right, Capron stated flatly, to become involved in assessing new technology' (Wright 1994). The third speaker, Roger Dworkin, a law professor at Indiana University, 'led the scientists out into the chilling atmosphere of legal liability. He described dangerous crevasses with names like proximate cause, negligence, and strict liability, and created courtroom scenes of litigation.... He discussed workers' compensation and regulatory agency involvement, including the roles of Occupational Safety and Health Administration' (Fredrickson 2001).

'Listening to the lawyers predict what might happen to the decisions made on the morrow, the scientists stiffened their resolve to close up ranks so that world would see that the scientific community was able to finish what it had began. More than all the others, the members of the organizing committee now realized that the product of their long last evening's work had to be definitive' (Fredrickson 2001).

Janet H Weinberg had recorded her observations in her own inimitable way on the proceedings of the third day of the conference (Weinberg 1975). 'Wednesday was another tense and frustrating day. More working and technical papers presented, more heated debate, no consensus. The group seemed to be caught in an argument as circular as a bacterial chromosome.' 'The central obstacle seemed to stem from deductive training, from the way scientists learn to think and analyse problems. They were taught that there is a set of facts for every situation from which one can deduce other pieces of information. In the absence of the facts, one cannot make an informed decision. Yet here they were being asked to balance the risks to society – before they could possibly be quantified – with the benefits to society – which they hoped would be large. And the conference deadline was pressing. Scientific minds, as the British researcher predicted, were probably the worst kind for making sociopolitical decisions'. 'The most potent force for stopping the deductive merry-go-round, though, was the constant and eloquent persuasion of scientists who felt the moral responsibility to protect society more pressing than their own needs for academic freedom and success or even their desires to improve the biological world. Conference organizers, like Berg, Brenner, Baltimore, and a couple dozen others from Europe and the United States, hinted, cajoled, and as time grew short, demanded that their peers become at long last voting members of the world community. Abstaining was no longer possible. They demanded from each other not just technical excellence but morality'.

'Whatever the conference decides, Brenner said, it will have no legal force. In the end, each scientist's judgment and behavior would have to guide him to use the powerful biotool with a sense of moral responsibility to the innocents inside and outside of the laboratory. These discussions, one graying researcher mused before the final session, will mark the end of the age of innocence for basic research. They are a catalyst, another said, a final logical step in the heightened awareness that scientists control people's lives to some extent, and must account for it'.

The final session started at 8:30 a.m. on Thursday, February 27. Paul Berg initiated the proceedings by stating the three objectives of the organizing committee of the conference: (a) to bring experts from all over the world to discuss the risks of recombinant DNA experiments, (b) to deliberate and arrive at consensus of opinion about the risks and prepare a statement based on it and (c) forward the statement to the National Academy of Science. The members of the organizing committee had worked all night and drafted a provisional statement, based on their perception of the majority views, a copy of which was given to all the participants. Some of the scientists found it difficult to accept that the statement has already been drafted without inputs from each of them and raised procedural questions. One thing was, however, obvious: people were worried about the wording, the grammar and the form, not about the substantive points. Someone asked whether the final version would be prepared by taking vote on each paragraph. There was confusion, when Sydney Brenner raised his voice and suggested 'that reaction to the following simple statement be tested, "Work should proceed, but with appropriate safeguards".' Hands flurried up, and the chairman said that he perceived an overwhelming consensus to the statement. The participants allowed the chairman to decree a general agreement that 'with reservations, some form of experiments should proceed, some, however, should not'. Upon entrance into the issues of actual levels of containment for experiments, the discussion deteriorated, but the chair gamely kept order.... Feelings ran high. There were numerous attempts to amend some definition of hazard from the floor. As the first lunch bell sounded, heralding the end of the conference, the discussion was jostling back and forth between the ultimate and penultimate paragraphs. The moment for the final question could no longer be delayed. Berg, making himself heard over the commotion, flatly said, 'All those in favor of this as a provisional statement, please raise your hands'. Almost all the participants raised their hands. Stanley Cohen protested loudly that 'he could not support something if he could not see the wording of it'. 'All those opposed to the statement', Berg now demanded. Richard Roberts counted 'somewhere about four hands'. Two of these belonged to Watson and Lederberg. Watson apparently said, 'They want to put me out of business for something (the risk) you can't measure'. A third hand was Cohen's.... Thus, the statement with which they began the morning – although frayed and variously patched along the way – had made it through, still holding to the framework fashioned by the organizing committee in their last night's vigil.... By 12:15 on February 27, Asilomar II had ended' (Fredrickson 2001).

A five-page provisional statement was issued in the same afternoon (see Chapter 39, Section 39.2).

A more detailed 'summary statement' of the report was submitted subsequently to the Assembly of Life

Sciences of the National Academy of Sciences and was approved by its Executive Committee on 20 May 1975. The summary statement has been published (Berg *et al.* 1975) and is reproduced in Section 18.12.

Janet H Weinberg's observations on the last day's proceedings were again in her own style, but brief and crisp (Weinberg 1975). 'The scientists and observers shuffled over the flagstone path to the chapel for the last time Thursday morning. Some of them had been up almost all night, drafting a final working paper to pull together the disparate opinion and capture a consensus. And within four hours, they had it. The international group decided, by majority vote, to end their self-induced ban but to replace it with general guidelines. The guidelines will allow some experiments, defer others until safer containment is available, and all but ban others considered highly dangerous. The guidelines will have only the force of peer pressure'. 'A very few of the scientists, still unconvinced, spun away like tops into deductive circles. The rest of them left Asilomar quietly, in groups of two or three, with their muted nonconformity and their corduroy jackets and their moral responsibility'.

18.12 Summary statement of the Asilomar Conference on Recombinant DNA Molecules

The following is the summary statement of the Asilomar Conference on Recombinant DNA Molecules.

18.12.1 Introduction and general conclusions

This meeting was organized to review scientific progress in research on recombinant DNA molecules and to discuss appropriate ways to deal with potential biohazards of this work. Impressive scientific achievements have already been made in this field and these techniques have a remarkable potential for furthering our understanding of fundamental biochemical processes in prokaryotic and eukaryotic cells. The use of recombinant DNA methodology promises to revolutionize the practice of molecular biology. While there has, as yet, been no practical applications of the new techniques, there is every reason to believe that they will have significant practical utility in the future.

Of particular concern to the participants at the meeting was the issue of whether the pause in certain aspects of research in this area, called for by the Committee on Recombinant DNA Molecules of the US National Academy of Sciences, in the letter published in July 1974, should end, and, if so, how the scientific work could be undertaken with minimal risks to workers in laboratories, to the public at large and to the animal and plant species sharing our ecosystems.

The new techniques, which permit combination of genetic information from very different organisms, place us in an area of biology with many unknowns. Even in the present, more limited conduct of research in this field, the evaluation of potential biohazards has proved to be extremely difficult. It is this ignorance that has compelled us to conclude that it would be wise to exercise considerable caution in performing this research. Nevertheless, the participants at the conference agreed that most of the work on construction of recombinant DNA molecules should proceed provided that appropriate safeguards, principally biological and physical barriers adequate to contain the newly created organisms, are employed. Moreover, the standards of protection should be greater at the beginning and modified as improvements in the methodology occur and assessments of the risks change. Furthermore, it was agreed that there are certain experiments in which the potential risks are of such a serious nature that they ought not to be done with presently available containment facilities. In the longer term, serious problems may arise in the large-scale applications of this methodology in industry, medicine and agriculture. But it was also recognized that future research and experience may show that many of the potential biohazards are less serious and/or less probable than we now suspect.

18.12.2 Principles guiding the recommendations and conclusions

Though our assessments of the risks involved with each of the various lines of research on recombinant DNA molecules may differ, few, if any, believe that this methodology is free from any risk. Reasonable principles for dealing with these potential risks are: (a) that containment be made an essential consideration in the experimental design and (b) that the effectiveness of the containment should match, as closely as possible, the estimated risk. Consequently, whatever scale of risks is agreed upon, there should be a commensurate scale of containment. Estimating the risks will be difficult and intuitive at first, but this will improve as we acquire additional knowledge; at each stage, we shall have to match the potential risk with an appropriate level of containment. Experiments requiring large-scale operations would seem to be riskier than equivalent done on a small scale and, therefore, require more stringent containment procedures. The use of cloning vehicles or vectors (plasmids, phages) and bacterial hosts with a restricted capacity to multiply outside of the laboratory would reduce the potential biohazard of a particular experiment. Thus, the ways in which potential biohazards and different levels of containment are matched may vary from time to time, particularly as the containment technology is improved. The means for assessing and balancing risks with appropriate levels of containment will need to be reexamined from time to time. Hopefully, through both formal and informal channels of information within and between the nations of the world, the way in which potential biohazards and levels of containment are matched would be consistent.

Containment of potentially biohazardous agents can be achieved in several ways. The most significant contribution to limiting the spread of the recombinant DNAs is the use of biological barriers. These barriers are of two types: (a) fastidious bacterial hosts unable to survive in natural environments and (b) non-transmissible and equally fastidious vectors (plasmids, bacteriophages or other viruses) able to grow only in specified hosts. Physical containments, exemplified by the use of suitable hosts, or where applicable, limited access or negative pressure laboratories provide an additional factor of safety. Particularly important is the strict adherence to good microbiological practices which, to a large measure, can limit the escape of organisms from the experimental situation and thereby increase the safety of operation. Consequently, education and training of all personnel involved in the experiments is essential to the effectiveness of all containment measures. In practice, these different means of containment will complement one another and documented substantial improvements in the ability to restrict the growth of bacterial hosts and vectors could permit modifications of the complementary physical containment requirements.

Stringent physical containment and rigorous laboratory procedures can reduce but not eliminate the possibility of spreading potentially hazardous agents. Therefore, investigators relying upon 'disarmed' hosts and vectors for additional safety must rigorously test the effectiveness of these agents before accepting their validity as biological barriers.

18.12.3 Specific recommendations for matching types of containment with types of experiments

No classification of experiments as to the risk and no set of containment procedures can anticipate all situations. Given our present uncertainties about the hazards, the parameters proposed here are broadly conceived and meant to provide provisional guidelines for investigators and agencies concerned with research on recombinant DNAs. However, each investigator bears responsibility for determining whether, in his particular case, special circumstances warrant a higher level of containment than is suggested here.

18.12.3.1 Types of containment

The various types of containment are as follows:

1. **Minimal risk:** This type of containment is intended for experiments in which biohazards may be accurately assessed and are expected to be minimal. Such containment can be achieved by following the operating procedures recommended for clinical microbiological laboratories. Essential features of such facilities are no drinking, eating or smoking in the laboratory, wearing laboratory coats in the work area, the use of cotton-plugged pipettes or preferably mechanical pipetting devices and prompt disinfection of contaminated materials.
2. **Low risk:** This level of containment is appropriate for experiments which generate novel biotypes but where the available information indicates that the recombinant DNA cannot alter appreciably the ecological behaviour of the recipient species, increase significantly the pathogenicity or prevent effective treatment of any resulting infections. The key feature of this containment (in addition to the minimal procedures mentioned above) are a prohibition on mouth pipetting, access limited to laboratory personnel and the use of biological safety cabinets for procedures likely to produce aerosols (e.g. blending and sonication). Though existing vectors may be used in conjunction with low-risk procedures, safer vectors and hosts should be adopted as they become available.
3. **Moderate risk:** Such containment facilities are intended for experiments in which there is a probability of generating an agent with a significant potential for pathogenicity or ecological disruption. The principal features of this level of containment, in addition to those of the two preceding classes, are that transfer operations should be carried out in biological safety cabinets (e.g. laminar flow hoods), gloves should be worn during the handling of infectious materials, vacuum lines must be protected by filters and negative pressure should be maintained in the limited-access laboratories. Moreover, experiments posing a moderate risk must be done only with vectors and hosts that have an appreciably impaired capacity to multiply outside the laboratory.
4. **High risk:** This level of containment is intended for experiments in which the potential for ecological disruption or pathogenicity of the modified organism could be severe and thereby pose a serious biohazard to laboratory personnel or the public. The main features of this type of facility, which was designed to contain highly infectious microbiological agents, are its isolation from other areas by air locks, a negative-pressure environment, a requirement for clothing

changes and showers for entering personnel and laboratories fitted with treatment systems to inactivate or remove biological agents that may be contaminants in exhaust air, liquid and solid wastes. All persons occupying these areas should wear protective laboratory clothing and shower at each exit from the containment facility. The handling of agents should be confined to biological safety cabinets in which the exhaust air is incinerated or passed through Hepa filters. High-risk containment includes, besides the physical and procedural features described above, the use of rigorously tested vectors and hosts whose growth can be confined to the laboratory.

18.12.3.2 Types of experiments

Accurate estimates of the risks associated with different types of experiments are difficult to obtain because of our ignorance of the probability that the anticipated dangers will manifest themselves. Nevertheless, experiments involving the construction and propagation of recombinant DNA molecules using DNAs from (a) prokaryotes, bacteriophages and other plasmids, (b) animal viruses and (c) eukaryotes have been characterized as minimal, low, moderate and high risks to guide investigators in their choice of the appropriate containment. These designations should be viewed as interim assignments, which will need to be revised upward or downward in the light of future experience.

The recombinant DNA molecules themselves, as distinct from cells carrying them, may be infectious to bacteria or higher organisms. DNA preparations from these experiments, particularly in large quantities, should be chemically inactivated before disposal:

1. **Prokaryotes, bacteriophages and bacterial plasmids:** Where the construction of recombinant DNA molecules and their propagation involves prokaryotic agents that are known to exchange genetic information naturally, the experiments can be performed in minimal-risk containment facilities. Where such experiments pose a potential hazard, more stringent containment may be warranted.

   Experiments involving the creation and propagation of recombinant DNA molecules from DNAs of species that ordinarily do not exchange genetic information generate novel biotypes. Because such experiments may pose biohazards greater than those associated with the original organisms, they should be performed, at least, in low-risk containment facilities. If the experiments involve either pathogenic organisms or genetic determinants that may increase the pathogenicity of the recipient species, or if the transferred DNA can confer upon the recipient organisms new metabolic activities not native to these species and thereby modify its relationship with the environment, then moderate- or high-risk containment should be used.

   Experiments extending the range of resistance of established human pathogens to therapeutically useful antibiotics or disinfectants should be undertaken only under moderate- or high-risk containment depending upon the virulence of the organism involved.
2. **Animal viruses** Experiments involving linkage of viral genomes or genome segments to prokaryotic vectors and their propagation in prokaryotic cells should be performed only with vector–host systems having demonstrably restricted growth capabilities outside the laboratory and with moderate-risk containment facilities. Rigorously purified and characterized segments of non-oncogenic viral genomes or of the demonstrably non-transforming regions of oncogenic viral DNAs can be attached to presently existing vectors and propagated in moderate-risk containment facilities; as safer vector–host systems become available, such experiments may be performed in low-risk facilities.

   Experiments designed to introduce or propagate DNA from non-viral or other low-risk agents in animal cells should use only low-risk animal DNAs as vectors (e.g. viral, mitochondrial) and manipulations should be confined to moderate-risk containment facilities.
3. **Eukaryotes:** The risks associated with joining random fragments of eukaryotic DNA to prokaryotic DNA vectors and the propagation of these recombinant DNAs in prokaryotic hosts are the most difficult to assess.

   A priori, the DNA from warm-blooded vertebrates is more likely to contain cryptic viral genomes potentially pathogenic for many than the DNA from other eukaryotes. Consequently, attempts to clone segments of DNA from such animals, and particularly primate genomes, should be performed only with vector–host systems having demonstrably restricted growth capabilities outside the laboratory and in a moderate-risk containment facility. Until cloned segments of warm-blooded vertebrate DNA are completely characterized, they should continue to be maintained in the most restricted vector–host system in moderate-risk containment laboratories; when such cloned segments are characterized, they may be propagated as suggested above for purified segments of virus genomes.

   Unless the organism makes a product known to be dangerous (e.g. toxin, virus), recombinant DNAs from cold-blooded vertebrates and all other lower eukaryotes can be constructed and propagated with the safest vector–host system available in low-risk containment facilities.

   Purified DNA from any source that performs known functions and can be judged to be non-toxic may be cloned with currently available vectors in low-risk containment facilities. (Toxic here includes potentially oncogenic products or substances that might perturb normal metabolism if produced in an animal or plant by a resident microorganism.)

4. **Experiments to be deferred:** There are feasible experiments which present such serious dangers that their performance should not be undertaken at this time with the currently available vector–host systems and the presently available containment capability. These include the cloning of recombinant DNAs derived from highly pathogenic organisms (i.e. class III, IV and V aetiologic agents as classified by the United States Department of Health, Education and Welfare), DNA containing toxin genes and large-scale experiments (more than 10 L of culture) using recombinant DNAs that are able to make products potentially harmful to man, animals or plants.

18.12.4 Implementation

In many countries, steps have already been taken by national bodies to formulate codes of practice for the conduct of experiments with known or potential biohazard. Until these are established, we urge individual scientists to use the proposal in this document (Summary statement of the Asilomar Conference on recombinant DNA molecules) as a guide. In addition, there are some recommendations which could be immediately and directly implemented by the scientific community.

18.12.4.1 Development of safer vectors and hosts

An important and encouraging accomplishment of the meeting was the realization that special bacteria and vectors can be constructed genetically which have a restricted capacity to multiply outside the laboratory, and that the use of these organisms could enhance the safety of recombinant DNA experiments by many orders of magnitude. Experiments along these lines are presently in progress and in the near future, variants of λ bacteriophage, non-transmissible plasmids and special strains of *E. coli* will become available. All of these vectors could reduce the potential biohazards by very large factors and improve the methodology as well. Other vector–host systems, particularly modified strains of *Bacillus subtilis* and their relevant bacteriophages and plasmids, may also be available for particular purposes. Quite possibly, safe and suitable vectors may be found for eukaryotic hosts such as yeasts and readily cultured plant and animal cells. There is likely to be a continuous development in this area and the participants at the meeting agreed that improved vector–host systems which reduce the biohazards of recombinant DNA research will be made freely available to all interested investigators.

18.12.4.2 Laboratory procedures

It is the clear responsibility of the principal investigator to inform the staff of the laboratory of the potential hazards of such experiments before they are initiated. Free and open discussion is necessary so that each individual participating in the experiment fully understands the nature of the experiment and any risk that might be involved. All workers must be properly trained in the containment procedures that are designed to control the hazard. It is also recommended that appropriate health surveillance of all personnel, including serological monitoring, be conducted periodically.

18.12.4.3 Education and reassessment

Research in this area will develop very quickly and the methods will be applied to many different biological problems. At any given time, it is impossible to foresee the entire range of all potential experiments and make judgements on them. Therefore, it is essential to undertake a continuing reassessment of the problems in the light of new scientific knowledge. This could be achieved by a series of annual workshops and meetings, some of which should be at the international level. There should also be courses to train individual in the relevant methods as it is likely that the work would be taken up by laboratories which may not have had extensive experience in this area. High priority should also be given to research that could improve and evaluate the containment effectiveness of new and existing vector–host systems.

18.12.5 New knowledge

This discussion represents our first assessment of the potential biohazards in the light of current knowledge. However, little is known about the survival of laboratory strains of bacteria and bacteriophages in different ecological niches in the outside world. Even less is known about whether recombinant DNA molecules will enhance or depress the survival of their vectors and hosts in nature. These questions are fundamental to the testing of any new organism that may be constructed. Research in this area needs to be undertaken and should be given high priority. In general, however, molecular biologists who may construct DNA recombinant molecules do not undertake these experiments and it will be necessary to facilitate collaborative research between them and groups skilled in the study of bacterial infection or ecological microbiology. Work should also be undertaken which would enable us to monitor the escape or dissemination of cloning vehicles and their hosts.

Nothing is known about the potential infectivity in higher organisms of phages or bacteria containing segments of eukaryotic DNA and very little is known about the infectivity of the DNA molecules themselves. Genetic transformation of bacteria does occur in animals, suggesting that recombinant DNA molecules can retain their biological potency in this environment. There are many questions in this area, the answers to which are essential for our assessment of the biohazards of experiments with recombinant DNA molecules. It will be necessary to ensure that this work will be planned and carried out.

And it will be particularly important to have this information before large-scale applications of the use of recombinant DNA molecules are attempted.

18.13 Comments of the science reporters who attended the conference

Altogether, 16 reporters attended the Asilomar Conference. These included reporters from daily newspapers such as the *New York Times*, *Washington Post*, *San Francisco Chronicle* and *Los Angeles Times* and financial papers such as the *Wall Street Journal*. Also included were reporters from scientific journals such as *New Scientist*, *JAMA*, *Nature*, *Science*, *Science News* and *Chemical and Engineering News*.

Nicholas Wade of *Science* wrote (Wade 1975), 'In a meeting that will probably rate at least a footnote in the history of science, an international group of biologists has voted in principle to lift the voluntary moratorium imposed last July on a new technique of genetic manipulation that involves constructing hybrid molecules of DNA'. 'Just as the moratorium seems to be unprecedented in the history of science, the action of the conference is a rare, if not unique, example of safety precautions imposed on a technical development before, instead of after, the first occurrence of the hazard being guarded against'. 'The control measures proposed by the conference are considerably stricter than many of those active in the field believed necessary and furthermore include a quite novel safety feature stipulating that the organisms involved in the experiments shall be biologically incapable of surviving outside the laboratory'. 'Paul Berg's committee was able to persuade a group of 140 competitive and strong willed individuals, most of them quite unaccounted to being told how to do their experiments, to reach a remarkable degree of consensus on how the problem should be resolved'. 'The decision reached on 27 February at Asilomar will be far reaching because it governs a technique which is widely expected both to revolutionize the study of molecular biology and to bring some of the wilder fantasies of genetic engineers into the realm of the possible'.

Colin Norman of *Nature* reported (Norman 1975), 'Future historians of science may well record that a highly significant event took place in a California state park between February 24 and 27, 1975. Meeting at the Asilomar Conference Center here for three days of intense – and at times heated – discussions, 140 scientists from 16 countries agreed that strict controls should be placed on an exciting new technique which enabled genes from one organism to be transplanted directly into another'. 'The attempt at self-regulation by a group of scientists is believed to be without direct precedent, for although restrictions have been imposed on research before, they have usually resulted from concern over the uses to which science will be put, or from public pressure. In this case, however, the call for controls has come from the very scientists most actively engaged in the research and it has come at the inception of a new field of study, before any known hazards have arisen'. 'It was remarkable that even a statement of general principles was approved by all but a tiny handful of participants at the conference, for at times it seemed inevitable that there would be a disastrous split between those who wanted to press ahead with their experiments and those who were arguing for restraint'. 'It was generally agreed that by using what one participant described as 'old fashioned steam genetic engineering', enfeebled strains could be developed and produced in quantity in a matter of weeks, so that even the gung-ho researchers were able to accept that it would be prudent to delay some of their riskier experiments until safer organisms are available', 'Coming as they do from the people who would be most affected by them, the proposed controls on such experiments may seem like the classic case of the fox being set to guard the chickens'. 'But the recommendations are much stricter than might be expected'. 'Whether this unprecedented attempt at self-regulation will work, however, remains to be seen'.

'The technique in question, according to Sydney Brenner of the MRC Laboratory for Molecular Biology, Cambridge, "is going to generate the most exciting period, I think, in biology, and it is going to last for at least 10 years, may be more".' 'In addition to that potentially rich scientific harvest from such genetic manipulation experiments, the technique may offer the more remote possibility of isolating genes which code for nitrogen fixation in leguminous plants and introducing them into the cells of crops such as cereals. If that could be accomplished, the resulting hybrid cereals would no longer need a constant supply of nitrogen fertilizer. Another widely touted possibility is that it may be possible to mass-produce the genes responsible for promoting the synthesis of insulin by growing them in bacterial cultures'.

Janet H Weinberg of *Science News* wrote (Weinberg 1975), 'A few young successful American molecular biologists called the international conference after it became clear that they and their colleagues had released a partly malevolent genie in the form of a technique for transplanting genetic information from one unrelated organisms to another. With it, man could manipulate evolution more surely than with breeding or medicines or pollution combined. Theoretically, in time, scientists could replace defective genes with working ones to prevent and cure diseases. And they could insert into animals and plants the genes for desirable characteristics, like nitrogen fixation in corn or insect resistance in apples or super weight gain in chickens'. 'But they knew the genie's malevolent side could not be ignored in the magic carpet scenario. Both by accident and by purposeful manipulation, genes for drug resistance or cancer or lethal toxin formation could be inserted into common organisms. Biological warfare agents and massive epidemics could be created too effectively this way. So the American scientists called their most prominent colleagues from around the world to ponder the hazards of this double-edged scimitar'. 'Before the sessions started, a British scientist had said, 'Any decision here is unlikely because scientists are probably the wrong people to make it.' The scientists spent the next three days proving that statement to be right over and over. And then, ultimately, they proved it wrong.

References

Berg P, Baltimore D, Boyer HB, Cohen SN, Davis RW, Hogness DS, *et al.* Potential biohazards of recombinant DNA molecules. *Science* 1974;185:303.

Berg P, Baltimore D, Brenner S, Roblin RO III, Singer MF. Asilomar conference on Recombinat DNA molecules. *Science* 1975;188:991–994.

Cell biology correspondent. A two-edged sword. *Nature* 1972;240:73.

Cohen SN, Chang ACY, Boyer HW, Helling RB. Construction of biologically functional bacterial plasmids *in vitro*. *Proc Natl Acad Sci USA* 1973;70:3240–3244.

Fredrickson DS. *The Recombinant DNA Controversy: A Memoir-Science, Politics, and the Public Interest 1974-1981.* Washington, DC: ASM Press; 2011. p. 11, 13, 14, 16, 17, 18, 19, 20, 21, 24, 25, 26.

Hotchkiss RD. Portents for a genetic engineering. *J Heredity* 1965;56:197–202.

Jackson DA, Symons RH, Berg P. Biochemical method for inserting new genetic information into DNA of Simian Virus 40 DNA molecules containing lambda phage genes and the galactose operon of *Escherichia coli*. *Proc Natl Acad Sci USA* 1972:69:2904–2909.

Kiss LR. The new biology: What price relieving man's estate. *Science* 1971;174:779–788.

Norman C. Berg conference favours use of weak strains. *Nature* 1975;254:6–7.

Shapiro J, Eron L, Beckwith J. More alarums and excursions. *Nature* 1969;224:1337–1337.

Shapiro J, Mac Hattie L, Eron L, Ihler G, Ippen K, Beckwith J. Isolation of pure lac operon DNA. *Nature* 1969;224:768–774.

Singer M, Soll D. Guidelines for DNA hybrid molecules. *Science* 1973;181:1114.

Wade N. Genetics: Conference sets strict controls to replace moratorium. *Science* 1975;187:931–934.

Weinberg JH. Decision at Asilomar. *Science News* 1975;107: 194–196.

Wright S. *Molecular Politics: Developing American and British Regulatory Policy for Genetic Engineering.* Chicago, IL: University of Chicago Press; 1994. p. 124, 129, 149, 153.

Ziff E. Benefits and hazards of manipulating DNA. *New Scientist* 1973;25 Oct:274–275.

19

Genetic engineering guidelines

H. K. Das

If the world scientific community was startled by the publication from the laboratory of Paul Berg (Jackson *et al.* 1972) and then by the other one from the laboratories of Stanley Cohen and Herbert Boyer (Cohen *et al.* 1973), they were put on high alert by the appeal to refrain from initiating recombinant DNA experiments (Berg *et al.* 1974) made by the Committee on Recombinant DNA appointed by the Assembly of Life Sciences of the United States (see Chapter 18, Section 18.9).

Many people felt that there should be some 'guidelines' that must be adhered to before undertaking any experiment using this new technology to ensure that the people in the laboratory and public in general are not exposed to any hitherto unknown hazard. Some, however, believed that imposing any restriction would tantamount to infringement on the scientists' right to free enquiry. Robert Sinsheimer of the California Institute of Technology had dwelt on this issue in a lecture to the Genetics Society of America. 'To impose any limit upon freedom of enquiry is specially bitter for the scientist whose life is one of enquiry; but science has become too potent. It is no longer enough to waive the flag of Galileo. Rights are not found in nature. Rights are conferred within a human society and for each there is expected a corresponding responsibility.... Would we wish to claim the right of individual scientists to be free to create novel self-perpetuating organisms likely to spread about the planet in an uncontrollable manner for better or worse? I think not. This does not mean we cannot advance our science or that we must doubt its ultimate beneficence. It simply means that we must be able to look as what we do in a mature way.... It is difficult for a scientist to conceive that there are certain matters best left unknown, at least for a time. But science is the major organ of inquiry for a society – and perhaps a society, like an organism, must follow a developmental program in which the genetic information is revealed in an orderly sequence'. (Wade 1975a).

19.1 Recommendations of the Working Party under the chairmanship of Lord Ashby of Cambridge, England

The Berg *et al.* recommendations were published in the 19 July 1974 issue of *Nature* and in the 26 July 1974 issue of *Science*. On July 26 itself, the Advisory Board to the Research Councils of the United Kingdom decided to appoint an expert Working Party under the chairmanship of Lord Ashby of Cambridge 'to assess the potential benefits and potential hazards of techniques which allow the experimental manipulation of the genetic composition of micro-organisms'. The Board decided 'that the membership of the Working Party should, as far as possible, not include those who were using the techniques and who might therefore be directly affected by the conclusions'. The 13 members of the Working Party were all established and renowned biologists and included Walter F Bodmer, HL Kornberg, JR Postgate and MHF Wilkins. The members of the Working Party, nevertheless, talked to as many as 28 experts; among them were Ephraim S Anderson, Sydney Brenner, Naomi Datta, Noreen and Kenneth Murray, H Williams Smith and Robert Williamson. The 23-page report of the Working Party (*Report of the Working Party on the Experimental Manipulation of the Genetic Composition of Micro-organisms*) was adopted by the Advisory Board to the Research Councils on 13 December 1974 and presented to the British Parliament in January 1975. It has been published by Her Majesty's Stationery Office (Cmnd. 5880; ISBN 0-10-158800-3). The Working Party acknowledged the enormous potential benefits of the applications of the techniques, but also took cognizance of the potential hazards that might arise from the use of the techniques.

The recommendations of the Working Party, reproduced next, contained the first enunciated 'guidelines' to be followed for performing experiments involving recombinant DNA technology:

a) It is desirable that anyone using these techniques, whether his experiments are judged to be hazardous or not, should have some experience in the techniques of handling pathogens (such as are used in a public health laboratory). He would then be familiar with the levels of containment which is practiced in such laboratories and the criteria of risk which medical microbiologists use.
b) It is essential that laboratories where these new techniques are being used should have the basic equipment necessary for the containment of pathogens, so that if experiments are judged to be hazardous, appropriate precautions can be taken.
c) The decision as to what precautions are appropriate has to remain a matter of judgment by workers experienced in using these techniques. We feel that funding bodies (research councils, universities, educational institutions and industry) should, as a matter of routine, ask both proposers and referees whether new research proposals, or extensions of existing ones, involve any biological hazard.
d) There is a scale of risks (although it cannot be spelt out), from obviously harmless experiments to obviously risky ones. We believe that for most purposes, the normal precautions used for dealing with ordinary pathogens would be sufficient. No smoking, eating or drinking is done in the laboratory. The workers wear gloves, which they remove before leaving. Liquids are drawn into pipettes mechanically, not by mouth. If anything is split or broken, it is washed with disinfectant. A common precaution is to do some operations in special cabinets or inoculation hoods. An essential feature in the discipline of containment is the disposal by sterilization of all contaminated material. No infected matter or equipment is allowed to leave the laboratory. Nothing is put down the sink until it has been sterilized. Waste material is incinerated or autoclaved. In institutions where experiments of this kind are going on or are proposed, someone should be designated as a biological safety officer, to give guidance on the precautions necessary to ensure containment for potentially hazardous experiments. His job would be more difficult than that of a radiation safety officer, but his role would be similar. As an initial step, a widely publicized advisory service, perhaps offered by public health laboratories, would help to safeguard the interests of the public and of those engaged in the experiment.
e) We have considered whether there should be some registration of workers using the techniques we have to assess. We decided that compulsory registration would be unreasonably difficult; it is practicable to register the use of a substance, such as radioactive isotopes; it is not practicable to register a technique which requires only standard laboratory equipment.
f) For some purposes, it is desirable to prepare comparatively large quantities of bacteria or virus or extracts from them, that is, by the ten-litre rather than by the millilitre. This greatly increases the risk from accidents and the difficulty of containment. We believe that large-scale operations, unless there are positive reasons for believing them to be harmless, should in the first instance be carried out in specially equipped laboratories with fail-safe arrangements for their confinement. We say 'in the first instance' because the prudent way to do research with unknown hazards is to conduct, in parallel with them, tests of pathogenicity on animals and to monitor all workers engaged in the research. This should give early warning of hazards, if these are substantial. In the absence of evidence of hazard, it would enable precautions to be cautiously relaxed.
g) We have made it clear that there are two categories of precautions: on the one hand, the well-known procedures for containment, familiar to medical microbiologists, as has been mentioned under (d); on the other hand, the use of vectors not likely to infect man or animals, such as bacteriophage lambda rather than plasmids to carry chromosome fragments; or mutations both in the plasmid and the host bacterium which require nutrients (like diaminopimelic acid) not found in the human gut, or which could not survive at body temperature, or restrict the host range of the fragment carried. Another strategy which would diminish hazards is to use an organism which does not occur in the body at all. One example of this—an organism whose genetical make-up is well understood—is *Bacillus subtilis*. But so much is known about the genetics of *E. coli* that research workers would be reluctant to give it up in favour of another bacterium. We realize this, but there might be circumstances in which it would be prudent to use an alternative host for DNA fragments. We think that this search for genetical safeguards should be pursued as an additional defence against possible hazards. These precautions are desirable because it would not be prudent, in our view, to assume that the commonly used K12 strain of *E. coli* is enfeebled and quite harmless.
h) Infection of animals with bacteria or viruses suspected of carrying disease may supply proof that there is a hazard, but cannot supply proof that there is no hazard. A more trustworthy test, although one that is necessarily retrospective and may take a long time, is to do epidemiological studies on the health records of people who work with (e.g.) oncogenic

viruses and other possibly dangerous organisms. We strongly recommend that epidemiological studies of this kind should be stimulated.

i) We recommend that further investigations should be sponsored forthwith to improve our knowledge of the viability in the human intestine of laboratory strains of *E. coli*, especially K12, and their capacity to transfer plasmids to resident strains.

j) We recommend that, subject to rigorous safeguards based on those set out above, these techniques should continue to be used because of the great benefits to which they may lead.

19.2 Interim guidelines of the NIH Recombinant DNA Program Advisory Committee

At 9 a.m. on 28 February 1975, a day after the conclusion of the Asilomar conference, the new NIH Recombinant DNA Program Advisory Committee comprising 12 members, including Edward E Adelberg, Roy Curtiss III, James E Darnell, Stanley Falkow, Donald R Helinski, David S Hogness, Jane K Setlow and Waclaw Szybalski, met for the first time in the Capri Room at the Bellevue Hotel in San Francisco and adopted the 'provisional statement' of the Asilomar conference as interim guidelines for federally supported laboratories in the United States. The text of the 5-page 'provisional statement, which was issued in the afternoon of the concluding day of the Asilomar conference and was adopted by the Recombinant DNA Program Advisory Committee, is reproduced next:

1. This meeting was organized to review scientific progress in the area of recombinant DNA molecules and to discuss appropriate ways to deal with the potential biohazards of this work. We have heard about the enormous scientific progress already achieved in this field, and have seen glimpses of the remarkable potential of these methods to accelerate the rate at which we may gain understanding of the fundamental processes occurring in eukaryotic cells. The use of recombinant DNA methodology promises to revolutionize the practice of molecular biology. While there has as yet been no practical applications of these techniques, there is every reason to believe that it will have significant impact in the future.

   The participants at the meeting agreed that the pause in research, called for in the July 1974 committee letter, ought not to be left unresolved. They considered whether there were ways in which the scientific work could be undertaken with minimal risks to the workers in the laboratories and to so society at large. It was emphasized that, in the longer term, even more difficult problems may arise in the probably large-scale applications of this working industry, medicine and agriculture. Even in the currently more limited area of the conduct of research in this field, the evaluation of potential biohazards has proved to be extremely difficult. The new techniques combining genetic information from very different organisms place us in an area of biology with many unknowns. It is this ignorance that has compelled us to conclude that it would be wise to exercise the utmost caution. Nevertheless, the work should proceed but with appropriate safeguards. Although future experience may dispel many fears, standards of protection should be set high at the beginning and each escalation, however small, should be carefully assessed.

2. Though our assessments of the risks involved with each of the various lines of research on recombinant DNA molecules may differ, few, if any, believe that this methodology is free of any risk. Reasonable principles for dealing with these risks are to adopt containment as a part of the experimental strategy and that the effectiveness of the containment should match the risk. Whatever scale of risks is ultimately devised, we shall need a commensurate scale of containment. Consequently, we must seek means for estimating the risks, perhaps subjectively at first but objectively as we acquire additional knowledge, and then to match that risk to the appropriate degree of containment. Experiments requiring large-scale operations would seem to be riskier than the equivalent experiment done on a small scale, and, therefore, require more stringent containment procedures. Improvements in the methodology, e.g., a 'disarmed' vector or host cell, could permit a reduction in the containment requirement. Quite possibly, the ways in which potential biohazards and different levels of containment are matched may vary from country to country; also, the ways could vary from time to time as the containment technology is improved. Thus, it is essential that there is a continued reassessment of the balancing of risk against the level of containment.

   Multiple factors define each level in the scale of containment. The most important factors, because they contribute most significantly to limiting the spread of the recombined plasmids, are biological barriers. These barriers are of two types: fastidious bacterial hosts unable to survive in natural environments and non-transmissible vectors designed to grow only in specified hosts. A second factor is adherence to good microbiological practices which, to a large measure, can limit the escape of organisms from the experimental situation. Physical containment, exemplified by the use of suitable hoods, or, where applicable, contained laboratories are an additional factor. Further, education and training of all personnel involved in the experiments is essential to the effectiveness of all of the above.

3. Specific Recommendations

These recommendations reflect the principle that a serious evaluation of biohazard potential and the adoption of appropriate biological and physical barriers are integral parts of experiments with recombinant DNA molecules. In this section, three levels of containment affording increasing protection are defined. Then, various types of experiments are evaluated for potential biohazards and matched with appropriate containment levels.

A. Types of Containment

The types of containment are an adaptation of the 'NCI Safety Standards'.

1. Low – This type of containment involves basic good medical microbiological techniques. The essential factors are wearing lab coats, using mechanical pipettes, no eating in the lab, and that sonication and other procedures which generate large aerosols should be done in biological safety cabinets.

   While existing vectors can be used for this level of containment, as safer vectors become available, their use is recommended.

2. Moderate – Both physical and biological containment enter into this containment system.

   The physical containment for moderate-risk agents was designed for handling moderate-risk oncogenic viruses. The main features are that transfer operations are to be carried out in biological safety cabinets, gloves are worn in addition to lab coats, vacuum lines are protected by filters, and negative pressure is maintained in limited-access laboratories.

   The physical containment procedures are recognized to provide significant but incomplete protection against the accidental spread of biological agents. The potential hazard of moderate-risk agents is such that a strong measure of biological containment is needed to ensure their safe handling. Therefore, experiments with such agents should only use prokaryotic vectors which have been designed with increased safety in mind. Such vectors are currently being designed and created, and should be available in near future so that these experiments can proceed.

3. High – This type of containment involves facilities which are isolated from other areas by air locks, clothing changes and shower rooms and which have treatment systems to inactivate or remove biological agents that may be contaminants in exhaust air, liquid and solid wastes. The handling of agents shall be confined to biological safety cabinets and all persons occupying these areas shall wear only laboratory protective clothing and shall shower at each exit from the containment facility. In addition, the containment facility shall be maintained under negative air pressure.

Again, only vectors designed for safety should be used.

B. Types of Experiments

The letter published by the Committee on Recombinant DNA Molecules in July 1974 requested that the scientific community join the Committee in deferring two types of experiments. The letter also advised caution regarding a third type of experiment. In the following assessments of potential hazards, the three original categories are maintained, but with some redefinition:

1. Prokaryotes – Potential biohazards of experiments involving genetic exchange among prokaryotic organisms can in general be accurately assessed.

   Experiments involving organisms that normally exchange genetic information involve no novel biotypes and pose no hazards that cannot be contained by the standard microbiological laboratory techniques appropriate for the handling of these organisms.

   Experiments involving the introduction of bacterial genes into species in which they have not been found to occur naturally result in novel biotypes and so pose increased potential biohazards. Such experiments involving genetic determinants affecting pathogenicity for man or other species or antibiotic resistance should be undertaken only under conditions of moderate or high containment; any large-scale industrial, commercial, agricultural or other applications should be deferred, pending the issuance of appropriate official guidelines by national scientific bodies.

2. Viruses – Experiments employing 'low-risk' animal viruses as vectors to introduce new genetic material into animal cells can be carried out under moderate physical containment conditions.

   Experiments involving linkage of viral genomes or genome segments to prokaryotic vectors and their introduction into prokaryotic cells should be carried out under moderate containment conditions including improved vectors. Rigorously purified fragments of the demonstrably non-transforming regions of oncogenic viral DNAs or regions of non-oncogenic viruses could be attached to plasmid DNAs and introduced into *E. coli* under moderate-risk conditions with existing vectors.

   Experiments involving high-risk viruses should be carried out under high containment.

3. Eukaryotic DNA – In the low-risk category are most experiments involving the fusion of

prokaryotic vectors with DNA from prokaryotes, lower eukaryotes, plants, invertebrates and cold-blooded vertebrates. Moderate-risk experiments involve the joining of DNA of warm-blooded vertebrates to prokaryotic vectors or the construction of recombinants between animal vectors and any DNA. Experiments with a high risk include the fusion of eukaryotic or prokaryotic genes to prokaryotic vectors when the resultant organism is likely to express a toxic or a pharmacologically active agent.

4. Implementation

It is clear that in many countries, steps are already being taken by national bodies to formulate and establish guidelines. Until such guidelines are established, we urge individual scientists to use the present document as a guide. There are in addition some recommendations which need not wait until that time and which the scientific community could implement directly.

A. Development of Vectors and Hosts

One of the most important and interesting accomplishments of the meeting was the beginning of the design and construction of bacteria and vectors which could radically improve the safety of these procedures by many orders of magnitude. It is certain that in the near future, there will become available special variants of λ phage, non-transmissible plasmids and special *E. coli* which will not only optimize safety but will also bring about considerable technical improvements in the methods themselves. Other bacterial systems, particularly suitably modified strains of *B. subtilis,* may also be specially useful for particular purposes. There is also the possibility that a suitable vector may be found for simple eukaryotic cells. We think that work which aims to improve the vectors or hosts should be given high priority.

B. Laboratory Procedures

Before any experiments of this kind are initiated, the research staff of the laboratory shall be informed of all of the hazards that might be associated with such experiments. Laboratory workers must also be trained in the containment procedures that are designed to control the hazards. This training must include emergency procedures that are to be performed in the event of an accident. It is also strongly recommended that appropriate health surveillance of all personnel, including the serological monitoring, be conducted periodically to establish a base for epidemiological analyses.

C. Education

A continuing reassessment of these problems to take into account and developing scientific knowledge is essential. This could be achieved by a series of annual courses, workshops and meetings which should also serve to train individuals in the relevant methods. Consideration should also be given to the establishment of a newsletter for the rapid dissemination of new information pertinent to the effectiveness of biological containment.

5. Proposed Model Containment Review Process

A review process should be established which would be able to determine whether a given laboratory has the appropriate containment facilities for a given type of experiment. As far as possible, the biohazard review process should not lengthen the time required for review of research proposals. The specific form of the review procedure in different countries for different scientific and industrial laboratories must depend on local circumstances. The following proposal is, therefore, presented as a model. The model is designed for universities in the United States but would have to be modified for other situations.

Each university or research institution should have a committee empowered and trained to grade the physical containment facilities of its laboratories (e.g., low, moderate or high according to established guidelines). The local committee would provide the laboratory head with a statement certifying the containment rating of the laboratory (subject to periodic reevaluation).

When an individual applies to an agency for funds to support work on recombinant DNA molecules, the certificate of containment rating would be appended. The group reviewing the grant would then determine whether the certified level of containment matched whatever biohazard might result from the proposed work. The biological barriers incorporated in the experiment, the magnitude of proposed growth of bacteria, the type of DNA to be cloned, etc., would all enter into the decision. If the reviewing group is satisfied, the grant would be processed for scientific merit in the usual fashion. If a question arises concerning the appropriateness of the certified containment level, the NIH Advisory Committee on Recombinant DNA Molecules or some other body would be asked for an opinion or ruling.

This procedure would not guarantee that all experiments would be performed under the required containment conditions but the investigators have reasonable good will, the system would generate widespread compliance.

6. New Knowledge

   This document represents our best assessment of the potential biohazards in the light of current knowledge. To improve this assessment, it will be important to have answers to the following questions.

   1. Are eukaryotic genes or viruses expressed in prokaryotic hosts and, if so, can they modify the biohazard potential of these cells?
   2. Can free DNA molecules infect animals or plants?
   3. Can prokaryote–eukaryote recombinant DNA molecules, either free or encapsulated as phage particles, infect animal or plant cells and be expressed there?
   4. Can mammalian cells in culture be genetically transformed by free homologous or heterologous DNA?
   5. Can hybrid animal virus DNA or virus-plasmid hybrids cause tumors in animals?
   6. Can methods be developed to monitor effectively the escape and dissemination of cloning vehicles?

In the first meeting of the Recombinant DNA Program Advisory Committee held in San Francisco on 28 February 1975, a subcommittee was formed under the chairmanship of David S Hogness to prepare a draft of the final guidelines.

The Hogness subcommittee recommended that each class of recombinant DNA experiment be assigned both a physical level and a biological level of containment. Four levels of physical containment designated P1–P4 were suggested. P1 consisted of good microbiological practice. P2 required some additional precautions such as avoidance of aerosol production giving transfers in an isolated chamber and treating all glass wares, etc. in that chamber itself with sterilizing solutions.

P3 involved putting the whole laboratory under negative pressure. A P4 laboratory would have the strictest containment facilities. The workers will have to do the experiments in glove chambers and can never come in direct contact with the experimental materials. They have to take a disinfecting shower before leaving the laboratory. Air locks have to be installed at the entry and exit points of the laboratory. All materials will go out of a P4 laboratory only after autoclaving.

Three levels of biological containment were recommended. In the least hazardous kind (low risk) of experiments, grouped under the category EK1, *Escherichia coli* strain K12 could be used. Low-risk experiments would be those with organisms that exchange genetic information in nature and most of those where prokaryotic cells would be used. The more hazardous kind (moderate risk) of experiments, grouped under the category EK2, could, however, be performed only with *E. coli* K12 strains that have been genetically incapacitated such that, on an average, only one bacterium in 100 million would be expected to survive in the environment outside the laboratory. Moderate-risk experiments were thought to be those where DNA from warm-blooded animals and certain cancer-causing animal viruses would be transferred to *E. coli*. Transfer of DNA from low-risk animal viruses to animal cells was also in this category. The most hazardous kind (high risk) of experiments were grouped under the category EK3 and these experiments could be performed with incapacitated *E. coli* K12 strains as mentioned under the EK2 category but only after the specified survival characteristic has been verified by feeding the incapacitated bacteria to animal hosts. High-risk experiments would be those in which DNA from high-risk viruses would be transferred to animal cells and also those involving eukaryotic genes that are likely to express any poison or pharmacologically active agent.

In addition to suggesting the above-mentioned general categorization, the Hogness subcommittee also assigned the level of both physical and biological containment for each class of experiment.

The draft prepared by the Hogness subcommittee was considered by the members of the Recombinant DNA Program Advisory Committee of the National Institutes of Health (NIH) in the meeting held in July 1975 at Woods Hole, MA. It was alleged that the draft was considerably 'watered down' in the Woods Hole meeting and a guideline much more lax than the one suggested in the Asilomar conference was adopted (Wade 1975a). The laxity was apparently with regard to assignment of containment levels of some experiments. This attracted widespread criticism, which was spearheaded by Richard Goldstein of Harvard Medical School and Harrison Echols of the University of California, Berkeley. Fifty biologists expressed their reservations and concern about the Woods Hole version in a petition to Dewitt Stetten of the NIH, who was the chairman of the Advisory Committee. Even Paul Berg, who was the chairman of the National Academy of Sciences committee that convened the Asilomar conference, criticized the Woods Hole guidelines. Dewitt Stetten formed a new subcommittee headed by Elizabeth Kutter, a biophysicist from Evergreen State College in Olympia, Washington. She was present as a special invitee in the Woods Hole meeting. Kutter's important qualification apparently was that she was a layperson or, at best, a semi-layperson in the area of recombinant DNA technology and would not be affected by the proposed guidelines. The other members of the new subcommittee were Stanely Falkow and Joe Sambrook. Kutter consulted a large number of persons, including David Hogness, Paul Berg and Joshua Lederberg, in an effort to 'get all the dissenting ideas and come up with compromises'. Her subcommittee met early in November 1975 and started to formulate the guidelines using the Hogness recommendation as their basic text. The Kutter subcommittee's recommendation was considered by the Recombinant DNA Program Advisory Committee of the NIH in their meeting held at La Jolla on 4 and 5 December 1975 and after tense arguments during these 2 days, the Committee came up with a guideline that was stricter than the Woods Hole version.

The following is the summary (Wade 1975b) of the guidelines issued by the NIH Committee for containment measures to be adopted for performing recombinant DNA experiments. Each class of experiment was assigned both a physical (P1–P4) and a biological (EK1–EK3) level of containment.

A. Shotgun experiments with *Escherichia coli* as host (use of recombinants to introduce undefined segments of an organism's genome into *E. coli*, classified by type of organism)
   (i) Eukaryotic DNA recombinants
   - Non-embryonic primates: P3 + EK3 or P4 + EK2
   - Embryonic primates: P3 + EK2
   - Other mammals: P3 + EK2
   - Birds: P3 + EK2
   - Cold-blooded vertebrates: P2 + EK2
   - Invertebrates and lower plants (ferns to algae): P2 + EK1
   - Higher plants: P2 + EK2, but P2 + EK1 if cells are taken from embryonic or germline tissue
   - Higher plants that produce pathogenic or toxic agents: P3 + EK2
   - Purification: If a cloned recombinant DNA can be made 99% pure on weight-for-weight basis, the P value of containment may be reduced by one level

   (ii) Prokaryotic DNA recombinants
   - Prokaryotes that naturally exchange genes with *E. coli*
     - Class 1 agents (as classified by the Center for Disease Control) such as enterobacteria: P1 + EK1
     - Class 2 agents such as *Salmonella typhi*: P2 + EK2
     - Class 3 and higher: Experiments banned
   - Prokaryotes that do not naturally exchange genes with *E. coli*
     - Non-pathogens: P2 + EK1
     - Pathogens: P3 + EK2 if of low pathogenicity
       P3 + EK3 if of moderate pathogenicity

B. Use of recombinants to insert genes from viruses, eukaryotic plasmids and organelles into *E. coli*
   - Animal viruses: P4 + EK2 or P3 + EK3
   - Plant viruses: P3 + EK1 or P2 + EK2
   - Eukaryotic plasmids or organelles: As for the shotgun categories, unless the recombinant DNA has been rendered 99% pure, in which case either the P or EK value may be reduced by one level

C. Use of animal virus vectors
   - Defective polyoma virus + class 1 virus or non-pathogens: P3
   - Defective polyoma virus + class 2 viruses: P4, but if the polyoma host range has not been changed and the virus segment can be proved harmless, then P3
   - Defective SV40 + class 1 virus or non-pathogens: P4
   - Defective SV40 + non-pathogenic and purified DNA, whether prokaryotic or eukaryotic: P3
   - Defective SV40 or defective polyoma (lacking late genes) + prokaryotic or eukaryotic DNA: P3 as long as no virus particles are produced by infected cells

It has, however, become clear today that many of the hazards feared in 1975 do not really exist. Bacterial strains used as hosts for recombinant plasmids cannot exchange DNA with human intestinal flora, nor can they survive at all in the human system. The guidelines have been revised every year and the containment levels, both physical and biological, prescribed in 1975 for various experiments have been relaxed considerably. Most experiments are now permitted to be done at, at least, one level, and sometimes at two levels below the original stipulation. Many experiments have been exempted from following any containment procedure. The 2013 version of the NIH Guidelines for Research Involving Recombinant DNA Molecules can be found on the website: http://osp.od.nih.gov/sites/default/files/NIH_Guidelines_0.pdf.

Today, each country has its own guidelines and regulations recommending the level of both physical and biological containments necessary for different kinds of recombinant DNA experiments. Incapacitated hosts and cloning vectors have been certified for suitability of use at different levels of biological containments.

19.3 Indian guidelines

The Indian guidelines were formulated by a committee constituted by the Department of Biotechnology, Ministry of Science and Technology, Government of India. The author was a member of this committee. Based on these formulations, a law has been enacted in 1986 under the Environment Protection Act (EPA). A set of Indian guidelines were also issued in 1990 (see Appendix 2), covering genetically engineered organisms, genetic transformation of plants and animals, containment measures at the laboratory level under different risk groups and modalities of implementation of biosafety guidelines. A set of revised guidelines have been issued in 1994. A separate 'Revised Guidelines for Research in Transgenic Plants and guidelines for Toxicity and Allergenicity for Evaluation of Transgenic Seeds, Plants and Plant Parts' have come into force in 1998. These can be found on the website http://dbtbiosafety.nic.in/Files/CD_IBSC/Files/transgenic.PDF.

References

Berg P, Baltimore D, Boyer HB, Cohen SN, Davis RW, Hogness DS, *et al.* Potential biohazards of recombinant DNA molecules. *Science* 1974;185:303.

Cohen SN, Chang ACY, Boyer HW, Helling RB. Construction of biologically functional bacterial plasmids *in vitro*. *Proc Natl Acad Sci USA* 1973;70:3240–3244.

Jackson DA, Symons RH, Berg P. Biochemical method for inserting new genetic information into DNA of Simian Virus 40 DNA molecules containing lambda phage genes and the galactose operon of *Escherichia coli*. *Proc Natl Acad Sci USA* 1972;69:2904–2909.

Wade N. Recombinant DNA: NIH group stirs storm by drafting laxer rules. *Science* 1975a;190:767–769.

Wade N. Recombinant DNA: NIH sets strict rules to launch new technology. *Science* 1975b;190:1175–1179.

20

Tools used in genetic engineering

H. K. Das

The general strategy adopted for cloning by genetic engineering is the following. The chosen vehicle DNA is cleaved at a specific site by a restriction endonuclease, preferably one that produces a staggered cut in the two strands. The piece of foreign DNA, which is proposed to be introduced into the vehicle, is also produced by the action of the same restriction enyzme, thus generating similar ends. The foreign DNA is then spliced to the vehicle DNA by using the enzyme DNA ligase. The vehicle containing the ligated foreign DNA is then introduced into the recipient organism by transformation or transfection.

20.1 Restriction endonucleases

Getting large-enough DNA segments containing whole genes is the first challenge in gene cloning. An appropriate cleaving device would be an enzyme which would not act on any phosphodiester linkage or next to one particular nucleotide (e.g. T1 RNase cleaves next to guanine), but would cleave only at a specific sequence of several (could be six) nucleotides. The discovery of such enzymes has been described in details in Chapter 17, Sections 17.1.2.2.1 to 17.1.2.2.8. A large number of such enzymes (restriction endonucleases, type II) have since been isolated from various microorganisms which recognize different base sequences as their cleavage sites. The recognition sequences of all these enzymes have a two-fold axis of symmetry, that is, the nucleotide sequence in a particular direction (say 5′→3′) is the same in both the strands. Some of these enzymes produce flush cut ends, while the others generate overhangs. There can be either a 5′-overhang (as in the case of *Hin*dIII) or a 3′-overhang (e.g. *Pst*I). The ends with an overhang can be used as cohesive ends and hence are preferred by genetic engineers. The two molecules to be joined are cleaved by the same enzyme and then annealed together. The cohesive ends (complementary overhangs) form hydrogen bonds and stick together.

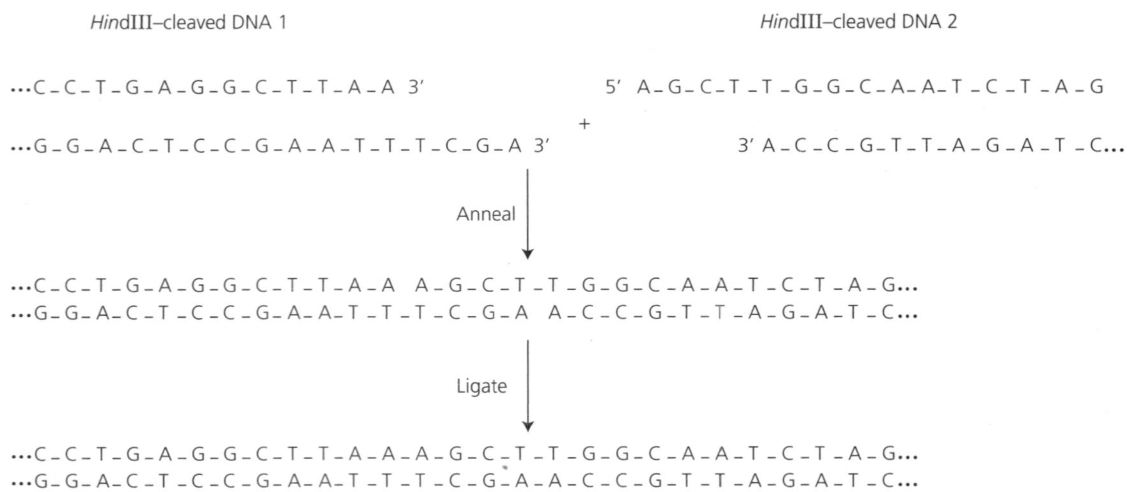

Figure 20.1 DNA fragments produced by cleavage with the same restriction endonuclease can be joined by DNA ligase.

Table 20.1 A few examples of type II restriction endonucleases

Recognition sequence		Fragment		Fragment
*Eco*RI isolated from *Escherichia coli* RY 13				
5′ GAATTC 3′	⟶	5′ G	+	AATTC 3′
3′ CTTAAG 5′		3′ CTTAA		G 5′
*Bam*HI isolated from *Bacillus amyloliquefaciens*				
5′ GGATCC 3′	⟶	5′ G	+	GATCC 3′
3′ CCTAGG 5′		3′ CCTAG		G 5′
*Pst*I isolated from *Providencia stuartii*				
5′ CTGCAG 3′	⟶	5′ CTGCA	+	G 3′
3′ GACGTC 5′		3′ G		ACGTC 5′
*Sal*I isolated from *Streptomyces albus*				
5′ GTCGAC 3′	⟶	5′ G	+	TCGAC 3″
3′ CAGCTG 5′		3′ CAGCT		G 5′
*Hae*III isolated from *Haemophilus aegyptius*				
5′ GGCC 3′	⟶	5′ GG	+	CC 3′
3′ CCGG 5′		3′ CC		GG 5′
*Hha*I isolated from *Haemophilus haemolyticus*				
5′ GCGC 3′	⟶	5′ GCG	+	C 3′
3′ CGCG 5′		3′ C		GCG 5′
*Sau*3AI isolated from *Staphylococcus aureus* 3A				
5′ NGATCN 3′	⟶	5′ N	+	GATCN 3′
3′ NCTAGN 5′		3′ NCTAG		N 5′
*Not*I isolated from *Nocardia otitidis-caviaram*				
5′ GCGGCCGC 3′	⟶	5′ GC	+	GGCCGC 3′
3′ CGCCGGCG 5′		3′ CGCCGG		CG 5′
*Pac*I isolated from *Pseudomonus alcaligenis*				
5′ TTAATTAA 3′	⟶	5′ TTAAT	+	TAA 3′
3′ AATTAATT 5′		3′ AAT		TAATT 5′
*Sfi*I isolated from *Streptomyces fimbriatus*				
5′ GGCCNNNNNGGCC 3′	⟶	5′ GGCCNNNN	+	NGGCC 3′
3′ CCGGNNNNNCCGG 5′		3′ CCGGN		NNNNCCGG 5′

The formation of the phosphodiester linkage is then achieved by the enzyme DNA ligase (Fig. 20.1).

Some more examples of type II restriction endonucleases are given in Table 20.1.

In addition to the restriction endonucleases listed in Table 20.1, many more of such enzymes are commercially available today. The genes for most of these enzymes have been cloned in *E. coli* and the enzymes are now isolated from *E. coli* strains harbouring the cloned genes. Under appropriate conditions, the recognition specificities of the restriction endonucleases are very rigid. A particular genome, when treated with a restriction endonuclease, would reproducibly yield the same DNA fragments over and over again, thus allowing the construction of the restriction map of the genome. Fig. 20.2 shows the photograph of an agarose gel with electrophoretically separated DNA restriction fragments, and Fig. 20.3 depicts the restriction map of the lambda genome derived from the sizes of the restriction fragments on the gel.

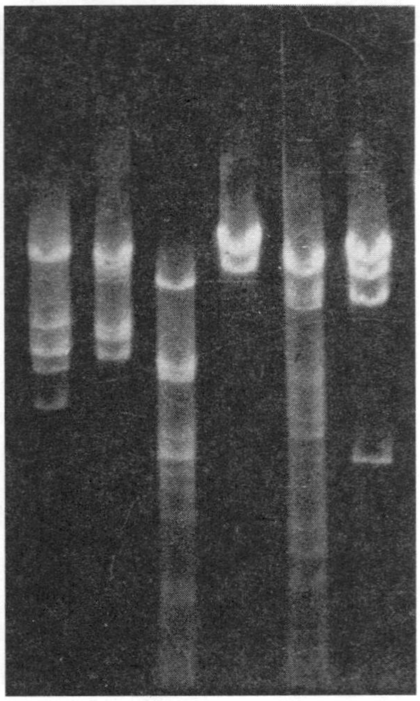

Figure 20.2 Photograph of an agarose gel with electrophoretically separated DNA fragments produced by the cleavage of the DNA from the bacterial virus lambda (λ) with several restriction endonucleases. The fragments of DNA were applied into slots in agarose gel at the top of the photograph and separated by electrophoresis. The mobility of the fragments (top to bottom in the photograph) is an inverse function of the length of the DNA fragments. Visualization of the fragments was achieved under ultraviolet rays as sharp fluorescent bands by staining with ethidium bromide. Right to left: slot 1, lambda (λ) DNA cleaved with the restriction endonuclease *Eco*RI; slot 2, lambda (λ) DNA cleaved with *Bam*HI; slot 3, lambda (λ) DNA cleaved with *Pst*I; slot 4, lambda (λ) DNA cleaved with *Sal*I: slot 5, lambda (λ) DNA cleaved with *Bgl*I; slot 6, lambda (λ) DNA cleaved with *Bgl*II.

All the above-mentioned restriction endonucleases belong to type II. Two other types of restriction endonucleases (type I and type III) have also been encountered. Examples of type I enzymes are *Eco* K and *Eco* B and of type III enzymes are *Eco*P1 and *Eco*P15. The following are the comparative characteristics of the three types of restriction endonucleases: the recognition site is often palindromic (GAATTC in *Eco*RI) in type II enzymes, but asymmetrical (TGAN$_8$TGCT for *Eco*B) in type I and type III enzymes; the cleavage site is the same as the recognition site in type II enzymes, while it is 24–26 bases away in type III enzymes and 1000 bases away in type I enzymes; ATP is not needed for type II enzymes, but is essential for type I and type III enzymes.

20.1.1 Modification functions associated with restriction endonucleases

Nucleotides in the genome are quite often modified after their incorporation into DNA. By far, the most common way of modification is methylation.. Three general methylation systems exist in *E. coli*. The *hsd* system methylates adenine in DNA and is thought to impart host specificity. The *dam* system also methylates adenine and thus distinguishes the strands of a newly replicated DNA. The *dem* system methylates cytosine. Similar general methylation systems have been detected in other prokaryotes too.

In contrast to the above-mentioned general methylation systems, the methylation processes associated with the DNA restriction systems are very specific. Restriction endonucleases in an organism appear to be a device for degrading any invading DNA molecule which

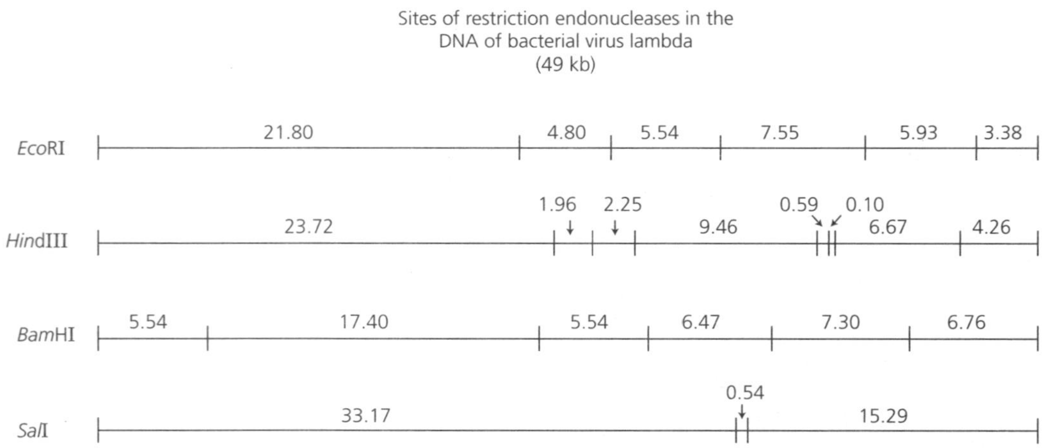

Figure 20.3 Positions of restriction sites in the lambda genome with respect to each individual endonuclease derived from the sizes of the restriction fragments on the gel shown in Fig. 20.2. The exact sizes of the DNA restriction fragments can be determined by comparison with DNA fragments of known lengths electrophoresed alongside several restriction endonucleases. The topmost line represents the restriction map of the lambda (λ) genome obtained with the restriction endonuclease *Eco*RI, the next with *Hin*dIII, the next with *Bam*HI and the bottommost with *Sal*I. The small vertical lines intersecting the horizontal lines represent the cleavage sites of the restriction endonucleases and the digits in between denote the sizes in kilobases of the fragments produced by cleavage with the enzymes.

obviously has to be distinguished from the host DNA. This is achieved through modification of the host DNA by methylation done by an enzyme activity associated with the restriction endonuclease. The site of methylation is the same sequence of bases that would have been recognized by the restriction endonuclease. The methylated sequence is not cleaved by the endonuclease. In contrast, the invading DNA is cleaved as the sites are not protected by methylation. If only one strand of the DNA is methylated, even then the cleavage by the endonuclease does not take place. Instead, the methylase component then methylates the second strand also. The characteristics of the restriction modification system are different depending on which type the restriction system belongs to Type I and type III are single enzymes having both endonuclease and methylase activities, while type II enzymes comprise two separate entities, one endonuclease and the other methylase; S-adenosyl methionine is essential for both types I and III modification systems, but is not needed for type II modification systems.

Modification enzymes associated with type II restriction endonucleases are made use of, when blunt-ended DNA fragments are to be cloned. The practice is to first blunt-end ligate a linker, which is a synthetic double-stranded oligonucleotide containing a specific restriction site, to the two ends of the DNA to be cloned. The specific restriction endonuclease is then used to liberate the staggered ends for subsequent ligation with complementary staggered ends of the vector. The problem arises when the DNA fragment to be cloned contains the same restriction site as is present in the linker. The restriction endonuclease would act on that internal site also and cleave the DNA fragment into subfragments, thus making it impossible to clone the intact fragment. The DNA fragment is, hence, treated with the specific modification enzyme prior to the ligation of the linkers to its two ends. The internal restriction site in the DNA fragment to be cloned would be methylated and render it unavailable to the specific restriction endonuclease when employed subsequently to make the staggered ends exposed.

20.2 DNA ligases

Two types of DNA ligases are used in recombinant DNA experiments. One type is typified the *E. coli* DNA ligase, which can only join staggered cuts but not blunt ends. The other class, represented by T4 DNA ligase, is capable of joining both staggered cut and blunt ends. The discovery of the DNA ligases and their characterization have been described in Chapter 17, Section 17.2.1.

20.3 *E. coli* DNA polymerase I

DNA polymerase I, discovered in 1958 by Arthur Kornberg and coworkers (Lehman *et al.* 1958) from the soluble fraction of the lysate of *E. coli*, could synthesize DNA *in vitro* using deoxyribonucleotide triphosphates as substrates. A DNA template was necessary for the reaction. With a natural DNA template, addition of all the four deoxyribonucleoside 5′-triphosphates was found essential for the polymerization reaction catalysed by the DNA polymerase. The diphosphates or the 3′-triphosphates did not serve as substrates. A DNA primer onto which the mononucleotides would actually be added by the formation of the phosphodiester linkage was also found to be a strong requirement. Conclusive evidence, that the nucleotides were added only to the 3′-end of the primer, was obtained by carrying out the polymerization reaction using a primer that was recessed both at the 5′-end and at the 3′-end with respect to the template (Fig. 20.4). The product produced by DNA polymerase exactly resembled DNA, as revealed by the analysis of base composition and base sequence.

20.3.1 Nuclease activities as integral parts of DNA polymerase I

DNA polymerase I has two nuclease activities as integral parts of it.

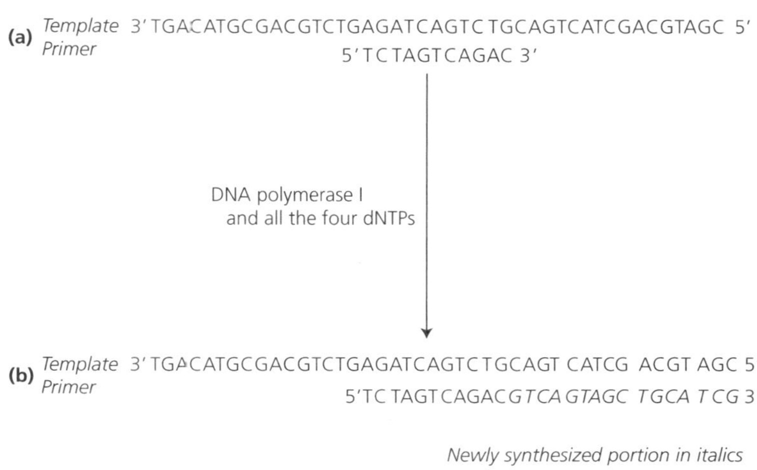

Figure 20.4 Addition of mononucleotides to the primer strand when the primer strand is recessed at both the 5′-end and the 3′-end. Thus, the template strand has both a 5′-overhang and a 3′-overhang. The newly synthesized portion of DNA is shown by the letters in italics. Addition took place only to the recessed 3′-end of the primer strand.

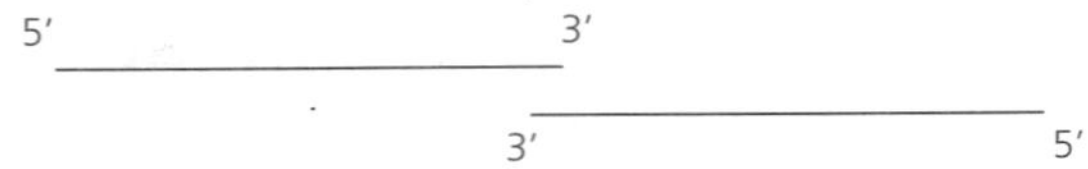

Figure 20.5 A double-stranded DNA molecule with 5′-overhangs on both sides serves as a template primer combination.

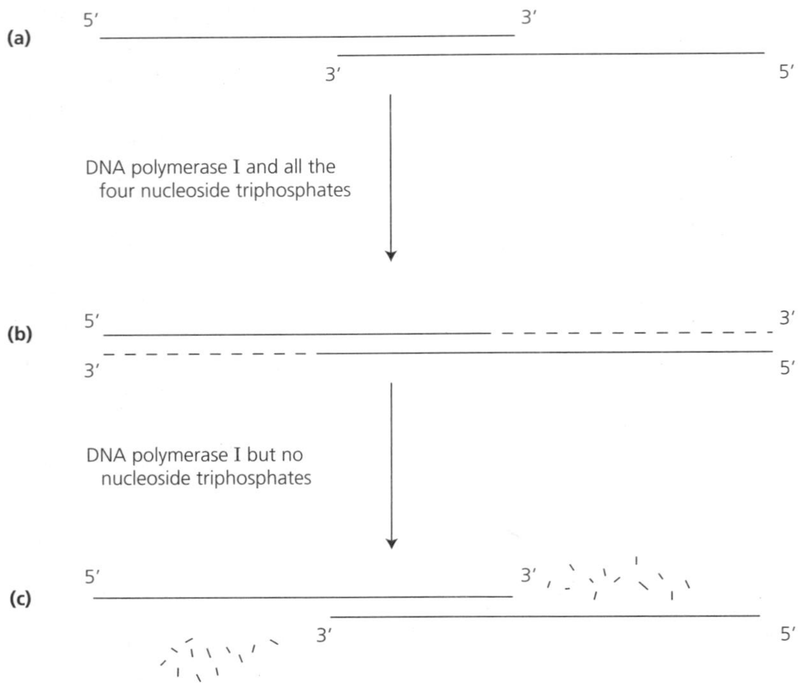

Figure 20.6 Schematic representation of a 3′→5′ exonuclease activity as an integral part of *Escherichia coli* DNA polymerase I. (a) Shown in solid lines is the template–primer combination used, which is similar to the molecule B in Fig. 9.1. (b) Completely double-stranded molecule obtained by employing DNA polymerase I and all the four ^{32}P-labelled nucleoside triphosphates. A part of the acid-soluble ^{32}P-radioactivity in the nucleoside triphosphates gets converted into acid-insoluble radioactivity. The newly synthesized ^{32}P-labelled portion of DNA is shown by broken lines. (c) Result of 3′→5′ exonuclease activity of DNA polymerase I, which can be apparent when the nucleoside triphosphates are withheld. Lot of acid-insoluble radioactivity is converted into acid-soluble radioactivity.

20.3.1.1 A 3′→5′ exonuclease activity

Completely purified homogeneous DNA polymerase I from *E. coli* contains as an integral part of the enzyme a 3′→5′ exonuclease activity (Lehman and Richardson 1964). This activity can be detected in the following manner. With a suitable template–primer combination, like that shown in Fig. 20.5, ^{32}P-labelled deoxyriboside triphosphates are used as substrates for the synthetic reaction and the newly synthesized DNA is purified free of the acid-soluble nucleotides by gel filtration. DNA polymerase I can now release labelled acid-soluble nucleoside monophosphates from the 3′-ends of the synthesized DNA strands if applied in the absence of the deoxynucleoside triphosphates (Fig. 20.6). This exonuclease activity releases only mononucleotides and degrades single-stranded DNA as well.

20.3.1.2 A 5′→3′ exonuclease activity

Purified DNA polymerase I from *E. coli* also contains as an integral part of the enzyme a 5′→3′ exonuclease activity (Klett *et al.* 1968), which has been detected in the following manner. A single-stranded polydAT copolymer was first synthesized from ^{32}P-labelled dATP and dTTP using DNA polymerase I from *E. coli* in the presence of small synthetic repeating dAdT oligomers. Incorporation of ^{32}P-labelled radioactivity in the acid-insoluble product ensured polymerization. The polymer was purified thoroughly from any ^{32}P-labelled dATP and dTTP by gel filtration. DNA polymerase I was again added, but this time along with ^{3}H-labelled d[DAP]TP (deoxyribose 2,6-diaminopurine triphosphate) and dTTP (Klett *et al.* 1968). Because of the possibility of the 3′-end of the dAT copolymer bending on to itself and forming a template–primer complex [Fig. 20.7(a)], a hairpin containing a hybrid double-stranded DNA product, as schematically shown in Fig. 20.7(b), was expected. This product was detected and purified by CsCl density gradient centrifugation, taking advantage of the difference in the buoyant density imparted by the incorporation of d[DAP]P. The buoyant density of pure dAT copolymer was found to be 1.679, while that of pure d[DAP]T copolymer was 1.718. The hybrid molecule banded with an intermediate buoyant density depending on the proportion of dA and d[DAP] in the product (Klett *et al.* 1968). While the dAT copolymer [Fig. 20.7(a)] contained ^{32}P-radioactivity, the hybrid dAT–d[DAP]T copolymer contained both ^{32}P- and ^{3}H-radioactivity. If this hybrid molecule was treated again with DNA polymerase I along with d[DAP]TP and dTTP, then it was exonucleolytically degraded from the 5′-end, apparently in a manner shown in Fig. 20.7(c). ^{32}P-labelled dAMP and dTMP were released

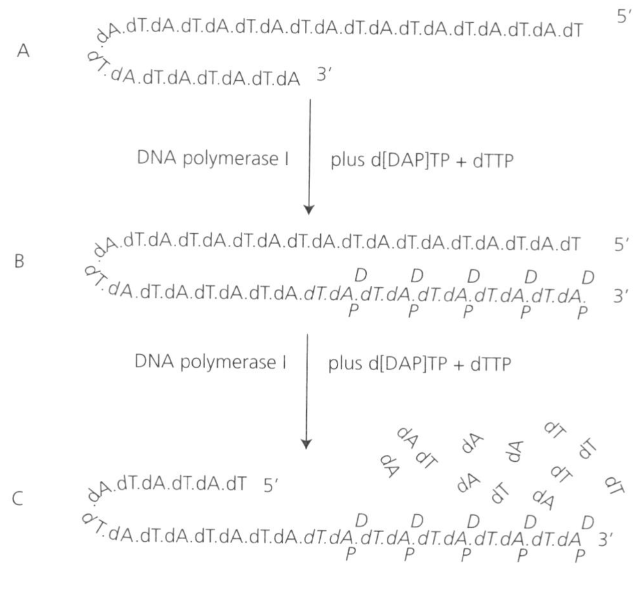

Figure 20.7 Schematic representation of a 5′→3′ exonucleolytic activity as an integral part of *Escherichia coli* DNA polymerase I. (a) A dAT copolymer, the 3′-end of which has bent on itself and formed a hydrogen-bonded hairpin-like structure, thus serving as a template–primer combination. (b) Addition of d[DAP]MP and dTMP to the 3′-end, forming a hybrid hairpin containing a double-stranded molecule. This molecule has dA–dT at the 5′-end and d[DAP]–dT at the 3′-end. (c) The dAMP and dTMP residues have been released from the 5′-end of the hybrid molecule due to the 5′-exonucleolytic activity of DNA polymerase I.

in the acid-soluble fraction and the buoyant density of the residual polymer gradually increased till it finally attained the value 1.718.

E. coli DNA polymerase I and its associated nuclease activities have been covered in greater detail in Chapter 9, Sections 9.2 to 9.5.

20.4 The Klenow fragment

Arthur Kornberg and his associates had observed that if a pure preparation of DNA polymerase I from *E. coli* was stored at 0°C for 4 weeks, it lost the 5′→3′ exonuclease activity, though the polymerase activity and the 3′→5′ exonuclease activity were unaffected (Brutlag *et al.* 1969). They also reported that if the freshly isolated enzyme was treated with an extract of *Bacillus subtilis*, a C-terminal fragment of 76,000 Da and an N-terminal fragment of 36,000 Da could be recovered. The larger fragment contained only the polymerase activity and the 3′→5′ exonuclease activity, while the smaller fragment contained the 5′→3′ activity (Brutlag *et al.* 1969). Soon afterwards, Klenow and Henningsen (1970) also reported the elimination of the 5′→3′ exonuclease activity by limited proteolysis with subtilisin. The larger fragment, which retains the polymerization fidelity of the holoenzyme, is now popularly known as the Klenow fragment. This is used more often to synthesize DNA, particularly when removal of the DNA strand complementary to the template strand is not necessary.

The Klenow fragment is used for second-strand synthesis of complementary DNA (cDNA), for other oligodeoxyribonucleotide synthetic processes and for DNA base sequencing by Sanger's dideoxy chain termination method.

20.5 T4 DNA polymerase

The T4 bacteriophage–coded DNA polymerase has been purified which functions as a holoenzyme of the core polymerase and three other phage-coded proteins (Goulian *et al.* 1968). The enzyme requires a template, a primer and deoxyribonucleoside 5′-triphosphates for replication that proceeds in the 5′→3′ direction. The T4 DNA polymerase has an 3′→5′ exonuclease activity which is much more active than that of *E. coli* DNA polymerase I (Bedinger and Alberts 1983), which possibly contributes to its fidelity (Sinha 1987). The exonuclease did remove *in vitro* 50–86% of alkyl mispairs (Khare and Eckert 2001). T4 DNA polymerase lacks the 5′→3′ exonuclease activity.

The T4 DNA polymerase is used to form blunt ends either by 3′-overhang removal or by 5′-overhang fill-in and second-strand synthesis in site-directed mutagenesis by the Kunkel procedure.

20.6 T7 DNA polymerase

The DNA polymerase from the bacteriophage T7 has been purified (Grippo and Richardson 1971) and its crystal structure has been elucidated (Doublie *et al.* 1998) The T7 DNA polymerase is similar to the Klenow fragment of the *E. coli* DNA polymerase I, in having an amino terminal 3′→5′ exonuclease domain and a carboxy terminal polymerase domain. The polymerase domain of the T7 DNA polymerase and the Klenow fragment of *E. coli* DNA polymerase I, both have a shape of a human right hand in which the fingers, thumb and palm subdomains form a DNA-binding groove that leads to the polymerase active site. This high fidelity polymerase has a stronger 3′→5′ exonuclease activity and a more rapid chain extension rate.

The T7 DNA polymerase is useful in copying long stretches of the DNA template.

20.7 Terminal deoxynucleotidyl transferase

The 32,360-Da enzyme can be purified from calf thymus gland (Chang and Bollum 1971). The enzyme uses deoxynucleoside triphosphates as substrate and adds on mononucleotides to the 3′-end of a DNA primer. Neither any template strand is necessary, nor the addition of nucleotides is directed by any template. The enzyme is not associated with any exonucleolytic activity (Bollum 1978). Another enzyme with similar activity but of a molecular weight of 79,000 Da has also been reported (Johnson and Morgan 1976). One of the isoforms of terminal deoxynucleotidyl transferase of mouse, which happens to be the isoform conserved across the vertebrate phylum, has been found to be unable to discriminate between ribonucleotides and deoxyribonucleotides for incorporation (Boule *et al.* 2001).

20.8 Reverse transcriptase

Two papers appeared in the 27 June 1970 issue of *Nature* that questioned the universality of the 'central dogma' – DNA → RNA → protein. These two papers by two independent groups of investegators convincingly demonstrated that the step RNA → DNA does operate in RNA tumour viruses mediated by an RNA-dependent DNA polymerase (Baltimore 1970; Temin and Mizutani 1970). This enzyme, which was also referred to as RNA-directed DNA polymerase or DNA polymerase of RNA tumour viruses, was finally named reverse transcriptase.

Purified virions of Rauscher mouse leukemia virus (R-MLV) and Rous sarcoma virus (RSV), though contained no DNA, could incorporate radioactivity into an acid-insoluble product when incubated with ^{3}H-TTP in the presence of dATP, dCTP and dGTP. Dithiothreotol, Mg^{2+}, Mn^{2+} and Na^+ were other requirements. The incorporation of radioactivity was stimulated if the virions were treated with a non-ionic detergent such as Nonidet P-40. The reaction product was unaffected by treatment with ribonuclease or by digestion with NaOH, but was hydrolysed by treatment with deoxyribonuclease or micrococcal nuclease. The purified virions did not incorporate any radioactivity when incubated with ^{3}H-UTP in the presence of ATP, CTP and GTP.

The enzyme reverse transcriptase has been purified after lysing the virions with non-ionic detergents such as Nonidet P-40 or Triton X-100. Reverse transcriptase can utilize both RNA and DNA as templates to synthesize cDNA. The substrate is deoxyribonucleotide 5′-triphophates. A primer is necessary to initiate incorporation of deoxyribonucleotides. A DNA primer is extended more efficiently, though an RNA primer is also acceptable. The primer can be as short as 4–8 nucleotides. Extension takes place at the 3′-end of the primer and the direction of synthesis is from 5′→3′. Reverse transcriptase from avian myeloblastosis virus (AMV) has been reported to incorpoate quite a large number of incorrectly paired bases (Verma 1977).

Purified reverse transcriptase exhibits, in addition to the synthetic activity, a degradative activity similar to ribonuclease H. A distinctive feature of this associated RNase H activity is that it is an exonuclease that needs free ends, while cellular RNase H is an endonuclease. Although both the synthetic and the degradative activities reside in the same polypeptide, the active sites are not common (Verma 1977).

20.9 T4 polynucleotide kinase

The bacteriophage T4–coded polynucleotide kinase was isolated from T4-infected *E. coli* lysates (Panet *et al.* 1973), and the enzyme transfers the γ-phosphate of ATP to the 5′-hydroxy terminal of polynucleotides, which could be single-stranded or double-stranded DNA or RNA. The enzyme also phosphorylates nucleoside 3′-monophosphates, but removes the 3′-phosphate group from polynucleotides and deoxynucleoside monophosphates and diphosphates (Richardson 1981).

T4 polynucleotide kinase is used for end-labelling of DNA and RNA for use as probes and for base sequencing by the Maxam and Gibert method. The enzyme is also used for the addition of 5′-phosphates to oligonucleotides for subsequent ligation and for removal of 3′-phosphates.

20.10 Calf intestinal alkaline phosphatase

Calf intestinal alkaline phosphatase removes 5′-phosphates from DNA, RNA and ribo- and deoxyribo triphosphates.

The enzyme is used to reduce self-ligation of DNA fragments and of cloning vectors and to prepare polynucleotides for subsequent 5′-phosphate labelling.

20.11 Exonuclease *Bal*31

An exonuclease activity was detected in the extracts of *Pseudomonas Bal*31 (Gray *et al.* 1975), which was further characterized to be active against linear double-stranded DNA (Legerski *et al.* 1978). The enzyme progressively shortens the DNA (Fig. 20.8) from both ends without producing any endonucleolytic cleavage. The series of progressively shortened pieces of DNA thus obtained would, on digestion with a site-specific restriction endonuclease and subsequent analysis by gel electrophoresis, permit establishment of the order of the restriction enzyme fragments through the original DNA molecule (Legerski *et al.* 1978).

*Bal*31 is used for shortening of DNA molecules by reduction from both ends and also for restriction mapping.

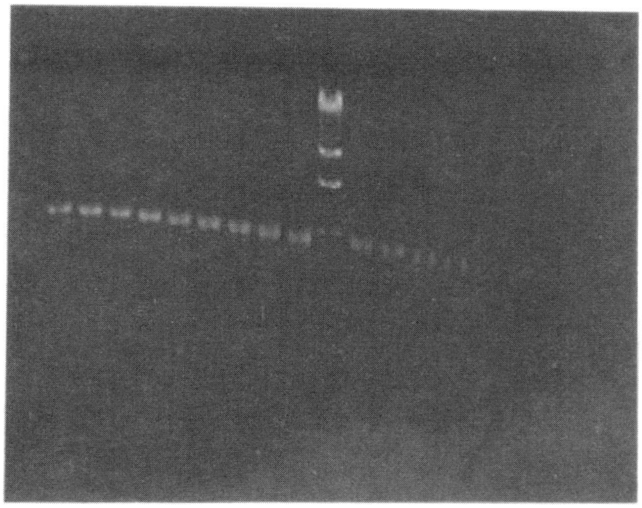

Figure 20.8 Photograph of an electrophoresed agarose gel showing DNA fragments obtained by periodic withdrawal of samples from a reaction mixture in which the DNA has been treated with Bal 31. The periodic samples have been applied to the slots in the gel sequentially from left to right. DNA markers were applied to the 10th slot from left.

20.12 Nuclease S1

S1 nuclease has been isolated from *Aspergillus oryzae* and cleaves endonucleolytically single-stranded DNA and RNA, releasing 5′-phosphoryl mono- or oligonucleotides. It is five times more active on DNA than on RNA. Although its primary substrate is single-stranded nucleic acid, at higher concentrations of the enzyme, it can also occasionally introduce single-stranded breaks in double-stranded DNA or RNA or DNA–RNA hybrids.

S1 nuclease is used to remove single-stranded overhangs of DNA fragments, to digest hair pin loops, to detect mismatches of nucleotides in hybrid structures, to map transcription initiation site (Berk and Sharp 1978) and to map intron–exon junction.

20.13 DNA and RNA markers

DNA molecules of known sizes are used as markers when the sizes of the DNA fragments produced by restriction endonucleases are desired to be found out. Such size markers are used when a restriction map of a genomic DNA fragment or a plasmid is to be drawn. The fragments produced by cleavage with restriction endonuclease are electrophoresed on an appropriate gel along with some DNA markers. The mobility of the newly produced DNA fragments is compared with the mobility of the markers and their sizes computed by extrapolation. As has been mentioned under the section 'Restriction Endonucleases', a particular restriction endonuclease would cleave a specific DNA molecule reproducibly at exactly the same sites. As many of the viral DNA genomes can easily be isolated in intact form, these molecules cleaved with a specific restriction endonuclease have been popular as DNA markers. A look at Fig. 20.3 would illustrate the point. In fact, the DNA fragments obtained by cleaving the genome of the bacterial virus lambda (λ) with the restriction endonuclease *Hin*dIII are very often used for determining the sizes of DNA fragments between 2 kb and 20 kb [Fig. 20.9(a)]. The same genome of the virus lambda (λ) when cleaved with the restriction endonuclease *Bst*EII can be used as a DNA marker between the size range of 700 bases to 8 kb [Fig. 20.9(b)]. Fragments obtained by cleaving the genome of the bacterial virus ϕx174 with the restriction endonuclease *Hae*III are used as markers for DNA fragments between 200 bases and 1.3 kb [Fig. 20.9(c)]. Restriction endonuclease digest of plasmid DNA is also used as a DNA size marker. Fig. 20.9(d) shows size markers in the range of 100 bases to 600 bases obtained by digestion of the plasmid pBR322 with *Msp*I. Step ladders of DNA fragments [Fig. 20.9(e)] are useful for precise size determination of small DNA fragments. At larger size ranges, intact chromosomes of the yeast *Saccharomyces cerevisiae* [Fig. 20.9(f)] are used, particularly in experiments involving pulse-field gel electrophoresis. A ladder of lambda (λ) genomes [Fig. 20.9(g)] is also useful for pulse-field gel electrophoresis experiments.

RNA markers are used for computing the sizes of single-stranded RNA molecules. RNA molecules from *E. coli* ribosomes (5S rRNA: 120 bases; 16S rRNA: 1541 bases; 23S rRNA: 2904 bases) are among the earliest known RNA markers. The intact chromosome of brome mosaic virus, a plant RNA virus, is used as an RNA marker in the 800 bases to 3.2 kb size range [Fig. 20.10(a)]. Mixtures of *in vitro* transcribed RNA molecules are also available in the molecular weight range of 500–9000 bases [Fig. 20.10(b)].

20.14 DNA linkers

A 'linker' is a short sequence of deoxynucleotides which includes bases specifying the recognition sequence of a restriction endonuclease. Thus, there could be a *Hin*dIII linker or a *Bam*HI linker and so on.

*Hin*dIII linker	*Bam*HI linker
*5′-CCC**AAGCTT**GGG-3′* *3′-GGG**TTCGAA**CCC-5′*	*5′-CCC**GGATCC**GGG-3′* *3′-GGG**CCTAGG**CCC-5′*

The recognition sequences of the restriction endonucleases in the above-stated linkers are shown in bold italics. The two complementary oligonucleotides of a linker are separately synthesized chemically and then

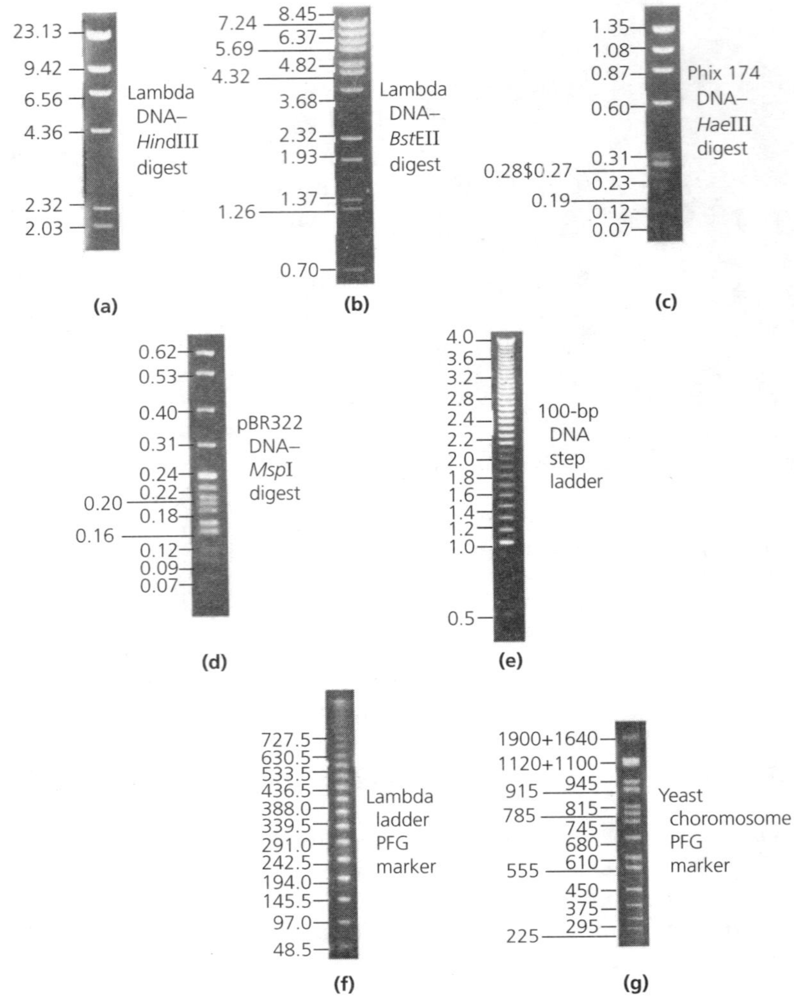

Figure 20.9 DNA size markers. Each band represents a DNA fragment. The numbers that appear on the left of each gel strip denote the size of the marker DNA fragments in kilobases (kb). The writings on the right of the strip describe the source of the markers. (a) The DNA of the bacterial virus lambda (λ) when digested with the restriction endonuclease *Hin*dIII yields DNA fragments of sizes between 2.0 kb and 23.1 kb. The figure shows fragments electrophoresed in 1.0% agarose gel, stained with ethidium bromide and visualized by fluorescence under ultraviolet ray. (b) The same DNA of the bacterial virus lambda (λ) when digested with the restriction endonuclease *Bst*EII yields DNA fragments of sizes between 0.7 kb and 8 kb. The figure shows fragments after electrophoresis in 1.0% agarose gel. (c) The DNA of the bacterial virus Φx 174 when digested with the restriction endonuclease *Hae*III yields fragments between 0.2 kb and 1.3 kb. The figure shows fragments electrophoresed in 1.7% agarose gel. (d) The DNA of the plasmid pBR322 when digested with the restriction endonuclease *Msp*I yields DNA fragments of sizes between 0.07 kb and 0.62 kb. The figure shows fragments electrophoresed in 1.8% agarose gel. (e) The figure shows a DNA step ladder formed by electrophoresis in 1.0% agarose gel of blunt-ended DNA fragments of 100-bp size differences between the range of 0.5 kb and 4 kb. (f) Another ladder, but of the genomes of the bacterial virus lambda (λ), is shown in this figure. Concatamers of the lambda (λ) DNA have been subjected to pulse-field electrophoresis in 1% agarose gel. This ladder serves as a marker for large DNA molecules separated by pulse-field gel electrophoresis. (g) Intact chromosomes of the yeast *Saccharomyces cerevisiae* separated by pulse-field electrophoresis in 1% agarose gel. This is also quite often used as marker for large DNA molecules.

annealed together. The linkers are attached to a double-stranded DNA by 'blunt-end ligation'.

20.15 DNA adapters

Adapters are used if subsequent to the initial cloning of the DNA, it has to be re-cloned into another vector through a different restriction endonuclease site(s). For example, a DNA fragment with *Hin*dIII ends may have to be cloned into a vector into its *Bam*HI sites. An adapter for such a purpose comprises two chemically synthesized DNA oligonucleotides which when annealed together produce a duplex DNA with two different ends that are suitable for fitting into two different staggered ends, each created by a separate restriction endonuclease.

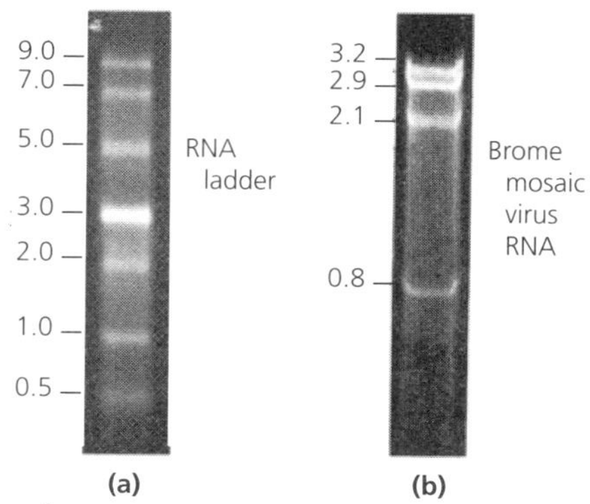

Figure 20.10 RNA size markers. Here also, the numbers that appear on the left of each of the two gel strips denote the size of the marker RNA fragments in kilobases (kb) and the writings on the right of the strips describe the source of the markers. Electrophoresis has been done in 1% agarose gel, staining was with ethidium bromide and visualization was against ultraviolet ray. (a) *In vitro* transcribed RNA molecules in the molecular weight range of 0.5 kb and 9 kb. (b) Intact chromosomes from the plant RNA virus, brome mosaic virus, which are in the range of 0.8 kb and 3.2 kb.

The following is a *Hin*dIII–*Bam*HI adapter:

5′-***AGCTT***GGGGG-3′

3′-ACCCCC***CCTAG***-5′

The 5′-overhang at the left of the top strand (bold italics) is designed to fit into a *Hin*dIII site, while the 5′-overhang at the right of the bottom strand (bold italics) is designed to fit into a *Bam*HI site. It is obvious that the adapter for connecting a restriction site with a 5′-overhang and a restriction site with a 3′-overhang would simply be a single-stranded oligonucleotide. The following is a *Hin*dIII–*Pst*I adapter with a *Hin*dIII-cleaved end at the left and a *Pst*I cleaved end at the right.

5′-A	5′-***AGCTTCTGCA***-3′	G-3′
3′-***TTCGA***-5′		3′-***ACGTC***-5′
*Hin*dIII overhang	*Hin*dIII–*Pst*I adapter	*Pst*I overhang

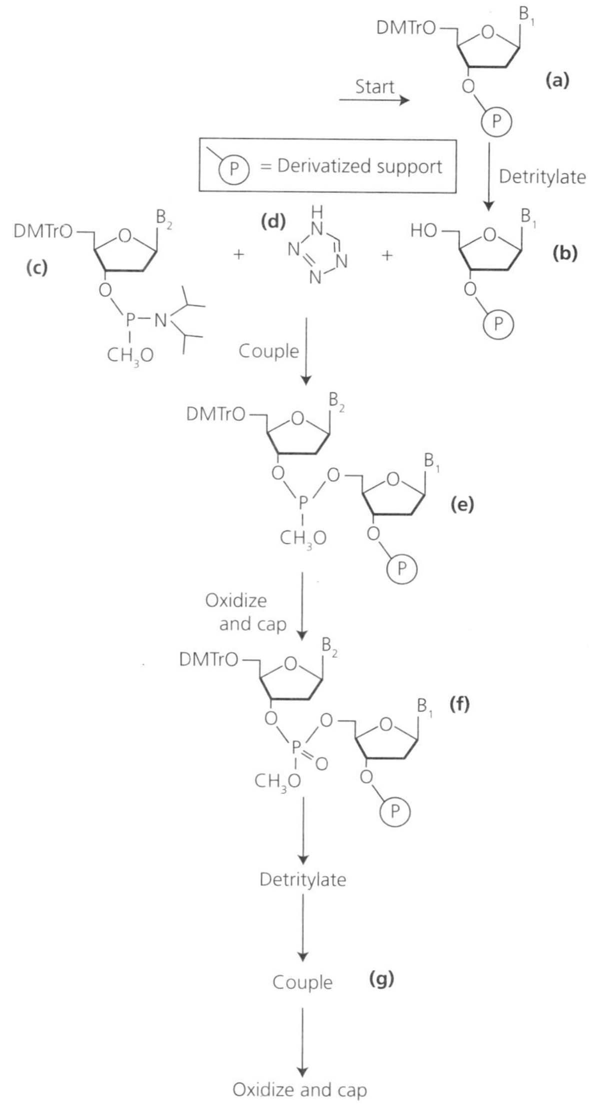

Figure 20.11 Chemical synthesis of oligonucleotides by the phosphoramidite method. (a) The dimethoxytrityl (DMT) derivative of the 3′-nucleotide hooked on to the solid support. (b) Exposed 5′-hydroxyl group of the 3′-nucleotide after treatment with trichloroacetic acid. (c) The phosphoramide derivative of the next nucleotide (second from the 3′-end in this case) with its 5′-OH group protected with the dimethoxytrityl group. (d) Tetrazole, which would create an intermediate with an activated 3′-OH group of the nucleotide at C. (e) The dinucleotide formed after the coupling reaction between the first and the second nucleotide from the 3′-end. (The capping of the unreacted free 5′-OH group, if any, of the first nucleotide by acetylation with acetic acid and 1-methylimidazole is done at this stage). (f) The dinucleotide after the linkage between the first and the second nucleotide has been converted from the phosphite to the more stable phosphotriester by using iodine as the oxidizing agent and water as the oxygen donor. (g) The next cycle of the removal of the DMT group on the second nucleotide, the coupling with the next nucleotide (third from the 3′-end in this case), is performed, followed by the usual capping of the unreacted 5′-OH group and conversion of the phosphite group into the more stable phosphotriester.

20.15.1 Chemical synthesis of adapters and linkers

Chemical synthesis of oligonucleotides is performed in the 3′→5′ direction by the phosphoramidite method, with the growing DNA chain attached to a solid support. The steps involved are described as follows (also illustrated in Fig. 20.11):

(i) **Detritylation:** The chosen derivatized solid support, with the required 3′-end nucleotide hooked on to it, is treated with trichloroacetic acid. This removes the dimethoxytrityl (DMT) group and exposes the 5′-OH group of the 3′-nucleotide for the subsequent reaction of coupling to the next nucleotide.

(ii) Coupling: The next nucleotide (second from the 3′-end in this case) is chosen as the phosphoramide derivative whose 5′-OH group is also protected with the DMT group. An intermediate with an activated 3′-OH group is created by adding tetrazole and the coupling reaction with the 3′-nucleotide (the first nucleotide from the 3′-end in this case) is performed.

(iii) Capping: There is always a possibility that some of the 3′-nucleotides (the first nucleotide from the 3′-end in this case) may have escaped the coupling reaction and still have a reactive free 5′-OH group. This must be inactivated (capped) to prevent it from reacting with the nucleotide next added, which is really intended to be the third nucleotide in the chain and not the second nucleotide. The capping is done by acetylation of the free 5′-OH group with acetic acid and 1-methylimidazole.

(iv) Oxidation: The linkage between the first and the second nucleotide is then converted from the phosphite to the more stable phosphotriester by using iodine as the oxidizing agent and water as the oxygen donor.

(v) Addition of the next nucleotide: The DMT group on the second nucleotide is then removed with trichloroacetic acid and the coupling with the next nucleotide (third from the 3′-end in this case) is performed as described under (ii) and operations as described under (iii) and (iv) are repeated.

(vi) Formation of the complete oligonucleotide: The cycles are repeated sequentially till complete synthesis of the desired oligonucleotide is achieved. After completion of the synthesis, the oligonucleotide is released from the solid support and purified. The whole operation described above has now been automated and can be performed in a DNA synthesizer.

20.16 Cloning vectors

Once the specific DNA fragment is available, it has to be ligated to a 'cloning vector' for its perpetuation and amplification. A cloning vector is a DNA molecule that replicates autonomously in a cell outside the main chromosome. Both plasmids and DNA viruses have been used as cloning vectors. Discovery of bacterial plasmids, their purification and characterization and also the construction of the first plasmid cloning vector pSC 101 have been described in Chapter 17, Section 17.3.2. The DNA of the animal virus SV40 was used as a cloning vector in the first ever-performed recombinant DNA experiment (Jackson *et al.* 1972) and has been described in Chapter 17, Section 17.3.1. As would be seen in subsequent chapters, the DNA of the bacterial viruses lambda (λ) and M 13 and their derivatives have also been used as cloning vectors.

20.17 Hosts for multiplication of clones

Once the specific DNA fragment is ligated to a cloning vector, the 'recombinant' vector has to be introduced into a host cell for its biological replication. Cloning vectors exhibit strict specificity with respect to their hosts. The replication signals on the vector have to conform to the replication machinery of the host. Vectors derived from Col E1 plasmid or lambda (λ) virus can be replicated only in *E. coli* and closely related bacteria as the host. Vectors having replication origins from plasmids such as RK2, RP1, RP4 or RSF1010 can, however, be replicated in other gram-negative bacteria (*e.g. Pseudomonas, Azotobacter, Azospirillum, Rhizobium,* etc.), in addition to *E. coli*. However, vectors derived from pUB110 can be replicated only in gram-positive bacteria such as *Staphylococcus aureus* and *Bacillus subtilis*. Vectors derived from the animal virus SV40 can be replicated only in mammalian cells. Thus, an appropriate host cell is essential for the biological perpetuation of the cloned fragment of DNA.

References

Baltimore D. Viral RNA-dependent DNA-polymerase. *Nature* 1970;226:1209–1211.

Bedinger P, Alberts BM. The 3′→5′ proofreading exonuclease of bacteriophage T4 DNA polymerase is stimulated by other T4 DNA replication proteins. *J Biol Chem* 1983;258:9649–9656.

Berk AJ, Sharp PA. Spliced early mRNAs of simian virus 40. *Proc Natl Acad Sci USA* 1978;75:1274–1278.

Bollum FJ. Terminal deoxynucleotidyl transferase: Biological studies. *Adv Enzymol* 1978;47:347–374

Boule J-B, Rougeon F, Papanicolaou C. Terminal deoxynucleotidyl transferase indiscriminately incorporates ribonucleotides and deoxyribonucleotides. *J Biol Chem* 2001;276:31388–31393.

Brutlag D, Atkinson MA, Setlow P, Kornberg A. An active fragment of DNA polymerase produced by proteolytic cleavage. *Biochem Biophys Res Commun* 1969;37:982–989.

Chang LMS, Bollum FJ. Deoxynucleotide-polymerizing enzymes of calf thymus gland. V. Homogeneous terminal deoxynucleotidyl transferase. *J Biol Chem* 1971;246:909–916.

Doublie S, Tabor S, Long AM, Richardson CC, Ellenberger T. Crystal structure of a bacteriophage T7 DNA replication complex 2.2 Å resolution. *Nature* 1998;391:251–258.

Goulian M, Lucas ZJ, Kornberg A. Enzymatic synthesis of DNA. XXV. Purification of properties of deoxyribonucleic acid polymerase induced by infection with phage T4. *J Biol Chem* 1968;243:627–638.

Gray HB Jr, Ostrander DA, Hodnett JL, Legerski RJ, Robberson DL. Extracellular nucleases of Pseudomonas BAL 31. I. Characterization of single strand-specific deoxyriboendonuclease and double-strand deoxyriboexonuclease activities. *Nucleic Acids Res* 1973;2:1459–1492.

Grippo P, Richardson CC. Deoxyribonucleic acid polymerase of bacteriophage T7. *J Biol Chem* 1971;246:6867–6873.

Jackson DA, Symons RH, Berg P. Biochemical method for inserting new genetic information into DNA of Simian Virus 40 DNA molecules containing lambda phage genes and the galactose operon of *Escherichia coli*. *Proc Natl Acad Sci USA* 1972;69: 2904–2909.

Johnson D, Morgan AR. The isolation of a high molecular weight terminal deoxynucleotidyl transferase from calf thymus. *Biochem Biophys Res Comm* 1976;72:840–849.

Khare V, Eckert KA. The 3′→5′ exonuclease of T4 DNA polymerase removes premutagenic alkyl mispairs and contrbutes to futile cycling at O^6-methylguanine lesions. *J Biol Chem* 2001;276:24286–24292.

Klenow H, Henningsen I. Selective elimination of the exonuclease activity of the deoxyribonucleic acid polymerase from *Escherichia coli* B by limited proteolysis. *Proc Natl Acad Sci USA* 1970;65:168–175.

Klett RP, Cerami A, Reich E. Exonuclease VI. A new nuclease activity associated with *E. coli* DNA polymerase. *Proc Natl Acad Sci USA* 1968;60:943–950.

Legerski RL, Hodnett JL, Gray HB Jr. Extracellular nucleases of Pseudomonas BAL 31. III. Use of the double-strand deoxyriboexonuclease activity as the basis of a convenient method for the mapping of fragments of DNA produced by cleavage with restriction enzymes. *Nucleic Acids Res* 1978;5:1445–1464.

Lehman IR, Bessman MJ, Simms ES, Kornberg A. Enzymatic synthesis of deoxyribonucleic acid. I. Preparation of substrates and partial purification of an enzyme from *Escherichia coli*. *J Biol Chem* 1958;233:163–170.

Lehman IR, Richardson CC. The DNases of *Escherichia coli*. IV. An exonuclease activity present in purified preparations of DNA polymerase. *J Biol Chem* 1964;239:233–241.

Panet A, van de Sande JH, Loewen PC, Khorana HG, Raae AJ, Lillehaug JR, *et al.* Physical characterization and simultaneous purification of bacteriophage T4 induced polynucleotide kinase, polynucleotide ligase and deoxyribonucleic acid polymerase. *Biochemistry* 1973;12:5045–5050.

Richardson CC. Polynucleotide kinase. In: Boyer PD, ed. *The Enzymes* (Volume 14). San Diego, CA: Academic Press; 1981. pp. 299–314.

Sinha NK. Specificity and efficiency of editing of mismatches involved in the formation of base-substitutions mutations by the 3′→5′ exonuclease activity of phage T4 DNA polymerase. *Proc Natl Acad Sci USA* 1987;84:915–919.

Temin HM, Mizutani S. RNA-dependent DNA polymerase in virions of Rous sarcoma virus. *Nature* 1970;226:1211–1213.

Verma IM. The reverse transcriptase. *Biochim Biophys Acta* 1977;473:1–38.

21

General strategies for expression of introduced genes in a foreign host

Shweta Sharma and Indranil Dasgupta

21.1 Vector engineering

The transcription efficiency of a gene largely determines its level of expression. Transcription of a gene initiates when the RNA polymerase complex interacts with the promoter sequences and proceeds from the 5′ 3′ direction of the gene, giving rise to an RNA transcript. In the next step, the RNA polymerase complex, after encountering the transcription termination signal, dissociates from the gene and thus releases the transcript for translation. To maximize the expression of heterologous genes, those genes which have been isolated from a particular organism but are being expressed in a different organism such as a bacterium, yeast, animal or plant cell, it is imperative to design a cloning vector which helps in an efficient transcription of genes. Such cloning vectors are known as expression vectors and the process of construction of such vectors is called vector engineering. Expression vectors are designed in such a manner that they include all the components required for optimal expression of the heterologous gene, that is, a strong promoter, which in some cases is induced by certain chemicals. The earliest vectors used for driving the expression of cloned genes in *E. coli* were constructed using the *lac* promoter. In 1976, Backman and coworkers and Gronenborn and coworkers, independently reported the construction of a plasmid in which the lac promoter was used to drive the expression of a repressor gene (*cI*) of bacteriophage λ. *E. coli* cells which carried these plasmids demonstrated overproduction of the repressor because transcription of *cI* originated at the *lac* promoter (Backman *et al.* 1976; Gronenborn *et al.* 1976). Building upon these results, Roberts and coworkers constructed a plasmid with the *lac* promoter driving the expression of *cro* gene of bacteriophage λ and reported synthesis of up to 1,90,000 *cro* protein monomers per cell in an *E. coli* host (Roberts *et al.* 1979). Guarente *et al.* in 1980 also reported construction of a plasmid that directs the synthesis of about 10,000–15,000 monomers per cell of rabbit β-globin (Guarente *et al.* 1980). Expression vectors are designed in such a manner that they include all the components required for maximum expression of the heterologous gene, that is, a strong promoter, which in some cases is induced either by certain chemicals or change in temperature, appropriate translation initiation signals, if required, and also components necessary for the easy purification of the expressed protein. For example, expression vectors designed for expression of heterologous genes in a bacterial system include a very common modification which consists of six consecutive histidine codons just before the start of the foreign gene. The foreign protein expressed using such expression vectors carries a tag of six histidine residues at the N-terminal end, which has very high affinity to metal ions such as nickel, which is included in the matrix used for separation. For isolation of such proteins which are expressed as histidine-tagged proteins, the total proteins isolated from such bacterial cells are allowed to pass through a matrix containing metal ions such as nickel. From the total proteins, only the histidine-tagged protein binds to the metal ions of the matrix, which can then be eluted out by high salt or temperature. The approach of vector engineering depends entirely upon the host system to be used for expression of the heterologous gene. In the case of prokaryotic hosts such as bacteria, a bacterial promoter such as *lac*, which is induced by the presence of a chemical such as isopropyl β-D-1-thiogalactopyranoside (IPTG) in the growth medium, would have to be used. In the case of certain vectors, bacteriophage promoters such as the P_L promoter derived from bacteriophage λ, which is inducible by high temperatures using a temperature-sensitive repressor, is used. The newly developed vectors available for bacterial expression systems utilize synthetic promoters such as *tac* derived from *trp* and *lac*, thus combining the high-expression properties of several promoters, which makes it capable of giving expression levels several fold higher than the individual native promoters. To obtain high expression of the heterologous genes in a eukaryotic system, the expression vectors would have to be based on the same principle of high-expression capability of specific promoters. But very few well-characterized promoters are available for eukaryotes as compared with prokaryotes. Most of the expression vectors available for the eukaryotic system contain viral promoters, derived from either animal viruses such as Simian Virus 40 (SV40), adenovirus or papillomavirus or plant viruses such as Cauliflower mosaic virus.

In expression vectors available for prokaryotic systems, the heterologous gene is expressed as a fusion protein either with a bacterial glutathione-S-transferase or with a maltose-binding protein, which allows a very easy affinity-based purification of the fusion protein by passing the bacterial extract through a column containing beads of glutathione-agarose or amylose resin. The fusion protein bound to the column is then eluted and cleaved with a protease called factor Xa, which is capable of cleaving any polypeptide between the amino acids arginine and valine. In these expression vectors, to provide cleavage sites for factor Xa, N-terminal amino acid sequences of the bacterial gene are followed by codons for arginine and valine, and for introduction of the heterologous gene to be expressed in the appropriate system, restriction enzyme sites are also included. These features make this system very efficient for the production of the heterologous protein as a fusion protein, from which the protein of interest can be obtained in a pure form.

Viruses are increasingly being used as vectors for expression of foreign genes in eukaryotic organisms. Such virus-derived vectors require significant modifications of the original virus, which can be achieved by means of vector engineering. Some of these modifications include reduction of the size of the derived vector to accommodate larger size of the foreign gene, removal of the symptom-determining genes to reduce or eliminate the disease symptoms caused by the virus, changing the tissue-specificity–determining viral genes to increase the likelihood of the virus reaching specific tissues or all the tissues, etc. All these modifications help in ensuring optimal expression of the foreign gene, which is being carried by the vector.

21.2 Codon optimization

To successfully achieve heterologous gene expression, a very important aspect to be considered is the phenomenon called codon bias. Except tryptophan and methionine, all other amino acids are encoded by two or more codons. In several organisms, all codons coding for one amino acid are not used equally, therefore showing a bias for one or more codons instead of others. For example, four codons, that is, GCA, GCC, GCG and GCT, code for alanine, out of which humans use GCC four times more than GCG for coding alanine. This phenomenon is known as codon bias. To ensure that correct codon bias is in use, appropriate changes in the coding sequence of the heterologous gene is required depending upon the system to be used. This procedure known as codon optimization becomes almost essential in those cases where heterologous transcript is required for translation at the maximum level. The concept of codon usage was pioneered by Grantham, who proposed the 'genome hypothesis', according to which there is a species-specific pattern to codon usage in organisms (Grantham *et al.* 1980; Grantham *et al.* 1981). Optimizations carried out to overcome this bias in codon usage are available. In one of the early reports, Robinson *et al.* demonstrated that the substitution of the rare arginine codon AGG with the *E. coli*-preferred CGU improved expression of the recombinant protein (Robinson *et al.* 1984). Brinkmann and coworkers demonstrated that co-expression of the *argU* (*dnaY*) gene coding for minor $tRNA^{Arg}$ (AGG/AGA), shown to be a limiting factor in the bacterial expression of several mammalian genes, resulted in high-level production of the target protein (Brinkmann *et al.* 1989). To confer resistance against insect pests, a bacterial gene for insecticidal crystal protein Cry was expressed in plants. The expression level of the *cry* gene was found to be very low because of the bacterial codon bias. However, the expression level of the *cry* gene in plants improved substantially following codon optimization, by changing the codon bias to a plant system from a bacterial one.

21.3 Host engineering

It is often seen that the expression level of heterologous genes is adversely affected by the genetic make-up of the host system. To achieve heterologous gene expression at the optimum level, these host factors need to be minimized by selecting a suitable mutant of the host to express the heterologous gene. This process of manipulating the host system to achieve optimum expression of the heterologous gene is known as host engineering. Basically, there are two host factors which affect the expression of the heterologous gene: the presence of the enzyme protease, which degrades the heterologous protein and high recombinogenic activity, bringing about the recombination or rearrangement of the introduced heterologous gene. To overcome the activity of the protease enzyme, an *E. coli* mutant, *lon*, which is deficient in protease activity, is generally used for the optimum expression of heterologous genes. This is very important in the case of an expression library, where each transformed cell of *E. coli* contains a different heterologous gene, which has the potential to produce different heterologous proteins. Hence, *lon* mutants are used to maximize the chances of such proteins remaining stable within cells. The *lon* gene of *E. coli* encodes an ATP-dependent protease (Chung and Goldberg 1981), which is involved in the degradation of unstable proteins. The *htpR* gene product controls the induction of the lon protease. In 1985, Buell and coworkers reported that the expression of a cloned gene product specific to human hormone somatomedin-C (SMC) in *lon, htpR* and *lon htpR* double mutants resulted in increased accumulation of SMC protein, presumably because of low lon protease activity (Buell *et al.* 1985). Similarly, Bishai *et al.* showed that proteolysis of the fusion protein consisting of N-terminal 485 amino acids of Diptheria toxin joined to the α-melanocyte–stimulating hormone was significantly reduced by growth in *lon* and *htpR* mutant strains of *E. coli* (Bishai *et al.* 1987).

To minimize the recombinogenic activity of the host *E. coli* system, various recombination-deficient mutants, such as *recA*, are used for the maintenance and expression of plasmids containing heterologous genes. RecA protein has been shown to be essential for recombination of DNA and, hence, *recA* mutants show less recombination of heterologous DNA. This results in a stable and improved expression of heterologous genes.

21.4 *In vitro* transcription and translation

The process of transcription takes place by the action of RNA polymerase on DNA inside the nucleus in a cell or close to the nucleoid in bacteria. Transcription can also be carried out outside the cell by mixing RNA polymerase and the four nucleotides to a cloned DNA in a tube under appropriate conditions. This process, known as *in vitro* transcription, results in RNA transcripts similar to the ones formed inside the cell. The enzyme RNA polymerase, obtained either from bacteria or from higher organisms, is a large enzyme containing several subunits. Purification of cellular RNA polymerase in an active form is extremely difficult. In contrast, most of the bacteriophages encode their own RNA polymerases, which are capable of transcribing genes at a high efficiency, are much simpler enzymes and are easy to purify – the reason being that they have evolved in such a way that they are capable of recognizing only certain bacteriophage promoters and transcribe whatever is downstream to that at a very high rate. *In vitro* transcription systems have been designed by utilizing this phenomenon. The commonly used bacteriophage RNA polymerases, which are used for *in vitro* transcription systems, include those from bacteriophage T3 and T7, which infect *E. coli* and SP6, which infects *Salmonella typhimurium*. Each of these RNA polymerases recognizes different promoter sequences, which can be as small as 21 bp in length. For *in vitro* transcription, these promoters are designed to be part of the special cloning vectors by placing them just upstream of the cloning sites for the DNA fragments to be transcribed. Transcripts specific for the cloned DNA are synthesized by incubating the cloned DNA, which is placed downstream of the above promoters, with the purified RNA polymerase from the bacteriophage and the precursor ribonucleotide triphosphates.

Melton and coworkers pioneered the technique of *in vitro* transcription in the year 1984. They constructed cloning vectors containing the promoter of the bacteriophage SP6 immediately upstream from a polylinker sequence. These SP6 cloning vectors were used under optimal conditions with bacteriophage SP6 RNA polymerase to achieve *in vitro* RNA synthesis. SP6 RNA polymerase, because of its specific recognition of the SP6 promoter, efficiently transcribes any sequence cloned downstream. They demonstrated the utility of the SP6-derived RNA as probes for nucleic acid blot hybridization. These probes are easy to prepare and significantly increase the sensitivity of nucleic acid hybridization methods. Furthermore, the SP6 transcription system can be used to prepare RNA substrates for studies on RNA processing and translation (Melton *et al.* 1984).

In a cell, ribosomes synthesize polypeptides by decoding the information carried by the mRNA through a process known as translation, which utilizes aminoacyl tRNAs and other accessory factors. The process of translation is not yet fully characterized biochemically as all the requirements for polypeptide synthesis are not known yet and very few of the components required for the process have actually been purified. Despite these limitations, mRNAs can be translated outside the cell, a process termed *in vitro* translation, if the given mRNA is incubated in a mixture of components, which supports translation. Generally, all the above components are isolated from *E. coli* cells (S-30 fraction), plants (wheat germ) or animals (rabbit reticulocytes). The first system for *in vitro* synthesis of proteins was developed in the laboratory of Paul Zamecnik in 1954 that contained amino acids, a donor molecule (ATP) for providing energy, microsomes and supernatant from high-speed ultracentrifugation of a raw cellular extract (Keller *et al.* 1954). Using this system, the investigators showed that peptide synthesis takes place on ribosomes and requires ATP, GTP and tRNA. Nirenberg and his collaborators, with the help of *in vitro* protein translation and using bacterial cell extracts, deciphered the genetic code for the first time (Matthaei *et al.* 1962). Shortly after the first complete protein, a coat protein of bacteriophage f2 was synthesized in a cell-free extract of *E. coli* by Nathans and coworkers (Nathans *et al.* 1962).

Irrespective of the *in vitro* translation system used, a given mRNA gets translated to the same extent and with the same efficiency owing to the universality of the genetic code. Synthesis of the new protein is monitored by adding radioactive amino acids in the translation mix as there are many types of protein molecules present in the translation mix. Therefore, radioactive labelling of all the newly synthesized proteins makes them easily detectable by autoradiography. Several systems having a combination of *in vitro* transcription and translation, which enables them to produce the encoded polypeptide directly from a given cloned gene, are commercially available now.

21.5 Expression in bacteria

Many of the strong and inducible promoters from bacterial systems have been well characterized and they have been used directly for the expression of heterologous genes in bacterial systems. Stanley Cohen and Herbert Boyer pioneered the technique of cloning and expressing genes in heterologous organisms with the help of recombinant DNA technology (Cohen *et al.* 1973). They ligated streptomycin-resistant plasmid RSF1010 from the bacterium *Salmonella typhimurium* with the tetracyclin-resistant plasmid pSC101 into *E. coli* and were able to isolate transformants, which were tolerant to both tetracycline and streptomycin (Cohen *et al.* 1973). A few years later, the biotechnology company Genentech successfully expressed the gene encoding human somatostatin in *E. coli*, which was the first report for the production of a human protein in *E. coli* (Itekura *et al.* 1977). Synthetic promoters, which are better suited for gene expression as compared with the natural ones, have also been created using some of the above promoters. One such example is the synthetic *tac* promoter, which utilizes the *lac* and *trp* promoters from *E. coli*, pL promoter from bacteriophage λ and gene 10 promoter from bacteriophage T7. The heterologous gene for bacterial expression should be free from introns, as bacteria lack the capacity to process them. These genes are cloned downstream of the appropriate promoter. To purify the expressed protein easily, the heterologous gene is cloned in such a way that it produces a fusion protein, which is affinity-purified, followed by cleavage with a protease to obtain the pure heterologous protein (Fig. 21.1). In the case of most bacterial expression systems, it is possible to obtain the heterologous protein at a concentration of 1–2% of the total soluble protein of the cell.

A large number of promoters, plasmids, transformation systems, mutant strains and selection systems are available for *E. coli*, a gram-negative bacterium, as it has been a favourite experimental organism with microbial geneticists and molecular biologists. Also, *E. coli* cells grow very easily, have a number of strongly induced promoters and a precisely controlled plasmid copy numbers, thus making it an excellent system for the expression of heterologous proteins. However, there are several drawbacks of the *E. coli* system, one of which being the sequestration of the expressed proteins in inclusion bodies within the cells, which are difficult to extract as they have low solubility. Also, proteins having disulphide linkages are difficult to obtain in their native form in *E. coli*. To overcome this problem, bacteria belonging to the genus *Bacillus*, which are gram-positive bacteria and have an efficient secretion system, are being exploited to obtain heterologous protein secreted into the growth media, thereby greatly simplifying the extraction of the desired protein.

Just like the gram-negative *E. coli*, gram-positive *Bacillus subtilis and B. licheniformis* are two species of bacteria whose complete genomes have been sequenced, and thus, a lot of information on several useful genes is available, which can be exploited to develop them into efficient hosts for heterologous expression of proteins. Apart from these two species, *B. brevis* and *B. megaterium* are other species, which are also popular systems

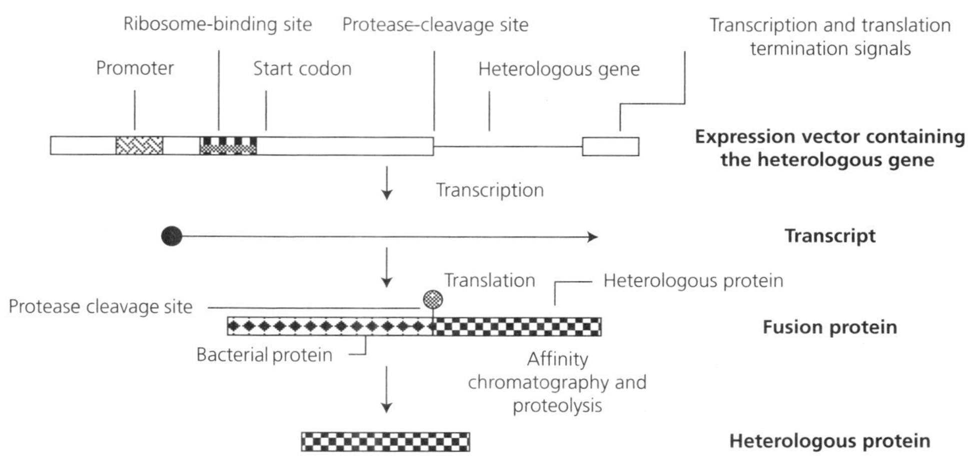

Figure 21.1 Expression of heterologous genes in bacterial systems and purification of the expressed protein. Bacterial expression vectors carrying a heterologous gene give rise to a fusion protein, consisting of bacterial polypeptide and the heterologous polypeptide. This fusion protein is purified from the bacterial extract by affinity chromatography. The fusion protein is then cleaved at the fusion site by a protease, whose recognition site is added to the bacterial polypeptide to obtain the heterologous protein.

for protein expression. However, a major problem with the *Bacillus* system is the secretion of extracellular proteases such as subtilisin, neutral protease, bacillopeptidase, etc. This potential problem has been dealt with by developing mutant strains, which secrete reduced levels of proteases.

21.6 Expression in yeast

In contrast to bacteria, a large number of post-translational modifications generally take place in most of the eukaryotic proteins, thus providing definite advantages to a eukaryotic expression system over a prokaryotic system. Along with their eukaryotic nature, yeast cells also have the advantage of being amenable to the application of powerful microbial genetic selection techniques on them. The first yeast to be used as a host for expression of foreign proteins was the baker's yeast *Saccharomyces cerevisiae*. Ronald Hitzeman's team pioneered hetrologous protein expression by successfully expressing the human leucocyte interferon protein (LeIF-D) in the *Saccharomyces cerevisiae* system (Hitzeman *et al.* 1981). This work resulted in a major patent titled 'Expression of polypeptide in yeast'. Since then, a host of foreign proteins have been successfully expressed in yeast. Several yeast strains such as *Saccharomyces cerevisiae, Kluvyeromyces lactis, Schizosaccharomyces pombe and Pichia pastoris* can be used for the expression of heterologous proteins. A shuttle expression vector, which is a plasmid capable of existing both in *E. coli* and in yeast cells, is used in most cases to achieve heterologous gene expression in yeast. These plasmids are created artificially by joining a small fragment of yeast DNA called 2-μm DNA, which can serve as the replication origin in yeast, into a bacterial plasmid, thus fulfilling the requirements of the cloning vector for replication in a yeast system. The heterologous gene is expressed under the control of a cloned yeast promoter like that of glyceraldehyde phosphate dehydrogenase. In addition, the expression vector also contains a yeast gene such as *leu*, which makes the transformed *leu*– strain of yeasts capable of growing on a leucine-deficient medium, thus providing a powerful method for selection of the transformed yeast cells. In certain cases, these plasmids get integrated into the yeast chromosome by homologous recombination, thereby ensuring the stable maintenance of the heterologous gene. Some yeasts such as *P. pastoris* are capable of growing by using only methanol as the source of carbon and energy, and thus can be grown very economically in large bioreactors. These can also be manipulated to secrete the heterologous protein into the medium, thus making them suitable for continuous purification of heterologous protein from the bioreactor.

21.7 Expression in insects

To achieve expression of heterologous genes in insects, the baculovirus system is very popular. Baculoviruses, which infect a large number of insects, are characterized by the high expression of the polyhedrin gene. The promoter of the polyhedrin gene, which becomes active 4–5 days after infection in the insect cells, can be used to express heterologous genes in a very efficient manner. *Autographa californica* multiple nuclear polyhedrosis virus (AcMNPV), which is a baculovirus, has been used extensively as an insect expression vector in both cultured cells and live larvae of the fall armyworm *Spodoptera frugiperda*. The pioneering work that resulted in the development of the baculovirus expression system for insects was carried out at Max Summers' laboratory in Texas, USA, in 1982. Smith and coworkers successfully expressed the human interleukin-2 protein using the AcNPV as an expression vector. By using specially constructed plasmids, the protein-coding sequences for interferon were linked to the AcNPV promoter for the gene-encoding polyhedrin. Using this system, they demonstrated that biologically active interferon was produced and greater than 95% was secreted from infected insect cells (Smith *et al.* 1983). Since then, more than 1000 prokaryotic and eukaryotic proteins have been expressed successfully using this system (Amer 2011). These expression vectors are based on the *E. coli* plasmid, which contains a fragment of AcMNPV to provide sites for homologous recombination, the polyhedrin promoter and the termination sequences. The heterologous genes, which are required to be expressed in an insect system, are inserted between the polyhedrin promoter and terminator sequences, followed by their introduction into insect cells or larvae, along with a full AcMNPV DNA. Homologous recombination between the AcMNPV sites present on the expression vectors and the full AcMNPV DNA results in the insertion of the heterologous gene into the AcMNPV DNA. The heterologous gene, which is now present between the polyhedrin expression signals following recombination, starts expressing at high levels in insect cells instead of the polyhedrin gene.

21.8 Expression in mammalian cells

To achieve optimum expression of the heterologous gene in a mammalian system, usually vectors derived from mammalian viruses such as SV40, polyomavirus, herpesvirus and papovavirus are used. Paul Berg and coworkers pioneered the expression of heterologous proteins in mammalian cells. SV40 was the first vector used for expression of a foreign gene in mammalian cells. Goff and Berg in 1976 prepared an SV40 expression system by replacing the entire late region of the viral DNA with λ-DNA fragment and used this hybrid virus to infect monkey kidney cell. To their disappointment, they observed that although the SV40-λ hybrid DNA replicated, but the λ-DNA sequences were not transcribed (Goff and Berg 1976). Building upon this result, the same group prepared another expression vector by replacing the major capsid protein (VP1) of SV40 by rabbit β-globin cDNA and used this recombinant genome for infecting CV1 monkey kidney cell cultures. They demonstrated that the globin-coding sequence was transcribed and cells produced substantial quantities of rabbit β-globin polypeptide (Mulligan *et al.* 1979). Human papovavirus BK, a double-stranded DNA virus that can exist as an extra-chromosomal element in cultured human cells, was also used early as an expression system for mammals. In these vectors, the pathogenicity genes had to be deleted, which is true for any vector derived from viruses, so that only the replication-associated genes and those for high copy number were retained. To drive strong expression of the transgene, a very strong promoter from mouse mammary tumour virus was incorporated in such vector systems. To incorporate a selection marker, the neomycin resistance gene, which provides resistance against the cytotoxic compound G418, was included in the vector system. These expression vectors can be used to express heterologous proteins in cultured mammalian cells. Although the levels of expression of heterologous genes in such vector systems are moderate, but in the case of proteins which can be correctly folded or processed only in human cells, it becomes important. Therefore, such vector systems find applications under special circumstances.

21.9 Expression in plants

To express heterologous proteins in plants, *Agrobacterium tumefaciens*, a soil bacterium with the unique capacity to introduce part of its resident plasmid into the plant genomic DNA, has been used widely. *Agrobacterium* causes a cancerous growth in the infected plant due to the introduction of a fragment of DNA called T-DNA derived from the plasmid called Ti plasmid. The Ti plasmid is modified by deleting the tumourogenic genes, which are responsible for causing tumours, from the T-DNA, thus making it incapable of causing tumours but still capable of transferring T-DNA to the plant. By inserting any heterologous DNA in the above plasmid, it can easily be introduced into the plant system. Strong promoters derived from plant viruses such as Cauliflower mosaic virus are utilized to obtain high expression levels of heterologous genes. Expression of heterologous genes in plants was first demonstrated in 1983, independently by three groups: Fraley *et al.* (1983), Bevan *et al.* (1983) and Herrera-Estrella *et al.* (1983), who expressed bacterial antibiotic resistance genes in plant cells by *in vitro* transformation technique using the *Agrobacterium*-mediated method. In the same year, Murai and coworkers reported expression of seed protein phaseolin from bean into sunflower and tobacco (Murai *et al.* 1983). A year later, Horsch *et al.* reported expression of the bacterial gene for neomycin phosphotransferase II (*npt II*) in tobacco (Horsch *et al.* 1984). In recent years, several heterologous genes have been expressed in plants to achieve either improved field performance of crops or inclusion of added qualities to the edible part, such as essential vitamins or even antigenic proteins, which can serve as vaccines against common diseases such as cholera. Expression of heterologous proteins in plants has become a very attractive proposition as the cost associated with growing such plants expressing heterologous proteins is quite low. Plant viruses are generally capable of infecting and multiplying in their host plant to a very high level in a very short time. Certain plant viruses such as Tobacco mosaic virus have been used for cloning heterologous genes by replacing some of the non-essential viral genes. These engineered viruses following infection of the host plants are capable of a very high-level expression of the heterologous gene in a matter of a few weeks in the infected plants. Introducing the gene to be expressed in a DNA-containing organelle is also possible. Plastids, which are present in 10–100 copies in any plant cell, with each having 10–100 copies of circular DNAs, is an example of one such organelle in plants. Incorporation of a heterologous gene into the plastid DNA results in a 'homoplastomic' condition, where all plastid DNA copies carry the heterologous gene. Such proteins are capable of showing very high expression levels, which can be close to 47% of the total soluble protein in a cell.

By using engineered plant viruses, in contrast to stable expression, proteins can also be expressed transiently in plant systems. In the case of most of the plant viruses, strong promoters, which bring about rapid expression of viral genes, are present in their small genomes. Such strong viral promoters present in the viral genomes are used for driving rapid expression of viral genes. These engineered viruses, following infection of the plant, trigger a transient production of heterologous proteins in the plant, sometimes to very high levels (10% of the leaf biomass) in a matter of weeks after inoculation. Several viruses such as Tobacco mosaic virus, Potato virus X, Cowpea mosaic virus and Tomato bushy stunt virus have been used to produce a variety of heterologous proteins in plants. In the case of certain viruses, for example, tomato bushy stunt virus, the heterologous protein can be 'displayed' on its capsid making it highly immunogenic. As a result, such systems are well suited for raising antibodies against the heterologous protein.

References

Amer HM. Baculovirus expression vector system: An efficient tool for the production of heterologous recombinant proteins. *Afr J Biotechnol* 2011;10:5927–5933.

Bevan MW, Flavell RB, Chilton M-D. A chimaeric antibiotic resistance gene as a selectable marker for plant cell transformation. *Nature* 1983;304:184–187.

Bishai WR, Rappuoli R, Murphy JR. High-level expression of a proteolytically sensitive diphtheria toxin fragment in *Escherichia coli*. *J Bacteriol* 1987;169:5140–5151.

Brinkmann U, Mattes RE, Buckel P. High-level expression of recombinant genes in *Escherichia coli* is dependent on the availability of the *dnaY* gene product. *Gene* 1989;85:109–114.

Buell G, Schulz M-F, Selzer G, Chollet A, Movva NR, Semon D, *et al.* Optimizing the expression in *E. coli* of a synthetic gene encoding somatomedin-C (IGF-I). *Nucleic Acids Res* 1985;13:1923–1938.

Chung CH, Goldberg AL. The product of the lon (capR) gene in *Escherichia coli* is the ATP dependent protease, protease La. *Proc Natl Acad Sci USA* 1981;78:4931–4935.

Cohen SN, Chang ACY, Boyer HW, Helling RB. Construction of biologically functional bacterial plasmids *in vitro*. *Proc Natl Acad Sci USA* 1973;70:3240–3244.

Fraley RT, Rogers SG, Horsch RB, Sanders PR, Flick JS, Adams SP, *et al.* Genetics expression of bacterial genes in plant cells. *Proc Natl Acad Sci USA* 1983;80:4803–4807.

Goff SP, Berg P. Construction of hybrid viruses containing SV40 and λ phage DNA segments and their propagation in cultured monkey cells. *Cell* 1976;9:695–705.

Grantham R, Gautier C, Gouy M, Jacobzone M, Mercier R. Codon catalogue usage is a genome strategy for genome expressivity. *Nucleic Acids Res* 1981;9:r43–r75.

Grantham R, Gautier C, Gouy M, Mercier R, Pave A. Codon catalogue usage and the genome hypothesis. *Nucleic Acids Res* 1980;8:r49–r62.

Guarente L, Lauer G, Roberts TM, Ptashne M. Improved methods for maximizing expression of a cloned gene: A bacterium that synthesizes rabbit β-globin. *Cell* 1980;20:543–553.

Hitzeman RA, Hagie FE, Levine HL, Goeddel DV, Ammerer G, Hall BD. Expression of a human gene for interferon in yeast. *Nature* 1981;293:717–722.

Horsch RB, Fraley RT, Rogers SG, Sanders PR, Lloyd A, Hoffman N. Inheritance of functional foreign genes in plants. *Science* 1984;223:496–498.

Itekura K, Hirose T, Cera R, Riggs AD, Heyneker HL, Boliver F, *et al.* Expression of *E. coli* of a chemically synthesized gene for the hormone somatostatin. *Science* 1977;198:1056–1063.

Keller EB, Zamecnik PC, Loftfield RB. The role of microsomes in the incorporation of amino acids into proteins. *J Histochem Cytochem* 1954;2:378–386.

Matthaei HJ, Jones OW, Martin RG, Nirenberg MW. Characteristics and composition of RNA coding units. *Proc Natl Acad Sci USA* 1962;48:666–677.

Melton DA, Krieg PA, Rebagliati MR, Maniatis T, Zinn K, Green MR. Efficient *in vitro* synthesis of biologically active RNA and RNA hybridization probes from plasmids containing a bacteriophage SP6 promoter. *Nucleic Acids Res* 1984;12:7035–7056.

Mulligan RC, Howard BH, Berg P. Synthesis of rabbit β-globin in cultured monkey kidney cells following infection with a SV40 β-globin recombinant genome. *Nature* 1979;277:108–114.

Murai NM, Kemp JD, Sutton DW, Murray MG, Slightorn JL, Merlo DJ, *et al.* Phaseolin gene from bean is expressed after transfer to sunflower via tumor-inducing plasmid vectors. *Science* 1985;222:476–482.

Nathans D, Notani G, Schwartz JH, Zinder ND. Biosynthesis of the coat protein of coliphage f2 by *E. coli* extracts. *Proc Natl Acad Sci USA* 1962;48:1424–1431.

Roberts TM, Kacich R, Ptashne M. A general method for maximizing the expression of a cloned gene. *Proc Natl Acad Sci USA* 1979;76:760–764.

Smith GE, Summers MD, Fraser MJ. Production of human β-interferon in insect cells infected with a baculovirus expression vector. *Mol Cell Biol* 1983;3:2156–2165.

22

Gram-negative bacterium *Escherichia coli* as host for recombinant DNA

H. K. Das

22.1 The host: *Escherichia coli*

The gram-negative enteric bacterium *Escherichia coli* has been the most commonly used host for recombinant DNA molecules.

22.1.1 History of *E. coli*

Joshua Lederberg, who won the Nobel prize in 1958 for his work on the genetics of *Escherichia coli* and related bacteria, has written a fascinating account of how *E. coli* gained the preeminent position it enjoys today (Lederberg 2004). 'Theodore Escherich, a German pediatrician, cultured "Bacterium coli" in 1885 from the faeces of healthy individuals, where it can be found almost universally in the large intestine, or colon, hence the "coli". It was renamed *Escherichia coli* in 1919 in a revision of bacteriological nomenclature, to lend more specificity to this form of Bacterium'.

'The strain K12 of *E. coli* was isolated at Stanford University in 1922 from human faeces, and was kept under that label for many years as a stock strain in the bacteriology department there. In the 1940s, Charles E. Clifton used it for studies of bacterial nitrogen metabolism, and his colleague Edward L. Tatum then borrowed it for his work on the biosynthesis of tryptophan from indole and serine. K12 entered the domain of genetics with Tatum's pioneering studies on the nutritionally deficient mutants in 1944'.

22.1.2 What is so special about *E. coli?*

Though some pathogenic strains of *E. coli* have been encountered, most *E. coli* strains have been found to be harmless, and in fact, human beings get benefited by their presence in the human gut. *E. coli* can be easily grown in synthetic media containing just glucose and salts with a doubling time of 50 min at 37°C. In rich media such as nutrient broth, *E. coli* would double in 20 min at 37°C. *E. coli* was being widely used by bacteriologists and biochemists during the early decades of the twentieth century as a favorite organism for metabolic studies.

Bacteria were then thought to be asexual and genetic studies involving recombination and segregation after cross-breeding of the kind being done with plants and animals were not possible. Lederberg discovered sexuality in *E. coli* K12 and demonstrated gene recombination and segregation after conjugation (Lederberg 1947). Sexuality is manifested through 'sex factor', which is nothing but a plasmid (see Chapter 17, Section 17.4.2.1). As Lederberg has stated, 'Soon the Mathew effect came into play: the very accumulation of knowledge, mostly concentrated on a single strain, "K12", made it more likely that it would be a prototype for still further studies.' Genetic markers in the form of gene mutations are essential for recombination and segregation studies. Tatum had produced nutritionally deficient mutants of *E. coli* K12 by using X-rays (Gray and Tatum 1944), but purification of the mutants was a laborious task. Lederberg devised the penicillin selection technique for purification of mutants (Lederberg and Zinder 1948). Lambda (λ) phage–mediated genetic transduction was subsequently demonstrated in *E. coli* K12 (Morse *et al.* 1956). This transduction was, however, limited to a 'cluster of genes for galactose fermentation' and is now referred to as 'specialized transduction'. It was found later that the bacteriophage P1 carries out 'generalized transduction' of *E. coli* K12 and can be used for unlimited transduction of diverse genes from around the genome.

Genetic transformation with purified DNA was first discovered in 1944 in the gram-positive bacteria *Streptococcus pneumoniae* and thus provided irrefutable evidence of DNA as the genetic material (Avery *et al.* 1944).

Such transformation has later been found to occur naturally in many gram-positive and gram-negative bacteria, but many other bacteria, including *E. coli,* were not amenable to such transformation. In 1970, however, it was discovered that a high amount of calcium ions could facilitate transformation by purified DNA (Mandel and Higa 1970) and this has now become a routine procedure for introducing recombinant plasmid DNA into *E. coli.*

As Lederberg has commented, 'Some of the most important of the scientific applications of K12 genetics have been in the field of gene regulation, and the elaboration of the concept of the 'operon', with work centered at the Pasteur Institute in Paris. The Nobel Prizes earned by Francois Jacob and Jacques Monod are but two of the dozen that by my account are affiliated with *E. coli'*. The overall scientific literature alluding to *E. coli* now encompass over 1,00,000 publications; Google reports almost 3 million hits with 'coli' on the World Wide Web.

Fred Neidhardt of the University of Michigan has aptly said, 'All cell biologists have at least two cells of interest, the one they are studying and *E. coli*'.

22.2 Plasmid cloning vectors for *E. coli*

The first report of expression in a foreign host of a cloned gene appeared in 1973. The cloned genes were of antibiotic resistance, the plasmid cloning vector was pSC101 and the foreign host was *E. coli* K12. The report has been described in Chapter 17, Sections 17.6.1 and 17.6.2. The plasmid cloning vector pSC101 has been described in Chapter 17, Section 17.4.2.7.

22.2.1 Desirable characteristics of a plasmid cloning vector

The following are considered to be the characteristics of a good plasmid cloning vector:

- The replication origin of the plasmid should be compatible with the host for which it is designed. It is advantageous to have the plasmid in multiple copies inside the host.
- There should be marker genes in the plasmid, such as those of resistance towards antibiotics, so that the cells which have been transformed with the plasmid can be easily selected.
- There should be unique cleavage sites for several restriction endonucleases on the plasmid, which could be used as cloning sites. The position of these cloning sites should be such that insertion of a segment of DNA in these sites would bring about a phenotypic change in the characteristics of the plasmid. This could be loss of resistance to an antibiotic or loss of expression of a gene whose product is an enzyme that normally carries out a reaction leading to an easily recognizable colour.

The plasmid pSC101, which was used as a cloning vector by Stanley Cohen (Chapter 17, Section 17.4.2.7) did have the first two characteristics, but the position of the unique *Eco*RI site which was used for cloning was not such that insertion of a segment of DNA in this site would bring about any phenotypic change in the characteristic of pSC101. They could detect insertion because they were cloning genes conferring resistance to new antibiotics.

One of the earliest cloning vectors that met all the essential criteria laid down above was pBR322 (Fig. 22.1). The plasmid pBR322 was constructed by Francisco Boliver and his colleagues in the laboratory of Herbert Boyer (Boliver *et al.* 1977b) and was derived from pBR313, which contained a ColE1-like replication origin (Boliver *et al.* 1977a). *E. coli* K12 and its derivatives could be transformed with pBR322 and the plasmid was found to exist in about 20 copies in *E. coli.* Conjugal transfer of pBR322 was undetectable. The genes for resistance towards ampicillin and tetracycline present in the plasmid pBR322 were sourced from

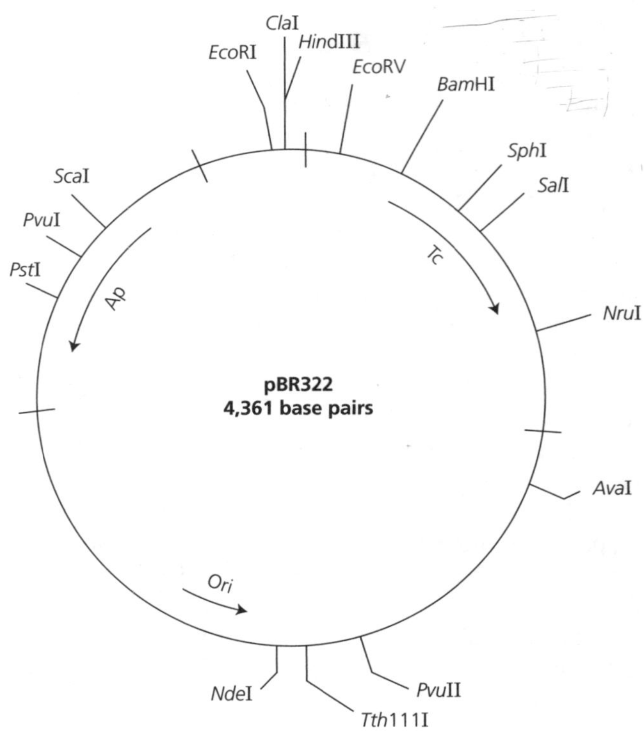

Figure 22.1 The plasmid vector pBR322. Positions of the genes responsible for resistance to ampicillin (Ap; taken from the plasmid pRSF2124) and tetracycline (Tc; taken from the plasmid pSC101), the replication origin (Ori; taken from the plasmid pMB1, which is similar to ColE1) and some unique restriction endonuclease sites are shown.

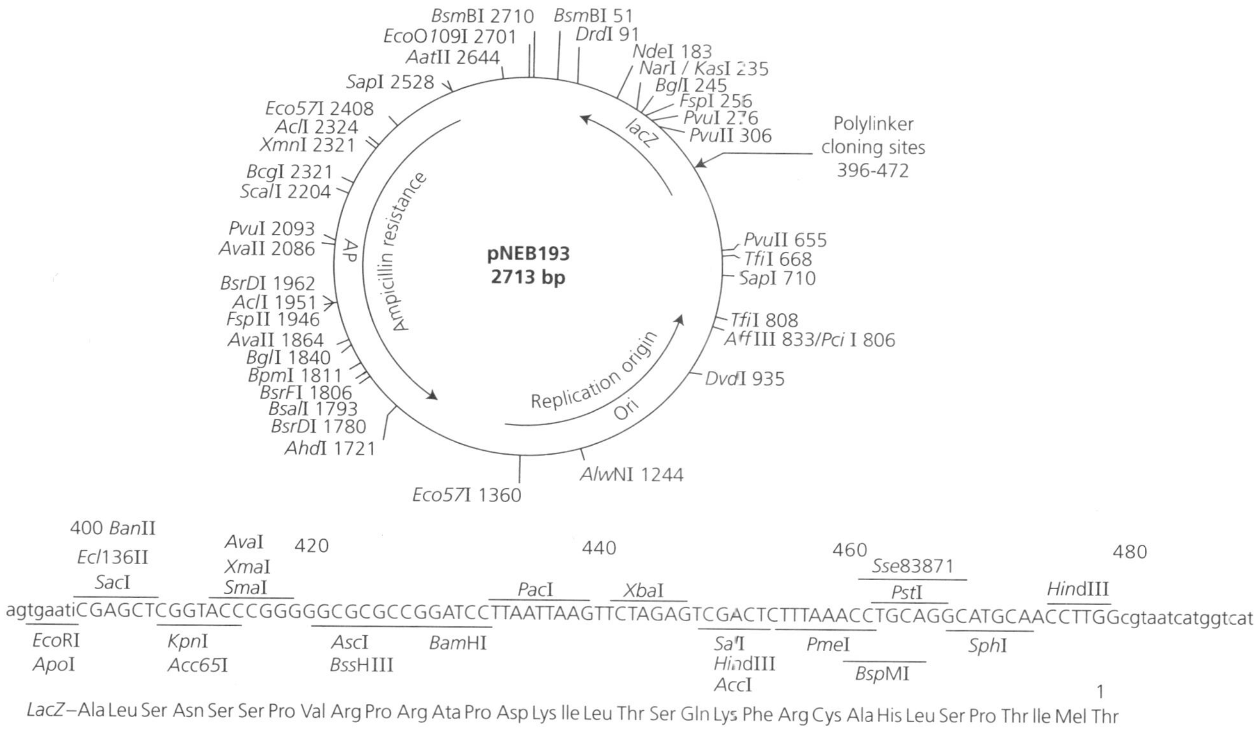

Figure 22.2 The plasmid vector pNEB193. The replication origin (Ori) and the gene for ampicilin resistance (Ap) have been taken from the plasmid pBR322. The gene for the enzyme β-galactosidase (*lacZ*) has been taken from *E. coli.* Some restriction endonuclease sites are also shown. The base sequence of the polylinker and the positions of the unique restriction endonuclease sites therein are shown at the bottom.

the plasmid pRSF 2124 (So *et al.* 1976) and the plasmid pSC 101 (Cohen *et al.* 1973), respectively. The plasmid vehicle pBR322 has several unique cleavage sites (as shown in Fig. 22.1), out of which those for *Pst*I, *Eco*RI, *Hin*dIII, *Bam*HI and *Sal*I are commonly used for cloning. The site for *Pst*I is in the middle of the β-lactamase gene, which confers ampicillin resistance. The site for *Hin*dIII is in the promoter region of the tetracycline resistance operon, while the sites for *Bam*HI and *Sal*I are in the structural genes that confer tetracycline resistance. An insertion in the *Pst*I site makes the cells sensitive to ampicillin and a similar phenomenon with respect to tetracycline occurs as a result of insertion in the *Bam*HI or *Sal*I site. A judicious test of resistance and sensitivity towards the two antibiotics after transformation with the plasmids would thus allow the selection of cells which have the plasmid cloning vector with a segment of foreign DNA as an insert in it.

22.2.2 Multicopy plasmids containing multiple cloning sites

A mutation in the copy number control region of the plasmid pBR322 has led to a class of plasmids that can exist in more than 1000 copies per cell of *E. coli.* Synthetic oligonucleotides representing the recognition sequences of various restriction endonucleases placed in succession have been introduced as multiple cloning sites (MCS). The MCS has been so designed as to form part of the β-galactosidase gene. Any insertion of a DNA fragment into any of the cloning sites would disrupt the β-galactosidase gene and can be detected by the lack of expression of the gene, which can be monitored by a colour reaction (blue to white). The pUC series of plasmids, developed by J Messing and coworkers (Messing *et al.* 1981; Yanish-Perron *et al.* 1985), belong to this class. The plasmid pNEB 193, a derivative of pUC 19, is an example (Fig. 22.2).

22.2.3 Compatibility of different plasmids among themselves

At times, it is desirable to study the *trans*-acting effect of the product of a gene on another gene. In such cases, it is advantageous to have one gene cloned in one plasmid vector and the other gene in a separate plasmid that can stably coexist with the first plasmid and is also not incompatible with it. Various plasmids have been classified into several incompatibility groups, implying that plasmids in the same incompatibility group cannot coexist. For example, ColE$_1$ and pMB1 are incompatible. However, the plasmid pSC101, the plasmids containing the P15A and F replicon are compatible with ColE1 and pMB1. Except for the F replicon, all other replicons so far mentioned are high copy number ones and all are suitable for use in *E. coli*.

22.2.4 Narrow host range versus broad host range plasmid vectors

The replicons mentioned above have a narrow host range and these are not stable in other gram-negative microorganisms such as *Pseudomonas, Azotobacter, Azospirillum, Rhizobium,* etc. Replication origins from broad host range plasmids such as RK2, RP1, RP4 and RSF1010 have been employed for the construction of vehicles meant to be used in such organisms. These replicons, nevertheless, are stable in *E. coli* and are compatible with each other and the replicons mentioned in the previous paragraph. The vector pRK404, which is derived from RK2, is an example (Fig. 22.3).

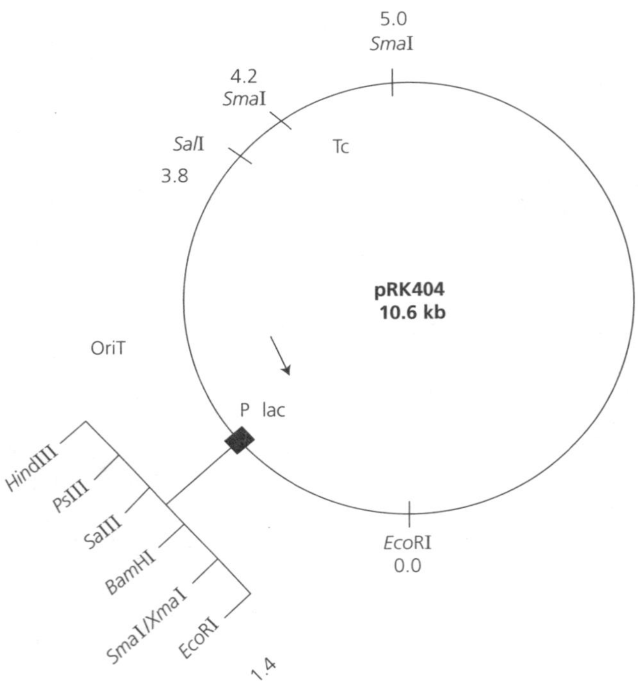

Figure 22.3 The wide host range cloning vector pRK404. The replication origin is from the plasmid RK2. *Hind*III, *Pst*I, *Sal*I and *Bam*HI are the unique cloning sites. Tc represents the tetracycline resistance gene complex.

22.3 Bacteriophages as cloning vectors

In addition to plasmids, bacteriophages have also been used as cloning vectors.

22.3.1 Cloning vectors based on the bacteriophage lambda (λ)

Derivatives of the genome of bacteriophage lambda have been constructed to serve as cloning vectors. Transfection or transduction is used to introduce such vectors into *E. coli*. Two properties of the lambda genome make it suitable for use as a vector:

- Only about 50% of the 50 or so genes of lambda are essential for its replication and for lysis of the host cell. Most of these non-essential genes are located together in a cluster around the middle of the genome (Fig. 22.4).
- Lambda genome is packaged inside the phage head by what is known as the 'head-full mechanism'. This means that not only there is an upper limit of the amount of DNA that is packaged inside the phage head, but also there is a lower limit. Effective packaging takes place only when a minimum amount of DNA, that is, about 35,000 base pairs (bp) (35 kilobases; kb), is available. The upper limit happens to be 53 kb.

22.3.1.1 Insertion-type lambda (λ) vectors

A lambda cloning vector can therefore be constructed by deletion (*in vivo* or *in vitro* by restriction deletion) of a part of the non-essential region such that the remainder is not less than 35 kb. Other mutations are also introduced such that restriction sites in the essential regions are eliminated. A segment of foreign DNA can be cloned in a unique restriction site in the non-essential region, the only condition being that the vector and the insert together would not be more than 53-kb long. Such vectors are termed 'insertion vectors'. Some of the Charon vectors are examples of this type of vector.

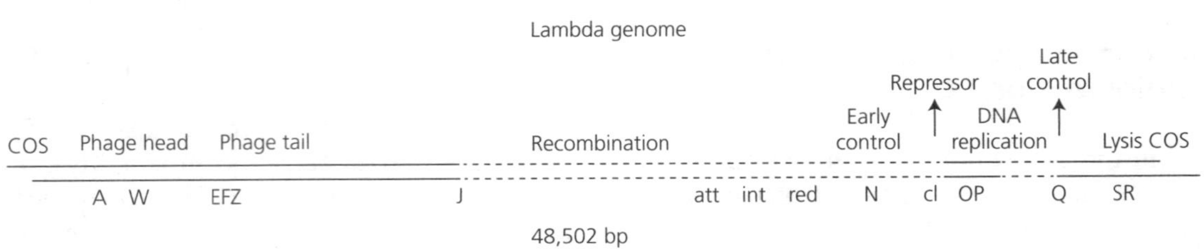

Figure 22.4 The bacteriophage lambda genome. The broken lines represent the portions that are not essential for the replication of the genome and lysis of the host cell.

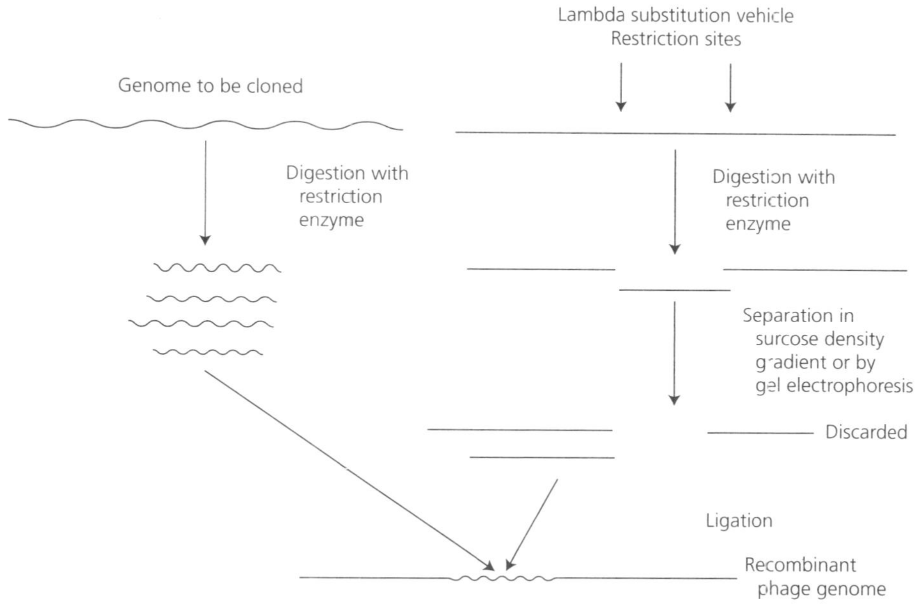

Figure 22.5 Scheme for cloning in a lambda substitution vehicle.

22.3.1.2 Substitution-type lambda (λ) vectors

The second type of lambda vector is of the substitution type, the examples being the lambda gt vectors and the EMBL vectors. These vectors have two *Eco*RI sites or two *Bam*HI sites in the non-essential region. On digestion with *Eco*RI at least three pieces are produced, two terminal ones containing the essential regions and a central piece(s) containing the non-essential genes. The central piece(s) is separated out by sucrose density gradient sedimentation and replaced by the foreign segment to be cloned (Fig. 22.5). The limit of the size of foreign DNA that can be cloned in the lambda gt vectors is 1–14 kb and in the Charon 4 vector, it is 8.2–22.2 kb. Such a replacement cloning vector has an advantage over the insertion vectors. The terminal pieces themselves if joined by DNA ligase do not make up 35 kb and hence cannot be packaged. Packaging occurs only when a segment of foreign DNA gets cloned between the two terminal pieces of the vector and hence no separate selection step for recombinant molecules is necessary.

22.3.2 Cloning vectors based on the bacteriophage M13

Double-stranded derivatives of the filamentous coliphage M13 have also found extensive use as cloning vectors. A series of such vectors (M13 mp19 is an example) have been developed by J Messing and coworkers (Messing *et al.* 1981; Yanish-Perron *et al.* 1985). Inside the *E. coli* host, the M13 DNA exists as double-stranded circles, but the released virions contain single-stranded circles. The DNA fragment is cloned in double-stranded M13 DNA isolated from inside the *E. coli* cell and is put back into *E. coli* by transformation. The virions containing their single-stranded DNA genome are then allowed to mature and get continuously secreted out of the *E. coli* cells without killing the cells. Single-stranded DNA can now be isolated from the virions in the supernatant and be directly subjected to base sequencing.

22.4 Phagemids as cloning vectors

A phagemid is a hybrid of a plasmid and a filamentous coliphage that can be propagated in either form. The coliphage could be either of the three virtually identical phages, M13, fd or f1. These are male-specific phages that contain single-stranded circular DNA as their genome. Upon infection of *E. coli* by the bacteriophage, double-stranded DNA is first formed as the replicative intermediate. Finally, single-stranded DNA is packaged into the virion. Both the replication origins of the plasmid and the coliphage are incorporated in the phagemid. The auxiliary replication functions necessary in *trans* for the coliphage replication are not, however, incorporated in the phagemid. Hence, replication from the coliphage origin can take place only in the presence of a helper phage. Otherwise, replication takes place from the plasmid replication origin. The pBLUESCRIPT phagemids have both the ColE1 (pMB 9–like) origin and the filamentous phage f1 origin (Fig. 22.6). The cloning is done in any of the MCSs in the double-stranded circular DNA of the plasmid form of the vector. This is introduced into *E. coli* by transformation and the synthesis of single-stranded DNA from the phage f1 origin is induced by superinfection with a helper phage. The single-stranded

DNA formed is packaged into the phage rods because of the presence of the phage-packaging signals in the phagemid. Direct base sequencing can be undertaken using the single-stranded DNA isolated from the virions secreted from the *E. coli* cells. Depending on the orientation of the f1 replication origin in the phagemid, either the (+) strand or the (–) strand of the phagemid is replicated in the presence of the helper phage. The pEMBL phagemids are similar to the pBLUESCRIPT phagemids. Phagemids that have replication origins of pUC plasmids and of the M13 bacteriophage have also been developed and are available under the trade name of LITMUS vectors. Replication in the circular double-stranded plasmid form uses the pUC replication origin. Cloning is done in the MCS located in the *lacZ* gene and the usual blue/white selection is available. Here also, replication of the single-stranded phage DNA form is induced by superinfecting with the M13 helper phage.

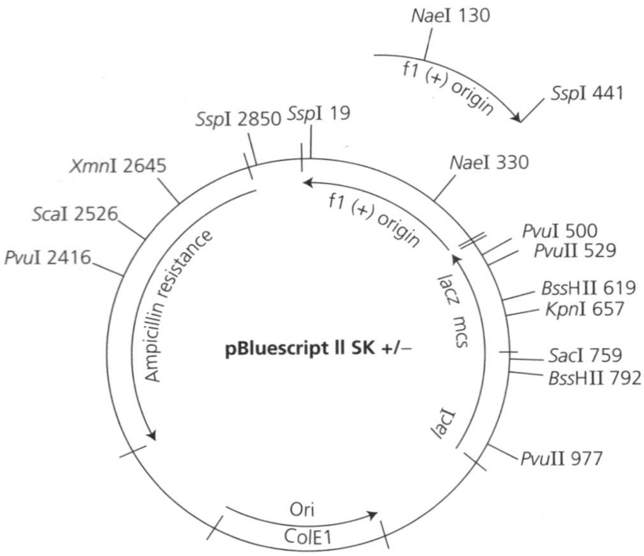

Figure 22.6 A pBLUESCRIPT phagemid. It contains the replication origin of bacteriophage f1 (both orientations shown) and also the replication origin of the plasmid ColE1. It also contains the genes for ampicillin resistance, *lac* repressor (*lacI*) and β-galactosidase (*lacZ*). The multiple cloning sites (MCSs) and positions of some restriction endonuclease sites are shown.

22.5 Cosmids as cloning vectors

Plasmid vectors are not suitable for cloning DNA fragments much larger than their own size, as the transformation frequency falls beyond acceptable limits and cloned fragments or their parts very often get deleted. Takagi and coworkers (Fukumaki *et al.* 1976; Umene *et al.* 1978) observed that the presence of the cohesive end site cos λ from the bacteriophage lambda DNA in a plasmid allows it to be packaged *in vivo* into virus

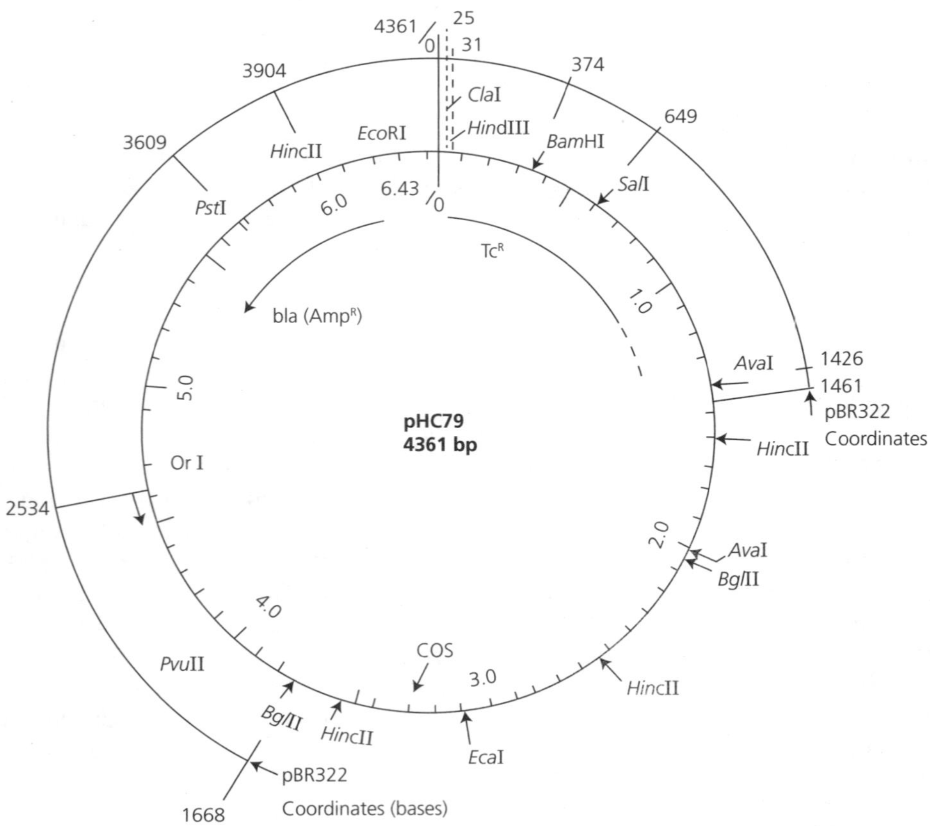

Figure 22.7 The cosmid cloning vector pHC79. Constructed by cloning the cohesive end site cosλ into pBR322. The coordinates of pBR322 are marked, as also those of the cohesive end site cosλ. The positions of the replication origin and of resistance functions against ampicillin and tetracycline can be seen.

particles. The interesting finding was that the *in vivo* packaging mechanism would select DNA molecules of the full size of the lambda genome (~48.5 kb). Making use of this finding, cosmid vectors were first developed in 1978 by J Collins and B Hohn (1978) to facilitate cloning of larger DNA fragments in plasmids. Extracts of lambda lysogens have been successfully used for *in vitro* packaging of lambda capsids. An example of a commonly used cosmid is pHC79 (Fig. 22.7), which is nothing but pBR322 containing the cohesive end site cos λ and which can accommodate up to 45-kb-sized inserts. A great advantage of such a cosmid vector is that gene libraries consisting of a smaller number of clone members can span the whole genome of an organism. For example, the genome of *E. coli* can be accommodated in just 120 cosmids. Other advantages are that large genes can be studied intact and genetic linkage studies can be carried out at the molecular level. An important practical advantage of a cosmid is that background molecules which do not have the inserts or have smaller inserts are eliminated during packaging. This is not possible to achieve with plasmid cloning vectors. Besides, the frequency of transduction of the lambda capsids with an *in vitro* packaging extract is much higher than the transformation frequency of plasmids. The following are the steps for construction of a cosmid library:

1. Cleavage of the genome by partial digestion with a restriction endonuclease
2. Sizing of the fragments by gel electrophoresis or velocity centrifugation
3. Cleavage of the cosmid vector and treatment with phosphatase to minimize polycosmid formation
4. Ligation of the genomic DNA and the cosmid DNA
5. Packaging the ligated DNA into infectious phage particles
6. Transduction into *E. coli*

As mentioned above, a cosmid vector can accommodate up to about 45 kb of an insert DNA. A complete cosmid library of the human genome would still require about 2,50,000 clones, which is indeed an awfully large number.

22.6 The PAC cloning system

The PAC cloning system (PAC) has been developed from the bacteriophage P1 by Nat Sternberg and coworkers (Sternberg 1990; Sternberg *et al.* 1994) in an effort to accommodate large DNA (90–100 kb) inserts. Bacteriophage P1 heads can accommodate 110 kb of DNA, in contrast to only 53 kb of DNA that can be accommodated in the bacteriophage lambda (λ) head. A 50,000-member human DNA library has been constructed using PACs (Sternberg *et al.* 1990). Cloning in a P1-derived vector is similar in principle to cloning in cosmids. The large DNA fragments to be cloned are ligated to the cleaved P1 vector that contains various genetic elements of the P1 bacteriophage and a plasmid replicon/partition system. The linear recombinant molecules are then packaged *in vitro* into P1 particles and *E. coli* cells are transfected. A commonly used P1 vector is pAd10sacBII, which contains two P1-encoded replicons, a positive selection system and an adenovirus stuffer fragment. A PAC vector has been constructed by deletion of the adenovirus stuffer and insertion of a pUC plasmid derivative in its place. Instead of packaging and transfection, the recombinant PAC vector is also introduced into *E. coli* by electroporation. Such a PAC vector can accommodate inserts of 130–150 kb. The recombinant PACs exhibit lower rates of rearrangement and chimerism than yeast artificial chromosomes (YACs). Recombinant PACs are also more tolerant of repetitive and palindromic DNA sequences than the YACs.

22.7 Bacterial artificial chromosomes

Bacterial artificial chromosomes (BACs) were developed by Mel Simon and coworkers (Shizuya *et al.* 1992) to accommodate ~300 kb of DNA and a human chromosomal library has been constructed using such vectors (Kim *et al.* 1996). BACs are based on the fertility factor (F factor) of *E. coli*. The F plasmid, a ~100-kb circular double-stranded DNA, is present in an *E. coli* cell in only 1–2 copies. The naturally occurring F2 factors consist of up to 25% of the *E. coli* genome integrated into the basic F factor and are very stable. This characteristic of the F factor contributes to the ability of BACs to accommodate very large amounts of external DNA, to the extent of 300 kb. Besides, BACs do not have any packaging requirement or limitations, as is the case with vectors derived from bacteriophages λ and P1. Synthetic BAC vectors, which are only ~7.5-kb double-stranded DNA circles, contain the replication origin oriS and the gene *repE* of the F plasmid, which are responsible for initiation and proper orientation of replication of the BAC vector, respectively. The *parA* and the *parB* genes of the F plasmid, which ensure efficient segregation of the F factor into the daughter *E. coli* cells after its replication, are also incorporated in BAC vectors. BAC vectors also contain MCSs, a selectable marker in the form of antibiotic resistance and colour-based identification (*lacZ* complementation system) of recombinants carrying inserts. Recombinant BACs have been found to exhibit a lower level of rearrangement and chimerism of the cloned DNA sequence than exhibited by YACs. The cloning of DNA in BACs is also done in the same way as in a plasmid, by linearizing the vector with a restriction endonuclease, treating with phosphatase and then ligating with the DNA fragments to be cloned. Here also, *E. coli* has to be transformed by electroporation because of the large size of the recombinant BAC.

22.8 Construction of a library of genes in *E. coli*

The most commonly used host for constructing a gene library has been *E. coli*.

22.8.1 Conversion of the DNA of the organism in question into small fragments: The first step in the construction of a library of genes

It has to be appreciated that there is no method available for conversion of a larger DNA into smaller fragments, which would always leave all the genes intact. The way to get around this problem is to create overlapping fragments, thus ensuring the availability of an array of fragments, so that each and every gene would be in the intact state in one or other fragment. The principle is illustrated in Fig. 22.8. The library would have all the cleavage products cloned, and hence, each and every intact gene would be available in some cloned form.

22.8.1.1 Cleavage of DNA by mechanical shearing

The technique that was used very early to attain such cleavage was mechanical shearing of the DNA (Maniatis *et al.* 1978). As has been mentioned in Chapter 17, Section 17.1, before 1959, people were at a loss to reconcile the wide difference in the size of DNA of bacteriophage T2 and T4 reported by different workers (Levinthal 1956; Butler *et al.* 1959). It was Peter F Davison who established conclusively for the first time that this inconsistency in the size of DNA was because naked DNA easily underwent double-strand breaks by hydrodynamic shear (1959) produced even by shaking the DNA solution [see Fig. 17.1(a)]. The reduction in the size of the DNA was indeed appreciable if the DNA solution was forced through a No. 27 needle of a syringe [see Fig. 17.1(b)]. This has now become a standard procedure for shearing large DNA molecules, the other procedures being subjecting the DNA solution to vortex mixing and to ultrasonic treatment. As PF Davison had observed (1959), 'The theoretical calculations show that the rupturing stress applied by the viscous forces is maximal in the center of the chain and increase as the square of the length of the molecules. The molecules should thus break roughly in half when the critical shear is applied, and a very long molecule should be halved repeatedly until a fairly uniform population of chains stable to the shear gradient is obtained'.

The only problem with shearing DNA mechanically is that the ends created are not cohesive.

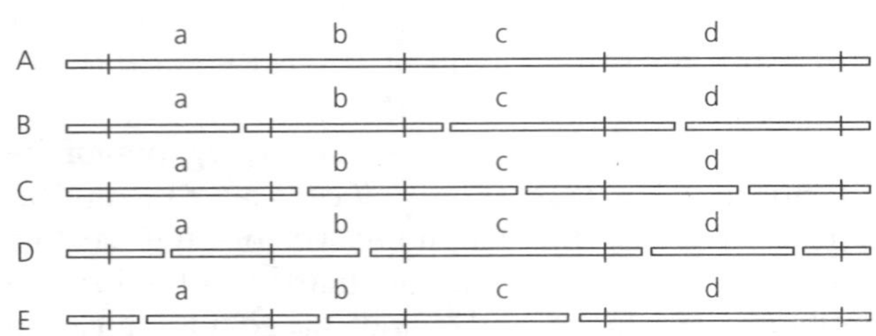

Figure 22.8 The principle of creating overlapping smaller fragments from a large DNA for ensuring the availability of some fragments which would have each and every gene in the intact state. The original DNA (A), which has four genes (a, b, c and d), is cleaved at variable sites, resulting in products B, C, D and E. Product B has only 'b' gene intact, product C has only 'a' gene intact, product D has only 'c' gene intact and product E has only 'd' gene intact.

22.8.1.2 Enzymatic addition of 'cohesive' ends to 'blunt' ends

'Cohesive' ends can be added to 'blunt' ends by the use of terminal deoxynucleotidyl transferase isolated from calf thymus (see Chapter 9, Section 9.8). This enzyme catalyses the successive addition of mononucleotides to the 3′-end of DNA from nucleoside triphosphates. Jensen *et al.* (1971) separately introduced several thymidylate residues to the 3′-ends of one batch of T7 DNA and deoxyadenylate residues to another batch of the same DNA. Equimolar amounts of the two batches of DNA were mixed and annealed. Catenanes measuring two or three times the length of normal T7 DNA were formed, as visualized by electron microscopy. The inference was that the polydA and polydT tails have served as 'cohesive' ends and hydrogen bonds have been formed between the complementary tails during annealing. DNA ligase (see Chapter 9, Section 9.7 and Chapter 17, Section 17.3.1) could not, however, join the hydrogen-bonded single strands of these catenanes. This was obviously because the lengths of the polydA and the polydT chains were not exactly the same; hence, there were gaps between the ends of the two tails after annealing (see Fig. 17.4 for a diagrammatic representation of the situation).

22.8.1.3 Cleavage of DNA by the action of restriction endonucleases

Restriction endonucleases cleave DNA at specific sequences (see Chapter 20, Section 20.1). As the sequence of bases in natural DNA is seldom random, sites of cleavage of restriction endonucleases are also not likely to be randomly distributed (*cf.* Fig. 20.3). This is particularly true for endonucleases having recognition sequences of six bases and more. Besides, a particular site of a particular restriction endonuclease might exist right inside the sequence of the gene of interest; hence, digestion with that endonuclease would disrupt the gene. These problems have been circumvented by only partially digesting the DNA with restriction endonuleases that have recognition sequences of only four bases, thus having sites of cleavage that are more frequent and are more random. The endonuclease most often used is *Sau*3AI, whose recognition sequence is 5′-GATC-3′. *Mbo*1 and *Dpn*II are isochizomers of *Sau*3AI, but these two are blocked by *dam* methylation. These three endonucleases cleave DNA of close to 50% GC content around every 250 bp (theoretical probability is once every 256 bp) if digestion is carried to completion. The principle operating behind achieving our objective by partial digestion with such an enzyme is illustrated in Fig. 22.9. Fragments cleaved by these enzymes can be cloned in vectors in restriction endonuclease sites *Bam*HI, *Bgl*II and *Bst*YI, which have six-base recognition sequences: 5′-GGATCC-3′, 5′-AGATCT-3′ and 5′-PuGATCPy-3′, respectively.

22.8.2 Desirable size of the DNA fragments meant for construction of a library of genes

An ideal library of genes would have all the genes of the organism in the intact state, but would not need to have too large a number of library members to be screened for obtaining a clone of a single-copy gene. It is obvious that the larger is the size of the DNA fragment used for construction of the library, the smaller would be the number of members necessary in the library to represent all the genes of the organism in question. Clarke and Carbon (1976) have derived a precise

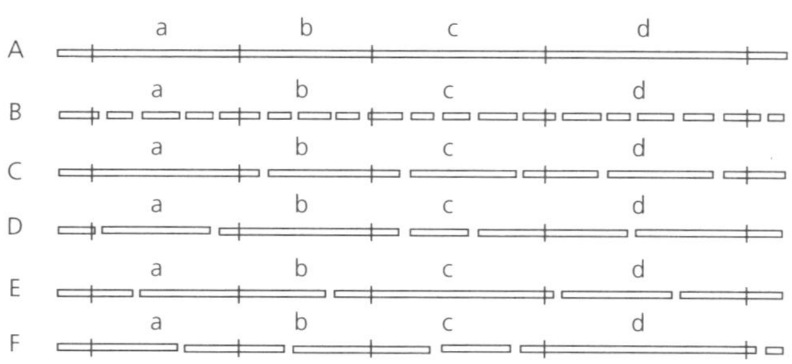

Figure 22.9 The principle of creating overlapping fragments by partial digestion with a restriction endonuclease which has more frequent recognition sites, thus ensuring the availability of some fragments which would have each and every gene in the intact state. The original DNA (A) has four genes (a, b, c and d) and has a large number of recognition sites of the restriction endonuclease. Complete digestion of the DNA with this endonuclease would yield product B, which consists of a large number of very small DNA fragments. The same DNA (A) would yield products C, D, E and F when partially digested with the same restriction endonuclease. Product C has 'a' gene intact, product D has 'b' gene intact, product E has 'c' gene intact and product F has 'd' gene intact.

relationship between the number of members (N) necessary in the library and the fractional proportion of the genome (f) in a single recombinant:

$$N = \ln(1 - P) / \ln(1 - f)$$

where P is the probability of a DNA sequence being present in the library. Thus, for a library of a 3×10^9-bp mammalian genome, if we want to limit the number of recombinants within, say, 5×10^4, the size of the DNA fragments to be cloned at 99% probability would be given by the following expression:

$$\ln(1 - f) = \ln(1 - P)/N$$

or $\ln(1 - [s/3 \times 10^9]) = \ln[(1 - 0.99)/5 \times 10^3]$ where s is the size of the DNA fragment

or $s = \sim 3 \times 10^5$ bp. Thus, we should aim at ~300-kb DNA fragments.

However, with the genome of a bacterium such as *E. coli*, which is only 4.6×10^6 bp, one can make do with $\sim 5 \times 10^3$ recombinants containing as small as 5-kb DNA fragments as inserts. Similarly, the number of recombinants needed with 45-kb DNA fragment inserts to represent the total genome of *E. coli* at 99% probability would be just about 533.

22.8.3 Choice of cloning vector for construction of the library

Various cloning vectors suitable for use in *E. coli* have been described in Sections 22.2 to 22.7. As has been mentioned in Section 22.5, ordinary plasmid vectors are not suitable for cloning DNA fragments much larger than their own size, as the transformation frequency falls beyond acceptable limits and cloned fragments or their parts very often get deleted. The size of the DNA fragments to be cloned is thus an important determining factor in the choice of the cloning vector. Phage lambda–based cloning vectors such as Charon 4 can be used to clone 8- to 22-kb-long DNA fragments, while cosmids can accommodate 45-kb-long DNA fragments. However, PAC vectors can accommodate about 100-kb DNA fragments and BAC vectors can be used to clone as large as 300-kb DNA fragments.

22.9 Non-fusion heterologous gene expression in *E. coli*

Eukaryotic genes have introns. Bacteria, however, lack the capacity to process introns. Hence, the heterologous gene, if from a eukaryotic source, for expression in a prokaryotic host, would have to be made free from

***lac*UV5 promoter**

−35 *Hpa*II −10 +1
CCAGGCTTTACACTTTATGCTTCCGGCTCGTATAATGTGTGGAATTGTG

***trp* promoter**

−35 *Taq*I −10 +1
GAGCTGTTGACAATTAATCATCGAACTAGTTAACTAGTACGCAAGTTCA

***tac*I promoter**

−35 *Taq–Hpa* −10 +1
GAGCTGTTGACAATTAATCATCGGCTCGTATAATGTGTGGAATTGTG

Concensus sequence of sigma70 promoter

TTGACA (16-18 bases) TATAAT

Figure 22.10 The *tac*I promoter has been obtained by fusing the −35 region of the promoter of the tryptophan operon (*trp* promoter) from *Escherichia coli* and the −10 region of the promoter of the lactose operon (*lac* promoter) from *E. coli.* The transcription initiation site is marked as +1. There is a *Hpa*II restriction site in between the − 10 region and the − 35 region of the *lac*UV 5 promoter, while there is a *Taq*I site in between the − 10 region and the − 35 region of the *trp* promoter. Both the restriction sites, *Hpa*II and *Taq*I, have a 5'-overhang of the bases GC, so ligation of these two restricted sites is straightforward.

introns. The cDNA is the answer and has been dealt with in Chapter 30. Such an intronless gene is then cloned downstream of the appropriate promoter. Several strong and inducible promoters from *E. coli* and its viruses have been used directly to express heterologous genes in *E. coli* without going through any fusion with any other gene or part of other gene. Examples are *lac* promoter from *E. coli, p*L promoter from bacteriophage lambda and gene 10 promoter from bacteriophage T7. Some of the promoters have also been used to create synthetic promoters, which are better suited for gene expression as compared with natural ones. For example, synthetic *tac* promoter utilizes the *trp* and *lac* promoters from *E. coli* (Boer *et al.* 1983). Nucleotides upstream of position −20 with respect to the transcriptional start site, including the −35 sequence of the *trp* promoter, was fused with nucleotides downstream of position −20, including the −10 sequence of the *lac* promoter (Fig. 22.10). The *tac* promoter is ~11 times and ~3 times more efficient than the *lac* promoter and the *trp* promoter, respectively. The *tac* promoter can be repressed by the *lac* repressor and can be derepressed by isopropyl thiogalactoside (IPTG).

22.10 Production of fusion proteins from heterologous genes in *E. coli*

The amount of heterologous proteins produced by *E. coli,* in spite of the strong promoter used, is not high, generally not more than 2% of the total soluble protein. Standard methods of protein purification involving multiple steps result in further reduction in the yield of the heterologous protein. The heterologous gene has, therefore been fused to a gene coding for a specific ligand so that a fused protein is produced. A single-step purification method has been developed based on affinity of the ligand towards a solid matrix. The affinity tag comprising six consecutive histidine residues, added either to the NH_2 terminus or the COOH terminus, is most often used in *E. coli* (Daabrowski *et al.* 2000; Busso *et al.* 2005). The hexa-histidine (His) tag has high affinity towards metals such as nickel and cobalt. The tagged protein is purified by immobilized metal affinity chromatography (IMAC), the most commonly used affinity column consists of nickel-agarose. The column-bound His-tagged recombinant protein is eluted at low pH or with imidazole (Loughran and Walls 2011). A less often used affinity tag is 218-residue glutathione S-transferase (GST) from *Schistosoma japonicum* (Smith and Johnson 1998), which has high and selective affinity for glutathione. The GST-tagged heterologous recombinant protein is eluted from the glutathione containing affinity resin, glutathione Sepharose 4B by reduced glutathione. Another affinity tag is the maltose-binding protein (MBP) coded by *malE* gene of *E. coli* (Maina *et al.* 1988). Amylose-agarose resins are used to bind the MBP-tagged heterologous recombinant protein and maltose is used to elute it. The fusions are engineered in such a way that there is a sequence Ile–Glu–Gly–Arg between the tag and the recombinant protein. The serine endopeptidase, factor Xa (Jenny *et al.* 2003) cleaves after Arg of the above-mentioned tetrapeptide sequence. The tagged recombinant protein, after purification through the affinity column, is treated with factor Xa to cleave the tag from the recombinant protein.

22.11 Vectors for non-fusion heterologous gene expression in *E. coli*

There are a large number of plasmid vectors available for non-fusion heterologous gene expression in *E. coli.* Some of these vectors are described next.

22.11.1 Marsh ptac-85 plasmid vector

An example of plasmid vectors for heterologous expression of non-fusion proteins in *E. coli* is the ptac-85 (Marsh 1986). This vector makes use of the *tac* promoter and a Shine/Dalgarno sequence from the *lacZ* gene. An unique *Nco*I site (CCATGG) in the plasmid includes the translation initiator ATG 10 bp downstream of the Shine/Dalgarno sequence. There is a *Bam*HI just upstream of the *Nco*I site and a *Sal*I site just downstream of it (Fig. 22.11). Any DNA fragment with a blunt end and lacking the promoter, the Shine/Dalgarno sequence and the initiator ATG can be joined to the plasmid DNA after cleavage with *Nco*I. The plasmid vector ptac-85 has been used successfully in the bacterial expression of human IgE fragments.

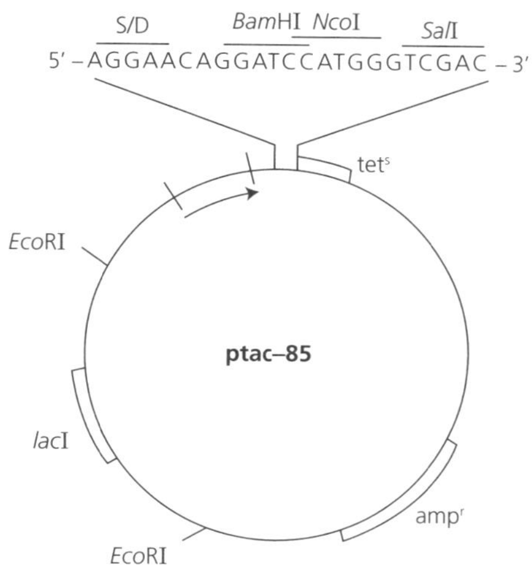

Figure 22.11 The plasmid vector ptac-85. The *tac* promoter and its polarity are shown by the arrow. The *lac*I repressor controls the *tac* promoter. Transcription termination occurs at the *tet* terminator in the 3′-segment of the *tet^r* gene, represented by *tet^s*. S/D represents the Shine/Dalgarno sequence and amp^r represents the β-lactamase gene.

22.11.2 Tabor and Richardson T7 promoter plasmid vector

Tabor and Richardson (1985) cloned the gene for phage T7 RNA polymerase in pBR322 under the control of the lambda P_L promoter and found that after induction, T7 RNA polymerase constituted 20% of the soluble protein of *E. coli*. This and the specificity of T7 RNA polymerase for its promoters prompted them to use the T7 RNA polymerase–promoter combination to develop a plasmid vector system for overproducing heterologous proteins. The system consisted of two compatible plasmids, pGP1-2 and pT7-1 (Fig. 22.12). The *kan*^{R} plasmid pGP1-2 has the P-15A replication origin and the gene for T7 RNA polymerase under the lambda P_L promoter along with the gene for the temperature-sensitive lambda repressor *c*I 857 under the *lac* promoter. The *amp*^{R} plasmid pT7-1 has the ColE1 replication origin and a multiple cloning site for inserting the heterologous gene to be expressed under the phage T7 promoter ϕ10. The plasmid pT7-7 is a later version of pT7-1.

22.11.3 Moffatt and Studier pET plasmid vectors based on T7 promoter

Moffatt and Studier (1986) had also designed an expression system based on bacteriophage T7 RNA polymerase. The gene for T7 RNA polymerase was inserted into the *E. coli* chromosome under the control of the *lac* promoter and the *lac* operator. The *lac*I gene was also incorporated into the chromosome. Production of T7 polymerase would be induced only when a compound like lactose or isopropyl thiogalactoside (IPTG) would be added to the *E. coli* cell culture. In addition to its own replication origin, the plasmid vector (pET vector) also contained the f1 replication origin. This made it possible to produce single-stranded DNA under appropriate conditions and direct determination of base sequence. A model pET plasmid vector is shown in Fig. 22.13.

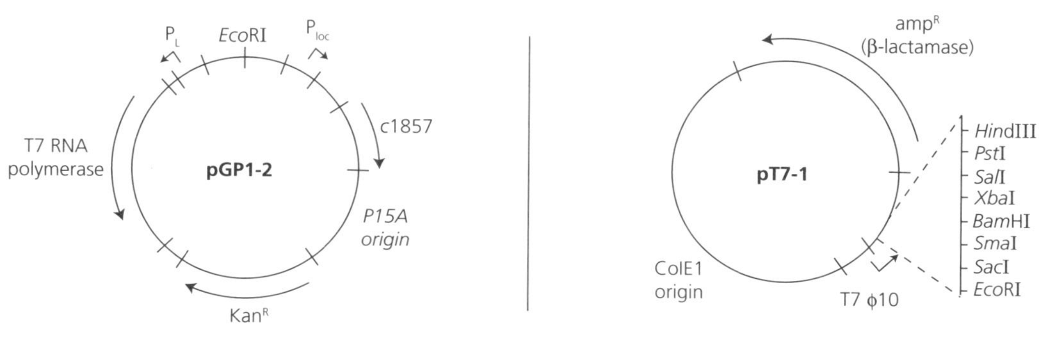

Figure 22.12 A two-plasmid T7 RNA polymerase expression system. The 7200-bp plasmid pGP1-2 (left) has the P15A replication origin and the neomycin phosphotransferase (kan^R) gene. The T7 RNA polymerase gene in it is under the lambda P_L promoter, which is controlled by the temperature-sensitive lambda *c*I857 repressor gene under the *lac* promoter. The 2400-bp plasmid pT7-1 has the compatible ColE1 replication origin and the β-lactamase (amp^R) gene. There is also a polylinker downstream of the T7 Φ10 promoter for cloning of the heterologous gene to be expressed.

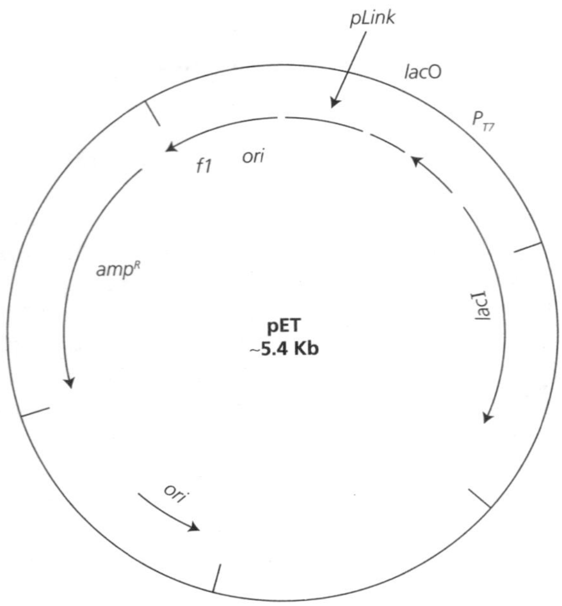

Figure 22.13 A model pET plasmid vector. The constituents of the vector are the f 1 replication origin, in addition to the plasmid's own replication origin, the β-lactamase gene as the ampicillin resistance marker, the T7 Φ10 promoter, the *lac* operator and the polylinker, in that order, and also the repressor *lac*I gene.

22.12 Gene fusion vectors for expression in *E. coli*

There are also many plasmid vectors available for fusion gene expression in *E. coli.* Description of some of these vectors follows.

22.12.1 Vectors designed for expression of proteins fused with glutathione S-transferase

These vectors (pGEX vectors; Fig. 22.14) express the cloned heterologous gene as a protein fused GST, which can be purified, in a one-step procedure, by glutathione Sepharose resin affinity chromatography. The column-bound fused protein is eluted with reduced glutathione. The *tac* promoter and the translation initiation signal of GST are used in pGEX vectors. pGEX vectors are available with a sequence coding for the recognition site of a specific protease to cleave the desired protein from GST. The recognition site could be of factor Xa or thrombin or PreScission Protease.

22.12.2 Vectors designed for expression of proteins fused with the maltose-binding protein

These vectors (pMAL vectors) express the cloned heterologous gene as a protein fused with MBP, which can be purified, in a one-step procedure, by amylase–agarose resin affinity chromatography. The column-bound fused protein is eluted with maltose. These vectors also contain the M13 origin of DNA replication to facilitate production of single-stranded DNA. The *tac* promoter and the translation initiation signal of MBP are used in pMAL vectors. These vectors have the additional feature that the insertion of the heterologous gene disrupts the *lacZα* gene, allowing blue-white selection in an X-gal plate. The *mal E* signal sequence has been deleted in the pMAL-c2 series of vectors, thus confining the fusion protein in the cytoplasm. The *mal E* signal sequence has been left unaffected in the pMAL-p2 series of vectors, thus allowing the fusion protein to pass through the cytoplasmic membrane. Here, the fusion protein can be purified from the periplasm. pMAL vectors are available with a sequence coding for the recognition site of a specific protease to cleave the desired protein from MBP. The recognition site could be of factor Xa, enterokinase or Genenase I.

22.12.3 Vectors designed for expression of proteins fused with the Staphylococcal protein A

These vectors express the cloned heterologous gene as a protein fused with the Staphylococcal A protein, which can be purified, in a one-step procedure, by IgG–Sepharose affinity chromatography. The column-bound

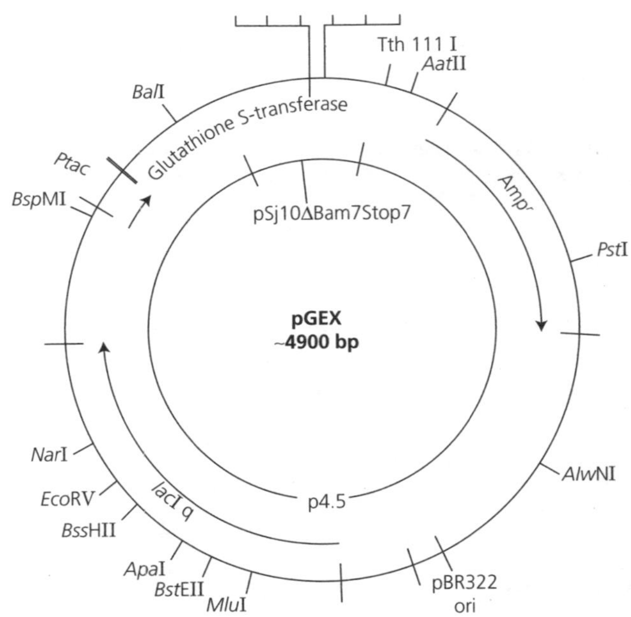

Figure 22.14 A model pGEX vector. The constituents of the vector are the pBR322 replication origin, the β-lactamase gene as the ampicillin resistance marker, the *tac* promoter, the *lac* operator, the glutathione S-transferase gene and the polylinker, in that order. The repressor *lac*I gene is also there.

fused protein is eluted with protein A. The plasmid vectors pRIT2 and pRIT5 are examples (Nilsoon *et al.* 1983; Uhlen *et al.* 1983).

22.12.4 Vectors designed for expression of proteins fused with the signal peptide of the ompA protein

The omp A protein is the major outer membrane protein of *E. coli.* The DNA fragment coding for the signal peptide of the omp A protein has been taken out from the *omp* A gene and has been inserted into the expression vector pIN III – A3 (Fig. 22.15) by Ghrayeb *et al.* (1984), thus creating a cloning vector that would secrete the expressed heterologous protein into the periplasmic space of *E. coli.* This vector has the β-lactamase gene (amp^R) and has both the *lpp* promoter and the *lac* promoter–operator in tandem. Regulation of expression is by the *lac* repressor *lacI*. Rentier-Delrue *et al.* (1988) improved upon this vector and made redundant the additional operation to remove the linker sequence between the ompA signal peptide sequence and the cloned heterologous gene. This improved secretion vector has been named pin III *omp A* Hind (Fig. 22.16).

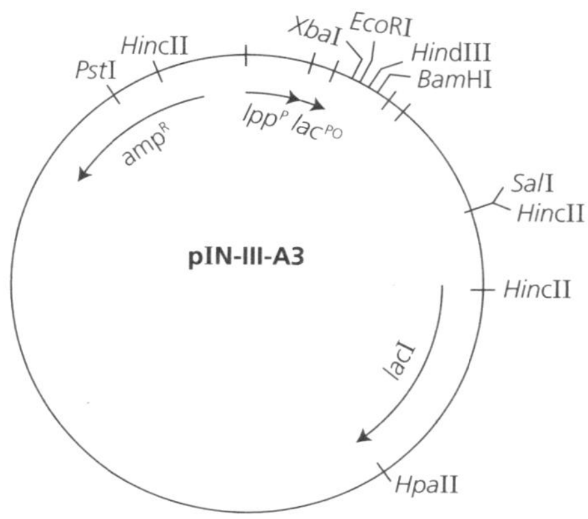

Figure 22.15 The pIN-III-A3 plasmid vector. The constituents of the vector are the β-lactamase gene as the ampicillin resistance marker, the lpp promoter (lpp^P) and the lac promoter – operator(lac^{PO}) in tandem and upstream of the polylinker cloning site. The repressor *lacI* gene is also there.

22.13 Codon optimization

Another important issue to be kept in mind while attempting heterologous gene expression is the phenomenon called codon bias. All amino acids except tryptophan and methionine are coded for by two or more codons. In most organisms, not all codons coding for one amino acid are used equally; there is bias towards using one or more codons instead of the others. For example, in *E. coli,* the codon GUU is used much more often than GUC in coding for valine, while in animals, the situation is the opposite. Again, in *E. coli,* the codon GCU is used much more often than GCC in coding for alanine, while in animals, GCC is used somewhat more often. Thus, appropriate changes in the coding sequence of the heterologous gene need to be made to ensure that the correct codon bias for *E. coli* is in place. This is called codon optimization and becomes

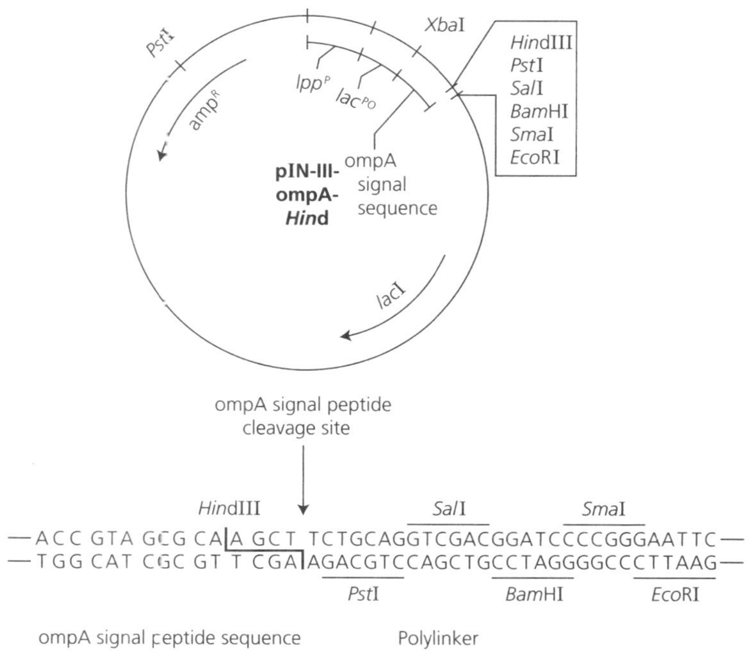

Figure 22.16 The pIN-III-ompA-*Hind* plasmid vector. The constituents of the vector are the β-lactamase gene as the ampicillin resistance marker, the lpp promoter (lpp^P) and the lac promoter – operator(*lac*^{PO}) in tandem and upstream of the omp A signal sequence, followed by the polylinker cloning site and the repressor *lacI* gene. The sequence of bases flanking the ompA signal peptide cleavage site is shown below the vector. The heterologous gene is to be cloned at any site between the *Hind*III site and any site in the polylinker.

essential if the heterologous transcript is to be translated at the maximum level.

22.14 Engineering the host *E. coli* for heterologous gene expression

A problem quite often encountered is that *E. coli* recognizes the foreign proteins and degrades it by proteolysis. A mutant deficient in protease activity, the *lon* mutant, is therefore generally used for the expression of heterologous genes in *E. coli*. The other problem is the rearrangement of the heterologous gene by recombinogenic activity in *E. coli*. This has been addressed by using the *recA* mutant of *E. coli*, which is deficient in the RecA protein. The RecA protein is essential for the recombination of DNA, resulting in a rearrangement of homologous regions of the cloned heterologous gene. *recA* mutants of *E. coli* indeed show less rearrangement of heterologous DNA, leading to an improved expression of the heterologous gene.

References

Avery OT, Macleod CM, McCarty M. Studies on the chemical nature of the substance inducing transformation of pneumococcal types. Induction of transformation by a deoxyribonucleic acid fraction isolated from *Pneumococcus* type III. *J Exp Med* 1944;79:137–158.

Boliver F, Rodriguez RL, Betlach MC, Boyer HW. Construction and characterization of new cloning vehicles. I. Ampicillin resistant derivative of the plasmid pMB 9. *Gene* 1977a;2:75–93.

Boliver F, Rodriguez RL, Greene PJ, Betlach MC, Heyneker HL, Boyer HW, *et al.* Construction and characterization of new cloning vehicles. II. A multipurpose cloning system. *Gene* 1977b;2:95–113.

Busso D, Delagoutte-Busso B, Moras D. Construction of a set gateway-based destination vectors for high-throughput cloning and expression screening in *Escherichia coli. Anal Biochem* 2005;343:313–321.

Butler JAV, Lawrence DJR, Robins AB, Shooter KV. Comparison of the molecular weights of deoxyribonucleic acids as determined from light scattering and from sedimentation and viscosity. *Proc Roy Soc Lond A* 1959;250:1–21.

Clarke L, Carbon J. A colony bank containing synthetic ColE1 hybrid plasmids representative of the entire *E. coli* genome. *Cell* 1976;9:91–100.

Cohen SN, Chang ACY, Boyer HW, Helling R. Construction of biologically functional plasmids *in vitro*. *Proc Natl Acad Sci USA* 1973;70:3240–3244.

Collins J, Hohn B. Cosmids: A type of plasmid gene-cloning vector that is packageable *in vitro* in bacteriophage λ heads. *Proc Natl Acad Sci USA* 1978;75:4242–4246.

Daabrowski S, Sobiewska G, Maciu ska J, Synowiecki J, Kur J. Cloning, expression, and purification of the His(6)-tagged thermostable beta-galactosidase from *Pyrococcus woesei* in *Escherichia coli* and some properties of the isolated enzyme. *Protein Expr Purif* 2000;19:107–112.

Davison PF. The effect of hydrodynamic shear on the deoxyribonucleic acid from T2 and T4 bacteriophages. *Proc Natl Acad Sci USA* 1959;45:1560–1568.

de Boer HA, Comstock LJ, Vasser M. The tac promoter: A functional hybrid derived from the trp and lac promoters. *Proc Natl Acad Sci USA* 1983;80:21–25.

Fukumaki Y, Shimada K, Takagi Y. Specialized transduction of colicin E1 DNA in *Escherichia coli* K-12 by phage lambda. *Proc Natl Acad Sci USA* 1976;73:3238–3242.

Ghrayeb J, Kimura H, Takahara M, Hsiung H, Masui Y, Inouye M. Secretion cloning vectors in *Escherichia coli. EMBO J* 1984; 3:2437–2442.

Jenny RJ, Mann KG, Lundblad RL. A critical review of the methods for cleavage of fusion proteins with thrombin and factor Xa. *Protein Expr Purif* 2003;31:1–11.

Jensen RH, Wodzinski RJ, Rogoff MH. Enzymatic addition of cohesive ends to T7 DNA. *Biochem Biophys Res Com* 1971;43:384–392.

Kim UJ, Birren BW, Slepak T, Mancino V, Boysen C, Kang HL, *et al.* Construction and characterization of a human bacterial chromosome library. *Genomics* 1996;34:213–218.

Lederberg J. Gene recombination and linked segregation in *E. coli*. *Genetics* 1947;32:505–525.

Lederberg J. *E. coli* K-12. *Microbiol Today* 2004;31:116.

Lederberg J, Zinder ND. Concentration of biochemical mutants of bacteria with penicillin. *J Amer Chem Soc* 1948;70:4267–4268.

Levinthal C. The mechanism of DNA replication and genetic recombination in phage. *Proc Natl Acad Sci USA* 1956;42:394–404.

Lobban PE, Kaiser AD. Enzymatic end-to-end joining of DNA molecules. *J Mol Biol* 1973;78:453–471.

Loughran ST, Walls D. Purification of poly-histidine-tagged proteins. *Methods Mol Biol* 2011;681:311–335.

Maina CV, Riggs PD, Grandea AG III, Slatko BE, Moran LS, Tagliamonte JA, *et al.* An *Escherichia coli* vector to express and purify foreign proteins by fusion to and separation from maltose-binding protein. *Gene* 1988;74:365–373.

Mandel M, Higa A. Calcium-dependent bacteriophage DNA infection. *J Mol Biol* 1970;53:159–162.

Maniatis T, Hardison RC, Lacy E, Lauer J, O'Connell C, Quon D, *et al.* The isolation of structural genes from libraries of eukaryotic DNA. *Cell* 1978;15:687–701.

Marsh P. ptac-85, an *E. coli* vector for expression of non-fusion proteins. *Nucleic Acids Res* 1986;14:3603.

Messing J, Crea R, Seeburg PH. A system for shotgun DNA sequencing. *Nucleic Acids Res* 1981;9:309–321.

Moffatt BA, Studier FW. Use of bacteriophage T-7 RNA polymerase to direct selective high-level expression of cloned genes. *J Mol Biol* 1986;189:113–130.

Morse L, Lederberg E, Lederberg J. Transduction in *E. coli* K12. *Genetics* 1956;41:142–156.

Nilsson B, Abrahmsén L, Uhlén M. Immobilization and purification of enzymes with staphylococcal protein A gene fusion vectors. *EMBO J* 1985;4:1075–1080.

Shizuya H, Birren B, Kim UJ, Mancino V, Slepak T, Tachiiri Y, *et al.* Cloning and stable maintenance of 300-kilobase-pair fragments of human DNA using an F-factor-based vector. *Proc Natl Acad Sci USA* 1992;89:8794–8797.

Smith DB, Johnson KS. Single-step purification of polypeptides expressed in *Escherichia coli* as fusions with glutathione *S*-transferase. *Gene* 1988;67:31–40.

So M, Gill R, Falkow S. The generation of a ColE1-Ap^r cloning vehicle which allows detection of inserted DNA. *Mol Gen Genet* 1976;142:239–249.

Sternberg N. Bacteriophage-P1 cloning system for the isolation, amplification and recovery of DNA fragments as large as 100 kilobase pairs. *Proc Natl Acad Sci USA* 1990;87:103–107.

Sternberg N, Ruether J, deRiel K. Generation of a 50,000-member human DNA library with an average insert size of 75-100 kbp in a bacteriophage P1 cloning vector. *New Biol* 1990;2:151–162.

Sternberg N, Smoller D, Braden T. Three new developments in P1 cloning. Increased cloning efficiency, improved clone recovery and a new P1 mouse library. *Genet Anal Tech Appl* 1994;11:171–180.

Tabor S, Richardson CC. A bacteriophage T7 RNA polymerase/promoter system for controlled exclusive expression of specific genes. *Proc Natl Acad Sci USA* 1985;82:1074–1078.

Uhlén M, Nilsson B, Guss B, Lindberg M, Gatenbeck S, Philipson L. Gene fusion vectors based on the gene for staphylococcal protein A. *Gene* 1983;23:369–378.

Umene K, Shimada K, Takagi Y. Packaging ColE1 DNA having a lambda phage cohesive end site. *Mol Gen Genet* 1978;159:39–45.

Yanish-Perron C, Vieira J, Messing J. Improved M13 phage cloning vectors and host strains: Nucleotide sequences of the M13mp18 and pUC19 vectors. *Gene* 1985;33:103–119.

23

Gram-positive bacterium *Bacillus subtilis* as host for recombinant DNA

H. K. Das

23.1 The host: *Bacillus subtilis*

Bacillus subtilis has been the second most popular bacterial host for recombinant DNA molecules only after *Escherichia coli*.

23.1.1 History of *B. subtilis*

Christian Gottfried Ehrenberg described *Vibrio subtilis* in 1835 as a rod-shaped bacterium found in soil and rotting material. Ferdinand Cohn renamed it as *Bacillus subtilis* in 1872. In 1941, the Nazi German Medical Corps isolated *B. subtilis* from the dung of horses and camels and discovered that a suspension of this bacterium on ingestion could cure soldiers suffering from acute and fatal dysentery. The story goes that the Germans had observed during their African campaign that the native Arabs would gulp fresh and warm dung droppings when they were afflicted with dysentery and alleviate their suffering almost overnight. Cultures of *B. subtilis* were sold in North America, Central America and Europe since 1946 till the late 1950s (one brand name was Bacti-Subtil) as a cure for dysentery and other intestinal ailments.

23.1.2 What is so special about *B. subtilis*?

Bacillus subtilis, a rod-shaped gram-positive bacterium, is the most studied prokaryote after *Escherichia coli*. It has a non-biased codon usage. Techniques of genetic manipulation of *B. subtilis* are highly developed. *B. subtilis* has a very high rate of protein synthesis and the cells can be grown to a very high density in a rather inexpensive medium. *B. subtilis* is non-pathogenic and does not elaborate any endotoxin. It has been given the GRAS (generally recognized as safe) status by the US Food and Drug Administration (FDA). This bacterium has been in use for years for the production of food additives and several industrial enzymes. The technology of large-scale culture of *B. subtilis* is very advanced. *B. subtilis* is naturally competent for transformation with DNA and its transformation has been in practice much before the methods for transformation of *E. coli* were developed. *B. subtilis* and other gram-positive bacteria have a single layer of cell membrane and transportation of proteins synthesized in the cytoplasm across the membrane to the culture medium is lot easier than it is in *E. coli*. *B. subtilis* is thus capable of excreting proteins directly into the culture medium. The problem of inclusion body formation is minimal and there is no need for cell disruption for recovering the desired protein. Downstream processing of proteins is a lot simpler. The yield of a secreted protein can be as high as 3 g/L of culture medium.

A major disadvantage of *B. subtilis* is the presence of high levels of intracellular and extracellular proteases, which tend to degrade foreign proteins. This problem has, however, been almost completely solved by constructing the strain *B. subtilis* WB 600, which is devoid of as many as six proteases (Wu *et al.* 1991). The chromosomal deletion of these six genes has been achieved through cloning, *in vitro* deletion and *in vivo* recombination of these deleted genes (Wu *et al.* 1991). The residual 0.32% of protease activity was inhibited by adding 2 mM phenylmethylsulphonyl fluoride to the culture medium. A strain (*B. subtilis* WB 800) has since been developed which is devoid of all the eight proteases (Murashima *et al.* 2002). The other disadvantage of *B. subtilis* is sporulation, but this problem has also been circumvented by developing asporogenic strains.

23.2 Plasmid cloning vectors for *B. subtilis*

Quite a number of plasmid vectors are available for cloning in *B. subtilis*.

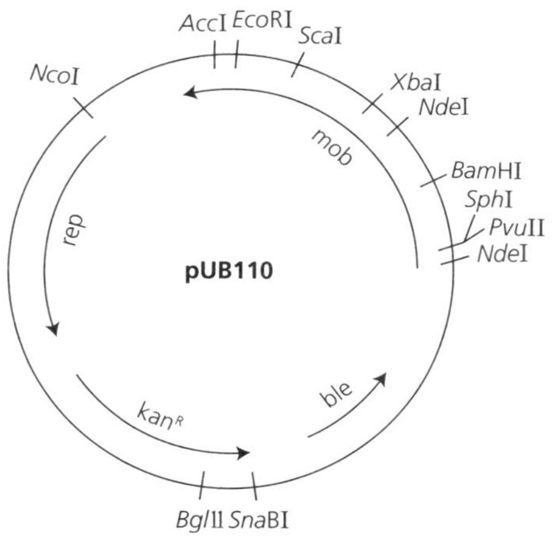

Figure 23.1 The plasmid pUB110, which was originally detected in *Staphylococcus aureus*, has been found to be stable in *Bacillus subtilis*.

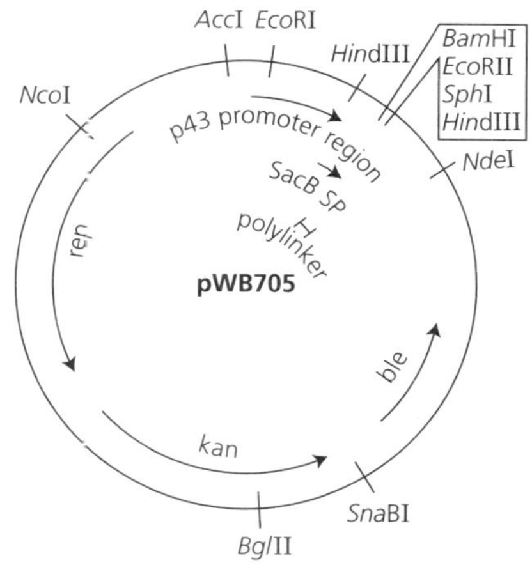

Figure 23.2 The plasmid pWB705. It is derived from pUB110, has a polylinker, a strong constitutive promoter, p43 and a levan sucrase signal sequence, making it a very good secretion vector for *Bacillus subtilis*.

23.2.1 Plasmids from *Staphylococcus aureus*

Most native plasmids of *Bacillus* are either cryptic or determine host functions and cannot be used as cloning vectors as such. However, certain plasmids carrying antibiotic resistance markers, which were originally detected in *Staphylococcus aureus*, have been found to be stable in transformed *B. subtilis*. One such plasmid that has found use as a cloning vector for *B. subtilis* is pUB110 (Km^r, ~4.5 kb) (Keggins *et al.* 1978), which has a copy number of 50–100, and whose restriction map given in Fig. 23.1 is adapted from Wu and Wong (1999). Various improved versions of pUB110 have been developed (Wong 1989; Behnke 1992; Wu and Wong 1999). The plasmid pWB705 is one such version, which has a polylinker, a strong constitutive promoter p43 and a levan sucrase signal sequence, making it a very good secretion vector [Fig. 23.2; the restriction map has been adapted from Wu and Wong (1999)]. A second derivative of pUB110, which has been named pWB980, has considerably higher expression levels than pWB705. Its restriction map, which has also been adapted from Wu and Wong (1999), is shown in Fig. 23.3. Another plasmid from *Staphylococcus aureus*, pE194 (~3.7 kb), which is compatible with pUB110, has also been used in *B. subtilis* as a component of a binary expression system co-producing accessory factors such as molecular chaperones and transcription activation factors (Wong *et al.* 1995; Wu *et al.* 1998). Sites of some restriction endonucleases in the plasmid pE 194 (Horinouchi

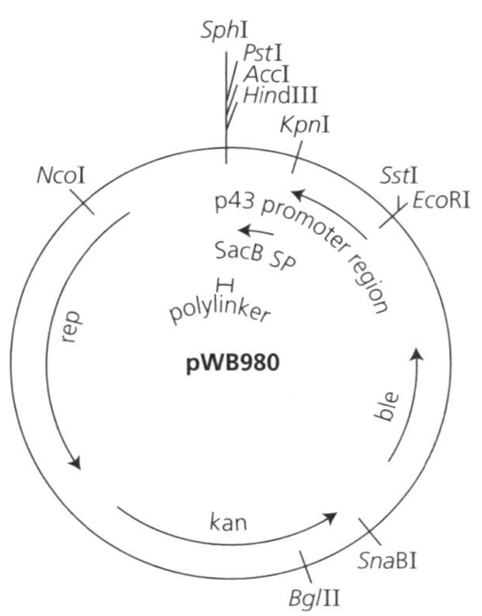

Figure 23.3 The plasmid pWB980. It is also derived from pUB110 and has considerably higher expression levels than pWB705.

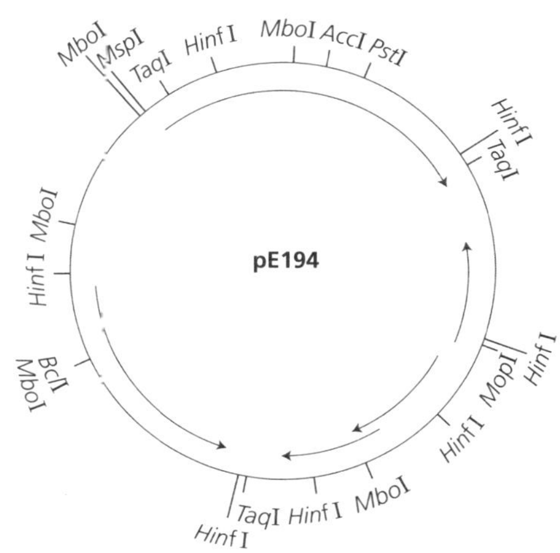

Figure 23.4 Sites of some restriction endonucleases in the *Staphylococcus aureus* plasmid pE194.

and Weisblum, 1982) are shown in Fig. 23.4. This plasmid confers resistance to macrolide, lincosamide and streptogramin type B. Though the original plasmid pE194 has a copy number of only 10, a mutant had subsequently been obtained, which has a copy number of 50–100 (Weissblum *et al.* 1979).

The plasmids mentioned above have, however, one drawback. Though by themselves these plasmids are quite stable in *B. subtilis,* sometimes instability arises or structural rearrangements take place subsequent to cloning of a foreign gene in these plasmids. The reason for this defect has been traced to the rolling circle mode of replication of these plasmids (Bron *et al.* 1991). As single-stranded DNA is formed as an intermediate of replication, short direct repeats, if present in the cloned DNA, sometimes result in the deletion of one of the repeats and the intervening DNA or in structural rearrangements.

23.2.2 Plasmid cloning vectors constructed using replication origin from plasmids native to *B. subtilis*

Cloning vectors have, therefore, been constructed using replication origins of comparatively large natural plasmids from *B. subtilis,* such as pAMβ1 (~26.5 kb) (Janniere *et al.* 1990) and pBS72 (>90 kb) (Titok *et al.*

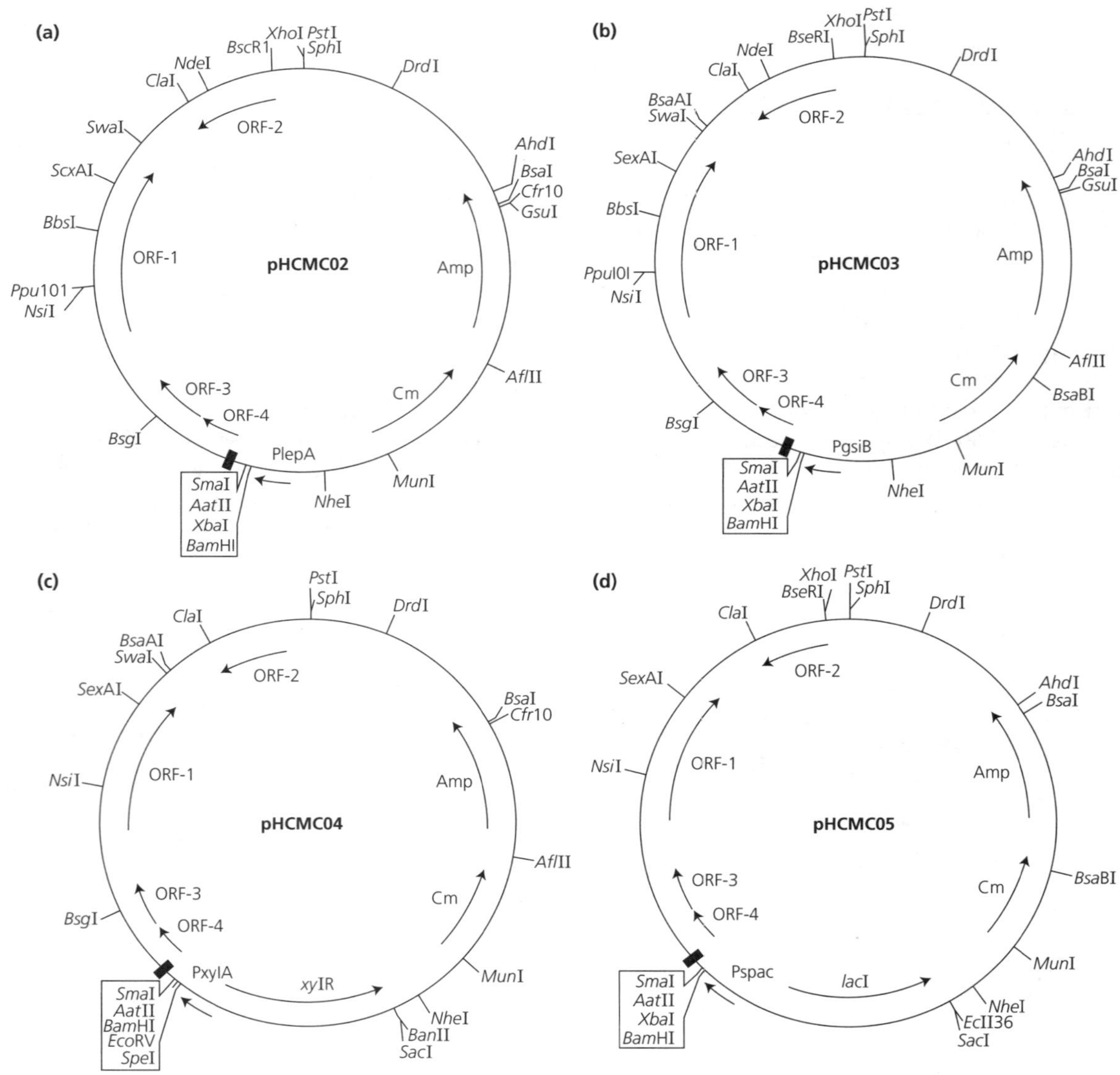

Figure 23.5 The pHCMC expression vectors for *Bacillus subtilis*. (a) pHCMC02, (b) pHCMC03, (c) pHCMC04 and (d) pHCMC05. All the pHCMC vectors contain the efficient transcriptional terminator *trpA* downstream from the promoters, as indicated by the solid bar perpendicular to the plasmid circle. The cloning sites, which are sandwiched between the promoter and the transcriptional terminator, are indicated.

2003), which replicate by the theta mode. The basic strategy has been to make a library of restriction fragments of these plasmids in pMTL21C (Chambers *et al.* 1988), which is a derivative of the *E. coli* plasmid vector pBR322, and then screen for the recombinant plasmids that stably replicate in *B. subtilis*. Examples of such cloning vectors, which can stably replicate in both *B. subtilis* and *E. coli*, are pMTLBS72 (Titok *et al.* 2003), the pHV vectors pHV1431 and pHV1435 (Janniere *et al.* 1990), the pHCMC series (Nguyen *et al.* 2005) and the pNDH series (Phan *et al.* 2006). The copy number of pHV1431 and pHV1435 in *B. subtilis* has been found to be 190, and stable cloning of large DNA fragments (20–40 kb) was possible in these vectors (Janniere *et al.* 1990). The different vectors of the pHCMC series, which have the chloramphenicol transacetylase gene, have a constitutive promoter; a promoter inducible by heat, acid shock, or ethanol; a promoter inducible by xylose; or a promoter inducible by IPTG. The four pHCMC plasmid vectors, the drawings of which have been adapted from Nguyen *et al.* (2005), are shown in Fig. 23.5. The expression vector pNDH33 (Phan *et al.* 2006) was constructed by replacing the promoter P_{spac} of pHCMC05 (see Fig. 23.5; Nguyen *et al.* 2005) by the *groE* promoter from *B. subtilis* fused to the *lac* operator, allowing its induction by IPTG. In pNDH37, the coding region of the signal peptide of the *amyQ* gene from *B. amyloliquiefasciens* was fused downstream of the P *groE–lacO* to ensure secretion of recombinant proteins (Phan *et al.* 2006). The drawings of the two pNDH vectors, adapted from Phan *et al.* (2006), can be seen in Fig. 23.6.

23.2.3 Plasmids from other bacilli

A tetracycline-resistant strain of a thermophilic bacillus isolated from river sludge at 55°C contained a 4.14-kb plasmid which could stably transform *B. subtilis* 168 and impart tetracycline resistance. This plasmid was named pAB124 (Bingham *et al.* 1979). Imanaka *et al.* (1981) isolated two plasmids, pTB19 (tetracycline and kanamycin resistance) and pTB20 (tetracycline resistance), from thermophilic bacilli, both of which could transform *B. subtilis*. Several restriction deletion derivates of pTB19 were constructed for use as vectors (Imanaka *et al.* 1984). One such derivative, pTB52 (tetracycline resistance), was further deleted to yield pTB522 (Imanaka *et al.* 1985). This theta mode–replicating plasmid has been used successfully as a vector by Itaya and coworkers.

23.3 Integration of foreign genes into the chromosome of *B. subtilis*

The basic principle is to design a plasmid that would not be able to replicate in *B. subtilis* but would have a stretch of DNA that would be homologous to some non-essential region in the *B. subtilis* chromosome. Of course, the plasmid would have to have (a) a replication origin compatible with another bacteria, which is generally *E. coli*, (b) an expression cassette positioned in the middle of the stretch of DNA that is homologous to

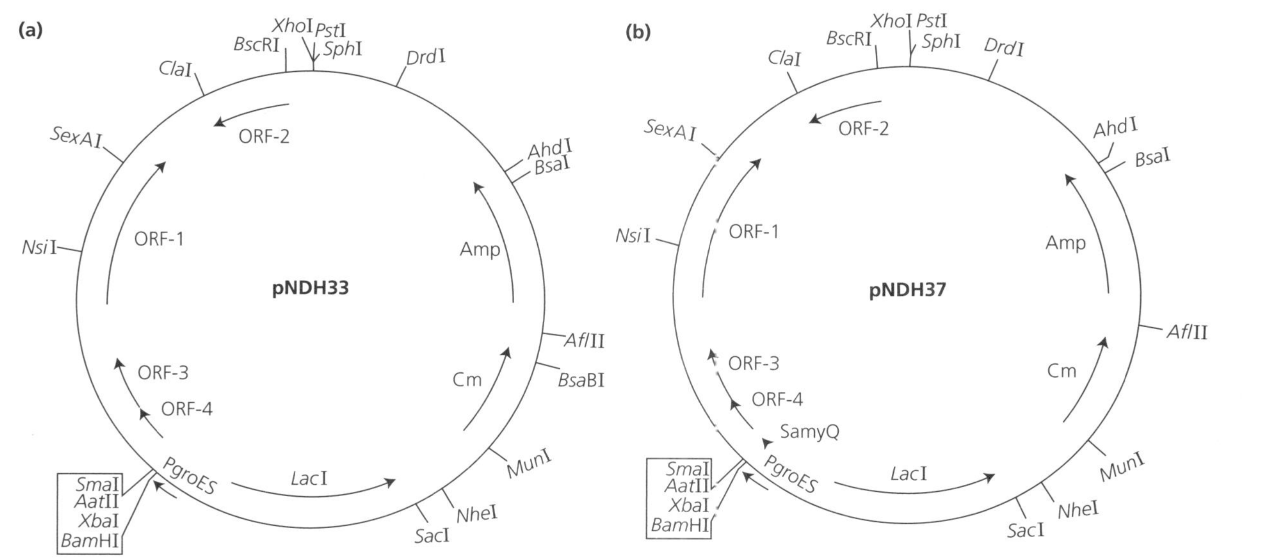

Figure 23.6 The pNDH expression vectors for *Bacillus subtilis*. (a) The expression vector pNDH33. (b) The expression-cum-secretion vector pNDH37. pNDH37 is the same as pNDH33, except that it has the coding region of the signal peptide of the *amyQ* gene from *B. amyloliquiefasciens* sandwiched between the ribosome-binding site and the *lac* operator.

the non-essential region in the *B. subtilis* chromosome, followed by a multiple cloning site and (c) an antibiotic resistance marker.

23.3.1 Identification of non-essential regions in the chromosome of *B. subtilis* and development of vectors for integration of foreign genes at such a site

One of the non-essential regions identified is the *amyE* gene, which codes for α-amylase (Shimotsu and Henner 1986). In the vector developed for this region, the expression cassette containing the *trp* promoter, the multiple cloning site and the antibiotic resistance marker (Cm^r) are sandwiched between the two halves of the *amyE* gene, which are designated as *amyE* front and *amyE* back. Once inside the *B. subtilis* cell, the two *amyE* halves recombine with the chromosomal *amyE* locus *via* a two-point crossover event, resulting in the integration of the DNA segment between the two *amyE* halves containing the cloned gene into the *B. subtilis* chromosome. A schematic representation of the phenomenon is given in Fig. 23.7. The two recombination regions, that is, the two *amyE* halves in this case, have later been named landing pad sequences (LPSs; Kaneko *et al.* 2003). The vector constructed by Shimotsu and Henner (1986) has *lacZ* gene cloned in it. The drawing of the vector, adapted from Shimotsu and Henner (1986), can be seen in Fig. 23.8. As expected, the transformed *B. subtilis* became Cm^r, *amyE*$^-$ *and lac*$^+$ and the level of β-galactosidase was regulated by exogenous tryptophan.

Two other integration sites, the *thr C* gene (Guerout-Fleury *et al.* 1996) and the *lacA* gene (Hartl *et al.* 2001), in the chromosome of *B. subtilis*, have been developed. Here also, the plasmid vectors are derivatives of pBR322. The specialty of the two vectors pAX01 and pA-spac of Hartl *et al.* (2001) is that the foreign gene integrated into the chromosome would be expressed only when induced. The drawings of the two vectors, adapted from Hartl *et al.* (2001), can be seen in Fig. 23.9. The vector, pAX01, features xylose induction, while the vector, pA-spac, features IPTG induction. Both the integration sites, the *amyE* locus and the *lacA* locus, can be used simultaneously. Middleton and Hofmeister (2004) have reported the development of three other integration sites, the *pyrD, gltA, and sacA* genes, in the chromosome of *B. subtilis*. The plasmid vectors constructed for these sites are all derivatives of pUC18. A schematic presentation of a representative vector (glt-kan integration vector), adapted from Middleton and Hofmeister (2004), can be seen in Fig. 23.10.

23.3.2 Integration of very large fragments of foreign DNA in the chromosome of *B. subtilis*

The large fragment of DNA was really not integrated as it was, but was indeed assembled in the chromosome

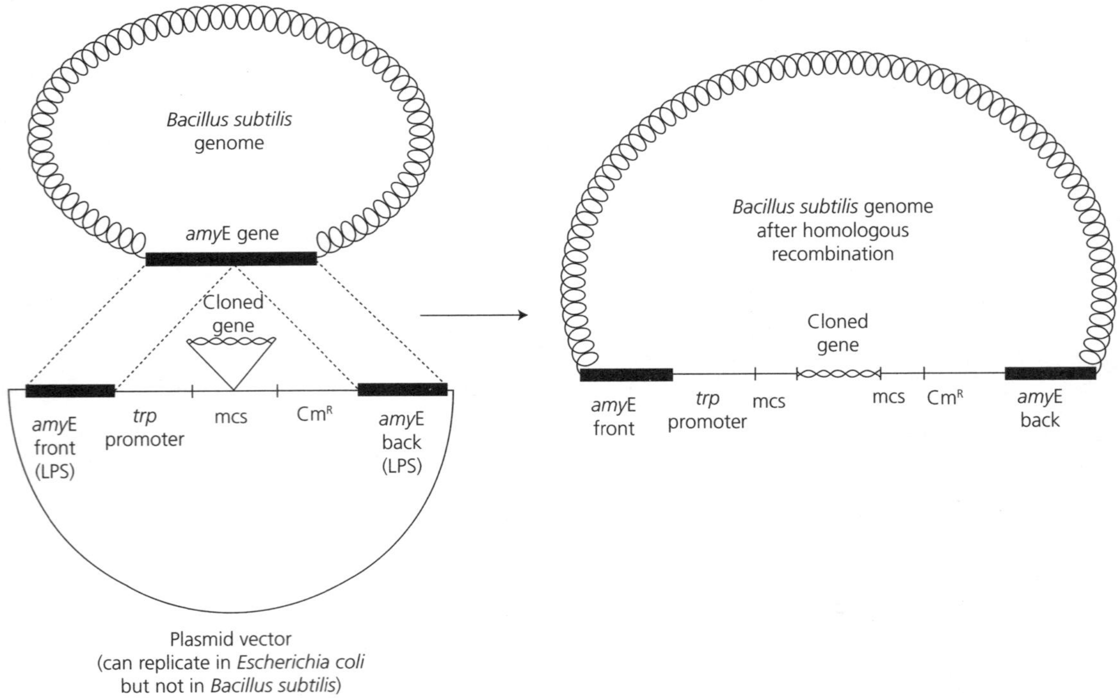

Figure 23.7 Schematic representation of the recombination phenomenon by which a cloned gene with the *trp* promoter upstream and the chloramphenicol resistance gene downstream is transferred from a plasmid in to the *amyE* gene in the chromosome of *Bacillus subtilis*.

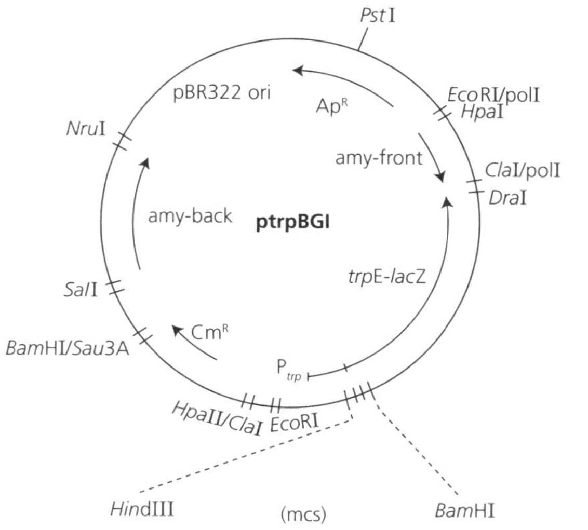

Figure 23.8 The plasmid ptrpBGI. The replication origin and the β-lactamase gene (ampicillin resistance) are from the *E. coli* plasmid pBR322; the *amyE* gene and the *trpE* promoter are from *B. subtilis*. The *amyE* gene is split in the plasmid. The *lacZ* gene from *E. coli* has been cloned in the multiple cloning site immediately downstream of the *trpE* promoter. *Bacillus subtilis* is transformed with ptrpBGI and selected for Cm^r and Amp^s character. The β-galactosidase gene and the chloramphenicol transacetylase gene get integrated into the chromosome of *B. subtilis* in the middle of the *amyE* gene.

of BEST2011. This is a derivative of *B. subtilis* 168 *trpC2*, which is now called a BGM vector, standing for *B.* Geno Me vector (Itaya *et al.* 2003). BEST 2011 has the linearized plasmid pBR322 inserted in the *Not* I site within the *degX* gene of *B. subtilis* 168 *trpC2*. The pBR 322 sequence in the bacterial chromosome would serve as LPSs. As the genes for ampicillin and tetracycline resistance present in pBR322 are ineffective in *B. subtilis*, the kanamycin/neomycin resistance gene cassette and the erythromycin resistance gene cassette were inserted in pBR322 into the *Pvu*II site and the *Bam*HI site, respectively. This plasmid was named pBASE2001. Four fragments of the whole 48.5-kb phage lambda DNA was designed to have overlapping regions on both sides and cloned separately in four derivatives of the plasmid pBR322, having four different antibiotic resistance markers. These plasmids would replicate in *E. coli* but not in *B. subtilis*. Assembly of the whole lambda DNA in the middle of the pBR322 sequence located in the chromosome of BEST2011 was achieved by successive transformation and recombination events with the four different plasmid constructs. The lambda DNA thus assembled was stable in the BEST2011 chromosome and replicated correctly as a part of the chromosome without any selection pressure (Itaya 1995). Subsequently, Itaya (1999) was successful in transferring contiguous DNA segments of up to 50 kb into the chromosome of BEST2011, though the transformation efficiency with large DNA segments was very low. Using this approach, a mouse genomic segment of ~120 kb has been assembled into the pBR322 locus in the chromosome of BEST2011 by using mouse DNA subsegments of 20–50 kb (Itaya *et al.* 2000). Later, Itaya *et al.* (2000) succeeded in using up to 77-kb DNA subsegments and assemble the whole 3573-kb genome of a cyanobacterium, *Synechocystis* sp. PCC 6803, in the chromosome of BEST2011. BEST2012 has a chloramphenicol acetyl transferase (CAT) gene cloned at the *Eco* RI site of the pBR322 sequence in BEST2011. Another BGM vector, BEST6016, is a derivative of BEST2011, which has an additional insert, a neomycin resistance gene under the control of the lambda Pr promoter, in the *Not* 1 site *yvfC–yveP* gene (Ytaya *et al.* 2000). BEST7003

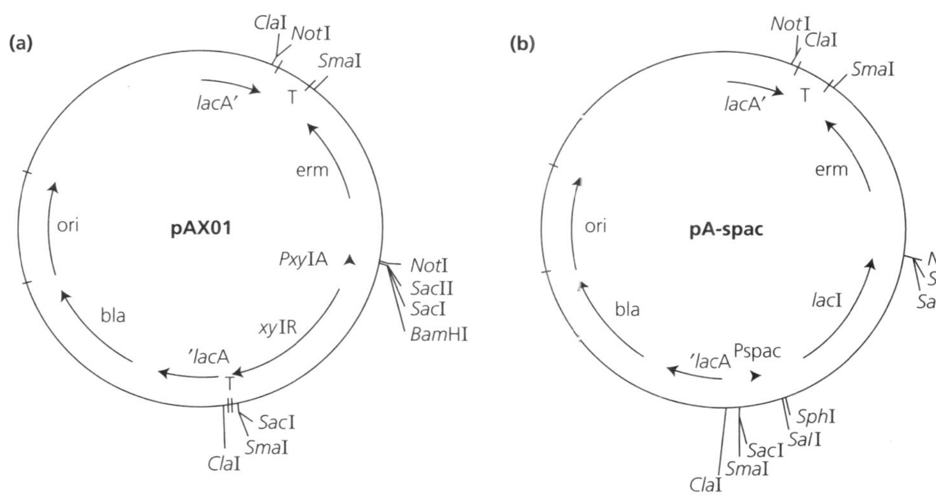

Figure 23.9 Delivery expression vectors. (a) pAX01. (b) pA-spac. Both vectors are derivatives of pBR322, which does not replicate in *Bacillus subtilis*. The two segments of *lacA* from *B. subtilis* allow integration of the DNA between them into the chromosome of *B. subtilis* via a double-crossover event. The selection is for Erm^r.

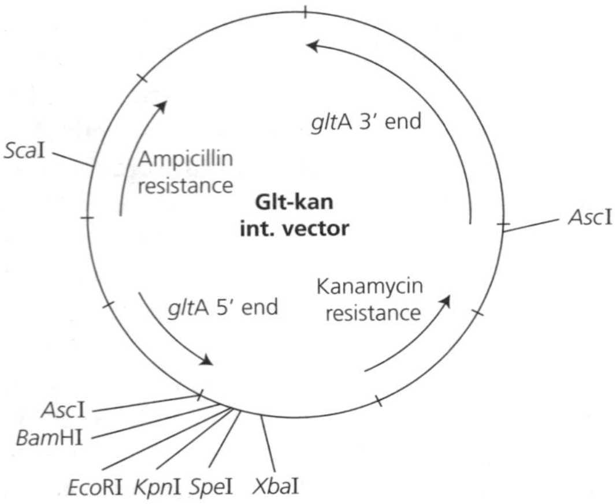

Figure 23.10 Schematic presentation of the Glt-kan integration vector. It contains a multiple cloning site and a kanamycin/neomycin resistance gene flanked by two halves of the *gltA* gene of *Bacillus subtilis*. The replication origin is from pUC18 (does not replicate in *B. subtilis*).

is a derivative of BEST6016 which has lost the BSU restriction modification system (Kaneko *et al.* 2003). BEST310 has been constructed from BEST7003 by providing two 3.5-kb guide sequences BAC [R] and BAC [L] from the BAC (bacterial artificial chromosome) vector in the pBR322 region so that BAC clones transferred to BEST310 can participate in a two-point crossover, leading to the direct integration of any BAC clone of around 100 kb into the BEST 310 genome (Kaneko *et al.* 2005). An 18-clone BAC library of the mitochondrial DNA of *Arabidopsis thaliana* was transferred to BEST310, thus constructing a very stable BGM library, which was unaffected even after sporulation of BEST310 and subsequent germination (Kaneko *et al.* 2005). The complete mitochondrial DNA of mouse was also transferred to BEST7003 (Yonemura *et al.* 2007).

23.3.3 Transfer of chromosomal inserts into plasmid

Transfer of a chromosomal insert into a plasmid is based on the observation that a linearized plasmid with a replication origin compatible with *B. subtilis*, when introduced into *B. subtilis* by transformation, can be circularized and stabilized only if it has sequences in it that have homology with sequences present in the *B. subtilis* chromosome (Tsuge and Itaya 2001). A recombination event occurs when the homologous regions pair and the sequence in the chromosome between the homologous regions gets transferred to the plasmid. This has been termed as the *Bacillus* recombinational transfer (BreT) phenomenon. A schematic presentation of this phenomenon, adapted from Tsuge and Itaya (2001), can be seen in Fig. 23.11. In contrast, a circular multimeric form of the same plasmid would be stable by itself inside the *B. subtilis* cell. Tsuge and Itaya (2001) used the plasmid vector pGETS103, which is derived from the theta mode–replicating plasmid pTB522 mentioned in Section 23.2.3, but also contains the linearized full-length *E. coli* plasmid pBR322. The drawing of the plasmid pGETS103, adapted from Tsuge and Itaya (2001), can be seen in Fig. 23.12. Transfer of previously cloned DNA sequence up to 90 kb from the chromosome to the plasmid has been demonstrated (Tsuge and Itaya 2001). A derivative of pGETS103 that has an additional replication origin compatible with *E. coli* is pGETS118. The latter plasmid has been used in shuttling of BAC clones to BGM (*B. subtilis* genome) and then to a plasmid that can replicate both in *B. subtilis* and in *E. coli* (Kaneko *et al.* 2005). For situations where transformation with a large DNA is inefficient, a conjugational plasmid-mediated DNA retrieval and transfer (CReT) system has been developed (Kuroki *et al.* 2007). The 65-kb plasmid pLS20, which can be conjugationally transferred between

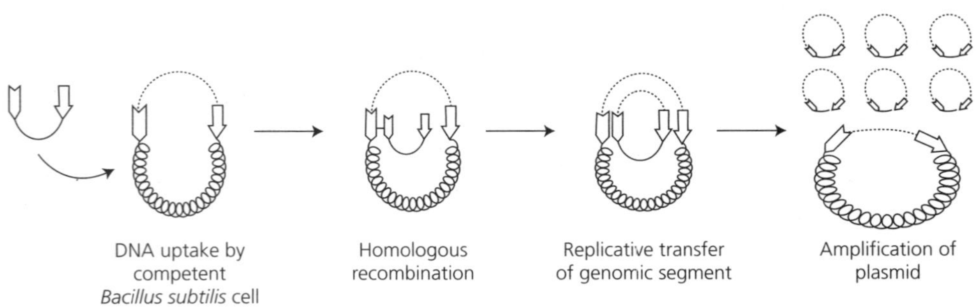

Figure 23.11 Schematic presentation of the replicative transfer of the DNA segment cloned in the chromosome of *Bacillus subtilis* (BGEM vector) to a linearized plasmid with a replication origin compatible with *B. subtilis*, which contains DNA sequences in it that have homology with DNA sequences present in the *B. subtilis* chromosome. The linearized plasmid that is used to transform *B. subtilis* (leftmost structure in the figure) has a cleaved pBR322 sequence (cleaved open arrow) attached to a sequence (solid black segment) of a plasmid that can replicate in *B. subtilis*. The *B. subtilis* genome already has the pBR322 sequence inserted into it (cleaved open arrow) (*cf.* genome vector, BEST2011) with the cloned large DNA fragment (grey segment) between the two pBR322 segments (second left structure in the figure). The next three structures in the figure show homologous recombination, replicative transfer of cloned genomic segment to the plasmid and amplification of the plasmid containing the transferred genomic segment.

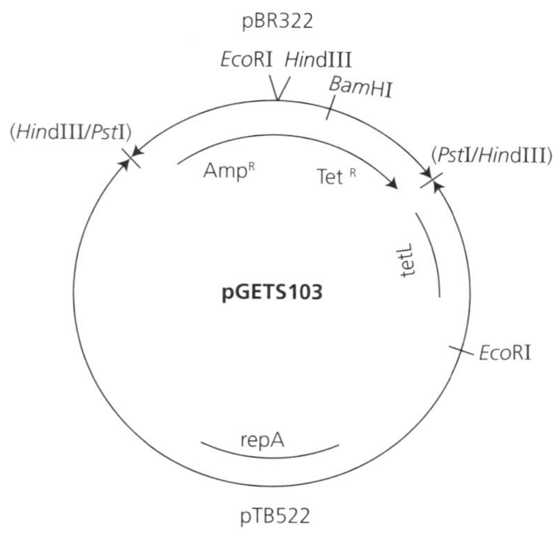

Figure 23.12 Drawing of the plasmid pGETS103. The plasmid pGETS103 is a fusion of pBR322 and pTB522, the latter having a replication origin compatible with *Bacillus subtilis*. The composite plasmid is linearized by cleavage at the *Hin*dIII site of pBR322 and introduced into *B. subtilis* cells containing BEST vectors.

different *B. subtilis* strains, has been modified to retrieve DNA cloned in the BGM vector by homologous recombination. Here also, retrieval of a 90-kb-long DNA has been demonstrated and no structural alteration has been encountered (Kuroki *et al.* 2007).

23.4 Bacteriophage cloning and expression system for *B. subtilis*

A derivative of the temperate bacteriophage ϕ105 of *B. subtilis* has been developed as a cloning and expression vector in *B. subtilis* (Thornewell *et al.* 1993; Liu *et al.* 2004). As has been pointed out by Thornewell *et al.* (1993), plasmid-based expression systems of *B. subtilis* have the problem of not very good stability in the absence of selection pressure. In contrast, prophage-based expression systems have the advantage of very good stability, even in the absence of selection pressure, being integrated into the chromosome of *B. subtitis.* A search for strong promoters in the genome of the phage o105 was undertaken by Thornewell *et al.* (1993) by shotgun ligation of a cartridge containing a promoterless *lacZ* gene followed by a *cat* (chloramphenicol acetyl transferase) gene in the opposite orientation into the prophage ϕ105 genome. A *B. subtilis* lysogen of ϕ105 was transformed with the shot gun ligation products and selected for chloramphenicol resistance (Fig. 23.13). Seven out of 300 transformant colonies showed β-galactosidase activity, presumably because in these, the promoterless *lacZ* gene was under a strong phage promoter. Phage induction was carried out in these transformed cells with mitomycin C, and based on β-galactosidase activity in liquid culture, the transformant containing the prophage designated ϕ105MU3 was selected. This transformant was found to be blocked in cell lysis (Thornewell *et al.* 1993), possibly because the insertion of the *lac–cat* casette into the prophage genome had disrupted a gene involved in lysis. The DNA from the prophage ϕ 105 MU3 was next used to transform the *B. subtilis* strain CU267 lysogenic for the temperature-inducible prophage ϕ105J119 (*cts*-52 mutant) to chloramphenicol resistance. A transformant was chosen which, 2 h after temperature induction, elicited 13,000 Miller units of β-galactosidase activity. The prophage in this transformed strain was designated ϕ105MU209. The strain CU267 (ϕ105MU209) was also lysis defective. Manipulation of the prophage ϕ105MU209 by transformation, with the aim of expression of other genes, failed, possibly because of poor competence for transformation. An *ind*⁻ background was, therefore, brought

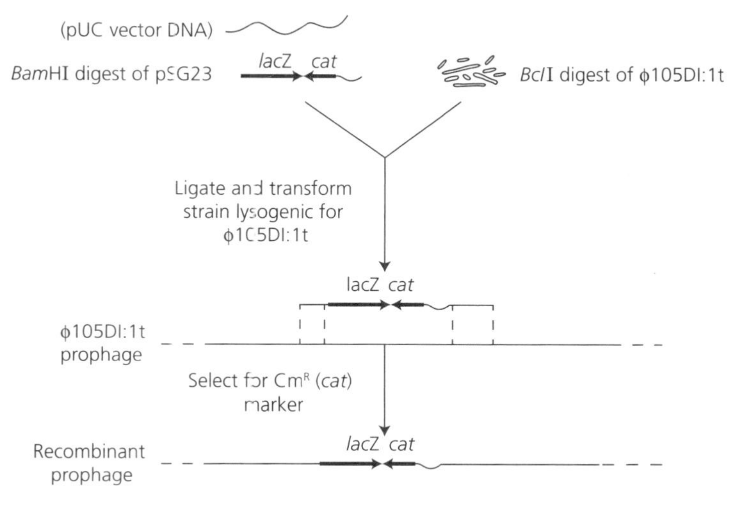

Figure 23.13 Schematic representation of random insertion of the *lacZ cat* (β-galactosidase–chloramphenicol acetyl transferase) gene cartridge into the ϕ105DI:1t prophage to identify strong phage promoters. A *Bam*HI DNA fragment containing the promoterless *lacZ* gene and the *cat* gene was ligated to the DNA fragments obtained by *Bcl* I digestion of the ϕ105DI:1t DNA. The *B. subtilis* strain CU 267 (ϕ105DI:1t) was then transformed with the ligation products and selected for Cm^R. The regions of homology are bounded by vertical broken lines, where homologous recombination is expected to take place.

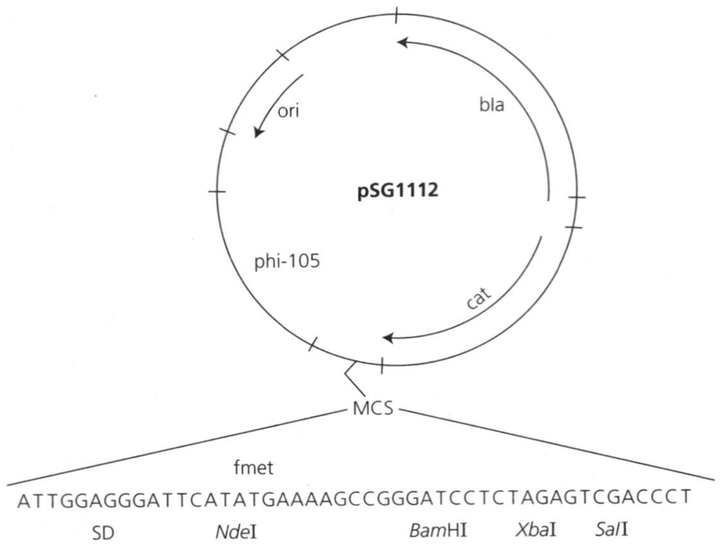

Figure 23.14 The integration plasmid vector pSG1112, which is a derivative of pSG703. The replication origin (ori) is from pBR322. It also has a ~600-bp ϕ105 phage DNA fragment containing the strong promoter and the Shine and Dalagarno–type ribosome-binding site followed by a multiple cloning site and also both chloramphenicol resistance (*cat* gene) and ampicillin resistance (*bla* gene) functions.

in to improve upon the competence. The resulting new prophage, which had useful properties such as strong induction of β-galactosidase with no host lysis and high transformability of the lysogen, was designated ϕ105MU330. For overexpression of heterologous genes other than β-galactosidase, the prophage ϕ105MU331 was subsequently constructed by deleting a DNA fragment comprising a part of the *lacZ* gene and a part of the *cat* gene from ϕ105MU330 and inserting the ErR (erythromycin resistance) gene there, rendering it Cm^S, Er^R. In addition, a 5.3-kb plasmid pSG703, which has a truncated *lacZ* gene followed by an multiple cloning site (MCS) and the chloramphenicol acetyl transferase gene, was also constructed ((Thornewell *et al.* 1993). Together, the prophage ϕ105MU331 and the plasmid pSG703 would comprise the heterologous gene expression system. To test the system, the *blaI* gene from *B. cereus* was cloned in the correct orientation in the MCS of pSG703 and the *B. subtilis* strain lA304 (ϕ105MU331) was transformed and Cm^R, Er^S colonies were selected. The prophage was induced at 50°C for 5 min and then growth was continued at 37°C. Production of β-lactamase I peaked after about 5 h and yields of 200–500 μg/mL were obtained (Thornewell *et al.* 1993). The plasmid pSG703 has since been replaced by the improved version 3.5-kbp pSG1112 (Liu *et al.* 2004), which has the MCS immediately next to the Shine and Dalgarno (SD)–type ribosome-binding site of prophage ϕ105MU331 and the initiator ATG in an *Nde*I site (Fig. 23.14).

References

Behnke D. Protein export and the development of secretion vectors. In: Doi RH, McGloughlin M, eds. *Biology of Bacilli: Applications to Industry*. Stoneham, MA: Butterworth-Heinemann; 1992. pp. 143–188.

Bingham AHA, Bruton CJ, Atkinson T. Isolation and partial characterization of four plasmids from antibiotic resistant thermophilic bacilli. *J Gen Microbiol* 1979;114:401–408.

Bron S, Holsappel S, Venema G, Peeters BP. Plasmid deletion formation between short direct repeats in *Bacillus subtilis* is stimulated by single-stranded rolling-circle replication intermediates. *Mol Gen Genet* 1991;226:88–96.

Chambers SP, Prior SE, Barstow DA, Minton NP. The pMTL *nic*- cloning vectors. I. Improved pUC polylinker regions to facilitate the use of sonicated DNA for nucleotide sequencing. *Gene* 1988;68:139–149.

Guerout-Fleury A-M, Frandsen N, Stragier P.Plasmids for ectopic integration in *Bacillus subtilis*. *Gene* 1996;180:57–61.

Härtl B, Wehrl W, Wiegert T, Homuth G, Schumann W. Development of a new integration site within the *Bacillus subtilis* chromosome and construction of compatible expression cassettes. *J Bacteriol* 2001;183:2696–2699.

Horinouchi S, Weisblum B. Nucleotide sequence and functional map of pC194, a plasmid that specifies inducible chloramphenicol resistance. *J Bacteriol* 1982;150:815–825.

Imanaka T, Fujii M, Aiba S. Isolation and characterization of antibiotic resistance plasmids from thermophilic bacilli and construction of deletion plasmids. *J Bacteriol* 1981;146:1091–1097.

Itaya M. Toward a bacterial genome technology: integration of the *Escherichia coil* prophage lambda genome into the *Bacillus subtilis* 168 chromosome. *Mol Gen Genet* 1995;248:9–16.

Itaya M. Genetic transfer of large DNA inserts to designated loci of the *Bacillus subtilis* 168 genome. *J Bacteriol* 1999;181:1045–1048.

Itaya M, Fujita K, Ikeuchi M, Maki M, Tsuge K. Stable positional cloning of long continuous DNA in the *Bacillus subtilis* genome vector. *J Biochem* 2003;134:513–519.

Itaya M, Nagata T, Shiroishi T, Fujita K, Tsuge K. Efficient cloning and engineering of giant DNAs in a novel *Bacillus subtilis* genome vector. *J Biochem* 2000;128:869–887.

Janniere L, Bruand C, Ehrlich SD. Structurally stable *Bacillus subtilis* cloning vectors. *Gene* 1990;87:53–61.

Kaneko S, Akioka M, Tsuge K, Itaya M. DNA shuttling between plasmid vectors and a genome vector: Systematic conversion and preservation of DNA libraries using the *Bacillus subtilis* genome vector. *J Mol Biol* 2005;349:1036–1044.

Kaneko S, Tsuge K, Takeuchi T, Itaya M. Conversion of sub-megasized DNA to desired structures using a novel *Bacillus subtilis* genome vector. *Nucl Acids Res* 2003;31:e112.

Keggins KM, Lovett PS, Duvall EJ. Molecular cloning of genetically active fragments of Bacillus DNA in *Bacillus subtilis* and properties of the vector plasmid pUB110. *Proc Natl Acad Sci USA* 1978;75:1423–1427.

Kuroki A, Ohtani N, Tsuge K, Tomita M, Itaya M. Conjugational transfer system to shuttle giant DNA cloned by *Bacillus subtilis* genome (BGM) vector. *Gene* 2007;399:72–80.

Liu H-B, Chui K-S, Chan C-L, Tsang C-W, Leung Y-C. An efficient heat-inducible *Bacillus subtilis* bacteriophage ϕ105 expression and secretion system for the production of the *Streptomyces clavuligerus* β-lactamase inhibitory protein (BLIP). *J Biotechnol* 2004;108:207–217.

Middleton R, Hofmeister A. New shuttle vectors for ectopic insertion of genes into *Bacillus subtilis*. *Plasmid* 2004;51:238–245.

Murashima K, Chen C-L, Kosugi A, Tamaru Y, Doi RH, Sui-Lam Wong S-L. Heterologous production of *Clostridium cellulovorans engB*, using protease-deficient *Bacillus subtilis*, and preparation of active recombinant cellulosomes. *J Bacteriol* 2002;184:76–81.

Nguyen HD, Nguyen QA, Ferreira RC, Ferreira LCS, Tran LT, Schumann W. Construction of plasmid-based expression vectors for *Bacillus subtilis* exhibiting full structural stability. *Plasmid* 2005;54:241–248.

Phan TTP, Nguyen HD, Schumann W. Novel plasmid-based expression vectors for intra- and extracellular production of recombinant proteins in *Bacillus subtilis*. *Protein Express Purific* 2006;45:189–195.

Shimotsu H, Henner DJ. Construction of a single-copy integration vector and its use in analysis of regulation of the trp operon of *Bacillus subtilis*. *Gene* 1986;43:85–94.

Thornewell SJ, East AK, Errington J. An efficient expression and secretion system based on *Bacillus subtilis* phage ϕ105 and its use for the production of *B. cereus* β-lactamase I. *Gene* 1993;133:47–53.

Titok MA, Chapuis J, Selezneva YV, Lagodich AV, Prokulevich VA, Ehrlich SD, *et al*. *Bacillus subtilis* soil isolates: Plasmid replicon analysis and construction of a new theta-replicating vector. *Plasmid* 2003;49:53–62.

Tsuge K, Itaya M. Recombinational transfer of 100-kilobase genomic DNA to plasmid in *Bacillus subtilis* 168. *J Bacteriol* 2001;183:5453–5458.

Weisblum B, Graham MY, Gryczan T, Dubnau D. Plasmid copy number control: Isolation and characterization of high-copy-number mutants of plasmid pE 194. *J Bacteriol* 1979;137:635–643.

Wong S-L. Development of an inducible and enhancible expression and secretion system in *Bacillus subtilis*. *Gene* 1989;83:215–223.

Wong S-L, Wu X-C, Yang L-P, Ng S-C, Hudson N. Production and purification of antibody using *Bacillus subtilis* as an expression host. In: Wang HY, Imanaka T, eds. *Antibody Expression and Engineering*. Washington, DC: American Chemical Society; 1995. pp. 100–107.

Wu S-C, Wong S-L. Development of improved pUB110-based vectors for expression and secretion studies in *Bacillus subtilis*. *J Biotechnol* 1999;72:185–195.

Wu S-C, Ye R, Wu X-C, Ng S-C, Wong S-L. Enhanced secretory production of a single-chain antibody fragment from *Bacillus subtilis* by coproduction of molecular chaperones. *J Bacteriol* 1998;180:2830–2835.

Wu X-C, Lee W, Tran L and Wong S-L. Engineering a *Bacillus subtilis* expression-secretion system with a strain deficient in six extracellular proteases. *J Bacteriol* 1991;173:4952–4958.

Yenomura I, Nakada K, Sato A, Hayashi J-I, Fujita K, Kanneko S, *et al*. Direct cloning of full-length mouse mitochondrial DNA using a *Bacillus subtilis* genome vector. *Gene* 2007;391:171–177.

24

Streptomyces as host for recombinant DNA

K. Dharmalingam and J. Jeya Maheshwari

24.1 Introduction

Gene manipulation in microbes began with the laboratory strain of *Escherichia coli*. Subsequently, several other organisms have been developed as cloning hosts for specific purposes, such as industrial production of commercially useful products. *Aspergillus niger*, a filamentous fungus, has been developed into a cell factory for the production of several enzymes. In the case of yeast *Saccharomyces cerevisiae*, a single-celled fungus, the driving force for gene manipulation is the need for the understanding of a model eukaryotic microbe. At present, several gram-positive bacilli are also used as cloning hosts. All these systems had the whole genome sequences of the host available very early on in public databases, and hence, the progress in the area of genetic manipulation was also rapid.

In the case of *Streptomyces*, gram-positive bacteria belonging to the group actinomyces, the driving force for developing these organisms as cloning hosts is their ability to produce a large number of antibiotics and other commercially useful products, including enzymes. In the words of Hopwood, father of *Streptomyces* genetics, '*Streptomyces* are microscopic chemists capable of production and modification of several chemical entities into potent antibiotics'. *Streptomyces* became famous and the study of *Streptomyces* biology took on a different path after the discovery of streptomycin, an antibiotic that kills *Mycobacterium tuberculosis*–another actinomyces that causes tuberculosis in humans.

In the early days, there were considerable difficulties in the classification of *Streptomyces*. Even their names were confusing. The word *Mycobacteria* means 'fungus bacterium' and *Streptomyces* means 'twisted fungus'. This created enough debates as to their true nature. However, their bacterial nature was confirmed with the start of the genetic studies on *Streptomyces coelicolor*. This strain contributed the most to the understanding of *Streptomyces* genetics and this bacterium produces colourful colonies, hence the name '*coelicolor*', meaning 'sky colour' or 'heavenly colour' in Latin.

24.2 Life cycle of *Streptomyces*

The life cycle of *Streptomyces* is complex, and the morphological changes that occur during the various stages of the life cycle closely reflect the physiological changes as well. The colonial complexity of *Streptomyces* increases when the colony grows out from a spore. After a suitable germination trigger, the germ tube emerges from the unigenomic spore and its cell wall grows mainly at the tip, at a rate that doubles with each replication of the chromosome in the initially septum-free germ tube. During the late vegetative growth, secondary metabolism is switched on and sporulation is initiated. One of the first steps in sporulation is the production of hydrophobic aerial hyphae that grow erect out of the plane of the agar. Aerial hyphae formation is dependent on the metabolic status, stress and a cascade of extracellular signalling molecules. Aerial hyphae become subdivided by the simultaneous in-growth of regularly spaced specialized double cross walls, necessitating the regular partitioning of DNA into each compartment. Once septation is complete, the compartments begin to separate from each other and eventually round up as their walls thicken to give chains of unigenomic spores (Manteca and Sanchez 2009).

The colonization ability of *Streptomyces* in the soil is greatly facilitated by vegetative hyphal mass and the spores that assist in its spread and persistence. The spores are a semi-dormant stage in the life cycle, which can survive in the soil for long periods. The spores impart resistance to low nutrient and water availability, whereas the mycelial stage is sensitive to drought. The germination of spores requires exogenous nutrients, water and Ca^{2+}, and the presence of nutrients in the growing site limits the extent of hyphal growth, differentiation into aerial hyphae and eventually spores.

Streptomyces produce many extracellular enzymes in the soil to decompose complex mixtures of polymers in dead plants, animal and fungal matter. *Streptomyces* aid in soil biodegradation by recycling the nutrients associated with recalcitrant polymers. The biodegradation capabilities of this genus have allowed characterization of various enzymes such as cellulases, xylanases, amylases, chitinases and maltases.

24.3 Discovery of antibiotics

Organisms belonging to the group *Streptomyces* are responsible for the production of several commercially useful antibiotics. As genetic manipulation has been used extensively in the analysis of antibiotic production, it is useful to have a brief account of antibiotics. Waxman defines antibiotics as follows, 'chemical substances that are produced by microorganisms and that have the capacity, in dilute solution, to selectively inhibit the growth of or even to destroy other microorganisms'.

Table 24.1 Number of chemical compounds produced by bacteria and fungi (data up to 2002)

Producer group	Antibiotics	Other bioactive compounds	Total compounds produced
Actinomycetes	8,700	1,400	10,100
Bacteria other than *Streptomyces*	2,900	900	3,800
Fungi	4,900	3,700	8,600
Total	16,500	6,000	22,500

Table 24.1 illustrates the number of antibiotics produced by bacteria and fungi. As of now, bacteria and fungi produce antibiotics whose commercial value is greater than 25 billion dollars.

24.4 *Streptomyces* genomics

The complete sequence of the genome of *S. avermitilis* was the first *Streptomyces* genome sequence to be published. Genome sequences of *S. coelicolor* and *S. griseus* were published subsequently. *Streptomyces* have a linear chromosome and this is very clear from the genome sequences reported till date. The linear nature of the *Streptomyces* chromosome was discovered in *S. lividans* and was later proven to be common to all the tested *Streptomyces*. Information pertaining to the whole genome sequence of three *Streptomyces* is summarized in Table 24.2. Terminal inverted repeats (TIRs) are present at both ends of the *Streptomyces* chromosome and are termed as telomeres as they bind proteins covalently. TIRs and the protein which binds to them are unique to *Streptomyces*. Bacteria such as *Agrobacterium tumefaciens* and *Borrelia burgdorferi* too harbour linear chromosomes but lack TIR sequences. Both *A. tumefaciens* and *B. burgdorferi* also lack the protein which binds to TIRs, which indicate that both TIRs and the protein can probably be unique to *Streptomyces* or other actinobacteria.

The origin of replication of the *Streptomyces* chromosome is located in the middle of the linear chromosome and replication proceeds bidirectionally towards the telomeres. The telomere-associated protein (Tap) plays an important role in the replication of telomeres by specifically binding to the telomeric regions. On an average, 43% of genes are common among *Streptomyces* and these proteins were identified in all three *Streptomyces*

Table 24.2 Properties of *Streptomyces* genome – a comparison

	S. coelicolor A3(2) M145	*S. avermitilis* ATCC31267	*S. griseus* IFO 13350
Chromosome topology	Linear	Linear	Linear
Length (bp)	86,67,507	90,25,608	85,45,929
Average GC content (%)	72.1	70.7	72.2
Plasmids	Strain M145 lacks SCP1 and SCP2	SAP1 and SAP2	Absent
Total ORFs	7,825	7,582	7,138
rRNA operons	6	6	6
No. of tRNA genes	63	68	66
Average CDS length (bp)	991	1,027	1.055
Coding density (%)	88.9	86.3	88.1
TIR length (bp)	21,653	49	1,32,910
oriC position	42,69,853 to 42,72,747	52,87,933 to 52,89,023	43,24,631 to 43,25,203

genomes sequenced. Comparison of genomes of *S. coelicolor*, *S. avermitilis* and *S. griseus* has revealed 6.4 Mb of the chromosome as an internal core region.

Most *Streptomyces* species are unstable for certain characters, the rate of spontaneous mutation of these genes being higher than 0.1% per colony-forming spore. The major alterations found in many of these mutants are the deletions in DNA. These DNA alterations are enhanced by treatments with acridines and ethidium bromide or by cold shock or protoplast generation. This phenomenon of genetic instability seems to be ubiquitous in the genus *Streptomyces*. Genetic instability affects several properties, including morphological differentiation, production of secondary metabolites and extracellular enzyme production, but rarely are genes for primary metabolism affected. The mutants generated by genetic instability in *S. lividans* are cml^S and arg^G. The cultures of *S. fradiae* and *S. lividans* treated with sub-inhibitory concentrations of antibiotics showed a loss of amplified DNA. Using this system, it has been shown that the loss of amplified DNA affected protease production and the onset of sporulation.

24.5 Genetic mapping in *Streptomyces*

Before the advent of chromosome manipulation using recombinant DNA techniques and physical techniques such as pulsed-field gel electrophoresis, the study of the genetics of *Streptomyces* was a rather cumbersome activity and only a few laboratories could successfully do a genetic cross, and what was still more difficult was to analyse the results of such a cross. The early studies using genetic cross in *S. coelicolor* implied that the chromosome of *Streptomyces* is circular. However, the analysis of the chromosome after restriction digestion, using rare cutting enzymes, and pulsed-field gel electrophoresis revealed that the chromosome of *Streptomyces* is linear. The introduction of genomic technologies thus replaced the classical bacterial genetic crosses to identify the linkage of markers. *In vivo* genetic analysis is now almost completely replaced by *in vitro* techniques. The genetic map and the physical map of the 8-Mb *S. coelicolor* genome can be seen in Redenbach *et al.* (1996).

24.6 *Streptomyces* in industrial production

Strategies for yield improvement in industrial processes have mainly been achieved in using media optimization and process development in the early days. For improving secondary metabolite production in *Streptomyces*, classical techniques such as mutation, selection and recombination protocols have been used for some time. However, these techniques do not provide information on the genes that are most important for the high-level production of secondary metabolites and those that relieve the rate-limiting steps in biosynthetic pathways. This information will enable the development of rational approaches to intervene in primary metabolism for a better supply of precursors and to influence the regulation and synthesis of secondary metabolites. In recent years, an increasing number of molecular genetics approaches have been developed and widely used to alleviate the problems encountered using classical techniques.

24.7 Gene cloning in *Streptomyces*

Most of the self-replicating plasmids in *Streptomyces*, which are used for constructing plasmid vectors, cause substantial reduction in secondary metabolite yields. The stability of cloned genes is also a major problem. The presence of a potent restriction system operating in many of the *Streptomyces* and actinomycetes interferes in the introduction of foreign DNA by conventional protoplast transformation. Successful cloning strategies evolved only when these problems were solved.

Basic steps involved in cloning genes from heterologous organisms or self-cloning (cloning genes from the same species or strain) in *Streptomyces* follow similar steps when compared to *E. coli*.

24.7.1 Isolation of passenger DNA

Each tissue source demands certain modifications in the isolation scheme of genomic DNA. Primary variations could be the growth in appropriate media and in the method used to break the cell wall. This step includes a detergent as well as a cell wall–digesting enzyme such as lysozyme. The next step is to separate proteins from nucleic acids. Chloroform, phenol and predominantly a mixture of both are used to denature and separate proteins. Upon centrifugation, denatured proteins will be in the interphase while nucleic acids will be in the aqueous phase. This step is usually followed by precipitation of DNA and RNA with ice-cold ethanol, polyethylene glycol (PEG) or isopropanol. This step concentrates nucleic acids and also removes the contaminating salts, detergents and phenol.

24.7.2 Isolation of plasmid vector DNA

Plasmid isolation methods developed for *E. coli* could be used for the isolation of plasmid DNA from *Streptomyces* also. The lysis procedure, such as alkaline lysis, removes most proteins. However, RNA contamination has to be cleared either by lithium chloride precipitation or RNAse treatment followed by isopropanol precipitation. A good preparation of plasmid DNA is essential for the succeeding steps.

24.7.3 Gene transfer procedures

The following procedures are currently being used for introducing foreign DNA into the *Streptomyces* mycelium.

24.7.3.1 Polyethylene glycol–mediated protoplast transformation

The conventional way of introducing plasmid DNA or recombinant plasmid DNA into *Streptomyces* is by protoplast transformation. The initial work on the formation, regeneration and transfection of protoplast led to the discovery that protoplast could be transformed by plasmid DNA at a very high frequency in the presence of PEG. A standard procedure for PEG-mediated transformation of *Streptomyces* was developed later. This protocol has been adopted with minor modifications for many different species. Several parameters influencing protoplast regeneration and transformation frequency were optimized for individual species. It has been reported that the methods used for protoplastation and transformation of *S. lividans* and *S. coelicolor* are ineffective for industrial *Streptomyces* strains that produce antibiotics.

A major limitation of this method is that most streptomycetes express a strong restriction system that can lower the efficiency of protoplast transformation and, in some cases, inhibit transformation completely. Apart from host restriction, it has been reported that protoplastation increases genetic instability in *Streptomyces.*

24.7.3.2 Transduction in *Streptomyces*

Bacteriophage-mediated transduction is another means of introducing DNA into *Streptomyces*. Transducing phages take advantage of introducing linear concatemers of DNA during physiological phases of cells that minimizes the expressions of restriction systems. However, this method is limited to self-replicating vectors derived from pIJ101 and its derivatives. Recently, generalized transducing phages were isolated and found to transduce markers or plasmids between derivatives of *S. coelicolor*.

24.7.3.3 Electroporation and electroduction

Electrotransformation of plasmid DNA into mycelium and germinating spores of *Streptomyces* can be used for electrotransformation. The procedures using fresh germinating spores are better than the other procedures.

Electroduction is a simple and fast method that allows the direct transfer of bifunctional plasmids from streptomycetes to *E. coli*. However, the transfer by electroduction from *E. coli* to *Streptomyces* has not been shown. This method depends on the formation of transient pores in cell membranes under conditions of electroporation, when *Streptomyces* mycelium carrying a shuttle plasmid is mixed with electroporation-competent *E. coli* cells; the plasmids are transferred to *E. coli* after an electrical pulse. This method is useful in rapid analysis of recombinant *Streptomyces* plasmids in *E. coli* and overcomes the difficulties associated with plasmid preparation from *Streptomyces.*

24.7.3.4 Conjugation in *Streptomyces*

Conjugation in *Streptomyces* is a very distinct process from genetic exchange in other bacteria not only in its molecular mechanism but also in its phenotypic appearance. Plasmid transfer results in the formation of a characteristic 'pock' structures when a plasmid-carrying donor is plated onto agar containing a plasmid-free recipient. The pocks are growth inhibition zones of 1–3 mm. Within this pock structure, the morphological differentiation of the recipient mycelium, which has newly acquired a plasmid, is temporarily retarded.

24.7.4 Cloning procedure

For cloning purposes, vector and passenger DNA are cleaved with restriction enzymes and an appropriate mix of cleaved passenger and vector DNA are mixed and ligated. One precaution to be observed in this step is the high G + C content of the *Streptomyces* chromosomal DNA. For example, complete digestion with *Eco*RI (recognition site GAATTC) leads to much larger-size DNA fragments. Another precaution needed is the amplification of certain sections of the chromosome. In extreme cases, more than 300 copies of up to a 5-kb segment may exist as tandem amplification. These will be over-represented in gene libraries if precaution is not taken to remove these before library construction. Some strains also have several copies of plasmid DNA, and some of these plasmids are linear and, hence, will also appear in chromosomal DNA preparation.

24.7.5 Cloning vectors

Plasmid and phage vectors are available for gene cloning in *Streptomyces*. There are several plasmid vectors constructed using the replicons of naturally occurring plasmids such as pIJ101, a multicopy vector. The host range of these vectors is good. However, the presence of restriction systems in different *Streptomyces* species makes introduction of DNA into industrial strains rather cumbersome. This limitation has to be kept in mind if one wishes to use particularly industrial strains as cloning hosts. The common host used for cloning purposes is *S. lividans*, as this strain does not have any restriction

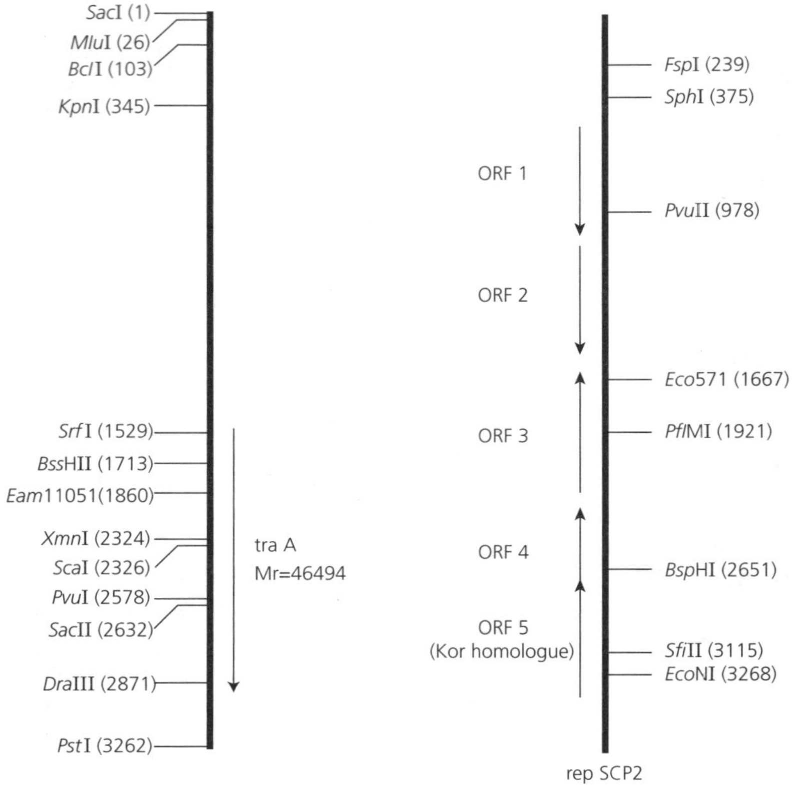

Figure 24.1 Map of the linear plasmid, SCP2.

system. This strain also has a high transformation efficiency and yields a large number of transformants.

24.7.5.1 Plasmid vectors

There is a wide choice of plasmid vectors for gene cloning in *Streptomyces*. The choice of a vector depends on the type of cloning experiment and the ultimate objective of the cloning exercises. Some of the common considerations are copy number of the cloned insert, size of the foreign DNA to be cloned and the expression of the cloned DNA. There are other special circumstances such as examination of promoter and other regulatory regions when one has to use special-purpose vectors.

Low-copy-number vectors

Low-copy-number vectors are derived from the naturally occurring linear conjugative plasmid SCP2 (Fig. 24.1). This plasmid is 31.4 kb in size and is linear. The derivatives of SCP2 such as pIJ920, however, are circular plasmids and can carry large segments of DNA. This vector was used for cloning the antibiotic biosynthetic gene cluster encoding the antibiotic actinorhodin. One of the major problems with this vector is the difficulty in the isolation of recombinant plasmid DNA due to its large size and low copy number.

High-copy-number vectors

These are the most popular vectors and most of them are derived from the naturally occurring plasmid pIJ101. The plasmid map of pIJ101 and its derivative vectors, pIJ702 and pIJ487, is shown in Fig. 24.2 with appropriate cloning sites. The copy number of these vectors could go up to 300 per cell. The naturally occurring plasmid pSG5 carries a temperature-sensitive replicon and will be lost when the cells are grown at 37°C. Useful vectors such as pGM9 and pGM160 are constructed using this plasmid.

24.8 Gene cloning using integrating vectors

Unlike the autonomously replicating plasmid vectors, integrating vectors have several advantages. The following subsections discuss some of the approaches and benefits of using integration vectors for gene cloning in *Streptomyces*.

24.8.1 Conjugative plasmids of *Streptomyces*

The ability to move from one cell to another germinating spore and to spread across the mycelial compartment makes conjugative plasmids a useful system for the construction of integrating vectors. These vectors also have the integration system from the *Streptomyces* phage ϕC31, which allows them to integrate stably in one or two specific sites. These vectors could move among phylogentically distant organisms also. Conjugal

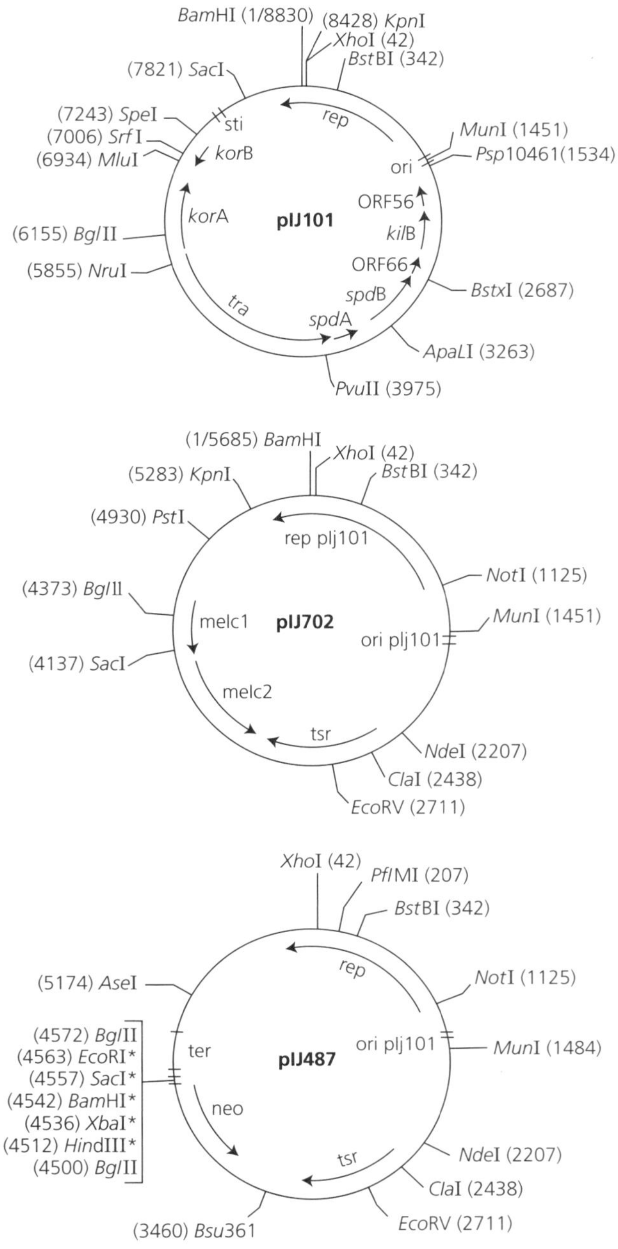

Figure 24.2 Physical map of pIJ101, pIJ702 and pIJ487.

transfer of these vectors from *E. coli* to *Streptomyces*, *Mycobacteria*, yeast as well as animal cells has been demonstrated.

24.8.1.1 Conjugation between *E. coli* and *Streptomyces*

Intergeneric conjugation is a convenient way to introduce DNA into *Streptomyces* strains. The protocol for transfer of DNA from *E. coli* to *Streptomyces* does not require any strain-specific optimizations of protoplasting conditions. Furthermore, this method circumvents the host restriction system operating in most of the *Streptomyces* strains. The frequency of conjugal transfer between *E. coli* and different gram-positive bacteria is very high.

Intergeneric conjugation protocol involves mixing of late-log phase *Streptomyces* mycelium or heat shock–given spores with the plasmid carrying an *E. coli* donor and plating onto agar plates. After 18 h, the plates are overlaid with appropriate antibiotics to select the transconjugants. Nalidixic acid was shown to be very effective in preventing outgrowth of the *E. coli* donor. Although conjugation with freshly germinated spores is convenient to do, mycelial fragments also could be used.

There are several advantages of using *E. coli* as a conjugative donor for introducing DNA into streptomycetes and other actinomycetes, and a variety of versatile *oriT* containing vectors are now available. These vectors replicate in *E. coli* and all the constructs can be made relatively easily in *E. coli* and transferred. The mechanism of this transfer has been studied extensively and the plasmid functions needed for this process is defined well.

E. coli donor S17-1, which carries a RP4 derivative integrated into the chromosome of a *res*/*mod E. coli* strain, is used extensively as donor. Plasmids containing the pIJ101 and pBR322 origins of replication could be transferred from *E. coli* S17-1 into *S. lividans*, *S. pristinaespiralis* and *S. viridochromogenes*.

24.8.1.2 *Streptomyces* as recipient

Spores as well as mycelial fragments of *Streptomyces* could act as recipients for intergeneric conjugation. Freshly germinated spores of *Streptomyces* were found to be the

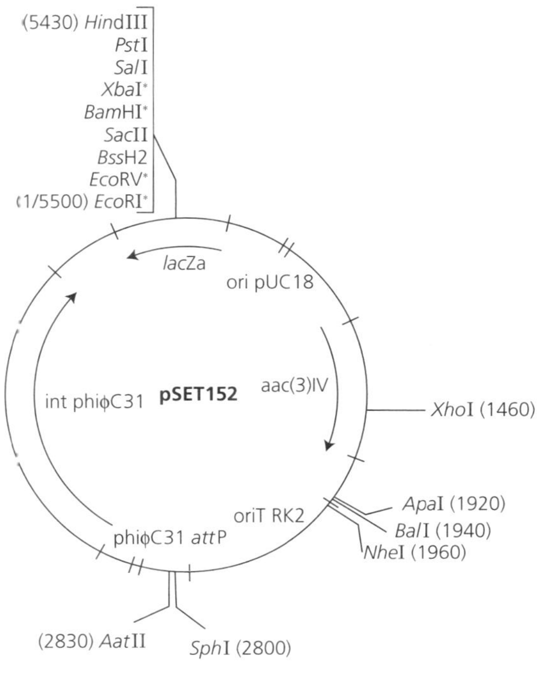

Figure 24.3 Physical map of the integrative vector, pSET152.

convenient recipient for intergeneric conjugation. The efficiency of exconjugant formation using germinating spores and mycelial fragments of *Streptomyces* indicated that mycelial fragments yielded a very high exconjugant frequency when compared with germinating spores.

24.8.1.3 ϕC31-based integrative vectors

A 4-kb segment of ϕC31 DNA (containing ϕC31 *int*/*att*) was sufficient to cause non-replicating vectors to integrate into the chromosome of *S ambofaciens*, *S. griseofuscus*, *S. fradiae*, *S. lividans*, *S. lipmanii* and *S. thermotolerans*.

Several vectors have been developed based on ϕC31 *int*/*att* functions and these vectors provide single-copy integration per genome and optimal segregational stability in *Streptomyces*. The derivatives of the integrative vector pSET152 (Fig. 24.3) have also been used for complementation and expression studies.

24.8.1.4 Effect of plasmid integration on cultural morphological properties and antibiotic production of the strain

The effect of a ϕC31 *int*/*att*–carrying integrating vector on development, differentiation and antibiotic production in exconjugants was also evaluated. The exconjugants of *S. aureofaciens* and *S. virginiae* did not show any alterations in cultural and morphological properties or antibiotic production.

Similarly, in the case of *Saccharopolyspora spinosa* and *Sacch. toyocaensis*, the plasmid pOJ436 could insert at the ϕC31 *attB* site without having any deleterious effect on secondary metabolite biosynthesis. No negative effect of plasmid integration (at ϕC31 *attB* site) on cultural and morphological properties of *S. albus*, *S. antibioticus*, *S. griseus*, *S. aureofaciens*, *S. bambergiensis*, *S. lavendulae*, *S. lividans*, *S. purpureus*, *S. rimosus*, *S. venezulae* and *S. virginiae* has been reported. Integration of pSET152, carrying inserts, at the ϕC31 *att*B site did not affect the expression of genes such as σ^{HrdB}, the essential sigma factor of *S. coelicolor* A3(2). However, it will be useful to ascertain this aspect before starting a cloning experiment.

24.8.2 *Streptomyces* integrative vector can be used for cloning in *Mycobacterium smegmatis*

The pSET152 and other integrating vectors based on the ϕC31 recombination system used widely in *Streptomyces* research have been exploited for cloning purposes in other organisms such as *Mycobacterium smegmatis* and *Rodococcus equi*. To illustrate the use of the *Streptomyces* integrative vector pSET152, cloning and expression of a gene encoding a small heat shock protein (sHSP18) from *Mycobacterium leprae* is described below. The complete coding region of this gene was cloned with its native promoter in pSET152. pSET152 is non-replicative in *Mycobacterium* as in *Streptomyces*, but can integrate

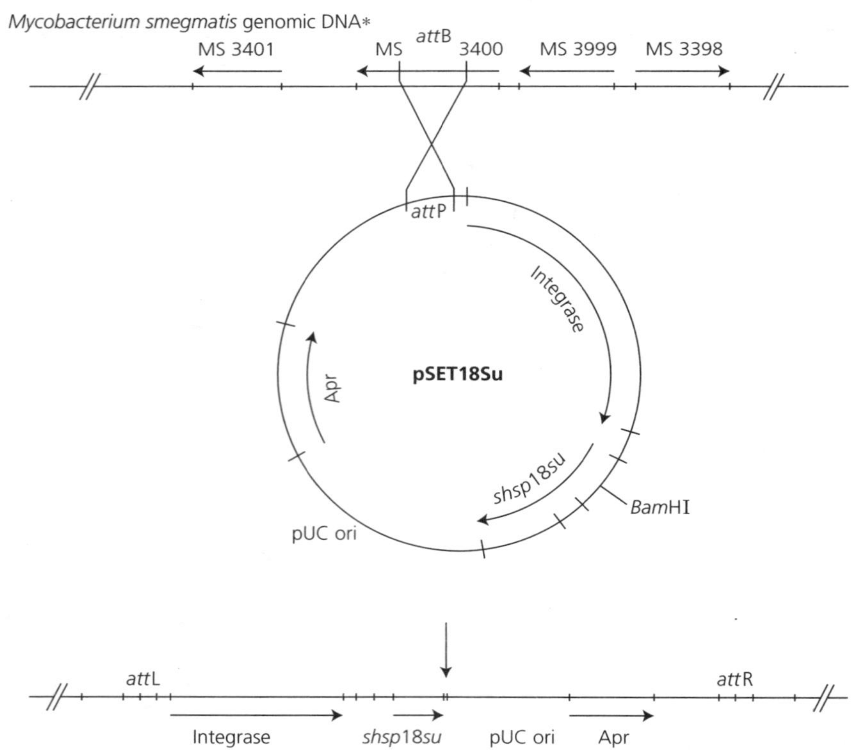

Figure 24.4 Schematic representation of the integration of pSET18Su in the genome of *M. smegmatis*. *For illustration purpose, only a part of the circular genome is shown. Genes are represented by their locus ID in the figure, where MS represents *M. smegmatis*.

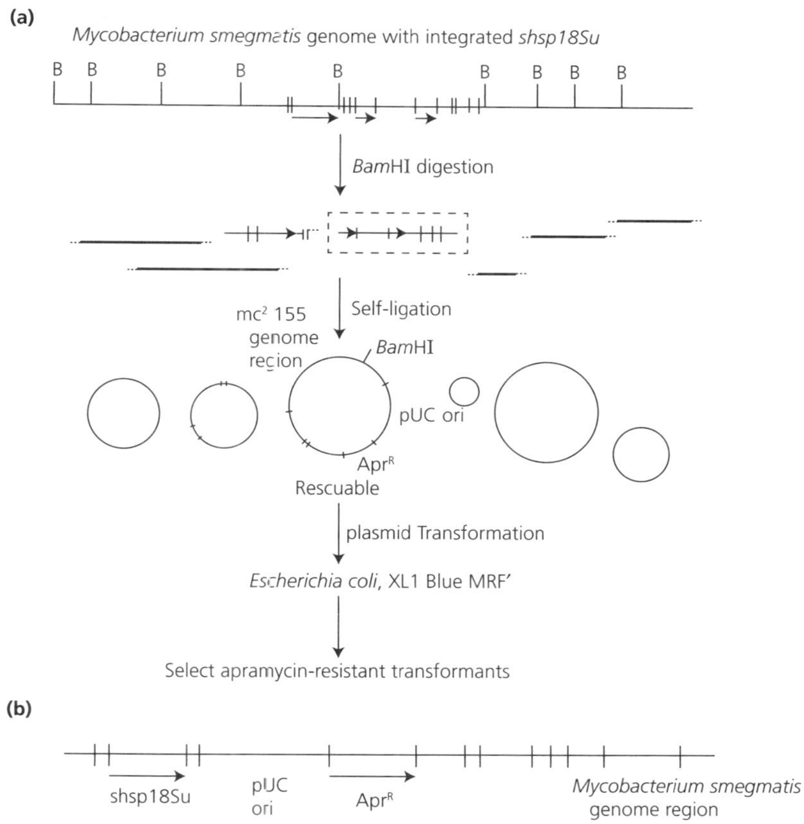

Figure 24.5 Plasmid rescue strategy. (a) Schematic representation of the steps involved in this strategy to confirm the integration of *shsp18Su* in *M. smegmatis* genome. (b) Details of the rescuable fragment.

into the chromosomal attachment site of bacteriophage ϕC31 (*attB*). An apramycin resistance gene serves as the selection marker. The recombinant plasmid, pSET18Su, was transformed in *E. coli*; XL1 Blue MRF and the transformants were selected based on apramycin resistance.

24.8.2.1 Integration of *shsp18* in the genome of *M. smegmatis*

A schematic representation on the integration of pSET18Su in *M. smegmatis* genome is shown in Fig. 24.4. The ϕC31 integrase acts on the *attP* site in the vector to bring about a site-specific recombination with the *attB* site in the genome. This results in the integration of the entire vector along with the cloned gene in an orientation as shown in Fig. 24.4. The *attB* and *attP* sites after integration form the *attL* and *attR* sites on either side of the integrated vector. *M. smegmatis*, which is inherently susceptible to apramycin, develops resistance after the integration event and, hence, apramycin-resistant transformants were selected.

24.8.2.2 Confirmation of the integration event

Integration of *shsp18Su* (the gene of interest cloned in pSET152) in *M. smegmatis* genome was confirmed by a plasmid rescue strategy. This strategy, illustrated in Fig. 24.5(a), exploits the fact that after the integration, digestion of genomic DNA with an enzyme in the multiple cloning site of the vector would generate a fragment with a segment of the integrated vector and this fragment can be rescued as a plasmid in *E. coli*. The enzyme chosen here was *Bam*HI, which is present only in the vector and not in the cloned gene, *shsp18Su*. Genomic DNA isolated from *M. smegmatis* transformant was digested completely with *BamH*I. This generates a large number of fragments from all possible *BamH*I sites. If the integration had taken place among these fragments, there will be one fragment carrying the *shsp18Su*, apramycin resistance gene, *E. coli* ori and *att*R along with a portion of the mycobacterial genome, as shown in Fig. 24.5(b). As all the fragments had *BamH*I cohesive ends, they were self-ligated and transformed to *E. coli*. Except the fragment shown in Fig. 24.5(b), carrying the pUC replicon, none of the other fragments would have a combination of *E. coli* ori and apramycin gene and hence would not be propagated in *E. coli* on apramycin selection. Apramycin-resistant *E. coli* transformants were obtained, indicating that the pSET152 vector was integrated in the genome of *M. smegmatis*, which was further confirmed by subsequent restriction analysis.

Many of these themes are common among *Streptomyces* and *Mycobacteria*, and hence, are useful in gene manipulation in an industrial microbe and a pathogen.

References

Anne J, Van Mellaert L. *Streptomyces lividans* as host for heterologous protein production. *FEMS Microbiol Lett* 1993;114:121–128.

Bierman M, Logan R, O'Brien K, Seno ET, Nagaraja Rao R, Schoner BE. Plasmid cloning vectors for the conjugal transfer of DNA from *Escherichia coli* to *Streptomyces* spp. *Gene* 1992;116:43–49.

Gusek TW, Kinsella JE. Review of the *Streptomyces lividans*/vector pIJ702 system for gene cloning. *Crit Rev Microbiol* 1992;18: 247–260.

Hopwood DA. *Streptomyces in Nature and Medicine: The Antibiotic Makers*. New York: Oxford University Press; 2007.

Hopwood DA, Bibb MJ, Bruton CJ, Chater KF, Fietlson JS, Gil J. Cloning *Streptomyces* genes for antibiotic production. *Trends Biotechnol* 1983;1:42-48.

Hutchinson CR. The impact of genetic engineering on the commercial production of antibiotics by *Streptomyces* and related bacteria. *Appl Biochem Biotechnol* 1987;16:169–190.

Kieser TB, Bibb MJ, Buttner MJ, Chater KF, Hopwood DA. *Practical Streptomyces Genetics*. Norwich: John Innes Foundation; 2000.

Kieser, TB, Hopwood DA, Wright HM, Thompson CJ. pIJ101, a multi-copy broad host-range *Streptomyces* plasmid: Functional analysis and development of DNA cloning vectors. *Mol Gen Genet* 1982;185:223–228.

Kuhstoss S, Richardson MA, Rao RN. Plasmid cloning vectors that integrate site-specifically in *Streptomyces* spp. *Gene* 1991;97:143–146.

Manteca A, Sanchez J (2009). *Streptomyces* development in colonies and soils. *Appl Environ Microbiol* 2009;75:2920–2924.

Paradkar A, Trefzer A, Chakraburtty R, Stassi D. *Streptomyces* genetics: A genomic perspective. *Crit Rev Biotechnol* 2003; 23:1–27.

Redenbach M, Kieser HM, Denapaite D, Eichner A, Cullum J, Kinashi H, *et al.* A set of ordered cosmids and a detailed genetic and physical map for the 8 Mb *Streptomyces coelicolor* A3(2) chromosome. *Mol Microbiol* 1996;21:77–96.

Tomich PK. *Streptomyces* cloning: Useful recombinant DNA systems and a summation of cloned genes. *Antimicrob Agents Chemother* 1988;32:1465–1471.

Wohlleben W, Hartmann V, Hillemann D, Krey K, Muth G, Nussbaumer B, *et al.* Transfer and establishment of DNA in *Streptomyces* (a brief review). *Acta Microbiol Immunol Hung* 1994;41:381–389.

25

Yeast *Pichia pastoris* as host for recombinant DNA

Upasana Arora and Navin Khanna

25.1 Introduction

Recombinant protein production has revolutionized the biotech industry and has helped in lowering the prices of several proteins and enzymes, making them affordable to people in the developing countries. Various hosts, such as bacteria, yeasts, insect and mammalian cell lines, have been evaluated for expression of different proteins. Interestingly, it has been found that two hosts, the bacterium *Escherichia coli* and the yeast *Pichia pastoris*, have been used for expression of 80% of the recombinant proteins reported in the literature in the last 15 years (Sørensen 2010). This chapter highlights the basic features of *P. pastoris* and the reasons behind the emergence and rise of this expression host.

As already stated, *P. pastoris* serves as an attractive eukaryotic host for recombinant protein expression. It was first used as a source of single-cell protein, but later, in 1980s, efforts were made to develop it as an expression host and since then its demand and popularity have grown over the years. Fig. 25.1 reflects the rise in the study and use of *P. pastoris* since 1990, as per the available literature in PubMed database (http://www.ncbi.nlm.nih.gov/pubmed/).

The potential of *P. pastoris* was largely exploited in the late 1990s and since then it has been able to grab and maintain the interest of both academicians and industrialists.

A study of publications on *P. pastoris* in 2013 reflects that it was largely used as a host for expression of heterologous proteins (Fig. 25.2).

With the increase in the usage of *P. pastoris*, publications related to various strategies, techniques and methods involved in cloning and expression have increased and constitute the second largest contributor to *P. pastoris* publications. To sort out problems associated with protein secretion and expression in *P. pastoris*, efforts are now being made to study the genome and proteome of *P. pastoris* strains in greater

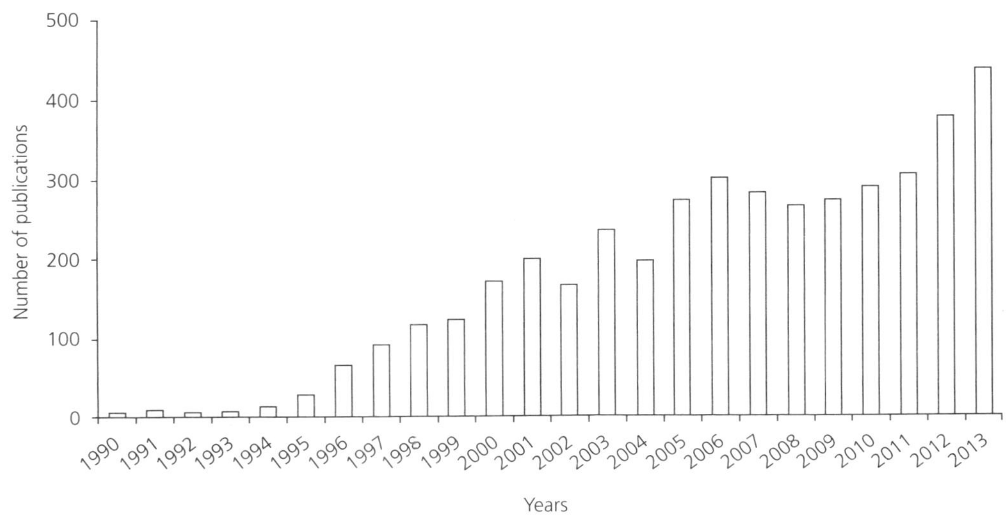

Figure 25.1 Increase in the number of publications (in PubMed) on *P. pastoris* (used as a key word) from 1990 to 2013.

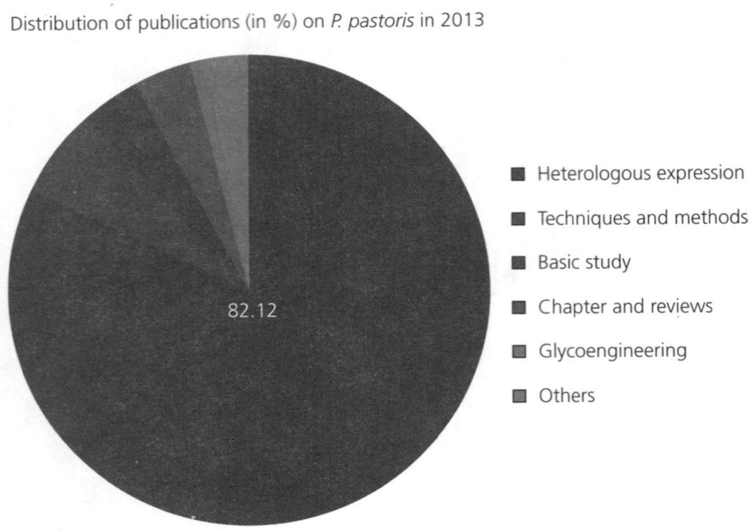

Figure 25.2 Sorting of publications (in PubMed), released on *P. pastoris* (used as a keyword) in 2013, according to the focus of the study. Most of the articles (82.12% of total publications) were related to the utilization of *P. pastoris* as an expression host for heterologous proteins. The other publications (clockwise: in descending order of their percent contribution to the total number of publications) focused on techniques and methods involved in cloning and expression, basic study of *P. pastoris* physiology, chapters and reviews, glycoengineering of *P. pastoris* and other publications. See Colour Plate 30 on page 521.

detail to identify host proteins which are naturally secreted. This would enable identification of other alternative secretion signals that can be used. Hence, the study of basic physiology of *P. pastoris* is also gaining importance. *P. pastoris*–expressed proteins are known to be over-glycosylated. Therefore, to express human proteins with glycosylation pattern similar to native proteins, efforts are being made to glycoengineer it to develop humanized *P. pastoris*. It is evident from Fig. 25.2 that publications on heterologous gene expression in *P. pastoris* outnumber the other spectrum of publications on this host, highlighting the keen interest in using it as an expression host.

25.2 Methylotrophic yeast

Methylotrophic yeasts were first described by Koichi Ogata (Cereghino and Cregg 2000). Methylotrophs are those organisms which are able to utilize either reduced one-carbon compounds (such as methanol or methane) or multi-carbon compounds lacking carbon–carbon bonds (such as dimethyl ether or dimethylamine) as sole carbon and energy source. These are obligate methylotrophs. Facultative methylotrophs can utilize a variety of other multi-carbon compounds. Methylotrophic yeasts predominantly use methanol as the one-carbon compound. The major factors which make methylotrophic yeasts highly competent for industrial application are the high cell densities to which they can grow and the low cost, abundance and purity of the carbon source such as methanol. *P. pastoris* is an example of such methylotrophic yeast. The other major genera included in this category are *Candida*, *Hansenula* and *Torulopsis* (Anthony 1982).

25.3 Emergence of *P. pastoris*

P. pastoris was identified as a potential source of single-cell protein (SCP) because of its ability to grow on an inexpensive, readily available waste product from oil industry as its carbon source. Growth media and protocols for continuous culture of *P. pastoris* on methanol at high cell densities (>130 g/dry cell weight) were developed for the first time by Phillips Petroleum Company (Cereghino and Cregg 2000). With the advent of oil crisis in 1970s, the market of *P. pastoris* as SCP fell because of increase in the cost of methane, and hence methanol. A decade later, *P. pastoris* was again explored by Philips Petroleum Company in association with Salk Institute Biotechnology/Industrial Associates, Inc. (SIBIA, La Jolla, CA); this time as a host for the expression of heterologous proteins (Cereghino and Cregg 2000; Macauley-Patrick *et al.* 2005). Fig. 25.3 gives a brief history of *P. pastoris* in a schematic manner.

This methylotroph was genetically manipulated to exploit its utilization for high levels of heterologous protein expression. After isolation of the gene and promoter of alcohol oxidase (*AOX*), vectors, strains and protocols for molecular manipulation of *P. pastoris* were developed. The combined advantage of its ability to grow at high cell density and high strength of the alcohol oxidase promoter, along with its regulated expression, resulted in surprisingly elevated levels of foreign protein production, which made it a highly attractive expression host (Cereghino and Cregg 2000; Macauley-Patrick *et al.* 2005).

The patent for *P. pastoris* expression system lies with Research Corporation Technologies (Tucson, AZ) since 1993 and this expression system can be obtained from Invitrogen Corporation (Carlsbad, CA) (Cereghino and Cregg 2000; Macauley-Patrick *et al.* 2005).

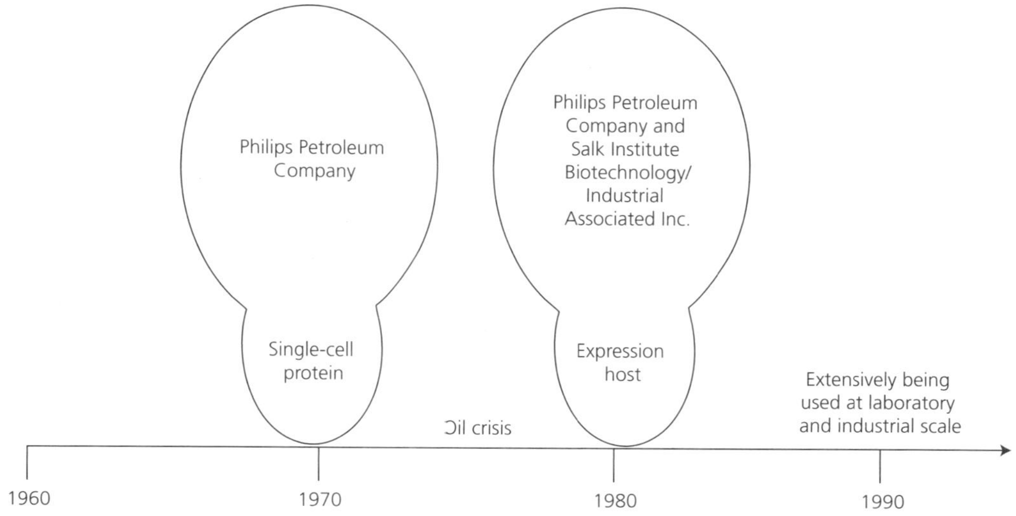

Figure 25.3 Emergence of *P. pastoris* on a time scale. It came into use, in the early 1970s, as a source of SCP, developed by Philips Petroleum Company. With the advent of oil crisis, in mid 1970s, its demand as SCP fell sharply. It emerged back again in 1980s, but this time as a eukaryotic expression host due to the combined effort of Philips Petroleum Company and Salk Institute Biotechnology/Industrial Associates, Inc. Since then, it has been used for the expression of a number of proteins.

25.4 Features

The widespread use of *P. pastoris* as an expression system is derived from its various characteristic features, namely its ability to grow on methanol, express heterologous proteins at high level and allow post-translational modifications.

25.4.1 *P. pastoris* vs. other expression hosts

Though *E. coli* has been the most desired expression host (Sørensen 2010), *P. pastoris* has been able to establish its own identity. The major disadvantage of *E. coli* as an expression host is that it is a prokaryote and hence does not have any membrane-bound organelles as found in eukaryotes. Therefore, it lacks the ability to carry out post-translational modifications, which may affect the functionality of the protein. To overcome this problem, eukaryotic systems such as mammalian, insect and yeast cells have been exploited as hosts for expression of eukaryotic proteins. Yeasts cells are the most desirable of them because of the combined advantage of a prokaryotic system such as the ease of genetic manipulation and rapid growth characteristics, and a eukaryotic system such as the presence of subcellular machinery for performing post-translational protein modification (Daly and Hearn 2005; Macauley-Patrick *et al.* 2005; Demain and Vaishnav 2009). Table 25.1 gives a relative comparison between prokaryotic, yeast, and insect and mammalian cell lines as host for heterologous protein production.

Table 25.1 Comparison between prokaryotes, yeast, and insect and mammalian cell lines

	Prokaryotes (*E. coli*)	Yeast	Insect and mammalian cell lines
Ease of genetic manipulation	✓	✓	×
Post-translational modification	×	✓	✓
Rapid growth	✓	✓	×
Growth to high cell densities	×	✓	×
Cost effective	✓	✓	×

Cell lines from multicellular organisms, such as S2 cells from *Drosophila melanogaster* and Chinese hamster ovary (CHO) cells, are also used as eukaryotic expression hosts. But *P. pastoris* is considered better than them because of the high yields and high densities (Fig. 25.4) that can be attained with it. Moreover, it requires simpler media as against complex rich media supplemented with growth factors required by eukaryotic cell lines (Demain and Vaishnav 2009). By virtue of these characteristics, *P. pastoris* is considered to be a more commercially viable host. A large number of scientists also use this host to express the proteins to be used in structural studies by nuclear magnetic resonance (NMR). The reason behind this growing trend is that it can grow on media containing a single carbon and nitrogen source, which can be radio labelled (Wood and Komives 1999; Demain and Vaishnav 2009).

One of the most well-studied yeast has been *Saccharomyces cerevisiae*, which has been used for the production of some eukaryotic proteins. However, with the use of this eukaryotic expression host, high expression levels are generally not attained. Moreover, expression of foreign proteins puts the cells under stress, leading to low growth rates. To overcome this problem, the growth

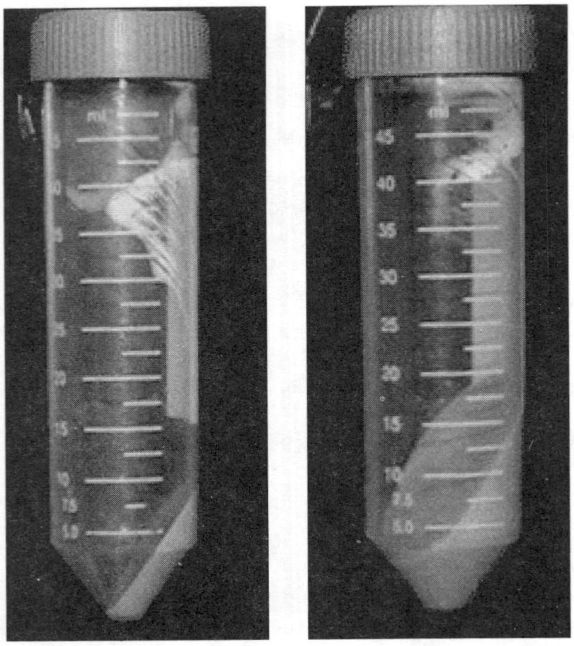

Figure 25.4 High cell density culture of *P. pastoris.* The *P. pastoris* culture in the two tubes depict shake flask culture of density of 1 OD_{600} unit on the left and high cell density culture of 130 g/L obtained in the fermenter (~500 OD_{600} units) on the right. See Colour Plate 31 on page 522.

and protein production phases were separated by using inducible plasmid promoters. This has also not been effective enough due to poor plasmid stability. Hence, it is not considered a very convenient system for scale-up to obtain high yields (Buckholz and Gleeson 1991). It has also been reported to hyper-glycosylate the expressed proteins, resulting in changes in their immunogenicity profile, reduced activity and shorter *in vivo* half-life of foreign proteins (Buckholz and Gleeson 1991; Demain and Vaishnav 2009; Damasceno *et al.* 2012).

To overcome these problems, other alternative yeasts such as *P. pastoris* were explored. *P. pastoris* has rapidly gained popularity as a suitable eukaryotic host because of several reasons. The protocols optimized for the growth of *S. cerevisiae* could be directly applied for *P. pastoris* growth, which reduced a lot of time and effort in developing this system as a suitable expression host (Higgins 1995). Higher and inducible expression levels could be achieved using *P. pastoris* as the host because of the presence of a strong, inducible promoter. As compared with *S. cerevisiae*, *P. pastoris* does not hyper-glycosylate its proteins and these modifications are in close resemblance to humans (Cregg *et al.* 1993; Hamilton and Gerngross 2007). Table 25.2 compares *S. cerevisiae* and *P. pastoris* with respect to different parameters.

P. pastoris has been used to express a large number of proteins on induction with methanol. The ability of *P. pastoris* to utilize methanol is attributed to peroxisomal enzymes, especially alcohol oxidase, the expression level of which is tightly regulated at the transcriptional level. In the presence of methanol, the expression level of alcohol oxidase increases to 30% of the total cellular protein, while in the presence of glucose or ethanol, alcohol oxidase is undetectable in cells. Hence, its promoter is most commonly used for inducible expression of heterologous proteins using methanol (Cregg *et al.* 1993; *Easy Select™ Pichia Expression Kit Manual* 2010).

There are two types of alcohol oxidases, AOX1 and AOX2, encoded by *AOX1* and *AOX2* genes. Though they are highly homologous, with 92% (Berardi 1997) and 97% (*Easy Select™ Pichia Expression Kit Manual* 2010) homology at the nucleotide and amino acid sequence level, respectively, they greatly differ in their methanol utilization capacity, with AOX1 playing the major role. Their promoters are believed to be responsible for the difference in the activity of the two enzymes (Berardi 1997). *AOX1* promoter is controlled by repression/de-repression and the induction mechanism. The presence or absence of an initial carbon source, such as glycerol, represses or de-represses transcription, respectively, and the addition of methanol induces the transcription of the gene under this promoter. *AOX2* gene promoter also functions similarly (Cereghino and Cregg 2000; Macauley-Patrick *et al.* 2005; Damasceno *et al.* 2012), but the difference in the activity of the two alcohol oxidases can be attributed to the difference in regulatory elements of the two genes. Several *cis*-acting elements in *AOX1* promoter, such as Regions D and E, have been identified (Xuan *et al.* 2009), as illustrated in Fig. 25.5. For instance, Regions D and E are positive *cis*-acting elements, located between positions –638 and –510 and –552 and –442, respectively. These regions have been identified through deletion analysis, where the deletion of either region resulted in the reduction of the transcription level by 90% and the methanol-induced expression level by 85%. Unlike Region E, Region D has been reported to form a DNA–protein complex with the cell extract under methanol-induced and glucose/methanol-repressed state. Hence, Region D is expected to contain the binding site for transcription regulator factor, which is responsive to methanol induction. Sequence analysis of Region D has revealed that it is an invert repeat-like sequence, GTGGGGTCAAATAGTTTCATGTTCCCCAA, similar to upstream activation sequence 1 (UAS1) of alcohol dehydrogenase 2 gene (*ADH2*) in *S. cerevisiae* (Xuan *et al.* 2009).

Utilization of methanol leads to the generation of formaldehyde and hydrogen peroxide. This reaction takes place in peroxisomes and all the key participating enzymes are compartmentalized there. The type of carbon source available for utilization also decides the number of peroxisomes present. Very few peroxisomes are present when cells are grown in the presence of glucose as substrate. On the contrary, in the presence of methanol as substrate, peroxisomes can take up to 80% of total cell volume (Gellissen 2000). Peroxisomes

Table 25.2 Comparison between *S. cerevisiae* and *P. pastoris*, highlighting the suitability of *P. pastoris* over *S. cerevisiae* as an expression host

	S. cerevisiae	*P. pastoris*	**Outcome**
Strong inducible promoter	×	✓	
High expression level	×	✓	
High cell density	×	✓	Selection of *P. pastoris* as a suitable expression host over *S. cerevisiae*
Easy to scale-up	×	✓	
Post-translational modification similar to mammalian proteins	×	✓	

constitute a very important part of cells for methanol utilization. Cells defective in peroxisome biogenesis have been found to be incapable of growing in methanol. Also, when cells grown in methanol are transferred to a glucose-containing medium, peroxisomes are eliminated by autophagic degradation in vacuoles (Glick 1996).

High expression levels with appropriate post-translational modification is always desired and preferred from an expression host. *P. pastoris* performs both *N*- and *O*-linked glycosylations. Though *O*-linked glycosylation, in most of the eukaryotes, preferably occurs on the hydroxyl group of threonine and serine, in *P. pastoris*, there is no preference for any amino acid. The type of *O*-linked saccharides is also different. In mammals, *O*-linked saccharides mainly consist of sialic acid, galactose and *N*-acetylgalactosamine, while in *P. pastoris*, mannose residues make up the *O*-linked saccharides (Damasceno *et al.* 2012).

Although *S. cerevisiae* can also produce glycosylated proteins, they are not preferred for use in therapeutics because of their highly antigenic nature. Hence, mammalian cell lines have been more commonly used for this purpose. It has been found that proteins expressed in *S. cerevisiae* are hyper-mannosylated as compared with those in *P. pastoris*. The length of these carbohydrate chains varies between the two yeasts. The longest carbohydrate chain in *P. pastoris* usually contains approximately 20 mannose residues (Demain and Vaishnav 2009). On the contrary, in *S. cerevisiae*, though these chains typically contain 50–150 residues (Demain and Vaishnav 2009), more than 200 residues (Hopkins *et al.* 2011, Striebeck *et al.* 2013) may also be incorporated. Unlike in *S. cerevisiae*, glycans in *P. pastoris* do not contain terminal α-1,3 linked mannose residues. Such linkages are catalysed by α-1,3-mannosyl transferase enzyme, which is absent in *P. pastoris* (Demain and Vaishnav 2009). These linkages are assumed to be responsible

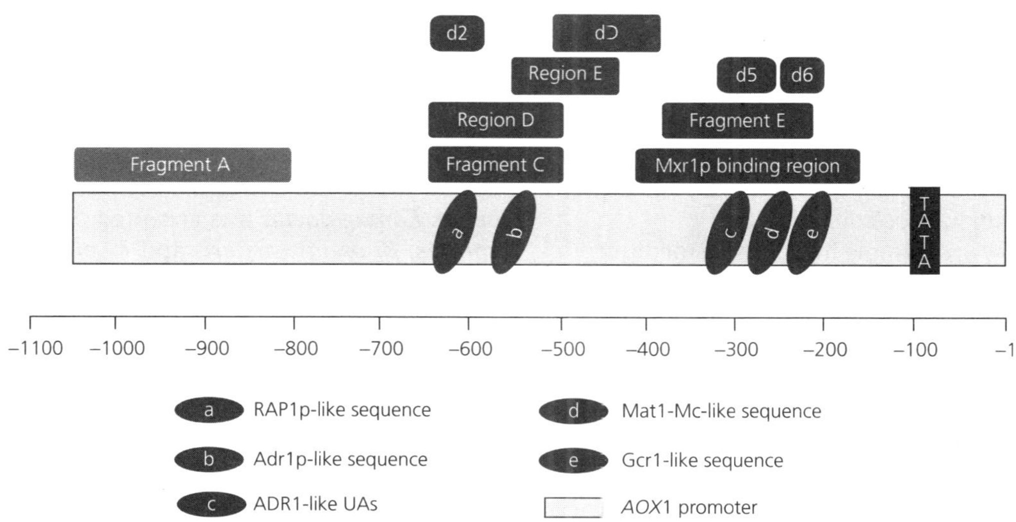

Figure 25.5 Schematic representation of the reported *cis*-acting elements in *AOX1* promoter (yellow horizontal block). Orange, blue and red rectangular blocks represent *cis*-acting elements, and blue and red ovals represent putative transcription factor binding sites; their relative position is indicated by the number line below. See Colour Plate 32 on page 523.

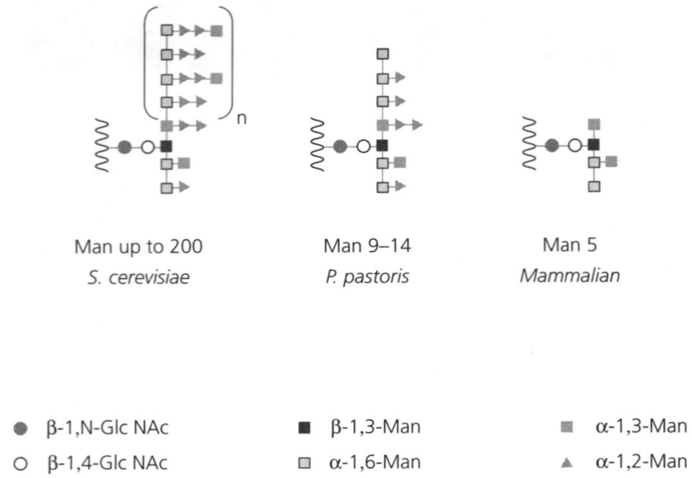

Figure 25.6 Generalized schematic representation of mannosylation in glycoproteins expressed in *S. cerevisiae*, *P. pastoris* and the mammalian system, indicating the hyper-mannosylated feature of *S. cerevisiae*–expressed proteins. Glycoproteins expressed in the mammalian system get further modified into complex glycans. See Colour Plate 33 on page 523.

for the highly antigenic nature of glycoproteins (Hopkins *et al.* 2011). Fig. 25.6 gives a schematic representation of the mannosylation pattern of glycoproteins expressed in *S. cerevisiae*, the mammalian system and *P. pastoris* (Cregg *et al.* 1993; Striebeck *et al.* 2013).

Heterologous proteins, in both the yeasts, can be made to be secreted out in the medium after introduction of a secretory signal sequence before the desired gene. Purification of secreted protein is easier in *P. pastoris* as compared with *S. cerevisiae* because very few endogenous proteins of *P. pastoris* are secreted, unlike in *S. cerevisiae*. Hence, *P. pastoris*–expressed secretory heterologous protein is relatively pure, thus simplifying the entire purification process (Macauley-Patrick *et al.* 2005).

Another reason which gives *P. pastoris* an edge over *S. cerevisiae* is that *P. pastoris* is a methylotroph and hence can utilize cheap sources of carbon and energy such as methanol. Also, it can grow to higher cell densities.

The following features make *P. pastoris* an attractive expression host (Macauley-Patrick *et al.* 2005; Demain and Vaishnav 2009; Damasceno *et al.* 2012):

- High growth rate
- High expression level
- Ability to attain high cell densities
- Ability to grow on a simple inexpensive medium
- Suitable for both small- and large-scale production
- Proteins expressed in *P. pastoris* are free of endotoxins, oncogenic and viral DNA

25.4.2 Taxonomical classification

Morphologically, *P. pastoris* cells are spherical or oval in shape (Fig. 25.7) and form hat-shaped ascospores during sexual reproduction. Asexual reproduction takes place through budding (Fig. 25.7). It has often been isolated from tree exudates such as chestnut tree (Kurtzman 1998). Tables 25.3 and 25.4 give taxonomic details (available at http://www.uniprot.org/taxonomy/4922) and a list of its synonymic names (fungal database of nomenclature and species bank, Mycobank; available at http://www.mycobank.org/Biolomics.aspx?Table=Mycobank&Page=200&ViewMode=Basic), respectively. This yeast, isolated from the French chestnut tree by Guilliermond in 1919, was initially classified as *Zygosaccharomyces pastori*, which was periodically re-classified with the increase in knowledge about this organism. Though for a long time, *pastori* and *pastoris* could be alternatively used as its species name, *pastoris* is now the scientifically accepted species name.

Yamada *et al.* (1995) proposed the formation of a distinct genus *Komagataella*, in which *P. pastoris* was placed. This was based on the study of partial sequences of 18S and 26S rRNAs of *P. pastoris* and other methanol-assimilating yeasts. Splitting of the genera *Pichia* was not accepted initially, but when this comparison was studied in greater depth by Kurtzman and Robnett (1998), it was concluded that *P. pastoris* was different phylogenetically from other methanol-assimilating yeasts and hence its inclusion under a distinct genus *Komagataella* was accepted. Further, two more species, *K. pseudopastoris* and *K. phaffii*, were added to this genus based on a phylogenetic study on nuclear large and small subunit ribosomal rDNA sequences (Kurtzman 2005).

Later, some of the biotechnological strains of *P. pastoris* were identified to be *K. phaffii* based on the phylogenetic study of nuclear genes such as large subunit rRNA, elongation factor (EF-1α), RNA polymerase 1 and mitochondrial small subunit rRNA (Kurtzman 2009).

Though its current scientific name is *Komagataella pastoris*, it is still commonly known as *P. pastoris*.

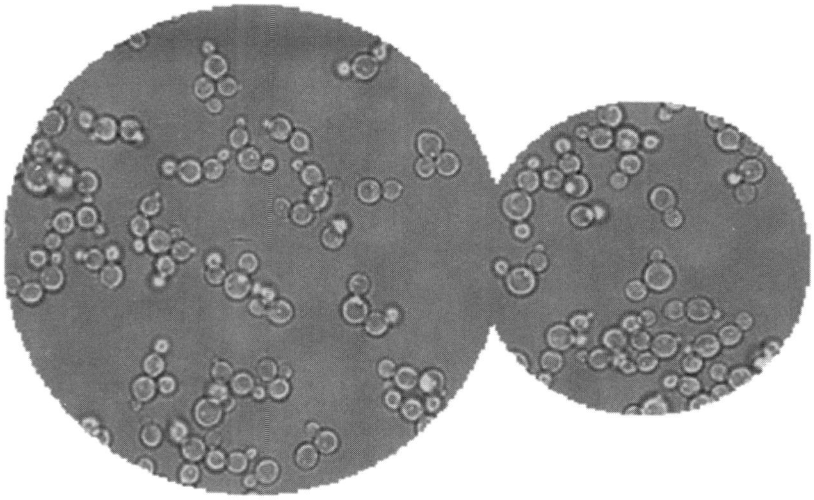

Figure 25.7 Two microscopic images of *P. pastoris* depicting their shape and asexual reproduction by budding, at x1000 magnification.

Table 25.3 Taxonomic classification of *P. pastoris*

Kingdom	Fungi
Subkingdom	Dikarya
Phylum	Ascomycota
Subphylum	Saccharomycotina
Class	Saccharomycetes
Order	Saccharomycetales
Family	Saccharomycetaceae
Genus	*Komagataella*
Species	*pastoris*

Table 25.4 Synonymic names of *P. pastoris*

Synonym	Taxonomic scrutiny
Komagataella pastoris	(Guilliermond) Y. Yamada, Matsuda, Maeda and Mikata (1995)
Zymopichia pastoris	(Guilliermond) Novák and Zsolt (1961)
Pichia pastori	(Guilliermond) Phaff (1956)
Zygowillia pastori	(Guilliermond) Kudryavtsev (1954)
Petasospora pastori	(Guilliermond) Boidin and Abadie (1954)
Saccharomyces pastori	(Guilliermond) Lodder and Kreger-van Rij (1952)
Endomyces pastori	(Guilliermond) Zender (1926)
Zygosaccharomyces pastori	Guilliermond A. (1919)

25.4.3 Methanol metabolism

The reactions involved in methanol metabolism take place in peroxisomes and the cytosol. During the induction phase, the peroxisomes swell and may constitute up to 80% of the cell's volume (Gellissen 2000). Methanol utilization begins with reactions in peroxisomes, harbouring three key enzymes – AOX, catalase and dihydroxyacetone synthase, which make these microbodies indispensable for growth. Alcohol oxidase and dihydroxyacetone synthase are present at very high levels (30% and 20%, respectively, of total cell proteins) in cells fed on methanol but are undetectable in cells growing on most other carbon sources such as glucose, glycerol and ethanol. Further reactions related to methanol assimilation and dissimilation take place in the cytosol (Gellissen 2000; Macauley-Patrick *et al.* 2005).

Fig. 25.8 gives a schematic representation of the methanol utilization pathway (De Schutter *et al.* 2009). Methanol utilization begins with the oxidation of methanol to formaldehyde, catalysed by AOX in peroxisomes. Hydrogen peroxide, formed as a byproduct, is degraded by catalase to oxygen and water. A portion of formaldehyde leaves the peroxisome as S-hydroxymethylglutathione and is oxidized to S-formylglutathione. This reaction is accompanied with the generation of NADH. In a cytoplasmic hydrolase-catalysed reaction, S-formylglutathione is further converted to formate. Formate is further oxidized to carbon dioxide by formaldehyde dehydrogenase, leading to the generation of NADH. The reactions which are accompanied with the generation of NADH are sources of energy for the cells growing on methanol (Cereghino and Cregg 2000; De Schutter *et al.* 2009).

The remaining portion of formaldehyde, generated in peroxisomes, is assimilated by a series of reactions to form cellular constituents and hence generate biomass. The third peroxisomal enzyme, dihydroxyacetone synthase, catalyses the condensation of formaldehyde and xylulose-5-phosphate to form dihydroxyacetone

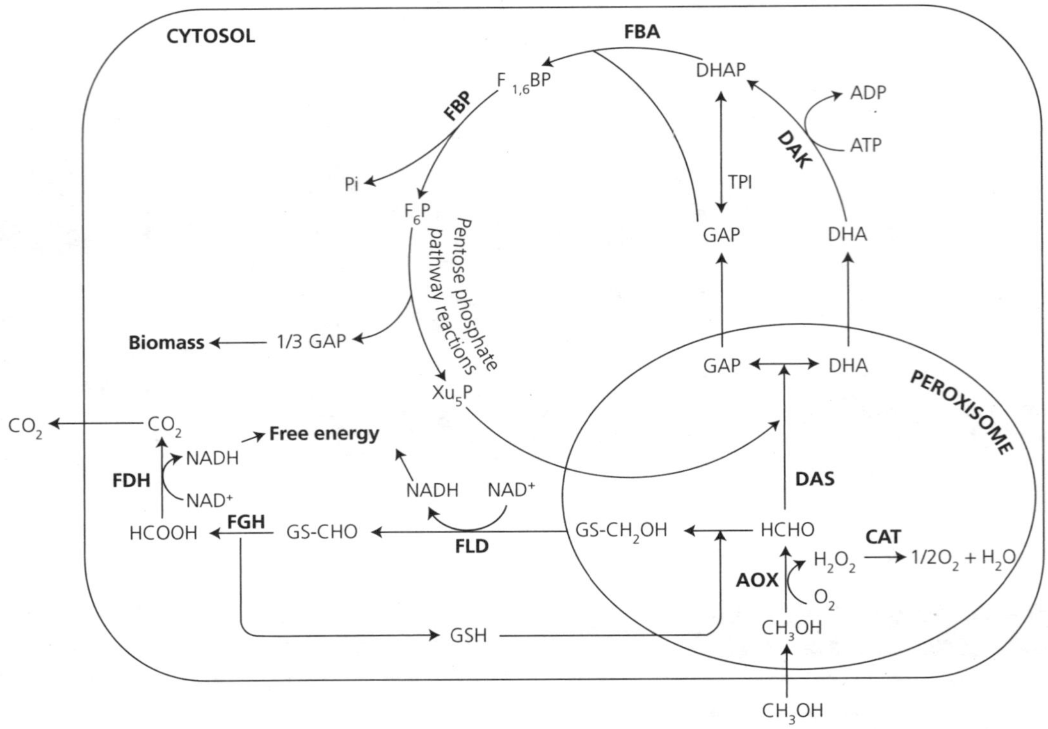

Figure 25.8 Reactions involved in methanol metabolism. Abbreviations: AOX, alcohol oxidase; CAT, catalase; DAS, dihydroxyacetone synthase; DHA, dihydroxyacetone; GAP, glyceraldehyde-3-phosphate; DAK, dihydroxyacetone kinase; DHAP, dihydroxyacetone phosphate; TPI, triosephosphate isomerase; FBA, fructose-1,6-bisphosphate aldolase; $F_{1,6}BP$, fructose-1,6-bisphosphate; FBP, fructose-1,6-bisphosphatase; FDH, formate dehydrogenase; F_6P, fructose-6-phosphate; Xu_5P, xylulose-5-phosphate; GSH, glutathione; GS-CH2OH, S-hydroxymethylglutathione; FLD, formaldehyde dehydrogenase; FGH, S-formylglutathione hydrolase; GS-CHO, S-formylglutathione.

phosphate and glyceraldehyde-3-phosphate, which enter the cytosol. In the cytosol, they get converted into xylulose-5-phosphate and glyceraldehyde-3-phosphate by a series of reactions. The stoichiometry of the reactions is in such a way that a set of three cycles regenerate three molecules of xylulose-5-phosphate and one molecule of glyceraldehyde-3-phosphate (Cereghino and Cregg 2000; De Schutter *et al.* 2009).

25.5 *P. pastoris* strains

The several strains of *P. pastoris* are broadly categorized according to their methanol utilization ability, as discussed next.

25.5.1 Methanol utilization phenotypes

The phenotype of a *P. pastoris* strain depends upon its rate of methanol utilization, which in turn depends on the presence of functional AOX1. Among the two alcohol oxidases, AOX1 and AOX2, AOX1 is more active in utilization of methanol than AOX2. On the basis of this, *P. pastoris* can be categorized into the following three methanol utilization phenotypes (Cereghino and Cregg 2000; Macauley-Patrick *et al.* 2005; *Easy Select™ Pichia Expression Kit Manual* 2010):

- **Mut^S (methanol utilization slow)**: Most of the cell's alcohol oxidase activity is lost with the disruption of the *AOX1* gene. Hence, its methanol utilization capacity largely relies on the weaker AOX2 activity, resulting in a strain that is phenotypically slow (Mut^S) in the growth and utilization of methanol. Mut^S cells utilize methanol at a very slow rate, resulting in poor growth on a methanol medium. KM71 and its derived strain KM71H are examples of Mut^S strains, where *AOX1* has been disrupted by replacement with *S. cerevisiae ARG4* gene.
- **Mut$^+$ (methanol utilization plus)**: This refers to the wild-type ability of strains to metabolize methanol as the sole carbon source due to the presence of a functional *AOX1* gene. Such cells require high methanol-feeding rates, for example, X-33, GS115.
- **Mut$^-$ (methanol utilization minus)**: This phenotype refers to the complete inability of the cells to metabolize methanol because of a lack of both *AOX1* and *AOX2* gene, for example, MC100-3.

25.5.2 Available strains

The common *P. pastoris* expression strains, listed in Table 25.5 (Cereghino and Cregg 2000; *Easy Select™ Pichia Expression Kit Manual* 2010), are all derivatives of the parent strain, NRRL-Y 11430 (Northern Regional Research Laboratories, Peoria, IL). Most of the strains differ from each other with respect to their auxotrophic mutations. All these non-transformed cells require supplementation of the respective nutrient in the media for their growth. After transformation with appropriately selected expression vector carrying the gene (as a selectable marker) which would allow a transformed cell to overcome its auxotrophy, cells grow on the same media without external supplementation of the specific nutrient. This enables selection of transformed cells over non-transformed cells (Cereghino and Cregg 2000).

The common auxotrophic mutations lie within the biosynthetic pathway genes like *HIS4*, *ARG4*, *ADE1* and *URA3*. Hence, these genes are common components of an expression vector as a selectable marker. *HIS4* gene encodes for histidinol dehydrogenase. With mutation in histidinol dehydrogenase gene (*his4*), cells are unable to synthesize histidine. Though such cells can grow on complex media, they require histidine to be externally supplemented for growth on minimal media. Similarly, a strain with mutation in argininosuccinate lyase gene (*arg4*) is unable to grow in the absence of arginine. Mutations in genes of PR-amidoimidazolesuccinocarboxamide synthase (*ade1*) and orotidine 5-phosphate decarboxylase (*ura3*) result in a failure of cells to grow in the absence of adenine and uracil, respectively (Cereghino and Cregg 2000; Macauley-Patrick *et al.* 2005).

Protease-deficient strains, such as SMD1163, 1165 and 1168, have been generated by mutating the *PEP4* and *PRB1* loci. The *PEP4* gene encodes a vacuolar protease, namely proteinase A. It is an aspartyl protease which further activates carboxypeptidase and partially activates proteinase B encoded by the *PRB1* gene. Hence, *pep4* mutants (SMD1168) are deficient in proteinase A and carboxypeptidase activity and partially lack proteinase B activity. In *prb1* mutants, such as SMD1165, only proteinase B activity is eliminated. The activity of all three proteases is substantially reduced in *pep4 prb1* double mutants, such as SMD1163 (Cereghino and Cregg 2000; Macauley-Patrick *et al.* 2005). A Kex1 protease-deficient variant of SMD1168 strain was also generated to prevent carboxy-terminal proteolysis of murine and human endostatin (Cereghino and Cregg 2000).

GS115 is the most commonly used strain. It is auxotrophic for histidine and allows selection of clone transformed with a vector-carrying *HIS4* gene (Higgins 1995). Strain X-33 is also often used. It has been created from GS115 after its transformation with wild-type *HIS4* gene (Blanchard *et al.* 2008). Hence, it does not require supplementation of histidine. Unlike X-33 and GS115,

Table 25.5 List of available *P. pastoris* strains

Strain	Genotype	Phenotype
GS115	*his4*	Mut^+, His^-
X-33	Wild type	Mut^+, His^+
GS190	*arg4*	Mut^+, Arg^-
JC220	*ade1*	Mut^+, $Ade1^-$
JC254	*ura3*	Mut^+, $Ura3^-$
GS200	*arg4 his4*	Mut^+, Arg^-, His^-
JC227	*ade1 arg4*	Mut^+, $Ade1^-$, Arg^-
JC304	*ade1 his4*	Mut^+, $Ade1^-$, His^-
JC305	*ade1 ura3*	Mut^+, $Ade1^-$, $Ura3^-$
JC306	*arg4 ura3*	Mut^+, $Ura3^-$, Arg^-
JC307	*his4 ura3*	Mut^+, $Ura3^-$, His^-
JC300	*ade1 arg4 his4*	Mut^+, $Ade1^-$, His^-, Arg^-
JC301	*ade1 his4 ura3*	Mut^+, $Ade1^-$, $Ura3^-$, His^-
JC302	*ade1 arg4 ura3*	Mut^+, $Ade1^-$, $Ura3^-$, Arg^-
JC303	*arg4 his4 ura3*	Mut^+, $Ura3^-$, Arg^-, His^-
JC308	*ade1 arg4 his4 ura3*	Mut^+, $Ade1^-$, $Ura3^-$, Arg^-, His^-
KM71	*his4 arg4 aox1::ARG4*	Mut^s, His^-
KM71H	*aox1::ARG4*	Mut^s
MC100-3	*aox1::SARG4 aox2::Phis4 his4 arg4*	Mut^-, His^-
SMD1168	*pep4::URA3 his4 ura3*	Mut^+, His^-, $pep4^-$
SMD1165	*prb1 his4*	Mut^+, His^-, $prb1^-$
SMD1163	*pep4 prb1 his4*	Mut^+, His^-, $pep4^-$, $prb1^-$
SMD1168 variant	*kex1::SUC2 pep4::URA3 his4 ura3*	Mut^+, His^-, $pep4^-$, $kex1^-$

which are Mut^+, KM71H is a Mut^S strain. *ARG4* gene was inserted into cloned, wild-type *AOX1* gene, resulting in deletion of codons 16 through 227 of *AOX1*, thus making AOX1 non-functional. This construct was transformed into KM71 (*arg4*, *his4*), the parent strain of KM71H, followed by the selection of Arg^+ transformant with Mut^S phenotype. These transformants were further transformed with the *HIS4* gene to develop KM71H strain (*Easy Select™ Pichia Expression Kit Manual* 2010). X-33, GS115 and KM71H, together, can be regarded as the most frequently used strains for heterologous protein expression.

25.6 Plasmid vectors

A number of plasmids are available for heterologous protein expression in *P. pastoris*, the key elements of which are discussed next.

25.6.1 Basic components

E. coli–P. pastoris shuttle vectors are used as plasmid vectors for heterologous expression in *P. pastoris*. They consist of an origin of replication and a selectable marker for plasmid maintenance and selection, respectively, in *E. coli*. They also carry a strong promoter and a transcription termination sequence, between which lies the multiple cloning site for insertion of the gene of interest. Fig. 25.9 gives a schematic representation of the basic components of a *P. pastoris* expression vector (Higgins 1995; Cereghino and Cregg 2000).

5′ P_{AOX1} is the methanol-inducible promoter of *AOX1* from *P. pastoris* which drives the expression of the gene of interest inserted into the multiple cloning site. The *AOX1* transcription termination (TT) sequence permits efficient 3′ mRNA processing, including polyadenylation, leading to an increase in mRNA stability. *P. pastoris* and *E. coli* selection markers allow the selection of transformants in *P. pastoris* and *E. coli*, respectively. An expression vector may also carry the 3′*AOX1* sequence, which lies 3′ to the transcription termination sequence. This sequence is utilized for integration of the vector into the host genome by gene replacement or by gene insertion 3′ in the chromosomal *AOX1* gene. The shuttle vector also carries the ColE1 origin of replication to allow replication of the plasmid in *E. coli* and *E. coli* selection marker ensures its maintenance (Higgins 1995; *Easy Select*™ *Pichia Expression Kit Manual* 2010).

Secretory expression vectors are also available, which carry an additional secretion signal to enable secretion of the foreign protein into the medium upon in-frame insertion of the foreign gene with the secretion signal. The commonly used secretion signals are *P. pastoris* acid phosphatase (PHO1) presequence or *S. cerevisiae* α-mating factor preprosequence (Domínguez *et al.* 1998; Cereghino and Cregg 2000; Damasceno *et al.* 2012).

25.6.2 Promoters

A number of promoters have been used for the heterologous expression of proteins in *P. pastoris*. Table 25.6 lists the different types of promoters and their key features (Cereghino and Cregg 2000; Macauley-Patrick *et al.* 2005).

Though a number of promoters are available, the most extensively used is the *AOX1* promoter because it enables high, inducible expression of foreign proteins. It allows tightly regulated transcription of the inserted gene, which is controlled by the repression/de-repression mechanism. When the expression of a foreign gene is under the *AOX1* promoter, the general cell cultivation practice is to grow the cells first on glycerol and then switch the medium to methanol for induction after sufficient cell density has been attained. The initial carbon source (glycerol) represses the expression of the foreign protein, thus ensuring good cell growth before induction of the foreign protein by methanol, which is achieved by replacement of the glycerol medium

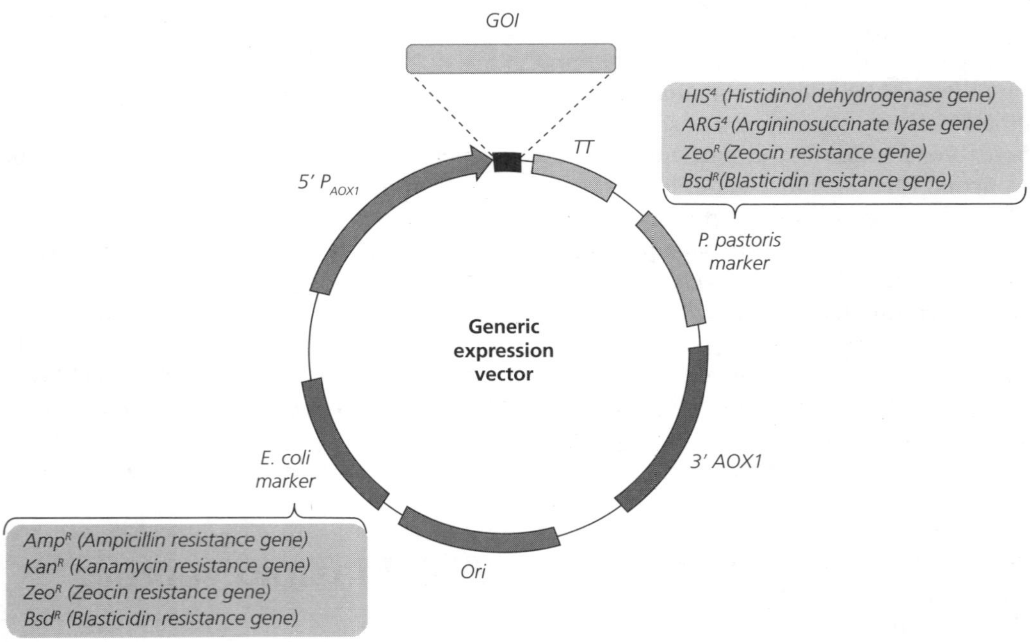

Figure 25.9 Schematic representation of a generic *P. pastoris* expression vector. P_{AOX1}, *GOI*, *TT*, *3′AOX1* and *Ori* denote *AOX1* promoter, gene of interest inserted at the multiple cloning site (black block), transcription terminator sequence, *AOX1* region 3′ to transcription termination sequence, and *E. coli* origin of replication, respectively. Commonly used *P. pastoris* and *E. coli* marker genes used for selection of the desired clone are listed in pink and green boxes, respectively. See Colour Plate 34 on page 524.

Table 25.6 Different promoters used to drive expression of heterologous proteins in *P. pastoris*, and their key features

Promoter	Expression	Advantages	Disadvantages
AOX1 gene promoter	Inducible on methanol	Strong and tightly regulated; high expression level	Requires careful monitoring of methanol; difficulty in methanol storage; methanol cannot be used for production of certain food products and additives; culture needs to be shifted from one carbon source during growth to the other during induction
Glyceraldehyde-3-phosphate dehydrogenase (*GAP*) gene promoter	Constitutive on glucose	Does not require methanol; can grow and express on the same carbon source and hence no need to shift culture	Cannot be used for production of toxic proteins as expression is constitutive
Formaldehyde dehydrogenase (*FLD1*) gene promoter	Inducible on methanol (as carbon source and ammonium sulphate as nitrogen source) or methylamine (as nitrogen source and glucose as carbon source)	High expression levels can be achieved; potential alternative to *AOX1* promoter	Needs to be explored more to make significant comparison with *AOX1* promoter
Peroxisomal matrix protein (*PEX8*) gene promoter	Inducible on methanol	Suitable for expression of those genes for which high expression levels have led to misfolded, unprocessed or mislocalized proteins	Moderate expression level
AOX2 gene promoter	Inducible on methanol	Inducible expression	Low expression level
GTPase gene promoter (*YPT1*)	Constitutive on either glucose, methanol or mannitol as carbon sources	Suitable for expression of those genes for which high expression levels have led to misfolded, unprocessed or mislocalized proteins	Low expression level
Isocitrate lyase (*ICL1*) gene promoter	Inducible on low levels of glucose or ethanol	Methanol-free expression	Needs to be studied more

with a methanol-containing medium. The other alternative promoters are used only when methanol storage is a hurdle or when methanol is not suitable for the production of a particular product such as some food products and additives (Cereghino and Cregg 2000). The *FLD1* promoter has been found to be an efficient alternative for the *AOX1* promoter. It allows methanol-free induction, where heterologous protein expression can be induced by the use of methylamine (Cos *et al.* 2005). Most commonly used *P. pastoris* expression vectors (commercially available vectors from Invitrogen; available at http://www.lifetechnologies.com/in/en/home.html) and their corresponding promoters are listed in Table 25.7.

25.6.3 Selectable markers

Selectable markers are extremely important in order to differentiate between transformed and non-transformed clones. Markers used are either involved in biosynthetic pathways or confer resistance to a specific antibiotic. Table 25.7 lists the most commonly used selection markers in commercially available vectors.

Expression vectors carrying biosynthetic markers such as *ARG4, HIS4, ADE1* and *URA3* are used to transform host strain of corresponding auxotrophy to enable selection of a positive clone. Apart from auxotrophic complementation, antibiotic resistance genes, such as Kan^R, Zeo^R and Bsd^R, are also used as selection markers (Cereghino and Cregg 2000; Macauley-Patrick *et al.* 2005). They are primarily used for selection of multicopy gene transformants on the basis that a greater degree of drug resistance is attained by transformants harbouring a greater copy number (Damasceno *et al.* 2012).

25.7 Cloning and expression

Heterologous expression of a protein in *P. pastoris* involves three basic steps:

1. Insertion of gene of interest into the desirable expression vector
2. Integration of the gene of interest into the genome of appropriate *P. pastoris* host strain
3. Screening of transformed clones to obtain the best expressing clone

Table 25.7 Promoter, secretion signal and selection marker gene related information of the various commercially available vectors from Invitrogen

Vector	Secretion signal	Promoter	Marker genes for selection of transformants	
			E. coli	*P. pastoris*
pPICZ	–	*AOX1*	Zeocin resistance gene	
pPICZα	α-Factor			
pGAPZ	–	*GAP*	Zeocin resistance gene	
pGAPZα	α-Factor			
pPIC6	–	*AOX1*	Blasticidin resistance gene	
pPIC6α	α-Factor			
pPIC9	α-Factor	*AOX1*	Ampicillin resistance gene	*HIS4* gene
pPIC9K				*HIS4* gene, Kanamycin gene – confers resistance to Geneticin
pPIC3.5	–	*AOX1*	Ampicillin resistance gene	*HIS4* gene
pPIC3.5K	–	*AOX1*	Ampicillin resistance gene	*HIS4* gene, Kanamycin gene – confers resistance to Geneticin
pAO815	–	*AOX1*	Ampicillin resistance gene	*HIS4* gene
pFLD	–	*FLD1*	Ampicillin and Zeocin resistance genes	Zeocin resistance gene
pFLD α	α-Factor			
pHIL-D2	–	*AOX1*	Ampicillin resistance gene	*HIS4* gene
pHIL-S1	PHO-1	*AOX1*	Ampicillin resistance gene	*HIS4* gene

Factors such as host strain availability and secretory or cytosolic expression are taken into consideration before selection of an expression vector. After insertion of the gene of interest into the vector, the recombinant plasmid is linearized to enable integration of the gene of interest into the genome of host strain via homologous recombination between the transforming DNA and regions of homology within the genome. Integration of gene into the genome maximizes the stability of expression, which occurs by either gene insertion (single crossover) or gene replacement (double crossover) events (Higgins 1995; *Easy Select*™ *Pichia Expression Kit Manual* 2010).

Gene insertion events (Fig. 25.10) take place at the *AOX1* (in X-33 and GS115) or *aox1::ARG4* (in KM71H) locus, resulting in the insertion of one or multiple copies of the vector upstream or downstream of the locus. It arises from a crossover event between the locus in the chromosome and any of the three *AOX1* regions on the vector (*AOX1* promoter, *AOX1 TT* or *3′AOX1*). The phenotype of such a transformant is Mut^+ (for X-33 or GS115) or Mut^S (for KM71H). Insertions are much more likely to happen than replacements. Multiple insertion events, as illustrated in Fig. 25.10, occur spontaneously, at about 1–10% of the single-insertion events (Higgins 1995; *Easy Select*™ *Pichia Expression Kit Manual* 2010).

A gene replacement (omega insertion) event (Fig. 25.11) arises from a double crossover between the *AOX1* promoter and 3′ *AOX1* regions of the vector and the genome. In the case of GS115 and X-33, it results in complete removal of the *AOX1* coding region, leading to conversion from Mut^+ to Mut^S phenotype of the transformed cells (Higgins 1995).

The clones are then screened by colony polymerase chain reaction (PCR) and expression of the protein of interest. Often, attempts are made to obtain multicopy transformants to attain high expression levels. One way to generate a multicopy strain is by integrating a multicopy expression cassette into the host genome. This method enables generation of a strain with known number of gene copies. A multicopy strain can also be generated by successive transformations. The post-transformation vector amplification method is also used for generation of multicopy strains. For example, strains transformed with zeocin resistance–based vector can be successively selected on increasing concentrations of zeocin to increase the gene copy number. It has been commonly observed that the protein expression level in relation to the gene copy number follows a bell-shaped curve. This indicates that the expression level increases with the increase in gene dosage till an optimum dosage is attained, after which there is a decrease in the expression probably due to increased endoplasmic reticulum (ER) stress. Hence, selecting the clone under optimum selection pressure is critical to obtain the best-expressing clone. The optimum gene dosage varies from protein to protein and has to be separately optimized in each case (Damasceno *et al.* 2012).

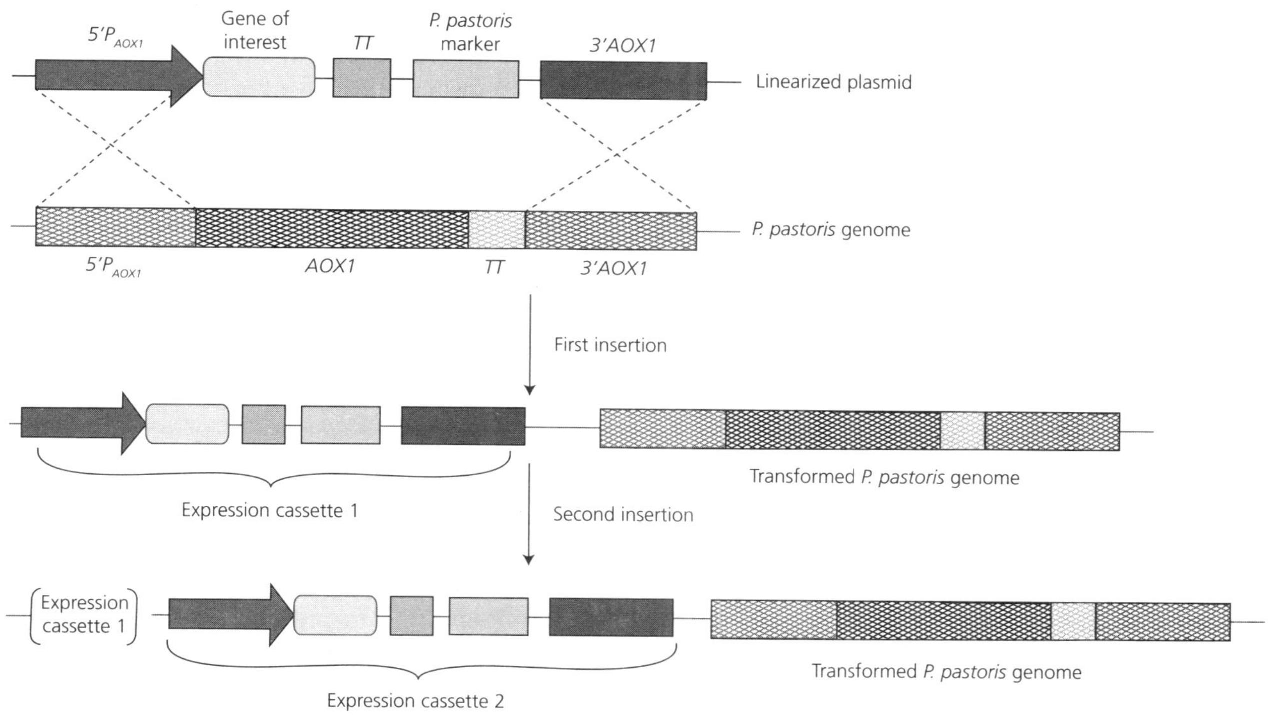

Figure 25.10 Schematic representation of two insertion events at *AOX1* promoter (5′ P_{AOX1}). TT denotes transcription terminator sequence. See Colour Plate 35 on page 524.

25.8 Glycoengineering of *P. pastoris*

Proteins as therapeutic compounds have gained importance since the success of the first recombinant therapeutic protein, insulin. Traditionally, *E. coli* and mammalian cells have been used for the production of such proteins. Lately, other alternative hosts are being explored because of some limitations with the use of these two hosts; first, lack of glycosylation in *E. coli*, second, the high production cost associated with protein purification from mammalian cell cultures and third, the differences in glycosylation pattern between native human proteins and those purified from mammalian cells. Though the glycosylation patterns are largely similar, these minor differences have been reported to be responsible for altered immune responses such as the presence of Hanganutiziu–Deicher antibodies. Thus,

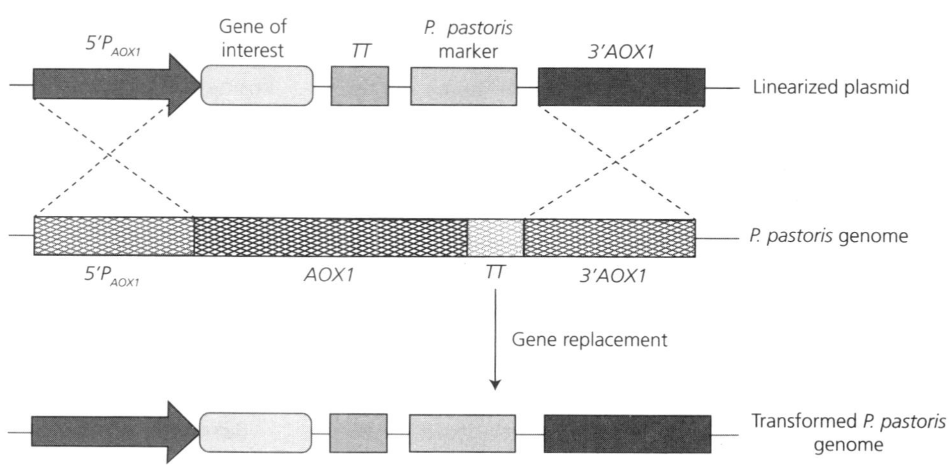

Figure 25.11 Schematic representation of gene replacement event, where the genomic region between the *AOX1* promoter (5′ P_{AOX1}) and the 3′ *AOX1* region is replaced with an expression cassette. See Colour Plate 36 on page 525.

other hosts are being explored, which can be used for the production of functional human-like proteins in a cost-effective manner (Hamilton and Gerngross 2007).

P. pastoris has proven to be the most promising host in achieving this objective. Its ability to grow to high cell densities on inexpensive media in the absence of exotic growth factors and to secrete a large amount of recombinant proteins in the culture media makes it an attractive host for the production of therapeutic proteins. Thus, attempts are now being made to engineer *P. pastoris* in such a way that it expresses human-like proteins with respect to the glycosylation pattern (mainly *N*-glycosylation; Hamilton and Gerngross 2007). The primary focus of glycoengineering is on *N*-linked glycosylation because the majority of the protein features such as protein folding, *in vivo* half-life and function have been found to be associated with *N*-glycosylation (Hopkins *et al.* 2011). Though the initial stages of *N*-glycosylation, occurring in the ER, are similar in both human cell and *P. pastoris*, the final modifications in Golgi differ significantly. The glycan entering Golgi in both the cases is Man_8-*N*-acetylglucosamine$_2$ ($Man_8GlcNAc_2$). This high mannose structure is met with different fates in human and yeast systems. In the human cell, this structure is trimmed in Golgi, whereas in yeasts, it is further extended (in the Golgi apparatus) by incorporation of mannose residues, though the number of residues incorporated varies from one host to the other. *P. pastoris* holds an edge over *S. cerevisiae* as it does not hyper-glycosylate its proteins. Excess of mannose residues can be detrimental for the recipient of that therapeutic protein as it can elicit altered immune response and can also decrease the half-life of the protein by binding to the mannose receptors on the macrophages, thus enhancing its clearance (Hopkins *et al.* 2011).

In the Golgi apparatus of yeasts, extension of the glycan chain begins with the formation of $Man_9GlcNAc_2$ by addition of α-1,6-mannose to the α-1,3-mannose arm of $Man_8GlcNAc_2$ through α-1,6-mannosyltransferase (Och1p) activity. $Man_9GlcNAc_2$ acts as a substrate for further mannosylation. Hence, in generation of humanized *P. pastoris*, removal of the *OCH1* gene becomes mandatory (Bretthauer 2003).

Currently, the various patterns and stages of glycosylation are being actively studied and various genes are being knocked out and introduced to generate human-like glycans. Vervecken *et al.* (2004) have been able to achieve humanization of *N*-glycosylation by preventing hyper-mannosylation through introduction of the required mannosidase (α-1,2-mannosidase) and glycosyltransferases (Kre2-GnTI and Kre2-β-1,4-galactosyltransferase). Hyper-mannosylation was also prevented by inactivation of the *OCH1* gene, which encodes for α-1,6-mannosyltransferase Och1p. Other groups have also been able to substitute humanized *P. pastoris* for mammalian cells for production of human-like proteins. Attempts to produce human-like recombinant erythropoietin, functionally similar to human erythropoietin, have been successful (Hamilton and Gerngross 2007). Mille *et al.* (2008) reported a family of four β-mannosyl transferase genes (*BMTs*) – *BMT1* to *BMT4* – to be responsible for β-Man glycan. The maximal BMT activity was reported to be associated with *BMT2* (Bmt2p). In the study conducted by Hopkins *et al.* (2011), recombinant human erythropoietin was expressed in glycoengineered *P. pastoris* strain devoid of *BMT2* gene expression to limit the extension of the mannose chain. The reactivity of this protein to host cell (lacking the gene of interest) antibodies (HCA) was checked and they were found to be recognizing the protein. It was believed that HCA were probably recognizing a glycan. This was later confirmed through α-1,2-mannosidase treatment, which eliminated the reactivity of protein with HCA. Glycan profile of the expressed protein revealed the presence of low amounts of α-mannosidase–resistant high-Man glycoforms. As a result, it was decided to knock out the other *BMT* genes in different combinations of double, triple and quadruple knockouts. It was observed that with triple knockout of *BMT1, 2* and *3* genes, formation of high-Man glycans, including those which were α-mannosidase resistant, could be prevented, thus increasing the control over the extent of glycosylation (Hopkins *et al.* 2011).

Humanized IgGs have also been produced from glycoengineered *P. pastoris* because the glycan profile of an antibody has a critical impact on the generation of antibody-dependent cell cytotoxicity (ADCC) and complement-dependent cytotoxicity (CDC) effects against the target antigen. It has been observed that deglycosylated antibodies are unable to mount the desired response. An anti-CD20 mAb, Rituximab, produced in glycoengineered *P. pastoris* has been reported to be more functional than that purified from CHO cell lines. Similar observations have been made for Trastuzumab. It is an anti-HER2 humanized mAb obtained from CHO cell lines. It is used for the treatment of breast cancer patients over-expressing HER2, a ligand-less tyrosine kinase receptor of the ErbB family (an epidermal growth factor receptor family). This mAb was produced in glycoengineered *P. pastoris* and was observed to be comparable to the commercial Trastuzumab in activity. Thus, *P. pastoris* is gaining popularity as a useful tool for the production of therapeutic antibodies, with improved glycan profile, in a cost-effective manner (Jiang *et al.* 2011).

While efforts are being made to humanize *P. pastoris*, it is important to ensure that the cells remain viable. The other challenging task is to ensure correct localization of the transformed glycosylation enzymes and ensure the availability of UDP-GlcNAc (Bretthauer 2003).

Research Corporation Technologies, in collaboration with Ghent University and Flanders Institute for

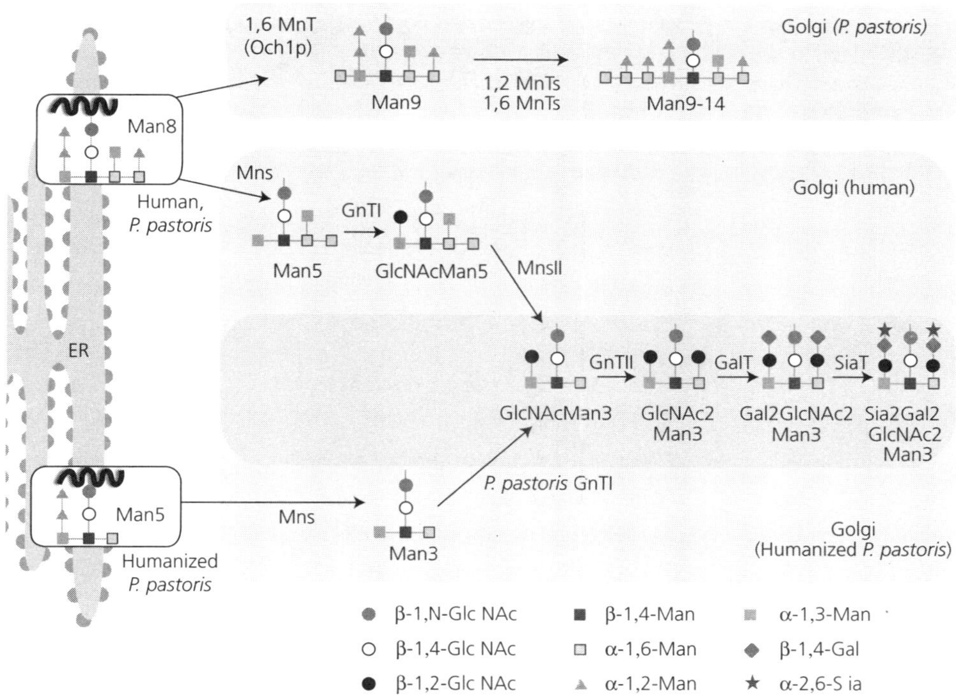

Figure 25.12 Schematic representation of the *N*-linked glycosylation pathway in *P. pastoris*, human cell and humanized *P. pastoris*. Abbreviations: MnT, mannosyltransferase; Mns, α-1,2-mannosidase; GnTI, β-1,2-*N*-acetylg ucosaminyltransferase I; MnsII, mannosidase II; GnTII, β-1,2-*N*-acetylglucosaminyltransferase II; GalT, β-1,4-galactosyltransferase; SiaT, α-2,6-sialyltransferase. Note that the two GlcNAc residues present at the reducing end of all glycans have been omitted from the nomenclature of glycans to keep it simple. See Colour Plate 37 on page 525.

Biotechnology in Belgium, has developed glycoengineered *P. pastoris* in the name of *Pichia* Glyco Switch®. A number of biologically important molecules have already been successfully produced using this *Pichia* platform (*Pichia* Glyco Switch®: available at http://www.rctech.com/licensing/gxt-pichia-glycoswitch.php).

Fig. 25.12 illustrates the differences in the glycosylation pattern of a protein in a human cell, *P. pastoris* and glycoengineered *P. pastoris* (Hamilton and Gerngross 2007).

25.9 *P. pastoris*–expressed proteins

Tucson-based Research Corporation Technologies, which holds the patent on the *P. pastoris* expression system, has licensed a number of biotechnological, pharmaceutical, animal health and food industries, and research organizations to study and use *P. pastoris* as a heterologous expression system.

The interest in use of *P. pastoris* as an expression system has grown at a tremendous pace over the past few years because of the already stated benefits of using this system. It has been used for the production of a wide variety of proteins. The major factor which makes this system highly popular for use in industries is the high expression level that can be achieved and the low production cost associated with this system, which makes the product economically feasible for production. Table 25.8 lists the proteins that have been expressed in *P. pastoris* at a level of >1 g/L.

25.10 Commercial success stories

P. pastoris has been successfully used for commercial production of a number of proteins, some of which are being discussed next.

25.10.1 Kalbitor

The first *P. pastoris*–expressed product to be approved for human use, in 2009, by the US Food and Drug Administration was ecallantide under the trade name KALBITOR® by Research Corporation Technologies licensee Dyax Corp (Walsh 2010). It is a very potent, selective and reversible plasma kallikrein inhibitor and has been approved for subcutaneous treatment of acute attacks of hereditary angioedema (HAE) by patients 16 years of age and older. HAE is a hereditary disease associated with a rise in activated plasma kallikrein levels, leading to fluid leakage from the veins, which gets accumulated in the tissues, causing painful swelling in the abdomen, face, hands, feet or other parts of the body (How Kalbitor works: available at

Table 25.8 List of proteins expressed in *P. pastoris* at a level >1 g/L. The proteins have been arranged in descending order of their expression level for each type of protein

Type of protein	Expression level (g/L)	Protein	References
Prokaryotic	12.0	*Clostridium tetani* tetanus toxin fragment C (l)	Clare *et al.* (1991)
	8.8	Staphylococcal protein A (s)	Hao *et al.* (2013)
	6.4	*Bacillus subtilis* β-propeller phytase (s)	Viader-Salvadó *et al.* (2013)
	6.4	*E. coli* phytase (s)	Chen *et al.* (2004)
	5.5	*Pseudomonas pseudoalcaligenes* organophosphorus hydrolase (s)	Chu *et al.* (2006)
	4.2	*Citrobacter amalonaticus* phytase (s)	Luo *et al.* (2007)
	4.0	Core streptavidin (s)	Casteluber *et al.* (2012)
	3.0	*Fibrobacter succinogens* glucanase (s)	Huang *et al.* (2008)
	3.0	*Rhodothermus marinus* xylanase (s)	Ramchuran *et al.* (2005)
	3.0	*Bordetella pertussis* pertactin (P69) (l)	Romanos *et al.* (1991)
	2.7	*Bacillus subtilis* β-mannanase (s)	Qiao *et al.* (2010)
	2.6	*Streptomyces sp.* S27 xylanase (s)	Li *et al.* (2009)
	2.5	*Bacillus licheniformis* α-amylase (s)	Paifer *et al.* (1994)
	2.5	*Streptomyces viridosporus* T7 peroxidase, endoglucanase (s)	Thomas and Crawford (1998)
	2.2	*Bacillus* sp. β-mannanase (s)	Zhu *et al.* (2011)
	1.3	*Talaromyces stipitatus* glucoamylase (s)	Xiao *et al.* (2014)
	1.1	*Bacillus licheniformis* rhamnogalacturonan I lyase (s)	Silva *et al.* (2011)
	1.0	*Thermus aquaticus* YT-1 pro-aqualysin I (s)	Olędzka *et al.* (2003)
Protozoan	1.4	*P. falciparum* merozoite surface protein 3 (s)	Wang *et al.* (2005)
Fungal	23.0	*Hansenula fabiani* phytase (s)	Watanabe *et al.* (2009)
	12.2	*Peniophora lycii* phytase (s)	Xiong *et al.* (2006)
	9.1	*Paecilomyces thermophila* β-glucanase (s)	Hua *et al.* (2010)
	8.4	*Paecilomyces* sp. FLH30 endoglucanase (s)	Hua *et al.* (2011)
	6.2	*Aspergillus oryzae* β-galactosidase (s)	Zhao *et al.* (2014)
	6.1	*Aspergillus niger* acid phytase (s)	Xiong *et al.* (2005)
	5.8	*Candida parapsilosis* lipase/acyltransferase (s)	Brunel *et al.* (2004)
	4.5	*Aspergillus niger* β-mannase (s)	Zheng *et al.* (2012)
	4.2	*Aspergillus niger* phytase (s)	Xiong *et al.* (2004)
	3.9	*Biospora* sp. MEY-1 endoxylanase (family 10) (s)	Luo *et al.* (2009)
	3.8	*Aspergillus sulphureus* β-xylanase (s)	Cao *et al.* (2007)
	3.6	*Irpex lacteus* rhamnogalacturonan hydrolase (s)	Normand *et al.* (2012)
	3.1	*Neosartorya fischeri* P1 α-galactosidase (s)	Wang *et al.* (2014)
	3.0	*Aspergillus fumigatus* endo-chitosanase (s)	Chen *et al.* (2012)
	3.0	*Candida antarctica* lipase B (s)	Yang *et al.* (2013)
	3.0	*Candida antarctica* lipase (s)	Liu *et al.* (2012)
	2.5	*Penicillium pinophilum* endo-1,3(4)-β-D-glucanase (s)	Chen *et al.* (2012)
	2.6	*Chaetomium thermophilum* superoxide dismutase (s)	Wakadkar *et al.* (2010)
	2.6	*Debaryomes castellii* phytase (s)	Ragon *et al.* (2008)
	2.5	*Penicillium notatum* glucose oxidase (s)	Gao *et al.* (2012)
	2.5	*S. cerevisiae* invertase (s)	Tschopp *et al.* (1987)
	2.3	*A. fumigatus* catalase L (s)	Calera *et al.* (1997)
	2.2	*Yarrowia lipolytica* lipase (s)	Wang *et al.* (2012)
	2.1	*Achaetomium* sp. Endo-polygalacturonase (s)	Tu *et al.* (2013)

... continued

Type of protein	Expression level (g/L)	Protein	References
	2.1	*Emericella nidulans* endogalactanase (s)	Michalak *et al.* (2012)
	2.0	*P. pastoris* GH45 endoglucanase (s)	Couturier *et al.* (2011)
	1.7	*Chaetomium thermophilum* cellobiohydrolase (s)	Li *et al.* (2009)
	1.5	*Chaetomium thermophilum* thermostable superoxide dismutase (s)	Zhang *et al.* (2011)
	1.5	*Neurospora crassa* cellobiose dehydrogenase (s)	Zhang *et al.* (2011)
	1.3	Plectasin-derived peptide (s)	Zhang *et al.* (2014)
	1.2	*Thermoascus aurantiacus* var. *levisporus* serine protease (s)	Li *et al.* (2011)
	1.2	*Candida antarctica* lipase B (s)	Eom *et al.* (2013)
	>1.0	*A. fumigatus* Z5 endoglucanases Egl2 and Egl3 (s)	Liu *et al.* (2011)
Viruses	7.0	Hepatitis B virus surface antigen (l)	Gurramkonda *et al.* (2009)
	3.0	SARS coronavirus N protein (s)	Cao *et al.* (2005)
	2.4	Hepatitis B virus surface antigen (l)	Bardiya (2006)
	2.0	Porcine reproductive and respiratory syndrome virus (PRRSV) membrane protein (s)	Qian *et al.* (2003)
	1.5	Rabbit hemorrhagic disease virus VP60 capsid (l)	Farnós *et al.* (2005)
	1.0	Foot and Mouth Disease Virus (FMDV) VP1 antigen (s)	Shi *et al.* (2006)
Plant	22.0	*Hevea brasiliensis* hydroxynitrile lyase (l)	Hasslacher *et al.* (1997)
	1.5	*Cynodon dactylon* (Bermuda grass) allergen Cyn d 1 (s)	Smith *et al.* (1996)
Invertebrates	5.0	Leech hirudin (s)	Xiao *et al.* (2006)
	2.0	*Nippostrongylus brasiliensis* acetylcholinesterase (s)	Richter *et al.* (2006)
	1.7	Sea anemone *Actinia equine* equistatin (s)	Outchkourov *et al.* (2002)
	1.5	*Boophilus microplus* gut glycoprotein Bm86 (l)	Rodríguez *et al.* (1994)
	1.5	Leech hirudin (s)	Rosenfeld *et al.* (1996)
	1.2	Sea anemone *Stichodactyla helianthus* BRTI-Kunitz protease inhibitor (s)	Gil *et al.* (2011)
Vertebrates (non-human)	14.8	Rat gelatin (s)	Werten *et al.* (1999)
	5.5	Chinese quail ovalbumin (s)	Yang *et al.* (2009)
	2.7	Duck interleukin 2 (s)	Du *et al.* (2011)
	1.8	Porcine interferon α (s)	Gao *et al.* (2013)
	1.4	Porcine interferon α (s)	Gao *et al.* (2012)
	1.3	Porcine interferon α (s)	Jin *et al.* (2011)
	>1.0	Bovine β-lactoglobulin (s)	Kim *et al.* (1997)
	1.0	Bovine interferon γ (s)	Shi *et al.* (2006)
	1.0	Bovine β-casein (l)	Choi *et al.* (1996)
Humans	10.0	Tumour necrosis factor α (l)	Sreekrishna *et al.* (1989)
	3.6	Insulin precursor protein (s)	Xie *et al.* (2008)
	3.0	Insulin precursor protein (s)	Gurramkonda *et al.* (2010)
	3.0	Serum albumin (s)	Barr *et al.* (1992)
	2.5	Truncated leukotactin 1 (s)	Woo *et al.* (2005)
	2.3	Proinsulin C peptide precursor (s)	Huang *et al.* (2006)
	1.5	Serum albumin (s)	Ohashi *et al.* (1999)
	1.5	Mini pro-insulin (s)	País *et al.* (2003)
	1.5	Insulin precursor protein (s)	Wang *et al.* (2001)

(Continued)...

... continued

Type of protein	Expression level (g/L)	Protein	References
	1.5	Human gelatin (s)	Olsen *et al.* (2005)
	1.4	Serum albumin (s)	Kobayashi *et al.* (2000)
	1.2	Consensus interferon α (s)	Hao *et al.* (2006)
	1.2	Consensus interferon α (s)	Wu *et al.* (2008)
	1.0	Keratinocyte growth factor 2 (s)	Wang *et al.* (2007)
	1.0	Staphylokinase (s)	Apte-Deshpnade *et al.* (2009)
	1.0	Kunitz-type protease inhibitor domain of protease nexin-2/amyloid β-protein precursor (s)	Wagner *et al.* (1992)
Humanized	4.9	Single chain Fv antibody fragment (s)	Damasceno *et al.* (2004)
	1.9	Monoclonal antibody (identity not revealed) (s)	Berdichevsky *et al.* (2011)
Chimeric proteins	>5.0	Anti CD3 immunotoxin (s)	Liu *et al.* (2005)
	3–4	Human serum albumin and human growth hormone fusion protein (s)	Wu *et al.* (2014)
	2.6	*P. falciparum* apical antigen domain III and C terminal fragment of merozoite surface protein (s)	Pan *et al.* (2004)
	1.5	*Neocallimastix patriciarum* cellulose 6A cellulose binding module and *Candida antarctica* lipase B (s)	Jahic *et al.* (2003)
	1.2	Red fluorescent protein in fusion with fungal cellulose binding domain (s)	Sugimoto *et al.* (2012)
	1.0	Bovine interferon γ in fusion with foot and mouth diseaase virus VP1 antigen (s)	Shi *et al.* (2006)
Custom designed	>3.0	Gelatin (s)	Werten *et al.* (2001)
	>2.1	Collagen-inspired gel-forming polymers (s)	Silva *et al.* (2011)

(s) and (I) indicate secretory and intracellular protein, respectively.

http://www.kalbitor.com/patient/why-kalbitor/how-kalbitor-works.html).

25.10.2 Hepatitis B surface antigen

The humble *P. pastoris*, which can survive on a single carbon source, such as methane, has joined the race for the ultimate eradication of hepatitis B virus (HBV) from humans. Despite having an effective vaccine against this virus, it continues to kill many people. It has been found to be 50–100 times more infectious than HIV and can survive outside the body for at least 7 days, unlike HIV (WHO hepatitis B fact sheet available at www.who.int/mediacentre/factsheets/fs204/en/). Moreover, virus-infected children, while playing, and licking the shared toy, have the risk of transmitting the virus through saliva.

The use of vaccine against the hepatitis B virus was the only way to effectively block the spread of this virus. The first-generation vaccine for hepatitis B became available in 1980s (Liaw and Chu 2009). This plasma-derived hepatitis B vaccine has been discontinued for safety-related issues of possible co-contamination of this vaccine from other unknown viruses of human origin. Moreover, this approach always required a cohort of hepatitis B virus–infected individuals, who would have to be willing to continuously donate their virus-infected plasma as a raw material for purification of hepatitis B surface antigen (HBsAg) virus-like particles, the main ingredient of the hepatitis B vaccine.

The cloning, expression and purification of HBsAg VLPs from recombinant *Saccharomyces cerevisiae* eliminated the use of plasma-derived hepatitis B vaccine (Hilleman 1987). But, this recombinant vaccine remained unaffordable in most developing countries, which bear the biggest burden of this virus. Despite repeated appeals to all nations by the World Health Organization, this expensive vaccine could not be introduced in the expanded programme of immunizations of >85 countries, including the most populated countries, India and China.

The discovery and development of *P. pastoris* as an expression system for the commercially viable production of recombinant proteins is changing this dreadful scenario. Now, Hepatitis B vaccine is included in the expanded programme of immunizations in several countries, including India and China, which together constitute 40% of the world population. The much higher expression levels of HBsAg that can be achieved from *P. pastoris* enabled the reduction in the production cost of this vaccine. Hence, it has been easier for the developing countries to include this low cost vaccine in their expanded programme of immunizations. HBsAg has been expressed in >7 g/L (Gurramkonda *et al.* 2009) of culture using *P. pastoris*, which is much higher than the expression levels achieved using *S. cerevisiae*. This methanol-utilizing yeast is contributing towards the WHO's objective of eliminating the hepatitis B virus from human population in the near future.

25.10.3 Products in the pipeline and those approved in different parts of the world

There have been a number of other products which have treaded the long path from the research laboratories to functional application in the market, but are awaiting approval for use in humans. Also, there are many products which have been approved for use in different parts of the world. Table 25.9 gives a list of such products (available at http://www.pichia.com/science-center/commercialized-products/).

25.11 Hurdles

As evident, *P. pastoris* has been used as a host for the expression of a number of proteins. This is a highly labour-intensive system due to the high cell densities to which it reached. Therefore, successful utilization of this system for product development is a highly appreciable achievement. Though this system is beneficial over *E. coli* for a number of reasons, there are factors under which *P. pastoris* scores lower than *E. coli*. For example, *E. coli* cells competent for transformation can

Table 25.9 *P. pastoris* products approved or awaiting approval

Product	Trade name	Company	Use	Status; Reference
Human neutrophil elastase inhibitor	Depelestat	Debiopharm S. A., Switzerland/Dyax Corporation, USA	Treatment of cystic fibrosis	Phase IIb; http://www.debiopharm.com/media/press-releases/53-other-news/2593-debiopharm-and-dyax-corp-restructure-development-agreement-for-depelestat-epi-hne4-or-dx-890-in-pulmonary-disorders.html
Ecallantide	Kalbitor	Dyax Corporation, USA	Hereditary angioedema (HAE)	On the market; http://www.kalbitor.com/index.html
Recombinant collagen		FibroGen, Inc., USA	Medical research reagents and dermal filler	On the market; http://www.fibrogen.com/recombinantcollagen
Recombinant allergens (house dust mite, cat, mouse and cockroach)		INDOOR Biotechnologies, USA	Research use	On the market; http://www.in-bio.com/UK/index.php?main_page=index&cPath=3_15
Nitrate reductase	Superior Stock Nitrate Reductase™	The Nitrate Elimination Co., USA	Enzyme-based kit for water testing and water treatment	On the market; http://nitrate.com/nr4.htm
Recombinant endotoxin-neutralizing protein		MP Biomedicals, USA	Detection and quantification of gram-negative bacterial endotoxins (research use)	On the market; http://www.mpbio.com/product_info.php?open=&cPath=&select tab=&family_key=0932353
Hepatitis B surface antigen	Shanvac®-B	Shantha Biotechnics Ltd., India	Hepatitis B vaccine	On the Indian market; http://www.shanthabiotech.com/files/Shanvac%20B%20Domestic%20Pack%20insert.pdf
Recombinant human insulin	Insugen®	Biocon, India	Diabetes	On the Indian market; http://www.biocon.com/insugen/insugen.html
Recombinant phytase	Optiphos®	Phytex LLC, USA	Animal feed additive	On the market; http://enzyvia.com/service/
Recombinant human cystatin C	P713-1	SCIPAC Ltd., UK	Laboratory use	On the European market; http://scipac.com/products/product/cystatin-c
Botulism vaccine	rBV A/B	DynPort Vaccine Company, USA	Botulism vaccine product	Phase II; http://assets1.csc.com/dvc/downloads/DVC_Botulinum_Vaccine_Case_Study_May_2010.pdf
Microplasmin	Ocriplasmin	ThromboGenics Ltd., Belgium	Treatment of symptomatic vitreomacular adhesion	Phase III; http://thrombogenics.com/wp-content/uploads/documents/10012010TG_company_presentation_OCT2010.pdf

be prepared, stored and used when required. In the case of *P. pastoris*, cells competent for transformation (by electroporation) need to be prepared fresh, which adds to the laborious feature of this expression host. Once the gene of interest has been transformed, a large number of clones need to be screened because of the high variation in the expression level between clones, which is not the case in *E. coli*. After selection of the desired clone, it is scaled up to shake the flask level for expression of the desired protein. It requires approximately 5–7 days in generating the induced biomass, as against *E. coli*, which requires 2–3 days. Hence, it is important to decide the expression host depending on the nature of the protein to be expressed. Striking a balance between the quality of protein and the ease of its preparation is highly critical for commercial success of the product.

References

Apte-Deshpnade A, Mandal G, Soorapaneni S, Prasad B, Kumar J, Padmanabhan S. High-level expression of non-glycosylated and active staphylokinase from *Pichia pastoris*. *Biotechnol Lett* 2009;31:811–817.

Anthony C. *The Biochemistry of Methylotrophs*. London: Academic Press; 1982.

Bardiya N. Expression in and purification of Hepatitis B surface antigen (S-protein) from methylotrophic yeast *Pichia pastoris*. *Anaerobe* 2006;12:194–203.

Barr KA, Hopkins SA, Sreekrishna K. Protocol for efficient secretion of HSA developed from *Pichia pastoris*. *Pharm Eng* 1992;12:48–51.

Berardi E. Genetics and molecular biology of methylotrophic yeasts. In: Spencer JFT, Spencer DM, eds. *Yeasts in Natural and Artificial Habitats*. Berlin: Springer; 1997. pp. 264–294.

Berdichevsky M, d'Anjou M, Mallem MR, Shaikh SS, Potgieter TI. Improved production of monoclonal antibodies through oxygen-limited cultivation of glycoengineered yeast. *J Biotechnol* 2011;155:217–224.

Blanchard V, Gadkari RA, George AV, Roy S, Gerwig GJ, Leeflang BR, *et al*. High-level expression of biologically active glycoprotein hormones in *Pichia pastoris* strains – Selection of strain GS115, and not X-33, for the production of biologically active N-glycosylated ^{15}N-labeled phCG. *Glycoconj J* 2008;25:245–257.

Bretthauer RK. Genetic engineering of *Pichia pastoris* to humanize *N*-glycosylation of proteins. *Trends Biotechnol* 2003;21:459–462.

Brunel L, Neugnot V, Landucci L, Boze H, Moulin G, Bigey F, Dubreucq E. High-level expression of *Candida parapsilosis* lipase/acyltransferase in *Pichia pastoris*. *J Biotechnol* 2004;111:41–50.

Buckholz RG, Gleeson MA. Yeast systems for the commercial production of heterologous proteins. *Biotechnology* 1991;9:1067–1072.

Calera JA, Paris S, Monod M, Hamilton AJ, Debeaupuis JP, Diaquin M, *et al*. Cloning and disruption of the antigenic catalase gene of *Aspergillus fumigatus*. *Infect Immun* 1997;65:4718–4724.

Cao S, Wang H, Luhur A, Wong SM. Yeast expression and characterization of SARS-CoV N protein. *J Virol Methods* 2005;130:83–88.

Cao Y, Qiao J, Li Y, Lu W. De novo synthesis, constitutive expression of *Aspergillus sulphureus* β-xylanase gene in *Pichia pastoris* and partial enzymic characterization. *Appl Microbiol Biotechnol* 2007;76:579–585.

Casteluber MC, Damasceno LM, da Silveira WB, Diniz RH, Passos FJ, Passos FM. Cloning and expression of a functional core streptavidin in *Pichia pastoris*: Strategies to increase yield. *Biotechnol Prog* 2012;28:1419–1425.

Cereghino JL, Cregg JM. Heterologous protein expression in the methylotrophic yeast *Pichia pastoris*. *FEMS Microbiol Rev* 2000;24:45–66.

Chen CC, Wu PH, Huang CT, Cheng KJ. A *Pichia pastoris* fermentation strategy for enhancing the heterologous expression of an *Escherichia coli* phytase. *Enzyme Microb Technol* 2004;35:315–320.

Chen X, Meng K, Shi P, Bai Y, Luo H, Huang H, Yuan T, *et al*. High-level expression of a novel *Penicillium* endo-1,3(4)-beta-D-glucanase with high specific activity in *Pichia pastoris*. *J Ind Microbiol Biotechnol* 2012;39:869–876.

Chen X, Zhai C, Kang L, Li C, Yan H, Zhou Y, *et al*. High-level expression and characterization of a highly thermostable chitosanase from *Aspergillus fumigatus* in *Pichia pastoris*. *Biotechnol Lett* 2012;34:689–694.

Choi BK, Jiménez-Flores R. Study of putative glycosylation sites in bovine β-casein introduced by PCR-based site-directed mutagenesis. *J Agric Food Chem* 1996;44:358–364.

Chu XY, Wu NF, Deng MJ, Tian J, Yao B, Fan YL. Expression of organophosphorus hydrolase OPHC2 in *Pichia pastoris*: Purification and characterization. *Protein Expr Purif* 2006;49:9–14.

Clare JJ, Rayment FB, Ballantine SP, Sreekrishna K, Romanos MA. High-level expression of tetanus toxin fragment C in *Pichia pastoris* strains containing multiple tandem integrations of the gene. *Biotechnology* 1991;9:455–460.

Cos O, Resina D, Ferrer P, Montesinos JL, Valero F. Heterologous production of *Rhizopus oryzae* lipase in *Pichia pastoris* using the alcohol oxidase and formaldehyde dehydrogenase promoters in batch and fed-batch cultures. *Biochem Eng J* 2005;26:86–94.

Couturier M, Feliu J, Haon M, Navarro D, Lesage-Meessen L, Coutinho PM, *et al*. A thermostable GH45 endoglucanase from yeast: Impact of its atypical multimodularity on activity. *Microb Cell Fact* 2011;10:103.

Cregg JM, Vedvick TS, Raschke WC. Recent advances in the expression of foreign genes in *Pichia pastoris*. *Biotechnology* 1993;11:905–910.

Daly R, Hearn MTW. Expression of heterologous proteins in *Pichia pastoris*: A useful experimental tool in protein engineering and production. *J Mol Recognit* 2005;18:119–138.

Damasceno LM, Huang CJ, Batt CA. Protein secretion in *Pichia pastoris* and advances in protein production. *Appl Microbiol Biotechnol* 2012;93:31–39.

Damasceno LM, Pla I, Chang HJ, Cohen L, Ritter G, Old LJ, *et al*. An optimized fermentation process for high-level production of a single-chain Fv antibody fragment in *Pichia pastoris*. *Protein Expr Purif* 2004;37:18–26.

De Schutter K, Lin YC, Tiels P, Van Hecke A, Glinka S, Weber-Lehmann J, *et al*. Genome sequence of the recombinant protein production host *Pichia pastoris. Nat Biotechnol* 2009;27: 561–566.

Demain AL, Vaishnav P. Production of recombinant proteins by microbes and higher organisms. *Biotechnol Adv* 2009;27: 297–306.

Domínguez A, Fermiñán E, Sánchez M, González FJ, Pérez-Campo FM, García S, *et al.* Non-conventional yeasts as hosts for heterologous protein production. *Int Microbiol* 1998;1: 131–142.

Du C, Han L, Xiao A, Cao M. Secretory expression and purification of the recombinant duck interleukin-2 in *Pichia pastoris. J Microbiol Biotechnol* 2011;21:1264–1269.

EasySelect™ Pichia Expression Kit Manual. Carlsbad, CA: Invitrogen; 2010. Accessed from: http://tools.invitrogen.com/content/sfs/manuals/easyselect_man.pdf

Eom GT, Lee SH, Song BK, Chung KW, Kim YW, Song JK. High-level extracellular production and characterization of *Candida antarctica* lipase B in *Pichia pastoris. J Biosci Bioeng* 2013;116:165–170.

Farnós O, Boué O, Parra F, Martín-Alonso JM, Valdés O, Joglar M, *et al.* High-level expression and immunogenic properties of the recombinant rabbit hemorrhagic disease virus VP60 capsid protein obtained in *Pichia pastoris. J Biotechnol* 2005;117: 215–224.

Gao MJ, Zhan XB, Zheng ZY, Wu JR, Dong SJ, Li Z, *et al.* Enhancing pIFN-α production and process stability in fed-batch culture of *Pichia pastoris* by controlling the methanol concentration and monitoring the responses of OUR/DO levels. *Appl Biochem Biotechnol* 2013;171:1262–1275.

Gao MJ, Zheng ZY, Wu JR, Dong SJ, Li Z, Jin H, *et al.* Improvement of specific growth rate of *Pichia pastoris* for effective porcine interferon-α production with an on-line model-based glycerol feeding strategy. *Appl Microbiol Biotechnol* 2012;93:1437–1445.

Gao Z, Li Z, Zhang Y, Huang H, Li M, Zhou L, *et al.* High-level expression of the *Penicillium notatum* glucose oxidase gene in *Pichia pastoris* using codon optimization. *Biotechnol Lett* 2012;34:507–514.

Gellissen G. Heterologous protein production in methylotrophic yeasts. *Appl Microbiol Biotechnol* 2000;54:741–750.

Gil DF, García-Fernández R, Alonso-del-Rivero M, Lamazares E, Pérez M, Varas L, *et al.* Recombinant expression of ShPI-1A, a non-specific BPTI-Kunitz-type inhibitor, and its protection effect on proteolytic degradation of recombinant human miniproinsulin expressed in *Pichia pastoris. FEMS Yeast Res* 2011;11: 575–586.

Glick BS. Cell biology: Alternatives to baker's yeast. *Curr Biol* 1996;6:1570–1572.

Gurramkonda C, Adnan A, Gäbel T, Lünsdorf H, Ross A, Nemani SK, *et al.* Simple high-cell density fed-batch technique for high-level recombinant protein production with *Pichia pastoris*: Application to intracellular production of Hepatitis B surface antigen. *Microb Cell Fact* 2009;8:13.

Gurramkonda C, Polez S, Skoko N, Adnan A, Gäbel T, Chugh D, *et al.* Application of simple fed-batch technique to high-level secretory production of insulin precursor using *Pichia pastoris* with subsequent purification and conversion to human insulin. *Microb Cell Fact* 2010;9:31.

Hamilton SR, Gerngross TU. Glycosylation engineering in yeast: The advent of fully humanized yeast. *Curr Opin Biotechnol* 2007;18:387–392.

Hao J, Xu L, He H, Du X, Jia L. High-level expression of Staphylococcal protein A in *Pichia pastoris* and purification and characterization of the recombinant protein. *Protein Expr Purif* 2013;90:178–185.

Hao Y, Chu J, Wang Y, Zhang S, Zhuang Y. Expression and aggregation of recombinant human consensus interferon-α mutant by *Pichia pastoris. Biotechnol Lett* 2006;28:905–909.

Hasslacher M, Schall M, Hayn M, Bona R, Rumbold K, Lückl J, *et al.* High-level intracellular expression of hydroxynitrile lyase from the tropical rubber tree *Hevea brasiliensis* in microbial hosts. *Protein Expr Purif* 1997;11:61–71.

Higgins DR. Overview of protein expression in *Pichia pastoris*, 5.7.1-5.7.16. In: Wingfield PT, ed. *Current Protocols in Protein Science, Supplement 2.* New York: Wiley; 1995.

Hilleman MR. Yeast recombinant hepatitis B vaccine. *Infection* 1987;15:3–7.

Hopkins D, Gomathinayagam S, Rittenhour AM, Du M, Hoyt E, Karaveg K, *et al.* Elimination of β-mannose glycan structures in *Pichia pastoris. Glycobiology* 2011;21:1616–1626.

Hua C, Yan Q, Jiang Z, Li Y, Katrolia P. High-level expression of a specific β-1,3-1,4-glucanase from the thermophilic fungus *Paecilomyces thermophila* in *Pichia pastoris. Appl Microbiol Biotechnol* 2010;88:509–518.

Hua C, Yi H, Jiao L. Cloning and expression of the endo-1,3(4)-β-glucanase gene from *Paecilomyces sp.* FLH30 and characterization of the recombinant enzyme. *Biosci Biotechnol Biochem* 2011;75:1807–1812.

Huang H, Yang P, Luo H, Tang H, Shao N, Yuan T, *et al.* High-level expression of a truncated 1,3-1,4-β-D-glucanase from *Fibrobacter succinogenes* in *Pichia pastoris* by optimization of codons and fermentation. *Appl Microbiol Biotechnol* 2008;78:95–103.

Huang YB, Li J, Gao X, Sun JR, Lu Y, Feng T, *et al.* Human proinsulin C-peptide from a precursor overexpressed in *Pichia pastoris. Acta Biochim Biophys Sin (Shanghai)* 2006;38:586–592.

Jahic M, Gustavsson M, Jansen AK, Martinelle M, Enfors SO. Analysis and control of proteolysis of a fusion protein in *Pichia pastoris* fed-batch processes. *J Biotechnol* 2003;102:45–53.

Jiang Y, Li F, Zha D, Potgieter TI, Mitchell T, Moore R, *et al.* Purification process development of a recombinant monoclonal antibody expressed in glycoengineered *Pichia pastoris. Protein Expr Purif* 2011;76:7–14.

Jin H, Liu G, Dai K, Wang H, Li Z, Shi Z. Improvement of porcine interferon-α production by recombinant *Pichia pastoris* via induction at low methanol concentration and low temperature. *Appl Biochem Biotechnol* 2011;165:559–571.

Kim TR, Goto Y, Hirota N, Kuwata K, Denton H, Wu SY, *et al.* High-level expression of bovine β-lactoglobulin in *Pichia pastoris* and characterization of its physical properties. *Protein Eng* 1997;10:1339–1345.

Kobayashi K, Kuwae S, Ohya T, Ohda T, Ohyama M, Ohi H, *et al.* High-level expression of recombinant human serum albumin from the methylotrophic yeast *Pichia pastoris* with minimal protease production and activation. *J Biosci Bioeng* 2000;89:55–61.

Kurtzman CP. *Pichia* E. C. Hansen emend. Kurtzman, In: Kurtzman CP, Fell JW, eds. *The yeasts, a taxonomic study.* Fourth revised and enlarged edition. Amsterdam: Elsevier Science BV; 1998. pp. 273–352.

Kurtzman CP. Description of *Komagataella phaffii* sp. nov. and the transfer of *Pichia pseudopastoris* to the methylotrophic yeast genus *Komagataella. Int J Syst Evol Microbiol* 2005;55: 973–976.

Kurtzman CP. Biotechnological strains of *Komagataella* (*Pichia*) *pastoris* are *Komagataella phaffii* as determined from multigene sequence analysis. *J Ind Microbiol Biotechnol* 2009;36: 1435–1438.

Kurtzman CP, Robnett CJ. Identification and phylogeny of ascomycetous yeasts from analysis of nuclear large subunit (26S) ribosomal DNA partial sequences. *Antonie Van Leeuwenhoek* 1998 73:331–371.

Li AN, Xie C, Zhang J, Zhang J, Li DC. Cloning, expression, and characterization of serine protease from thermophilic fungus *Thermoascus aurantiacus* var. *levisporus*. *J Microbiol* 2011;49:121–129.

Li N, Shi P, Yang P, Wang Y, Luo H, Bai Y, *et al.* Cloning, expression, and characterization of a new *Streptomyces* sp. S27 xylanase for which xylobiose is the main hydrolysis product. *Appl Biochem Biotechnol* 2009;159:521–531.

Li YL, Li H, Li AN, Li DC. Cloning of a gene encoding thermostable cellobiohydrolase from the thermophilic fungus *Chaetomium thermophilum* and its expression in *Pichia pastoris*. *J Appl Microbiol* 2009;106:1867–1875.

Liaw YF, Chu CM. Hepatitis B virus infection. *Lancet* 2009;373: 582–592.

Liu D, Zhang R, Yang X, Xu Y, Tang Z, Tian W, *et al.* Expression, purification and characterization of two thermostable endoglucanases cloned from a lignocellulosic decomposing fungi *Aspergillus fumigatus* Z5 isolated from compost. *Protein Expr Purif* 2011;79:176–186.

Liu YY, Woo JH, Neville DM Jr. Overexpression of an anti-CD3 immunotoxin increases expression and secretion of molecular chaperone BiP/Kar2p by *Pichia pastoris*. *Appl Environ Microbiol* 2005;71:5332–5340.

Liu ZQ, Zheng XB, Zhang SP, Zheng YG. Cloning, expression and characterization of a lipase gene from the *Candida antarctica* ZJB09193 and its application in biosynthesis of vitamin A esters. *Microbiol Res* 2012;167:452–460.

Luo H, Huang H, Yang P, Wang Y, Yuan T, Wu N, *et al.* A novel phytase appA from *Citrobacter amalonaticus* CGMCC 1696: Gene cloning and overexpression in *Pichia pastoris*. *Curr Microbiol* 2007;55:185–192.

Luo H, Li J, Yang J, Wang H, Yang Y, Huang H, *et al.* A thermophilic and acid stable family-10 xylanase from the acidophilic fungus *Bispora* sp. MEY-1. *Extremophiles* 2009;13:849–857.

Macauley-Patrick S, Fazenda ML, McNeil B, Harvey LM. Heterologous protein production using the *Pichia pastoris* expression system. *Yeast* 2005;22:249–270.

Michalak M, Thomassen LV, Roytio H, Ouwehand AC, Meyer AS, Mikkelsen JD. Expression and characterization of an endo-1,4-β-galactanase from *Emericella nidulans* in *Pichia pastoris* for enzymatic design of potentially prebiotic oligosaccharides from potato galactans. *Enzyme Microb Technol* 2012;50: 121–129.

Mille C, Bobrowicz P, Trinel PA, Li H, Maes E, Guerardel Y, *et al.* Identification of a new family of genes involved in β-1,2-mannosylation of glycans in *Pichia pastoris* and *Candida albicans*. *J Biol Chem* 2008;283:9724–9736.

Normand J, Bonnin E, Delavault P. Cloning and expression in *Pichia pastoris* of an *Irpex lacteus* rhamnogalacturonan hydrolase tolerant to acetylated rhamnogalacturonan. *Appl Microbiol Biotechnol* 2012;94:1543–1552.

Ohashi R, Mochizuki E, Suzuki T. A mini-scale mass production and separation system for secretory heterologous proteins by perfusion culture of recombinant *Pichia pastoris* using a shaken ceramic membrane flask. *J Biosci Bioeng* 1999;87:655–660.

Olędzka G, Dabrowski S, Kur J. High-level expression, secretion, and purification of the thermostable aqualysin I from *Thermus aquaticus* YT-1 in *Pichia pastoris*. *Protein Expr Purif* 29: 223–229.

Olsen D, Jiang J, Chang R, Duffy R, Sakaguchi M, Leigh S, *et al.* Expression and characterization of a low molecular weight recombinant human gelatin: development of a substitute for animal-derived gelatin with superior features. *Protein Expr Purif* 2005;40:346–357.

Outchkourov NS, Stiekema WJ, Jongsma MA. Optimization of the expression of equistatin in *Pichia pastoris*. *Protein Expr Purif* 2002;24:18–24.

Paifer E, Margolles E, Cremata J, Montesino R, Herrera L, Delgado JM. Efficient expression and secretion of recombinant alpha amylase in *Pichia pastoris* using two different signal sequences. *Yeast* 1994;10:1415–1419.

País JM, Varas L, Valdés J, Cabello C, Rodríguez L, Mansur M. Modeling of mini-proinsulin production in *Pichia pastoris* using the AOX promoter. *Biotechnol Lett* 2003;25:251–255.

Pan W, Huang D, Zhang Q, Qu L, Zhang D, Zhang X, *et al.* Fusion of two malaria vaccine candidate antigens enhances product yield, immunogenicity, and antibody-mediated inhibition of parasite growth *in vitro*. *J Immunol* 2004;172:6167–6174.

Qian P, Li X, Tong G, Chen H. High-level expression of the ORF6 gene of porcine reproductive and respiratory syndrome virus (PRRSV) in *Pichia pastoris*. *Virus Genes* 2003;27:189–196.

Qiao J, Rao Z, Dong B, Cao Y. Expression of *Bacillus subtilis* MA139 beta-mannanase in *Pichia pastoris* and the enzyme characterization. *Appl Biochem Biotechnol* 2010;160: 1362–1370.

Ragon M, Neugnot-Roux V, Chemardin P, Moulin G, Boze H. Molecular gene cloning and overexpression of the phytase from *Debaryomyces castellii* CBS 2923. *Protein Expr Purif* 2008;58:275–283.

Ramchuran SO, Mateus B, Holst O, Karlsson EN. The methylotrophic yeast *Pichia pastoris* as a host for the expression and production of thermostable xylanase from the bacterium *Rhodothermus marinus*. *FEMS Yeast Res* 2005;5:839–850.

Richter S, Nieveler J, Schulze H, Bachmann TT, Schmid RD. High yield production of a mutant *Nippostrongylus brasiliensis* acetylcholinesterase in *Pichia pastoris* and its purification. *Biotechnol Bioeng* 2006;93:1017–1022.

Rodríguez M, Rubiera R, Penichet M, Montesinos R, Cremata J, Falcón V, *et al.* High level expression of the *B. microplus* Bm86 antigen in the yeast *Pichia pastoris* forming highly immunogenic particles for cattle. *J Biotechnol* 1994;33:135–146.

Romanos MA, Clare JJ, Beesley KM, Rayment FB, Ballantine SP, Makoff AJ, *et al.* Recombinant *Bordetella pertussis* pertactin (P69) from the yeast *Pichia pastoris*: High-level production and immunological properties. *Vaccine* 1991;9:901–906.

Rosenfeld SA, Nadeau D, Tirado J, Hollis GF, Knabb RM, Jia S. Production and purification of recombinant hirudin expressed in the methylotrophic yeast *Pichia pastoris*. *Protein Expr Purif* 1996;8:476–482.

Shi XJ, Wang B, Zhang C, Wang M. Expressions of bovine IFN-γ and foot-and-mouth disease VP1 antigen in *P. pastoris* and their effects on mouse immune response to FMD antigens. *Vaccine* 2006;24:82–89.

Silva CI, Teles H, Moers AP, Eggink G, de Wolf FA, Werten MW. Secreted production of collagen-inspired gel-forming polymers with high thermal stability in *Pichia pastoris*. *Biotechnol Bioeng* 2011;108:2517–2525.

Silva IR, Larsen DM, Meyer AS, Mikkelsen JD. Identification, expression, and characterization of a novel bacterial RGI lyase enzyme for the production of bio-functional fibers. *Enzyme Microb Technol* 2011;49:160–166.

Smith PM, Suphioglu C, Griffith IJ, Theriault K, Knox RB, Singh MB. Cloning and expression in yeast *Pichia pastoris* of a biologically active form of Cyn d 1, the major allergen of Bermuda grass pollen. *J Allergy Clin Immunol* 1996;98:331–343.

Sørensen HP. Towards universal systems for recombinant gene expression. *Microb Cell Fact* 2010;9:27.

Sreekrishna K, Nelles L, Potenz R, Cruze J, Mazzaferro P, Fish W, *et al.* High-level expression, purification, and characterization of recombinant human tumor necrosis factor synthesized in the methylotrophic yeast *Pichia pastoris*. *Biochemistry* 1989;28: 4117–4125.

Striebeck A, Robinson DA, Schüttelkopf AW, van Aalten DM. Yeast Mnn9 is both a priming glycosyltransferase and an allosteric activator of mannan biosynthesis. *Open Biol* 2013; 3:130022.

Sugimoto N, Igarashi K, Samejima M. Cellulose affinity purification of fusion proteins tagged with fungal family 1 cellulose-binding domain. *Protein Expr Purif* 2012;82:290–296.

Thomas L, Crawford DL. Cloning of clustered *Streptomyces viridosporus* T7A lignocellulose catabolism genes encoding peroxidase and endoglucanase and their extracellular expression in *Pichia pastoris*. *Can J Microbiol* 1998;44:364–372.

Tschopp JF, Sverlow G, Kosson R, Craig W, Grinna L. High-level secretion of glycosylated invertase in the methylotrophic yeast, *Pichia pastoris*. *Biotechnology* 1987;5:1305–1308.

Tu T, Meng K, Bai Y, Shi P, Luo H, Wang Y, *et al.* High-yield production of a low-temperature-active polygalacturonase for papaya juice clarification. *Food Chem* 2013;141:2974–2981.

Vervecken W, Kaigorodov V, Callewaert N, Geysens S, De Vusser K, Contreras R. *In vivo* synthesis of mammalian-like, hybrid-type N-glycans in *Pichia pastoris*. *Appl Environ Microbiol* 2004;70:2639–2646.

Viader-Salvadó JM, Castillo-Galván M, Fuentes-Garibay JA, Iracheta-Cárdenas MM, Guerrero-Olazarán M. Optimization of five environmental factors to increase β-propeller phytase production in *Pichia pastoris* and impact on the physiological response of the host. *Biotechnol Prog* 2013;29:1377–1385.

Wagner SL, Siegel RS, Vedvick TS, Raschke WC, Van Nostrand WE. High level expression, purification, and characterization of the Kunitz-type protease inhibitor domain of protease nexin-2/amyloid β-protein precursor. *Biochem Biophys Res Commun* 1992;186:1138–1145.

Wakadkar S, Zhang LQ, Li DC, Haikarainen T, Dhavala P, Papageorgiou AC. Expression, purification and crystallization of *Chaetomium thermophilum* Cu, Zn superoxide dismutase. *Acta Crystallogr Sect F Struct Biol Cryst Commun* 2010;66: 1089–1092.

Walsh G. Biopharmaceutical benchmarks 2010. *Nat Biotechnol* 2010;28:917–924.

Wang J, Nguyen V, Glen J, Henderson B, Saul A, Miller LH. Improved yield of recombinant merozoite surface protein 3 (MSP3) from *Pichia pastoris* using chemically defined media. *Biotechnol Bioeng* 2005;90:838–847.

Wang H, Shi P, Luo H, Huang H, Yang P, Yao B. A thermophilic α-galactosidase from *Neosartorya fischeri* P1 with high specific activity, broad substrate specificity and significant hydrolysis ability of soymilk. *Bioresour Technol* 2014;153:361–364.

Wang X, Sun Y, Ke F, Zhao H, Liu T, Xu L, *et al.* Constitutive expression of *Yarrowia lipolytica* lipase LIP2 in *Pichia pastoris* using GAP as promoter. *Appl Biochem Biotechnol* 2012;166:1355–1367.

Wang Y, Liang ZH, Zhang YS, Yao SY, Xu YG, Tang YH, *et al.* Human insulin from a precursor overexpressed in the methylotrophic yeast *Pichia pastoris* and a simple procedure for purifying the expression product. *Biotechnol Bioeng* 2001;73:74–79.

Wang Y, Yuan S, Wang P, Liu X, Zhan D, Zhang Z. Expression, purification, and characterization of recombinant human keratinocyte growth factor-2 in *Pichia pastoris*. *J Biotechnol* 2007;132:44–48.

Watanabe T, Ikeda H, Masaki K, Fujii T, Iefuji H. Cloning and characterization of a novel phytase from wastewater treatment yeast *Hansenula fabianii* J640 and expression in *Pichia pastoris*. *J Biosci Bioeng* 2009;108:225–230.

Werten MW, van den Bosch TJ, Wind RD, Mooibroek H, de Wolf FA. High-yield secretion of recombinant gelatins by *Pichia pastoris*. *Yeast* 1999;15:1087–1096.

Werten MW, Wisselink WH, Jansen-van den Bosch TJ, de Bruin EC, de Wolf FA. Secreted production of a custom-designed, highly hydrophilic gelatin in *Pichia pastoris*. *Protein Eng* 2001;14:447–454.

Wood MJ, Komives EA. Production of large quantities of isotopically labeled protein in *Pichia pastoris* by fermentation. *J Biomol NMR* 1999;13:149–159.

Woo SH, Park SH, Lim HK, Jung KH. Extended operation of a pressurized 75-L bioreactor for shLkn-1 production by *Pichia pastoris* using dissolved oxygen profile control. *J Ind Microbiol Biotechnol* 2005;32:474–480.

Wu D, Hao YY, Chu J, Zhuang YP, Zhang SL.. Inhibition of degradation and aggregation of recombinant human consensus interferon-α mutant expressed in *Pichia pastoris* with complex medium in bioreactor. *Appl Microbiol Biotechnol* 2008;80:1063–1071.

Wu M, Liu W, Yang G, Yu D, Lin D, Sun H, Chen S. Engineering of a *Pichia pastoris* expression system for high-level secretion of HSA/GH fusion protein. Appl Biochem Biotechnol 2014;172:2400–2411.

Xiao A, Zhou X, Zhou L, Zhang Y. Improvement of cell viability and hirudin production by ascorbic acid in *Pichia pastoris* fermentation. *Appl Microbiol Biotechnol* 2006;72:837–844.

Xiao Z, Wu M, Grosse S, Beauchemin M, Lévesque M, Lau PC. Genome mining for new α-amylase and glucoamylase encoding sequences and high level expression of a glucoamylase from *Talaromyces stipitatus* for potential raw starch hydrolysis. Appl Biochem Biotechnol 2014;172:73–86.

Xie T, Liu Q, Xie F, Liu H, Zhang Y. Secretory expression of insulin precursor in *Pichia pastoris* and simple procedure for producing recombinant human insulin. *Prep Biochem Biotechnol* 2008;38:308–317.

Xiong AS, Yao QH, Peng RH, Han PL, Cheng ZM, Li Y. High level expression of a recombinant acid phytase gene in *Pichia pastoris*. *J Appl Microbiol* 2005;98:418–428.

Xiong AS, Yao QH, Peng RH, Li X, Fan HQ, Guo MJ, *et al.* Isolation, characterization, and molecular cloning of the cDNA encoding a novel phytase from *Aspergillus niger* 113 and high expression in *Pichia pastoris*. *J Biochem Mol Biol* 2004;37:282–291.

Xiong AS, Yao QH, Peng RH, Zhang Z, Xu F, Liu JG, *et al.* High level expression of a synthetic gene encoding *Peniophora lycii* phytase in methylotrophic yeast *Pichia pastoris*. *Appl Microbiol Biotechnol* 2006;72:1039–1047.

Xuan Y, Zhou X, Zhang W, Zhang X, Song Z, Zhang Y. An upstream activation sequence controls the expression of *AOX1* gene in *Pichia pastoris*. *FEMS Yeast Res* 2009;9:1271–1282.

Yamada Y, Matsuda M, Maeda K, Mikata K. The phylogenetic relationships of methanol-assimilating yeasts based on the partial sequences of 18S and 26S ribosomal RNAs: The proposal of *Komagataella* gen. nov. (Saccharomycetaceae). *Biosci Biotechnol Biochem* 1995;59:439–444.

Yang JK, Liu LY, Dai JH, Li Q. De novo design and synthesis of *Candida antarctica* lipase B gene and α-factor leads to high-level expression in *Pichia pastoris*. PLoS ONE 2013;8:e53939.

Yang S, Chen G, Yu X, Li M, Wang J. Cloning of a novel ovalbumin gene from quail oviduct and its heterologous expression in *Pichia pastoris*. *J Basic Microbiol* 2009;49:S73–S78.

Zhang LQ, Guo FX, Xian HQ, Wang XJ, Li AN, Li DC. Expression of a novel thermostable Cu, Zn-superoxide dismutase from

Chaetomium thermophilum in *Pichia pastoris* and its antioxidant properties. *Biotechnol Lett* 2011;33:1127–1132.

Zhang R, Fan Z, Kasuga T. Expression of cellobiose dehydrogenase from *Neurospora crassa* in *Pichia pastoris* and its purification and characterization. *Protein Expr Purif* 2011;75:63–69.

Zhang Y, Teng D, Mao R, Wang X, Xi D, Hu X, *et al.* High expression of a plectasin-derived peptide NZ2114 in *Pichia pastoris* and its pharmacodynamics, postantibiotic and synergy against *Staphylococcus aureus*. *Appl Microbiol Biotechnol* 2014;98:681–694.

Zhao Q, Liu F, Hou Z, Yuan C, Zhu X. High level production of β-galactosidase exhibiting excellent milk-lactose degradation ability from *Aspergillus oryzae* by codon and fermentation optimization. Appl Biochem Biotechnol 2014;172: 2787–2799.

Zheng J, Zhao W, Guo N, Lin F, Tian J, Wu L, Zhou H. Development of an industrial medium and a novel fed-batch strategy for high-level expression of recombinant β-mananase by *Pichia pastoris*. *Bioresour Technol* 2012;118:257–264.

Zhu T, You L, Gong F, Xie M, Xue Y, Li Y, Ma Y. Combinatorial strategy of sorbitol feeding and low-temperature induction leads to high-level production of alkaline β-mannanase in *Pichia pastoris*. *Enzyme Microb Technol* 2011;49:407–412.

26

Insect cells as host for recombinant DNA

Krishnaveni Mohareer and Seyed Ehtesham Hasnain

26.1 Introduction

The knowledge of baculovirus-associated pathogenesis dates back to the sixteenth century when silkworm jaundice was widespread and the aetiology had remained a mystery (Miller 1993). It was only during the 1940s that the agent for the causation of the wilting disease of the silkworm was known and baculovirus was thus exploited as a biopesticide (Benz 1986). With the advent of recombinant DNA technology, the *Escherichia coli* expression system had become the universal host of choice for expression of heterologous proteins. Realizing their limited capacities in terms of post-translational modifications (PTMs) (phosphorylation, acetylation, glycosylation, mannosylation, etc.) for foreign gene expression, interest in eukaryotic expression systems has gained importance in both academia and industry. Eukaryotic systems of gene expression are in use for other reasons, besides post-translational modifications, such as protein folding, expression of higher-molecular-sized proteins, codon bias and synthesis of endotoxin-free protein, which overcome the limitations of prokaryotic expression systems. Yeast-, baculovirus- and adenovirus-based systems are majorly used as vector systems for expression of eukaryotic proteins. Of late, newer systems such as *Leishmania* (Basile 2009) are emerging. Wheat germ lysates and rabbit reticulolysates are typically used to address specific academic questions and demand the use of radioactive isotopes and are not of industrial importance. The protein of interest is expressed in the choice of the system depending on the ultimate use of the target protein. Eukaryotic expression systems are, however, limited by factors such as the time required for expression, the yield and the expenditure. In transition between prokaryotic expression systems (which are fast, easy and inexpensive, but constrained by PTMs) and the advanced mammalian expression systems (which are more expensive and

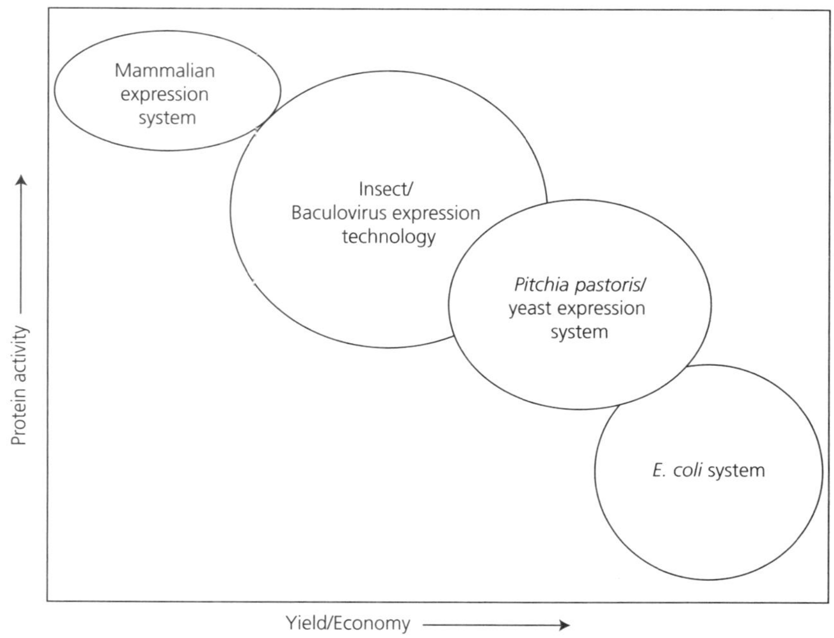

Figure 26.1 Comparative graph of yield versus activity in different expression systems. Comparative features of different expression systems in terms of yield and protein activity. It can be noted that as protein activity of the expression system increases, the yield decreases. Among the several expression systems, the BEVS offers most of the advantages and occupies the central focus of heterologous gene expression.

Table 26.1 Comparative features of different expression systems

Feature	*E. coli*	Yeast	Sf9	CHO	Adeno	Rabbit reticulo-lysate (*in vitro*)
Inducible	Yes	Yes	Temporal	Constitutive	Constitutive	Yes
Time of expression	3–12 h	1 day	3–5 days	3–4 days	3–4 days	3 h
Extracellular expression	Periplasm	Secreted into media	Secreted into media	Secreted into media	Secreted into media	Cell free
Protein folding	Restricted	Refolding may be required	Properly folded	Properly folded	Properly folded	Properly folded
PTMs						
N-linked glycosylation	No	High mannose	Simple; no sialic acid	Yes	Yes	Yes
O-linked glycosylation	No	Yes	Yes	Yes	Yes	Yes
Phosphorylation	No	Yes	Yes	Yes	Yes	Yes
Acetylation	No	Yes	Yes	Yes	Yes	Yes
Acylation	No	Yes	Yes	Yes	Yes	Yes
Gamma-carboxylation	No	No	No	Yes	Yes	Yes
Protein limit	Difficult to express and purify higher–molecular-size proteins	Higher molecular size than *E. coli* but less than insect and other mammalian systems	Easy to express and purify higher-molecular-size proteins	Easy to express and purify higher–molecular-size proteins	Easy to express and purify higher-molecular-size proteins	Can express higher-molecular-size proteins
Yield	Very good	Good	Good	Less	less	Very less
Ease of handling	Simple	Have to take care of bacterial contamination and cross contamination	Simple, with basic tissue culture precautions	Simple, with basic tissue culture precautions	To be handled with care	To be handled with care involves radioactivity
Scale up	Yes	Yes	Yes	Yes, but expensive	Yes, but expensive	No; can be used for academic purpose only
Safety	Yes	Yes	Yes	Yes	To be handled with care	To be handled with care; involves radioactivity

time-consuming but achieve functional, correctly folded proteins with PTMs) is the insect expression system with an appealing performance and is usually sufficient for many biotechnological processes and expression of functional folded proteins carrying out a majority of PTMs (Fig. 26.1). A comparative table of different expression parameters among the different expression systems has been listed in Table 26.1. With the advanced and user-friendly methods for baculovirus expression, this technology, which was once limited to researchers in insect biology, is now within the reach of an average molecular biologist. This chapter focuses on the current trends in the baculovirus expression vector (BEVS) system and future prospects.

26.2 Baculovirus technology: An overview

Baculoviruses specifically infect insect hosts belonging to the phylum Arthropoda (Miller 1997). The different members of the Baculoviridae infect distinct hosts from Lepidoptera, Hymenoptera and Diptera, multiply and ultimately lyse the host cell, liberating the progeny virus to infect more insects. It has consequently been used as a biopesticide. Like other animal viruses, gene expression in baculoviruses occurs in a cascade (O'Reilly *et al.* 1994). The natural life cycle of the virus is carried out in 3–4 days, during which there is co-ordinated gene expression. Based on gene expression, the baculovirus infection can be divided into three phases: the early (subdivided into immediate early and delayed early), the late and the very late phases. During early gene expression, genes encoding proteins required in late gene transcription as well as DNA replication are expressed. The major proteins expressed by the baculovirus during the late and very late phase are the surface proteins P10 and Polh, constituting the capsid, and are non-essential. The promoter driving the *polh* gene is immensely strong and expresses about 50% of total cellular proteins at a given time; hence, it has become the promoter of choice for driving recombinant protein expression. Besides the *polh* gene, another gene, *p10,* is also expressed in high levels during the late phase, but lesser than that of *polh.* The *polh* and *p10* comprise the workhorse promoters of the BEVS (as reviewed in Ramchandran *et al.* 2001). The foreign gene of interest is cloned downstream of the *polh* promoter and the recombinant baculovirus expressing the foreign gene is allowed to infect a host cell line (usually *Sf*9/*Sf*21) at a ratio of 10 virus particles per host cell. The *polh* promoter expresses at a late stage of infection and the expression rises dramatically from 48 h onwards. The infection cycle is allowed to proceed until the very late stage to allow maximal expression of foreign genes. The cells are then harvested and used for protein purification. A schematic representation of the BEVS is given in Fig. 26.2.

The advantages of the BEVS are as follows:

- The protein expressed through BEVS is endotoxin free.
- Glycosylation and other post-translational modifications are possible.

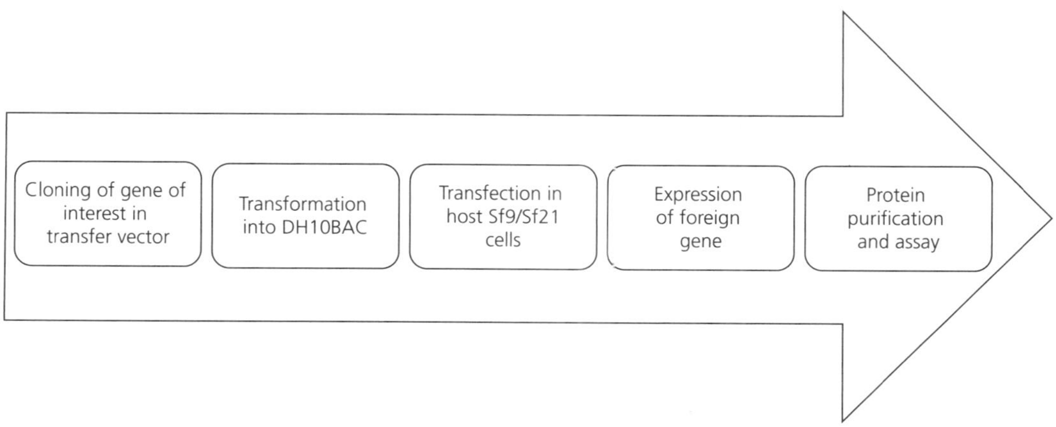

Figure 26.2 Overview of baculovirus expression. Schematic representation of the BEVS. The gene of interest is initially cloned in a suitable transfer vector such as pFastBac (Invitrogen). The construct is then transformed nto DH10Bac competent cells and selected by blue/white screening. White colonies indicate successful transposition. The colonies are further confirmed by PCR using primers flanking the cloning site. An expected size of 350 bp is obtained for a non-transposed colony and 2420 + *n* bp for a transposed one, where *n* is the size of the gene cloned using commercially available primers (pUC M13 F/R) from Invitrogen. The confirmed clone is then transfected into *Sf*9 cells and allowed to bud the virus for 3–5 days, the titer of which is very low. The resulting culture supernatant is used to infect a fresh lot of *Sf*9 cells. The infection process is repeated till a titer of 10^8–10^9 pfu is obtained. The titer of the recombinant virus is done by using antibodies specific to baculovirus such as gp-64/gp-120 or neutral red staining.

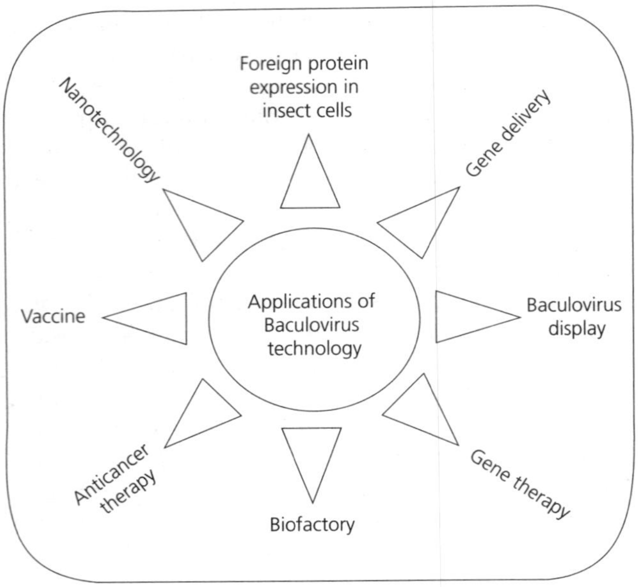

Figure 26.3 Schematic representation of different applications of the baculovirus technology. The baculovirus expression technology can be used for various applications as depicted. It offers maximum protein expression among the various expression systems and is very safe and easy to handle and has hence become the choice of expression system for heterologous gene expression.

- The recombinant protein is functionally active.
- It can be made to secrete into the culture medium, easing the process of purification.
- It can be easily scaled up in fermentation, reaching high cell density (up to 10 million cells/mL).
- It is less expensive as compared with mammalian expression systems.
- Insect larvae can be used to express recombinant proteins, rendering them as 'bio-factories'.

The limitations of the BEVS are as follows:

- Complex post-translational modifications cannot be done.
- It is expensive as compared with yeast-based expression systems.

A schematic representation of the applications of the baculovirus technology is given in Fig. 26.3.

26.3 Transcription from the polyhedrin promoter

Baculovirus polyhedrin promoter is apparently the strongest eukaryotic promoter next to the SV40 promoter. Polyhedrin is expressed from 48 h of infection in host cells and continues till the very late stage where only viral proteins are synthesized. During this stage, polyhedrin accounts for >50% of total cellular protein content. Several molecular elements of the virus and the host have been elucidated to be essential in driving the transcription from the *polh* promoter. The polyhedrin gene expression is driven by alpha-amanitin–resistant viral RNA polymerase. A library of 18 late expression factor (*lef*) genes has been established as essential for full transactivation of late and very late genes, including polyhedrin (Hefferon and Miller 2002). These factors are expressed early during infection and *trans*-activate late gene expression. However, two of the 18 factors (IE-1, IE-2) also *trans*-activate early genes. Some of the *lefs* are also involved in viral DNA multiplication. Several *cis*-elements are involved in *polh* and other late/very late gene transcription. The 'burst sequence (TAAG)'(Rankin *et al.* 1988) distinguishes late and very late genes. A very late factor, vlf-1, binds to the sequence encompassing the burst sequence. An increase in the number of burst sequences further augments the transcription (Manohar *et al.* 2010). Gene knockout of *vlf-1* causes infection-defective virus similar to that of non-infectious control virus (Vanarsdall *et al.* 2006). However, overexpression of vlf-1 stimulated the expression of *polh* and *p10*. Other important *cis*-elements include enhancer elements, termed as homologous regions (*hr*), which enhance the *polh* transcription manifold. They are characterized by repetitive palindromic repeats. There are 8 *hr* elements in *Autographa californica* nuclear polyhedrosis virus (*Ac*NPV), which is the prototype baculovirus (*hr1, hr1a, hr2, hr3, hr4a, hr4b* and *hr5)* (Lu *et al.* 1997), which also function as enhancers in heterologous systems such as in mammalian cells (Vishwanathan *et al.* 2003). They serve as binding sites for *trans*-acting factors such as IE-1 (Rodems and Friesen 1995), which function as *trans*-activators of late genes. The *hr* elements have been used to *trans*-activate early baculovirus promoters (Friesen and Rueckert 1984). Apart from transcription, *hr* elements are involved in DNA replication by functioning as the origin of replication (Leisy *et al.* 1995). Besides viral *trans*-activators, host proteins are also involved in transcription. The *polh* promoter has been an extensively studied late promoter. It harbours binding sites for several host transcription factors such as polyhedrin promoter binding protein (PPBP) (Ghosh *et al.* 1998; Mukherjee *et al.* 1995), hr1-BP (Habib and Hasnain 1996), Sp-1 (Ramachandran *et al.* 2001), Oct-1 and heat shock elements (Kumar *et al.* 2009), which are known to influence transcription from the *polh* promoter. Deletion of these cognate motifs resulted in decreased luciferase reporter activity. A schematic representation of transcription from the polyhedrin promoter is given in Fig. 26.4.

26.4 Other promoters used in baculovirus expression

Although the *polh* promoter has been extensively explored, other promoters of baculovirus have also been studied for various reasons – early gene expression to enhance post-translational modifications (PTMs), minimize protein aggregation, targeted secretion, use in dual expression systems, etc. Some of the other promoters and their advantages over that of *polh* have been described next.

26.4.1 The promoter *p10*

The *p10* promoter, similar to *polh,* belongs to the late phase of viral gene expression. The *p10* promoter has also been in use in BEVS. The *p10* promoter competes with the *polh* promoter, and in the absence of *p10*

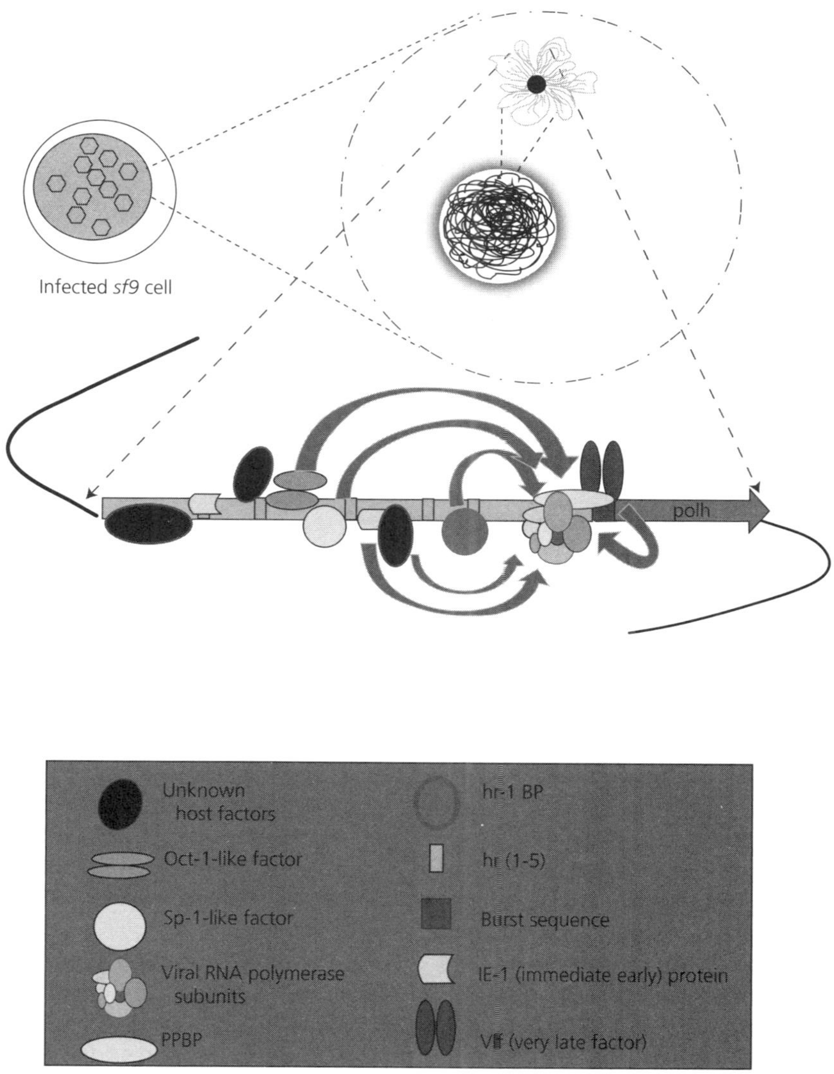

Figure 26.4 Transcription from polyhedrin promoter. Schematic illustration of polyhedrin gene transcription and related factors. The polyhedrin gene is expressed in the late phase of infection. The late gene promoters have a characteristic motif, termed as burst sequence, to which a very late factor (vlf-1) binds and enhances transcription initiated by viral polymerase. In addition, there are *hr* elements, which serve as binding sites for several transcription activators, including host (hr-1 BP) and viral factors (IE1), which results in tremendous activation of the polyhedrin gene expression. The host factor PPBP binds to the initiator promoter and serves to recruit several viral proteins constituting the viral RNA polymerase, which is influenced by both viral and host factors such as SP-1, Oct-1, heat shock elements, etc. See Colour Plate 38 on page 526.

promoter elements, the *polh* promoter induces the reporter gene expression much stronger, implying the competitiveness of *p10* promoter. However, when the reverse happens, the *p10* expression is not enhanced, implying *polh* does not compete with *p10* (Chaabihi *et al.* 1993). The same could be used to enhance gene expression from the *polh* promoter. Certain dual vectors such as pFASTBacDual use two different promoters, including *polh* and *p10*.

26.4.2 The promoter *ie-1*

Immediate early (*ie*) genes (*ie-0, ie-1, ie-2*) are expressed very early in infection and deploy host RNA polymerase and are alpha-amanitin sensitive. Their expression levels are 10–50 times lower than that of the *polh* promoter. However, their expression levels can be enhanced by the addition of *hr*3 of *Bm*NPV, as demonstrated by Chen and colleagues (Chen *et al.* 2004). It would be of great advantage as it would reduce the time of expression and also improve secretion and also efficient PTMs, which are compromised during the late stage of infection.

26.4.3 The promoter *vp-39*

vp-39 is expressed in the late stage of baculovirus infection, much before polyhedrin and with much lesser promoter activity. Expression under the *vp-39* promoter, in the context of an additional *Bm*NPV *hr*3, causes enhanced protein expression as compared with *polh*-driven expression with minimal aggregation of the recombinant protein, when protein quality control mechanisms are much more effective (Ishiyama and Ikeda 2010).

26.4.4 *Bombyx*-specific promoters

Bombyx mori is a large domestic insect used in sericulture and is the most studied insect in terms of physiology, biochemistry, pathology and genetics. *Bm*NPV infects and successfully replicates in *B. mori* but not *Ac*MNPV for reasons discussed elsewhere in the chapter. *B. mori* has the potential to be used as biofactory for expressing foreign proteins with all the advantages of BEVS discussed earlier and at minimal cost. The *hr3* enhancer of *Bm*NPV has been reported to function as a super-enhancer (Chen *et al.* 2004). Recombination of elements from *Ac*MNPV and *Bm*NPV could be exploited to increase expression levels favourably.

26.4.4.1 The sericin promoter

Sericin is one of the major silk proteins along with fibroin. Silk is secreted by the last instar larva in the cocoon stage of *B. mori*. A reporter gene, cloned under sericin promoter, which is expressed specifically in the silk gland, can be secreted along with silk proteins that are water soluble (Guo *et al.* 2005). A recombinant protein with 'ser' signal peptide could be targeted for secretion, making the purification of the recombinant protein much easier.

26.5 Process of heterologous gene expression in the baculovirus system

The process of heterologous gene expression in the baculovirus system (summarized in Fig. 26.2) includes the following steps.

26.5.1 Molecular cloning of foreign genes in the baculovirus system

The gene encoding the protein of interest is cloned in a choice of vector depending on the downstream application and nature of the recombinant protein. The choice of host cells for expression again varies on the downstream protein application and purpose. The range of available vectors, host, process and their advantages over others have been described next.

26.5.1.1 Choice of vectors

Transfer vectors are commercially available for cloning the gene of interest within the range of restriction enzyme sites in a multiple cloning site (MCS). Table 26.2 gives a list of commercially available vectors. The foreign gene can be fused to a tag such as His or GST. The protein can also be fused to a secretory signal and secreted out. The gene is cloned into the vector of choice by using appropriate enzyme sites.

26.5.1.2 Transposition in DH10Bac cells

After cloning the foreign gene in the MCS of the transfer vector of choice, the construct is transformed into DH10Bac cells, which carries the entire baculovirus genome and two plasmids, one of which encodes a transposase that transposes the gene of interest into the '*att*' sites engineered in the flanking regions of the MCS in the bacmid. Transposition of the cloned gene takes place by virtue of the transposase expressed in DH10Bac cells. The recombinants are selected by blue-white screening (MCS is flanked by the X-gal ORF, which is disrupted by the cloning process). The white colonies are screened and confirmed by PCR.

Table 26.2 List of commercially available transfer vectors for expression of foreign genes using the baculovirus system

S. No.	Vector	Company	Remark
1	pFastBac	Invitrogen	Has no tag
2	pFastBacHT(a/b/c)	Invitrogen	Has His tag and is available in three frames
3	pFastBacDual	Invitrogen	Can be used for expression of two proteins in the same cell
4	pPolh-FLAG™-1	Sigma	Flag tag at N-terminal
5	pPolh-FLAG™-2	Sigma	Flag tag at C-terminal
6	pPolh-MAT™-1	Sigma	Metal affinity tag at N-terminal
7	pPolh-MAT™-2	Sigma	Metal affinity tag at C-terminal
8	pQE-TriSystem	Qiagen	The same construct can be used to express protein in *E. coli*/ insect/mammalian system
9	pOET1	BioNovus	Uses polh promoter, multiple restriction sites, smaller size of vector
10	pOET2	BioNovus	Similar to pOET1 but with MCS in reverse orientation
11	pOET3	BioNovus	Uses p6.9 promoter, early expression as compared to polh, optimal for extensive post-translational modifications
12	pOET4	BioNovus	Similar to pOET3 with reverse MCS
13	pOET1C	BioNovus	Similar to pOET1 with C-terminal 6X His tag
14	pOET1N	BioNovus	Similar to pOET1 with N-terminal 6X His tag
15	pOET2C	BioNovus	Similar to pOET2 with C-terminal 6X His tag
16	pOET2N/2C	BioNovus	Similar to pOET2 with N-terminal 6X His tag if stop codon is included in reverse primer and both if not included
17	pTriEx	Novagen/Merck	Single clone can be expressed in bacterial, baculo- and mammalian expression systems

26.5.2 Generation of recombinant virus

The PCR-confirmed bacmid preparations are then transfected into *Sf*9 or any other suitable insect host by simple transfection. Expression of the recombinant protein in the transfected cells is monitored by Western blot analysis using anti-His antibody. After confirmation of expression, the recombinant virus is allowed to accumulate in the culture supernatant till the time the cells begin to lyse. The process is reiterated using uninfected cells up to a few cycles till the viral titer reaches at least 10^8 pfu/mL. For every cycle, the viral titer is expected to increase 10 fold.

26.5.3 Choice of host cells

Several insect cell lines have been used as host, but *Sf*9/*Sf*21 cells are often used with the baculovirus system. Other host cell lines include *Tni*, Hifive (Invitrogen), *Bombyx mori*, etc. The cell line can be chosen based on the target protein requirements. The dipteran cell line S2 (*Drosophila*) has also been used for expression using the baculovirus system (Lee *et al.* 2000). For ease of protein purification and economy of production, whole larvae (*B. mori* or *S. litura*) can be used as biofactories (Palhan *et al.* 1995, Hasnain *et al.* 2003), reducing the economic burden of cell culture. *T. ni* cell line has been reported to perform better in terms of PTMs (Davis *et al.* 1993).

26.5.4 Host specificity and host range

Baculoviruses have a limited host range. This raises crucial issues while considering the choice of the cell line and vector. Different host determinants have been shown to be factors involved at various levels in the viral life cycle (McClintock *et al.* 1986; Wickham *et al.* 1992; Clem 1993; Morris and Miller 1994; Mukherjee *et al.* 1995). *Ac*MNPV and *Bm*NPV share 90% identity at the nucleotide level, yet their host specificity is different. *Ac*MNPV cannot amplify in *B. mori* cells due to defective DNA replication. The host determinant in *Ac*MNPV was later determined to be p143, which performs helicase activity during DNA replication. Two single mutations in two amino acids of this gene make the non-permissive *B. mori* cells permissive and result in their subsequent killing (Argaud *et al.* 1998). Hcf-1, gp-64 and inhibitor of apoptosis (*Ac*MNPVP35, IAPs) factors also contribute to host range determination (Miller 1997).

Homologous recombination of different baculoviruses can lead to the expansion of the host range (Iwanaga *et al.* 2004). Of the several baculoviruses known, *Ac*MNPV and *Bm*NPV are most widely used in BEVS. Although they propagate in their specific host, they can still infect cells of different species, such as mammalian cells (Smolewski and Robak 2011). However, they do not replicate, but do induce anti-viral effects. This property of baculovirus has been exploited as a gene delivery vehicle.

26.6 Advances in the BEVS

BEVS has been explored as the choice of eukaryotic protein expression system. There have been several biotechnological advances which address some of the present limitations of BEVS described earlier so as to enhance both protein quality and quantity. Not only as a protein expression system, the use of the baculovirus technology as a gene delivery vehicle for gene therapy and in vaccine therapy and tumour immunity has been explored. Some of the advances in this direction have been described next.

26.6.1 Expression of multi-protein complexes

Structures of many proteins which exist in multimeric forms cannot be isolated from *E. coli* in the native folded state. Multiple recombinant baculoviruses expressing individual desired genes could be co-infected to express all the desired proteins, which can be assembled to form the multi-protein complex. Insect expression system offers a suitable choice where multiple proteins could be expressed under same/different baculoviral promoters provided in the transfer vector (pFastBacDual, pTri-Ex, etc). Recently, a MultiBac system has been developed which can express four proteins under four different promoters in two transfer vectors based on cre-lox p recombination and Tn7 transposition (Berger *et al.* 2004). The production of multi-protein complexes is very demanding in structural studies where the structure of complex interacting partners is difficult to achieve.

26.6.2 Modification of genome for enhanced protein expression

To generate enhanced levels of recombinant protein using baculovirus, there have been various modifications in the baculoviral genome so as to increase the expression. A tremendous increase in foreign gene expression has been obtained by introducing single (Venkaiah *et al.* 2004) or multiple copies of *hr1*. For instance, Hitchman *et al.* have deleted the non-essential genes, *p26*, *p10* and *p74*, and replaced them with antibiotic selection markers to facilitate screening of the recombinant virus which enhance the protein expression and yield (Hitchman *et al.* 2010). As described earlier, burst sequences are indispensable motifs for late gene expression and markedly enhance transcription from the *polh* promoter. The transcription from the *polh* promoter has been enhanced by increasing the number of burst sequences and, consequently, increased reporter gene expression (Manohar *et al.* 2010). Integration of *hr1* into the bacmid-based transfer vector also resulted in enhanced yield of the recombinant proteins, protein kinase B and CYP-1A2 (Tiwari *et al.* 2010). Insertion of *Ac*MNPV *FP25k* gene into *Bm*MPV increases expression of the foreign gene in cultured cells and hastens the secretion of the foreign protein from the larval nymph (Nakanishi *et al.* 2010). Large-scale production of proteins in silkworm larvae would be economic and easy (Palhan *et al.* 1995). However, the *Bm*NPV expression is limited, which has been overcome by genetic modification, including the deletion of proteases and co-expression of chaperones for assisting in folding and expression. Generation of stable transformants of chaperones would facilitate increased yield of foreign protein (Kato *et al.* 2005). Translation of proteins from other promoters driving transcription of foreign genes has been reported to be enhanced by the addition of 5′-UTR (Iizuka *et al.* 2008). Metabolic engineering of host cells by RNAi of v-cath has resulted in an approximately threefold increase in the protein yield (Kim *et al.* 2007). Table 26.3 gives a representative list of proteins expressed in the baculovirus system for various applications (references therein).

26.6.3 Baculovirus display technology

The phage display system is an extremely useful molecular biology tool that has been used to display and screen for peptides/proteins among a library for a specific biological activity (Smith 1985). The foreign protein, termed as 'passenger protein', is fused to the phage membrane protein, termed 'anchor protein', which displays the foreign protein stably on the surface. Phages have been used to generate a library of genes that are expressed in a certain context. Based on a similar principle, microbial cell surface display systems have been developed followed by yeast surface display. They have been used as a live bacterial vaccine delivery vehicle, as novel microbial biocatalysts with immobilized heterologous enzymes (Samuelson *et al.* 2002). However, bacterial systems have constraints in the expression of relatively complex eukaryotic proteins. Overcoming the problems in prokaryotic display systems, the baculovirus surface display of foreign proteins has emerged as an alternative tool for screening libraries and in many other functions associated with surface display such as vaccines, viral protein–mediated receptor endocytosis and gene therapy (Ernst *et al.* 2006). Often, the eukaryotic/mammalian proteins that are displayed on the membrane are glycosylated and are not, therefore,

Table 26.3 List of representative recombinant proteins produced in baculovirus

S. No.	Recombinant protein	Reference	Remark
1	**Vaccines**		
	Influenza	Treanour *et al.* (2011)	Marketed as FluBlok® from baculovirus-expressing trivalent influenza virus haemagglutinin (rHA0).
	H5N1	Lin *et al.* (2011)	Trimeric recombinant HA (rHA) proteins from two HPAI H5N1 viruses were purified in the baculovirus system
	Western equine encephalitis virus (WEEV)	Toth *et al.* (2011)	Full-length E1, the E1 ectodomain, an E26KE1 polyprotein precursor and an artificial, secretable E2E1 chimera under immediate early (ie1), late (p6.9) and very late (polh) promoters, respectively, were expressed and purified
2	**Interleukins/Interferons**		
	Interleukin-2 (IL-2)	Shin *et al.* (2003)	Produced in stable *Drosophila* S2 cells as secreted protein
	Interleukin-3	Ding *et al.* (2003)	Highly active, soluble and stable protein was purified in single step
	Interleukin-4	Nuntaprasert *et al.* (2005)	Swine IL-4 was expressed in insect cells in a biologically active form
	Interleukin-6	Metz *et al.* (2007)	IL-6 inhibitor was developed as a fusion protein with the ligand-binding domain of IL-6 and gp130 (also termed as cytokine trap)
	Interleukin-7 (human)	Mirzaei *et al.* (2009)	hIL-7 is a valuable supplementary agent for immunotherapeutical treatments in patients with immunodeficiency
	Interleukin-18	Lu *et al.* (2005)	Purified recombinant protein induced similar levels of IFN-gamma as native IL-18 protein in human PBMC and can also specifically bind to EGFR, which can be further developed as novel tumour therapeutics
	Interferon-γ (Human, bovine, canine, equine, chicken, etc.)	Chen *et al.* (2011)	rhIFN-γ protein was expressed in *T. ni* that had potent antiviral activity
	Interleukin-12 (human, porcine, bovine, mouse, etc.)	Shen *et al.* (1998)	Recombinant viruses AcNPV-hp40 and AcNPV-hp35 were co-infected to generate active IL-12
3	**Hormones**		
	CTB insulin (Cholera toxin B conjugated insulin)	Meng *et al.* (2011)	The conjugate protein induces tolerance and protects against diabetes type1 (auto-immune) involving $CD4^+CD25^+Foxp3^+$Treg cells
	Adiponectin	Do Carmo *et al.* (2008)	Recombinant adiponectin can be used to treat deficiencies of this hormone which results in type II diabetes and cardiovascular disease
	Cecropin B-binding site of luteinizing hormone-releasing hormone (CBLHRH')	Li *et al.* (2007)	The anticancer effects of the CB-LHRH' protein was shown through the LHRH receptor, and it is a potential candidate as peptide drug for targeted cancer therapy
	Human calcitonin	Yang *et al.* (2002)	Shows a hypercalcaemic effect in rats
	Human TSH glycosylated	Cundiff *et al.* (2001)	Absorbs antibodies to TSH and blocks hyper-activation of thyroid (Graves disease)
	Human prolactin	Strokovskaya *et al.* (2001)	Substitute for pituitary-derived prolactin
	Human chorionic gonadotropin-luteinizing hormone receptor ectodomain complex	Fralish *et al.* (2001)	High levels of protein that bind LHR with high affinity have been generated
4	**Growth factors**		
	Human growth hormone (GH)	Lan *et al.* (2010)	Investigated the activity of human GH in rats, which showed a significant gain in body weight
	Human FSH	Hou *et al.* (2004)	hFSH is essential to stimulate the ovary to release the egg, which is used in *in vitro* fertilization and also in fertility treatment.
	EGFL7	Picuric *et al.* (2009)	EGFL7 helps in the recruitment of endothelial and smooth muscle cells in the process of angiogenesis

Continued ...

... continued

S. No.	Recombinant protein	Reference	Remark
5	**Others**		
	Aminopeptidase A	Ishii *et al.* (2008)	Aminopeptidase A can be used to treat hypertension, which degrades circulating angiotensin II before it binds to its receptor
	Human dipeptidyl peptidase IV/ CD26	Dobers *et al.* (2002)	Protein used for crystallographic studies
	OATP2B1	Tschantz *et al.* (2008)	OATP2B1 is an important member of the organic anion-transporting polypeptides and is implicated in the intestinal and hepatic disposition of endo- and xenobiotics
	Human bradykinin B2 receptor	Reyes-Cruz *et al.* (2000)	The protein expressed in Sf21 demonstrated that tyrosine kinase(s) regulates the activity of the human B2 receptor by affecting its coupling to G proteins and its phosphorylation

presented in their native context on prokaryotic display systems. A library of proteins can be screened for a specific biological activity. This system allows combination of the genotype with the phenotype, wherein the foreign eukaryotic proteins on the viral envelope or capsid protein such as GP-64 are displayed. Several expression vectors of surface display have been developed for various applications.

The baculovirus surface display holds immense promise as an emerging molecular tool both to study fundamental cell biology and in industry for screening peptide libraries, mutational studies, receptor-mediated biological activities, as vaccines, biocatalysts, in bioremediation processes (Grabherr *et al.* 2010) and so on. This technology is in its inception and further research into optimizing the strategies for insertion and selection need to be worked out. Table 26.4 gives a list of the major applications using the baculovirus display (references therein).

26.6.4 Baculovirus as gene delivery vehicle for gene therapy and vaccine

Baculovirus has the potential to infect non-permissive cells, such as mammalian cells, although it fails to replicate within. This property has been exploited for its potential as a gene delivery vehicle (Ghosh *et al.* 2000). Expression of the foreign protein under the host promoter facilitates their expression within the cell. Besides, there are several advantages of using baculoviruses, such as ease of amplification, safety and ability to transduce both dividing and non-dividing cells.

The bacmam virus, which carries a mammalian promoter and flanking baculoviral nucleotide sequences, has been developed as a transfer vector for expression in mammalian cells. It has a high transduction efficiency up to 70%, making gene delivery into difficult-to-transfect cells exceedingly simple and efficient (O'Grady *et al.* 2011). The bacmam virus can be used to deliver multiple genes at a time and has no restriction on the gene size. Besides gene delivery, transfer vectors can be used to generate stable cell lines by either random integration (Condreay *et al.* 1999; Merrihew *et al.* 2001) or site-specific recombination using the *rep* gene of adeno-associated virus (Zeng *et al.* 2007). In cases of protein dysfunction, gene delivery into tissues under specific promoters can solve the problems. Table 26.5 gives a list of gene therapy candidates using the baculovirus technology (references therein).

These approaches, however, have certain limitations. As the proteins are produced in insect cells, there are certain glycoproteins that are displayed on the viral surface, which could increase immunogenic responses and are consequently inactivated by the complement system. The safety issues have been studied and they show no signs of toxicity (Ignoffo and Heimpel 1965; Heimpel and Buchanan 1967). That the baculoviruses avoid the complement system has been demonstrated by using complement-protecting factors (Hofmann *et al.* 1999; Huser and Rudolph 2001; Pieroni and Maione 2001), avoiding exposure to the complement (Airenne *et al.* 2000; Sandig *et al.* 1996), or by using the virus in immune-privileged areas such as the eye (Haeseleer *et al.* 2001) and the brain (Sarkis *et al.* 2000; Lehtolainen *et al.* 2002). Baculovirus-mediated gene delivery has offered most promising results *in vivo* in the central nervous system (CNS).

26.6.5 Baculovirus for induction of antitumour immunity

It is now clear that baculovirus does enter mammalian cells, although it does not replicate. It has also been later observed and demonstrated that baculovirus infection stimulates the immune responses such as secretion of pro-inflammatory cytokines, expression of co-stimulatory molecules, CD80, CD86, MHC class I and class II

Table 26.4 Use of baculovirus display in several biotechnological applications

S. No.	Baculovirus display use	Protein expressed	Remarks	Reference
1	Production of antibodies	Mammalian claudin-1	Overcomes tedious purification protocols	Hotzel *et al.* (2011)
2	Biotracking/Imaging	His-tagged ZnO inorganic binding peptide	The binding of nano-sized ZnO powders to the virus capsid was visualized by transmission electron microscopy (TEM)	Song *et al.* (2010)
3	Site-specific gene delivery	Avidin	Avidin-displaying baculovirus showed a fivefold increase in transduction efficiency in rat malignant glioma cells (BT4C) and a 26-fold increase in rabbit aortic smooth muscle (RAASMC) cells compared with wild-type baculovirus	Raty *et al.* (2004)
4	**Vaccine/Immunization**			
	Human Enterovirus 71	VP1	After two immunizations in mice, Bac-Pie1-gp64-VP1 elicited neutralization antibody against EV71 (subgenogroup C4) and showed high cross-neutralization activities against all 11 subgenogroup EV71 strains	Meng *et al.* (2011)
	African horse sickness	VP7	Soluble VP7 fusion proteins, including a fusion with a 36-amino acid insert that overlaps important immunological domains on protein VP1 of foot and mouth disease virus (FMDV) as well as a 110-amino-acid peptide derived from AHSV VP2.	Rutkowska *et al.* (2011)
	Japanese encephalitis virus	E envelope glycoprotein	BacSC-E successfully induced neutralizing antibody response and protective immunity towards a lethal challenge of the JEV in mouse and swine models	Xu *et al.* (2011)
	Influenza	Haemagglutinin	HA-DBV can stimulate strong humoral as well as cellular immune responses	Tang *et al.* (2010)
	Malaria	Pfs25	Both intranasal and intramuscular immunizations of mice with AcNPV-Pfs25surf induced high levels of Pfs25-specific antibodies, which strongly reacted with ookinetes of transgenic *Plasmodium berghei* expressing Pfs25	Mlambo *et al.* (2010)
5	Screening of binding partners	Catalytic domain of human tumour necrosis factor-alpha (TNF-alpha)–converting enzyme (TACE)	Fluorimetric monitoring of activity of TNF-α can be screened in a high-throughput system in the presence of inhibitors	Chen *et al.* (2002)
6	Functional reconstitution of membrane-associated proteins		G-protein–coupled receptor (GPCR) signalling proteins were reconstituted	Sakihama *et al.* (2008)
7	Tumour homing peptide	LyP-1, F3 and CGKRK tumour homing peptides	LyP-1 peptide inhibits viral binding to MDA-MB-435 more specifically than CGKRK and F3 peptides	Makela *et al.* (2006)

molecules, which can be a favourable outcome for activating immune responses. This property of baculovirus has, therefore, been exploited as a vehicle for vaccine delivery. Their role in inducing immune responses favourable to anti-cancer immunity has been studied by Suzuki *et al.* (2010) and they have reported the use of baculovirus-infected dendritic cells against lung cancer and melanoma in mouse models. *Ac*MNPV infection has also been reported to activate NK cells and prolong their survival in liver metastasis model equivalent to five treatments with IL-12 protein (Kitajima *et al.* 2008). T-cell co-stimulation plays a significant role in the

Table 26.5 List of gene therapy candidates under trial in baculovirus

S. No.	Candidate gene delivered	Disease under therapy	Remarks	Reference
1	*Ang-1*	Myocardial infarction	Successful in rat model	Paul *et al.* (2011)
2	*VEGF*	Potential replacement angioplasty, stenting or bypass surgery	Induces angiogenesis in rabbit models	Heikura *et al.* (2011)
3	Monster green fluorescent protein –factorial design	Colorectal carcinoma	Baculovirus shows highest transduction in colorectal cancer cell lines	Paul *et al.* (2010)
4	*SiRNA*	Influenza A and B	Bispecific short-hairpin siRNA constructs containing an 8-nucleotide intervening spacer, targeted against influenza virus A or influenza virus B, have been shown to inhibit the production of both types of virus in infected cell lines	Suzuki *et al.* (2009a) and (2009b)
5	*ShRNA*	Anti-HCV	Baculovirus-based shRNA-expressing vectors that contain the latent viral protein Epstein–Barr nuclear antigen 1 (EBNA1) and the origin of latent viral DNA replication (OriP) sequences	Suzuki *et al.* (2009)
6	*hSox9*	Intervertebral disc degeneration	Provides a novel expression vector system for basic research and clinical treatment of intervertebral degenerative disc disease	Liu *et al.* (2007)
7	*hEA*	Prostrate cancer/ ovarian tumour/ cancer metastasis	With sustained hEA expression, Bac-SB-T2hEA/w repressed the angiogenesis *in vivo*, hindered the growth of two different tumours, including prostate tumour allografts and human ovarian tumour xenografts in mice	Luo *et al.* (2011)
8	Herpes simplex virus thymidine kinase suicide gene	Cancer/Glioma	Utilize tissue tropic neural /mesenchymal stem cells transduced with herpes thymidine kinase gene to migrate to distant tumours	Bak *et al.* (2010); Zhao *et al.* (2011)
9	*p53*	Glioblastoma	Combination therapy, including p53 gene transduction as well as histone deacetylase inhibitor	Guo *et al.* (2011)
10	*NIS*	Colon tumour	Shows synergistic effect with sodium butyrate, hdac inhibitor	Yin *et al.* (2010)
11	*Pdcd4*	Tumour	Successful tumour regression with folate/PEG/baculovirus transduction in mouse xenograft model	Kim *et al.* (2010)
12	Apoptin	Murine hepatoma	p53-independent, bcl-2-insenstive apoptin protein has the ability to specifically induce apoptosis in tumours or transformed cells	Pan *et al.* (2010)

induction of antitumour immunity (Schlom and Hodse 1999). Table 26.6 lists the various baculovirus-based vaccines against different types of cancer (references therein).

26.6.6 Baculovirus to produce virus-like particles

Virus-like particles (VLPs) are non-infectious particles that contain self-assembled virion structural proteins without any genetic material. Several VLPs have been created in the BEVS which are used as vaccines for protection against various diseases (Table 26.7; references therein). Chimeric VLPs have been generated consisting of both L1 capsid protein and E2/E7 protein of HPV that provide additional viral targets for cell-mediated anti-immune response (Greenstone 1998). This system acts as an adjuvant in itself and can function as prophylactic vaccines.

26.7 Post-translational modifications: Advances and limitations in BEVS

Eukaryotic proteins are usually distinguished from prokaryotic proteins by PTMs that are carried out in eukaryotic cells. There are about 100 protein modifications, of which glycosylation is the most frequent PTM. PTMs are affected by high levels of expression. In the case of BEVS, PTM quality is compromised in the very late stage of infection. However, methods to overcome

Table 26.6 List of vaccines generated through the baculovirus system

S. No.	Vaccine	Remarks	Reference
1	Japanese encephalitis	Among E, polyprotein and NS1 proteins of Japanese encephalitis virus, E and polyprotein induced protective immunity up to 70% in mice but not NS1	Cown *et al.* (1990)
2	Bursal disease in chicken	VP2-pseudotyped baculovirus vaccine (BacSC-VP2) was used for vaccination in chickens, which significantly elicited higher levels of VP2-specific antibodies and neutralizing antibodies	Xu *et al.* (2011)
3	Malaria	PbCSP was expressed by a dual promoter that consists of tandemly arranged baculovirus-polyhedrin and mammal cytomegalovirus promoters, which elicited PbCSP-specific antibody titers, stimulating both Th1 and Th2 responses in mice	Yoshida *et al.* (2009)
4	H5N1 influenza	H5N1 influenza virus haemagglutinin proteins (BV-HAs) have been used to immunize mice, which conferred 100% protection	Wu *et al.* (2011)
5	Avian reo virus	SigmaC and sigmaB proteins were expressed on the envelope of baculovirus, and upon immunization in mice, displayed high levels of neutralizing antibodies	Lin *et al.* (2008)
6	Classical swine fever virus (CSFV)	The E2 glycoprotein expressed on baculovirus envelope protein gp64 was evaluated in mice for protection against CSFV	Xu *et al.* (2008)
7	Human papillomavirus type 45	Human papillomavirus type 45 virus-like particles were generated using baculovirus expression system, wherein VLP were assembled by the major capsid protein L1	Touze *et al.* (1996)
8	Newcastle disease virus	F protein was expressed using the baculovirus system in silkworm. The resulting virus particles were used as vaccine in chickens, which exhibited protection against NDV	Mori *et al.* (1994)

Table 26.7 Generation of different VLPs using baculovirus*

S. No.	VLP	Remark	Utility	Reference
1	HIV-1	Assembly of gag and vpr into HIV-1–like virions in the baculovirus system	Design of future inhibitors of HIV particle assembly	Gonzalez *et al.* (2011)
2	Influenza	BacMam BV, in which three CMV-promoters drive the haemagglutinin, neuraminidase and a matrix of influenza virus and, subsequently, influenza-like virus particles were generated	As vaccine	Tang *et al.* (2011)
3	Severe acute respiratory syndrome coronavirus (SARS-CoV)	S-Glycoprotein and influenza matrix 1 efficiently form virus-like particles	As vaccine	Liu *et al.* (2011)
4	Human papillomavirus (HPV)	HPV-16 and HPV-18 L1 proteins were expressed in the baculovirus system, which formed HPV-like particles	As prophylactic vaccine	Deschuyteneer *et al.* (2010)
		Also covers HPV type 58		Zhang *et al.* (2010)
		Chimeric L1-L2		Schellenbacher *et al.* (2009)
5	H1N1 influenza	Recombinant influenza virus-like particle (VLP) vaccine consisting of haemagglutinin, neuraminidase, and matrix proteins were generated in the baculovirus system	As vaccine	Pushko *et al.* (2010)
	H5N1 influenza	Haemagglutinin, neuraminidase and matrix 1 proteins were expressed in the baculovirus system		Mahmood *et al.* (2008)
6	Foot-and-mouth disease type	VP1-2A-VP3-VP0 serotype was used to express VLP in baculovirus	As vaccine	Cao *et al.* (2010)
7	Norovirus	The baculovirus system expressing a NoV VP1 capsid protein was used to generate VLF	As vaccine	Ji-Young *et al.* (2011)

Continued ...

... continued

S. No.	VLP	Remark	Utility	Reference
8	Papilloma	Two copies of the L1 gene were expressed from independently controlled cassettes	As vaccine	Senger *et al.* (2009)
9	Mouse polyoma virus	VP1 tagged was with a secretion signal for targeting to the extracellular medium	Cell-free preparation of VLP; optimal for large-scale preparation	Ng *et al.* (2007)
10	Hepatitis E (HEV)	HEV genes ORF2 (112-660) and ORF2 (112-608) were expressed in the baculovirus system using papilloma VLP, conferring protection against both HEV and papilloma viruses	As vaccine	Renoux *et al.* (2008)
11	Feline calcivirus	Recombinant baculovirus containing the capsid gene of feline calicivirus strain F9 was used to generate VLP, which elicited neutralizing antibodies against several strains of the virus	As vaccine	Di Martino *et al.* (2007)
12	Ebola virus	VP40 (rBV-VP40) and GP (rBV-GP) proteins were expressed in the baculovirus system	As vaccine	Ye *et al.* (2006)
13	SIV	SIV envelope expressed and packaged in insect/baculovirus system induces both mucosal and systemic immunity	As vaccine	Yao *et al.* (2002)
14	Rotavirus	VP2, VP6 and VP7 of rotavirus were co-expressed in insect cells, which assembled into virus-like particles	As vaccine	Jiang *et al.* (1998)
15	Hantaan virus	Small segment (S), which encodes the viral nucleocapsid protein, was expressed and assembled into VLP	As vaccine	Betenbaugh *et al.* (1995)

these problems have been addressed in terms of modifications in either transfer vectors or host cells expressing additional chaperones and knocking out proteases. The use of a late promoter instead of the *polh* promoter also would help in getting more amount of post-translationally modified proteins. Advanced PTMs such as N-linked glycosylations that do not occur in insect cells may be the greatest limitation of BEVS. About 50% of all human proteins are glycosylated, and 1–2% of these proteins contribute to N-glycosylation (Walsh 2009). Often, the glycosylated component contributes to biological activity. Sometimes, the glycosylation also increases the shelf-life in the serum. N-glycosylation patterns in insect and mammalian systems differ in the final steps of modification such as transfer of monosaccharides to the trimmed N-glycan. There is a subtle difference in the processing of N-glycan, where there is a proteolytic cleavage of α-1,3 instead of α-1,6 branch mannose. In mammalian systems, there is an addition of residues on the α-1,3 carbon, which is cleaved off in insect systems. As a result of N-linked glycosylation, there is greater mobility of mammalian proteins expressed in insect systems in acrylamide gel electrophoresis as compared with proteins expressed in mammalian systems (Davis *et al.* 1993). This is responsible for the failure of complete N-linked glycosylation by BEVS and the ensuing failure of its activity *in vivo*. Therefore, protein biopharmaceuticals for enzyme replacement cannot be raised through BEVS. Table 26.8 gives a list of differences in glycosylation patterns of insect and mammalian systems.

26.8 Prospects and future directions

Ever since the inception of BEVS by Summers in 1983, there has been a revolution in the developments in both basic and applied research in BEVS. BEVS has almost emerged as an alternative to the *E. coli* expression system and the first system of choice for eukaryotic gene expression among others. There have been limitations and constraints for the expression of foreign proteins in BEVS, which have been successfully addressed currently, as described earlier. There have been several improvements in the screening and expression of foreign proteins in insect cells. A recent development in the BEVS is the introduction of the FlashBAC system, where the there is no possibility of non-recombinant viruses.

Several emerging areas in BEVS have been depicted in Fig. 26.2. The potential areas in BEVS development are the use of silkworm and insect larvae as bio-factory,

Table 26.8 List of differences in glycosylated patterns in insect and mammalian systems

S. No.	Feature	Baculovirus system	Mammalian system	Outcome
1	N-terminal sialic acids	Absent	Present	Activity loss
2	A-1,3-linked fucose	Present	--	Allergenicity
3	Mannosyl terminal glycans	Present	--	More efficient interactions
4	N-glycan structure	Simple: mono- and bi-antennary pauci mannose lacking sialic acid	Complex, multi-branched, siaylated	May result in low protein activity of recombinant proteins
5	Addition of residues to N-glycan	No	Yes	May result in low protein activity of recombinant proteins
6	Removal of N-aceytl glucosamine	Yes	No	May result in low protein activity of recombinant proteins

which can simplify the whole method of protein expression and purification. Besides, silkworms can be easily maintained, and they can produce a huge amount of the protein almost more than 100 times the active protein than the *Sf*9 system and 100 times than CHO cells. In this system, the researchers have employed the *vp-39* promoter and protease-deficient silkworm, which enhances the yield and activity of the protein. The purification process has been made easy by use of Dock-tag as N-terminal fusion, which binds to Ca^{2+} ion attached to the cohesion domain covalently attached to a resin, and this could be used several times very efficiently (Usami *et al.* 2010). In terms of glycosylation patterns, there have been some limitations, which could be overcome by transfection of glycosyltransferase into the silkworm. They have been able to express around 5000 foreign proteins in the silkworm system in a fully functional form (Kato *et al.* 2010).

Several applications of the BEVS have been examined and the most promising application in terms of human health care is the use of BEVS in targeted gene therapy for cancer, cardiac problems, as vaccine, etc. BEVS platform has generated three products that have progressed through phase III clinical development, including Flu-Blok (seasonal influenza vaccine) by Protein Sciences Corporation, Diamyd (therapeutic vaccine for type I diabetes) by Diamyd Medical AB and Glybera (gene therapy for lipoprotein lipase deficiency) by AMT. Also, prophylactic vaccines for the prevention of infection by influenza virus (seasonal and pandemic), human papillomavirus, emerging pathogens (SARS) and biodefense (Ebola) are in progress using BEVS. Therapeutic vaccines for prostate and breast cancer and diabetes are also in progress. While gene therapy for muscular dystrophy, genetically based lipid disorders, CNS disorders (Alzheimer's and Parkinson's diseases) have been successful using rAAV (recombinant adeno-associated virus), problems of wound healing, tissue repair, antivirals (hepatitis C) and anti-inflammatory diseases/arthritis are being pursued using BEVS.

Some of the anti-apoptotic genes of baculovirus are of great importance for academic research and also as a therapeutic agent for gene therapy in cells where unwarranted apoptosis leads to pathogenic conditions such as neuro-degenerative diseases, myocardial infarction, cataract, etc. Extensive research on one of the anti-apoptotic proteins of baculovirus, P35, has revealed that P35 can inhibit apoptosis at different conserved levels in the apoptotic pathway (Bump *et al.* 1995; Sah *et al.* 1998; Hasnain *et al.* 2003; Mohan *et al.* 2003; Sahdev *et al.* 2003) and is a potential candidate for gene therapy of several disease conditions mentioned earlier (Sahdev *et al.* 2010). As the virus also stimulates anti-viral properties in mammalian cells and that the safety issues are clear, they can go a long way in human health management. Academically and industrially, BEVS offers several avenues for further research and application.

References

Airenne KJ, Hiltunen MO, Turunen MP, Turunen AM, Laitinen OH, Kulomaa MS, *et al.* Baculovirus-mediated periadventitial gene transfer to rabbit carotid artery. *Gene Ther* 2000;7:1499–1504.

Argaud O, Croizier L, López-Ferber M, Croizier G. Two key mutations in the host-range specificity domain of the p143 gene of *Autographa californica* nucleopolyhedrovirus are required to kill *Bombyx mori* larvae. *J Gen Virol* 1998;79:931–935.

Bak XY, Yang J, Wang S. Baculovirus-transduced bone marrow mesenchymal stem cells for systemic cancer therapy. *Cancer Gene Ther* 2010;17:721–729.

Basile GPM. Recombinant protein expression in *Leishmania tarentolae*. *Mol Biotechnol* 2009;43:273–278.

Benz GA. Introduction: Historical perspectives. In: Granados RR, Federici BA, eds. *Biology of Baculoviruses*. Boca Raton, FL: CRC Press; 1986. p. 1235.

Berger I, Fitzgerald DJ, Richmond TJ. Baculovirus expression system for heterologous multiprotein complexes. *Nat Biotechnol* 2004;22:1583–1587.

Betenbaugh M, Yu M, Kuehl K, White J, Pennock D, Spik K, *et al.* Nucleocapsid- and virus-like particles assemble in cells infected with recombinant baculoviruses or vaccinia viruses expressing the M and the S segments of Hantaan virus. *Virus Res* 1995;38:111–124.

Bump NJ, Hackett M, Hugunin S, Seshagiri K, Brady P, Chen C, *et al.* Inhibition of the ICE family proteases by baculovirus anti apoptotic protein p35. *Science* 1995;269:1885–1888.

Cao Y, Sun P, Fu Y, Bai X, Tian F, Liu X, *et al.* Formation of virus-like particles from O-type foot-and-mouth disease virus in insect cells using codon-optimized synthetic genes. *Biotechnol Lett* 2010;32:1223–1229.

Chaabihi H, Ogliastro MH, Martin M, Giraud C, Devauchelle G, Cerutti M. Competition between baculovirus polyhedrin and p10 gene expression during infection of insect cells. *J Virol* 1993;67:2664–2671.

Chen WS, Villaflores OB, Jinn TR, Chan MT, Chang YC, Wu TY. Expression of recombinant human interferon-γ with antiviral activity in the bi-cistronic baculovirus-insect/larval system. *Biosci Biotechnol Biochem* 2011;75:1342–1348.

Chen Y, Diener K, Patel IR, Kawooya JK, Martin GA, Yamdagni P, *et al.* Phage display of functional human TNF-alpha converting enzyme catalytic domain: A rapid method for the production of stabilized proteolytic proteins for assay development and high-throughput screening. *J Biomol Screen* 2002;7:433–440.

Chen Y, Yao B, Zhu Z, Yi Y, Lin X, Zhang Z, *et al.* A constitutive super-enhancer: Homologous region 3 of *Bombyx mori* nucleopolyhedrovirus. *Biochem Biophys Res Commun* 2004;318:1039–1044.

Clem RJ, Miller LK. Apoptosis reduces both the *in vitro* replication and *in vivo* infectivity of a baculovirus. *J Virol* 1993;67:3730–3738.

Cundiff JG, Kaithamana S, Seetharamaiah GS, Baker JR Jr, Prabhakar BS. Studies using recombinant fragments of human TSH receptor reveal apparent diversity in the binding specificities of antibodies that block TSH binding to its receptor or stimulate thyroid hormone production. *J Clin Endocrinol Metab* 2001;86: 4254–4260.

Condreay JP, Witherspoon SM, Clay WC, Kost TA. Transient and stable gene expression in mammalian cells transduced with a recombinant baculovirus vector. *Proc Natl Acad Sci USA* 1999;96:127–132.

Davis T, Wickham T, McKenna K, Granados R, Shuler M, Wood H. Comparative recombinant protein production of eight insect cell lines. *In Vitro Cell Dev Biol Anim* 1993;29:388–390.

Deschuyteneer M, Elouahabi A, Plainchamp D, Plisnier M, Soete D, Corazza Y, *et al.* Molecular and structural characterization of the L1 virus-like particles that are used as vaccine antigens in Cervarix™, the AS04-adjuvanted HPV-16 and -18 cervical cancer vaccine. *Hum Vaccin* 2010;6:407–419.

Di Martino B, Marsilio F, Roy P. Assembly of feline calicivirus-like particle and its immunogenicity. *Vet Microbiol* 2007;120: 173–178.

Ding H, Griesel C, Nimtz M, Conradt HS, Weich HA, Jäger V. Molecular cloning, expression, purification, and characterization of soluble full-length, human interleukin-3 with a baculovirus-insect cell expression system. *Protein Expr Purif* 2003;31:34–41.

Dobers J, Zimmermann-Kordmann M, Leddermann M, Schewe T, Reutter W, Fan H. Expression, purification, and characterization of human dipeptidyl peptidase IV/CD26 in Sf9 insect cells. *Protein Expr Purif* 2002;25:527–532.

Ernst W, Schinko T, Spenger A, Oker-Blom C, Grabherr R. Improving baculovirus transduction of mammalian cells by surface display of a RGD-motif. *J Biotechnol* 2006;126:237–240.

Fralish GB, Narayan P, Puett D. High-level expression of a functional single-chain human chorionic gonadotropin-luteinizing hormone receptor ectodomain complex in insect cells. *Endocrinology* 2001;142:1517–1524.

Friesen PD, Rueckert RR. Early and late functions in a bipartite RNA virus: evidence for translational control by competition between viral mRNAs. *J Virol* 1984;49:116–124.

Ghosh S, Jain A, Mukherjee B, Habib S, Hasnain SE. The host factor polyhedrin promoter binding protein (PPBP) is involved in transcription from the baculovirus polyhedrin gene promoter. *J Virol* 1998;72:7484–7493.

Ghosh S, Parvez MK, Banerjee K, Sarin SK, Hasnain SE. Baculovirus as mammalian cell expression vector for gene therapy: An emerging strategy. *Mol Ther* 2002;6:5–11.

Gonzalez G, DaFonseca S, Errazuriz E, Coric P, Souquet F, Turcaud S, *et al.* Characterization of a novel type of HIV-1 particle assembly inhibitor using a quantitative luciferase-Vpr packaging-based assay. *PLoS ONE* 2011;6:e27234.

Grabherr R, Ernst W. Baculovirus for eukaryotic protein display. *Curr Gene Ther* 2010;10:195–200.

Greenstone HL. Chimeric papillomavirus virus-like particles elicit antitumor immunity against the E7 oncoprotein in an HPV16 tumor model. *Proc Natl Acad Sci USA* 1998;95:1800–1805.

Guo H, Choudhury Y, Yang J, Chen C, Tay FC, Lim TM, *et al.* Antiglioma effects of combined use of a baculoviral vector expressing wild-type p53 and sodium butyrate. *J Gene Med* 2011;13:26–36.

Guo XY, Guo TQ, Wang SP, Wang JY, Lu CD. Silk gland specific secretory expression of egfp gene in silkworm *Bombyx mori* with r*Ac*MNPV system. *Arch Virol* 2005;150:1151–1160.

Habib S, Hasnain SE. A 38-kDa host factor interacts with functionally important motifs within the *Autographa californica* Multinucleocapsid Nuclear Polyhedrosis virus homologous region (hr1) DNA sequence. *J Biol Chem* 1996;271:28250–28258.

Haeseleer F, Imanishi Y, Saperstein DA, Palczewski K. Gene transfer mediated by recombinant baculovirus into mouse eye. *Invest Ophthalmol Vis Sci* 2001;42:3294–3300.

Hasnain SE, Begum R, Ramaiah KV, Sahdev S, Shajil EM, Taneja TK, *et al.* Host-pathogen interactions during apoptosis. *J Biosci* 2003;28:349–358.

Hefferon KL, Miller LK. Reconstructing the replication complex of *Ac*MNPV. *Eur J Biochem* 2002;269:6233–6240.

Heikura T, Nieminen T, Roschier MM, Karvinen H, Kaikkonen MU, Mähönen AJ, *et al.* Baculovirus-mediated vascular endothelial growth factor-D(ΔNΔC) gene transfer induces angiogenesis in rabbit skeletal muscle. *J Gene Med* 2012;14:35–43.

Heimpel AM, Buchanan LK. Human feeding tests using a nuclear-polyhedrosis virus of *Heliothis zea*. *J Invertebr Pathol* 1967;9: 55–57.

Hitchman RB, Possee RD, Crombie AT, Chambers A, Ho K, Siaterli E, *et al.* Genetic modification of a baculovirus vector for increased expression in insect cells. *Cell Biol Toxicol* 2010;26:57–68.

Hofmann C, Hüser A, Lehnert W, Strauss M. Protection of baculovirus-vectors against complement-mediated inactivation by recombinant soluble complement receptor type 1. *Biol Chem* 1999;380:393–395.

Hötzel I, Chiang V, Diao J, Pantua H, Maun HR, Kapadia SB. Efficient production of antibodies against a mammalian integral membrane protein by phage display. *Protein Eng Des Sel* 2011;24:679–689.

Huser A, Rudolph M, Hofmann C. Incorporation of decay-accelerating factor into the baculovirus envelope generates complement-resistant gene transfer vectors. *Nat Biotechnol* 2001;19:451–455.

Ignoffo CM, Heimpel AM. The nuclear-polyhedrosis virus of *Heliothis zea* (Boddie) and *Heliothis virescens* (Fabricius): V. Toxicity-pathogenicity of virus to white mice and guinea pigs. *J Invert Pathol* 1965;7:329–340.

Iizuka M, Tomita M, Shimizu K, Kikuchi Y, Yoshizato K. Translational enhancement of recombinant protein synthesis in transgenic silkworms by a 5'-untranslated region of polyhedrin gene of *Bombyx mori* Nucleopolyhedrovirus. *J Biosci Bioeng* 2008;105:595–603.

Ishii M, Hattori A, Numaguchi Y, Tsujimoto M, Ishiura S, Kobayashi H, *et al.* The effect of recombinant aminopeptidase A on hypertension in spontaneously hypertensive rats: Its effect in comparison with candesartan. *Horm Metab Res* 2008;40: 887–891.

Ishiyama S, Ikeda M. High-level expression and improved folding of proteins by using the vp39 late promoter enhanced with homologous DNA regions. *Biotechnol Lett* 2010;32:1637–1647.

Iwanaga M, Takaya K, Katsuma S, Ote M, Tanaka S, Kamita SG, *et al.* Expression profiling of baculovirus genes in permissive and nonpermissive cell lines. *Biochem Biophys Res Commun* 2004;323:599–614.

Jiang B, Barniak V, Smith RP, Sharma R, Corsaro B, Hu B, *et al.* Synthesis of rotavirus-like particles in insect cells: Comparative and quantitative analysis. *Biotechnol Bioeng* 1998;60:369–374.

Kato T, Kajikawa M, Maenaka K, Park EY. Silkworm expression system as a platform technology in life science. *Appl Microbiol Biot* 2010;85:459–470.

Kato T, Murata T, Usui T, Park EY. Improvement of the production of GFPuv-beta1,3-N-acetylglucosaminyltransferase 2 fusion protein using a molecular chaperone-assisted insect-cell-based expression system. *Biotechnol Bioeng* 2005;89:424–433.

Kim EJ, Kramer SF, Hebert CG, Valdes JJ, Bentley WE. Metabolic engineering of the baculovirus-expression system via inverse 'shotgun' genomic analysis and RNA interference (dsRNA) increases product yield and cell longevity. *Biotechnol Bioeng* 2007;98:645–654.

Kim YK, Kwon JT, Choi JY, Jiang HL, Arote R, Jere D, *et al.* Suppression of tumor growth in xenograft model mice by programmed cell death 4 gene delivery using folate-PEG-baculovirus. *Cancer Gene Ther* 2010;17:751–760.

Kitajima M, Abe T, Miyano-Kurosaki N, Taniguchi M, Nakayama T, Takaku H. Induction of natural killer cell-dependent antitumor immunity by the *Autographa californica* multiple nuclear polyhedrosis virus. *Mol Ther* 2008;16:261–268.

Kumar MS, Ramachandran A, Hasnain SE, Bashyam MD. Octamer and heat shock elements regulate transcription from the *Ac*MNPV polyhedrin gene promoter. *Arch Virol* 2009;154:445–456.

Lan H, Nie Z, Liu Y, Lv Z, Liu Y, Quan Y, et al. *In vivo* bioassay of recombinant human growth hormone synthesized in *B. mori* pupae. *J Biomed Biotechnol* 2010;2010:Article no. 306462.

Lee DF, Chen CC, Hsu TA, Juang JL. A baculovirus superinfection system: Efficient vehicle for gene transfer into *Drosophila* S2 cells. *J Virol* 2000;74:11873–11880.

Lehtolainen P, Tyynela K, Kannasto J, Airenne KJ, Yla-Herttuala S. Baculoviruses exhibit restricted cell type specificity in rat brain: a comparison of baculovirus- and adenovirus-mediated intracerebral gene transfer *in vivo*. *Gene Ther* 2002;9:1693–1699.

Leisy DJ, Rasmussen C, Kim HT, Rohrmann GF. The *Autographa californica* nuclear polyhedrosis virus homologous region 1a: Identical sequences are essential for DNA replication activity and transcriptional enhancer function. *Virology* 1995;208:742–752.

Li XY, Li HL, Zheng GY. Construction and expression of recombinant cecropin B-binding site of luteinizing hormone releasing hormone gene and its anticancer function. *Zhonghua Fu Chan Ke Za Zhi* 2007;42:477–481.

Lin YH, Lee LH, Shih WL, Hu YC, Liu HJ. Baculovirus surface display of sigmaC and sigmaB proteins of avian reovirus and immunogenicity of the displayed proteins in a mouse model. *Vaccine* 2008;26:6361–6367.

Liu XY, Yang SH, Liang CY, Song JH, Li KH, Chen XW. Construction of recombinant baculovirus Ac-CMV-hSox9 for gene therapy of intervertebral disc degeneration. *Chin J Traumatol* 2007;10: 94–100.

Liu YV, Massare MJ, Barnard DL, Kort T, Nathan M, Wang L, *et al.* Chimeric severe acute respiratory syndrome coronavirus (SARS-CoV) S glycoprotein and influenza matrix 1 efficiently form virus-like particles (VLPs) that protect mice against challenge with SARS-CoV. *Vaccine* 2011;29:6606–6613.

Lu JX, Peng Y, Meng ZF, Jin LQ, Lu YS, Guan MX. Rational design of an EGF-IL18 fusion protein: Implication for developing tumor therapeutics. *Biochem Biophys Res Commun* 2005;334:157–161.

Lu M, Farrell PJ, Johnson R, Iatrou K. A baculovirus (*Bombyx mori* nuclear polyhedrosis virus) repeat element functions as a powerful constitutive enhancer in transfected insect cells. *J Biol Chem* 1997;272:30724–30728.

Luo WY, Shih YS, Lo WH, Chen HR, Wang SC, Wang CH, Chien CH, Chiang CS, Chuang YJ, Hu YC. (2011). Baculovirus vectors for antiangiogenesis-based cancer gene therapy. *Cancer Gene Ther*. 18:637–645.

Mahmood K, Bright RA, Mytle N, Carter DM, Crevar CJ, Achenbach JE, *et al.* H5N1 VLP vaccine induced protection in ferrets against lethal challenge with highly pathogenic H5N1 influenza viruses. *Vaccine* 2008;26:5393–5399.

Mäkelä AR, Matilainen H, White DJ, Ruoslahti E, Oker-Blom C. Enhanced baculovirus-mediated transduction of human cancer cells by tumor-homing peptides. *J Virol* 2006;80:6603–6611.

Manohar SL, Kanamasa S, Nishina T, Kato T, Park EY. Enhanced gene expression in insect cells and silkworm larva by modified polyhedrin promoter using repeated burst sequence and very late transcriptional factor-1. *Biotechnol Bioeng* 2010;107: 909–916.

McClintock JT, Dougherty EM, Weiner R. Semipermissive replication of a nuclear polyhedrosis virus of *Autographa californica* in a gypsy moth cell line. *J Virol* 1986;57:197–204.

McCown J, Cochran M, Putnak R, Feighny R, Burrous J, Henchal E, *et al.* Protection of mice against lethal Japanese encephalitis with a recombinant baculovirus vaccine. *Am J Trop Med Hyg* 1990;42:491–499.

Meng Q, Wang W, Shi X, Jin Y, Zhang Y. Protection against autoimmune diabetes by silkworm-produced GFP-tagged CTB-insulin fusion protein. *Clin Dev Immunol* 2011;2011:831704.

Meng T, Kolpe AB, Kiener TK, Chow VT, Kwang J. Display of VP1 on the surface of baculovirus and its immunogenicity against heterologous human enterovirus 71 strains in mice. *PLoS ONE* 2011;6:e21757.

Merrihew RV, Clay WC, Condreay JP, Witherspoon SM, Dallas WS, Kost TA. Chromosomal integration of transduced recombinant baculovirus DNA in mammalian cells. *J Virol* 2001;75:903–909.

Metz S, Wiesinger M, Vogt M, Lauks H, Schmalzing G, Heinrich PC, *et al.* Characterization of the interleukin (IL)-6 inhibitor IL-6-RFP: Fused receptor domains act as high affinity cytokine-binding proteins. *J Biol Chem* 2007;282:1238–1248.

Miller LK. Baculoviruses: High-level expression in insect cells. *Curr Opin Genet Dev* 1993;3:97–101.

Miller LK. *The Baculoviruses*. New York: Plenum Press; 1997.

Mirzaei M, Xu Y, Elias CB, Prakash S. Nonviral production of human interleukin-7 in spodoptera frugiperda insect cells as a soluble recombinant protein. *J Biomed Biotechnol* 2009;2009:Article no. 637942.

Mlambo G, Kumar N, Yoshida S. Functional immunogenicity of baculovirus expressing Pfs25, a human malaria transmission-blocking vaccine candidate antigen. *Vaccine* 2010;28:7025–7029.

Mohan M, Taneja TK, Sahdev S, Mohareer K, Begum R, Athar M, *et al.* Antioxidants prevent UV-induced apoptosis by inhibiting mitochondrial cytochrome c release and caspase activation in *Spodoptera frugiperda* (*Sf*9) cells. *Cell Biol Int* 2003;27:483–490.

Mori H, Tawara H, Nakazawa H, Sumida M, Matsubara F, Aoyama S, *et al.* Expression of the Newcastle disease virus (NDV) fusion glycoprotein and vaccination against NDV challenge with a recombinant baculovirus. *Avian Dis* 1994;38:772–777.

Morris TD, Miller LK. Mutational analysis of a baculovirus major late promoter. *Gene* 1994;140:147–153.

Mukherjee B, Burma S, Hasnain SE. The 30-kDa protein binding to the 'initiator' of the baculovirus polyhedrin promoter also binds specifically to the coding strand. *J Biol Chem* 1995;270:4405–4411.

Nakanishi T, Goto C, Kobayashi M, Kang W, Suzuki T, Dohmae N, *et al.* Comparative studies of lepidopteran baculovirus-specific protein FP25K: Development of a novel *Bombyx mori* nucleopolyhedrovirus-based vector with a modified fp25K gene. *J Virol* 2010;84:5191–5200.

Ng J, Koechlin O, Ramalho M, Raman D, Krauzewicz N. Extracellular self-assembly of virus-like particles from secreted recombinant polyoma virus major coat protein. *Protein Eng Des Sel* 2007;20:591–598.

Nuntaprasert A, Mori Y, Fujita K, Yoneda M, Miura R, Tsukiyama-Kohara K, *et al.* Expression and characterization of the recombinant swine interleukin-6. *Comp Immunol Microbiol Infect Dis* 2005;28:103–120.

O'Grady M, Batchelor RH, Scheyhing K, Kemp CW, Hanson GT, Lakshmipathy U. BacMam-mediated gene delivery into multipotent mesenchymal stromal cells. *Methods Mol Biol* 2011;698:485–504.

O'Reilly DR, Miller LK, Luckov VA. *Baculovirus Expression Vectors. A Laboratory Manual* (2nd Edition). New York: Oxford University Press; 1994.

Palhan VB, Sumathy S, Gopinathan KP. Baculovirus mediated high-level expression of luciferase in silkworm cells and larvae. *Biotechniques* 1995;19:97–104.

Pan Y, Fang L, Fan H, Luo R, Zhao Q, Chen H, *et al.* Antitumor effects of a recombinant pseudotype baculovirus expressing apoptin *in vitro* and *in vivo*. *Int J Cancer* 2010;126:2741–2751.

Paul A, Binsalamah ZM, Khan AA, Abbasia S, Elias CB, Shum-Tim D, *et al.* A nanobiohybrid complex of recombinant baculovirus and Tat/DNA nanoparticles for delivery of Ang-1 transgene in myocardial infarction therapy. *Biomaterials* 2011;32:8304–8318.

Paul A, Jardin BA, Kulamarva A, Malhotra M, Elias CB, Prakash S. Recombinant baculovirus as a highly potent vector for gene therapy of human colorectal carcinoma: Molecular cloning, expression, and *in vitro* characterization. *Mol Biotechnol* 2010;45:129–139.

Picuric S, Friedrich M, Oess S. Expression and purification of recombinant human EGFL7 protein. *Protein Expr Purif* 2009;68:1–6.

Pieroni L, Maione D, La Monica N. *In vivo* gene transfer in mouse skeletal muscle mediated by baculovirus vectors. *Hum Gene Ther* 2001;12:871–881.

Pushko P, Kort T, Nathan M, Pearce MB, Smith G, Tumpey TM. Recombinant H1N1 virus-like particle vaccine elicits protective immunity in ferrets against the 2009 pandemic H1N1 influenza virus. *Vaccine* 2010;28:4771–4776.

Ramchandran A, Bashyam MD, Viswanathan P, Ghosh S, Kumar MS, Hasnain SE. The bountiful and baffling baculovirus: The story of polyhedrin transcription. *Curr Sci India* 2001;81:998–1010.

Ramachandran A, Jain A, Arora P, Bashyam MD, Chatterjee U, Ghosh S, *et al.* Novel Sp family-like transcription factors are present in adult insect cells and are involved in transcription from the polyhedrin gene initiator promoter. *J Biol Chem* 2001;276:23440–23449.

Rankin C, Ooi BG, Miller LK. Eight base pairs encompassing the transcriptional start point are the major determinant for baculovirus polyhedrin gene expression. *Gene* 1988;70:39–49.

Räty JK, Airenne KJ, Marttila AT, Marjomäki V, Hytönen VP, Lehtolainen P, *et al.* Enhanced gene delivery by avidin-displaying baculovirus. *Mol Ther* 2004;9:282–291.

Renoux VM, Fleury MJ, Bousarghin L, Gaitan J, Sizaret PY, Touzé A, *et al.* Induction of antibody response against hepatitis E virus (HEV) with recombinant human papillomavirus pseudoviruses expressing truncated HEV capsid proteins in mice. *Vaccine* 2008;26:6602–6607.

Reyes-Cruz G, Vázquez-Prado J, Müller-Esterl W, Vaca L. Regulation of the human bradykinin B2 receptor expressed in sf21 insect cells: A possible role for tyrosine kinases. *J Cell Biochem* 2000;76:658–673.

Rodems SM, Friesen PD. The hr5 transcriptional enhancer stimulates early expression from the *Autographa californica* nuclear polyhedrosis virus genome but is not required for virus replication. *J Virol* 1993;67:5776–5785.

Rutkowska DA, Meyer QC, Maree F, Vosloo W, Fick W, Huismans H. The use of soluble African horse sickness viral protein 7 as an antigen delivery and presentation system. *Virus Res* 2011;156:35–48.

Sah NK, Taneja TK, Pathak N, Begum R, Athar M, Hasnain SE. The baculovirus antiapoptotic p35 gene also functions via an oxidant-dependent pathway. *Proc Natl Acad Sci USA* 1999;96:4838–4843.

Sahdev S, Saini KS, Hasnain SE. Baculovirus P35 protein: An overview of its applications across multiple therapeutic and biotechnological arenas. *Biotechnol Prog* 2010;26:301–312.

Sahdev S, Taneja TK, Mohan M, Sah NK, Khar AK, Hasnain SE, *et al.* Baculoviral p35 inhibits oxidant-induced activation of mitochondrial apoptotic pathway. *Biochem Biophys Res Commun* 2003;307:483–490.

Sakihama T, Masuda K, Sato T, Doi T, Kodama T, Hamakubo T. Functional reconstitution of G-protein-coupled receptor-mediated adenylyl cyclase activation by a baculoviral co-display system. *J Biotechnol* 2008;135:28–33.

Habib S, Hasnain SE. A 38-kDa host factor interacts with functionally important motifs within the *Autographa californica* multinucleocapsid nuclear polyhedrosis virus homologous region (hr1) DNA sequence. *J Biol Chem* 1996;271:28250–28258.

Samuelson P, Gunneriusson E, Nygren PÅ, Ståhl S. Display of proteins on bacteria. *J Biotech* 2002;96:129–154.

Sandig V, Hofmann C, Steinert S, Jennings G, Schlag P, Strauss M. Gene transfer into hepatocytes and human liver tissue by baculovirus vectors. *Hum Gene Ther* 1996;7:1937–1945.

Sarkis C, Serguera C, Petres S, Buchet D, Ridet JL, Edelman L, *et al.* Efficient transduction of neural cells *in vitro* and *in vivo* by a baculovirus-derived vector. *Proc Natl Acad Sci USA* 2000;97:14638–14643.

Schellenbacher C, Roden R, Kirnbauer R. Chimeric L1-L2 virus-like particles as potential broad-spectrum human papillomavirus vaccines. *J Virol* 2009;83:10085–10095.

Schlom J, Hodse JW. The diversity of T-cell co-stimulation in the induction of antitumor immunity. *Immunolog Rev* 1999;170:73–84.

Senger T, Schädlich L, Gissmann L, Müller M. Enhanced papillomavirus-like particle production in insect cells. *Virology* 2009;388:344–353.

Shen H, Li C, Lin M, Zhang Z, Shen Q. Expression of human interleukin 12 (hIL-12) in insect cells. *Chin J Biotechnol* 1998;14:205–212.

Shin HS, Lim HJ, Cha HJ. Quantitative monitoring for secreted production of human interleukin-2 in stable insect *Drosophila* S2 cells using a green fluorescent protein fusion partner. *Biotechnol Prog* 2003;19:152–157.

Smith GP. Filamentous fusion phage: Novel expression vectors that display cloned antigens on the virion surface. *Science* 1985;228:1315–1317.

Smolewski P, Robak T. Inhibitors of apoptosis proteins (IAPS) as potential molecular targets for therapy of hematological malignancies. *Curr Mol Med* 2011;11:633–649.

Song L, Liu Y, Chen J. Baculoviral capsid display of His-tagged ZnO inorganic binding peptide. *Cytotechnology* 2010a;62:133–141.

Song L, Liu Y, Chen J. Inorganic binding peptide-mediated immobilization based on baculovirus surface display system. *J Basic Microbiol* 2010b;50:457–464.

Strokovskaya L, Bartoszewicz Z, Szolajska E, Kikhno I, Solomko A, Michalik J. Expression and one-step purification of intracellular human prolactin in insect cells. *Protein Expr Purif* 2001;22: 242–248.

Summers MD. Baculovirus-directed foreign gene expression. In: Menn J, Kelly T, Mosler E, eds. *Insect Neuropeptides*. Washington, DC: American Chemical Society; 1991. pp. 237–251.

Suzuki H, Kaneko H, Tamai N, Miyano-Kurosaki N, Hashimoto K, Shimotohno K, *et al*. Suppression of HCV RNA replication by baculovirus-mediated shRNA expression. *Nucleic Acids Symp Ser (Oxf)* 2005;49:339–340.

Suzuki H, Matsumoto N, Suzuki T, Chang MO, Takaku H. Stable replication of the EBNA1/OriP-mediated baculovirus vector and its application to anti-HCV gene therapy. *Virol J* 2009;6:156.

Suzuki H, Saitoh H, Suzuki T, Takaku H. Inhibition of influenza virus by baculovirus-mediated shRNA. *Nucleic Acids Symp Ser (Oxf)* 2009;53:287–288.

Suzuki T, Oo Chang M, Kitajima M, Takaku H. Induction of antitumor immunity against mouse carcinoma by baculovirus-infected dendritic cells. *Cell Mol Immunol* 2010;7:440–446.

Tang XC, Lu HR, Ross TM. Hemagglutinin displayed baculovirus protects against highly pathogenic influenza. *Vaccine* 2010;28:6821–6831.

Tang XC, Lu HR, Ross TM. Baculovirus-produced influenza virus-like particles in mammalian cells protect mice from lethal influenza challenge. *Viral Immunol* 2011;24:311–319.

Tiwari P, Saini S, Upmanyu S, Benjamin B, Tandon R, Saini KS, Sahdev S. Enhanced expression of recombinant proteins utilizing a modified baculovirus expression vector. *Mol Biotechnol* 2010;46:80–89.

Touzé A, Dupuy C, Chabaud M, Le Cann P, Coursaget P. Production of human papillomavirus type 45 virus-like particles in insect cells using a recombinant baculovirus. *FEMS Microbiol Lett* 1996;141:111–116.

Tschantz WR, Pfeifer ND, Meade CL, Wang L, Lanzetti A, Kamath AV, *et al*. Expression, purification and characterization of the human membrane transporter protein OATP2B1 from Sf9 insect cells. *Protein Expr Purif* 2008;57:163–171.

Usami A, Suzuki T, Nagaya H, Kaki H, Ishiyama S. Silkworm as a host of baculovirus expression. *Curr Pharm Biotechnol* 2010;11:246–250.

Vanarsdall AL, Okano K, Rohrmann GF. Characterization of the role of very late expression factor 1 in baculovirus capsid structure and DNA processing. *J Virol* 2006;80:1724–1733.

Venkaiah B, Viswanathan P, Habib S, Hasnain SE. An additional copy of the homologous region (hr1) sequence in the *Autographa californica* multinucleocapsid polyhedrosis virus genome promotes hyperexpression of foreign genes. *Biochemistry* 2004;43: 8143–8151.

Viswanathan P, Venkaiah B, Kumar MS, Rasheedi S, Vrati S, Bashyam MD, *et al*. The homologous region sequence (hr1) of *Autographa californica* multinucleocapsid polyhedrosis virus can enhance transcription from non-baculoviral promoters in mammalian cells. *J Biol Chem* 2003;278:52564–52571.

Walsh G. Post-translational modifications in the context of therapeutic proteins: An introductory overview. In: Walsh G, ed. *Post-translational Modification of Protein Biopharmaceuticals*. Weinheim: Wiley VCH Verlag GmbH & Co. KGaA; 2009. pp. 1–76.

Wickham TJ, Davis T, Granados RR, Shuler ML, Wood HA. Screening of insect cell lines for the production of recombinant proteins and infectious virus in the baculovirus expression system. *Biotechnol Progress* 1992;8:391–396.

Wu Q, Xiao S, Fan H, Li Y, Xu J, Li Z, Lu W, Su X, Zou W, Jin M, Chen H, Fang L. (2011). Protective immunity elicited by a pseudotyped baculovirus-mediated bivalent H5N1 influenza vaccine. Antiviral Res. 92(3):493-6.

Xu XG, Liu HJ. Baculovirus surface display of E2 envelope glycoprotein of classical swine fever virus and immunogenicity of the displayed proteins in a mouse model. *Vaccine* 2008;26: 5455–5460.

Xu XG, Tong DW, Wang ZS, Zhang Q, Li ZC, Zhang K, *et al*. Baculovirus virions displaying infectious bursal disease virus VP2 protein protect chickens against infectious bursal disease virus infection. *Avian Dis* 2011;55:223–229.

Xu XG, Wang ZS, Zhang Q, Li ZC, Zhao HN, Li W, *et al*. Baculovirus surface display of E envelope glycoprotein of Japanese encephalitis virus and its immunogenicity of the displayed proteins in mouse and swine models. *Vaccine* 2011;29:636–643.

Yang GZ, Chen ZZ, Cui DF, Li BL, Wu XF. Production of recombinant human calcitonin from silkworm (*B. mori*) larvae infected by baculovirus. *Protein Pept Lett* 2002;9:323–329.

Yao Q, Vuong V, Li M, Compans RW. Intranasal immunization with SIV virus-like particles (VLPs) elicits systemic and mucosal immunity. *Vaccine* 2001;20:2537–2545.

Ye L, Lin J, Sun Y, Bennouna S, Lo M, Wu Q, *et al*. Ebola virus-like particles produced in insect cells exhibit dendritic cell stimulating activity and induce neutralizing antibodies. *Virology* 2006;351: 260–270.

Yin HY Zhou X, Wu HF, Li B, Zhang YF. Baculovirus vector-mediated transfer of NIS gene into colon tumor cells for radionuclide therapy. *World J Gastroenterol* 2010;16:5367–5374.

Yoshida S, Kawasaki M, Hariguchi N, Hirota K and Matsumoto M. (2009). A baculovirus dual expression system-based malaria vaccine induces strong protection against *Plasmodium berghei* sporozoite challenge in mice. Infect Immun. 77(5):1782-9.

Zeng J, Du J, Zhao Y, Palanisamy N, Wang S. Baculoviral vector-mediated transient and stable transgene expression in human embryonic stem cells. *Stem Cells* 2007;25:1055–1061.

Zhang T, Xu Y, Qiao L, Wang Y, Wu X, Fan D, *et al*. Trivalent Human Papillomavirus (HPV) VLP vaccine covering HPV type 58 can elicit high level of humoral immunity but also induce immune interference among component types. *Vaccine* 2010;28:3479–3487.

Zhao Y, Lam DH, Yang J, Lin J, Tham CK, Ng WH, *et al*. Targeted suicide gene therapy for glioma using human embryonic stem cell-derived neural stem cells genetically modified by baculoviral vectors. *Gene Ther* 2012;19:189–200.

27

Animal cells as host for recombinant DNA

Vijay Kumar

27.1 Introduction

Recombinant DNA technology or genetic engineering was developed in the 1970s to express mammalian genes in bacteria. However, it soon became apparent that large complex proteins, especially those of therapeutic value, could not be produced in bacteria, as they do not have the appropriate cellular machinery to make appropriate post-translational modifications (PTMs) in these proteins. Therefore, genetically engineered animal cells were developed for large-scale commercial production of such important proteins. Nowadays, animal cells are used such as bacteria or yeast to produce a variety of recombinant proteins of human origin, including secreted proteins such as monoclonal antibodies (mAb) and growth factors, intracellular proteins such as enzymes, and membrane proteins such as growth factor receptors. The mammalian cells offer a distinct advantage of making recombinant proteins with fully humanized PTMs which drastically improve their biological activity and minimize their immunogenicity. Such proteins can remain in body circulation for an extended period to produce desirable pharmacological effects.

Animal cell culture is a process of growing animal cells *in vitro* in a flask, dish or fermenter under a strictly controlled environment. Because maintaining the purity of cell cultures and preventing their contamination constitute the greatest challenge in animal cell culture, the principles of aseptic technique followed in cell culture are of paramount importance. Optimized animal cell culture has the ability to deliver sufficient quantities of recombinant products, and thus, it constitutes an important activity of the present-day biopharmaceutical industries. The range of commercially available recombinant pharmaceuticals produced by cell culture technologies has increased rapidly over the past few years. This is particularly so for therapeutic proteins synthesized from selected or genetically engineered mammalian cells. These are needed in large quantities and hence require careful study of the underlying principles of the production processes. Biopharmaceutical industries are currently facing the challenge of producing biological recombinant medical products in a quick and cost-effective manner, while complying with the safety requirements set by the regulatory agencies. Mammalian cell lines, particularly the Chinese Hamster Ovary (CHO) cell line, are the eukaryotic systems of choice for many biopharmaceutical industries for large-scale production of recombinant proteins and mAbs due to their ability to grow in high-density suspension cultures and to produce proteins with human-like PTMs. However, the lengthy, expensive and tortuous path for approval of a new therapeutic candidate has a significant bearing on the timeline and economics of process development. Nowadays, animal cell biotechnology has advanced to the stage where it is used in a variety of manufacturing processes. In fact, products derived from animal cell cultures are reported to generate nearly half of the revenues from the sale of modern biotechnology products.

This chapter provides a historical perspective of animal cell culture and describes the commonly used methods, equipment, supplies and reagents in this process. The key steps required to maintain adherent cell lines and suspension-grown cells, their long-term cryogenic storage and revival from frozen stocks are outlined. More important, different methods of gene transfer in animal cells, their selection and scale-up culture for the production of recombinant proteins and methods used in their quality control are elaborated.

27.2 Some milestones in the development of tissue and cell cultures and their use in recombinant DNA technology

The concept of maintaining live animal cell lines separated from their original tissue source was discovered in the early 20th century and animal cell culture became a routine laboratory technique by 1950s. The earliest tissue culture experiment dates back to 19th century when Sydney Ringer – an English physiologist – developed 'salt solutions' containing chlorides of sodium, potassium, calcium and magnesium for maintaining isolated animal hearts outside the body in culture. In 1885, Wilhelm Roux – a German embryologist – introduced 'saline culture' by successfully cultivating the medullary plate of a chick embryo in a warm saline solution for several days and, thus, opened the roads for tissue culture. Between 1907 and 1910, Ross Harrison from Johns Hopkins Medical School published a series of papers describing the methods of tissue culture. He used frog embryonic tissue to grow nerve fibers in the presence of frog lymph by the hanging-drop method. In 1911, Alexis Carrel and Montrose Burrows used horse plasma to grow explants of chick heart tissue and introduced strict aseptic techniques so that cells could be cultured for long periods. The development of antibiotics during World War II simplified long-term animal cell culture by minimizing the problems of microbial contamination in cell culture. Peyton Rous and FS Jones (1916) introduced the proteolytic enzyme trypsin for the subculture of adherent cells. In 1950s, Wilton Earle and his colleagues standardized the technique to dissociate cells from growing chick embryo with the help of trypsin. He observed that when a single-cell suspension was mixed with plasma and embryo extract and transferred in a sterile glass container, the cells adhered to the glass and divided to form a monolayer of primary culture that covered the entire bottom surface of the vessel and then stopped dividing. The primary culture that contained a variety of cell types could then be re-dispersed with trypsin and planted in new culture vessels containing fresh medium. Often, these secondary cultures were composed entirely of spindle-shaped cells called fibroblasts because of their resemblance to cultured connective tissue. Earle also isolated mouse L fibroblasts, which formed clones from single cells. In 1952, George Gey and colleagues established a cell line derived from a human cervical carcinoma, which later became the well-known HeLa cell line, while the first successful suspension cultures, lymphoblastic MBIII cells, were grown by Owens, Gey and Gey (1953). In 1981, the teams led by Martin Evans and Gail Martin first derived embryonic stem (ES) cells from mouse embryos, which laid the foundation of 'gene knock-out technology'. Subsequently, the teams led by James Thomson and John Gearhart established human ES cells in 1998 that are extremely useful in regenerative medicine.

In 1955, Harry Eagle made the first systematic investigation of the essential nutritional requirements of cultured cells and reported that animal cells can propagate in a defined mixture of small molecules supplemented with a small proportion of serum protein. He began by showing that HeLa and mouse L cells would grow in a mixture of salts, amino acids, vitamins and cofactors, carbohydrates and horse serum. By eliminating one component at a time, he then determined which nutrients were essential for cell growth. His minimum essential medium (MEM) contains 13 amino acids (human tissue *in vivo* requires only eight), eight vitamins and cofactors, glucose as an energy source and a physiological salt solution that is isotonic to the cell. The pH is maintained at 7.2–7.4 by $NaHCO_3$ in equilibrium with CO_2. The pH indicator phenol red is usually incorporated into the medium; it turns red-purple if the medium is basic, yellow if the medium is acidic, and remains red-orange if the pH is in the right range. Serum in concentrations of 1–10% must be added to the medium to provide the cells with additional poorly defined factors, without which most cells will not grow. Most mammalian cells are incubated at 37°C. However, the avian, reptilian and arthropod cells may grow optimally at higher or lower temperatures. CMRL-1066 – a chemically defined medium – was developed in the late 1950s by Connaught Medical Research Laboratories. Later, Richard Ham (1965) introduced a defined, serum-free medium able to support the clonal growth of certain mammalian cells. Cherry and Hull (1956) used a magnetic stir bar suspended from a fishing swivel to keep cells in suspension in a round bottom flask so that they could study the effects of varying the medium, speed and seeding densities on cell growth. McLimans and his group (1957) took the magnetic stir bar and attached it to a sliding wire so that it could be raised or lowered to match the medium volume. They added a side port for sampling and called it a 'spinner flask', which was used to grow 400 mL suspension cultures of HeLa and L cells. They adapted this approach to work in commercially available 5-L microbial fermenter with baffles and spargers driven by an overhead impeller and showed that suspension animal cell culture could be successfully scaled up to grow products at commercial level. As this method was not suitable for anchorage-dependent cell cultures, AL van Wezel (1967) resolved this issue by using small porous beads called microcarriers (200–250-μm diameter) as substances for cell attachment. These particles are made from a variety of materials, including polystyrene, dextran, glass and cross-linked collagen. The use of these beads in culture required the use of more gentle stirring techniques, and by the 1980s, spinner vessels of different designs were available to minimize the damage caused by bead–bead collisions and sheer effects. Today, a number of therapeutic protein drugs and vaccines are produced from cells grown in suspension on microcarriers.

Frank Graham and Alex van der Eb (1973) developed the famous calcium–phosphate method of DNA transfection into mammalian cells. The viral method of gene delivery was introduced by Paul Berg (1976), who used a modified SV40 virus containing DNA from the bacteriophage lambda to infect monkey kidney cells in culture. Manual microinjection of DNA into nuclei of mammalian cells was successfully tried by Graessman and coworkers (1978) to introduce genes into mammalian cells and study their expression and functions. The method of electroporation to introduce DNA into fibroblasts was established by Tai-Kin Wong and Eberhard Neumann (1982), while the cationic liposome-based DNA transfection was initiated by Philip Felgner and coworkers (1987). Today, cationic liposome method is one of the most popular methods of mammalian cell transfection.

In the early days, the vaccine industry used animals such as artificially infected rabbits for rabies vaccine or cows for vaccines against smallpox as a source of some of the required viruses for producing viral vaccines. Bacteria or their toxins were also used as the basis for bacterial vaccines. Between 1920 and 1950, virus and bacterial vaccines against a number of diseases such as typhoid, diphtheria, tuberculosis, tetanus, cholera, pertussis, influenza and yellow fever were developed and marketed. However, the major advancement made in animal cell culture during 1940s and 1950s due to major epidemics of polio during these years prompted a lot of effort to develop polio vaccine through cell culture. Growing the virus in cell culture paved the way for manufacturing effective viral vaccines. The polio vaccine was made possible by the incessant efforts of John Enders, Thomas Weller and Frederick Robbins, who shared the Nobel Prize in 1954 for their discovery to grow poliomyelitis viruses in tissue culture. In fact, the polio vaccine developed by Jonas Salk was the first product to be manufactured in mass scale using animal cell culture techniques. The Salk vaccine changed the medical history by preventing many thousands of cases of crippling illness and saving thousands of lives. Another important landmark in the animal cell culture was the development of 'hybridoma technology' by Ceaser Milstein and George Kohler in 1975, which allowed the continuous production of a single type of antibody or mAb. The diagnostic and therapeutic potential of mAbs is evident from their production in kilogram quantities from large-scale cultures of hybridomas.

In short, the development of science of animal culture had a tremendous impact on both basic and applied research as well as contributed immensely to the area of biotechnology (Box 27.1). Its application ranges from basic studies such as the cell cycle control mechanisms, molecular biology of cancer, discovery of new viruses, creation of new disciplines such as genomics and transcriptomics, etc. to anticancer drug discovery, production of recombinant pharmaceuticals and stem cell therapy. The recombinant DNA and hybridoma techniques boosted the use of animal cell technology for the production of vaccines and natural therapeutic proteins. Today, the list of products resulting from these methods is extensive and includes drugs for the treatment of cardiovascular diseases (tissue plasminogen activator; tPA), dwarfism (human growth hormone; hGH), haemophilia (coagulation factors VIII and IX), anaemia (erythropoietin; EPO), cancer and viral infections (interferons). The list also includes many mAbs used in diagnostics. These examples illustrate the tremendous impact of animal cell biotechnology on modern medicine.

Box 27.1 Major discoveries from cell culture

- Identification and isolation of almost all animal/human viruses
- Discovery and study of bacteria associated with cells such as *Chlamydia*
- Growth and attenuation of viruses in culture for vaccines, e.g. rubella, polio, chicken pox, measles, mumps, rabies, yellow fever, etc.
- Studies of the genetic basis of cancer, immunology, neuroscience, embryology and *in vitro* fertilization
- Genomic revolution: Human genomic DNA sequencing made possible
- Characterization of expressed genes by DNA microarrays (transcriptomics) and pattern of protein expression under a defined condition (proteomics)
- Production of mAbs from hybridomas
- Screening/evaluation of drugs against viruses and cancer
- Development of gene therapy/stem cell therapy

27.3 Characteristic features of animal cell culture

Animal cells, just as plant cells, when removed from tissues will continue to grow in a favourable artificial environment having appropriate nutrients and conditions. Therefore, cell culture is a technique of growing and maintaining the cells outside of the multicellular organism in specially designed vessels and which are incubated under conditions to mimic precise environmental conditions such as temperature, humidity and nutrition and contamination-free conditions that were present in that organism. The cells may be removed from the tissue directly and disaggregated by mechanical or enzymatic means before culturing, or they may be derived from an established cell line or cell strain. The culture process allows single cells to act as independent units, much like a microorganism. These cells are capable of dividing as they increase in size or in a batch culture and can continue to grow until limited by nutrient depletion or other culture variables. Further, the cells in culture may be genetically identical (homogenous population) or may exhibit some genetic variation (heterogeneous population). A homogenous population of cells is derived from a single parental cell and are therefore genetically identical and referred to as 'clone'.

27.3.1 Purpose of cell culture

Animal cell culture has proven to be of immense value in biomedical research and biotechnology industry (Box 27.2). It permits the study of cells under controlled conditions and allows researchers to investigate the normal physiology and biochemistry of cells, cell metabolism and the cell cycle. It also allows us to examine the effects of specific conditions and mutations on cell physiology and evaluate the effect such as the toxicity of chemical compounds or drugs on a given cell type. Most important, it allows us to do high-throughput screening of anticancer agents.

27.3.2 Missing features in cell culture

Cells are often but not always in contact with other cells because in cultures, cells are never left to complete or 100% confluency. This is a major problem as normal cells (primary cells) need cell-to-cell contacts and show 'contact inhibition'. This means that when cells grow and reach the walls of the container (i.e. reach confluency), they stop growing further. Transformed cells, on the other hand, can grow and divide quite well in the absence of cell-to-cell contacts and form multilayers. The following important features are missed in animal cell culture:

- No original tissue organization and three-dimensional (3D) structure
- Poor or no cell–cell and cell–matrix interaction
- No original blood circulation
- Altered hormonal and nutritional environment. Further, hormones are usually added at high concentrations which may not be physiological

Box 27.2 Importance of animal cell culture

Advantages

- Has a homogenous population of cells
- Is grown in controlled physicochemical environment
- Has the flexibility to add genes (transfection) or regulate protein levels (RNAi)
- Generates sufficient number of cells for doing biochemistry
- Used in production of biopharmaceuticals
- Is ethical to use and with fewer regulatory controls
- Involves low-cost screening/assays

Disadvantages

- Requires sensitive techniques to detect changes in tiny cells
- Has a challenging scale-up
- Does not necessarily represent the original living systems

27.3.3 Limitations of cell culture

Cell culture is referred to as an *ex vivo* study of the cellular milieu. This is a serious limitation because the cell is not in its normal physiological and original environment. Cell culture is simply an attempt to provide a simulated environment. Thus, the study of cells in cell culture is akin to studying animal behaviour in captivity. With the exception of some tumour-derived cells, most primary cell cultures have a limited lifespan. After a certain number of doubling, cells undergo the process of senescence and stop dividing, but may retain their viability. This phenomenon is known as the Hayflick limit (Hayflick and Moorhead 1961). Also a problem in cell culture is the usage of cells which are transformed. For example, HepG2 cells are derived from human hepatoma or liver cancer and, therefore, do not represent normal human hepatocytes. These cells differ markedly from normal cells as they have altered or lost specific cell functions due to mutations. Alternatives to using transformed cells are the cultures of primary cells or *immortalized cells* which are untransformed. Immortalized cells can grow continuously in culture, but they are unique in the sense that they do not form tumours in xenograft models (see Section 27.4).

27.3.4 Problems with animal cell culture

Cell culture has many problems, including cell culture contamination by bacteria, mycoplasma and fungus. Cells must be subcultured frequently to prevent overcrowding of cells, leading to a change in their properties. Further, the culture media must be changed regularly to prevent the build-up of contaminants, metabolites and toxins and to provide fresh nutrients as the nutritional requirements of fast-growing transformed cells are higher.

27.4 Origin of cell culture, cell types and cell culture systems

The term 'tissue culture' was originally used to denote the explants of tissue grown in the plasma. The term subsequently became associated with the culture of cells and is now obsolete in its original sense. Cell culture refers to tissue dissociated into a single-cell suspension, usually with the help of trypsin. After a brief wash, cells are counted, diluted in a growth medium and allowed to settle onto the flat-bottom culture vessels. Freshly isolated cultures from mammalian tissues are known as primary culture until subcultured (Fig. 27.1). At this stage, cells are usually heterogeneous but still closely represent the parent cell types as well

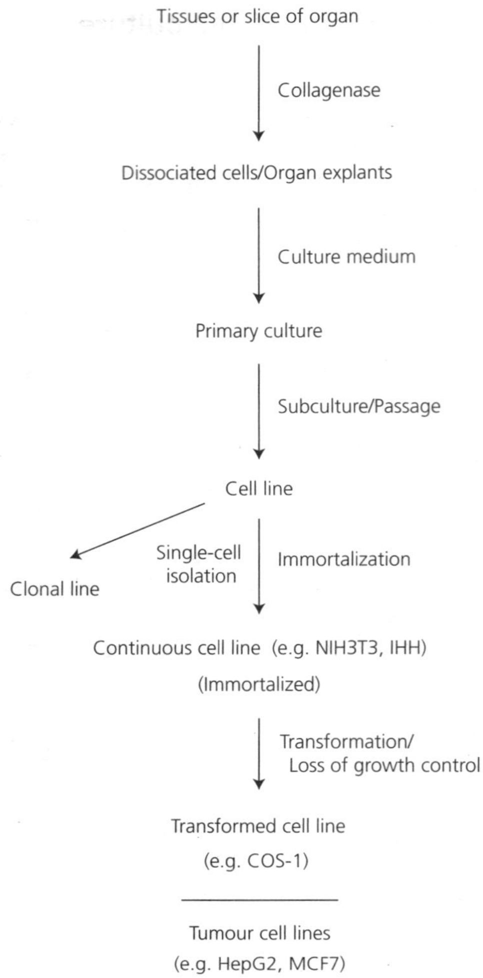

Figure 27.1 Scheme showing the possible modes of origin of cell lines [modified from Ian Freshney (2010)].

as the expression of tissue-specific properties. When these cells are subcultured in new containers with fresh medium, this is called a secondary culture. After several subcultures onto fresh media, the cell line will either die out or 'transform' to become a continuous cell line. Cells can also be virally transformed. Such cell lines show many alterations from primary cultures, including changes in morphology, chromosomal variations and an increase in the capacity to form colonies in soft agar (colony formation assay) or give rise to tumours when implanted in immune-deficient hosts (xenograft models; Fig. 27.2).

27.4.1 Types of cells

Cultured cells are usually described based on their morphology or their functional characteristics. Animal cells exhibit three basic morphologies (Fig. 27.3):

- **Epithelial like:** cells which are attached to a substrate and appear flattened and polygonal in shape
- **Fibroblast like:** cells which are attached to a substrate and appear bipolar and elongated, frequently forming swirls in heavy cultures
- **Lymphoblast like:** cells which do not attach normally to a substrate but remain in suspension and have a spherical shape.

27.4.2 Types of cultured cells

There are three main types of cultured cells depending on the number of times the cells can divide:

- **Primary cell cultures:** When cells are taken freshly from animal tissue and placed in culture, the culture consists of a wide variety of cell types, most of which are capable of very limited growth *in vitro*, usually fewer than 10 divisions. These cells retain their diploid karyotype, that is, they have the chromosome number and morphology of their tissues of origin. They also retain some of the differentiated characteristics that they possessed *in vivo*. For this reason, these cells support the replication of a wide range of viruses. Primary cultures derived from monkey kidneys, mouse foetuses and chick embryos are commonly used for many laboratory experiments and for diagnostic purposes.
- **Diploid cell strains:** Some primary cells can be passed through secondary and several subsequent subcultures while retaining their original morphological characteristics and karyotype. Subcultures will have fewer cell types than primary cultures. After 20–50 passages *in vitro*, these diploid cell strains usually undergo a crisis in which their growth rate slows and they eventually die out. Diploid strains of fibroblasts derived from human foetal tissue are widely used in diagnostic virology and vaccine production.
- **Continuous cell lines:** Certain cultured cells, notably mouse foetal fibroblasts, kidney cells from various mammalian species and human carcinoma cells, are able to survive the growth crisis and undergo indefinite propagation *in vitro*. After several passages, the growth rate of the culture slows down; then isolated colonies of cells begin to grow more rapidly than diploid cells, their karyotype becomes abnormal (aneuploid), their morphology changes and other poorly understood changes take place that make the cells immortal. The cells are now 'dedifferentiated', having lost the specialized morphology and biochemical abilities they possessed as differentiated cells *in vivo*. Continuous cell lines such as KB and HeLa, both derived from human carcinomas, support the growth of many viruses.

27.4.3 Cell culture systems

For growing cells, two culture systems are used. They are mainly based on the ability of the cells to either

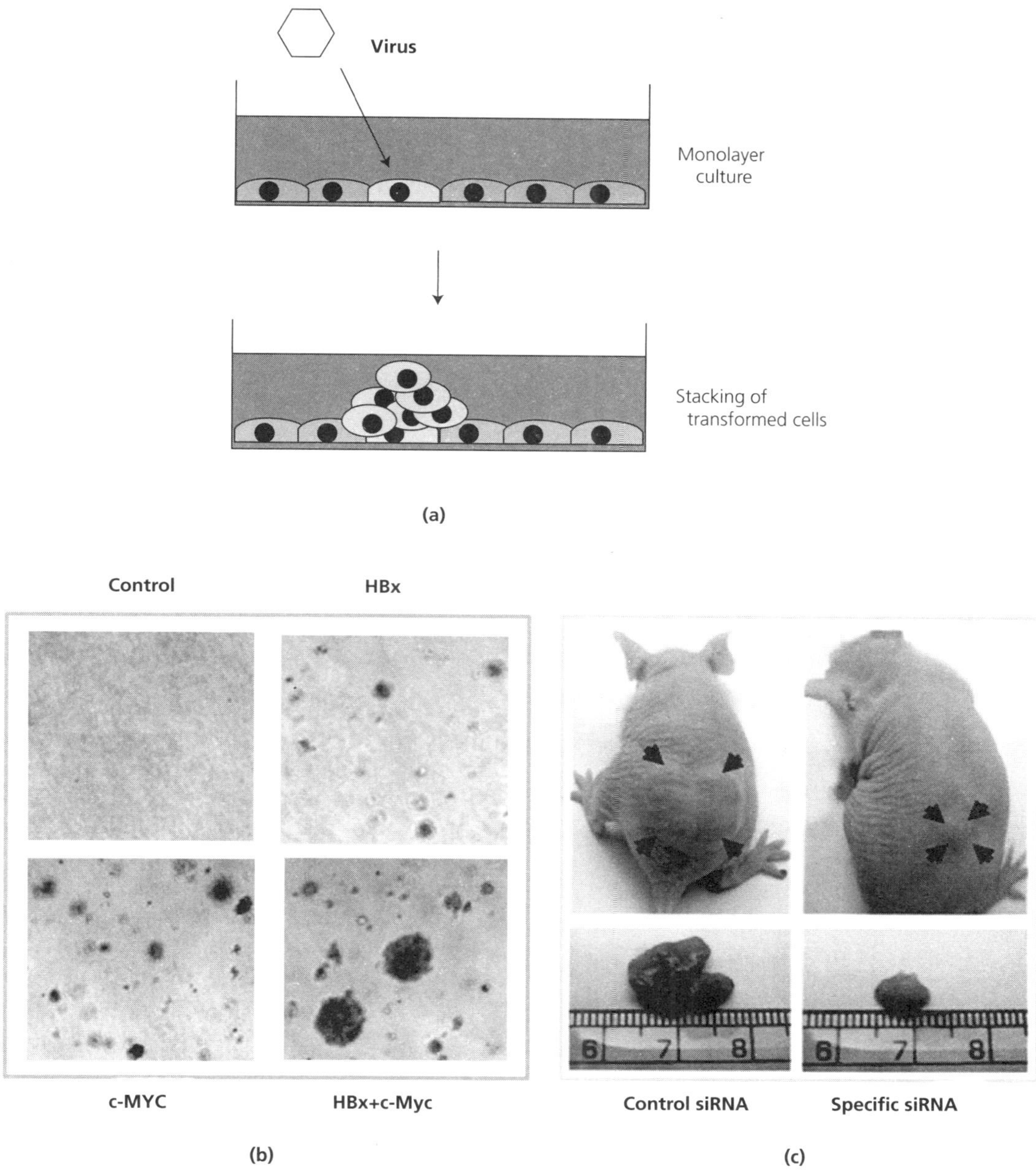

Figure 27.2 Animal cell transformation and its assays. (a) Viral transformation of cells. (b) Colony formation assay. (c) Xenograft model (*Courtsey*: Dr Anil Suri, National Institute of Immunology, New Delhi).

attach to a glass or treated plastic substrate, known as monolayer culture or 'anchorage-dependent system', and the floating free in the culture medium is known as suspension culture or 'anchorage-independent system'. Monolayer cultures are usually grown in tissue culture–treated dishes, T-flasks, roller bottles, petri plates or multiple well plates; the choice should be based on the number of cells needed, the nature of the culture, personal preference and cost. While the suspension cultures are usually grown either in magnetically rotated spinners or in shaken Erlenmeyer flasks, where the cells are keep in a fully suspended medium. The advantages and limitations of using monolayer and suspension cultures are summarized in Table 27.1.

27.4.3.1 Monolayer culture

Adherent cell lines will grow *in vitro* until they have covered the surface area available or the medium is depleted of nutrients. Before this point, the cell lines should be subcultured to prevent the culture from

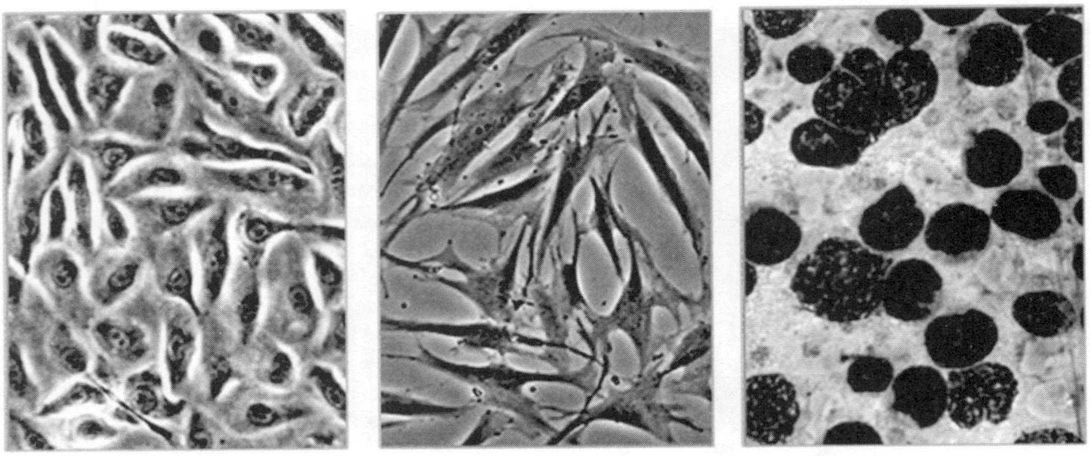

(a) (b) (c)

Figure 27.3 Types of mammalian cell lines. (a) Human mammary epithelial cells. (b) Human fibroblast cells. (c) Human large lymphoid cells.

Table 27.1 Monolayer versus suspension cell culture

Monolayer culture	Suspension cell culture
Appropriate for most cell types, including primary cultures	Appropriate for cells adapted to suspension culture or cells of hematopoietic origin
Requires periodic passaging	Easier to passage
Easy visual inspection under microscope	Requires daily cell counts and viability determination
Cells are dissociated enzymatically (e.g. trypsin) or mechanically	Does not require enzymatic or mechanical dissociation
Growth is limited by surface area	Growth is limited by cell density in the medium
Limits product yield	Allows easy scale-up
Requires cell culture–compatible vessels	Can be maintained in any sterile culture vessels
No agitation is required	Requires agitation for gas exchange
Good for cytology, many research applications, harvesting of products continously	Good for bulk protein production and batch harvesting, many research applications

dying. For subculture, the cells need to be brought into suspension. The degree of adhesion varies from cell line to cell line, but in most cases, proteases, for example, trypsin, are used to release the adhered cells from the culture vessel. Note however that trypsin treatment sometimes could be harmful to cells as this may remove membrane markers or receptors of interest. In these cases, cells should be brought into the suspension into a small volume of medium with the help of cell scrapers/rubber policeman.

27.4.3.2 Suspension culture

Some cells, particularly those derived from haematopoietic or certain tumour tissues, are anchorage independent and grow in suspension. Propagation of such cells has several advantages over propagation in monolayer as this is just a matter of dilution. Thus, there is no growth lag after splitting of cells as there is no trauma associated with proteolytic enzyme treatment. Further, the scale-up of suspension cultures is straightforward and the cells can be easily grown in bioreactors. The seeding densities range from 2×10^4 to 5×10^5 viable cells/mL.

27.4.4 Procedure for cell culture

After cultured cells have formed a confluent monolayer on the surface of their culture vessel, they may be removed from the surface, diluted and seeded into new vessels. If the initial culture was primary, the new cultures are called secondary and are likely to consist of fewer cell types. Removal of cells from glass or plastic surfaces may be by either physical methods such as scraping with a sterile rubber policeman or enzymatic methods using proteolytic enzymes such as trypsin, dispase or collagenase or chelating agents or a combination of the two. After removal, cells are pipetted up and down against the bottom of the flask to break up clumps, diluted and counted. Primary cultures can usually be diluted 1:2 or 1:3 for secondary culturing, and after one becomes familiar with the growth characteristics of a certain cell type, counting can usually be dispensed with. The same procedure can be used to transfer both primary cells and a continuous cell line after removing the cells from flasks with a mixture of trypsin and EDTA in physiological saline (TES, i.e.Tris-EDTA-Saline with 10 mM Tris chloride, pH 7.4, 150 mM NaCl, 1 mM EDTA).

27.4.5 Culture conditions

Culture conditions vary widely for each cell type, but the artificial environment in which the cells are cultured invariably consists of a suitable vessel containing the following:

- A medium that supplies the essential nutrients (amino acids, carbohydrates, vitamins, minerals)
- Growth factors
- Hormones
- Gases (O_2, CO_2)
- A regulated physicochemical environment (pH, osmotic pressure, temperature)

Cells can also be adapted to different culture environments (e.g. different nutrients, temperatures, salt concentrations, etc.) by varying the activities of their enzymes. Most cells are anchorage dependent and must be cultured while attached to a solid or semi-solid substrate (adherent or monolayer culture), while others can be grown floating in the culture medium (suspension culture):

- **Incubation conditions:** Most mammalian cells are grown and maintained at an appropriate temperature and gas mixture (typically at 37°C with 5% CO_2 for mammalian cells) in a CO_2 incubator. This temperature is selected because it is the core temperature of the human body. Besides, most cells derived from other warm-blooded animals grow well at 37°C.
- **pH:** The extracellular and intracellular pHs are critical for the survival of mammalian cells. They help in maintaining the appropriate ion balance and optimal function of cellular enzymes, hormones and growth factors. Fluctuations in pH adversely affect cell metabolism, which can lead to cell death. Most media strive to achieve and maintain the pH between 7 and 7.4. Most media use the bicarbonate–CO_2 buffering system. The interaction of CO_2 derived from cells or the atmosphere with water leads to a drop in the pH, while the bicarbonate content of the medium neutralizes the effect of increased CO_2 until equilibrium is reached at pH 7.4. This kind of system is called an open system and can be described by the following equation:

$$H_2O + CO_2 \rightarrow\leftrightarrows\rightarrow H_2CO_3 \rightarrow\leftrightarrows\rightarrow H^+ + HCO_3^-$$

- **Osmolality:** The osmolality of the culture medium also plays a crucial role in cell growth and function by maintaining the outside and inside osmotic pressure of the cell. Osmolality of the medium used is determined by the media formulation. Salt and glucose are the major contributors to the osmolality of the medium, but amino acids are equally important Most commercial media are formulated to have a final osmolality of ~300 mOsm, which can be checked with the help of an osmometer.
- **Culture medium:** Apart from temperature and CO_2, the most common variable in culture systems is the growth medium. Recipes for growth media can vary in pH, glucose concentration, growth factors and the presence of other nutrients. Glucose (0.8 g to >5 g/L) is commonly used as an energy source, which helps in maintaining osmolarity of the medium. Many of the media contain phenol red as a pH indicator, which is very helpful in monitoring the pH of the culture medium in a CO_2 incubator. Highly acidic conditions turn the phenol red into yellow, while highly alkaline conditions turn the phenol red into pink. The growth factors used to supplement media are often derived from animal blood, such as foetal bovine serum (FBS). One complication of these blood-derived ingredients is the potential for contamination of the culture with viruses, mycoplasmas or prions and shows variation in growth factor and hormone contents. Therefore, current practice is to minimize or eliminate the use of these ingredients wherever possible and use chemically defined media. Commercially available media are sterile and ready-to-use liquids in a concentrated liquid or powdered form. Instead of providing nutrients for growing cells, the medium is generally supplemented with fungicides and antibiotics, or both, to inhibit contamination.
- Dulbecco's Modified Eagle Medium (DMEM) is a basal medium consisting of amino acids, vitamins, glucose, salts and a pH indicator which contains no proteins or growth-promoting agents. Therefore, it needs supplementation to be a 'complete' medium. It is commonly supplemented with 5–10% FBS. DMEM uses a sodium bicarbonate buffer system (3.7 g/L) and therefore requires artificial levels of CO_2 to maintain the required pH 7, but many researchers successfully use it in 5% CO_2.
- **Serum and antibiotics:** Serum is one of the most important components of animal cell culture as it supports cell proliferation. It is the source of peptide hormones and hormone-like growth factors that promote a healthy growth of cells. Serum is also a source of various amino acids, hormones, lipids, vitamins, polyamines and salts of calcium, potassium and iron, etc However, one of the key limitations of FBS usage is the variation in its growth factor contents and chances of its contamination with viruses or prions that may adversely affect the culture. Therefore, current practice is to minimize the use of FBS in cell culture and use serum-free medium instead. Although not essential for cell growth, antibiotics such as penicillin and

streptomycin are often used in the culture medium to control microbial growth.

- **Seeding density:** Plating density (the number of cells per volume of the culture medium) plays a critical role for some cell types. For example, HepG2 cells when seeded at lower density do not grow optimally in culture vessels.
- **Mode of culturing:** Cells can be grown either in suspension or adherent cultures. Some cells naturally live in suspension, without being attached to a surface, such as cells that exist in the bloodstream. There are also cell lines that have been modified to be able to survive in suspension cultures so they can be grown to a higher density than adherent conditions would allow. Adherent cells require a surface, such as tissue culture plastic or microcarrier, which may be coated with extracellular matrix components to increase adhesion properties and provide other signals needed for growth and differentiation. Most cells derived from solid tissues are adherent. Another type of adherent culture is organotypic culture, which involves growing cells in a 3D environment as opposed to two-dimensional culture dishes. This 3D culture system is biochemically and physiologically more similar to *in vivo* tissue, but is technically challenging to maintain because of many factors (e.g. diffusion). Cultures should be examined daily for their morphology, colour of the medium and density of cells.

27.4.6 Typical experimental protocol for cell culture

A typical experimental protocol for cell culture is as follows:

1. Examine the cells growing in a 25-cm^2 flask with an inverted microscope to see if they have formed a confluent (~80%) cell monolayer. If there are sufficient cells, pour off the medium.
2. Wash the monolayer with 2 mL of phosphate-buffered saline. Rinse well without shaking the flask (shaking produces bubbles) and pour off. Repeat.
3. Add 1.0 mL of TES to the flask and incubate at 37°C for 2–10 minutes with TES covering the cells. Observe periodically to determine when cells are loosened from the plastic. (Note: TES should contain a pH indicator. Below pH 7, trypsin is inactive. However, above pH 8, trypsin is damaging to cells.)
4. When cells are seen to detach from the surface upon shaking or jarring against the heel of one's hand (this can be checked under the microscope), add 4 mL of fresh growth medium with 10% serum and suspend cells by pipetting up and down a few times. Count cells in a haemacytometer, calculate the volume of additional medium needed to bring the cell concentration to ~5 × 10^5 cells/mL and add this volume to the cell suspension.
5. Seed an appropriate volume of cell suspension for the size of the flask to get the desired cell density.

27.4.7 Cell counting

Cell counting is done in the following steps with the help of a haemocytometer (Fig. 27.4):

1. Clean the chamber and cover slip with alcohol. Dry and fix the coverslip in position.
2. Harvest the cells. Add 10 µL of the cells to the haemacytometer.
3. Place the chamber under the inverted microscope under a 10× objective. May use phase contrast to distinguish the cells.
4. Count the cells in four 16-square grids (marked 1 to 4) and take their average.

The following example illustrates how to convert your cell count to the number of cells in the original suspension. Assume a 1:10 dilution is being used to count cell numbers; if you count an average of 12 cells in the four 16-square grid (1 mm^2), then the cell numbers in a given suspension can be derived as follows:

$$12 \times 10^4 \times 10 = 1.2 \times 10^6 \text{ cells/mL}$$

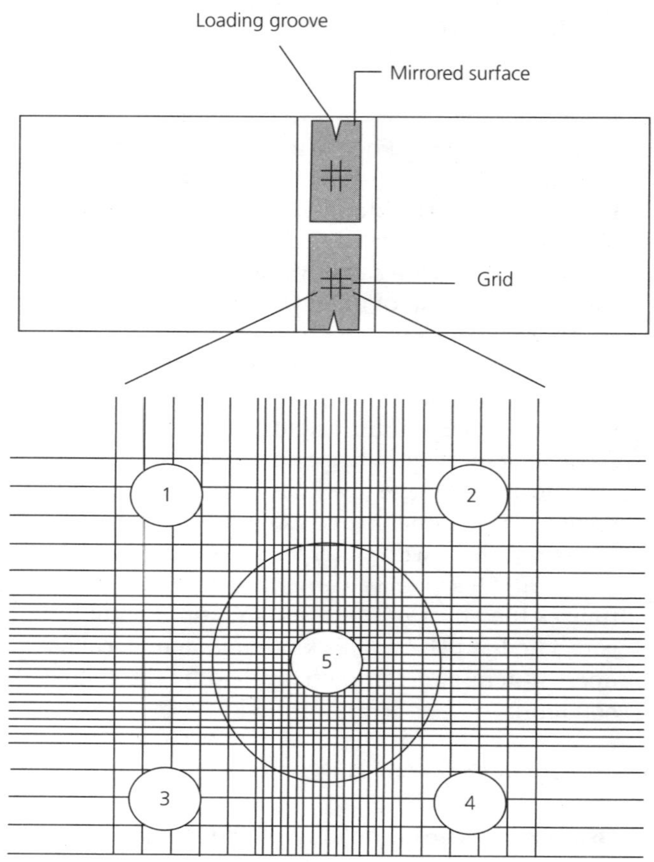

Figure 27.4 Haemocytometer along with an enlarged view of the counting chamber or grids.

where 12 is the average number of cells in four grids, × 10^4 converts to cells per millilitre, × 10 is the dilution factor and 1.2 × 10^6 cells/mL is the number of cells in the original suspension.

27.4.8 Cell viability

Viability measures the number of viable cells in a population and provides an accurate indication of the health of cells in culture. These assays rely upon the integrity of the cell membrane as an indicator of cell viability. Stains such as trypan blue are actively excluded by viable cells, while they are taken up and retained by dead cells that lack an intact membrane.

27.4.9 Cryopreservation (freezing, storage and revival) of cells

Cell culture has many problems, including culture contamination by bacteria, mycoplasma and fungus. Also, cells must be subcultured frequently to prevent overcrowding of cells. Furthermore, media must often be changed to prevent the build-up of contaminants and toxins and to provide fresh nutrients as nutrients are used up quickly by rapidly proliferating cells. Therefore, as soon as a small surplus of cells becomes available from subculturing, they should be frozen as a seed stock, protected and should not be made available for general laboratory use. Working stocks can be prepared and replenished from frozen seed stocks. If the seed stocks become depleted, cryo preserved working stocks can then serve as a source for preparing a fresh seed stock with a minimum increase in the generation number from the initial freezing. Cell lines in continuous culture are also prone to genetic drift and are fated for senescence, and even the best-run laboratories can experience equipment failure. As an established cell line is a valuable resource and its replacement is expensive and time-consuming, it is vitally important that it is cryopreserved for long-term storage.

27.4.9.1 Storage by freezing

Viability of viruses and bacteria is well preserved during freezing, but original attempts to preserve animal cells by freezing resulted in cell death. This was first thought to be due to laceration of cell plasma membranes by ice crystals, but more recent evidence suggests that the cause may be osmotic changes during freezing which give rise to irreversible changes in lipoprotein complexes in intracellular membranes. In any case, mammalian cells are cryopreserved to avoid loss by contamination, to minimize genetic change in continuous cell lines and to avoid aging and transformation in finite cell lines. Before cryopreservation, one need to ensure that cells are healthy (>90% viability) and growing in the log phase. Further, the culture should be characterized and checked for contamination. There are several common media used to freeze cells.

For serum-containing medium, the constituents may be as follows:

- Complete medium containing 10% glycerol
- Complete medium containing 10% DMSO (dimethylsulphoxide)
- 50% cell-conditioned medium with 50% fresh medium containing 10% glycerol or 10% DMSO.

Cells are stored in sterile cryogenic storage vials or cryovials that have a screw-top and a rubber seal to keep the nitrogen out. Cryoprotective agents such as DMSO help in reducing the freezing point and facilitate a slow cooling rate. Gradual freezing decreases the risk of cell damage and ice crystal formation. Ideally, a cooling rate of 1°C/min is recommended. Afterwards, cells must be stored at –70°C or lower (ideally in liquid nitrogen at –196°C). One must ensure that the liquid nitrogen containers carrying the canisters of cryovials are relatively full of nitrogen during the period of storage.

27.4.9.2 Resuscitation of frozen cell lines

When cryopreserved cells are needed for study, it is vital to thaw cells correctly to maintain the viability of cells and enable the culture to recover quickly. Some cryoprotectants, such as DMSO, are toxic above 4°C; therefore, it is essential that cultures are thawed quickly and diluted in culture medium to reduce their toxic effects. With careful handling, 50–80% of the cells of a healthy culture will survive freezing.

The following steps are recommended for getting best results:

1. Thaw frozen cells rapidly (<1 min) in a 37°C water bath.
2. Dilute the thawed cells slowly using a pre-warmed growth medium.
3. Plate thawed cells at a high density to optimize recovery.

27.4.10 Commonly used equipment in animal cell culture

A typical cell culture laboratory should have the following main equipment for doing animal cell culture work: cell culture hood, CO_2 incubator, inverted microscope, centrifuge, etc.

27.4.10.1 Cell culture hood

All cell culture manipulations must be performed aseptically in laminar air flow hoods to keep the work area free from any bacterial or fungal contamination. Therefore, the cell culture hoods should never be used for bacterial or fungal work. A laminar hood has a dual role in cell culture:

- It protects the tissue culture from the users (by providing a sterile environment).
- It protects the user from the tissue culture (from possible infection risk).

There are two types of laminar air flow hoods – vertical and horizontal (Fig. 27.5). In a vertical hood, the filtered air blows down from the top of the cabinet; in a horizontal hood, the filtered air blows out at the user in a horizontal fashion. The vertical hood is also known as a biology safety cabinet (BSC) and is appropriate for working with infectious agents. Here, the infectious aerosols generated in the hood are filtered out before they are released into the surrounding environment. Depending on the nature of the infective agent being used in cell culture, the BSCs are designated as class I, II or III. The horizontal hoods are unsuitable for working with hazardous agents as it blows air directly at users, but are good for cultures. Both vertical and horizontal hoods have continuous air flow that passes through a high efficiency particle air (HEPA) filter to remove particulate matter from the air. The hoods are also equipped with a short-wave ultra violet (UV) light source which needs be turned on for a few minutes before use to sterilize the work surfaces of the hood. Also, the hood should be turned on about 10–20 min before being used. It is important to keep the hood free of all clutter as they interfere with the laminar flow air pattern. Wiping of all work surfaces with 70% ethanol inside the hood before and after each use is a good practice.

27.4.10.2 CO_2 incubator

The CO_2 incubator is designed to maintain sterility inside the chamber and keep a constant temperature, high relative humidity and an atmosphere with a fixed level of CO_2. Thus, it reproduces an environmental

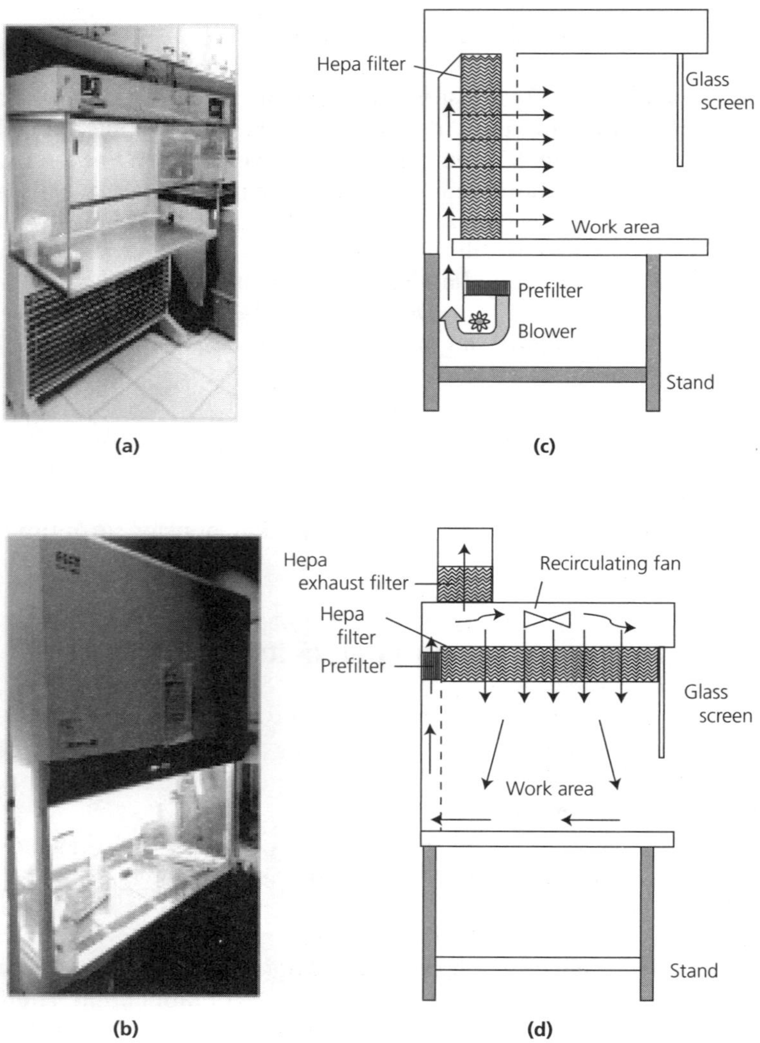

Figure 27.5 Cell culture laminar hoods. (a) Horizontal laminar hood. (b) Vertical laminar hood. (c) Horizontal laminar flow. (d) Vertical laminar flow.

(a) (b)

Figure 27.6 Cell culture equipment. (a) CO_2 incubator. (b) Inverted microscope.

condition very close to that of living cells. The incubator chamber has an airtight silicone gasket on the inner door which keeps it isolated from the external environment [Fig. 27.6(a)]. The animal cells are maintained in an atmosphere of 5–10% CO_2, which keeps the medium buffered with sodium bicarbonate/carbonic acid. Culture vessels have vents to allow for sufficient gas exchange. A pan of water is kept at all times in the incubator chamber to maintain a high relative humidity, prevent desiccation and maintain the correct osmolarity of the culture medium. The incubator doors should not be left open for very long during use.

27.4.10.3 Inverted microscope

In tissue culture vessels such as petri dish, the cells are present at the bottom, with the culture medium above. The inverted microscope allows viewing of cells from the bottom because its optical system is at the bottom, with the light source on top [Fig. 27.6(b)]. Observation of cultures in this way gives an immediate idea of the health and growth of cells. The lights of microscope should be turned off when not in use and should be kept covered to protect the optical parts from dust.

27.4.10.4 Centrifuge

A low-speed centrifuge is required for animal cell pelleting during the subculturing step. In most cases, cells are centrifuged at ambient temperature. Nevertheless, a low operation temperature is desirable to minimize exposure of cells to uncontrolled higher temperatures.

27.4.11 Commonly used culture wares

The need for sterility also applies to the glassware or plastic ware used to prepare media and reagents and to grow cells. Metal and glass devices can be sterilized by autoclaving while disposable plastic culture flasks and other plastic ware are sterilized by gamma radiation. Sterilization methods have the goal of killing contaminating organisms without affecting the surface of the flask or vessel. The animal cells are usually grown and maintained in petri dishes, culture flasks or multi-well plates of various shapes and sizes (Fig. 27.7) at an appropriate temperature and gas mixture (typically, 37°C, 5% CO_2 for mammalian cells) in an incubator. There are numerous suppliers of cell culture plastic ware such as BD Biosciences, Corning, Falcon and Nunc International. Note that

Figure 27.7 Cell culture plastic wares and accessories.

the most expensive cell culture plastic ware does not always mean the best choice for your cells. A plastic ware check is always worth doing with primary cells as their requirements are more specific. However, for cell lines, one may start with the brand currently used in the laboratory.

27.5 Gene delivery in cells

Animal cell transfection is a commonly used method to deliver exogenous DNA or RNA into cells to express them in cell culture. It offers an important mechanism for advancing the knowledge about the structure and functions of animal genomes and understanding their new traits through cell culture. There are several different ways to transfect mammalian cells, depending on the cell line characteristics, desired effect and downstream applications. These methods can be broadly divided into two categories: those used to generate transient transfection and those used to generate stable cell lines. Table 27.2 summarizes the important aspects of stable and transient transfection methods.

27.5.1 Transient versus stable transfection

Transient transfections are most commonly used to investigate the short-term impact of gene and/or protein expression on the cell cycle, cell physiology and metabolism or even reporter genes. Plasmid DNA, messenger RNA (mRNA), short-interfering RNA (siRNA), short-hairpin RNA (shRNA) and microRNA (miRNA) are commonly used in transfection experiments. During transient expression, however, the nucleic acid sequence is not integrated into the host cell genome. Therefore, the effect on target gene expression is temporary (24–72 h for RNA probes, 48–96 h for DNA probes).

Table 27.2 Transient versus stable expression of genes

Transient expression	Stable expression
Short-term expression (usually 48 h)	Sustained expression on a long-term basis
Transfection by chemical methods or electroporation	Transfection by chemical methods or viral vectors followed by selection for antibiotic or drug resistance
Useful for evaluating gene activity or their regulation	Useful for continuous source of a gene product or gain/loss of a gene function
No genomic integration of transfected DNA	Non-specific integration of transfected DNA in host cell genome

For long-term gene expression, the transfected gene is integrated into the host cell genome to make permanent cell lines. A selection marker is used to identify cells that have successfully integrated the gene sequence of interest. The selection agent enriches and enables the growth of a subpopulation of cells where the exogenous genetic material has been incorporated into the genome.

Transfection of DNA or RNA molecules into cultured mammalian cells can be accomplished using a variety of methods and reagents. These methods include physical (electroporation, microinjection), chemical (calcium phosphate method, lipofection) and virus-based (adenovirus, retrovirus) delivery systems. The important parameters that affect the outcome of cell transfection include: (i) cell type, (ii) cell density, (iii) amount of DNA to be transfected, (iv) ratio between the transfection reagent and the DNA, (v) incubation period of cells with the DNA complex, (vi) post-transfection incubation time of cells, etc.

27.5.2 Methods of transient transfection

The various methods of transient transfection are as follows.

27.5.2.1 Electroporation

The use of high-voltage pulses or electroporation to introduce DNA into cultured cells was first established by Wong and Neumann (1982) for fibroblasts. It is a highly efficient technique for delivering exogenous nucleic acids to suspension cells and non-adherent primary cells. This technique uses electricity to create transient pores (electropores) in the cellular membrane to enable the uptake of charged nucleic acid molecules into the target cells [Fig. 27.8(a)]. A high cell mortality in this case is a major drawback.

27.5.2.2 Calcium phosphate method

This method was first used by Graham and van der Eb in 1973 to introduce adenovirus DNA into mammalian cells. Here HEPES-buffered saline solution containing phosphate ions is mixed with a calcium chloride solution containing the DNA to be transfected. When the two solutions are combined, a fine precipitate of the positively charged calcium and the negatively charged phosphate is formed, binding the DNA to be transfected on its surface [Fig. 27.8(b)]. The suspension of the precipitate is then overlaid on the cells to be transfected. The resulting calcium phosphate–DNA complexes adhere to the cell membrane and enter the cytoplasm by endocytosis. This method is extremely simple to handle but not good for all cell types. Transient hypertonic shock of the transfected cells with 10% glycerol or DMSO improves the transfection efficiency.

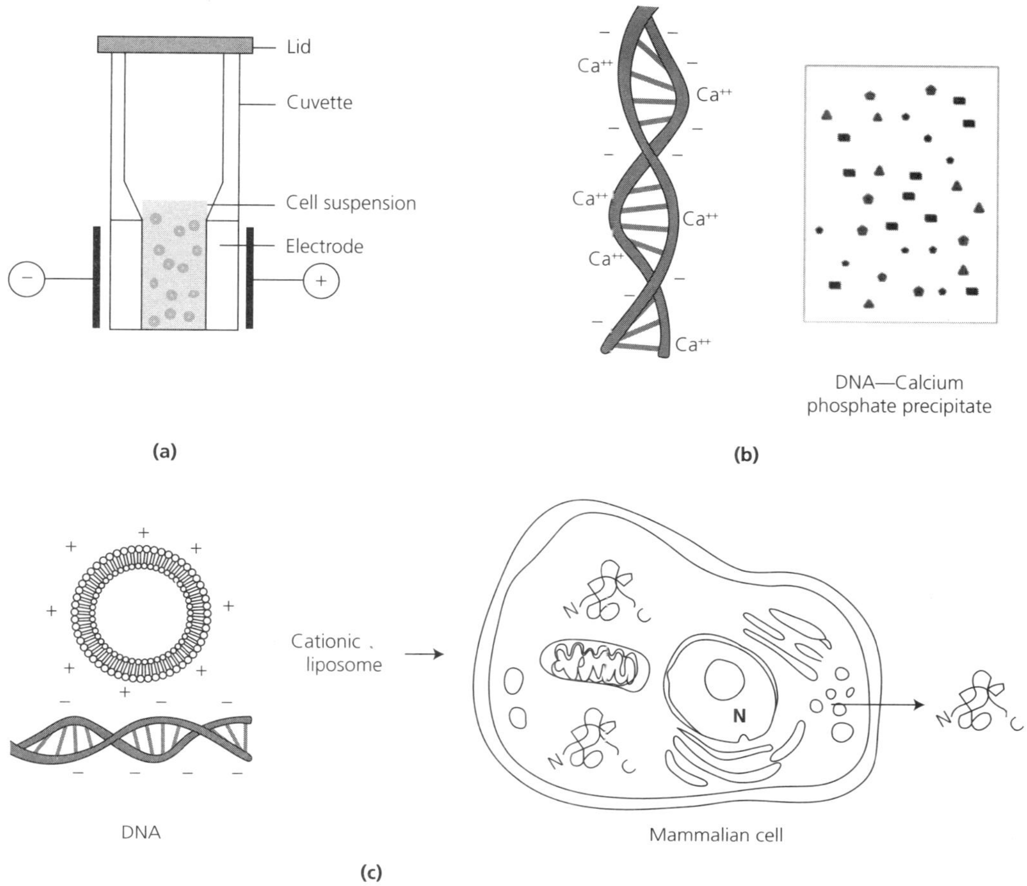

Figure 27.8 Methods of transient transfection. (a) Electroporation. (b) Calcium phosphate method. (c) Lipofection.

27.5.2.3 Liposome-mediated transfection

Liposomes are synthetic analogues of the phospholipid bilayer, the building block of the cellular membrane. These transfection compounds share a number of characteristics with their natural counterparts, including the presence of hydrophobic and hydrophilic regions of each molecule, which allow the formation of spheroid liposomes under aqueous conditions. In the presence of free DNA or RNA, liposomes encapsulate the nucleic acids to create an efficient delivery system. The charge, composition and structure of the liposome define the affinity of the complex for the cellular membrane. The cationic liposomes were first introduced in 1987 by Felgner and coworkers. The liposomes currently in use typically contain a mixture of cationic and neutral lipids organized into lipid bilayer structures. Transfection complex formation is based on the interaction of the positively charged liposome with the negatively charged phosphate groups of the nucleic acid. Liposomes fuse with the cell membrane and deliver their cargo inside [Fig. 27.8(c)]. The method is quite simple as well as reproducible, and the transfection efficiency is reasonably high. However, cell toxicity could be a problem sometimes.

27.5.3 Methods of stable transfection

The various methods of stable tranfection are as follows.

27.5.3.1 Microinjection

Microinjection delivers the genetic material directly into the cell nucleus (Fig. 27.9). Using a light microscope and a fine needle (micromanipulator) guided into the nucleus, a small amount of DNA or RNA is injected. However, this method is labour intensive and requires trained personnel. Nevertheless, this method is extremely useful in the case of animal transgenesis and gene knock-in and knock-out technologies.

27.5.3.2 Virus-mediated gene delivery (transduction)

Viral vector is the most effective means of gene transfer to modify specific cell type or tissue and can be manipulated to express heterologous genes. Several virus types are currently being investigated for use to deliver genes to cells to provide either transient or permanent transgene expression. The viral vectors in current use are based on different RNA and DNA viruses

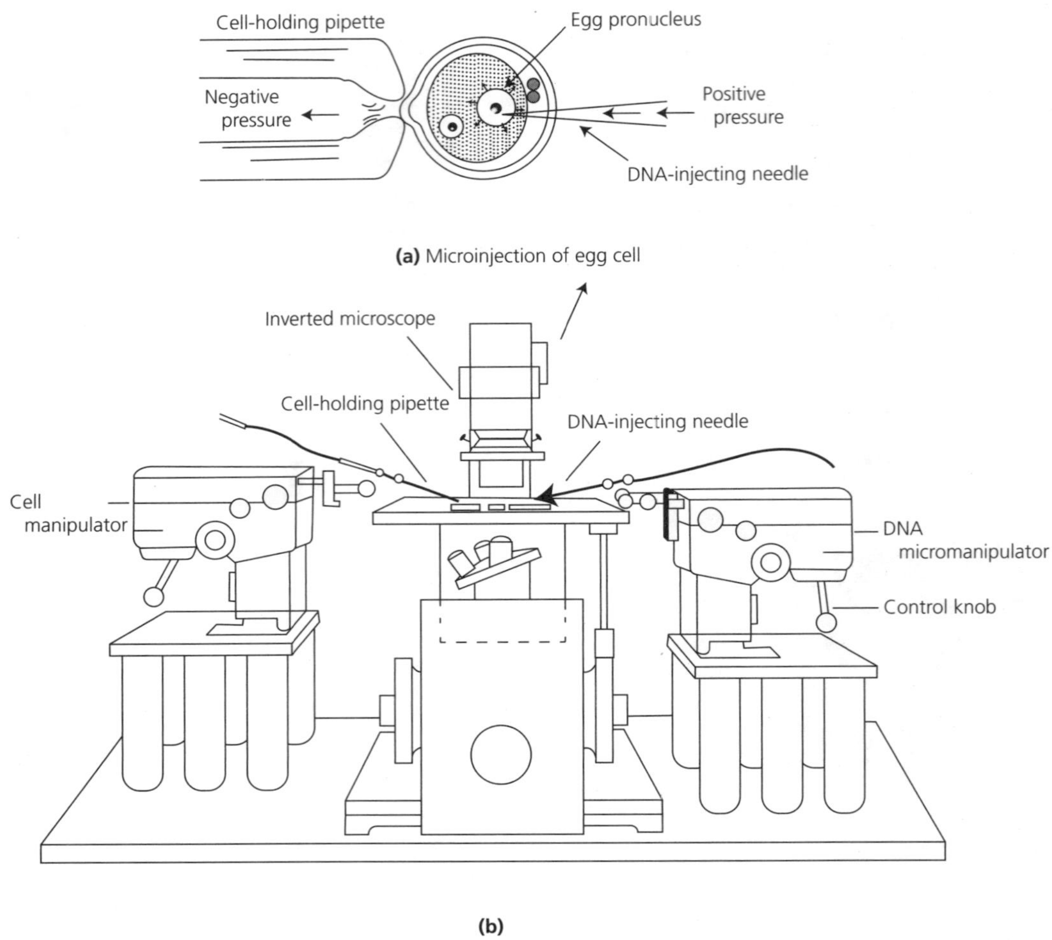

Figure 27.9 Method of microinjection. (a) Microinjection of egg cell. (b) Micromanipulator.

that possess very different genomic structures and host ranges. Among viral vectors developed so far, retroviral vectors represent the most prominent delivery system, as these vectors have high gene transfer efficiency and mediate high expression of heterologous genes. Members of the DNA virus family such as adenovirus, adeno-associated virus or herpesvirus have also become attractive for efficient gene delivery. The viral genomes are modified to act as viral vectors to introduce foreign DNA into mammalian cells. The viruses have their own mechanism for moving nucleic acids into cells and transduction by this route is usually very efficient. The viral vectors are selected as gene delivery vehicles based on their capacities to carry foreign genes and their ability to efficiently deliver these genes combined with their efficient gene expression and integration into the target cell genome with the help of the viral machinery. While in adeniviral vectors, the early *E*1 and *E*3 genes can be replaced, the entire *gag*, *pol* and *env* genes of the retroviral genome can be replaced with foreign DNA. The lost viral gene functions can be provided in *trans* using appropriate packaging cell lines. The simplified maps of adenoviral and retroviral vectors are shown in Fig. 27.10. Very often, a selection marker (either antibiotic based or fluorescent protein based) is used to select for cells that have been successfully transduced with the virus. Once the genetic material is incorporated into the host genome, it relies on the host transcriptional machinery for expression, while in the case of shRNA/miRNA probes, the host RNAi machinery is required. However, viral vectors have limited laboratory uses because the viral vector integrates randomly into the genome and could yield different results.

27.5.4 Development of stable cell line

The episomal DNA stability is often limited, resulting in a gradual loss of transfected DNA vector from the cells. As DNA integration into host chromosomes is a rare event, stably transfected cells need to be selected and cultured in different ways. A variety of systems for selecting transfected cells exist, including resistance to antibiotics or expression of metabolic enzymes. After gene transfer, cells are cultivated in a medium containing the selective agent. Only those cells which have

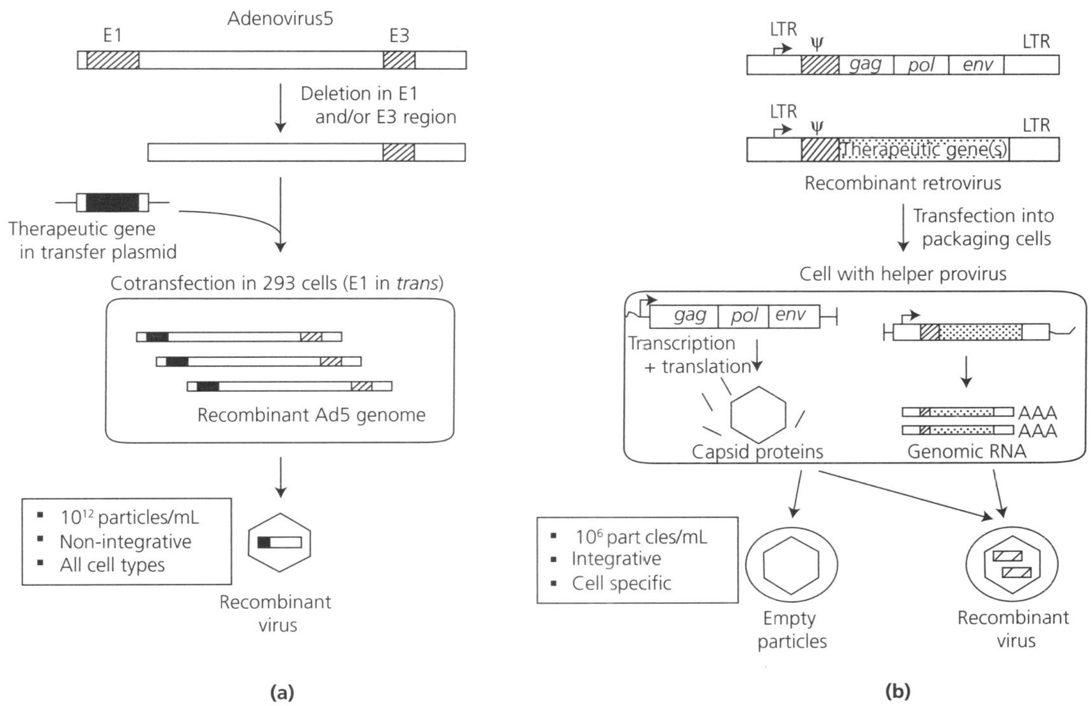

Figure 27.10 Scheme of viral vectors for gene delivery. (a) Adenoviral vector. (b) Retroviral vector.

integrated the plasmid survive, containing the selection marker gene. Both dominant and recessive selection marker genes are used for establishing permanent cell lines:

- **Dominant selection markers:** These marker genes are usually of microbial origin and confer resistance to the toxic effects of a drug or other pharmacologically active compounds used for selection. Such markers are used to confer an advantage to mammalian cells that have been transfected with exogenous DNA constructs. Dominance refers to the fact that a single copy of the marker is sufficient to confer the resistant phenotype. For example, G418 is an aminoglycoside antibiotic produced by *Micromonospora rhodorangea* that blocks polypeptide synthesis by inhibiting the elongation step in both prokaryotic and eukaryotic cells. Resistance to G418 is conferred by the neomycin resistance gene (*neo*) from Tn5, encoding an aminoglycoside 3'-phosphotransferase (Fig. 27.11). Selection in mammalian cells is usually with G418 concentrations ranging from 0.2 to 2 mg/mL.

  Puromycin is another commonly used antibiotic for making permanent cell lines. It is produced by the bacterium *Streptomyces alboniger*. As puromycin is an aminonucleoside and resembles an aminoacyl-tRNA, it can inhibit protein synthesis by disrupting peptide transfer on ribosomes, causing premature chain termination during translation. It is also a potent translational inhibitor in both prokaryotic and eukaryotic cells. Resistance to puromycin is conferred by the puromycin N-acetyl-transferase gene (*pac*) from *Streptomyces*. Puromycin has a fast mode of action, causing rapid cell death at low antibiotic concentrations. Adherent mammalian cells are sensitive to concentrations of 2–5 μg/mL and stable mammalian cell lines can be generated within 1 week.
- **Recessive selection markers:** A number of mammalian cells that are deficient for enzymes of key metabolic pathways are suitable for developing stable cell lines. The cells are transiently transfected with plasmids carrying genes for deficient enzymes and then selected using appropriate substrate analogues. For example, dihydrofolate reductase (DHFR) is a ubiquitous cytosolic enzyme that catalyses the reduction of 5,6-dihydrofolate to 5,6,7,8-tetrahydrofolate (THF). THF is an essential co-factor in several metabolic pathways, including purine and thymidylate biosynthesis. Methotrexate is a folate analogue that acts as a slow, tight-binding competitive inhibitor of DHFR. It acts by inhibiting cellular THF synthesis and, in turn, cellular proliferation. Therefore, only those cells that express the *DHFR* gene survive, while other cells die.

  Popular hosts for stable expression are Chinese hamster ovary (CHO) cells, baby hamster kidney (BHK-21) cells, myeloma cells and human embryonic kidney (HEK) 293 cells.

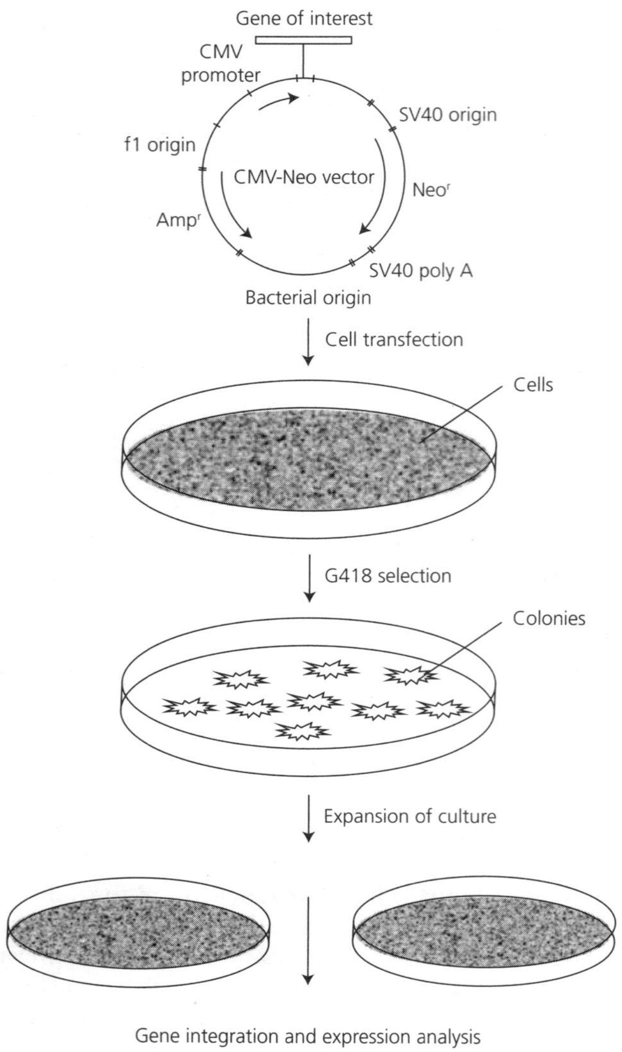

Figure 27.11 Development of stable cell lines by neomycin selection.

27.6 Scale-up of animal cell culture process

Mammalian cells are cultured by a variety of methods at a range of volumes for the production of therapeutic and diagnostic proteins. For commercial exploitation, the cells are grown in a bioreactor or fermentation vessels where physicochemical and biological factors are maintained at optimum levels. The most suitable bioreactor used is a compact-loop bioreactor with marine impellers. Here, glucose and glutamine are the main carbon and energy source and the metabolic products that affect cell growth are monitored by gas chromatography or high-performance liquid chromatography (HPLC). However, for batch cultures, mainly roller bottles with micro-carrier beads (for adherent cells) and spinner flasks (for suspension cultures) are used in scale-up of animal cell culture process:

- **Roller bottles:** Roller bottles provide total curved surface area of the micro-carrier beads for growth. This system offers the following advantages over the static monolayer culture: (i) provides an increase in the surface area, (ii) provides constant gentle agitation of the medium and (iii) provides an increased ratio of surface area of medium to its volume, which allows better gaseous exchange. Roller bottles are cylindrical vessels that revolve slowly (between 5 and 60 revolutions per hour). These tissue culture bottles can be used in specialized CO_2 incubators with attachments that rotate the bottles along the long axis [Fig. 27.12(a)]. After each complete rotation of the bottle, the entire cell monolayer has transiently been exposed to the medium. The volume of medium need only be sufficient to provide a shallow covering over the monolayer. Typically, a surface area of 750–1500 cm^2 with 200–500 mL medium will yield $1–2 \times 10^8$ cells.
- **Micro-carrier beads:** Micro-carrier beads (~90–300 μm diameter) can be used in any application where anchorage-dependent cells are to be cultured. These are either dextran or glass based, and are available in a range of densities and sizes. They are primarily used to increase the surface area for cell growth, and thus, the number of adherent cells in a culture vessel is very high. For example, micro-carrier beads usually provide 0.24 m^2 area for every 100 mL of a culture flask. As the beads are buoyant, these can be easily used with a spinner culture vessel [Fig. 27.12(b)]. Further, adherent cells can be grown to high densities with the help of micro-carrier beads.
- **Spinner cultures:** Spinner-type vessels are used for scaling up the production of suspension cells or anchorage-dependent cells attached to micro-carrier beads. Spinner is a type of bioreactor having a suspended Teflon impeller, stirrer or similar device to agitate the medium when placed on a magnetic stirrer. The vessel is usually made of glass or stainless steel, with port holes to accommodate sensors, medium input and gas flow [Fig. 27.12(b)]. Commercial versions incorporate one or more side arms for sampling and/or decantation. As cells are not allowed to settle to the bottom of the flask, they can be grown to very high densities. Further, stirring of the medium improves gas exchange.

27.7 Quality control of recombinant products

It is extremely important to do analytical and pharmacokinetic characterization of protein biopharmaceuticals produced through rDNA technology before these could be marketed. Specific assays have been designed

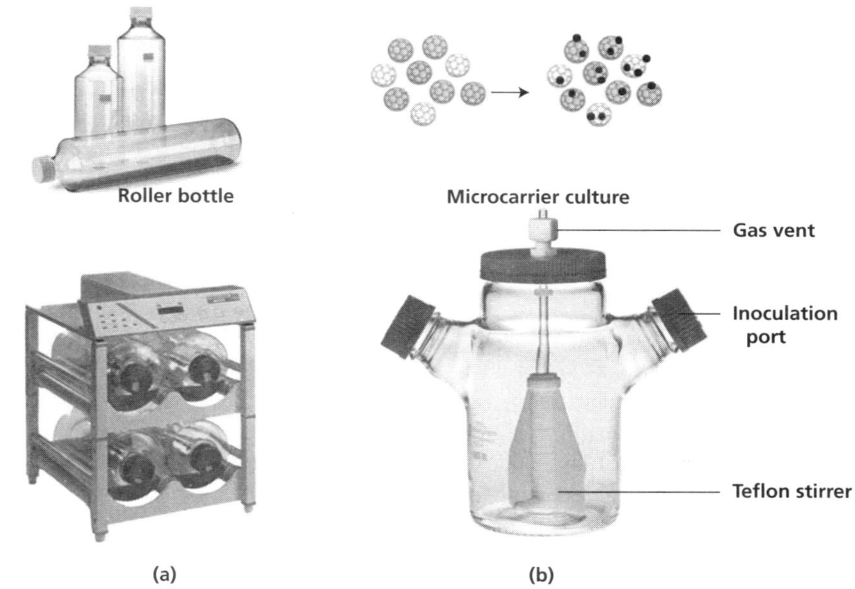

Figure 27.12 Scale-up culture of animal cells. (a) Roller incubator. (b) Spinner flask.

to establish the identity, quality, purity and quantity of recombinant proteins, mAbs, vaccines, antigens, allergens and biosimilars and perform pharmacokinetic and toxicokinetic studies.

27.7.1 Analytical procedures

Combinations of specific analytical procedures are performed to ascertain different properties of bioproducts. For example, the antibody characterization involves:

- Heavy- and light-chain molecular weight determination by mass spectrometry (MS)
- Peptide mapping by MS of heavy and light chains
- Sequencing of variable complementary-determining regions
- N- and C-terminal sequencing of heavy and light chains by matrix-assisted laser desorption ionization–in-source decay (MALDI-ISD)
- Edman N-terminal sequencing
- Electrophoretic analysis by 1D and 2D polyacrylamide gel electrophoresis (PAGE)
- Analysis of common post-translational modifications–deamidation, oxidation, pyroglutamate, N-glycosylation
- Quantification by liquid chromatography–mass spectrometry (LC-MS)/MS
- Host cell protein identification assay

Likewise, antigen and vaccine characterization involves establishing the identity and purity of the antigen in question using the following methods:

- Mass spectrometric peptide mapping with high sequence coverage
- N-terminal Edman sequencing
- C-terminal sequencing by MS
- Molecular weight determination of intact protein isoforms
- Western blot analysis by specific antibodies
- Quantitative amino acid analysis
- HPLC profiling
- Quantification by LC-MS/MS with isotope dilution

The analytical characterization of antigens is even more relevant in the case of subunit or complex protein-based vaccines. Further, it is important to do MS and proteomics-based analysis for the identity, potency and stability of each preparation, combined with documentation of antigen concentrations and comparison of different batch preparations.

27.7.2 Pharmacokinetic analysis

Just for any other drug, pharmacokinetic studies are an integral part of human clinical studies to evaluate the efficacy and safety of recombinant protein biopharmaceuticals. This is required to ascertain the relationship between administered dose, the observed biological fluid concentration of the drug and time. These methods are used to measure the bio-product concentration in biological fluids (serum, plasma, urine) or target tissue collected at different timepoints. These studies are performed in compliance with good laboratory practice (GLP).

27.7.3 Toxicokinetic analysis

Toxicokinetic studies are performed to understand the relationship between administered drug dose and toxicity. It is a regulatory requirement done in accordance with OECD (Organization for Economic

Co-operation and Development) guidelines to assess systemic exposure in toxicology studies, that is, in pre-clinical animal studies for setting human plasma concentration limits and safety margins. Quantification of systemic exposure is often represented by plasma, serum or blood concentration of the compound. Toxicokinetic studies also need to be performed in consistence with GLP.

27.7.4 Purity analysis

When human proteins are expressed recombinantly in another organism, the purified protein always contains trace amounts of contaminating host proteins from the expression organism. Such process-related impurities are identified and evaluated qualitatively and quantitatively using procedures such as 1D sodium dodecyl sulphate polyacrylamide gel electrophoresis (SDS-PAGE), combined with silver staining of the gel, followed by protein identification using mass spectrometric peptide mapping. The twin procedures offer an attractive supplement to the enzyme-linked immunosorbent assay (ELISA)-based protein assay as these are less time-consuming and do not involve development of specific antibodies. The 1D SDS-PAGE is helpful in visualizing the individual host protein bands and allows batch-to-batch comparison of the relative amounts. The protein bands are then cut out from the gel, digested with trypsin and the peptide mixture is analysed by MS. Then, the Mascot database is used to establish the identity of the individual host protein bands. This database includes protein/peptide sequences from different host systems, such as yeast, *E. coli*, sf9 insect cells, rice and CHO cells.

27.8 Applications of animal cell technology

Cell culture is one of the major tools used in cellular and molecular biology, providing excellent model systems to study the normal physiology and biochemistry of cells (e.g. cell cycle, metabolic studies), the effects of drugs and toxic compounds on the cells, and mutagenesis and carcinogenesis (e.g. apoptosis, genetoxicity). It is also used in drug screening and development, and large-scale manufacturing of biological compounds (e.g. therapeutic proteins and vaccines). The main advantage of using animal cell culture in the above applications is the consistency and reproducibility of results obtained through batch culture of clonal cells. However, some changes in the characteristics of these cells after a period of continuous growth are an important concern.

When a biopharmaceutical is developed, the institution or company will apply for a patent. Then exclusive or non-exclusive rights are granted to manufacturing companies, followed by its marketing. This is the primary means by which the developer of the drug can recover the investment cost for development of the biopharmaceutical. The first therapeutic recombinant biopharmaceutical for human use was 'human' insulin. The recombinant human insulin (rHI), also known as Humulin, was developed by the US pharmaceutical giant Genentech and licensed to Eli Lilly for manufacturing and marketing in 1982. The global biopharmaceutical industry has come a long way since the development of Humulin. Today, more than 300 biopharmaceuticals have already been approved and many more are in the late stage of clinical development, including blood factors, colony-stimulating factors, enzymes, growth factors, growth hormones, immunoglobulins, insulin, interferons, interleukins and mAbs. Some important milestones in the therapeutic use of recombinant protein biopharmaceuticals are given in Box 27.3, while those produced using animal cell culture and rDNA technology are listed in Table 27.3. These drugs have not only advanced the prevention and treatment of a number of life-threatening diseases, but have also provided the thrust for the continued success of the pharmaceutical industry. The industry's US$ 92 billion figures and double-digit growth rate in recent years is a testimony of the higher approval rate for investors. According to a global biopharmaceutical market report (International Market Analysis Research and Consulting's group report, 2010–2015), the sale of recombinant protein pharmaceuticals is likely to touch US$ 167 billion by 2015, including a share of $79 billion for mAbs. Next are some examples of recombinant therapeutics that are currently available in the market.

27.8.1 Blood coagulation factors

Haemophilia A is a common heritable genetic disorder where the body lacks the ability to produce factor VIII, required for blood clotting, while haemophilia B or Christmas disease is the second most common type of bleeding disorder due to the deficiency of factor IX – a

Box 27.3 History of therapeutic use of recombinant biopharmaceuticals

1982: Human insulin became the first recombinant protein to be licensed.

1985: Human growth hormone was produced from recombinant bacteria.

1986: Lymphoblastoid gamma-interferon (IFN) was licensed.

1987: Tissue-type plasminogen activator (tPA) from recombinant animal cells became commercially available.

1989: Recombinant erythropoietin came into clincial trial.

1990: Recombinant products came into clinical trial (HBsAg, factor VIII, HIV gp120, CD4, GM-CSF, EGF, mAbs, IL-2).

Table 27.3 Protein pharmaceuticals produced from animal cell culture

Protein	Animal cell line used	Therapeutic use
Erythropoietin (EPO)	CHO cells	Anaemia
Factor VIII	CHO cells	Haemophilia A
Factor IX	CHO cells	Haemophilia B
Follicle stimulating hormone (FSH)	CHO cells	Infertility
Human growth hormone (hGH)	CHO cells	GH deficiency
Interleukin 2 (IL2)	CHO cells	Cancer therapy
Tissue plasminogenactivator (tPA)	CHO cells	Stroke
Monoclonal antibodies (mAbs)	Hybridoma cells	Cancer therapy and autoimmune diseases

serine protease of the blood coagulation system. Many patients with haemophilia are treated with factor VIII or IX. Traditionally, factors VIII and IX were isolated from the plasma. However, due to the short supply of plasma, increased blood testing and quality demands on donor's blood in the light of the potential danger from HIV, hepatitis and some unknown viruses, it increasingly became difficult to produce these factors from human blood. Further, some of the patients treated with conventional factor VIII or IX products developed inhibitors against these drugs. The production of factors VIII and IX using animal cell technology will ensure sufficient virus-free supplies.

In principle, another coagulation factor, factor VII, could also be used for treating haemophilia patients. As factor VII is available only in minute quantities in our body fluids, it is a non-viable proposition to isolate it from human blood. However, with the help of modern recombinant DNA technology, it has become possible to produce this factor in mammalian cells. The recombinant factor VII is expected to reach the market soon for human use.

27.8.2 Erythropoietin

Erythropoietin (EPO) is a glycoprotein hormone essential for RBC production and wound healing. It stimulates the bone marrow to produce more red cells and, thus, increases the oxygen-carrying capacity of the blood. Its requirement increases in the case of anaemia. Therefore, EPO is useful in the treatment of anaemia caused due to cancer, chronic renal failure and even AIDS. Earlier, 2500 L of urine from patients with aplastic anaemia had to be processed to obtain small quantities of EPO. Now, with the aid of genetically modified cells, it is possible to produce sufficient EPO to treat many thousands of patients with anaemia associated with renal failure. The recombinant human EPO (r-HuEPO) is now produced using CHO cell lines. The use of r-HuEPO is advantageous over blood transfusion as it does not require donors or blood transfusion facilities and is free of risk for transfusion-associated diseases.

27.8.3 Follicle-stimulating hormone

Follicle-stimulating hormone (FSH) is a glycoprotein hormone produced by the anterior pituitary gland. It regulates the development, growth, pubertal maturation and reproductive processes of the human body along with luteinizing hormone. Low levels of FSH may result in failure of gonadal functions. Recombinant FSH produced in CHO cell lines is used commonly in infertility therapy to stimulate follicular development, notably in *in vitro* fertilization (IVF) therapy, as well as with interuterine insemination.

27.8.4 Growth hormone

Human growth hormone (hGH) is a peptide hormone produced by the anterior pituitary gland and responsible for regulating normal body growth by stimulating cell division and regeneration in the body. GH deficiency is linked with growth failure and short stature in children and delayed sexual maturity. hGH is used for the treatment of dwarfism. Initially, hGH was obtained from the pituitary glands of deceased humans, before it could be produced from genetically modified animal cells. The quantity of hormone produced by human pituitaries was only sufficient to treat some, but not all, patients. In addition, this production method involved a risk of transferring other infectious agents. The current hGH production method using genetically modified animal cells enables the treatment of all patients and is a much safer product.

27.8.5 Interleukin-2

Interleukin-2 (IL-2) is a type of cytokine produced by lymphocytes as a natural immune response in the body. It is secreted in response to microbial infections to increase growth and activity of other T- and B-cells and activate maturation of the immune response. The recombinant form of IL-2 is used in the case of cancers, chronic viral infections and adjuvant for vaccines.

27.8.6 Tissue plasminogen activator

Tissue plasminogen activator (tPA) is a thrombolytic (clot-dissolving) agent produced by the endothelial lining of blood vessels. It is a serine protease that catalyses the conversion of plasminogen to plasmin, which triggers the dissolving blood clots. It is quite effective in patients of pulmonary embolism and myocardial infarction, provided it is given within a few hours after symptoms begin. In different multicentric clinical trials, it has been shown that the impact of such thrombolytic therapy is significant. tPA is another product that could not be previously made in sufficient quantities to treat these patients. However, through biotechnological production of recombinant tPA in animal cell cultures, it has been made possible to produce sufficient quantities of tPA for clinical treatment of patients. So far, more than 5,00,000 patients worldwide have been treated with recombinant tPA. The benefits for the patients are higher survival rates and higher reperfusion. There are fewer side effects of tPA as compared with other thrombolytic agents or invasive methods.

27.8.7 Monoclonal antibodies

mAbs antibodies are antibodies with a single specificity that can be produced *in vitro* by culturing the hybridoma cells or *in vivo* as mouse ascites tumours. Because of their specificity, mAbs are most suited as diagnostic agents for laboratory tests and were among the first products of the modern biotechnology to enter the market. At present, hundreds of such products are available. Their application ranges from detecting pregnancy to blood group typing prior to blood transfusion as well as detection of hormone deficiencies. The therapeutic applications of mAbs are just starting, and many applications are being evaluated in the clinic. Developments with respect to therapeutic mAbs have already made significant advancements in the treatment of cancer, AIDS and other diseases. Some products are already on the market, for example, mAbs for the treatment of transplant rejection and for the diagnosis of some cancers. Another example of the therapeutic use of mAbs is the 'magic bullet' concept for treating cancer using a toxin or an enzyme pro-drug coupled to a specific antibody. The specific antibody binds to specific antigens on cell surface, leading to the accumulation of toxin or pro-drug in tumour cells and their eventual death.

27.9 Concluding remarks

Animal cell culture is a powerful tool in the hands of biologists. It has not only allowed us to do fundamental studies such as understanding of cell metabolism and cell cycle control but is also very useful in the testing effects of drugs and toxins, and production of many important diagnostics and therapeutics. The animal cells can be cultured and scaled up to provide large quantities of the product of interest. Some of the therapeutics would never have been available for clinical use because they could not be extracted from the natural sources in good quantities. Besides, mammalian cells offer special advantage of producing biopharmaceuticals with correct post-translational modifications, making them most suitable to get desired pharmacological effects without being immunogenic. Biopharmaceutical industries are currently facing the challenge of producing biological recombinant medical products in a cost-effective manner, while meeting the safety standards formulated by the regulatory agencies. Recent new developments in the area such as better methods of gene delivery into cells and better protocols for establishing stable cell lines combined with new tools for scale-up culture have given a major thrust to animal cell biotechnology. The protein biopharmaceutical industry has been growing at an amazing pace in recent years, suggesting that it holds a bright future.

Web resources

http://www.atcc.org/

http://flashserver.sac.ac.uk/biology_sac/cellbiol_index.htm

http://www.scribd.com/doc/427679/Introduction-to-Cell-and-Tissue-Culture

References

Carrel A, Burrows MT. Cultivation of tissues *in vitro* and its technique. *J Exp Med* 1911;13:387–396.

Earle WR, Schilling EL, Bryant JC, Evans VJ. The growth of pure strain L cells in fluid-suspension cultures. *J Nat Cancer Inst* 1954;14:1159–1171.

Eagle, H. Nutritional needs of mammalian cells in tissue culture. *Science* 1955;122:501–514.

Evans MJ, Kaufman MH. Establishment in culture of pluripotential cells from mouse embryos. *Nature* 1981;292:154–156.

Felgner PL, Gadek TR, Holm M, Roman R, Chan HW, Wenz M, *et al.*Lipofection: A highly efficient, lipid-mediated DNA-transfection procedure. *Proc Natl Acad Sci USA* 1987;84:7413–7417.

Goff S, Berg P. Construction of hybrid viruses containing SV40 and lambda phage DNA segments and their propagation in cultured monkey cells. *Cell* 1976;9:695–705.

Graham FL, van der Eb AJ. A new technique for the assay of infectivity of human adenovirus 5 DNA. *Virology* 1973;52: 288–293.

Ham RG. Clonal growth of mammalian cells in a chemically defined synthetic medium. *Proc Natl Acad Sci USA* 1965;53:7413–7417.

Harrison RG. Observations on the living developing nerve fiber. *Proc Soc Exp Biol* 1907;4:140–143.

Harrison RG. The outgrowth of the nerve fiber as a mode of protoplasmic extension. *J Exp Zool* 1910;9:787–846.

Hayflick L, Moorhead PS. The serial cultivation of human diploid cell strains. *Exp Cell Res* 1961;25:585–621.

Ian Freshney R. *Culture of Animal Cells: A Manual of Basic Technique and Specialized Applications*. Hoboken, NJ: Wiley-Blackwell; 2010

Lewis MR, Lewis WH. The cultivation of tissues from chick embryos in solutions of NaCl, $CaCl_2$, KCl and $NaHCO_3$. *Anat Rec* 1911;5: 277–293.

Martin GR. Isolation of a pluripotent cell line from early mouse embryos cultured in medium conditioned by teratocarcinoma stem cells. *Proc Natl Acad Sci USA* 1981;78:7634–7638.

McLimans WF, Davis EV, Glover FL, Rake GW. The submerged culture of mammalian cells: The spinner culture. *J Immunol* 1957;79:428–433.

McLimans WF, Giardinello FE, Davis EV, Kuchera CJ, Rake GW. The submerged culture of mammalian cells: The five-liter fermentor. *J Bacteriol* 1957;74:768–774.

Mueller G, Graessmann A, Graessmann G. Mapping of early SV40 specific functions by microinjection of different early viral DNA fragments. *Cell* 1978;15:579–585.

Rous P, Jones FS. A method for obtaining suspensions of living cells from the fixed tissues, and for the plating out of individual cells. *J Exp Med* 1916;23:549–555.

Shamblott MJ, Axelman J, Wang S, Bugg EM, Littlefield JW, Donovan PF, *et al.* Derivation of pluripotent stem cells from cultured human primordial germ cells. *Proc Natl Acad Sci USA* 1998;95 13726–13731.

Thomson JA, Iskovitz-Eldor J, Shapiro SS, Waknitz MA, Swiergiel JJ, Marshall VS, *et al.* Embryonic stem cell lines derived from human blastocysts. *Science* 1998;282:1445–1447.

van Wezel AL. Growth of cell-strains and primary cells on micro-carriers in homogeneous culture. *Nature* 1967;216:64–65.

Wong TK, Neumann E. Electric field mediated gene transfer. *Biochem Biophys Res Commun* 1982;107:584–587.

28

Plants as host for recombinant DNA

Arun Kumar Sharma and Manoj K Sharma

28.1 Introduction

Plants are our best friends since our existence and are backbone of all the life on earth, as they convert sun's energy to chemical energy that feeds the world. Thousands of years ago, human societies began to understand gathering by agriculture. In ancient times, human beings could not read or write, even then they started a scientific discipline 'plant domestication', that is, genetic modification of plants to create new forms of plants, which transformed human life. Today, thousands of plant species are used for food, feed and medicines to cure acute and chronic health problems. Advances in plant genetics and genomics have revolutionized our understanding of plant biology considerably. We are now beginning to use this knowledge to expedite the improvement of existing agricultural commodities or modification of specialized constituents of plants such as carbohydrates, proteins, oils, fats or vitamins. A relatively new bioscience 'plant molecular farming' has emerged and is gaining momentum. It involves the genetic modification of the host plant through the purposeful addition of a gene or group of genes (that hold the information for synthesizing a biomolecule) to the plant genome. For this purpose, genes of interest and nucleotide sequences required for its expression are stitched together as transcriptional units and inserted in target genome. The concept of molecular farming was first recognized during late 1980s, when tobacco plants were genetically engineered to be used as bioreactors. Hiatt and colleagues (1989) reported the production of immunoglobulin fragments and their assembly in tobacco, followed by a large number of proof-of-principle studies that reported the ability of several plant species to make recombinant proteins (Fischer *et al.* 1999). The promising research outcomes soon led to great commercial interests (Ma *et al.* 2003). A large number of products are currently being produced in plants that include bioactive peptides, vaccines, antigens, antibodies, diagnostic proteins, nutritional supplements, enzymes and biodegradable plastics. The economic drivers that are burgeoning this industry include the ability of plants to synthesize animal proteins, cost benefits, safety issues (lack of pathogen contamination) and easy scale-up (Fischer *et al.* 2004; Sharma and Sharma 2009; Egelkrout *et al.* 2012). Apart from many advantages, there are several challenges associated with using plants as bioreactors that include differences in glycosylation patterns in plants and humans, low expression level of recombinant biomolecules and environmental contamination. More extensive research efforts are required to optimize plant-based production systems and address the concerns. The hierarchy of events involved in the process of using plants as bioreactors include (i) choice of the host species and expression system, (ii) sequence optimization for enhanced expression levels of the biomolecule, (iii) engineering strategies, (iv) methods for efficient downstream processing, (v) plant transformation and regeneration of plants expressing the desired protein, (vi) identification and stabilization of the plant lines for commercial production and (vii) purification and characterization of recombinant proteins. Recombinant DNA has been engineered in plants for a variety of purposes that include germplasm improvement and molecular farming (Fig. 28.1).

This chapter discusses the recent developments, opportunities and challenges involved in using plants and plant cells for molecular farming.

28.2 Transgene expression strategies

Transgene is any DNA sequence that is transferred from one organism to another that does not carry it naturally. For the expression of a transgene, it should be provided with all the appropriate elements that are critical for expression such as promoter, coding sequence, transcriptional stop sequence and translational start codon. Transgenes may be integrated into the host genome and inherited by next generations or can be transferred to plant cells where they use the host's machinery to express themselves without integrating into the host

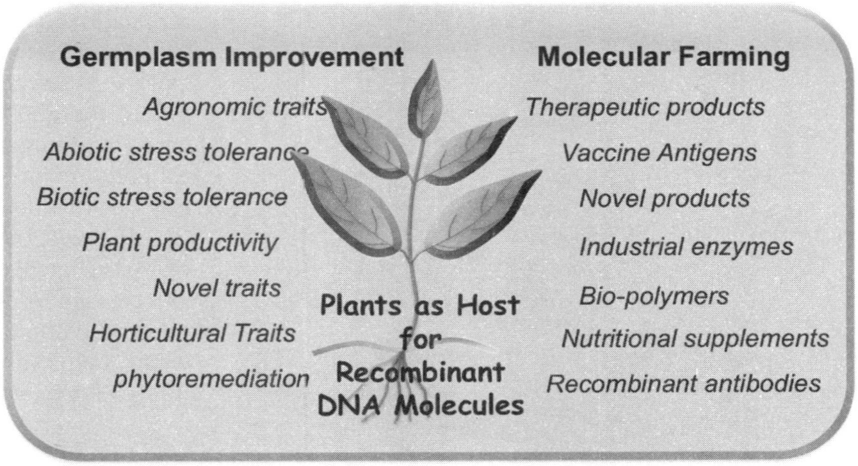

Figure 28.1 Plants as host for various types of recombinant DNA molecules.

genome and, therefore, are not inherited by the next generation.

28.2.1 Stable expression

Stable plant transformation involves the integration of foreign DNA into the nuclear or organellar genome to generate stable transgenic plants, transplastomic plants or plant cell lines. It involves tissue culture that facilitates regeneration of whole plants from a genetically modified single cell in the presence of hormones and selectable markers such as antibiotic resistance genes. Research findings in early 1980s identified Ti-plasmid of *Agrobacterium* as a vehicle to carry foreign DNA to plant cells. Soon after, first demonstration of plant transformation and inheritance of the transgene was reported (Herrera-Estrella *et al.* 1982; Horsch *et al.* 1984). Since then, several gene transfer methods have been developed for stable integration of foreign DNA into the plant genome. These include *Agrobacterium*-mediated transformation, microprojectile bombardment, microinjection, electroporation-mediated transformation, pollen tube pathway and chemical treatment of protoplasts (polyethylene glycol [PEG]/liposome mediated or silicon carbide mediated). However, *Agrobacterium*-mediated genetic transformation is the most popular and routinely used method to generate transgenic plants. The whole process of *Agrobacterium*-mediated transformation involves several stages (Fig. 28.2), and various strategies have been employed to optimize transformation efficiency. In most of such attempts, either the exposure time of plant cells to *Agrobacterium* was optimized or the tissue was wounded prior to inoculation either by bombarding with small particles, coating *Agrobacterium* onto microprojectiles before bombardment or sonication of embryogenic cell cultures. The floral dip method, developed for *Arabidopsis* transformation, bypasses the time-consuming tissue culture. Young flowering shoots are dipped in *Agrobacterium* solution and then grown to yield seeds. Because the gynoecium is open in *Arabidopsis* flowers, *Agrobacterium* colonizes in the interior of the gynoecium and transforms cell lineages that develop into the female gametophyte. Seeds are grown and transgenic seedlings are selected.

Although crops of major interest have been successfully transformed, many others remain recalcitrant to transformation using *Agrobacterium*. Optimization of transformation technologies is required to establish a routine transformation protocol for these crops. Microprojectile bombardment is the method of choice when the target species is resistant to *Agrobacterium* infections. Recently, Broothaerts and colleagues (2005) showed that several other species of bacteria other than *Agrobacterium* could be modified to mediate gene transfer in plants. A typical plant cell has two types of genomes, that is, nuclear genome and organellar genome. The foreign DNA can be inserted in either of them.

28.2.1.1 Nuclear transformation

Nuclear transformation is the most common method of plant transformation that involves integration of foreign DNA into the nuclear genome and its subsequent transfer to next generations as a native trait. *Agrobacterium*-mediated transformation is the most common method used to generate transgenic plants by nuclear transformation. Expression from recombinant DNA is directly influenced by the genomic environment of the host plant and may vary depending upon the site of integration of T-DNA. Several methodologies are developed to minimize the variation in transgene expression in nuclear-transformed plants (Butaye *et al.* 2005). Significant progress has been made in optimizing transcription or translation of transgenes as well as post-translational modification of the recombinant products and this has been discussed in Section 28.4.

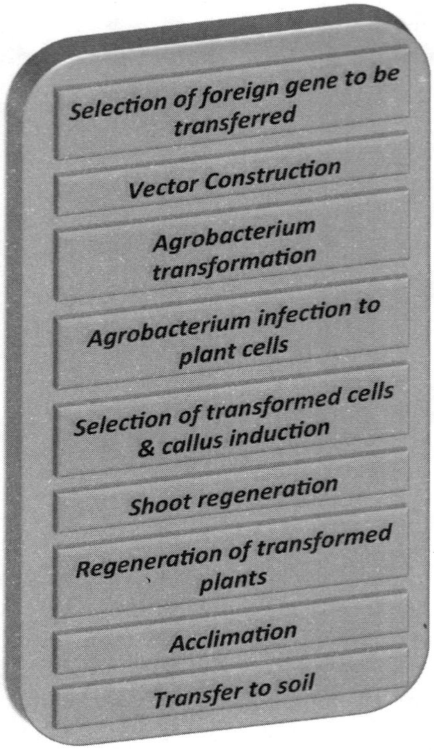

Figure 28.2 Steps for transfer of foreign genes into plants.

Nuclear-encoded genes are translated in the cytoplasm, but the protein may be targeted to various subcellular compartments or secreted out of the cell.

28.2.1.2 Plastid transformation

Plastids are semi-autonomous organelles and their various categories include chloroplasts, chromoplasts and leucoplasts. They have their own genome as well as transcription–translation machinery. Expression of recombinant DNA from the plastid genome offers an attractive alternative to nuclear integration of transgenes. Plastids transformation was reported for the first time during early 1990s when Svab and colleagues (1990) transferred recombinant DNA to tobacco plastids using particle bombardment. Microprojectile bombardment is the most common method for plant transformation, though protoplast-mediated plastid transformation has also been achieved. The vectors used for plastid transformation are *E. coli* plasmid derivatives that carry the gene expression cassette and targeting sequences flanking the gene expression cassette. These flanking sequences are used for integration into the plastid genome through homologous recombination, at a predetermined location in its genome (Verma and Daniell 2007). If any transformation vector lacks the plastid replication origin, it is lost with cell division, but if it has the plastid-based origin of replication, it is maintained in plastids as an extra chromosomal element and is called a shuttle vector. Practically, shuttle vectors are not very successful as they are lost in the absence of selection pressures. Plants that carry foreign gene sequences integrated to their plastid genome are called transplastomic plants and offer several advantages over conventional transgenic plants (Cardi *et al.* 2010). One of the most important advantages is that plastids provide natural containment for the transgene. Further, each plant cell carries several plastids (>50), and each plastid has several genomes, which helps to achieve an exceptionally high level of foreign protein accumulation.

28.2.2 Transient expression

Transient expression refers to the expression of a gene in a heterologous system without its integration into the host chromosomal DNA. Expression of transgenes from these extragenomic transcriptional cassettes provides a rapid alternative and recombinant product that can be detected as early as 3 h from its delivery (Jones *et al.* 2009; Shepherd 2009). Generally, the expression level reaches to its maximum by the second day and may persist for more than a week. During the last decade, the use of transient expression systems gained momentum and the production of several biomolecules was reported (Komarova *et al.* 2010). The major advantages of transient expression systems over stable transgenic plants are convenience to work, reduced overall cost of production and significantly reduced time to recover the recombinant product. Because the transgene is not integrated into the host chromosomes, it does not have to face undesirable consequences of stable integration such as position effects. Further, they have been demonstrated to be safe and environment friendly. Transient expression is achieved either through agroinfiltration or plant virus–based vectors and is described here.

28.2.2.1 Agroinfiltration method

The *Agrobacterium*-based transient expression system involves infiltration of recombinant *Agrobacterium tumefaciens* into live tissues/cells that facilitate the transfer of T-DNA to cells. Janssen and Gardner (1990) reported *Agrobacterium*-based transient expression in plant cells for the first time and also found that transient expression was several fold higher than expression from a transgene stably integrated into the host genome. Binary vectors are used for this purpose. A binary vector has a T-DNA segment that carries the expression cassette, including the gene of interest, and is delimited by specialized border sequences. The region of binary vector other than T-DNA comprises the vector backbone that carries the sequences responsible for its own replication and selectable marker genes (Lee and Gelvin

2008). Agroinfiltration of plant leaves is the most widely exploited approach for transient expression. *Agrobacterium*-based T-DNA delivery to plant roots for transient expression of transgenes has also been reported (Skarjinskaia *et al.* 2008; Yang *et al.* 2008) and is seen as advantageous over leaf infiltration (Komarova *et al.* 2010). Several factors can influence the yield and quantity of the recombinant product, for example, silencing effects of the host genome on the transgene, host specificity of codons used in the transgene, activity of the promoter used to drive transgene expression and, finally, stability of the recombinant product in a heterologous system (for details, see Section 28.3).

28.2.2.2 Plant virus–based infection method

With technical advances in the field, viral-based transient expression systems are emerging as a wonderful platform for the production of recombinant biomolecules. The unique characteristics of viruses, including strong infectivity, ease of manipulation, exceptionally high level of amplification, etc., are being utilized for the development of robust transient expression systems. Viruses with positive-sense RNA-type genomes are of specific interest as they are directly translated upon entry into plant cells. The DNA sequence of interest is inserted into plants as part of the viral vector that is replicated and systemically transported throughout the plant. It results in the production of a very high level of recombinant products. Several viruses belonging to different virus families have been modified to carry genes of interest and overexpress them in plants (Komarova *et al.* 2010). These include tobacco mosaic virus, tomato mosaic virus, cauliflower mosaic virus, potato virus X, tobacco etch virus, cucumber mosaic virus, cowpea mosaic virus, bean yellow dwarf virus, etc. Based upon the strategies used to design, viral vectors are generally categorized into substitution, insertion, deconstructed and peptide display vectors (Lico *et al.* 2008). A cauliflower mosaic virus (CaMV)–based substitution vector has been used successfully to express foreign proteins in plants. The strategy used for the construction of substitution vectors is to replace existing non-essential viral genes with foreign genes of interest. For example, the gene coding for the coat protein (CP) has been replaced with a heterologous gene of interest (Scholthof *et al.* 1993). However, insertion vectors are constructed by adding foreign genes to the viral genome without deleting any native gene. These vectors behave as functional viruses and are capable of cell-to-cell and systemic movement as well as infection. The viral gene encoding for CP has also been modified to fuse coding sequences of foreign genes of interest for their heterologous expression (Johnson *et al.* 1997; Porta and Lomonossoff 1998; Wu *et al.* 2003; Yang *et al.* 2007) and foreign proteins are expressed as chimeric proteins with CP. CP may or may not be required for cell-to-cell and systemic movement. Therefore, substituting the CP with the gene of interest may result in defects in systemic movement. For the successful expression of the heterologous protein, it should not interfere with viral assembly or movement. This method is improved by making CP stop codon to be leaky or inserting a small peptide from foot and mouth disease virus between the CP gene and the epitope-encoding sequence (Hamamoto *et al.* 1993; Turpen *et al.* 1995; Borovsky *et al.* 2006). Cowpea mosaic virus was first to be investigated for peptide display (Usha *et al.* 1993). Since then, several other viruses have been used to develop the epitope/peptide presentation system (Canizares *et al.* 2005; Lico *et al.* 2008).

Research has unfolded several mysteries about viral genomes, and several strategies have been developed to boost the expression of foreign genes using viral vectors. Various genomic components of a viral genome can be broken down into several units, maintaining their infectivity when they operate together. These include components involved in replication, cell-to-cell movement, systemic spread, assembly of viral particles and components required for the infectious process. This fact that these viral sequences need not be together in a single piece of DNA opened another avenue of opportunities and led to development of modular/deconstructed vectors. During the construction of these vectors, various components of a viral genome are inserted into binary vectors. The replicative portion of the viral genome is separated, and because of its small size, it can accommodate large segments of foreign DNA. Other components of the viral genome required for the infection can be provided in *trans*. CP helps the virus to move systemically in the plant. Magnifection, a combination of transient expression methods, has been developed where the CP-encoding gene is removed and viral vectors are delivered to the entire plant using *Agrobacterium* in mild vacuum (Gleba *et al.* 2005, 2007; Lico *et al.* 2008). *Agrobacterium* strains carrying these binary plasmids are mixed together and used for infection on plants. Magnifection relies on transient amplification of systematically delivered viral vectors (Gleba *et al.* 2005). This technique combines the vector efficiency and systemic delivery capability of *Agrobacterium* and rapid accumulation and high-level expression of viral vector–based transiently expressed genes. It involves the use of modified plant virus vectors where putative cryptic splice sites were removed and multiple plant introns were inserted (Komarova *et al.* 2010). In the process, the whole plant is infected with a diluted suspension of *Agrobacteria* carrying T-DNAs, which encode RNA replicons. Bacteria initiate primary infection and systematic movement throughout the plant, and the viral vector provides cell-to-cell spread, amplification and high-level expression. The magnifection-based expression system has been tested by expressing several biomolecules that include interferons, growth hormones, monoclonal antibodies (mAbs), single-chain antibodies, enzymes, bacterial and viral antigens (Gleba *et al.* 2005). Scale-up is simple and requires a simple apparatus for high-throughput vacuum infiltration of whole plants. Recently, the process has been used to produce full-size mAbs (Giritch *et al.* 2006) and immunogenic fusion protein F1-V from *Yersinia pestis* (Rosales-Mendoza *et al.* 2010). To express hetero-oligomeric proteins, Giritch and colleagues exploited the capability of viral vectors from different taxonomic groups to replicate within the same cell. They coexpressed two polypeptides antibody chains using two different non-competing viral vectors. Target recombinant protein was assembled into functional hetero-oligomeric units and expression levels of up to 100-fold were achieved.

28.3 Search for a suitable host

A large number of plant species are amenable to genetic manipulation and have been used for the production of recombinant biomolecules (Rybicki 2009; Sharma and Sharma 2009). These include rice, wheat, soybean, maize, *Arabidopsis*, tobacco, tomato, potato, banana, beans, pea, carrot, lettuce, spinach, alfalfa, duckweed, strawberries, aloe, single-celled algae, etc. The performance of a plant host depends upon several biological as well as geographical factors. Therefore, each host needs to be evaluated separately. The choice of the host species is directly associated with the type of recombinant product in question, life cycle of the host plant, transformation efficiency, availability of species-specific promoter sequences, biomass yields, containment and scale-up investments. Success largely depends upon the understanding of biological environments, that is, species- or tissue-specific factors that affect the accumulation of recombinant products and their quality. Generally, domesticated species are preferred over the wild species for molecular farming as they were adapted to a wide range of environmental conditions, availability of transformation and regeneration methods, more predictive and synchronous development and higher yields. But wild species can be more advantageous as they will not be mistaken for food crops and also address the issue of food crop mix-up with transgenic material. Further, the self-pollinating species are advantageous for the containment of transgenic crops compared with cross-pollinating crops. The issue can also be addressed by using plants that can be grown in containment, for example, alfalfa, which can be grown in green houses, and plant cell cultures that can be grown on an industrial scale using fermenters. For the proteins that need to be purified, downstream processing may affect the overall production cost to a large extent. Therefore, hosts from which extraction and purification of the recombinant protein is easy and economic have an edge over other species. To further optimize the expression system, a deeper understanding of the biochemical nature of target product is required.

28.3.1 Food crops

Most of the earlier work towards plant-based commercial production of recombinant proteins has been done in food crops, mostly because the availability of efficient transformation procedures and more options for targeting and delivery of the recombinant proteins make food crops alluring. The first transgenic plant–derived recombinant product to reach the market place was a diagnostic enzyme (Hood *et al.* 1997). Maize-produced bacterial beta-glucuronidase and egg white avidin were commercialized in late 1990s by collaborative efforts of biotech firms Prodigene and Sigma-Aldrich. Currently, major crops used for this purpose include maize, rice, barley, wheat, tomato, potato, soybean, peas, carrots and oilseed. Rice and maize are easy to transform and their production scale-up is easy. However, with large biomass yields, maize is still considered cost-efficient, while rice, in spite of the lower yield, is considered as a better choice due to self-pollination. Ventria Biosciences has developed a transgenic rice grain–based production platform and is making human lysozyme and lactoferin. Recently reported preclinical trial results showed that the recombinant product is safe and well tolerated. As targeting recombinant proteins to oil bodies helps in easy downstream purification of the recombinant products and may reduce the overall product cost significantly, oil crop safflower is also promoted for this purpose. It is a minor food crop and is easily transformable.

Although there are several arguments in favour of using food crops, but a large group of people are concerned about the contamination of the food or feed chain with genetically modified (GM) crops. Public opposition grew specifically after Prodi Gene's accidental contamination incident in 2002. This company tested its GM maize to make trypsin in the field. After harvest of the GM crop, this land was used to grow conventional soybean for human consumption. GM maize seed left in the field grew with soybean and contaminated soy crop. This led to the imposition of additional regulation by regulatory agencies for any release or field trials of plant-made recombinant product (http://www.ventria.com/content/news/549/; accessed on 9 July 2012).

28.3.2 Non-food/Non-cultivated crops

The use of non-food crops such as tobacco and non-cultivated crops such as algae, mosses, duckweed, etc. for molecular farming will provide containment advantages and address the issue of food chain contamination. Challenges are far more than opportunities, as we do not have established cultivation practices for these crops. Some of them produce toxins and we do not know about the genetics of the non-cultivated species. Tobacco, which is easy to transform and manipulate, is the system of choice for the production of recombinant products among the non-food crops category. It produces a very high biomass. Planet Biotechnology has developed a tobacco-based production platform for developing antibodies (http://www.planetbiotechnology.com/; accessed on 9 July 2012).

Greenovation BioTech is using a *Physcomitrella*-based production platform for the commercial production of recombinant proteins (http://www.greenovation.com/bryotechnology.html; accessed on 9 July 2012). *Physcomitrella* can be easily grown and has a high growth rate. Being easy to transform, *Lemna* (Duckweed) has been used as the model organism for several years. *Lemna* is a free-floating aquatic plant and grows very fast. Under ideal conditions for its growth, *Lemna* can double its biomass as early as 48 h. It has been engineered to humanize recombinant proteins, that is, to remove plant-specific glycans and add human-specific protein modifications. Biolex Therapeutics developed the *Lemna*-based LEX platform and has produced several mAbs and interferons. Recently, Synthon, a pharmaceutical company, acquired the LEX-based system from Biolex Therapeutics (http://www.synthon.com/en/nieuwsarchief/synthon-acquires-lex-systemsm-from-biolex-therapeutics.aspx; accessed on 28 August 2012). Being a flowering plant, *Lemna* produces pollen and, therefore, strict containment systems are required to rule out the contamination of natural environments. Having a short generation time and a rapid growth rate, algae have also been explored for the production of biopharmaceutical proteins (Rasala *et al.* 2010; Specht *et al.* 2010). Photosynthetic single-cell algae, *Chlamydomonas reinhardtii*, has been transformed to produce mammalian proteins and has been demonstrated as a promising platform for the production of recombinant biomolecules.

28.3.3 Cell suspension cultures

The use of plant cell cultures for the production of foreign proteins is gaining momentum and this is a step towards making plant molecular farming a safer industry. Plant cell culture lines have been established for a few plants such as tobacco, *Arabidopsis*, rice, carrot, maize and wheat and offer a plant-based cheaper alternative to mammalian cells. The main advantage is containment, that is, plant cell suspension cultures can be grown in self-contained production systems, thereby addressing the issue of transgene flow in natural ecosystems and contamination of food/feed chain. The maintenance of plant cell cultures is very cheap as compared with their mammalian counterpart because plant cells need a relatively simple growth medium. Further, the capability to secrete the recombinant product into the medium helps in the purification of the product and reduces the cost of downstream processing as compared with whole plant production systems. There are challenges that need to be addressed to make them viable at commercial scale. The genetic instability of the recombinant product and its low yield are the major concerns. Various designs for bioreactors offer an optimal environment for effective cell growth and synthesis of foreign biomolecules. Several bioreactor systems have been designed that include reusable, disposable, moss and hairy root bioreactor systems (Huang and McDonald 2012). With new technological developments, different systems are being optimized and the production of several foreign proteins, including growth factors, antibodies, hormones, etc., in plant cell cultures has been demonstrated in recent years. Recently, the US Food and Drug Administration has approved 'Elelyso', an enzyme produced in genetically engineered carrot cells, developed by Protalix Bio-Therapeutics (Fox 2012). This enzyme is used to treat Gaucher disease in humans. It is the first plant-made human biologics that is approved for human use. Earlier, the first commercially approved plant-made pharmaceutical produce 'Poultry Vaccine' was also produced in plant cell cultures (Obembe *et al.* 2011). Plant cell culture technology is making progress, but the need of the hour is to put more effort to develop and characterize more cell lines from various plant species.

28.4 Optimizing the expression

While several plants have been demonstrated to support the expression of heterologous proteins, their expression levels are not enough for commercialization. There are several parameters that can influence protein expression levels such as strength of the promoter, destabilization sequences, cryptic splice sites, translational regulation, etc. The following sections discuss these parameters under five categories, including transcriptional regulation, optimization of the genetic code and transcript stability, translational regulation, protein targeting, and novel expression methodologies.

28.4.1 Transcriptional regulation

Transcription is the primary level of gene regulation and several constituents of its machinery affect the transcription rate. Therefore, it is necessary to optimize the gene of interest towards transcriptional requirements of the host plant. Various factors that influence transcription are discussed here.

28.4.1.1 Selection of promoter

In any transcriptional unit, promoter is a key regulatory element that is required to initiate transcription, and its *cis*-elements directly affect the rate of transcription.

Therefore, promoter biology has received considerable attention from the research community, which has led to the identification of a wide selection of ubiquitous or species-specific promoters. In general, promoters can be classified into the following categories based on their activity and specificity.

Constitutive promoters

The promoters that induce the expression of genes throughout the system irrespective of the environment or developmental factors are called as constitutive promoters. They are generally used for the production of recombinant proteins in all the tissues of a plant. The cauliflower mosaic virus *35S* promoter (Odell *et al.* 1985) has been used extensively for this purpose and high levels of expression of recombinant proteins have been achieved (Gutierrez-Ortega *et al.* 2005). It is more effective in dicots than monocots, probably because of differences in the quality/quantity of regulatory factors. Several other prokaryote- or virus-derived promoters such as nopaline synthase gene promoter, tobacco cryptic constitutive promoter and mac promoter (hybrid of mannopine synthase promoter and CaMV promoter) have been identified. Several plant-derived promoters such as ubiquitin promoter from maize, actin 1 promoter from rice and alcohol dehydrogenase 1 promoter from maize are commonly used plant-derived constitutive promoters.

Tissue/Organ-specific promoters

These promoters drive gene expression either in a tissue or in a developmental stage–specific manner. Recombinant products may have unintended effects on the host plant. The use of tissue-specific promoters limits the presence of recombinant proteins to specific tissues such as seeds, leaving all other parts of the host plant unaffected. Jefferson and colleagues (1990) identified patatin promoter as tuber specific and used it to express several biomolecules in potato tubers. Several promoters that drive the expression of seed storage proteins are seed specific. The earliest identified seed-specific promoters include globulin-1 and zein promoters from maize, arcelin promoter from common bean and P-conglycinin a1-subunit promoter from soybean. The arcelin 5-1 promoter of common bean is a strong constitutive promoter, and when used to drive the expression of the single-chain variable fragment (ScFv), it resulted in the accumulation of ScFv up to 36% of total soluble protein (TSP) (De Jaeger *et al.* 2002). Several promoters, including the promoter of globulin, glutelin and prolamin of rice, hordein genes from barley and zein protein genes from maize, drive gene expression in an endosperm-specific manner and have been used to express several biomolecules. Several promoters that drive gene expression in leaves, flowers or floral parts, mesophyll and bundle sheath cells, epidermis, vascular tissues, fruits, and roots have been described (Egelkrout *et al.* 2012). To express a recombinant product in chloroplasts, the most commonly used promoters are the 16S ribosomal RNA promoter (*Prrn*) and *psbA* gene promoter.

Inducible promoters

Promoters whose transcriptional activity is directly modulated either by chemical compounds or by environmental factors belong to this category. The use of these promoters is a powerful tool in genetic engineering as they can be turned on/off as needed. Several promoters that either respond to chemicals such as alcohol, tetracycline and steroids or environmental factors such as desiccation, salt stress, temperature and light have been identified. A chemically inducible promoter system offers high inducibility and specificity of regulated gene expression and is an attractive alternative to constitutive expression of foreign biomolecules that interfere with the growth and development of the host. For a highly efficient system, an inducer should have the features of high specificity, non-toxic nature, quick on/off response and easy applicability and should not be present in the plant naturally. These promoters are quiescent in the absence of an inducer. Uses of chemical-responsive transcription factors (TFs) with an inducible promoter can further restrict the production of foreign biomolecules in specific organs, tissues or even cell types (Zuo and Chua 2000). A lot of effort has been invested to identify inducible promoters and several inducible systems, including tetracycline and ecdysone-, oestrogen-, glucocorticoid- or ethanol-based systems, have been identified and used in plants (Padidam *et al.* 2003). In general, an inducible system consists of two transcription units. The promoter of the first transcription unit responds to the chemical inducer and initiates the synthesis of a TF that recognizes a specific sequence in the second transcriptional unit and activates transcription. Several inducible promoter systems have been shown to work in plants. These include stress-inducible rd29-based, sucrose starvation–inducible and mechanical stress–inducible promoters. A versatile two-component system was developed to have a precise control of expression of target gene with respect to tissue type and time. For this, a chimeric TF, XVE, was constructed using six different promoters/enhancers, each of which had a different tissue specificity. In this system, XVE is used to efficiently and faithfully activate target gene in a variety of cell types. Using this system, a conditional transactivation was achieved in a predictable tissue-specific manner (Brand *et al.* 2006). Selection of the promoter will depend upon the production platform and recombinant product in question.

Artificial promoters

In addition to indispensable elements of a promoter such as TATA box, transcription start site, etc., promoters have a diverse range of elements that either up- or down regulate its activity. With demonstration that promoter regulatory elements such as enhancers or suppressors can regulate the expression of heterologous promoters, the idea of synthesizing promoters artificially came into existence and is a promising strategy to achieve higher expression levels. Artificial promoters have been synthesized in the past by combinatorial engineering of *cis*-elements, which include enhancers, activators or repressors, upstream to core promoters. The CaMV *35S* promoter, engineered to carry more than one copy of its enhancer element, boosts up the expression level of target gene and has been used to produce several foreign biomolecules in plants (Sharma and Sharma 2009). Initially, synthetic promoters were designed to characterize enhancer–promoter interactions or understand metabolic regulatory complexities. However, their ability to amplify target gene expression by several folds has made them a promising choice to drive recombinant DNA expression. Galvin and associates tested many combinations for regulatory sequences from octopine synthase and mannopine synthase gene promoters, and a hybrid promoter '(Aocs)3Amas Pmas or super promoter' was constructed by combining a triple repeat of *ocs* activator sequence, *mas* activator element and *mas* promoter (Ni *et al.* 1995). Super promoter has been used for the expression of SARS Coronavirus spike protein in tomato as well as tobacco (Pogrebnyak *et al.* 2005). High-throughput sequencing methods such as RNA-seq, microarray, etc. are very helpful to identify new promoters having constitutive expression or ones that express in specific tissues or specific stages of development.

Bidirectional promoters

Promoters that can regulate the expression of two flanking genes simultaneously are known as bidirectional promoters. They are capable of transcribing a gene upstream as well as downstream. The construction of a bidirectional promoter was reported for the first time in 1995, when Baron and colleagues used seven *tet* operator sequences flanked by two minimal promoters to drive expression of two reporter genes. However, in plants, the first ever use of bidirectional promoters was reported by Xie and colleagues (2001). Another bidirectional promoter was synthesized that carries multiple *cis*-regulatory DNA sequence elements arranged to construct a transcription initiation module (Chaturvedi *et al.* 2006). Recently, Dhadi and colleagues (2009) have identified several natural bidirectional promoters from rice, *Arabidopsis* and *Populus* by computational sequence analysis. Very little is known about the organization, regulatory mechanisms and biological significance of bidirectional promoters in plants. A deeper understanding of the regulatory motifs and their organization in bidirectional promoters will open a whole new avenue for development and engineering of new and highly efficient bidirectional promoters in plants.

28.4.1.2 Transgene copy number

Research efforts have been put to generate transgenic plants carrying multiple transgene copies under the control of same or different promoters and an increase in transgene expression levels has been reported (Egelkrout *et al.* 2012). An increase in the number of transcriptional units carrying the same transgene does not always work additively and may result in homology-dependent gene silencing. Several methods have been used to increase the expression levels of transgenes by increasing their copy number. Particle bombardment–mediated transformation often generates transgenic plants with multiple copies of the transgene, either at the same or at different locations. A more effective way to increase transgene copy number is to cross homozygous stable transgenic lines generated from independent transformation events. Although in late 1990s, it was shown that selfing heterozygous plants can increase transgene copies, recently, Choi and colleagues (2009) at UC Berkeley crossed independently transformed homozygous plants. This resulted in a progeny expressing multiple transgene copies obtained from parents with an overall additive effect on expression levels. Further chances to boost the expression level are higher as multiple copies of the transgene would be stably expressing in the parents used for crossing. Moreover, stable transgenic plants can be retransformed using the same construct (single/multicopy construct), thereby increasing the transgene copy number and expression levels. Overall, the concept of engineering several copies of transgenes to increase expression levels appears to be promising.

28.4.1.3 Transcription factor–binding sites

As discussed in Section 28.4.1.1, transcription factors can be used to fine-tune transgene expression. Because the presence of engineered TF activates the production of recombinant proteins, levels of the recombinant product directly depend upon the availability of that TF, its amount and timing of expression. If the engineered TF recognizes host promoters and binds to them, its availability to target promoter may become limiting and results in a low level of transgene expression. Therefore, selection of TF that does not recognize host promoters can direct higher level of foreign protein accumulation. The transcription unit encoding the TF can be transferred to host plant with the gene of interest using one construct or can be supplied in *trans* also. Kasuga and colleagues (2004) overexpressed DREB1A, which interacts with dehydration-responsive promoter elements and induces expression of the gene derived by that promoter. They used the stress-inducible rd29A promoter and reported overexpression of DREB1A-controlled stress-inducible genes in response to environmental stress. Engineering artificial transcription factors is another interesting approach to control the expression of transgenes. In 1990s, it was clear that the regulatory domain of a transcription factor can be separated from its DNA binding domain and fused to the other DNA binding domain to regulate gene targets of the new binding domain. It led to the development of artificial transcription factors. Although this is more common in animals, the strategy has also been demonstrated to work with plant systems. Stege and colleagues used a synthetic Zn finger DNA binding domain fused to VP16 activation domain to regulate target endogenous genes (Stege *et al.* 2002).

28.4.1.4 Transcriptional gene silencing

Transcriptional gene silencing (TGS) is very common occurrence that may result from transgene methylation. There are several mechanisms that trigger transgene methylation. Although the sequence of events in transgene methylation is poorly understood, most generalized reasons include the presence of dsRNA complementary to the promoter, pairing of transgene copies and hairpin loops. Therefore, potential hairpin loops should be avoided while designing the constructs. Further cryptic splice sites formed by AU-rich regions should not be included. Scaffolds/matrix attachment regions are the DNA sequences that are thought to influence gene expression and their use with transgenes has been shown to stabilize transgene expression. They are supposed to place the surrounding loci in the regions of the genome that are suitable for recruitment of TF to the promoter. Further, they bind specifically to the nucleoskeleton and have been found to increase transformation efficiency and reduce variations in expression (Vain *et al.* 2002). Based upon the knowledge gained from naturally occurring matrix attachment regions (MAR), artificial MARs have been synthesized and tested for their effect on transgene expression with regard to their orientation, promoter type and different species. Post-transcriptional gene silencing is a surveillance mechanism that evolved in eukaryotes to fight against viral infections and may result in transgene silencing. Some viruses have evolved and developed ways to overcome this silencing effect. They deploy suppressors for this purpose. Engineering of these viral suppressors along with the gene of interest can be an effective alternative to overcome post-transcriptional transgene silencing. Voinnet *et al.* (2003) described a transient expression system where co-expression of silencing suppressor protein p19 of Tomato Bushy Stunt Virus (TBSV) could enhance the expression of several proteins by 50 folds or more.

28.4.1.5 Use of enhancers

Enhancers are DNA sequences that have been evolved to augment the rate of transcription from a promoter or provide tissue specificity to its target. They can function from a long distance and activate target transcription independent of their orientation; that is, they may be located either upstream or downstream of the gene they regulate. Because of this orientation-independent nature, they can provoke non-specific enhancer–promoter interactions, specifically in the constructs carrying multiple genes or promoters. This cross talk may alter the specificity and strength of the promoters (Blackwood and Kadonaga 1998; Singer *et al.* 2010a). Enhancer-insulating sequences provide a viable alternative to negate these non-specific effects. A 1-kb sequence of bacteriophage lambda has been identified as an enhancer-insulating sequence and can potentially be utilized to minimize enhancer–promoter interferences (Singer *et al.* 2010b).

28.4.2 Genetic code optimization and mRNA stability

Because of the extensive third-position degeneracy of the genetic code, a species-specific codon bias has evolved during the course of evolution. Therefore, synonymous codons used for any given amino acid may differ between different species. Similarly, tRNAs have also evolved to serve existing codons in a species, and therefore, if the codon usage of the transgene differs from the heterologous host, it may not be translated efficiently. Optimization of the coding sequence according to the host plant is another parameter to enhance recombinant protein expression and is now routinely used for transgene expression studies using heterologous systems. It has been reported that the presence of rare codons may lead to the formation of the secondary structure and slow down or pause ribosome movement through the region. As a strong correlation exists between the frequency of codon usage and the presence of their cognate tRNA in a species, a cluster of rare codons may block the translation completely as all the tRNAs are trapped in the protein complexes. Replacement of such undesired signals to fit the tRNA availability dramatically increases the expression of transgenes in plants. With the availability of whole-genome sequences, the codon usage information is available for several species or can be generated easily using software programs such as CoGe Pedia (http://genomevolution.org/CoGe/; accessed on 28 August 2012). When codons of the gene encoding cholera toxin B subunit were optimized according to tobacco codon usage, it resulted in a 15-fold improvement in expression level compared with non-optimized codons in the same conditions (Kang *et al.* 2004). Further, sequence within the coding region may have a regulatory role and any change during codon optimization may have profound effects on its regulation. Therefore, any changes made for codon optimization should be examined carefully. A sequence within the coding region may be important for post-transcriptional or post-translational processing and any change in this region may contribute negatively, resulting in unintended consequences. In plants, signals that target the protein to vacuoles where it is degraded because of lytic activity have been identified in the coding region. Such motifs may be present anywhere in the pro-peptides. Several such sequence elements are known, but there is no consensus sequence that represents the vacuolar sorting signal (Vitale and Hinz 2005).

Stability of mRNA directly affects recombinant product accumulation. Typically, the lifespan of mRNA ranges from a few minutes to a few days. Both 5′- and 3′-untranslated regions (UTRs) of mRNA are important for its stability and also play an important role in translation. Monocot- or dicot-specific 5′-UTRs may function as translational enhancers (Sugio *et al.* 2008). Several upstream reading frames that have key translational regulatory properties are known. The first ATG codon in optimal context is the exclusive site of initiation even when the second initiation codon is positioned just a few bases downstream. This was true even when the second initiation was in the same favourable context. Further, the sequence surrounding the initiation codon is important for translation initiation. In most of the high-expressing plant genes, alanine follows N-terminal methionine. When used after methionine in a transgene, it augmented the expression of the reporter gene in an incremental fashion with the insertion of each successive codon in tobacco, and also increased the stability of the reporter protein by two folds. With the construct having Met–Ala–Ser–Ser–GUS, a 30- to 40-fold increase in GUS activity was reported (Sawant *et al.* 2001). Plant-preferred translation initiation context has been used for the expression of foreign biomolecules to optimize transgene expression in plants (Stomp *et al.* 2004; Mishra *et al.* 2006; Sharma *et al.* 2008a; 2008b).

28.4.3 Post-translational modifications

A higher rate of transcript accumulation may not result in a higher amount of target protein. After translation, a protein may undergo several modifications that include glycosylation, phosphorylation, methylation, etc. and is prone to degradation by proteases or ubiquitin-mediated proteolytic activity. This can be a major hurdle for recombinant protein accumulation. The development of protease-deficient lines can be an alternative, but in plants, there are hundreds of proteins that can hydrolyse peptide bonds and it is almost impossible to generate protease-deficient plant lines. Alternatively, the use of plants expressing protease inhibitors can improve recombinant protein accumulation (Rivard *et al.* 2006). Fusion to thermostable proteins is another attractive recombinant protein–stabilizing strategy because the higher thermal stability of a protein is directly correlated with its resistance to proteolysis. Several studies showed increased accumulation of target protein as a fusion partner. HIV p25 fused to human IgA heavy-chain constant region accumulated 13 folds as compared with when it was expressed alone.

For recombinant protein synthesis, several heterologous systems are used such as bacteria, yeast, insect cells, mammalian cells or plants. These expression systems differ in their post-translational machinery, which has its own advantages and disadvantages. Prokaryotes lack the glycosylation machinery and, therefore, cannot be used to synthesize eukaryotic proteins that require glycan residues for their activity. In eukaryotes, glycosylation occurs in the endoplasmic reticulum (ER) and Golgi apparatus. The glycosylation machinery of plants is very similar to mammals except that they do not add galactose and terminal sialic acids but add plant-specific $\alpha(1,3)$-fucose and $\beta(1,6)$-xylose residues, which are not desirable. Several plant engineering strategies are used to humanize the plant glycosylation machinery and are discussed in Section 28.7.

28.4.4 Targeting of the protein

The majority of proteins are synthesized in the cytosol and then directed to subcellular compartments such as the ER, Golgi complex and hydrolytic compartments of the cell or secreted out from the cell. Depending upon the advantages and disadvantages in terms of yield, purification and biosafety, there are several options available for protein targeting. The ER, protein storage vacuoles and oil bodies are the desired targets for recombinant protein accumulation in transgenic plants.

28.4.4.1 Endoplasmic reticulum

The ER of plant cells can tolerate unusually high accumulation of proteins without compromising plant development and reproduction. The first report of successful targeting of any cytosolic protein to ER appeared in late 1990s when Iturriaga *et al.* (1989) used a signal peptide of patatin from potato to direct reporter protein β-glucuronidase. Analysis of several ER resident proteins determined that a C-terminal tetrapeptide (KDEL/HDEL) was responsible for retaining these proteins in the ER. Proteins having these C-terminal tetrapeptides are recognized by a receptor located in the Golgi complex and are directed back to the ER (Gomord *et al.* 1997). This property of ER retention is often used to accumulate proteins in the ER. To target protein to the ER, a signal peptide at the N-terminal end that directs that protein to the ER after its synthesis in the cytosol and a C-terminal retention signal to retain the protein in the ER are added. An increased accumulation of foreign proteins in the ER indicates that the ER compartment has low hydrolytic activity and is very plastic. Prolamins are also ER resident proteins but lack the

KDEL/HDEL tetrapeptide. Another ER retention signal was identified from γ-zein, a prolamin from maize. The retention sequence of γ-zein is highly efficient and its use increased the accumulation of the foreign protein by approximately seven folds compared with protein accumulation using the KDEL retention signal. Zein protein induces the formation of protein bodies, which may help in a higher level of protein accumulation (see Section 28.5.2).

28.4.4.2 Oil bodies

These are small intracellular organelles that store oil (triglycerides) and are surrounded by a phospholipid monolayer that is in direct contact with the lipid content. The phospholipid membrane is equipped with an abundant amount of a specialized lipophilic protein, oleosin. The potential of oleosins to carry recombinant proteins was investigated in mid-1990s, for the first time, when Gijs and colleagues (1995) demonstrated the expression of beta-glucuronidase–fused oleosin in seeds of *Brassica napus*. Fusion proteins expressed using translational fusions with oleosins remained tightly associated with the oil bodies. Recombinant oil bodies facilitate the separation of the recombinant protein from other cellular proteins by flotation centrifugation, taking advantage of their low density. To separate the fused protein from oleosins, a specific protease cleavage site can be inserted between oleosin and the desired recombinant protein. The engineered oil bodies with their associated proteins can be used as affinity matrices for the selective, non-covalent binding of desired target molecules. For this, the oil body proteins may be genetically fused to a ligand having specificity for the desired target molecule. The expression of the recombinant protein as translational fusion with oleosin protein exposes the recombinant protein in the cytosol, but at the same time, it protects the foreign protein from cytosolic degradation. Because the fusion protein is not exposed to the environment of the ER lumen, as an added advantage, it avoids the post-translational modifications. The use of C-terminal hydrophobic region of tail-anchored (TA) proteins also targets the recombinant protein to the membrane system without passing through the ER lumen, thereby avoiding post-translational modifications in the ER lumen. TA proteins lack the N-terminal signal peptide and are synthesized on free ribosomes in the cytosol. As the C-terminal region emerges from the ribosome only upon translation termination, it is very unlikely that the C-terminal region interacts with the signal recognition particle that binds only signal anchors as long as they are part of a nascent polypeptide (Maggio *et al.* 2007). The TA domain of mammalian cytochrome b5 was used to express recombinant antigens. The results indicate that these molecules are more stable than the cytosolic counterparts and may provide a good strategy for preventing the rapid degradation of recombinant proteins, thus improving the yield (Vitale and Pedrazzini 2005).

28.4.4.3 Plastids

As described in Section 28.2.1.2, plastids are transformable and the expression of genes in plastids has several advantages over nuclear-based protein expression. Each cell contains a large number of plastids, ~100 chloroplasts per cell, and each chloroplast contains about 100 genomes. Therefore, their transformation permits the introduction of thousands of copies of transgenes per plant cell. It dramatically enhances the protein production in the cell. Several proteins have been expressed in plastids and an increased foreign protein expression has been reported and reviewed by Scotti *et al.* (2012). Exceptionally high levels of accumulation are reported for several heterologous proteins that include cardiotrophin-1 (4.8%), endolysin Pal (30%), insulin-like growth factor-1 (32%), *Bt* Cry2Aa2 (46%), fusion product of cholera toxin B subunit and proinsulin (53%) and endolysin PlyGBS (70%).

The use of synthetic genes with plastid-optimized codons has played an important role in enhancing protein expression levels. Plastoglobules, the subchloroplastic compartment of the chloroplast is another destination to target recombinant proteins. They are low-density lipoprotein particles that are attached to the thylakoid membrane (Vidi *et al.* 2007). Like the oil bodies, they are light in weight and contain only a few proteins, which is advantageous for purification. The major disadvantage of transplastomic-based molecular farming is that the protein synthesis machinery in the chloroplast does not add sugar residues to the recombinant protein backbone, which is required for the activity of several mammalian proteins. Transplastomic plants have been generated in the past to engineer agricultural traits (McBride *et al.* 1995; Liu *et al.* 2008), metabolic pathways (Hasunuma *et al.* 2008; Apel *and* Bock 2009) and expression of recombinant products (Daniell *et al.* 2009; Scotti *et al.* 2012). Recently, research efforts have been put towards plastid transformation in algae also (Franklin *et al.* 2002). Because of the very high level of recombinant proteins in transplastomic plants, they can be an attractive platform for the production of industrial enzymes and can play a crucial role to make ligno-cellulosic biofuels an economically viable alternative to fossil fuels.

28.4.4.4 Plant vacuoles

Vacuoles in plant cells are multifunctional membrane-bound organelles within the cytoplasm. They not only maintain the cell turgor but also store proteins and secondary metabolites. Vacuoles in the leaf cells are generally lytic type and can degrade the recombinant protein. Generally, vacuolar sorting in plants is mediated by short amino acid sequences and recombinant proteins engineered to use a plant's secretory pathway are prone to vacuolar delivery, leading to their degradation. There is no canonical consensus and the signals are very diverse; therefore, it is difficult to make predictions. However, seeds of higher plants accumulate a large amount of proteins in specialized vacuoles that are called protein storage vacuoles (PSVs). Two pathways have been suggested for the storage proteins to reach into PSVs. Storage proteins can form protein bodies on ER itself that are transported to PSVs, as in cereals, or storage proteins are transported in soluble form to PSV by vesicle traffic mediated by the Golgi apparatus, where they make protein bodies, as is suggested for legumes. PSVs are an attractive target for recombinant protein accumulation. Signal sequences that are responsible for targeting proteins to protein-storing vacuoles have been identified, but no consensus protein-sorting signal has been optimized. Rice has both ER-stored and vacuole-stored proteins. Therefore, it is possible that the higher accumulation of these proteins has a dominant effect on the localization of foreign proteins expressed in the secretory pathway (Vitale and Pedrazzini 2005). The recombinant lysozyme when expressed in rice endosperm and recombinant human serum albumin when expressed in wheat endosperm were accumulated in PSVs (Yang *et al.* 2003; Arcalis *et al.* 2004). The mechanisms or signals required to store recombinant proteins in vacuoles need further exploration before designing strategies to target recombinant proteins to PSVs.

28.4.5 Novel expression systems

Because of the heterogeneous chromatin structure, transcriptional activity is not uniform throughout the eukaryotic genomes. Therefore, the same transgene can show variable transgene expression under the influence of the chromatin environment at the site of integration. Several strategies are being developed to integrate transgenes as a single copy at a predetermined site of the genome. Recombinase-mediated transgene integration has been used to produce transgenic plants having single-copy inserts at precise locations in genomic regions. To achieve this, a unique single recombinase recognition site is used to target T-DNA. Two plasmids (one carrying a recombinase expression cassette and the other carrying T-DNA having a recognition site) are co-transformed. Other than Cre, Flp and R, recently, some novel recombination systems have also been identified for this purpose that include R4, PhiC31, TP901-1 and Bxb1.

Targeted DNA double-strand breaks and their homology-dependent repair is another strategy. Zinc finger nucleases (ZFNs) are used to produce site-specific cleavage in DNA. DNA segments carrying multiple ZFN sites can be transferred to plants, and transgenic lines carrying T-DNA in transcriptionally active regions are identified. Using ZFNs that recognize the engineered binding sites, T-DNA containing the gene of interest can be integrated to this location. ZFNs are designed to bind predetermined gene sequences and they are used in combination with T-DNA for targeted transgene integration. Cai and colleagues (2009) used an endochitinase gene as target for T-DNA integration using ZFNs. Although most of the integration events occurred randomly, but in ~10% of the integration events, the endochitinase gene was disrupted. Identification of genomic regions for best transcription activity and screening for ZFNs that specifically recognize these regions can have a tremendous impact on transgene expression.

Engineering *trans*-acting factors can also be used to enhance expression levels of transgenes. These factors either bind directly to the promoter that drives the expression of target gene or interact with other factors, recruiting them to target promoter. Yang *et al.* (2001) engineered bZIP TF, REB (rice endosperm bZIP), along with the human lysozyme gene driven by the rice globulin promoter with the REB upstream activation sequence. The expression of REB enhanced the lysozyme expression levels. Similarly, prokaryotic TFs can be supplied in *trans* to enhance the expression of the transgene targeted to plastids.

Bacteriophage T7 RNA polymerase (T7 RNAP) is a DNA-dependent RNA polymerase that has an exceptionally high specificity. T7 RNAP-based transcription does not require additional host factors for promoter recognition. It has also been demonstrated to direct foreign protein production in plants and significantly high expression levels are achieved (Nguyen *et al.* 2004). Further, transgene expression levels were found to be uniform among independent transformation events. T7-RNAP-based expression system can be optimized further for tissue specific and inducible expression of foreign proteins.

28.5 Strategies for efficient downstream processing

Recovery of recombinant proteins from plant-based systems is another challenge and it is believed that about 70–80% of the total production cost is contributed by downstream processing of the recombinant product. Therefore, along with a high level of transgene expression to provide good yields in plant-based production systems, efficient recovery of the recombinant proteins must also be optimized. There are several undesired biomolecules (proteins, oils, phenolics, etc.) that need to be removed during purification. The processing of plant tissues for the recovery of recombinant proteins generally includes fractionation of plant tissues, extraction of recombinant proteins in an aqueous medium and purification. Various strategies used to improve upon the efficiency of downstream processing include targeting the recombinant protein to oil bodies, protein bodies and the secretory system, and the use of affinity tags and phase transition, etc.

28.5.1 Oil bodies

As described in Section 28.4.4.2, oil bodies are small intracellular organelles that store oil and their membrane contain an abundant protein, oleosin. Oleosins can accumulate up to 10% of total seed protein content. The open reading frame of the recombinant protein–encoding gene can be fused with the open reading frame of oleosin and the product is expressed as a translational fusion to oleosin protein. Oil bodies can be readily separated from other protein-rich subcellular compartments by flotation centrifugation, thereby reducing the cost as well as the time required for purification.

28.5.2 Protein bodies

Plants store large reservoirs of proteins in specialized intracellular ER-derived compartments, protein bodies (PBs). This unique ability of plants has been used for the expression of foreign biomolecules. This encapsulation of heterologous proteins into stable intracellular organelles may exclude them from normal physiological turnover, a component of the protein quality control system. It is hypothesized that PBs not only protect proteins from proteolytic degradation but also protect the cell from the toxic effects associated with a high accumulation of heterologous proteins. Because protein bodies are insoluble, it has an added advantage, that is, purification by centrifugation, thereby reducing the downstream processing cost (Conley *et al.* 2011).

The major storage protein of maize, γ-zein, is capable of inducing the formation of ER-derived protein bodies. Phaseolin is a major storage protein of beans and accumulates in storage vacuoles. Mainieri and colleagues synthesized a chimeric protein, zeolin, from the N-terminal half of γ-zein and phaseolin protein. When expressed in tobacco, it accumulated in PBs up to 3.5% of TSP (Mainieri *et al.* 2004). This unique ability of γ-zein (Zara^R, from ERA Biotech, Barcelona, Spain) and zeolin has been shown to improve expression levels of several biomolecules, which include human growth hormone, epidermal growth factor, F1-v antigen of *Yersinia pestis*, etc. To compare γ-zein and zeolin, a highly unstable HIV antigen Nef was fused for plant expression. It was found that γ-zein–Nef could not escape from ER quality control and was degraded; however, zeolin–Nef formed PBs and the fusion protein accumulated up to 1.5% of TSP. Packaging of recombinant proteins in dense PBs also helps in pre-purification enrichment by centrifugation methods. By sucrose density centrifugation, PBs can be separated from cell components to up to 85% purity in a single step. ERA Biotech is developing low-speed direct centrifugation method for substantially concentrating PBs (Torrent *et al.* 2009).

The use of an elastin-like polypeptide (ELP) tag also induces the formation of PBs (Conley *et al.* 2009b). ELPs are derived from mammalian elastin proteins and composed of a repeating pentapeptide 'VPGXG'. Upon increase in temperature, ELPs undergo a reversible phase transition from the soluble form into insoluble hydrophobic aggregates. It makes the downstream purification of recombinant biomolecules rapid, scalable and inexpensive. A minimum number of 20 pentapeptide repeats are required for the ELP tag to selectively enrich the fusion protein. In general, ELP tags having >80 pentapeptide repeats result in the highest recovery of the recombinant protein (Conley *et al.* 2009a). In addition to purification benefits, ELP tags significantly enhance recombinant protein accumulation as has been recorded for scFv antibody, spider silk proteins, interleukin-4, etc. The orientation of the ELP tag is also crucial for its activity. The C-terminal placement of the ELP tag results in significantly higher levels of recombinant protein accumulation as compared with its N-terminal placement. Further, ELPs may protect the recombinant protein against irreversible aggregation and denaturation. ELPs are tolerant to hydrolysis and protease cleavage (Zhang *et al.* 1996; Raucher and Chilkoti 2001); therefore, they may stabilize the recombinant protein attached to it. Recently, another class of proteins 'hydrophobins' was hypothesized to induce the formation of protein bodies in plants. Hydrophobins are a class of small surface-active proteins produced by filamentous fungi.

28.5.3 Phase transition

Aqueous two-phase systems have received considerable attention for separation and purification of proteins. These are formed of either water-soluble two structurally distinct polymers or a polymer and a phase-forming salt. Partitioning behaviour is affected by several factors that include type and concentration of polymer/salt, temperature and pH of the system. Polymer-salt-based systems are less expensive, have shorter separation times, handle large biomass loads and are suitable for large scale-up operations. Shorter separation time is crucial for the quality of recombinant proteins, as it will reduce the time for which recombinant proteins are exposed to deleterious effects of plant extracts. Therefore, they are widely used for bioseparation of proteins. PEG is a commonly used polymer. It can be used for pre-purification or concentrating the recombinant proteins by removing phenolics, alkaloids and plant debris. Aqueous two-phase extraction (ATPE) has been used to purify recombinant proteins from plants successfully. Recently, Ross and Zhang (2010) investigated its potential for purification of recombinant acidic β-glucuronidase from transgenic tobacco plants. This study resulted in 74% recovery of rGUS from the top PEG-rich phase and more than 90% of the native tobacco proteins were removed. Earlier, Balasubramaniam and colleagues (2003) used ATPE to separate recombinant lysozyme produced in tobacco and reported 79% recovery of the lysozyme.

Hydrophobins, a class of small surface-active proteins, are found in filamentous fungi. These proteins are characterized by the presence of a hydrophobic patch and presence of eight cysteine residues. As a fusion partner, they are capable of altering the hydrophobicity of the fusion partner and thus are being exploited to develop a rapid and inexpensive surfactant-based aqueous two-phase purification system. Joensuu and colleagues (2010) fused GFP to hydrophobin 1 (HFB1) from *Trichoderma reesei* and expressed it in plant cell extracts. Using ATPS, they could recover up to 91% of the expressed protein. Plants expressing HFB1-GFP survived longer than those expressing GFP, thereby indicating that fusion protects the cells from toxicity of highly expressed proteins. Lee and Forciniti (2010) used ATPS to recover recombinant mAbs produced in corn seeds in a three-stage process, where the first two stages were typical phase partitioning with the target of protein enrichment in the aqueous salt phase, followed by precipitation at the two-phase interphase. Using optimized conditions of composition, pH and ionic strength, product purity up to 72% and a yield of 49% were achieved. Overall, the aqueous two-phase partitioning system is a promising strategy for efficient downstream processing.

28.5.4 Use of affinity tags

The use of protein/peptide-based affinity tags is a powerful and highly efficient tool to recover recombinant proteins from crude extracts. Several affinity tags have been described for recombinant protein purification (Li 2011). Ideally, affinity tags should be small and able to purify with high purity and yield. In general, there are two classes of affinity tags. The first class includes epitope tags, which have the affinity to bind to small ligand molecules, and the second class includes peptide tags, which can bind to another protein. Poly-histidine tag is the most commonly used tag for recombinant protein purification. While in microbial systems, it can purify recombinant proteins with high purity, in plants, it is not very successful. The use of Strep II tag is reported to successfully purify recombinant proteins to homogeneity from plants. White and colleagues (2004) compared three affinity tags, that is, eight amino acid tag Strep II, His_6 and 181-amino acid tandem affinity purification (TAP) tag, for their potential to purify its fusion partner. The protein purified using His_6 tag was of low purity, whereas the recombinant proteins having TAP or StrepII tag were purified to homogeneity. While Strep II tag purification achieved yield and high purity that was comparable to that obtained with TAP tag, it was considerably easier and faster. The possibilities of using histamine, tryptamine, phenylamine and tryamine as affinity tags for affinity purification of recombinant proteins have also been evaluated (Platis and Labrou 2009). The use of histamine exhibited a high degree (up to 95%) of purity and >90% recovery. Similarly, HA tag, Myc tag, Flag tag and GST tag are also used for purification purposes.

Another group of fusion tags is self-cleaving tags. These are the protein modules that possess inducible proteolytic activities. When fused to recombinant proteins, they can be triggered to undergo site-specific self-cleavage and separation of the tag from target recombinant protein can be achieved in a single step. Available self-cleaving fusion tags includes inteins, sortase A, N-terminal protease from pestivirus, FrpC and cysteine protease domain (Li 2011). Inteins, the protein-splicing elements, can catalyse their self-cleavage and have been explored to purify recombinant proteins from plants. Target protein can be fused to either the N- or the C-terminal end of the intein, though the C-terminal is more successful. SMAP 29, a mammalian antimicrobial peptide, was expressed with the intein fusion tag in tobacco plants and was purified using a chitin column (Morassutti *et al.* 2002). Recently, the self-cleavage feature of intein was used with ELP tag, which is used to purify proteins via temperature-sensitive phase transition (Section 28.5.3). ELP–intein tag (Ei-tag) was fused to plant lectin (PAL) and rice seeds were used to produce ELP–intein–PAL. Using the ELP–intein system, target protein was purified up to 1.1 mg/g of total seed protein without any complicated instrumentation (Tian and Sun 2011). A major drawback of the system is uncontrolled *in vivo* cleavage, reported in a few cases, which can leave most of the fusion product cleaved during expression itself. Further, the steric hindrance caused by the fusion partner can lead to failure of cleavage even after induction.

28.5.5 Secretory systems

Secretory systems for the production of recombinant proteins in plants can be an effective strategy to reduce overall production cost. When secreted outside the plant cell, recombinant proteins have better accessibility and are easy to purify. The purification can be achieved from cell suspension cultures, root exudates or guttation fluid. Secretion of recombinant proteins outside the cell was reported in early 1990s in tobacco transgenic plants as well as cell suspension cultures (Firek *et al.* 1993). Plant cell suspension cultures are attractive targets for recombinant protein secretion into the culture medium, and several proteins have been investigated targeting them outside the cell. Instability and degradation of the protein in the culture medium, leading to low productivity, is the major obstacle that limits the use of this platform. Various approaches have been investigated in the past to improve the stability and effectiveness of recombinant proteins, which includes the use of protease inhibitors, protein fusions, etc. Xu and colleagues (2007) used a hydroxyproline (Hyp)-rich repetitive peptide tag that directs extensive proline hydroxylation and, subsequently, Hyp-O-glycosylation. The peptide tag is composed of X–Hyp–X–Hyp–X–Hyp repeats, where X can be either Ser or Ala. Hyp-O-glycosylation increases the molecular mass of the recombinant protein and stabilizes it. The authors used this tag to express interferon alpha 2b in tobacco cells and the fusion protein was secreted out in the culture medium. The fusion product accumulated up to 1400 folds compared with the untagged secreted product. Later, they (Xu *et al.* 2010) expressed recombinant growth hormone with this tag and a yield up to 500 folds was achieved in the culture medium.

A natural rhizosecretion-based phenomenon was investigated to secrete recombinant proteins during late 1990s. Green fluorescent protein and human placental-secreted alkaline phosphatase (SEAP) were targeted through rhizosecretion in a hydroponic medium and accumulated up to 20 and 1 μg/g dry root weight/24 h, respectively. Drake and colleagues (2009) expressed mAbs and microbicide polypeptide cyanovirin-N and targeted them through rhizosecretion. It was found that mAbs accumulated up to 58 μg/g root dry weight/24 h and cyanovirin-N accumulated up to 766 μg/g dry root weight/24 h. Further, they found that the application of growth hormone IAA, BAP and kinetin resulted in increased accumulation of the mAbs. Whereas NAA and IAA increased the root dry weight, BAP and kinetin increased the rhizosecretion of the recombinant protein without affecting the dry weight of roots. Other strategies used include co-expression of protease inhibitors and the use of a root-specific promoter and improved accumulation has been achieved using these strategies (Paul and Ma 2011; Paul *et al.* 2011).

The phenomenon of loss of water with dissolved materials from uninjured plant organs is called guttation and can be used to accumulate recombinant proteins. It is labour intensive to collect this fluid, but because it can be collected throughout the plant's life, it provides a continuous source of recombinant proteins. Komarnytsky and colleagues (2000) expressed human placental alkaline phosphatase in tobacco plants, targeting it to the guttation fluid. The recombinant alkaline phosphatase accumulated up to 3% of TSP in the guttation fluid. Tobacco is not an ideal plant for guttation fluid production and other plants that have a higher production of the guttation fluid can be better alternatives for this purpose.

28.6 Multi-gene engineering

Technological developments have made it possible to transfer several genes simultaneously into a plant's genome. It is now possible to manipulate the entire pathway or engineer an entirely new pathway in a single engineering event. Co-transformation allows simultaneous introduction of multiple genes into plants. Transgenes are often integrated at the same locus and, therefore, prevent segregation of these transgenes in progenies. The genes to be transferred to plants are either linked together in one plasmid or are placed on separate plasmids. In both cases, co-transformation works efficiently. Standard binary vectors can transfer up to 20 kb of T-DNA efficiently, but it is challenging to transfer larger T-DNAs. Specialized vectors have been developed to transfer larger fragments of T-DNA to plant genomes. Transformation-competent artificial chromosome–based vectors can transfer up to 80 kb of DNA. Similarly, binary bacterial artificial chromosome–based vectors can transfer up to 150-kb-long DNA fragments (Hamilton *et al.* 1996).

As chloroplasts have evolved from bacteria and have preserved the operon system for gene regulation, they are an attractive target for expression of several genes. De Cosa and colleagues (2001) transformed tobacco plastids with three *cry2Aa2* gene operons from *Bacillus thuringiensis*. The open reading frame upstream to Cry2Aa2 codes for a putative chaperonin component that facilitates its folding. The recombinant protein accumulated up to 45% of TSP. Three genes encoding for enzymes of the β-carotene biosynthetic pathway were introduced in plastids of rice endosperm that resulted in an increase in the β-carotene content. To increase the production of very long-chain polyunsaturated fatty acids such as arachidonic acid or eicosapentaenoic acid, Wu and colleagues (2005) introduced three genes encoding for Δ5-, Δ6 fatty acid desaturase and Δ6 fatty acid elongase into *Brassica*. New fatty acids were detected in transgenic plants. The transfer of these genes improved the overall elongation rate. For the production of n-3 fatty acids, the gene encoding ω3 desaturase was introduced, which resulted in a significant increase in the EPA content. To produce DHA in AA- and EPA-producing transgenic plants, three genes encoding for elongase, Δ4 desaturase and putative lysophosphatidic acid acyltransferase were introduced. There are several instances of successful transfer of multi-genes either by step-wise engineering or by co-transformation (Naqvi *et al.* 2010). Synthesizing mini-chromosomes is another strategy to engineer several genes together. Wada and colleagues (2009) reported a bioactive bead-based (calcium alginate microbeads) efficient transformation method to deliver large DNA fragments.

28.7 Humanizing protein production in plants

Post-translational modification, especially glycosylation, is crucial for the protein activity and its immunogenicity. The process of adding sugar moieties (monomer/oligomers) to the protein backbone is called glycosylation. In general, it is classified into two main categories: N-glycosylation and O-glycosylation. While N-glycans are attached to the amide nitrogen of asparagine (Asn) residues, O-glycans are linked to amino acids containing an OH group, that is, serine, threonine, hydroxylysine and hydroxyproline. In both animals and plants, the glycosylation machinery is distributed in the ER compartment and Golgi apparatus. Plant cells share essentially same initial steps of glycosylation in the ER compartment and add similar glycans that belong to the oligomannose category, that is, high-mannose glycans to the same Asn residues (Chen *et al.* 2005). Oligosaccharides undergo several maturation steps that involve removal of glucose or mannose residues and addition of new residues, thereby synthesizing complex N-glycans, which differ in plants and mammals. Certain plant-specific glycan residues (β-1,2-xylose and α-1,3-fucose) that are absent on human proteins may cause immunological reactions that are of major concern regarding the use of plants as bioreactors for pharmaceutical proteins. Further, plant glycoproteins lack terminal sialic acid, an important component in human glycoproteins, and this may make these proteins short lived and ineffective. To address these concerns, several approaches have been developed and rapid progress has been made towards humanizing plant glycosylation patterns. Because the ER compartment of plants and humans add similar types of glycan structures, retaining glycoproteins in the ER compartment (see Section 28.4.4.1) can avoid further processing and recombinant proteins may lack plant-specific glycans. The other option is to modify the enzymatic machinery of the Golgi apparatus. It can be done either by knocking out genes related to β(1,2)-xylosylation and α(1,3)-fucosylation and/or by adding new glycosyltransferase to modify the processing of glycan structures. Strasser and colleagues (2004) reported the generation of knockout *Arabidopsis* plants with the complete deficiency of β(1,2)-xylosyltransferase (XylT) and α(1,3)-fucosyltranferase. Interestingly, these plants were able to produce N-glycans with two β-N-acetylglucosamine residues but lacked β(1,2)-linked xylose and α(1,3)-linked fucose. RNA interference technology–based shutdown of genes encoding for plant-specific endogenous β-1,2-xylosyltranferase and α-1,3-fucosyltransferase enzymes is an effective and attractive strategy. Another impressive and successful example is the production of MDX-060 mAb in *Lemna minor*, where glycosylation was optimized by co-expression of a RNAi transcript designed to silence endogenous α(1,3)-fucosyltransferase and β(1,2)-xylosyltransferase activities. The mAb with optimized glycosylation had a single major N-glycan species without detectable plant-specific N-glycans. Plant-produced mAb showed better antibody-dependent cell-mediated cytotoxicity and was indistinguishable from the mAb produced in Chinese hamster ovary (CHO) cells (Cox *et al.* 2006). RNAi technology was used to downregulate the endogenous XylT and FucT genes in *Nicotiana benthamiana*, which were used for the expression of anti-HIV monoclonal antibody 2G12 (Strasser *et al.* 2008). *Nicotiana*-produced recombinant mAb showed antigen binding and HIV neutralization activity similar to the mammalian-derived mAb.

Plants do not add terminal galactose, which contributes to the immunological functions and the correct folding of the antibodies. Further, plant glycoproteins lack terminal sialic acids. Therefore, plant-made proteins may have reduced clinical efficiency and may be cleared rapidly. It is possible to make glycoproteins in plants that carry terminal galactose and sialic acid residues by introducing human N-glycan machinery. Earlier, the human β(1,4)-galactosyltranferase (GalT) gene was transferred into *Nicotiana tabacum* and the production of galactosylated N-glycans was reported (Palacpac *et al.* 1999). Recently, Castilho and colleagues (2011) used glycol-engineered *Nicotiana benthamiana*, which lacks plant-specific sugars and expressed erythropoietin and serum transferrin genes. They co-expressed the genes that code for human β 1,4-mannosyl1-β 1,4-N-acetylglucosaminyltransferase (GnTIII,), α1,3-mannosyl-β1,4-N-acetylglucosaminyltrajsferese (GnTIV) and α 1,6-mannosyl-β1,6-N-acetyglucosaminlytransfererase (GnTV) transiently, which led to the generation of bisected tetra-antennary complex N-glycans. Endogenous sialylated glycol conjugates have been found in plants, which means that the sialylation pathway exists in plants. However, Neu5Ac, the major sialic acid present in humans, is not synthesized in plants in detectable amounts. Several sialic acid synthesis pathway components have been successfully engineered in plants producing biologically active recombinant proteins. These include mammalian α-2,6-sialtransferase, a glycosyltransferase of the mammalian trans-Golgi cisternae expressed in *Arabidopsis* plants, human CMP-N acetylneuraminic acid synthetase and CMP-sialic acid transporter in tobacco suspension-cultured cells, genes encoding Neu5Ac lyase from *Escherichia coli* and Neu5Ac synthase (neuB2) from *Campylobacter jejuni* in BY2 cells as well as alfalfa (Sharma and Sharma 2009). Castilho and colleagues (2010) have introduced the entire mammalian sialation biosynthetic pathway in *N. benthamiana*. Six mammalian proteins from the sialation biosynthetic pathway that are required for biosynthesis, activation, transport and transfer of Neu5Ac to the terminal galactose were identified for expression.

Co-expression of these genes with human mAb in *N. benthamiana* plants deficient in endogenous xylosylation and fucosylation resulted in the production of mAb 2G12 carrying a biantennary complex type N-glycan at the Fc domain. The recombinant mAb showed efficient neutralization activity. Interestingly, plant-produced mAbs had a higher level of Fc sialylation as compared with CHO cell line–derived mAbs.

28.8 Biomolecules produced in plants

Some of the biomolecules which have been expressed in plants in high amount have been described in following subsections.

28.8.1 Vaccine antigens

During mid-1990s, the technology of expressing antigenic determinants in plants was tapped for the production of edible vaccines and local production of engineered edible plants for consumption was proposed. Later, advancement in our understanding about the pros and cons of the system and technological developments made it clear that the concept of edible vaccines is far from reality as two plants grown in the same environment can express vaccine antigens at different levels. To define doses, the idea of processing a large amount of produce with the help of healthcare workers is practically possible. Therefore, the concept of 'edible vaccines' is now more popular as 'plant vaccines'. Although the amount of protein required for oral delivery will vary among various antigens, but generally, oral delivery needs large doses of the antigens. For defining vaccine doses, powdered formulations can be produced from seeds or freeze-dried fruits and leaves using standard, inexpensive food industry milling and processing techniques. Large-scale batches of powdered plant materials can be tested to ensure consistency of vaccine dose and adequate quality control. This material may also be supplemented with mucosal adjuvant, vitamins and/or other vaccines to make it more effective. If vaccines are supplied in powdered formulations, it may also improve the acceptability of vaccines in countries that are against growing GM crops. The idea of plant vaccines was conceived about two decades ago. Research developments and commercial achievements in the field during this time have brought the idea of plant vaccine closer to reality. In recent years, a large number of antigens have been produced in plants and shown to activate the immune response against the antigen in the animal model. The first ever-regulatory approval for a plant-based vaccine was given to Dow Agro Sciences in 2006. Despite regulatory approval, this vaccine against Newcastle disease is not commercialized yet, possibly because of low product yields, which is commercially unviable (http://en.wikipedia.org/wiki/Dow_AgroSciences; accessed on 23 June 2014).

Several antigenic determinants belonging to various pathogens, causing a variety of diseases, including Alzheimer disease, anthrax, bacterial and viral diarrhoea, bovine pneumonia, bursal disease, cancer, cytomegalovirus infections, diphtheria, foot and mouth disease of cattle, gastroenteritis, goat plague, haemorrhagic disease, HIV, malaria, measles, parvoviral infections of dogs, pertussis, tetanus, rabies, rinderpest virus, respiratory syndrome, SARS and tuberculosis, have been produced in plants (Rybicki 2009; Sharma and Sharma 2009; Ling *et al.* 2010; Lossl and Waheed 2011). Several of these plant-made vaccines have been through phase I clinical trial in humans and more are scheduled.

The efficiency of various antigens to target the mucosal system varies; therefore, antigens are often co-administered with adjuvants to boost the immune response. Several adjuvants are identified that enhance immunity by facilitating antigen transport from the gut lumen to the gut-associated lymphoid tissues. Adjuvants such as cholera toxin B subunit, heat-labile enterotoxin B subunit or interleukin can be used with other antigens for the production of multicomponent vaccines. Rotavirus enterotoxin, accessory colonization factor A or P4/P6 epitopes of TCPA from *Vibrio cholera*, HIV antigens, VP2 of parvovirus, anthrax lethal factor and several others have been engineered as chimeric products where one of the partners is either CTB or heat-labile enterotoxin B subunit (LTB). Chimeric antigens could assemble in functionally active oligomers. The oral administration of CTB linked to antigens can trigger acceptable peripheral and mucosal immune responses that were otherwise unattainable when the antigen is given alone. The assembly of these chimeric proteins into biologically active oligomers in transgenic plants demonstrates the feasibility of using plants for the synthesis of various antigenic epitopes fused with CTB/LTB. Further, a recent example of plant-produced influenza A H1N1 vaccine is a very impressive demonstration of developing plant-based vaccines. Researchers could produce up to 800 mg of the recombinant protein in less than a month via transient expression in tobacco (Shoji *et al.* 2009; D'Aoust *et al.* 2010), which is one of the fastest production time recorded. It appears that plants are not only good for conventional vaccine production but can also be very viable for a rapid response during pandemic situations.

28.8.2 Therapeutic products

The production of the first therapeutic product using transgenic plants was reported in late 1980s, when genes encoding immunoglobulin fragments were engineered in tobacco. Since then, several technological milestones have been achieved and the concept of molecular farming has expanded rapidly. Today, a large number of therapeutic products are being produced in plants, which include diagnostic proteins (antibodies, enzymes), replacement proteins (factor VIII for haemophilic patients, insulin for diabetic patients), immune system stimulator/suppressants (interleukins, interferons, colony-stimulating factors), biopolymers and adhesive proteins for surgical purposes or growth factors (Sharma and Sharma 2009). The engineering of recombinant antibodies and their successful expression in plants have opened up new opportunities, not only for the medical sciences but also for applied and fundamental agronomic research. Therapeutic antibodies or their fragments derived from plants are called as 'plantibodies'. Purified plantibodies can be used as diagnostic tools for immunochromatography or in medical therapy. Plantibodies have several *in planta* applications, which include studying the function of antigen or even the epitope of an antigen in plants, change in the agronomic traits, immunization of plants against pathogen infection or providing resistance against herbicides (De Jaeger *et al.* 2002). Single-chain variable fragments (scFvs) have a binding affinity that is similar to complete antibodies but are easy to produce in plants. Recombinant antibodies can be produced in plants in several forms such as full-size recombinant antibody, chimeric antibody, secretory antibody, scFvs, scFv fusion, bispecific scFv, antibody fragments or heavy-chain variable domains (Ma *et al.* 2003). Sugar residues present on therapeutic antibodies are required for their activity, and therefore, the choice of the plant system and the cell compartment for their expression has a direct impact on their functionality. Several methods are being developed to make the glycosylation patterns of plant-produced proteins human friendly (Section 28.7). The goal of these attempts is to remove plant-specific sugar residues that can be immunogenic in humans and add animal-specific residues that are absent in plants. Full-size Abs are generally unstable in the cytosol and are engineered through the secretory pathway for improved accumulation. Selection of the final compartment also depends upon the ultimate goal. For example, recombinant antibodies are targeted to the cytosol when the goal is to engineer pathogen resistance or to modulate the metabolic pathway. Several pharmaceutical proteins of mammalian origin are being produced in plants that include blood products such as human serum albumin, for which there is an annual demand of more than 500 tons, enzymes such as gastric or pancreatic lipases, structural proteins, cytokines and other signalling molecules that are required in much smaller amounts. For the production of proteins, which are required in large quantities and where huge investment in other production systems is required, plants offer a huge competitive advantage. For example, anti-HIV mAbs are required in milligram amounts for effective neutralization and one would need these antibodies on the scale of tons to meet the global demand. Similarly, human glucocerebrosidase (hGC) is one of the most expensive drugs and is required at a large scale for enzyme replacement therapy. The dose required for a single person per year may cost thousands of dollars. The US Food and Drug Administration has approved plant-made glucocerebrosidase for treating type 1 Gaucher disease. This is the first plant-made drug which is approved for human use. Protalix Biotherapeutics has produced this enzyme by employing carrot cells cultures. Human somatotropin (hST), which is used to treat hypopituitary dwarfism in children, Turner syndrome and chronic renal failure, has been produced in tobacco chloroplasts. Other pharmaceutical proteins produced include trypsin, aprotonin, lactoferrin, human glucocerebrosidase (GCD), gastric lipase, proinsulin and alpha interferon 2b. One of the major benefits of plantibodies is the economy of scale-up. This is specifically suitable for antibodies that are required in very large doses. Safflower seed oil–based recombinant insulin, SBS-$_{1000}$, has progressed through phase I/II clinical trials (Paul and Ma 2011). Another very impressive example is *Lemna*-based production of human plasmin (BLX-301) by Biolex Therapeutics (Paul and Ma 2011). Human plasmin is considered as a good anticoagulant, but because of difficulties in purification of active plasmin, no therapeutic version has been developed. The recombinant human plasmin accumulated in *Lemna* in active conformation and at high yields. It is being tested in preclinical trials and looks very promising.

28.8.3 Nutritional products

The advances in our understanding of plant genetics and development of tools to engineer genetic modifications have made it possible to enhance the nutritional quality of plants, too. Suitable targets are macronutrients such as proteins (e.g. it may be desirable to increase levels of essential amino acids), fatty acids, micronutrients such as vitamins A, B, C, etc., carotenoids such as lycopene, flavonoids and minerals such as iron, zinc, etc. The modulation or redirection of flux through metabolic engineering involved initial attempts towards expression or silencing of the single gene. Flux redirection can lead to either elevation of existing compounds or synthesis of novel products. The last two decades have seen a lot of developments in the field of bio-fortification, where plants were engineered to increase the accumulation of β-carotene, lycopene, α-tocopherols or vitamin E, flavonoids, resveratrol, polyamines, amino acids, nutritional proteins, minerals, fatty acids and carbohydrates (Sharma and Sharma 2009). Plants have been engineered in several aspects such as uptake of micronutrients from soil, translocation from root to shoot, phloem loading, remobilization, appropriate localization and bioavailability. Engineering at the level of any single process may not be sufficient to achieve the desired results and simultaneous enhancement of several physiological processes is required. The development of golden rice was a major breakthrough and has overcome several hurdles since the initial proof of concept studies. Golden rice has been bred into local adapted rice varieties and has been incorporated into breeding programmes in India, the Philippines, Bangladesh, Indonesia and Vietnam. Several other successful stories include lysine-rich corn, iron-rich lettuce, lycopene-enhanced tomatoes, etc. and have been reviewed extensively (Zhu *et al.* 2007; Sperotto *et al.* 2012). Naqvi and colleagues (2009) engineered three separate metabolic pathways in corn simultaneously to develop enhanced levels of β-carotene, ascorbate and folate. Recently, the iodine content has been enhanced in *Arabidopsis* by overexpression of human sodium–iodide symporter, which facilitates the uptake of iodine or by knocking-out HOL-1, a halide methyltransferese that catalyses methylation of halides to produce volatile methyl iodide (Landini *et al.* 2012). Before actual engineering to improve upon nutritional fortification, there is a need to develop an integrated understanding of the genetic networks and biochemical and molecular processes. Therefore, it is very important to complement breeding programmes with functional genomic technologies to understand expression networks and protein and metabolite profiles, and ultimately, integrate technological approaches for trait improvement. The capability of multi-gene engineering and efficient transfer methods have made it possible to develop nutritionally improved crops by stacking multiple traits together.

28.8.4 Additional products

Several non-pharmaceutical or non-therapeutic biomolecules are also produced in plants. These include spider silk proteins; collagen for the cosmetic industry; food/feed additives, industrial enzymes such as trypsin, laccase, β-glucuronidase; research and diagnostic agents such as avidin, aprotinin, lysozyme and lactoferrin; cell wall deconstruction enzymes for biofuel industry; and polyhydroxyalkanoates (PHAs) and poly(3-hydroxybutyrate) (PHBs) to make biodegradable plastics. Plant-produced recombinant enzymes β-glucuronidase and avidin were first to reach the market. PHAs are biodegradable polymers which occur naturally in plants. Plants contain thioesters, which are precursor molecules for PHA/PHB synthesis. Transgenic plants have been generated to attempt PHA/PHB production in various cell compartments. For the economic feasibility of transgenic plants–derived biodegradable plastics, accumulation of at least 15% of the tissue dry weight is required. Various approaches used to improve the production levels have been reviewed elsewhere (Suriyamongkol *et al.* 2007). To increase the expression level of biodegradable plastic-like compounds in plants, plastids have also been targeted and expression levels ranging up to 40% of dry weight have been obtained (Bohmert *et al.* 2000). However, defects in plant growth and metabolism were associated with their expression. Alternative strategies have been used to reduce the associated negative effects. Attempts have also been made to produce PHB from plastid-encoded genes and expression levels up to 18.8% of dry weight of the sample have been achieved (Bohmert-Tatarev *et al.* 2011). As thiol esters are abundant in peroxisomes of plant cells, *Pha* gene products were targeted to peroxisomes and accumulation of PHBs up to 2% dry weight was reported.

The biofuel industry needs tons of enzymes to meet its goals of green fuels. Several attempts have been made towards the production of cellulolytic enzymes in plants. The capability of plants to make cellulases was investigated in late 1990s. E2 and E3 cellulases were produced in potato, tobacco and alfalfa. E1 cellulase, an endo-β-1,4-glucanase, has been produced in transgenic maize and recovered at levels >16% of TSP (Hood *et al.* 2007). The heterologous expression of many enzymes results in deleterious phenotypes, as is observed in the case of expression of xylanases in monocots. In another study, several enzymes, which include endoglucanase, exoglucanase and lipase from bacteria and pectate lyases, cutinase, endoglucanases, swollenin, xylanase, acetyl xylan esterase and beta-glucosidase from fungi, were produced in chloroplasts and a cocktail of these plant-produced enzymes was used to degrade lignocellulose material (Verma *et al.* 2010). Recently, a 2A self-cleaving oligopeptide from foot and mouth disease virus was used to fuse genes encoding for β-glucosidase, endoglucanase, xylanase and exoglucanase and these fusion products were successfully expressed as polyproteins in tobacco chloroplasts (Lee *et al.* 2012).

28.9 Containment and risk management of GM crops

A large number of published reports are available that discuss possible risks associated with GM crops. Concerns have been raised about their safety in relation to environment and human health. Although there is no scientific evidence that current modified foods involve any new or magnified risks, but certain environmentalists and consumers are still not convinced. To some extent, a negative opinion for transgenic plants already exists among the general public, and this notion that genetic engineering is against nature, makes an impact on regulators. Public opposition grew specifically after

ProdiGene's accidental contamination incident of 2002, where conventional soybean was found contaminated with GM maize. Regulations are getting clearer, but still there are several challenges ahead for the regulating agencies. Because of opposition from environmental and consumer organizations, the regulations for pharmaceutical and food industry are strict and turn out to be very costly. The regulatory challenges that molecular farming faces and how they are different from those for first-generation transgenic crops have been reviewed recently (Spok 2007; Spok and Karner 2008). The most important issue is to segregate GM crops from non-GM crops to prevent intermixing. A variety of approaches, including physical containment as well as genetic strategies such as seed sterility, maternal inheritance, male sterility, selective elimination by engineering sensitivity to chemicals, etc., have been postulated to address this question. The question of liability in the case of accidental contamination is another key issue. Even after stringent confinements, it is very difficult to maintain a complete segregation for GM and non-GM crops in the open field set-ups and, therefore, it is difficult to follow a zero-tolerance policy. Who would be responsible or liable for economic damage caused by low-level residue contamination from pharma crops? For this reason, threshold limits of accidental contamination have been suggested. Molecular farming crops need to be evaluated on a case-by-case basis as different biomolecules that are targets for their production in plants would pose different levels of risks. For the plant-made bio-molecules other than pharmaceuticals that do not pose hazardous risks, the threshold limit for contamination of non-GM crops is 0.9% (Spok 2007).

Another important issue is the labelling of GM products. To empower the consumer to select between GM and non-GM products, labelling could be an option. But the mandatory labelling might not be economically justifiable in every situation and would not provide the consumer with the required information. Alternatively, information domains can be built that can provide the essential information related to the GM content of the product to the consumer. Also, a system that could trace the product in the market to its source and a good strategy for post-market monitoring and surveillance would be practical (Smyth and Phillips 2003).

It has been estimated that the annual tonnage of any biopharmaceuticals very rarely exceeds 1000 kg/year, except certain high-dose drugs, where the annual requirement may touch up to 5,00,000 kg. Therefore, for most biopharmaceuticals, pharma crops grown on about 1000 acres can satisfy the world's needs (Spok and Karner 2008). The present insulin demand is about 16,000 kg and safflower growing on about 4000 hectares of land can produce that. As more and more pharma crops enter commercial production, regulatory regimes regarding coexistence and segregation are likely to become more complex. Overall, the future of plant molecular farming will depend upon several factors, including policy decisions on the use of food/feed crops, economic viability, technological developments, regulatory burden on the industry and how plant-made pharmaceuticals impact human health, society and environment.

28.10 Commercialization scenario

The plant molecular farming–based industry is growing very fast. Because of accidental contamination of non-GM crops with GM residues and resulting economic damages, the public as well as the food industry is campaigning for stricter rules. Several transgenic plant-derived products are undergoing clinical trials or have reached at an advanced stage of development; some have already been commercialized (Sharma and Sharma 2009). Prodi Gene, Inc. was the first to develop and commercialize products from transgenic plants. GUS (β-glucuronidase) was the first commercialized product. The transgenic corn–derived GUS protein was commercialized and sold in the market by Sigma-Aldrich. Later, plant-produced human therapeutic protein, aprotonin, was commercialized under the trade name Aproli Zean™. Avidin, which is a diagnostic reagent, was produced in transgenic corn and commercialized by Sigma-Aldrich. Recently, the US Food and Drug Administration approved the first plant-made human biologics, Elelyso. Elelyso is an enzyme that is used to treat Gaucher disease and has been produced in carrot cell cultures. Ventria Bioscience has developed Lacromin™ (recombinant human lactoferrin) and Lysobac™ (recombinant human lysozyme). Plant-derived lysozyme is four times more active than chicken lysozyme for lysing *E. coli*.

28.11 Concluding remarks

Though more than two decades have passed since the first report of expression of foreign biomolecules in plants, the pharmaceutical industry is still reluctant to integrate plant expression platforms into the production strategy. However, an increasing number of major drug companies are utilizing transgenic plants as an essential component of their business model. The last decade has seen a lot of technological developments, but some technical problems remain to be solved, which include plant-specific glycan patterns, low product yield, processing costs and biosafety. Facing a looming crisis in the production capacity for mAbs and other protein-based therapeutics, the industry is concentrating to solve the technical problems associated with molecular farming. In the near future, it is expected to provide a captive new market for many agricultural products. The key to success of transgenic plant–derived products in the future will be the level of expression. It is very important with regard to economics, because it affects the cost of growing, processing, extraction, purification and waste disposal. If the technical hurdles can be overcome, soon it might be possible to make protein-based pharmaceuticals available to the needy at an affordable cost. The full realization and impact of the aforementioned developments, however, depend not only on consistent, successful and innovative research and developmental activities, but also on a favourable regulatory climate and public acceptance. Overall, public perception, eventual acceptance and trust on the regulatory regime will shape up further developments in the field of plant molecular farming.

References

Apel W, Bock R. Enhancement of carotenoid biosynthesis in transplastomic tomatoes by induced lycopene-to-provitamin A conversion. *Plant Physiol* 2009;151:59–66.

Arcalis E, Marcel S, Altmann F, Kolarich D, Drakakaki G, Fischer R, *et al.* Unexpected deposition patterns of recombinant proteins in post-endoplasmic reticulum compartments of wheat endosperm. *Plant Physiol* 2004;136:3457–3466.

Balasubramaniam D, Wilkinson C, Van Cott K, Zhang C. Tobacco protein separation by aqueous two-phase extraction. *J Chromatograph A* 2003;989:119–129.

Baron U, Freundlieb S, Gossen M, Bujard H. Co-regulation of two gene activities by tetracycline via a bidirectional promoter. *Nucleic Acids Res* 1995;23:3605–3606.

Blackwood EM, Kadonaga JT. Going the distance: A current view of enhancer action. *Science* 1998;281:60–63.

Bohmert K, Balbo I, Kopka J, Mittendorf V, Nawrath C, Poirier Y, *et al.* Transgenic *Arabidopsis* plants can accumulate polyhydroxybutyrate to up to 4% of their fresh weight. *Planta* 2000;211:841–845.

Bohmert-Tatarev K, McAvoy S, Daughtry S, Peoples OP, Snell KD. High levels of bioplastic are produced in fertile transplastomic tobacco plants engineered with a synthetic operon for the production of polyhydroxybutyrate. *Plant Physiol* 2011;155:1690–1708.

Borovsky D, Rabindran S, Dawson WO, Powell CA, Iannotti DA, Morris TJ, *et al.* Expression of *Aedes* trypsin-modulating oostatic factor on the virion of TMV: A potential larvicide. *Proc Natl Acad Sci USA* 2006;103:18963–18968.

Brand L, Horler M, Nuesch E, Vassalli S, Barrell P, Yang W, *et al.* A versatile and reliable two-component system for tissue-specific gene induction in *Arabidopsis*. *Plant Physiol* 2006;141:1194–1204.

Broothaerts W, Mitchell HJ, Weir B, Kaines S, Smith LM, Yang W, *et al.* Gene transfer to plants by diverse species of bacteria. *Nature* 2005;433:629–633.

Butaye K, Cammue B, Delaure S, De Bolle M. Approaches to minimize variation of transgene expression in plants. *Mol Breeding* 2005;16:79–91.

Cai CQ, Doyon Y, Ainley WM, Miller JC, Dekelver RC, Moehle EA, *et al.* Targeted transgene integration in plant cells using designed zinc finger nucleases. *Plant Mol Bio* 2009;69:699–709.

Canizares MC, Nicholson L, Lomonossoff GP. Use of viral vectors for vaccine production in plants. *Immunol Cell Biol* 2005;83: 263–270.

Cardi T, Lenzi P, Maliga P. Chloroplasts as expression platforms for plant-produced vaccines. *Expert Rev Vaccines* 2010;9:893–911.

Castilho A, Gattinger P, Grass J, Jez J, Pabst M, Altmann F, *et al.* N-glycosylation engineering of plants for the biosynthesis of glycoproteins with bisected and branched complex N-glycans. *Glycobiology* 2011;21:813–823.

Castilho A, Strasser R, Stadlmann J, Grass J, Jez J, Gattinger P, *et al.* In planta protein sialylation through overexpression of the respective mammalian pathway. *J Biol Chem* 2010;285: 15923–15930.

Chaturvedi CP, Sawant SV, Kiran K, Mehrotra R, Lodhi N, Ansari SA, *et al.* Analysis of polarity in the expression from a multifactorial bidirectional promoter designed for high-level expression of transgenes in plants. *J Biotechnol* 2006;123:1–12.

Chen M, Liu X, Wang Z, Song J, Qi Q, Wang PG. Modification of plant N-glycans processing: the future of producing therapeutic protein by transgenic plants. *Med Res Rev* 2005; 25:343–360.

Choi HW, Yu XH, Lemaux PG, Cho MJ. Stability and inheritance of endosperm-specific expression of two transgenes in progeny from crossing independently transformed barley plants. *Plant Cell Rep* 2009;28:1265–1272.

Conley AJ, Joensuu JJ, Jevnikar AM, Menassa R, Brandle JE. Optimization of elastin-like polypeptide fusions for expression and purification of recombinant proteins in plants. *Biotechnol Bioeng* 2009a;103:562–573.

Conley AJ, Joensuu JJ, Menassa R, Brandle JE. Induction of protein body formation in plant leaves by elastin-like polypeptide fusions. *BMC Biol* 2009b;7:48.

Conley AJ, Joensuu JJ, Richman A, Menassa R. Protein body-inducing fusions for high-level production and purification of recombinant proteins in plants. *Plant Biotechnol J* 2011;9: 419–433.

Cox KM, Sterling JD, Regan JT, Gasdaska JR, Frantz KK, Peele CG, *et al.* Glycan optimization of a human monoclonal antibody in the aquatic plant *Lemna minor*. *Nat Biotechnol* 2006;24:1591–1597.

D'Aoust MA, Couture MM, Charland N, Trepanier S, Landry N, Ors F, *et al.* The production of hemagglutinin-based virus-like particles in plants: A rapid, efficient and safe response to pandemic influenza. *Plant Biotechnol J* 2010;8:607–619.

Daniell H, Singh ND, Mason H, Streatfield SJ. Plant-made vaccine antigens and biopharmaceuticals. *Trends Plant Sci* 2009;14:669–679.

De Cosa B, Moar W, Lee SB, Miller M, Daniell H. Overexpression of the Bt cry2Aa2 operon in chloroplasts leads to formation of insecticidal crystals. *Nat Biotechnol* 2001;19:71–74.

De Jaeger G, Scheffer S, Jacobs A, Zambre M, Zobell O, Goossens A, *et al.* Boosting heterologous protein production in transgenic dicotyledonous seeds using *Phaseolus vulgaris* regulatory sequences. *Nat Biotechnol* 2002;20:1265–1268.

Dhadi SR, Krom N, Ramakrishna W. Genome-wide comparative analysis of putative bidirectional promoters from rice, *Arabidopsis* and *Populus*. *Gene* 2009;429:65–73.

Drake PM, Barbi T, Sexton A, McGowan E, Stadlmann J, Navarre C, *et al.* Development of rhizosecretion as a production system for recombinant proteins from hydroponic cultivated tobacco. *FASEB J* 2009;23:3581–3589.

Egelkrout E, Rajan V, Howard JA. Overproduction of recombinant proteins in plants. *Plant Sci* 2012;184:83–101.

Firek S, Draper J, Owen MR, Gandecha A, Cockburn B, Whitelam GC. Secretion of a functional single-chain Fv protein in transgenic tobacco plants and cell suspension cultures. *Plant Mol Biol* 1993;23:861–870.

Fischer R, Drossard J, Commandeur U, Schillberg S, Emans N. Towards molecular farming in the future: Moving from diagnostic protein and antibody production in microbes to plants. *Biotechnol Appl Biochem* 1999;30:101–108.

Fischer R, Stoger E, Schillberg S, Christou P, Twyman RM. Plant-based production of biopharmaceuticals. *Curr Opin Plant Biol* 2004;7:152–158.

Fox JL. First plant-made biologic approved. *Nat Biotechnol* 2012;30:472.

Franklin S, Ngo B, Efuet E, Mayfield SP. Development of a GFP reporter gene for *Chlamydomonas reinhardtii* chloroplast. *Plant J* 2002;30:733–744.

Gijs JH, Rooijen V, Motoney MM. Plant seed oil-bodies as carriers for foreign proteins. *Nat Biotechnol* 1995;13:72–77.

Giritch A, Marillonnet S, Engler C, van Eldik G, Botterman J, Klimyuk V, *et al.* Rapid high-yield expression of full-size IgG antibodies in plants coinfected with noncompeting viral vectors. *Proc Natl Acad Sci USA* 2006;103:14701–14706.

Gleba Y, Klimyuk V, Marillonnet S. Magnifection – a new platform for expressing recombinant vaccines in plants. *Vaccine* 2005;23: 2042–2048.

Gleba Y, Klimyuk V, Marillonnet S. Viral vectors for the expression of proteins in plants. *Curr Opin Biotechnol* 2007;18:134–141.

Gomord V, Denmat LA, Fitchette-Laine AC, Satiat-Jeunemaitre B, Hawes C, Faye L. The C-terminal HDEL sequence is sufficient for retention of secretory proteins in the endoplasmic reticulum (ER) but promotes vacuolar targeting of proteins that escape the ER. *Plant J* 1997;11:313–325.

Gutierrez-Ortega A, Sandoval-Montes C, de Olivera-Flores TJ, Santos-Argumedo L, Gomez-Lim M. Expression of functional interleukin-12 from mouse in transgenic tomato plants. *Transgenic Res* 2005;14:877–885.

Hamamoto H, Sugiyama Y, Nakagawa N, Hashida E, Matsunaga Y, Takemoto S, *et al.* A new tobacco mosaic virus vector and its use for the systemic production of angiotensin-I-converting enzyme inhibitor in transgenic tobacco and tomato. Biotechnology 1993;11:930–932.

Hamilton CM, Frary A, Lewis C, Tanksley SD. Stable transfer of intact high molecular weight DNA into plant chromosomes. *Proc Natl Acad Sci USA* 1996;93:9975–9979.

Hasunuma T, Miyazawa S, Yoshimura S, Shinzaki Y, Tomizawa K, Shindo K, *et al.* Biosynthesis of astaxanthin in tobacco leaves by transplastomic engineering. *Plant J* 2008;55:857–868.

Herrera-Estrella L, Depicker A, Van Montagu M, Schell J. Expression of chimaeric genes transferred into plant cells using a Ti-plasmid-derived vector. *Biotechnology* 1982;24:377–381.

Hiatt A, Cafferkey R, Bowdish K. Production of antibodies in transgenic plants. *Nature* 1989;342:76–78.

Hood EE, Love R, Lane J, Bray J, Clough R, Pappu K, *et al.* Subcellular targeting is a key condition for high-level accumulation of cellulase protein in transgenic maize seed. *Plant Biotechnol J* 2007;5:709–719.

Hood EE, Witcher DR, Maddock S, Meyer T, Baszczynski C, Bailey M, *et al.* Commercial production of avidin from transgenic maize: characterization of transformant, production, processing, extraction and purification. *Mol Breeding* 1997;3:291–306.

Horsch RB, Fraley RT, Rogers SG, Sanders PR, Lloyd A, Hoffmann N. Inheritance of functional foreign genes in plants. *Science* 1984;223:496–498.

Huang TK, McDonald KA. Bioreactor systems for *in vitro* production of foreign proteins using plant cell cultures. *Biotechnol Adv* 2012;30:398–409.

Iturriaga G, Jefferson RA, Bevan MW. Endoplasmic reticulum targeting and glycosylation of hybrid proteins in transgenic tobacco. *Plant Cell* 1989;1:381–390.

Janssen BJ, Gardner RC. Localized transient expression of GUS in leaf discs following cocultivation with *Agrobacterium*. *Plant Mol Bio* 1990;14:61–72.

Jefferson R, Goldsbrough A, Bevan M. Transcriptional regulation of a patatin-1 gene in potato. *Plant Mol Biol* 1990;14:995–1006.

Joensuu JJ, Conley AJ, Lienemann M, Brandle JE, Linder MB, Menassa R. Hydrophobin fusions for high-level transient protein expression and purification in *Nicotiana benthamiana*. *Plant Physiol* 2010;152:622–633.

Johnson J, Lin T, Lomonossoff G. Presentation of heterologous peptides on plant viruses: Genetics, structure, and function. *Ann Rev Phytopathol* 1997;35:67–86.

Jones HD, Doherty A, Sparks CA. Transient transformation of plants. *Methods Mol Biol* 2009;513:131–152.

Kang TJ, Loc N, Jang M, Yang M. Modification of the cholera toxin B subunit coding sequence to enhance expression in plants. *Mol Breeding* 2004;13:143–153.

Kasuga M, Miura S, Shinozaki K, Yamaguchi-Shinozaki K. A combination of the *Arabidopsis* DREB1A gene and stress-inducible rd29A promoter improved drought- and low-temperature stress tolerance in tobacco by gene transfer. *Plant Cell Physiol* 2004;45:346–350.

Komarnytsky S, Borisjuk NV, Borisjuk LG, Alam MZ, Raskin I. Production of recombinant proteins in tobacco guttation fluid. *Plant Physiol* 2000;124:927–934.

Komarova TV, Baschieri S, Donini M, Marusic C, Benvenuto E, Dorokhov YL. Transient expression systems for plant-derived biopharmaceuticals. *Expert Rev Vaccines* 2010;9:859–876.

Landini M, Gonzali S, Kiferle C, Tonacchera M, Agretti P, Dimida A, *et al.* Metabolic engineering of the iodine content in *Arabidopsis*. *Scientific Rep* 2012;2:338.

Lee DS, Lee KH, Jung S, Jo EJ, Han KH, Bae HJ. Synergistic effects of 2A-mediated polyproteins on the production of lignocellulose degradation enzymes in tobacco plants. *J Exp Botany* 2012;63:4797–4810.

Lee JW, Forciniti D. Purification of human antibodies from transgenic corn using aqueous two-phase systems. *Biotechnol Prog* 2010;26:159–167.

Lee LY, Gelvin SB. T-DNA binary vectors and systems. *Plant Physiol* 2008;146:325–332.

Li Y. Self-cleaving fusion tags for recombinant protein production. *Biotechnol Lett* 2011;33:869–881.

Lico C, Chen Q, Santi L. Viral vectors for production of recombinant proteins in plants. *J Cellular Physiol* 2008;216:366–377.

Ling HY, Pelosi A, Walmsley AM. Current status of plant-made vaccines for veterinary purposes. *Expert Rev Vaccines* 2010;9:971–982.

Liu CW, Lin CC, Yiu JC, Chen JJ, Tseng MJ. Expression of a *Bacillus thuringiensis* toxin (cry1Ab) gene in cabbage (*Brassica oleracea* L. var. capitata L.) chloroplasts confers high insecticidal efficacy against *Plutella xylostella*. *Theoretic Appl Genet* 2008;117:75–88.

Lossl AG, Waheed MT. Chloroplast-derived vaccines against human diseases: Achievements, challenges and scopes. *Plant Biotechnol J* 2011;9:527–539.

Ma JK, Drake PM, Christou P. The production of recombinant pharmaceutical proteins in plants. *Nat Rev Genet* 2003;4:794–805.

Maggio C, Barbante A, Ferro F, Frigerio L, Pedrazzini E. Intracellular sorting of the tail-anchored protein cytochrome b5 in plants: A comparative study using different isoforms from rabbit and *Arabidopsis*. *J Exp Botany* 2007;58:1365–1379.

Mainieri D, Rossi M, Archinti M, Bellucci M, De Marchis F, Vavassori S, *et al.* Zeolin. A new recombinant storage protein constructed using maize gamma-zein and bean phaseolin. *Plant Physiol* 2004;136:3447–3456.

McBride KE, Svab Z, Schaaf DJ, Hogan PS, Stalker DM, Maliga P. Amplification of a chimeric *Bacillus* gene in chloroplasts leads to an extraordinary level of an insecticidal protein in tobacco. *Biotechnology* 1995;13:362–365.

Mishra S, Yadav DK, Tuli R. Ubiquitin fusion enhances cholera toxin B subunit expression in transgenic plants and the plant-expressed protein binds GM1 receptors more efficiently. *J Biotechnol* 2006;127:95–108.

Morassutti C, De Amicis F, Skerlavaj B, Zanetti M, Marchetti S. Production of a recombinant antimicrobial peptide in transgenic plants using a modified VMA intein expression system. *FEBS Lett* 20002;519:141–146.

Naqvi S, Farre G, Sanahuja G, Capell T, Zhu C, Christou P. When more is better: Multigene engineering in plants. *Trends Plant Sci* 2010;15:48–56.

Naqvi S, Zhu C, Farre G, Ramessar K, Bassie L, Breitenbach J, *et al.* Transgenic multivitamin corn through biofortification of endosperm with three vitamins representing three distinct metabolic pathways. *Proc Natl Acad Sci USA* 2009;106:7762–7767.

Nguyen HT, Leelavathi S, Reddy VS. Bacteriophage T7 RNA polymerase-directed, inducible and tissue-specific over-expression of foreign genes in transgenic plants. *Plant Biotechnol J* 2004;2:301–310.

Ni M, Cui D, Einstein J, Narasimhulu S, Vergara CE, Gelvin SB. Strength and tissue specificity of chimeric promoters derived from the octopine and mannopine synthase genes. *Plant J* 1995;7:661–676.

Obembe OO, Popoola JO, Leelavathi S, Reddy SV. Advances in plant molecular farming. *Biotechnol Advance* 2011;29:210–222.

Odell JT, Nagy F, Chua NH. Identification of DNA sequences required for activity of the cauliflower mosaic virus 35S promoter. *Nature* 1985;313:810–812.

Padidam M, Gore M, Lu DL, Smirnova O. Chemical-inducible, ecdysone receptor-based gene expression system for plants. *Transgenic Res* 2003;12:101–109.

Palacpac NQ, Yoshida S, Sakai H, Kimura Y, Fujiyama K, Yoshida T, *et al.* Stable expression of human beta1,4-galactosyltransferase in plant cells modifies N-linked glycosylation patterns. *Proc Natl Acad Sci USA* 1999;96:4692–4697.

Paul M, Ma JK. Plant-made pharmaceuticals: leading products and production platforms. *Biotechnol Appl Chem* 2011;58: 58–67.

Paul M, van Dolleweerd C, Drake PM, Reljic R, Thangaraj H, Barbi T, *et al.* Molecular pharming: Future targets and aspirations. *Human Vaccine* 2011;7:375–382.

Platis D, Labrou NE. Application of a PEG/salt aqueous two-phase partition system for the recovery of monoclonal antibodies from unclarified transgenic tobacco extract. *Biotechnol J* 2009;4: 1320–1327.

Pogrebnyak N, Golovkin M, Andrianov V, Spitsin S, Smirnov Y, Egolf R, *et al.* Severe acute respiratory syndrome (SARS) S protein production in plants: development of recombinant vaccine. *Proc Natl Acad Sci USA* 2005;102:9062–9067.

Porta C, Lomonossoff GP. Scope for using plant viruses to present epitopes from animal pathogens. *Rev Med Virol* 1998;8:25–41.

Rasala BA, Muto M, Lee PA, Jager M, Cardoso RM, Behnke CA, *et al.* Production of therapeutic proteins in algae, analysis of expression of seven human proteins in the chloroplast of *Chlamydomonas reinhardtii*. *Plant Biotechnol J* 2010;8: 719–733.

Raucher D, Chilkoti A. Enhanced uptake of a thermally responsive polypeptide by tumor cells in response to its hyperthermia-mediated phase transition. *Cancer Res* 2001;61:7163–7170.

Rivard D, Anguenot R, Brunelle F, Le VQ, Vezina LP, Trepanier S, *et al.* An in-built proteinase inhibitor system for the protection of recombinant proteins recovered from transgenic plants. *Plant Biotechnol J* 2006;4:359–368.

Rosales-Mendoza S, Soria-Guerra RE, Moreno-Fierros L, Alpuche-Solis AG, Martinez-Gonzalez L, Korban SS. Expression of an immunogenic F1-V fusion protein in lettuce as a plant-based vaccine against plague. *Planta* 2010;232:409–416.

Ross KC, Zhang C. Separation of recombinant β-glucuronidase from transgenic tobacco by aqueous two-phase extraction. *Biochemical Eng J* 2010;49:343–350.

Rybicki EP. Plant-produced vaccines: Promise and reality. *Drug Discov Today* 2009;14:16–24.

Sawant SV, Kiran K, Singh PK, Tuli R. Sequence architecture downstream of the initiator codon enhances gene expression and protein stability in plants. *Plant Physiol* 2001;126: 1630–1636.

Scholthof HB, Morris TJ, Jackson AO. The capsid protein gene of Tomato bushy stunt virus is dispensable for systemic movement and can be replaced for localized expression of foreign genes. *Mol Plant Microbe Interact* 1993;6:309–322.

Scotti N, Rigano MM, Cardi T. Production of foreign proteins using plastid transformation. *Biotechnol Advance* 2012;30:387–397.

Sharma AK, Sharma MK. Plants as bioreactors: Recent developments and emerging opportunities. *Biotechnol Advance* 2009;27:811–832.

Sharma MK, Jani D, Thungapathra M, Gautam JK, Meena LS, Singh Y, *et al.* Expression of accessory colonization factor subunit A (ACFA) of *Vibrio cholerae* and ACFA fused to cholera toxin B subunit in transgenic tomato (*Solanum lycopersicum*). *J Biotechnol* 2008a;135:22–27.

Sharma MK, Singh NK, Jani D, Sisodia R, Thungapathra M, Gautam JK, *et al.* Expression of toxin co-regulated pilus subunit A (TCPA) of *Vibrio cholerae* and its immunogenic epitopes fused to cholera toxin B subunit in transgenic tomato (*Solanum lycopersicum*). *Plant Cell Rep* 2008b;27:307–318.

Shepherd CT. Transient expression of GFP in immature seed tissues. *Methods Mol Biol* 2009;526:23–28.

Shoji Y, Bi H, Musiychuk K, Rhee A, Horsey A, Roy G, *et al.* Plant-derived hemagglutinin protects ferrets against challenge infection with the A/Indonesia/05/05 strain of avian influenza. *Vaccine* 2009;27:1087–1092.

Singer SD, Cox KD, Liu Z. Both the constitutive cauliflower mosaic virus 35S and tissue-specific AGAMOUS enhancer activate transcription autonomously in *Arabidopsis thaliana*. *Plant Mol Biol* 2010a;74:293–305.

Singer SD, Hily J, Liu Z. A 1-kb bacteriophage lambda fragment functions as an insulator to effectively block enhancer-promoter interactions in *Arabidopsis thaliana*. *Plant Mol Biol Rep* 2010b;28:69–76.

Skarjinskaia M, Karl J, Araujo A, Ruby K, Rabindran S, Streatfield SJ, *et al.* Production of recombinant proteins in clonal root cultures using episomal expression vectors. *Biotechnol Bioeng* 2008;100:814–819.

Smyth S, Phillips PW. Labeling to manage marketing of GM foods. *Trends Biotechnol* 2003;21:389–393.

Specht E, Miyake-Stoner S, Mayfield S. Micro-algae come of age as a platform for recombinant protein production. *Biotechnology Lett* 2010;32:1373–1383.

Sperotto RA, Ricachenevsky FK, Waldow Vde A, Fett JP. Iron biofortification in rice: It's a long way to the top. *Plant Sci* 2012;190:24–39.

Spok A. Molecular farming on the rise--GMO regulators still walking a tightrope. *Trends Biotechnol* 2007;25:74–82.

Spok A, Karner S. *Plant Molecular Farming: Opportunities and Challenges*). Seville, Spain: Joint Research Center, Institute for Prospective Technolocal Studies; 2008.

Stege JT, Guan X, Ho T, Beachy RN, Barbas CF III. Controlling gene expression in plants using synthetic zinc finger transcription factors. *Plant J* 2002;32:1077–1086.

Stomp A, Dickey L, Gasdaska JR. Expression of biologically active polypeptide in duckweed. United States Patent no. 6815184; 2004.

Strasser R, Altmann F, Mach L, Glossl J, Steinkellner H. Generation of *Arabidopsis thaliana* plants with complex N-glycans lacking beta1,2-linked xylose and core alpha1,3-linked fucose. *FEBS Lett* 2004;561:132–136.

Strasser R, Stadlmann J, Schahs M, Stiegler G, Quendler H, Mach L, *et al.* Generation of glyco-engineered *Nicotiana benthamiana* for the production of monoclonal antibodies with a homogeneous human-like N-glycan structure. *Plant Biotechnol J* 2008;6: 392–402.

Sugio T, Satoh J, Matsuura H, Shinmyo A, Kato K. The 5′-untranslated region of the *Oryza sativa* alcohol dehydrogenase gene functions as a translational enhancer in monocotyledonous plant cells. *J Biosci Bioeng* 2008;105:300–302.

Suriyamongkol P, Weselake R, Narine S, Moloney M, Shah S. Biotechnological approaches for the production of polyhydroxyalkanoates in microorganisms and plants – a review. *Biotechnol Advance* 2007;25:148–175.

Svab Z, Hajdukiewicz P, Maliga P. Stable transformation of plastids in higher plants. *Proc Natl Acad Sci USA* 1990;87:8526–8530.

Tian L, Sun SS. A cost-effective ELP-intein coupling system for recombinant protein purification from plant production platform. *PloS ONE* 2011;6:e24183.

Torrent M, Llompart B, Lasserre-Ramassamy S, Llop-Tous I, Bastida M, Marzabal P, *et al.* Eukaryotic protein production in designed storage organelles. *BMC Biol* 2009;7:5.

Turpen TH, Reinl SJ, Charoenvit Y, Hoffman SL, Fallarme V, Grill LK. Malarial epitopes expressed on the surface of recombinant tobacco mosaic virus. *Biotechnology* 1995;13:53–57.

Usha R, Rohll JB, Spall VE, Shanks M, Maule AJ, Johnson JE, *et al.* Expression of an animal virus antigenic site on the surface of a plant virus particle. *Virology* 1993;197:366–374.

Vain P, Worland B, Kohli A, Snape JW, Christou P, Allen GC, *et al.* Matrix attachment regions increase transgene expression levels and stability in transgenic rice plants and their progeny. *Plant J* 2002;18:233–242.

Verma D, Daniell H. Chloroplast vector systems for biotechnology applications. *Plant Physiol* 2007;145:1129–1143.

Verma D, Kanagaraj A, Jin S, Singh ND, Kolattukudy PE, Daniell H. Chloroplast-derived enzyme cocktails hydrolyse lignocellulosic biomass and release fermentable sugars. *Plant Biotechnol J* 2010;8:332–350.

Vidi PA, Kessler F, Brehelin C. Plastoglobules: A new address for targeting recombinant proteins in the chloroplast. *BMC Biotechnol* 2007;7:4.

Vitale A, Hinz G. Sorting of proteins to storage vacuoles: How many mechanisms? *Trends Plant Sci* 2005;10:316–323.

Vitale A, Pedrazzini E. Recombinant pharmaceuticals from plants: The plant endomembrane system as bioreactor. *Molecular Interv* 2005;5:216–225.

Voinnet O, Rivas S, Mestre P, Baulcombe D. An enhanced transient expression system in plants based on suppression of gene silencing by the p19 protein of tomato bushy stunt virus. *Plant J* 2003;33:949–956.

Wada N, Kajiyama S, Akiyama Y, Kawakami S, No D, Uchiyama S, *et al.* Bioactive beads-mediated transformation of rice with large DNA fragments containing *Aegilops tauschii* genes. *Plant Cell Rep* 2009;28:759–768.

Witte CP, Noel LD, Gielbert J, Parker JE, Romeis T. Rapid one-step protein purification from plant material using the eight-amino acid StrepII epitope. *Plant Mol Biol* 2004;55:135–147.

Wu G, Truksa M, Datla N, Vrinten P, Bauer J, Zank T, *et al.* Stepwise engineering to produce high yields of very long-chain polyunsaturated fatty acids in plants. *Nat Biotechnol* 2005;23:1013–1017.

Wu L, Jiang L, Zhou Z, Fan J, Zhang Q, Zhu H, *et al.* Expression of foot-and-mouth disease virus epitopes in tobacco by a tobacco mosaic virus-based vector. *Vaccine* 2003;21:4390–4398.

Xie M, He Y, Gan S. Bidirectionalization of polar promoters in plants. *Nat Biotechnol* 2001;19:677–679.

Xu J, Okada S, Tan L, Goodrum KJ, Kopchick JJ, Kieliszewski MJ. Human growth hormone expressed in tobacco cells as an arabinogalactan-protein fusion glycoprotein has a prolonged serum life. *Transgenic Res* 2010;19:849–867.

Xu J, Tan L, Goodrum KJ, Kieliszewski MJ. High-yields and extended serum half-life of human interferon alpha2b expressed in tobacco cells as arabinogalactan-protein fusions. *Biotechnol Bioeng* 2007;97:997–1008.

Yang CD, Liao JT, Lai CY, Jong MH, Liang CM, Lin YL, *et al.* Induction of protective immunity in swine by recombinant bamboo mosaic virus expressing foot-and-mouth disease virus epitopes. *BMC Biotechnol* 2007;7:62.

Yang D, Guo F, Liu B, Huang N, Watkins SC. Expression and localization of human lysozyme in the endosperm of transgenic rice. *Planta* 2003;216:597–603.

Yang D, Wu L, Hwang YS, Chen L, Huang N. Expression of the REB transcriptional activator in rice grains improves the yield of recombinant proteins whose genes are controlled by a Reb-responsive promoter. *Proc Natl Acad Sci USA* 2001;98: 11438–11443.

Yang L, Wang H, Liu J, Li L, Fan Y, Wang X, *et al.* A simple and effective system for foreign gene expression in plants via root absorption of agrobacterial suspension. *J Biotechnol* 2008;134:320–324.

Zhang X, Urry DW, Daniell H. Expression of an environmentally friendly synthetic protein-based polymer gene in transgenic tobacco plants. *Plant Cell Rep* 1996;16:174–179.

Zhu C, Naqvi S, Gomez-Galera S, Pelacho AM, Capell T, Christou P. Transgenic strategies for the nutritional enhancement of plants. *Trends Plant Sci* 2007;12:548–555.

Zuo J, Chua NH. Chemical-inducible systems for regulated expression of plant genes. *Curr Opin Biotechnol* 2000;11:146–151.

29

Chemical and enzymatic characterization of a cloned DNA fragment

H. K. Das

29.1 Restriction mapping of DNA fragments and map construction

A restriction map reveals the positions of the cleavage sites of restriction endonucleases on a DNA fragment or a gene or a genome. As an example, the restriction map of the genome of the lambda bacteriophage with respect to several restriction endonucleases is shown in Fig. 29.1. Such a map is very useful for a base sequencing programme to be initiated. It is obvious that the restriction map and the genetic map are complementary to each other. Proper superimposition of the restriction map and the genetic map of a genomic region or the whole genome makes it possible to locate the positions of genes in absolute physical terms and their eventual isolation. The following are some of the approaches for the construction of a restriction map:

29.1.1 Treatment with a number of single restriction endonucleases followed by simultaneous treatment with pairs of restriction endonucleases

For the sake of easy illustration, a very simplified situation is considered here. Assume that each of the restriction endonucleases, *Bam*HI, *Bgl*II, *Eco*RI, *Hin*dIII and *Pst*I, has a single cleavage site in a 10-kb DNA fragment. Hence, each endonuclease would produce two subfragments from the 10-kb DNA fragment. The results of such treatments are given in Fig. 29.2. It can be seen that *Bam*HI produces two fragments of 1 and 9 kb, while *Bgl*II produces two fragments of 3 and 7 kb. Either of the possibilities shown in Fig. 29.2(a) or (b) could be correct. Simultaneous treatment with *Bam*HI and *Bgl*II, however, produces three fragments of 1, 2 and 7 kb. Hence, possibility (a) has to be the correct one and not possibility (b). Again, it can be seen that *Hin*dIII produces two fragments of 4 and 6 kb. So, between

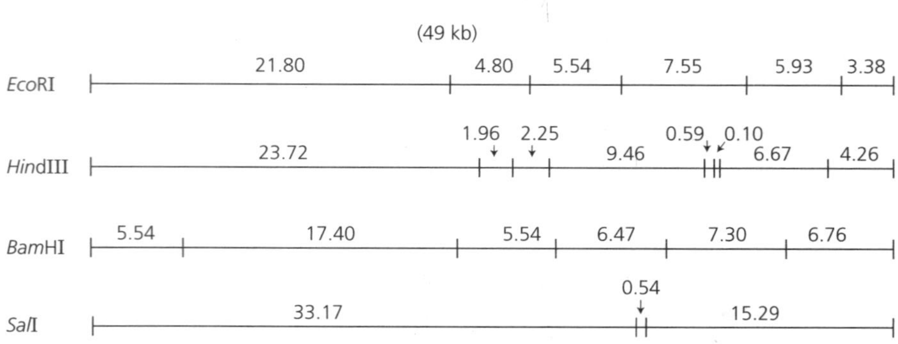

Figure 29.1 Restriction maps of the bacteriophage lambda genome. The topmost line represents the restriction map of the lambda (λ) genome obtained with the restriction endonuclease *Eco*RI, the next line below with *Hin*dIII, the next line below with *Bam*HI and the bottommost line with *Sal*I. The small vertical lines intersecting the horizontal lines represent the cleavage sites of the restriction endonucleases and the digits in between denote the sizes in kilobases of the fragments produced by cleavage with the enzymes.

*Bam*HI and *Hin*dIII, either (c) or (d) as shown in Fig. 29.2 could be correct, but as simultaneous treatment with these two endonucleases produces fragments of 1, 4 and 5 kb, possibility (d) has to be the correct one. The locations of the sites for *Eco*RI and *Pst*I could also be determined in a similar manner [Fig. 29.2(e)–(j)] and the complete restriction map could finally be determined [Fig. 29.2(k)].

29.1.2 Interrupted treatment with restriction endonucleases

Interrupted treatment or partial digestion of a DNA fragment with a restriction endonuclease could reveal information on the location of all the restriction sites on the fragment when multiple sites exist. On partial digestion, each DNA molecule does not suffer cleavage at every restriction site on it. Consider a situation when complete digestion of a 10-kb DNA fragment with the restriction endonuclease *Bam*HI produces, let us say, subfragments of 1, 3 and 6 kb, suggesting that there are two BamHI sites at locations shown in (a) or (b) or (c) in Fig. 29.3. Partial digestion with the same restriction endonuclease yields DNA subfragments of 1, 3, 4, 6 and 9 kb. It is obvious that possibility (b) would never generate on partial digestion, a subfragment of 4 kb, while possibility (c) cannot generate a subfragment of 9 kb. Possibility (a), on the other hand, can generate on partial digestion, all the subfragments found. Hence, possibility (a) is the correct one. The same 10-kb DNA fragment yields on complete digestion with *Bgl*II, subfragments of 2, 3 and 5 kb, suggesting that possibilities (d–f) exist. Partial digestion with the same enzyme, however, produces subfragments of 2, 3, 5, 7 and 8 kb. Now, possibility (d) can in no way produce on partial digestion, a subfragment of 7 kb, while possibility (f) would never generate a subfragment of 8 kb. Again, possibility (e) is capable of yielding all the subfragments mentioned, and thus has to be the correct one. The same kind of

10-kb fragment

Enzyme(s) used	Fragment size (kb)	Enzymes used	Fragment size (kb)
*Bam*HI	1, 9	*Bgl*II + *Eco*RI	2, 3, 5
*Bgl*H	3, 7	*Bgl*II + *Hin*dIII	3, 3, 4
*Eco*RI	5, 5	*Bgl*II + *Pst*I	2, 3, 5
*Hin*dIII	4, 6		
*Pst*I	2, 8	*Eco*RI + *Hin*dIII	1, 4, 5
		*Eco*RI + *Pst*I	2, 3, 5
*Bam*HI + *Bgl*II	1, 2, 7	*Hin*dIII + *Pst*I	2, 2, 6
*Bam*HI + *Eco*RI	1, 4, 5		
*Bam*HI + *Hin*dIII	1, 4, 5		
*Bam*HI + *Pst*I	1, 2, 7		

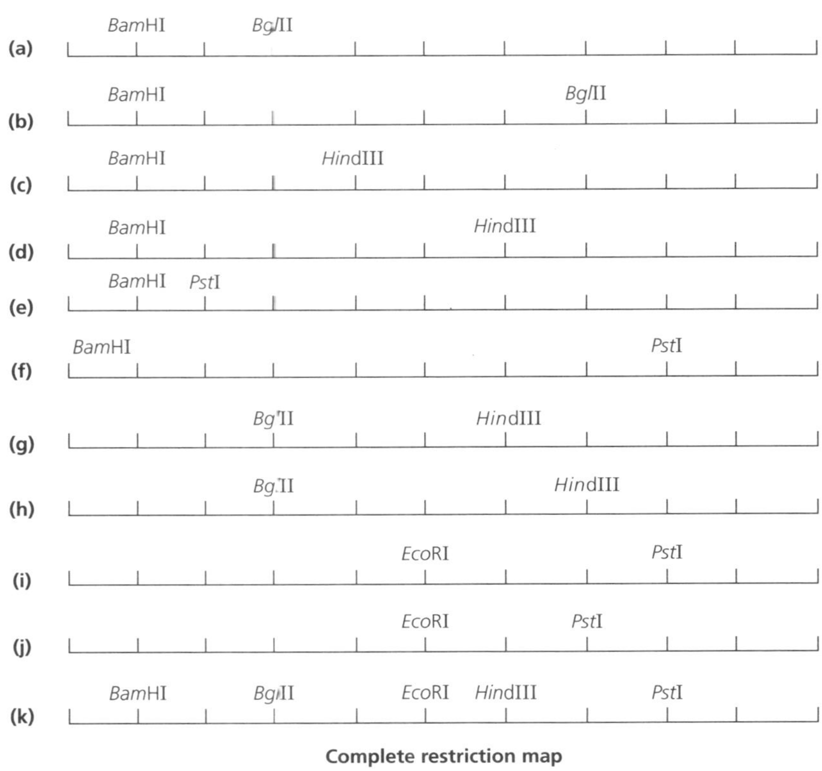

Figure 29.2 Construction of a restriction map by treatment with a number of single restriction endonucleaes followed by simultaneous treatment with pairs of restriction endonucleases. The 10-kb DNA fragment contains one site each for the restriction endonucleases *Bam*HI, *Bgl*II, *Eco*RI, *Hin*dIII and *Pst*I. Treatments with single enzyme and pairs of enzymes have been used.

experimentation and argument would enable one to locate the *Hin*dIII sites. Finally, double digestion with pairs of endonucleases would allow confirmation of the relative locations of the sites of all the three endonucleases as shown in Fig. 29.3. The complete restriction map deduced from the observations is shown in Fig. 29.3(g). Drawing inference from this kind of experimentation can be made more certain if one end of the DNA is radioactively labelled. One could label the 5′-end with ATP-γ-^{32}P using polynucleotide kinase.

29.1.3 Interrupted treatment with an exonuclease

The order of multiple sites of a restriction endonuclease in a DNA fragment can also be determined by incomplete treatment with an exonuclease. First, the DNA fragment is completely digested with the specific restriction endonuclease and the sizes of the subfragments are recorded. Let us say the subfragments obtained from a 21-kb DNA fragment are of 1, 2, 3, 4, 5 and 6 kb. How does one order these subfragments? The DNA fragment is subjected to digestion with the exonuclease *Bal*31, which cleaves sequentially from both ends. Samples are withdrawn periodically and *Bal*31 treatment is interrupted. Each sample is then subjected to complete digestion with the restriction endonuclease and the sizes of the subfragments produced are determined. Let us say in sample 1, the subfragments obtained are of sizes 0.5, 2, 3, 3.5, 5 and 6 kb. As the subfragments of 2, 3, 5 and 6 kb are unaffected, these must not be at the termini. So, we start constructing the map; the subfragments of 1 and 4 kb are at the two termini. Assume that sample 2 would have subfragments of 1.5, 2.5, 3, 5 and 6 kb. Now, the original subfragment of 1 kb is completely degraded and the subfragment that is newly affected is of

Complete digestion of 10-kb DNA fragment

Enzyme used	Subfragments formed (kb)	Enzyme used	Subfragments formed (kb)
*Bam*HI	1, 3, 6	*Bam*HI + *Bgl*II	1, 1, 2, 3, 3
*Bgl*II	2, 3, 5	*Bam*HI + *Hin*dIII	1, 1, 1, 3, 4
*Hin*dIII	1, 4, 5	*Bgl*II + *Hin*dIII	1, 1, 2, 3, 3

Partial digestion

Enzyme used	Subfragments formed (kb)
*Bam*HI	1, 3, 4, 6, 9
*Bgl*II	2, 3, 5, 7, 8
*Hin*dIII	1, 4, 5, 5, 6

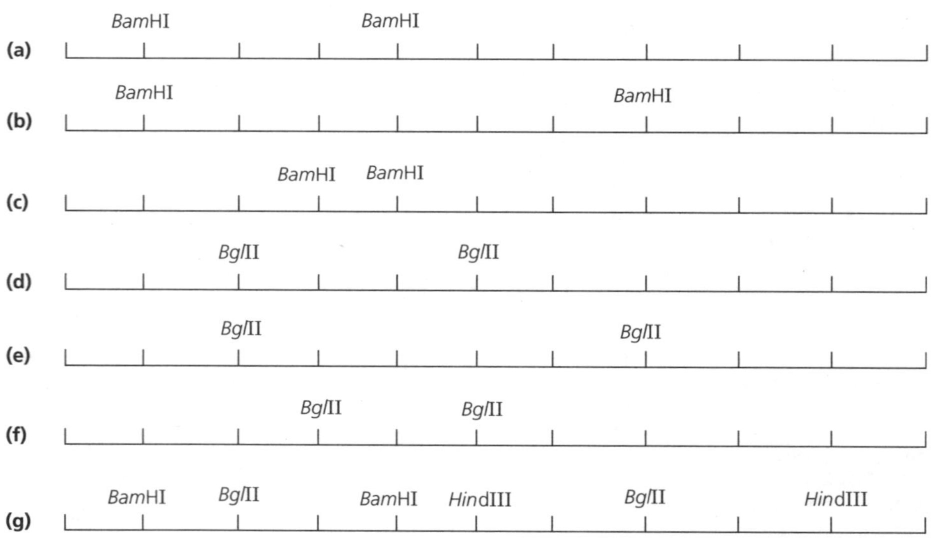

Complete restriction map

Figure 29.3 Construction of a restriction map by interrupted treatment with restriction endonucleases. The 10-kb DNA fragment contains two sites each for the restriction endonucleases *Bam*HI, *Bgl*I and *Hin*dIII. Both complete treatment and interrupted treatment with restriction endonucleases have been used.

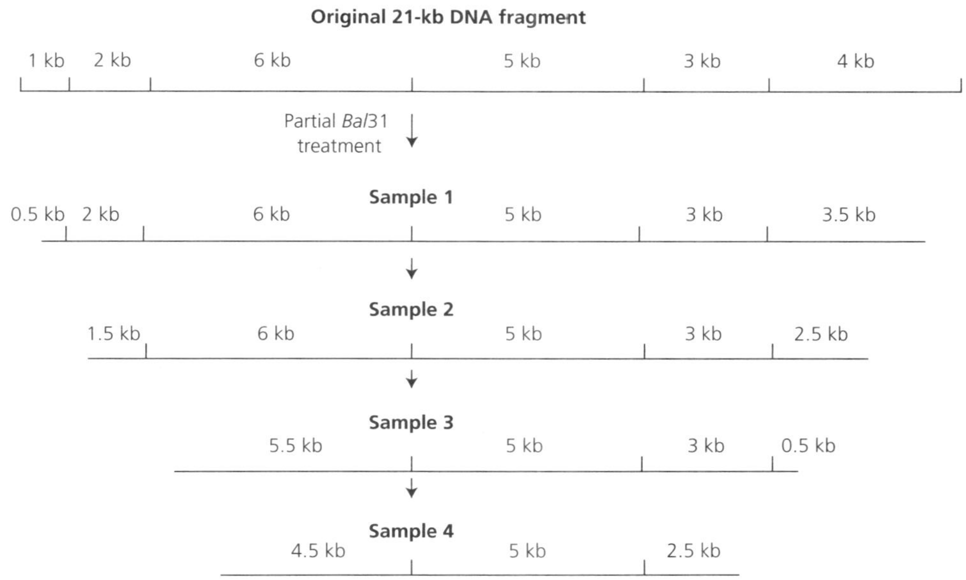

Figure 29.4 Construction of a restriction map by interrupted treatment with an exonuclease. The 21-kb DNA fragment contains five sites of one restriction endonuclease. The exonuclease *Bal*31 was used of to determine the order of the location of the six subfragments. First, the fragment was subjected to complete digestion with the restriction endonuclease and the sizes of the six subfragments were noted. Then, the whole 21-kb fragment was subjected to partial *Bal*31 treatment, samples were withdrawn periodically and *Bal*31 digestion was interrupted in the withdrawn samples. DNA of each sample was purified and then subjected to complete digestion with the restriction endonuclease. The sizes of the subfragments produced were determined. Sample 1: sizes of the subfragments of 1 kb and 4 kb were found smaller and hence were allocated terminal positions. Sample 2: the 2-kb subfragment has started getting reduced in size, and hence allocated a position next to the 1-kb subfragment. Sample 3: the 6-kb subfragment has been affected, and is hence placed next to the 2-kb subfragment. Sample 4: the original 4-kb subfragment was completely digested and the 3-kb subfragment was reduced in size; hence, the 3-kb subfragment is placed next to the 4-kb subfragment. The 5-kb subfragment then would be between the 6-kb and the 3-kb subfragment.

2 kb. The subfragments of 3, 5 and 6 kb are still untouched. The original subfragment of 4 kb continues to be digested (now 2.5 kb). Hence, we place the 2-kb subfragment next to the 1-kb subfragment. Sample 3 could have subfragments of 0.5, 3, 5 and 5.5 kb. The subfragments of 3 and 5 kb are still unaffected. The original subfragment of 4 kb is still being degraded (now 0.5 kb), while the degradation of the subfragment of 6 kb has just been initiated (now 5.5 kb). This would let us place the subfragment of 6 kb next to the 2-kb subfragment. Sample 4 may have subfragments of 2.5, 4.5 and 5 kb. Thus, the original 4-kb subfragment is completely gone, while the original 6-kb subfragment continues to be degraded (now 4.5 kb). This time, only the subfragment of 3 kb has been newly affected; hence, we can allot it a place next to the original 4-kb subfragment. This lets us complete the map construction; the subfragments are in the order of 1, 2, 6, 5, 3 and 4 kb. Fig. 29.4 illustrates the course of the *Bal*31 reaction.

29.2 DNA size markers

We have just discussed about the determination of restriction maps. We have assumed that the sizes of the subfragments produced by a restriction endonuclease are known to us. Let us now talk about how one finds the sizes of the DNA fragments produced by a restriction endonuclease. The subfragments produced by cleavage with the restriction endonuclease are electrophoresed on an appropriate gel. Some DNA fragments of known molecular weights are applied to a well nearby and simultaneously electrophoresed. Such DNA fragments of known molecular weights are used as size markers to determine the sizes of the newly produced DNA fragments by extrapolation. A particular restriction endonuclease would cleave a specific DNA molecule reproducibly at exactly the same sites. As many of the viral DNA genomes can easily be isolated in intact form, these molecules cleaved with a specific restriction endonuclease have been popular as DNA markers. A look at Fig. 29.5 would illustrate the point. In fact, the DNA fragments obtained by cleaving the genome of the bacterial virus lambda (λ) with the restriction endonuclease *Hin*dIII are very often used for determining the sizes of DNA fragments between 2 kb and 20 kb [Fig. 29.5(a)]. The same genome of the virus lambda (λ) when cleaved with the restriction endonuclease *Bst*EII can be used as a DNA marker between the size range of 700 bases and 8 kb [Fig. 29.5(b)]. Fragments obtained by cleaving the genome of the bacterial virus ϕx174 with

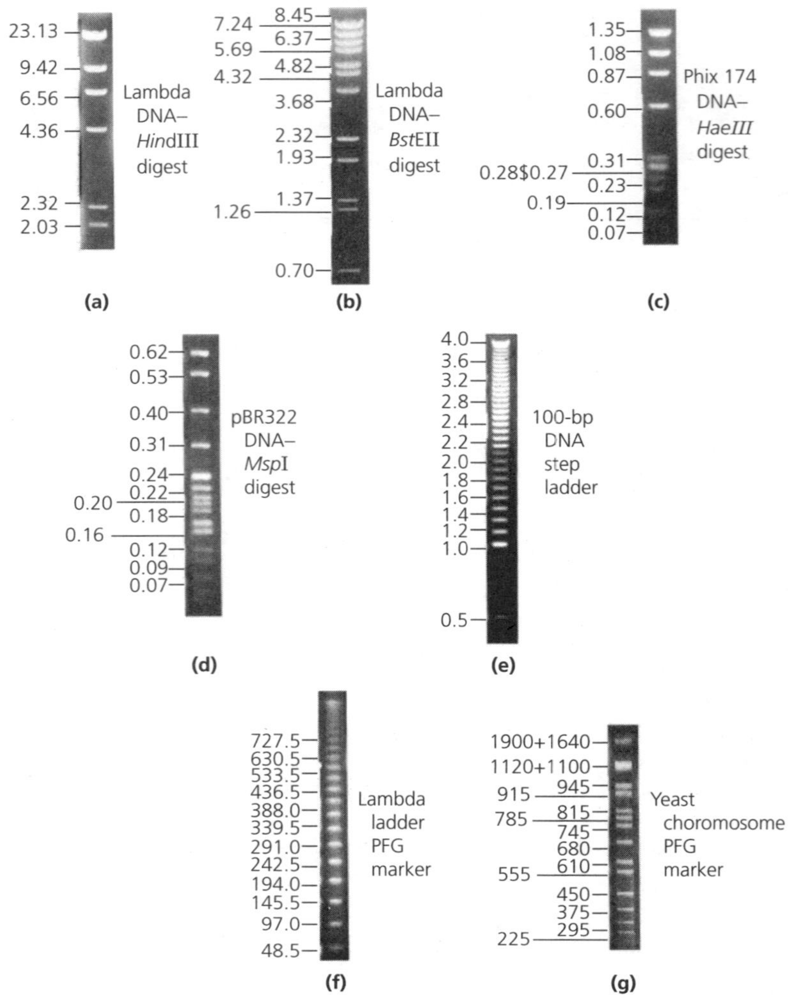

Figure 29.5 DNA size markers. Each band represents a DNA fragment. The numbers that appear on the left of each gel strip denote the size of the marker DNA fragments in kilobases (kb). The writings on the right of the strip describe the source of the markers. (a) The DNA of the bacterial virus lambda (λ) when digested with the restriction endonuclease *Hin*dIII yields DNA fragments of sizes between 2.0 kb and 23.1 kb. The figure shows fragments electrophoresed in 1% agarose gel, stained with ethidium bromide and visualized by fluorescence under ultraviolet rays. (b) The same DNA of the bacterial virus lambda (λ) when digested with the restriction endonuclease *Bst*E II yields DNA fragments of sizes between 0.7 kb and 8 kb. The figure shows fragments after electrophoresis in 1% agarose gel. (c) The DNA of the bacterial virus φx174 when digested with the restriction endonuclease *Hae*III yields fragments between 0.2 kb and 1.3 kb. The figure shows fragments electrophoresed in 1.7% agarose gel. (d) The DNA of the plasmid pBR322 when digested with the restriction endonuclease *Msp*I yields DNA fragments of sizes between 0.07 kb and 0.62 kb. The figure shows fragments electrophoresed in 1.8% agarose gel. (e) The figure shows a DNA step ladder formed by electrophoresis in 1% agarose gel of blunt-ended DNA fragments of 100-bp size differences between the range of 0.5 kb and 4 kb. (f) Another ladder, but of the genomes of the bacterial virus lambda (λ), is shown in this figure. Concatamers of the lambda (λ) DNA have been subjected to pulsed-field gel (PFG) electrophoresis in 1% agarose gel. This ladder serves as a marker for large DNA molecules separated by PFG electrophoresis. (g) Intact chromosomes of the yeast *Saccharomyces cerevisiae* separated by PFG electrophoresis in 1% agarose gel. This is also quite often used as marker for large DNA molecules.

the restriction endonuclease *Hae*III are used as markers for DNA fragments between 200 bases and 1.3 kb [Fig. 29.5(c)]. Restriction endonuclease digest of plasmid DNA is also used as DNA size markers. Fig. 29.5(d) shows size markers in the range of 100 and 600 bases obtained by digestion of the plasmid pBR322 with *Msp*I. Step ladders of DNA fragments [Fig. 29.5(e)] are useful for precise size determination of small DNA fragments. A ladder of lambda (λ) genomes [Fig. 29.5(f)] is useful for pulsed-field gel (PFG) electrophoresis experiments. At larger size ranges, intact chromosomes of the yeast *Saccharomyces cerevisiae* [Fig. 29.5(g)] are used in experiments involving PFG electrophoresis.

29.3 Determination of the sequence of bases in DNA

A number of methods are available for the determination of base sequence of DNA, some of which are discussed next.

29.3.1 Sequencing by specific chemical cleavage of phosphodiester linkages

Allan Maxam and Walter Gilbert introduced the method of chemical sequencing of DNA (Maxam and Gilbert 1977, 1980). The guanine residues in DNA can be methylated at the position N7 by dimethyl sulphate. These methylated guanines are eliminated by subsequent treatment with piperidine, thereby cleaving the DNA at those points. The reactions are shown in Fig. 29.6. Some adenines also get methylated at position N3 by dimethyl sulphate, but piperidine does not eliminate the methylated adenines. Treatment with formic acid (adjusted to pH 2 with pyridine) weakens the glycosidic bonds of both adenine and guanine by protonating the purine ring nitrogens, which are subsequently breached by alkali. Piperidine can then eliminate these bases. Hydrazine reacts with the thymine and cytosine residues in DNA at positions C4 and C6 and subsequently forms a new five-membered ring with hydrazine. Piperidine reacts with these hydrazine rings, resulting in the elimination of the thymine and cytosine residues. The sequence of reactions with thymine residues is shown in Fig. 29.7. If molar sodium chloride is present during the hydrazine reaction, only cytosine is affected. The four reactions are, thus, specific for (i) guanine residues, (ii) guanine plus adenine residues, (iii) thymine plus cytosine residues and (iv) cytosine reactions.

Before initiating the cleavage reaction, either the 5′-end or the 3′-end of the DNA strands is labelled with ^{32}P. Polynucleotide kinase and γ-^{32}P–labelled ATP are used to label the 5′-end. DNA polymerase I and α-^{32}P–labelled nucleoside triphosphates are used to label the 3′-end.

The rationale of the procedure is that if the cleavage reactions are not allowed to proceed to completion and breaks in the DNA strand (labelled at one end) are allowed to take place only at a limited number of sites

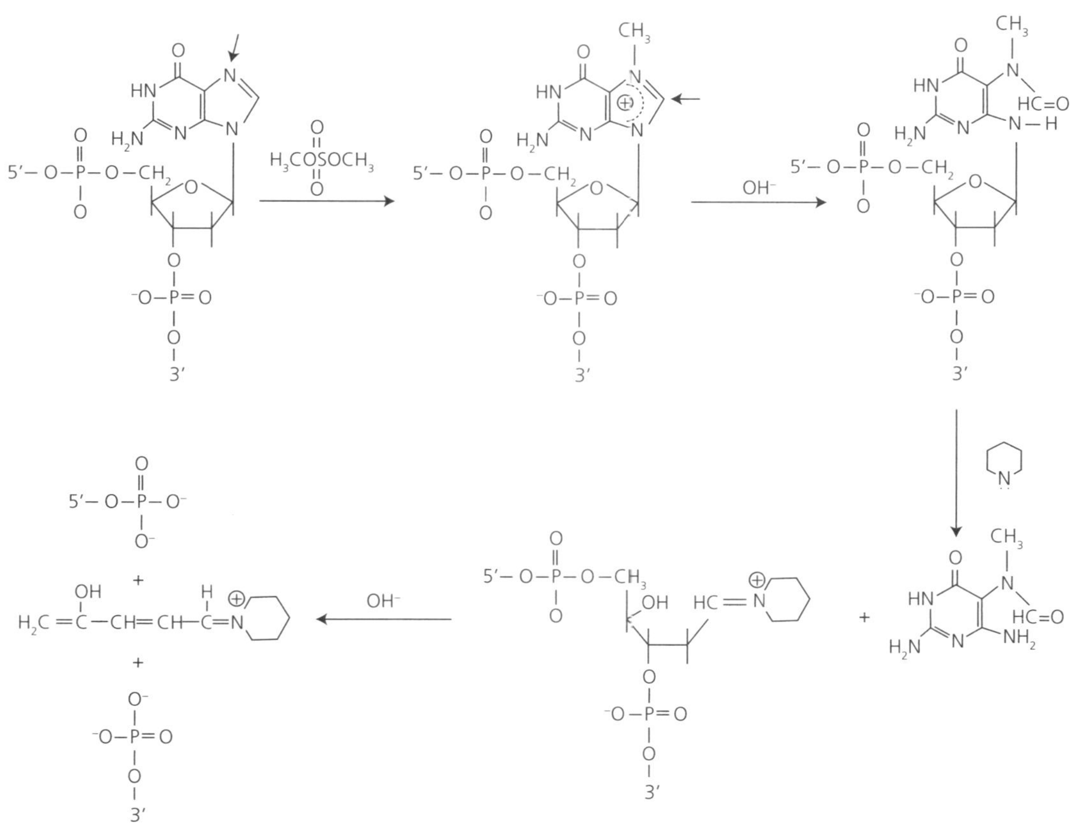

Figure 29.6 Reaction of dimethyl sulphate and piperidine with guanine.

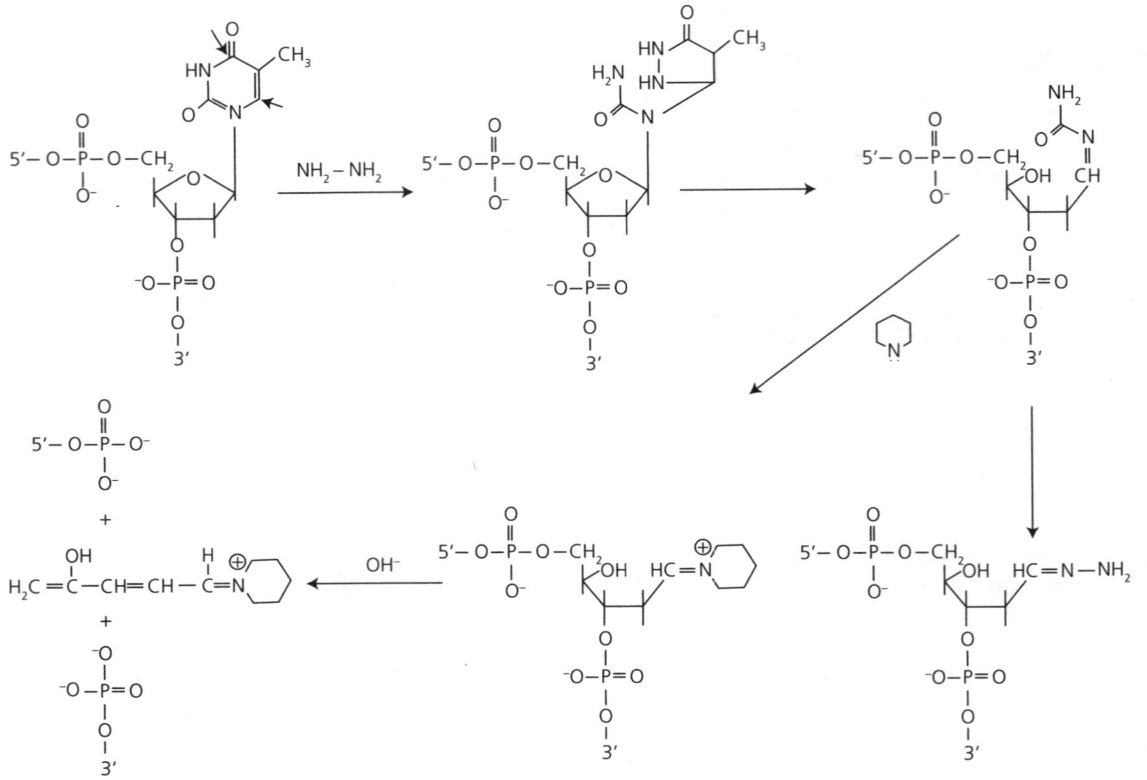

Figure 29.7 Reaction of hydrazine and piperidine with thymine.

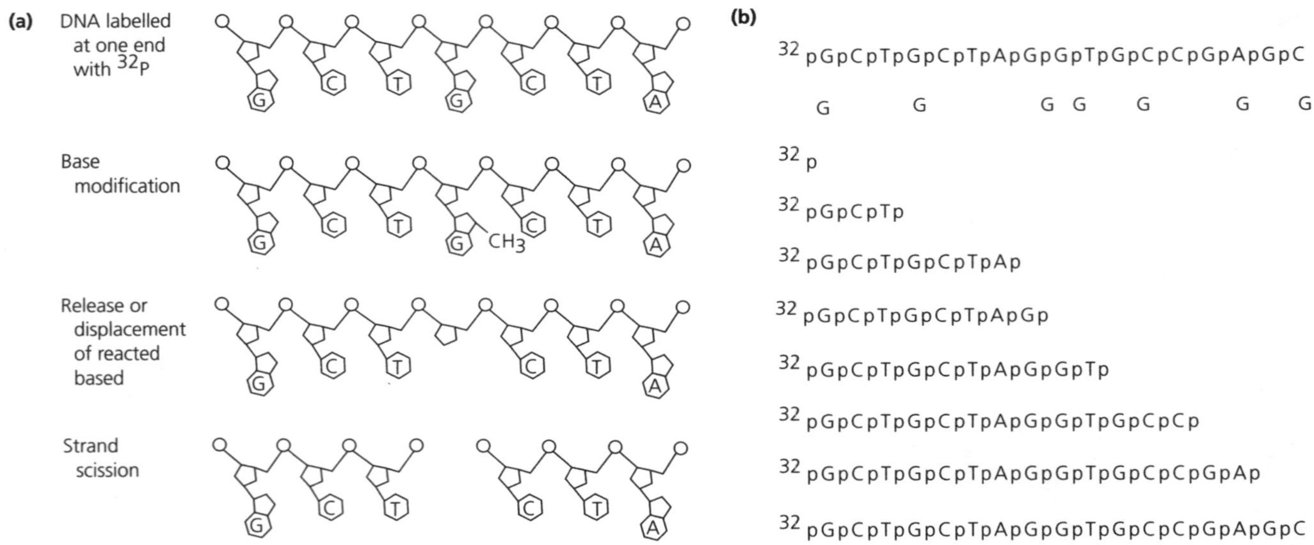

Figure 29.8 Cleavage of guanine residues (a) in DNA and formation of fragments of assorted lengths (b).

at random, then one would very likely obtain a complete assortment of lengths of labelled DNA fragments (Fig. 29.8). If all these fragments can be separated by electrophoresis on the basis of length, then the positions of any individual bases in the DNA can be determined. The resolution of the electrophoretic separation process has to be such that two fragments of DNA differing in length by only one base would also be resolved. The fragments obtained by the treatment with dimethyl sulphate would indicate the positions of guanine residues in the DNA strand. The products obtained from the four reactions mentioned above are electrophoresed in four adjacent lanes in a very thin urea–polyacrylamide gel held upright between two glass plates. One glass plate is removed and the gel on the other glass plate is exposed to an X-ray film. The X-ray film is developed to reveal the

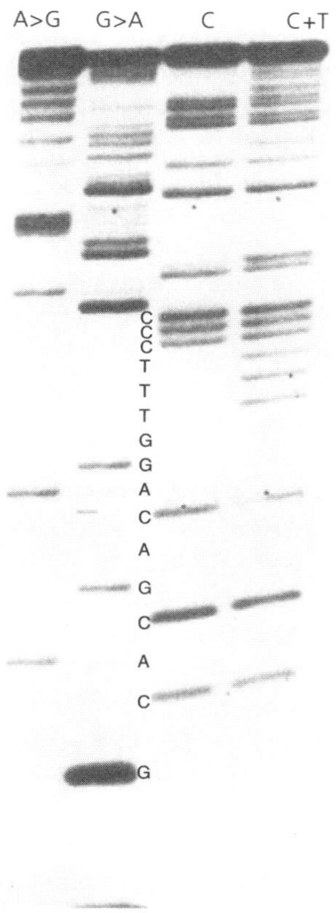

Figure 29.9 A sample X-ray plate showing separation of labelled fragments of DNA for base sequence analysis done by the chemical cleavage method of Maxam and Gilbert. *Source:* Maxam and Gilbert, *Proc Natl Acad Sci USA* 1977;74:563.

sharp black bands. The base sequence can be worked out from the positions of the bands in the four lanes. A sample X-ray plate is shown in Fig. 29.9. As the distance traversed by all the fragments would be according to the length from one specific end (the end which has been labelled), the polarity of the bases would also be obvious.

29.3.2 Sequencing by enzymatic synthesis using chain-terminating derivatives of nucleotides

Frederick Sanger introduced the use of *E. coli* DNA polymerase I for DNA base sequencing (Sanger *et al.* 1973). Chapter 9, Section 9.2 contains a detailed treatment of the synthesis *in vitro* of DNA complementary to a given DNA template using DNA polymerase I from *E. coli* and the 5′-triphosphates of deoxyguanosine, deoxyadenosine, deoxycytidine and thymidine. Sanger's 'plus and minus' method (Sanger and Coulson 1975) was itself a big advancement in this area. In 1977, he used chain-terminating derivatives of nucleotides to randomly terminate the reaction (Sanger *et al.* 1977). If the 3′-hydroxyl of one of the four deoxynucleotide triphosphates (say of guanosine) is removed (the compound is called 2′, 3′-dideoxy derivative), the chain would be terminated at the next guanine residue, as no subsequent 5′–3′-phosphodiester linkage can be formed. If a mixture of the deoxynucleotide (2′-monodeoxy) triphosphate (the normal substrate) and 2′, 3′-dideoxy derivative of guanosine is used along with the other three deoxynucleotide (2′-monodeoxy) triphosphates, termination would not always take place at the first guanine residue. Termination would actually take place at random at all possible guanine residues, thus creating a complete collection of DNA fragments ending with guanine. In practice, each of the four dideoxy derivatives is used in four separate reactions that are initiated with ^{32}P-labelled primers. The labelled DNA fragments of all sizes are then electrophoresed in a thin (~1 mm) polyacrylamide gel containing 8 M urea and the sequence is determined as described for the chemical cleavage method. A sample X-ray plate is shown in Fig. 29.10.

29.3.3 Automation of DNA base sequencing

DNA base sequencing by the Sanger method was automated by the replacement of the ^{32}P-labelled deoxyribonucleotides with fluorescent dye–tagged deoxyribonucleotides during the synthesis of the primer used for DNA replication. Tetramethylrhodamine iodoacetamide was the reagent that was chosen to be conjugated with sulphydryl-containing M13 sequencing primers by Wilhelm Ansorge and coworkers at the

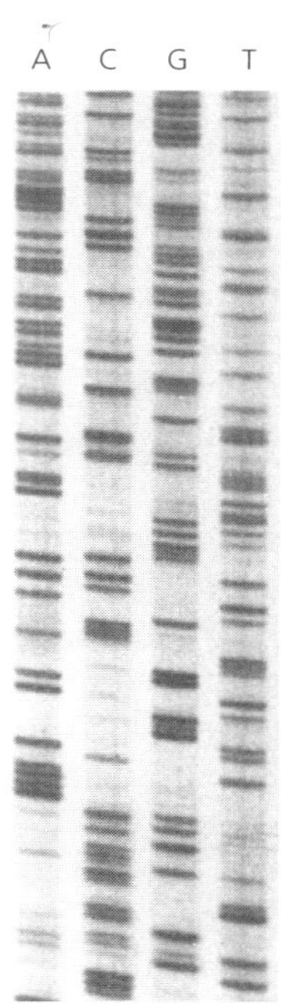

Figure 29.10 A sample X-ray plate showing separation of labelled fragments of DNA for base sequence analysis done by the enzymatic synthesis method of Sanger and coworkers.

European Molecular Biology Laboratory in Heidelberg, Germany, in 1986. Otherwise, the basic Sanger procedure of enzymatic synthesis using chain-terminating derivatives of nucleotides was followed and electrophoresis was done in four separate lanes. The fluorescent bands in the four lanes of the gel were excited by a laser and the sequences from the four lanes were transferred directly into a computer (Ansorge *et al.* 1986, 1987). This technology was adopted for commercialization by Pharmacia-Amersham.

Six months ahead of the first Ansorge paper, Leroy Hood and coworkers from the California Institute of Technology in Pasadena, CA, reported an automated procedure of DNA sequencing using four different fluorescent tagging dyes, which would emit distinct and non-overlapping spectra on excitation with laser. Four different synthetic reactions were carried out with *E. coli* DNA polymerase I with the four different chain-terminating dideoxyribonucleotides, as in the Sanger procedure, but the four primers used for the four different reactions were labelled with four different tagging dyes. The tagging dyes used were NBD aminohexanoic acid (NBD), fluorescein isothiocyanate (FITC), tetramethylrhodamine isothiocyanate (TM) and Texas Red (TEX) (Smith *et al.* 1986). Electrophoresis in four separate lanes was unnecessary and the products of the four different reactions were combined and electrophoresed together in a single gel tube. Differently coloured fluorescent bands were detected by laser excitation near the bottom of the tube and this sequence information was recorded directly by a computer. The Hood technology was adopted for commercialization by Applied Biosystems. With time, there was considerable improvement in the sensitivity and resolution of the detection system. The separation of the DNA fragments was carried out then in a capillary tube of short length. Higher voltage could now be applied, as heat dissipation from the air-cooled capillary was faster, thus drastically reducing the time of electrophoretic separation.

29.3.4 Capillary array DNA sequencer

The Human Genome Project demanded simultaneous automatic sequencing of a large number of DNA samples. This was achieved by going back to slab gel electrophoresis and applying each sample in each lane and carrying out base sequencing by the Hood technology. A scanning laser beam irradiation method, where the laser beam was aimed at different lanes on the slab gel sequentially, was used. The advantages of capillary gel electrophoresis were, however, lost.

Dr Hideki Kambara of the Hitachi Laboratories in Japan went back to capillary gel electrophoresis and developed an automated capillary array DNA sequencer where an array of as many as 96 capillaries were employed in a single sequencer, each capillary analysing one DNA sample. A 'side entry multiple focusing laser irradiation method' of individual capillaries was used, in which disturbing reflection and refraction from capillaries were minimized by making each capillary behave like a convex lens through adjustment of the inner and outer diameters of the capillaries. Fluorescence emission from every individual capillary was simultaneously detected by a CCD camera and recorded in a computer (Kambara 1998). The technology was commercialized in 2000 by Applied Biosystems and subsequently by Molecular Dynamics/Amersham, offering up to 384 capillaries in one sequencer.

29.3.5 Sequencing by monitoring pyrophosphate release during DNA synthesis: No tagging, no chain terminators, no gels, no electrophoresis

E. coli DNA polymerase I catalyses the *in vitro* synthesis of DNA using the four deoxyribonucleoside triphosphates as substrates. The deoxyribonucleoside monophosphates are added to the primer sequentially, releasing an inorganic pyrophosphate molecule at each addition step (Chapter 9, Section 9.2 contains more details). If the triphosphate of only one deoxyribonucleoside is added at a time into the reaction mixture containing the DNA template, the primer, DNA polymerase I and other necessary factors and if the release of pyrophosphate or the lack of release can be monitored, then the identity of the deoxyribonucleoside actually being added to the primer can be determined. Repetition of the steps would thus furnish base sequence information of the DNA template. Pal Nyren and A Lundin had reported from Stockholm (Nyren and Lundin 1985) a sensitive method of continuous monitoring of the release of inorganic pyrophosphate. Adenosine 5′-phosphosulphate and ATP sulphurylase were used to generate ATP from inorganic pyrophosphate. The ATP thus generated was acted upon by oxyluciferin and firefly luciferase and the light emitted was monitored. Inorganic phosphate did not interfere with this assay. The assay procedure was adapted by ED Hyman to report on a new method of sequencing of DNA (Hyman 1988). Pal Nyren and coworkers (Ronaghi *et al.* 1996) had further improved upon the procedure. Roche Applied Science adapted the technology and developed the commercial version 454 GenomeSequencer FLX instrument in 2005. In the commercial version, the DNA fragment to be sequenced is ligated to a specific adapter that binds it to a bead present in a water droplet. The DNA molecule is amplified by emulsion polymerase chain reaction (PCR), with the PCR reagents immersed in oil. The bead containing the amplified and denatured DNA is placed at the top end of an etched glass fibre in an optical fibre chip, the bottom end of the fibre facing a sensitive CCD camera.

The chip contains several hundred thousand fibres and each fibre would carry a bead containing a DNA sample, with each bead sitting on an addressable position. The DNA replication process is next initiated with a primer, DNA polymerase and other factors. Only one deoxyribonucleotide triphosphate (say dCTP) is added to all the beads. Pyrophosphate would be released (which is detected by the emitted light) only on those beads where the particular deoxyribonucleotide (dC in this example) would enter into a phosphodiester linkage. This step is repeated for the other three deoxyribonucleotide triphosphates (dA, dG and T). As the identity of the nucleotide added in each step is known, the appearance or non-appearance of the light signal would indicate whether that specific nucleotide was incorporated or not in the DNA sample on that particular bead. Homopolar stretches of identical bases could, however, be a source of inaccuracy in this procedure.

29.3.6 Sequencing by synthesis: Use of reversible chain terminators tagged with four different fluorescent dyes

Solexa in 2006 commercialized a sequencing procedure based on synthesis using reversible terminator nucleotides, each tagged with a different fluorescent dye. The DNA fragments are ligated on both sides to adapters that are immobilized on a solid support. Amplification by PCR and denaturation follow, resulting in single-stranded clusters of about 1000 copies of each DNA fragment. These clusters are termed DNA 'polonies', which resemble cell colonies adhering to the solid support. DNA polymerase I, the primer and the four reversible terminator nucleotides, each tagged with a different fluorescent dye, are added subsequently to initiate sequencing by synthesis. The position of the 'polony' into which a specific terminator nucleotide is incorporated is identified by a CCD camera through its specific fluorescent dye. The terminator groups and the fluorescent dyes are then removed from the 3′-ends of the DNA fragments and another round of incorporation of reversible terminator nucleotides is carried out. About 35 successive nucleotides can thus be incorporated into each DNA fragment. Simultaneously, 40 million such polonies can thus be handled.

29.3.7 Sequencing by just hybridization and ligation, no synthesis

Applied Biosystems in their ABI SOLiD system have introduced a sequencing process which comprises just hybridization and ligation, but requires no synthesis. The DNA fragments are ligated to oligonucleotide adapters and then allowed to bind to beads, each bead containing one DNA fragment. Emulsion PCR to amplify the experimental DNA fragment is followed by denaturation and hybridization of primers complementary to the adapters. A set of four random dinucleotides labelled with four different fluorescent tags are now allowed to anneal with the DNA. A ligation reaction is carried out next. If the ligation is successful, as monitored by the presence of the bound fluorescent tag, the identity of the first two nucleotides in the DNA is revealed, being complementary to the sequence of the specific fluorescent dinucleotide probe. All the 16 dinucleotide probes are used sequentially. The probes are cleaved after successful ligation and multiple cycles of annealing, ligation, detection and cleavage are subsequently carried out. About 36 cycles are possible, each cycle allowing reading of two bases. The extension product along with the primer is removed after possible ligation cycles and a new primer containing $n - 1$ nucleotides is hybridized with the DNA fragment. The cycle of annealing, ligation, detection and cleavage is carried out to confirm the base sequence determined in the first round. A total of four confirmatory rounds are run for the same DNA with primers containing n to $n - 3$ bases. More than 50 million bead clusters can be employed at a time, thus allowing a throughput of gigabases per run.

29.3.8 Sequencing by synthesis using a single DNA molecule as template

PCR amplification of the DNA molecule has been an indispensible step for intensifying the fluorescence signal for any procedure dependent on detection by CCD cameras. PCR amplification can, however, introduce errors or specifically copy certain sequences in preference to others. Stephen R Quake and coworkers (Braslavsky *et al.* 2003) from the California Institute of Technology reported a procedure in which no amplification of DNA was necessary. Monitoring was done by a fluorescence microscope with total internal reflection illumination that minimized background fluorescence. Successive incorporation of fluorescently labelled nucleotides could be detected at single-base resolution during DNA polymerase I–catalysed synthesis on single DNA molecule templates. The denatured DNA fragments were immobilized on a quartz slide, which was pretreated to specifically bind polynucleotides, but not free mononucleotides. The concentration of the fragments was kept low to resolve single DNA molecules. Fluorescently tagged primers were annealed to the immobilized DNA molecules and the position of each DNA molecule was recorded. The tag was then photobleached and the DNA synthesis reaction was initiated by adding DNA polymerase I and only one fluorescent dye–tagged deoxyribonucleoside triphosphate. The incorporation or lack of incorporation of the specific nucleotide was recorded and the tag was photobleached and washed out. The next fluorescently labelled deoxyribonucleotide triphosphate was then added along with DNA polymerase I and the cycle was repeated. An average 'read length' of 23 bases could be achieved. Millions of single DNA fragments were sequenced simultaneously, resulting in a gigabase sequence throughput. Quake's group has used this procedure to determine the base sequence of an individual human's genome (Pushkarev *et al.* 2009). Helicos Biosciences commercialized the procedure in 2007.

29.3.9 Massive parallel sequencing

The massive parallel sequencing platforms differ in engineering configurations and sequencing chemistry. The common thing is that chain termination chemistry is not used. Sequencing is achieved by synthesis by addition of nucleotides to the complementary strand. The spatially segregated, amplified DNA templates are sequenced simultaneously in a massively parallel fashion without the requirement for a physical separation step. Depending on the specific platform, the chemistry may be monitoring pyrophosphate release (Section 28.3.5), or using reversible dye terminator (Section 28.3.6), or oligonucleotide probe ligation (Section 28.3.7) or by incorporation of fluorescently labeled nucleotides (Section 28.3.8). The review article by Tucker *et al.* (2009) is informative. The genome of James D Watson has been sequenced by massive parallel sequencing to 7.4-fold redundancy in 2 months (Wheeler *et al.* 2008).

29.3.10 Other emerging sequencing technologies based on synthesis

A DNA polymerase which has a fluorescent dye tag close to the active site of the enzyme has been developed by VisiGen Biotechnologies. In addition, the four nucleotides also have four different dye tags. A fluorescent resonant energy transfer (FRET) signal emanates as energy and is transferred from the polymerase dye to the nucleotide dye when the correct nucleotide reaches the active site. The dye tag on the nucleotide determines the frequency of this signal, which reveals the base that is incorporated. The speed of the polymerase reaction is usually a few hundred bases per second. The base sequencing can, therefore, be done at that speed by the VisiGen procedure.

Zero-mode waveguides (ZMW) have been developed at the Cornell University Nanobiotechnology Centre. These are nanometer-scale aperture chambers in a 100-nm metal film. The measurement of fluorescence of nucleotides being incorporated into a single growing chain was possible by reducing the observation volume to only 20 zL (zeptolitres). Read lengths are about 1500 bases and about 3000 ZMWs can be handled at a time. Pacific Biosciences has adapted this technology.

29.3.11 Single-molecule, motion-based DNA sequencing using RNA polymerase

The principle of a method for determining the base sequence of DNA that relied on the motion of a single RNA polymerase molecule has been demonstrated in 2006 by Stanford University (Greenleaf and Block 2006). An RNA polymerase molecules pauses when the required ribonucleotide triphosphate substrate is limiting. An optical trapping device has been developed which was capable of base pair resolution and this device has been used to monitor transcription under limiting amounts of each of the four ribonucleotide triphosphates. The base sequence of the DNA template has been correctly inferred from the patterns of pauses recorded. Thus, the motion of a processive nucleic acid polymerase can reveal base sequence information of DNA.

29.3.12 Determination of the sequence of bases of DNA during translocation through a nanopore

As a DNA molecule travels through a nanopore, ionic current is generated, which is a signature of the diameter, length and conformation of the monodeoxynucleotides. The base sequence of DNA can thus be determined without recourse to any biochemical reaction (Trepagnier *et al.* 2007). The normal speed of travel of DNA through artificial nanopores is quite high, but was reduced about 200 folds using optical tweezers. Besides, optical tweezers made it possible 'to "floss" single polymers back and forth through the pore', thus making it possible to gather complete information about the bases. Oxford Nanopore Technologies, BioNanomatrix and Complete Genomics have come together to commercialize the technology, with a goal to sequence the human genome in 8 h at a cost of less than US $100.

References

Ansorge W, Sproat BS, Stegemann J, Schwager C. A non-radioactive automated method for DNA sequence determination. *J Biochem Biophys Meth* 1986;13:315–323.

Ansorge W, Sproat BS, Stegemann J, Schwager C, Zenke M. Automated DNA sequencing: Ultrasensitive detection of fluorescent bands during electrophoresis. *Nucl Acids Res* 1987;15:4593–4602

Braslavsky I, Hebert B, Kartalov E, Quake SR. Sequence information can be obtained from single DNA molecules. *Proc Natl Acad Sci USA* 2003;100:3960–3964.

Greenleaf WJ, Block SM. Single-molecule, motion-based DNA sequencing using RNA polymerase. *Science* 2006;313:801–801.

Hyman ED. A new method of DNA sequencing. *Anal Biochem* 1988;174:423–436.

Kambara H. Capillary array electrophoresis system. United States of America Patent no. 5,730,850; dated 24 March 1998.

Maxam A, Gilbert W. A new method for sequencing DNA. *Proc Natl Acad Sci USA* 1977;74:560–564.

Maxam A, Gilbert W. Sequencing end-labelled DNA with base-specific chemical cleavages. *Methods Enzymol* 1980;65:499–560.

Nyren P, Lundin A. Enzymatic method for continuous monitoring of inorganic pyrophosphate synthesis. *Anal Biochem* 1985;151: 504–509.

Ronaghi M, Karamohamed S, Pettersson B, Uhlén M, Nyrén P. Real-time DNA sequencing using detection of pyrophosphate release. *Anal Biochem* 1996;242:84–89.

Sanger F. Determination of nucleotide sequences in DNA. *Science* 1981;214:1205–1210.

Sanger F, Coulson AR. A rapid method for determining sequences in DNA by primed synthesis with DNA polymerase. *J Mol Biol* 1975;94:441–448.

Sanger F, Donelson JE, Coulson AR, Kössel H, Fischer D. Use of DNA polymerase I primed by a synthetic oligonucleotide to determine a nucleotide sequence in phage f1 DNA. *Proc Natl Acad Sci USA* 1973;70:1209–1213.

Sanger F, Nicklen S, Coulson AR. DNA sequencing with chain-terminating inhibitors, *Proc Natl Acad Sci USA* 1977;74: 5463–5467.

Smith LM, Sanders JZ, Kaiser RJ, Hughes P, Dodd C, Connell CR, *et al.* Fluorescence detection in automated DNA sequence analysis. *Nature* 1986;321:674–679.

Trepagnier EH, Radenovic A, Sivak D, Geissler P, Liphardt J. Controlling DNA capture and propagation through artificial nanopores. *Nano Lett* 2007;7:2824–2830.

Tucker T, Marra M, Friedman JM. Massively parallel sequencing: The next big thing in genetic medicine. *Am J Hum Genet* 2009;85:142–154.

Wheeler DA, Srinivasan M, Egholm M, Shen Y, Chen L, McGuire A, *et al.* The complete genome of an individual by massively parallel DNA sequencing. *Nature* 2008;452:872–876.

30

Reverse transcription and cDNA cloning

H. K. Das

30.1 What is cDNA and why is it cloned?

cDNA stands for complementary DNA. It is a polydeoxyribonucleotide duplex, the sequence of one strand of which is complementary to an RNA molecule, and has been obtained by the use of the enzyme reverse transcriptase on the RNA as template (Baltimore 1970; Temin *et al.* 1970). Sometimes, it is easier to isolate an mRNA molecule than the gene itself, because of selective expression of a particular gene, resulting in the relative abundance of that mRNA. Insulin mRNA in the islet cells of pancreas and ovalbumin mRNA in the hen oviduct on treatment with oestrogens are examples. In such situations, cDNA cloning is the choice of the experimenter. As the cDNA is derived from the mRNA, there is no question of any intervening sequence in it (cf. Bretnach *et al.* 1977; Chambon 1981), and hence, its transcription and translation in bacteria such as *Escherichia coli* would be straightforward. The other important point is that only about 1–10% of the total DNA in a plant or animal cell is expressed or is functional genetically. A cDNA library, which would thus represent only the active genes, would be much smaller than a gene library and hence would be much simpler to construct and handle. Besides, there are RNA viruses, such as influenza virus and reovirus, which do not replicate through a DNA intermediate. Their cDNA can only be cloned.

30.2 Reverse transcription: Its initiation on DNA primers

The first step in cDNA synthesis is the production of a hybrid mRNA–DNA duplex catalysed by the enzyme reverse transcriptase. The deoxyribonucleotides are added on to the 3′-end of an exogenous primer. The primer needs to have a sequence of bases complementary to the 3′-end of the mRNA, which has to be reverse transcribed. Most eukaryotic mRNAs contain a poly-A tail at the 3′-end. The primer used has, therefore, almost invariably been oligo dT (Efstratiadis and Villa-Komaroff 1979; Okayama and Berg 1982). In the next step, the RNA is eliminated by digestion with alkali and the cDNA single strand is converted into a double-stranded DNA duplex by the use of reverse transcriptase or *E. coli* DNA polymerase I. The cDNA itself serves as the primer, a loop is formed and then a duplex DNA. The loop between the two strands is digested by the single-strand–specific nuclease S1. A cDNA duplex does not have any cohesive ends; hence, such structures are added to the ends before it is cloned in a vehicle. This is achieved by either of the following two ways.

30.3 Use of linkers in cDNA cloning

The addition of synthetic 'linkers' through 'blunt-end ligation' (Bahl *et al.* 1976; Kurtz and Nicidemus 1981) is one way of adding cohesive ends. A 'linker' is a short sequence of deoxynucleotides which includes bases specifying the recognition sequence of a restriction endonuclease. A *Hin*dIII linker and a *Bam*HI linker are shown as follows:

*Hin*dIII linker	*Bam*HI linker
5′ CCC***AAGCTT***GGG 3′	5′ CCC***GGATCC***GGG 3′
3′ GGG***TTCGAA***CCC 5′	3′ GGG***CCTAGG***CCC 5′

Letters in bold italics represent the recognition sequences of the two restriction endonucleases. The

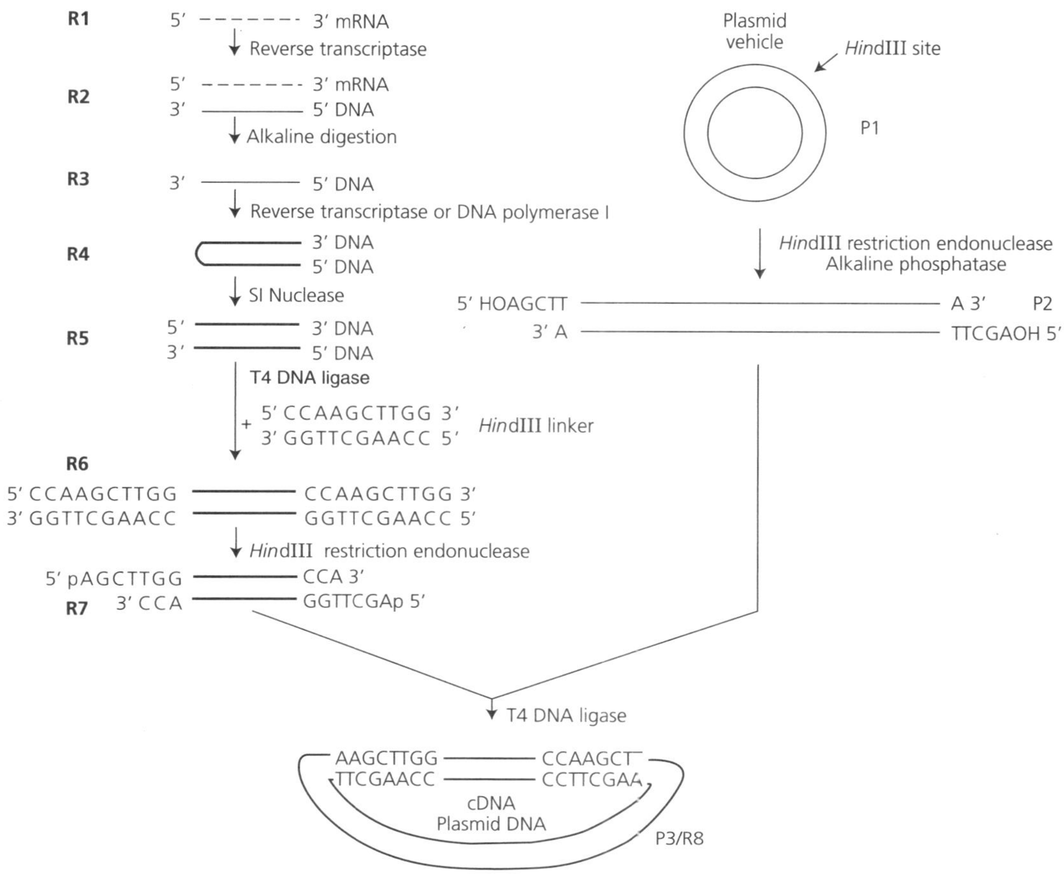

Figure 30.1 Steps involved in cDNA cloning. P1, plasmid cloning vector. P2, linearized DNA with staggered ends having 5′-overhangs after digestion of the plasmid with the restriction endonuclease *Hin*dIII and treatment with alkaline phosphatase. R1, the mRNA with poly-A tail. R2, oligo dT–primed reverse transcribed RNA–DNA hybrid. R3, complementary single-stranded DNA left over after alkali digestion of the RNA–DNA hybrid. R4, looped double-stranded DNA formed by replication of the complementary DNA by reverse transcriptase or DNA polymerase I. R5, removal of the single-stranded looped portion of DNA by S1 nuclease. R6, treatment with *Hin*dIII methylase, followed by joining of *Hin*dIII linkers to both ends by blunt-end ligation with T4 DNA ligase. R7, production of staggered ends by digestion with *Hin*dIII. P3/R8, cloning of the cDNA in the plasmid by joining with T4 DNA ligase.

two complementary oligonucleotides of a linker are separately synthesized chemically and then annealed together. 'Blunt-end ligation' of the linkers to the cDNA duplex is carried out by using T4 DNA ligase. It may, however, be recognized that the cDNA duplex itself may already contain a site for the restriction endonuclease of the very same restriction endonuclease, the site for which is embedded in the linker. Hence, before the ligation of the linker, the cDNA must be treated with the modifying enzyme relevant to the restriction endonuclease, whose site is present in the linker, so as to methylate the particular bases. After ligation of the linkers to both ends of the cDNA duplex, cohesive ends are generated by digesting with the specific restriction endonuclease. The preceding methylation step is essential to ensure that the cDNA is not cleaved into pieces while the cohesive ends are being liberated. It can then be ligated to the cloning vehicle. A cDNA library is constructed by using the total mRNA of the cell as template for reverse transcription. Fig. 30.1 illustrates the steps involved in cDNA cloning through the use of linkers.

30.4 Use of adapters

Adapters are used if subsequent to the initial cloning of the DNA (cDNA or otherwise), the DNA has to be recloned into another vector through different restriction endonuclease sites (Bahl and Wu 1978). For example, a DNA fragment with *Hin*dIII ends may have to be cloned into a vector in its *Bam*HI sites. An adapter for such a purpose comprises two chemically synthesized DNA oligonucleotides which when annealed together produce a duplex DNA with two different ends that are suitable for fitting into two different staggered ends, each created by a separate restriction endonuclease. A *Hin*dIII–*Bam*HI adapter is shown as follows:

5′ ***AAGCTT***GGGGG 3′
3′ ACCCCC***CCTAGG*** 5′
*Hin*dIII–*Bam*HI adapter

The 5′-overhang at the left of the top strand (bold italics) is designed to fit into a *Hin*dIII site, while the 5′-overhang at the right of the bottom strand (bold italics) is designed to fit into a *Bam*HI site. It is obvious that the adapter for connecting a restriction site with a 5′-overhang and a restriction site with a 3′-overhang would simply be a single-stranded oligonucleotide. A *Hin*dIII–*Pst*I adapter with a *Hin*dIII cleaved end at the left and a *Pst*I cleaved end at the right is shown as follows:

5′ A 5′	***AGCTTCTGCA*** 3′	G 3′
3′ ***TTCGA*** 5′		3′ ***ACGTC*** 5′
*Hin*dIII overhang	*Hin*dIII–*Pst*I adapter	*Pst*I overhang

30.5 Homopolymer tailing in cDNA cloning

The other method of adding cohesive ends to a blunt-ended DNA duplex is by homopolymer addition (Jackson *et al.* 1972; Lobban and Kaiser 1973; Wesnik *et al.* 1974), illustrated in Fig. 30.2. The homopolymers polydG and polydC are preferred to the homopolymers polydA and polydT, because of greater stability of the

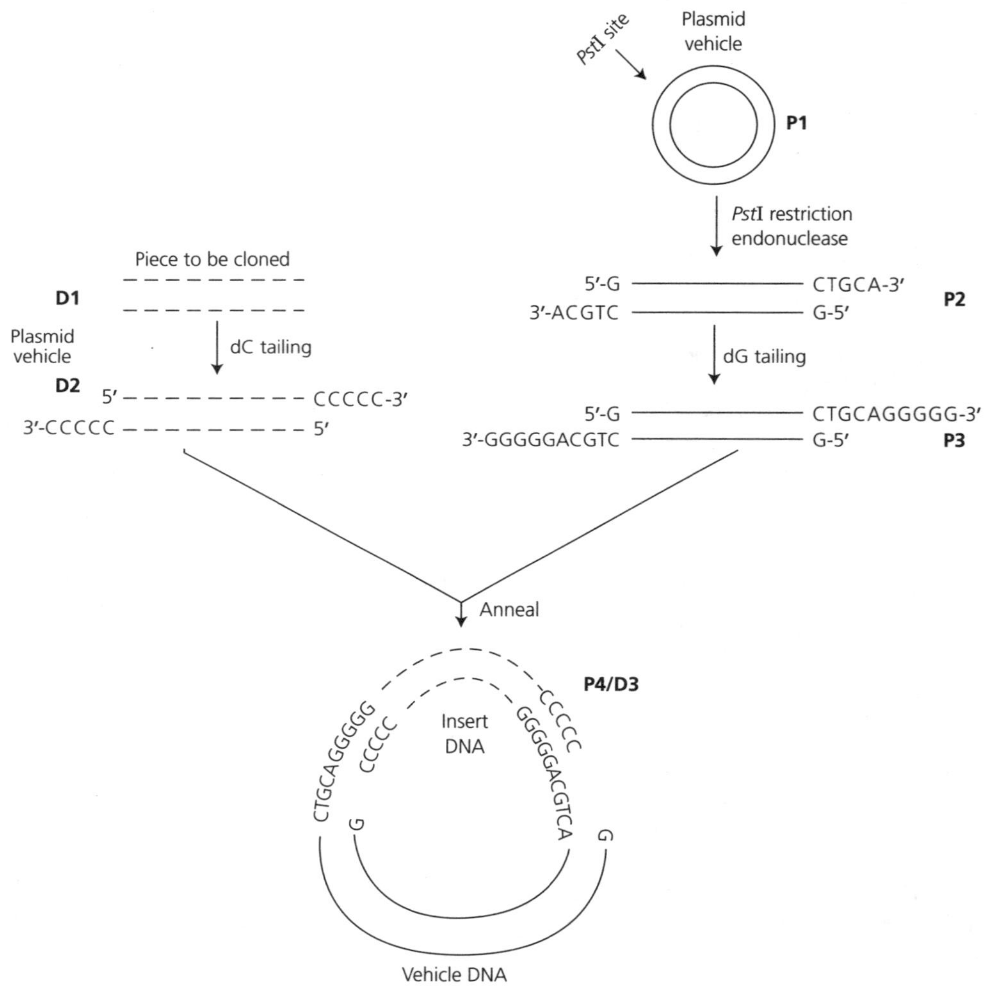

Figure 30.2 Cloning of a DNA fragment by homopolymer tailing. P1, plasmid cloning vector. P2, linearized DNA with staggered ends having 3′-overhangs after digestion of the plasmid with the restriction endonuclease *Pst*I. P3, dG-tailed linear plasmid DNA obtained by using dGTP and terminal transferase. D1, double-stranded DNA to be cloned with blunt (flush) ends. D2, DNA to be cloned dC-tailed using dCTP and terminal transferase. P4/D3, annealing of the dG-tailed linear plasmid DNA and the dC-tailed DNA to be cloned.

polydG–polydC duplex. The enzyme terminal transferase and the specific deoxyribonucleotide triphosphate are used when the homopolymers are added to the 3′-ends. About 10–20 nucleotide-long tails are sufficient. For effective terminal transferase reaction, the cloning plasmid vector would have to be cleaved with a restriction endonuclease that produces either a blunt end or a 3′-overhang, but not a 5′-overhang. *PstI* is an endonuclease that generates a 3′-overhang. In the presence of Co^{2+} ions, however, the terminal transferase reaction has been shown to proceed with recessed 3′-terminii also. If polydG tails are added to the cloning vector, polydC tails would be added to the DNA to be cloned. The tailed vector DNA and the DNA to be cloned are then annealed and the bacterial host is transformed with the annealed product. The gaps are filled up and ligated *in vivo* and the recombinant plasmid clone then gets replicated.

30.6 Chemical synthesis of adapters, linkers and other oligonucleotides

Oligonucleotides are chemically synthesized by joining phosphoramide derivatives of mononucleotides in the 3′→5′ direction (Itakura *et al.* 1984; Crauthers 1985). The following are the steps of chemical synthesis (illustrated in Fig. 30.3):

1. **Detritylation:** The first nucleotide from the 3′-end with its 3′-phosphate group and 5′-dimethoxytrityl group is allowed to bind to the derivatized solid support. This is then treated with trichloroacetic acid. This treatment removes the dimethoxytrityl (DMT) group and exposes an active 5′-OH group of the first nucleotide from the 3′-end for the subsequent reaction of coupling to the second nucleotide from the 3′-end.
2. **Coupling:** The would-be second nucleotide from the 3′-end with its 3′-phosphoramide group and 5′-dimethoxytrityl group is treated with tetrazole to liberate an activated 3′-OH group, which is allowed to undergo coupling with the active 5′-OH group of the first nucleotide.
3. **Capping:** There is some possibility that a few of the first nucleotides from the 3′-end might have escaped the coupling reaction in spite of the presence of the reactive free 5′-OH group. These reactive 5′-OH groups of the first nucleotide from the 3′-end must be capped to prevent these from reacting with the would-be third nucleotide from the 3′-end, which would be next incoming and would be meant to react only with the second nucleotide from

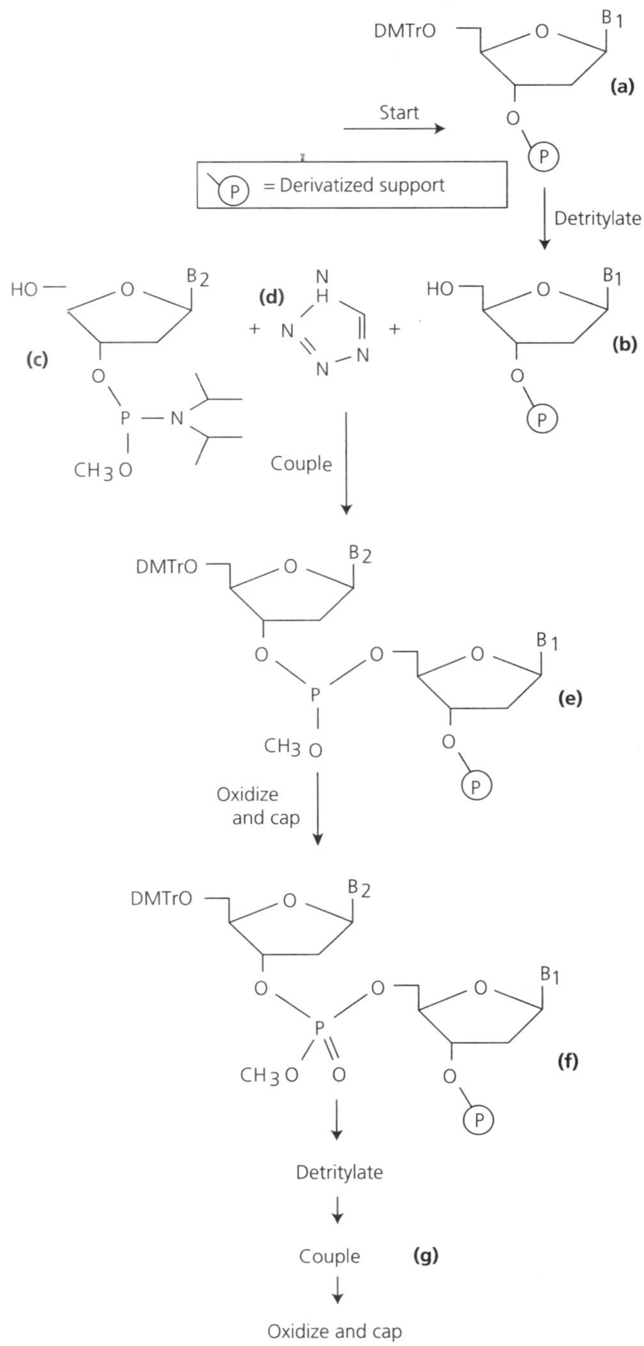

Figure 30.3 Use of the phosphoramidite method for chemical synthesis of oligonucleotides. (a) The dimethoxytrityl derivative of the first nucleotide from the 3′-end bound to the solid support. (b) Liberated active 5′-OH group of the first nucleotide from the 3′-end after treatment with trichloroacetic acid. (c) The phosphoramide derivative of the would-be second nucleotide from the 3′-end with its 5′-OH group protected with the dimethoxytrityl group. (d) Tetrazole for the liberation of an intermediate with a free active 3′-OH group of the would-be second nucleotide from the 3′-end. (e) The dinucleotide formed as a result of the coupling reaction between the first and the second nucleotide from the 3′-end. The subsequent capping of the unreacted free active 5′-OH group, if any, of the first nucleotide from the 3′-end by acetylation with acetic acid and 1-methylimidazole. (f) The dinucleotide after the linkage between the first and the second nucleotide from the 3′-end is converted from the phosphite to the more stable phosphotriester by using iodine as the oxidizing agent and water as the oxygen donor. (g) The next cycle of the removal of the dimethoxytrityl group on the second nucleotide, the coupling with the would-be third nucleotide from the 3′-end is performed, followed by the usual capping of the unreacted active 5′-OH group and conversion of the phosphite group to the more stable phosphotriester.

the 3′-end. The capping of the free active 5′-OH groups is performed by acetylation using acetic acid and 1-methylimidazole.

4. **Oxidation:** The linkage between the first and the second nucleotide from the 3′-end is then converted from the phosphite to the more stable phosphotriester by using iodine as the oxidizing agent and water as the oxygen donor.
5. **Addition of the would-be third nucleotide from the 3′-end:** The dimethoxytrityl group on the second nucleotide from the 3′-end is then removed with the usual trichloroacetic acid and the coupling with the would-be third nucleotide from the 3′-end is performed as described under step 2 and operations as described under steps 3 and 4 are repeated.
6. **Synthesis of the complete oligonucleotide:** The cycles are repeated sequentially till complete synthesis of the desired oligonucleotide is achieved. After completion of the synthesis, the oligonucleotide is released from the solid support and purified.

An automated DNA synthesizer is now available in which the whole operation described above can be performed.

30.7 Construction of a cDNA library

A cDNA library is similar to a gene library except that the genomic fragments are replaced by a collection of diverse cDNAs derived from all the mRNA transcripts of the genome of an organism or a particular tissue at the point of time when the RNA was isolated. The cDNA library is constructed the same way as an individual cDNA, as described in the preceding subsections, but taking the total mRNA population instead of an individual enriched mRNA. A cDNA library in a plasmid vehicle is stored in a host microorganism, quite often *E. coli*, after transformation of the host with the recombinant plasmids. The library can thus be multiplied at will by letting the host *E. coli* cells grow. The recombinant plasmids are then isolated from the cultured cells. A phage cDNA library is stored as bacteriophage particles. Such a library is multiplied by reinfecting the host cells with the recombinant phages and isolation of the virus particles after cell lysis.

30.8 Screening of a library for the isolation of a particular clone

A specific cDNA or a specific gene can be identified and isolated from a library in several ways. Some of the more common ways are described as follows.

30.8.1 Functional complementation is one way of identifying a clone

Functional complementation is feasible to adopt only when the host organism has a gene whose product has the same function as the product of the cloned cDNA or gene that has to be selected from the library. It is obvious that if the host is a bacterium such as *E. coli* or *Bacillus subtilis* and the clone is from a eukaryote, only cDNAs are likely to work, as transcripts of eukaryotic genes with introns cannot be processed by bacteria. In addition, adequate expression of the cloned cDNA from a eukaryote or the gene from a prokaryote would be a prerequisite. In practice, a strain of the host (let us say *E. coli*) is chosen which has a mutation, preferably deletion in the gene, which is functionally equivalent to the gene or cDNA to be selected. An example could be tryptophan synthetase 'A' protein cDNA from the cDNA library from humans. The *E. coli* tryptophan synthetase 'A' protein mutant is then transformed with the cDNA library in a plasmid and a selection is done on the basis of complementation. The selection medium would lack tryptophan and only those *E. coli* cells would grow that would have the recombinant plasmid containing the human tryptophan synthetase 'A' protein cDNA. The plasmids may be isolated from these cells and the cDNA released by cleavage with suitable restriction endonuclease(s). It need not be mentioned that this approach would not work with a library constructed in a lytic lambda bacteriophage vehicle, since the *E. coli* host would be lysed once the bacteriophage grows. Libraries in cosmid vehicles (Collins and Hohn 1978; Hohn and Collins 1980) should work, as these are really plasmids and only contain the 'cos' region of the lambda genome. The *recA* gene and the genes of the leucine biosynthetic pathway of *Azotobacter vinelandii* have been successfully cloned by functional complementation of *E. coli* mutants (Venkatesh *et al.* 1990; Manna and Das 1997).

30.8.2 Immunological method is another way of identifying a clone

All cDNAs, even if expressed well, may not be possible to be identified by functional complementation. It is essential that the translated product of the cDNA must be of some relevance to the host organism. For example, insulin cDNA can never be identified by functional complementation in the host *E. coli*. Immunological methods could be of use for identification of cDNA of these kinds. The translated protein product such as insulin is purified to homogeneity and an antibody is raised against it. Bacterial colonies containing recombinant plasmids or plaques of recombinant bateriophages are fixed on nitrocellulose membranes, lysed and the nucleic acids are washed out after digestion with nucleases. The proteins released from the colonies or plaques remain bound to the membranes. The membranes are then treated with the antibody that has been labelled with radioactive iodine and then washed. The antibody–antigen complexes stay stuck to the membranes. The membranes are next kept in contact with X-ray plates. Silver grains appear where the labelled antibody has formed a complex with the antigen. A replica membrane prepared before fixation allows tracing of the colony or plaque that contains the desired clone. Antibodies conjugated with horse radish peroxidase are also used. The bound antibody is detected here by colour-producing reagents such as 4-chloro-1-naphthol. Glucosamine 6-phosphate deaminase cDNA of *Candida albicans* has been cloned using this approach (Natarajan and Datta 1993). It is obvious that this technique can be used only when the pure protein product of the cDNA to be selected is available beforehand for raising the antibody. Higher sensitivity has been achieved by using solid matrix–bound antibodies and also by employing ^{125}I-labelled *Staphylococcus* protein 'A' that specifically binds to the Fc portion of the antibody. Better results have also been obtained by using lon mutants of the *E. coli* host that are defective in protein degradation. High-frequency lysogeny mutants (*hfl* A) of lambda phage expression vehicles have been developed. With these mutants, no lysis takes place at 32°C and normal *E. coli* colonies form on the membrane. The temperature is then raised to 42°C, when the phage is induced and a large number of the viral genome and copies of the cloned cDNA are formed and are expressed, yielding the protein product of the cloned cDNA in huge quantities. Immunological detection has been found to be very efficient in such situations.

30.8.3 Hybridization with a labelled DNA or RNA probe could be yet another way of identifying a clone

If a DNA fragment from a related organism or RNA from the same organism or a related one is available which has homology with the desired cDNA clone, then that can be used as a probe. Instead of the usual plating on agar, here the plating is done on a nylon or nitrocellulose membrane, which has been placed on the agar medium containing all the nutrients. Bacterial colonies or a bacterial lawn with bacteriophage plaques would form on the membrane. A second sterile membrane is gently pressed on the first membrane and placed on another agar plate containing all the nutrients. This would serve as a replica of the first membrane. The first membrane would be saved as the reference membrane. The replica membrane is then bathed in alkali so that the cells of the colonies or the bacteriophages of the plaques would lyse, releasing and denaturing the DNA, which would remain stuck to the membrane. The replica membrane is then washed and air dried. The replica membrane is next bathed in a solution of a labelled RNA or DNA probe under conditions that favour nucleic acid hybridization. The labelled DNA/RNA probe hybridizes to the DNA of only those colonies or plaques which have some sequence complementarity with the probe. The membrane is washed, dried and kept in contact with an X-ray film. Only those colonies/plaques would fog the X-ray film whose DNA has complementarity with the probe. The positions of these colonies/plaques that fog the X-ray film are compared with the colonies/plaques on the reference membrane to identify the desired colony or plaque. The bacteria in the particular colony or the bacteriophage in the particular plaque are then multiplied. Fig. 30.4 shows a schematic representation of the procedure. The advantage of the procedure is that it is independent of any expression of any gene. Non-radioactive labelling protocols are now available, but their sensitivity may not be comparable.

Barack and Tonegawa (1977) and Lai *et al.* (1978) have used labelled probes to identify the immunoglobulin light chain gene and the ovalbumin gene, respectively. Synthetic oligodeoxynucleotide probes, several from different stretches of the gene if desired, can be designed using a sequence database comprising entire nucleotide sequences of a large number of bacteria and several eukaryotic organisms available today. For example, a gene is known in *E. coli* or *S. cerevisiae* and a homologous gene is suspected to exist in, say, humans. The human DNA database is searched through a myriad of available software for sequences homologous to that of the *E. coli* or *S. cerevisiae* gene and the putative human gene is located. Probes are designed based on the sequence of the putative human gene and the cDNA in question is screened from the human cDNA library. The cDNA that codes for the human DNA polymerase η has been cloned by this approach using the sequence of the RAD30 gene of *S. cerevisiae* (Johnson *et al.* 1999). The cDNA for human DNA polymerase κ (which was earlier named as DNA polymerase θ) has been obtained similarly following homology with the gene for *E. coli* DNA polymerase IV (Johnson *et al.* 2000).

Deoxyoligonucleotide probes have also been derived from partial amino acid sequences of the protein product of the gene in question. Imagine that the partial amino acid sequence of a protein is the following:

Tryptophan—phenyl alanine—lysine—serine—methionine

The following eight oligodeoxyribonucleotides would code for the above-mentioned amino acid sequence, taking into consideration the degeneracy of the code:

1. TGG TTTAAAAGT ATG
2. TGG TTCAAGAGC ATG

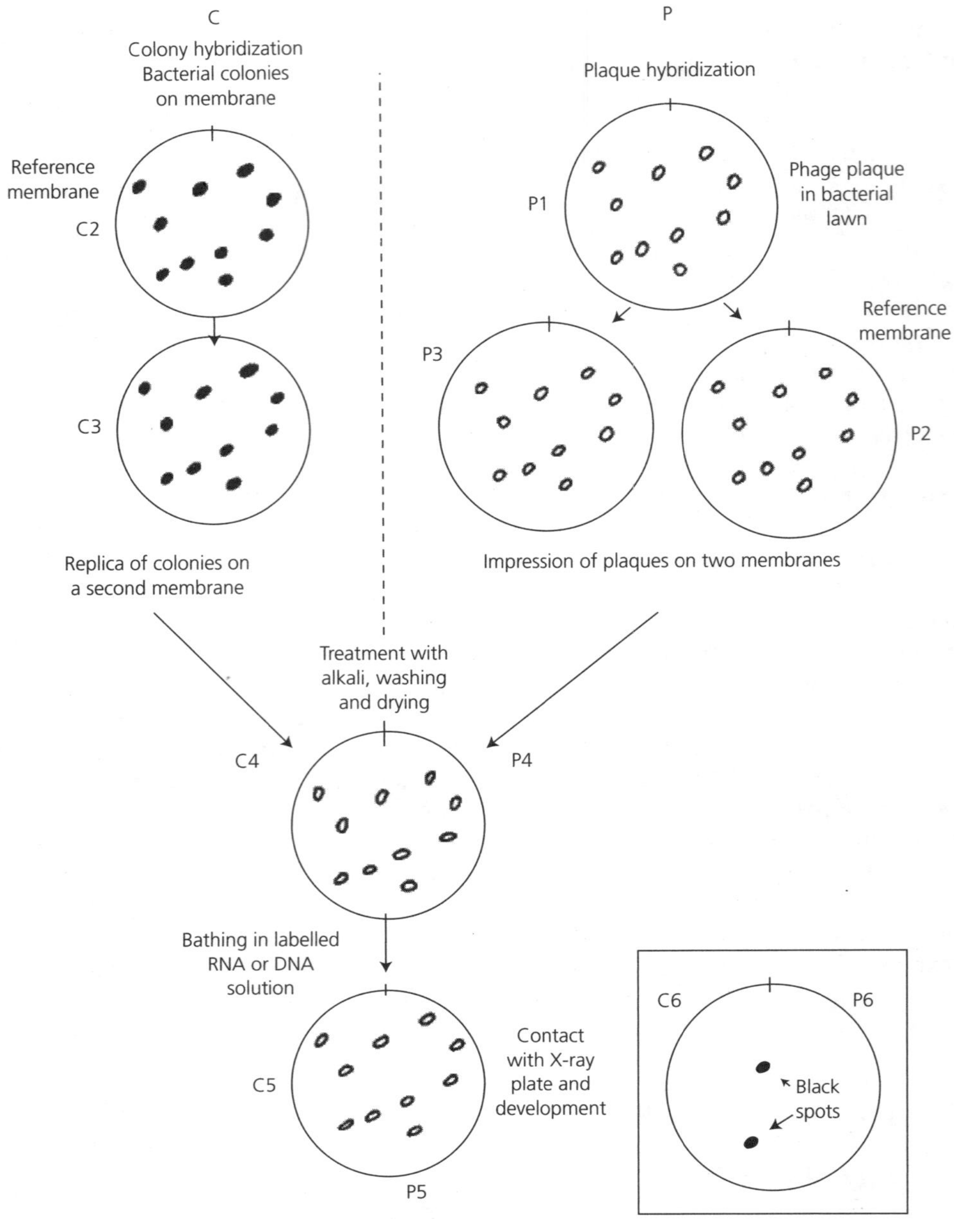

Figure 30.4 Schematic representation of the procedure for identification of a clone in a library by hybridization with a labelled DNA or RNA probe. P, bacteriophage plaque. C, bacterial colony. P1, bacteriophage plaques in the bacterial lawn. P2, reference membrane containing bacteriophage plaques. C2, reference membrane containing bacterial colonies. P3, replica of the reference membrane containing impression of bacteriophage plaques. C3, replica of the reference membrane containing impression of bacterial colonies. C4/P4, replica membrane containing impression of colonies/plaques in a bath of labelled RNA or DNA solution. C5/P5, washed and dried replica membrane kept in contact with an X-ray plate. C6/P6, the developed X-ray plate reveals black spots, indicating the location of hybridization with the labelled probe.

3. TGG TTTAAAAGT ATG
4. TGG TTCAAGAGC ATG
5. TGG TTTAAAAGT ATG
6. TGG TTCAAGAGC ATG
7. TGG TTTAAAAGT ATG
8. TGG TTCAAGAGC ATG

All these eight oligodeoxyribonucleotides would be synthesized chemically from labelled precursors and used as probes to select the cDNA in question from a library. Feredoxin I gene from *Azotobacter vinelandii* has been cloned using synthetic oligonucleotide probes (Dimri *et al.* 1988).

References

Bahl CP, Marians KJ, Wu R, Slawinsky J, Narang SA. A general method for inserting specific DNA sequences. *Gene* 1976;1: 81–92.

Bahl CP, Wu R. Cloned seventeen-nucleotide-long synthetic lactose operator is biologically active. *Gene* 1978;3:123–124.

Baltimore D. Viral RNA dependent DNA polymerase. *Nature* 1970;226:1209–1211.

Brack C, Tonegawa S. Variable and constant parts of the immunoglobulin light chain gene of a mouse myeloma cell are 1250 nontranslated bases apart. *Proc Natl Acad Sci USA* 1977;74:5652–5656.

Breathnach R, Mandel JL, Chambon P. Ovalbumin gene is split in chicken DNA. *Nature* 1977;270:314–319.

Chambon P. Split genes. *Sci Amer* 1981;244:60–71.

Collins J, Hohn B. Cosmids: A type of plasmid gene cloning vector that is packageable *in vitro* in bacteriophage heads. *Proc Natl Acad Sci USA* 1978;75:4242–4246.

Crauthers MH. Gene synthesis machines: DNA chemistry and its uses. *Science* 1985;230:281–285.

Dimri GP, Roy KB, Das HK. Cloning of feredoxin I gene from *Azotobacter vinelandii* using synthetic oligonucleotide probes. *J Biosci* 1988;13:323–327.

Efstratiadis A, Villa-Komaroff L. Cloning of double stranded DNA. In: Setlow JK, Hollander A, eds. *Genetic Engineering* (Volume 1). New York: Plenum; 1979. pp. 15–27.

Hohn B, Collins J. A small cosmid for efficient cloning of large DNA fragments. *Gene* 1980;11:241–248.

Itakura K, Rossi JJ, Wallace RB. Synthesis and use of synthetic oligonucleotides. *Ann Rev Biochem* 1984;53:323–356.

Jackson DA, Symons RH, Berg P. Biochemical method for inserting new genetic information into DNA of simian virus 40: Circular SV40 DNA molecules containing lambda phage genes and the galactose operon of *Escherichia coli*. *Proc Natl Acad Sci USA* 1972;69:2904–2909.

Johnson RE, Kondratick CM, Prakash S, Prakash L. *hRAD30* mutations in the variant form of *Xeroderma pigmentosum*. *Science* 1999;285:263–265.

Johnson RE, Prakash S, Prakash L. The human DINB1 gene encodes the DNA polymerase Pol θ. *Proc Natl Acad Sci USA* 2000;97:3838–3843.

Kurtz DT, Nicodemus CF. Cloning of α2μ globulin DNA using a high efficiency technique for the cloning of trace messenger RNAs. *Gene* 1981;13:145–155.

Lai EC, Woo SLC, Dugaiczyk A, Catertall JF, O'Mally B. The ovalbumin gene: Structural sequences in native chicken DNA are not contiguous. *Proc Natl Acad Sci USA* 1978;75:2205–2209.

Lobbar PE, Kaiser AD. Enzymatic end-to-end joining of DNA molecules. *J Mol Biol* 1973;78:453–471.

Manna A, Das HK. Characterization and mutation of the leucine biosynthetic genes of *Azotobacter vinelandii*: An analysis of the rarity of amino acid auxotrophs. *Mol Gen Genet* 1997;254: 207–217.

Natarajan K, Datta A. Molecular cloning and analysis of the NAG1 cDNA coding for glucosamine 6-phosphate deaminase from *Candida albicans*. *J Biol Chem* 1993; 268:9206–9214.

Okayama H, Berg P. High-efficiency cloning of full-length cDNA. *Mol Cell Biol* 1982;2:161–170.

Temin HM, Mizutani S. Viral RNA dependent DNA polymerase. *Nature* 1970; 226:1211–1213.

Venkatesh TV, Reddy MA, Das HK. Cloning and characterization of the *recA* gene of *Azotobacter vinelandii* and construction of a *recA* deletion mutant. *Mol Gen Genet* 1990;224:482–486.

Wensink PC, Finnegan DJ, Donelson KE, Hogness DS. A system for mapping DNA sequences in the chromosomes of *Drosophila melanogaster*. *Cell* 1974;3:315–325.

31

Nucleic acid amplification and its applications

H. K. Das

31.1 Amplification of plasmid DNA *in vivo*

A very important observation on the replication of the plasmid Col E_1, the colicinogenic factor in *Escherichia coli*, was made in 1972 by Don B Clewell and Donald R Helinski.

The rate of synthesis of the Col E_1 plasmid DNA gradually increased and within 2–4 h, reached a rate about eight times the original rate when protein synthesis was inhibited by the addition of the antibiotic chloramphenicol or by starvation of an amino acid, even though the replication of chromosomal DNA gradually came to a complete stop within one generation time. The replication of the plasmid DNA continued till 12–16 h after the inhibition of protein synthesis, and the number of copies of the plasmid per *E. coli* cell reached about 3000. This was in contrast to the normal 24 copies of the plasmid per cell. As, in the presence of ethidium bromide, the supercoiled plasmid has a higher buoyant density than the linear fragments formed as a result of shearing of the chromosomal DNA, these two species of DNA can be easily separated by equilibrium centrifugation in a caesium chloride density gradient. Fig. 31.1 shows such a separation and the remarkable progressive amplification of plasmid DNA after addition of chloramphenicol. For many years, it was believed that plasmid amplification occurs only when the host bacteria are grown in a minimal medium. It was later found that sufficiently high amplification could also be obtained in lysogeny broth (LB) medium. Such amplification, however, occurs only with $ColE_1$ or other closely related plasmids whose replication is mediated by DNA polymerase I. In practice, the DNA fragment to be amplified is first cloned in a $ColE_1$ or a closely related plasmid, *E. coli* cells are transformed with the recombinant plasmid and then the plasmid is amplified in *E. coli* after the addition of chloramphenicol at the appropriate growth phase of the bacteria. The recombinant plasmid is then isolated and the DNA fragment is purified after digestion of the recombinant plasmid with the appropriate restriction endonuclease. It is noteworthy that no base sequence information is necessary for achieving amplification of a DNA fragment by this process.

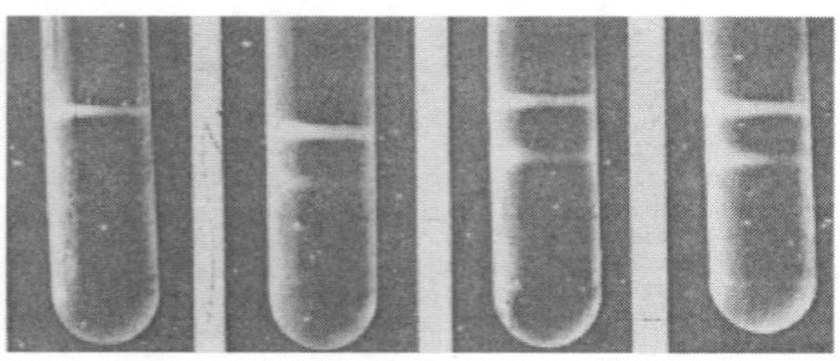

(a) (b) (c) (d)

Figure 31.1 Time course of amplification of $ColE_1$ plasmid DNA on addition of chloramphenicol to the *E. coli* cells harbouring the plasmid. Isolated total DNA was subjected to equilibrium centrifugation in a caesium chloride–ethidium bromide density gradient and the bands were visualized under ultraviolet rays. The upper band contains the chromosomal DNA, while the lower band comprises the $ColE_1$ plasmid DNA. (a) DNA was isolated from cells that were not treated with chloramphenicol. (b) DNA isolated after 4 h of addition of chloramphenicol. (c) DNA isolated after 8 h of addition of chloramphenicol. (d) DNA isolated after 16 h of addition of chloramphenicol.

31.2 Amplification of DNA fragments *in vitro* by polymerase chain reaction

Polymerase chain reaction (PCR) is the name given by Kary Banks Mullis (1985) to a procedure by which he was able to amplify *in vitro* a very small amount of any DNA

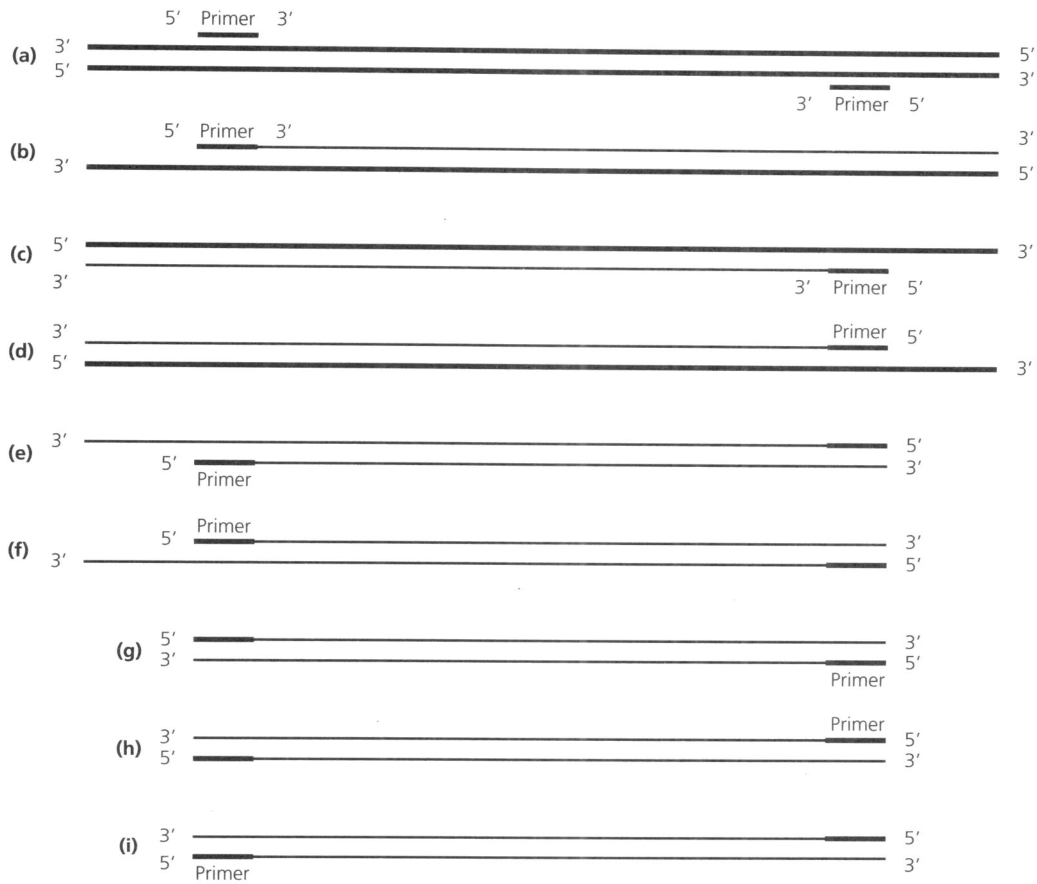

Figure 31.2 Schematic representation of polymerase chain reaction. (a) The DNA to be amplified is annealed with two primers (small lengths of complementary DNA) to the two strands. The replication process would be initiated by the addition of nucleotides to the two primers. The original DNA and the primers are shown in thick lines. The two strands of the original DNA get separated and two daughter DNA duplexes are formed as a result of DNA replication by the semiconservative mechanism. The newly synthesized DNA strands are shown in thin lines. In the next cycle, the primers are annealed to both the newly synthesized duplexes (b) and (c) and the replication process is repeated with either fresh *E. coli* polymerase or thermostable polymerase. In the diagram, only the replication of the duplex (c) has been shown. The two strands of (c) get separated as the two new daughter duplexes (d and e) are formed. In the next cycle, all the steps are repeated. Again, the replication of only the duplex (e) has been shown. The daughter duplexes (f) and (g) are formed. In the next cycle, the duplex (g) has been shown in the figure, forming the daughter duplexes (h) and (i). It may be noted that the duplexes (g) onwards are all of the same length bounded by the two primers. The PCR product would thus be of a fixed size, exhibiting a specific band of DNA on electrophoresis (this figure is the same as Fig. 9.32).

sequence to a very large amount. After 20 cycles of the reaction catalysed by DNA polymerase, he could achieve an amplification of 2,00,000 folds. No transformation of bacterial cells was required; only base sequence information of the two terminal regions of the DNA fragment was necessary (Saiki *et al.* 1985, 1988; Mullis *et al.* 1986). Mullis has been awarded the Nobel Prize for developing this procedure. The protocol he followed was as follows. The DNA to be amplified was heat denatured. Added to this in large excess, were two oligodeoxynucleotide primers that have sequence complementary to the flanking 3′-regions of the two strands of the DNA to be amplified. Annealing was done by slow cooling till 37°C. DNA polymerase I from *E. coli* was added along with all the four deoxyribonucleotides for DNA replication to proceed.Mononucleotides were added at the 3′-ends of the two primers and both the strands of DNA were replicated in the 5′ → 3′ direction. This completed the first cycle. The mixture was heated to denature the two newly formed DNA duplexes and then cooled so that the primers, which were in excess, could again anneal, but this time to both the two original DNA strands and the two newly synthesized ones. Once the mixture was cooled to 37°C, DNA polymerase I from *E. coli* was again added and replication was allowed to proceed. The second cycle was now complete. The steps were repeated for the third and subsequent cycles. At the end of 20 cycles,an amplification of 2^{20} folds was theoretically expected, which was equivalent to more than 10^6. The logic of the amplification steps is schematically represented in Fig. 31.2.

The subsequent isolation of the thermostable DNA polymerase from *Thermus aquaticus* has rendered the amplification procedure simpler. It is not necessary any more to add fresh polymerase in the beginning of every cycle and the reaction mixture does not have to be cooled to 37°C for the polymerization reaction. After the annealing is carried out at 40°C or so, the temperature is raised to 70°C, the optimum temperature for polymerization by the *Thermus aquaticus* DNA polymerase. At this temperature, all poorly matched primer–template hybrids dissociate and only highly complementary hybrids survive, thus imparting strict specificity and hence purer polymerized products. The reaction rate is very high and hence a greater yield is obtained in much less time.

31.3 Applications of DNA amplification

Applications of the PCR are many and varied. Some of these are described in the following sections.

31.3.1 Cloning of a gene

Any gene of any organism for which sequence information is available can be cloned now without having to first obtain a gene library. One needs to obtain a few cells or a tiny amount of the tissue of the organism, isolate total DNA and carry out the PCR using the appropriate primers. Only the gene in question would be amplified from the total DNA. One can also incorporate appropriate restriction endonuclease sites in the primer for ease of subsequent cloning of the gene.

31.3.2 Detection of bacterial or viral infecting agent

At an early stage of infection, the population of bacteria or virus in a biological sample can be very small. For example, at the very early stage of infection of the lungs by *Mycobacterium tuberculosis*, the sputum is likely to contain only a few *Mycobacterium* cells. As a result,the slide of the sputum may fail to reveal any acid-fast bacillus under the microscope. When, however, DNA is isolated from the sputum sample and a PCR is performed with primers specific for unique regions of the genome of *M. tuberculosis*, a sharp PCR band of the expected size is obtained, thus confirming the presence of the infecting *M. tuberculosis* (Eisenach *et al.* 1991). Fig. 31.3 is from the laboratory of Dr Pradeep Seth. It is possible to detect by PCR the presence of a single copy of the DNA of the infectious agent, whether bacteria or virus. A combination of *in situ* PCR and immunohistochemistry has allowed detection of cellular reservoirs of HIV-1 in the central nervous system of infected individuals (Bagasra *et al.* 1996). Plant pathogenic *Mycoplasma*-like organisms have also been detected by PCR (Ahrens 1992).

31.3.3 Paternity testing and individual identification by PCR

The usual technique of individual identification or paternity testing by DNA finger printing involves Southern blotting and the use of radioactive materials and requires several days' time. Some of the variable-number tandem-repeat (VNTR) DNA sequences of the human genome show considerable variations in the number of repeats within the population. For example, the VNTR locus D1S58(defined by the probe pMCTll8) [16-base pair repeating unit containing the core sequence (GNNGTGGG)] revealed heterozygosity of 78% and more than 10 alleles among 100 Caucasians (Nakamura *et al.* 1988). Just 2 ng DNA isolated from human blood revealed, after amplification by PCR of the MCTll8 locus and electrophoresis on polyacrylamide gel, polymorphic bands on ethidium bromide staining. The sizes of the bands from individuals were reproducible and varied between only 387 and 723 bases. Good resolution of individual alleles was possible even when the individual alleles differed by only one repeat unit (Kasai *et al.* 1990). Another VNTR locus D1S80 has also been amplified by

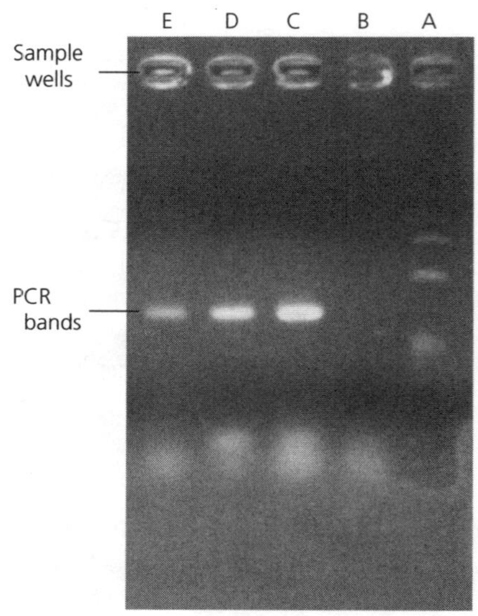

Figure 31.3 Agarose gel electrophoresis of the polymerase chain reaction product. Specific primers have been used to select a segment of DNA from *M. tuberculosis* out of total DNA isolated from human sputum. Lane A: molecular weight marker - ϕx 174 replicative form DNA digested with the restriction endonuclease *Hae*III. Lane B: negative control PCR undertaken with DNA isolated from the sputum of a person who has not been afflicted with *M. tuberculosis*. Lane C: positive control PCR undertaken with DNA isolated from the sputum of a person who has been confirmed to be suffering from tuberculosis. Lanes D and E: PCR undertaken with DNA isolated from the sputum of two persons who had tested sputum negative; that is, no acid fast bacilli could be detected in their sputum, but still they were diagnosed to have been afflicted with *M. tuberculosis*, because they have tested positive on the basis of the PCR. *Source:* Courtesy Dr Pradeep Seth.

PCR and used for forensic analysis (Sajantila *et al.* 1992; Kloosterman *et al.* 1993).

31.3.4 Detection of infecting agent with RNA genome

A reverse transcription reaction has to precede the PCR (RT-PCR) if the genome of the infecting agent is not DNA but RNA. An example is blood products (platelets) tainted with hepatitis C virus. An RT-PCR test could detect hepatitis C virus in a sample even when the enzyme-linked immunosorbent assay (ELISA) test had failed. Sometimes, the sensitivity of the test is enhanced by carrying out two successive amplifications, first using two outer primers and then employing two inner primers. RT-PCR is also useful in determining whether or not a particular gene is being transcribed. Dengue viruses have been detected and typed by PCR after reverse transcription (Lanciotti *et al.* 1992). Based on PCR of specific regions in the C gene of 44 hepatitis C virus samples, four distinct types of the virus have been postulated (Sugiyama *et al.* 1992).Response to interferon-alpha therapy on hepatitis C virus in relation to genotypes of hepatitis C virus has also been studied by PCR (Yoshioka *et al.* 1992).Quantitative RT-PCR has been used to study mRNA decay (Schmittgen *et al.* 2000).

31.3.5 Quantification of a gene or an infecting agent by competitive or differential PCR

The principle is to amplify the DNA in question by PCR through a limited number of cycles and then quantify the PCR product; the logic being that the amount of product after a limited fixed number of cycles of PCR would be a measure of the amount of the initial amount of DNA under conditions of abundant substrates. The problem is that PCR is extremely sensitive, so even a very small variation in the reaction conditions would result in significant differences in the amount of the PCR product formed. Hence, there is a need for an internal reference. This is achieved by adding a known amount of reference DNA into the sample, carrying out the PCR on the combined DNA and ascertaining the amounts of products formed from both the reference and the experimental DNA. This process is known as competitive PCR or differential PCR. It is essential that the reference DNA is designed in such a way that the size of the reference DNA be the same as the experimental DNA and the same PCR primers can also serve the reference DNA. The product of the reference DNA, obviously, would have to be distinguished from the product of the target DNA and both would have to be assayed separately. A standard procedure is to incorporate a unique restriction site in the reference DNA and treat the total product with the enzyme. The product of the experimental DNA would be unaffected, while the product of the reference DNA would be cleaved. The products can, therefore, be separated electrophoretically after the PCR reaction and assayed separately. An important application of this technique has been to determine the presence of amplified oncogenes involved in breast tumour metabolism. Some workers use radioactively labelled precursors during the PCR and determine radioactivity of the products. A gene transcript or an RNA virus can also be quantified by performing RT-PCR and then following the above-mentioned protocol. Competitive PCR has been made use of for the quantitative analysis of the expression of two cytokines, granulocyte-macrophage colony-stimulating factor (GM-CSF) and interleukin 3 (IL-3),in MLA-144 cells before and after induction with phorbol 12-myristate 13-acetate (Gilliland *et al.* 1990). Quantitation of mRNA has also been done by PCR by using an internal standard (Wang *et al.* 1989). A variation of this method is the PCR-aided transcript titration assay (PATTY) developed by Becker-André and Hahlbrock (1989). Here, co-amplification is done of an *in vitro*–generated transcript differing by a single base exchange from the experimental mRNA, which creates a novel restriction endonuclease cleavage site in the cDNA produced from the externally added RNA. The PCR product is digested with the specific restriction endonuclease and the target cDNA is separated by electrophoresis from the standard cDNA.

31.3.6 Quantification of a gene or an infecting agent by real-time PCR

Another method of quantification of the number of copies of a gene or genome in a sample is real-time PCR. In this method, the course of the PCR process is followed rather than determining the amount of the end product. The sensitivity of the detection process is enhanced by replacing ethidium bromide with SYBR Green as the staining dye. The fluorescence of SYBR Green, when bound to double-stranded DNA, is 11 times more than when bound to single-stranded DNA and 1000 times more than when in a free state (Zipper *et al.* 2004). A laser detector is used to continuously follow the increase in fluorescence, which is plotted against the number of PCR cycles as more and more double-stranded DNA is formed with more and more cycles of PCR. The plot is known as the 'amplification plot' and looks very similar to melting curves of DNA (Fig. 5.16), but instead of absorbance at 260 mμ versus temperature, here, it is fluorescence versus number of cycles of PCR (Fig. 31.4). Tracings 1, 2 and 3 represent real-time PCR of samples containing DNA in decreasing amounts. A threshold is decided arbitrarily depending on the sensitivity of the detection mechanism. The number of cycles at which the fluorescence crosses the threshold is termed as the cycle threshold (Ct) (somewhat analogous to melting temperature,Tm), which is an inverse

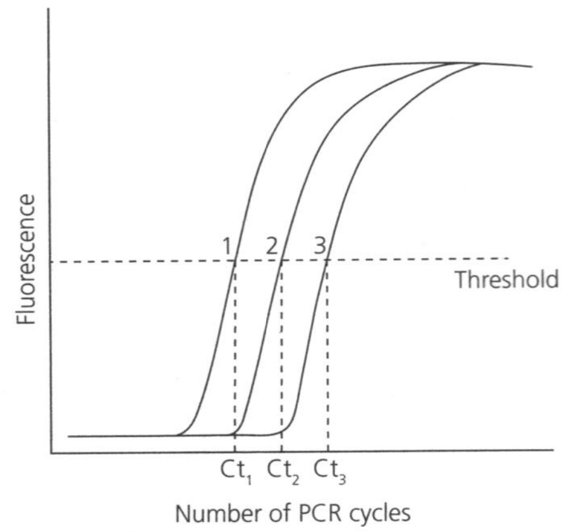

Figure 31.4 Real-time PCR. The fluorescence emitted by SYBR Green (bound to double-stranded DNA) is monitored continuously as the polymerase chain reaction continues and is plotted against the number of PCR cycles completed. Curves 1, 2 and 3 represent plots of three samples with amounts of DNA in sample 1 > in sample 2 > in sample 3. Ct_1, Ct_2 and Ct_3 are cycle threshold values of sample 1, 2 and 3, respectively.

measure of the amount of DNA present in the sample. The higher the value of Ct, the lower is the amount of DNA present in the sample. The quantitation is done by either the standard curve method or the comparative Ct method using a reference gene.

Expression levels of mRNA can also be monitored by real-time PCR by introducing an initial step of reverse transcription. In addition to the standard curve method and the comparative Ct method, an assay based on simulation of PCR kinetics has also been developed (Liu and Saint 2002). Levels of mRNA of atrial natriuretic peptide and c-fos have been determined in an *in vitro* model of cardiac hypertrophy (Winer *et al.* 1999). By a variation of this method,expression kinetics of cyclins A, B1, D1 and E and of the immediate-early genes c-fos, c-myc and -actin had earlier been studied (Wong *et al.* 1994).

31.3.7 PCR with nested primers

Nested primers are used to minimize the chance of amplification of an unwanted region of the genome because of the chance possibility of partial homology with the wanted region. Let us consider a situation where in a particular genome, the following two sequences exist in two distantly located loci. The dotted lines represent some unrelated long sequence.

Sequence 1:
5′ ATCGCGCTTTCCGC**ACCAT**CACCGGC
TACGGCAGCGAGGA.................
.......................GTCGTCGAGCTGCCCTTC
TC**GGCCT**CACCTCTAGAGTCGA 3′

Sequence 2:
5′ GTCAACCGCCGCAAG**ACCAT**GACCCTGT
ACCTGGCCGAAC..................
.................CACGTAGACACCGACTATC
G**GGCCT**GCGAAGGCTGCCGGAT 3′

Let us say the following pair of primers are chosen for PCR:

5′ ATCGCGCTTTCCGC**ACCAT** 3′ and 3′**CCGGA**CGCT TCCGACGGCCTA 5′

Depending on the annealing temperature chosen, there is a distinct possibility that sequence 2 would also be amplified in addition to sequence 1. Now, if after a couple of cycles of PCR, the following pair of 'nested' primers are used,

5′CACCGGCTACGGCAGCGAGGA3′and3′CAGCAGCT CGACGGGAAGAG 5′, then there is no chance of amplification of any wrong sequence. An example of the application of the use of nested primers for PCR is the diagnosis of herpes simplex encephalitis by detection of herpes simplex virus DNA in the cerebrospinal fluid (CSF) (Aurelius *et al.* 1991).

31.3.8 PCR coupled to DNA sequencing

Mutation(s) in gene(s) can lead to genetic diseases; several such mutations in humans is known. Base sequencing of such genes offers confirmatory evidence of such a genetic disease. Quite often, the amount of tissue available is very small, so the isolated DNA has to be first amplified by PCR and then subjected to base sequencing. For routine sequencing, it is always advantageous to employ single-stranded DNA templates. There are several ways of obtaining single-stranded DNA from a PCR product.

31.3.8.1 PCR conducted asymmetrically

Instead of using the two amplification primers in equal amounts, grossly unequal amounts are used. This causes overproduction of a single-stranded DNA [Fig. 31.5(a)]. Gyllensten and Erlich (1988) applied this method to directly determine the base sequence of the *HLA-DQS* locus and study allelic diversity by using allele-specific oligonucleotides as primers.

31.3.8.2 Lambda (λ) exonuclease treatment of the PCR product

A unique characteristic of lambda (λ) exonuclease is that it specifically cleaves a DNA strand that has a 5′-phosphate. If a particular primer, one of the two PCR primers, is treated with T4 polynucleotide kinase, then that primer would have a 5′-phosphate. The DNA strand that would be generated by the kinase-treated

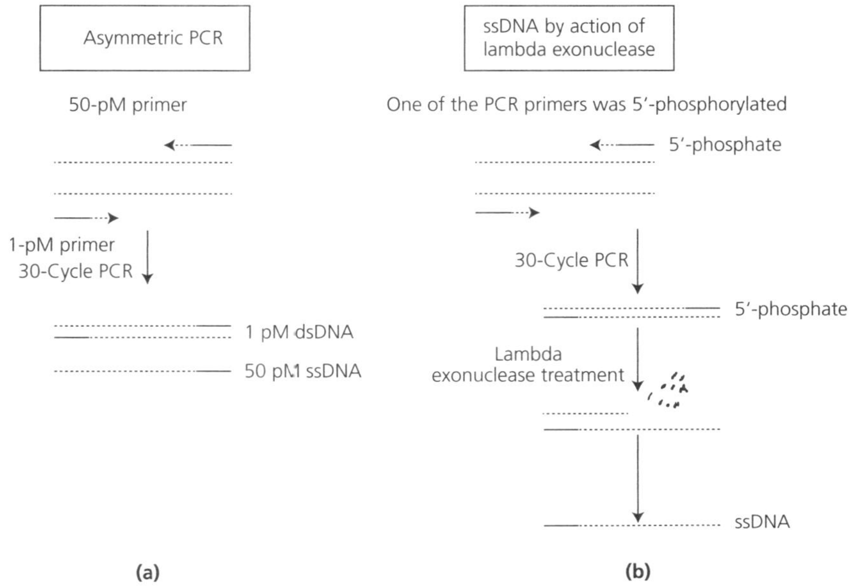

Figure 31.5 Isolation of single-stranded DNA from the product of polymerase chain reaction (PCR). (a) PCR conducted asymmetrically. Of the two primers, one primer was used in a disproportionately large excess, resulting in a large excess of the specific single-stranded DNA product. (b) Lambda exonuclease treatment to obtain a single-stranded DNA product. Of the two primers, one primer was designed to have a 5′-phosphate. This end of the double-stranded PCR product was specifically attacked by lambda exonuclease, leaving a single-stranded DNA.

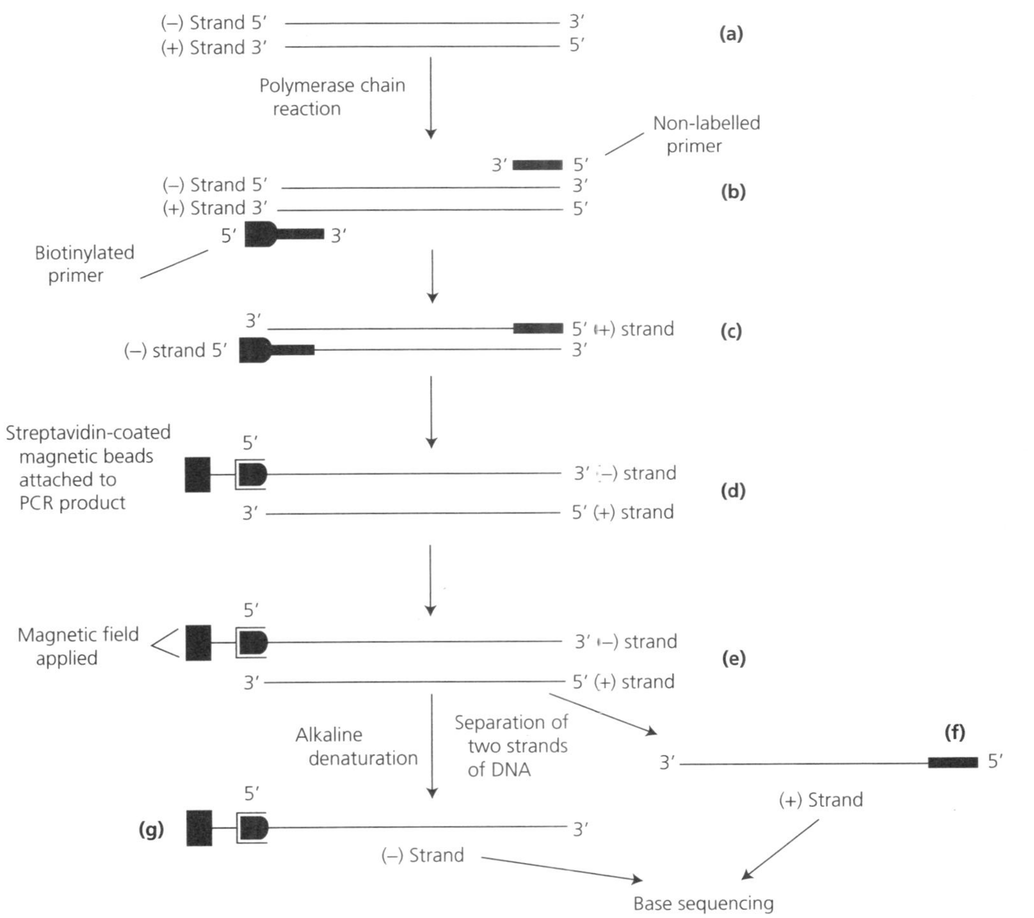

Figure 31.6 Separation of the two strands of DNA by a magnetic field. (a) The target DNA; one strand has been arbitrarily designated (–) strand and the other the (+) strand. (b) The 5′-biotinylated primer has been annealed to the (+) strand, while the primer annealed to the (–) strand is non-labelled. (c) The PCR product, thus, has the 5′-end of the (–) strand biotinylated. (d) Stretavidin-coated magnetic beads bind specifically to the 5′-biotinylated end of the (–) strand. (e) A magnetic field immobilizes the (–) DNA strand. (f) On alkaline denaturation, the (+) strand is released, but the (–) strand is held back. (g) The (–) strand is released on withdrawal of the magnetic field. Either or both the separated strands can now be individually subjected to base sequencing.

primer would also have a 5′-phosphate.That particular strand of the PCR product would then be degraded by the lambda (λ) exonuclease treatment, while the other strand would be spared [Fig. 31.5(b)].

31.3.8.3 Separation of the two strands of DNA by a magnetic field

Of the two PCR primers, one is chosen to have a 5′-biotin label. Hence, only one strand of the product of the PCR would have biotin in it. The product is next subjected to treatment with streptavidin-coated magnetic beads, denatured and subjected to a magnetic field. The two strands of DNA are thus separated (Fig. 31.6).

31.3.9 Site-directed mutagenesis by PCR

There are several ways of using PCR as a tool for achieving site-directed mutagenesis, some of which are discussed next.

31.3.9.1 Site-directed mutagenesis *in vitro* by PCR of a plasmid containing the DNA fragment to be mutagenized

The basic principle here is to incorporate the mutation in the primers to be used in the PCR. This implies that the base sequence of the region of the DNA fragment where the mutation has to be introduced needs to be known. The primers are required to be 25–45 bases long, with the mutation in the middle and 10–15 correct bases on both sides of the mutated bases. The DNA fragment to be mutagenized is cloned in a plasmid, which is isolated from a *dam* (+) *E. coli*. The two primers are designed to anneal on opposite strands of the same stretch of the DNA where the mutation is desired.

The PCR generates a plasmid containing staggered nicks close to the mutated bases in the relevant DNA stretch (Fig. 31.7). The nicks can be joined *in vitro* by using DNA ligase or *in vivo* by transforming *E. coli.*

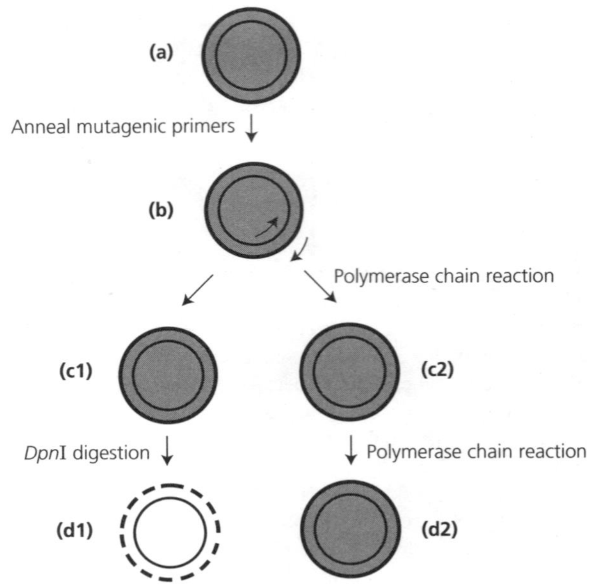

Figure 31.7 Site-directed mutagenesis *in vitro* by PCR of a plasmid containing the DNA fragment to be mutagenized. (a) The plasmid containing the fragment of DNA to be mutagenized. The plasmid has been isolated from a *dam* (+) *E. coli* cell. (b) The plasmid (a) has been denatured and the two mutagenic primers have been annealed. PCR has been carried out. (c1) The parental plasmid template. (c2) The product of PCR containing the mutated (base substituted or base deleted or base added) bases. The parental plasmids and the PCR products are digested with the restriction endonuclease *DpnI*. (d1) The original template DNA is degraded. (d2) The PCR product (mutated) is unaffected.

31.3.9.2 Site-directed mutagenesis by overlap extension of a DNA fragment by PCR

This is another method of site-directed mutagenesis. Here, the DNA fragment need not be cloned in a plasmid. In fact, by this method, a site-directed mutagenesis can be done in a specific DNA fragment present in total DNA isolated from a tissue as long as the base sequence of the fragment is known. The principle is illustrated in Fig. 31.8. The species (a) is to be mutagenized. The mutation can be substitution, or addition or deletion of bases. Two separate PCR operations are done, first using primers P_1 and P_2. Primer P_1 is at one end, while primer P_2 carries the mutation and is in the middle of the fragment (a). The next PCR operation is done by using primers P_3 and P_4. Primer P_4 is at the other end of the fragment (a), while primer P_3 is in the middle of the fragment and carries the mutation. Primers P_2 and P_3 have complementary base sequences. There is thus an overlapping region in the two PCR products. The two PCR products, species (b) and (c), are purified from species (a) on the basis of size by gel electrophoresis. The next step is to mix the two PCR products, denature them and then renature. Some of the molecules formed by renaturation would be as shown by species (d) in Fig. 31.8. Species (d) is converted to species (e) by treatment with *E. coli* DNA polymerase I and all the four deoxyribonucleoside triphosphates. Species (e) is amplified by PCR using primers P_1 and P_4. This amplified product has the desired mutation. This method was proposed by Ho *et al.* (1989) and they had generated three variants of a mouse major histocompatibility complex class I gene using this method. A variation of this method has been described by Urban *et al.* (1997).

31.3.9.3 Site-directed mutagenesis using one mutagenic primer and two universal primers

Two universal sequencing primers flanking the region to be mutated along with just one mutagenic primer are all

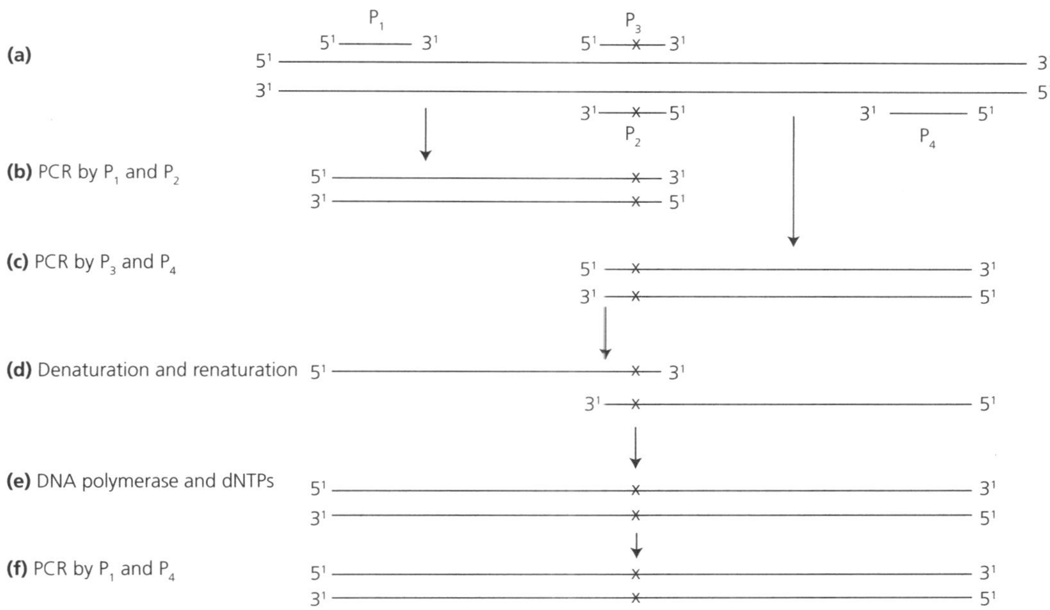

Figure 31.8 Site-directed mutagenesis *in vitro* by overlap extension of a DNA fragment by PCR. The species (a) has to be mutagenized. Four primers are designed. Primers P_1 and P_4 are at the two ends and primers P_2 and P_3 are in the middle of the DNA fragment (a). Primers P_2 and P_3 carry the mutation and are complementary to each other. Species (b) is the product of PCR using primers P_1 and P_2. Species (c) is the product of PCR using primers P_3 and P_4. Species (d) is one of the products of denaturing species (b) and (c) and renaturing together. Species (e) has been obtained by treating species (d) with *E. coli* DNA polymerase I and all the four deoxyribonucleotide triphosphates. Species (f) is the PCR-amplified species (e).

that is needed for the PCR in this method. Two different mutants of the RNAse T1 gene have been constructed and have been found to be mutated in 12 out of 12 cases (Landt *et al.* 1990).

31.3.9.4 Site-directed mutagenesis by the megaprimer method

Two rounds of PCRs are performed. Two oligonucleotide primers are used for the first round of PCR. A third oligonucleotide and the product of the first round of PCR reaction (megaprimer) are used as primers for the second round of PCR. Different variants of the megaprimer method have been suggested (Sarkar and Sommer 1990; Barik 1993; Barettino *et al.* 1994; Ke and Madison 1997; Brons-Poulsen *et al.* 1998).

31.3.10 Ligation-mediated PCR

Unlike RNA polymerase, DNA polymerase cannot initiate synthesis; it can only add nucleotides to an existing primer. Hence, primers are essential for PCR.

To design the primers one needs,knowledge of the base sequence of the two flanking regions of the DNA fragment to be amplified by PCR is thus essential. It might, however, happen that the sequence information of only one of the two flanking regions is available. Such a situation is encountered during genomic sequencing and also during *in vivo* foot printing. The method to be adopted then is ligation-mediated PCR (LMPCR; Fig. 31.9). Step 1 is to generate a 5′-phosphate end by creating a nick in the experimental DNA. This could be achieved by acting with DNase I, restriction endonucleases or the Maxam–Gilbert sequencing reagent DMS/piperidine. A primer is next designed so that it would be complementary to the flanking region of the nicked DNA strand but the 5′-end of the primer would be recessed and would not be flush against the 3′-end of the template strand. The target DNA is denatured,the primer is annealed after denaturing the experimental DNA and the first strand synthesis is allowed. The DNA synthesis terminates at the nick in the template strand, creating a duplex flush end. The new duplex DNA produced would thus have one staggered end and one flush end. A unidirectional linker is next ligated to the experimental DNA. This linker consists of two oligomeric unequal complementary strands of known base sequence with one flush end and one staggered end. Because of the specific nature of the ends of the experimental DNA and the linker, their ligation can take place only through their flush (blunt) ends. Conventional PCR can now be carried out, as the sequence information of both the flanking regions is available.

31.3.11 Aribitrarily primed PCR

This procedure is useful when the objective is to fingerprint a whole genome for determination of

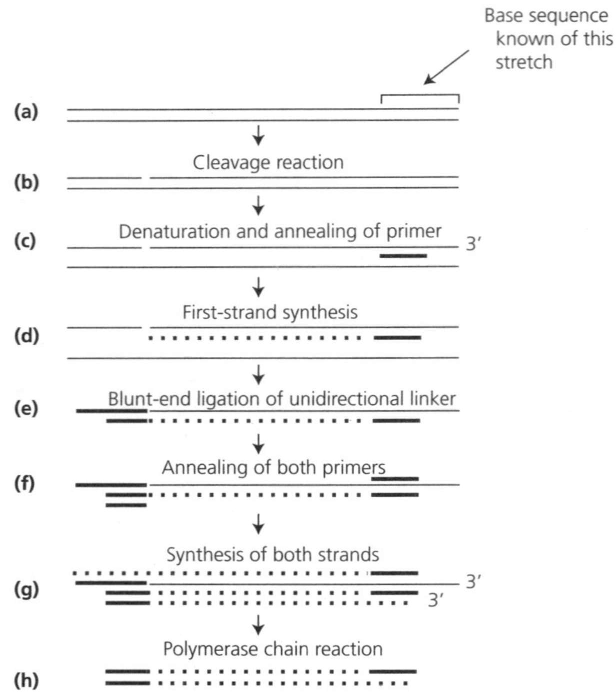

Figure 31.9 Ligation-mediated PCR (LMPCR). (a) The segment of DNA of which the base sequence of only one end is known. (b) The DNA after the cleavage reaction. (c) Annealed primer designed on the basis of the sequence of one end such that the 5′-end of the primer is recessed with respect to the 3′-end of the target strand. (d) The duplex formed after first-strand synthesis has one flush end and one staggered end. (e) The duplex DNA after blunt-end ligation of the unidirectional linker. (f) The second primer designed on the basis of the sequence of the unidirectional linker annealed as also the first primer. (g) Both DNA strands replicated. (h) The product of PCR.

relatedness with other genomes, and not for the usual amplification of DNA. The primers used in this procedure do not have base sequences complementary to the two flanking regions of any specific stretch of DNA. The base sequences of the primers are quite arbitrary; consequently, the annealing conditions used are of low stringency. As a result, the primers would be able to anneal almost non-specifically at several loci, even if the complementarity is somewhat imperfect. High-stringency conditions are finally imposed, but only after running a few low-stringency cycles. The pattern of DNA bands visible on gel electrophoresis after the completion of the amplification process is quite reproducible as long as the arbitrary primers used are unchanged and the experimental conditions are the same. The pattern of bands in the gel is indeed a fingerprint of the particular genome. Relatedness of two genomes can be determined from such fingerprints, or pattern of bands, provided that the same two primers and the same set of experimental conditions have been employed.Genetic alterations in cancer cells have also been evaluated by comparing the fingerprint of cancer cells with that of the fingerprint of normal cells of the same person.

References

Ahrens U. Detection of DNA of plant pathogenic *Mycoplasma*-like organisms by a polymerase chain reaction that amplifies a sequence of the 16S rRNA gene. *Phytopathology* 1992;82:828–832.

Aurelius E, Johansson B, Staland A, Forsgren M, Aurelius E, Skoldenberg B. Rapid diagnosis of herpes simplex encephalitis by nested polymerase chain reaction assay of cerebrospinal fluid. *Lancet* 1991;337:189–192.

Bagasra O, Lavi E, Bobroski L, Khalili K, Pestaner JP, Tawadros R, Pomerantz RJ. Cellular reservoirs of HIV-1 in the central nervous system of infected individuals: Identification by the combination of in situ polymerase chain reaction and immunohistochemistry. *AIDS* 1996;10:573–585.

Barettino D, Feigenbutz M, Valcarcel R, Stunnenberg HG. Improved method for PCR-mediated site-directed mutagenesis. *Nucleic Acids Res* 1994;22:541–542.

Barik S. Site-directed mutagenesis by double polymerase chain reaction: Megaprimer method. *Methods Mol Biol* 1993;15:277–286.

Becker-André M, Hahlbrock K. Absolute mRNA quantification using the polymerase chain reaction (PCR). A novel approach by a PCR aided transcript titration assay (PATTY). *Nucleic Acids Res* 1989;17:9437–9446.

Brøns-Poulsen J, Petersen NE, Horder M, Kristiansen K. An improved PCR-based method for site directed mutagenesis using megaprimers. *Mol Cell Probes* 1998;12: 345–348.

Clewell DB, Helinski DR. Effect of growth conditions on the formation of the relaxation complex of supercoiled ColE1 deoxyribonucleic acid and protein in *Escherichia coli.* J Bacteriol 1972;110: 1135–1146.

Eisenach KD, Sifford MD, Cave MD, Bates JH, Crawford JT. Detection of *Mycobacterium tuberculosis* in sputum samples using a polymerase chain reaction. *Am J Respir Crit Care Med* 1991;144:1160–1163.

Gilliland G, Perrin S, Blanchard K, Bunn HF. Analysis of cytokine mRNA and DNA: Detection and quantitation by competitive polymerase chain reaction. *Proc Natl Acad Sci USA* 1990;87: 2725–2729.

Gyllensten UB, Erlich HA. Generation of single stranded DNA by the polymerase chain reaction and its application to direct sequencing of the HLA-DQA locus. *Proc Natl Acad Sci USA* 1988;85:7652–7656.

Ho NS, Hunt HD, Horton RM, Pullen JK, Pease LR. Site-directed mutagenesis by overlap extension using the polymerase chain reaction. *Gene* 1989;77:51–59.

Kasai K, Nakamura Y, White R. Amplification of a variable number of tandem repeats (VNTR) locus (pMCTII8) by the polymerase chain reaction (PCR) and its application to forensic science. *J Forensic Sci* 1990;35:1196–1200.

Ke S-H, Madison EL. Rapid and efficient site-directed mutagenesis by single-tube 'megaprimer' PCR method. *Nucleic Acids Res* 1997;25:3371–3372.

Kloosterman AD, Budowle B, Daselaar P. PCR-amplification and detection of the human D1S80 VNTR locus. Amplification conditions, population genetics and application in forensic analysis. *Int J Leg Med* 1993;105:257–264.

Lanciotti RS, Calisher CH, Gubler DJ, Chang GJ, Vorndam AV. Rapid detection and typing of dengue viruses from clinical samples by using reverse transcriptase- polymerase chain reaction. *J Clin Microbial* 1992;30:545–555.

Landt O, Grunert H-P, Hahn U. A general method for rapid site-directed mutagenesis using the polymerase chain reaction. *Gene* 1990;96:125–128.

Liu W, Saint DA. A new quantitative method of real time reverse transcription polymerase chain reaction assay based on simulation of polymerase chain reaction kinetics. *Anal Biochem* 2002; 302:52–59.

Mullis K, Faloona F, Scharf S, Saiki R, Horn G, Erlich H. Specific enzymatic amplification of DNA *in vitro*: The polymerase chain reaction. *Cold Spring Harbor Symp Quant Biol* 1986;51: 263–273.

Nakamura Y, Carlson M, Krapcho K, White R. Isolation and mapping of a polymorphic DNA sequence, pMCTII8, on chromosome tp *(D1S58). Nucleic Acids Res* 1988;16:9364.

Okamoto H, Sugiyama Y, Okada S, Kurai K, Akahane Y, Sugai Y, *et al.* Typing hepatitis C virus by polymerase chain reaction with type-specific primers: Application to clinical surveys and tracing infectious sources. *J Gen Virol* 1992;73:673–679.

Saiki R, Gelfand D, Stoffel S, Scharf S, Higuchi R, Horn G, *et al.*. Primer-directed enzymatic amplification of DNA with a thermostable DNA polymerase. *Science* 1988;239:487–491.

Saiki R, Scharf S, Faloona F, Mullis K, Horn G, Erlich H, *et al.* Enzymatic amplification of beta-globin genomic sequences and restriction site analysis for diagnosis of sickle cell anemia. *Science* 1985;230:1350–1354.

Sajantila A, Budowle B, Ström M, Johnsson V, Lukka M, Peltonen L, *et al.* PCR amplification of alleles at the DIS80 locus: Comparison of a Finnish and a North American Caucasian population sample, and forensic casework evaluation. *Am J Hum Genet* 1992;50:816–825.

Sarkar G, Sommer SS. The 'megaprimer' method of site-directed mutagenesis. *BioTechniques* 1990;8:404–407.

Schmittgen TD, Zakrajsek BA, Mills AG, Gorn V, Singer MJ, Reed MW.

Quantitative reverse transcription-polymerase chain reaction to study mRNA decay: Comparison of endpoint and real-time methods. *Anal Biochem* 2000;285:194–204.

Urban A, Neukirchen S, Jaeger KE. A rapid and efficient method for site-directed mutagenesis using one-step overlap extension PCR. *Nucleic Acids Res* 1997;25:2227–2228.

Wang AM, Doyle MV, Mark DF. Quantitation of mRNA by the polymerase chain reaction. *Proc Natl Acad Sci USA* 1989;86:9717–9721.

Winer J, Jung CK, Shackel I, Williams PM. Development and validation of real-time quantitative reverse transcriptase-polymerase chain reaction for monitoring gene expression in cardiac myocytes *in vitro. Anal Biochem* 1999;270:41–49.

Wong H, Anderson WD, Cheng T, Riabowol KT. Monitoring mRNA expression by polymerase chain reaction: The 'primer-dropping' method. *Anal Biochem* 1994;223: 251–258.

Yoshioka K, Kakumu S, Wakita T, Ishikawa T, Itoh Y, Takayanagi M, *et al.* Detection of hepatitis C virus by polymerase chain reaction and response to interferon-alpha therapy: Relationship to genotypes of hepatitis C virus. *Hepatology* 1992;16:293–299.

Zipper H, Brunner H, Bernhagen J,Vitzthum F. Investigations on DNA intercalation and surface binding by SYBR Green I, its structure determination and methodological implications. *Nucleic Acids Res* 2004;32:e103.

32

Alternative strategies of gene cloning

Indranil Dasgupta

32.1 Cloning interacting genes: Yeast one-hybrid, two-hybrid and three-hybrid systems

Intermolecular interactions of macromolecules drive the life processes. These interactions usually result in the formation of functional complexes of large molecules such as those involved in nucleic acid replication, transcription of messenger RNAs or the components constituting signalling processes, which help the various cellular compartments to communicate with each other or with the outside environment. The interactions could be of three types: (a) between a protein and DNA, (b) between two proteins and (c) between a protein and RNA. The DNA fragment, which participates in an interaction with a protein, can be cloned using a method known as the yeast one-hybrid system. Similarly, when two proteins interact with each other, their corresponding genes are known as interacting genes and can be cloned using a method known as the yeast two-hybrid system. Similarly, interactions between proteins and RNA molecules can be investigated and the RNA molecules can be cloned using a method known as the yeast three-hybrid system.

32.1.1 Yeast one-hybrid system

The yeast one-hybrid system is used for cloning genes encoding a DNA-binding protein specific for a particular DNA sequence. Most transcription factors have a DNA-binding domain and an activation domain, which can be separated from each other. The yeast one-hybrid system is based on the principle that an activation domain alone, when fused to a specific DNA-binding protein, can activate transcription of a reporter gene if its upstream region contains the specific binding sites for that protein. In actual practice, first, the DNA sequence, for which protein binding has to be studied, is cloned upstream of a gene, whose product is essential for the survival of yeast in the presence of a growth inhibitor or antibiotic (Fig. 32.1). Next, the yeast cells are transformed with a library consisting of a large number of DNA molecules, each expressing a fusion of a test protein and the activation domain of the transcription factor. If the gene encodes a DNA-binding protein for the DNA sequence used, its encoded protein will bind to the binding site present upstream of the essential gene, which will be transcribed, resulting in the growth of the yeast cell to form a colony on the selection medium. Thus, the gene encoding the DNA-binding protein for a particular DNA sequence can be cloned. For example, Kim *et al.* (2002) could isolate transcription factors, which interacted with and bound to DNA elements located upstream to genes expressing abundantly in the embryos of carrot. In other words, it could be said that the transcription factors controlled the expression of these genes by binding specifically to the promoter region. For this work, the investigators screened 25 million yeast transformants and, on the basis of growth in the selection medium, could isolate nine clones coding for transcription factors.

32.1.2 Yeast two-hybrid system

The yeast two-hybrid system is a method of studying interaction between two proteins or to find out unknown interacting partners of a given protein. The two-hybrid method is based on the observation that most proteins involved in promoting transcription of a gene (transcription factors) contain two distinct parts or domains. These domains can be physically separated from each other, in which case the protein loses its activity, and can be re-activated once they are brought

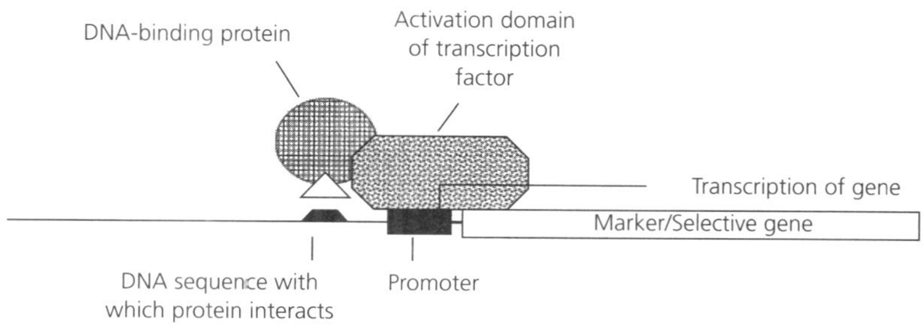

Figure 32.1 Schematic diagram of the yeast one-hybrid system. The activation domain of a yeast transcription factor is fused with a test protein, which can bind to a specific DNA sequence. The binding sequence is placed directly upstream of the promoter of a reporter or a selectable gene. If the test protein binds to the sequence, the transcription of the gene is activated.

together. One of the two domains recognizes the promoter of the gene to be activated, known as the DNA-binding domain, and the other interacts with the RNA polymerase enzyme or a general transcription factor, which regulates transcription and performs the activation of transcription, known as the activation domain. If the above two domains are fused individually to genes whose products interact, then the domains are brought together by virtue of the above interaction and can be used to transcribe a gene essential for growth in yeast (Fig. 32.2). Only those yeast cells where the two interacting genes have actually come together grow, because the two domains of the transcription factor become functional by virtue of the 'bridging' action of the products of the interacting genes. This provides a very powerful system for the selection of such genes. One of the interacting genes, known as 'bait', is fused to the DNA-binding domain of the transcription factor and its interacting pair, known as 'prey', is selected from a c-DNA expression library designed to form a fusion protein with the activation domain of the same transcription factor. Thus, given a bait gene, its interacting pair, the prey, can be cloned using a yeast two-hybrid system.

An example of the use of a yeast two-hybrid analysis to investigate protein–protein interaction is illustrated by the work of Li and Fields (1993). The interacting proteins used were the tumour suppressor protein p53 and its interacting partner SV40 T antigen. The investigators used the yeast two-hybrid system to study the interaction process. Using mutants of p53, which were defective in binding to T antigen, they concluded that the yeast two-hybrid method of studying the interacting proteins was superior to other methods such as immunoprecipitation. Using this system, the investigators could identify a large number of mutants of p53, which showed compromised binding to T antigen.

32.1.3 Yeast three-hybrid system

Protein–RNA interactions form a cornerstone of life processes, such as the assembly and function of ribosomes, the control of expression of genes in RNA viruses, mRNA stability and expression, etc. The yeast three-hybrid system provides a means of study of these interactions. In this system, a DNA-binding domain of a transcription factor, for example, the *E. coli* Lex A protein, is expressed as a fusion product with another protein having a strong RNA-binding property, such as the *E. coli* RNA phage MS2 capsid protein. This is then allowed to interact with a hybrid RNA, consisting of a part of MS2 and a part of the specific RNA (X), whose interaction with proteins needs to be studied. An interacting protein, if fused with the activation domain of a transcription factor, can cause the activation of a detectable gene in yeast (Fig. 32.3). Because of the presence of two proteins and a single RNA, the method is called the three-hybrid system. Sengupta *et al.* (1999) used this system to study the RNA-binding partners of the yeast protein Snp1, a homologue of the human U1-70K protein. The analysis indicated that the strongest interacting partner to Snp1 was

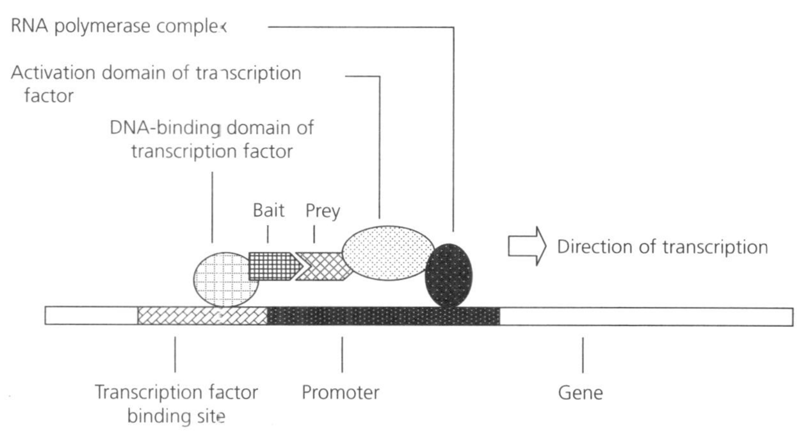

Figure 32.2 Schematic diagram of the yeast two-hybrid system. The activation domain of the yeast transcription factor is fused with a polypeptide called 'prey'. Prey interacts with another polypeptide, 'bait', which in turn is fused with the DNA-binding domain of the same transcription factor. Successful transcription of an essential gene is possible only when bait and prey interact, bringing together the DNA-binding and the activation domains of the transcription factor.

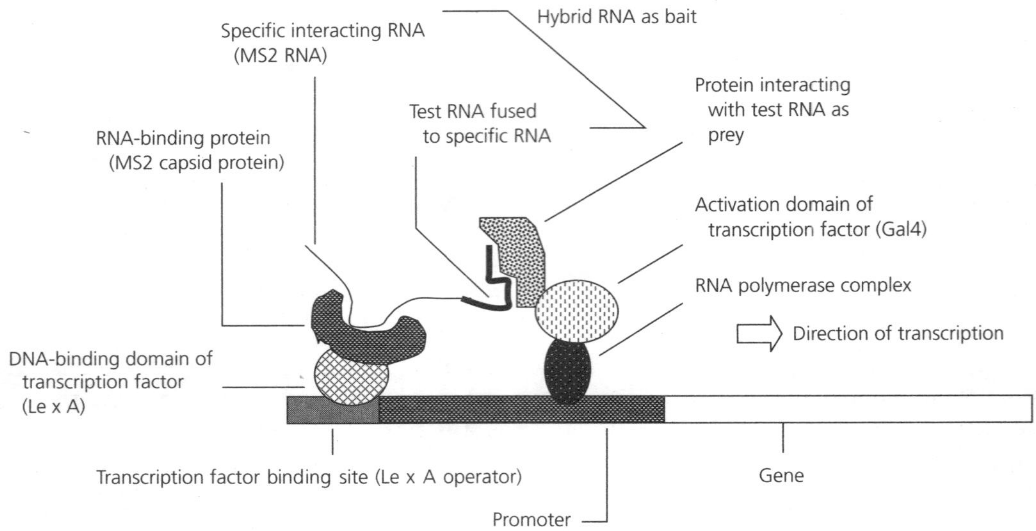

Figure 32.3 Schematic diagram of the yeast three-hybrid system. The DNA-binding domain of the yeast transcription factor is fused with a bacteriophage capsid protein MS2, which specifically binds to a specific RNA (MS2 RNA). Test RNA is fused with MS2 RNA, which acts as the bait. Protein interacting with the test RNA (prey), fused to the activation domain of the same transcription factor, can initiate the transcription of an essential gene.

the fragment of U1 RNA-containing loop 1. Four other RNA ligands were also identified having weaker affinities to Snp1. Hence, the yeast three-hybrid system presents a very effective method to perform a fine analysis of RNA–protein interactions.

32.2 Using phage display to clone genes of interacting proteins

Several types of bacteriophages can infect *E. coli*. Bacteriophages are useful in contributing strong and inducible promoters and RNA polymerases for heterologous gene expression in bacterial systems. Here, we will see how some of them can be useful in helping in the selection and cloning of interacting genes. Bacteriophages such as M13 and fl are filamentous in shape, having thousands of capsid protein subunits of different types surrounding their single-stranded DNA genomic material in a particular fashion to give rise to a thread-like filamentous particle. Some of the capsid protein subunit types, like that of protein III or protein VIII of M13 and f1, which are not crucial to the architecture of the particle can be used to form hybrid proteins if a heterologous fragment of DNA is inserted adjacent to its gene. In most such cases, the above protein can still be included in the infectious particles. Such particles will 'display' the heterologous protein as their part. Such a system of exposing the expressed heterologous proteins on bacteriophage particles is known as 'phage display' and was first described by George Smith in 1985. The power of the technique lies in the fact that a large number of such recombinant phage particles can be generated to give a library of high complexity, which means that the phage library can display a large collection of peptides. Usually, a large collection of phage particles, commonly known as a 'phage display library', is produced, where each particle expresses a different heterologous protein depending upon the fragment of DNA cloned in it. Such a system gives an opportunity to easily select and clone genes corresponding even to interacting proteins of low abundance. The protein to which interacting proteins have to be selected is immobilized in a microtitre plate or a suitable surface or on a column matrix, which is then exposed to the phage display library. Interacting phages to the immobilized protein interact with the protein and are in turn immobilized on the surface or column, from where they can be released and their corresponding inserted DNA can be obtained (Fig. 32.4). As the interacting peptide is encoded by the heterologous gene, which is part of the phage DNA, the gene can be multiplied and characterized.

Phage display has proved to be very important in studying protein–protein interactions, in searching agonists and antagonists to particular active molecules and in drug design. For example, Burton *et al.* (1991) used this method to screen for and select highly interactive antibody fragments against the surface antigen gp120 of HIV from a library of random antibody fragments. The library was prepared from about 5 mL of bone marrow from an HIV-positive individual who has been asymptomatic for several years. The antibodies selected have both high affinities to the antigen as well as diversity in their structures. Such antibody collections help in designing an effective therapy against AIDS.

32.3 Using gene tagging to clone genes

Gene tagging is a method of inserting a recognizable DNA fragment within a gene such that the function of the gene is disrupted and the gene can be easily

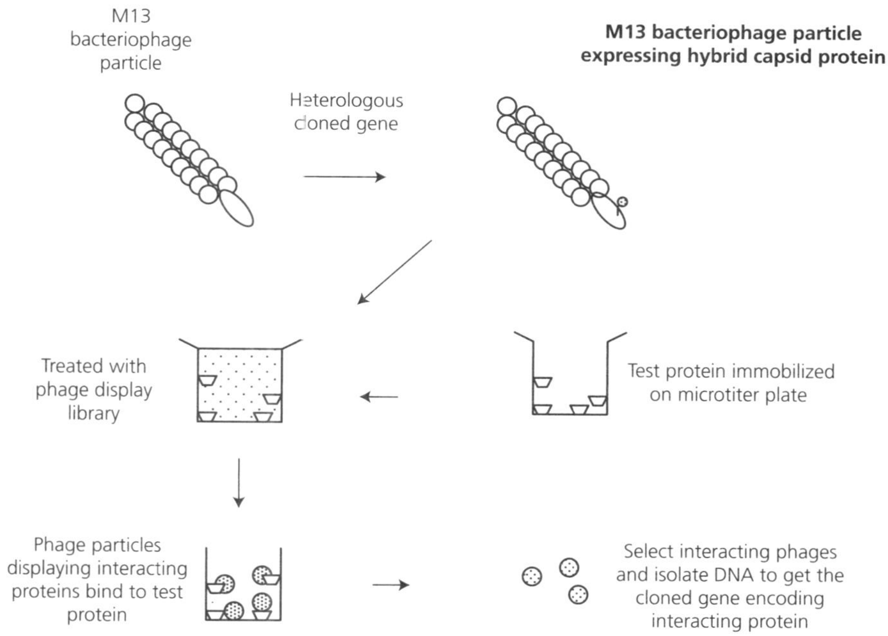

Figure 32.4 Phage display to clone interacting genes. Bacteriophage M13 particles, displaying heterologous proteins as part of their capsids, are allowed to hybridize with test proteins. Phage particles interacting with the test proteins display interacting proteins in their capsids and are then purified to obtain the corresponding gene.

recognized by virtue of the inserted fragment. The disruption of function can be used as a means to locate the gene by looking for the presence of the known inserted fragment that has caused the gene disruption (Fig. 32.5). Transposable elements are pieces of DNA which are mobile and insert themselves, mostly at random locations, in the genome. Hence, transposable elements, which have already been characterized by sequencing, are ideally suited for gene tagging. Another example of random insertion of a DNA fragment in the genome is the T-DNA, the piece of DNA transferred by the soil bacterium *Agrobacterium tumefaciens,* which is randomly inserted into the DNA of many plants. Gene tagging helps in locating a gene by using the disruption of its function as the starting point and to assign the function disrupted as the normal function of the gene.

In the genome of the tagged mutated individual, the gene responsible for the mutation can be located by searching for the inserted transposon or T-DNA. The location of the transposon points to the gene which has been disrupted and it can be cloned using the tag as an identifiable marker. The gene can also be cloned by using a variant of the method of PCR by using the tag sequences as primers for the amplification reaction. These methods have been successfully used in cloning of several genes, both from bacteria as well as from higher organisms. In large-scale gene identification projects, transposons are inserted into the genome of organisms and allowed to transpose at random locations in the genome. Mutants are selected which have resulted from the transposition events. Genomic libraries are then prepared from each mutant and clones

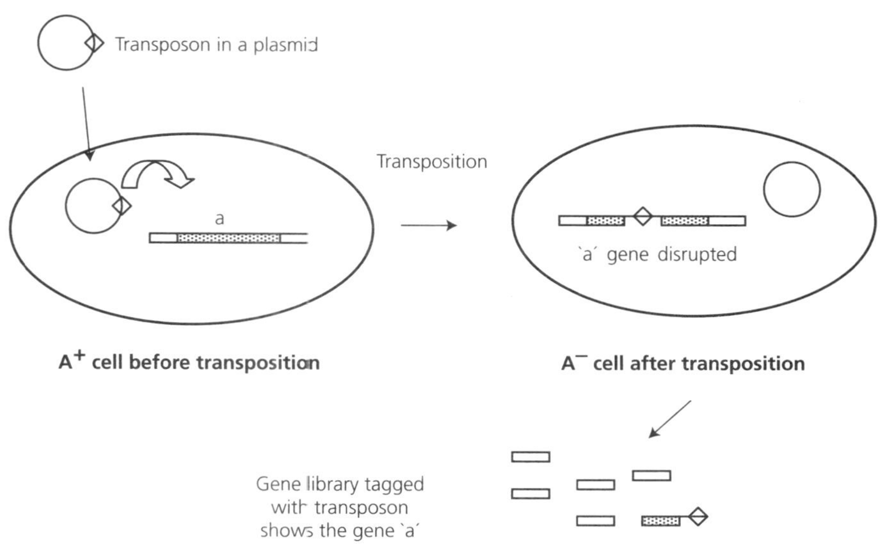

Figure 32.5 Transposon tagging of genes. A transposon inserted into a gene 'a' produces a mutant (A⁻). The mutant is selected by its phenotype, and a gene library, prepared from the mutant, is screened for the presence of the transposon. As the transposon is inserted into gene a, its presence indicates the position of gene a.

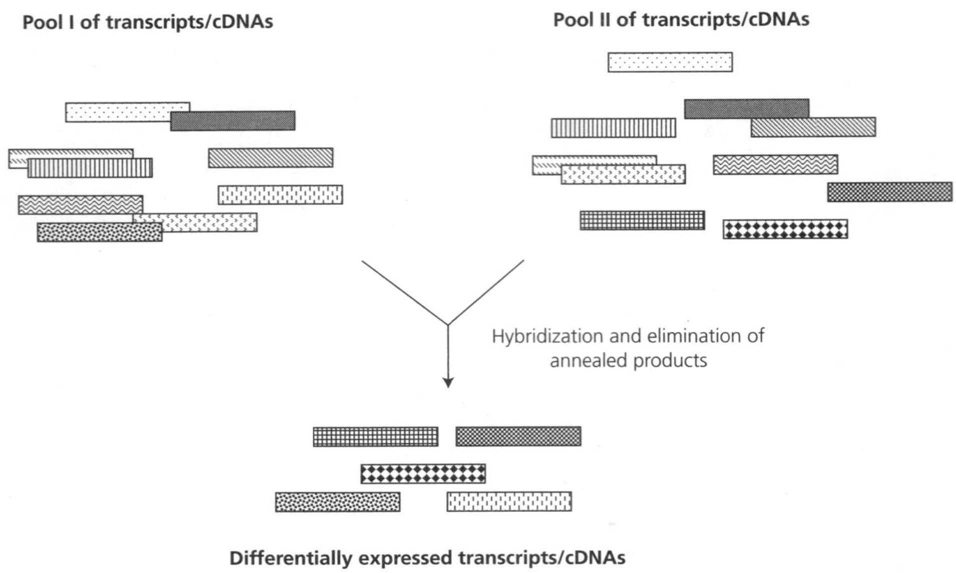

Figure 32.6 Schematic diagram of subtractive hybridization. RNA/cDNA derived from tissues under two different treatments are hybridized with each other. Same molecules present in both the pools anneal with each other and can be removed. Non-identical molecules or those present in differing quantities survive and can then be specifically cloned or amplified.

containing the transposons are selected by hybridization with a DNA probe from the transposon itself. Genes present on either side of the inserted transposon represent those whose disrupted function has resulted in the mutation. Thus, genes responsible for specific functions can be cloned and identified.

An example of the use of transposon tagging was the report of the functional analysis of a large number of yeast genes by Ross-Macdonald *et al.* (1999). The investigators used a mini-transposon derived from the bacterial transposon Tn3 to create the mutations. Transposon mutagenesis resulted in the generation of almost 11,000 individual mutants, each having a mutation caused by the insertion of the mini-transposon within a region of the genome expressed during the vegetative or sporulation stage of growth. Almost 2000 of a total of 6000 genes of yeast were affected by these insertions. The investigators used this collection of mutants to determine the disruption phenotypes of nearly 8000 yeast strains using 20 different growth conditions. By selecting for the transposable element, the genes mutated could be cloned and the corresponding function of the gene could be determined by the mutant phenotype.

32.4 Cloning differentially expressed genes

Some genes are expressed more or less at the same level in all tissues and under different environmental or biological conditions of the organism. These are generally the housekeeping genes and are known to express constitutively. There are many genes in the organism, however, which are expressed at varying levels under different conditions or in different tissues. These are known as differentially expressed genes and represent those required by the organism only under certain conditions. Such genes can be cloned using a variety of methods, all of which are based on the principle of 'subtraction'. In subtraction, two populations of cDNA are synthesized, representing all the transcripts found under two different conditions in the cell. To enrich for differentially expressed transcripts, the two populations are reverse transcribed to double-stranded cDNAs and are allowed to hybridize with each other. Same molecules in the two populations hybridize with each other and the unhybridized transcripts represent those which are differentially expressed. In some cases, a cloned cDNA library is hybridized with a labelled RNA extracted from the tissue grown under a different condition. All those clones which do not hybridize represent cDNA which are synthesized from differentially expressed transcripts. Such approaches are sometimes termed 'subtractive hybridization' (Fig. 32.6). In recent years, considerable improvements have been incorporated into such techniques. In one such clever approach, known as suppression subtractive hybridization, differentially expressed cDNAs are specifically amplified by PCR as they are linked to special oligonucleotide linkers, allowing for the amplification of only the differentially expressed cDNAs. These can then be cloned in suitable cloning vectors.

References

Burton DR, Barbas CF, Persson MA, Koenig S, Chanock RM, Lerner RA. A large array of human monoclonal antibodies to type 1 human immunodeficiency virus from combinatorial libraries of asymptomatic seropositive individuals. *Proc Natl Acad Sci USA* 1991;88:10134–10137.

Kim SY, Chung H-J, Thomas TL. Isolation of a novel class of bZIP transcription factors that interact with ABA-responsive and embryo-specification elements in the *Dc3* promoter using a modified yeast one-hybrid system. *Plant J* 1997;11:1237–1251.

Li B, Fields S. Identification of mutations in p53 that affects its bnding to SV40 large T antigen by using yeast two-hybrid system. *FASEB J* 1993;7:957–963.

Ross-Macdonald P, Coelho PS, Roemer T, Agarwal S, Kumar A, Jansen R, *et al.* Large-scale analysis of yeast genome by transposon tagging and gene disruption. *Nature* 1999;402:413–418.

Sengupta DJ, Wickens M, Fields S. Identification of RNAs which bind to a specific protein using the yeast three hybrid system. *RNA* 1999;5:596–601.

33

Site-directed *in vitro* mutagenesis and protein engineering

Shweta Sharma and Indranil Dasgupta

33.1 Site-directed *in vitro* mutagenesis using primer–template system of DNA polymerase reaction

Induced mutations are mostly random in nature as the precise nucleotide of a gene, which is going to be mutated during mutagenesis, cannot be predicted. However, site-directed mutagenesis is a technique by which the position of the mutation in a gene can be controlled accurately. The primer–template system of DNA polymerase reaction [Fig 33.1(a) and (b)] is at the core of the *in vitro* mutagenesis process developed in several laboratories. If a primer is, however, chosen such that most of the bases in it are complementary to the template but one or two are not [Fig. 33.1(c)], then the newly formed DNA strand synthesized by DNA polymerase would have those one or two wrong or mismatched bases [Fig. 33.1(d)]. During subsequent replication, either *in vitro* or *in vivo*, 50% of the duplexes formed would have the wrong bases on both the strands [Fig. 33.1(e)]; hence, the mutation would have been achieved at the desired site. The problem, however, is to enrich or select for the mutants in preference to the wild-type DNA duplexes [Fig. 33.1(f)]. In 1985, TA Kunkel devised a rapid and efficient site-specific mutagenesis process that could enrich the mutant without any phenotypic selection. Kunkel's method provided a very strong selection against the non-mutagenized strand of a double-stranded DNA. When DNA is synthesized in a *dut, ung* double-mutant of *Escherichia coli*, the newly formed DNA strand carries a number of uracils in thymine positions. This is because while the *dut* mutation inactivates the enzyme dUTPase and results in high intracellular levels of dUTP, the *ung* mutation inactivates the enzyme uracil N-glycosylase, which would have otherwise removed the incorporated uracil in DNA. In practice, the DNA to be mutagenized is cloned in an M13 phage vector and *dut ung E. coli* is transformed with it. The virions of M13 that come out of the host *E. coli* contain a single-stranded DNA as their genetic material, which in this case contains uracil. This uracil-containing single strand is then used as a template for the *in vitro* synthesis of a complementary strand primed by an oligonucleotide containing the desired mutation. When the resulting double-stranded DNA is introduced into an *E. coli* cell having a proficient uracil N-glycosylase, the uracil-containing strand is degraded *in vivo* with high efficiency, leaving the non-uracil–containing survivor strand with the mutated bases to replicate *in vivo*. This results in a considerable increase in the proportion of the mutated DNA. Fig. 33.2 shows the steps involved in *in vitro* mutagenesis using an M13 bacteriophage vector.

33.2 Site-directed *in vitro* mutagenesis by PCR

This is generally achieved by designing two oligonucleotide primers from the site of the cloned DNA which needs to be mutagenized. The primers for the polymerase chain reaction (PCR) are designed such that: (i) the two primers would anneal to the same stretch of DNA on opposite strands of the plasmid where the mutation is desired and (ii) the primers are 25–45 bases

(a) Primer 5′ ACTGTACGCTGCAGA 3′
Template 3′ TGACATGCGACGTCTGAGATCAGTCTGCAGTCATCGACGTAGC 5′

Newly synthesized portion in italics

(b) 5′ ACTGTACGCTGCAGACT *CTA GTCAGA CGTCA GTAGC TGCA TCG* 3′
Template 3′ TGACATGCGACGTCTGA GAT CAGTCT GCAGT CATCG ACGTAGC 5′

(c) Primer 5′ ACTGTA[TAT]TGCAGA 3′
Template 3′ TGACATGCGACGTCTGAGATCAGTCTGCAGTCATCGACGTAGC 5′

Newly synthesized portion in italics

(d) 5′ ACTGTA[TAT]TGCAGA*CTCTA GTCAGA CGTCA GTAGC TGCATCG* 3′
Template 3′ TGACATGCGACGTCTGAGAT CAGTCT GCAGT CATCGACGTAGC 5′

(e) Template 5′ ACTG[TAT]ATTGCAGACTCTAGTCAGACGTAGT AGCTGCATCG 3′
3′ *TGACA* [*TAT*] *AACGTCTGAGA TCAGTCTGCAGTCA TCGACGTAGC* 5′

Newly synthesized strand in italics

Newly synthesized strand in italics

(f) 5′ *ACTGTACGCTGCAGACTCTAGTCAGACGTCAGTAGCTGCATCG* 3′
Template 3′ TGACATGCGACGTCTGAGATCAGTCTGCAGTCATCGACGTAGC 5′

Figure 33.1 The principle of *in vitro* mutagenesis. (a) The standard primer– template system for *in vitro* DNA synthesis. All the bases of the primer are exactly complementary to those of the template. (b) Duplex DNA formed as a result of extension of the primer by the addition of deoxynucleotides to its 3′-end catalysed by DNA polymerase I from *E. coli*. The newly synthesized portion is shown in italics. (c) The three boxed bases of the primer are changed during its chemical synthesis and are different from the corresponding bases in (a). These three bases are not complementary to the bases in the template. (d) Duplex DNA formed by extension of the primer containing the changed (mutated) bases. (e) and (f) Semiconservative replication *in vivo* of the duplex shown in (d) results in two daughter duplexes (e and f). The duplex shown in (e) has mutated bases (shown within boxes) in both the strands. The duplex shown in (f) has no mutated base.

long, with the mutation in the middle and 10–15 correct bases on both sides of the mutated bases.

The overlapping oligonucleotide primers, because of their partial self-complementarity, are known as 'abutting primers'. The DNA fragment to be mutated is ligated into a plasmid isolated from a *dam* (+) *E. coli.* A PCR amplification reaction, which is performed using these abutting primers, results in the original gene carrying a mutated residue at the predetermined position. In the next step, the non-mutagenized parental molecules are removed by digestion with a restriction enzyme (e.g. *Dpn* 1), whose activity is seen only when there are methylated nucleotides at the recognition sequence. The parental DNA templates contain methylated residues, as the plasmid DNA isolated from a *dam* (+) *E. coli* strain is methylated and is therefore susceptible to digestion with such restriction enzymes. However, the mutated DNA molecules, synthesized following PCR amplification, survive the above digestion because they lack methylated residues. This is followed by cloning of the product obtained after the amplification reaction, which in turn represents the mutated gene carrying a mutation in the desired location (Fig. 33.3). The effect of such mutations can be tested by reintroducing this mutated gene in the organism and studying the function of the mutated gene. Using site-directed mutagenesis, the effect of mutation of individual amino acid residues in a protein could be studied on the overall functions performed by the protein. In addition, by changing a particular codon into a stop codon, which leads to premature termination of the protein synthesis, this technique can also be used to produce truncated proteins.

Site-directed mutagenesis using oligodeoxyribonucleotides was first shown by Michael Smith and his collaborators in 1978. Hutchinson *et al.* (1978) mutated nucleotide G at position 587 to nucleotide A of the viral DNA strand of the bacteriophage phiX174 using oligodeoxyribonucleotides complementary to this region. As a result, the phages obtained using this method contained A in place of G at position 587 of the viral strand and were found to be mutants of gene E, which is present in this region of DNA. Therefore, the mutated *am* phages were found to show similarity with the *am3* mutants, where the change G to A at position 587 is sufficient to produce a defective E function. For his contribution in the development of site-directed mutagenesis technique, Michael Smith shared the Nobel Prize in Chemistry in 1993 with Karry Mullis, who invented PCR.

33.3 Protein engineering

The use of site-directed mutagenesis resulting into precise mutation of single nucleotide of a gene led to the concept of protein engineering. Protein engineering is

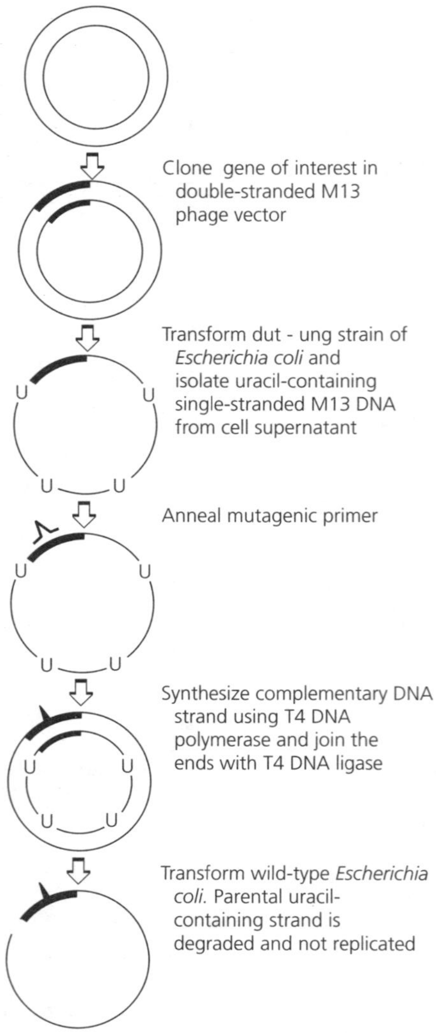

Figure 33.2 Steps involved in *in vitro* mutagenesis by Kunkel's method using an M13 bacteriophage vector.

a method using which the function of proteins can be improved by making precise changes in their amino acid sequence, which in turn can be achieved by accurate mutations in their encoded genes. The successful use of protein engineering to modify the functions of known proteins was reported for the first time in 1982 for enzymes tyrosyl-transfer RNA synthetase (TyrRS) (Winter *et al.* 1982) and β-lactamase (Dalbadie-Mcfarland *et al.* 1982; Sigal *et al.* 1982). In 1982, Michael Smith and his collaborators incorporated specific changes in the gene coding for enzyme tyrosyl-transfer RNA synthetase of *Bacillus stearothermophilus*, which is involved in the aminoacylation of $tRNA^{tyr}$ with tyrosine. Using oligonucleotide-based site-directed mutagenesis, the cysteine residue (Cys35) present at the catalytic site of TyrRS, which forms a putative hydrogen bond with the ribose portion of the ATP and is conserved in other aminoacyl-tRNA synthetases (Brannigan and Wilkinson 2002), was mutated into a serine residue. The mutated enzyme showed reduced activity as compared with the wild-type enzyme, which confirmed the important role of Cys35 for the catalytic action of this enzyme. At the end of 1982, Dalbadie-Mcfarland *et al.* also used oligonucleotide-directed mutagenesis to incorporate specific changes in the β-lactamase gene of the plasmid pBR322, which confers resistance to the antibiotic ampicillin. Using a chemically synthesized 16-base oligodeoxyribonucleotide, they inverted the serine–threonine dyad, which is present at the active site of this enzyme, to the threonine–serine dyad. This inversion resulted in a mutant plasmid with an ampicillin-sensitive phenotype.

Monoclonal antibodies bind to a wide spectrum of antigens with high affinity and selectivity, but as they originate in mice, they can give rise to complications in patients requiring chronic therapy as they often elicit immunogenic response. To overcome, the antigen recognition domain and surfaces (complementarity-determining regions; CDRs) of rodent origin were grafted onto the framework of human antibody (Jones *et al.* 1986). The humanized antibodies thus obtained retained the antigen-binding characteristics of the rodent antibody but eliminated the problems of immunogenicity. These antibodies, which work more efficiently, are now being used in the treatment of

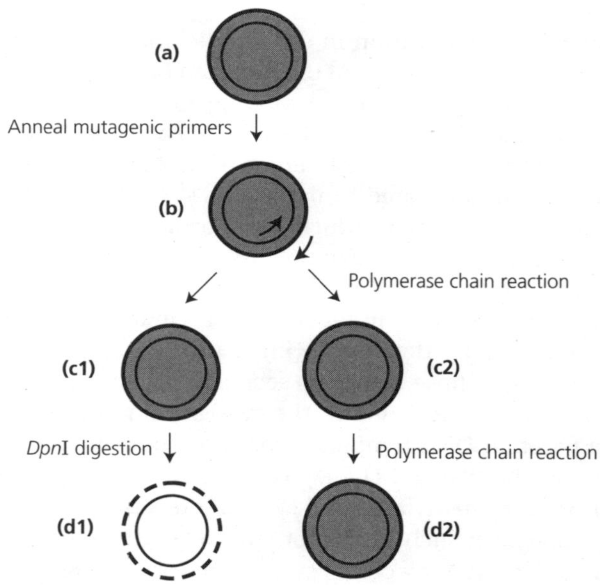

Figure 33.3 Site-directed mutagenesis by PCR. (a) The plasmid containing the fragment of DNA to be mutagenized. The plasmid has been isolated from a dam (+) *E. coli* cell. (b) The plasmid (a) has been denatured and the two mutagenic primers have been annealed. PCR has been carried out. (c1) The parental plasmid template. (c2) The product of PCR containing the mutated (base substituted or base deleted or base added) bases. The parental plasmids and the PCR products are digested with the restriction endonuclease *Dpn*I. (d1) The original template DNA is degraded. (d2) The PCR product (mutated) is unaffected.

cancer and transplant rejections (Holliger and Bohlen 1999; McCafferty and Glover 2000).

Another example of the successful use of protein engineering is the development of improved insulin for patients of diabetes mellitus. One of the major challenges in the use of traditional insulin preparation was the delay in the absorption of subcutaneously injected insulin in the bloodstream (Brange and Vølund 1999; Vajo and Duckworth 2000). Moreover, it was observed that in concentrated preparations, insulin self-associates to form dimers and hexamers, which suggested that lowering the tendency to self-associate might result in its quicker absorption. To achieve this, interfaces on the surface of insulin involved in self-assembly, which were identified using the crystal structure of its hexamer, were disrupted successfully by protein engineering (Blundell *et al.* 1972). Mutations that either juxtapose negative charges, such as replacement of proline 28 in B chain (P28B) by an asparagine residue or by destabilizing the intermolecular β-sheet interaction by exchanging ProB28 and LysB29, were incorporated to disrupt self-assembly. These improved insulins retained their biological activity as monomers (Brange *et al.* 1988).

In addition, the first example of a tailored enzyme being introduced in a crop plant was reported in 1999. Transgenic brassica plants expressing a modified Garm FatA1, an acyl–acyl carrier protein (ACP) thioesterase isolated from *Garcinia mangostana,* were developed. Several mutants that showed approximately a 13-fold increase in the specific activity of enzyme thioesterase were generated using site-directed mutagenesis. The seed-specific expression of one of the mutated thioesterase enzymes in *Brassica* plants resulted in transgenic plants that accumulate 55–68% more stearate than plants expressing the wild-type enzyme, which resulted in an increased proportion of unsaturated fatty acids in seed oil (Facciotti *et al.* 1999).

One of the successful examples has been reported in the case of protein T4 lysozyme. Stability of T4 lysozyme is dependent upon the presence of disulphide linkages between two cysteine residues present in the polypeptide subunits, which help them to achieve a particular secondary structure. A greater stability of this protein was achieved by replacing certain amino acids in the polypeptide with cysteine, which resulted in an overall increase in the number of cysteine residues. However, certain amino acids such as asparagine and glutamine are unstable and thus adversely affect protein functions, as they tend to undergo deamination at higher temperatures. Based on this information, replacing asparagines residues by threonine or isoleucine in triosephosphate isomerase enzyme of yeast resulted in a more thermostable enzyme. Using the concept of protein engineering, a significant improvement in the binding ability of the enzyme to its substrate has been reported in a few cases. Moreover, some enzymes have also been engineered to bind to altogether new substrates, such as specific DNA sequences. Despite these successful attempts, in the absence of knowledge regarding the functions of most proteins, the field of protein engineering is still in its infancy.

References

Blundell T, Dodson G, Hodgkin D, Mercola D. Insulin: The structure in the crystal and its reflection in chemistry and biology. *Adv Protein Chem* 1972;26:279–402.

Brange J, Ribel U, Hansen JF, Dodson G, Hansen MT, Havelund S, *et al.* Monomeric insulins obtained by protein engineering and their medical implications. *Nature* 1988;333:679–682.

Brange J, Vølund, A. Insulin analogs with improved pharmacokinetic profiles. *Adv Drug Deliv Rev* 1999;35:307–335.

Brannigan JA, Wilkinson AJ. Protein engineering 20 years on. *Nat Rev Mol Cell Biol* 2002;3:964–970.

Dalbadie-MacFarland G, Cohen LW, Riggs AD, Morino C, Itakura K, Richards JH. Oligonucleotide-directed mutagenesis as a powerful method for studies of protein function. *Proc Natl Acad Sci USA* 1982;79:6409–6413.

Facciotti MT, Bertain PB, Yuan L. Improved stearate phenotype in transgenic canola expressing a modified acyl-acyl carrier protein thioesterase. *Nat Biotechnol* 1990;17:593–597.

Holliger P, Bohlen H. Engineering antibodies for the clinic. *Cancer Metastasis Rev* 1999;18:411–419.

Hutchison CA, Phillips S, Edgell MH, Gillam S, Jhanke D, Smith M. Mutagenesis at a specific position in a DNA sequence. *J Biol Chem* 1978;253:6551–6560.

Jones PT, Dear PH, Foote J, Neuberger MS, Winter G. Replacing the complementarity-determining regions in a human antibody with those from a mouse. *Nature* 1980;321:522–525.

Kunkel TA. Rapid and efficient site specific mutagenesis without phenotypic selection. *Proc Natl Acad Sci USA* 1985;82:488–492.

McCafferty J, Glover DR. Engineering therapeutic proteins. *Curr Opin Struct Biol* 2000;10:417–420.

Sigal IS, Harwood BG, Arentzen R. Thiol β-lactamase: replacement of the active site serine of RTEM β-lactamase by a cysteine residue. *Proc Natl Acad Sci USA* 1982;79:7157–7160.

Vajo Z, Duckworth WC. Genetically engineered insulin analogues: Diabetes in the new millennium. *Pharmacol Rev* 2000;52:1–10.

Winter G, Fersht AR, Wilkinson AJ, Zoller M, Smith M. Redesigning enzyme structure by site-directed mutagenesis: Tyrosyl tRNA synthetase and ATP binding. *Nature* 1982;299:756–758.

34

Processing of recombinant proteins

Vikash Kumar Dubey and Suman Kundu

34.1 Introduction

Remarkable advancement in recombinant DNA technology has empowered protein science beyond the limits of imagination, enabling one to clone any gene in a suitable vector (of which there are several options) and subsequently express the gene product (protein) in a desired host system (Bradley 1990; Das 1990), which is generally an organism other than the one to which the gene originally belonged (heterologous expression). The host system most frequently and widely used is the bacterial expression system (*Escherichia coli*), because of its ease of handling and fast doubling time (Baneyx 1999; Jonasson *et al.* 2002). However, this simpler expression system suffers from disadvantages such as its inherent inability to post-translationally modify the expressed protein or produce a protein with multiple disulphide bridges. Other host systems, which overcome such shortcomings, include yeast, insect and mammalian cell lines, plant, etc. (Cregg *et al.* 2000; Kost *et al.* 2005; Rosser *et al.* 2005; Olczak and Olczak 2006; Lackner *et al.* 2008; Dyring 2011; Malys *et al.* 2011; McCarroll and King 1997; Werner *et al.* 2011; Gellert *et al.* 2012). These latter host systems, however, provide proteins in much lower yields than the bacterial expression system (Yin *et al.* 2007) and are not suitable where large quantities of recombinant proteins are required.

Recombinant proteins produced by a heterologous expression system need to be processed for downstream applications (Jonasson *et al.* 2002). Such processing, while routine in most laboratory applications, could also present significant challenges, especially where protein production needs to be scaled up. Proteolysis of the expressed protein, aggregation and poor solubility or inclusion body formation constitute some problems often faced in processing recombinant proteins (Sørensen and Mortensen 2005a, 2005b). The various steps in processing recombinant proteins are thus outlined.

34.2 Recovery of expressed recombinant protein

Host cells that express the recombinant protein in a suitable media containing necessary nutrients are harvested by centrifugation at the vegetative phase (Conn 2003). The harvested cells are suspended and homogenized in a solution containing a physiological buffer, a protease inhibitor cocktail, often a reducing agent such as dithiothreitol (DTT) and other additives that enhance protein stability and solubility (Deutscher 1990). The suspended cells are subsequently lysed to release the cell contents (Cull and McHenry 1990). There are several methods available for cell lysis which include sonication and pressure lysis (French press). However, if proteins get denatured during sonication or pressure lysis, enzymatic lysis by lysozyme is used to digest the cell wall. Another round of centrifugation at this stage segregates the recombinant protein either in the supernatant in soluble form or in the precipitate in insoluble form (Jonasson *et al.* 2002; Sørensen and Mortensen 2005b).

34.2.1 Soluble protein

Soluble recombinant proteins obtained as above are prone to degradation by proteases released by host cells and present a major hindrance in the recovery of functional proteins (Jonasson *et al.* 2002). Thus, an inhibitor cocktail (a mixture of small molecules and/or synthetic peptides that inhibit a diverse class of proteases) is employed at the time of cell lysis and processing is performed at lower temperatures (Jonasson *et al.* 2002; Sørensen and Mortensen 2005b), both strategies favouring minimal proteolysis. Purification of soluble proteins may be initiated directly. Mostly, the first step of purification is ammonium sulphate precipitation (salting out) or fractionation, where the bulk undesired material (proteins, nucleic acid, cell debris) is removed, resulting in a clear solution (Englard and Seifter 1990). Once partial purification is achieved, protein-specific strategies for purification are employed (Linn 1990). However, while designing a purification strategy, stability of the recombinant protein should also be considered (protein should remain in active form after purification). A harsh purification strategy may result in the isolation of an inactive or aggregated protein. If the target recombinant protein is expressed in conjugation with a fusion tag, then affinity chromatography can be initiated directly with the clarified lysate (Jonasson *et al.* 2002). With the advent of recombinant DNA technology, the use of fusion tags is a common practice, and as such, affinity chromatography (Wilchek 1972; Jonasson *et al.* 2002) has remained the method of choice for protein purification.

34.2.2 Insoluble protein

Proper folding of synthesized linear polypeptide into its correct tertiary structure is a pre-requisite for a protein to assume its functional role (Ptitsyn 1992; Ahluwalia *et al.* 2012). Often, the expressed recombinant protein does not assume the proper folded state and tends to aggregate. Although the complex process of protein folding depends mainly upon its primary amino acid sequence, under different physicochemical environments, the same protein can fold differently (Kaur *et al.* 2012). In the case of bacterial recombinant system, for example, a lower expression temperature or/and a low concentration of the inducer (viz. isopropyl β-D-1-thiogalactopyranoside [IPTG]) may help in correct folding, as slow synthesis of recombinant proteins allows more time for them to fold in the native form (Jonasson *et al.* 2002; Sørensen and Mortensen 2005b). Once a protein aggregates, however, it has to be solubilized and refolded to the desired native, functional state (Rudolph and Lilie 1996; Lilie *et al.* 1998; Middelberg 2002). Mostly, addition of chaotropes improves the recovery of the insoluble recombinant protein (Chow *et al.* 2006). Thus, insoluble proteins are usually refolded following solubilization and denaturation with mild denaturants (a low concentration of urea, guanidine hydrochloride and a suitable pH; Rudolph and Lilie 1996; Middelberg 2002; Chow *et al.* 2006). The denaturants are subsequently removed by dialysis in the presence of a suitable buffer system containing various reagents and additives that help in the refolding process (Jonasson *et al.* 2002; Sørensen and Mortensen 2005b). One arrives at the right combination of 'refolders' by an iterative trial-and-error process (Chow *et al.* 2006). Several commercial kits that provide such solutions for refolding are currently available in the market as well (Pierce protein refolding kit, Thermo Scientific; QuickFold, AthenaES; iFold, Merck Millipore; Protein refolding kit, Novagen; BioAssay, US Biological; QuickFold, Molecular Dimensions).

Further, to avoid misfolding of an expressed protein in a heterologous system, mainly due to the absence of the requisite folding machinery in the host cell, it is prudent to use a system similar to that of the original organism. Thus, many eukaryotic proteins cannot be properly obtained in the active form in prokaryotes and requires a eukaryotic expression system (Yin *et al.* 2007; Dyring 2011; Malys *et al.* 2011; Werner *et al.* 2011; Gellert *et al.* 2012). Recombinant proteins are also often expressed in fusion with proteins (tags) that enhance solubility such as SUMO (small ubiquitin-like modifier proteins) or GST (glutathione S-transferase) (Kimple and Sondek 2004; Walls and Loughran 2011). Although these fusion constructs help in the solubility of proteins and, in some cases, assist folding, mostly these tags need to be removed for the optimum activity of the recombinant protein.

Oxidative refolding of denatured proteins is another important method (Heyehan and Clark 1997; Clark 1998; Bessette *et al.* 1999; de Marco 2009, 2012). There are two requirements for an oxidative refolding buffer: (a) presence of a redox couple and (b) alkaline pH. The redox couple (GSH/GSSG, i.e. reduced glutathione/oxidized glutathione) is necessary to activate the cysteine residues, while the alkaline pH facilitates the nucleophilic attack–dependent correct disulphide bond formation. Generally, correct disulphide bond is formed in an oxidizing environment as the interaction is thermodynamically driven to yield the lowest-energy native state of the protein.

Bacterial expression systems (*E. coli*) have chaperones (mainly DnaK/DnaJ and GroEL/GroES systems) that assist in the folding of cellular proteins under normal physiological conditions (Schroder *et al.* 1993; Horwich and Fenton 2009; Hoffman *et al.* 2010). However, during overexpression of a recombinant protein, these chaperones may not be sufficient. Hence, an alternative is to coexpress a recombinant protein with chaperones (Kagawa 2011; Folwarczna *et al.* 2012) or with enzymes such as PDI (protein disulphide isomerase) that help in the rearrangement of disulphide bridges (Wang and Tsou 1993; Humphreys *et al.* 1996; Hatahet and Ruddock 2009). The coexpressed chaperones help in refolding of the desired recombinant protein. Once refolded under proper conditions, insoluble proteins can be processed further for downstream applications in a manner similar to soluble proteins discussed above.

34.3 Purification of recombinant protein

The level of expression of most heterologous proteins is rather low – typically less than 2% of the total soluble protein. Classical chromatographic methods of purification (ion exchange chromatography, hydrophobic interaction chromatography, size exclusion chromatography) result in loss of yield at every step. Thus, specific methods of purification are generally employed to obtain the expressed protein in a pure form in minimal number of steps. Many bacterial expression vectors are designed to express the heterologous protein as part of a bacterial protein or in fusion to small tags (such as 6X His Tag), which can be purified using the specificity and extremely high efficiency of affinity chromatography (Terpe 2003). The bacterial proteins, which are commonly used for this purpose, are β-galactosidase, maltose-binding protein, glutathione-S-transferase, etc. (Kimple and Sondek 2004). The cell extract containing the fusion protein is passed over a column containing the affinity group (the ligand) that binds the fusion protein. After the column is thoroughly washed to remove any unbound protein, the bound fusion protein is eluted out of the column in a pure form. The heterologous protein thus purified in a single chromatographic step is cleaved off from the fused bacterial protein or the small tag using a specific protease such as factor Xa. Following cleavage, the mixture (bacterial protein used for fusion and the target heterologous protein or the heterologous protein and the small tag) is once again passed over the same affinity column, which now binds only the bacterial protein or the tag releasing the heterologous protein in a pure form. As a polishing step, to get rid of any other minor impurities or oligomerized heterogenous population, size exclusion chromatography of the recombinant protein often follows the affinity chromatography step (Kimple and Sondek 2004). With the availability of fast, reliable and efficient systems for size exclusion chromatography under high pressure (Fast protein liquid chromatography [FPLC], high-performance liquid chromatography [HPLC]), protein purification has witnessed unprecedented quality and success in the recent times. Such an exercise also helps to exchange the buffer used in the affinity chromatography step to one which is required for downstream applications or for stabilization of the pure protein. In fact, size exclusion chromatography can be used in combination with affinity chromatography to remove cleaved fusion tags as well, especially when the tags are much smaller in size than the target recombinant protein (Walls and Loughran 2011). Affinity chromatography for insoluble proteins can be performed in the presence of chaotropes such as urea or guanidine hydrochloride, which help to keep the protein in solution (Walls and Loughran 2011). The purified, denatured protein can subsequently be refolded as outlined above.

34.4 Stabilization of proteins

Protein stabilization in a heterologous system sometimes becomes important as it is prone to proteolytic degradation by the host proteases or self-aggregation. Many expression systems have been designed to use hosts deficient in proteases. This is especially common in the case of bacterial expression systems, where many of the *E. coli* hosts used for expression carry the *lon* mutation, making them deficient in proteases (Lan *et al.* 2007). This increases the general stability of the expressed heterologous proteins. Stabilizing reagents such as sucrose, trehalose, carboxylic acid salts, etc. can also be employed to enhance stability and prevent aggregation (Schultz *et al.* 2007). Many enzymes and receptors are stable in the presence of their substrates or ligands or substrate analogues. Some proteins may require divalent metal ions for stability (Sørensen and Mortensen 2005a).

34.5 Overcoming other challenges

The success of a recombinant protein expression in *E. coli* is impeded by unproductive interaction, which generally leads to insoluble aggregation and/or inactive recombinant protein. Several strategies have been reported for avoiding protein aggregation such as co-expressing molecular chaperone and/or interacting subunits of larger complexes (de Marco and De Marco 2004). Disulphide bond formation, a post-translational modification, is important for the folding and stability of most membrane and secreted proteins (Hatahet *et al.* 2010; Clark 1998). The correct pairing of cysteine residues for disulphide bond formation is another key issue that needs to be taken care of in heterologous expression of proteins. The formation of native disulphide bonds involves a complex pathway that takes place in the endoplasmic reticulum (ER) of eukaryotes, the intermembrane space of mitochondria and the periplasm of prokaryotic organisms. As there is no ER in prokaryotes, eukaryotic protein expression in bacteria sometime needs to be translocated to the periplasm by tagging the bacterial signal sequence with the protein sequence (de Marco 2012). It is important to note that the periplasmic compartment of a bacterial cell contains enzymes that catalyse the formation of disulphide bonds and subsequent rearrangement or isomerization of incorrect or non-native disulphide bonds. The native disulphide bond formation is often the rate-limiting step in protein folding *in vitro* and *in vivo* (Hatahet *et al.* 2010).

34.5.1 Mechanism of protein oxidation in periplasm

Disulphide bond formation in the periplasmic space is enzymatically catalysed (Fig. 34.1). Periplasmic protein oxidation is controlled by five members of Dsb protein systems (DsbA, DsbB, DsbC, DsbD and DsbG). With the exception of DsbB, other Dsb's share a thioredoxin fold

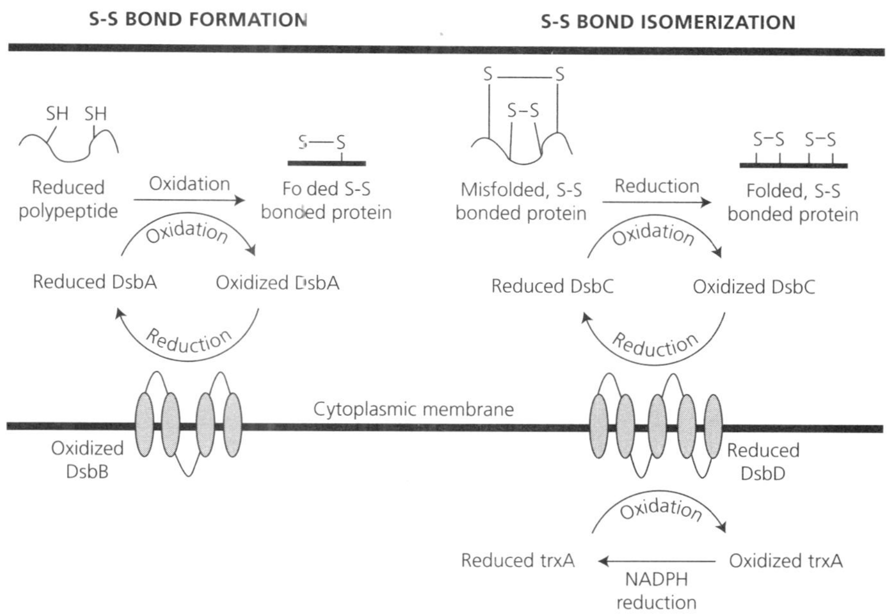

Figure 34.1 Dsb-dependent protein disulphide formation and rearrangement machinery in the bacterial periplasm. DsbB is involved in disulphide bond formation as well as in correct disulphide bond rearrangement, and DsbA donates its disulphide bond to newly synthesized polypeptides for its folding in native confirmation.

with a conserved Cys–Xxx–Xxx–Cys motif (de Marco 2009). These proteins constitute an important quality control machinery, where DsbA catalyses disulphide bond formation. In the process of disulphide bond formation, DsbA is reduced and further recycled by DsbB for another cycle of reaction.

Another important quality control requirement is to ensure correct disulphide bond formation in the protein. If a wrong non-native disulphide is formed, the protein will not be in the lowest-energy native state. In a reducing environment, incorrectly formed disulphide will open up and correct disulphide will be formed (de Marco 2009).

The process will be energetically favourable as correct disulphide–bonded protein will have lower energy. DsbC and DsbD perform these functions. DsbC reduces incorrect disulphide bonds and the reduced Cys residues are subsequently disulphide bonded by DsbA (de Marco 2009). Periplasmic proteins may be recovered by disruption of the outer membrane by osmotic shock. There are some chemical permeabilization schemes also available which involve the use of detergents and/or denaturants. These methods have the advantage of limiting contamination with cytoplasmic material, making purification easier.

34.5.2 Expression of disulphide-rich protein in cytoplasm

The cytoplasm of *E. coli* has a reducing environment, which is not suitable for disulphide bond formation (Hatahet *et al.* 2010). The environment of the cytoplasm of most prokaryotes not only lacks components that catalyse the formation of disulphide bonds, but also glutathione and thioredoxins keep the disulphide bond in the reduced form. Thus, the production of proteins that contain disulphide bonds is difficult in wild-type prokaryotes (Khushoo *et al.* 2005). When disulphide-rich proteins are expressed in wild-type prokaryotes such as *E. coli*, they are unable to fold in the native form and get precipitated as insoluble aggregates, known as inclusion bodies (Lilie *et al.* 1998). Using recombinant DNA technology, the cytoplasm of *E. coli* may be converted into an oxidizing environment by modifying the redox state. Several transgenic *E. coli* strains have been developed to overcome the problem associated with the production of disulphide bond formation in the cytoplasm of *E. coli* (Hatahet *et al.* 2010). These strains have altered pathways, ensuring an oxidizing environment of the cytoplasm (Fig. 34.2). This was achieved by knocking out two NADPH-dependent reductase genes *trxB* and *gor* (Hatahet *et al.* 2010).

34.6 Characterization of heterologous protein

To confirm whether the heterologous protein is properly synthesized in the host organism, specific tests need to be performed. If the protein is an enzyme not present in the organism, its activity can be assayed and the quantity of the protein can be estimated. For example, the bacterial enzyme β-glucuronidase expressed in plants can be estimated by assaying its activity as it is totally absent in plants. Other proteins such as green fluorescent protein (GFP) from jellyfish (Nojima *et al.* 2012) expressed in plants can be assayed by virtue of

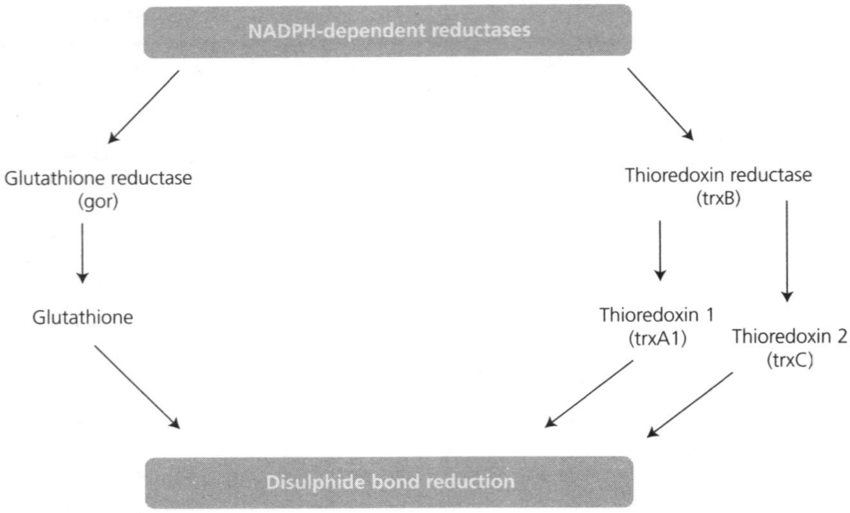

Figure 34.2 Depiction of disulphide bond reduction in the cytoplasm of *E. coli*. The pathway involves mainly two NADPH-dependent enzymes, gor and trxB, which are involved in maintaining an oxidizing environment in the cytoplasm of *E. coli*.

its property of fluorescing in the presence of ultraviolet light. However, for most heterologous proteins, no such specific tests are available. In such cases, the inducibility and size of the heterologous protein are estimated by electrophoretic means. This can be performed by comparing the electrophoretically separated protein profile of the uninduced and the induced cell extracts. If the expected protein is being formed, then there should be induction of a fusion protein of the estimated size under the appropriate inducing conditions. If a specific antibody against the heterologous protein is available, immunological detection tools such as enzyme-linked immunosorbent assay (ELISA) (Friguet *et al.* 1989) can be used to test its presence. For further confirmation, a Western blot can also be performed. However, in both the above cases, there should be sufficient immunological distinctiveness between the heterologous protein from any of the proteins already present in the host. If the heterologous protein is fused to a tag, then affinity binding of the same to its corresponding ligand also verifies target protein production. If one observes a protein of expected molecular weight in SDS-PAGE (sodium dodecyl sulphate polyacrylamide gel electrophoresis), the corresponding band can be excised from the gel, subjected to in-gel trypsin digestion and the product can be analysed by mass spectrometry (such as MALDI TOF [matrix-assisted laser desorption/ionization time-of-flight mass spectrometer]) (Conn 2003) to verify its identity. Such observations help the researcher to draw conclusions regarding the authenticity of the fusion protein. Other biophysical characteristics such as circular dichroism (Sears and Beychok 1973) or fluorescence spectrum (Ptitsyn 1992) of the recombinant protein can be measured to develop an idea as to whether the processed protein has been obtained in native functional form.

References

Ahluwalia U, Katyal N, Deep S. Models of protein folding. *J Protein Proteomics* 2012;3:85–93.

Baneyx F. Recombinant protein expression in *Escherichia coli*. *Curr Opin Biotechnol* 1999;10:411–421.

Bessette PH, Åslund F, Beckwith J, Georgiou G. Efficient folding of proteins with multiple disulfide bonds in the *Escherichia coli* cytoplasm. *Proc Natl Acad Sci USA* 1999:96;13703–13708.

Bradley MK. Overexpression of proteins in eukaryotes. In: Deutscher MP, ed. *Guide to Protein Purification*. San Diego, CA: Academic Press; 1990. pp. 112–132.

Chow MK, Amin AA, Fulton KF, Fernando T, Kamau L, Batty C, *et al.* The REFOLD database: A tool for the optimization of protein expression and refolding. *Nucleic Acids Res* 2006;34:D207–D212.

Clark EDB. Refolding of recombinant proteins. *Curr Opin Biotechnol* 1998;7:157–163.

Conn PM. *Handbook of Proteomic Methods*. Totowa, NJ: Humana Press; 2003.

Cregg JM, Cereghino JL, Shi J, Higgins DR. Recombinant protein expression in *Pichia pastoris*. *Mol Biotechnol* 2000;16: 23–52.

Cull M, McHenry CS. Preparation of extracts from prokaryotes. In: Deutscher MP, ed. *Guide to Protein Purification*. San Diego, CA: Academic Press; 1990. pp. 147–153.

Das A. Overproduction of proteins in *Escherichia coli*: Vectors, hosts, and strategies. In: Deutscher MP, ed. *Guide to Protein Purification*. San Diego, CA: Academic Press; 1990. pp. 93–112.

de Marco A. Strategies for successful recombinant expression of disulfide bond-dependent proteins in *Escherichia coli*. *Microb Cell Fact* 2009;8:1–18.

de Marco A. Recent contributions in the field of the recombinant expression of disulfide bonded proteins in bacteria. *Microb Cell Fact* 2012;11:129.

de Marco A, De Marco V. Bacteria co-transformed with recombinant proteins and chaperones cloned in independent plasmids are suitable for expression tuning. *J Biotechnol* 2004;109:45–52.

Deutscher MP. Maintaining protein stability. In: Deutscher MP, ed. *Guide to Protein Purification*. San Diego, CA: Academic Press; 1990. pp. 83–89.

Dyring C. Optimising the *Drosophila* S2 expression system for production of therapeutic vaccines. *Bioprocess J* 2011;10: 28–35.

Englard S, Seifter S. Precipitation techniques. In: Deutscher MP, ed. *Guide to Protein Purification*. San Diego, CA: Academic Press; 1990. pp. 285–300.

Folwarczna J, Moravec T, Plchova H, Hoffmeisterova H, Cerovska N. Efficient expression of human papillomavirus 16 E7 oncoprotein fused to C-terminus of Tobacco mosaic virus (TMV) coat protein using molecular chaperones in *Escherichia coli*. *Protein Expr Purif* 2012;85:152–157.

Friguet B, Djavadi-Ohaniance L, Goldberg ME. Immunochemical analysis of protein conformation. In: Creighton TE, ed. *Protein Structure—A Practical Approach*. Oxford: IRL Press; 1989. pp. 287–310.

Gellert A, Salanki K, Tombacz K, Tuboly T, Balazs E. A cucumber mosaic virus based expression system for the production of porcine circovirus specific vaccines. *PLoS ONE* 2012;7:e52688.

Hatahet F, Nguyen VD, Salo KEH, Ruddock LW. Disruption of reducing pathways is not essential for efficient disulfide bond formation in the cytoplasm of *E. coli*. *Microb Cell Fact* 2010:9:67.

Hatahet F, Ruddock LW. Protein disulfide isomerase: A critical evaluation of its function in disulfide bond formation. *Antioxid Redox Signal* 2009;11:2807–2850.

Hevehan DL, Clark EDB. Oxidative renaturation of lysozyme at high concentrations. *Biotechnol Bioeng* 1997;54:221–230.

Hoffmann A, Bukau B, Kramer G. Structure and function of the molecular chaperone *Trigger Factor Biochim Biophys Acta* 2010;1803:650–661.

Horwich AL, Fenton WA. Chaperonin-mediated protein folding: Using a central cavity to kinetically assist polypeptide chain folding. *Q Rev Biophys* 2009;42:83–116.

Humphreys DP, Weir N, Lawson A, Mountain A, Lund PA. Co-expression of human protein disulphide isomerase (PDI) can increase the yield of an antibody Fab′ fragment expressed in *Escherichia coli*. *FEBS Lett* 1996;380:194–197.

Jonasson P, Liljeqvist S, Nygren P-A, Stahl S. Genetic design for facilitated production and recovery of recombinant proteins in *Escherichia coli*. *Biotechnol Appl Biochem* 2002;35:91–105.

Kagawa N. Efficient expression of human aromatase (CYP19) in *E. coli*. *Methods Mol Biol* 2011;705:109–122.

Kaur I, Pillai MG, Singh N. An overview of biologics: science behind protein therapeutics. *J Protein Proteom* 2012;3:229–237.

Khushoo A, Pal Y, Mukherjee KJ. Optimization of extracellular production of recombinant asparaginase in *Escherichia coli* in shake-flask and bioreactor. *Appl Microbiol Biotechnol* 2005;68:189–197.

Kimple ME, Sondek J. Overview of affinity tags for protein purification. Curr Protoc Protein Sci 2004;73:Unit 9.9.

Kost TA, Condreay JP, Jarvis DL. Baculovirus as versatile vectors for protein expression in insect and mammalian cells. Nat Biotechnol 2005;23:567–575.

Lackner A, Genta K, Koppensteiner H, Herbacek I, Holzmann K, Spiegl-Kreinecker S, *et al.* A bicistronic baculovirus vector for transient and stable protein expression in mammalian cells. *Anal Biochem* 2008;380:146–148.

Lan L, Deng X, Xiao Y, Zhou JM, Tang X. Mutation of Lon protease differentially affects the expression of *Pseudomonas syringae* type III secretion system genes in rich and minimal media and reduces pathogenicity. *Mol Plant Microbe Interact* 2007;20: 682–696.

Lilie H, Schwarz E, Rudolph R. Advances in refolding of proteins produced in *E. coli*. *Curr Opin Biotechnol* 1998;9:497–501.

Linn S. Strategies and considerations for protein purifications. In: Deutscher MP, ed. *Guide to Protein Purification*. San Diego, CA: Academic Press; 1990. pp. 9–15.

Malys N, Wishart JA, Oliver SG, McCarthy JE. Protein production in *Saccharomyces cerevisiae* for systems biology studies. *Methods Enzymol* 2011;500:197–212.

McCarroll L, King LA. Stable insect culture for recombinant protein production. *Curr Opin Biotechnol* 1997;8:590–594.

Middelberg A. Preparative protein refolding. *Trends Biotechnol* 2002;20:437–443.

Nojima T, Kaneda S, Kimura H, Yamamoto T, Fujii T. Application of cell-free expression of GFP for evaluation of microsystems. *Front Biosci* 2012;17:1931–1939.

Olczak M, Olczak T. Comparison of different signal peptides for protein secretion in nonlytic insect cell system. *Anal Biochem* 2006;359:45–53.

Ptitsyn OB. The molten globule state. In: Creighton TE, ed. *Protein Folding*. New York: WH Freeman; 1992. Pp. 243–300.

Rosser MP, Xia W, Hartsell S, McCaman M, Zhu Y, Wang S, *et al.* Transient transfection of CHO-K1-S using serum-free medium in suspension: A rapid mammalian protein expression system. *Protein Expr Purif* 2005;40:237–243.

Rudolph R, Lilie H. *In vitro* folding of inclusion body proteins. *FASEB J* 1996;10:49–56.

Schröder H, Langer T, Hartl FU, Bukau B. DnaK, DnaJ and GrpE form a cellular chaperone machinery capable of repairing heat-induced protein damage. *EMBO J* 1993;12:4137–4144.

Schultz T, Liu J, Capasso P, de Marco A. The solubility of recombinant proteins expressed in *Escherichia coli* is increased by otsA and otsB co-transformation. *Biochem Biophys Res Commun* 2007;355:234–239.

Sears DW, Beychok S. Circular dichroism. In: Leach SJ, ed. *Physical Properties and Techniques of Protein Chemistry. Part C*. New York: Academic Press; 1973. pp. 445–593.

Sørensen HP, Mortensen KK. Advanced genetic strategies for recombinant protein expression in *Escherichia coli*. *J Biotechnol* 2005a;115:113–128.

Sørensen HP, Mortensen KK. Souble expression of recombinant proteins in the cytoplasm of *Escherichia coli*. *Microb Cell Fact* 2005b;4:1.

Terpe K. Overview of tag protein fusions: From molecular and biochemical fundamentals to commercial systems. *Appl Microbiol Biotechnol* 2003;60:523–533.

Walls D, Loughran ST. Tagging recombinant proteins to enhance solubility and aid purification. *Methods Mol Biol* 2011;681:151–175.

Wang CC, Tsou CL. Protein disulfide isomerase is both an enzyme and a chaperone. *FASEB J* 1993;7:1515–1517.

Werner S, Breus O, Symonenko Y, Marillonnet S, Gleba Y. High-level recombinant protein expression in transgenic plants by using a double-inducible viral vector. *Proc Natl Acad Sci USA* 2011;108:14061–14066.

Wilchek M. Affinity chromatography: A tool in protein chemistry. *Biochem J* 1972;127:7P–9P.

Yin J, Li G, Ren X, Herrler G. Select what you need: A comparative evaluation of the advantages and limitations of frequently used expression systems for foreign genes. *J Biotechnol* 2007;127: 335–347.

35

Techniques for the study of regulation of a cloned gene

Shweta Sharma and Indranil Dasgupta

35.1 Introduction

Regulation of genes is studied mostly at the level of transcription, a step which involves the formation of a transcript from a given gene. The specific mRNA copy produced from a particular gene with the help of the enzyme RNA polymerase is known as a transcript. The basic information required about the transcription process involves determining the size of the transcript, its start and finish points on the gene and its abundance at a given point of time within particular tissues or cells. Another point of interest for investigators is the presence of introns in a gene, their number and their location. As the activity of the promoter driving a particular gene largely determines the transcription of that gene, transcriptional control can be investigated by fusing an easily assayable reporter gene downstream of the promoter. By determining the amount of the reporter gene product being formed under defined conditions, the expression patterns of the promoter and hence the gene under investigation could be studied in a better way. Some of the methods that are usually employed to answer the questions raised above are dealt with in this chapter.

35.2 DNA transfection

To study the transcription of a cloned DNA, it is required to be reintroduced inside a cell. The process where a cloned DNA is reintroduced inside a cell is termed as DNA transfection. The history of DNA transfection goes back to 1928 when Frederick Griffith transformed non-pathogenic pneumococcus bacteria *Streptococcus pneumoniae* into a virulent variety by mixing non-pathogenic pneumococcus with heat-killed pathogenic bacteria (Griffith 1928). This was followed by the seminal work of Avery, Macleod and McCarty in 1944, which showed that DNA is the transforming factor (Avery *et al.* 1944). The term 'transfection' (derived from transformation and infection) was first coined by Födes and Troutner in 1964 when they introduced the phage SP8 DNA into *Bacillus subtilis*. They suggested that this term refers to the production of the complete virus in cells infected with isolated nucleic acid from that virus. Vaheri and Pagano first described the diethyl-amino-ethyl dextran (DEAE-dextran)–mediated cell transfection in 1965 in which positively charged DEAE-dextran binds to the negatively charged phosphate groups of the DNA, forming aggregates. These complexes, when applied to cells, subsequently bind to the negatively charged plasma membrane. Cellular uptake is mediated by endocytosis, which is further assisted by an osmotic shock. If eukaryotic cells, that is, cells of higher multicellular organisms such as animals and plants, are used for reintroduction of the cloned DNA, then the process is known as transfection. However, if bacterial cells are used, then the process is generally termed as 'transformation', though the basic principle behind both remains the same. A cell is surrounded by a membrane, which is made up of a lipid bilayer embedded with a number of proteins and is normally not permeable to large molecules such as DNA. Transfection of a cloned DNA through the non-permeable membrane into a cell can essentially be achieved by increasing the permeability of the cell membrane towards large molecules. Several methods, which can be broadly divided into physical and chemical, are being used routinely to create such permeable conditions of the membrane. In some cases, viruses are also used to deliver the cloned DNA inside a cell. Among the physical methods, electroporation and biolistics are two very widely used methods. Electroporation is a method where electric pulses are applied across cell membranes, as a result of which transient pores are formed in the membrane, which close soon after. Through these transient pores, some molecules of DNA are able to diffuse inside the cell, which brings about transfection. In the case of biolistics, tiny metal particles, usually of tungsten or gold, coated with the cloned DNA on their surface are shot directly into the cell. The target cells can be taken either as intact tissues or as a suspension. The problem is that this process kills many of the cells thus targeted. However, those cells that survive the physical damage caused by the entry of the metal particles receive the DNA and thus transfection takes place. The other category of methods, that is, chemical methods uses either lipids or poly-cations such as DEAE-dextran for transfection. In this process, DNA is mixed with lipids or poly-cations, which form globules containing the DNA. The lipid part of the globules fuses with the lipid bilayer of the cell membrane, thereby releasing the DNA inside the cell. Transfection achieved specifically using lipids is sometimes called as lipofection. A similar mechanism is also believed to take place in the case of poly-cations.

35.3 Southern blot

The technique which is used to detect the presence of a known fragment of DNA in a given DNA preparation is known as Southern blot. This technique is used mainly to check the presence and the copy number of a foreign DNA in a genetically modified organism or that of an endogenous gene in the organism's genome. This technique has been named after Edwin Southern, who pioneered the technique in 1975, prior to which similar studies were based on annealing data of two DNA strands, which were difficult to reproduce and were less informative. Southern developed a method of transferring DNA fragments from agarose gel to cellulose nitrate filters (Southern 1975). The transferred fragments were then hybridized with radioactively labelled DNA and the hybrids were detected by autoradiography or flouorography. He demonstrated the efficacy of this method by analysing the restriction fragments complementary to the ribosomal RNA from *Escherichia coli, Xenopus laevis* and several mammals. In this method, the DNA preparation under investigation is first digested with a particular restriction enzyme; the resulting fragments thus obtained are then separated according to their sizes by gel electrophoresis. In the next step, the gel containing the restricted fragments is subjected to alkali treatment to denature the DNA fragments, which are then transferred to special membranes either by capillary action or under the influence of an electric field. To bind the DNA fragments tightly to these membranes, they are either baked at a particular temperature or treated with ultraviolet rays. The DNA fragment of interest, the presence of which has to be detected in the DNA preparation, is labelled using either radioactive or non-radioactive methods. The labelled DNA fragment thus obtained is known as a probe and is allowed to hybridize to the test DNA immobilized on the membrane. During the hybridization reaction, the probe molecules bind specifically to the complementary DNA sequences under appropriate temperature and salt conditions. The unhybridized/excess probe is removed by washing the membrane under suitable conditions and the hybridized probe is detected by either exposing the membrane to an X-ray film, in the case of radioactive probes, or using a suitable detection reagent for the probe, if it is non-radioactive. Thus, the presence of DNA sequences complementary to the probe used is confirmed by the signal obtained (Fig. 35.1).

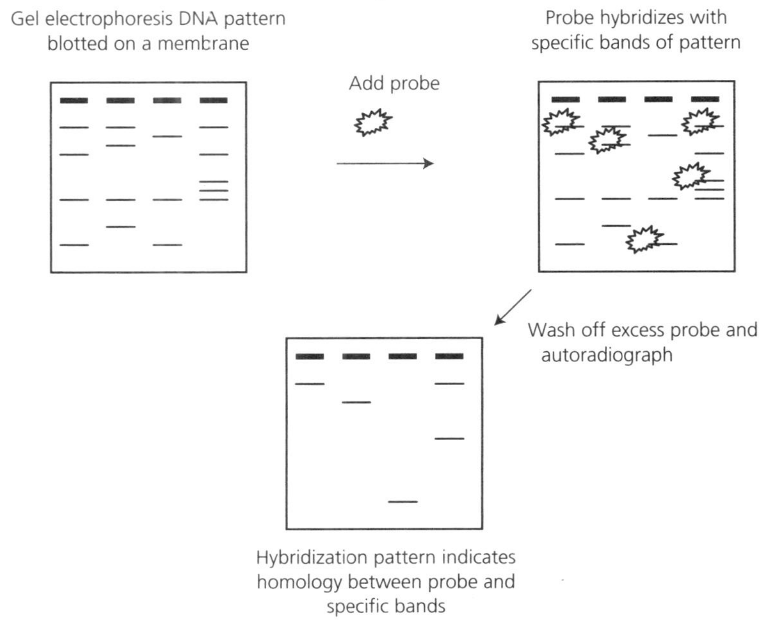

Figure 35.1 Schematic diagram of Southern hybridization. DNA fragments are first separated by electrophoresis and blotted on a membrane. The DNA to be used as a probe is labelled appropriately and allowed to hybridize with the blotted DNA on the membrane. The hybridized probe is then detected by autoradiography or any other suitable method and its position indicates the size of the homologous DNA fragment to which it has hybridized.

35.4 Northern blot

The method of estimating the size and abundance of a particular transcript is called as northern blot, based on its similarity to the Southern blot method described earlier. The only difference between northern and Southern blot lies in the use of RNA instead of DNA for gel electrophoresis and immobilization to the membrane. The technique of northern blot was first used by Alwine *et al.* in 1977 at Stanford University to investigate the RNA of the fruitfly *Drosophila melanogaster*. The procedure involved transferring electrophoretically separated total RNA of *D. melanogaster* from agarose gel to special paper strips to immobilize them on that support. RNA, which normally does not bind to nitrocellulose paper, was coupled covalently to specially developed DBM (diazobenzyloxymethyl) paper. After transfer and appropriate treatment of the paper to remove excess of the diazo group, specific sequences of the RNA were detected with high sensitivity using ^{32}P-labelled probes. In this method, total RNA extracted either from a tissue or a suspension or its poly(A)–enriched fraction is subjected to electrophoresis in an agarose or polyacrylamide gel, followed by its immobilization to the membrane, as described for Southern blot. This is followed by hybridization of the membrane with radioactively labelled probes and the signals thus obtained indicate the position of the transcript on the membrane. It could potentially also give a fairly good idea of the size of the transcript if compared with the positions of known fragments of DNA or RNA. During hybridization, the labelled probe is usually used in excess, with the quantity of the target transcript immobilized on the membrane determining the extent of hybridization of the probe. By comparing the intensity of the signal obtained by hybridization of the labelled probe to the transcript extracted from various tissues, a rough estimate of the abundance of the transcript in those tissues can be determined. Northern blot analysis could thus be very informative for studying the relative levels of transcripts in various tissues of an organism, at various stages of growth and development or at various time intervals following treatment with a pathogen or stress-inducing agent.

35.5 Nuclease S1 mapping

Nuclease S1 mapping is a widely used method based on the principle of degradation of single-stranded regions of hybrid nucleic acids, followed by further analysis based on size separation. Nuclease S1, which is obtained from the fungus *Aspergillus oryzae*, is an enzyme which can specifically degrade single-stranded nucleic acids. Berk and Sharp in 1977 devised this simple and sensitive technique for detecting, sizing and mapping RNA transcripts from viral or cloned DNA. In their procedure, RNA was first hybridized in solution with the DNA probe; the specific DNA complement is then generated by digestion with the single-strand–specific nuclease S1 and the size determined by alkaline agarose gel electrophoresis. They used this method to examine the cytoplasmic transcripts produced during the early phase of Adenovirus 2 (Ad2) infection of human cell lines (HeLa). This technique allowed mapping of all the early Ad2 transcripts and showed that they clustered in four regions of the viral DNA. The technique has also been used to map the position of introns, those regions of the pre-mRNA that are absent from a mature mRNA. Hybridization of the mature transcript with the corresponding gene results in the formation of loops, which represents the introns in the gene. These looped-out portions stay single stranded because of the absence of the corresponding regions in the mature transcripts. The DNA–RNA hybrids thus formed can be cleaved at the introns upon treatment with nuclease S1. The fragments resulting after nuclease S1 treatment of the hybridized molecule can be separated by gel electrophoresis using special conditions such as an alkaline running buffer (Fig. 35.2).

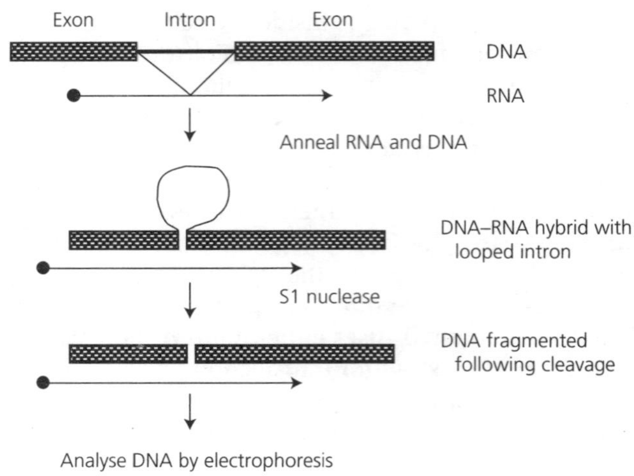

Figure 35.2 Schematic diagram of mapping of introns by digestion with S1 nuclease. RNA is hybridized with its corresponding genomic DNA, causing the intron, which is not represented in the RNA, to loop out. The looped-out intron is specifically digested with S1 nuclease, which degrades single-stranded nucleic acids. Analysis of the fragmented DNA by gel electrophoresis indicates the number and presence of introns.

35.6 Nuclease protection assay

A variant of the nuclease S1 mapping where the unlabelled test transcript is hybridized with the *in vitro* synthesized radioactively labelled RNA or DNA molecules obtained usually from a cloned gene is

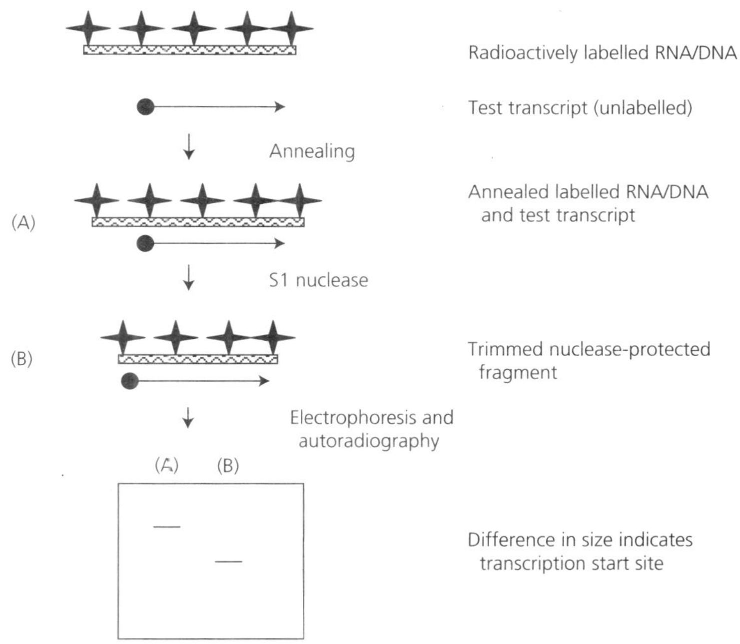

Figure 35.3 Schematic diagram of mapping transcription start site by nuclease protection assay. Radioactively labelled RNA/DNA is annealed to a test transcript whose transcription start site has to be mapped. The unprotected labelled molecule is reduced in size by S1 nuclease digestion depending upon the amount of protection by the annealed transcript. The difference in the size of the protected molecule from that of the unprotected molecule can be used to map the site of transcription initiation.

known as nuclease protection assay. This method is a modification of the S1 nuclease mapping developed by Berk and Sharp (1977, 1978). This modified method was used by for the first time by Weaver and Weissman in 1979 for mapping the exon–intron boundaries, which involved the use of 5′ and 3′ terminally labelled probes. The authors determined that the 5′ termini of the mouse 15S beta-globin precursor and the mature mRNA have the same map coordinates. In this method, the unlabelled test transcript is allowed to hybridize with an excess of the *in vitro* synthesized radioactively labelled transcript. The hybridized transcripts are then subjected to digestion with nuclease S1 or pancreatic ribonuclease, which degrades single-stranded nucleic acids. The required information regarding the presence of introns, transcriptional termini and the relative abundance of the test transcript can be obtained by the difference in the sizes of nuclease-treated and untreated hybridized molecules, as determined by gel electrophoresis (Fig. 35.3).

35.7 Primer extension

Primer extension is another reliable method for determining the 5′-end of transcripts. This method was first used to detect the transcript of Herpes Simplex virus thymidine kinase gene that had been injected into *Xenopus laevis* oocytes (Mcknight and Gavis 1980; Mcknight and Kingsbury 1982). By using this method, the investigators were able to delimit the functional boundaries corresponding to the 5′ and 3′ termini of the gene. In this method, a preparation of RNA is allowed to hybridize to a synthetic 5′-labelled oligonucleotide primer which has bases complementary to a few hundred bases downstream to that of the expected 5′-end of the test transcript. In the next step, the enzyme reverse transcriptase is allowed to extend the primer on the template, which is provided by the transcript itself. This step usually synthesizes an extended primer till it reaches the 5′-end of the test transcript, with the size of the resulting cDNA representing the distance between the 5′-end of the primer and the 5′-end of the transcript.

The length of this cDNA can be estimated accurately by electrophoresis in a sequencing gel, which in turn helps in accurate estimation of the 5′-end of a transcript. In this method, usually two different oligonucleotide primers, having a known number of bases separating them on the transcript, are used for carrying out two separate primer extension reactions on the transcript. These two separate primers also act as a negative control as the length of the cDNA obtained from the distant primer should be longer than that of the nearer primer by the distance separating them on the transcript. Taking into consideration both sets of data, the transcription start point can also be located on the transcript (Fig. 35.4).

35.8 Rapid amplification of cDNA ends (RACE)

The method which is used to map the 5′- or 3′-ends of transcripts by utilizing polymerase chain reaction (PCR) is known as 5′-RACE (rapid amplification of

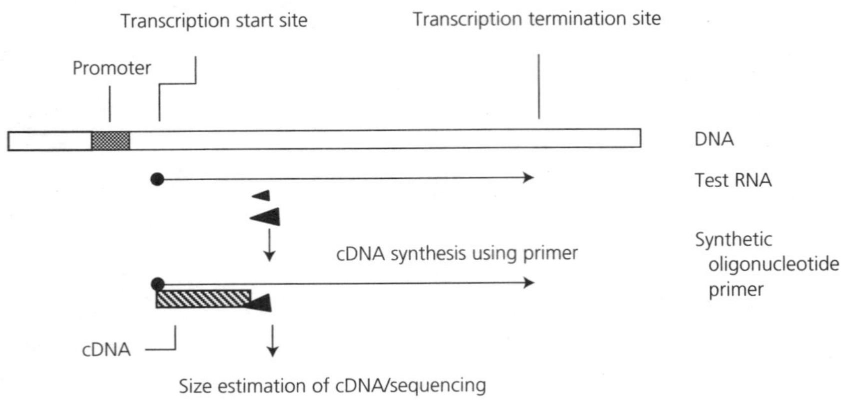

Figure 35.4 Schematic diagram of mapping transcription start site by primer extension. An oligonucleotide primer is used to synthesize cDNA using the test RNA as template. The cDNA represents the length between the primer site and the 5′-end of the transcript. Sequence analysis of the cDNA gives the accurate transcription start site of the test RNA.

cDNA ends) or 3′-RACE, depending upon the end to be mapped. Frohman *et al.* in 1988 devised this efficient technique that achieves amplification and cloning of the region between a single short sequence in a cDNA molecule and its unknown 5′- or 3′-end. The efficacy of this method was demonstrated by isolating a full-length cDNA clone of mRNA from *int*-2, a mouse gene that expresses four transcripts at low abundance in the early embryo. The initial steps of 5′-RACE are similar to that of primer extension; that is, the enzyme reverse transcriptase extends a synthetic oligonucleotide primer which is complementary to the region within a few hundred nucleotides of the estimated 5′-end of the transcript. In the next step, a string of deoxyadenosines is added to the 3′-end of the newly synthesized cDNA by the enzyme terminal transferase. This is followed by amplification of the resulting cDNA having 3′-adenosine residues using a primer pair consisting of an original primer and an oligodeoxythimidine primer (designed to anneal to the added 3′-adenosine residues of the cDNA) and a Taq DNA polymerase or any other suitable thermostable enzyme. By using the combination of the above two primers, the cDNA representing the portion between the 5′-end of the original primer and the 5′-end of the transcript gets amplified, whose size can then be accurately determined. In the case of 3′-RACE, cDNA synthesis is initiated on the transcript whose 3′-end has to be mapped using an oligothymidine primer. In

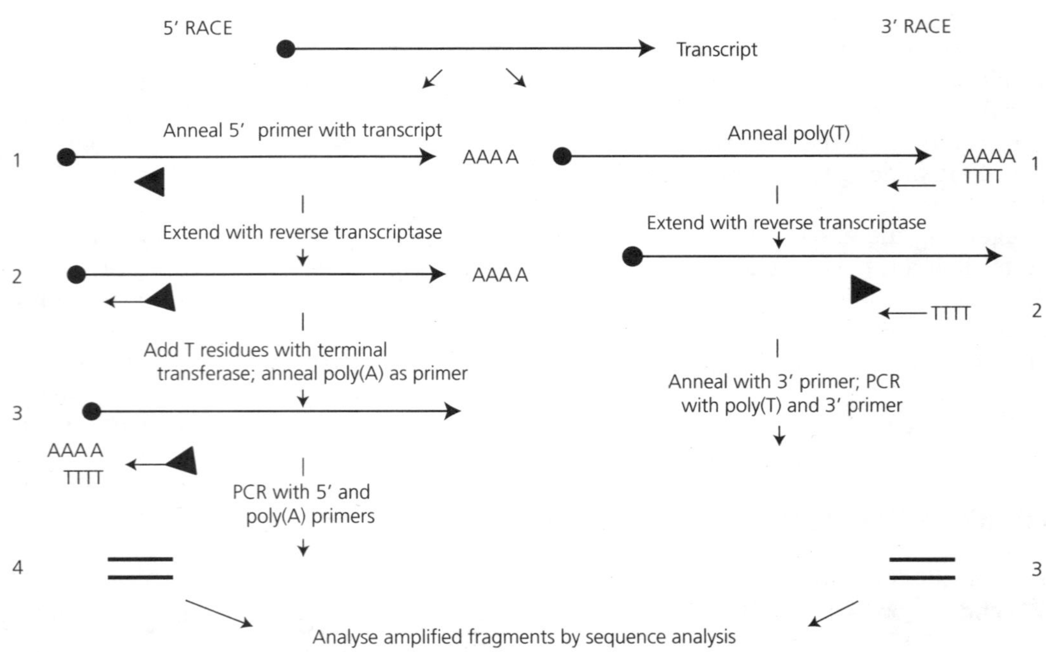

Figure 35.5 Schematic diagram of RACE for the mapping of the 5′-end and 3′-end of a transcript. For mapping the 5′-end, cDNA is synthesized on the test transcript using a 5′-primer. The cDNA is extended by adding poly(T) residues using the enzyme terminal transferase. The cDNA is then amplified by using poly(A) and the 5′-primer, which is then sequenced to obtain the 5′-end of the transcript. For the 3′-end mapping, a poly(T) primer is used for cDNA synthesis, which is then amplified by a 3′-end primer and poly(T). Sequence analysis indicates the 3′-end of the transcript.

combination with the oligothymidine primer, another oligonucleotide primer, based on the sequences of a region a few hundred bases from the expected 3′-end of the transcript, is used to amplify the cDNA representing the fragment between the above oligonucleotide primer and the 3′-end of the transcript. Similar to the 5′-RACE method, the position of the 3′-end of the transcript can be determined by accurate estimation of the size of the amplified fragment (Fig. 35.5). The method is highly efficient and rapid, illustrated by the results obtained by Frohman *et al.* (1988), where working with the low abundance *int-2* mRNA of mouse, 29 independent *int-2* clones were isolated after screening just 0.05% of all cDNAs, all clones showing accurate 3′- and 5′-ends.

35.9 Reporter assays

One of the widely used methods of determining the expression patterns of a gene within the cell is fusion of its promoter to a reporter gene and transfecting it inside a living cell, a method known as reporter assay. Those genes which give rise to easily detectable and quantifiable protein products are known as reporter genes. Chloramphenicol acetyl transferase (Gorman *et al.* 1982; DeBlock *et al.* 1984), β-galactosidase (Hall *et al.* 1983; Sanes *et al.* 1986), β-glucuronidase (Jefferson *et al.* 1986, 1987), luciferase (de Wet *et al.* 1987; Bronstein *et al.* 1994) and green fluorescent protein (GFP; Chalfie *et al.* 1994; Flach *et al.* 1994) are some of the commonly used reporter proteins whose enzymatic activity or mere presence reflects the activity of the promoter under investigation. The reporter genes chloramphenicol acetyl transferase, β-galactosidase and β-glucuronidase, all obtained from the bacterium *E. coli*, are used for transfected mammalian cells, bacteria and plant cells, respectively. Luciferase, obtained from the firefly *Photinus pyralis*, and GFP, obtained from the jellyfish *Aequorea victoria*, have been used for a variety of organisms. Each reporter gene has its own set of advantages and disadvantages, and the user has to carefully select the type of reported gene to be used depending upon the requirements of the experiment. For example, luciferase assay, which is extensively used to investigate promoter activity in a variety of animal cells, is considered to be 30–1000 times more sensitive than chloramphenicol acetyl transferase assay (de Wet *et al.* 1987). GFP can be detected by exposing the intact tissue to ultraviolet light, resulting in a green fluorescence, and hence the expression of the promoter in question can be studied even in intact tissues and cells just by observing for green fluorescence under ultraviolet illumination.

Broadly, there are two types of requirements of reporter genes: the first is to check whether the promoter under investigation is active or not in a particular tissue of a multicellular organism, and the second is to determine its level of expression, depending on which the assays can be quite different. To check whether the promoter under study is active in a given tissue or not, the tissue is subjected to histochemical staining using a substance which produces a coloured dye in the presence of the reporter gene product. To confirm whether the promoter is active or not in that particular tissue, a histochemical examination of the tissue is carried out to check the presence of the coloured dye. However, the level of the promoter activity is determined by quantifying the amount of reporter gene product present in the tissue concerned or in the organism.

References

Alwine JC, Kemp DJ, Stark GR. Method for detection of specific RNAs in agarose gels by transfer to diazobenzyloxymethyl-paper and hybridization with DNA probes. *Proc Natl Acad Sci USA* 1977;74:5350–5354.

Avery OT, MacLeod CM, McCarty M. Studies on the chemical nature of the substance inducing transformation of pneumococcal types: Induction of transformation by a DNA fraction isolated from *Pneumococcus* type III. *J Exp Med* 1944;79:137–158.

Berk AJ, Sharp PA. Sizing and mapping of early adenovirus mRNAs by gel electrophoresis of S1 endonuclease digested hybrids. *Cell* 1977;12:721–732.

Berk AJ, Sharp PA. Spliced early mRNAs of simian virus 40. *Proc Natl Acad Sci USA* 1978;75:1274–1278.

Bronstein I, Fortin J, Stanley PE, Stewart GSAB, Kricka LJ. Chemiluminescent and bioluminescent reporter gene assays. *Anal Biochem* 1994;219:169–181.

Chalfie M, Tu Y, Euskirchen G, Ward WW, Prasher DC. Green fluorescent protein as a marker for gene expression. *Science* 1994;263:802–805.

DeBlock M, Herrera-Estrella L, Van Montagu M, Schell J, Zambryski P. Expression of foreign genes in regenerated plants and in their progeny. *EMBO J.* 1984;3:1681–1689.

deWet JR, Wood KV, DeLuca M, Helinski DR, Subramani S. Firefly luciferase gene: Structure and expression in mammalian cells. *Mol Cell Biol* 1987;7:725–737.

Flach J, Bossie M, Vogel J, Corbett A, Jinks T, Willins DA, Silver PA. A yeast RNA-binding protein shuttles between nucleus and cytoplasm. *Mol Cell Biol* 1994;14:8399–8407.

Földes J, Trautner TA. Infectious DNA from a newly isolated *B. subtilis* phage. *Z VererbLebre* 1964;95:57–63.

Frohman MA, Dush MK, Martin GR. Rapid production of full-length cDNAs from rare transcripts: Amplification using a single gene-specific oligonucleotide primer. *Proc Natl Acad Sci USA* 1988;85:8998–9002.

Gorman CM, Moffat LM, Howard BH. Recombinant genomes which express chloramphenicol acetyltransferase in mammalian cells. *Mol Cell Biol* 1982;2:1044–1105.

Griffith F. The significance of pneumococcal types. *J Hygiene* 1928;27:113–159.

Hall CV, Jacob PE, Ringold GM, Lee F. Expression and regulation of *E. coli lacZ* gene fusion in mammalian cells. *J Appl Mol Genet* 1983;2:101–109.

Jefferson RA, Burgess SM, Hirsh D. 3-glucuronidase from *Escherichia coil* as a gene-fusion marker. *Proc Natl Acad Sci USA* 1986;83:8447-8451.

Jefferson RA, Kavanagh TA, Bevan MW. GUS fusions: 3-Glucuronidase as a sensitive and versatile gene fusion marker in higher plants. *EMBO J* 1987;6:3901–3907.

Sanes JR, Rubenstein JLR, Nicolas J-F. Use of a recombinant retrovirus to study post-implantation cell lineage in mouse embyo. *EMBO J* 1986;5:3133–3142.

Southern EM. Detection of specific sequences among DNA fragments separated by gel electrophoresis. *J Mol Biol* 1975; 98:503–517.

Weaver RF, Weissman C. Mapping of RNA by Berk–Sharp procedure: The 5′ termini of the 15S mRNA of β-globin mRNA and mature β-globin 10S mRNA have identical map co-ordinates. *Nucleic Acids Res* 1979;7:1175–1193.

36

Transgenic and gene knockout technologies

Indranil Dasgupta

36.1 Introduction

Transgenic organisms are those which contain a combination of genes not possible by the normal reproductive process. In nature, reproductive barriers exist, which make it impossible to have genetic exchange between two species. Genetic engineering can join any two molecules of DNA, irrespective of their origins, and when introduced in organisms by the process of genetic transformation, can result in new combinations of genes not found in nature. Site-directed changes in genes can be performed and new genes which never existed in nature before can be synthesized. Such organisms that possess new and useful properties by intelligent design of gene combinations are called transgenic organisms. They are being used in delivering more nutritious food products, improving the environment and in scientific research.

36.2 Transgenic animals

Transgenic animals have been produced keeping four broad uses in mind: (a) to make them grow bigger than normal to get increased meat products, (b) to make pharmaceutically useful products available in an easily deliverable form (such as milk), (c) to create animals which develop human diseases to help understand the process of disease development better and to test the efficacy of new drugs before being tested on humans and (d) to create animals which can serve as organ donors for humans.

36.2.1 Transgenic animals which grow bigger than normal to serve as better sources of meat

Pigs were first such animals to be made, bearing genes expressing increased levels of the growth hormone. These animals, called 'Superpig', contained the growth hormone gene from cows, pigs and rats. In the first such report, researchers produced transgenic pigs expressing the growth hormone gene under a metallothionein promoter. The promoter was activated by feeding the animals with a diet supplemented with zinc. This caused an increased production of the growth hormone and resulted in the pigs gaining more weight than normal pigs. Using the same approach, transgenic fishes (tilapia and salmon) were created, making them bigger in size. The results were particularly impressive in salmon, with the growth hormone–over-producing fishes growing 17 folds bigger than non-transgenic ones within 14 months of fertilization.

36.2.2 Transgenic animals (mammals) producing pharmaceutically important products in their milk

Pharmaceutically important proteins, if available in the milk of transgenic mammals, can be very efficiently delivered to humans. However, the production of transgenic dairy animals, such as cow or goat, is an extremely inefficient process, and thus, such attempts are preceded by first testing the experimental approach in mice. The first pharmaceutical product, which have been thus been expressed in mice was the blood clot-dissolving tissue plasminogen activator by Gordon *et al.* (1987). Human α-lactalbumin, used for the treatment of emphysema, was expressed in the mammary glands of transgenic rats, producing the protein in the milk at concentrations between 2.0 and 4.3 mg/mL. The coagulation factor (factor IX) from humans was used to produce transgenic sheep, expressing it in the milk. Human lactoferrin, a protein absent in cows' milk but required by human babies as a source of iron, was produced in transgenic cows (Krimpenfort *et al.* 1991). Human anti-thrombin III, a type of anticoagulant, was expressed in the milk of transgenic goats (Baguisi *et al.* 1999).

36.2.3 Transgenic animals as disease models

Transgenic animals have been developed which contain a heterologous gene, making them prone to a particular disease. Stewart *et al.* (1984) were the first to develop such a transgenic animal, the 'Oncomouse', which carried a viral oncogene, causing the animal to develop carcinomas. Games *et al.* (1995) developed an 'Alzheimer Mouse', which carried a heterologous gene producing a protein responsible for the development of characteristic amyloid plaques in the brain, a hallmark of the disease in humans. These animals displayed signs similar to Alzheimer's disease and were used for the testing of drugs. Reid *et al.* (2001) developed the HIV mouse, which contained a portion of the HIV genome as the heterologous gene. All the above animals serve as models to study the pathological changes during the disease and for testing of potential drugs.

36.2.4 Transgenic animals which can serve as organ donors for human patients

Shortage of donors for patients requiring organ transplants is a worldwide problem. The problem mainly arises due to the difficulty in finding a compatible donor whose organs will not be rejected by the patient's immune system. Rejection of an organ after transplantation takes place due to recognition of various antigens present on the surface of the organ tissue by the patient's immune system. To find a solution to this problem, transgenic animals can be produced, which have reduced levels of surface antigens. Such animals, called xenotransplanters, can be potentially raised as sources of organ donors for patients. Lai *et al.* (2002) have raised transgenic pigs with reduced levels of a surface glycoprotein, which is otherwise recognized by the human immune system, as the first step in this direction.

36.3 Techniques employed to produce transgenic animals and plants

Transgenic animals are produced using two main methods: by microinjection of DNA into the pronucleus of a fertilized egg or by using viruses to infect embryos or embryonic cells (Fig. 36.1). Apart from embryos, certain stages of early development of animals are characterized by the presence of cells known as 'embryonic stem cells', which have the capability to differentiate into various types of tissues. To ensure optimal expression, the DNA to be introduced into the animal is cloned so as to be under the control of a desirable promoter and transcription termination sequences.

For microinjection, the male pronucleus is generally preferred. This is mainly due its larger size and proximity to the surface. The animal to be used as the donor is hormonally induced for super-ovulation so that sufficient quantities of eggs are obtained. Eggs obtained are cleaned of adhering tissues and DNA is microinjected into the nucleus using micropipettes.

Embryonic stem cells are generally obtained from the inner cell mass of 3-day post-fertilization blastocyst embryos in mice. Once obtained, the cells can be kept in culture indefinitely under carefully controlled conditions. Viruses can be used as a means to transfer genes to embryonic stem cells. The first such transfer was performed by Jaenisch and Minz (1974), where Simian Virus 40, a DNA virus infecting monkeys, was used to inject blastocysts of mice. Irrespective of the method used (microinjection or virus-mediated transfer) for gene transfer, the embryonic cells which have taken up the transgene are generally selected by checking for growth in media supplemented with antibiotics, which will otherwise prevent the growth of embryos not having the transgene. Needless to say, the antibiotic resistance genes are part of the transgene construct.

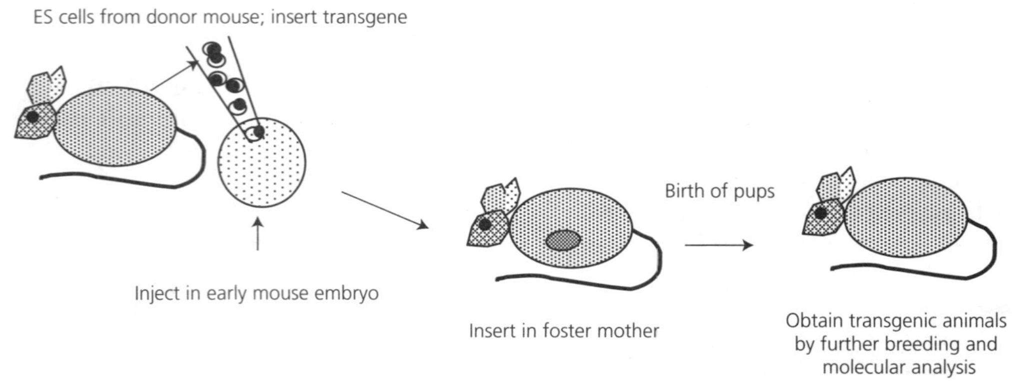

Figure 36.1 Methods for the production of transgenic animals.

Selected embryos, or transfected stem cells which are embedded in embryos, containing the transgene are then placed in oviducts of recipient female animals. The offspring born within 3 weeks of embryo transfer in mice are screened for the presence of the transgene by molecular tools, such as PCR or Southern hybridization. They are then mated with their transgenic siblings to obtain homozygous transgenic animals. Similar procedure is also followed to obtain transgenic birds and fish, for example, chicken and salmon.

DNA transfer to plant cells for producing transgenic plants is generally done by direct transfer or through the mediation of the soil-dwelling bacteria, *Agrobacterium tumefaciens* (Fig. 36.2). In the direct transfer method, naked DNA is introduced into plant cells which have been specially treated with enzymes that remove the thick cell wall, resulting in the formation of spherical cells without cell walls, known as protoplasts. Treatment with polyethylene glycol (PEG) promotes the transfer of DNA into protoplasts. The transgene, under the control of appropriate expression signals (promoter and transcription terminator signals), can get integrated in the protoplast genome. Similar to the process of selection mentioned for animals, protoplasts which have taken up the transgene can be selected for by growing them on a medium containing appropriate antibiotics or other toxic agents, resistance genes for which are a part of the transgene construct.

Another direct method to transfer DNA to plant cells is by using a biolistic device, commonly known as 'gene gun'. In this method, the transgene is coated onto tiny metal particles (usually gold or tungsten) and is propelled at a high speed into the plant tissue. The metal particles penetrate the cells and hence transfer the DNA within the nucleus, where the transgene can integrate in the genomic DNA.

The indirect method, which is in fact used by most plant biotechnologists, is to use the natural inter-kingdom gene transfer phenomenon occurring between the soil bacterium *Agrobacterium tumefaciens* through the transfer of a plasmid from the bacteria to plant cells. The plasmid, known as the tumor-inducing (Ti) plasmid, transfers part of itself (T-DNA), encoding various plant growth substances into the plant genome, which results in enhanced plant cell division in that region, leading to the formation of a tumour. Transgenes are introduced as part of the T-DNA from which the tumour-causing genes have been removed and replaced with appropriate gene expression signals to bring about the expression of the transgene in the transformed cell. A good example to illustrate this method of transformation for rice is the one described by Hiei *et al.* (1994), in which calli derived from scutella of rice seeds were used for transformation. The authors used a DNA construct which contained the reporter gene *Gus* and the gene encoding for resistance to the antibiotic hygromycin. The use of certain *Agrobacterium* strains resulted in a transformation efficiency of 28% in rice, considered to be a difficult plant species for transformation. The efficiencies reported for rice were very similar to the efficiencies reported for plants highly amenable for *Agrobacterium*-mediated transformation, such as tobacco.

One major difference between animal and plant cells with respect to the potential to develop into a complete organism is that this property is bestowed only to stem cells in animals, whereas plant cells are more elastic in this respect, meaning that most plant cells can be coaxed to regenerate into a mature plant, a property known as totipotency. To use the property of totipotency to obtain complete plants from a few cells, the plant tissue is removed from the plant and placed in a nutrient medium containing a mixture of plant hormones under specific light regimes, which prompts the tissue to form a clump of undifferentiated tissue known as the callus. The T-DNA also contains a gene which gives resistance against antibiotics. Thus, callus cells which

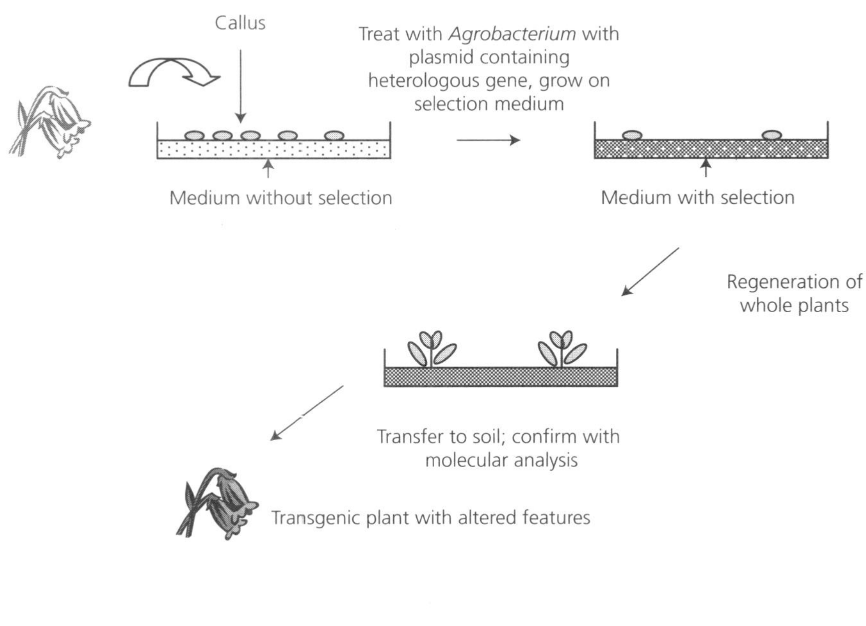

Figure 36.2 Methods for the production of transgenic plants.

have received the T-DNA containing the transgene are selected by virtue of their survival in the presence of an antibiotic and then grown into mature plants by providing a careful mixture of plant hormones which allow dedifferentiation of the calli into mature plants.

36.4 Targeted gene replacement

During the production of a transgenic organism, the transgene can be integrated at any location of the recipient genome. Gene targeting or gene replacement is only possible if the transgene integration event can be targeted to a predetermined location of the genome. This was done initially by utilizing the phenomenon of homologous recombination, which depends upon the sequence homology between two similar stretches of DNA. If segments of DNA homologous to the target region in the genome are included at either sides of a transgene, homologous recombination can insert the transgene in the recipient genome at the site of homology. To take an example in mouse, the heterologous gene (Hg) to be introduced in the vector DNA is flanked on either side by a DNA sequence homologous to a mouse gene (M), which allows homologous recombination to take place between these sequences and the corresponding gene present in the chromosome. Flanking the homologous gene is placed a copy each of the *tk* gene, which can be used for counter-selection, a method used to negatively select/eliminate events not desirable under the experimental strategy. In addition, there are also genes coding for antibiotic resistance to allow the transformed cells to survive in the presence of the antibiotic, a positive selection step. In the case where homologous recombination takes place, the transgene is introduced in between the two homologous stretches of DNA. On the other hand, in the case where spurious or non-homologous recombination takes place, the *tk* gene also gets integrated, giving rise to counter-selection and elimination of such events. Cells in which the above vector containing the transgene has been integrated are then treated with antibiotics and a compound known as gancyclovir, which kills all cells expressing *tk* gene. Such a selection enriches the homologous recombination events in the selected cells (Fig. 36.3). In *Drosophila*, a similar approach was used to replace a defective *y* gene with the y^+ allele, a 8-kb DNA fragment with a rate of 1/800 gametes. Also, a wild-type gene was disrupted with a marker gene at the rate of 1/380 gametes (Gong and Golic 2003).

More recently, gene targeting has been made more efficient by the use of artificial site-specific nucleases, termed the zinc-finger nucleases. These nucleases are designed to recognize a predetermined DNA sequence, to which they bind and cleave, resulting in a double-stranded break. Such a break allows integration of a heterologous DNA fragment to be inserted at the cleavage site.

36.5 Chromosome engineering

Chromosome engineering is a broad term encompassing techniques available from changing the mode of inheritance of chromosomes to the production of

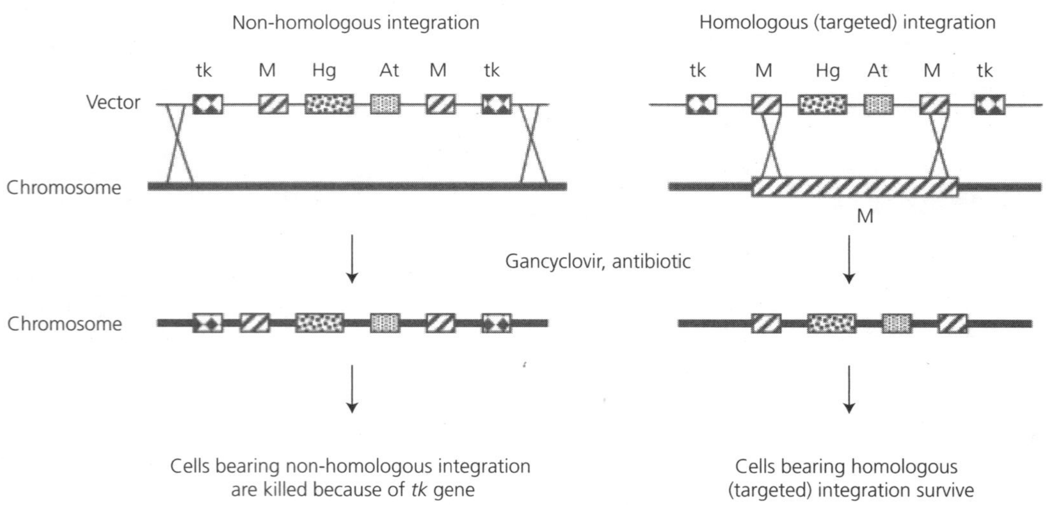

Figure 36.3 Schematic diagram of targeted gene replacement. A vector containing the gene for thymidine kinase (*tk*), a homologous host gene (M), heterologous gene (Hg) and an antibiotic resistance gene (At) is allowed to recombine with the host chromosome and is treated with gancyclovir. Cells which have undergone non-homologous recombination are killed because of the presence of *tk* gene in the chromosome. Cells having undergone homologous recombination at the M gene survive after gancyclovir treatment. Such cells are further selected by growing in the presence of antibiotics.

artificial chromosomes, which can be used to clone large DNA fragments. In plant breeding, this technology has been used to change the nature of the kinetochore proteins, which are responsible for chromosomal segregation during mitosis and meiosis. Such a change can result in homozygous offspring in plants, which can help in 'fixing' a beneficial character in the offspring. Chromosome engineering can also be used to convert meiosis to mitosis, thus producing diploid gametes, resulting in clonal propagation of organisms, a key step in the phenomenon of apomixis, the clonal propagation of plants through seeds (Chan 2010).

Cloning large fragments of DNA has always been a problem. Most plasmid-cloning vectors cannot accept DNA fragments more than a few kilobases (kb). The availability of bacteriophage-based cloning vectors have increased the upper size limit of cloning large DNA fragments, which has further increased with the advent of bacterial artificial chromosomes. However, designing a full artificial chromosome promises to give a powerful tool towards cloning very large DNA fragments, complete with genes and flanking control elements. An artificial chromosome should be an independently segregating entity, and because it can accommodate very large pieces of DNA, heterologous gene expression, with its own promoter, is expected to be more stable. The methods being adopted towards this goal can be termed 'chromosome engineering'. Conceptually, an artificial chromosome should have three elements: a telomere to control the replication at the ends, replication origins to initiate bidirectional replication and a centromere to help the artificial chromosome to segregate during mitosis. Once these three elements can be assembled, an artificial chromosome can be built up. The attempts to construct an artificial chromosome can be divided into either the 'bottom–up' or the 'top–down' approach. In the bottom–up approach, the three elements can be arranged such that they form an artificial chromosome. The identities of DNA sequences constituting telomeres and replication origins are already known in detail and can be assembled. However, centromeric sequences are much less defined and known to be rich in alpha satellite sequences (Willard 1998).

The first human artificial chromosome (HAC) was reported by Harrington *et al.* (1997), who assembled HACs by joining together synthetic alpha satellite DNAs fragments and random pieces of DNA (to provide the replication origin). These HACs were stably maintained in cultured cells. The top–down approach in chromosome engineering involves removing most of the DNA of chromosome arms. Such artificial chromosomes could attain a size of 60–400 Mb in size.

References

Baguisi, Behboodi E, Melican DT, Pollock JS, Destrempes MM, Cammuso C, *et al.* Production of goats by somatic cell nuclear transfer. *Nat Biotechnol* 1999;17:456–461.

Chan SWL. Chromosome engineering: Power tools for plant genetics. *Trends Biotechnol* 2010;12:605–610.

Games D, Adams D, Alessandrini R, Barbour R, Berthelette P, Blackwell C, *et al.* Alzheimer-type neuropathology in transgenic mice overexpressing V717F beta amyloid precursor protein. *Nature* 1995;373:523–527.

Gordon JW, Scangos GA, Plotkin DJ, Barbosa JA, Ruddle FH. Genetic transformation of mouse embryos by microinjection of purified DNA. *Proc Natl Acad Sci USA* 1980;77:7380–7384.

Gordon K, Lee E, Vitale J, Smith AE, Westphal H, Henninghausen L. Production of human tPA in transgenic mouse milk. *Biotechnology* 1987;5:1183–1187.

Harrington JJ, Van Bokkelen G, Mays RW, Gustashaw K, Willard HF. Formation of *de novo* centromeres and construction of first generation human artificial chromosomes. *Nat Genet* 1997;15:345–355.

Krimpenfort P, Rademakers A, Eyestone W, van der Schans A, van den Broek S, Kooiman P, *et al.* Generation of transgenic dairy cattle using "in vitro" embryo production. *Biotechnology* 1991;9:844–847.

Lai L, Kolber-Simonds D, Park KW, Cheong HT, Greenstein JL, Im GS, *et al.* Production alpha 1-3 galactosyl transferase knockout pigs by nuclear transfer cloning. *Science* 2002;295:1089–1092.

Miskin R, Masos T, Yahav S, Shinder D, Globerson A. Alpha-MUPA mice: A transgenic model for increased lifespan. *Neurobiol Aging* 1999;20:555–564.

Reid W, Sadowska M, Denaro F, Rao S, Foulke J Jr, Hayes N, *et al.* An HIV-1 transgenic rat that develops HIV-related pathology and immunology dysfunction. *Proc Natl Acad Sci USA* 2001;98:9271–9276.

Stewart TA, Pattengale PK, Leder P. Spontaneous mammary adenocarcinomas in transgenic mice that carry and express MTV/myc fusion genes. *Cell* 1984;38:627–637.

Tang YP, Shimizu E, Dube GR, Rampon C, Kerchner GA, Zhuo M, *et al.* Genetic enhancement of learning and memory in mice. *Nature* 1999;401:63–69.

Willard HF. Centromeres: The missing link in the development of human artificial chromosomes. *Curr Opin Gene Dev* 1998;8:219–225.

37

Gene therapy

Uttam Pati

37.1 Gene transfer technology: Concept

The characteristics of a species depend both upon the genes it inherits and the environment it lives in. However, in genetic disorders, which are phenotypically very heterogeneous, an individual either inherits defective gene(s) from parents or the gene defect has been acquired due to a change in the lifestyle or due to an unfriendly environment. Both these events could lead to diseases such as diabetes, cancer, coronary artery diseases, autoimmune disease, cystic fibrosis (CF) and muscular dystrophy and diseases related to protein aggregation. From its earliest conception, gene therapy held the promise of correcting inherited diseases by inserting a normal copy of the relevant gene into somatic cells (Friedmann and Roblin 1972).

Thousands of rare diseases are caused by a Mendelian genetic error, and more than 1800 genes associated with rare diseases have been identified. Gene therapy strategies, in general, include addition of a normal copy of the mutated gene for loss-of-function mutations, modification of messenger RNA to avoid the consequences of mutation (in cases where the mutated exon is not indispensable), inhibition of expression of a mutated gene to prevent the expression of a gain-of-function protein or to inhibit a cryptic splice site, and site-specific repair of the mutated gene (Bachtarzi *et al.* 2008). An appropriate strategy is then determined to execute a suitable protocol. The invention of recombinant DNA technology consequently led to the immediate inception of engineered gene transfer into human cells, aiming at reversing a cellular dysfunction or creating a new cellular function. As a result, the concept of direct therapeutic benefit based on a gene defect correction in human cells is coined as gene therapy (Denefle 2011). 'Gene therapy' is a term that is often used as a technique in which 'the genes causing a defect are themselves substituted by correct genes in the patients to cure a disease' (Macer and Eubios Ethics Institute 1990).

37.1.1 Background

In gene delivery protocols, DNA that encodes a therapeutic protein is packaged within a 'vector' and is then delivered inside cells within the body. Inside the cell, the DNA is subsequently expressed by the host cell's machinery, resulting in the production of a therapeutic protein, which in turn treats the patient's disease. It was thus assumed that DNA could be used as a pharmaceutical agent to treat disease; the most common approach has been to utilize DNA either to supplement or alter genes within an individual's cells as a means to eradicate the disease. Two categories of gene therapy protocol are commonly used based upon the classification of the source of cells: somatic gene therapy (SGT) and germ line gene therapy (GGT); the majority of the cells in our body are somatic cells and are not passed on to future generations, whereas germ line cells are present in the egg or sperm. The majority of the gene therapy that is practised today is of the somatic variety because there are a lot more cells to work with and any therapy through germ line in human is prohibited on ethical grounds as germ line therapy not only affects the individual but their potential offspring as well. An extended version of the gene therapy technology called 'preventive therapy', where a gene could be repaired prior to the existence of a medical condition or preventing the gene expression *in utero* before the child is born, has also successfully been used (Blaese *et al.* 1995).

In 1979, Martin Cline had made a first early and certainly premature human gene therapy attempt in treating severe thalassaemia patients through an *ex vivo* β-globin gene transfer protocol in the bone marrow of two patients in Italy and Israel. As the protocol did not have the prior approval of regulatory bodies, the study was promptly terminated and Cline was forced to resign from his job from the University of California and had lost research grants (Joyce 1982). One year later in 1980, a recombinant DNA Advisory Committee (RAC) at the National Institute of Health (NIH) was constituted to expand its regulatory function beyond recombinant DNA experiments so as to include human gene therapy studies. In 1982, in a seminar that was held at the Branbury Conference Center of Cold Spring Harbor Laboratory, a group of scientists, led by Ted Friedmann and Paul Berg, came together to build the foundations of gene therapy and its future. As an outcome, the first book on gene therapy (Friedmann 1983) was and is still a landmark reference to this field. In 1989, Rosenberg initiated the first RAC-approved gene therapy clinical trial as a 'gene-labelling' study in which a neomycin resistance gene was transferred in to tumour-infiltrating lymphocytes using a retroviral construct for the treatment of metastatic melanoma with interleukin-2 (Rosenberg *et al.* 1988).

37.2 Gene delivery systems

A few types of gene delivery systems are discussed as follows.

37.2.1 Direct delivery

The delivery of a copy of a full-length cDNA encoding the therapeutic gene to a patient is generally achieved through different pathways: physical and chemical methods and through viral vectors. As nucleic acid has a poor cell penetration capacity, the primary goal of all the three methods is to encapsulate a copy of the therapeutic gene in delivering to the defective cells in patients. As the production of plasmid DNA that carries a copy of the therapeutic gene is an easy process, the idea of using plasmid DNA directly in the patient through physical approach was earlier explored. The advantage in favour of a direct plasmid DNA delivery was based upon the fact that plasmids, unlike viruses, are non-immunogenic and do not encode accessory proteins. As plasmids lack the innate ability to enter cells and incorporate into the genome, their delivery to cells is generally dependent on physical methods, which enable them to penetrate the cell membrane.

The problem is that the expression of plasmid DNA in most cells is transient, continuing for up to a week due to the fact that DNA is lost during cell division or is degraded. The injection of naked DNA into joints through gene transfer into synovial cells by intra-articular administration of plasmid DNA had low-level transient gene expression (Yovandich *et al.* 1995). In another attempt, a naked dystrophin gene–carrying plasmid was injected into the radialis muscle of nine Duchenne muscular dystrophy (DMD)/Becker muscular dystrophy (BMD) patients; although the expression of dystrophin was low, no adverse side effect was observed after the treatment (Romero *et al.* 2004).

However, a direct use of plasmid DNA from a bacterial source has a certain disadvantage in which unmethylated CpG islands of the bacterial DNA could elicit strong host immune responses (Krieg 2002). In addition, it was further reported that plasmid DNA injection could potentially exacerbate autoimmunity through the development of antibodies to double-stranded DNA autoantigens (Mor *et al.* 1997). As the cell wall is made of fatty lipids and liposome ion can move across the membrane directly into the cell, another idea was to pack the gene with liposome through chemical methods. Chemical vectors include lipoplexes, polyplexes and inorganic nanoparticles (Tani *et al.* 2011). The liposome-encapsulated DNA could then be delivered into the cell, although it might not target a specific cell. In CF, where the patient has breathing problem because of chloride ion built-up in the respiratory tract, inhaling a liposome that is coated with genetically altered genes can help the patient considerably without any side effect. Electroporation and some emerging physical delivery methods such as ultrasound and magnetofection have also shown encouraging results. Direct electroporation into mouse lung enhances reporter gene expression (Pringle 2004), in ovine (Emerson *et al.* 2003) and in both rodent and larger animal models, such as sheep (Xenariou *et al.* 2005). Electroporation has since been shown to enhance gene transfer in other tissues, including synovium (Ohashi *et al.* 2002), spinal cord (Lin *et al.* 2002) and skin (Zhang *et al.* 2002). In a physical method, specifically in mice, a helium gun could fire very small gold bullets coated with genetically altered genes (Gould and Favorov 2003). Genes coding for cytokines that activate immune cells were used successfully in regression of tumours.

37.2.2 Delivery through viral vectors

In the last few decades, viral gene delivery has been the method of choice to develop the nucleic acid–based biomimetic vehicles. However, the human defence system might also play a reverse role in rejecting the delivered gene. These defence systems, which are mostly restricted to higher primates, are of concern in the context of a gene therapy clinical development plan, especially when it comes to the use of a natural human-derived virus. Other hurdles are the active virus loads and the amount of virus particles to be used to achieve therapeutic effects, which combined with the administration route, are very difficult to predict in terms of clinical pharmacology and drug safety (Denefle 2011). When choosing a vector for gene therapy, there are some fundamental differences between the vectors available in terms of capacity of transgene cassette, duration of transgene expression, immunogenicity, ability to target quiescent or dividing cells, and extra chromosomal or genomic integration of a transgene as well as basic safety issues to be considered (Gould and Favorov 2003).

Both DNA and RNA viruses are used for gene delivery: DNA viruses such as adenovirus (ADV), herpes simplex virus (HSV), and Epstein–Barr virus (EPV) provide non-integrative means of transferring therapeutic gene, as the genetic material of these viruses does not usually integrate into cells and is not replicated at cell division. Their restricted use is limited to the infection of post-mitotic cells (Verma and Weitzman 2005). However, RNA viruses, which include retroviruses, spuma viruses and lenti viruses, are used to mediate integrative gene transfer. Retroviral vectors integrate with a high frequency near CpG islands, which suggests that its long terminal repeat (LTR) region, which controls site-specific integration, can also target active gene albeit without selective tropism to the transcription start site (Wu *et al.* 2003; Mitchell *et al.* 2004; Bushman 2007).

The advantage of a virus delivery system is that in a virus, the target gene could be wrapped in a protein coat that could easily bind the surface of the target cells, with retroviruses being the first choice as they can change the cells' defective genes with the new therapeutic gene. Retroviruses are effective in dividing cells because the pre-integration complex cannot cross the nuclear membrane. Retroviruses introduce the gene they carry into the target cell genome, while ADVs introduce the gene into the target cell's nucleus without incorporating it into the target cell's genome. The disadvantage of these vectors is the possibility of their random insertion into the host genome, which causes serious adverse effects (SAEs) (Tsurumi *et al.* 1996). The remedial approach has been to design 'non-integrative vectors', which are mainly used to transfer DNA into quiescent cells but which will be lost after a few replication cycles in dividing cells (e.g. ADV or adeno-associated virus [AAV] vectors). The nature of target tissue/cell and the length of the desired therapeutic effect have, therefore, to be taken into consideration in the gene therapy project charter (Denefle 2011).

Retroviruses are a group of RNA containing viruses and are lipid-enveloped nucleoprotein particles which contain a single-stranded linear RNA molecule (7–11 kb long). The RNA template after entering the cell is reverse transcribed into DNA in the cytoplasm, which is then imported to the nucleus in a mitosis-dependent fashion and then randomly integrated into the host cell genome. Wild-type retroviruses are responsible for chronic non-pathological infections. The ability of retroviruses to integrate into the genome, combined with their relatively low immunogenicity, is an advantageous feature for gene therapy applications (Walther and Stein 2000). Retroviral infection is initiated by attachment to specific cell surface molecules, the retroviral receptors (Hawley 1996). For example, the retroviral genome, such as that of Moloney murine leukaemia virus (MMLV), contains two LTRs at its ends. LTRs and their neighbouring regions contain *cis*-signals employed during reverse transcription, integration and transcription. Besides, there are three genes that encode structural proteins (*gag*), enzymes of nucleic acid metabolism (*pol*), and a surface envelope glycoprotein (*env*). As most of the *cis*-acting elements are located in the terminal region, it has been proved possible to engineer retroviral vectors for gene transfer by removal or inactivation of all protein-coding regions of the retroviral genome. Vector plasmids, containing the gene of interest flanked by LTRs and *cis*-elements, can be assembled into retroviral particles by their transfection into packaging cells, producing in *trans* all viral proteins. To avoid genetic recombination of replication competent viruses, viral genes in the packaging cell genome were divided between two separate genomic regions. This method allows preparation of viral stocks with a titre up to 10^7 transduction units (TU) per millilitre.

ADVs infect both dividing and non-dividing cells with high efficiency and the capacity to produce high-titre adenovirus (Ad) stocks has greatly aided their application. Two adenoviral serotypes Ad2 and Ad5 are most commonly used as vectors for gene therapy; they bind to a 46-kDa glycoprotein receptor called Coxsackie virus and ADV receptor (CAR). After the initial adsorption, secondary interactions between RGD (arginine–glycine–aspartic acid) motifs of the Ad penton base and avb3 and avb5 integrins on the host cell surface then follow, which enable internalization of the capsid (Shayakhmetov *et al.* 2005). The infection efficiency is dependent upon the expression level of the primary and secondary receptors on the cell surface. Ad capsids are icosahedral particles composed of three major protein components – the hexon, penton base and fibre. Ad infection is initially mediated by binding of the 'knob' domain of the fibre protein to specific receptors on the cell surface. It may be essential that the immune system may be weakened before the therapy starts.

AAV is a small (less than 5 kb) single-stranded DNA non-enveloped parvovirus. AAV infection does not require host cell proliferation, although expression from AAV vectors may exhibit a relative preference for actively dividing cells. Wild-type AAV and modified AAV vectors tend to persist in infected cells for prolonged periods of time, without any significant adverse consequences for the host. Wild-type AAV integrates into a specific region of chromosome 19, whereas rep-deleted AAV vectors integrate in a less specific fashion in the host cell genome and may also persist in an episomal state (Flotte and Carter 1995). The natural route of infection of wild-type AAV is the upper respiratory tract. For productive infection to occur, AAV requires co-infection with Ad, which allows the viral genome to replicate episomally and leads to the synthesis of AAV proteins. A helper virus is essential for the AAV genome to integrate into the host cell chromosome 19 (19q13.3-qter). Site-specific integration requires enzyme activity of replication (Rep) proteins (Rep68 and/or Rep78) (Xiao *et al.* 1996; Surosky *et al.* 1997; Lamartina *et al.* 1998). The clinical application of recombinant AAV (rAAV) is potentially limited by the fact that humans are the natural host of AAV2 and seropositive antibodies are detected in up to 96% of the population. Virtually all the AAV vectors developed for gene therapy application are based on AAV serotype 2 (AAV2), which interacts with target cells by binding to membrane-expressed heparan sulphate (Summerford and Samulski 1998). To date, at least 12 serotypes (AAV1–12) have been identified, each with different tissue tropisms (Li). rAAV has been prepared by deleting almost the entire AAV genome with the exception of inverted terminal repeats, which are required for packaging, integration and DNA replication (Samulski *et al.* 1989). Production of rAAV is achieved by different approaches, but in each case, it is dependent upon *rep* and capsid (*cap*) genes being supplied in *trans*, along with a source of helper virus gene sequences, to direct replication and packaging of the vector (Monahan and Samulski 2000).

37.2.3 Avoiding silence zone during gene delivery

During gene delivery through viral vectors, the delivered gene is introduced into the cells of patients in an indiscriminate fashion; it may be integrated into the genome in the 'silence zone', where the genes are not expressed at all. This happens as the newly introduced therapeutic gene lands randomly in the genome. In a novel approach, to make sure that the β-globin gene that is stitched to lentiviral vectors (LVVs) is delivered to the active site of the genome, the researchers hooked an 'ankyrin insulator' to the globin gene. The ankyrin insulator creates an active area in the genome where the new gene can work efficiently irrespective of its landing area (Breda *et al.* 2012). By treating eight cell specimens taken from sickle cell anaemia patients, it was discovered that attaching the ankyrin insulator to a normal globin gene increases the amount of normal β-globin protein while reducing the level of sickle protein.

37.3 Gene therapy in disease

Gene therapy is being used in treating various diseases, some of which are as follows.

37.3.1 Severe combined immunodeficiency

Although the idea of gene therapy was first conceptualized in 1972, the first ever Food and Drug Administration (FDA)–approved gene therapy experiment occurred in 1990, at the US National Cancer Institute (NCI), where a 4-year-old patient was treated for ADA-SCID (adenosine deaminase-severe combined immunodeficiency) deficiency; the effects were temporary but successful (Sheridan 2011). SCID is due to the lack of ADA and is a fatal disorder of purine metabolism and immunodeficiency. A normal *ADA* gene produces adenosine deaminase; the patient with ADA deficiency is susceptible to most minor illnesses (Laino 2006). In 1992, Claudio Bordignon performed the first procedure of gene therapy using haematopoietic stem cells as delivery vectors to correct SCID (Abbott 1992). In 1993, in a modified 'preventive approach', a patient with SCID could be treated before birth; stem cells from the placental blood containing retrovirus–ADA gene was injected into patient's vein. For four years, T-cells produced by stem cells made the ADA enzyme (Blaese *et al.* 1995). Further, a network of European groups led by A. Fischer in France, A. Trascher in the United Kingdom, and M. Roncarolo in Italy initiated similar protocols for the treatment of SCID. The success of SCID gene therapy got a setback in 2002 when two of the first 10 children developed SAEs and died of a leukaemia-like condition in a clinical trial in Paris (Gore 2003). Two more deaths too had occurred in the United States, in 1999, after which the United States cancelled clinical trials, pending the re-evaluation of ethical and procedural practice. After a detailed regulatory review of the process, the trial resumed (Cavazzana-Calvo *et al.* 2004). The cause of SAEs was determined to be the insertion of the therapeutic DNA into the patient cells next to one specific locus LMO2 (Lim domain only 2 is a member of LMO, LIM only, class of transcription factors that contain a zinc-binding finger-like motif termed the LIM domain only) (Rabbitts *et al.* 1997; Nam and Rabbitts 2006; Pike-Overzet *et al.* 2006). Although the SCID gene therapy trials were then stopped, due to lack of alternative treatment, ADA and SCID-X1 trials still continued under strict monitoring. Between 1999 and 2007, gene therapy had restored the immune systems of at least 26 children with two forms (ADA-SCID [9 children] and SCID-X1 [10 children]) of the disorder, although 4 of the 10 SCID-X1 patients had developed leukaemia-related SAEs (Deichmann *et al.* 2007). Out of 20 children who were treated, four developed leukaemia-like adverse effects and one patient died from leukaemia. Today, gene therapy, combined with reduced-intensity conditioning, is a safe and effective treatment for SCID in patients with ADA deficiency. In a recent study, autologous CD34$^+$ bone marrow cells transduced with a retroviral vector containing the *ADA* gene was infused into 10 children with SCID, who lacked an HLA (human leucocyte antigen)-identical sibling donor, without any adverse effect such as leukaemic transformation (Aiuti *et al.* 2009). Although 18 of 20 treated infants with X-linked SCID are alive and well with restored immunity, a T-cell lymphoproliferative syndrome developed within 2–5 years after the procedure in five children; one of these children died as a consequence of complications of the syndrome despite therapy. Investigations have implicated insertional oncogenesis in the pathogenesis of leukaemia-like illness, in which the insertion of the corrective retroviral vector may activate expression of cellular proto-oncogenes near the integration site.

37.3.2 X-linked adrenoleucodystrophy

X-linked adrenoleucodystrophy (ALD) is a severe brain demyelinating disease in boys that is caused by a deficiency of the ALD protein, an adenosine triphosphate–binding cassette transporter encoded by the *ABCD1* gene. A gene therapy for X-linked ALD was successfully conducted. Autologous $CD34^+$ cells were removed from patients, genetically corrected *ex vivo* with an LVV encoding wild-type *ABCD1*, and then re-enfused into patients (Cartier *et al.* 2009). ALD progression can be halted by allogeneic haematopoietic cell transplantation (HCT). Beginning 14–16 months after infusion of the genetically corrected cells, progressive cerebral demyelination in the two patients was prevented. This result could be well compared with a clinical outcome by allogeneic HCT.

37.3.3 Parkinson and Alzheimer disease

Most small-molecule drugs do not cross the blood brain barrier (BBB). Second- and third-generation three-plasmid vector systems, termed FELIX, which were constructed from FIV (feline immunodeficiency virus), offer high efficiency and stable delivery of genes to numerous non-dividing and primary cell types, opening new avenues for biological inquiry into normal human cells (Curran *et al.* 2000). A breakthrough was achieved in 2003 when scientists transferred a gene into the brain; it was significant because viral vectors are normally too big to get across the BBB. Molecular Trojan horses are genetically engineered proteins that cross the BBB via endogenous receptor-mediated transport processes (Pardridge 2006). Although this procedure opened up a new dimension in treating Parkinson and Alzheimer disease, unexpectedly, immune rejection after the newly introduced gene, such as organ transplantation, has been a problem since gene therapy was practised as the new gene was perceived by the host as foreign. In a novel approach, scientists rationalized that micro RNA could potentially be used to selectively turn off the identity of the newly introduced gene in cells of the immune system (Brown *et al.* 2006). When mice were injected with a gene containing an immune cell micro RNA target sequence, the mice did not reject the gene.

37.3.4 Leber's congenital amaurosis

Leber's congenital amaurosis (LCA) is an inherited retinal disease that causes blindness because of mutations in the *RPE65* gene. In 2008, three independent groups in the United Kingdom and the United States conducted gene therapy trials through subretinal delivery of recombinant AAV in patients carrying the *RPE65* gene who recovered functional vision without any side effects. Human gene therapy with an rAAV2 vector was performed for the RPE65 form of childhood blindness; the procedure was deemed safe and there was evidence of visual gain in the short term (Cideciyan *et al.* 2009). In 2009, a similar approach successfully restored trichromatic vision in squirrel monkey, a hopeful precursor to the treatment for colour blindness in humans (Mancuso *et al.* 2009). Further breakthrough was reported in 2010, for treating achromatopsia, or complete colour blindness, directed towards cone photoreceptors in dogs (Komaromy *et al.* 2010).

37.3.5 Skeletal muscle

The AAV vector of various serotypes provides superior transduction efficiency to the skeletal muscle (Mingozzi and High 2007; Cooper *et al.* 2009; Wang *et al.* 2010). Although AAV vectors were used for treating CF, their usage has dropped in the last decade due to poor transduction efficiency in human airway epithelial cells and the inability to be readministered. In addition, anti-ADV immune responses, in particular cytotoxic T-lymphocyte–mediated (CTL) responses and major histocompatibility complex class I antigen (MHC-I) presentation, were enhanced if the host has a pre-existing *Pseudomonas* infection (Tosi *et al.* 2004). However, Lai *et al.* (2005) have recently shown that the efficiency of AAV *trans*-splicing can be greatly improved through rational vector design and may, therefore, allow the CFTR cDNA to be split between two viral vectors. So far, two human gene therapy phase I/II protocols have been undertaken with incremental and repeat doses of AAV, up to 2×10^{12} and 2×10^{13} DNase-resistant particles, respectively (Flotte Zeitlin *et al.* 2003; Moss *et al.* 2004). In both studies, viral shedding and increases in neutralizing antibodies were observed, but no SAE (Tsurumi *et al.* 1996) was associated to the virus administration. Importantly, a significant reduction in sputum IL-8 and some improvement in lung function were noted after the first administration. Studies are currently underway to assess the feasibility of repeated administration of lentivirus-based vectors into airways by several groups (Medina *et al.* 2003; Sinn *et al.* 2005), and further data will be needed before the relevance of such viruses for CF gene therapy can be decided. However, various pseudotyped AAVs (Grimm and Kay 2003) were later on used as a vehicle to deliver either truncated versions of the gene (mini- or microdystrophin) or an exon-skipping RNA structure, both thought to achieve truncated albeit functional dystrophin protein expression (Alexander *et al.* 2007; Rodino-Klapac *et al.* 2007).

37.3.6 Chronic pain

Gene therapy approaches to treat chronic pain due to cancer, infectious diseases, autoimmune-related syndromes and surgery have been reported in numerous preclinical studies that demonstrate pain control at the level of the spinal cord. In a phase-I human trial in which a replication-defective herpes simplex virus (HSV) vector was used to deliver the human pre-proenkephalin (hPPE) gene, encoding the natural opioid peptides met- and leu-enkephalin (ENK), to cancer patients with intractable pain resulting from bone metastases with excellent result in a month (Fink *et al.* 2011). Patients who received higher doses of the therapeutic vector experienced a substantial reduction in their overall pain scores for up to a month post vector injection. These exciting early clinical results await further patient testing to demonstrate treatment efficacy and will likely pave the way for other gene therapies to treat chronic pain (Goins *et al.* 2012).

37.3.7 Cancer

One must realize that successful therapeutic protocols in gene delivery system have been restricted to single genes. However, the protocol is still in the beginning stage in treating complex lifestyle diseases such as cancer, diabetes, and coronary artery disease (CAD), in which multiple genes might be involved. In 2006, an international group of scientists announced the successful use of gene therapy to treat two adult patients for a disease affecting myeloid cells (Ott *et al.* 2006).

Immunotherapy uses genetically modified cells and viral particles to stimulate the immune system to destroy cancer cells. Recent clinical trials of second- and third-generation vaccines have shown encouraging results with a wide range of cancers, including lung cancer, pancreatic cancer, prostate cancer and malignant melanoma. Oncolytic virotherapy, which uses viral particles that replicate within the cancer cell to cause cell death, is an emerging treatment modality that shows great promise, particularly with metastatic cancers (Cross and Burmester 2006). Initial phase I trials for several vectors have generated excitement over the potential power of this technique. Gene transfer is a new treatment modality that introduces new genes into a cancerous cell or the surrounding tissue to cause cell death or slow down the growth of the cancer. This treatment technique is very flexible, and a wide range of genes and vectors are being used in clinical trials with successful outcomes. As these therapies mature, they may be used alone or in combination with current treatments to help make cancer a manageable disease (Cross and Burmester 2006). It was also demonstrated that metastatic melanoma could be treated using killer T cells genetically retargeted to attack cancer cells (Morgan *et al.* 2006).

In 2011, a pilot study had also confirmed that chronic lymphocytic leukaemia (CLL) could be tackled through genetically modified T cells to fight the disease; two of the three subjects were confirmed to be cured (Kalos *et al.* 2011; Porter *et al.* 2011). Gene therapy has also been used to produce recombinant cancer vaccines. Unlike vaccines for infectious agents, these vaccines do not prevent disease; they may contain it by training the patient's immune system to recognize the cancer cells by presenting it with highly antigenic and immune-stimulatory cellular debris. Cancer cells are grown *in vitro* after harvesting them either from the patient (autologous cells) or from established cancer cell lines (allogeneic); they are engineered to be more recognizable to the immune system of the patients. Cytokine genes that produce proinflammatory immune-stimulating molecules are preferred. These altered cells are grown *in vitro* and killed, and the cellular contents are incorporated into a vaccine (Kowalczyk *et al.* 2003). Since 2006, continuous breakthroughs in gene therapy protocol have opened up a new cure for cancer (Morgan *et al.* 2006), specifically in the myeloid system (Ott *et al.* 2006). In a phase I clinical trial (Goins *et al.* 2012), a replication-defective HSV vector was recently used to deliver the human pre-proenkephalin (hPKE) gene, encoding natural opioid peptides met- and leu-enkephalin (ENK), to cancer patients with intractable pain resulting from bone metastasis (Fink *et al.* 2011). Tumour necrosis factor-α (TNF-α) is a cytokine with potent anticancer activity and high systemic toxicity; TNF-α gene therapy could be used in cancer patients through intra-tumoural injections and a promoter that is further activated by radiation therapy (Rasmussen *et al.* 2002).

37.3.8 Cardiovascular disease

Clinical trials to transfer genes as a therapeutic approach to treat cardiac diseases have recently been standardized in studying key factors such as vector type, injected dose, delivery method and targeted cardiac disease, for which large animal studies are best suited for comprehensive evaluation at the preclinical stages of therapeutic development to ensure safe and effective gene transfer (Ishikawa *et al.* 2012). In addition, the generation of pluripotent stem cells and transdifferentiation, two emerging strategies for the production of cardiac myocytes from human cells and for the investigation of human diseases, have been reported. Normal rhythms originate in the sino-atrial node, a specialized cardiac tissue consisting of only a few thousands of nodal pacemaker cells; malfunction of pacemaker cells due to diseases or aging leads to rhythm generation disorders (bradycardias and sick sinus syndrome), which often necessitate the implantation of electronic pacemakers. Various gene- and cell-based approaches, with a particular emphasis placed on the use of pluripotent stem cells and the hyperpolarization-activated cyclic nucleotide–gated encoded pacemaker gene family, have been pursued in the past decade to reconstruct bio-artificials (Li 2012). LVVs have also been successfully manipulated for cardiomyocyte differentiation and function (Di Pasquale *et al.* 2012).

Adenoviral and adeno-associated viral vectors have been the preferred system for myocardial gene therapy applications for many years; however, their popularity has been diminished by immunogenicity issues, their innate tropism for some human tissues and their inability to integrate into the host genome as well as their short duration of transgene expression. LVVs were shown to potentially overcome such limitations, and for this reason, they are now widely used in biological research (Wilson 2002). They resemble retroviral vectors, and so are able to stably integrate into the genome of the target cell, allowing for persistent expression of the gene of interest and can also transduce both replicating and non-replicating cells with high efficiency, making them suitable for use in terminally differentiated cells such as neurons, hepatocytes and cardiomyocytes (Naldini 1998; Di Pasquale *et al.* 2012). Treatment for CAD and repair of the damage after myocardial infarction have been under progress through human growth factor (HGF) plasmid DNA therapy of cardiomyocytes (Yellon and Hausenloy 2007; Yang *et al.* 2009). Indeed, AAVs present an option for studying long-term *in vivo* genetic modifications and have been used in clinical trials for heart failure (Giacca and Baker 2011; Tilemann *et al.* 2012).

37.3.9 Diabetes

The alarming worldwide prevalence of diabetes mellitus by the year 2011 is projected to reach 347 million (Danaei 2011), 97% of these having type 2 diabetes, based on estimates of changes in lifestyle, economic development and population figures. Diabetes mellitus is usually classified as type 1 or type 2 diabetes. Type 1 results from a β-cell defect, often due to an autoimmune process. Type 2 diabetes is characterized by insulin resistance, which is often combined with an insulin secretory defect.

The gene therapy for diabetes might be a risky affair if the number of gene-modified insulin-producing cells and their insulin release are not under perfect control; the result could be severe hypoglycaemia (Welsh 2000). In principle, non-insulin–producing cells could be turned into insulin-producing β-cells using a suitable promoter and an insulin gene construct, thus restoring insulin production in type 1 and some type 2 diabetics. Cell lines such as fibroblasts, hepatocytes, myocytes, pituitary and exocrine cells have also been made into artificial insulin-producing cells, although none of these cells respond to glucose with physiological secretion of insulin. Another gene therapy approach aims at genetically manipulating β-cells so that they produce a local β-cell protection factor, thus raising the possibility of targeted therapy for autoimmunity. Transplantation of haematopoietic stem cells modified to express a protective form of MHC class II corrects a defect in central tolerance. This method contrasts with other targeted therapies that attempt to modify peripheral tolerance, which is also defective in type 1 diabetes mellitus (Creusot and Fathman 2004).

37.3.10 HIV

The highly active antiretroviral therapy (HAART) can suppress HIV replication to undetectable levels in the plasma, but it is unlikely to eradicate cellular reservoirs of the virus. HIV-specific CD4(+) and CD8(+) T cells bear a chimeric T-cell receptor (CD4zeta) composed of extracellular and transmembrane domains of human CD4 (which binds HIVgp120) linked to the intracellular zeta signalling chain of the CD3 T-cell receptor. CD4, zeta-modified T cells can inhibit viral replication, kill HIV-infected cells *in vitro* and survive for prolonged periods *in vivo*. In a phase II randomized highly active antiretroviral therapy (HAART) trial of CD4 (zeta signalling chain of CD3-T-cell receptor) gene–modified versus gene–unmodified T cells in 40 HIV-infected subjects, with plasma viral loads <50 copies/mL, HIV replication was suppressed to undetectable levels in the plasma (Deeks *et al.* 2002).

37.4 VEGF gene therapy

Vascular endothelial growth factor (VEGF) indeed refers to a family of five mammalian factors (VEGF-A, -B, -C, -D and placenta growth factor) and to additional, related proteins of viral (VEGF-E) and snake venom. The proteins encoded by these genes act as homo- or heterodimers through their interaction with three structurally related tyrosine kinase receptors (Olsson *et al.* 2006; Adams and Alitalo 2007; Carmeliet and Jain 2011; Potente *et al.* 2011) of (VEGF-Fs) origin. The induction of new blood vessel formation by delivering angiogenic genes to ischaemic tissues continues to appear as a promising, alternative strategy to currently available therapies. VEGF family members exert multiple, non-angiogenic effects on various cell types, including neurons, skeletal muscle and cardiac cells. As a result, VEGF-based therapies have already reached clinical experimentation and highlight the pleiotropic activities of VEGF factors for therapeutic application. However, the very characteristics may warrant caution for unwanted, non-vascular toxicity upon administration of VEGF-based gene therapies (Giacca and Zacchigna 2012).

37.5 RNAi therapy

As short, non-coding RNA molecules affect protein expression levels, RNA interference, or RNAi, could be used to reduce the levels of a specific protein to reverse its deleterious effect. When such RNAs were injected into the mouse tail vein, hepatitis could be prevented (McCaffrey *et al.* 2002, 2003). Viral vectors could also deliver RNA sequences into the cell, thus preventing neurodegeneration in spino-cerebellar ataxia (Xia *et al.* 2004). Lentiviral-mediated silencing of SOD1 through the RNAi technology was shown to retard progression of amyotrophic lateral sclerosis (ALS) in mouse mode capsid (cap) genes being supplied in *trans*, along with a source of helper virus gene sequences to direct replication and packaging (Ralph *et al.* 2005; Raoul *et al.* 2005). Administration of RNAs through nasal spray for inhibition of respiratory infection through respiratory syncytical virus (RSV) and parainfluenza virus (PIV) has also been successfully used in mice (Bitko *et al.* 2005).

37.6 Current scenario

At the present time, a large number of clinical trials across the globe are in progress; 65 trials are declared to be in the late stage (phase II–III) (Denefle 2011). Since 2002, approximately 1700 trials have been conducted using a number of techniques (Gene Therapy Clinical Trials Database: http://www.genetherapynet.com/clinical-trials.html). In all the examples that are described above, the expression of a deficient gene is induced in the patient's cells. In cases where a gene is over-expressed, thus creating a pathogenic condition, a reasonable approach has been to knock out the gene through specific gene knockout therapy. This has been possible through engineered zinc finger nucleases, I-CreI homing endonucleases and nucleases from TAL effectors. Knocking out the *CCR5* gene in T cells has been effective in controlling HIV infection (Jouvenot *et al.* 2005) and is being used in human clinical trials. However, in the last 5–7 years (Foster *et al.* 2006), unrivalled progress in efficient systemic delivery of synthetic and chemically modified oligonucleotides has made it possible to enforce mutated exon splicing ('t Hoen 2008). This progress has led to several more clinical trials, which are labelled as 'small-molecule' trials, that is, outside the boundaries of gene therapy. The most advanced clinical trial, led by a company called Prosensa in the Netherlands, is completing its phase II.

37.7 Regulation and ethics

In practice, before starting a gene therapy study, a balance of benefits and risks is weighed upon; it is important to consider the safety and efficacy of the proposed research and the availability of alternatives (Resnik 2001). Schematic representation of the prevalence of mutation types available from the Human Genome Data Base (HGDB) shows that more than 60% of 73,411 catalogued mutations responsible for human diseases are caused by single-nucleotide polymorphisms (SNPs) (Seyhan 2011). After the discovery that different individuals with different genotypes respond more or less positively to targeted drugs (Fleck 2012), the idea of personalized medicine would definitely be the choice of future. Thus, it is suggested that a triad synergy must be outlined based upon pharmacogenomics, biotechnology and regulatory issues (Mesko *et al.* 2012). Further, in the age of genomic medicine, genetic testing and gene therapy will appropriately be integrated without any choice. The situation demands that gene therapy be an integral part of the human health-care system in the 21st century. The rapid progress in translational medicine, including emerging gene therapy protocols, brings a new set of ethics that has to be strictly followed.

In the United States, there is a three-tier system for execution of gene therapy: the RAC, Office of Biotechnology Activities (OBA) and US FDA. The RAC was constituted in 1974 in response to public concerns regarding the safety of manipulating genetic material through the use of recombinant DNA techniques. Human gene transfer trials conducted at, or sponsored by, institutions receiving NIH (National Institutes of Health) funding for recombinant DNA research are registered with the OBA and reviewed by the RAC. The OBA is responsible for the NIH system of oversight of recombinant DNA research. The FDA is the primary government agency charged with protecting the health of US citizens by ensuring that drugs, medical devices and biological products are safe and effective before they are used by doctors, nurses, other health-care professionals, hospitals and consumers. FDA's Center for Biologics Evaluation and Research (CBER) regulates human gene therapies, which fall under the legal definition of a 'biologic'. Manufacturers of gene therapy products must test their products extensively and meet FDA requirements for safety, purity and potency before they can be sold in the United States. The FDA may tell the manufacturer to change the study or stop it altogether. The FDA has not yet approved for sale any human gene therapy product.

There should be a distinction between gene therapy for disease genes and gene therapy to enhance desired traits. However, there is a thin line between disease and traits in regard to the large number of mutations called SNPs. Removing a disease state through gene therapy to ease the suffering of a patient is quite different from opting for an enhancement of a desired trait through an introduction or deletion of gene mutants. In the past, the world has witnessed the ills of eugenics under the sponsorship of states that include elimination of mentally feeble people across nations. Besides, the current scenario of 'SGT only' may change when the protocols for germ line therapy improve with time; future research may then fail to distinguish between the correction of disease gene and enhancement of a normal gene as an advanced gene therapy protocol. Gene therapy in human beings brings a variety of important ethical, legal and social issues. In spite of a wide approval of SGT in treating or preventing diseases in individuals, there are serious reservations concerning the safety and efficacy of treatments and protection of human research subjects. GGT has been much more controversial than SGT, because it creates risks not only to patients but also to future generations, and because it may lead to genetic enhancement. Genetic enhancement raises a number of difficult issues, including the ethics of changing human traits. It may bring risks to future generations through parental control, social diktats and inequalities, exacerbation of discrimination and new eugenics (Glover 2006; Harris and William 2007). A global approach in regard to cultural variations, individual priorities and maximum medical benefit should be the prime criterion to implement new technologies, which should invariably include the genome, ethics, law and policy.

References

Abbott A. Gene therapy. Italians first to use stem cells. *Nature* 1992;356:465.

Adams RH, Alitalo K. Molecular regulation of angiogenesis and lymphangiogenesis. *Nat Rev Mol Cell Biol* 2007;8:464–478.

Aiuti A, Cattaneo F, Galimberti S, Benninghoff U, Cassani B, Callegaro L, *et al.* Gene therapy for immunodeficiency due to adenosine deaminase deficiency. *N Engl J Med* 2009;360: 447–458.

Alexander BL, Ali RR, Alton EWF, Bainbridge JW, Braun S, Cheng SH, *et al.* Progress and prospects: gene therapy clinical trials (part 1). *Gene Ther* 2007;14:1439–1447.

Bachtarzi H, Stevenson M, Fisher K. Cancer gene therapy with targeted adenoviruses. *Expert Opin Drug Deliv* 2008;5:1231–1240.

Bitko V, Musiyenko A, Shulyayeva O, Barik S. Inhibition of respiratory viruses by nasally administered siRNA. *Nat Med* 2005;11:50–55.

Blaese RM, Culver KW, Miller AD, Carter CS, Fleisher T, Clerici M, *et al.* T lymphocyte-directed gene therapy for ADA-SCID: Initial trial results after 4 years. *Science* 1995;270:475–480.

Breda L, Casu C Gardenghi S, Bianchi N, Cartegni L, Narla M, *et al.* Therapeutic hemoglobin levels after gene transfer in beta-thalassemia mice and in hematopoietic cells of beta-thalassemia and sickle cells disease patients. *PLoS ONE* 2012;7:e32345.

Brown BD, Venneri MA, Zingale A, Sergi L, Naldini L. Endogenous microRNA regulation suppresses transgene expression in hematopoietic lineages and enables stable gene transfer. *Nat Med* 2006;12:585–591.

Bushman FD. Retroviral integration and human gene therapy. *J Clin Invest* 2007;117: 2083–2086.

Carmeliet P, Jain RK. Molecular mechanisms and clinical applications of angiogenesis. *Nature* 2011;473:298–307.

Cartier N, Hacein-Bey-Abina S, Bartholomae CC, Veres G, Schmidt M, Kutschera I, *et al.* Hematopoietic stem cell gene therapy with a lentiviral vector in X-linked adrenoleukodystrophy. *Science* 2009;326:818–823.

Cavazzana-Calvo M, Thrasher A, Mavilio F. The future of gene therapy. *Nature* 2004;427:779–781.

Cideciyan AV, Hauswirth WW, Aleman TS, Kaushal S, Schwartz SB, Boye SL, *et al.* Vision 1 year after gene therapy for Leber's congenital amaurosis. *N Engl J Med* 2009;361:725–727.

Cooper M, Nayak S, Hoffman BE, Terhorst C, Cao O, Herzog RW. Improved induction of immune tolerance to factor IX by hepatic AAV-8 gene transfer. *Hum Gene Ther* 2009;20:767–776.

Creusot RJ, Fathman CG. Gene therapy for type 1 diabetes: A novel approach for targeted treatment of autoimmunity. *J Clin Invest* 2004;114:892–894.

Cross D, Burmester JK. Gene therapy for cancer treatment: Past, present and future. *Clin Med Res* 2006;4:218–227.

Curran MA, Kaiser SM, Achacoso PL, Nolan GP. Efficient transduction of nondividing cells by optimized feline immunodeficiency virus vectors. *Mol Ther* 2000;1:31–38.

Danaei G, Finucane MM, Lu Y, Singh GM, Cowan MJ, Paciorek CJ, *et al.* National, regional, and global trends in fasting plasma glucose and diabetes prevalence since 1980: Systematic analysis of health examination surveys and epidemiological studies with 370 country-years and 2.7 million participants. *Lancet* 2011;378:31–40.

Deeks SG, Wagner B, Anton PA, Mitsuyasu RT, Scadden DT, Huang C, *et al.* A phase II randomized study of HIV-specific T-cell gene therapy in subjects with undetectable plasma viremia on combination antiretroviral therapy. *Mol Ther* 2002;5:788–797.

Deichmann A, Hacein-Bey-Abina S, Schmidt M, Garrigue A, Brugman MH, Hu J, *et al.* Vector integration is nonrandom and clustered and influences the fate of lymphopoiesis in SCID-X1 gene therapy. *J Clin Invest* 2007;117:2225–2232.

Denefle PP. Introduction to gene therapy: A clinical aftermath. *Methods Mol Biol* 2011;737:27–44.

Di Pasquale E, Latronico MV, Jotti GS, Condorelli G. Lentiviral vectors and cardiovascular diseases: A genetic tool for manipulating cardiomyocyte differentiation and function. *Gene Ther* 2012;19:642–648.

Emerson M, Renwick L, Tate S, Rhind S, Milne E, Painter HA, *et al.* Transfection efficiency and toxicity following delivery of naked plasmid DNA and cationic lipid-DNA complexes to ovine lung segments. *Mol Ther* 2003;8:646–653.

Fink DJ, Wechuck J, Mata M, Glorioso JC, Goss J, Krisky D, *et al.* Gene therapy for pain: Results of a phase I clinical trial. *Ann Neurol* 2011;70:207–212.

Fleck LM. Pharmacogenomics and personalized medicine: Wicked problems, ragged edges and ethical precipices. *N Biotechnol* 2012;29:757–768.

Flotte TR, Carter BJ. Adeno-associated virus vectors for gene therapy. *Gene Ther* 1995;2:357–362.

Flotte TR, Zeitlin PL, Reynolds TC, Heald AE, Pedersen P, Beck S, *et al.* Phase I trial of intranasal and endobronchial administration of a recombinant adeno-associated virus serotype 2 (rAAV2)-CFTR vector in adult cystic fibrosis patients: A two-part clinical study. *Hum Gene Ther* 2003;14:1079–1088.

Foster K, Foster H, Dickson JG. Gene therapy progress and prospects: Duchenne muscular dystrophy. *Gene Ther* 2006;13:1677–1685.

Friedmann T. *Gene Therapy: Fact and Fiction in Biology's New Approaches to Disease.* Cold Spring Harbor, NY: Cold Spring Harbor Laboratory; 1994.

Friedmann T, Roblin R. Gene therapy for human genetic disease? *Science* 1972;175:949–955.

Giacca M, Baker AH. Heartening results: The CUPID gene therapy trial for heart failure. *Mol Ther* 2011;19:1181–1182.

Giacca M, Zacchigna S. VEGF gene therapy: Therapeutic angiogenesis in the clinic and beyond. *Gene Ther* 2012;19: 622–629.

Glover J. *Choosing Children: Genes, Disability, and Design.* Oxford: Clarendon Press; 2006.

Goins WF, Cohen JB, Glorioso JC. Gene therapy for the treatment of chronic peripheral nervous system pain. *Neurobiol Dis* 2012;48:255–270.

Gore ME. Adverse effects of gene therapy: Gene therapy can cause leukaemia: No shock, mild horror but a probe. *Gene Ther* 2003;10:4–4.

Gould DJ, Favorov P. Vectors for the treatment of autoimmune disease. *Gene Ther* 2003;10:912–927.

Grimm D, Kay MA. From virus evolution to vector revolution: Use of naturally occurring serotypes of adeno-associated virus (AAV) as novel vectors for human gene therapy. *Curr Gene Ther* 2003;3:281–304.

Harris J, William B. Enhancing Evolution: The Ethical Case for Making Better People. Princeton, NJ: Princeton University Press; 2007.

Hawley RG. Therapeutic potential of retroviral vectors. *Transfusion Sci* 1996;17:7–14.

Ishikawa K, Tilemann L, Ladage D, Aguero J, Leonardson L, Fish K, *et al.* Cardiac gene therapy in large animals: Bridge from bench to bedside. *Gene Ther* 2012;19:670–677.

Jouvenot Y, Perez E, Urnov FD, Beausejour C, Rock J, Lee Y-L, *et al.* Towards gene knock out therapy for AIDS/HIV: Targeted disruption of CCR5 using engineered zinc finger protein nucleases (ZFNs). *Mol Ther* 2005;11:S376–S376.

Joyce C. New moves to control the splice of life. *New Sci* 1982; 96:486.

Kalos M, Levine BL, Porter DL, Katz S, Grupp SA, Bagg A, *et al.* T cells with chimeric antigen receptors have potent antitumor effects and can establish memory in patients with advanced leukemia. *Sci Transl Med* 2011;3:95ra73.

Komaromy AM, Alexander JJ, Rowlan JS, Garcia MM, Chiodo VA, Kaya A, *et al.* Gene therapy rescues cone function in congenital achromatopsia. *Hum Mol Genet* 2010;19:2581–2593.

Kowalczyk DW, Wysocki PJ, Mackiewicz A. Cancer immunotherapy using cells modified with cytokine genes. *Acta Biochim Pol* 2003;50:613–624.

Krieg AM. CpG motifs in bacterial DNA and their immune effects. *Annu Rev Immunol* 2002;20:709–760.

Lai Y, Yue Y, Liu M, Ghosh A, Engelhardt JF, Chamberlain JS, *et al.* Efficient *in vivo* gene expression by trans-splicing adeno-associated viral vectors. *Nat Biotechnol* 2005;23:1435–1439.

Laino C. Racing toward gene therapy: After years of dashed hopes, researchers are poised to start clinical trials for muscular dystrophy. *Neurology Now* 2006;2:22–30.

Lamartina S, Roscilli G, Rinaudo D, Delmastro P, Toniatti C. Lipofection of purified adeno-associated virus Rep68 protein: Toward a chromosome-targeting nonviral particle. J *Virol* 1998;72:7653–7658.

Li RA. Gene and cell-based bio-artificial pacemaker: What basic and translational lessons have we learned? *Gene Ther* 2012;19:588–595.

Lin CR, Yang LC, Lee TH, Lee CT, Huang HT, Sun WZ, *et al.* Electroporation-mediated pain-killer gene therapy for mononeuropathic rats. *Gene Ther* 2002;9:1247–1253.

Macer DRJ; Eubios Ethics Institute. Shaping Genes: Ethics, Law and Science of Using New Genetic Technology in Medicine and Agriculture. Christchurch: Eubios Ethics Institute; 1990.

Mancuso K, Hauswirth WW, Li Q, Connor TB, Kuchenbecker JA, Mauck MC, *et al.* Gene therapy for red-green colour blindness in adult primates. *Nature* 2009;461:784–787.

McCaffrey AP, Meuse L, Pham TT, Conklin DS, Hannon GJ, Kay MA. RNA interference in adult mice. *Nature* 2002;418:38–39.

McCaffrey AP, Nakai H, Pandey K, Huang Z, Salazar FH, Xu H, *et al.* Inhibition of hepatitis B virus in mice by RNA interference. *Nat Biotechnol* 2003;21:639–644.

Medina MF, Kobinger GP, Rux J, Gasmi M, Looney DJ, Bates P, *et al.* Lentiviral vectors pseudotyped with minimal filovirus envelopes increased gene transfer in murine lung. *Mol Ther* 2003;8:777–789.

Mesko B, Zahuczky G, Nagy L. The triad of success in personalised medicine: pharmacogenomics, biotechnology and regulatory issues from a Central European perspective. *Nat Biotechnol* 2012;29:741–750.

Mingozzi F, High KA. Immune responses to AAV in clinical trials. *Curr Gene Ther* 2007;7:316–324.

Mitchell RS, Beitzel BF, Schroder AR, Shinn P, Chen H, Berry CC, *et al.* Retroviral DNA integration: ASLV, HIV and MLV show distinct target site preferences. *PLoS Biol* 2004;2:E234.

Monahan PE, Samulski RJ. Adeno-associated virus vectors for gene therapy: More pros than cons? *Mol Med Today* 2000;6:433–440.

Mor G, Singla M, Steinberg AD, Hoffman SL, Okuda K, Klinman D. Do DNA vaccines induce autoimmune disease? *Hum Gene Ther* 1997;8 293–300.

Morgan RA, Dudley ME, Wunderlich JR, Hughes MS, Yang JC, Sherry RM, *et al.* Cancer regression in patients after transfer of genetically engineered lymphocytes. *Science* 2006;314:126–129.

Moss FB, Rodman D, Spencer LT, Aitken ML, Zeitlin PL, Waltz D, *et al.* Repeated adeno-associated virus serotype 2 aerosol-mediated cystic fibrosis transmembrane regulator gene transfer to the lungs of patients with cystic fibrosis: A multicenter, double-blind, placebo-controlled trial. *Chest* 2004;125:509–521.

Naldini L. Lentiviruses as gene transfer agents for delivery to non-dividing cells. *Curr Opin Biotechnol* 1998;9:457–463.

Nam CH, Rabbitts TH. The role of LMO2 in development and in T cell leukemia after chromosomal translocation or retroviral insertion. *Mol Ther* 2006;13:15–25.

Ohashi S, Kubo T, Kishida T, Ikeda T, Takahashi K, Arai Y, et al. Successful genetic transduction *in vivo* into synovium by means of electroporation. *Biochem Biophys Res Commun* 2002;293:1530–1535.

Olsson AK, Dimberg A, Kreuger J, Claesson-Welsh L. VEGF receptor signalling - in control of vascular function. *Nat Rev Mol Cell Biol* 2006;7:359–371.

Ott MG Schmidt M, Schwarzwaelder K, Stein S, Siler U, Koehl U, et al. Correction of X-linked chronic granulomatous disease by gene therapy, augmented by insertional activation of MDS1-EVI1, PRDM16 or SETBP1. *Nat Med* 2006;12:401–409.

Pardridge WM. Molecular Trojan horses for blood-brain barrier drug delivery. *Curr Opin Pharmacol* 2006;6:494–500.

Pike-Overzet K, de Ridder D, Weerkamp F, Baert MRM, Verstegen MM, Brugman MH, *et al.* Gene therapy: Is IL2RG oncogenic in T-cell development? *Nature* 2006;443:E5–E5.

Porter DL, Levine BL, Kalos M, Bagg A, June CH. Chimeric antigen receptor-modified T cells in chronic lymphoid leukemia. *N Engl J Med* 2011;365:725–733.

Potente M, Gerhardt H, Carmeliet P. Basic and therapeutic aspects of angiogenesis. *Cell* 2011;146:873–887.

Pringle IA, Davies LA, McLachlan G, Collie DDS, Gill DR, Hyde SC. Duration of Reporter gene expression from naked pDNA in the mouse lung following direct electroporation and development of wire electrodes for sheep lung electroporation studies. *Mol Ther* 2004;9:S56.

Rabbitts TH, Axelson H, Forster A, Grutz G, Lavenir I, Larson R, *et al.* Chromosomal translocations and leukaemia: A role for LMO2 in T cell acute leukaemia, in transcription and in erythropoiesis. *Leukemia* 1997;11:271–272.

Ralph GS, Radcliffe PA, Day DM, Carthy JM, Leroux MA, Lee DC, *et al.* Silencing mutant SOD1 using RNAi protects against neurodegeneration and extends survival in an ALS model. *Nat Med* 2005;11:429–433.

Raoul C, Abbas-Terki T, Bensadoun JC, Guillot S, Haase G, Szulc J, *et al.* Lentiviral-mediated silencing of SOD1 through RNA interference retards disease onset and progression in a mouse model of ALS. *Nat Med* 2005;11:423–428.

Rasmussen H, Rasmussen C, Lempicki M, Durham R, Brough D, King CR, *et al.* TN Ferade biologic: Preclinical toxicology of a novel adenovector with a radiation-inducible promoter, carrying the human tumor necrosis factor alpha gene. *Cancer Gene Ther* 2002;9:951–957.

Resnik DB. *Bioethics of Gene Therapy.* eLS, John Wiley & Sons; 2001.

Rodino-Klapac LR, Janssen PM, Montgomery CL, Coley BD, Chicoine LG, Clark KR, *et al.* A translational approach for limb vascular delivery of the micro-dystrophin gene without high volume or high pressure for treatment of Duchenne muscular dystrophy. *J Transl Med* 2007;5:45.

Romero NB, Braun S, Benveniste O, Leturcq F, Hogrel JY, Morris GE, *et al.* Phase I study of dystrophin plasmid-based gene therapy in Duchenne/Becker muscular dystrophy. *Hum Gene Ther* 2004;15:1065–1076.

Rosenberg SA, Packard BS, Aebersold PM, Solomon , Topalian SL, Toy ST, *et al.* Use of tumor-infiltrating lymphocytes and interleukin-2 in the immunotherapy of patients with metastatic melanoma. A preliminary report. *N Engl J Med* 1988;319:1676–1680.

Samulski RJ, Chang LS, Shenk T. Helper-free stocks of recombinant adeno-associated viruses: Normal integration does not require viral gene expression. *J Virol* 1989;63:3822–3828.

Seyhan AA. RNAi: A potential new class of therapeutic for human genetic disease. *Hum Genet* 2011;130:583–605.

Shayakhmetov DM, Eberly AM, Li ZY, Lieber A. Deletion of penton RGD motifs affects the efficiency of both the internalization and the endosome escape of viral particles containing adenovirus serotype 5 or 35 fiber knobs. *J Virol* 2005;79:1053–1061.

Sheridan C. Gene therapy finds its niche. *Nat Biotechnol* 2011;29:121–128.

Sinn PL, Shah AJ, Donovan MD, McCray PB. Viscoelastic gel formulations enhance airway epithelial gene transfer with viral vectors. *Am J Respir Cell Mol Biol* 2005;32:404–410.

Summerford C, Samulski RJ. Membrane-associated heparan sulfate proteoglycan is a receptor for adeno-associated virus type 2 virions. *J Virol* 1998;72:1438–1445.

Surosky RT, Urabe M, Godwin SG, McQuiston SA, Kurtzman GJ, Ozawa K, *et al.* Adeno-associated virus Rep proteins target DNA sequences to a unique locus in the human genome. *J Virol* 1997;71:7951–7959.

't Hoen PA, de Meijer EJ, Boer JM, Vossen RH, Turk R, Maatman RG, *et al.* Generation and characterization of transgenic mice with the full-length human DMD gene. *J Biol Chem* 2008;283: 5899–5907.

Tani J, Faustine, Sufian JT. Updates on current advances in gene therapy. *West Indian Med J* 2011;60:188–194.

Tilemann L, Ishikawa K, Weber T, Hajjar RJ. Gene therapy for heart failure. *Circ Res* 2012;110:777–793.

Tosi MF, van Heeckeren A, Ferkol TW, Askew D, Harding CV, Kaplan JM. Effect of *Pseudomonas*-induced chronic lung inflammation on specific cytotoxic T-cell responses to adenoviral vectors in mice. *Gene Ther* 2004;11:1427–1433.

Tsurumi C, Shimizu Y, Saeki M, Kato S, Demartino GN, Slaughter CA, *et al.* cDNA cloning and functional analysis of the p97 subunit of the 26S proteasome, a polypeptide identical to the type-1 tumor-necrosis-factor-receptor-associated protein-2/55.11. *Eur J Biochem* 1996;239:912–921.

Verma IM, Weitzman MD. Gene therapy: Twenty-first century medicine. *Annu Rev Biochem* 2005;74:711–738.

Walther W, Stein U. Viral vectors for gene transfer: A review of their use in the treatment of human diseases. *Drugs* 2000;60:249–271.

Wang Z, Storb R, Lee D, Kushmerick MJ, Chu B, Berger C, *et al.* Immune responses to AAV in canine muscle monitored by cellular assays and noninvasive imaging. *Mol Ther* 2010;18:617–624.

Welsh N. Prospects for gene therapy of diabetes mellitus. *Gene Ther* 2000;7:181–182.

Wilson JF. Murine gene therapy corrects symptoms of sickle cell disease. *The Scientist*; 18 March 2002.

Wu X, Li Y, Crise B, Burgess SM. Transcription start regions in the human genome are favored targets for MLV integration. *Science* 2003;300:1749–1751.

Xenariou S, Liang H, Griesenbach U, Farley R, Somerton L, Singh C, *et al.* Sonoporation increases non-viral gene transfer to the murine lung. *Mol Ther* 2005;11:S139–S139.

Xia H, Mao Q, Eliason SL, Harper SQ, Martins IH, Orr HT, *et al.* RNAi suppresses polyglutamine-induced neurodegeneration in a model of spinocerebellar ataxia. *Nat Med* 2004;10:816–820.

Xiao W, Chow BL, Rathgeber L. The repair of DNA methylation damage in *Saccharomyces cerevisiae*. *Curr Genet* 1996;30:461–468.

Yang ZJ, Zhang YR, Chen B, Zhang SL, Jia EZ, Wang LS, *et al.* Phase I clinical trial on intracoronary administration of Ad-hHGF treating severe coronary artery disease. *Mol Biol Rep* 2009;36:1323–1329.

Yellon DM, Hausenloy DJ. Myocardial reperfusion injury. *N Engl J Med* 2007;357:1121–1135.

Yovandich J, O'Malley B Jr, Sikes M, Ledley FD. Gene transfer to synovial cells by intra-articular administration of plasmid DNA. *Hum Gene Ther* 1995;6:603–610.

Zhang L, Nolan E, Kreitschitz S, Rabussay DP. Enhanced delivery of naked DNA to the skin by non-invasive *in vivo* electropora tion. *Biochim Biophys Acta* 2002;1572:1–9.

Colour plates

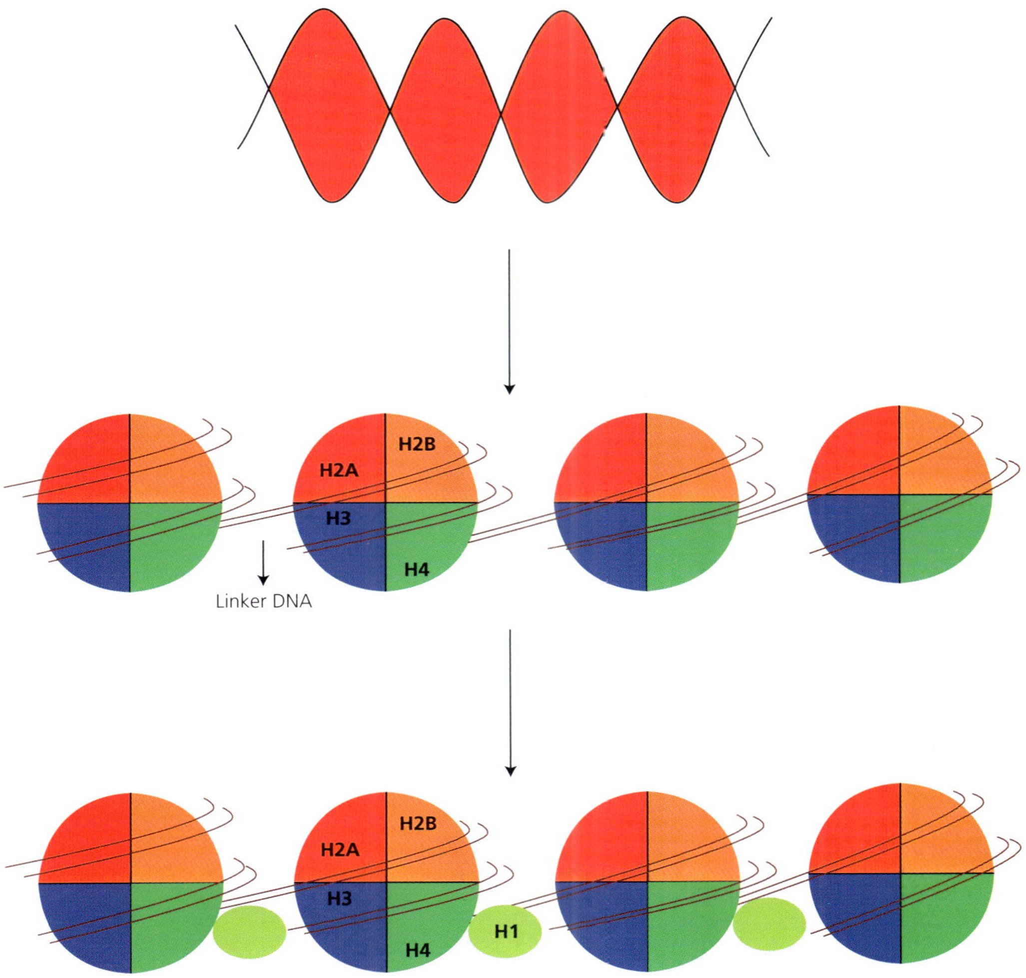

Plate 1 Wrapping of the DNA around the nucleosome to create the bead-on-the-string structure.

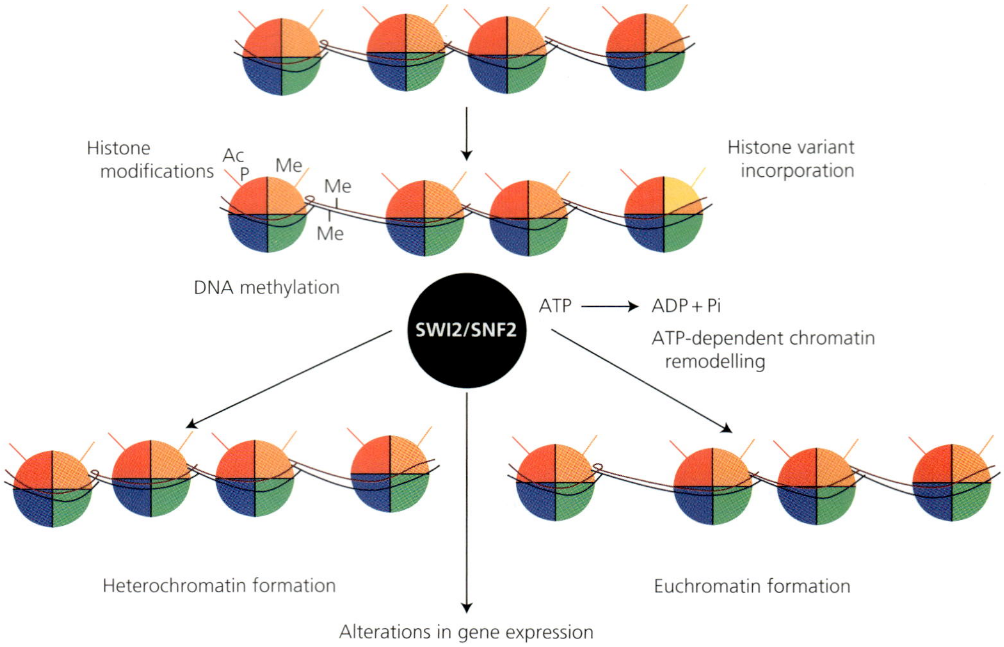

Plate 2 Chromatin remodelling involves the concerted action of histone modifications, histone variant incorporation, DNA methylation and ATP-dependent chromatin remodelling, leading to alterations in gene expression.

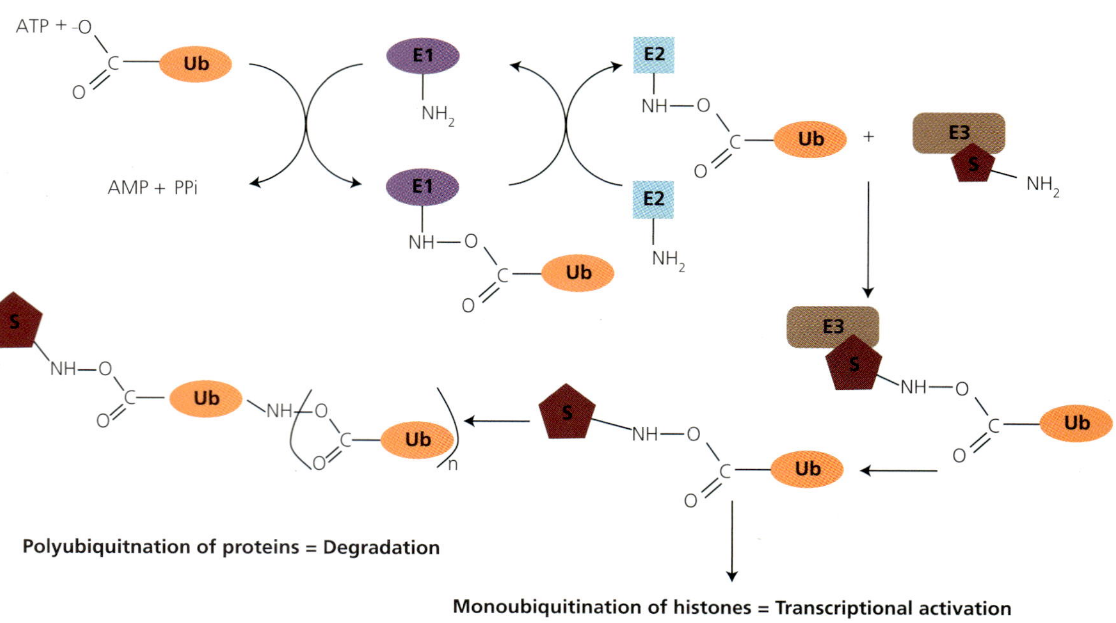

Plate 3 Histone ubiquitination mediates transcription activation.

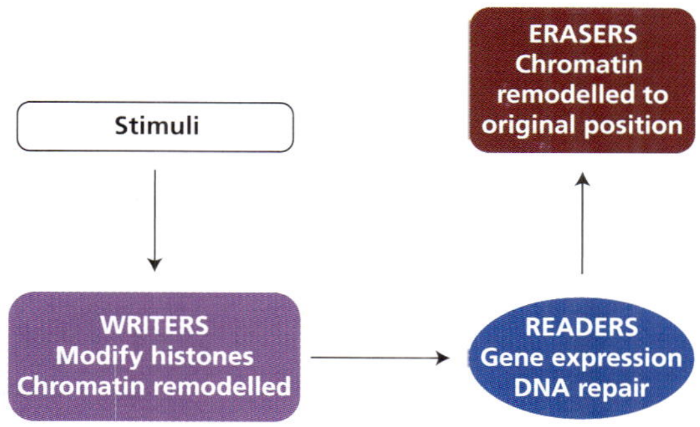

Plate 4 Histone code involves three components: writers, readers, and erasers.

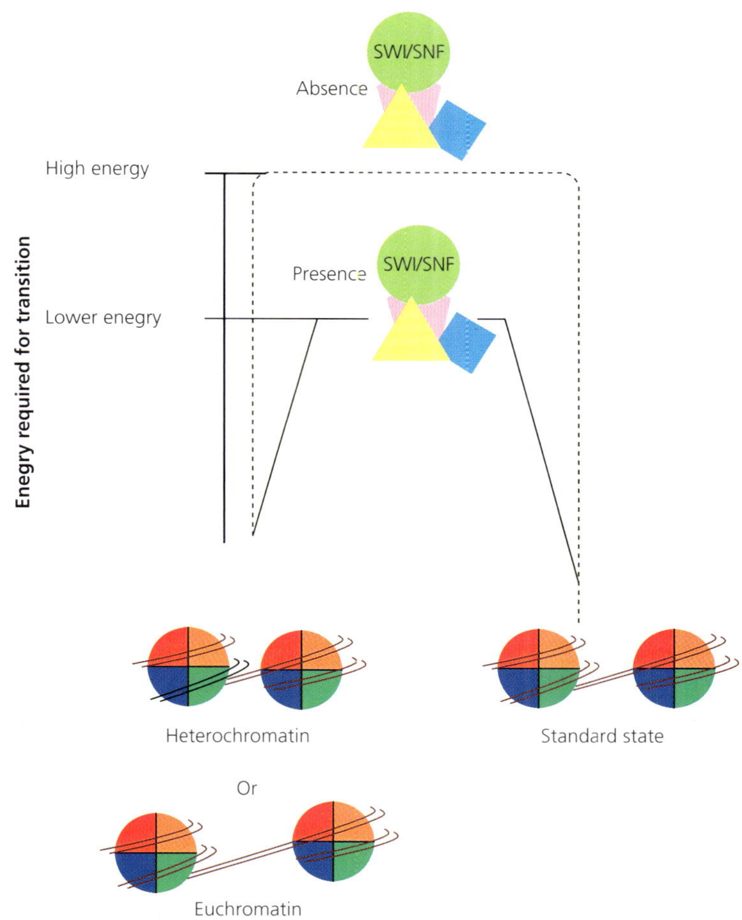

Plate 5 ATP-dependent chromatin remodellers catalyse the formation of both the euchromatin and the heterochromatin state.

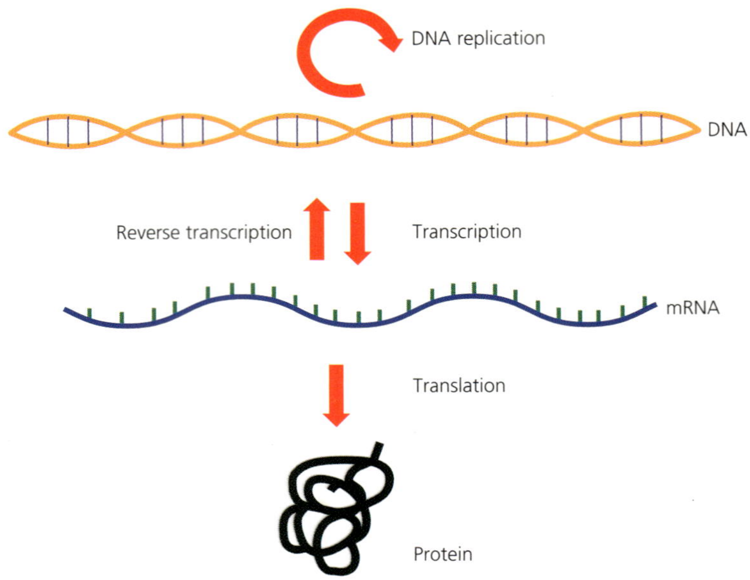

Plate 6 Central dogma of molecular biology: Schematic representation. The genetic information flows from DNA to RNA to protein, with the exception of RNA viruses. In such viruses that contain RNA as the genetic material, RNA is reverse transcribed to DNA and then it follows the general scheme.

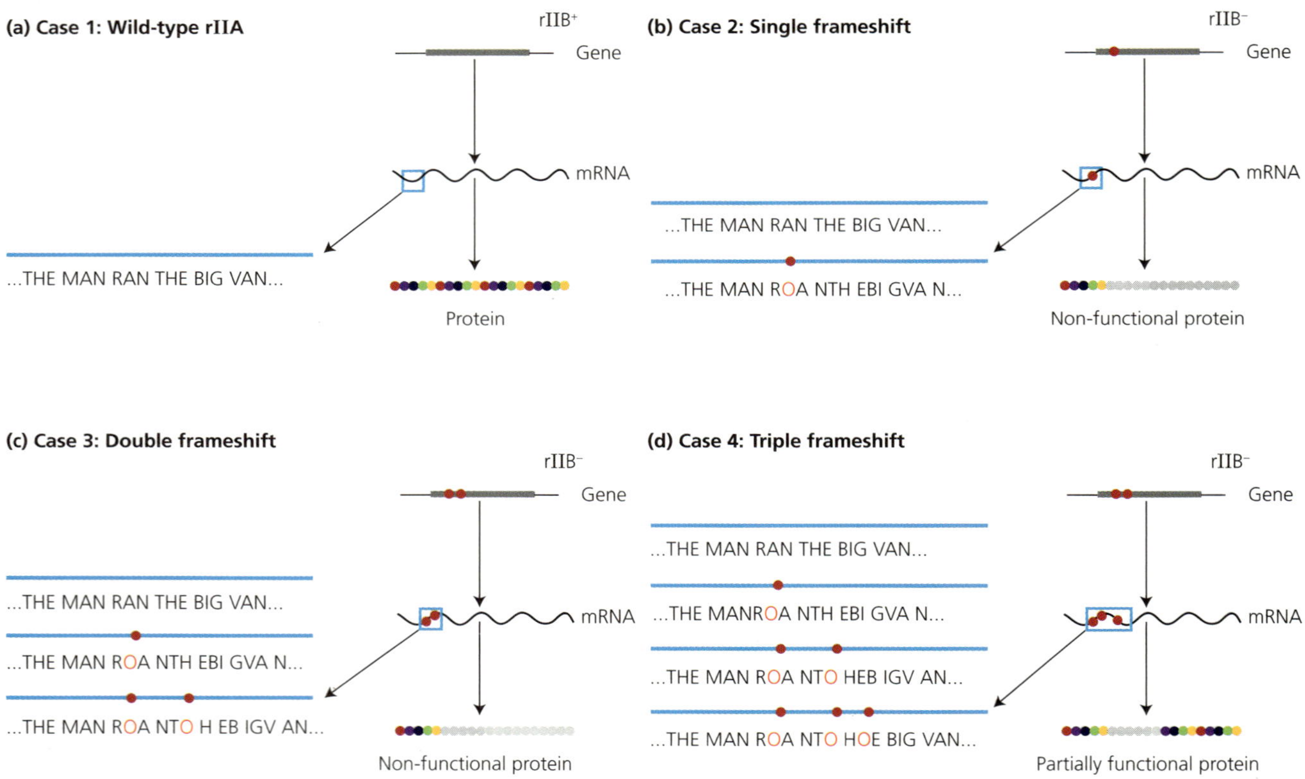

Plate 7 Graphical representation of proflavine-induced genetic mutation experiments in T4 bacteriophage that led to the concept of reading frame and triplet codons. (a) Case 1: Wild-type T4 bacteriophage with rII B gene. In this case, the functional protein is produced from the mRNA. (For better understanding, the message in the mRNA is denoted by the phrase 'THE MAN RAN THE BIG VAN', with each word representing a codon.). (b) Case 2: T4 bacteriophage with rII B mutant (denoted as rII B–) with a single frameshift mutation leads to the message being misread as 'THE MAN ROA NTH EBI GVA N' and thus a non-functional protein is produced. (c) Case 3: T4 bacteriophage with rII B mutant with a double frameshift mutation. Second mutation in the above rII B mutant results in the message being misread as 'THE MAN ROA NTO HEB IGV AN'. (d) Case 4: T4 bacteriophage with rII B mutant with a triple frameshift mutation. The third mutation in the above rII B mutant results in the message being partially misread as 'THE MAN ROA NTO HEO BIG VAN'. Thus, a partially functional protein is produced.

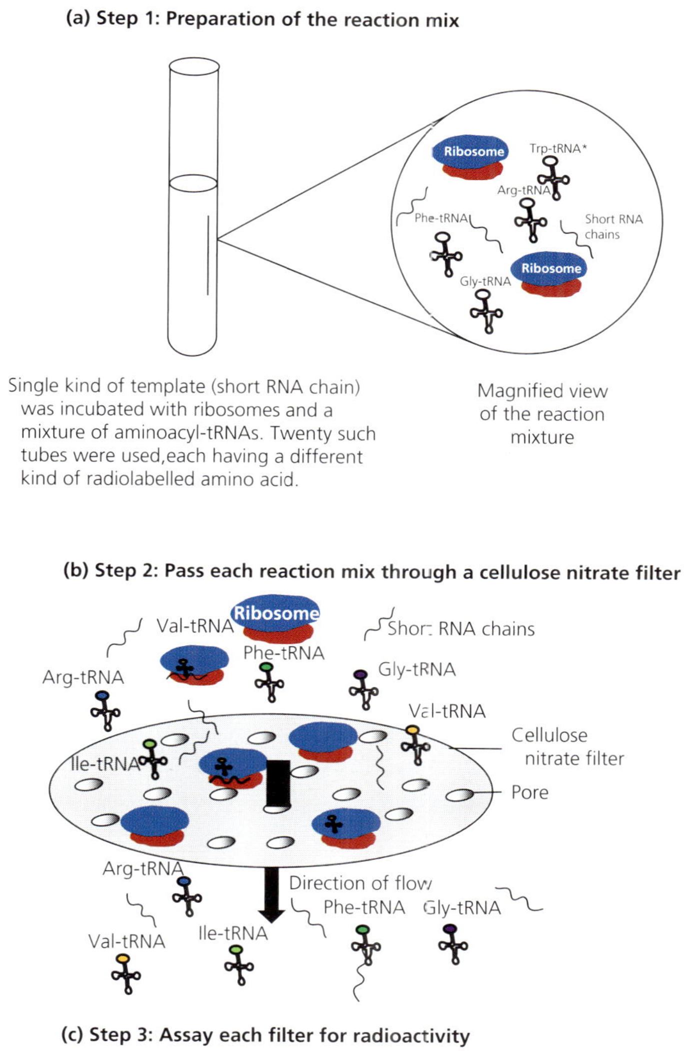

Plate 8 Graphical representation of the filter-binding assay done by Nirenberg and Leder (1964). (a) Step 1: Preparation of the reaction mix – short RNA chains were incubated with ribosomes, a mixture of aminoacyl-tRNAs and a single-kind template. Twenty such tubes were used, each having a different kind of radiolabelled amino acid. (b) Step 2: Pass each reaction mix through a cellulose nitrate filter.
(c) Step 3: Assay each filter for radioactivity.

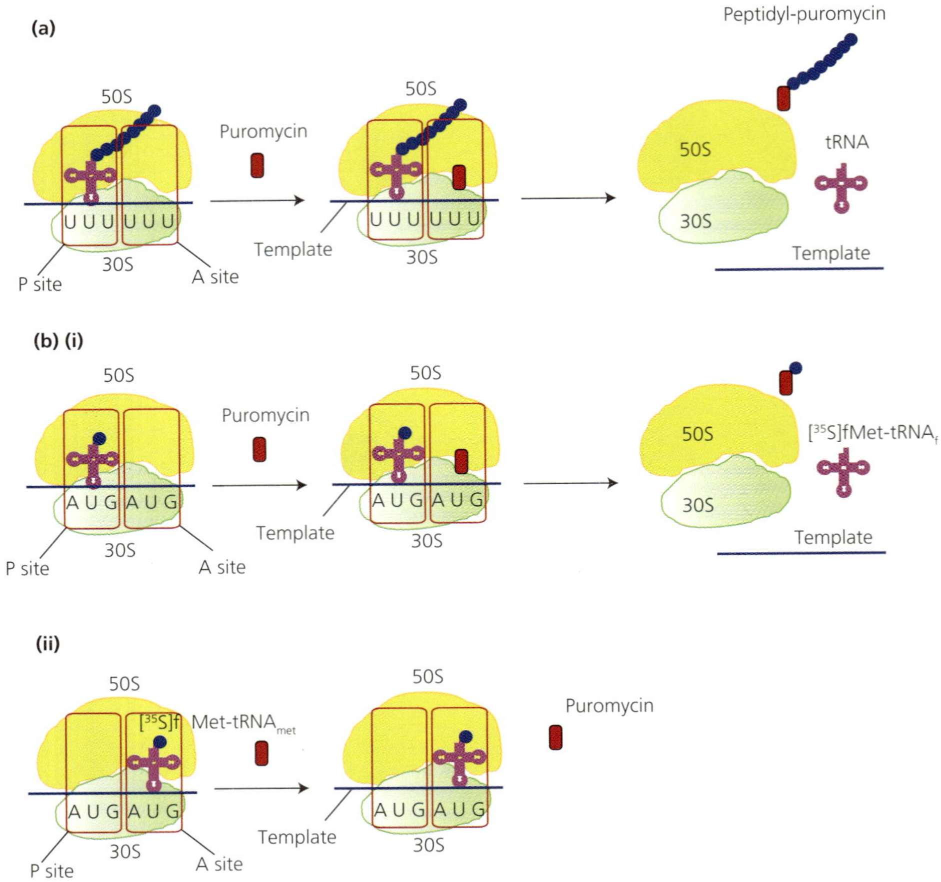

Plate 9 (a) Puromycin release assay. A synthetic template such as poly(U) is incubated with ribosomes, tRNA and other buffer components. Such ribosomes are isolated and incubated with puromycin. Puromycin attaches to the growing peptide chain and causes the release of the peptide from the ribosome. This peptidyl puromycin is acid soluble; hence, it can be isolated and characterized. (b) Puromycin release assay as employed by Bretscher and Marcker. They utilized an AUG triplet template in the above reaction along with (i) [35S]fMet-tRNAf or (ii) [35S] Met-tRNAmet. In the first reaction, [35S]fMet-tRNAf occupies the P site; hence, puromycin is able to enter the A site and terminate the peptide chain. In the second case, [35S] Met-tRNAmet occupies the A site; hence, puromycin is not able to enter. As a result, chain termination does not occur.

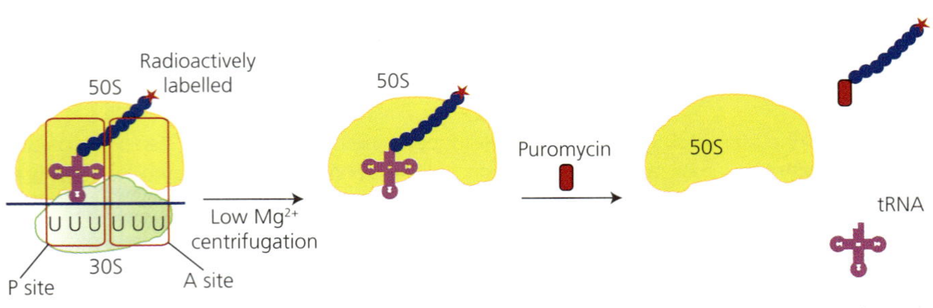

Plate 10 Peptidyl transferase assay. A synthetic poly(U) template is incubated with ribosomes, tRNA and other buffer components. Such ribosomes are isolated and centrifuged under low Mg^{2+} concentration. Pure 50S subunit complexed with poly(phe)-tRNA in the P site is isolated by repeatedly washing the fraction with salt solutions to remove traces of initiation or elongation factors in it. Such primed 50S subunit was utilized for puromycin release assay [see Fig. 14.8(a)].

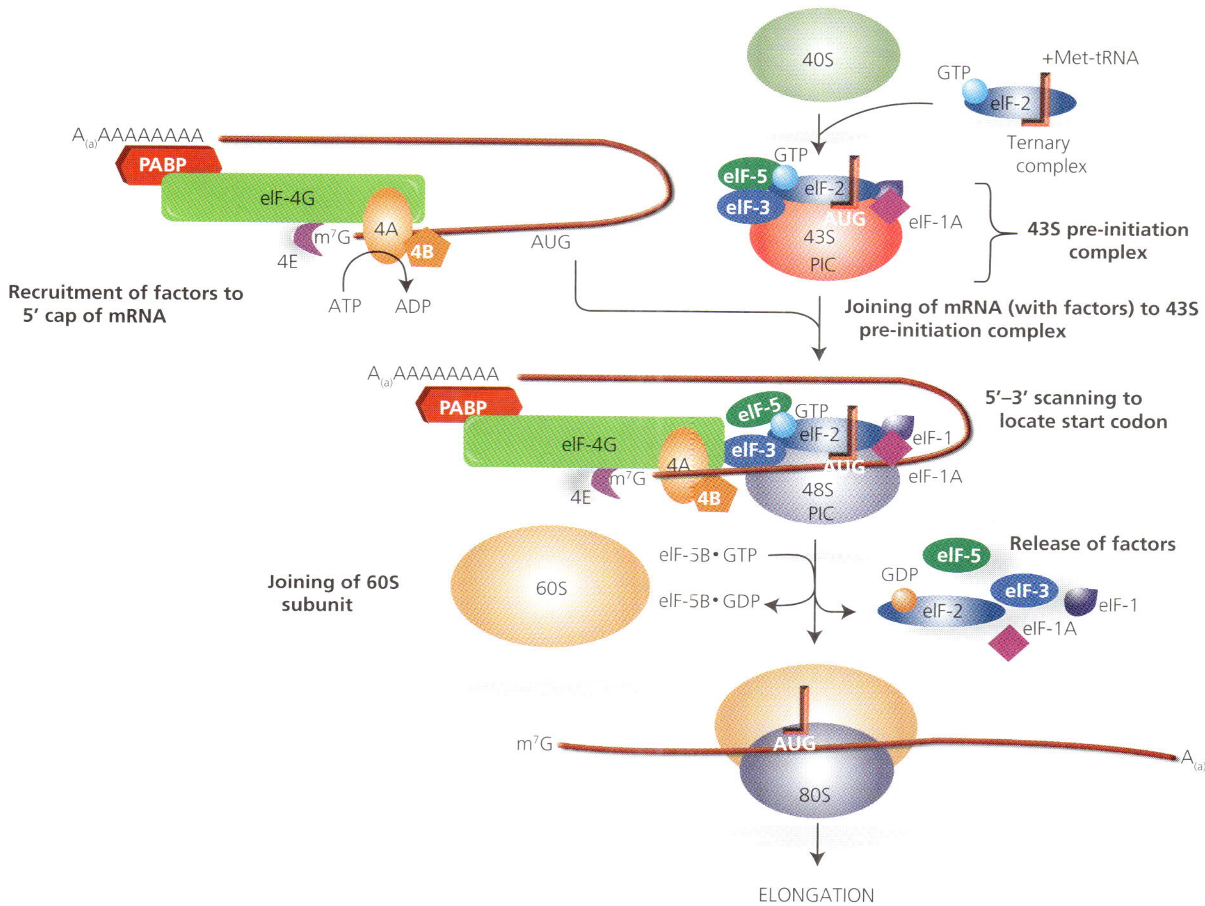

Plate 11 Summary of initiation of eukaryotic protein synthesis. There are four major steps: (1) Formation of the pre-initiation complex (PIC) containing 40S ribosome, eIF2.GTP.Met-tRNA and other initiation factors (eIF1A, -3 and -5). (2) Preparation of mRNA with several initiation factors, including (eIF-4E) (cap-binding protein), -4A, -4B, -4G and possibly PABP (polyA binding protein), which links 5′- and 3′-ends of mRNA. (3) Interaction of PIC with such mRNA, followed by scanning for AUG by the 40S ribosome. (4) Joining of the 60S ribosome to form the 80S initiation complex along with eIF5.

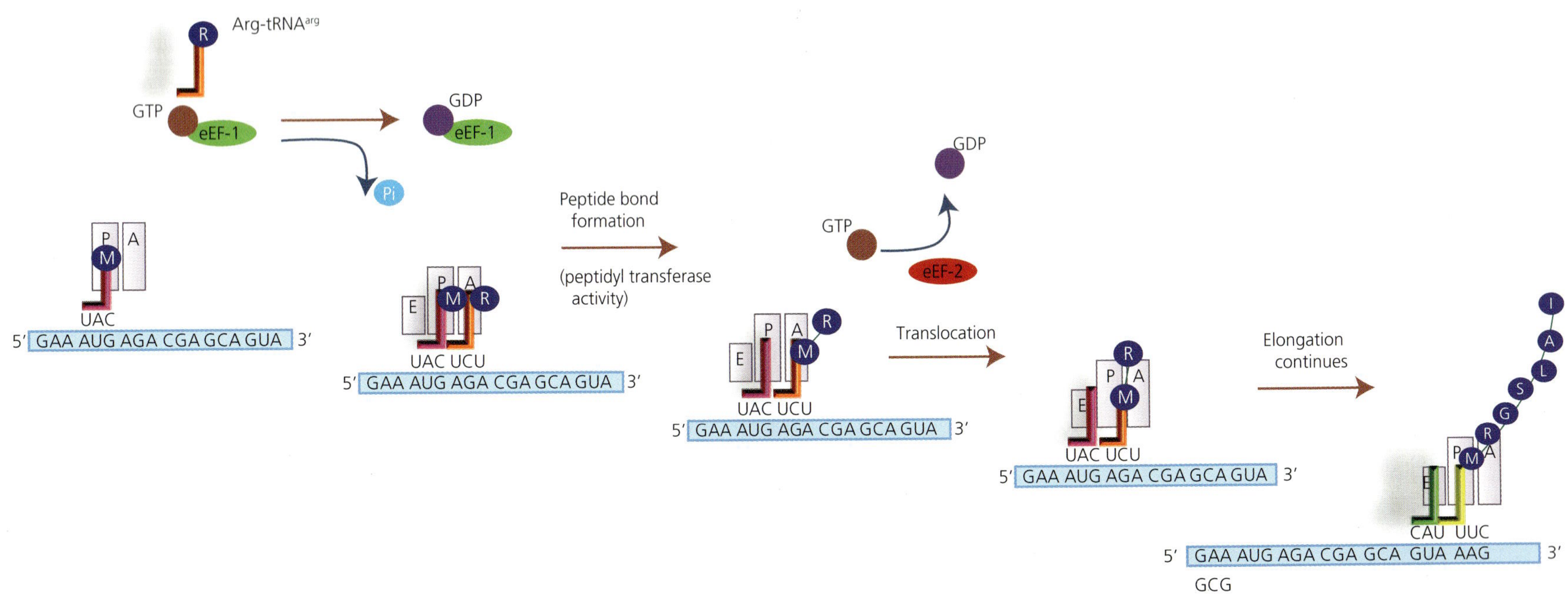

Plate 12 Elongation during translation. First, the initiator tRNA binds at the P site in the ribosome. In the second step, the next aminoacyl tRNA binds at the A site, followed by peptide bond formation. Translocation of the ribosomes shifts the vacant tRNA to the E site and the aminoayl tRNA to the P site. Then, the next aminoacyl-tRNA comes at the A site and the process is repeated till the termination codon is reached.

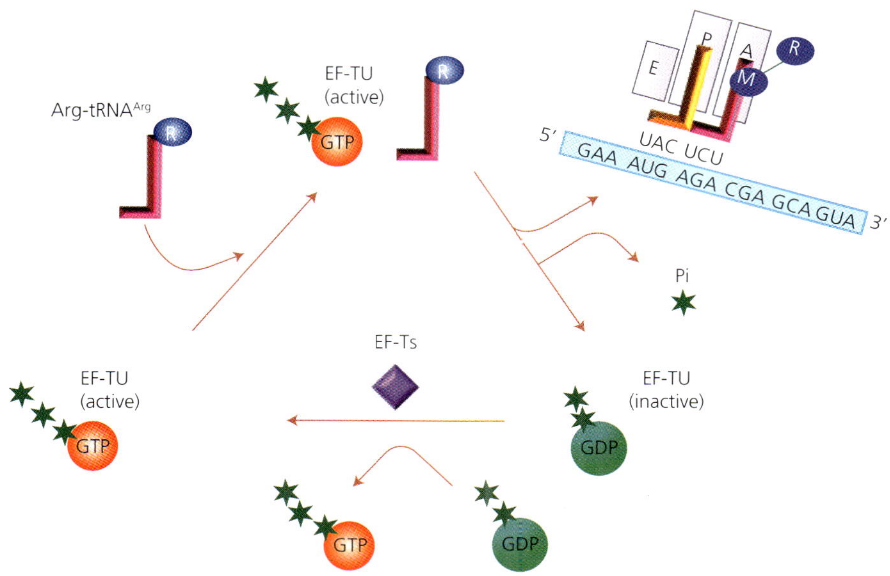

Plate 13 Recycling of EF-Tu/GTP during elongation of translation. During elongation, EF-Tu-GTP binds to the aminoacyl-tRNA. On binding of the correct aminoacyl-tRNA to the codon, EF-Tu-GTP gets hydrolysed and EF-Tu-GDP is released. The inactive EF-Tu-GDP complex is converted to EF-Tu-GTP with the help of EF-Ts, making the EF-Tu-GTP ready for binding to the next aminoacyl-tRNA.

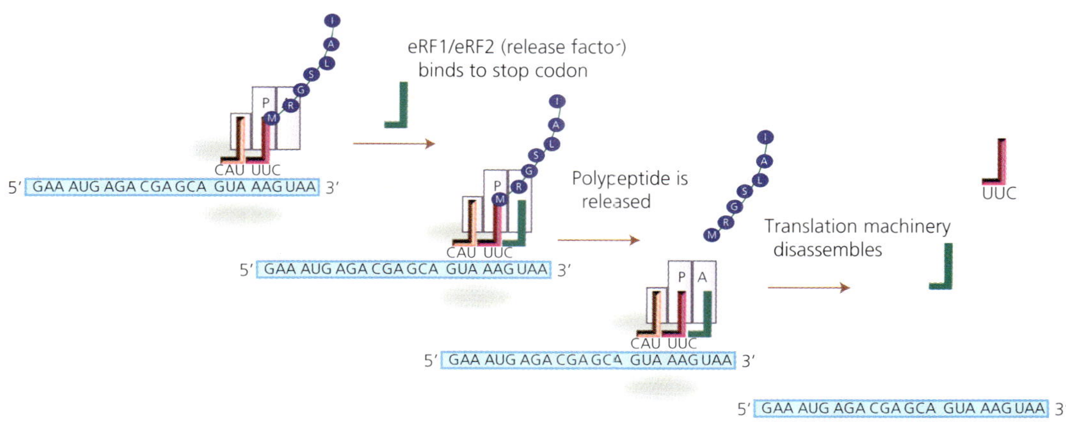

Plate 14 Termination of translation. On reaching the stop codon, the release factor binds to it (UAA, UGA and UAG). After the binding of the release factor, the translational machinery disassembles as the ribosome dissociates and the polypeptide and the tRNA are released.

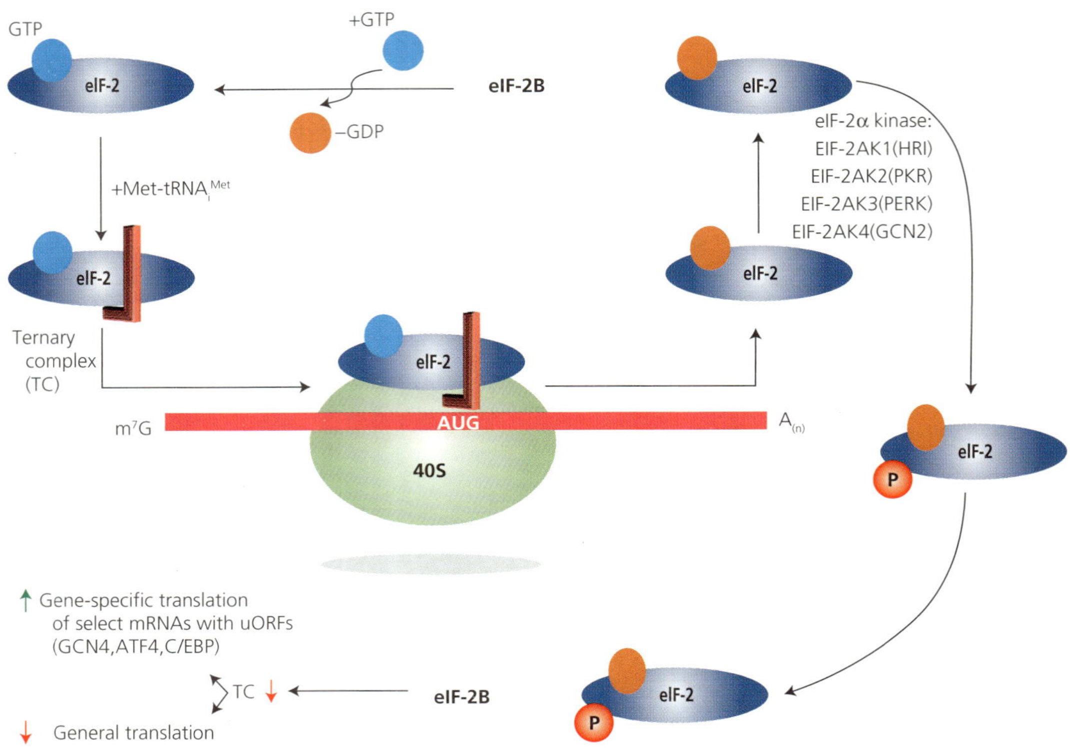

Plate 15 Schematic diagram showing the role of eIF-2α kinases in modulating initiation of protein synthesis through phosphorylation of eIF-2α. eIF-2 is rate limiting in a cell and thus is required to be recycled for continuation of initiation of protein synthesis. eIF-2 is liberated after one round of initiation as a non-functional eIF2.GDP binary complex. This is acted upon by eIF-2B, a guanine nucleotide exchange factor which converts eIF-2-GDP into a functional complex eIF-2-GTP. eIF-2-GTP binds to Met $tRNA_i$ and forms a ternary complex eIF-2-GTP-Met $tRNA_i$, which is utilized for initiation. If a member of the family of eIF-2α kinases (HRI, PKR, PERK and GCN2) gets activated, it phosphorylates eIF-2α. Phosphorylated eIF-2α cannot be recycled by eIF-2B, resulting in the inhibition of initiation of protein synthesis. However, in contrast to this, mRNAs of some specific genes such as *GCN4*, *ATF4* and *C/EBP* with uORFs are favourably translated under this condition.

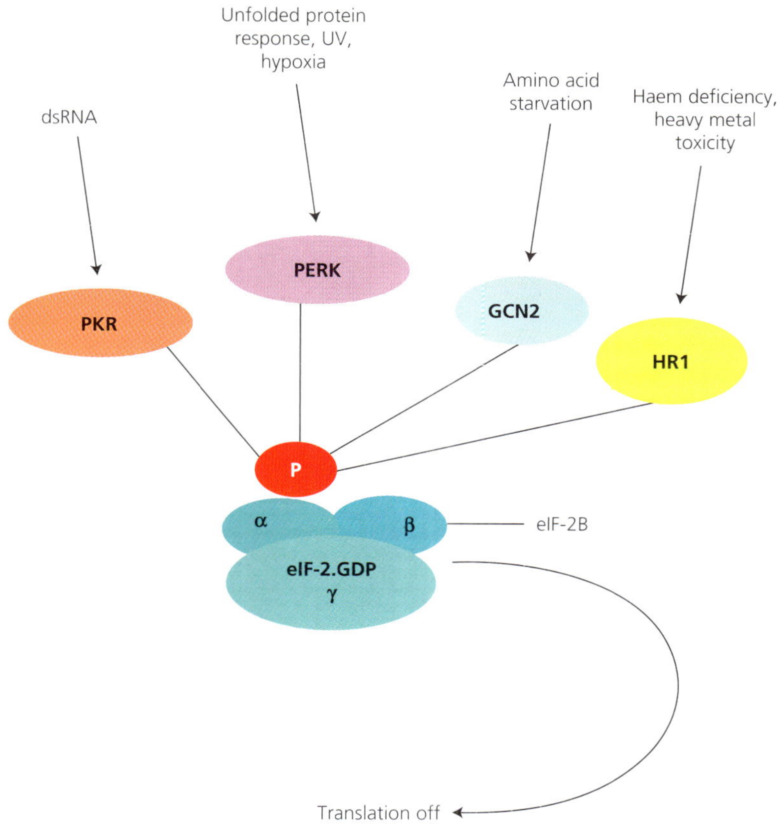

Plate 16 Differential regulation of eIF-2α kinases in mammals. Although each of these eIF-2α kinases phosphorylates the same Ser 51 residue of the substrate eIF-2α, each gets activated differently. The major modes of activation are indicated on the top of each kinase.

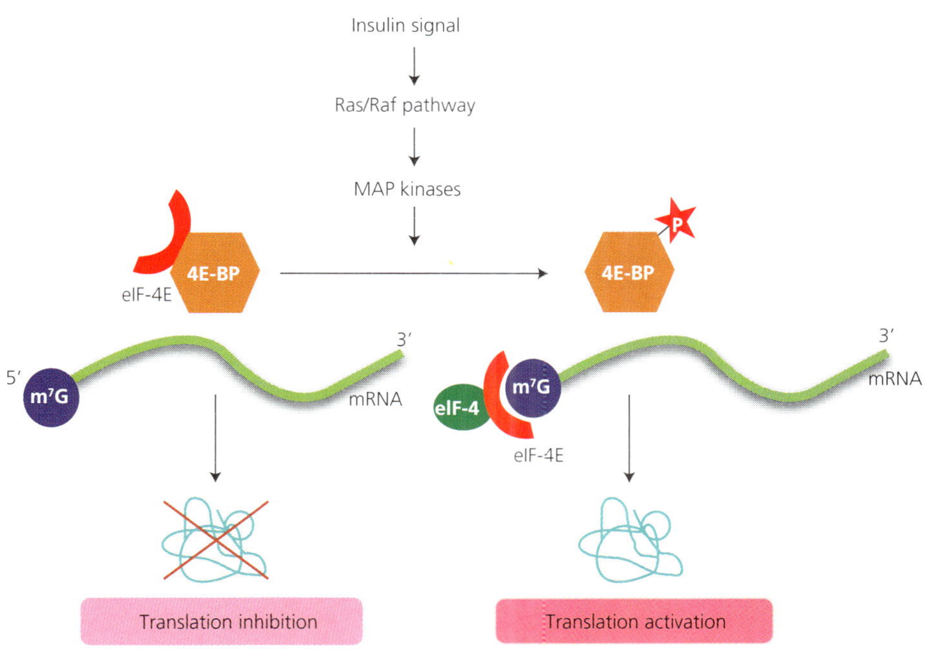

Plate 17 Schematic diagram showing stimulation of translation initiation (exposed to insulin) with the availability of increased quantity of eIF-4E. The cap-binding protein – eIF-4E – is crucial for cap-dependent initiation of translation. eIF-4E-BPs binds to eIF-4E and inhibits its cap-binding activity. During insulin signalling, MAP kinases phosphorylate eIF-4E-BPs. The phosphorylated eIF-4E-BPs fall off eIF-4E and thus more free eIF-4E is available for translation.

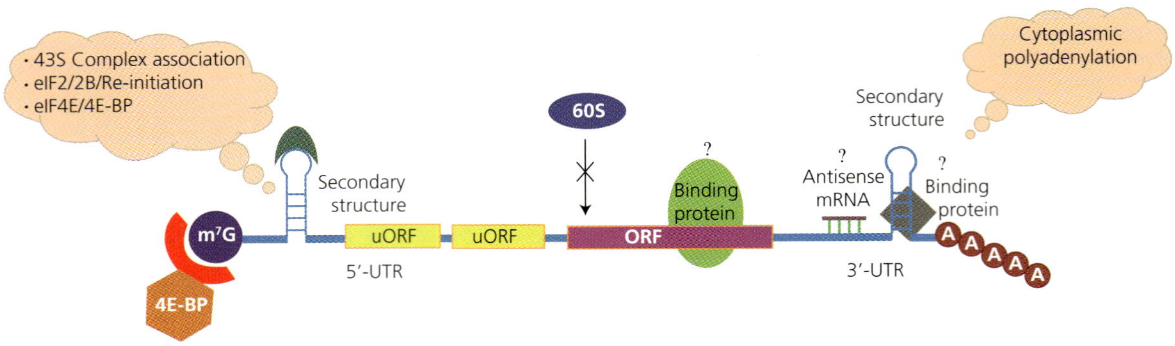

Plate 18 Summarized scheme of translation regulation in eukaryotic cells. Interaction of various *cis* elements present at the 5′- and 3′-UTR (untranslated regions) in the mRNA with specific *trans*-acting factors present in the cytoplasm determines the regulation of its translation. Some possible such *cis-trans* regulations that are still to be unravelled are indicated with notes of interrogation.

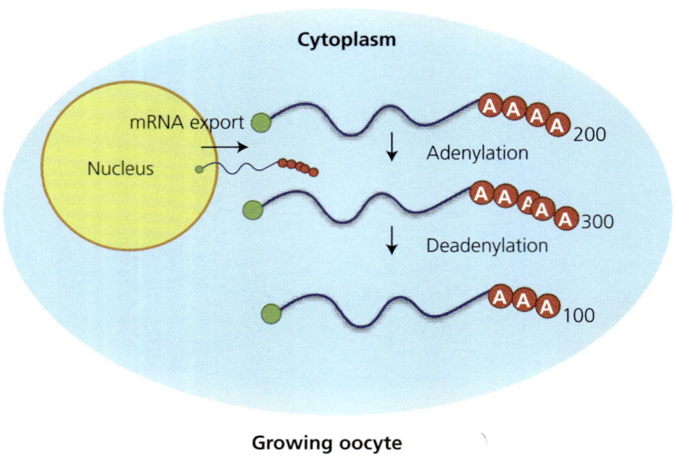

Plate 19 Translation regulation by changing the length of the poly(A) tail of mRNAs. In oocytes, the length of the poly(A) tail is altered by adenylation or deadenylation by cytoplasmic enzymes, resulting in the regulation of its translation.

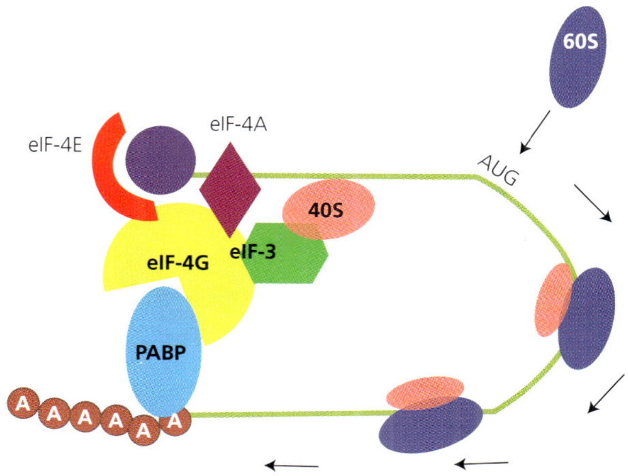

Plate 20 Translation regulation by eIF-4G and poly(A) binding protein (PABP) through circularization of mRNA. The circular form generated due to the interaction of eIF-4G (as a part of the cap-binding complex at the 5′-end of an mRNA) and PABP at the 3′-end is stabilized by proteins which are bound to the 5′- and 3′-end of the mRNA. The ribosome which dissociates from the 3′-end of mRNA can easily find the neighbouring ribosomal subunits for reassembling at the 5′-end. This further ensures translation of only full-length mRNAs.

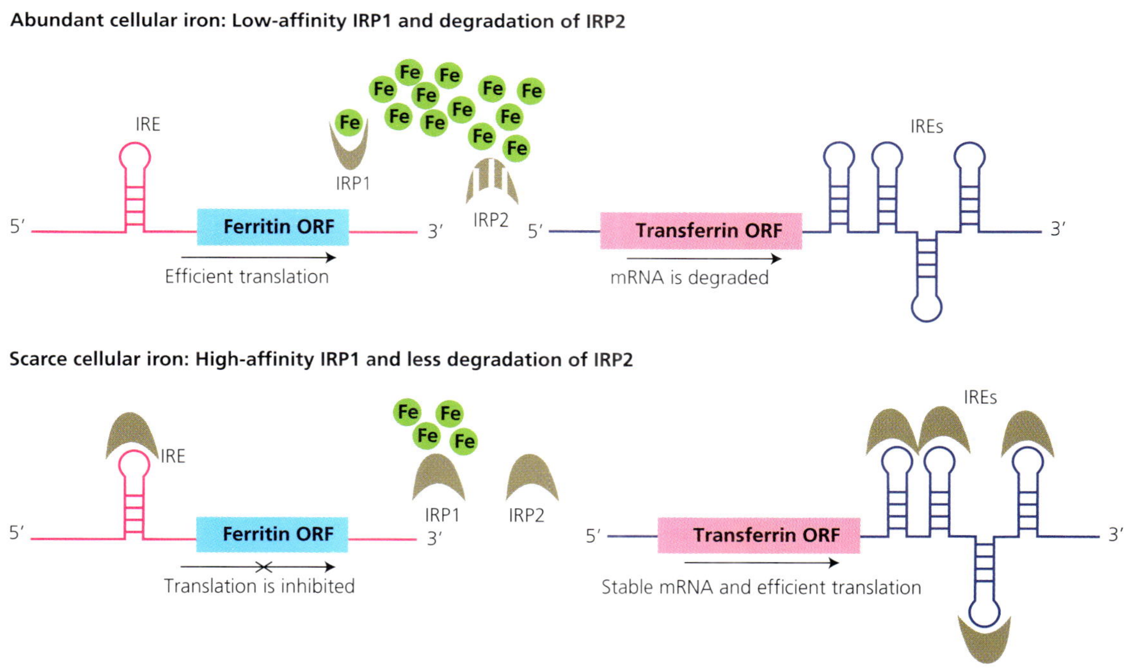

Plate 21 Translation regulation of ferritin and transferrin receptor mRNAs. 5′-UTR of ferritin mRNAs have IREs (iron-responsive elements). The IRP (iron regulatory protein) binds to IRE and inhibits translation of the mRNA. When iron is present in excess, iron binds to IRP; thus, IRP cannot bind to IRE and translation of ferritin mRNA takes place (a). The reverse is true in the case of iron-deficient condition (b). 3′-UTR of transferrin receptor mRNA also contains IREs. During iron deficiency, IRPs stabilize the mRNA by binding to it and thus facilitate its efficient translation (b). But when iron is present in excess, IRPs do not bind to IREs and the transferrin mRNA gets degraded (a).

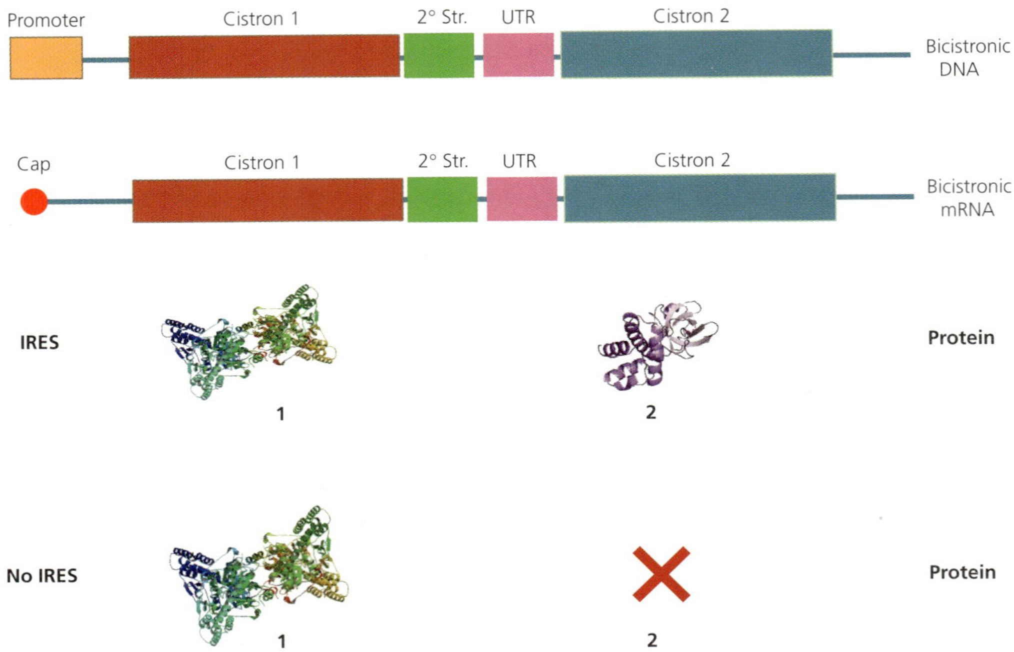

Plate 22 Schematic representation of bicistronic assay. A bicistronic construct is made in such a way that the transcribed mRNA from the first cistron is translated in a cap-dependent manner, while the mRNA of the second cistron is translated in a cap-independent manner. The 5′-UTR of the prospective IRES-containing mRNA (to be tested) is cloned next to a highly structured sequence in the UTR of the second cistron. If this prospective UTR contains an IRES, due to internal ribosome entry, it will be translated, as can be observed by expression of the specific reporter gene (second cistron).

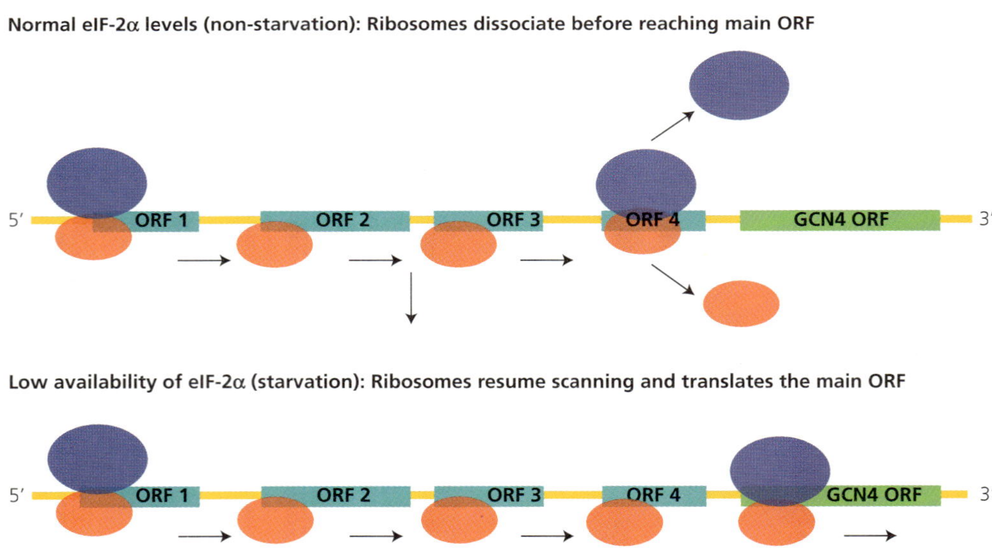

Plate 23 Translation regulation of GCN4 mRNA in yeast. GCN4 mRNA contains 4 uORFs apart from the actual ORF. Sequences downstream of the ORF 1 allow the ribosomes to resume scanning and to initiate at downstream ORFs. During normal conditions, ribosomes are released after translating one of these ORFs. On amino acid starvation, the ribosomes resume scanning and re-initiate at the actual ORF (coding sequence), thus allowing its translation.

Plate 24 Artist's representation of petunia flowers that led to the discovery of cc-suppression.

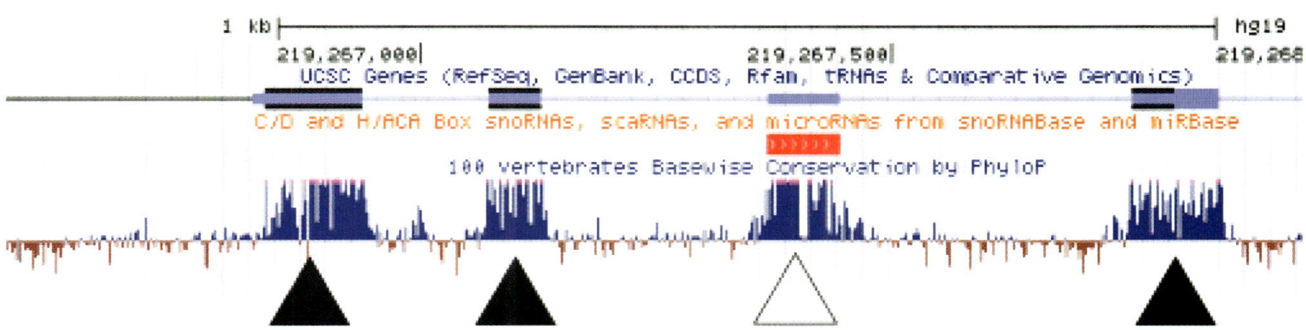

Plate 25 A region of human chromosome 2 that harbours the protein-coding gene *CTDSP1* and within its intron, the microRNA (miRNA) hsa-miR-26b as displayed in the UCSC Genome Browser (www.genome.ucsc.edu). The black arrowheads highlight the conservation in protein-coding exons, while the white arrowhead shows the comparable conservation in the miRNA region.

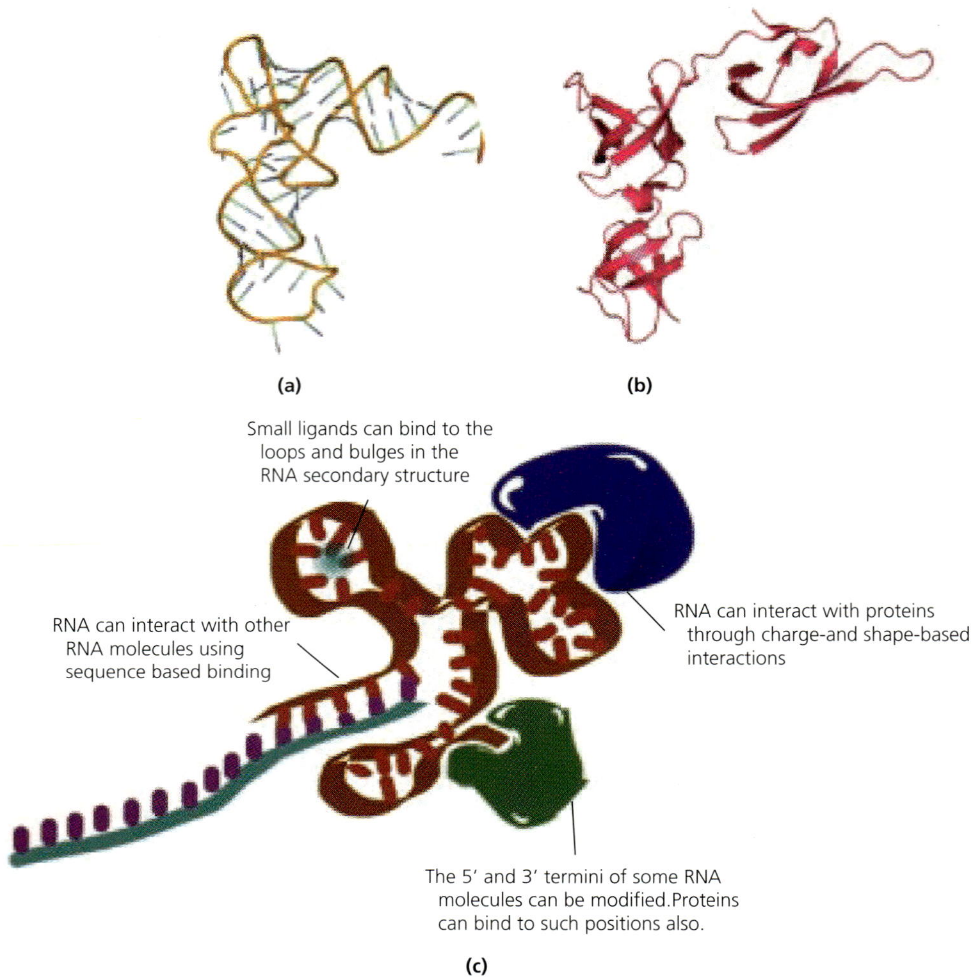

Plate 26 The three-dimensional L-shaped structure of the non-coding transfer RNA (tRNA) (a) is mimicked by some proteins (b) that modulate translation. The tRNA coordinates from chain R of PDB ID 1ASY and EF-P coordinates from chain D of PDB ID. 3A5Z were used to make figures in Pymol (image by Bhupesh Taneja, CSIR–Institute of Genomics and Integrative Biology, New Delhi). Schematic representation of the types of interactions RNA can show with small molecules, other RNA, DNA and proteins (c).

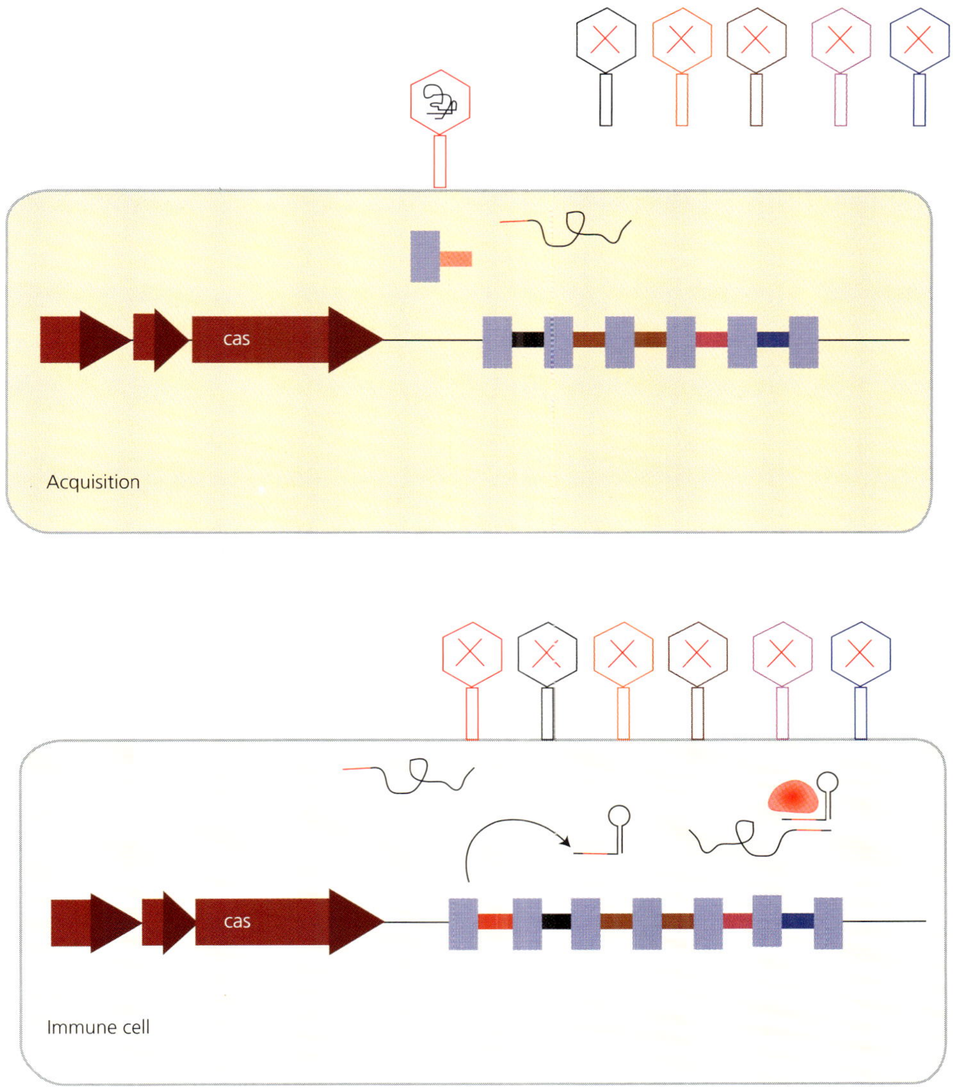

Plate 27 The CRISPR locus consists of the CAS genes, followed by a leader sequence and several repeats (vertical boxes) separated by spacers (horizontal boxes). A new spacer, with an additional repeat, is acquired when a phage infects the bacterial cell. The transcripts from the CRISPR locus are processed (curved arrow) to produce the ncRNA, which binds to the infecting phage genetic material during subsequent infections and directs its cleavage by protein factors. This provides acquired immunity against the phage. The hexagons represent phages. Phages against which the cell is immune are represented with a cross within the hexagon.

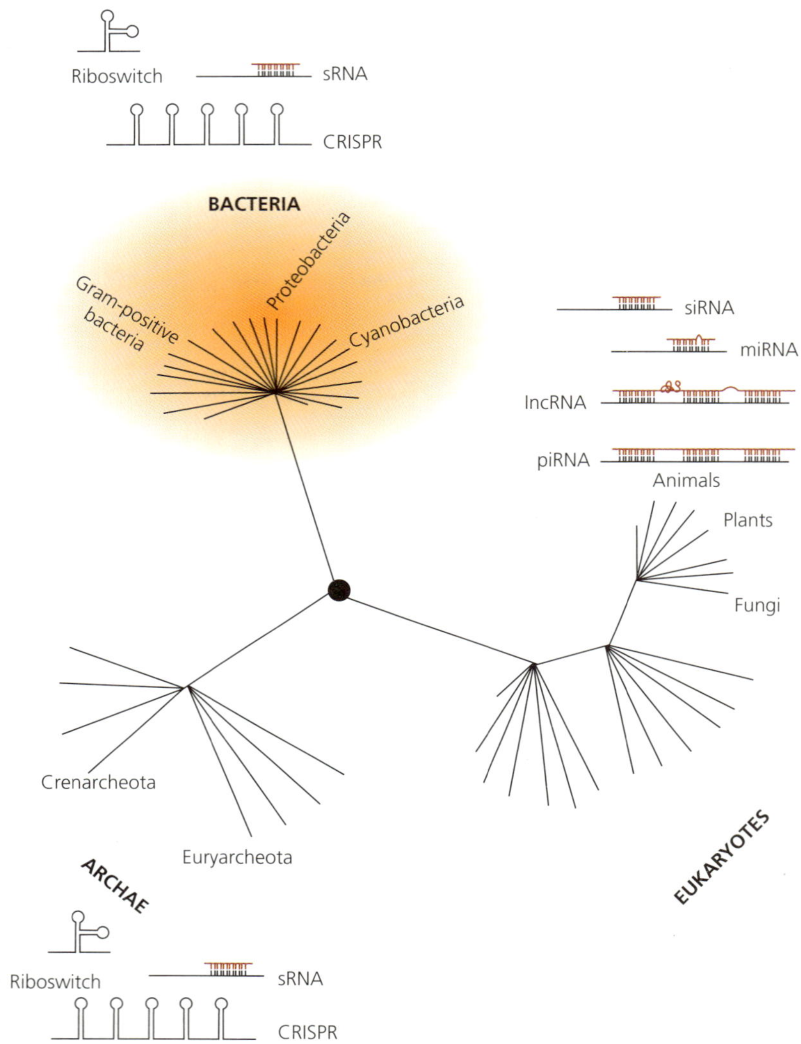

Plate 28 The major types of non-coding RNA found in different classes of organisms.

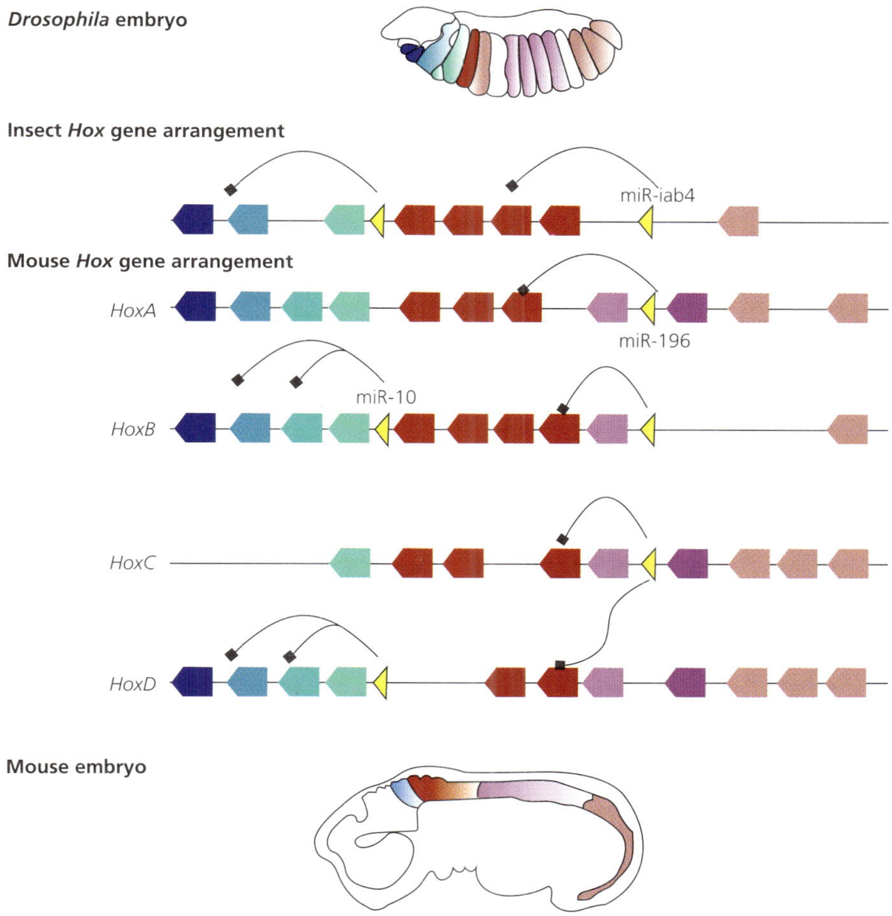

Plate 29 The mammalian and fly Hox clusters contain genes that are arranged in the spatial order of their expression along the anterior–posterior axis. Two microRNAs found within these clusters can repress the more anterior *Hox* genes in the cluster. As microRNAs act in *trans*, they can target *Hox* genes from distant clusters.

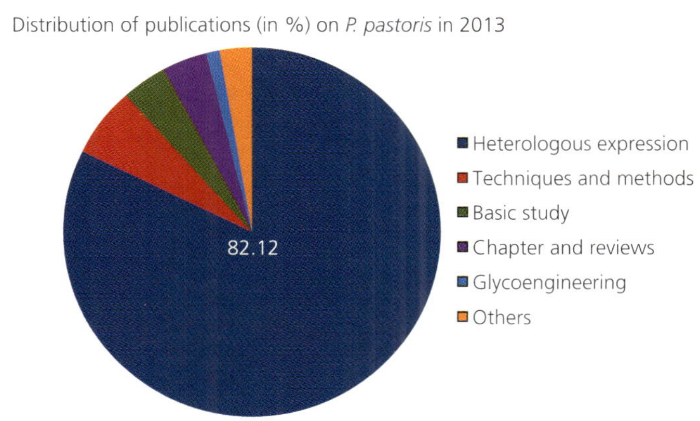

Plate 30 Sorting of publications (in PubMed), released on *P. pastoris* (used as a keyword) in 2013, according to the focus of the study. Most of the articles (82.12% of total publications) were related to the utilization of *P. pastoris* as an expression host for heterologous proteins. The other publications (clockwise: in descending order of their percent contribution to the total number of publications) focused on techniques and methods involved in cloning and expression, basic study of *P. pastoris* physiology, chapters and reviews, glycoengineering of *P. pastoris* and other publications.

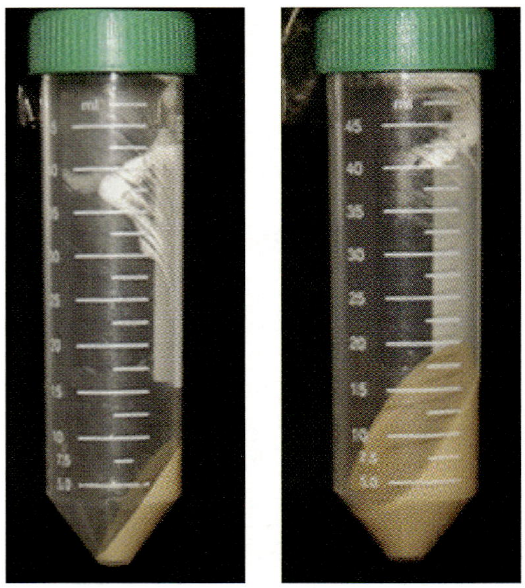

Plate 31 High cell density culture of *P. pastoris*. The *P. pastoris* culture in the two tubes depict shake flask culture of density of 1 OD_{600} unit on the left and high cell density culture of 130 g/L obtained in the fermenter (~500 OD_{600} units) on the right.

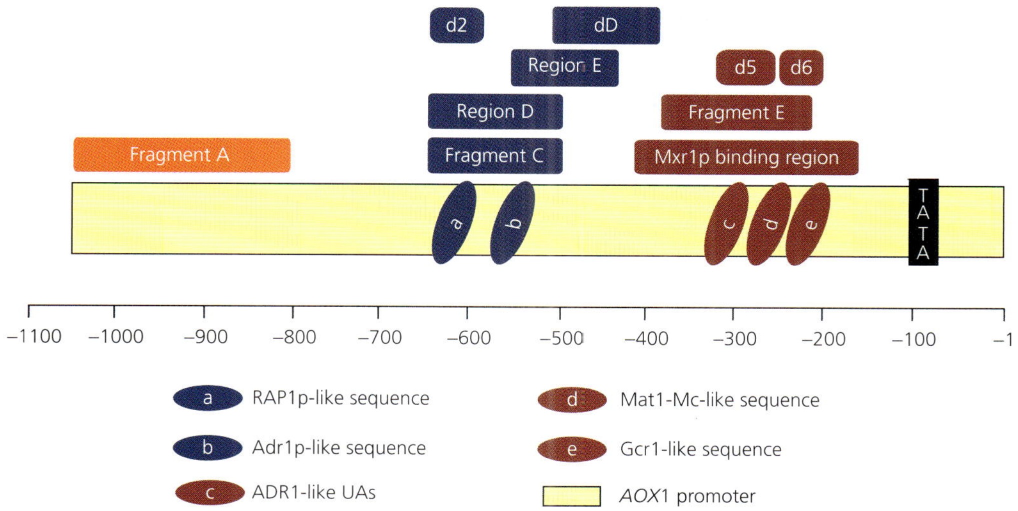

Plate 32 Schematic representation of the reported *cis*-acting elements in *AOX1* promoter (yellow horizontal block). Orange, blue and red rectangular blocks represent *cis*-acting elements, and blue and red ovals represent putative transcription factor binding sites; their relative position is indicated by the number line below.

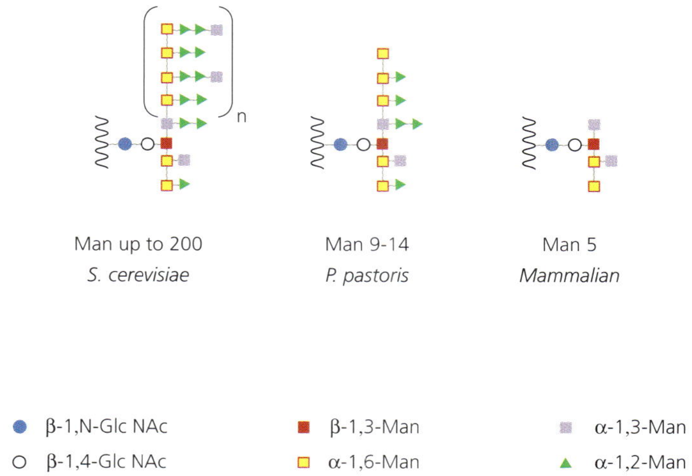

Plate 33 Generalized schematic representation of mannosylation in glycoproteins expressed in *S. cerevisiae, P. pastoris* and the mammalian system, indicating the hyper-mannosylated feature of *S. cerevisiae*-expressed proteins. Glycoproteins expressed in the mammalian system get further modified into complex glycans.

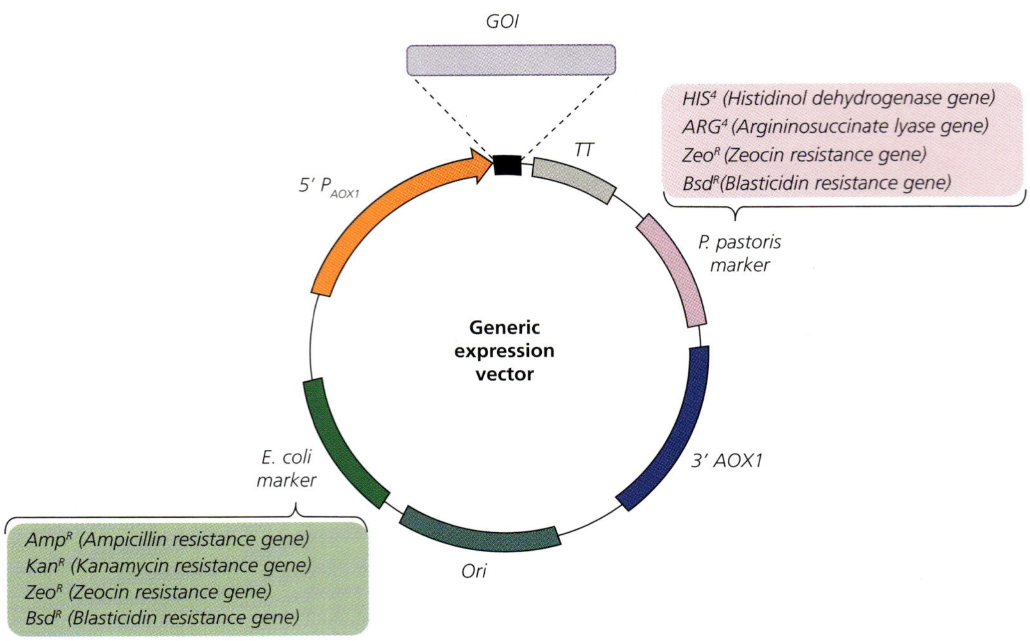

Plate 34 Schematic representation of a generic *P. pastoris* expression vector. P_{AOX1}, *GOI*, *TT*, *3′ AOX1* and *Ori* denote *AOX1* promoter, gene of interest inserted at the multiple cloning site (black block), transcription terminator sequence, *AOX1* region 3′ to transcription termination sequence, and *E. coli* origin of replication, respectively. Commonly used *P. pastoris* and *E. coli* marker genes used for selection of the desired clone are listed in pink and green boxes, respectively.

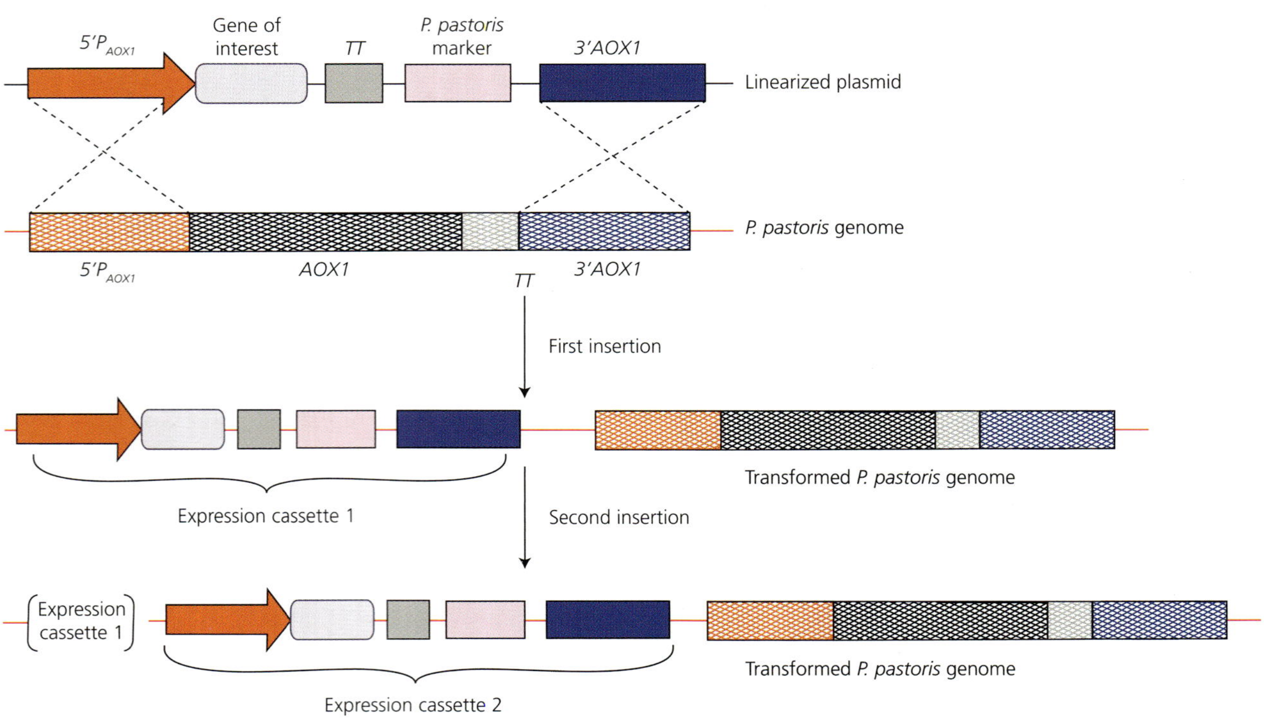

Plate 35 Schematic representation of two insertion events at *AOX1* promoter (5′ P_{AOX1}). TT denotes transcription terminator sequence.

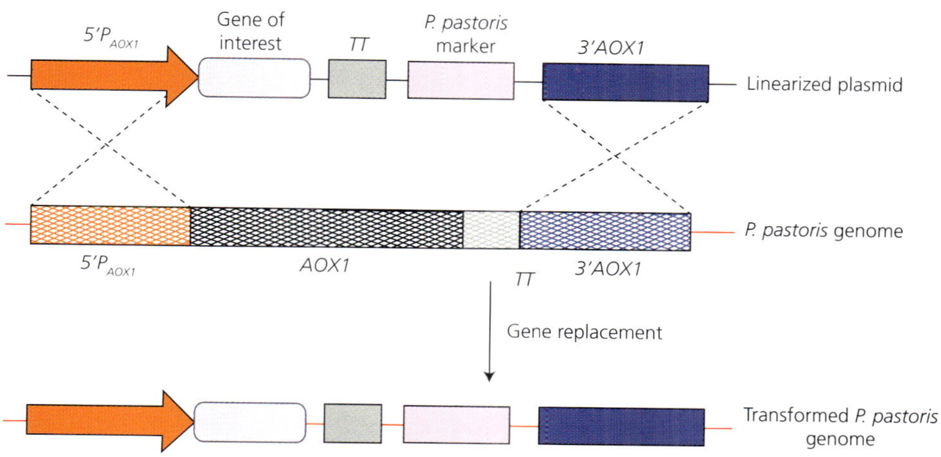

Plate 36 Schematic representation of gene replacement event, where the genomic region between the *AOX1* promoter (5′ P_{AOX1}) and the *3′AOX1* region is replaced with an expression cassette.

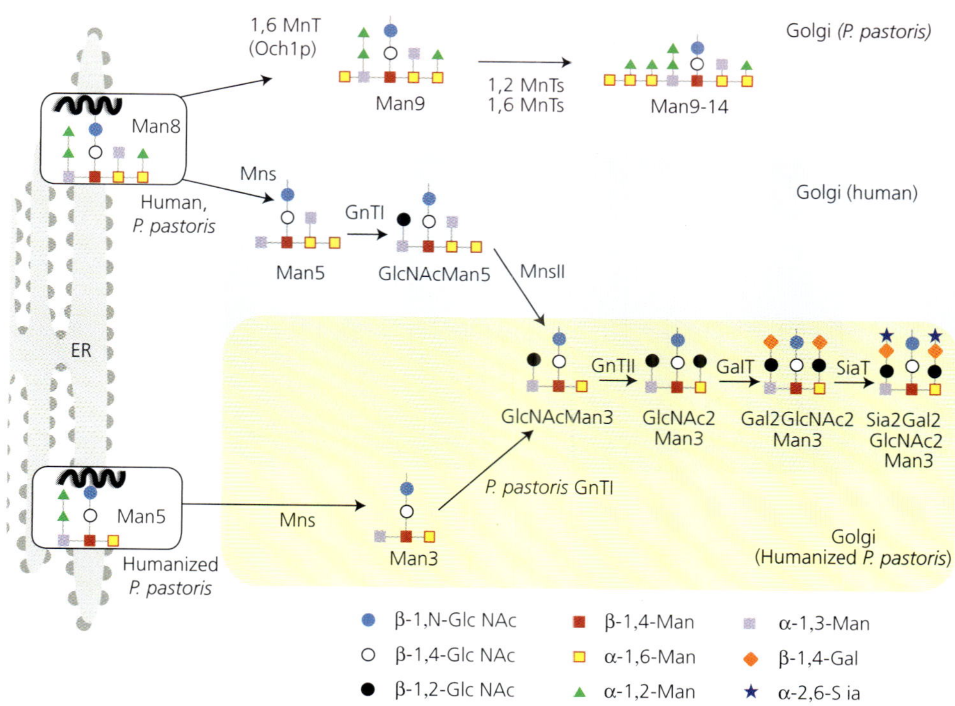

Plate 37 Schematic representation of the *N*-linked glycosylation pathway in *P. pastoris*, human cell and humanized *P. pastoris*. Abbreviations: MnT, mannosyltransferase; Mns, α-1,2-mannosidase; GnTI, β-1,2-*N*-acetylglucosaminyltransferase I; MnsII, mannosidase II; GnTII, β-1,2-*N*-acetylglucosaminyltransferase II; GalT, β-1,4-galactosyltransferase; SiaT, α-2,6-sialyltransferase. Note that the two GlcNAc residues present at the reducing end of all glycans have been omitted from the nomenclature of glycans to keep it simple.

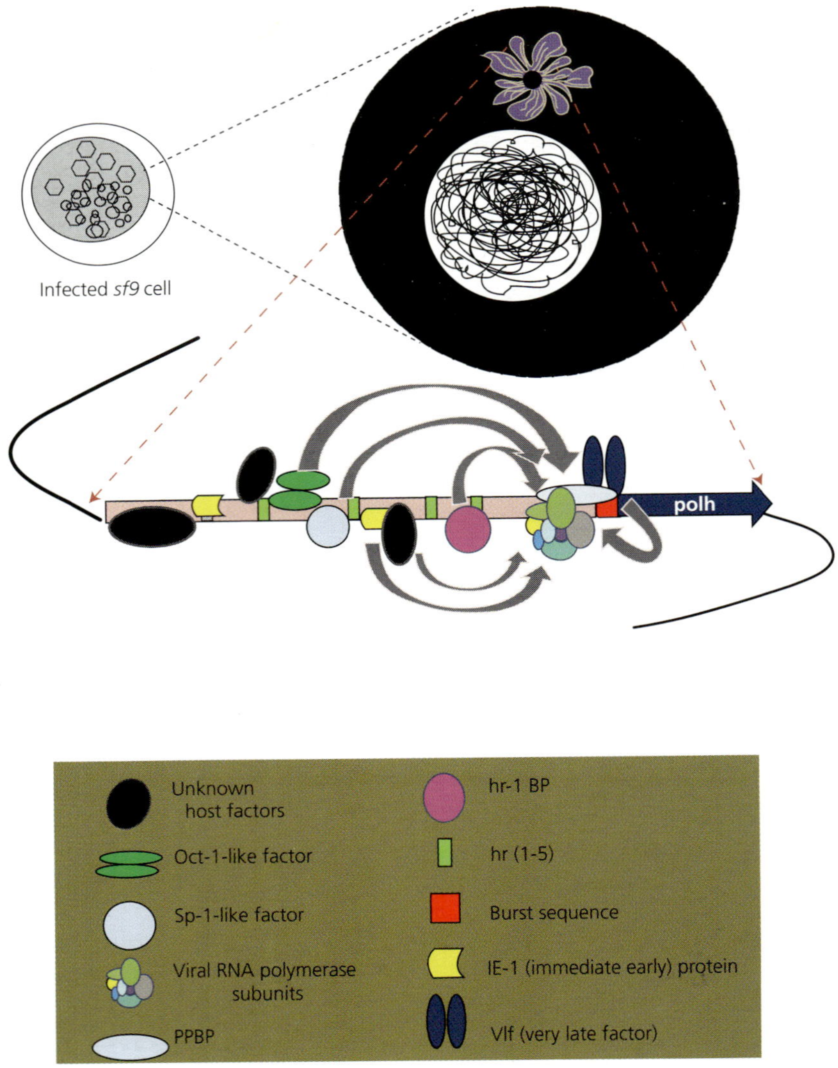

Plate 38 Transcription from polyhedrin promoter. Schematic illustration of polyhedrin gene transcription and related factors. The polyhedrin gene is expressed in the late phase of infection. The late gene promoters have a characteristic motif, termed as burst sequence, to which a very late factor (vlf-1) binds and enhances transcription initiated by viral polymerase. In addition, there are *hr* elements, which serve as binding sites for several transcription activators, including host (hr-1 BP) and viral factors (IE1), which results in tremendous activation of the polyhedrin gene expression. The host factor PPBP binds to the initiator promoter and serves to recruit several viral proteins constituting the viral RNA polymerase, which is influenced by both viral and host factors such as SP-1, Oct-1, heat shock elements, etc.

Index

(The subjects are shown against the section, subsection or subsubsection they appear in.)

A

B

C

D

E

F

G

H

I

J

K

L

M

N

O

P

Q

R

S

T

U

V

W

X

Y

Z

Acknowledgements

For Chapter 14, the authors wish to thank Ms Shilpa Rao, Ms Varsha Bhavnani and Dr Sunil Berwal for their help in preparing a few figures included in this chapter. Financial help in the form of research grants from the Department of Atomic Energy, Government of India, Bhabha Atomic Research Centre, Mumbai (BRNS grant no. 2007/37166/BRNS) and from the Department of Science and Technology (DST), Government of India, under the DST PURSE Programme is duly acknowledged.

For Chapter 15, the authors wish to thank Ms Shilpa Rao for her help in preparing a few figures included in this chapter. Financial help in the form of research grants from the Department of Atomic Energy, Govt. of India, Bhabha Atomic Research Centre, Mumbai (BRNS grant no. 2007/37166/BRNS) and from the Department of Science and Technology (DST), Government of India, under the DST PURSE Programme is duly acknowledged.

For Chapter 28, the author Arun Kumar Sharma is thankful to the Department of Biotechnology, Government of India, which provided financial support for the work related to this field.

Appendix 1

United States Patent 4,237,224

TITLE:

Process for producing biologically functional molecular chimeras

US PATENT:	4237224	SERIES CODE:	6
APPL. NO.:	001021	PATENT TYPE:	1
ART REC. NO.:	172	APP FILL. DT:	19790104
ISSUE DATE:	19801202	NO.OF CLAIMS:	14
EXEMPL.CLAIM:	1	PRI.EXAMINER:	Tanenholtz; Alvin E.

INVENTORS

NAME	CITY:	STATE:
Cohen; Stanley N.	**Portola Valley**	**CA**
Boyer; Herbert W.	**Mill Valley**	**CA**

ASSIGNEE

NAME	**Board of Trustees of the Leland Stanford Jr. University**
CITY:	**Stanford**
STATE:	**CA**
ASS/REIS CD:	**02**

RELATED U.S. APPLICATION DATA

ASS/REIS CD:	APPL. NO.:	APP FILL. DT:	PRNT ST CODE:
72	959288	19781109	0
82	687430	19760517	0
82	520691	19741104	0

CLASS

ORG US CLS:	**435/68**	XREF CLASS:	**435207**
XREF CLASS:	**435172**	XREF CLASS:	**2601125S**
XREF CLASS:	**435231**	XREF CLASS:	**260/27**
XREF CLASS:	**435183**	XREF CLASS:	**435212**
XREF CLASS:	**435317**	INTL. CLASS:	**C12P/2100**
XREF CLASS:	**435849**	FLD SRCH CLS:	**195**
XREF CLASS:	**435820**	FLD SRCH SUB:	**1;28 N;28 R;112;78;79**
XREF CLASS:	**435 91**	FLD SRCH CLS:	**435**
FLD SRCH SUB:	**68;172;231;183**		

REFERENCES CITED

U.S. PATENT DOCUMENTS

PATENT NO.:	ISSUE DATE:	NAME:	ORG US CLS:
3813316	**19740500**	**Chakrabarty**	**195/28R**

OTHER PUBLICATIONS

Morrow et al., Proc. Nat. Acad. Sci. USA, vol. 69, pp. 3365-3369, Nov. 1972.
Morrow et al., Proc. Nat. Acad. Sci. USA, vol. 71, pp. 1743-1747, May 1974.
Hershfield et al., Proc. Nat. Acad. Sci. USA, vol. 71, pp. 3455 et seq. (1974).
Jackson et al., Proc. Nat. Acad. Sci. USA, vol. 69, pp. 2904-2909, Oct. 1972.
Mertz et al., Proc. Nat. Acad. Sci. USA, vol. 69, pp. 3370-3374, Nov. 1972.
Cohen, et al., Proc. Nat. Acad. Sci. USA, vol. 70, pp. 1293-1297, May 1973.
Cohen et al., Proc. Nat. Acad. Sci. USA, vol. 70, pp. 3240-3244, Nov. 1973.
Chang et al., Proc. Nat. Acad. Sci, USA, vol. 71, pp. 1030-1034, Apr. 1974.
Ullrich et al., Science vol. 196, pp. 1313-1319, Jun. 1977.
Singer et al., Science vol. 181, p. 1114 (1973).
Itakura et al., Science vol. 198, pp. 1056-1063 Dec. 1977.
Komaroff et al., Proc. Nat. Acad. Sci. USA, vol. 75, pp. 3727-3731, Aug. 1978.
Chemical and Engineering News, p. 4, May 30, 1977.
Chemical and Engineering News, p. 6, Sep. 11, 1978.

LEGAL REPRESENTATION

Attorney, Agent, or Firm - Bertram I. Rowland

ABSTRACT

Method and compositions are provided for replication and expression of exogenous genes in microorganisms. Plasmids or virus DNA are cleaved to provide linear DNA having ligatable termini to which is inserted a gene having complementary termini, to provide a biologically functional replicon with a desired phenotypical property. The replicon is inserted into a microorganism cell by transformation. Isolation of the transformants provides cells for replication and expression of the DNA molecules present in the modified plasmid. The method provides a convenient and efficient way to introduce genetic capability into microorganisms for the production of nucleic acids and proteins, such as medically or commercially useful enzymes, which may have direct usefulness, or may find expression in the production of drugs, such as hormones, antibiotics, or the like, fixation of nitrogen, fermentation, utilization of specific feedstocks, or the like.

GOVERNMENT SUPPORT

The invention was supported by generous grants of NIH, NSF and the American Cancer Society.

CROSS-REFERENCE TO RELATED APPLICATIONS

This application is a continuatin-in-part of applicatin Ser. No. 959,288, filed Nov. 9, 1978, which is a continuation of application Ser. No. 687,430 filed May 17, 1976, now abandoned, which was a continuation-in-part of application Ser. No. 520,691, filed Nov. 4, 1974, now abandoned.

BACKGROUND OF THE INVENTION

1. Field of the Invention

Although transfer of plasmids among strains of *E. coli* and other Enterobacteriaceae has long been accomplished by conjugation and/or transduction, it has not been previously possible to selectively introduce particular species of plasmid DNA into these bacterial hosts or other microorganisms. Since microorganisms that have been transformed with plasmid DNA contain autonomously replicating extrachromosomal DNA species having the genetic and molecular characteristics of the parent plasmid, transformation has enabled the selective cloning and amplification of particular plasmid genes.

The ability of genes derived from totally different biological classes to replicate and be expressed in a particular microorganism permits the attainment of interspecies genetic recombination. Thus, it becomes practical to introduce into a particular microorganism, genes specifying such metabolic or synthetic functions as nitrogen fixation, photosynthesis, antibiotic production, hormone synthesis, protein synthesis, e.g. enzymes or antibodies, or the like—functions which are indigenous to other classes of organisms—by linking the foreign genes to a particular plasmid or viral replicon.

BRIEF DESCRIPTION OF THE PRIOR ART

References which relate to the subject invention are Cohen, et al., Proc. Nat. Acad, Sci., USA, 69, 2110 (1972); ibid, 70, 1293 (1973); ibid, 70, 3240 (1973); ibid, 71, 1030 (1974); Morrow, et al., Proc. Nat. Acad. Sci., 71, 1743 (1974); Novick, Bacteriological Rev., 33, 210 (1969); and Hershfeld, et al., Proc. Nat. Acad. Sci., in press; Jackson, et al., ibid, 69, 2904 (1972);

SUMMARY OF THE INVENTION

Methods and compositions are provided for genetically transforming microorganisms, particularly bacteria, to provide diverse genotypical capability and producing recombinant plasmids. A plasmid or viral DNA is modified to form a linear segment having ligatable termini which is joined to DNA having at least one intact gene and complementary ligatable termini. The termini are then bound together to form a "hybrid" plasmid molecule which is used to transform susceptible and compatible microorganisms. After transformation, the cells are grown and the transformants harvested. The newly functionalized microorganisms may then be employed to carry out their new function; for example, by producing proteins which are the desired end product, or metabolities of enzymic conversion, or be lysed and the desired nucleic acids or proteins recovered.

DESCRIPTION OF THE SPECIFIC EMBODIMENTS

The process of this invention employs novel plasmids, which are formed by inserting DNAhaving one or more intact genes into a plasmid in such a location as to permit retention of an intact replicator locus and system (replicon) to provide a recombinant plasmid molecule. The recombinant plasmid molecule will be referred to as a "hybrid" plasmid or plasmid "chimera." The plasmid chimera contains genes that are capable of expressing at least one phenotypical property. The plasmmid chimera is used to transform a susceptible and competent microorganism under conditions where transformation occurs. The microorganism is then grown under conditions which allow for separation and harvesting of transformants that contain the plasmid chimera.

The process of this invention will be divided into the following stages:

I. preparation of the recombinant plasmid or plasmid chimera;
II. transformation or preparation of transformants; and
III. replication and transcription of the recombinant plasmid in transformed bacteria.

Preparation of Plasmid Chimera

In order to prepare the plasmid chimera, it is necessary to have a DNA vector, such as a plasmid or phage, which can be cleaved to provide an intact replicator locus and system (replicon), where the linear segment has ligatable termini or is capable of being modified to introduce ligatable termini. Of particular interest are those plasmids which have a phenotypical property, which allow for ready separation of transformants from the parent microorganism. The plasmid will be capable of replicating in a microorganism, particularly a bacterium which is susceptible to transformation. Various unicellular microorganisms can be transformed, such as bacteria, fungii and algae. That is, those unicellular organisms which are capable of being grown in cultures of fermentation. Since bacteria are for the most part the most convenient organisms to work with, bacteria will be hereinafter referred to as exemplary of the other unicellular organisms. Bacteria, which are susceptible to transformation, include members of the Enterobacteriaceae, such as strains of *Escherichia coli*; Salmonella; Bacillaceae, such as *Bacillus subtilis*; Pneumococcus; Streptococcus, and *Haemophilus influenzae*.

A wide variety of plasmids may be employed of greatly varying molecular weight. Normally, the plasmids employed will have molecular weights in the range of about 1 X 10^6 to 50 X 10^6d, more usually from about 1 to 20 X 10^6 d, and preferably, from about 1 to 10 X 10^6 d. The desirable plasmid size is determined by a number of factors. First, the plasmid must be able to accommodate a replicator locus and one or more genes that are capable of allowing replication of the plasmid. Secondly, the plasmid should be of a size which provides for a reasonable probability of recircularization with the foreign gene(s) to form the recombinant plasmid chimera. Desirably, a restriction enzyme should be available, which will cleave the plasmid without inactivating the replicator locus and system associated with the replicator locus. Also, means must be provided for providing ligatable termini for the plasmid, which are complementary to the termini of the foreign gene(s) to allow fusion of the two DNA segments.

Another consideration for the recombinant plasmid is that it be compatible with the bacterium to be transformed. Therefore, the original plasmid will usually be derived from a member of the family to which the bacterium belongs.

The original plasmid should desirably have a phenotypical property which allows for the separation of transformant bacteria from parent bacteria. Particularly useful is a gene, which provides for survival selection. Survival selection can be achieved by providing resistance to a growth inhibiting substance or providing a growth factor capability to a bacterium deficient in such capability.

Conveniently, genes are available, which provide for antibiotic or heavy metal resistance or polypeptide resistance, e.g. colicin. Therefore, by growing the bacteria on a medium containing a bacteriostatic or bacteriocidal substance, such as an antibiotic, only the transformants having the antibiotic resistance will survive. Illustrative antibiotics include tetracycline, streptomycin, sulfa drugs, such as sulfonamide, kanamycin, neomycin, penicillin, chloramphenicol, or the like.

Growth factors include the synthesis of amino acids, the isomerization of substrates to forms which can be metabolized or the like. By growing the bacteria on a medium which lacks the appropriate growth factor, only the bacteria which have been transformed and have the growth factor capability will clone.

One plasmid of interest derived from *E. coli* is referred to as pSC101 and is described in Cohen, et al., Proc. Nat. Acad. Sci., USA, 70, 1293 (1972), (referred to in that article as Tc6-5). Further description of this particular plasmid and its use is found in the other articles previously referred to.

The plasmid pSC101 has a molecular weight of about 5.8 X 10^6 d and provides tetracycline resistance.

Another plasmid of interest is colicinogenic factor EI (ColE1), which has a molecular weight of 4.2 X 10^6 d, and is also derived from *E. coli*. The plasmid has a single EcoRI substrate site and provides immunity to colicin E1.

In preparing the plasmid for joining with the exogenous gene, a wide variety of techniques can be provided, including the formation of or introduction of cohesive termini. Flush ends can be joined. Alternatively, the plasmid and gene may be cleaved in such a manner that the two chains are cleaved at different sites to leave extensions at each end which serve as cohesive termini. Cohesive termini may also be introduced by removing nucleic acids from the opposite ends of the two chains or alternatively, introducing nucleic acids at opposite ends of the two chains.

To illustrate, a plasmid can be cleaved with a restriction endonuclease or other DNA cleaving enzyme. The restriction enzyme can provide square ends, which are then modified to provide cohesive termini or can cleave in a staggered manner at different, but adjacent, sites on the two strands, so as to provide cohesive termini directly.

Where square ends are formed such as, for example, by HIN (Haemophilus influenzae RII) or pancreatic DNAse, one can ligate the square ends or alternatively one can modify the square ends by chewing back, adding particular nucleic acids, or a combination of the two. For example, one can employ appropriate transferases to add a nucleic acid to the 5′ and 3′ ends of the DNA. Alternatively, one can chew back with an enzyme, such as a λ-exonuclease, and it is found that there is a high probability that cohesive termini will be achieved in this manner.

An alternative way to achieve a linear segment of the plasmid with cohesive termini is to employ an endonuclease such as EcoRI. The endonuclease cleaves the two strands at different adjacent sites providing cohesive termini directly.

With flush ended molecules, a T_4 ligase may be employed for linking the termini. See, for example, Scaramella and Khorana, J. Mol. Biol. 72: 427-444 (1972) and Scaramella, DNAS 69: 3389 (1972), whose disclosure is incorporated herein by reference.

Another way to provide ligatable termini is to leave employing DNAse and Mn^{++} as reported by Lai and Nathans, J. Mol. Biol, 89: 179 (1975). The plasmid, which has the replicator locus, and serves as the vehicle for introduction of a foreign gene into the bacterial cell, will hereafter be referred to as "the plasmid vehicle."

It is not necessary to use plasmid, but any molecule capable of replication in bacteria can be employed. Therefore, instead of plasmid, viruses may be employed, which will be treated in substantially the same manner as the plasmid, to provide the ligatable termini for joining to the foreign gene.

If production of cohesive termini is by restriction endonuclease cleavage, the DNA containing the foreign gene(s) to be bound to the plasmid vehicle will be cleaved in the same manner as the plasmid vehicle. If the cohesive termini are produced by a different technique, an analogous technique will normally be employed with the foreign gene. (By foreign gene is intended a gene derived from a source other than the transformant strain.) In this way, the foreign gene(s) will have ligatable termini, so as to be able to covalently bonded to the termini of the plasmid vehicle. One can carry out the cleavage or digest of the plasmids together in the same medium or separately, combine the plasmids and recircularize the plasmids to form the plasmid chimera in the absence of active restriction enzyme capable of cleaving the plasmids.

Descriptions of methods of cleavage with restriction enzymes may be found in the following articles: Greene, et al., *Methods in Molecular Biology*, Vol. 9, ed. Wickner, R. B., (Marcel Dekker, Inc., New York), "DNA Replication and Biosynthesis"; Mertz and Davis, 69, Proc. Nat. Acad. Sci., USA, 69, 3370 (1972);

The cleavage and non-covalent joining of the plasmid vehicle and the foreign DNA can be readily carried out with a restriction endonuclease, with the plasmid vehicle and foreign DNA in the same or different vessels. Depending on the number of fragments, which are obtained from the DNA endonuclease digestion, as well as the genetic properties of the various fragments, digestion of the foreign DNA may be carried out separately and the fragments separated by centrifugation in an appropriate gradient. Where the desired DNA fragment has a phenotypical property, which allows for the ready isolation of its transformant, a separation step can usually be avoided.

Endonuclease digestion will normally be carried out at moderate temperatures, normally in the range of 10° to 40° C. in an appropriately buffered aqueous medium, usually at a pH of about 6.5 to 8.5. Weight percent of total DNA in the reaction mixture will generally be about 1 to 20 weight percent. Time for the reaction will vary, generaly being from 0.1 to 2 hours. The amount of endonuclease employed is normally in excess of that required, normally being from about 1 to 5 units per 10 μg of DNA.

Where cleavage into a plurality of DNA fragments results, the course of the reaction can be readily followed by electrophoresis. Once the digestion has gone to the desired degree, the endonuclease is inactivated by heating above about 60° C. for five minutes. The digestion mixture may be worked up by dialysis, gradient separation, or the like, or used directly.

After preparation of the two double stranded DNA sequences, the foreign gene and vector are combined for annealing and/or ligation to provide for a functional recombinant DNA structure. With plasmids, the annealing involves the hydrogen bonding together of the cohesive ends of the vector and the foreign gene to form a circular plasmid which has cleavage sites. The cleavage sites are then normally ligated to form the complete closed and circularized plasmid.

The annealing, and as appropriate, recircularization can be performed in whole or in part in vitro or in vivo. Preferably, the annealing is performed in vitro. The annealing requires an appropriate buffered medium containing the DNA fragments. The temperature employed initially for annealing will be about 40° to 70° C., followed by a period at lower temperature, generaly from about 10° to 30° C. The molar ratio of the two segments will generally be in the range of about 1-5:5-1. The particular temperature for annealing will depend upon the binding strength of the cohesive termi.

While 0.5 hr to 2 or more days may be employed for annealing, it is believed that a period of 0.5 to 6 hrs may be sufficient. The time employed for the annealing will vary with the temperature employed, the nature of the salt solution, as well as the nature of the cohesive termini.

The ligation, when in vitro, can be achieved in conventional ways employing DNA ligase. Ligation is conveniently carried out in an aqueous solution (pH 6-8) at temperatures in the range of about 5° to 40° C. The concentration of the DNA will generally be from about 10 to 100 g/ml. A sufficient amount of the DNA ligase or other ligating agent e.g. T_4 ligase, is employed to provide a convenient rate of reaction, generally ranging from about 5 to 50 U/ml. A small amount of a protein e.g. albumin, may be added at concentrations of about 10 to 200 g/ml. The ligation with DNA ligase is carried out in the presence of magnesium at about 1-10 mM.

At the completion of the annealing or ligation, the solution may be chilled and is ready for use in transformation.

It is not necessary to ligate the recircularized plasmid prior to transformation, since it is found that this function can be performed by the bacterial host. However, in some situations ligation prior to transformation may be desirable.

The foreign DNA can be derived from a wide variety of sources. The DNA may be derived from eukaryotic or prokaryotic cells, viruses, and bacteriophage. The fragments employed will generally have molecular weights in the range of about 0.5 to 20 X 10^6 d, usually in the range of 1 to 10 X 10^6 d. The DNA fragment may include one or more genes or one or more operons.

Desirably, if the plasmid vehicle does not have a phenotypical property which allows for isolation of the transformants, the foreign DNA fragment should have such property. Also, an intact promoter and base sequences coding for initiation and termination sites should be present for gene expression.

In accordance with the subject invention, plasmids may be prepared which have replicons and genes which could be present in bacteria as a result of normal mating of bacteria. However, the subject invention provides a technique, whereby a replicon and gene can coexist in a plasmid, which is capable of being introduced into a unicellular organism, which could not exist in nature. The first type of plasmid which cannot exist in nature is a plasmid which derives its replicon from one organism and the exogenous gene from another organism, where the two organisms do not exchange genetic information. In this situation, the two organisms will either be eukaryotic or prokaryotic. Those organisms which are able to exchange genetic information by mating are well known. Thus, prior to this invention, plasmids having a replicon and one or more genes from two sources which do not exchange genetic information would not have existed in nature. This is true, even in the event of mutations, and induced combinations of genes from different strains of the same species. For the natural formation of plasmids formed from a replicon and genes from different microorganisms it is necessary that the microorganisms be capable of mating and exchanging genetic information.

In the situation, where the replicon comes from a eukaryotic or prokaryotic cell, and at least one gene comes from the other type of cell, this plasmid heretofore could not have existed in nature. Thus, the subject invention provides new plasmids which cannot naturally occur and can be used for transformation of unicellular organisms to introduce genes from other unicellular organisms, where the replicon and gene could not previously naturally coexist in a plasmid.

Besides naturally occurring genes, it is feasible to provide synthetic genes, where fragments of DNA may be joined by various techniques known in the art. Thus, the exogenous gene may be obtained from natural sources or from synthetic sources.

The plasmid chimera contains a replicon which is compatible with a bacterium susceptible of transformation and at least one foreign gene which is directly or indirectly bonded through deoxynucleotides to the replicon to form the circularized plasmid structure. As indicated previously, the foreign gene normally provides a phenotypical property, which is absent in the parent bacterium. The foreign gene may come from another bacterial strain, species or family, or from a plant or animal cell. The original plasmid chimera will have been formed by in vitro covalent bonding between the replicon and foreign gene. Once the originally formed plasmid chimera has been used to prepare transformants, the plasmid chimera will be replicated by the bacterial cell and cloned in vivo by growing the bacteria in an appropriate growth medium. The bacterial cells may be lysed and the DNA isolated by conventional means or the bacteria continually reproduced and allowed to express the genotypical property of the foreign DNA.

Once a bacterium has been transformed, it is no longer necessary to repeat the in vitro preparation of the plasmid chimera or isolate the plasmid chimera from the transformant progeny. Bacterial cells can be repeatedly multiplied which will express the genotypical property of the foreign gene.

One method of distinguishing between a plasmid which originates in vivo from a plasmid chimera which originates in vitro is the formation of homoduplexes between an in vitro prepared plasmid chimera and the plasmid formed in vivo. It will be an extremely rare event where a plasmid which originates in vivo will be the same as a plasmid chimera and will form homoduplexes with plasmid chimeras. For a discussion of homoduplexes, see Sharp, Cohen and Davidson, J. Mol. Biol., 75, 235 (1973), and Sharp, et al, ibid, 71, 471 (1972).

The plasmid derived from molecular cloning need not homoduplex with the in vitro plasmid originally employed for transformation of the bacterium. The bacterium may carry out modification processes, which will not affect the portion of the replicon introduced which is necessary for replication nor the portion of the exogenous DNA which contains the gene providing the genotypical trait. Thus, nucleotides may be introduced or excised and, in accordance with naturally occurring mating and transduction, additional genes may be introduced. In addition, for one or more reasons, the plasmids may be modified in vitro by techniques which are known in the art. However, the plasmids obtained by molecular cloning will homoduplex as to those parts which relate to the original replicon and the exogenous gene.

II. Transformation

After the recombinant plasmid or plasmid chimera has been prepared, it may then be used for the transformation of bacteria. It should be noted that the annealing and ligation process not only results in the formation of the recombinant plasmid, but also in the recircularization of the plasmid vehicle. Therefore, a mixture is obtained of the original plasmid, the recombinant plasmid, and the foreign DNA. Only the original plasmid and the DNA chimera consisting of the plasmid vehicle and linked foreign DNA will normally be capable of replication. When the mixture is employed for transformation of the bacteria, replication of both the plasmid vehicle genotype and the foreign genotype will occur with both genotypes being replicated in those cells having the recombinant plasmid.

Various techniques exist for transformation of a bacterial cell with plasmid DNA. A technique, which is particularly useful with *Escherichia coli*, is described in Cohen, et al., ibid, 69, 2110 (1972). The bacterialcells are grown in an appropriate medium to a predetermined opticaldensity. For example, with *E. coli* strain C600, the optical density was 0.85 at 590 nm. The cells are concentrated by chilling, sedimentation and washing with a dilute salt solution. After centrifugation, the cells are resuspended in a calcium chloride solution at reduced temperatures (approx. 5°-15° C.), sedimented, resuspended in a smaller volume of a calcium chloride solution and the cells combined with the DNA in an appropriately buffered calcium chloride solution and incubated at reduced temperatures. The concentration of Ca^{++} will generally be about 0.01 to 0.1 M. After a sufficient incubation period, generally from about 0.5-3.0 hours, the bacteria are subjected to a heat pulse generally in the range of 35° to 45° C. for a short period of time; namely from about 0.5 to 5 minutes. The transformed cells are then chilled and may be transferred to a growth medium, whereby the transformed cells having the foreign genotype may be isolated.

An alternative transformation technique may be found in Lederberg and Cohen, I. Bacteriol., 119, 1072 (1974), whose disclosure is incorporated herein by reference.

III. Replication and Transcription of the Plasmid

The bacterial cells, which are employed, will be of such species as to allow replication of the plasmid vehicle. A number of different bacteria which can be employed, have been indicated previously. Strains which lack indigenous modification and restriction enzymes are particularly desirable for the cloning of DNA derived from foreign sources.

The transformation of the bacterial cells will result in a mixture of bacterial cells, the dominant proportion of which will not be transformed. Of the fraction of cells which are transformed, some significant proportion, but normally a minor proportion, will have been transformed by recombinant plasmid. Therefore, only a very small fraction of the total number of cells which are present will have the desired phenotypical characteristics.

In order to enhance the ability to separate the desired bacterial clones, the bacterial cells, which have beeen subjected to transformation, will first be grown in a solution medium, so as to amplify the absolute number of the desired cells. The bacterial cells may then be harvested and streaked on an appropriate agar medium. Where the recombinant plasmid has a phenotype, which allows for ready separation of the transformed cells from the parent cells, this will aid in the ready separation of the two types of cells. As previously indicated, where the genotype provides resistance to a growth inhibiting material, such as an antibiotic or heavy metal, the cells can be grown on an agar medium containing the growth inhibiting substance. Only available cells having the resistant genotype will survive. If the foreign gene does not provide a phenotypical property, which allows for distinction between the cells transformed by the plasmid vehicle and the cells transformed by the plasmid chimera, a further step is necessary to isolate the replicated plasmid chimera from the replicated plasmid vehicle. The steps include lysing of the cells and isolation and separation of the DNA by conventional means or random selection of transformed bacteria and characterization of DNA from such transformants to determine which cells contain molecular chimeras. This is accomplished by physically characterizing the DNA by electrophoresis, gradient centrifugation or electron microscopy.

Cells from various clones may be harvested and the plasmid DNA isolated from these transformants. The plasmid DNA may then be analyzed in a variety of ways. One way is to treat the plasmid with an appropriate restriction enzyme and analyze the resulting fragments for the presence of the foreign gene. Other techniques have been indicated above.

Once the recombinant plasmid has been replicated in a cell and isolated, the cells may be grown and multiplied and the recombinant plasmid employed for transformation of the same or different bacterial strain.

The subject process provides a technique for introducing into a bacterial strain a foreign capability which is genetically mediated. A wide variety of genes may be employed as the foreign genes from a wide variety of sources. Any intact gene may be employed which can be bonded to the plasmid vehicle. The source of the gene can be other bacterial cells, mammalian cells, plant cells, etc. The process is generally applicable to bacterial cells capable of transformation. A plasmid must be available, which can be cleaved to provide a linear segment having ligatable termini, and an interact replicator locus and system, preferably a system including a gene which provides a phenotypical property which allows for easy separation of the transformants. The linear segment may then be annealed with a linear segment of DNA having one or more genes and the resulting recombinant plasmid employed for transformation of the bacteria.

By introducing one or more exogeneous genes into a unicellular organism, the organism will be able to produce polypeptides and proteins ("poly(amino acids)") which the organism could not previously produce. In some instances the poly(amino acids) will have utility in themselves, while in other situations, particularly with enzymes, the enzymatic product(s) will either be useful in itself or useful to produce a desirable product.

One group of poly(amino acids) which are directly useful are hormones. Illustrative hormones include parathyroid hormone, growth hormone, gonadotropins (FSH, luteinizing hormone, chorionogonadatropin, and glycoproteins), insulin, ACTH, somatostatin, prolactin, placental lactogen, melanocyte stimulating hormone, thyrotropin, parathyroid hormone, calcitonin, enkephalin, and angiotensin.

Other poly(amino acids) of interest include serum proteins, fibrinogin, prothrombin, thromboplastin, globulin e.g. gamma-globulins or antibodies, heparin, antihemophilia protein, oxytocin, albumins, actin, myosin, hemoglobin, ferritin, cytochrome, myoglobin, lactoglobulin, histones, avidin, thyroglobulin, interferin, kinins and transcortin. Where the genes or genes produce one or more enzymes, the enzymes may be used for fulfilling a wide variety of functions. Included in these functions are nitrogen fixation, production of amino acids, e.g. polyiodothyronine, particularly thyroxine, vitamins, both water and fat soluble vitamins, antimicrobial drugs, chemotheropeutic agents e.g. antitumor drugs, polypeptides and proteins e.g. enzymes from apoenzymes and hormones from prohormones, diagnostic reagents, energy producing combinations e.g. photosynthesis and hydrogen production, prostaglandins, steroids, cardiac glycosides, coenzymes, and the like.

The enzymes may be individually useful as agents separate from the cell for commercial applications, e.g. in detergents, synthetic transformations, diagnostic agents and the like. Enzymes are classified by the I.U.B. under the classifications as I. Oxidoreductases; II. Transferases; III. Hydrolases; IV. Lyases; V. Isomerases; and VI. Ligases.

EXPERIMENTAL

In order to demonstrate the subject invention, the following experiments were carried out with a variety of foreign genes.

(All temperatures not otherwise indicated are Centrigrade. All percents not otherwise indicated are percents by weight.)

EXAMPLE A

A. Preparation of pSC101 Plasmid

Covalently closed R6-5 DNA was sheared with a Virtis stainless steel microshaft in a one milliliter cup. The R6-5 DNA was sheared at 2,000 r.p.m. for 30 minutes in TEN buffer solution (0.02 M Tris-HCl (pH 8.0)-1 mM EDTA (pH 8.0)-0.02 M NaCl), while chilled at 0°-4°.

The sheared DNA sample was subjected to sucrose gradient sedimentation at 39,500 r.p.m. in a Spinco SW 50.1 rotor at 20°. A 0.12 mil fraction was collected on a 2.3 cm diameter circle of Whatman No. 3 filter paper, dried for 20 minutes and precipitated by immersion of the disc in cold 5% trichloroacetic acid, containing 100 μg/ml thymidine. The precipitate was filtered and then washed once with 5% trichloroacetic acid, twice with 99% ethanol and dried. pSC101 was the 27S species having a calculated molecular weight of 5.8.times.10^6 d.

B. Generalized Transformation Procedure

E. coli strain C600 was grown at 37° in H1 medium to an optical density of 0.85 at 590 nm. At this point the cells were chilled quickly, sedimented and washed once in 0.5 volume 10 nM

NaCl. After centrifugation, the bacteria was resuspended in half the original volume of chilled 0.03 M calcium chloride, kept at 0° for 20 minutes, sedimented, and then resuspended in 0.1 of the original volume of 0.03 M of calcium chloride solution. Chilled DNA samples in TEN buffer were supplemented with 0.1 M calcium chloride to a final concentration of 0.03 M. 0.2 ml of competent cells treated with calcium chloride was added to 0.1 ml of DNA solution with chilled pipets and an additional incubation was done for 60 minutes at 0°. The bacteria were then subjected to a heat pulse at 42° for two minutes, chilled, and then either placed directly onto nutrient agar containing appropriate antibiotics or, where indicated, diluted 10 times in L-broth and incubated at 37° before plating. The cell survival is greater than 50% after calcium chloride treatment and heat pulse. Drug resistance was assayed on nutrient agar plates with the antibiotics indicated in specific experiments.

EXAMPLE I: Construction of Biologically Functional Bacterial Plasmids in vitro

A. Covalently closed R6-5 plasmid DNA was cleaved by incubation at 37° for 15 minutes in a 0.2 ml reaction mixture containing DNA (40 μg/ml, 100 mM Tris.HCl (pH 7.4)), 5 mM $MgCl_2$, 50 mM NaCl, and excess (2 U) EcoRI endonuclease in 1 μl volume. An additional incubation at 60° for 5 minutes was employed to inactivate the endonuclease.

The resulting mixture of plasmid fragments was employed for transformation of *E. coli* strain C600 in accordance with the procedure previously described. A single clone was examined further which was selected for resistance to kanamycin and was also found to carry resistance to neomycin and sulfonamide, but not to tetracycline, chloramphenicol, or streptomycin after transformation of *E. coli* by EcoRI generated DNA fragments of R6-5. Closed circular DNA obtakined from this isolate (plasmid designation pSC102) by CsCl-ethidium bromide gradient centrifugation had an S value of 39.5 in neutral surcrose gradients.

Treatment of pSC102 plasmid DNA with EcoRI resistriction endonuclease in accordance with the above-described procedure resulted in the formation of 3 fragments that were separable by electrophoresis in agarose gels. Intact pSC102 plasmid DNA and pSC101 plasmid DNA, which had been separately purified by dye-buoyant density centrifugation, were treated with EcoRI endonuclease followed by annealing at 0°-2° for about six hours. The mixture was then subjected to ligation with pSC101 and pSC102 in a ratio of 1:1 respectively, by ligating for 6 hours at 14° in 0.2 ml reaction mixtures containing 5 mM $MgCl^2$, 0.1 mM NAD, 100 μg/ml of bovine-serum albumin (BSA), 10 mM ammonium sulphate (pH 7.0), and 18 U/ml of DNA ligase. (J. Mertz and Davis, Proc. Nat. Acad. Sci., USA, 69, 3370 (1972); and Modrich, et al., J. Biol. Chem., 248, 7495 (1973). Ligated mixtures were incubated at 37° for 5 minutes and then chilled in ice water. Aliquots containing 3.3-6.5 μg/ml of total DNA were used directly for transformation.

Transformation of *E. coli* strain C600 was carried out as previouslydescribed. For comparison purposes, transformation was also carried outwith a mixture of pSC101 and pSC102 plasmid DNA, which had been subjectedto EcoRI endonuclease, but not DNA ligase. The antibiotics used for selection were tetracycline (10 μg/ml) and kanamycin (25 μg/ml). The results are reported as transformants per microgram of DNA. The following table indicates the results.

Kanamycin resistance in the R65 plasmid is a result of the presence of the enzyme kanamycin monophosphotransferase. The enzyme can be isolated from the bacteria by known procedures and employed in an assay for kanamycin in accordance with the procedure described in Smith, et al., New England J. Medicine, 286, 583 (1972).

In the preparation for the enzyme extracts, the *E. coli* are grown in ML-broth and harvested in a late logarithm phase of growth. The cells are osmotically shocked (see Nossal, et al., J. Biol. Chem. 241, 3055 (1966), washed twice at room temperature with 10 ml 0.01 M Tris and 0.03 M NaCl, pH 7.3, and the pellet suspended in 10 ml 20% sucrose, 3×10^3 MEDTA and 0.033 M Tris (pH 7.5), stirred for 10 minues at room temperatureand centrifuged at 16,000 g for 5 minutes. The pellet is then suspended in 2 ml of cold 5×10^{-4} M M_9Cl_2, stirred for 10 minutes at 2° and centrifuged at 26,000 g for 10 minutes to yield a supernatant fluid referred to as the osmotic shockate. The solution should be stored at −20° or lower. (See Benveneste, et al., FEBS Leters, 14 293 (1971).

The osmotic shockate may then be used in accordance with the procedure of Smith, et al., supra.

EXAMPLE II: Genome Construction between Bacterial Species in vitro: Replication and Expression of Staphylococcus Plasmid Genes in *E. coli*

S. aureus strain 8325 contains the plasmid pI258, which expresses resistance to penicillin, erythromycin, cadmium and mercury. (Lindberg, et al., J. Bacteriol., 115, 139 (1973)). Covalently closed circular pSC101 and pI258 plasmid DNA were separately cleaved by incubation at 37° for 15 minutes in 0.2 ml reaction mixtures by EcoRI endonuclease in accordance with the procedure described previously. Aliquots of the two cleaved species were mixed in a ratio of 3 μg of pI258:1 μg of pSC101 and annealed at 2°–4° for 48 hours. Subsequent ligation was carried out for six hours at 14% as described previously and aliquots containing 3.3–6.5 μg/ml of total DNA were used directly in the transformation as described previously.

Other transformations were carried out employing the two plasmids independently and a mixture of the two plasmids. Selection of transformants was carried out at antibiotic concentrations for tetracycline (Tc, 25 μg/ml) or pencillin (Pc, 25 OU/ml). The transformation was carried out with *E. coli* strain C600 r_K-m_K-. The following table indicates the results.

The above table demonstrates that bacteria can be formed which have both tetracycline resistance and penicillin resistance. Thus, one can provide the phenotypical property penicillin resistance in bacteria from DNA, which is indigenous to another biological organism. One can thus use *E. coli* for the production of the enzyme, which imparts penicillin resistance

Table I Transformation of *E. coli* C600 by a mixture of pSC101 and pSC102 DNA

	Transformation frequency for antibiotic resistence markers		
Treatment of DNA	Tetracycline	Kanamycin	Tetracycline + kanamycin
	2 × 10	1 × 10	2 × 10
*Eco*RI	1×10^4	1.1×10^3	7×10^1
*Eco*RI+ DNA ligase	1.2×10^4	1.3×10^3	5.7×10^2

to bacteria, and assay for penicillin in a manner similar to that employed for kanamycin. Penicillinase is used for destroying penicillin in blood serum of patients treated with penicillin in order to determine whether pathogenic organisms whose growth is inhibited by penicillin may be present.

EXAMPLE III: Replication and Transcription of Eukaryotic DNA in *E. coli*

The amplified ribosomal DNA (rDNA) codeing for 18S and 28S ribsomal RNA of the South African toad, *Xenopus laevis* was used as a source of eukaryotic DNA for these experiments. Dawid, et al., J. Mol. Biol., 51, 341 (1970). *E. coli-X. laevis* recombinant plasmids were constructed in vitro as follows:

The reaction mixture (60 μl) contained 100 mM Tris.HCl (pH 7.5) 50 mM NaCl, 5 mM $MgCl_2$, 1.0 μg of pSC101 plasmid DNA and 2.5 μg of *X. laevis* rDNA, and excess EcoRI restriction endonuclease (1 μl, 2 U). After a 15 minute incubation at 37°, the reaction mixture was placed at 63° for 5 minutes to inactivate EcoRI endonuclease. The product was then refrigerated at 0.5° for 24 hours, to allow association of the short cohesive termini.

The reaction mixture for ligation of phosphodiester bonds was adjusted to a total volume of 100 μl and contained in addition to the components of the endonuclease reaction, 30 mM Tris.HCl (pH 8.1), 1 mM sodium EDTA, 5 mM $MgCl_2$, 3.2 nM NAD, 10 mM ammonium sulphate, 5 μg BSA, and 9 U of *E. coli* DNA ligase. All components were chilled to 5° before their addition to the reaction mixture. The ligase reaction mixture was incubated at 14° for 45 minutes, and then at 0.5° for 48 hours. Additional NAD and ligase were added and the mixture incubated at 15° for 30 minutes and then for 15 minutes at 37°. The ligated DNA was used directly in the plasmid transformation procedure previously described. The DNA was used to transform *E. coli* strain C600 r_K-m_k- and tetracycline resistant transformants (3.3×10^3 /μg of pSC101 DNA) were selected and numbered consecutively CD1, CD2, etc. Plasmid DNA was isolated from a number of the transformants.

^{32}P-labeled 18 S and 28 S *X. laevis* rRNA were hybridized with DNA obtained from the plasmids CD4, CD18, CD30, and CD42. CD4 DNA annealed almost equally with both the 18 S and 28 S rRNA species. CD18 plasmid DNA hybridized principally with 28 S *X. laevis* rRNA, while the DNA of plasmids CD30 and CD42 annealed primarily with 18 S rRNA. These data indicate that portions of the *X. laevis* rDNA were, in fact, incorporated into a plasmid recombinant with pSC101, which was capable of transforming *E. coli*, so as to be capable of replicating *X. laevis* rDNA.

Transcription of *X. laevis* DNA was also carried out in *E. coli* minicells. The minicell producing *E. coli* strain P678-54 was transformed with plasmid DNA isolated from *E. coli* strain C600 r_K-m_k- containing CD4, CD18, or CD42. Many cells containing the plasmids were isolated and incubated with [^{3}H] uridine; RNA purified from such minicells was hybridized with *X. laevis* rDNA immobilized on nitrocellulose membranes in order to determine whether the *X. laevis* rDNA linked to the pSC101 replicon is transcribed in *E. coli*. The results in the following table show that RNA species capable of annealing with purified *X. laevis* rDNA are synthesized in *E. coli* minicells carrying the recombinant plasmids, CD4, CD18, and CD42, but not by minicells carrying the pSC101 plasmid alone.

Minicells containing plasmids were isolated as described by Cohen, et al., Nature New Biol., 231, 249 (1971). They were incubated with [^{3}H] uridine (50 μCi/ml, 30 Ci/mol) as described by Roozen, et al., J. Bacteriol., 107, 21 (1971) for 10 minutes at 37°. Minicells collected by centrifugation were resuspended in Tris.HCl (20 mM, pH 7.5)-5 mM $MgCl_2$-1 mM EDTA pH 8.0 and rapidly frozen and thawed 3 times. RNA was extracted as described in Cohen, et al., J. Mol. Biol., 37, 387 (1968). Hybridization assays were carried out in nitrocellulose membranes as described in Cohen, et al., ibid, at saturating levels of pSC101 DNA. Hybridizations involving *X. laevis* DNA were not performed at DNA excess. Counts bound to blank filters (5-10 c.p.m.) were substracted from experimentally determined values. 3 H count eluted from filters containing *X. laevis*

Table II Transformation of C600 rK mK by pSC101 and pl258 plasmid DNA

	Transformants/μg DNA	
DNA	Tc	Pc
pSC101 closed circular	1 × 106	<3
pl258 closed circular	<3.6	<3.6
pSC101 + pl258 untreated	9.1×10^5	<5
pSC101 + pl258 *Eco*RI-treated	4.7×10^3	10

DNA were rendered acid soluble by ribonuclease A 20 μg/ml, 0.30 M NaCl-0.030 M sodium citrate, 1 hour, 37°. The following table indicates the results.

EXAMPLE IV: Plasmid ColEl as a Molecular Vehicle for Cloning and Amplification of Trp Operon

In a volume of 200 μl (100 mM Tris.HCl (pH 7.5)-5 mM $MgCl_2$ -50 mM NaCl), 5.7 μg of ColEl (*E. coli* JC411Thy-/ColEl) (Clewell, et al., Proc. Nat. Acad. Sci., USA, 62, 1159 (1969) and 6.0 μg DNA from bacteriophage .phi.80pt190 (Deeb, et al., Virology, 31, 289 (1967) were digested to completion with homogeneously purified EcoRI endonuclease, monitoring the digestion by electrophoresis of the fragments in an agarose gel. The endonuclease was inactivated by heating at 65° for 5 minutes, the digest dialyzed overnight against 5 mM Tris.HCl, pH 7.5, and the sample concentrated to 50 μl. The fragments were ligated as described in Dugaiczyk, et al., Biochemistry, 13, 503 (1974) at a concentration of 75 pmoles/ml of fragments.

Transformation was carried out as previously described except that the cells were grown to A_{590}=0.600 and following exposure to DNA were incubated in L-broth for 90 minutes. The cells were collected and resuspended in 10 mM NaCl before plating. Cells employed as recipients for the transformations were *E. coli* strains C600 trpR', δtrpE5(MVl), C600 trpR- trpE 10220 recA(MV2), C600 δtrpE5(MV10) and C600 5trpE5 recA(MV12). (trpR- is the structural gene for the trp repressor and δtrpE5 is a trp operon deletion entirely within trpE and removing most of the gene.) Approximately 2 μg of the DNA was used to transform the cells.

Cultures were plated on Vogel-Bonner agar supplemented with 50 μg/ml of the non-selective amino acids, 0.2% glucose and 5 μg/ml of required vitamins. Transformants to colicin immunity were initially selected on a lawn of a culture of a mutant strain carrying ColE1. Clones were then selected for their ability to grow in the absence of tryptophan. Cells capable of producing tryptophan were isolated, which could be used for the production of exogenous tryptophan. The subject example demonstrates the introduction of a complete operon from foreign DNA to provide a transformant capable of replicating the operon and transcribing and translating to produce enzymes capable of producing an aromatic amino acid.

EX. V: Cloning of Synthetic Somatostatin Gene

The deoxyribonucleotide sequence for the somatostatin gene was prepared in accordance with conventional procedures. (Itakura et al, Science, 198 1056 (1977)). To prepare the recombinant plasmid, plasmid pBR 322 was digested with Eco RI. The reaction was terminated by extraction with a mixture of phenol and chloroform, the DNA precipitated with ethanol and resuspended in 50 μl of T_4 DNA polymerase buffer. The reaction was started by the addition of 2 units of T_4 DNA polymerase. After 30 min at 37°, the mixture was extracted with phenol and chloroform and the DNA precipitated with ethanol. The λplac5 DNA (3 μg) was digested with the endonuclease Hae III and the digested pBR 322 DNA blunt end ligated with the Hae III-digested λplac5 DNA in a final volume of 30 μl with T_4 DNA ligase (hydroxylopatite fraction) in 20 mM tris-HCl pH 7.6), 10 mM $MgCl_2$, 10 mM dithiothreitol and 0.5 mM ATP for 12 hrs at 12°. The ligated DNA mixture was dialyzed against 10 mM tris-HCl (pH 7.6) and used to transform *E. coli* strain RR1. Transformants were selected for tetracycline resistance and ampicillin resistance on antibiotic (20 μg/ml) X-gal (40 μg/ml) medium. Colonies constitutive for the synthesis of .beta.-galactosiodase were identified by their blue color and of 45 colonies so identified, 3 of them were found to contain plasmids with 2 Eco R1 sites separated by ~200 base pairs.

The plasmid so obtained pBH10 was modified to eliminate the Eco R1 site distal to the lac operator and plasmid pBH20 was obtained.

Plasmid pBH20 (10 μg) was digested with endonucleases Eco R1 and Bam HI and treated with bacterial alkaline phosphatase (0.1 unit of BAPF, Worthington) and incubation was continued for 10 min at 65°. After extract with a phenol-chloroform mixture, the DNA was precipiated with ethanol. Somatostatin DNA (50 μl containing 4 μg/ml) was ligated with the Bam HI-Eco R1 alkaline phosphatase=treated pBH20 DNA in a total volume of 50 μl with 4 units of T_4

DNA ligase for 2 hrs at 22° and the recombinant plasmid used to transform *E. coli* RR1. Of the Tcr transformants isolated (10), four plasmids has Eco R1 and Bam HI sites. Base sequence analysis indicated that the plasmid pSOM1 had the desired somatostatin DNA fragment inserted. Because of the failure to detect somatostatin activity from cultures carrying

Table III [³H] RNA synthesized by *E. coli* minicells

Plasmid carried by minicells	Input cpm	[³H] RNA counts hybridized to *X. laevis* rDNA 0.2 μg	0.4 μg	pSC101 DNA 18 μg
CD42	4810	905(19%)	1436 (30%)	961 (20%)
CD18	3780	389 (10%)	--	1277 (34%)
CD4	5220	789 (15%)	--	1015 (19%)
pSC101	4170	0 (0%)	--	1500 (36%)

plasmid pSOM1, a plasmid was constructed in which the somatostatin gene could be located at the COOH-terminus of the .beta.-galactosidase gene, keeping the translation in phase. For the construction of such a plasmid, pSOMI (50 μg) was digested with restriction enzymes Eco RI and Pst I. A preparative 5 percent polyacrylamide gel was used to separate the large Pst I-Eco RI fragment that carries the somatostatin gene from the small fragment carrying the lac control elements (12). In a similar way plasmid pBR322 DNA (50 μg) was digested with Pst I and Eco RI restriction endonucleases, and the two resulting DNA fragments were purified by preparative electrophoresis on a 5 percent polyacrylamide gel. The small Pst I-Eco RI fragment from pBR322 (1 μg) was ligated with the large PstI-Eco RI DNA fragment (5 μg) from pSOM1. The ligated mixture was used to transform *E. coli* RR1, and transformants were selected for Ap^r on X-gal medium. Almost all the Ap^r transformants (95 percent) gave white colonies (no lac operator) on X-gal indicator plates. The resulting plasmid, pSOM11, was used in the construction of plasmid pSOM11-3. A mixture of 5 μg of pSOM11 DNA and 5 μg of λplac5 DNA was digested with Eco RI. The DNA was extracted with a mixture of phenol and chloroform; the extract was precipitated by ethanol, and the precipitate was resuspended in T4 DNA ligase buffer (50 μl) in the presence of T4 DNA ligase (1 unit). The ligated mixture was used to transform *E. coli* strain RR1.

Transformants were selected for Ap^r on X-gal plates containing ampicillin aidn screened for constitutive.beta.-galactosidase production. Approximately 2 percent of the colonies were blue (such as pSOM11-1 and 11-2). Restriction enzyme analysis of plasmid DNA obtained from these clones revealed that all the plasmids carried a new Eco RI fragment of approximately 4.4 megadaltons, which carries the lac operon control sites and most of the .beta.-galactosidase gene (13, 14). Two orientations of the Eco RI fragment are possible, and the asymmetric location of a Hind III restriction in this fragment can indicate which plasmids had transcription proceeding into the somatostatin gene. The clones carrying plasmids SOM11-3, pSOM11-5, pSOM11-6, and pSOM11-7 contained the Eco RI fragment in this orientation.

It is evident from the above results, that both DNA from a eukaryotic source and RNA transcribed from the eukaryotic DNA can be formed in a bacterial cell and isolated. Thus, the subject process provides a simple technique for producing large amounts of eukaryotic DNA and/or RNA without requiring the reproduction and maintenance of the eukaryotic organism or cells. The employment of DNA for production of ribosomal RNA is merely illustrative of using a genome from a eukaryotic cell for formation of a recombinant plasmid for replication in a bacteria. Genomes from a eukaryotic cell for formation of genotypical properties, such as the production of enzymes, could have equivalently been used. As evidenced by the transformation with DNA from a bacteriophage, and entire operon can be introduced into a bacterial cell and the cell becomes capable of its transcription, translation, and production of a functional gene product. Thus, a wide variety of auxotrophic properties can be introduced into a bacterial cell.

In accordance with the subject invention, DNA vehicles are provided, which are covalently closed circular extrachromosomal replicons or genetic elements, including plasmids and viral DNA. The vehicles generally will have molecular weights in the range of about 1 to $20X10^6$ and are characterized by having an intact replicon, which includes a replicator locus and gene. The vehicle is capable of clevage by a restriction enzyme to provide a linear segment having an intact replicon and cohesive termini, which may be directly obtained by the cleavage or by subsequent modification of the termini of the linear segment. The vehicle will be capable of transforming a bacterial cell and to that extent is compatible with the cell which will provide replication and translation. Preferably, the vehicle will have a phenotypical property which will allow for segregation of the transformant cells. Phenotypical properties include resistance to growth inhibiting materials, such as antibiotics, peptides and heavy metals, morphological properties, color, or the like, and production of growth factors, e.g. amino acids.

The vehicle is combined with DNA indigenous to a biological organism other than the cell which provides replication and provides a genotypical or phenotypical property which is alien to the cell. The source of the DNA can be prokaryotic or eukaryotic, thus including bacteria, fungi, vertebrates, e.g. mammals, and the like.

The plasmid vehicle and the alien DNA having complementary cohesive termini can be annealed together and covalently linked to provide a recombinant plasmid, which is capable of transforming a bacterial cell, so as to be capable of replication, transcription, and translation. As a result, a wide variety of unique capabilities can be readily introduced into bacteria, so as to provide convenient ways to obtain nucleic acids and to study nucleic acids from a foreign host. Thus, the method provides the ability to obtain large amounts of a foreign nucleic acid from bacteria in order to be able to study the function and nature of the nucleic acid. In addition, the subject method provides means for preparing enzymes and enzymic products from bacteria where the natural host is not as convenient or efficient a source of such product. Particularly, bacteria may allow for more ready isolation of particular enzymes, uncontaminated by undersirable contaminants, which are present in the original host. In addition, the products of the enzymic reactions may be more readily isolated and more efficiently produced by a transformant than by the original host. Besides enzymes, other proteins can be produced such as antibodies, antigens, albumins, globulins, glycoproteins, and the like.

Although the foregoing invention has been described in some detail by way of illustration and example for purposes of clarity of understanding, it will be obvious that certain changes and modifications may be practiced within the scope of the appended claims.

CLAIMS

We claim:

1. A method for replicating a biologically functional DNA, which comprises:

transforming under transforming conditions compatible unicellular organisms with biologically functional DNA to form transformants; said biologically functional DNA prepared in vitro by the method of:

(a) cleaving a viral or circular plasmid DNA compatible with said unicellular organism to provide a first linear segment having an intact replicon and termini of a predetermined character;

(b) combining said first linear segment with a second linear DNA segment, having at least one intact gene and foreign to said unicellular organism and having termini ligatable to said termini of said first linear segment,wherein at least one of said first and second linear DNA segments has a gene for a phenotypical trait, under joining conditions where the termini of said first and second segments join to provide a functional DNA capable of replication and transcription in said unicellular organism; growing said unicellular organisms under appropriate nutrient conditions; and isolating said transformants from parent unicellular organisms by means of said phenotypical trait imparted by said biologically functional DNA.

2. A method according to claim 1, wherein said unicellular organisms are bacteria.

3. A method according to claim 2, wherein said transformation is carried out in the presence of calcium chloride.

4. A method according to claim 3, wherein said phenotypical trait is resistance to growth inhibiting substance, and said growth is carried out in the presence of a sufficient amount of said growth inhibiting substance to inhibit the growth of parent unicellular organisms, but insufficient to inhibit the growth of transformants.

5. A method according to claim 1, wherein said unicellular organism is *E. coli.*

6. A method according to claim 1, wherein said predetermined termini are staggered and cohesive.

7. A method according to claim 6, wherein said joining conditions includes enzymatic ligation.

8. A method according to claim 6, wherein said cohesive ends are formed by staggered cleavage of said viral or circular plasmid DNA and a source of said second segment with a restriction enzyme.

9. A method acording to claim 6 wherein said cohesive termini are formed by addition of nucleotides.

10. A method according to claim 1, wherein said predetermined termini are blunt end and said joining conditions include enzymatic ligation.

11. A method for replicating a biologically functional DNA comprising a replicon compatible with a host unicellular organism joined to a gene derived from a source which does not exchange genetic information with said host organism, said method comprising:

isolating said biologically functional DNA from transformants prepared in accordance with claim 1;

transforming unicellular microorganisms with which said replicon is compatible with said isolated DNA to provide second transformants; and

growing said second transformants under appropriate nutrient conditions to replicate said biologically functional DNA.

12. A method for producing a protein foreign to a unicellular organism by means of expression of a gene by said unicellular organism, wherein said gene is derived from a source which does not exchange genetic information with said organism, said method comprising:

growing transformants prepared in accordance with any of claims 1 and 11 under appropriate nutrient conditions, whereby said organism expresses said foreign gene and produces said protein.

13. A method according to claim 12, wherein said protein is an enzyme.

14. A method according to claim 11, wherein said method is repeated substituting said biologically functional DNA from transformants prepared in accordance with claim 1 with second or subsequent transformants to produce additional transformants.

* * * * *

Appendix 2

Recombinant DNA Safety Guidelines and Regulations, January 1990

Department of Biotechnology, Ministry of Science and Technology, Government of India

1. Introduction

The new capabilities to manipulate the genetic material present tremendous potential and find use in many novel experiments and applications. These developments have generated a sense of concern among scientists working in biological areas and others to find ways as to how safely the research in the field should be carried out and also possible means to regulate the work involving pathogenic microorganisms and virulent genes. With the safety considerations in view, the Department of Biotechnology is mandated to evolve the Recombinant DNA Safety Guidelines. The Department has set up the Recombinant DNA Advisory Committee (RDAC) for this purpose. On the basis of current scientific information a document on Recombinant DNA Safety Guidelines has been brought out on the use of this technique in the area of research, manufacture and applications. The booklet brings forth the salient features in the document as a guide towards observance of the guidelines in research and also to meet the regulatory requirements by those in production, testing and use of genetically modified organisms and products.

2. Research

The guidelines cover areas of research involving

i) Genetically engineered organisms,
ii) Genetic transformation of green plants, animals,
iii) rDNA technology in vaccine development and
iv) Large scale production and deliberate/accidental release of organisms, plants, animals & products derived by rDNA technology.

The guidelines prescribe 4 levels of risk, while carrying out experiments with microorganisms. Classification of organisms within these levels is based on pathogenicity, local prevalence of disease and epidemic causing strains in India.

3. Notification

Related to the incremental risk involved in the use of microorganisms, plants, animals in experiments, the corresponding biosafety practices are required. Accordingly the notification procedures are defined under 3 categories.

i) Exempt category—(self cloning experiments)
ii) Intimation to competent authority—(e.g. experiments involving nonpathogenic DNA vector systems)
iii) Review and approval of competent authority—(e.g. Toxin gene cloning, antibiotic resistant genes, etc.)

4. Containment

Containment facilities are recommended for necessary safeguards while carrying out experiments involving organisms belonging to 4 different biosafety levels.

a) PHYSICAL C5ONTAINMENT
Physical containment is to limit the spread of dangerous microorganisms by way of
i) Technique (Good Laboratory Practice)
ii) Safety equipments
iii) Laboratory design and facilities

b) BIOLOGICAL CONTAINMENT
Biological containment involves the use of the combination of vector and host in such a way so that it
i) can limit the infectivity of vector to specific hosts and
ii) control host vector survival in the environment

The growth of whole plants will however require special environmental conditions which may be achieved by using glasshouse containment.

Glasshouse containment A is appropriate to plant experiments involving no plant pathogen and would be suitable for experiments involving non-pathogen DNA vector systems and regeneration from single cells. (Notification to competent authority is needed).

Glasshouse containment B is recommended for experiments involving i) genetically manipulated plant pathogens including plant viruses such as the propagation of genetically manipulated organisms in plant and ii) the growth of plants regenerated from cells transformed by genetically manipulated pathogen vector systems which still contain the pathogen. (Approval of competent authority is required before commencement of activity).

5. Recognition of facility

Application for recognition of research facility to carry out genetic manipulation should be made to the Department of Environment before the commencement of work in the prescribed proforma as per 7(2) of the rules on hazardous microorganisms/genetically engineered organisms notified under the EPA 1986.

6. Competent authority

a) INSTITUTIONAL BIOSAFETY COMMITTEE (IBSC) (IN R&D CENTRES, UNIVERSITY ETC.)

   i) The IBSC shall be the nodal point for interaction within the institution for implementation of the guidelines. As such, in the first instance it is necessary that the institutions intending to carry out research activities involving genetic manipulation of microorganisms, plants or animals should constitute the IBSC. Any research project which is likely to have biohazard potential (as envisaged in the guidelines) during the execution stage or which involve the production of either microorganisms or biologically active molecules that might cause bio-hazard should be notified to IBSC. The on-site emergency plan should be prepared by the institution or occupier for each of the above activities with the help of IBSC. IBSC will allow genetic engineering activity on classified organisms only at places where such work should be performed as per guidelines. The prescribed form-Notification of intention to carry on genetic manipulation—Part B should be submitted to funding agency with comments of IBSC for financial support on experiments falling under categories II, III & IV. Support on approved projects will be withdrawn in case of deliberate violation or avoidable negligence of rDNA guidelines. In addition, it may attract action under EPA.

   ii) IBSC will provide half yearly reports on the on-going projects to RCGM.

   iii) Authorisation for inter-state exchange of etiologic agents, diagnostic specimens and biological products will be done by IBSC on standard proforma evolved for this purpose.

   iv) Manipulation of plants under containment would be performed under the regulatory clearance of IBSC. Development of organisms for agricultural and environmental applications should be conducted in a step-wise fashion, moving where appropriate from the laboratory to the growth chamber and green-house under containment conditions and good laboratory practice.

b) REVIEW COMMITTEE ON GENETIC MANIPULATION (RCGM)

   The RCGM will have the function :

   i) To review the reports in all approved ongoing research projects involving high risk category and controlled field experiments.

   ii) To visit site of experimental facilities periodically where projects with biohazard potential are being pursued and also at a time prior to the commencement of the activity to ensure that adequate safety measures are taken as per the guidelines.

   iii) To issue authorisation for import and receipt of etiologic agents and vectors, germ plasms, organelle, etc needed for experimental work/training and research. (Applications should be made in prescribed proforma).

7. Large scale experiments and manufacture

Regulations of large scale production and field testing of engineered organisms and products including environmental release has been laid down under statutory provisions of Environmental Protection Act 1986.

Experiments beyond 20 litres capacity for research as well as industrial purposes are included in the category of large scale experimentation/operations. For such activities it is recommended that one should seek approval of the competent authority (GEAC) on furnishing the relevant details in a prescribed format (To be issued by GEAC with Department of Environment).

The following safety criteria are to be complied :

i) The host organism should not be a pathogen, should have extended history of safe use and built in environmental limitations.

ii) The vector/insert should be well characterised and free from known harmful sequences, limited in size as much as possible (with insert less than 10 base pair or below 30,000 M.W.) should be poorly mobilisable, should not transfer any resistance markers. In cases, where the insert sequence exceeds the above limit, toxicity screening should be made. The constructs should have 3 markers for bioprocess monitoring.

iii) The genetically manipulated organism should not be a pathogen.

For large scale operations, measures such as proper engineering for containment, quality control, personnel protection, medical surveillance are recommended.

Offsite contingency plans in event of unanticipated effects of novel organisms/products on accidental release are to be worked out and appropriate control measures are to be developed in consultation with the competent authority (State and District level coordination committees of GEAC) to meet any exigency.

8. Biologicals produced by rDNA technology

The general regulations normally applicable for biologicals are relevant to the recombinant DNA products. The specific relevant aspects to a particular product should be discussed with the appropriate Govt, agency (Drugs controller) on a case by case basis.

i) A new licence for a product drug application whould be required on products made of recombinant DNA technology even if the product is considered to be chemically and physically similar to the naturally occurring substance or previously approved product produced in conventional system.

ii) A recombinant DNA product demonstrated to be identical to normally occurring substance would not require toxicological and pharmacological data if the information is already available at the dose levels intended for specific use. Otherwise the data on the above would be necessary on such products.

9. Release to the environment & field experiments

i) Depending on the type of organism handled & the assessment of potential risks involved appropriate containment facilities must be provided to ensure safety and to prevent unwanted release in the environment.

ii) It is important to evaluate rDNA modified organism for potential risk, prior to application in agriculture and environment. Prior to introduction of micro-organisms, properties of the organisms, the possible interaction with other disease causing agents and the infected wild plant species should be evaluated. An independent review of potential risks should be conducted on a case by case basis to application.

iii) Pre-release tests of genetically engineered organisms in Agricultural applications should include elucidation of genetic markers, host range, requirements for vegatative growth, persistence and stability in small plots and experimental field trials for 2 year. Soil samples in experiments under controlled containment conditions should be tested for the absence of viable cells before disposal into the environment.

iv) Biowastes resulting from laboratory experiments, industrial operations should be properly treated, so organisms are either destroyed or rendered harmless before disposal into the environment.

10. Genetic Engineering Approval Committee (GEAC)

i) Testing of genetically altered organisms, transgenic animals, plant material tested against pathogens & products in the environment should follow regulatory guidelines seeking field use permit from GEAC in the prescribed proforma (to be evolved by GEAC).

ii) The large scale planned release of organisms into the environment both for environmental and agricultural applications should be done under licence. Applications to this effect should be made to GEAC on the prescribed proforma which will be examined on case by case basis (to be evolved by GEAC). The validity of approval is for a period of 4 years.

iii) Import for large scale use, export, manufacture, process, sell, use of any genetically engineered substances or cells including food stuffs and additives that contains rDNA products are subject to regulatory control by GEAC. Prior approval to this effect should be obtained on applications in the prescribed proforma (to be evolved-by GEAC).

11. A copy of the Recombinant DNA Safety Guidelines and the rules under EPA 1986 for the manufacture, use, import and storage of hazardous microorganisms/genetically engineered organisms or cells may be obtained on request.

CONTACT ADDRESSES

Research

Dr. K. Narayanaswami
Director,
Deptt. of Biotechnology,
Block 2, 7th Floor,
C.G.O. Complex, Lodi Road,
New Delhi-110003
Tel. 363012/237

Large Scale Production, Environmental applications & Use

Dr. (Mrs.) I. Chandrasekaran
Scientist 'SE' (HSMD),
Ministry of Environment & Forests,
Paryavaran Bhavan, C.G.O. Complex,
Lodi Road, New Delhi-110003
Tel. 360488

Good laboratory practice

- Never do direct mouth pipetting of infectious or toxic fluids; use a pipettor.
- Plug pipettes with cotton.
- Do not blow infectious material out of pipettes.
- Do not prepare mixtures of infectious material by bubbling expiratory air through the liquid with a pipette.
- Use an alcohol—moistened pledget around the stopper and needle when removing a syringe and needle from a rubber stoppered vaccine bottle.
- Use only needle-locking hypodermic syringes. Avoid using syringes whenever possible.
- Expel excess fluid and bubbles from a syringe vertically into cotton pledget moistened with disinfectant, or into a small bottle of cotton.
- Before and after infecting an animal, swab the site of injection with a disinfectant.
- Sterilize discarded pipettes and syringes in pan where they were first placed after use.
- Before centrifuging, inspect tubes for cracks. Inspect the inside of the trunnion cup for rough walls caused by erosion or adhering matter. Carefully remove all bits of glass from the rubber cushion. A germicidal solution added between the tube and the trunnion cup not only disinfects the surfaces of both these, but also provides an excellent cushion against shocks that otherwise might break the tube.
- Use centrifuge trunnion cups with screw caps or equivalent.
- Avoid decanting centrifuge tubes; if you must do so, afterwards wipe off the outer rim with a disinfectant. Avoid filling the tube to the point that the rim ever becomes wet with culture.
- Wrap a lyophilized culture vial with disinfectant-wetted cotton before breaking. Wear gloves.
- Never leave a discarded tray of infected material unattended.
- Sterilize all contaminated discarded material.
- Periodically, clean out deep-freeze and dry-ice chests in which cultures are stored to remove broken ampoules or tubes. Use rubber gloves and respiratory protection during the cleaning.
- Handle diagnostic serum specimens carrying a risk of infectious hepatitis with rubber gloves.
- Develop the habit of keeping your hands away from your mouth, nose, eyes and face. This may prevent self-inoculation.
- Avoid smoking, eating, and drinking in the laboratory.
- Make special precautionary arrangements for respiratory, oral, intranasal, and intratracheal inoculation of infectious material.
- Give preference to operating room gowns that fasten at the back.
- Evaluate the extent to which the hands may become contaminated with some agents and operations, forceps or rubber gloves are available.
- Wear only clean laboratory clothing in the dining room, library and other nonlaboratory areas.
- Shake broth cultures in a manner that avoids wetting the plug or cap.

2. Aerosol minimization

Because of their insidious nature, aerosols pose special problems in that the laboratory worker may be unwillingly exposed. Procedures which can produce aerosols include:

- grinding
- blending
- sonicating
- resuspending packed cells or viruses
- inserting a hot loop into a culture
- centrifugation
- flaming an inoculation loop so that it splatters
- forceful ejection of fluid from a pipette or syringe
- opening a tube containing a lyophilized agent
- releasing the vaccum on a freeze dryer
- opening a tube within which the air pressure may differ from that of the room, such as may occur when the tube is opened at a temperature different from that which it was sealed.

Glasshouse containment conditions for plant experiments

1. **Glasshouse Containment A** are:
   i) Plants should be grown in a designated glasshouse or compartment, clearly marked with a bio-hazard sign indicating "glasshouse containment A"
   ii) Any other plants grown in the designated glasshouse or compartment must be handled under conditions appropriate for the experimental plants.
   iii) Plants should be managed by suitably trained personnel with the principles of good glasshouse hygiene.
   iv) The IBSC should consider whether any additional factors such as pest control, screening to prevent ingress by vermin, birds and insects and destruction of surplus plants and seed are relevant to the particular experiment.
2. **Glasshouse containment B** conditions will be specified by the committee (RCGM) and will vary with the pathogen, being particulary dependant on its mode of dispersal, host range and pathogenicity and they are to be worked out on case by case basis.
   Special conditions may be needed in addition to those given under 'A' to prevent dissemination of the genetically manipulated plant pathogen especially during transfer between glasshouse and laboratory, during disposal of plants and equipment and through survival of pollen, seeds or other biological vectors.

Table1 1* Summary of recommended biosafety levels for infectious agents

Biosafety level	Practice and techniques	Safety	Facilities
1.	Standard microbiological practices	None : primary containment provided by adherence to standard laboratory practices.	Basic
2.	Level 1 practices plus: Laboratory coats; decontamination of all infectious wastes limited access; protective gioves and biohazard warning signs as indicated.	Partial containment equipment (i.e. Class I or II Biological Safety Cabinets) used to conduct mechanical and manipulative procedures that have aerosol potential that may increase the risk of exposure to personnel	Basic
3.	Level 2 practice plus: special laboratory clothing; controlled access.	Partial containment equipment used for all manipulations of infectious material	Containment
4.	Level 3 practices plus: entrance through change room where street clothing is removed and laboratory clothing is put on; shower on exit; all wastes are decontaminated on exit from the facility.	Maximum containment equipment (i.e. Class III biological safety Containcabinet or partial containment equipment in combination withfull body air-supplied, positive pressure personnel suit used for all procedures and activities.	Maximum containment

* Biosafety in Microbiological and Biomedical Labs. (US Deptt. of Health & Human Services) 1st. Ed., 1984.

Examples of containment approaches for large scale industrial applications other than GLSP (good industrial large scale practice)

	Containment categories		
Specifications	**1**	**2**	**3**
1. Viable organisms should be handled in a system which physically separates the process from the environment (closed system)	Yes	Yes	Yes
2. Exhaust gases from the closed system should be treated so as to	Minimise release	Prevent release	Prevent release
3. Sample collection, addition of materials to a closed system and transfer of viable organisms to another closed system, should be performed so as to	Minimise release	Prevent release	Prevent release
4. Bulk culture fluids should not be removed from the closed system unless the viable organism has been	Inactived by validated means	Inactived by validated means	Inactived by validated means
5. Seals should be designed so as to	Minimise release	Prevent release	Prevent release
6. Closed system should be located within a controlled area	Optional	Optional	Yes, and purpose built
a) Biohazard signs should be Posted	Optional	Yes	Yes
b) Access should be restricted to nominated personnel only	Optional	Yes	Yes, via an airlock
c) Personnel should wear protective clothing	Yes	Yes	A complete change
d) Decontamination and washing facilities should be provided for personnel.	Yes	Yes	Yes
e) Personnel should shower before leaving the controlled area	No	Optional	Yes
f) Effluent from sinks and showers should be collected and inactivated.	No	Optional	Yes
g) The controlled area should be adequately ventilated to minimise air contamination.	Optional	Optional	Yes
h) The controlled area should be maintained at an air pressure negative to atmosphere.	No	Optional	Yes
i) Input air and extract air to the controlled area should be HEPA filtered.	No	Optional	Yes

Continued ...

... continued			
j) The controlled area should be designed to contain spillage of the entire contents of the closed system.	No	Optional	Yes
k) The controlled area should be sealable to permit fumigation.	NO	Optional	Yes
l) Effluent treatment before final discharge	Inactivated by validated means.	Inactivated by validated chemical or physical means.	Inactivated by validated chemical or physical means.

a) Need for negative pressure and air filtration double doors etc. in cases where airborne dispersal is a potential hazard.

b) Need for effluent treatment plant where water borne dispersal is a hazard.

c) Need for suitable construction of glasshouse (floors, dwarfwalls, threshold at door etc.) in cases where waterborne or soilborne dispersal are potential hazards.

d) Need to prevent pollination and seeding, or to contain pollen and seed in cases where pollen and seed-borne dispersal is a potential hazard.

e) Need for measures either to prevent contamination of, or to decontaminate the clothing of personnel or tools, pots, equipment etc., where mechanical transmissions is an above average hazard.

f) Need to limit the growing of host plants in the vicinity of the containment facility and to provide monitoring for escape.

Inspection of a 'Glasshouse Containment B' facility by IBSC will be required before approval.